*Presented to the Santa Maria Public Library
by the Santa Maria Valley Chapter of 99s
International Organization of Women Pilots
September, 1998*

THE NINETY-NINES
Yesterday – Today – Tomorrow

TURNER PUBLISHING COMPANY

TURNER PUBLISHING COMPANY
412 Broadway, P.O. Box 3101
Paducah, KY 42002-3101
Phone: (502) 443-0121

Copyright © 1996. Turner Publishing Company.
All rights reserved.

Turner Publishing Company Staff:
Editor: Julie Agnew Thomas
Designer: Herbert C. Banks II

Ninety-Nines History Book Committee:
Lu Hollander
Gene Nora Jessen
Verna West

This book or any part thereof may not be reproduced without the written consent of Turner Publishing Company and The Ninety-Nines, Inc.

This publication was produced using available material. Turner Publishing Company regrets they cannot assume liability for errors or omissions.

Library of Congress Catalog Card Number: 96-60849
ISBN: 1-56311-203-5

Additional copies may be purchased directly from Turner Publishing Company or The Ninety-Nines, Inc.

Printed in the U.S.A.
Limited Edition.

On the cover:
Painting by Douglas Ettridge provided by Betty Basar Robin.

Title Page:
Amelia Earhart, Charter Member and First President. (Courtesy of Schlesinger Library, Radcliffe College.)

Early pilot Mathilde Moisant (right) models appropriate flight attire for women of her day. In 1995, her nephew John Weyl (above) presented The Ninety-Nines with much of her memorabilia, including the snappy flying costume on the mannequin at his side.

Jacqueline Cochran and Amey Otis Earhart (Amelia's mother) visit during the second All Woman Air Show, June 5-6, 1948, in Miami, Florida.

Mrs. Thomas A. Edison christens the new plane owned by Opal Kunz July 13, 1929. The affair being a formal one, even the pilot wore an evening dress.

FOREWORD

It's been nearly 20 years since an extensive history book about 99s was assembled, and here we are again — trying our best to capture the activities of a living, vital organization in a collection of words on pages, and in the activities of people frozen in still photographs.

Not an easy task, but what an exciting one. There are so many interesting stories within the pages of this book, and so many more ongoing — ready to be collected for the next time a book like this one is published.

Meanwhile, enjoy reading about over 1,900 dynamic women who have submitted their personal stories, review the history of The Ninety-Nines and relish the part each of you has played in producing this airborne tapestry of women's contributions to the world of flight.

ACKNOWLEDGMENTS

Every 99 who has been, or is, a member of this organization

The Headquarters staff who searched for photographs and otherwise assisted with research

Others who contributed their time, personal photographs, articles and other documents to make this book more complete

HISTORY BOOK COMMITTEE

Lu Hollander
Gene Nora Jessen
Verna West

MISSION STATEMENT

Promote fellowship through flight

Provide working and scholarship opportunities for women and aviation education in the community

Preserve the unique history of women in aviation

CONTENTS

International Headquarters 4
Past Presidents 6
Charter Members 7
Ninety-Nines History 8
Ninety-Nines Members 42
Ninety-Nines Roster 236
Index 263

INTERNATIONAL HEADQUARTERS

In 1995, Palms Chapter members Claire Walters, C.J. Strawn and Gail Kass flew to Oklahoma City to present The Ninety-Nines with the Mathilde Moisant Collection of aviation memorabilia. The collection had to be stored because there was no space to display the items.

The three flew home and told their chapter members about the need to complete the second floor of Headquarters so that the Moisant Collection, and others being stored, could be displayed. Palms members decided to take the money they had set aside to build a cabinet for the Moisant items and use it as seed money for the second-floor completion.

Claire Walters, chapter founder, became the fund-raising chairman, and C.J. Strawn, a professional movie set designer, took the assignment of designing the space.

Plans for the second floor include a library; a gallery; a wall of flyers featuring past presidents, charter members and special contributors; a theatre area; a Wall of Merit; and reception and seating areas.

By the end of September 1996, more than half the estimated $130,000 needed for the completion had been raised, and initial construction begun. Completion is scheduled for June 1997.

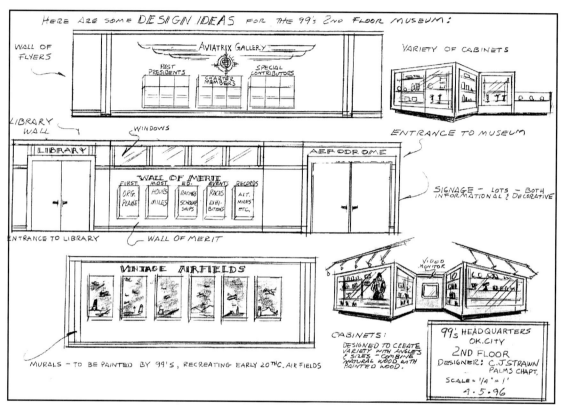

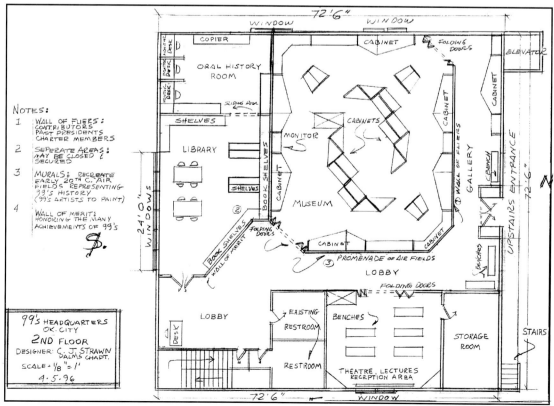

The new two-story, 20,000-square-foot Headquarters/Resource Center in Oklahoma City was completed and dedicated in 1988, prior to the annual Convention. The Ninety-Nines currently are fund-raising to complete the second-floor space. Ultimately, the two buildings (the original single-story structure is at left) will be connected by a glass atrium.

Several large display cases contain memorabilia from early women pilots. In the background (left) is a display of women astronauts.

First-floor space at Headquarters includes a board meeting room (background) complete with portraits of all past presidents. The display case at left contains memorabilia from women airline pilots, and the quilt in the background was made by Idaho Chapter members. A bust of Amelia occupies a prominent space in the exhibit area.

Along one wall is a viewing table and storage for a number of early scrapbooks.

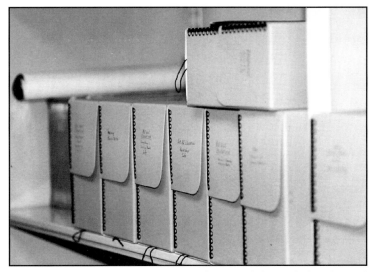

Acid-free archival storage cartons line the wall-to-wall shelves in the humidity- and temperature-controlled archives storage room at Headquarters.

PAST PRESIDENTS

Amelia Earhart*
1931-1933

Margaret Manser*
1933-1935

Mabel Britton*
1935-1937

Daisy Kirkpatrick*
1937-1939

Betty Huyler Gillies
1939-1941

Jacqueline Cochran*
1941-1943

Ethel Sheehy*
1943-1945

Jeannete Sovereign*
1945-1947

Belle Hetzel*
1947-1948

Blanche Noyes*
1948-1950

Kay Menges Brick*
1950-1951

Alice H. Hammond*
1951-1953

Geraldine Mickelsen*
1953-1955

Edna Gardner Whyte*
1955-1957

Broneta Davis Evans*
1957-1959

Eugenia R. Helse
1959-1961

Louise M. Smith
1961-1963

Ruth Deerman
1963-1965

Alice Roberts
1965-1967

Donna T. Myers
1967-1969

Bernice T. Steadman
1969-1970

Betty W. McNabb
1970-1972

Elizabeth Sewell
1972-1974

Patricia Z. McEwen*
1974-1976

Lois Feigenbaum
1976-1978

Thon Griffith
1978-1980

Janet Green
1980-1982

Marilyn Copeland
1982-1984

Hazel Jones*
1984-1986

Barbara Sestito
1986-1988

Gene Nora Jessen
1988-1990

Marie Christensen
1990-1992

Lu Hollander
1992-1994

Joyce Wells
1994-1996

Lois Erickson
1996-1998

*Indicates Deceased

LIVING CHARTER MEMBERS

Barbara W. *Bancroft*
Phyllis *Fleet* Nelson Crary
Jean *Davidson*
Achsa B. *Peacock* Donnels
Betty *Huyler* Gillies
Margery Louise *Doig* Greenberg
Candis I. *Hall* Hitzig Gullino
Ruth E. *Halliburton*
Alberta B. *Worley* Homan
Mary H. *Goodrich* Jensen
Frances *Ferguson* Leitch Leistikow
Ila *Fox* Loetscher
Ethel *Lovelace*
Olivia "Keet" *Mathews* Maugham
Lillian *Porter* Metcalf
Anges A. *Mills*
Margaret Fzandee *O'Mara* Murphy
Bernice C. *Blake* Perry
Elizabeth F. *Place*
Mathilda J. *Ray*
Madeleine B. *Kelly* Royle
Nancy *Hopkins* Tier
Evelyn Bobbi *Trout*
Josephine C. *Wood* Wallingford
Wilma L. *Walsh*
Margaret *Thomas* Warren
E. Ruth *Webb*
Fay *Gillis* Wells
Bonnie *Chittenden* Whitman

DECEASED CHARTER MEMBERS

Adeline *Fisit* Anderson, 1992
Mary Ellen *Bacon*, 1936
Melba *Gorby* Beard, 1987
Helen V. *Cox* Clohecy Bikle, 1992
Bernice C. *Blake* Perry
Ruth T. *Bridewell*
Margery H. *Brown*, 1961
Myrtle *Brown*, 1934
Vera *Brown*, 1976
Myrtle R. *Caldwell*, 1992
Irene J. *Chassey* Greene, 1991
Marion *Clark* Clendaniel, 1992
Retha *McCulloch* Crittenden, 1993
Joan Faye *Shankle* Davis, 1951
Jane *Dodge*
Amelia *Earhart*, 1937
Sarah S. *Fenno*
Jean *LaRene* Foote, 1960
Viola *Gentry*, 1988
Thelma *Elliot* Giesen, 1991
Lady Mary *Heath*, 1935
Mary C. *Alexander* Held, 1955
Jean Davis *Hoyt*, 1988
Katherine F. *Johnson*, 1967
Thelma R. *Burleigh* Johnston, 1982
Angela L. *Joseph*, 1930
Cecelia Roy *Kenny*, 1980
Cecil W. "Teddy" *Kenyon*, 1985
Ruth *Elder* Camp King, 1977
Florence E. *Klingensmith*, 1933
Opal Logan *Kunz*, 1967
Eva Mae *Lange*, 1963
Dorothea Backenstoe *Leh*, 1955
Geraldine *Grey* Loffredo
Lola L. *Lutz*, 1968
Mildred Chase *MacDonald*, 1982
Margaret Cooper *Perry* Manser, 1951
Sasha *Hall* Martin, 1992
Frances E. *Harrell* Marsalis, 1934
Helen *Manning* Matthews, 1963
Jessie Maude *Miller*, 1973
Sylvia Anthony *Nelson*, 1984
Ruth Rowland *Nichols*, 1960
Mary Webb *Nicholson*, 1943
Blanche *Noyes*, 1981
Gertrude C. *Ruland* Oberlander, 1972
Gladys *O'Donnell*, 1973
Phoebe Fairgrave *Omlie*, 1975
Neva *Paris*, 1930
Peggy J. *Paxon*, 1992
Phyllis M. *Goddard* Penfield, 1984
Thea *Rasche*, 1971
Eleanore B. *Lay* Ross, 1981
Meta *Rothholz*, 1974
Margaret *Willis* Smith, 1971
Hazel Mark *Spanagle*, 1992
Edith *Foltz* Stearns, 1956
Ruth W. *Stewart*, 1932
Mildred *Stinaff*, 1931
Marjorie C. *Stinson*, 1975
Dorothy L. *Stocker*
Louise McPhetridge *Thaden*, 1979
Edwina *McConnell* Thro, 1992
Marjorie May *Lesser* Van Antwerp, 1969
Mary E. *Von Mach*, 1980
Esther *Combes* Vance, 1983
Vera Dawn *Walker*, 1978
Betsy *Kelly* Weeks, 1985
Nora Alma *White*, 1931
Nellie Zabel *Willhite*, 1991
Mildred E. *Kauffman* Workman, 1932

NOTE: Charter Member name is shown in italics

CHARTER MEMBERS Top, L to R: Blanche Noyes, Mary H. Goodrich Jensen, Barbara Bancroft. **Middle, L to R:** Nellie Zabel Willhite, Betty Huyler Gillies, Phyllis Fleet, Nelson Crary. **Bottom, L to R:** Ila Fox Loetscher, Edwyna McConnell Thro, Retha McCulloch Crittenden.

Ninety-Nines History

CHARTER MEMBERS Top, L to R: *Irene J. Chassey Greene, Madeleine B. Kelly Royle, Edith Foltz Stearns.* **Bottom, L to R:** *Thelma Elliot Giesin, Bonnie Chittenden Whittman.*

THE NINETY-NINES 1929-1979

by Gene Nora Jessen
Originally published in 1979 as part of *History of The Ninety-Nines, Inc.*

With special appreciation to Melba Beard, Glenn Buffington, Virginia Thomas, Ruth Rueckert and many other 99s and friends who contributed so generously in researching the early years.

Vivacious French Baroness Raymonde de Laroche had painted and sculpted, acted upon the stage, driven racing cars and made flights in balloons. What was left for the adventuresome woman of 1909?

She tried flying — a Voisin Biplane. She learned to manipulate the unstable and unpredictable machine and became the first woman licensed to fly by the Federation Aeronautic Internationale. When asked about the danger of handling one of those early-type planes subject to engine failure and even structural collapse, she talked of fate and fear.

"Most of us spread the perils of a lifetime over a number of years. Others may pack them into a matter of only a few hours. In any case, whatever is to happen will happen — it may well be that I shall tempt Fate once too often. Who knows? But it is to the air that I have dedicated myself, and I fly always without the slightest fear."

Fate did catch up with her, as it sought out so many early fliers, and she was killed in an airplane crash in 1919.

Other women flew even before there were airplanes to ride aloft. In 1784, before America even had its first president, Madame Thible ascended as a balloon passenger in Lyons, France.

Tiny Broadwick, in 1908 at age 15, became the first person to make a parachute jump, doing so from a hot air balloon. She made close to a thousand jumps from balloons while traveling with a carnival, then later demonstrated parachuting from aircraft to the U.S. Army. Today (1979) she's a spry 85, her lifetime having traversed pre-aircraft to space.

"First" is a slippery word and will be avoided whenever possible here. America's first woman pilot? Well, there are many nuances to first. Bessica Raiche built her own airplane in her living room then flew it on Sept. 16, 1910. She later observed, "The only good thing about the old days was the splendid spirit and unquenchable enthusiasm of the experimenters. I do not say fliers, for so few ever achieved that breathless moment when their dream ship actually rose into the air. My complete flying instruction consisted of a mechanic placing the wheel in my hands and saying: 'Pull it this way to go up and that way to come down.' As a throttle control was unheard of, a group of volunteers held fast to the wings while the motor was driven at top speed. Then they let go and theoretically one flew."

The intrepid lady later became a physician and settled in Southern California.

Blanche Stuart Scott reveled in firsts. She became a trick bicycle rider, then drove an automobile across the United States at a time when there were only 216 miles of paved road in the country. She took up flying with the famous Glenn Curtiss, becoming a member of his exhibition team. Curtiss himself declared her America's first aviatrix on Sept. 6, 1910, though she never did obtain a pilot's license. She later tested planes for Glenn Martin and retired from piloting in 1913.

America's first licensed woman pilot? Harriet Quimby was officially the first with a license obtained from the Moisant School in 1911. Fellow student Mathilde Moisant became number two and later established an altitude record of 1,500 feet. Harriet Quimby traveled to France where she acquired a Bleriot monoplane to attempt a flight across the English Channel. She had no opportunity to flight-test the aircraft and handled a compass for the first time above the fog in the Channel. She returned to America successful and triumphant. Only three months later she was to die in the unstable Bleriot over Boston harbor.

Katherine Stinson learned to fly in 1912, becoming famous for precision acrobatics and skywriting. Subsequently, her younger sister Marjorie learned to fly and taught at the family school in San Antonio. The much-admired Ruth Law was a cohort of Katherine's, becoming an expert exhibition flyer and record holder. She established an American long-distance record flying solo from Chicago to New York, and later carried the mail.

And a black woman, against staggering odds, gained renown in aviation. One of 13 children, she had picked cotton to earn money for school. As a black and as a woman Bessie Coleman found the door locked at flying schools. So she studied French and sailed for Paris to learn to fly. She returned in 1921, the world's first licensed black pilot. "Brave Bessie," as she became known, became a popular attraction on the air show circuit.

For some unknown reason, at an air show in 1926, Bessie had failed to wear her parachute and hadn't strapped herself in. A wrench jammed her controls. As the airplane rolled over, Bessie fell to her death.

The early years exacted a horrendous toll on aviation's pioneers. We can hardly comprehend the machines they called airplanes. But nevertheless, more eager aviators followed and aviation grew up with women pilots competitive and yet supportive of each other.

From the Margery Brown Collection is her helmet, covered with signatures of well-known pilots of the day, and a photo of her wearing it.

Mathilde Moisant and Harriet Quimby.

Eight of the 1929 National Women's Air Derby contestants with "ground escorts" at San Bernardino, the first stop in the historic air race. Front row: Vera Dawn Walker, Louise Thaden, Jessie Maude Keith-Miller, Ruth Elder, Edith Foltz. Back row: Thea Rasche, Margaret Perry, Neva Paris.

By 1929 there were over 100 American women, and numerous women in other countries, licensed to fly. A Women's Air Derby was launched that year in conjunction with the Cleveland Air Races. The rules were straightforward. Whoever got to Cleveland in the least time won!

The entries were divided into two classes, depending upon size of engine. The starting point was Santa Monica, then the route proceeded by way of Santa Monica, San Bernardino, Yuma, Phoenix, Douglas, El Paso, Pecos, Midland, Abilene, Fort Worth, Tulsa, Wichita, Kansas City, East St. Louis, Terre Haute, Cincinnati, Columbus and Cleveland. The most famous names of the day were in the lineup, with Will Rogers on hand at the takeoff. He prophetically remarked that the start looked like a "powder puff derby."

The contest was not without irregularities. Amelia Earhart nosed over on a landing at Yuma, necessitating replacement of her prop (and those who were there stopped for three hours while a new prop was flown in); Margaret Perry dropped out at Fort Worth with typhoid fever; Florence "Pancho" Barnes made an unscheduled stop in Mexico; Blanche Noyes landed in the desert to successfully extinguish a fire, then continued; Ruth Elder landed in a pasture; and, tragically, Marvel Crosson crashed and was killed, probably due to carbon monoxide poisoning.

Despite their obstacles, the majority of the entrants finished. Louise Thaden claimed first in the faster aircraft class with her Travel Air J-5, and Phoebe Omlie was first in the second division.

Despite the competitive nature of these talented women, but probably more because of it, they felt their camaraderie called for a more formalized bond. After all had arrived in Cleveland, Amelia Earhart, Gladys O'Donnell, Ruth Nichols, Blanche Noyes, Phoebe Omlie and Louise Thaden gathered under the grandstand and, at the suggestion of Phoebe Omlie, considered forming some kind of organization just for women pilots. Louise Thaden was supported for president.

Others were thinking along the same lines, and Clara Studer, not a pilot herself but in public relations at Curtiss-Wright, galvanized some of the East Coast pilots into action. An organizational letter went out to all the licensed pilots in the United States signed by Frances Harrell, Neva Paris, Margery Brown and Fay Gillis (Wells). Of the 117 licensed women pilots in the country, 86 responded to the call. Others later regretted they had missed being a charterite, but some were traveling and received the invitational letter too late to respond.

On Nov. 2, 1929, 26 women gathered at Curtiss Airport, Valley Stream, NY. The weather wasn't favorable and most drove in or came by train. The first order of business was selection of Neva Paris as temporary chairman, then the presentation of flowers to Viola Gentry, who was recovering from a crash following an endurance record attempt. The women conducted their business in a hangar above the din of a Curtiss Challenger engine running up as the work of the mechanics proceeded around them. Tea was served from a tool box wagon on wheels.

Club eligibility and purpose were quickly decided upon. Membership would be open to any woman with a pilot's license, and the purpose was "good fellowship, jobs and a central office and files on women in aviation."

Choosing a name was a little harder. Some offerings were The Climbing Vines, Noisy Birdwomen, Homing Pigeons and Gadflies. Amelia Earhart and Jean Davis Hoyt put a stop to the nonsense, proposing the name be taken from the sum total of charter members. Thus the group was momentarily the 86s, then the 97s and finally the

First Air Derby entrants, August 1929, Clover Field, Santa Monica, CA. Louise Thaden, Bobbi Trout, Patty Willis, Marvel Crosson, Blanche Noyes, Vera Dawn Walker, Amelia Earhart, Marjorie Crawford, Ruth Elder and Pancho Barnes. The Gypsy Moth in the background belongs to Thea Rasche. (Crawford and Willis did not start.)

This teapot is the one used at the Nov. 2, 1929, organizational meeting of The 99s.

First National Women's Air Derby, Santa Monica, CA, to Cleveland, OH, Aug. 18 - 26, 1929.

Ninety-Nines. The name/number stopped at 99, but the membership thereafter grew worldwide.

Leadership problems arose since these kinds of ladies were of strong will or they wouldn't have been flying in that day and age. Opal Kunz became acting president until a formal election could be held. However, Neva Paris, who was in charge of the election, was killed while enroute to air races in Florida. Members pulled in opposing directions and officers were not finally selected, leaving the club informally structured. Louise Thaden served as secretary for several years and Blanche Noyes as treasurer, until Amelia Earhart became the group's first elected president in 1931.

Other female pilot groups sprang up also. Five of the '29 Air Derby participants met at the home of Gladys O'Donnell in Long Beach and formed the Skylarks. Opal Kunz, disappointed that the 99s had not gone in the direction of her desire for a tightly-disciplined group dedicated to national defense, formed the Betsy Ross Corps in 1931. The objective was to be of service to the country through humanitarian relief work. Later, at the outbreak of WWII, Opal Kunz once again formed a women's defense program, this time called the Women Fliers of America. The group was devoted to training women pilots to replace men going off to war.

An offshoot of the Betsy Ross Corps came through Florence "Pancho" Barnes, who envisioned an unofficial division of the U.S. Army Air Corps. The Women's Air Reserve evolved under Pancho's strong leadership and was quite successful. The Reserve served in the disastrous earthquake, and later the Montrose Flood. They were undoubtedly ahead of their time, and aspects of their service later could be seen in the WASPs and the Civil Air Patrol.

But what of the women, the 99? Each was an outstanding person in her own right. They made contributions to aviation totally out of proportion to their numbers. Now, 50 years later (1979), many are still actively contributing and some are still current pilots. We "new" pilots take huge pride in the heritage with which they've endowed us.

Of the four who signed the original letter of invitation to form a women pilot's organization, Fay Gillis Wells has continued a lifetime involvement in aviation and service to her beloved 99s. Fay Gillis just happened to be living in Russia when her friend Wiley Post decided to fly solo around the world. Fay supervised his refueling in Siberia, no

The 1929 organizational meeting of The 99s found 26 women gathered in a hangar in Valley Stream, Long Island. Front row: Viola Gentry, Teddy Kenyon, Wilma Walsh, Frances Harrell Marsalis, Meta Rothholz. Back row: Neva Paris, Mary Alexander, Betty Huyler, Opal Logan Kunz, Jean Davis Hoyt, Jessie Keith-Miller, Amelia Earhart, Marjorie May Lesser, Sylvia A. Nelson, unidentified, Dorothea Leh, Margaret F. O'Mara, Margery Brown, Mary Goodrich, Irene Chassey, "Keet" Mathews, E. Ruth Webb and Fay Gillis.

Two of the items on display at International Headquarters are the license granted to Amelia Earhart, first president of The 99s, by the Federation Aeronautique Internationale (left) and her president's pin (above).

mean feat, contributing to his world record. Post later invited Fay to accompany him on another record attempt, and Fay was forced to decide whether to fly with Wiley Post or honeymoon with Linton Wells in Africa. The world-famous humorist Will Rogers took Fay's place on the flight, and he and Post were killed in Alaska.

On another occasion Fay Gillis Wells proved that hers was a charmed life. The day after her first solo, Fay was flying an experimental airplane with her instructor. It seems the craft was overpowered and they literally tore it apart. Both fell out, and some 400 feet above the ground Fay figured out how to pull her parachute's ripcord for a safe landing. Thus Fay qualified for membership in the Caterpillar Club, open only to those who have bailed out of an airplane to save their lives.

Fay's primary vocation through the years has been journalism, and we have watched her on television at presidential press conferences where she served Storer Broadcasting as a White House correspondent. She was one of three women to accompany President Richard M. Nixon on his historic visit to China. Fay's dynamic and creative projects for The Ninety-Nines culminated in her SEE THE USA "Friendship Through Flying," and the International Forest of Friendship. The Forest was The Ninety-Nines' project for America's Bicentennial celebration, commemorating our organization's international scope, and is located at the birthplace of Amelia Earhart in Atchison, Kansas.

Frances Harrell Marsalis was another signer of the invitational letter. She had invested a small legacy in learning to fly at the Curtiss School then stayed on to work for Curtiss Flying Service. She was a bubbly, popular flier, and teamed up with Louise Thaden on an endurance flight. The two spent eight days, four hours and five minutes airborne in a Curtiss Thrush, engraving their names in the record books.

The racers were the early women of the hour — those marvelous women and their flying machines. Amelia Earhart was to become the most famous woman pilot of all time, but in 1929 she was but one of a dozen glamorous, daring female aviators.

Amelia had flown the Atlantic as a passenger, gaining fame and adulation. In 1932, she realized her dream of crossing the Atlantic alone, for which she reaped international honors, and other record flights followed. Amelia was a strong advocate of awakening women's potential. She encouraged young girls to dream big, and said about women and aviation: "The more women fly, the more who become pilots, the quicker will we be recognized as an important factor in aviation."

Ninety-Nines who knew her remember her as a warm and feminine person and a catalyst for achievement. Her parting words to Louise Thaden were, "If I should bop off, it'll be doing the thing that I've always most wanted to do."

Another sparkling star of the day was Louise Thaden, who had convinced Walter Beech that she

> "If I should bop off, it'll be doing the thing that I've always most wanted to do."

Fay Gillis Wells

Ruth Elder takes the physical exam for her pilot's license.

Louise Thaden meets the press after being presented with the Bendix Trophy.

should help him gain recognition for his airplanes. She promptly gained an altitude record, an endurance record and then speed record in the Beech Travel Air. More records fell to this whirlwind, then in 1936 the all-male Bendix Trophy Race was opened to women. Louise, along with Blanche Noyes, flew to first place in a Staggerwing Beech. That year Louise Thaden was awarded the Harmon Trophy as the world's outstanding flyer.

Though she doesn't appear in our records as a former president, she served in all but title during the first two years and remained all her life a stalwart supporter of The 99s and all women in aviation.

The late 1920s were aviation's adolescence; a time to prove oneself and shout to the world, "Here I am!" Air races, endurance flights, altitude and speed records were the challenges. Engine failures and off-airport landings were expected. Aviators were colorful and adored, and Ruth Elder was a heroine.

Five months after Lindbergh's epic flight, Ruth Elder and George Haldeman took off for Paris in a Stinson monoplane named "The American Girl." Encountering storms over the Atlantic, they flew for 28 hours and made it to within 360 miles of the Azores when an oil leak forced them to land in the water. The pair was rescued by a Dutch tanker and a tumultuous welcome awaited them in Paris and then back home. The beautiful aviatrix went on to a successful Hollywood acting career.

The Ninety-Nines has always been international. Jessie Keith Miller, an Australian, competed in the '29 Women's Air Derby and attended the first 99s organizational meeting on Long Island. A German, charter 99 Thea Rasche, was for several years the only woman in her country with a pilot's license and a plane of her own.

Quotable Thea offers the sage advice, "Flying is more thrilling than love for a man, and far less dangerous."

Another German woman, 99 Hanna Reitsch, is recognized even today as the world's first and foremost female test pilot. The tiny 88-pound dynamo started out in gliders but conquered everything she touched. The firsts boggle — first woman to fly jet planes, rocket planes and helicopters, the first pilot to fly a glider over the Alps, and the only woman to fly a robot V-1, commonly known as a buzz bomb, modified for pilot control. During WWII, Hanna tested all types of military planes for the Luftwaffe.

An intriguing chapter of her life concerns the

Test pilot Hanna Reitsch.

final days of the war. It has been reported that Hanna Reitsch flew Hitler to safety in a helicopter and that he fled to South America in a submarine, but she says that is not true. She has described her last flight over Berlin and meeting the demented Fuerher in his bunker. On the last day of Hitler's life, Hanna escaped the bunker in a tank amidst falling buildings and thundering guns. Into the 1970s, this remarkable woman remained in the top ranks of glider and helicopter pilots.

Other international 99s have gained worldwide fame. Nancy Bird Walton barnstormed Australia, then operated a charter and aerial ambulance service in Queensland. Lady Marie Casey contributed through the years to Australian aviation. Nancy Ellis Leebold built up time and experience ferrying war surplus aircraft across Australia after WWII. Then she made a 12,000-mile flight from Britain to Australia in a single-engine Miles Messenger. The 145-hp aircraft had only one four-channel, short-range VHF transceiver, so the flight was accomplished by dead reckoning with a war surplus magnetic compass. Nancy's forté was aeronautical engineering, and she worked for Rolls Royce on the Dart turbo-prop and Avon jet engines.

Brazil produced Ada Rogato, who in 1951 flew her 90-hp Cessna 25,000 miles solo across the

Nancy Bird Walton, 1933 (left), and 1996 (right).

Andes, to Alaska and across Canada and the United States. The Dean of South America's flyers, Anesia Pinheiro Machado, also made a grand tour in 1951. On a goodwill flight from New York to Rio de Janeiro, she visited Mexico and all of Central and South America. The much-honored pilot had learned to fly in 1922 in a French Caudron C-3 with a rotary engine burning castor oil. The aircraft had no brakes, idle cutoff or ailerons. In this primitive craft, she made her first cross-country from Sao Paulo to Rio following the railroad tracks, since there were no charts. As a stunt pilot, Anesia flew a German

Bucker Jungman. Mrs. Machado was a pioneer pilot of Brazil, and has been recognized by dozens of governments and aviation organizations.

Britain produced daring and colorful women pilots in the early years, exemplified by Amy Johnson Mollison, who flew solo from Britain to Australia in 1930 in an open cockpit biplane, taking 19 days for the task. In 1966 Sheila Scott became Britain's first pilot to fly around the world solo, doing so in a Piper Comanche 260. Sheila had been an actress before her addiction to flight. After her round-the-world flight, Sheila went on to other record-breaking dashes. She set a number of new records flying from London to Cape Town. For these flights she was awarded the Britannia Trophy, the highest award of the Royal Aero Club of the United Kingdom.

Sheila had been preceded by a colorful Irish lass, Lady Mary Heath, a charter 99. She had soloed her Avro Avian on the Cape Town to London route, a treacherous flight in 1928. Then later she flew 70,000 miles with KLM as second pilot in Fokker Tri-motors.

Lady Mary Heath fought the bureaucratic battle in Britain for women who followed. She had passed her "A" license without difficulty, but when it came to the "B," permitting the carriage of passengers, the English Air Ministry was skeptical about entrusting passengers' lives to a woman. Finally, a committee studied the question of physical tests for women pilots, and the athletic Lady Heath accomplished the breakthrough.

American socialite Ruth Nichols became a flying addict flying dirigibles, gliders, autogyros, seaplanes, amphibians and four-engine aircraft. She held three different world records at one time and an early transport license. Not content with transcontinental record flights, Ruth aspired, along with Laura Ingalls and Amelia Earhart, to solo the Atlantic. Amelia got off first as Ruth Nichols wrecked her plane in New Brunswick and was seriously injured. Ruth, along with two partners, organized flying clubs across the country called Aviation Country Clubs. She promoted this venture doing a 12,000-mile air tour. Later, as thoughts of war couldn't be avoided, Ruth became convinced that airplanes could play a prominent role as air ambulances. Ruth founded Relief Wings, whose purpose was a humanitarian air service in case of either civilian disasters or war. After the outbreak

Sheila Scott was England's first woman pilot to fly solo around the world in 1966.

From the Manila Davis Talley Collection at International Headquarters: her ground school text.

Anesia Pinheiro Machado, pioneer pilot from Brazil.

Melba Gorby Beard after her solo, Long Beach, 1929.

of WWII, Ruth turned the basic structure over to the Civil Air Patrol.

Another woman who involved herself in the war effort through aviation was Phoebe Omlie. Her entré to aviation had been as a parachute jumper with flying circuses in 1921. She soon bought a Curtiss Jenny, was taught to fly by Vernon Omlie, married her instructor and set out on a honeymoon exhibition tour of the country. Phoebe obtained the first (female) transport license, No. 199, and also earned an aircraft and engine license.

At the outbreak of WWII the Tennessee Bureau of Aeronautics decided to aid the war effort by training a select group of women as flight instructors, replacing the men gone to war. They borrowed former Tennessean Phoebe Omlie from the CAA to set up and supervise the program. Housing and staffing were the immediate problems. Phoebe found an old colonial home "with a fireplace in every room, dirt in every closet and not much else." Phoebe rolled up her sleeves, cleaned house, midnight-requisitioned supplies, and found faculty. From 235 applicants, 10 girls were chosen, trained and graduated to teach flying. Phoebe Omlie was never accused of being just an onlooker.

Bobbi Trout was another racer and charter 99.

Betty Gillies

In the early years all the pilots knew each other and swapped stories and advice. The women pilots were close, but the California girls had a special bond. Bobbi was (and is) a Californian. After learning to fly in 1928, Bobbi got a job demonstrating the Golden Eagle Monoplane, then made the first solo endurance flight for women. Later flights with Elinor Smith Sullivan and Edna Mae Cooper were once again endurance refueling flights, the latter going to 122 hours until the engine gave up. Bobbi tells of the refueling ship dumping fuel on her head and of dropping reusable cans overboard in tiny parachutes which became entangled in the tail. Those flights were anything but smooth. Bobbi always had an inventive bent, and during the war she realized that thousands of dollars worth of rivets were being discarded so she promptly formed the Aero Reclaiming Company. Today she is still an inventor, and enjoys prospecting too.

Vera Dawn Walker and Gladys O'Donnell were fellow Californians. Vera Dawn was working in the movies as a stand-in on a Tom Mix picture when all the company went for airplane rides. Vera Dawn didn't want to go but Tom Mix dared her so she did and fell in love with flying. The tiny girl got a job putting time on the Panther McClatchie powerplant. Vera Dawn commented, "I became the unofficial forced-landing champion of the world!"

After an adventuresome flight to Guatemala, tuberculosis forced her retirement from aviation. Even in fighting for her health Vera Dawn Walker showed the adventurous spirit of an aviator. Not content to rest out her illness in a sanitorium, the lady set up a tent south of Tucson and started placer mining! Her four-year recovery period brought in substantial gold.

Gladys O'Donnell started out in business. Since her husband was a pilot, she learned to fly and they started a flying school. Gladys competed in the 1929 Women's Air Derby with a grand total of 46 hours in her log book (strictly contrary to the rules), taking a second place only to Louise Thaden, an amazing achievement. The next year she won. And four events at the National Air Races, too.

The Californians gravitated to Jim and Clema Granger's flight school in Santa Monica where Hoot Gibson, Wallace Beery and Ruth Elder learned to fly. Pancho Barnes was involved with the movies too, flying in early aviation pictures. Pancho could only be called a character; a staunch supporter of a person's individual rights and freedoms, generous to a fault, certainly an activist. When WWII shut

down flying on the West Coast, Pancho purchased considerable acreage on the Mojave Desert near the future Edwards Air Force Base. She established a guest ranch with a landing strip, accommodations, working ranch and stables (Pancho herself was an accomplished equestrian), provocatively called Happy Bottom Riding Club. Sadly, Pancho's later years were spent doing battle with the air force base, leading her to study law to better defend her maze of legal entanglements.

Melba Gorby Beard was in the thick of the California activities, having learned to fly in a wooden International biplane with a 90-hp OX5 engine, just in time to be a charter 99. She had quit her art studies to go to work so that she could earn enough to pay for a half hour's dual a week. After achieving her transport license, Melba started instructing though she was not yet licensed to drive a car.

Melba met her husband Bill when he showed up at the flight school with his Bird biplane, seeking some dual in spins. The Bird became Melba's as a wedding present, and both the airplane and the marriage seemed to take well. Forty-five years later, same husband, still flying a Bird. Melba has had six Birds since the original one, using the surplus to cannibalize for parts. She is a mechanic and still does all the work on the '28 Bird herself.

Melba competed and conquered in the early air races, served the 99s in office, and enthusiastically supported all 99 activities. Fortunately for the 99s, Melba is a saver and has accumulated priceless records and photographs of the early years of aviation.

The East Coast 99s were every bit as active as the westerners. After all, the organization had officially started on Long Island, and Valley Stream remained a focus for the women pilots.

Betty Huyler was the one who was drawn to aviation partly because of an article she read by Amelia Earhart, but then just as much because she was in love with a naval aviator and wanted to enter his world. Soon, her nurse's training fell by the wayside, and she flew professionally under her new name of Gillies.

Betty started in sales for Curtiss, then came association with other aviation companies and three babies along the way. For the years 1939-1941 this charter member served the 99s as president. Under her direction, along with Alma Harwood, the Amelia Earhart Memorial Scholarship Fund was established, and The 99s waged a long, hard battle with the CAA over regulations regarding pregnancy. It seemed that pregnancy was considered an illness and regaining a license after "recovery" entailed both rewriting and reflying the examinations. The regulations were successfully modified.

In 1942 Betty became one of the original group of 25 women forming the WAFS (Women's Auxiliary Ferrying Squadron) organized by Nancy Harkness Love. They ferried aircraft for the Army Air Corps within the continental limits of the United States. Betty became WAFS commander at New Castle Army Air Base; the organization later became the WASPs. During the war years, Betty ferried fighters and bombers, transports, cargo and utility aircraft.

In 1945, the family moved to California and Betty flew for the Ryan Aeronautical Co., giving instrument instruction to their test pilots. She checked out in the Ryan Fireball, a Navy fighter with propeller engine in front and jet engine in the rear, in anticipation of becoming a production test pilot. VJ Day curtailed production and test flying for Betty. She utilized her flying and management skills for the next nine years directing the Powder Puff Derby, and is an active pilot today (1979).

Teddy Kenyon and Nancy Hopkins Tier are two other charter members who have remained

Phoebe Omlie acquaints participants with the facilities at the Tennessee Research Instructor School for Women.

active pilots these 50 years. Teddy, while going to school in Boston, met aviation enthusiast Ted Kenyon studying engineering at MIT. That was the beginning of an enduring flying team. Teddy soloed in an Arrow Sport, then in 1933 won the title National Champion Non-professional Woman Pilot. During the years she assisted her husband with the instruments he designed for the Sperry Gyroscope Co.

Nancy Hopkins Tier attended that first meeting, too. She had learned to fly across the Potomac from Washington, D.C., with her progress chronicled by the famous Ernie Pyle. With 16 hours solo and her private license in hand, Nancy applied for a job in the operations at Roosevelt Field on Long Island and got it. She lived on Campbell's Soup to save money for more training.

Robert Gross, the future president of Lockheed, hired Nancy to demonstrate the new Kittyhawk airplane. She entered the sixth Ford Reliability Tour with it, crossing Canada, flying down the Rockies then back to Detroit. She flies a mint Cessna 170 today (1979) and was named New England Section's Woman of the Year in 1976.

Mildred H. Chase MacDonald flew down to Valley Stream from Boston for the 99 organizational meeting, and explained why she wasn't in the famous picture of those pioneers. It seems that the night before, at a party at Opal Kunz' house, she had met Jack Donaldson, the fourth-ranking ace of WWI. The next day, during a lull in the meeting, Donaldson invited Mildred to go for a spin in Opal's plane. They wrung it out and Mildred missed the photograph. Mildred's instructor was August Pabst of the Milwaukee brewing family, a stern taskmaster. Mildred did all her own engine work.

The name Fleet peppers any account of early flying, for Major R. H. Fleet was the founder of Consolidated Aircraft, whose planes bore his name. His daughter Phyllis was a charter 99, though she never attended a meeting.

Phyllis Fleet Crary's introduction to flying was by Barney Oldfield in Texas, who showed her every known stunt, and Phyllis was hooked. She returned to Buffalo, where the Fleet airplanes were made, and learned to fly. At her father's urging, Phyllis studied engineering at Cornell. Phyllis said the reason she didn't attend any 99 meetings from Buffalo was that they were in New York City, and in the age of chaperones, "My mother didn't want me running around New York unchaperoned." It was to be 36 years before she met other chapter 99s.

Viola Gentry learned to fly from Curtiss Field, but had quite a time getting to the 99 organizational meeting in November 1929. It seems that she had been attempting to establish a refueling endurance flight record with Jack Ashcraft when ground fog rolled in around 2 a.m. By 6 a.m., the two were out of fuel and had to make a forced landing through the fog. They hit a big tree across from Curtiss Field, with Jack being killed instantly, and Viola spending 22 months in the hospital. When the invitation came for the meeting at Valley Stream, Viola's doctor said she could go, encased in a body cast, if she were accompanied by a nurse. They were flown to the meeting, the nurse becoming thoroughly airsick.

Viola was born in 1900 but says she isn't that old, and those who know her believe it.

Annette Gipson Way was a compadre of Viola's and an early 99. An air racing enthusiast, she started the Annette Gipson All Women Air Races held annually for several years. She loved pylon racing and once snap-rolled under the Brooklyn Bridge. Performing over Macy's Thanksgiving Day Parade in New York almost led to disaster, as her airplane became entangled with a floating animal balloon. Annette collected over 100 trophies for her flying, and later ran the Fort Lauderdale Airport.

Charter member Nancy Hopkins Tier flies First Day Covers from Atchison, KS, to Hartford, CT, and Teterboro, NJ, on July 24, 1976.

There is country between the coasts, and there were women flying inland, too. A woman who flew commercially for over 60 years is Edna Gardner Whyte. Somehow she missed being a charter 99, but she has missed little else in aviation. A disproportionate number of pilots seem to have come from nursing's ranks, and that is Edna Whyte's background, too. Even today her pride in nursing shows as she signs her letters "R.N."

Edna started flying in 1926, and logged more than 30,000 hours. She has taught 6,000 people to fly, and her greatest pleasure today is greeting airline pilots who come back to see their first instructor. Edna tells the story of the student who said she had taught his instructor's instructor and asked what relation that made them. Edna told him he was her great-grand pilot!

Edna's passion always has been competition, and she has over 100 trophies attesting to her piloting skills. Prior to the war she ran the New Orleans Air College, later owned and operated Aero-Valley Airport near Fort Worth, still instructing daily. She served The Ninety-Nines as president in the mid-50s and continues to take great interest in job breakthroughs for women pilots.

Ohio always has been a strong aviation state, and provided well for The Ninety-Nines. Cleveland's Blanche Wilcox was a leading lady in the theater and movies, a career she abandoned to marry airmail pilot Dewey Noyes. Dewey taught her to fly in December 1928, making her the first woman pilot in Ohio. Her participation in the 1929 Women's Air Derby already has been described. In 1931 Blanche flew a 300-hp Pitcairn autogyro for Standard Oil Co. of Ohio. It was the forerunner of the helicopter.

In 1935, 99s Helen Richey, Louise Thaden, Helen MacCloskey, Nancy Love and Blanche Noyes were appointed as airmarking specialists for the CAA. Their assignment was to produce a marking sign every 15 miles along a given route. Through attrition the group diminished to one, Blanche, and the job was reversed during the war; signs were to come down to thwart any enemy aircraft. Civilian aviation came back, and so did airmarking. Blanche headed up airmarking for 35 years for the CAA and FAA, providing for over 75,000 markers across the country. Blanche Noyes served The 99s as president in the late 1940s.

Blanche had another kid to fly with on the

Teresa James soloed in 1933 and flew as a WASP. She also flew a commemorative mail flight between Wilkinsburg Airport to Allegheny County Airport in May 1938.

Cleveland Airport, 16-year-old Abby Dill (Haddaway), who learned to fly in 1929 in an American Eagle. Shortly after starting her flying, the youngster had a speech assignment in her English class. The teacher pointed out that if you knew your subject, you'd make a better speech. So, naturally, Abbie made a parachute jump the next Thursday night so she could make a speech about it in English class on Friday. She later became the CAA's first woman air traffic controller, in the Southwest Region.

Arlene Davis learned to fly when her husband ordered an airplane, then could hardly keep her out of it. Once Arlene was qualified to fly four-engine airplanes, she was said to have all the possible ratings to fly anything on land or water. A woman of means, Arlene's philanthropy became well known and she was ever ready to give a boost to Scouts, college flyers (the Arlene Davis Memorial Trophy is still presented annually at the National Intercollegiate Flying Meet today), and provided aeronautical engineering scholarships to women.

Arlene Davis became interested in racing, finishing fifth in the 1939 Bendix race and was, in fact, the only woman to finish that year. She taught instrument flying to Army cadets during the war.

Ohio even had a woman airport manager. Lauretta Schimmoler learned to fly in Akron in 1930, then developed and managed the first airport at Bucyrus, Ohio, serving 99s as Section Governor at the same time. Lauretta founded the Aerial Nurse Corps in Cleveland, which was the forerunner of the flight nurses. After moving to California, she helped make the motion picture "Parachute Nurse" for Columbia. She also founded the Amelia Earhart Post of the American Legion at Glendale, being its first commander. Though Lauretta has retired in California, much of her memorabilia has returned to Ohio and is housed at the Bucyrus Historical Society.

The Northwest had its pioneers, too. Edith Foltz Stearns may have been born a Texan, but she learned to fly in Portland, OR. Her husband was president of Oregon Airways, for which Edith directed public relations. A charter 99, Edith served as her section's governor. Prior to WWII, she taught primary CPT at Salem, then served in England as a first officer in the Air Transport Auxiliary.

Mary (Mae) Haizlip, winner of the 1932 World's Speed Record for Women, is presented her trophy by Maj. James Doolittle, left. With them is James Wedell, who designed the aircraft Haizlip flew.

Fashionably-dressed early-day air markers include Blanche Noyes, Helen Richey, Helen MacCloskey and Louise Thaden.

Edith designed a clever flying outfit, trademarked "Folzup" flying togs. The skirt pulled up and became a jacket, then knickers were worn beneath for climbing into the open cockpits.

Another Northwesterner, Gladys Buroker, started flying in 1932 near Bellingham, WA. Soon both she and her husband were teaching students at Port Angeles and Olympia, and Gladys got into gliding at the Snohomish Airport. When war broke out, all flying was shut down on the West Coast, so the Burokers moved their Cubs to Coeur d'Alene, Idaho.

Gladys herded her students over in the airplanes without incident, but her husband Herb got into a little more weather than he had bargained for. Crossing the Cascades, he picked up carburetor ice and sank down into the fog below. As power decreased, he spiraled down until he saw a big pine limb go by, pulled all the way back on the stick to stall the airplane, and hooked one wheel over a tree limb. He hung in the tree 40 feet above the ground with nary a scratch and shinnied down the tree by the aileron cables. Ah, flying was ADVENTURE.

Gladys had a few unusual experiences herself. In later years she became intrigued with balloons. On the day of her checkride, she had been out flying in airplanes and was conscious of the wind being out of the north. The flight examiner was sure it would switch at altitude and Gladys was sure he was wrong, but she wasn't about to argue with an examiner. They took off and, sure enough, they picked up a north wind and drifted south over Coeur d'Alene Lake.

Since dark was coming on, they descended toward a beach and yelled at some boaters to tow them in, getting neither the envelope nor gondola (let alone feet) wet. The boater turned out to be a student of Gladys' from years before, and while her ground crew was still out searching for the elusive aeronauts, Gladys and her student and examiner toasted her new license.

A melancholy chapter of the early years was that of Laura Ingalls. Another Roosevelt Field, Long Island, pilot (and another former nurse), she was the darling of aviation. In 1920, she broke both men's and women's records for the barrel roll, doing 714. She made 980 consecutive loops. She was the first woman to make an aerial circuit of South and Central America, a 17,000-mile flight, and the first woman to fly across the Andes. In 1934, she won the Harmon Trophy. Two years later, she won second place in the Bendix race.

Then in 1939 Laura Ingalls bombed the Capitol in Washington, D.C., with anti-war pamphlets. In her defense, few pilots were aware of restrictions on where they could fly, and Laura maintained that she had done it out of love for her country. But in February 1942, she was convicted of being the unregistered representative of German agents from whom she was receiving pay. She was released from prison in 1943. Pilot friends were greatly saddened by her misguided involvement.

War clouds were gathering and the first to respond with a call for pilots was England. The British were desperate for civilian pilots to ferry aircraft from factories to aerodromes to free the military pilots to fly combat. The Air Transport Auxiliary was formed, soon cautiously utilizing eight highly experienced women pilots on a limited basis. As the needs grew, so did the service, with additional women soon flying all 120 types of aircraft flown by the RAF. Hazards were very real. All aircraft observed radio silence so en route weather was not available; weather minimums were 800-foot ceilings and a mile visibility. Dodging artillery ranges, barrage balloons and training aircraft made navigation circuitous. The country's famed pilot, Amy Johnson Mollison, was killed flying for the ATA. She was seen parachuting through the clouds, but landed in water. Her body was never recovered, and her husband was convinced that she had been shot down.

Male and female pilots from other countries flew for the ATA, and Jacqueline Cochran recruited approximately 25 American girls to serve. One of the first was Helen Richey, who later resigned, then joined the WASP.

Nancy Miller (Livingston) from California flew from July 1942 - July 1945. Nancy flew 50 different types of British aircraft across England, into Europe, Scotland and Ireland.

Each woman flew as captain with no co-pilot

Edna Gardner Whyte with a Swallow — 1927.

Nancy Harkness Love, first woman to fly for the U.S. Air Force in 1952, founded the original Women's Auxiliary Ferry Service and commanded the WAFS during WWII.

A photo of a young Jacqueline Cochran form the Gerry Mickelsen Collection in the 99s Resource Center adds to the store of information about the famous pilot.

since there just weren't any extra pilots, the exception being the presence of a flight engineer in a twin when the pilot couldn't reach the emergency gear. After the war, Nancy married and she and her husband Arlo ran a helicopter service for a number of years in Juneau, AK.

During the '30s, aviation sponsored by governments was growing all over the world. As many as 65,000 German men were being trained as pilots and mechanics with nearly 200,000 in glider training. Japan had 51,000 pilots. Russia was training 600,000 in aeronautics. The United States had 23,000 civilian pilots, the majority of them rank amateurs.

On Dec. 27, 1938, President Roosevelt authorized the CAA to train 20,000 private pilots a year in the nation's colleges. The program was called the Civilian Pilot Training program, and its purpose was twofold: to stimulate aviation's growth and to build up a reserve of pilots to which the military might turn. The CPT program was a **huge** success, and many women pilots learned to fly under the auspices of CPT, or taught in the program.

During this time, two prominent women aviators, Jacqueline Cochran and Nancy Harkness Love, had recognized the role that women could assume in time of war, and each pursued independently the military use of civilian women pilots. On Sept. 10, 1942, the Air Transport Command announced a plan to utilize women pilots would be called WAFS, Women's Auxiliary Ferrying Squadron. Nancy Love was appointed squadron commander.

The first group of 25 went to Wilmington, DE, and consisted of experienced pilots who would need only to transition into the military airplanes they would be flying. Betty Gillies became the first WAFS member; Cornelia Fort its second.

Cornelia had been a flight instructor teaching a student in Honolulu on Dec. 7, 1941. She described what happened.

"Coming in just before the last landing, I looked casually around and saw a military plane coming directly toward me. I jerked the controls away from my student and jammed the throttle wide open to pull above the oncoming plane. He passed so close under us that our celluloid windows rattled violently, and I looked down to see what kind of plane it was.

"The painted red balls on the tops of the wings shone brightly in the sun. I looked again with complete and utter disbelief. Honolulu was familiar with the emblem of the Rising Sun on passenger ships but not on airplanes.

"I looked quickly at Pearl Harbor, and my spine tingled when I saw billowing black smoke. Still, I thought hollowly it might be some kind of coincidence or maneuvers, it might be, it must be. For surely, dear God . . .

"Then I looked way up and saw the formations of silver bombers riding in. Something detached itself from an airplane and down, down and even with knowledge pounding in my mind, my heart turned convulsively when the bomb exploded in the middle of the harbor. I knew the air was not the place for my little baby airplane, and I set about landing as quickly as ever I could. A few seconds later a shadow passed over me and simultaneously bullets all around me.

"Suddenly that little wedge of sky above Hickam Field and Pearl Harbor was the busiest, fullest piece of sky I ever saw.

"We counted anxiously as our little civilian planes came flying home to roost. Two never came back. They were washed ashore weeks later on the windward side of the island, bullet-riddled. Not a pretty way for the brave little yellow Cubs and their pilots to go down to death."

As a WAFS pilot, Cornelia Fort was killed in a mid-air collision in Texas while ferrying a bomber.

As a nucleus of experienced pilots started flying successfully, Jacqueline Cochran's idea of training lower-time pilots and introducing them into the WAFS was adopted. As young women graduated from the training program, they would join the ranks of the Ferry Command. All had civil service status and a salary of $250 per month ($50 less than the male pilots in the Air Transport Command) with a $6 per diem on ferrying trips. Miss Cochran's training program commenced in Houston and was designated the Women's Flying Training Detachment or WFTD (they called themselves "Woffteddies").

Four classes went through training, contending with such unmilitary problems as housing (they lived in boarding houses all over town), transportation (they called the buses "cattle trucks"), no mess hall (which often meant just one meal a day), and clothing (that was no problem — they each wore just what they happened to have). They had come to Houston to fly for their country and they worked hard.

Halfway through the fourth training class, the whole operation was moved to Sweetwater, TX. The WAFS and WFTD were merged into one organization called WASP, Women's Airforce Service Pilots. Miss Cochran was named Director

Nancy Miller Livingston ferries a Barracuda torpedo bomber while attached to the British Air Transport Auxiliary.

WASP Dora (Dougherty) Strother climbs out of the cockpit of an A-25 Helldiver she used to tow targets for anti-aircraft practice in 1944.

Dr. Dora (Dougherty) Strother is congratulated by Bell Helicopter's chief production pilot and the company's president just after breaking the women's record for rotorcraft altitude.

Jacqueline Cochran, here with Gen. Chuck Yeager, established more world records than any other pilot. Her last flying opportunity was at a Paris air show in 1971; she was grounded with a heart condition and flew to a new horizon Aug. 9, 1980.

of Women Pilots. Mrs. Love continued as WASP Executive of the Ferrying division of the ATC.

Dependence upon the WASPs grew, and they flew everything the U.S. built. By 1944, WASPs were the only ferry pilots flying the P-47 Thunderbolt out of the Republic factory.

Little people such as Betty Gillies did have some problems which they worked out on their own. As she transitioned out of military aircraft and into the fighters, Betty had some wooden blocks made so she could reach the rudder pedals. She had a set for the AT-6, then for the P-47. "I had a set for the Lockheed P-38, which was really necessary, because if I had put enough cushions behind me to reach the rudder, my nose would smash right into the gun panel."

Sometimes the women were used to demonstrate the safety of aircraft. After all, if a woman can fly it, anybody.... Nancy Love and Betty Gillies checked out as first pilot on Boeing's Flying Fortress, the B-17. They were cleared to deliver one to Scotland. As they were awaiting clearance at the end of the runway at Goose Bay, Labrador, for the final leg, Gen. Arnold stopped the flight and restricted the WASPs thereafter to domestic service.

In a similar demonstration, WASPs Dorothea Moorman and Dora Dougherty checked out in the B-29 Superfortress to demonstrate it to reluctant male pilots. Lt. Col. Paul Tibbets Jr., later famed pilot of the "Enola Gay" atomic bomb flight, checked them out.

They flew the maligned "beast" to Alamogordo, NM, to demonstrate it to flight crews there. Dora reported that "Flight crews, their male egos challenged, approached the aircraft with renewed enthusiasm."

Was the WASP program successful? By any measure it was. Safety and reliability were tops, and the purpose of freeing male pilots for other flying was surely accomplished. Civilians Barbara Erickson (London) and Nancy Love received the Air Medal, and Jacqueline Cochran received the Distinguished Service Medal.

Whatever happened to these women after the war? Many left aviation, as did their male counterparts, since the pilot market was saturated. Others became aviation's leaders.

Dora Dougherty Strother picked up an ATP and Ph.D. and became chief of the human factors engineering group at Bell Helicopter Co. Her records and honors could fill the rest of this book.

Jacqueline Cochran's credentials and fame had been established solidly before the war. Yet she went on to fly a Northrop T-38 jet and break every speed, altitude and distance record for women. In 1962, she established over 30 speed records in a Lockheed Jetstar. Later she flew 1,429 mph in a Lockheed F-104G Starfighter.

The astonishing thing about Jackie Cochran's achievements has to do with her origins. Orphaned at an early age, the little girl's schooling ended at the third grade. She had worked in a beauty parlor and then founded her own successful cosmetics business. She learned to fly at Roosevelt Field, Long Island, and fell to the competition fever. She won the Bendix Race, set innumerable speed records and has been a multiple Harmon Trophy winner. She served the 99s as president.

The war years ended aviation's youth. Postwar meant growing up, with all the attendant growing pains. The Ninety-Nines moved on.

ACTIVITIES

by Page Shamburger

"The War," they called it, and thought World War II was the war to end all wars, was over; Ninety-Nines got back their airplanes and could fly again on civilian wings.

And fly they did. In 1947, Jeannette Lempke Sovereign was president of The Ninety-Nines, and down in Florida came a dream. Why not an All-Women Air Show? So the Florida Chapter of the Southeast Section put their shoulders to the wheel and developed, put on, flew in and promoted the First All-Women Air Show in March 1947. Scheduled for two days, March 15-16, they were faced with the same thing that has plagued airshows always — weather. The Air Show became a one-day event.

Likely the major spectacular of the show was the opening bell. Carolyn West and Bea Medes flew an Ercoupe in from Palm Springs. That took 21 hours and 45 minutes, a near record for the time.

It was reported that over 13,000 spectators watched as Marge Hurlburt set a new international women's speed record of 337 mph in her clip-wing Corsair. Kaddy Landry, a former WASP, took first place in the aerobatic competition flying a Stearman, and second place in the Military Pilots' Association 15-mile closed-circuit race.

Second to Kaddy in the aerobatic competition was a youngster, then a new name to aviation, Betty Skelton, who also won that first free-for-all closed-circuit race. Jane Page of Chicago won the Military Pilots' Association race, and Ruth Hubert placed third in that race and in aerobatics.

That air show's 65-hp Handicap Race was won by Carolyn Cullen of Pittsfield, MA, and Anne Shields placed second.

Caro Bayley thrilled the crowd with sailplane flying; Jean Broadhead surprised them by flying in a 1910 Curtiss Pusher; Duke Caldwell made them laugh when she flew her "broken-back" Cub.

Gladys Pennington was chairman of the First All-Women Air Show. Marion Bertram was contest chairman and Verna Burke announced. It was a rousing success.

The reason for listing so many names is that these ladies, winners in 1947, continued taking all top honors for several years. Some are yet flying in air races. And winning!

The Florida Ninety-Nines were pleased with the results of their first effort and made the decision for a Second Annual All Women Air Show of the World, this to be staged June 1-6, 1948. The location was changed from Peter O'Knight Airport, Tampa, to the Opa Locka U.S. Naval Air Station north of Miami, named Amelia Earhart Field for the event. That one started with a Los Angeles-to-Miami transcontinental air race offering a total purse of $1,500. Interesting — "gas stops and route were left up to pilot's option."

In June 1949, came the third. Lauretta Foy won the Women's Transcontinental Air Race with Betty Gillies second. The International leg from Montreal to Miami was won by Peggy Lennox; Helen Greinke was second, Fran Nolde third, and Kay Brick came in fourth.

In the '49 event, Helen McBride took top honors in the 65-hp contest; Anne Shields was second to Peggy Lennox in the 75-125 hp competition. The AT6 Race went to Kaddy Landry, Caro Bayley second; Helen McBride was third and Mary Tracy placed fourth.

In the meantime, Miami's All-American Air Maneuvers ignited, and with a female division. Some of the stalwarts of the All Women's Air shows jumped into bigger headlines. For instance, in July 1949, Betty Skelton and her "Little Stinker Too," a Pitts Special, sailed for England. She had been invited to put on an exhibition in London and in Belfast, Ireland, the first American to be so invited. At that time, Betty, 22, held many records, including the altitude record for light planes and the International Aerobatic Title for Women.

On June 17-18, 1950, the Florida Ninety-Nines gave it one more shot: the All Women Air Maneuvers, then the fourth annual, at West Palm Beach. Blanche Noyes was Ninety-Nines president and Ann Ross was chairman of the Florida Chapter. Reading that program, you see an aerobatic demonstration by Mary Tracy — yes, her married name is Gaffaney and aerobatics, you'd better BELIEVE! Another aerobatic demonstration was flown by Caro Bayley, who was an upcoming international champion, too, in her black Pitts Special. Fran Nolde was the NAA representative.

From the beginning, Florida Ninety-Nines sponsored international air races, and they fostered aerobatic championships that still lure dozens of Ninety-Nines, many still in Pitts Specials and some in helicopters. Proceeds from the first all-women's air shows went to the Amelia Earhart Scholarship Fund.

The Ninety-Nines president at that time, Blanche Noyes, has quite a background in aviation but when you say Noyes, don't you think airmarking?

The Air Marking Program had a very timid birth in 1935. In those days, pilots didn't have OMNI, or ADFs, or DME. Even their charts were doubtful. Where a pilot was could be a major problem. Water tanks, warehouse roofs, drag strips, airports all have felt the furor of 99 paint brushes. But the start came from the pushing of Blanche Noyes, Helen Richey, Helen Rough, Louise Thaden, Amelia Earhart, Phoebe Omlie and Nancy Love. Many of those were charter members of The Ninety-Nines. They started airmarking through government channels and Blanche remained as the FAA's chairman of air marking until the '70s.

If you haven't wielded a paint brush or roller on an airmarking exercise, you're either a new Ninety-Nine or a lazy one!

But Ninety-Nines have many special interests. Only a very few have remained vital for most of the organization's life. Of these few, only one has been financially able to grow into more productivity each year. The Amelia Earhart Scholarship is that extremely rare entity.

For the purpose of the AE Scholarship, what better to quote than "Thirty Sky-Blue Years" (1959): "As a living memorial to the first President, Amelia Earhart, The Ninety-Nines selected a scholarship through which all members could participate in carrying on her enthusiastic and unselfish aims... most particularly that of strengthening and cementing women's permanent place in aviation."

How, though, did the AE Scholarship start? A study committee chaired by Ruth Nichols in 1939 came up with the idea. The scholarship is to assist deserving Ninety-Nines to further their aviation aims.

The Ways and Means of such a purpose was created as the Amelia Earhart Memorial Scholarship on April 7, 1940. The governing body is a Board of Trustees, two of whom are permanent, three of whom are elected by The Ninety-Nines. From the beginning, they've had the responsibility

Just one example of 99s' airmarking style.

of acquiring and investing the monies to continue the trust, and through the efforts of outside judges, dispersing awards to the most deserving. At no time do the trustees even receive expenses for the hour after dedicated hour they give to the future of women in aviation.

The first two permanent trustees were Alma Hendricks Harwood and Margaret Cooper Manser. Mrs. Manser, on her death in 1951, was replaced by Jeannette Lempke Sovereign, who served until she resigned in 1958. Marjorie Fauth succeeded her as chairman and, in 1959, Alice Hammond succeeded Alma Harwood.

Perhaps you'll notice that many of the trustees are former Ninety-Nines officers? That's true. The bylaws of the AE Trust document prohibit an officer of The Ninety-Nines from holding office on the AE Board of Trustees (rightly, of course). Former presidents have continued efforts for the organization, though, by taking on the duties of the AE Fund.

The first AE Scholarship winner was Pat Gladney in 1941. Pat received $150. When she was awarded the prestigious award, she was instructing at the University of California. The winner's name was announced on the fourth anniversary of the Amelia Earhart "lost" flight across the Pacific. Pat is a former WASP.

From the $150 start, the care and conservative nature of the AE Trustees and the ever-loyal contributions of individuals, chapters and sections of The 99s have made possible more than a single award annually. The awards are in the neighborhood of $1,000 each, depending on the amount contributed by The Ninety-Nines and the monies realized from interest on the trust fund. The trustees now consider and have made awards to other than U.S. fliers, and have added a research scholar category.

The Ninety-Nines throughout the world understand the importance of the AE Scholarship. They understand the need of contributing as much

In 1984, International Librarian Dorothy Niekamp spends free time at International Headquarters cataloging books and periodicals in the Resource Center. Niekamp also was a two-time recipient of the AE Research Scholarship to create and update a bibliography of women in aviation.

as possible, both as individuals and as groups. They have sponsored a variety of events to benefit the fund like, for instance, the First All Women Air Show. Rummage sales, bake sales, concession stands, hangar dances, plane washes, poker flights, and Pennies-a-Pound flights have led the list.

Pennies-a-Pound flights — with inflation, 2 cents a pound or more — can net the sponsoring group upwards of $2,000, with a little luck. Bad weather or a lack of publicity can bring nothing but frustration. But there's always next year. Sponsoring Ninety-Nines (and their families) have given freely of their time, their airplanes and even, oft times, bought their own fuel for such events.

The main value, though, is that a group will

AMELIA EARHART MEMORIAL SCHOLARSHIP FUND

Members who conceived the Amelia Earhart Memorial Scholarship Program in the late '30s could hardly have envisioned how it would grow over time, but grow and strengthen it did.

From a single scholarship awarded in 1941, the first year, to a record 21 scholarships awarded at the 1995 International Convention in Halifax, Nova Scotia, 99s have benefited from the vision of those early members.

In 1992, the first contribution from an outside entity, United Parcel Service, was received. The same year United Airlines began their relationship with the program by awarding a Flight Engineer Scholarship to Laurel Cameron. United then presented 737 Type ratings to Jenny Beatty and Patricia Turney in 1995, and to Kelly Sue Hamilton and Janet Patton in 1996.

Also in 1996, the largest scholarship amount distributed from internally-generated funds was awarded to Ann Leininger to obtain her Type rating.

In addition, eight Research Scholar Grants have been awarded to the following women:

1978 - Dorothy Niekamp to develop an annotated bibliography of women in aviation
1983 - Shirley Render to research and write a scholarly paper on Canadian women pilots
1984 - Gail J. Vail to correlate pilot error with social/psychological factors research
1985 - Claire B. Koop, Ph.D., to study spatial orientation abilities of general aviation pilots
1986 - Anita B. Crockett to study motion sickness susceptibility, focus of control and personality factors in women pilots and non-pilots
1989 - Dorothy Niekamp to update her bibliography on women in aviation
1993 - Susie Sewell to establish an indexing program for *The 99 NEWS*

The following list includes all 302 AE Scholarship winners:
1941-Patricia Thomas Gladney
1942-Dorothy Broadfield Monahan
1947-Elizabeth Sewell
1948-Jean Hixson
1949-Virginia Sweet
1950-Amalie Ward Stone
1951-Jean Swartwood
1952-Janet Dietrich
1953-Donna Evans
1954-Nancy Leebold, Dorothy Woodham
1955-Lucile Cheetham
1956-Shirley Mahan
1957-Georgianna McConnell
1958-Anne Shields
1959-Cora McDonald Fraser
1960-Ann Piggott Mentzer
1961-Harriet Wladyke
1962-Ruth Wikander
1963-Velta Benn, Nancy E. Brumlow
1964-Mary Creason, Jill McCormick, Florence Toney
1965-Janet Ferguson, Carol Rayburn, Ruby St. Onge
1966-Mary Frances Blair, Jean Reynolds, Christine Winzer
1967-Wendy Blanchard, Evelyn Braese, Yvonne Pope, Ardyth Trenholm
1968-Helen Kelton, Hazel Jones, Martha Graham, Theresse Pirrung
1969-Yvonne van den Dool, Donna Flaum, Mary Reindl, Carol Wright
1970-Carolina Luhta, Evelyn Sedivy, Lorna deBlicquy
1971-Joyce Jones, Charlotte Parker, Helen Sheffer, Ann White
1972-Ann Esselburne, Karen Walker Harris, Ruth Hurst Jefford, Connie Jo-Ellen Jones, Jean Schiffman
1973-Peggy Bolton Husby, Ora Merk, Patricia Domas, Dorothy Tuller, Frances Sargent
1974-Carol Borgerding, Mary Kochanek, Barbara Goetz, Gloria Homes, Helen McGee
1975-Nancy Fairbanks, Orene Hirth, Linda Hooker, Margaret Stanford, Mary Ward
1976-Lynn Palmer Cary, Susan Linsley, Mary Elaine Anglin, Elizabeth Schermerhorn, Nicole Radecki, Amy L. Pilkinton
1977-Ursula Davidson, Elizabeth Dinan, Angela Izzo, Carole Sutton, Roberta Taylor
1978-Lane Joan Basler, Joanruth Baumann, Linda Hollowell, Rita Reo, Evelyn Snow,
Verene Trubey, Rene Wicks
1979-Jacqueline Breeden, Joan DíAmico, Marijane N. Howard, Anitra Doss Ross
Janice Orr Young, Terry Zeidler, Ann McNamara
1980-Gray Gordon Bower, Rosalie Burchette, Carol DePue, Donna Miller, Mary Murphy Monterubio, Patricia Rockwell, Virginia Unger, Mary Jo Voss
1981-Carolyn Clarke, Evelyn June Craik, Theresa Donner, Charlene Giebe, Mary S. Jablonski, June Perry, Carol Ann Phelps
1982-Eileen Anderson, Joan DíAmico, Judy Graham, Glenda Martlew, Virginia Mitchell, Gabrielle Thorp, Melissa Vreeland, Mary Lou Westmoreland
1983-îMikeî Alexander, Pat Bizzoso, Bonnie Carr, Candi Chamberlin, Marcy Glasermann, Michelle Miller, Carol Nielsen, Melody Rich, Sylvia Rickett
1984-Martha Bailey, Suzanne Batz, Dorothy Dickerhoof, Georgina Harris, Loretta Haskell, Karen Johnson, Rosemary S. Jones, Evelyn Kropp, Teresa Ludtke, Susan DeEtta Maule, Constance McConnell, Lawanna Steel
1985-Cynthia Bergstrom, Diane Dwelle, Bessie Hensley, Deborah Kaeder, Mary Kelly, Estelle Kirkpatrick, Cecelia Stratford, Elizabeth Wieben
1986-JoAnne Carpenter, Linda DeForest, Cathy Fraser, Joan Jones, Enid Kaspar, Aimee Kuprash, Jean Pickering, Catherine Shanahan, Patricia Tormey, Mary Trusler, Karen Winters
1987-Janet Bartos, Louise DeMore, Juanne L. Hoelscher, Delores Jewett, Linda Mattingly, Madeleine Monaco, Laurie Peterson, Lynette Renneke-West, Lynn Schug, Barbara Stott
1988-Deborah Cunningham, Shari Egan, Barbara Goodwin, Jessica Hatfield, Leslie Highleyman, Carol Landefeld, A. Lee Orr, Kathy Osborne, Mary Rutherford, Anna Scott, Valerie Suberg
1989-Linda Barker, Marion Bullington, Teresa Evans, Holly Friedman, Vicki Grandy, Joanne Hodges, Dee Ramachandran, Karen Rutledge, Patricia Thomas, Laura Warman, Gail Schroeder, Ann Marie Schorsch
1990-Holly Brenneman, Stacy Burger, Margaret Doyle, Lesa K. Grider, Pamela Hengsteler, Cynthia Jean Huffman, Judy Lanning, Gail LaPook, Carolyn Martell, Nancy McGinnis, Angela Wilkerson, Laura Winkelmann
1991-Janet Andersen, Nancy Clinton, Audrey Cook, Lorraine Jones, Janet Lewis, Denise C. Louth, Carol MacNeil, Linda Marshall, Beverly Roediger, Julia Schmitt, Anna Scholten, Yvonne M. Souza, Linda Thomas, Jessica Waltz, Evie Washington, Cynthia Wright
1992-Evelyn Ambrose, Katherine Bauer, Lisa Cotham, Anne English, Patricia Haley, Mary Henig, Adele McDonald, Pamela Parask, Katherine Price, Margaret Puckette, Christine St. Onge, Donna Stevens, Tina Thomas, Melinda Walton, Laura Warman, Gay Zena Williams
1993-Gladys Bowditch, Candace Covington, Tara Anne Donn, Kimberly Eggert, Denise Eggleston, Amy A. Ellsworth, Peggy Sue Figley, Karen Foster, Kathy Gardella,
Sandra Gordley, Wendy Grimm, L. Bernadette Hayward, Christine Hettenbach, Linda Maloney, Marie Miller, Wendy Paver, Jane Phillips, Gayle Conklin Prichard, Anneliese Rene Stark, Susie Sewell-Research School
1994-Belinda Allen, Gail Allison, Robin Andersen, Susan Bailey, Susan Barber, Carol Church, Mia Donnelly, Linda Friedman, Amy Hoover, Andrea Lende, Laurel Lippert, Karen Monteith, Sandra Reagan, Tiffany Tokar-Vlasek, Celia Vanderpool, Aileen Watkins, Michele Yarbrough
1995-Katherine Anderson, Annemaire Brainerd, Heather Brown, Patricia Compton, Kimberly Coonce, Jan Currie, Teresa DeGraaff, Linda Draper, Christine Hettenbach, Cathleen Jameson, Stephanie Martin, Jennifer McCann, Jessica McMillan, Donna Moore, Jo Ellen Peters, Carol Renneisen, Merav Schwartz, Linda Scully, Margaret Shaffer, Laura Smith, Anneliese Rene Stark
1996-Phyllis Berry, Coleen Campbell, Nohema Fernandez, Karen Diane Helly, Ann Marie Leininger, Tracy Leonard, Donna Jeanne Miller, Danuta Pronczuk, Susan Gene Thomas, Janice M. Welch

and does work so hard for a single, and in the case of the AE Fund, worthy cause.

Air races have been of supreme interest to many Ninety-Nines ever since the beginning: endurance races, then-big ones like the Bendix or Cleveland and the transcontinental air races; small ones staged to section meetings and conventions; the international one, the All Women International Air Race — the Angel Derby, which indeed does have many, many enthusiastic boosters.

The 1978 race was the 28th AWIAR. Remember we said it started during those first All Women Air Shows?

Over the years, the international competition has covered routes from Canada and Cuba to the Bahamas, El Salvador, Nicaragua, and once even from Canada, over the U.S. and ending in the Bahamas. The second race, in 1951, between Orlando and Windsor, Ontario, was won by Margaret Carso, a Canadian, the first to be won by a pilot from outside the United States.

In 1961, the Florida Women Pilots Association was formed to assume management of the race. How did it get the name "Angel Derby"? Credit that to a Mexican newspaper writer in 1964.

Both the All Women's International Air Race and the Angel Derby were incorporated in 1973 and, since that time, the race has been conducted under the direction of All Women's International Air Race, Inc.

The Angel Derby has been won by many distinguished pilots who are members of The Ninety-Nines. B Steadman, a past President and former chairman of the Board of the International Women's Air and Space Museum, is a two-time winner. Edna Gardner Whyte has won the race four times and is one of aviation's most notable personalities. Around-the-world pilots Jerry Mock and Joan Merriam Smith have flown in this international competition. And Pat Arnold, a winner, later organized the Women's Pylon Racing Association. Angel Derby four-time winner Judy Wagner also tore up the winning trail in those pylon races, placing first in Frederick, MD, Fort Worth, Reno, Las Vegas and Cleveland.

Other races the gals have originated, developed and flown in are Formula 1, many proficiency air races, the Kachina Doll Air Race in Arizona, the Indiana Fairladies Air Races (which saw action for 17 consecutive years), the ever-popular Palms to Pines Air Race, and likely the largest, and oldest, proficiency race, the Michigan SMALL Race. There are and have been dozens of others. Some in specific areas, like the New England Air Race, have drawn competitors from many states and from Canada.

Jerrie Mock

Betty Miller

Past president and charter member Kay Brick and Doris Renninger pose with Capt. Kyung O. Kim, the only female in Korea with the opportunity to fly. Kim was the recipient of an organized S&H Green Stamp savings program among 99s that netted her a Colt airplane. The Colt went to Korea in 1963 to aid other women who wanted to fly.

Women's Advisory Committee on Aviation, 1974

NATIONAL INTERCOLLEGIATE FLYING ASSOCIATION AND THE 99s

by Pat Ward

NIFA needs no introduction to 99s in the U.S. Providing ground personnel, judging staff and financial assistance to the colleges and universities that compete in the National Intercollegiate Flying Association's annual Safety Conferences (SAFECON) are exciting activities for individual 99s and chapters across the country.

Intercollegiate competition in air games had its humble beginning in 1911 when AERO magazine reported "the first intercollegiate glider meet" at which Cornell, Harvard, Pennsylvania, Swarthmore, Tufts, Volkman and a college simply called "Technology" met to compete in gliders weighing from 60 to 150 pounds. Volunteers from the flying colleges who were represented in that initial contest went on to design and pilot gliders during WWI.

In 1929, an associate of Wilbur Wright, holder of the first master's degree in aeronautics, and a staunch supporter of collegiate aviation programs, Grover Loening designed a sterling silver trophy to reward leadership achievement in college flying. Today, Delta Air Lines perpetuates Loening's contribution to collegiate aviation by sponsoring the Loening Trophy, joining a host of other aviation industry benefactors.

Women's Achievement Award winners at the 1987 NIFA competition include Ann Marie Wychelewski, Southern Illinois University; Jennifer Boyle, Mount San Antonio College; and Donna Heilig, also of SIU. The winners are flanked by past presidents Hazel Jones and Lois Feigenbaum, who presented their medals and checks.

A meeting in 1935 brought 74 men and women interested in college aviation programs together to organize college flying competitions. They called their new organization the National Collegiate Flying Club and operated under the auspices of the National Aeronautic Association. Early participants included Thomas Watson Jr., future chairman of IBM, and J.B. "Doc" Hartranft, founder of the Aircraft Owners and Pilots Association.

The first regional competition was held in Northampton, ME, in May 1935, with eight schools competing; by early the next year, 50 collegiate flying clubs were enrolled in NIFC.

The first known 99 to compete in an NIFC event was Caroline Etheridge (Hembel), who flew with the University of South Carolina team in 1940. Arlene Davis, All-Ohio Chapter, championed the concept of the intercollegiate competition and may have been responsible for involving the 99s organization. The chapter continues to award the Arlene Davis Trophy at annual SAFECONs in her honor, and the international organization also recognizes the Top Woman Pilot each year.

NIFC, the club, became NIFA the association in 1972 when as many as 60 schools and over 700 competitors entered the annual competition. A more manageable qualification system was developed, and regional meets were established to narrow the number of national competitors to no more than 26 schools.

In 1994, 99s Hazel Jones, Pat Roberts, Carole Sue Wheeler and Jan Maxwell were responsible for helping to develop the judges' manual used in today's events. Jody McCarrell subsequently redesigned the navigation flying event, which has become the benchmark by which excellence in precision navigation is judged.

Today's competitive air events include precision landings, a complex navigation exercise, a message drop using a carefully-designed object and a precision flight. In addition, ground events cover computer accuracy, aircraft preflight inspection, aircraft recognition, electronic flight computer tests and a simulated comprehensive aircraft navigation exam. In all events, safety, educational skills and sportsmanship are of primary concern.

The majority of the judges for NIFA events are 99s; also, many former NIFA competitors return to volunteer their services in a variety of roles after graduation. Individual 99s, as well as chapters and sections, provide financial assistance to the NIFA Foundation, the NIFA Council and to member schools. In 1996, three members of The 99s served on the NIFA Council: Pat Roberts, Jody McCarrell and Kelli Hughes-Lager.

A sequence of photos provide a flight-line's-eye view of a precision flight team member's landing effort over a barrier.

Other races outside the United States have drawn 99s' support, both as workers and competitors, particularly in Australia and in Canada.

We're not ignoring the AWTAR — the Powder Puff Derby. It rates a section all its own.

Another type of competition which has drawn Ninety-Nines, both as competitors and workers, is the National Intercollegiate Flying Association. We have reports of the 1940 college "nationals" with a 99 contestant, Caroline Etheridge (Hembel) from the University of South Carolina. Caroline still flies mostly Cessnas; her daughter, Bunny, flies airplanes in air races, and helicopters at home. Caroline's husband, Les, is very active in aviation and his helicopter school has turned out many 99 Whirly Girls (helicopter pilots).

But for NIFA, likely the first Ninety-Nine really to become a booster was Ohio's Arlene Davis. The All-Ohio Chapter continues the tradition of awarding a magnificent Arlene Davis trophy at the annual college meet. From the roles of female college fliers have come many Ninety-Nines and on the roles of workers for NIFA are many Ninety-Nines. The International budget contributes money to the trust fund for the college fliers; officers have worked as judges and scorekeepers. But individual Ninety-Nines have worked hard for the group. They have seen the value of advice, money, labor — this is the future of aviation and Ninety-Nines are proud to be a part of it.

Contributions for NIFA have come in from chapters and sections. Fairly new is the opportunity for each Ninety-Nine to become an associate member of NIFA; for that $10 in membership dues, the Ninety-Nine receives the college fliers' newsletter, and the college flier receives $10 worth of financial aid to keep costs within reason. The Ninety-Nines has an overall NIFA chairman who tries to keep the various aims centered towards the most needed slots.

Individual Ninety-Nines have brought reputation to the organization. Perhaps one of the best is a quiet, unassuming lady named Betty Miller. She learned to fly in 1950 and then married her flight instructor, Chuck Miller. They owned and operated a large flight school — flight training only — in Santa Monica. She became well known as an instrument and helicopter instructor.

She was awarded the 1964 Harmon Trophy. For what? Being the first woman to fly the Pacific solo, that's what. When an opportunity to fly an Apache to Australia came along, she says she "thought it over for two minutes." The flight took 54 hours and 8 minutes. That was in May 1963 and she took off from Oakland, made four landings, then arrived at Brisbane on May 12 — 7,400 miles later.

Later, she delivered a Comanche to Holland. Both oceans! If you ask her about the flights, likely she'll say something like, "Well, airplanes needed to be delivered."

The next year was one of several attempts — and two successful around the world flights. Solo. Jerrie Mock of Columbus, Ohio, in her 11-year-old Cessna took off on March 19, 1964, from Columbus and landed back there on April 17. At the time, Jerrie had only 750 hours flying time.

By completing this globe-circling flight, Jerrie set three aviation records: the first woman to solo the globe in a single-engine plane, first to fly the Pacific Ocean solo west to east; and first to solo it either way in a single-engine plane.

Jerrie continued to set FAI/NAA world records for distance and speed; she flew in some of the air races.

Two days prior to Jerrie's Columbus take-off, Joan Merriam Smith left Oakland for around the world, the Amelia Earhart route. Joan's route was longer, yes, but too she was plagued by bad weather and equipment problems. Joan flew an Apache.

She completed her flight on May 12, 25 days behind Jerrie. Jerrie Mock was awarded the FAA Gold Medal for exceptional service.

The ladies were competitive, of course, or they would never have tried to circle the globe. Each claimed she knew nothing of the other's plans. The press wouldn't have it that way and it became a headline battle — unfairly. Both ladies completed their flights — around the world — and no mean feat that is!

Round-the-world continued as a dream for a few years. Ann Pellegrino, an experienced pilot and history buff, dreamed of the mystery of Amelia Earhart's disappearance over the Pacific. A friend, Lee Koepke, owned a restored Lockheed 10 — very similar to AE's Electra. In 1967, the 30th anniversary year of Amelia's last flight, Ann, Lee Koepke and co-pilot Bill Payne flew the Lockheed on the AE route — they found the Howland Island that Amelia missed — and completed their round-the-world dream.

When over the ocean flying is talked about, then the name of the maestro must lead the topic. Pennsylvania's Louise Sacchi has delivered over 300 general aviation planes to Europe, South and East Africa, Australia and the Philippines. Not big airplanes, either — Sierras, for instance, to Holland. Her comments on one Sierra delivery are typical. After a delay necessitated by lack of a Dutch license, she arrived May 2 instead of the planned April, explaining, "This proved perfect for me because I have been trying to get to Holland in tulip time."

She estimates that about 70 to 75 percent of her ocean-crossing deliveries have been in single-engine planes. She likes flying the Atlantic better — the Pacific, she says, is a "bore." She's the boss of Sacchi Air Ferry Enterprises in Philadelphia. She first flew the Atlantic as a navigator in 1962; Beech awarded her the Distinguished Service Medal for 100 ocean crossings in their airplanes. In 1971, she set a record from New York to London in a Bonanza.

Another name for ocean-flier has GOT to be Louise Sacchi!

Over the ocean AIRPLANE was another major project for The Ninety-Nines. In 1959, a Korean (ROK) Air Force captain in the reserves joined the organization. Captain Kyung 0. Kim — the name means Beautiful Golden Tree Castle — had 119 missions in L-19's as a liaison pilot in the Korean War. But she was the only female in Korea with the opportunity to fly. If she only had an airplane ...

Kay Brick, then International president, and her following officer, Louise Smith, said yes, The 99s would help. Doris Renninger spearheaded the drive on Long Island and it spread like wild fire. S&H Green Stamps — about 2,500 books — bought a Colt for Kim. The Colt went to Korea in 1963 and now there indeed are other women, besides Beautiful Golden Tree Castle, flying there.

U.S. President Lyndon B. Johnson announced the formation of the Federal Aviation Agency's Women's Advisory Committee on Aviation on May 4, 1964. Jane Hart was designated chairman; Jean Ross Howard served as co-chairman. Both are 99s. Most of the 27 non-government members and five government members were. The original committee members were elected to a two-year term of office; later, the project of the committee, then called WACOA, was voted to continue. The first members rotated off as of June 30, 1967, and replacements came in at that time.

WACOA was proposed to the president by then-FAA Administrator N. E. Halaby. The name of the committee was changed to Citizens Advisory Committee on Aviation on Jan. 23, 1975, and terminated on Jan. 23, 1977.

In 1961, Jerrie Cobb, later a member of WACOA, was the first and the only female to pass all three phases of the Mercury Astronaut Program. She was appointed a consultant, but, unfortunately was 17 years too early to become what she had worked for so hard — an astronaut.

Twelve other 99s passed the series of 75 exhaustive physical competence tests, x-ray and

Mary Jo Oliver Knouff brings aviation to life for youngsters.

Lt. Rosemary Bryant Conatser (Mariner) became one of the first female Naval aviators in 1973. In 1975, she became the first woman to fly a tactical jet aircraft as a designated military aviator. Commander Mariner received the 1993 Award of Merit from The 99s to recognize her achievements for women in aviation.

Oklahom native Jerrie Cobb began flying at the age of 12 and went on to set numerous world records in speed, altitude and distance. The first woman to successfully complete the Mercury Astronaut testing program, she currently flies medical missions to the Amazon.

Emily Warner, the first woman jet airline pilot in the United States, left her captain's position flying a B727 to join the Denver office of the Federal Aviation Administration in 1991.

Italian Fiorenza de Bernardi flying a DC-8 for Aerol.

laboratory tests: Jane B. Hart, Rhea Hurrle Allison, Mary Wallace Funk, B. Steadman, Sarah Gorelick, Myrtle Cagle, Gene Nora Jessen, twins Jan and Marion Dietrich, Jerrie Sloan, Irene Leverton and Jean Hixson.

Not because American women, particularly American Ninety-Nines, didn't try — but once again, the Russians achieved a first — the first female in space was Russian.

Jerrie Cobb was deeply discouraged by the failure of NASA to put a female in space. She had learned to fly at 12, holds four world aviation records and has won numerous awards for her flying. Perhaps her most notable record is a solo non-stop between Guatemala and Oklahoma. She worked in a corporate position with Aero Commander in Oklahoma, her home state.

Her strong religious convictions led her to help the world's unfortunate through her ability as a pilot. For more than 30 years, she has given her airplane, her personal financial resources, all her efforts and time as a jungle pilot in Amazonia. She flies in doctors, missionaries, anthropologists, medicines and supplies to Indian tribes in unexplored parts of six countries. She flies out sick and injured Indians. Hers is a completely unselfish way of life.

Mary Tracy Gaffaney was co-owner and chief instructor at a flying school near Miami and has made a life of winning championships! She has won the U.S. National Women's World Championship at Salon de Provence, France, in July 1972 — the first such American victory. Her airplane? "Fair Play" is the name — a Pitts Special. Mary likely had every possible license and rating and has many firsts to her credit. She was the first women's helicopter instructor in Florida, the first woman skywriter in the nation. She flew in races, both closed-course and cross-country, and was chosen for pilot's position of the U.S. Helicopter Team at Middle Wallop, England, in July 1973. She's Whirly-Girl # 33.

In 1974, Mary Tracy Gaffaney was awarded the Lady Hay Drummond Hay - Jessie A. Chamberlin Memorial Trophy by the Women's International Association of Aeronautics. Later, she became president of the Whirly-Girls Scholarship Fund.

Now, exactly what is that? Or who are they, The Whirly-Girls? They're licensed helicopter pilots, female-type. The Whirly-Girls organization was dreamed up and is kept in order by Jean Ross Howard, a Washington 99 when she earned her own 'copter rating in 1953. The first gathering, now called "hovering," was in 1955. At that time, only 13 women had 'copter ratings in the world! Hanna Reitsch is #1 Whirly-Girl. She flew a helicopter INSIDE Berlin's Deutchland Halle in 1938; the first American member, Whirly-Girl #2, is Ann Shaw Carter. There are now 1,000 Whirly-Girls in 25 countries with many 99 members in their rolls.

The Whirly-Girls started a scholarship in memory of Doris Mullen, a member of both organizations, with a $500 grant towards helping the applicant earn a 'copter rating. Most of those winners have been 99s. Some are now with the airlines. The 1978 scholarship was international for the first time. A second scholarship, the Whirly-Girls/Enstrom Helicopter Corp. Scholarship, is designated for a U.S. applicant only.

The Whirly-Girls and The Ninety-Nines are close perhaps because they have so many mutual members.

It was Clara Trenckmann Studer, first editor under Amelia Earhart of *The 99 News*, the *Airwoman*, who through years of dogged, persistent persuasion, prodded the U.S. Post Office into issuing the Amelia Earhart 8-cent commemorative airmail stamp on her birthday, July 24, 1963. Before that, the rules said that honorees had to be dead for 25 years.

And it was only another hard campaign that persuaded the powers-that-be to issue the first day cover in Atchison, KS, her birthplace. A 99s Flyaway, delivering covers to all 50 states, and commercial Friendship Specials flown to cities in European countries where Amelia Earhart was honored — Rome, London, Paris and Brussels — were a major part of world-wide memorial activities arranged by Fay Gillis Wells. Charter 99s, flying covers out of Atchison, reached all rendezvous points on schedule despite tornadoes and turbulent weather. Clara also arranged Fifth Avenue window displays, an entire bilingual magazine issue, ceremonies in foreign cities, and sales at philatelic centers. The Coast Guard reactivated the Amelia Earhart beacon on Howland Island. Many VIPs attended a July 24 ceremony at the Amelia Earhart Memorial Stadium in Atchison.

Special hand-stamped 99 covers have cleared over $45,000 for the Amelia Earhart Scholarship Fund. Since 1963, all special occasions have been commemorated with a consecutively-numbered issue of 100: the Mount Amelia issue, for instance; the Smith and Scott around-the-world flights; the dedication of the International Forest of Friendship, and others.

On the 50th anniversary of Charles A. Lindbergh's transatlantic solo flight, Ann Morrow

THE POWDER PUFF DERBY AND OTHER RACES

There are several books and articles about the famous Powder Puff Derby which we don't have space to include. So that there can be at least some coverage of this race for the new readers of this history, we have excerpted segments of the Air Racing section from "History of The Ninety-Nines, Inc." First Edition, published in 1979.

The annual All-Woman Transcontinental Air Race (AWTAR), flown in summer and best known by its trademark as the Powder Puff Derby (PPD), was the oldest, longest and largest speed air race for women in the world.

While the race was conducted under the sporting code of the world body, the Federation Aeronautique Internationale, the six- to nine-woman Board of Directors annually reevaluated and revised the specific rules under which the race was run.

A route was laid out between two cities which had bid for and won the start and terminus of the race in cooperation with their local Ninety-Nines chapters. Over the years, The Ninety-Nines (organization) endorsed this race only.

Eligible aircraft, single- and multi-engine, were between 145 and 450 hp, extended to 600 hp for the 25th Jubilee race and 30th Commemorative only. All were handicapped for fair competition.

Designated stops along the route permitted racers to refuel, wait out weather which was not VFR (visual flight rules), and to RON (remain over night), since only sunup to sundown flying was permitted.

Qualified timers, approved by the National Aeronautic Association and the FAI representative in the United States (under whose sanction the race was run) clocked the contestants in and out of designated airports so that time on the ground was not counted in the scoring.

Harbinger of the now famous PPD was the first "race" in 1947 from Palm Springs, CA, to Tampa, FL. The Florida Chapter of The 99s, laden with eager post-war WASPs, created the Florida All-Woman Air Show and invited the race to terminate in Tampa as part of the show.

"Sun-up and five-ish on a nippy desert Palm Springs morning," as Mardo Crane, pioneer chairman, recalls; she, Dee Thurmond, Irma "Babe" Story and Helen Hooper enthusiastically directed Californians Caroline West with copilot Bea Medes in Caroline's Ercoupe to the starting line to await Dianna Bixby in her military A-26 bomber.

After great delay the "timers" waved the Ercoupe off for its 21-hour, 45-minute flight to Tampa. It was all "honor system" for time and no authorized stops. Not until they reached Tampa did the Ercoupe team realize that Dianna had never started due to engine trouble. In that they raced against their own advertised air speed, this "'Amelia Earhart Memorial Race' could be called a contest," related Mardo, "and the possibility of an annual all-woman transcontinental light plane race was now more than just an exciting idea."

In 1948, with permission of the Florida Chapter, Mardo's committee set about organizing the second race into Miami, the site of the second All-Woman Air Show. Basic rules were drawn up for the six planes that entered... planes up to 300 hp, daylight VFR flying only, minimum of a private pilot's license with 25 hours in type of plane flown, females only, and still on the honor system for timing "verified" to the racers by tower operators.

Now the need for money asserted itself — telephone calls, wires, travel, publicity and most important, prize money. Charter member Jacqueline Cochran rallied to the need and the 1948 and 1949 races were called the "Jacqueline Cochran All-Woman Trophy Race."

About this time, the "honor system" for timing was challenged and "time clocks" at check-in airports were suggested. Locating and setting these up created the "Trail Blazing" flights by race officials.

In 1950, the "Ninety-Nines Transcontinental Air Race" drew 33 entries, and the purse was provided by Odessa, TX, and Mrs. Olive Ann Beech (Beech Aircraft). Late in 1950, the race became the "All-Woman Transcontinental Air Race, Inc.," a non-profit corporation with a four-member Board of Directors: chairman Mardo Crane, Betty Gillies, Ethel Sheehy and Arlene Davis.

As the race grew, it became a year-round job to keep it winging across the United States. The route must be set and surveyed, rules revised, data processing and personnel for tabulation of scores secured, funds raised, official timers appointed and instructed, an official race program produced, awards and trophies secured, airplanes test-flown and handicaps assigned, race aircraft inspected at start and finish, promotion and publicity beneficial to all race sponsors conducted, navigation services arranged... the list went on, and not the least effort was coordinating the entire project with every official body involved.

In 1951 and 1952, during the Korean War, the race operated as a training mission, the objective to "provide stimulation as a refresher course in cross-country flying for women whose services as pilots might once again be needed by their country." The race also was opened to all women pilots, not just 99s.

As early as 1950, the look-alike dress trend started. Many teams had complete wardrobes of matched outfits specifically designed for the race.

The first mother-daughter team to race, in 1950, was Betty Gillies with college daughter Pat, 16, who had 200 hours at the controls of the Navion this team flew. Subsequently, any passenger under 16 was ruled out as "not contributing to a racing endeavor." Later, only pilots could participate, making the youngest possible age 17. Eight mother-daughter teams flew the race in 1966, and 16 in 1976, indicating that air education begins at home.

While some women raced almost annually, each year the dream of flying the PPD came true for the over 40 percent of entries who had never flown it before.

The number of entries continued to climb, and soon a system of drawing for entries was established. In 1971, for example, 143 entries were postmarked April 15, and within four days entries were closed, with 16 stand-bys closely following.

Marion Andrews designed the unique gold PPD pin, a latticed half-globe encrested with a map of the United States and the words "Powder Puff Derby," which may be worn by only those who have flown this challenging race.

The 20th anniversary race was documented in a 50-minute color TV film by WCBS-TV; in 1972, a 25-minute documentary was filmed by a German company for worldwide distribution.

Many famous people participated in the race as honorary starters, official greeters, banquet speakers, etc. In 1969, artist Milton Caniff, in his syndicated comic strip "Steve Canyon," entered his character pilot Bitsy Beekman and her "Bug" in the PPD for an eight-week run.

The greatest of all honors paid the racers and officials occurred in July 1969, when Pat Nixon, wife of U.S. President Richard M. Nixon, graciously entertained them at a reception and tea at the White House.

In 1973, a national emergency -- a severe fuel shortage — captured the cooperation of PPD officials, who canceled the 1974 race. Since there was no race, it seemed a good time to compile the "Powder Puff Derby Commemorative Album" — 178 pages capturing 27 years, 2,700 racers and 166 stops in 41 states.

The 1975 race, with 102 entries, was not without its denouement, for a discrimination suit filed by a male applicant who had been denied entry as unqualified by the rules of 28 years. The day before takeoff, U.S. District Judge Lawrence T. Lydick ruled the race was on, for women only.

Pauline Glasson, Gini Richardson, and Pat Gladney are the three who have flown the PPD "the mostest" --22 times. Gini won in 1971 on her 19th try, and Pauline holds the record for providing an opportunity to fly the race for the most different co-pilots --18 different women (as of 1979).

THE AIR RACE CLASSIC

excerpted from an article by Anne Honer which appeared in the July/August 1996 issue of the International Women Pilots/99 News *magazine*

Air racing could not dissolve, it seemed. Too many women — and aviation itself — had gained too much momentum to stop. With the blessing and assistance of the Board members in the old organization, the Air Race Classic (ARC) organization was created in 1977. It continues today with two of the original board members still active.

As in many pioneering endeavors, the AWTAR spawned many procedures pilots live with today, for example, regular inspections of aircraft for airworthiness and up-to-date documentation, and the use of oxygen above 10,000. The ARC would continue to test pilots and their equipment and expertise.

Over the years the ARC has been to 39 states; in 1996, the race, sponsored in its entirety by Embry-Riddle Aeronautical University, began at its Prescott, AZ, campus and terminated at its Florida campus in Daytona Beach, a total of 2,381 statue miles. Pilots flew in VFR weather the entire route, a first, but endured record-breaking heat throughout the length of the race.

Veteran racer Mardell Haskins offered this advice to pilots considering flying in this type of race: "Always keep going as if you're winning, regardless of how badly you think you're doing. You don't know how anyone else is doing."

ARC President Pauline Glasson flew the 1996 race with one of her students as copilot and another as a passenger. Pauline, who flew 24 of the 29 PPDs, has flown all of the Air Race Classics with a greater variety of copilots than any other race pilot.

Ann Savage flags off Racer #2 during the 1993 Air Race Classic. Conrad Camden monitors an air-to-ground radio.

First for 1975 included the first Japanese team, Yae Nozoki and Chiyoko Murakami; the first female U.S. Navy team; the first grandmother/granddaughter team, Dell Hinn and Gail Champlin; and the first Sopwith Camel to enter the race, flown by the character Marci from the Charles Schulz comic strip "Peanuts."

Later that year, a difficult decision was reached; the 1976 race would be the last, for several reasons, including the increasing difficulty of planning the course due to escalating Terminal Control Areas, as well as the constant search for funds and accommodations for hundreds of racers. The aviation world was shocked.

Anticipating a deluge of entries, arrangements were made to accommodate 150 teams; on opening day, 235 properly postmarked entries had applied. Finally, 200 were permitted to enter, coming from as far away as South Africa, Alaska, the Bahamas, Canada and Rhodesia.

The final race, coinciding with the bicentennial of the American Declaration of Independence, would be the longest route — 2,926 miles between Sacramento, CA, and Wilmington, DE. The start was adjusted to allow faster planes to depart first, and weather provided a challenge along the route.

Late in 1976, AWTAR officials Marian Banks, Wanda Cummings and Kay Brick visited the new, unfinished National Air and Space Museum in Washington, D.C., to check the updating of winners' names on the PPD trophy ensconced there, and to determine disposition of memorabilia. Jack Whitelaw, deputy director of the museum, declared, "It's a shame to end the Derby at 29 years. Why not round it out to 30, following the first route? We could consider it our July 1977 'Milestone of Flight.'"

Thus the 30th Commemorative Flight was born. A complete overhaul of the rules permitted a broader spectrum of pilots and planes, and passengers were allowed if they ever held pilot status. Entry response was overwhelming, prompting the original 99 planned to be increased to 150 aircraft. Participants included 52 grandmothers; seven FAA flight examiners; three flying the race for the 24th time — Pauline Glasson, Pat Gladney and Gini Richardson; and all members of the International Board of Directors.

Jacqueline Cochran braved the 127-degree temperature at Palm Springs to flag off TAR #1. Wishing blue skies to all contestants were the first PPD racer, Caroline West; first Derby winner Clema Granger; and cartoonist Milton Caniff. At the Tampa terminus were celebrities including Charter 99s Viola Gentry, Louise Thaden and Blanche Noyes.

Why did participants do it? As contestant Helen Shropshire tersely summarized it, desire to compete in the Derby is:

"To TOUCH the past and thus preserve the traditions of our pioneers. To SHAKE the boundaries of earth and view the beauty of my country. To COMPETE with my contemporaries and accept the challenge of the sky."

FAA Administrator Langhorne Bond bestowed a baccalaureate on the Derby with these words: "The Powder Puff Derby was a grand event carried off with unusual expertise and professionalism, and most assuredly will rank with those aviation feats of our biplane pioneers."

PALMS TO PINES AIR RACE

excerpted from a copyrighted article by Betty McMillen Loufek

On this Friday morning in August, Ann Savage stands at the edge of the Santa Monica Airport runway, her starter's flag raised. Behind Savage stands race official Jeffrey Weldon, radio in hand. Engines purring, 44 planes flow slowly forward on the taxiway while the one on the runway holds.

The control tower signals "go." The word is relayed, the flag flashes. Brakes released, the lead plane gathers speed. Others follow at 45-second intervals, and the 1996 27th Annual Palms to Pines Air Race for women is on its way.

The 1996 race is 750 miles, from Santa Monica, CA, to Bend, OR, with timing or stops at Modesto and Red Bluff, and an overnight at Redding, CA. The race is open to all women with current private or commercial certificates. Passengers also must be female and at least 16 years old.

Friends race together year after year, as do family members. There have been teams of mother, daughter and grandmother; many mother-daughter teams; sister teams; aunt-niece teams; and grandmother-granddaughter teams. Awards are given each year to each "family" team.

The race operates with labor volunteered by 99s and their families. Entry fees pay race costs, trophies and prize money, which is limited to $500 divided among the first five places. Prize money is deliberately kept low so that the focus remains on the fun of flying.

The Palms to Pines race began 27 years ago when John Koich of Independence, OR, asked his former flight instructor Claire Walters at Santa Monica to run a women's air race to his town. Claire and several 99s members flew to Independence and found a grass field with few planes, no hangars and no motels. The town had purchased land and given it to the state for an airport, but nothing had been developed.

The air race changed all that. Claire became chairman of the race committee. Scorers, timers and handicappers were found, and race stops were designated. Four months later, 34 aircraft flew in the first race. And because of the popularity of the annual event, the state began constructing taxiways, runways, lights and hangars.

For 12 years the race ended at Independence, even though the Willamette Valley weather was often marginal. The 13th year, weather blocked flying beyond Klamath Falls, OR, and the race ended there. The next year the race went to Sunriver, OR, where fog delayed the finish. Bend, 15 miles north, enthusiastically welcomed the race, and has ever since.

Ask an air racer and she will tell you she'd like to see more information about any and all air races and rallies, large or small, in this second volume of history about 99s. We've included only three, but these races, particularly the PPD, spawned many races around the world — from chapter-size proficiency events like the Nutmeg Air Rally, to the Zimbabwe Sun Air Rally, to races the size and longevity of the Michigan SMALL race and the Mile High Derby. Perhaps it's time for a history of air racing.

Lindbergh, a gentle, very private person, made her second appearance at a public function in nearly 40 years, when she accepted the Amelia Earhart Medal at the Lindbergh Memorial Fund Dinner at the Waldorf Astoria in New York.

Ruth Dobrescu presented it on behalf of The 99s, for her "achievements as an aviation pioneer... accomplishments as an author... and shining example of gracious womanhood." As long-time collectors of first-day covers, Ruth and her husband Charles were a strong force behind issuance of the Charles A. Lindbergh commemorative stamp. As Charles Dobrescu put it, a stamp is the greatest honor that can be given a person since it lives forever, in albums all over the world.

Anyone who pictures women pilots as a bunch of lightweights who want to get into a man's world for the thrill of it simply is a generation (or two) behind the times. The 99s' humanitarian work is limitless: Happy Flyers, Flying Samaritans, Blood Flights, and medical airlifts are good examples.

Janie Postlethwaite, receiver of her chapter's Pilot of the Year Award in 1976, co-founded the Happy Flyers, an international organization of ham radio operators and pilots, with her husband Hartley. For the first time, through development of new techniques and inexpensive special equipment for ELT monitoring and DF radio location, rescuers can be led to a crash site accurately and quickly, and the vast number of false alarms can be greatly reduced. One life already has been saved, a severely injured 10-year-old girl in a plane crash in Colorado, found alive on Dec. 29, 1977, two days after her grandparents were killed, and after severe weather had hampered other rescue operations.

Aileen Saunders is another honors recipient who flies for others. A Powder Puff Derby winner, she was at the controls of a plane weathered in El Rosario, Mexico, in 1961. She and the 99s with her found a desperately poor but hospitable village in need of food, clothing and medical care. Their first pre-Christmas airlift included a doctor, and from his observations grew the year-round, bi-weekly airlifts of the Flying Samaritans, bringing volunteer doctors, nurses and technicians, donated equipment and medicine for an empty government-built hospital.

Rosella Bjornson made aviation history in 1973 when she became the first woman to fly for a major scheduled airline in Western Canada.

Blood flights, carrying donated blood from outlying towns to city processing centers, have spread to 99 chapters all over the country from their beginning in 1975 with the Minnesota Chapter. Blood processing must begin within four hours of its drawing, an impossibility without airlift. A mutually-beneficial arrangement had the Red Cross paying for gas and oil, the 99s pilots building time and updating ratings, and recipients getting their red cells, platelets, etc., while still fresh and useable.

Ninety-Nines also have set up, through DRF, an informal transportation of medicines across the country, eventually going into Mexico. A 99 flies her own aircraft full of medical supplies to another 99 who flies the next leg. Pat McEwen, for instance, former International president, at one time used her hangar as a way station loaded for donated medical supplies to be ferried out west.

All military services are now open to women,

Karen Kahn

including positions as jet pilots. Army women are flying helicopters and Naval aviators include Ensign Rosemary Conatser and Lt. Judith Ann Neuffer.

Many women such as Lorraine Jenick and Jan Dietrich have been flying the more sophisticated planes as corporate pilots. Lorraine flies a Jetstar for Xerox Corp. Jan flew for Golden Pacific Airlines at San Francisco, then went into corporate flying on Corvairs and DC-7s. In 1968, Jan was the first U.S. woman to receive an ATP in four-engine jets.

Among women who have made special contributions in various fields is Mary Jo Oliver Knouff in aviation education, first in Montana schools, then at Cessna Aircraft Co., then with the FAA. Past president Betty McNabb has a missionary fever about aviation education and Civil Air Patrol.

Margaret Mead has a double-barreled talent (not in anthropology), selling over a million dollars in Piper airplanes in 1969, and as a winner of numerous air races. She wound up tutoring in racing clinics, and selling corporate jets.

Another 99 taking the early astronaut tests was Wally Funk, later an accident investigator for the National Transportation Safety Board. Joyce Case, who started her dual when she could barely reach the pedals, was three tines the women's national aerobatics champion, holds an ATP, directed training programs at Cessna, became a production test pilot for Beech Aircraft Corp and then joined the FAA.

Forty years ago, Amelia Earhart told reporters, "Treat me as a human being, a person, rather than a woman." Today, with men hired as flight attendants or stewards, and over 1,000 women as pilots on every major U.S. airline, equality of opportunity is at least on the way. More than 15 other countries are known to have women pilots on airlines, and there may be more.

There are still, however, occasional remarks heard like, "Who's the broad in the cockpit?" and "I didn't know the captain had a secretary." A sense of humor helps.

One sign of change in the air was the January 1978 report that upcoming pilot labor contracts were adding six-month maternity-leave clauses. What ever happened to the old complaint that it wasn't fair, since men don't get pregnant? Perhaps that belongs in the same category as the decision of certain city fathers not to change references from "man-holes" to "person-holes."

A pioneer pilot and long-time flight training

With a 1941 Vultee BT13, a basic trainer for WWII pilots, are past president Janet Green (in cockpit) and Louise Pfoutz.

A member of the Long Island Chapter since 1971, Ida Van Smith earned a slot in the Smithsonian's Air and Space Museum for her work in providing children with exposure to aviation through the Ida Van Smith Flight Clubs

simulator operator for United Airlines was Doris M. Langher who, with her accomplishments and honors, paved the way for acceptance of other women pilots by major airlines, though she herself did not hold such a position. Joining United in 1935, Doris taught virtually every United pilot in the big simulators. An ATP, she piloted Sen. Estes Kefauver in his 1956 presidential campaign, and in 1966 was appointed to the President's Women's Advisory Committee on Aviation.

Captain Emily Warner earned her four stripes with Frontier in 1976, in less than four years: She has flown the Boeing 737, Corvair 580, and deHaviland Twin Otter. Before going to Frontier, she was Clinton Aviation Co.'s flight school manager and chief pilot in Denver. Emily flew the DC-8 for United Parcel Service, then returned to Denver with the FAA. Her uniform is on display in the Smithsonian Air and Space Museum.

She earned the Amelia Earhart Award in 1973 as outstanding woman in U.S. aviation, the Wright Brothers Memorial Trophy in Colorado, and in 1974 was given membership in the Air Line Pilots Association. Nothing came to her on a silver platter — she was persistent, and a hard worker.

Rosella Bjornson, Winnipeg, Canada, flew as an infant on her father's knee; he was a Flying Farmer. Her playhouse was an old Anson Mark V from WWII, which she "flew," sans wings, with her dolls and sisters. Rosella stepped into the majors flying for Transair Limited in a Fokker F28 twin jet (resembling a DC-9), one of only two flying in North America.

Claudia Jones, second officer of a Boeing 727 jet for Continental Airlines, is also a helicopter pilot and president of the Whirly Girls. As a singer and dancer, and accomplished on 19 musical instruments, she has entertained all over the country. It was through her transportation difficulties that she got into aviation; someone in the troupe had to learn to fly and she was elected. She and her husband Hal also own Oases Aviation in Las Vegas. A little old lady once saw her in the cockpit and asked, "And how many of those do you have?"

"About 1,100," answered the stewardess — the number of Continental pilots at that time.

An interesting father-daughter commercial pilot team has been that of YS-11 First Officer Denise Blankenship, flying for Piedmont, and DC-9 Captain Clyde D. Blankenship for Eastern Airlines, both based in Atlanta. Though at 5'11" Denise is taller than many pilots, she had been warned she'd never make it — she doesn't smoke, drink, or chase women! Pilot requirements do change.

Turi Wideroe of Norway flies for SAS over the Arctic Circle. She worked her way up the hard way, though her father does own a small airline: eight years of bush flying and competition with 200 men applicants for the same job. We're not sure we agree fully, though, with her offhand comment: "A modern aircraft is built so simply that a child could fly it." As recipient of the 1970 Harmon Trophy, Turi is not exactly a child.

Another 99, Yvonne Cunha of Antwerp, Belgium, progressed from throwing her dolls out the window to see them fly, through glider school, to powered flight. As a second officer for Trans European Airways on charter flights in Europe, Africa and the Middle East, training in Boeing 707-720 in Ireland at Aer Lingus, her work required fluency in English, French and Flemish, and designing her own uniform.

Mary Hirsch taught flying in Virginia, Okinawa and Hawaii, and ferried planes across the continent. She also served as chairman of the Women's Advisory Committee on Aviation for the FAA, then as a first officer for Continental Airlines. She sees no difference between men and women in the cockpit. She equates her ATP ticket to a Ph.D.: not many have earned either.

Another Continental pilot is Karen Kahn. She finds acceptance of women pilots today good, both in the cockpit and among passengers. Among her considerable past flying experiences was a stretch for a film production company.

Terry London Rinehart seemed destined to fly. Her mother, well-known 99 Barbara London, was a WASP; Terry flew for Western Airlines, her husband for United.

Beverly Bass, with American Airlines, got her private certificate at Meacham Field in Texas and was hooked. Her first part-time flying job, still in college, was transporting corpses in an aging Bonanza. She found flying charters a big help in building time toward her ATP. Her most valuable experience was flying freight at night for an air taxi company in a twin-engine plane, and in a single-pilot operation. She feels her achievement of an airline pilot's job in 5 1/2 years of flying is due to being in the right place at the right time.

All of these women, of course, have built on a foundation laid by pilots like Edith Foltz Stearns, Ruth Nichols, Helen Richey, Nancy Ellis Leebold of Australia, and England's Yvonne Pope.

Central Airlines broke precedent to hire Helen Richey as copilot on their route between Detroit and Washington, over vehement objections of fellow pilots — all male. In 1934 she held with Frances Marsalis the world's refueling endurance record for women, nearly ten days. By 1940 she had over 10,000 flying hours in her log. At 5'4", the best other pilots found to say about her was, "She flies like a man!"

Today, 99s in commercial slots need no better compliment than, "She's a good pilot. I'd fly with her anywhere."

REACHING FOR THE EIGHTIES AND BEYOND

Amelia Earhart said, "If enough of us keep trying, we'll get someplace." By the 50th anniversary of the 99s (1979), the WWII Women's Airforce Service Pilots (WASP), gained veteran status for their wartime efforts. Women took their places in the cockpits of national airlines. With the exception of the Marine Corps, the military services opened their ranks to women pilots. Women were selected as astronaut candidates, held key positions in aerospace advertising and marketing and sales, and worked as engineers, scientists, lawyers and leaders in important aviation research and development projects.

Women also held one-of-a-kind flying jobs through the world. In a typical day, Dr. Anne Spoerry of the East African Flying Doctors Service flew over rugged deserts, mountains and deserted shorelines, often holding clinics directly on the airfield.

Women became employed in highly technical civilian and government positions. By 1979, more than 5 percent of air traffic controllers were women. Women also were employed as Air Safety Investigators with the National Transportation Safety Board.

In the 1930s, Helen Richey was the first woman airline pilot. By 1979, 110 of the 45,000 airline pilots were women — perhaps unimpressive on the large scale, but a giant step for women in the industry. At that time, Kim Goodwin, age 22, was the youngest female member of a major air carrier flight crew.

Leading the way to space in 1963 was Valenhtina Tereshkova, a Russian woman. For U.S. women, 1979 proved to be a banner year in NASA's space program. From a field of 35 highly qualified female astronaut candidates, six women were selected by NASA as mission specialists for assignments on future crews. Duties of the mission specialists were to coordinate with the shuttle commander on space shuttle operations in the area of crew activity planning, consumable usage and activities affecting experiment operations. At that time, though, none of the candidates were interviewed for astronaut pilot.

Among these extraordinary women were 99s, adding to achievements in aviation as well as being noted as history-makers within the organization.

Two years prior to shuttle candidates being chosen, Maude H. Oldershaw of Bakersfield, CA, became the first woman to fly the Gossamer Condor. "The Gossamer Condor Hanger Mother," as she was nicknamed, told us about the strange, see-through craft with a wingspan of 96 feet, or longer than a DC-9.

"The craft is built from piano wire, aluminum tubing, and a covering of clear mylar. The low, front-positioned canard wing assisted with turning control, and the aircraft was peddled into the air like a bicycle, with the pilot seated in a cocoon-like, wind-resistant enclosure positioned at a 45-degree angle. The 12-foot plastic prop, located aft of the wing's trailing edge, pushed the 70-lb. craft aloft."

On Nov. 2, 1979, the 99s celebrated their golden jubilee in New York. Over 600 members attended the festivities. In addition to the aforementioned, women still were breaking records and achieving in other ways. Earlier that year Kathleen Snaper, a Las Vegas flight instructor, asked and received approval from the FAA, National Aeronautical Association and National Park Service to set an official low-altitude endurance record at Death Valley. In her words, "That's one record that no one has tried before."

On Jan. 16, 1979, Kathleen set two new world records. One was for low-altitude flying and the other was for covering the longest distance in a closed course at low altitudes. Kathleen flew for four hours at an altitude of 25 to 30 feet off the desert floor below sea level in dangerous, unpredictable air currents to achieve her goals.

The 99s also were celebrating other aviation achievements. Susan Horstman, Kansas Chapter, started classes to be the first woman to fly for National Airlines. Angela Masson of the Golden Triangle Chapter, who was flying for American Airlines, was in the process of upgrading to a DC-10. She was the first woman assigned to a jumbo jet.

That banner year, two 99s received the Bishop Wright Air Industry Award. Jerrie Cobb, the first woman to complete Lovelace Clinic's astronaut tests, received the International Harmon Trophy. The trophy, presented by the president of the United States, was awarded for her previous 14 years of dedication, service and humanitarian flights to the people of the Amazon basin.

Ida Van Smith, founder of the Ida Van Smith Flight Clubs, was given the award for her dedication to children in the Long Island area. For 12 years, Ida cultivated youngsters' interest in flying while encouraging high scholastic and ethical standards in the field of aviation.

The impressive list of aviation accomplishments continued. Mardo Crane received the Barnstormers Trophy for her contributions to aviation, including having served as first chairman of the Powder Puff Derby. Lois Feigenbaum received the Laurence P. Sharples award for her outstanding contribution for the advancement of general aviation. Joyce Case became the first female production test pilot for Beech Aircraft, testing everything from the King Air down the line. Jean Ferrell, a DC-10 flight operations instructor for United Airlines, received a merit award from the company. Sadly, not more than a year later, The 99s would lose Jean and another accomplished member, Marion Barnick, on an ill-fated flight of an Air New Zealand DC-10 that crashed during a sight-seeing flight over Antarctica. A scholarship in Marion's name has helped many young women since.

By 1979, The 99s had grown to 164 chapters, including two in South Africa. Membership had grown to over 5,000 members.

All 152 hours of television coverage of the Moscow Summer Olympics were canceled after the U.S. withdrew from the Games in 1980 to protest the Soviet invasion of Afghanistan. At last count, 300 million fans in 57 countries shared a common obsession, "Who shot J.R. Ewing?" of television's nighttime soap "Dallas." While Mount St. Helen's volcano erupted, women also were exploding into the world of important aviation positions.

The 99s launched the new decade of the '80s with an organization-wide media day. Each chapter was encouraged to spread the word about the mission and work of The 99s. In addition, the organization became a charter member of the World Aerospace Education Organization.

Members continued to introduce others to the exciting realm of aviation. The Wyoming Chapter took deaf children for thrilling airplane rides. California State Rep. Carol Hallett used her aircraft for transportation to meetings and political events across the state. The first all-women's aerobatic team, consisting of Betty Stewart, Paula Moore and Patti Johnson, swept the World Aerobatic Championships. Winners of the 1980 Air Race Classic, Texans Pat Jetton and Elinor Johnson, collected $3,000 for averaging 248 miles per hour over their handicap in a Beech 33A.

Each year brought another year of "firsts" to the history pages. Another 99 aviation pioneer, Janice L. Brown, Bakersfield Chapter, set the first solar-powered endurance records in the Solar Penguin in August 1980, while Emily Warner, the first modern female airline pilot in 1973, was honored with the 1980 Annual Achievement Award sponsored by the International Northwest Aviation Council.

Sherry Knight of the Santa Rosa Chapter was the first woman to fly for the California National Guard and the first to complete the Guard's short course for pilots, transitioning from fixed-wing to rotorcraft. Betty Rogers became the first female airworthiness inspector for the FAA. (*The 99 NEWS* saw its first advertisement for epaulet shirts especially tailored to a woman's figure.) And Hallie McGonigal, Monterey Chapter, told 99s about her job as a flight instructor and charter pilot. "I love every minute of it," she explained, "and they pay me for it. This is a dream job."

While her husband, John, held the family Bible, Sandra Day O'Connor was sworn in as America's first female justice of the Supreme Court by Chief Justice Warren Burger. This historic event was one of many to take place during the decade of the 80s that placed women in the public eye. Women increasingly captured significant roles traditionally held by men.

The space shuttle Columbia roared off the pad at Cape Canaveral on its historic first flight in April 1981.

Outside the Washington Hilton Hotel, John Hinkley Jr. made an unsuccessful assassination attempt on President Ronald Reagan's life on March 30, 1981. Two months later, a similar assassination attempt was made on Pope John Paul II at St. Peter's Square in Rome, Italy.

The world watched England's Prince Charles marry Lady Diana in the century's grandest wedding. With a congregation of 2,500 under the great painted dome of St. Paul's Cathedral, more than 75 technicians manning 21 cameras dazzled an estimated television audience of 750 million. At the movies, the endlessly-talented George Lucas crafted a real cliffhanger with the enormously popular "Raiders Of The Lost Ark."

In the exciting world of women making strides in aviation, the new year brought the first ever mother-daughter airline pilots. Claudia Jones worked for Continental Airlines, and her daughter Cathy for Western Airlines. Cessna was the high bidder for *The 99 NEWS* back cover, sporting a Stationaire 6 on floats, Cessna 172s and the Hawk HP. The future looked bright for aviation. No one could conceive of the idea that Cessna ever would cease production of their single-engine models.

Muriel Earhart Morrissey, Amelia's sister, and President Marilyn Copeland flank the number-one bronze casting of a bust of Amelia made by artist Don Weigand to recognize her contributions to aviation. The bust was commissioned by Debbie and Jack Scharr and presented at a Headquarters reception March 25, 1983.

Three 99s made the 1984 U.S. Aerobatic Team. Members include (standing) Gene Beggs, Harold Chappell, Henry Haigh, Kermit Weeks, Alan Bush, (kneeling) 99s Debby Rihn, Brigitte de St. Phalle and Julie Pfile, and Linda Meyers.

PROJECT AIR BEAR

In January 1987, the Chicago Area Chapter worked with the Illinois Department of Transportation Bureau of Aviation Education and Safety to begin presenting Project Air Bear to school children in the Chicago area.

Phase I, "Air Bear Goes to the Classroom," worked on the premise that kindergarten and first grade children are especially receptive to stimuli to which they are exposed. To have the maximum impact in familiarizing future citizens with aviation, 99s felt it should be introduced to this particular age group.

In the Air Bear Program, everyone is a participant. All of the children have job assignments revolving around making a commercial flight, complete with props to make it more real to them. Air Bear 1 takes an imaginary trip to Disney World.

After the flight is completed, all the children have the opportunity to be the pilot and receive a pilot certificate. Comments from teachers and parents included, "They really enjoyed acting out the various jobs, especially the pilot."

Chicago area teacher-presenters included Polly Gilkison, Bev Greenhill, Connie Miller-Grubermann, Marge Krupa, Ruth Rockcastle, Sharon Ann Schorsch and Pat Thomas.

Phase II, "An Airport Field Trip," begins with a visit to the airport ramp. The children will be given the opportunity to sit in the pilot's seat of an airplane. They also will visit a hangar and have a short lesson in aviation language. A snapshot of the group with an airplane will also be taken.

In 1981, Janice L. Brown set distance, time airborne and altitude records in a strange sun-powered aircraft called the Solar Challenger.

STS-7 Mission Specialist Astronaut Sally K. Ride strides away from a T-38 aircraft at Ellington Air Force Base following a flight.

The library shelves at Headquarters grew from an armful to over 300 volumes. The Houston Chapter vowed to keep the library growing. The valuable Archives took on a representation of the past, present and future of the organization.

Janice Brown was in the news again for piloting the solar-powered Solar Challenger over the Arizona desert to an altitude of 3,500 feet. She set records for altitude, distance flown and time airborne. The Solar Challenger was a craft 29 feet long with an 11-foot propeller, weighing 175 lbs. The experimental aircraft cruised at 20 to 30 mph, using a 2.47 hp electric motor, powered by 15,000 photovoltaic cells located on its 47-foot wingspan.

And more firsts for the 99s: In 1981, Carolyn Pilaar, Greenville, SC, won a spot on the U.S. Precision Flight Team. Juanita Blumberg and Bonnie Quenzler won the Air Race Classic. Karen Cox was the first FAA-certified female flight navigator. As a hurricane-hunter flying for the NOAA, Karen took her checkride in a C-130 over the Gulf of Mexico between Miami and San Antonio. Jerrie Cobb, missionary-pilot to the Amazon, was nominated for the Nobel Peace Prize. Olive Ann Beech, known to many as "the first lady of aviation," was inducted into the Aviation Hall of Fame in Dayton, OH.

The 99s commemorated the 40th anniversary of the Amelia Earhart Scholarship program. The Frank G. Brewer Award was given to Margaret Goldeman and Jeanne McElhatton for outstanding achievement in aerospace education. Judy Wagner won the Angel Derby, another long distance race, in her 33C Bonanza. CPT Leah Moser was one of the first three women to graduate as a pilot in the Canadian Armed Forces. Bonnie Tiburzi, Greater New York Chapter, was recognized for her significant achievements in aviation.

Another humanitarian, Central Illinois 99 Wanda Whitsitt, founded a non-profit organization of private pilots named Lifeline. The organization was established to fly medical supplies, blood, organs and health personnel where needed, as well as to assist in disaster relief operations.

The 99s learned about Safe Air Taxi Inc., located at Fort Lauderdale-Hollywood International Airport. Each pilot was a veteran with numerous ratings and considerable experience. The all-female air taxi company operated a fleet of aircraft from a Twin Beech model 18 to single-engine aircraft. Besides frequent trips to the Bahamas, the staff marketed their skills by offering the public a mystery dinner flight. Based on the couple's dining preference, they were whisked off to a mystery location within an hour's flight of Fort Lauderdale for dinner, then returned.

That year the North Central Section celebrated its golden anniversary with the slogan, "50 years in the making, 24 months in the planning and three days for the celebration."

The computer was touted as the machine of the year in 1982. By the millions, personal computers beeped their way into schools, homes and businesses around the world.

Though American women took stock and stepped up the pace by holding jobs in politics, education and science-related positions, they still were fighting for equal pay. In 1972, women still wondered hard about the possibility of having a family and career and being able to manage both. In 1982, more women — including some of the daughters of the past generation — take all this as a birthright.

The issue of disabled pilots was on the minds of 99s during the spring of 1982. Alverna Williams Bennett, a double-leg amputee, overcame scrutiny, several FAA checkrides and a court battle before receiving her medical certificate. In the end she bravely set a precedent for physical medical waivers, paving the way for other disabled pilots who followed. A photo of Alverna sitting on the wing of her Ercoupe is proudly displayed at the Smithsonian Air and Space Museum in Washington D.C.

During the years, The 99s placed an emphasis on a "seek and keep" program, encouraging existing members to recruit and keep female student pilots, termed the "66" Program. Education, legislation and community service were ongoing 99 contributions.

The pages of *The 99 NEWS* included continuous airmarking events, while other chapters held high flying density clinics. Ninety-Nines hosted GENAVAC (the General Aviation Council) to discuss problems of mutual interest to all involved with the future of general aviation. Others participated in the first U.S. Proficiency Flight Team event held that year in New Orleans. The competition measured navigation, flight planning and landing skills.

The 99 Minnesota Lady Lifeguards, who flew much-needed blood to various locations, participated in Red Cross recognition day for Lifeguard flights. Hundreds of miles to the east, Patricia Blum, Greater New York Chapter, was impacting other lives in a positive way. She founded the Corporate Angel Network, helping to set up a means for cancer patients to fill an empty seat on corporate flights. By June 1984, she received a volunteer action award from President Reagan.

The 99s were honored with the Crown Circle Award at the National Congress on Aerospace Education for their work promoting and educating in the areas of aviation and space technology fields. That year, the international convention was held in St. Louis and Sally K. Ride, future shuttle astronaut, was one of the newest members.

The year of 1983 saw the 200th anniversary of manned flight, and the first space flight of a U.S. woman astronaut, Sally K. Ride. Though many 99s were making tremendous strides in aviation, events of 1983 were overshadowed by Ride's ride. A member of the South Central Section, Ride was a former schoolgirl tennis star with a Ph.D. in physics. She was cool, witty, attractive and possessed as much of the "right stuff" as any man preceding her into space. At the launch site, 99s watched with great interest and excitement. As the space shuttle lifted off the launch pad, 99s shared a surge of feelings fraught with pride, apprehension, anticipation and joy, along with millions of people around the world.

Consumers celebrated a rebound year of turnaround in the economy. Growth continued to surge, unemployment fell, inflation stayed low, and businesses slimmed down and shaped up. Just as music lovers thought they had assembled the best audio system, a new technological innovation, compact discs or CDs were introduced to the marketplace, making expensive turntables and libraries of LPs as out of date as Edison's first talking machine.

In the U.S. the breakup of AT&T caused a communications upheaval, and Trivial Pursuit became the country's hottest board game.

The 99 president, Marylin Copeland, stated priorities for the upcoming year. Members were encouraged to participate in safety seminars, airmarking events, sponsor a USPFT, NIFA or other flying activity, conduct flying companion seminars, safety education projects, maintenance and mountain flying clinics. In addition, she strongly encouraged members to achieve by acquiring new ratings, participating in air race activities and membership meetings, and contributing to the AE Scholarship Fund.

The 99 Memorial Project Fund was established at headquarters for building expansion, scholarship programs and a resource center to reflect on the organization's original purpose, "to provide a close relationship among women pilots and to unite them in any movement that may be for their benefit or for that of aviation in general."

Canadian 99s enlightened the entire organization with history about the famous "Flying Seven," a group of seven Canadian women pilots from the 1930s. The women established a training school to teach other women ground school, fabric work and parachute packing. They also stated that by WW II, aviation fuel sold for an exorbitant 90 cents a gallon. Additionally, a Gypsy Moth trainer was $17 per hour, but $15 per hour solo. Flight instructors were considered well-paid at $3.50 per hour!

From historical events to the present, 99s learned more specifics about their plucky Canadian sisters. Rosella Bjornson became Canada's first woman airline pilot. Elizabeth Webster, a petite mother of four, flew a deHavilland Beaver on floats into the wilds of Northwestern Ontario.

Mimi Tompkins, a member, was profiled in an article about women pilots, Hawaiian style. A year later she would be recognized on a national scale for her heroism and fine pilot skills during an Aloha Airlines emergency. Her Boeing 737 experienced a structural failure, blowing a section of the airliner's top off.

Ninety-Nines continued to break world records and make major contributions to the aviation world. Los Angeles businesswoman Brooke Knapp set a new around-the-world speed record in a Lear 35, circling the globe in 50 hours, 22 minutes, 22 seconds, shattering the previous record held by Hank Beard. The Air Force's first all-female transatlantic flight crew flew in a C-141 Starlifter. Eight women competed in the USPF nationals and 99 Pat Dennehy, Palisades Chapter, completed a special celebration of her 30th birthday by setting five world-class speed-over-distance records in her 30-year-old Cessna 170. For the second time in as many years, a 99 was named flight instructor of the year.

Adventurous 99s gave others tips about flying to the scenic Bahamas, into the remoteness of Southwest Africa, Kenya and "Down Under" to Australia. Alaska 99s invited others to visit their majestic, rugged state, while other 99s participated in FAA-sponsored survival courses.

1983 proved to be another year of celebrations. At headquarters in Oklahoma City, International Marilyn Copeland and Muriel Earhart Morrissey, Amelia's sister, unveiled a stainless steel bust of Amelia Earhart during a memorable ceremony. As part of the International Forest of Friendship celebration, 99 Lucile M. Wright honored Hideko Yohoyama, the first licensed woman pilot in Japan. The ceremony was held in Tokyo upon the completion of United Airlines' inaugural flight to Japan.

By March 1983, 48 chapters had completed 65 airmarkings, and the 99s had grown to 5,900 members. In June, The 99s recognized Terry McCullough of the newly-formed Heart of Texas Chapter as the 6,000th member.

Wanting to be associated with an outbreak of peace, Sarajevo opened its impressive mountain peaks and passes to host the 1984 winter Olympic games. The summer games, held in Los Angeles, were boycotted by the Soviet Union and were dominated by Americans, including 4-foot, 9-inch gymnastics dynamo Mary Lou Retton. South African Bishop Desmond Mpilo Tutu was awarded the Nobel Peace prize, and Donald Duck turned 50. Along with the discussion of Ronald Reagan's "Star Wars" defense system came the theory that if used, such a system could cause a nuclear winter. Blasts from such a war would reportedly send 10-24 million tons of dust into the stratosphere, blanketing 99 percent of the sky with plumes of thick dust and smoke that would blot out the sun and cause frigid, grim, climatic consequences.

In 1984, one of the largest 99s convention was held in Anchorage, AK. Many 99s, including members from Finland, flew their own planes to Anchorage. Past President Broneta Evans was enshrined in the Oklahoma Aviation Hall of Fame and Patrice Francise Clark became the first female commercial pilot in the Bahamas. Later she would be employed with United Parcel Service, and eventually become the first black female captain working for a major air carrier.

In 1985, Charter Member Ila Fox Loetscher, known as the "Turtle Lady," works with one of her charges. Loetscher's organization, Sea Turtle, Inc., works to save endangered sea turtles.

In 1984, the National Aviation Hall of Fame in Dayton, OH, presented the Spirit of Flight Award to The 99s. With the award are Dr. John Copeland, then-International President Marilyn Copeland, presenter Anne Sawyer Green, Mrs. Wally Schirra and retired astronaut Wally Schirra.

ASTRONAUT RESEARCH GROUP REUNITES

from an article by Gene Nora Jessen that appeared in the July-August 1994 issue of the International Women Pilots/99 News magazine

The original Mercury astronauts' hijinks while undergoing physical exams at Lovelace Clinic in Albuquerque, NM, could best be described as "The Right Stuff." A group of women also endured those tests in the early years of the astronaut program, but they were all business — not at all sure of the outcome of their segment of the program and unwilling to chance frivolity.

Twenty-five American women were invited to tackle the astronaut physical exams, a six-day endurance test at our country's premier center for space medicine. They were all commercial pilots holding second-class medicals, and also had at least 1,000 total flying hours and a college degree.

These qualifications were not common coming out of the 1950s. Jerrie Cobb had been Dr. Lovelace's first "subject," passing the physical with flying colors and subsequently, the psychological, jet orientation and just about all the state-of-the-art could conjure up at the time. Of the 24 additional women taking the physical exams in 1961, 12 passed, making a base group of 13 women (including Jerrie).

As each candidate prepared to undergo further testing, some had to quit their jobs in order to participate or were fired. Not to worry, all were enthusiastic about seeing how far they could go toward becoming an astronaut. In the fall of 1961, the ax fell and the program was canceled. As the individuals were departing for Pensacola, NASA declared that there was no need for further testing.

Subsequently, Jerrie Cobb and Janey Hart (whose husband was a U.S. senator) testified at a congressional hearing on, roughly, "Is NASA prejudiced against women astronauts?" Of course it was, but it helped little to point that out. John Glenn and the Mercury astronauts testified that women were not qualified to become astronauts because they were not military test pilots. Congress agreed.

Some 20 years later, women astronauts were to join the corps. They included highly-qualified scientists and physicians, thus sidestepping the issue of military test pilot school. Our group had at least set physical parameters for the women to follow.

Most recently, the "Lovelace Class of '61" or FLATS (Fellow Lady Astronaut Trainees) as our leader Jerrie Cobb called us, took on the lighthearted "Mercury 13" tag coined by Jim Cross. Whatever we're called, we had never all met each other — having gone through the testing individually or in pairs, and for a long time didn't even know each other's names.

Even now, no one knows the names of the rest of the 25 who originally took the physical exams. Two of the group of 13, Marion Dietrich and Jean Hixson, are deceased. Two others, Jan Deitrich and Janey Hart, could not join the reunion, but the other nine gathered at the 99s museum in Oklahoma City in May — the first time all had met each other. It was not to be a subdued event.

Lt. Col. Eileen Collins, most recently an Air Force test pilot and now a NASA astronaut shuttle pilot, joined the pioneer group in Oklahoma City for reminiscing, much laughter and lively discovery. James Cross, a Hollywood film producer who was instrumental in bringing the group together, interviewed each member in depth over a period of three days for a television documentary.

Members of the "Mercury 13" group at the launch site include Gene Nora Jessen, Wally Funk, Jerrie Cobb, Jerri Truhill, Sarah Ratley, K. Cagle and B. Steadman.

This historic film will be placed permanently in the 99s Resource Center archives for the use of historians and researchers. The nine in attendance took on the demeanor of proud mothers as they became acquainted with the bright and talented shuttle pilot.

Closure happened for a group of women pilots who had been frustrated by the dearth of female spaceship drivers. Now, 33 years later, it is happening with the shuttle. The group planned to be on hand with a capacity cargo of inspiration, good will and love when astronaut Collins flew in February 1995.

And, surely they were. Collins blasted off Feb. 3 at 12:22 a.m., as the first woman pilot in command of a space shuttle, cheered on by members of the astronaut research group.

Women were continuing to make major strides in aviation. Emily Howell Warner made captain with Frontier Airlines in May 1984. That year, three 99s held positions on the U.S. Aerobatic Team and the Michigan Chapter celebrated 50 years.

The 99s were recognized at the national Aviation Hall of Fame in Dayton, OH, for 55 years of aviation contributions in the areas of general aviation and safety. The Hall of Fame, a public foundation, is the only one of its kind recognized by the U.S. Congress.

The 99s hosted the World Precision Flying Championship at Kissimmee, FL, in August, a first for the U.S. In November 1985, DC-8 jet pilot Manuel Cervero was flying cargo from Miami across the Andes to the Colombian capital city of Bogota. At the very moment he crossed the 17,715 foot-high, long-dormant volcano known as Nevado del Ruiz, it became thunderously alive.

"First came a reddish illumination that shot up to about 26,000 feet," the pilot recalled. Then came a shower of ash that covered the aircraft and leaving no forward visibility. The cockpit filled with smoke, heat and the smell of sulfur. The blast charred the nose of the DC-8 and turned the aircraft's windows white. The pilot successfully diverted to the city of Cali and pushed open one of the cockpit windows in order to see the runway lights on final approach. That eruption of Nevado del Ruiz would rate as one of the deadliest volcanic eruptions in all of recorded history.

U. S. President Reagan and Soviet President Mikhail Gorbachev met in Geneva, Switzerland. The "fireside summit" was the first meeting between leaders of the two countries in six years.

Chrysler President Lee Iacocca's book parked itself on the bestseller list for 36 weeks. Only a few books, such as *Gone With The Wind* and *Jonathan Livingston Seagull* exceeded his mark of 2 million books off the press. Tom Clancy, author of *The Hunt For Red October*, watched gleefully while the White House and the Soviet embassy in Washington displayed a high-level interest in his captivating submarine thriller.

While some 99s took up tandem skydiving, others made progress on plans to expand headquarters. Mary Jo Knouff received the Frank Brewer Award for significant contributions of enduring value to aerospace education. The 99s marveled at the operation of a new technological aviation wonder — the stormscope, an avionics instrument able to detect electrical activity within clouds. Members continued to educate themselves as well as others. In the forefront was a re-examination of aircraft noise levels and subsequent hearing lose due to prolonged exposure. *The 99 NEWS* gave chapters explicit instructions on the ins and outs of airmarking.

Victoria "Vicky" Katherine Wingett was upgraded to captain for Southwest Airlines, while Pat Jenkins used her canary yellow Hughes 300-C, nicknamed "Woodstock," to herd cattle on her 100,000-acre Oregon ranch.

Halley's Comet passed close enough to Earth to be photographed Feb. 24, 1986. Voyager II returned images of distant planets to Earth. The nuclear power plant at Chernobyl experienced a type of core meltdown in its reactor, sending plumes of radioactive gases and particles over the skies of Eastern Europe and Scandinavia.

In London, the air was sweet with jubilation as the world watched the wedding ceremony of toothy young lieutenant Prince Andrew and bonny redhead, Sarah Ferguson.

In January, the space shuttle Challenger exploded 73 seconds into its flight from Cape Canaveral, FL, killing all aboard. It was 32 months before another successful U.S. shuttle mission would be completed. In October 1988, Discovery carried the nation's hopes aloft again on a thundering pillar of fire.

For the 99s, the sadness of the Challenger explosion overshadowed other significant aviation events. The 99s held their convention in balmy, sun-soaked Hawaii, and members wrote about daring air safaris in Kenya and Nairobi. In 1986 the Amelia Earhart Birthplace Museum was opened to the public for tours.

The year 1987 was clouded with a steady stream of U.S. congressional hearings on the Iran-Contra arms scandal, of war threats in the Persian Gulf, huge budgetary and trade deficits, a declining dollar and a crashing stock market.

In aviation, 99 Jeana Yeager and Dick Rutan

WORLD PRECISION FLYING CHAMPIONSHIPS

excerpted from a historical report by Verna West

The Ninety-Nines invited the world to Kissimmee, FL, to compete in the first-ever-in-the-U.S. World Precision Flying Championship, Aug. 11-18, 1985.

The invitation was issued in early 1983, with the formal presentation being made by USPFT Council members Lois Feigenbaum, Hazel Jones, Janet Green and me, Verna West, in Norway during that year's competition.

We came away impressed, enthusiastic and well aware of what we were facing. The Council was expanded to include Jody McCarrell, Pat Roberts and Marie Christensen. Linda Dickerson joined as meeting planner, and we went to work.

The dual challenge for The 99s was to set up a structure under which to hold rallies throughout the U.S. to select a team, and to prepare for a world championship in Florida. Both goals were met.

Sixty-five pilot contestants participated from as far away as New Zealand, Argentina, South Africa and Finland. Teams brought family members, friends, a coach, a team manager, mechanics, etc., as well as two international judges.

Rules for the competition were established by the Federation Aeronautique Internationale (FAI), and judges were selected from a list approved by the Commission Internationale d' Aviation Generale (CIAG).

The navigation contest consisted of a route approximately 100 nautical miles long with 11 legs, 10 turning points, four secret check points and six photographs to be identified and located on the chart with a pin prick. Pilots were expected to execute four different landings, one of each type: normal with choice of power and flaps; forced without power, flaps as desired; forced without power and no flaps; and approach and landing over a barrier two meters high placed 50 meters before the touchdown (0) line, flaps and power as on a normal landing.

Only basic, minimum navigation equipment is used during the competition. Two-way radio equipment is required; additional navigation equipment is taped and sealed against use by the pilot during competition, except in case of disorientation.

As with any volunteer project, those who make it happen are the key to success; in the case of The 99s, the list is too long and the risk of overlooking someone's efforts too great to risk. Suffice to say, everyone who came to help, from near and far, contributed to the success of the first all-woman sponsored Championship.

To quote Verna, "I truly wish all Ninety-Nines could have been there. We proved that women can indeed plan and execute a professional-level World Championship. We deeply appreciate the men who came to help, from Washington, D.C., to California, from Michigan to Texas, and many places in between. A special 'thank you' to our husbands who kept things on an even keel at home so we could be there.

"There is a feeling that comes with the shared experience of pilots from all over the world that transcends all the hard work that went on to make the Championship possible. They had a good time and were appreciative of all we had done to make them welcome. We are happy they could come!"

And come again, they will. In 1991, Jody McCarrell proposed to the International Board of Directors that The Ninety-Nines again bid for a world championship to be hosted in the United States. They agreed, and the bid was presented and accepted by CIAG/FAI. The 1996 World Precision Flying Championship was scheduled at Meacham Field, Fort Worth, TX, Sept. 28-Oct. 5, 1996. And once again, The 99s will pull it off!

NOTE: The folowing is an excerpt from an article by Jody McCarrell in the January/February 1995 issue of 99NEWS/International Women Pilots *magazine.*

For the first time in the history of the Championships, women were appointed as International Judges. The first four were all 99s — Hazel Jones, Pat Roberts, Carole Sue Wheeler and me, Jody McCarrell. Now, several countries have women international judges and team managers, as well as workers — and we take pride in having led the way. Today, additional 99s judges include Jerry Anne Jurenka, Jan Maxwell and Pat Ward.

How to fund an international event of this scope? At the 1993 annual meeting in Portland, OR, members ratified hosting the event and adding $1 to member dues for 1994-95 to help fund the event.

As of August 1996, the Championship are just around the corner, and a number of 99s are committed to helping with judging and otherwise staffing navigation and landing competitions, serving as greeters for the estimated 22 teams from around the world, producing a daily newsletter, selling advertising and other fund-raising for the event, acquiring the 75-100 aircraft needed by the competitors, etc.

And have 99s come through? Only history will tell the full story; based on past experience, they'll come through with flying colors!

The 1987 U.S. Precision Flying Team arrives in Finland for the world championship with high hopes, having performed well in two previous world events. Poor quality rental aircraft contributed in part to a ninth-place finish.

---

completed a record-setting round-the-world journey of Voyager, traveling, 25,012 miles without refueling. The flight began and ended at Edwards AFB in the Mojave desert. Made of stiffened paper and plastic, the Voyager at times carried more than five times its weight in fuel. It took 9 days, 3 minutes and 44 seconds to complete the global flight.

On Dec. 26, 1986, American Airlines Flight 412 departed with an all-female flight crew. Dr. Angela Masson was the first woman to fly as captain in the left seat of a Boeing 747.

In the education arena, Air Bear was introduced to chapters as a way to encourage young children to develop an interest in aviation. In addition, the Young Astronaut program was launched. While 99s showcased women military aviators, the search for any evidence of Amelia Earhart, her plane or remains, continued.

The coveted 1987 Flight Instructor of the Year Award was given to 99 June Nonesteel of Phoenix, AZ. Also in the southwest, the 99s flew relief missions, carrying supplies that would ultimately be sent to areas of the world stricken by drought, famine or war. (Wings for Direct Relief Foundation was founded by 99 Dell Hinn in 1970.)

At headquarters, the oral history program was successfully implemented by member Judy Logue, who conducted a special training session.

The environment continued to be a hot issuein 1988, with concerns over global warming. The world called attention to the thick industrial wastes being produced in the USSR as being a danger to local residents as well as the rest of the planet. Other critical topics included overpopulation; a future ban of ozone-destroying chlorofluorocarbons (CFC's for short, found primarily in refrigerants and spray cans); the destruction of Brazil's rain forests; and toxic waste disposal.

Punk demon "Beetlejuice," played by sprightly actor Michael Keaton, was one of 1988's best feel-weird cinema picks. Food trendies looked for safe culinary havens at moderately-priced Italian trattorias and American bistros. The U.S. watched popular talk show host Oprah Winfrey publicly slim down using a liquid diet, and looked on as German teenager Steffi Graf apologized to the crowd at the U.S. Open for her quick, slam-dunk success over Argentine rival Gabriela Sabatini. Michael Jackson exceeded Bill Cosby as the year's highest paid entertainer, earning $60 million.

Airmarkings, beautification, safety programs, community involvement on a local level and junior astronaut programs continued to be high on the list.

Also a priority was the National Intercollegiate Flying Association at both regional and national levels, and the U.S. Precision Flying Team. The 99s were recognized as a group of impatient, innovative, courageous women united by their love of flight.

On the mind of international President Barbara Sestito was membership. She suggested members consider establishing honorary membership or associate membership programs and solicited member opinion. Should the Amelia Earhart Scholarship Funds be allowed to widen their investment choices? What is the conscience regarding long-term operational and organizational expenses?

That summer the International Forest of Friendship celebration coincided with the 60th anniversary of Amelia Earhart's first flight across the Atlantic.

The 99s moved into a new two-story headquarters building that housed the executive offices and museum facility, which included a premiere archive and resource library located adjacent to Will Rogers World Airport in Oklahoma City. Orchestrating the move was Loretta Grag, executive director, who coincidentally was celebrating 25 years of service with the 99s. Airline captain and 99 Lori Griffith created a new, permanent display

Fear of Flying Clinic co-founders Fran Grant and Jeanne McElhatton established the program after presenting it at a 1977 educational workshop in Oklahoma City.

June 17, 1985, was a red-letter day for 99 Victoria Wingett, when she was upgraded to captain for Southwest Airlines. Vickie became one of the first women pilots in the Air Force program when their doors were opened to women in 1976.

President Barbara Sestito presents June Bonesteel with the 1987 FAA Certified Flight Instructor of the Year Award.

commemorating women airline pilots at Headquarters.

Hurricane Hugo, with a 2,300-mile arc of destruction from the Caribbean to the Carolinas, was marked as one of the most fierce storms of the decade. George Bush was sworn in as president on Jan. 20, 1989. Douglas Wilder became America's first elected black governor, serving the state of Virginia. With no warning, a major earthquake measuring 6.9 on the Richter scale occurred in the San Francisco Bay area in October 1989, causing loss of life and widespread damage.

For the decade, technological wonders included the introduction of the boom box, and increased use of personal stereos, fax machines, cellular phones, home video games, compact discs, videocassette recorder-players and voice mail.

A 99 crew member was present on the first-ever Category IIIa manual approach. First officer Sandy Simons made aviation and 99 history on Alaska Airlines flight 93 by accomplishing the first manual Cat IIIa approach, using a heads-up guidance system and landing in a Part 121 operation on SEA runway 16R with a visibility of 800 Runway Visual Range (RVR).

Ninety-Nines continued to achieve: Carol Rayburn, a 99 since 1963, was the first woman in the FAA to enter the Senior Executive Services (SES). Carol was named manager, Flight Standards Division, New England Region, and with this move became the first in the SES ranks to return "to the field." Hazel Jane Raines was posthumously inducted into the Georgia Aviation Hall of Fame, and 99 Suzie Azar was elected mayor of El Paso, TX. Two 99s, Dr. Peggy Baty and Pat Church, were honored with the FAA prestigious Administrator's Championship Awards for excellence in aviation. Fran Grant was honored by the FAA for her clinics that help adults conquer the fear of flying. The first all-woman flight crew in an Indian Airlines Boeing 737 took off from Bombay to Dabolim (GOA) and returned with two 99s in command of the aircraft.

In chapter and section news, the Arabian Section was formed in October 1989 with 12 charter members. Members of the North Central Section welcomed spring by participating with the American Cancer Society in the Daffodil Day Flower Drop.

The Kansas Historical Society's sites review board approved a $7,500 allocation of federal Historic Preservation Funds to draw up architectural plans and specifications for the restoration of the Amelia Earhart Birthplace in Atchison, KS. Several years later, in 1991, Southwest Section 99s would be honored by the AOPA for their earthquake airlift operation. Members assisted with an airlift of approximately 350,000 pounds of emergency supplies, from food and clothing to teddy bears, to help hard-hit victims of the devastating Bay area earthquake.

At summer's end, 99s shared in celebrating the 60th anniversary of the 1929 First Women's Air Derby with a commemorative flight of Louise Thaden's winning Travel Air from Santa Monica, CA, to Cleveland, OH. The restored 1929 Travel Air was flown by Airborne Captain Susan Dusenberry.

By 1990 women were achieving greater recognition for their choices of non-traditional careers and other diverse aspects such as sports or hobby interests. Women were holding positions as chiefs of police, church leaders, AIDS activists, fashion tycoons, jazz virtuosos, choreographers, Indian chiefs and winning world-class rock climbers.

Women were making dramatic inroads into occupations previously dominated by men. By 1990 nearly 18 percent of U.S. doctors were women, as were 22 percent of the lawyers, 32 percent of the computer system analysts, and half of the accountants and auditors.

Major companies, including airlines, were re-examining the ways they found and recruited new talent, the incentives and benefits they offered and how they organized work from the top down. Wishing for women to be a greater part of their workforce, employers had to reconsider ways of helping working mothers balance job and family responsibilities. According to statistics and popular assumption, women did not quit their careers to tend the hearth. By 1990, women owned more than twice as many companies as they had 10 years earlier.

Jean Kaye Tinsley Atherton, CA, a 30-year-plus Bay Cities 99 and Whirly Girl, was the first woman in the world to fly the XV-15 Tilt-Rotor aircraft. Honoring her personal contribution to aviation, Achsa Barnwell Peacock Donnels received the coveted Katharine B. Wright Memorial Trophy. Cameron Park 99 Julie Clark was awarded the FAA Certificate of Appreciation, and Operation Skywatch was established and added to the International Committees list. The program, started by the First Canadian Chapter in the province of Ontario, was organized to identify illegal pollution of the environment.

The war in the Persian Gulf was the top story in 1991. The U.S. closely examined and attempted to understand Saddam Hussein's war. The world, concerned about the environmental impact of bombs and marauding armies, was doubly troubled with Saddam's senseless dumping of millions of gallon of oil into the gulf. Equally bothersome was some 650 oil-well fires that spewed untold tons of smoke into the air. Also on the environmental concern list was the Antarctic ozone hole.

1991 saw 99s doing what they did the best — aviation community involvement, performing such tasks as ferrying medical supplies, conducting flying companion seminars, giving Scouts airport tours, and communicating with the public by conducting safety-proficiency seminars to encourage safe flying.

Overseas, LT Manja Blok of the Royal Netherlands Air Force became the first female F-16 combat pilot. Making more aviation history: For the first time, women raced in all four classes at the Reno Air Races — formula one, biplane, unlimited and AT-6. Julie Clark was in the news again for receiving the Bill Barber Award for showmanship in aerobatic flying maneuvers.

Efforts by 99s continued and their hard work and dedication were recognized with a variety of awards. Bobbi Trout and Marie McMillian were selected as Elder Statesmen of Aviation by the National Aeronautic Association of the United States for significant contributions to aeronautics and reflecting credit upon America and themselves.

AMELIA EARHART BIRTHPLACE MUSEUM

In a simple ceremony on Nov. 19, 1984, the Amelia Earhart Birthplace was presented to The Ninety-Nines, Inc., by 102-year-young Dr. Eugene J. Bribach of Atchison, KS.

Through the previous efforts of local resident Evah Cray and 99s past presidents Marie Christensen and Marilyn Copeland, the 99s International Board of Directors had decided it would be willing to assume the ownership, management and restoration of the birthplace. In addition to gifting the birthplace (purchased for $90,000) to The 99s, Dr. Bribach donated $10,000 for maintenance and upkeep.

The first Board of Trustees — 99s Pat Roberts, Janet Green, Thon Griffith and Marie Christensen — was joined by three Atchisonians — Evah Cray, Joe Carrigan and Catherine Ryan — in formulating plans and funding for restoration and ongoing maintenance of the property.

In a June 1985 report in *The 99 NEWS*, Christensen indicated the electrical system had been brought up to acceptable standards, and the first floor had been re-plumbed. The roof was temporarily repaired, a new caretaker's kitchen was completed and a security system installed. The home was opened for limited viewing; however, for safety concerns, it later had to be closed until additional funds became available.

Honorary grant deeds were sold to help generate funds, and grants also were sought. By mid-1987, a caretaker was living in the home and conducting tours by appointment. Then, in April 1989, a sweepstakes fund-raiser was introduced to raise money for restoration. An Avid Flyer kit plane was the top prize, and members were urged to sell tickets for it and many other prizes. That year, grants were received from the Kansas State Historical Society for an architectural study and from the Atchison County Commissioners for ongoing maintenance.

"This is the first time I've won anything," declared 99 Edith Geneva McNamee, Bakersfield, CA, when she won the Avid Flyer kit, valued at over $12,000. Over $10,000 in sweepstakes tickets were sold to generate funds to restore the birthplace to its condition in the time when Amelia lived there.

Concern was expressed by some members regarding the appropriateness of The 99s owning the home and assuming the attendant expenses/obligations. In 1991, an ad hoc committee including 99s Pam Mahonchak, Carole Sutton, Janet Green, Hilda Devereux, Alexis Ewanchew and 99 supporter Glenn Buffington studied the issues.

While significant funds for restoration were slow to accrue, the committee found the birthplace to be self-sustaining in terms of normal maintenance and operational expenses. It also determined ownership of the birthplace suited the "educational" criteria of the organization's non-profit tax status.

In January 1991, an NBC camera crew arrived to shoot scenes for an upcoming segment of the television series, "Untold Stories." In addition, a grant from Sterling Savings & Loan was received, and other grants were submitted for consideration. The Birthplace continued to be the leading public attraction in Atchison.

A member survey in 1992 indicated a majority wished to retain ownership of the Birthplace, and substantial funding for restoration was still being sought.

Publicity about the ongoing search for Amelia brought many visitors to the doorstep in 1993, and signatures in the guest book continued to grow. An expanded Board of Administrators continued their efforts to raise funds for restoration.

In 1994, the southwest porch was fully restored utilizing accumulated funds. In addition, a representative from U.S. Sen. Robert Dole's (R-KS) office visited the site and met with 99s President Lu Hollander and representatives from the City of Atchison. The possibility of pursuing funding for an 18-month National Park Service study to determine whether the birthplace met the criteria to become a National Park was discussed.

Reorganization of the AEBM Resolution and the Board of Trustees took place in 1994, and was subsequently approved by the 99s' International Board of Directors. The new AEBM Board of Trustees consisted of four 99s and three Atchisonians, including Chairman Marilyn Copeland, Co-Chairman Jim Taylor, Secretary Carole Sutton, Treasurer Dick Senecal, and Trustees Joan Adam, Alexis Ewanchew and Linda Marshall.

In mid-1995, a successful open house honored charter members of The 99s; the Birthplace also sported a brand-new wood shingle pitched roof and copper flat roofs, thanks in great part to contributions from the Cray Foundation of Atchison.

Subsequent meetings with architect Dean Graves and representatives from the Kansas Heritage Trust led to a successful grant proposal which, in 1996, awarded $75,000 to restore the exterior of the Birthplace. At this writing, members of The 99s and the citizens of Atchison have more than matched the $20,000 required to fulfill the Kansas Heritage Trust grant.

More good news — once the old aluminum siding was removed from exterior walls, only about 20 percent of the original wood siding had to be repaired or replaced. In addition, an unused portion of the second floor was renovated to create sparkling new caretaker's quarters.

Now that the exterior walls of the Birthplace have been restored, the focus will move to restoration of the north porch, interiors and the development of educational outreach programs.

The Birthplace received a $25,000 grant from the Courtney Turner Trust, thanks to AEBM Trustee Dick Senecal and Turner Trust representative Dick Cray. The Zonta organization, led by Carolyn Mohler, has raised over $7,000 selling bricks for the sidewalks, while the Bookmark mailing to 99s has netted over $5,000.

At this writing, AEBM Trustees and the City of Atchison also are developing plans for the 1997 celebration to commemorate Amelia's 100th birthday.

Scaffolding on the north side of the Amelia Earhart Birthplace Museum signals the exterior restoration is moving forward. A Kansas Heritage Trust $75,000 grant, combined with matching funds from 99s and Atchison residents, made the work possible.

Three generations of Maule women get together at the 1989 Forest of Friendship celebration. With 99 Susan are mother Rathgunde and grandmother June.

A ride in a Tiger Moth in the early '70s introduced New Zealand 99 Pam Collings to aerobatics. A pilot since 1964, she went on to become a member of her country's team, placing 12th during the 1976 world championships in Kiev, Russia.

Alma Smith was recognized for 50 years in aviation. Combining her interest in aviation along with her newspaper work, Alma spent her life promoting her hometown airport of Laconia, NH, and the realm of aviation.

In August 1991, Evelyn Bryan Johnson was honored at the Forest of Friendship and celebrated over 50,000 hours in the air. She was given the Kitty Hawk Award by the FAA for her contributions to aviation. She logged her time in the air as a flight instructor and FAA designee, working tirelessly 12 hours a day, seven days a week, 52 weeks a year.

Jerrie Cobb was inducted into the Oklahoma Aviation Hall of Fame. Several states to the east, Charlote Frye was inducted into the Georgia Hall of Fame. She learned to fly in 1931, when there were only six women pilots in the entire state. Carolyn Pilaar, a member of the U.S. Precision Flight Team, received the Top Woman of the Year Award during the ninth annual International Precision Championship event held in Rio Cuarto, Argentina.

Chapters continued to be busy with community service events and programs. In Louisiana, the New Orleans Chapter developed and instituted an aerospace program to present in area schools. Members discussed careers in aviation, history, Civil Air Patrol programs and emphasized the importance of a high moral codes of ethics and standards.

During Operation Desert Storm, the war in the Persian Gulf, the Arabian 99s gave unselfishly of their time by entertaining troops in their homes, raising funds for MASH units and helping recruit volunteers for the Red Cross. And there was an additional, sad note to add regarding the war in the Gulf -- MAJ Marie Rossi was the first U.S. female pilot to fly into Iraq during the conflict; she crashed while piloting a CH-47 Chinook helicopter in a non-combat related mission.

Ninety-Nines and their companion 49 1/2s worked together to make aviation history. Husband-and-wife team Rich Gritter and MayCay Beeler, Kitty Hawk Chapter, broke more than three aviation records in the time-to-climb category for light piston-engine powered aircraft, for a total of nine records in their Questair Venture experimental aircraft. Meanwhile, 99 Laverne Lawerence built the first fiberglass Pulsar, an experimental homebuilt design, in six months.

In a symbolic move, astronaut Dr. Linda Goodwin carried Louise Thaden's flying helmet into space. But Dr. Goodwin was not the only woman in space this time. Rhea Seddon, Memphis Chapter, was aboard space shuttle Columbia, accompanied by 29 rats, 2,500 jellyfish, and her other NASA crew members.

The annual historian's report captured some of the highlights from 1991 — the U.S. recession deepened, aviation saw the bankruptcy of Piper, TWA and America West, and the closure of Pan Am and Midway airlines. The Amelia Earhart Birthplace had more visitors the first nine months of 1991 than all of 1990. The 99s worked with the FAA to establish private pilot ground schools and adult education classes. Additionally, 99s worked closely with the FAA developing cockpit resource management programs and co-sponsored the Air Bear program.

The torch was passed as Arkansas Gov. Bill Clinton celebrated his election to the U.S. presidency in 1992. Johnny Carson retired as TV's late-night talk-show king.

In science, evidence was found to support the Big Bang theory; the Hubble space telescope revolutionized astronomy with its long-distance vision, and the number of Americans who smoked reached a record low, as nonsmokers outnumbered smokers 3 to 1. 1992 would go down as the "year of clear." Hoping that consumers would equate clear with clean, manufacturers introduced clear dishwashing detergent and see-through sodas.

The list of firsts continued for 99s and women achieving in aviation — Patty Wagstaff was the first woman pilot to win the prestigious title of U.S. National Aerobatic Champion 260. For Patty, this would be one of many titles to her credit. MAJ Eileen Collins was selected by NASA as the first female shuttle pilot. The Alabama Aviation Hall of Fame inducted Melba Iris Harris for her "Fantastic Flight" program, the aviation-awareness program designed by Melba to stimulate and inspire elementary school students, a model for schools nationwide.

Doris Lockness received the Certificate of Honor from the National Aeronautic Association and was then the only pilot in the world qualified for membership in all five exclusive organizations — the Whirly Girls, OX Aviation Pioneers, WASPs, United Flying Octogenarians and The 99s.

Ninety-Nines were members of the U.S. helicopter team during an international fly-off and competition in Swindon, England. The all-female helicopter team was headed by Dorothy Cummins of San Antonio, TX.

The 99s participated in the first joint rally of the New Zealand Airwoman's Association and the Australian Woman's Pilot Association, held in Christchurch, New Zealand. In addition, Pam Collings, a New Zealand 99, won the Nancy Bird-Walton Trophy for her tremendous contribution to aviation in that country.

Continuing to break records and remind others that The 99s have an insatiable appetite for adventure, 99 Joann Osterud broke Dorthy Hester Stenzel's record by completing an astounding 208 outside loops in over four hours of flight time in her unlimited stunt plane, the Ultimate 10-300S. Air racers Marion Jayne and Sue Nealy related tales of hazardous weather, navigation challenges and pilot stress loads during the first around-the-world air race, which encompassed 16,500 miles.

Awards were numerous during 1992 — Marjorie M. Gray was inducted into the New Jersey

Julie Clark began flying in 1967, using college book money to buy flying lessons. As a DC-9 captain, Clark is also a sought-after airshow performer in a restored T-34 aircraft.

Astronaut Eileen Collins and charter member Fay Gillis Wells dedicate a wreath in memory of Challenger crew members during 1996 Forest of Friendship ceremonies.

Evelyn Bryan Johnson

Aviation Hall of Fame. Awards of Merit presented by The 99s were given to Lotfia-El Nadi, Thon Griffith and Evelyn (Bobbi) Trout. The Katharine B. Wright Award was given to Betty Pfister, and Grace Harris won the NAA, EAA and 99 awards at the 99s international convention that year.

Florenza de Bernardi of Rome, Italy, recognized as the sixth female airline pilot in the world, then retired captain of Aeral, received the award of Outstanding Leadership and Support of Aerospace Education from the World Aerospace Education Organization (WAEO). The Museum of Flight in Seattle, WA, recognized Barbara Erickson London as an aviation pioneer. Ruth Jefford, one of Alaska's most celebrated female pilots, was named to the National Aviation Hall of Fame.

The spring and summer of 1993 brought torrential rains which caused the worst flooding ever recorded in the Midwest. United Airline employees offered to buy the company, while researchers duplicated a human embryo, provoking cries that technology had gone too far. As of 1994, space probes had flown by more than 60 planets, moons and other heavenly bodies without turning up any evidence of life.

Eyes were focused above that year, as 99s watched Suzanne Asbury-Oliver paint smiley faces and "Pepsi" as the Pepsi Skywriter in her vintage 1929 Travel Air D4D.

Ninety-Nines were once again in the news, accepting awards and distinctive honors across the country and world. Charter member Nancy Hopkins Tier, a Connecticut 99, was inducted into the WIAC aviation Pioneer Hall of Fame. Kalina Yeda Cox de Barros, a 99 from Rio de Janeiro, was the first woman to graduate from Varig Brazilian Airlines' highly competitive flying course. The 737 co-pilot was the airline's first female in their 65-year existence.

Patty Wagstaff, Alaska Chapter, made history again in 1993 when she won the title of U.S. National Aerobatic Champion for the second year. She went on to place third at the International Aerobatic Club championships.

Evelyn Sharp, a charter member of the Nebraska 99s and an original member of the Women's Auxiliary Ferry Squadron (WAFS), became the first woman to be inducted into the Nebraska Aviation Hall of Fame.

That year the U.S. Defense Department changed its policy, allowing women to fly combat missions. Former WASPs reflected on non-combat status.

A 1990 photo of British Section 99s includes Eileen Egan, Connie Fricker, Barbara Cannon, Jill Honisett, Eve Saunders, Naomi Christy and Gwen Bellew.

Susan Dusenbury and her Beech Travel Air grace the cover of the September-October 1989 issue of The Ninety-Nine News *magazine. Her cross-country flight commemorated Louise Thaden's historic flight.*

COUNCIL OF GOVERNORS

At the 1990 International Convention in Las Vegas, a group of Section governors met and agreed to serve the organization as an additional voice of the membership to the International Board of Directors. A resolution was presented on the floor of the 1990 annual meeting and was unanimously adopted by the delegates.

The resolution pledged the service of the Council of Governors to enhance the communication process between members and the Board of Directors by attending meetings of the board at their own expense. The board was bound to keep the Council of Governors advised of all issues under consideration and permit the council to present their views to the board.

Members of the first council included Pat Ward, South Central Section, who wrote the resolution and served as the first council spokesperson; Lois Erickson, Southwest Section; A. Lee Orr, Southeast Section; Dodie Jewett, North Central Section; Shirley Ludington, New York-New Jersey Section; Marie Oswald, West Canada Section; Sue Ehrlander, East Canada Section; Betty Erickson, New England Section; Gayl Henze, Middle East (now Mid-Atlantic) Section; and Eileen Egan, British Section.

Over time, Sections recognized the value of having their governors attend board meetings and voted to budget funds to cover these expenses. A closer relationship among the various governors also resulted from the additional interaction at these meetings, leading to more focus on attending each other's section meetings as well as planning joint section meetings.

The function of the council has evolved to include serving on Board/Governors ad hoc committees and frequent participation in board discussions as they occur. Council members were asked to help draft Standard Operating Procedures for, and serve on, the organization's Grievance Committee, as well as assist in locating member resources within their own sections for specific projects.

Governors were able to take back to their members a better understanding of the decision-making process at the international level.

In 1995, governors Cathie Mayr, North Central Section, and Joy Parker-Blackwood, East Canada Section, offered to organize and conduct a series of long-range planning sessions, beginning with a survey sent to all members to provide initial input. An unprecedented response from the membership led to the development of a mission statement and initiation of an ongoing long range planning program.

In 1996, delegates at the annual business meeting voted to place the Council of Governors in the bylaws under its own article.

The year 1994 welcomed the passage of the General Aviation Revitalization Act in Washington, D.C., restricting the limits of liability on lawsuits claiming product defects in private planes to 18 years.

Shortly before dawn in January, a devastating earthquake centered north of Los Angeles caused many deaths and widespread damage to the entire Southern California area. Genetically-altered vegetables, modified to increase shelf-life and be more pest-resistant, were approved by the U.S. government. Former President Ronald Reagan announced in an open letter to the public that he had Alzheimer's disease. The economy turned from manufacturing to a more service-based focus.

Netscape quickly became the navigator of choice for dedicated surfers on the Internet. Services on the Net included the opportunity to obtain information as well as obtain goods through many on-line shopping centers and catalogs.

Two-thirds of the households in the U.S. watched the oddest car chase in TV history: O.J Simpson's slow-speed flight along the Los Angeles freeways, ending at his Brentwood home.

The Comet Shoemaker-Levy 9 slammed into Jupiter, creating a 2,000-mile-high fireball that was visible from most backyard telescopes.

The U.S. Post Office honored another sister of the skies when it issued a 32-cent stamp in honor of Bessie Coleman, the first African-American to earn a pilot's certificate. Bessie joined the ranks of Amelia Earhart and Harriet Quimby in being so celebrated.

Seeing a need to ease the "Good Ol' Boy" job network existing in aviation, The 99s started a Career Data Bank. The data bank allowed members to network exclusively with other 99s that held high-level aviation-related positions. That year, The 99s signed a partnership with EAA to participate in the Young Eagles Program. "The goal of the Young Eagles program is to fly with one million young people by the year 2003, the dawn of aviation's second century."

Honors and awards were bestowed upon more-than-deserving 99s. Evelyn Greenblastt Howren, North Georgia Chapter, was the fourth woman to be inducted into the Georgia Hall of Fame. Lucile Bledsoe was inducted into Colorado's Aviation Hall of Fame. Jean Ross Howard, founder of the Whirly Girls, received the prestigious Elder Statesman of Aviation Award.

Making 99 and U.S. military history, 1stLt Jeannie Flynn became the Air Force's first female pilot. She flew the most advanced tactical fighter aircraft, the Strike Eagle F-15E. 2ndLT Sarah Deal was selected as the first female Marine pilot. Naval Officer Shanon Workman became the first aircraft carrier pilot. CPT Kathy McDonald of the Texas Air National Guard because the first female fighter pilot to graduate from the Guard's six-month F-16 air-combat training school, at Kingsley Field, Klamath Falls, OR.

Setting records, hot air balloonist Jetta Schantz, Florida First Coast Chapter, achieved an altitude record. Jetta exceeded the previous record of 32,572 feet by 1,272 feet, and brought her total national records to 18. Her interest in balloons started simply enough, with a balloon ride for a birthday present. The interest led to a hobby, then a passion and finally a full-time profession.

The year 1995 began with a roar as Eileen Collins, the first female shuttle pilot, took off from Kennedy Space Center for a joint American-Russian space mission. Watching the launch were the 13 women who had passed the astronaut physical exams 34 years before. Sarah Ratley said, "Eileen is carrying out our dreams and wishes. We're happy she's finally been accepted and the sex barrier has been broken."

In addition to the technological wonders incorporated into the space shuttle, 99s took advantage of a fairly new computer advancement by networking via the Internet. In 1995, The 99s went "on-line," using CompuServe to reach other 99s for forums and sharing specific 99-related information between members.

The Amelia Earhart birthplace received a $75,000 grant from the Kansas Historical Society requiring a 20-percent match.

Awards and honors to 99s were plentiful in 1995. Audrey Poberenzy, wife of EAA founder Paul Poberenzy, was presented with the Katharine Wright Memorial Award at the 99s annual convention, held that year in Halifax, Nova Scotia. The award is presented annually by the 99s and the National Aeronautic Association, to a woman who has made a personal contribution to the advancement of the art, science, and/or sport of aviation and space flight over an extended period of time. The award also may be given to an individual who has provided encouragement, support or inspiration.

Betty Pfister, Aspen Chapter, received the FAI Rotorcraft Gold Metal Award for more than 50 outstanding years as an aviatrix and aviation contributor locally, nationally, and internationally.

Canadian 99s were inducted into Canada's Aviation Hall of Fame. The group was presented with the Belt of Orion Award of Excellence for advancement and contributions to aviation.

Doris Lockness was named one of six persons to receive the Elder Statesman Aviation Award. Mickey Axton, Kansas Chapter, received the OX-5 Aviation Pioneers Aviation Historian of the year award. Mickey received the award because she demonstrated qualities toward the preservation of the true aviation pioneering spirit.

Shelley Breedon, First Canadian Chapter, won Canada's Webster Memorial Trophy Competition and was declared Canada's "Top Amateur Pilot."

To commemorate the 20th anniversary of the International Women's Year, the Washington, D.C., Chapter and Zonta International honored several women as pioneers in the fields of aviation and aerospace. Included were Jean Ross Howard, founder of Whirly Girls; Sheri Coin Marshall, a right-arm amputee, for achieving professional pilot status, earning a CFII and ATP, and for assisting other physically-impaired individuals with their flying goals; Ida Van Smith for her long-time efforts encouraging young people to pursue aviation and aerospace interests; Mary Feik for her aviation designs and contributions to aviation; Christine Fox for her career as a Naval tactical analyst; Carolyn Shoemaker for her discovery of more that 800 asteroids and 32 comets including the Shoemaker-Levy 9; Ella D. Williams, who formed her own aviation-related engineering and computer-research consulting firm; and Moya Lear, who took charge of Lear Jet Inc. who upon the death of her husband, Bill.

Certainly The 99s' high visibility at the community level, working often quietly and without much fanfare, has helped women reach their aviation-related goals. It's been those honorable tasks such as airmarkings, pinch-hitting courses for non-pilot passengers, and educating both children and adults, that 99s have been recognized for their efforts.

JUNE 1995 - 1996

Excerpted from the annual report by Lynn Houston, International Historian, and from reports in The International Women Pilots Magazine/99 News

Women aviators experienced both triumph and tragedy in a year that encouraged the same extreme range of events around the world. Despite the presence of UN peacekeepers, the conflict in Bosnian continued. From this desperate arena, a true American hero emerged in the person of Air Force Capt Scott F. O'Grady when his F-16 jet was shot down by the Bosnian Serbs. For six days, O'Grady evaded Bosnian troops by using skills he learned in training school, until his rescue by U.S. Marines backed by NATO air support.

Twenty-year-old Shannon Faulkner made history by being the first female admitted to the South Carolina college The Citadel. Five days later, after years of court battles and personal ordeals to accomplish her goal, Faulkner quit the school.

Positive steps towards peace in the Middle East seemed certain when Israeli Prime Minister Yitzhak Rabin shook the hand of Palestinian leader Yasser Arafat on the White House lawn. Months later, however, Rabin was assassinated in Israel by a man opposed to his peace policies. The subsequent election of Benjamin Netanyahu has left the world wondering where the next steps in the Middle East will lead.

In the United States, Senate Majority Leader Robert Dole of Kansas emerged from the primaries as the Republican presidential candidate. President Bill Clinton ran uncontested in the Democratic race. Nearly a million men marched on Washington in a statement of pride for African-American fathers, husbands, sons and brothers. And much of America and the world watched the televised criminal trial of former football star O.J. Simpson for the murders of ex-wife Nicole Brown Simpson and her friend Ronald Goldman. In a controversial verdict following daily courthouse controversy, Simpson was acquitted.

For women in aviation, this was a year to both look back in remembrance and look forward as women became more visible in the skies and beyond to space.

August 1995 saw several events commemorating the triumphs, the sacrifices and the heroism during World War II. To mark the 50th anniversary of VJ Day, 120 planes flew across the U.S. to salute air veterans of WWII. Dubbed Freedom Flight America, the event included 260 pilots — many veterans or their sons or daughters — flying antique planes from Long Beach to New York City. Organized by Morey Darzniek, a refugee who fled Latvia in 1944, the flight started at the Queen Mary — used as a troop carrier in WWII — and ended with a Statue of Liberty flyover on August 15, VJ Day. The Pentagon's WWII Commemoration Committee recognized Freedom Flight America as an official anniversary event.

As part of the honors, Barbara London, the only woman awarded the Air Medal during WWII, was one of 60 women honored at a USO dinner and dance in Long Beach Hangar. The commander of a squadron of women ferrying new planes to airfields nationwide, she won the medal after making four transcontinental flights in five days.

Also in August, the Planes of Fame Museum in Chino, CA, honored Women Airforce Service Pilots (WASPs) who were drafted into civilian service to ferry war planes east to be shipped overseas. They flew an aeronautical arsenal that included P-51's, B-25's, Lockheed P-38's and Northrop P-61's. American women pilots flew more than 60 million miles during the war. More than 30 women pilots were killed, yet it wasn't until 1977 that they were finally granted veteran status and benefits.

January 29, 1996, marked another anniversary — it had been 10 years since the space shuttle Challenger blew up soon after launch, killing all seven astronauts on board, including Judith Resnick and Christa McCauliffe.

The best tribute to the Challenger is the continued growth of the space shuttle program, as evidenced during the year. Since Discovery blasted into orbit in February 1995 with a woman in the pilot's seat for the first time in NASA history — Lt. Eileen M. Collins — women have continued to become more and more prominent in space.

In March, Shannon Lucid became an official crew member of the Russian Mir space station. Lucid, a 53-year-old biochemist, is the first woman

A PASSAGE TO INDIA

Under the auspices of the Aero Club of India, the India Section of 99s assembled an impressive array of speakers and dignitaries for the first World Aviation Education and Safety Congress in New Delhi Feb. 22-26, 1986.

Prime Minister Rajiv Gandhi inaugurated the event, and other speakers included Sheila Scott, a British record-setting pilot; Australian Nancy-Bird Walton; Wally Funk, U.S. pilot retired from the National Transportation Safety Board; Dr. V.S. Arunachalam, then president of the Aeronautical Society of India; and John Baker, then president of the Aircraft Owners and Pilots Association.

Organizers Chanda Budhabhatti, Mohini Shroff and other India Section members worked hard to produce a meaningful international aviation event — and according to 99s who attended, they succeeded.

So much so that a second Congress was scheduled for March 1993, but had to be moved to the same time in 1994 due to unrest in India. The India Section persevered, and the second event took place in Bombay March 14-19, 1994.

Several hundred delegates from around the world, representing virtually every aspect of aviation, attended the presentation of papers on 27 topics. Key speakers included Eileen Egan, British Section 99s; Susan Darcy and Rose Loper, Boeing Aircraft; Lt. Manja Blok, Netherlands, the world's first woman F-16 combat pilot; and several prominent figures from Indian aviation circles.

Truly an international event hosted again by the India Section 99s, the second Congress brought this comment from North Central Section 99 Cathie Mayr, who likened attending the Congress to "living a *National Geographic* article."

Dr. Sunila Bhajekar and Mohini Shroff, India Section 99s, pose with Wally Funk during the World Aviation Safety Congress in New Delhi in 1986.

NATIONAL CONGRESS ON AVIATION AND SPACE EDUCATION

excerpted from the July-August 1996 issue of The International Women Pilots Magazine/99 News

With nearly 700 other participants, 99s from many states attended the 1996 National Congress on Aviation and Space Education (NCASE) in Little Rock, AR.

Ninety-Nines have participated in this important activity for many years as attendees, presenters and recipients of the Brewer Trophy for excellence and dedication to aerospace education. The organization also received the Crown Circle Award from NCASE in 1982.

NCASE is an excellent opportunity to meet a variety of representatives of the aviation community, especially those in the aviation organizations and associations who support aviation education.

LCDR Lori Tanner, FA-18 test pilot based at China Lake, CA, discussed her military flying career, and Mary Feik, pioneer woman engineer and an active Potomac Chapter member, provided insights into women's early roles in the aerospace industry.

A number of teacher workshops were held, providing educators with ideas for their classrooms and material for assisting students with career options.

to fly in space five times and, after her scheduled six-month Mir stay, is expected to become the first American to ever have spent so much time in orbit. The Mir missions are the precursors to an international space station, an orbiting research lab to be jointly built and owned by the United States, Russia, the European Space Agency, Japan and Canada. If all goes well, the first components will be launched in November 1997, and by June 2002 the station will be ready for permanent human habitation.

Closer to earth, Lt. Sarah Deal became the first female Marine Corps pilot commanding a combat helicopter, the Super Stallion CH53E.

And even as women soar to greater achievements, the year held sadness for some of those achievers. Former 99 Candalyn Kubeck, 35, became the first woman commercial jetliner captain to perish in a U.S. crash when the ValuJet she was piloting plunged into the Florida Everglades in May. She is survived by her husband Roger, a pilot for America West.

In perhaps one of the saddest aviation stories, 7-year-old Jessica Dubroff, accompanied by flight instructor Joseph Reid and her father, died in Cheyenne, WY, on April 11. The initial National Transportation Safety Board report stated that the "Cessna 177B, registered to and being flown by a commercial pilot/flight instructor, was destroyed during the collision with the terrain following a loss of control during takeoff/initial climb from the Cheyenne Airport."

Jessica, trying to become the youngest pilot to fly across the continental U.S., was not listed or recognized as a student pilot or crew member in the crash report since she was not of legal age to be either.

The 1995 annual meeting took place in Halifax, Nova Scotia, Canada, in conjunction with The 99s' International convention. As with the 1994 meeting, this was divided into two morning sessions. The awards banquet featured the presentation of the Katharine Wright Memorial Award to Audrey Poberenzy, wife of Experimental Aircraft Association founder Paul Poberenzy. Since she was unable to attend the banquet, 99s President Joyce Wells made the presentation in Oshkosh on July 30.

A motion was passed to add the words "Organization of" to the name of The Ninety-Nines, Inc., in the Articles of Incorporation. Completed, the name will read "The Ninety-Nines, Inc., International Organization of Women Pilots."

At the annual meeting, 270 delegates (232 voting/39 nonvoting) holding 1,301 votes and representing 14 sections, and 89 guests, represented 87.6 percent of the membership.

A record number of 21 Amelia Earhart Career Scholarships were awarded to 99s at the convention. In addition, two jet type ratings donated by United Airlines were presented.

Homecoming 1996, the 1996 convention, was held in Oklahoma City, so members could take the opportunity to visit International Headquarters.

At the Fall 1995 Board of Directors meeting in Oklahoma City, the Board and Council of Governors developed the following mission statement:
• Promote world fellowship through flight
• Provide networking and scholarship opportunities for women and aviation education in the community, and
• Preserve the unique history of women in aviation.

The Bessie Coleman Commemorative stamp is the 18th in the U.S. Postal Service Black Heritage series and was presented in ceremonies in June 1996.

Born in Texas in 1893 and unable to attain a U.S. license, Coleman earned her license from the Federation Aeronautique Internationale. She returned to the United States and began teaching other women how to fly, giving lectures and performing in flying exhibitions. As she gained fame as a barnstormer, she became known as "Queen Bessie." Queen Bessie died April 30, 1936, while practicing for an airshow in Florida.

Bobbi Trout, Charter member, was presented the 17th annual Howard Hughes Memorial Award in impressive ceremonies Jan. 18, 1996, in Malibu, CA. She was the first woman to receive this award, given by the Aero Club of Southern California.

Shelley Breedon won Canada's Webster Memorial Trophy Competition and was declared Canada's "Top Amateur Pilot." Mickey Axton received the OX-5 Aviation Pioneers Aviation Historian of the Year Award. Doris Lockness was selected by the National Aeronautic Association as one of six people to receive the Elder Statesman of Aviation Award for 1995. Janeen Kocelan became the first female DC-8 captain for Airborne Express. Jetta Schantz was honored by the U.S. National Air and Space Museum in March 1996 for establishing the world record for altitude in a hot air balloon — 32,572 feet.

LOCATIONS OF INTERNATIONAL CONVENTIONS

1930 Drake Hotel, Chicago, IL
1931 Scarlet Hotel, Cleveland, OH
1932 Westlake Hotel, Cleveland
1933 Grand Hotel, Santa Monica, CA
1934 Scarlet Hotel, Cleveland
1935 Carrer Hotel, Cleveland
1936 Clark Hotel, Los Angeles, CA
1937 Los Angeles
1938 Los Angeles
1939 Los Angeles
1940 Brown Hotel, Denver, CO
1941 Alvarado Hotel, Albuquerque, NM
1942 Due to war, no annual meeting was held
1943 Due to war, no annual meeting was held
1944 Ambassador Hotel, New York, NY
1945 No annual meeting
1946 Cleveland
1947 Troutdale-in-the-Pines, Evergreen, CO
1948 Muelebach Hotel, Kansas City, MO
1949 Waldorf-Astoria Hotel, New York
1950 Fort Clark Ranch, Bracketville, TX
1951 Grand Hotel, Mackinac Island, MI
1952 Boston, MA
1953 San Diego Hotel Manor, San Diego, CA
1954 Ashville, NC
1955 Sheraton-Kimball Hotel, Springfield, MA
1956 Ramona Park Hotel, Harbor Springs, MI
1957 Casa de Palmas, McAllen, TX
1958 Jefferson Davis Hotel, Montgomery, AL
1959 Davenport Hotel, Spokane, WA
1960 Hotel Dupont, Wilmington, DE
1961 El Cortez Hotel, San Diego
1962 Nassau Inn, Princeton, NJ
1963 Skirvin Hotel, Oklahoma City, OK
1964 Sheraton-Gibson Hotel, Cincinnati, OH
1965 Read house and Motor Inn, Chattanooga, TN
1966 Olympic Hotel, Seattle, WA
1967 Shoreham, Washington, DC
1968 Statler Hilton, Los Angeles
1969 Waldorf-Astoria, New York
1970 Mr. Washington Hotel, Brenton Woods, NH
1971 Regal Inn, Wichita, KS
1972 Hyatt Regency, Toronto, Canada
1973 Pfister Hotel, Milwaukee, WI
1974 El Conquistador, San Juan, Puerto Rico
1975 Coeur d'Alene, ID
1976 Benjamin Franklin Hotel, Philadelphia, PA
1977 Hyatt Regency, San Francisco, CA
1978 Lakeside International Hotel, Canberra, Australia
1979 Turri Inn, Albany, NY
1980 Mark Hotel, Vail, CO
1981 Copley Plaza, Boston
1982 Marriett Pavalion, St. Louis, MO
1983 New Orleans Marriott, New Orleans, LA
1984 Sheraton, Anchorage, AK
1985 Hyatt Regency Hotel, Baltimore, MD
1986 Hilton Hawaiian Village, Honolulu, HI
1987 Hotel Vancouver, Vancouver, B.C., Canada
1988 Shangri-la Resort, Afton, OK
1989 New York Marriott Marquisk, New York
1990 Las Vegas Hilton, Las Vegas, NV
1991 Stouffer Hotel, Orlando, FL
1992 Crown Plaza, Kansas City
1993 Red Lion Inn, Portland, OR
1994 Norfolk Waterside Marriott, Norfolk, VA
1995 Sheraton-Halifax, Halifax, Nova Scotia
1996 Marriott Hotel, Oklahoma City

CANADIAN 99S HONORED

by Mary Oswald, Alberta Chapter

On June 1, 1995, Canadian Ninety-Nines were inducted into Canada's Aviation Hall of Fame. The 99s were presented with the Belt of Orion Award for Excellence in an impressive ceremony held at Government House in Edmonton, Alberta.

This award was founded with the purpose of honoring organizations or groups which have made outstanding contributions to the advancement of aviation in Canada. The award consists of a round, engraved plaque, which was presented to Joan Lynum, governor of the West Canada Section, and Joy Parker-Blackwood, governor of the East Canada Section, by the Honourable Gordon Towers, lieutenant governor of the province of Alberta.

The awards ceremony was held in the beautifully-restored historical building which once served as the residence of Alberta's representatives of the Crown. A reception followed in the main salon, where 99s visited with many present members of the Hall, their families and members of the Board. Afterwards, guests moved out to the gardens to watch a fly-past of four distinctive Harvards, flown by members of the Western Canada Warbirds.

It was a very exciting day. We are so pleased to have this kind of attention drawn to our organization.

AWARD OF MERIT

The 99s Award of Merit was established in 1990 to recognize individuals who have made significant contributions to aviation, aviation education, science, aviation history, or The Ninety-Nines, Inc.

General criteria for the award include achievements which have occurred in the present or prior years and that meet the general objectives of the organization. Recipients may not necessarily be members of The 99s, may be living or dead, male or female, and not limited to individuals in the United States.

Recipients include:
1990-Pat and Sheldon Roberts, Muriel Earhart Morrissey, Jeana Yeager, Mimi Tompkins, Gaby Kennard
1991- Hazel Jones, John Baker, Nancy Bird-Walton, Olive Ann Beech, Alice Hammond
1992-Thon Griffith, Lotfia El-Nadi, Bobbi Trout
1993-Glenn Buffington, Lorna deBlicquy, Commander Rosemary Mariner, WASPs
1994-Evelyn Bryan Johnson, William Kershner
1996-Pauline Glasson

Recipients Margaret Goldman, Jeanne McElhatton and Fran Grant display the national Frank G. Brewer Award for Outstanding Achievement in Aerospace Education awarded during the 1981 National Congress on Aerospace Education.

CHARTER MEMBERS Top, L to R: *Fay Gillis Wells, Frances E. Harrell Marsalis.* **Middle, L to R:** *Nancy Hopkins Tier, Helen Manning Matthews, Cecil W. "Teddy" Kenyon.* **Bottom, L to R:** *Margaret Perry Cooper Manser and Dorothea Backenstoe Leh.*

Ninety-Nines Members

CHARTER MEMBERS Top, L to R: Jessie Maude Miller, Florence E. Klingensmith, Mary Webb Nicholson. **Middle, L to R:** Thea Rasche, Louise McPhetridge Thaden. **Bottom, L to R:** Phoebe Fairgrave Omlie, Ruth W. Stewart, Eleanore B. Lay Ross.

DORIS ABBATE's 35-year love of flying started when her 49 1/2 Ron, challenged her after obtaining his private license. In 1959, with an infant and three toddlers at home, she soloed on Long Island at Zahns Airport, Amityville, NY, in a J3 Piper Cub. She earned her private in 8801C, her own Tripacer, followed by her commercial, instrument and instrument ground instructor ratings.

Doris has competed in air rallies including the Air Race Classic and Garden State 300; took first place in the USPFT New England Regional, advancing to USPFT National Finalist in 1985. In 1985 she served as a judge for the World Precision Flight Championship and the US Precision Flight Team.

The life member joined the Greater New York Chapter in 1961 and became a charter member of her Long Island Chapter in 1965. Doris' most visible service to the 99s was on the International board of directors as secretary 1988-90, International director 1992-94, and NY - NJ section governor 1986-88. She was recognized as an International Forest of Friendship honoree in 1982 and by New York State for Aerospace Community Service in 1987.

Internationally, between 1982-94, Doris also held a number of offices. Doris served on the USPFT Council as national and regional coordinator. Beyond her service on the International board, Doris was a candidate for president in 1994.

However, other projects have made her, what others have called, "99s Extraordinaire." With her 49 1/2, Doris focuses on youngsters, giving school programs and taking classrooms of kids for hands-on tours to the airport and into her own airplane. She also has held many chapter and section offices and committee chairs, including governor, vice governor; Amelia Earhart Memorial Scholarship, International Convention committees, Bylaws and Standing Rules committees. She wrote the 1986 New York - New Jersey bylaws and standing rules, chapter chairman's manual and 50th Anniversary History. During her tenure as governor, Doris initiated a New York - New Jersey Newsletter, Annual Pilot of Year Award, the 1987 honorary celebration for 26 charter members and past New York - New Jersey governors, and the first International Joint Section meeting between New York - New Jersey and East Canada Sections.

Doris has served on the International Forest of Friendship Committee since 1990. Memberships include Women in Aviation International, AOPA, EAA, International Women's Air & Space Museum, New Zealand Women Pilots Assoc., Silver Wings, World Aerospace Education Organization, Grasshoppers, American Assoc. of Parliamentarians, National Assoc. of Parliamentarians, American Institute of Parliamentarians of Long Island and Parliamentarian of Great New York.

Doris retired from her salaried career as corporate secretary/treasurer at Exhibit Corp. of America and Ronnie Exhibits. Doris' family includes four children: Gregory, Lorraine, Carol and Vivian and, to date, 11 grandchildren.

"The 99s and flying," says Doris "had a major influence on my life. I feel part of the worldwide project to help each other advance women pilots. I love having friends and colleagues in many countries around the world; best of all I love the reunions at conventions."

MARY ANN ABBOTT, Akron, OH; retired account executive; Skyhawk - 172 at Akron, Fulton Airport; private rating; solo date: Nov. 7, 1994; license date: September 1995. Birthdate: May 5, 1946.

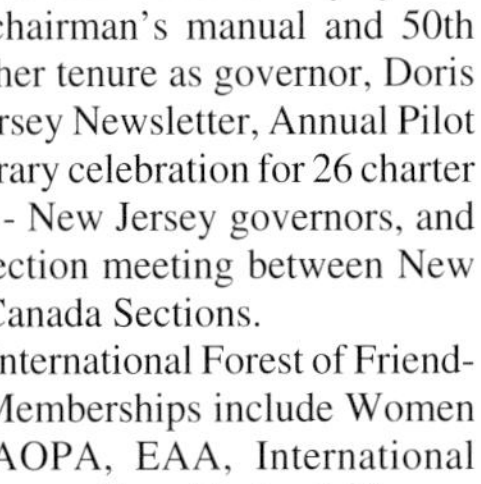

It was a simple notice in the newspaper that led her to the first event of what was to become a series of huge changes in her life. The notice had been placed by a local chapter of the 99s, and invited women interested in aviation or in learning to fly to attend one of their meetings.

Actually, it was her husband, Richard who first noticed the article; and before she knew what was happening she was whisked off and escorted to their local airport. After opening a door and nudging her through it, he disappeared to go hanger talkin'.

From that first contact with the 99s a whole new world has opened up to her. Her next step was to attend a Flying Companion Seminar. From then on, she's wanted to be part of the world of aviation.

She will always be grateful to her local Women With Wings Chapter for all their support, encouragement, and for her new friends; also to her husband for all his support, love and understanding.

"Flying gives me a feeling of personal achievement. Learning is a challenge, but the rewards are many: To feel free and almost touch heaven; to climb straight towards a rising full moon; to watch a sunset showering gold all over the western horizon ... all kinds of memories to last a lifetime!"

DELLA ABERNATHY, born May 9, in Herford, TX. Started as "Rosie the Riveter" for Douglas Aircraft, El Segundo, CA. Later worked on Nike-Zeus missile program at Pacific Missile Test Center, Point Mugu.

Trained and licensed in family Swift 145 Continental. Joined 99s in 1960.

Worked as flying exchange realtor over 30 years. Owned Comanche 250, Cessna 172 and 182, and two Swift 145s.

Married to Bob 55 years. Has two sons, eight grandchildren and two great-grandchildren.

Involved and flew in many air races. Held chapter offices, and currently flying copilot on Angel Flights.

Work with Republican women, Eastern Star and church.

Two days after receiving her private license in their Swift, she landed at Santa Paula Airport with lots of other planes in the pattern. She was met by several 99s from San Fernando and other chapters having a spot landing contest. They congratulated her on being the closest, and asked what chapter she was from. Having never heard of the 99s, she didn't get the prize. She then joined the happy gang, thinking she'd keep the prize next time.

SUE FEIGENBAUM-ACKLEY, born Feb. 15, 1949, St. Louis, MO. She has flown 172, Cherokee, Apache and Aztec planes. She began flying in 1964 at the age of 15 with her mother as her flight instructor. She has a private and multi-engine and joined the 99s at 18 years old. Her parents would let her take the twin Apache to college for football season, but having a car was too dangerous! She served as chapter chairman of the Cape Girardeau Area Chapter, flew the Powder Puff Derby (while five months pregnant) in 1971 with her mother, Lois Feigenbaum, and enjoyed working on four NIFA meets. She is married to a TWA MD80 Captain Gene A. Ackley, has two children, Jacqueline and Jeffrey, and one grandson, Chandler Johnson. (His mother being the pregnancy she carried during the 1971 PPD.)

MEIGS KIRKENDALL ADAMS, born Feb. 19, 1934, Kokomo, IN. Military Service included Civil Air Patrol, USAF Auxiliary (joined 1974); major. Soled at Welcome Field, Northfield, OH, in September 1968 (sod, 2400'). She has flown PA140, Cherokee 6, Cessna 310, and has owned PA28-180 N4815L for 23 years.

She is employed as a speech pathologist and RN Her daughter Paula M. Adams is a CPA and her son Scott M. died in 1973. Adams moved to Ohio in 1937, when her dad joined Taylor Craft Aircraft Co. She has more than 2,000 hours logged, flying Cleveland to Los Angeles to visit her two brothers there. Adams is an active 99, Lake Erie Chapter, having held positions up to and including chairman. Section Meetings, International Conventions and Australian Women Pilots 45th Annual Meeting, including New Zealand WPA have enabled her to make friends worldwide. Father, George W. Kirkendall, US Army Flying Corps, 1925-26; test flew Taylor Cub first time off ground. Her aviation affiliations include 99s, Silver Wings, EAA, OX-5, AOPA, ECOPA, CAP, Piper Cub, Women in Aviation and Forest of Friendship.

RITA V. ADAMS, born Chicago, IL. Received private license in 1967 and joined the 99s Chicago Area Chapter in 1968. Chapter chairman 1984-86. Currently vice-chairman, worked on almost all committees ranging from nominating to airmarking, etc. Member of Illi-Nines Air Derby Race Board for 11 years. Flew Illi-Nines and Chapter Air Meet. Recruited new members and greeted 99s at International Tent EAA Oshkosh for 14 years. Volunteer at Glenview Navy Air Shows, DuPage Air Shows and Springfield Rendezvous. Assisted Girl Scouts with aviation badges. Most memorable experience was taking the flight test when five months pregnant. Flying for "the fun of it" included many vacations locally and to Mexico and Canada. Proud to be a 99! She has flown Cherokee, Comanche and Warrior planes.

Adams is employed as office manager of an animal hospital. She has two children, Debbie, 27 and Carol, 20.

VANECIA ADDERSON, inspired to fly because of her husband's wish to fulfill his boyhood desire. They purchased a 172, then 182. Vanecia got her private license in February 1957. She joined the 99s in August 1957 and later became a life member. Her main flight instructor turned out to be a gentleman that was a childhood friend from elementary school. She flew for the March of Dimes airlift a number of times, as pilot in three races and two as copilot.

Vanecia held all offices on chapter level-treasurer, vice governor twice and governor two years for Northwest Section. She was proud to have received the Section Achievement Award twice and Chapter Achievement once. Vanecia has been very active in airmarking and flying companion seminars. She is also a volunteer for the Museum of Flight since its inception. Vanecia was co-founder and co-director of the Fear of Flying Clinic in Seattle. She has attended many International and section conventions and one Canadian Section.

Vanecia has spoken about 99s to various clubs and organizations. She is proud to be a 99 and will continue to be active always.

BONITA JOAN ADES, born Oct. 17, 1941, Beatrice, NE, into an aviation family. Her father and mother started with United Airlines in 1942. Her father was a machinist and her mother was "Rosie the Riveter." Her brother also works for United. Bonita has been a flight attendant for United for 33 years. She currently flies international routes out of San Francisco. Her career as a pilot started with a Pinch Hitter course and along the way she became "addicted." She continued to get her private license, ASEL rating in 1984. Bonita has been a member of the 99s since 1986. She started working with the Flight Without Fear program that the Colorado Chapter sponsors and was coordinator for six years. She has flown and worked for the Mile High Air Derby many years. She has also participated in the Air Bear program. She has held the office of vice-chairman and chairman of the Colorado Chapter. She is ways and means co-chairman South Central Section and initiated several successful fund-raising programs. She currently is a member of the International Ways and Means Committee. She is a NIFA judge and plans to judge the Nationals in 1995. She is also participating in the planning for the 1996 WPFC.

She has flown 150, 152, 172, 182 RG; simulator 727, DC10, 767, 747 planes. She and her husband Jon have two children, Aaron, 22 and Rebecca, 17.

CHERI WINE ADKINS, born in Long Island, NY, on Nov. 27, 1950. Being a dependent of an Air Force fighter pilot, Cheri became accustomed to living like "gypsies," as her Irish grandmother would say, and was fortunate to experience living in a variety of places, including the Far East. Cheri grew up with the sound of aircraft engines and the sights of runway lights, but her best aviation memories are when her dad would talk about his flying experiences. His love of flying was somehow passed on to Cheri and became a dream that she thought would never come true. But the dream did come true, and when Cheri soloed in May 1994, she finally understood why her father felt like he did about aviation. She received her private pilot's rating on Jan. 18, 1995. Her only regret is that she didn't pursue her license sooner, before her father passed away. Sharing flying experiences would have been really special.

Cheri is currently residing in Bristol, VA, with her husband, Mike. She is a senior programmer analyst, and Mike is a computer operations supervisor. Her ultimate aviation goal is to obtain a CFII rating. And she promotes women in aviation every chance she gets, telling them that it's never too late to get started.

GINETTE CHERVONAZ AELONY, born Aug. 12, 1943, in Nice, France; she learned to fly in Torrance, CA, as a present from her husband for her 40th birthday. A whole new world (literally) opened for her and a new understanding of California and real America started a love affair with the flying community from EAA, AOPA, and the valiant 99s, and all the restoration projects – specially the project Tomahawk on Zamperini field ... all these crazy wonderful people and their flying machines. After a dreadful encounter with an unmarked power line and trying to obtain legislation for an Adopt-A Wire program, which would work similarly to Adopt-a-Highway and "hold harmless" anyone who had marked a line with a big orange ball. She is flying her Ercoupe. She is married to Yossef who is not a pilot but an internationally recognized pulmonologist. She has two children and all sorts of friends.

CHARL AGIZA, born Sept. 13, 1962, Lake Village, AR. She fulfilled a dream in 1986 when she first soloed and shortly thereafter joined the 99s. At 300 hours of flight time she apprenticed for two months to learn about low-level flying and became a patrol pilot, flying weekly over AT&T fiber optic cable from Ft. Worth, TX, to El Paso. After several years at that commercial flying job, she moved to freight-hauling with a 135 operator based in Albuquerque, NM. She flies singles, mostly, but holds a multi-engine ATP with more than 6,000 hours flight time. Flying has brought many unexpected gifts to her life.

"Besides the adventure and challenge of it all, flying has given me the insight to see our land and the forces of nature from a different perspective. For me, it has opened up the path of sacredness with regards to Mother Earth.

"The camaraderie of commercial flying and the precious moments we share together with loved ones have become other lessons to treasure. The reality that flight can be a high-risk occupation heightens the importance of living purposefully and meaningfully.

"I never planned on the occupation I've chosen opening windows of the soul to me, but the surprises have been delightful. My advice to someone dreaming of flight – Do It, and always, always, always help anyone you possibly can."

Agiza has flown Cessna 402, 310, 210, 182, 172RG, 172, 152; Citabria, Stinson Voyager, Interstate Cadet, V-tall Bonanza.

LYNN AHRENS was bitten by the flying bug in 1948 after a ride in a J-3 Cub; however, it wasn't until 1967 that time and money coincided for flying lessons. She now holds a CFII rating and has flown about 4,500 hours, mostly in Cessnas from 150 to T-210. Lynn is an active member of the Civil Air Patrol as a search pilot, check pilot and mountain flying instructor.

She is a proud member of the 99s, Inc., Aircraft owners and Pilots Assoc. and the Idaho Aviation Assoc. Air races flown include the Powder Puff Derby, Palms to Pines, Pacific Air Race and Shamrock Derby. She has participated in Flying Companion Seminars and has presented several FAA safety seminars.

Ahren's educational background includes a BA in education and an MA in physical therapy. She is a member of the American Physical Therapy Assoc. and several local medical organizations.

BETTY ALAIR, a 22-year member of the Sacramento Valley 99s since learning to fly in 1973. Flight training was in the Beech Debonair owned by Betty and husband, Neil. Total time to date is more than 2,000 hours. Betty is a member of AOPA.

She is a native Californian, living in Sacramento since 1950. Membership in the 99s has brought many exciting flying experiences, including flying to an International Convention in Anchorage, AK.

Betty has participated in education flight seminars, community speaking programs on aviation, chairing a regional convention and putting together the first "Air Fair" at Sacramento Executive Airport. She has flown a number of noteworthy passengers, ranging from visiting government officials to the mayor of Sacramento.

She is the mother of two daughters and a pilot son. There is also a "troop" of three grandchildren. Betty flies for the feeling of freedom and pure joy of being aloft ...

ANITA P. ALBERT, born Sept. 26, 1936, in Peru, IL. Joined the 99s in 1971. Earned her private license on Oct. 6, 1970. Flown Cessna 152, Cessna 182 and Skylane planes. Her husband Donald K. is also a private pilot.

They have three children: Greta L. Sellers, 33; Charles A. Albert, 31; and Gail A. Buccafurri, 29. They also have four grandchildren. Flew Illi-Nines Air Derby with daughter, Greta, as copilot; she has private and instrument certificates. There have been several pilots in the Albert family from many years back.

R. MAXCINE ALBRIGHT, born March 8, 1936, Stratford, OK. Oklahoma Chapter started flying in 1966 in Aeronca Champs. An engineer husband moved the family with three small children to Oklahoma, Texas and Illinois, where lessons were continued. She earned her license in 1969. She was fortunate to live in an airport community until April 1995 and share an ownership in a Cessna 172 with her son, Stan. She also has a son, Randy, a daughter, Suzanne, and three grandsons.

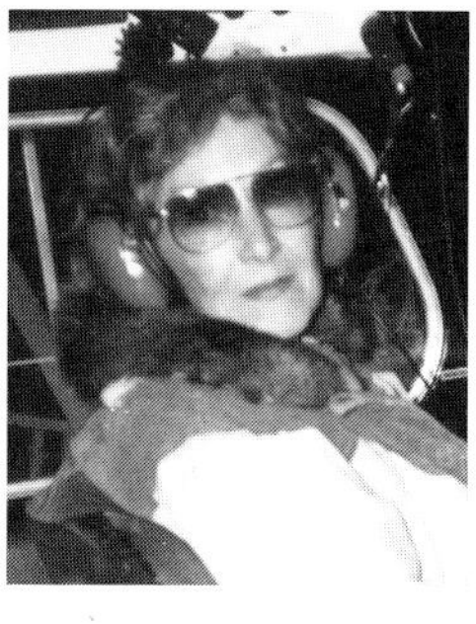

A high point as a 99 was the privilege of working at the Must Stop in Oklahoma City during the last Powder Puff Derby.

Retirement with husband, Gene, is back in Ada, OK, where she is looking forward to being active with Oklahoma 99 friends.

MARILYN ALDERMAN, graduated from Oberlin College as a chemistry major in 1961 and moved to the Delaware valley to work for Hercules Inc. She met her future husband, George, there and after the wedding helped him build his sports car repair and sports car racing-related businesses. She began another career in 1984 by taking flight lessons with a flight instructor friend. She quickly earned her pilot license, adding instrument and commercial ratings soon after. She and her flight instructor became partners in a Cherokee 140. Later the same year she began to transport her husband, a race car driver, and his race crew to race tracks around the country. After a couple of years, she advanced to multi-engine and later ATP. She has participated in two Air Race Classics with her partner in an Aerospatiale Trinidad, helped local Pennies-a-Pound activities, flying companion seminars, and other local aviation events. She is a member of Delaware Chapter 99s, Aircraft Owners and Pilots Assoc., Experimental Aircraft Assoc., Delaware Aviation Support Inc., and the Sports Car Club of America (regional and national). She is married to a race car driver and former Nissan dealer. She has one daughter living in Chicago, and one son who is working in the automotive machine shop and is also a race car driver.

NANCY WELZ ALDRICH, took her first flying lesson on Mother's Day, 1977, at the age of 37. She quickly realized her heart was in aviation. Spending all her time, energy and money on flying, she was able to acquire commercial, instrument and flight instructor ratings by fall of 1978, and became a traveling ground school instructor, working for King Accelerated Ground Schools.

Nancy joined the Colorado Chapter of the 99s in 1979. She participates in many activities, especially the Flight Without Fear classes for people who are afraid to fly. She has also conducted many Safety Seminars around the Denver area, and was designated an accident prevention counselor by the FAA.

Her dream was fulfilled when Nancy was hired by United Airlines. She flew as a flight engineer on the DC-10s, as First Officer on the Boeing 737-300, 767, 757 and DC-8s. She became a captain in October 1991, and is rated in DC-10s, Boeing 727, 737, 757 and 767s.

Nancy has two adult children and two grandsons. She is dedicated to sharing her love of flying with her grandsons, as well as with almost everyone she meets!

CHARLOTTE TUDOR ALEXANDER, born Sept. 20, 1960, Santa Barbara, CA. Joined Greater Seattle 99s 1995; commercial, instrument rated. Trained at Galvin Flying Service, Boeing Field, Seattle, WA. Flown Cessna 152, 172, 172 RG, 182, Piper Seneca and Navajo Pitts S2B and experimental A/C. After 10 years of procrastinating, Charlotte began flight training in June 1994 and is currently working on her CFI. She hopes to begin instructing by 1996 and adding on CFII multi and ATP ratings. After flying the Pitts S2B, she would like to continue aerobatic training and perhaps compete in the future. Her mom, also a pilot, has been her biggest fan and role model

ROSAMOND TUDOR ALEXANDER, October 1940, first flight. 50 hp J-3 Piper Cub at E.W. Wiggins Airport, Norwood, MA. November 1940, first solo, J-3 Cub on skis. September 1941, Ryan School of Aeronautics in San Diego. On Dec. 7, 1941, school closed forever on Pearl Harbor Day. March 1942 Dallas, TX, Aviation School. June 1942, private flight test, 100 hp B-5 Kinner-powered Fleet Biplane. August 1942, commercial flight, Fleet Biplane #798V. November 1942, instructor's rating. Hired by DAS to teach civilian students. Flew Aeroncas, Fleets, Waco UPF7s, Fairchild 24s, Stinson (SR8B and SR7) Reliants, among others. Also had ground instructor rating in CAR before it became FAR.

August 1943, instrument rating test, 220-hp Lycoming Stinson 'O' #13817. A unique aircraft, the only open cockpit aircraft Stinson ever built, designed originally as an observation plane for the government of Honduras. Taught instrument flight air work and radio navigation to War Training Service student cadets. Fairchild 24s and Stinson 10As. In free time, hopped passengers for a small time FBO at $5 a flight, $2 for her.

March 1944, Avenger Field, Sweetwater, TX, the 318th Army Air Force Flight Training Detachment, taught Army Primary to WASP trainees in Stearman PT-17s.

Postwar, flew sporadically in assorted modern aircraft. 1960 - joined Santa Barbara Chapter of the 99s. Has been inactive for the last 10 years, but – who knows?

Aircraft flown at one time or another include: J-2 Cub,

J-3 Cub, Cub Cruiser, Aeronca, Aeronca Chief, Taylorcraft, Porterfield, OX5 Curtis Robin, Ford Tri-motor (not solo), Gypsy Moth, Travelair, AT-6, PT-19, PT-17, JR Speedmail "Bull" Stearman, Stinson "O", Stinson Reliant, Stinson 10A, Ryan ST-A, Waco UPF7, Fairchild 24, Fleet biplanes from model "1" through Model "8", Tri-Pacer, Cessnas 150 and 172, and Hughes Helicopters.

PAULA JEAN ALGER, born Oct. 17, 1961, Jonesboro, AR, USA. Joined the 99s in 1980. Flight training: approximately 200 hours, private pilot. Metropolitan State College graduate 1985. She has flown C152, C172, C172XP, PA-38-112, B19, C23, AASA, PA 24-180. She has been a New York Life Insurance agent for eight years. She has two children, Randall Steven, born Nov. 24, 1982, and Dustin Paul, born March 9, 1984. She has enjoyed flying with Metro State College (Denver, CO) and going to NIFA competitions.

ANGELA GAIL ALLEN began the fulfillment of a lifelong dream when she had her first flying lesson in March 1978. Having already established herself as a professional musician performing with the Atlanta Symphony Orchestra, learning to fly was, at first, only a hobby, but one which she had wanted to pursue for many years. Angela soon realized that she wanted to make flying more than a hobby. She quickly obtained her instrument, commercial, multi-engine, and CFI in order to gain more flight experience. While working full-time as a musician, she also worked as a flight instructor and flew canceled checks at night. Her first big break came in May 1982, when she was hired as a First Officer with Atlantic Southeast Airlines (ASA). Only a year and a half later, she was hired by Federal Express to begin training as a B-727 second officer. Currently, she flies as a DC-10 captain for FEDEX and is also type-rated in the B-727.

Angela has been a member of the North Georgia Chapter of the 99s since 1981. She is also an active member of ISA+21 (International Society of Women Airline Pilots) and has served on its executive council.

JEAN ALLEN, born March 5, 1935, Toledo, OH. Joined the 99s in 1984. Flight training: private pilot, instrument rating, hot air balloon rating. She owns her own Cherokee Six. She is a registered nurse working with AME doing flight physicals for 30 years.

She is a widow and has two children, Christine, 40, and Robert, 35. Also has a 4-year-old grandchild, Alison. Logged 800 hours, flown over 200 children, young Eagles. Has been flying for 12 years.

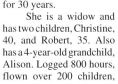

MARIE BARCLAY ALLEN, born Dec. 14, 1921, Napoleon, OH. Joined the 99s in 1970. Received her flight training at Lima, OH, Airport. She has flown Cessna 182, C150, C172, C177, C411, C206, Cherokee 6.

She and her husband Woodrow D. have one son, W. Dale Allen. Their grandchildren are Margot, 27; Gretchen, 23; and Molly, 19.

Member of International Organization of Women Pilots and the 99s. She has participated in several air races, Ohio Air Race, Michigan SMALL Race and a local precision air race. Has worked on the Dayton, OH, Air Show.

SHIRLEY KATHLEEN ALLEN, First Canadian Chapter, learned to fly and joined the 99s in 1967. Was appointed Canada's first airmarking chairman in 1968. Held a commercial rating with multi-engine rating in 1971.

Was co-chairman for the first International Convention held outside the US at the Hyatt Regency Hotel in Toronto, Canada - 1972.

Appointed International Public Relations chairman by the president of the 99s for 1973-74. Wrote an International "Hot-Line" for *The 99 News* magazine about all the overseas sections and did a PR trip across Canada and Europe.

Chairman of First Canadian Chapter in 1976-77. Was Canadian news reporter for the 99s "Up and Away" column in *Canadian Flight* magazine in 1980-81 and 1985-87.

Program chairman and organizer of the six-week series of Aviation Safety Seminars from 1975-81. These seminars earned the chapter an FAI Diploma in the year 1981.

Chairman and coordinator for the beginning of Operation Skywatch with the Ministry of Environment in 1978.

Taught copilot courses at Humber College, Toronto, and #666 Squadron Air Cadets for three years. Also a member of the British Women Pilots Assoc.

DONNA J. ALLEY, born Oct. 30, 1947, Pontiac, MI. She began her pursuit of the thrill of flight in 1988 at Grosse Ile Municipal Airport (previously Grosse Ile NAS) on the island known as Grosse Ile, MI, where, during WWII, President George Bush trained in fighters. She joined the Greater Detroit Area Chapter 99s as a "66" in the spring of 1989 and received her private pilot license on June 9, 1990. Shortly after beginning flight training, her husband, Alex, surprised her on their anniversary with a Cessna 150 so she would always have an airplane available when she wanted to fly! Her husband soon discovered that they had outgrown the little Cessna and the next "surprise" was a Piper Cherokee 140. Their first trip to Mackinac Island, MI, was fantastic. As they flew over the beautiful Mackinac Bridge, Donna said to her husband, "I still can't believe we can actually fly." To this day, she still feels that way every time the wheels leave the ground!

GAIL ALLINSON made her first solo in 1988 at NAS Glenview in a Cessna 150. Among the planes she now flies are the Cessna 172RG and the three-place Schweizer 2-32 sailplane.

Presently employed by Racine Soaring in Sturtevant, WI, Gail has flown for Windy City Soaring and Stone Mountain Soaring. She is a volunteer glider instructor for the Civil Air Patrol and the Gary Flight Academy. A Chicago Area 99s Chapter member, she is also an active member of the Soaring Society of America, the Chicago Glider Club, and the Chicagoland Glider Council. A 1994 Amelia Earhart Scholarship winner, she was recently appointed an accident prevention counselor by the DuPage FSDO. Her writing has appeared in *Soaring* magazine and the "Sailplane Safety" newsletter.

Gail has two sons and lives in Brookfield, IL. She is a Lyons Township High School graduate and holds a BA in art from Western Illinois University.

JOSEPHINE LOUISE ALLISON, born Nov. 8, 1906, Ft. Worth, TX. Learned to fly at the Old White Rock Airport in Dallas, TX, in 1941. She received her license in March 1942. She became a member of the 99s in 1946 and is a life member. She held the offices of chairman of the Texas Chapter, secretary and treasure of the Dallas Chapter of the 99s.

Josephine is the mother of three, grandmother of nine and great-grandmother of 19. She was known as Texas' Flying Grandmother.

In October 1960 Josephine as copilot and Jerry Sloan as pilot, won the 4th Annual Dallas Doll Derby.

Josephine and copilot Doris Weller flew 12 All-Texas Air Tours that covered different areas of Texas and about 3,000 miles each tour.

She was the deputy wing commander for the Texas Wing of the Civil Air Patrol, holding the rank of lieutenant colonel.

Now retired, Josephine fondly remembers the excitement and challenges presented to those pilots who literally "flew by the seat of their pants."

JULIA ROSE AMARAL learned to fly in just three months during 1974 after visiting a real estate agent in the San Joaquin Valley to look at some property and noticing that he had three planes in his back yard. If he could do that, why couldn't she? She was licensed in a Cherokee 180 and immediately purchased a 1/9th share in that plane with eight men. At that time, she didn't know any women pilots. In 1976 they sold that plane and six of the nine bought a Cherokee 6/300. After years of gradual attrition, Julia achieved sole ownership in 1992 and continues to use the plane in her real estate business to visit properties.

Julia married a fellow pilot in 1993 and together they bought a home in Grass Valley, CA, on the Nevada County Airpark and finally built a hangar in their backyard. They also own a ranch in Mendocino County where they recently built an airstrip. There is nothing like flying door-to-door.

A member of the 99s since the early 80s, Julia has enjoyed many flying adventures with other 99s. With its large cargo space, the "6" will hold two full-size bicycles plus lots of luggage. Thus, she and her husband, Mark, are instantly mobile wherever they land.

Julia earned her instrument ticket in 1991, but has not been able to keep it current. Her future goals include regaining instrument currency and making flying adventures to Alaska and across the country to Oshkosh, hopefully with other 99s.

JAN AMUNDSON has been active in flying activities since becoming a commercial pilot in 1965. She and five other Willamette Valley 99s have been making annual trips in three small airplanes, flying from Oregon to Alaska, on three different routes, to New England, New York, Louisiana, California, trans-Canada, and many other destinations between. They have all found the long-distance cross-countries have increased their flying skills, as well as improved their geography.

Jan was appointed to the Eugene, Oregon, airport commission in 1975 and served on that commission for 13 years. She was chairman twice during those years. She was a charter member of Willamette Valley Chapter of 99s and has been active in offices and projects of the chapter since that time.

AILEEN RODA ANDERSON, born and educated in New York state, started flying as a hostess with TWA in 1948. She flew domestic and international flights. Her TWA contract forced her to quit when she married Dick Anderson in 1954.

She was a school teacher, mother of two, a Girl Scout troop leader, and a business woman. When their children became independent, so did Aileen and Dick. They bought a Cessna Hawk XP and started taking flying lessons. Aileen joined the 99s and began planning Girl Scout Aviation Career Days with the Northern New England Chapter. This program has been a great success. She became Aerospace Education Chapter chairman and later Aerospace Education Section chairman.

Other activities include CAP mission pilot and membership in the World Aerospace Education Organization.

DOROTHY J. "DOTTIE" ANDERSON wanted to learn to fly from the first grade and learned to fly as a teenager in a Piper J-3. It was only after soloing an aircraft that her father let her drive the family car. Since that time, she has accumulated in excess of 40,000 hours, still instructs daily and is a flight examiner for private, commercial and instrument certificates. She built and flew her own Pitt Special, is an active amateur radio operator (W80VV) and has participated in the Powder Puff Derby, The Women's International Air Race and the Air Race Classic, where she serves as a member of the board. She holds a BS degree in education from Bowling Green State University, has worked closely with the Air Scouts and with the Flying Farmer Landit Program, and has acted as an Aviation Safety Counselor. She has twice been selected as District Flight Instructor of the Year. She is single but shares her home with two possessive calico cats!

LILLIAN KAY ANDERSON became a charter member of the Bay Cities Chapter March 2, 1932, and today lives in Hinckley, MN. She recalls a clear sunny day in May some years ago when she was a dietitian in Brooklyn, NY. News had come that a young man named Lindbergh had landed in Paris and he was from a town in Minnesota just a stone's throw from her own hometown. That day she decided that she, too, would learn to fly.

In 1929 Lillian took a mechanic's course in Nebraska. She still has that ticket though she never used it. She soloed in 1930 in a Travel Air OX-5 and finished her license (#20.300) at the Oakland Airport. She paid for her flying lessons and a DeSoto car she kept for 23 years by selling "confession stories" revolving around airplanes.

Lillian first heard about the 99s in 1931, and when the Bay Cities Chapter was formed she was proud to be a member. "So what did I get from my contact with aviation? The answer is very clear. It is meeting and knowing the women I met in those flying days. They are to me a race apart."

MARJORIE RAGLIN ANDERSON, at the age of 10, announced she would someday learn to fly an airplane, which she did in 1945 with money earned from her very first job. Earning a private license the same year at Galesburg, IL, she continued working for flight time there, during summers ferrying aircraft across the Middle West. In 1950, she attained her commercial rating and seaplane ratings and joined the 99s.

Marjorie has served as chairman of airmarking and scholarship committees, secretary, treasurer, and chairman of the Chicago Area Chapter, and vice-governor of the North Central Section. Although her professional career has been in teaching high school English in the US and in Europe, flying and camaraderie with the 99s have been a 50-year hobby. In 1987 she was inducted into the International Forest of Friendship and presently serves on the International Forest Board.

Marjorie and her 49 1/2 husband, Arthur, have one son, daughter-in-law, and three grandchildren, ages 8, 5, and 3.

BABETTE ANDRE's credentials include almost 5,000 hours of flight time, an Airline Transport Pilot certificate, FAA-designated accident prevention counselor, Gold Seal instrument and multi-engine flight instructor. Combine that experience with her real estate license, her assistant professorship at Metro State College of Denver, her founding and selling of an aviation magazine and one thing becomes clear: If there's an offshoot occupation in aviation, the entrepreneurial André will find it.

If her down-to-earth side launched her teaching career in the 1960s, teaching English in the Peace Corps in Bafousam, Cameroun, and substitute teaching French and Afro-Asian studies, then her propensity for adventure drew her into aviation. That and a new boyfriend with an airplane, that is.

But an incident over the Utah desert determined her aviation career.

"One day, the engine quit and I thought, 'Oh my God.' It was a simple thing like changing the fuel tanks, but I didn't know that. It just got real quiet real fast." She decided to learn to fly.

From the start, however, she was determined to become a flight instructor.

In 1978 she struck out on her own as a "self-employed flight instructor," although she also held a part-time faculty position in the Aerospace Science Department at Metro State College of Denver – a post she held from 1975-80, and then picked up again in 1991. She also did quick stints as airborne traffic reporter on KOA Air One in Denver in 1978 and KHOW Sky Spy in 1982.

She's an FAA certified ATP and a commercial certified multi-engine-rated pilot. She taught flying at P.C. Flyers at Centennial Airport in Englewood, CO, and is a member of many local and national professional organizations.

André's been listed in the *Who's Who of American Women* since 1979; *Who's Who in the World in 1984-85*; and *Who's Who in Aerospace, 1983*.

In the interim, she turned to a new venture. And, in 1985 launched *Wings West* magazine, a quarterly publication on western flying. She presently is associate editor of *Plane Bargains,* and writes and consults on other local and national publications.

Her contributions earned her the 1990 Aviation Journalist of the Year Award from the National Air Transportation Assoc. and in 1990, she was inducted into the Colorado Aviation Hall of Fame by the Colorado Aviation Historical Society.

JUDYANNE ANDRESS, a native Californian, began flight training in 1989. She earned her private certificate in 1990, her instrument rating in 1993, her commercial rating in 1995 and is currently at work on her flight instructor rating. She has had the opportunity to fly a Pitts S2B, the Otis Spunkmeyer DC 3, a Lake Buchanneer and soled in the summer of 1994 in a Grob 103 glider in Truckee, CA. JudyAnne has done the bulk of her flight training in the T-41 and PA28RT as a member of Civil Air Patrol and the Beale AFB Aero Club in California. She is a member of the Grass Valley Chapter of the 99s, Women in Aviation, AOPA and EAA. JudyAnne believes that the discipline and self-sufficiency that flying demands has enriched her entire life and has greatly expanded her horizons.

MARION ANDREWS, Greater New York Chapter, free-lance artist/calligrapher, a 99s since 1955 licensed 1953 - SEL. Learned to fly to overcome fear of air travel brought on by the death of a long-time girlfriend, when the wing came off of a Taylorcraft in which a pilot was practicing aerobatics for which the aircraft was not stressed.

Governor of the New York - New Jersey Section 1960-62; member Powder Puff Derby Board of Directors 1963-72; chief judge 1975-77; chairman of several Powder Puff terminuses and one 99 Convention. Flew PPD Races 1958, 59, 61, New England Air Race 1952.

She designed the official Powder Puff Derby pin, first-day covers and cachets. Designed the official Air Race program covers for 18 years and many items for the 99s, including the award-winning booklet "Thirty Sky Blue Years" and the 50th anniversary 99 insignia.

Flew on a training flight with the "Blue Angels" when commissioned by the US Navy to do a series of paintings of the flying team.

A graduate of Worcester Art Museum School, with a BS in education from Massachusetts College of Art and was honored by that college in 1993 with an honorary degree in fine arts.

She served 1988-92 as president of the Pen and Brush Club in New York City, the oldest professional woman's club in the US; now on the board of directors of the Salmagundi Club, NYC. Listed in *Who's Who of American Women, Who's Who of the East* and *World Who's Who of Women.*

"My closest friends are 99s."

HELENA BIRGITTA ANDTBACKA, born in Bergö, Finland, July 17, 1940. She is now living in Kronoby. She started flying training in October 1988, soloed in February 1989 and got her private pilot license April 4, 1989. Joined 99s in April 1993 and is a member of the local flight club in Kokkola, the only active lady in the club of 90 members.

Flying during wintertime on the 64th latitude includes lots of extra preparation, such as warming up the engine electrically and waiting for proper weather during the short period of daylight. Therefore is controlled VFR night flight training already scheduled.

Planes flown: Cessna 150 and 172, Piper Cherokee PA28, Piper Arrow 2, Fournier RF 5.

Helena is a school teacher with 32 active years teaching 7-8 year old children to read, write and count. She is married since 1963 to 49 1/2 Hans-Erik. She has two boys, Anders and Mats, and two grandchildren, Tobias, born 1993, and Michaela, born 1995.

CLAIRE ZIMMERMAN ANGELINI, enjoyed her first flight in a seaplane in 1947, Highgate Springs, VT. She knew after that first exciting ride that she would be a pilot some day.

In 1968 after finishing college and enjoying her two children, Kenneth and Virginia, she set out to learn to fly.

She received her private pilot license in September 1968 and joined the 99s the same month. She now holds her multi-engine and seaplane ratings. She has been the proud owner of two planes: a Piper Cherokee 180 and her Christmas present last year, a Bellanca Super Viking. She and husband George "take turns" piloting.

Claire has been a member of the Garden State Chapter of the 99s since 1968. Service to the 99s includes: chapter chairman, chapter committees, nominating chairman, APT chairman, membership chairman, publicity chairman, flying activities chairman, NIFA chairman, aerospace education chairman, and vice governor New York - New Jersey Section, 1976-78, governor of New York - New Jersey Section 1979-81. Section membership chairman and air age education chairman.

Races flown include the New England Air Derby, the Angel Derby, the Power Puff Derby, the Empire State 300 and three Garden State 300 races.

Highlights in aviation were many: Official timer for the London Daily Mail Trans-Atlantic Air Race 1969, Hospitality Committee for the 40th anniversary of the Intentional Convention of the 99s July 1969. Committee for the Power Puff Derby Terminus, hospitality and registration chairman 1972, banquet chairman for the Garden State 300 Race 1973-79.

Claire was on the committee to choose the flight instructor of the year by Teterboro, NJ GADO, 1973 and 1974. She has participated as hostess for many safety seminars with that group.

She is a graduate of Southern Seminary, VA, and Kean College, NJ. She is presently teaching in the Howell, NJ, school system.

KELLY O'ROURKE ANGOTT, born Oct. 1, 1969, Southfield, MI. Flight training: Cochise College private, comm., instructor, multi, CFI, CFII, ATP, C172 182, Beech Baron, Piper Aerostar, Cheyenne II, DC-3, C401. Started flying directly out of high school at the age of 18; by the age of 20 she was already a flight instructor at Cochise College, Douglas, AZ, the school she attended. At Cochise she obtained her private, commercial, instrument, CFI, CFII and multi-engine ratings; later she obtained her ATP certificate. She returned to Detroit where she taught at Detroit City Airport and also had a ground school at Troy Continuing Education. Her next career move was to fly corporate for Dunlop Golf, in which she flew a Piper Aerostar 60IP. Other airplanes she flew include DC-3, Cheyenne II and Cessna 401. In July 1995 she married Jack Angott. Her career aspirations are to fly corporate jets.

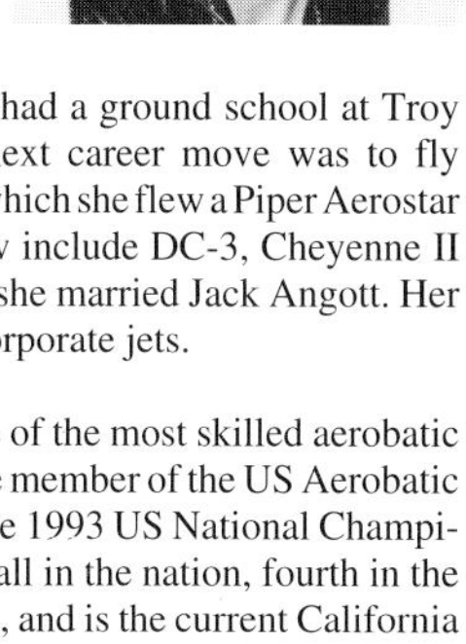

CECILIA ARAGON, one of the most skilled aerobatic pilots flying today. A two-time member of the US Aerobatic Team, she was a medalist at the 1993 US National Championships, is ranked fourth overall in the nation, fourth in the world in the women's division, and is the current California State Champion.

She won the second contest she ever entered, defeating 18 other competitors in the hotly-contested Sportsman category. She rapidly ascended to the top Unlimited category, where again, she won the second contest she entered, winning over competitors with many years of experience.

On her first try for the US Aerobatic Team in 1991, less than six years after she first soloed an airplane, Cecilia won one of the coveted slots. No other aerobatic pilot has equaled her record.

A native of Indiana, Cecilia graduated from the California Institute of Technology with a degree in mathematics and English literature. She moved to the San Francisco Bay Area to study computer science at the University of California at Berkeley, then stayed to take a job as a computer programmer.

But in 1985, a co-worker offered her a ride in a Piper Archer. Cecilia, who almost turned down the offer because she had always considered small planes dangerous, had her first taste of something that would eventually change her life.

"I was in heaven," she recalls. "I said, 'This is my dream, this is it.'"

Immediately, she started taking flying lessons and working towards her dream. Cecilia had to work two jobs to support the flying, often putting in 80 to 100 hours a week, and squeezing the practice in on the side.

Today, with more than 4,000 cockpit hours in her logbook, Cecilia flies in regional, national, and international competitions, performs in airshows, and instructs students at the aerobatic school she founded in Livermore, CA.

ZADA F. ARENTZ, born Zada Fillmore, Nov. 15, 1918, Rigby, ID, descendant of Idaho pioneer family. Attended the University of Idaho. While working in the state house in Boise she was invited to go flying with friends in their planes and was thus introduced to the ease, practicality and fun of flying and so was bitten by the flying bug. She started her flying instruction with Bill Woods at his Floating Feather Airport in Boise, 1942.

At some point she was in the Civil Air Patrol as observer, and later as pilot, till after the end of the war. She acquired the rank of captain.

She was on the waiting list for the WASP program which was disbanded before she was inducted. In 1945 she was licensed in Ogden, UT, and joined the 99s in 1947 in Salt Lake City. She enjoys the camaraderie of other pilots. It's great!

Her flying experience has been transportation, sightseeing and pleasure. Married Bob Arentz, photojournalist, pilot and safety engineer in 1944. They have four children: Robert, 47; David, 42; Christine, 46; Cathy, 36; and grandsons, Jason, 17 and Sean, 12 (who each have their own logbooks).

ANNE C. ARMSTRONG, born in Washington, DC, in 1961. Joined the AF on an AFROTC scholarship at Vanderbilt University. Second lieutenant at Columbus AFB for flight training. Flew C-12F at Langley AFB upgraded to flight examiner and moved to C-5s at Dover AFB, DE. Flew C-5 in Operation Desert Shield/Storm. Left active duty in January 1992. Flew with Reserves at Dover in C-5 till mid 1992 then joined the DC Air National Guard to fly B-727s.

Since the airline world is too unpredictable, she decided to go to law school. Will finish that next May. Plans on going into aviation law for a living and continue flying F2's with the Guard until she needs a walker to get into the cockpit!

MARGARET ARMSTRONG, B.Mus., M.Ed., A.T.C.L., was born Nov. 14, 1946, in Liverpool, England, of Scottish parents. She trained at the Rockcliffe Flying Club in Ottawa, Ontario, Canada, in 1993.

Just three weeks prior to her PPL flight test Margaret had an engine failure at 3,000 feet, caused by a cracked piston. Thanks to excellent training and a cool head, she brought the aircraft to a safe forced landing at an airfield within gliding range.

Her article about this experience was subsequently published in the nationally circulated Canadian DOT Safety bulletin.

Margaret flies Cessna 150s and 172s and holds a night VFR rating. She is a proud member of the 99s Eastern Ontario Chapter and is particularly interested in promoting the Young Eagles program at the high school level, where she teaches music.

Margaret's spouse, Ted Paul, is also a pilot and an electrical engineer. They have two boys but, as yet, no grandchildren.

CAROLYN J. ARNOLD, born Dec. 21, 1937, Spring Valley Township, WI. Joined the 99s in 1968. She has flown Cherokee 140, 180, Cessna 150, 170, 172, 182, Cardinal RG, Rockwell Commander, Beech Musketeer. Her love was the 1949 Cessna 170.

She and her husband, Al Arnold, have two children, Suzanne M. Masterson, 36, and Sandra C. Hoadley, 32. Their grandchildren Joshua, 9; Mikki, 5; and Meaghan, 1 1/2.

She learned to fly in 1968 and joined the 99s in 1969. She was very active in the group and was Wisconsin Chapter chairman in 1973. She was also treasurer at the 1973 International Convention in Milwaukee, WI. She flew in the 1976 Powder Puff Derby with Peggy Mayo and the 1977 Commemorative flight with Joan McArthur. She no longer flies, but enjoy trips with her husband, who is chief pilot for outboard Marine Corp.

CONNIE TAKSEL ARNOLD, began her affiliation with the 99s in the 60s at the age of 4, when her mother, Ruth Taksel Benedict, joined the Greater St. Louis Chapter and continued it in the 70s when Ruth transferred her membership to the Maryland Chapter. Connie accompanied her mother to many 99s functions, anxiously awaiting her day to officially join the organization.

In the 1970s, Connie became active in the CAP. In 1978 she earned her private license just in time to attend her obligatory three meetings and become a 99 in November before moving to Arizona the following January.

Connie studied aviation technology at both Southern Illinois and Arizona State universities. During this time, she earned her instrument, commercial and multi-engine ratings. In December 1982 Connie graduated from ASU with a BS in aviation technology. She also transferred her 99 membership to the Phoenix Chapter, became a charter member of the Sundance Chapter in the 1980s and flew with her mother as the "Benedict/Arnold" team in the Kachina Doll Air Rally.

Following her graduation, Connie landed a job flying copilot on a King Air for Circle K Corp. and worked as a flight instructor. Later she worked at America West Airlines, and is now employed in the safety/environmental field for the city of Phoenix.

Connie married David Arnold in 1979 and has two sons, Matthew and Nicholas. She credits her mother for instilling within her a love for flying, and counts her mother as her greatest influence.

VERA ARNOLD, began flying in 1967 at the age of 47, as a student of 99 Betty Hicks. She has been active in Santa Clara Valley Chapter activities, serving as vice-chairman, membership chairman, airmarking chairman, and fly-in chairman. She holds commercial, ASES, ASEL, instrument ratings and has more than 3,000 hours. Vera flew many races, including the AWTAR, the Palms to Pines, and the Angel Derby. Vera was also active as a NIFA judge both for local USPFT competitions and at the national level.

She and her husband, Sid, have owned many airplanes, including a C182 with a Canadian maple leaf on the tail, and then another C182 with a red 99 on the tail which took them to Alaska, where Vera got her floatplane rating in Ketchikan in 1977.

She has two sons and is both a grandmother and a great-grandmother. She lives in San Jose.

CAMILLE MICHELLE ARSENAULT, born in Chatham, Ontario, Canada, and now resides in Ottawa, where she flies with the Ottawa Flying Club, situated at the MacDonald-Cartier International Airport. She joined the Eastern Ontario Chapter of the 99s in June 1995.

Camille's interest in flying began when she enrolled in the Royal Canadian Air Cadets in 1981 and was able to participate in their glider program. She graduated from the University of Ottawa with a bachelor of arts degree in June 1990. She then started her flight training Oct. 6, 1990, soloed Feb. 23, 1991, and received her private pilot license Jan. 22, 1992. During the following three years, she obtained her night rating, and accumulated enough pilot-in-command and cross-country hours for her commercial rating, which she will obtain this year (she has passed the flight test and has only the written part remaining).

Camille currently flies Cessna 150s, 172s, 172 RGs and Piper Arrows (RG). She plans on obtaining her certified flight instructor rating by the spring of 1996, (she hopes to teach on a part-time basis), and her instrument, multi-engine and multi-engine instrument certifications by the end of 1996. Her ultimate goal is to obtain her air transport pilot license and fly for one of the major airline companies. This she hopes to accomplish before the turn of the century.

ANN ASH, joined the Cleveland, OH, Chapter in 1945. During WWII, she obtained her commercial rating while flying with the Cleveland Civil Air Patrol. After the war, she opened an airport at Willard, OH, offered training in Stearmans, BT-13s and a Cessna UC78 and had one of the largest VA flight schools in Ohio. She was one of the first distributors for Bonanzas and Piper aircraft. A highlight was placing second in a Pylon Race at Columbus, OH, with five other male pilots in 1948.

She married Maj. William Ash, commander of the Air National Guard Fighter SQ at Mansfield, OH, in 1949. They have three sons and two grandchildren. Later, his Air Force career gave her the opportunity to join the Wilmington, DE, and Washington, DC chapters. While stationed in Japan in 1954, she chartered the first Japanese chapter. Col. Ash retired in 1972 and they moved to San Antonio, TX, where she is active with that chapter and Ann Ash, Realtors.

ROMAINE JOHNS AUSMAN, born Aug. 14, 1932, Macon County, IL. Joined the 99s in 1975. Received her flight training at Boulder, CO, Western Flight. She has flown Cessnas 150s - 210.

She is married to E.L. Ausman, Jr. They have five children: Carol Ausman, 39; Evan, 36; S. Jordan, 34; Edmund, 32; and Ellen, 29.

She would have liked to learn to fly at University of Illinois in 1950s. Following surgery for cancer of the thyroid (20 years ago) husband gave her flying instruction (lessons) as a Christmas gift. She's flown to all 48 states and to Alaska for pleasure and pursuing her business as designer and creator of bridal gowns.

ALICIA AUSTIN, learned to fly in Moses Lake, WA, in 1987, with a part 141 flight school. She has earned her CFII with 325 hours. She moved to Burlingame, CA, and joined the Santa Clara Valley Chapter in 1991.

Alicia works in the corporate accounting department at Hewlett Packard and has two children. Alicia's hobbies are sewing and skiing.

DIANA PAYNE AUSTIN, born Oct. 1, 1946, Lafayette, IN. Private pilot, learned to fly at Monticello, IN, at age 22 (ASEL). She has flown Piper 140 and 180.

Employed as primary teacher for 26 years in Valparaiso, IN. She has two children B.J., age 19, and Brett, 15 1/2. Has served in every chapter office of 99s, served as international student pilot chairman, spoken in several classrooms regarding aviation history, chairman of Air Bear Program for Indiana. Learned to fly after experiencing "snow at night" in the mountains during December! Her former husband was a pilot, and she thought she'd just learn how to land a plane in case something happened. After she soloed, she was hooked! Has been in the 99s since January 1971. Flew two IN Air Rallies, placing second in 1981. Has flown Indiana Dunes Air Rallies. Placed first in 1982. Flew a congressional candidate, a lady, to nine airports in one day for her to campaign. Presently secretary and membership chairman for Indiana Dunes Chapter. Profession: teacher of second grade.

LEE GRELL WINFIELD AVERMAN, born in Minneapolis, April 6, 19—. Attended University of Redlands. Four children: Edmund J. III, Barbara Lee (Mrs. George Harris), Lawrence D., Linda Ann. Secretary-treasure/co-owner Regina Airlines 1950-59; president American Air Taxi 1961-62; assistant secretary Southeast Airlines 1962-75; various executive positions in several small airlines (financial departments) from 1976 to 1990. Recipient of the Key to the City of Miami Springs, 1963; awarded Florida Air Pilots Associate plaque for Outstanding Contribution to Private Aviation 1962; listed in *Who's Who of American Women,* 1972-73.

She received her private pilot license July 14, 1958, and was the proud owner of a Piper Tri-pacer (N821A). Her happiest days were during the 60s when she was active in the 99s Florida Air Pilots Assoc., Florida Women Pilots Assoc. and Florida Aero Club. The highlight of her aviation career was in 1963 when, as copilot with Virginia Britt, they won the Powder Puff Derby. She was an official in several Angel Derbys and as such visited many foreign countries and met many interesting and famous people.

In 1976 she moved to Tucson, AZ, and although she was active in the 99s for a year or two, she no longer had the opportunity to keep up with her flying. She retired in 1991 and has not been an active 99 for several years. She does keep up her membership and through the newsletter stays up-to-date with old friends.

JOAN F. AXINN, in May purchased a 1975 Cessna 182P and flew it from Opalocka, FL, to Islip, NY.

In Vermont, she flies a Piper Super Cub from an airstrip on her farm (the private airport is listed on the Montreal and New York sectionals).

Last summer, she and her husband, Donald Everett Axinn, flew their 414W to the Canadian Northwest and to Alaska.

Joan also flies with her husband in his Stearman N2S3 biplane, a WWII US Navy trainer.

Next summer, she is extending an invitation for fellow members to fly to Spruce Creek (Daytona, FL), an airport community where the Axinns are building a new home.

Joan earned a JD from Hofstra University Law School and ran for US Congress in 1992.

ENA E. AYERS, received her private license in August 1941. Joined the Bay Cities Chapter of 99s in August 1946. Flew to her first International Convention in Cleveland, OH, in September 1946 with Marjorie Fauth. Was chapter secretary for one year prior to moving to New York in 1950. Served on committees for five International Conventions and the start or terminus of six AWTARS. Was copilot to Dottie Gable Bock in the first New England Air Race, they placed sixth. Returned to San Francisco in 1963, entered Interior Design School, graduating in 1966. Worked in that field, and as hobby enjoys water color painting. Is an active member of Bay Cities Chapter.

SUZANNE L. "SUZIE" AZAR, born April 29, 1946, in Bay City, MI. Joined the 99s in 1980. Flight training: Piper's "Blue Sky Special" in a Tomahawk. Ratings: Comm., CFII, SEL, MEL, SES, instrument glider. She has flown Cessna 152, 172, 172XP Hawk, T41B, 180, 182, 206, 210, 310, 340, 414, Cessna 150 float plane, pressurized Skymaster P337, Beech Sundowner, Sierra, Bonanza, King Air, Duke, Duchess, Queen Air, Aero Commander (turbo prop), Maule MX7, Grob 109B, Schweitzer 233A glider, Aeronca Champ, C305 A Bird Dog, BD 4, Pitts Special, Piper Tomahawk, Warrior, Cherokee, Arrow, Aerostar, Lance, Supercub, and Aztec. Helicopters: Brantley, Hughes 300 and Jet Ranger. Military: T-34 Navy Aerobatic trainer and F-15 Fighter Eagle (a thrill of a lifetime).

Suzie Azar has enjoyed sharing flying adventures with her husband and daughters. They have flown all over Mexico, to many parts of Canada, and most of the US. Suzie and her youngest, Michelle, won best first-timer and fastest mother-daughter team in the 1994 Palms to Pines Air Race. Joined by her husband in 1995, they placed in the money racing through the Bahamas in the Great Southern Air Race. Suzie will challenge a World Speed Aviation Record, circumnavigating the globe in the fall of 1995. This quest will be joined by her husband, Richard, and their friend, Charlie Justiz.

Suzie has been an instructor since 1984, has flown radar tracking missions for Stanford Research Institute, flown aerial photography work, flown some charter trips and bought, sold, leased, traded and enjoyed a lot of aircraft.

Through her participation at 15 in Civil Air Patrol, Azar always talked about taking flying lessons. But was truly inspired when she met the 99s who flew through El Paso in the Powder Puff Derby. "When I finally decided that I deserved to fly, it was easy to find the money and time to do it," said Azar.

ADELE MAY BACHMAN, born May 1, 1926, in Albany, CA. Joined the 99s in 1977. Received her flight training at Buchanan Field, Concord Airport, Concord, CA. She has flown Cessna 150, Cessna 172, purchased and owned Cessna 172 for three years. Flew Palms to Pines Air Race from Santa Monica, CA, to Independence, OR, two times.

She is married to Julius A. Bachman. They have five children: Kenneth, 48; Mark, 46; Todd, 43; Gregg, deceased at 39; and Alain, 38. They also have three grandchildren: Jay, 20; Courtney, 18; and Alana, 6. At the age of more than 50 years, she started flying lessons and received private pilot, single engine, land rating on Jan. 8, 1977.

ANNE BRIDGE BADDOUR, began her aviation career in 1953 at Revere Aviation, Revere, MA. She further trained at the Learjet training school in Connecticut and at Flight Safety, Wichita, KS. She holds an airline transport pilot license with SEL, MEL, SES ratings and an AAS from Pine Manor College.

Anne has flown more than 40 different types of aircraft, from helicopters to fixed-wing, prop, turboprop, jets, including an F3D and an F/A 18. She is qualified in 32 different aircraft and has more than 2,700 hours of flight time.

Anne currently works as a research pilot for the Massachusetts Institute of Technology Lincoln Laboratory Flight Test Facility, doing airborne research for the Department of Defense and the Federal Aviation Administration. The aircraft she flies there include a DeHavilland Twin Otter, Beechcraft King Air, Twin Cessna 421, Beechcraft Bonanza, Cessna 172, Navajo Chieftain and Beechcraft 1900 Airliner. She has been a pilot with MIT Lincoln Laboratory for more than 17 years, and is their first woman pilot.

Anne has established a total of 27 NAA/FAI world speed records. The records were set during two separate transatlantic flights from Hanscom Field, Bedford, MA, to Europe and one trans-Canada flight, all with another woman as copilot. The aircraft used were a Mooney 252 in 1985, a pressurized Beechcraft Baron 58 in 1988 and a single-engine Beechcraft Sierra in 1991. She has also participated and placed in 16 regional, transcontinental and international air races.

Anne was appointed a Massachusetts Aeronautics Commissioner from 1979-85. Chairman, FAA New England Regional Women's Advisory Committee, 1984-87. FAA Judge; Aerospace Education Awards for New England, 1989 and 1991; flight dispatcher, ferry pilot, Comerford Flight School, Hanscom Field; administrative assistant, ferry pilot, Jenney Beechcraft Corp., Hanscom Field; manager, pilot, Baltimore Airways, Inc., Hanscom Field, 1976-77; aviation consultant, corporate pilot, Energy Resources, Inc., Cambridge, MA, 1974-84.

Anne was an instructor in basic aerodynamics, meteorology, aviation medicine, airline codes, navigation and other aviation courses at J.R. Power's Aerospace School, Boston, MA, 1966. Director of Aero Club of New England from 1977 to present. Chairman, Aero Club of New England education committee and founder of aviation scholarship program and fund raising annual auction, and has established five annual scholarships for young people in need of financial assistance in advance pilot training, aeronautical science, airframe and powerplant mechanic and aeronautical engineering. She has also established a scholarship for young women in advanced pilot training for 99s Eastern New England Chapter.

Anne has held the following offices in the 99s; New England Section; vice-governor, secretary, national news reporter, New England Section news reporter, co-chairman of All-Woman's Transcontinental Air Race (The Powder Puff Derby) and program chairman.

Anne has been a featured speaker at numerous aviation and travelers clubs, also for the FAA and USAF.

In England in 1983, Anne was received by the commanding officer of Biggin Hill RAF Station and special arrangements made for her to fly over city of London at a special assigned altitude of 1,500 feet, an honor rarely given. In 1992, the US Navy invited Anne to fly an orientation flight in their top-gun fighter; the F/A 18 Hornet at Cecil Field NAS, Jacksonville, FL.

In 1993, Anne was recipient of the Exemplary Service Award from the Aero Club of New England. In 1992, recipient of the Amelia Earhart Medal, Honor Award of the Year, naming her "Pilot of the Year" from the 99s Inc., New England Section. In 1991, an honoree in The International Aviation Forest of Friendship, Atchison, KS, in recognition of "Outstanding Contributions to Aviation." In 1990, the New England Federal Aviation Administration director presented her with a special award in "Recognition of Outstanding Aviation Achievements." In 1988, winner of the International Clifford B. Harmon Trophy, Aviatrix. This award is presented annually to a recipient for "the most outstanding international achievement in the art/science of aeronautics for the preceding year with the art of flying receiving first consideration." In 1986, Anne received the Eastern New England 99s Flight Safety Award.

Among other memberships, Anne is a Life Member of the 99s, Inc., a member of the board of The Aero Club of New England, an associate member in The Society of Experimental Test Pilots, Federation Aeronautique International, US Sea Plane Pilots Assoc., National Aeronautics Assoc., Daughters of The American Revolution, and Airplane Owners and Pilots Assoc. and a trustee of Daniel Webster College, Nashua, NH.

Anne is married to Professor Raymond F. Baddour, chemical engineer/entrepreneur. They have three children, all married; Cynthia Anne, Frederick Raymond and Jean Bridge. Anne and Raymond have five grandchildren, Frederick Baddour, Miles and Serena Ryan, Hannah and Edward Nardi.

BARBARA B. BAER, born Dec. 1, 1928, in Rockford, IL. She started taking flying lessons off the grass runways of Rockford, Illinois' Machesney Airport when she was 17; soloed on April 9, 1946; and received her license on Sept. 3 of that year.

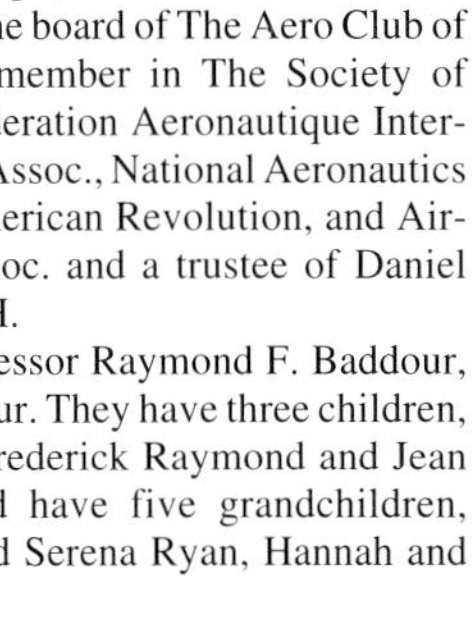

As a member of the UCLA Bruins Flying Club, between 1946 and 1950, she flew off of her first paved runways at Van Nuys Airport and Santa Monica's Clover Field. Barbara took part in the intercollegiate air meet in Ft. Worth, TX, which UCLA won, and checked out pontoon sea planes at Lake Elsinore.

In 1950, she had to forego piloting due to illness, but she has continued over the years to fly with friends and sit in the copilot's seat when in Alaska and near New Zealand's Mt. Cook. She's now returned to her first love as a proud member of the 99s.

Barbara has two children, Daniel, 41 and Suzanne, 38; she and her husband Dan have 11 grandchildren: John, 12; David, 10; Matt, 7; Chana, 13; Shmuel, 13; Rachael, 11; and Ariela, 6 and a half.

She has flown the following planes: Cubs, Aeronca Champ, Cessna 140, 160, 172, Beechcraft Bonanza, five passenger Stinson.

DEANNA DITTNOCK BAER, started learning to fly in 1963, shortly after the birth of her fourth child. She learned to fly while living on a ranch in Winnemucca, NV, in a Taylorcraft. She was husband, Rick's first student. Private license 1964, as fifth woman pilot in area was promptly recruited to form the Fallon Chapter of 99s (now Reno Area). By 1967 earned commercial instrument, MEL, flight instructor A&I, ground instructor and began working as charter pilot flight instructor. Past chairman Reno Area Chapter 99s. Participant Kachina Air Race, worked air race stops and Reno Air Races.

Great flying experiences include teaching son Marc to fly, helping him through ratings to become an airline pilot, and teaching daughter Sherri, a charter pilot and flight instructor, to fly. Family includes husband Rick, three daughters, one son, five grandchildren. Currently own and fly a Citabria for fun.

DOROTHY H. BAER, born Feb. 3, 1923. Joined the Reno Chapter of the 99s in 1964. Has held the office of chairman. Her current rating is ASEL - instrument.

SYLVIA BAHR, raised in Girard, OH, and moved to Cleveland after earning a BS in biology from John Carroll University. She then worked in a hospital. However, she learned to fly in 1983 and currently holds an airline transport license and flies a Cessna Citation and Beechcraft King Air.

She began her aviation career flight instructing and eventually began flying for local corporation. She especially enjoys giving career workshops to community schools and would like to have more women in aviation.

Bahr is a member of the 99s, Inc. and the Aircraft Owners and Pilots Assoc.

A rare professional opportunity came when she and her husband, Bob (now a pilot for America West Airlines) flew as a cockpit crew in a Citation for a corporation in 1990.

ELEANOR J. BAILEY, more than 40 years of flying for fun from Alaska to the Bahamas, Quebec to Mexico, Eleanor and her blue Comanche have covered the continent. Aviation activities have included involvement in Flying Farmers, Canadian Owners and Pilots Assoc. Board representing 99s, Calgary Airport Authority, Western Warbirds and of course, 99s. She has served as Alberta Chapter chairman, International Resolutions chairman, and co-chaired the 1971 Powder Puff Derby Start in Calgary.

Eleanor has flown a variety of single engine aircraft, and she and husband Bill, an ATR pilot, have owned a wide variety of planes from Super Cubs to many warbirds, including Eleanor's 1939 deHavilland Tiger Moth and Bill's P-51 Mustang, which they operated from their own farm airstrip.

Eleanor and Bill have two children and three grandchildren. Their oldest daughter, Karen, is a former AE Scholarship winner.

JUANITA PRITCHARD BAILEY, started flying at Betti's Airport near McKeesport, PA, in 1940. During the war flew with Civil Air Patrol. After the war, started delivering new Pipers for the Piper distributors, from the Piper Factory at Lock Haven, PA.

Twelve Pipers to Alaska, one on floats. Thirty-five to Panama, two Bellancas, one 450 hp Stinson, the rest were Pipers. Two Tri-Pacers to Bogota, Columbia. Flew own plane to Lima, Peru. Flew AWTAR in 1950, placed sixth.

Flew 50 Pipers to Wichita, KS, on week-ends, had a beauty shop in hometown, Clairton, PA. Good customers and operators let her enjoy flying. Flew two Pipers on floats to Portland, OR. She has more than 6,000 hours flying time. Has been a member of 99s since 1945, Sacramento Valley Chapter 99s since 1960. Has Silver Wings, OX5, land, sea and glider ratings.

KAREN JOAN BAILEY, born Aug. 20, 1957, Lethbridge, Canada. She made her first cross-country trip (1,500 miles return) at the age of six weeks. Most of the trip was spent sleeping behind the back seat in a Super Cub. Both her parents are pilots, so learning to fly was a natural thing to do. She learned to fly in a Citabria and was licensed in November 1976 with 37 hours in her log book.

She joined the 99s in 1977. Since then, she has been active at chapter and International levels. She received an Amelia Earhart Scholarship in 1989 to obtain her instrument rating. She was International nominating committee chairman 1990-92.

She now holds a commercial multi-engine instrument rating and has 575 hours in 10 types. When not busy flying, she practices obstetrics and gynecology in Glendale, AZ.

AUDREY L. BAIRD, born July 22, 1919, in Minneapolis, MN. Joined the Minneapolis Chapter of 99s in 1949. Received her flight training at Robbinsdale Airport (now Crystal) Minneapolis/Dickinson Airport/North Dakota. Has flown most time in J3 Piper Cub, Cessna 182, Cessna 210 and Bonanza. Flies occasional charter for Dickinson Air Service. Married LaRoy Baird in 1948, widowed in 1969. They had four children: LaRoy III, 46; John, 42; David, 38; and Ellen, 30. They also had grandchildren: Jesse, 20; Ondine, 19; Stephen, 10; and Anna, 5.

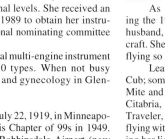

The winter of 1945 in Minneapolis, she flew her first solo in a J3 Piper Cub on snow skis. In 1946 she obtained her private license at Daytona Beach, FL, in a Piper J3-Seaplane. Part of the preparation was to wear a parachute, go solo, and practice spins over the water.

In 1948 she put her flying on hold for eight years while she raised her three sons. In 1957 she moved from Minneapolis to Dickinson, ND, where she resumed her flying in a Cessna 120.

In October 1964 their Chapter flew "Penny a Pound" rides. She flew her Cessna 182, still pregnant. Another 99 member flew her 182, and she had a nice, big, white cast on her leg. When taking up passengers they did not dare to get out of their planes, for fear that no one would dare to enter. Her daughter was born in November.

In 1971 she flew her Cessna 182 in the Powder Puff Derby - TAR #60. And in subsequent years she flew charter and ambulance trips in a Cessna 210 for Dickinson Air Service. In 1993 she completed 19 years of serving on their local board for the airport authority.

She is a charter member of the Minnesota 99s and the North Dakota 99s, former board member and treasurer of the Northwest Section and secretary of North Dakota Chapter. She has attended Northwest Sectionals and International Conventions, also North Central Sectionals.

Through the years she has flown and logged time in a J3 Piper Cub, a Taylorcraft, a Piper PA18, Cessna Cardinal, a Cessna 182, a Mooney 21, Cessna 210, and presently her Bonanza. She has currently logged 2,600 hours.

Her biggest thrill was flying her Bonanza to Alaska in 1993, with her youngest son who is an air traffic controller, a private pilot, and the only one out of her four children that has followed her enthusiasm of flying.

Her husband was an attorney and a judge; he was not a pilot, but always stayed as her mentor with much encouragement.

In 1969 she became a widow, and president of her husband's corporation of abstracting and title insurance, which at the present time she is still operating.

SHIRLEY GLASSLOCK BAIRD, born Sept. 12, 1935, in Virginia. Joined the 99s in 1990. Received her flight training private pilot SEL 1990, 200 hours presently. She has flown C-152 and C-172.

She is married to Bryant W. Baird, Jr. They have two children, Beverly B. Crowe, 35, and Scott B. Baird, 32.

JOYE SMITH BAKER, Colorado Chapter member since 1971; 1975-76 Chapter chairman; 1975 chairman SCS Section, Denver; 1980 A.E. Luncheon Chairman, Vial, CO; Jan Gammell's copilot 1975 Illi-Nines Air Derby and copilot for Anita Hessin in the 1978 Angel Derby, Dallas to Freeport, Bahamas.

After a 14 year hiatus, some spent sailing a Nor'Sea 27, Joye is again flying interesting aircraft, a 1961 Piper Colt and a 1949 Bellanca Cruisair Senior.

As an avocation during the 1970s Joye and her husband, John, enjoyed purchasing and refurbishing aircraft. She considers herself very lucky to have experienced flying so many different aircraft in only 550 hours TT.

Learning to fly in a 1961 Skylane and a 1946 Piper Cub; some of her other favorites have been a 65 hp Mooney Mite and a 450 hp DeHavilland Beaver; Bellanca Viking, Citabria, Decathlon; Piper Comanche, Arrow; Grumman Traveler, Cheetah, Tiger; Aeronca L-3; Globe Swift; and flying with her husband in a Stinson Gullwing before obtaining her license.

A memorable flight was a delivery trip in the Beaver to Alaska in 1975. Camping and fishing along the way was a thrill for all the family including 16-year-old Jill and John Art, 14. Joye likes to say that the trip took 35 hours by Beaver and five by Boeing!

The mid 1990s find Joye looking forward to taking her grandchildren on their first flight.

TRICIA BAKER, current ratings: ASEL, AMEL, instrument. She has flown Cessna 414, 303, 172, 150, PA 140, Bonanza A36 and J35, Piper Tomahawk. Tricia and her husband, Gary, learned to fly in 1982 and bought their first airplane, an A36 Bonanza, as a wedding present to one another. Since then, she has owned a Cessna 303 Crusader and a 414 Chancellor, but has returned to her favorite the A36. For three years, she and Gary had two Bonanzas and delighted in practicing formation flying.

Instrument rated, Tricia has logged more than a thousand hours using her plane in her work and play. With a BS in medical technology she has flown for a major laboratory network as a sales representative and with the Aeromedicos in Santa Barbara providing medical aid to the Yaqui Indians in Guaymas, Mexico.

Baker has been a member of the 99s since 1986, staring with the San Fernando Valley Chapter, becoming a charter member of the Santa Maria 99s, transferring to the Reno Area for three years, and finally back to Santa Maria Chapter.

Together Tricia and Gary own Performance Aero, a Beechcraft accessories business which advertises internationally. She belongs to Aircraft Owners and Pilots Assoc., American Beechcraft Society, World Beechcraft Society, Pacific Beechcraft Society, and is currently seeking membership in Santa Barbara Sheriff's Aero Squadron.

BEV. BAKTI, born Aug. 17, 1950, in Ontario. Joined the 99s in 1987.

Received her private license in 1986. She has flown 172 and Cherokee 140. Married to John.

"I've met some very dear friends through the 99s. Because of this association I'm flying today; and have flown trips to Edmonton and Florida, with future plans to the Bahamas and Halifax."

MARGE BALAZS, grew up in the Midwest, majored in music at the University of Wisconsin, and was a police officer before learning to fly while in the Navy in 1981. She served as a Naval aircraft maintenance officer until 1987, when she transferred to the Army Reserves to fly helicopters. Marge was designated an Army aviator in 1987.

She received her MA in aeronautical science from Embry-Riddle University while flight instructing at the Quantico Flying Club and working on the Navy's HH-60 helicopter acquisition program. Marge then flight instructed for the Air Force Flight Screening Program, concluding the military services.

Marge is currently a pilot for United Airlines, flying the B-767 and B-757. Her activities include flight instruction, EAA, Antique Classic and 99s member. Marge holds an ATP rating for the B-767 and B-757, is a CFI (ASEL, AMEL, instrument) and commercial rotorcraft - helicopter with instrument rating.

CONNIE BALL, born June 23, 1944, in Somerset, PA. Now a Maryland resident, she learned to fly in 1969. She flies a 1948 Swift that was completely rebuilt by her late husband, James Ball. Ball took one airplane ride in 1968 and said, "this is for me." She called a local flight school, took an intro flight, and was on her way. Since then she has earned her instrument and multi-engine ratings, SEL/MEL commercial rating, CFI, and ground instructor ratings. Ball also took aerobatics training in a 1946 J3 Cub and has 35 hours in a helicopter. Ball is a proud member of the 99s,

Inc., Aircraft Owners and Pilots Assoc., and the International Swift Assoc. Ball's educational background includes a BA in business and is presently a student in graduate school working for a MBA in management. Ball's late husband, James, was also a pilot. She has a daughter, Carolyn, 34, and a son, Roger, 30; and two grandchildren, Randy, 4, and Taylor Marie, 2.

She has flown the following planes: Piper Cherokee, Aztec, Navajo, J3 Cub, Warrior and 180, Swift, Cessna 152 and 172, Beechcraft/Duke. Employed as a commercial pilot for Bal-Mar Aviation and Air Hearse Service.

KAREN J. BALLARD, born Aug. 26, 1946, in Chicago. Joined the 99s in 1992. Holds a private pilot license. Karen really enjoys flying to various gatherings around the country with husband Gene in their Grumman Tiger. Always interested in astronomy and space flight, she grew up following the achievements of the astronauts and finally learned to fly. In addition to the 99s, she is a member of the Aircraft Owners and Pilots Assoc., American Yankee Assoc. and Grumman Gatherers. She is especially proud of winning first place in the "Egg Drop Contest" at Rough River, KY, Fly-In 1994.

Educational background includes a BS in dietetics and nutrition from Northern Illinois University with internship at Northwestern. As a registered dietitian, she is a member of various nutrition associations. She has a son, Dale, a civil/environmental engineer; four step-children and is waiting for grandchildren.

MARY LOU BALOGH, born June 3, 1937, in New York City. Joined the 99s in May 1971. She began flying in 1967. Flight training was in an Aeronca Seaplane, Cessna 150 and Cherokee 180. Also acquired simulator time in the Falcon 50. Received license in 1971 and joined the 99s. Educational background includes a JD from Rutgers University. Presently a real estate broker, involved in property management. Married to Louis Balogh, also a pilot. One of the greatest thrills was formation flying with her husband when they could not legally fly together. Many vacations were present flying their Cherokee 180 to Oshkosh, FL, and weekends to Cape Cod and the Islands. Her mom, now 86 years old, also loved to fly and joined them on many trips.

CAROL JOHNSTON BANEY, started flying out of Willoughby while stationed at Perry, OH, with CAA 1943. She was transferred to CV, MOT, DHN, ATL. Was vice chairman when they organized a ND 99s chapter in 1954. (Her aunt Thelma Burliegh was a charter member.) She organized and taught a CAP cadet group, later was squadron commander, then on group staff coordinating search and rescue.

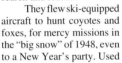

They flew ski-equipped aircraft to hunt coyotes and foxes, for mercy missions in the "big snow" of 1948, even to a New Year's party. Used county road maps to deliver repair parts to farmers and for visiting around the area.

She got her commercial rating just prior to flying the 1959 AWTAR.

Taught basic instruments to helicopter pilots at Fort Rucker in the link trainer and retired in ATL after 25 years with ATC.

DENISE BANSEMER, born Jan. 21, 1955, in Harrisonburg, VA. Joined the 99s in 1993. Flight training included commercial hot air balloon and private ASEL. She has flown Adams AX-7 Balloons, Cessna 150/152, 172, and Piper Warrior. She is divorced and has two children, Lauren, 10 and Rachael, 8.

She can remember as a child in southeastern Florida, how school and all other activities stopped to watch the rockets launch and fly over from the Cape. One of the first papers she wrote was on John Glenn and his historic flight. She also remembers

driving to Miami to watch the "jets" come in and of hearing the roar of Pratt & Whitney testing their engines. Caught up in all that, she wanted to be an astronaut. In 1981, she married a hot-air balloon pilot and learned to fly commercially. They published a book entitled *The Art of Hot Air Ballooning* and went as far as Australia to fly (painting a balloon mural on a restaurant at Sydney Harbor). After her divorce, against the odds, she finally got her ASEL certificate. She loves flying whether she is a passenger, an airport visitor, a viewer of a space launch or as a pilot. She is continuing her certificates, is considering aviation management or flight service as a career and is a member of the Florida Suncoast Chapter of the 99s. She has two girls that were introduced to flight before they were born. She is a fiber artist, depicting environmental and social issues and solutions. She also teaches fiber arts and personal growth classes.

SUSAN ELIZABETH BARBER, born May 28, 1955 and raised in Seattle, WA, she became interested in flying in 1988 when the challenge of flight presented itself. Her first flight was in an Aeronca Champ. Since then she has accumulated more than 1,200 hours and her credentials are impressive. She has flown five airshows in a Cessna 140 Aerobat including the 1990 NW EAA Arlington Airshow, holds commercial and certified flight instructor certificates with instrument, multi-engine and seaplane ratings. She was also the recipient of the 1994 Amelia Earhart Award Scholarship presented

by the 99s. Her long term goal in aviation is to do accident investigation for the Federal Aviation Administration (FAA) or the National Transportation Safety Board (NTSB).

She has flown the following planes: Cessna 150, 150 Aerobat, 152, 172, 172XP, 182, 206, 414; Piper Apache, Arrow, Warrior; Mooney, Bonanza, Citabria, etc. Employment as a pilot includes: Chuckanut Aviation, flight and ground instruction, and Auburn Valley Fliers, maintenance officer.

COURTNEY DENIESE BARGERHUFF, born Dec. 21, 1948, in Joplin, MO. An honors student, flying was not part of her life while she was growing up.

Years later, her husband talked excitedly of wanting to get his pilot's license. Earl asked Courtney to go to ground school with him "just to keep him company." Then he suggested she take the written test with him "just for fun." She passed with flying colors. When Earl earned his license in 1984, Courtney decided maybe she had better learn to land the plane, just in case.

When she had learned to land, Courtney figured out that she had all the tough parts behind her so she might as well get in on the fun parts, too. She earned her license in 1985 in South Bend, IN, in the Piper Cherokee she and Earl purchased. Now Courtney also owns and flies a 26 mph powered parachute, which is in the ultralight category.

Since joining the 99s in 1985, Courtney has been the North Central Section Ways and Means Chairman, working with the chapters to develop and improve various fund-raising activities. She is also active in fund-raising for the Amelia Earhart Memorial Scholarship Fund and promoting the International Forest of Friendship.

Professionally, Courtney worked 20 years in a male maximum security correctional institution. She says she has seen the worst of society so she really appreciates the best that is so common in the 99s.

Now she keeps busy with her custom-imprinted clothing business and with quilting, sewing, crafts and collecting books about aviation.

DOROTHY BARKER, shortly after learning to fly in 1970, she became a 99 and held many offices, including Kansas Chapter chairman for 1983/84.

Her flying experience includes island-hopping the Bahamas, trips to New York, Washington, DC, California and Alaska.

In Alaska spiraling up to catch a glimpse of cloud-shrouded McKinley and also flying on top of Katmai volcano and then flying out in the Bering Strait over Big and Little Diomede Islands. After landing at King Salmon, AK, they met WWII, Ace LT

Taylor, who was having his radio worked on. He was the first ace in WWII (Pearl Harbor, HI).

She has flown in (as a passenger) a sailplane, helicopter, ultralight, KC135 tanker on a refueling mission, sea plane and this year she finally took a hot air balloon ride over the Arizona desert on Friday the 13th.

She owned a Piper Cherokee for many years. Her flying was for pleasure.

Some of the other activities of her life before retiring in 1990 were: a secretary for 30 years, singing with the Wichita Sweet Adelines for 20 years, gospel singer throughout the Midwest for several years, and also active in church functions.

Her husband was self-employed in the industrial heating business in Wichita for many years before retiring in 1990.

They raised three daughters and have seven grandchildren and three great-grandchildren.

They miss their flying (her husband and she flew together a lot). Their life is still exciting. They took a pontoon houseboat down from Paducah, KY, clear to New Orleans then up the Tom Bigby Waterway back to Paducah, 21 days and 2,000 miles, what an experience! Life is still fun!

JUDITH H. BARKER, born Nov. 25, 1942, Vancouver, WA. Joined the 99s in 1977. Flight training included first solo on July 11, 1975, and single engine land private pilot June 2, 1976. Flew Cessna 150, 152, 172, 175, 182RG, 210 and 310 planes. She was employed as a waitress and cake decorator.

She is married to Noel T. Barker and has two children, Nora Flores and Valerie Wright. Their grandchildren are Robert, 10; Alex, 10; Matt, 9; Chazz, 7; Lauren, 7; and a baby girl on the way. Noel is a CFII. They have enjoyed many vacations, flying to and from, with their daughters when they were young and now with their grandchildren.

MARY L. BARKER, currently residing in New York, has been flying since February 1975. She is an ATP AMEL (Nord 262, Gulfstream 1159), CFIA AMEL, and FEJ (B-727, C-141). She is working on her FET (C-130).

She is a FAA aviation safety inspector at the Albany Flight Standards Office, NY. As part of her duties she inspects and certifies charter and commuter air carrier operations. She evaluates pilot training, designated examiners, and conducts certification for flight instructors and type ratings. Additionally she conducts airshows and accident investigations.

Barker has logged more than 10,000 hours as an Air Force/Reserve flight engineer, flight instructor, charter and commuter pilot. Flight experiences include military/passenger and cargo movement, airdrop, aerial refueling, medical evacuation, and humanitarian relief operations. She has flown throughout the US, Canada, Greenland, Europe, the Caribbean, South America, Latin America, Middle East, and Africa. In 1995, she started flying for the Air National Guard on the C-130. Flights involve passenger and cargo operations to support scientific research stations in Antarctica, Greenland and the Arctic.

Barker received a BS in psychology from the University of the State of New York. She flies for the CAP. She is a member of the 99s, AOPA, CAP, Empire State Aerosciences Museum, International Society of Air Safety Investigators, Technical Women's Organization, and Women in Military Aviation.

JUNE BARNES, Phoenix 99s, born in Phoenix, AZ, June 2, 1956, one of four girls. She left home at age 17 and was successful on her own. She dated the son of a man who had two airplanes, and discovered flying. Somewhere in her early "roaring" 20s she arrived in Hawaii and indulged in parachuting and hang-gliding. She insists that is the place to do both where you have soft sand in case things go wrong.

She returned to Phoenix and in 1978 earned her wings at Precision Aviation at the old Glendale Airport. Using her license, she flew her father all over the state checking out power plants he had worked on. This was rather unusual because she never used a chart, and she always made it back.

June joined the 99s in 1991, has served as corresponding secretary for two years and until recently was in charge of distributing their newsletter. She did disclose she is fascinated by her fellow members.

She worked for a while at Deer Valley making aircraft interiors, but now has her own company making pillows and doing interior decorating. Her 8-year-old son is being increasingly exposed to aviation. He has had a Young Eagle flight and she is checking out getting him a "hands on" flying ride in the near future.

MARION BARNICK, died 1979, was an active 99 in both the Bay Cities and then the Santa Clara Valley Chapters. She served as Southwest Section governor from 1974-76. She received primary training from the pre-WWII Civil Pilot Training Program in Idaho, 1939-40. She eventually held all ratings through ATP and flew as a flight instructor until her death. Marion won the first Hayward-Tuscon Air Race, and flew many other races, including copiloting the London to Victoria, BC, race in 1971, in a Britton-Norman Islander.

Marion and her husband, Herman, were part owners of San Jose Airport, southeast of Reid-Hillview Airport, from 1946 until it was sold in 1958. Marion was a diligent advocate for general aviation and very active in aerospace education. She and her daughter, Kay, were killed when their Air New Zealand flight crashed on a scenic tour of Antarctica in 1979.

LEILA J. BAROODY, Connecticut Chapter, started flying in June 1976, receiving her private pilot's license three months later. Earned commercial, instrument, and flight instructor certificates in the following years. Currently vice chairman, she has held several offices at the Connecticut Chapter level as well as various positions within the New England Section. While having learned to fly in Cessnas, most of her flight time has been in Piper aircraft, including her own Piper Cherokee (1978 - 1983). Leila has a strong interest in 35mm aerial photography, including aerial photographic research for a master's thesis at Cornell University during 1978-79.

In addition to the 99s, Leila's other aviation organization affiliations include the International Women's Air and Space Museum, and AOPA. Her flying and traveling interests often intersect, including wonderful visits with 99s in New Zealand and Australia, Alaskan glacier flying, and Icelandic volcano and waterfall sightseeing by air. Leila also flew her Cherokee from Massachusetts to a family home in St. Barthelemy, French West Indies, during the 1980s.

Currently a portfolio manager providing investment management services, Leila continues with her flying activities out of Great Barrington, MA.

AGNES BARR, born Aug. 2, 1922, lifetime resident of Newell, IA. Joined the 99s in 1990. Did her flight training in a Cherokee 150 and has also flown Comanche 180. Married to Art and has four children, two daughters and two sons. They have 13 grandchildren and one great-grandchild. When she married Art in 1982, she had never been in a small plane. Art was retired, owned a Comanche, and had been a pilot for more than 30 years. She enjoyed flying and knew it would be safer and more enjoyable for both of them if she got her license, which she did in 1988. She has been copilot and navigator on many trips in the US, into Canada, the Bahamas and Central America.

She is very happy to be a member of the 99s, and is currently serving as chairman of Aerospace and Aviation for the Iowa Chapter, which she enjoys very much.

She represents the 99s on the IA Aviation Promotion Committee, which sponsors the Annual "Fly IA" Aviation Fair.

She is also active in local flying clubs, as well as International Comanche Society.

She is a public speaker and loves to tell them her story ... "to never let go of their dreams, go for it, because you can do it."

NORMA MCELVAIN BARR, born April 29, 1921, in Canton, OH. She joined the 99s in 1965. Flight training included Alamo Flyers, Stinson Field San Antonio, Piper Super Cub Taildragger. She has flown Piper Super Cub, Cessna 172 and Navion planes. She has a BS from Temple in medical technology and worked several years in basic research.

She was first married to Col. Wilbert H. McElvain, MD, pilot-physician, USAF, 1943-69. After his death she married Col. (Ret.) John E. Barr, USMA '31 Pilot USAF, who died in 1993.

She has two children, Col. (Ret.) Kevin L. McElvain, 50, and Maj. (Ret.) Scott C. McElvain, 44. Her grandchildren are, Lt. Andrea McElvain Williams, 24, and Tyson K. McElvain, 21. She belonged to military aero clubs and flew mostly copilot. When she landed from her checkride for a license in January 1965 her first words were "How do I become a 99!" She has been an admirer of Amelia Earhart since her childhood.

PATTY BARRERA, knew at a very young age that she wanted to become an airline pilot. At the age of 17, after receiving her private pilot license, she set a course to achieve her goal. She began her flying career in 1985 as a flight instructor at Oakland, CA, and thereafter gained flight time and experience flying cargo, air ambulance, and commuter airline passengers. In 1990 her dreams came true when she became an airline pilot for USAir. In 1991 after being furloughed from USAir, Barrera started flying for United Airlines. At present she is a First Officer flying the Boeing 727 throughout the US and Canada, and she is living in her home-town area of the San Francisco Bay Area.

Barrera has been a member of the 99s since 1983, serving in various positions on the Chapter level, including chapter chairman. She was the recipient of many flight-training scholarships, including the Amelia Earhart Scholarship and the Marion Barnick Scholarship. She is also a proud member of ISA + 21, the International Society of Women Airline Pilots.

Her educational background includes a BA in liberal studies from California State University Hayward.

She is married to Jose Barrera, who is an accountant. In her free time, Patty enjoys speaking at schools, encouraging young people to set goals, believe in themselves, and pursue their dreams.

ADA MITCHEL BARRETT, deceased 1989, joined the 99s in 1949. She started flying in 1928, when she took two lessons from Eddie Rickenbacker. She got her private pilot rating in 1931, but renewed and got a new license in 1943. In a total 3,000 hours, she earned many more ratings, which were canceled in 1968 due to a serious heart operation. When asked what aircraft she was checked out in her answer was "You name it and I've checked out in it."

Ada retired from the FBI after 25 years of service, and during those years she added to her education by studying communications and radio, stage direction, stage and public speaking, beauty culture, law school and accounting.

She married Bill Barrett in August 1970, who was in aviation also. He was one of the first private airmail pilots and was a stunt flyer for the first air circus. She originally was a member of the Washington, DC, Chapter and was instrumental in getting the Maryland Chapter (1963) and the Southern Maryland Chapter (1976) started, she served in all capacities at chapter and section level. Maryland Chapter honored Ada by entering her in the Forest of Friendship in 1984. She was a life member of the 99s and was instrumental in getting the 49 1/2 initiation started. She loved to tell flying stories, especially the one when she had a forced landing at a nudist colony! *Submitted by Doris Jacobson.*

BARBARA BARRETT, the highest ranking woman at the US Civil Aeronautics Board (vice chairman) and at the Federal Aviation Administration (deputy administrator). While a DACOWITS advisor to Defense Secretary Dick Cheney, she appeared on network television shows and in Congress to urge opening fighter and bomber aircraft to women pilots. A private pilot and a 99, she is reportedly the first civilian woman to land in an F/A-18 Hornet on an aircraft carrier.

Internationally, she headed or participated in aviation negotiations in China, the United Kingdom, Ireland, Poland, Singapore, Peru, the Philippines, Hong Kong and many other sites. She chaired Airshow Canada Symposium in 1989 and 1991.

She served on the board of the National Air and Space Museum in Washington, DC, Embry-Riddle Aeronautical University, Phoenix Sky Harbor Airport and two general aviation airports. Her passions include increasing young girls' math and science skills to open career opportunities.

BERNICE M. BARRIS, learned to fly at Willoughby Airport in Willoughby, OH, where both her brothers flew. "Anything they could do, little sister could do better," she said.

When WWII broke out, her brothers became part of the Civil Air Patrol. At that time several of Barris' friends applied to the WASPS and eventually she did, too, and was accepted. While waiting to be called into the WASPS training program, she became an active member of the CAP in South Carolina and Florida. However, when the European war ended, the military had enough trained pilots to meet its needs and the WASPS program was discontinued.

Marriage, babies and money caused her to quit flying for a time, with only an occasional ride to feel the controls. Her son became interested in flying, and on the day he soloed she got into a Cessna 150 and wondered if she could remember how to fly. While she was sitting there, an instructor climbed aboard and said "Let's go!" and she experienced the freedom of flight again.

Since then, many instructors have climbed into the cockpit with her and helped her attain her commercial rating, instrument rating multi-engine rating, single engine sea, ground instructor's rating multi-engine sea and instrument instructor's rating.

She has flown in numerous races, including the Michigan Little Race, Illi-Nines, Buckeye Air Rally, Powder Puff Derby, Angel Derby, Air Race Classic, and finally the Great Southern Air Race where she came in first with Marion Jayne flying with her as pilot.

She has served as chairman of the Lake Erie Air Derby for five years, when the Lake Erie 99s put on an air race in connection with the world renowned Cleveland National Air Show.

She chaired the final stop at Cleveland for the 99s Angel Derby, as well as being chairman for the final stop of the 60th anniversary of the First Women's Air Derby, with Susan Dusenberry flying the restored Travelaire aircraft originally flown by Louise Thaden, who had won the first race.

As a former 99s chapter chairman, vice-chairman, secretary, treasurer and several committee chairman, also with her current involvement as a lieutenant/commander in the CAP, and with the support of her sister Zonta members, she will continue to support aviation and aerospace education and teach all who will listen about the joy of freedom of flight.

Her 99s chapter has given her the Achievement Award and she was named pilot-of-the-year twice.

She is an RN, and a first aid/CPR instructor. She volunteers for the American Red Cross on a monthly basis.

She's a member of the Silver Wings. Recently she has been asked to again chair a General Aviation Race in connection with the well known Cleveland National Air Races for Cleveland's Bicentennial year.

In 1994 she was awarded Outstanding Accident Prevention Counselor of the Year by the FAA, flew Coast Guard Shore Patrol.

DIANE RUTH ARMOUR BARTELS, earned her private pilot license Aug. 12, 1966. A member of Nebraska Chapter since June 3, 1968, Diane has held nearly all chapter offices, her efforts often recognized at sectionals for chapter accomplishments. Under her leadership, the Nebraska Chapter received the South Central Governor's Annual Chapter Achievement Award in 1976 and 1977.

Programs and projects initiated by Diane include aviation heritage and preservation, statewide poster contest, flight plan signs, statewide membership drive, state of Nebraska proclamation signings, safety clinics, and aviation education programs, including FAA ground school instruction at a local high school.

Diane has served as chairperson of Powder Puff Derby stops, air races/poker games, safety clinics, General Aviation Week, 50th Anniversary of the Nebraska Chapter, Nebraska's International Women's Year Historical/Contemporary Hall of Fame, and dedicatory tributes to Amelia Earhart, Belle Hetzel and Evelyn Sharp. In 1991 the National Endowment for the Humanities selected Diane as Nebraska's Teacher-Scholar, making it possible for her to write the biography of Evelyn Sharp, a WWII ferry pilot from Nebraska.

Diane, who is currently aerospace education chair, a member of Civil Air Patrol, and a Friend of WASP, teaches fifth grade in the Lincoln public schools. She has a son, Steven Scott and a daughter, Kaye Diane.

JUNE BASILE BARTELT, earned pilot's license and joined Chicago Area 99s in 1957. Took commercial and instrument tickets in 1961.

Held corresponding and recording secretary post as well as scrapbook chairman. Wrote newsletter reports for two years. Actively involved in many chapter committees and also on the Section level.

Flew as pilot in several Chicago Area Air meets and Illi-Nines Air Derbies. Flew as pilot in three Michigan Small Races. After the 1959 Michigan Small Race, was initiated into the Paul Bunyan Clan. Flew in Powder Puff Derby as pilot; as copilot in last one.

Awarded first place Chicago Area Achievement Award two consecutive years, 1960 and 1961. Received second place Achievement Award in 1959. Retired from active flying but holds life membership as a 99.

HAZEL BARTOLET, born March 27, 1913, in Huntington, PA. Joined the 99s in 1963. Flight training included private pilot license. She has flown Piper Comanche. She and her husband, William, have twins, Deborah and Donna, age 39. Their grandchildren are Jennie, 12; Chad, 10; Sarah, 4; and Karri, 2.

She organized the Central Pennsylvania Chapter of the 99s in 1965. Held every office in the chapter, also vice-governor and governor of the Mid-Atlantic Section. Flew the Powder Puff Derby in 1970. Chairman of the different stops for the Powder Puff Derby and the Air Race Classic. Scout leader from Brownies to Seniors. Volunteer work for the Red Cross and the Meals on Wheels. Also chairman of Camp Barree for Scouts for 10 years. They were an oriented aviation family, husband William and daughter Deborah were also pilots.

ELSIE WAHRER BASCOMB, took 21 years from the time she became a 99 to fly her first Powder Puff Derby race (AWTAR), but that was her crowning achievement of being a pilot. It's now 40 years since she joined and she never ceases to be proud of being a 99 and is forever grateful for the experiences and friendships she's made for life. She is a member of the Chicago Area Chapter of 99s.

RHEA REED BASTIAN, born May 8, 1957, in Bridgeton, NJ, and moved to Oregon in 1972. Her lifelong interest in aviation took wing when husband Richard gave her a gift certificate to begin flight training. Rhea earned her private pilot certificate in a Tomahawk in 1989 and Richard followed suit in 1990. They've enjoyed many trips throughout the western US in their Cherokee 140. She's also flown a 172, 182, 182RG and a Yak 18, and has participated in the Palms to Pines Air Race.

Bastian added an instrument rating in 1993 and has logged 450 hours. She's currently working on her commercial, with a goal to be a CFII.

Rhea is a member of AOPA, is a charter member and currently chairman of the 99's Crater Lake Flyers Chapter and currently aerospace education chairman for the Northwest Section.

A graphic artist, Rhea owns a screen-printing business that specializes in producing aviation designs for garments.

JEAN F. BATCHELDER, born April 13, 1922, in New Hampshire. She has flown the following planes Aeronca Coupe, Cessna 172, Tri-pacer, Cherokee, Arrow, Grumman Cougar and Apache.

She became interested in learning to fly and joining the 99s through her flight instructor, Alma Gallagher Smith, a member.

She joined the organization in 1964, went on to get a commercial rating with SEL&S, MEL and IFR ratings and logged 1,800 hours. She flew in connection with business as office manager of several plants and as an aviation columnist/writer.

Jean has been active in the 99s, Northern New England Chapter chairman, New England Section governor, NE Air Race and Mahn Scholarship committees - the Grasshoppers, Aviation Assoc. of New Hampshire (past president) and FAA Women's Advisory Committee on Aviation (New England).

She flew several Angel Derbies, proficiency races and as copilot in the 1976 Powder Puff.

Born in New Hampshire, where she lived before retiring to Florida with husband Christopher, she attended Dresser Business School, was a member of the CAP during WWII and has been active in business and women's organizations. Now writing a history of aviation in New Hampshire.

RHODA BATSON, born May 6, 1936, Baton Rouge, LA. Joined the 99s in 1994. Received her private license in Flagstaff, AZ, PIC in Utah. She has flown Cessna 172, Cessna 182, Piper Arrow and Glasair III planes. She and her husband, Ray, have three children: Beverly White, Fritz Batson and Tom Batson (all adults, all college grads). They also have grandchildren: Jessica White, 11; Danielle Batson, 10; Jennifer Batson, 7; and Douglas Batson, 5.

Her special story regarding flight experiences is not one of achievements or careers; rather it is one of personal accomplishment. Her first flight as a passenger was in a T-craft BC12D in 1956. Her second flight as a passenger was over the Rocky Mountains in a Taylorcraft with her husband as pilot. She was in awe with the landscape but her physical well-being was not one of joy.

Over the years she flew with her husband over beautiful country in various airplanes. Only when she became a grandmother did she toy with the idea of learning to fly. She took a "Pinch Hitter" course and upon its completion she thought that if faced with emergency that maybe the best she could manage would be a controlled crash. Further instruction seemed advisable. She hired two flight instructors, one military, one civilian, and each week alternated days of instruction. She bought audio tapes, several books, asked lots of questions, and taught herself ground school. She passed the written with a perfect score and actually had fun taking a hard and long flight exam. She received her private license April 1992 and now has more than 600 hours; mostly in their 182 and their Glasair III.

Her professional career was in the public schools as a teacher, and as a supervisor she takes joy in combining these skills with the continuing learning process involved with flying and passing this knowledge onto her grandchildren.

ANN BATTY, born June 27. She is a retired station agent for American Airlines. She joined the 99s in 1982 and the Reno Chapter in 1982. She became involved in aviation in 1954. Employed with United Air Lines, reservations. Got first lesson in a Piper Cub, but did not complete training. Finally got back into instruction in 1980 and was certified in 1981. Was active until 1991 with approximately 300 hours.

53

Current rating is SEL Private. Background information (professional, educational, interests, hobbies): has been around pilots most of her life. Was raised in an Air Force town in Missouri. Spent her entire working career with commercial air carriers in ground and office positions. She is married to a (now retired) commercial pilot. Made several cross-country trips to Missouri and Illinois when they were active in their Cessna 172. Still love to travel and go every chance they get. She likes to bowl and pitch horseshoes. She is a member of RV groups.

VIRGINIA BATZEL, since she wrote of her beginnings of flying 2 years ago in the last history book, she thought she would contribute her fond experiences with "Betsy." She was born in 1965 and they adopted her in 1970. She was red and white, had two feet, two wings, a spinner for a nose, and a big tail section. She was beautiful and she could fly! They flew their Cessna 182 up and down the East Coast and trips out West. She was well equipped with instruments, so she got a lot of IFR time flying from the right seat. She completed ground school, but never got her license. With husband and son having their IFRs, they were well covered! "Betsy" and she used to fly for parts, for their family owned construction company when needed. Even when they felt like a bunch of good crabs in Crisfield, she and "Betsy" would hop on down, with one stop at Laurel to air out! "Betsy" gave many Pennies-A-Pound rides, flew on CAP missions, and generally any reason to poke holes in the sky! After 15 years of pleasure, "Betsy" is no longer with them. She still remains very active with the 99s. She has held the following offices: Delaware Chapter Charter Member, chapter chairmen, treasurer, secretary, chapter and section safety chairmen and aerospace chairmen.

JUDY BAWCOM, when growing up in Algoma, WI, she never dreamed or had a goal of becoming a pilot. Quite unexpectedly, this opportunity came when she met Harry in 1983.

Prior to this point, she'd known little else besides being a mother. In 1968, when her seven children were ages 1-9, she was faced with the inevitable reality of becoming a widow. The 1972 move to Tempe, AZ, and the climate there, did help extend her husband's life. Nine long years later, he was laid to rest. Alone, she faced and somehow survived raising the children.

Harry gave her her very first small airplane ride in his 1962 Cessna 150. As their friendship and relationship grew closer, it was suggested she learn to fly in order to safely land the plane, should something happen to Harry. Loaned his plane to her and she covered gas plus instruction costs.

Many times during instruction and solo times, the fear level elevated for this respectable grandmother, and she wondered what she was doing up there. Although it took her longer than most, she did earn her pilot's license.

In 1985, Harry and she did marry. His big day was Sept. 19, 1987, when he did first flight in the LongEZ he built, which has now flown 1,100 hours.

Although they don't fly the Cessna very often, they still own it. She remains current and has been a member of the Arizona 99s Sundance Chapter for nine years.

DEBORAHA´ GAYLE BEACH, learned to fly and spread her wings to solo on Nov. 1, 1992, in a Dream (hang glider). She was towed behind a boat to 1,500 feet. Since then, she and her hang-glider have been towed behind an ultralight, released and soared at altitudes of up to 7,000 feet and stayed aloft for one hour and 16 minutes with no aid of power or instruments, "just feeling the thermal. This is what first got me in the air."

May 5, 1994, at the age of 45, she soloed a Cessna 152. July 1, 1995, she got her private. That very same day

she joined the 99s. Three goals achieved. Since then she flies a Tomahawk and has her tail wheel sign off.

Deboraha´ is also a welder, having graduated from Alabama Aviation Technical College with an associate degree in welding. Deboraha', currently a flight attendant with USAir Airlines Inc., lives in Clearwater, FL, with her 49 1/2, Curt Morehouse, a captain with USAir Airlines Inc. Her 22-year-old son, Andrew Long, is an air traffic controller with the US Air Force. Thea, her 24-year-old daughter, is married to Arthur Agnew and has three children: Deboraha´ calls her heartbeats Erikka', Nickolas and her namesake Deboraha´. "I can't wait to take my grandchildren up and show them what makes Gamma sooooo happy."

Currently she's taking aerobatic lessons in a Citabria. She plans to compete in aerobatics against her 49 1/2, and beat him with "no Problem."

Deboraha´ plans to continue with her flying training at Aviator's Flying Club in St. Petersburg, FL, and obtain her instrument, commercial and certified flight instructor certifications. Then she hopes to instruct students and give a little passion of flying back to aviation that she has received with hard work and humor.

MILDRED BEAMISH, 85, Marshall, Saskatchewan, was already a grandmother when she learned to fly in the 1950s, has held a flying license for more than 25 years. Beamish flew from Marshall, near Lloydminster, to Atchison, KS, a flight of about six hours, by herself, July 22, 1983. She took part in the celebrations before returning July 24, in the flyaway. While in Atchison she was asked to bring four trees back to North and South Dakota and Manitoba, in addition to Saskatchewan.

Beamish says women pilots from all over the world took part in the flyaway. Earhardt's sister, who is in her 80s attended the celebrations, and although she is not a pilot she did take a hot air balloon ride.

Beamish says when she began flying there were not many women with a private pilot's license and even fewer who flew alone. Most were passengers in their husband's planes. That's how Beamish caught the flying bug.

"Oh, I just fell in love with flying. My husband was a pilot and I was navigating and just fell in love with flying. I thought, well, I'll get my landed. Then when I got my landed I just went right on … you don't stop then."

And she didn't. It wasn't long before she bought a plane.

"I decided I'd buy my own plane. I had some money and I thought well that's my old-age money, (but) I might just as well put it into a plane. So I bought my own plane and that's the best thing I ever did. I've had some lovely flying and some wonderful experiences."

Beamish has flown all over Canada, the States, and through the north as far as Inuvik in the Northwest Territories. She has mostly flown solo and alone, which she prefers.

MELBA GORBY BEARD, charter member, who made flying her business. Between 1929 - 1933, she managed an airport, did charter flying and was one of the earliest women flight instructors in the West. She was a consistent winner in Precision Flying Meets. Melba also held an airframe and engine mechanic's license which enabled her to pursue her hobby, which was preserving and restoring planes of yesteryear, and just flying for fun.

ELLOUISE SKINNER BEATTY, born on March 26, 1928, in Yakima, WA. A graduate of the University of Denver with graduate degrees from Union Theological Seminary in New York City and the University of Wisconsin-Madison, she is a longtime resident of Madison, WI. She and Marvin Beatty were married in 1956, and they have four daughters and two grandchildren.

Growing up in a family of aviation enthusiasts in Yakima, Ellouise soloed on Aug. 16, 1944, and earned the private pilot license in 1945. She and her mother, Dora Davis Skinner, were an early mother-daughter pilot team in the 99s. Ellouise's greatest thrill was flying aerobatics in the Stearman biplane.

In 1977, after many years away from flying, Ellouise became current again to take each of her family members aloft. This sparked her daughter Jenny Beatty's interest, and in 1981 Jenny became the third generation of women pilots and 99s in the family.

The photo is Ellouise in her parachute next to a Taylorcraft, upon achieving the Private Pilot certificate, 1945.

JENNY TAY BEATTY, born on June 2, 1961, in Madison, WI, attended one year of high school in Bucaramanga, Colombia, as an exchange student, and is a graduate of the University of Wisconsin-Madison.

Jenny represents the third generation of women pilots and 99s in her family. First experiencing flight as a passenger on a commercial airliner at age 24 months, Jenny heard only stories about the flying exploits of her mother, grandmother and grandfather until 1977, when her mother Ellouise Skinner Beatty gave her a taste of flying light airplanes. Jenny knew then what her life's passion would be, and soloed in Santa Fe, NM, on July 25, 1981, the 97th birthday of her flying grandmother Dora Davis Skinner.

An active member of the 99s since obtaining the private pilot license in 1981, Jenny is a past chairman of the Albuquerque Chapter and presently participates in the 99s South Central Section, as well as in other aviation organizations. Her efforts are oriented towards encouraging other career-minded women pilots through networking and mentoring projects.

Jenny has worked her way up the general aviation and regional airline ranks to achieve the airline transport pilot certificate, in addition to the flight instructor, ground instructor, and aircraft dispatcher certificates. A soaring and aerobatics enthusiast, she is rated in multi-engine and single-engine airplanes, seaplanes, gliders and is type-rated in the Beechcraft 1900 airliner. Jenny's aviation skill and fluency in Spanish are utilized in her position as captain, instructor and check airman for a commuter airline based in Tucson, AZ, with scheduled passenger airline service in the Southwestern US and northern Mexico.

The photo is Jenny with a Beechcraft 1900 airliner she commands for a commuter airline, 1995.

SUSAN CHARETTE BEAUREGARD, was born April 1, 1948, in Fort Sill, OK, but has lived in many parts of the world as an Army brat. Massachusetts has been her home for many years.

Susan holds a BS in computer science and a BA and MA in education. As an educator, Susan in involved with after school programs on aviation and in general can be found talking to kids and teachers about flying. She is also active in the Young Eagles program.

Some of her other interests are computer programming, "ham" radio, church organist and quiltmaker.

Susan did not start flying until 1993, but is trying hard to make up for lost time. In 1994 she got her instrument rating and is now pursuing taildraggers and aerobatics, just for the fun of it. She flies a Piper Archer II and a Super Decathlon.

Susan and her husband, James, also an educator, started flying together. They have a son, James Jr., and a daughter, Lisa.

RUTH ENSLEY BECHERUCCI, Canton, OH, has been flying her own single-engine Cessna 172 as a recreational hobby since 1978, learning to fly at age 50. Her greatest personal satisfaction came the day she received her VFR pilot license. She is presently secretary of the Lake Chapter of the 99s. Also an active member of the East Central Ohio Pilots Assoc., and a member of the National Council for Women in Aviation. In 1988, she went to Russia with a delegation of women pilots from the 99s, on an aviation education exchange program, and gave several talks of this trip, along with a write-up in the Canton

paper. Then in 1995 she went to Brazil with a group of women pilots from the National Council for Women in Aviation, on an aviation exchange. These two trips, covering both Russia and Brazil extensively, were the highlights of her life. She has been privileged to have many opportunities and fortunate enough to have the ability and motivation to do many things.

NANCY PARKER BECKUM, born Nancy Parker, April 29, 1956, in Agana, Guam. Nancy is married to an airline captain (Chuck) and is the mother of two sons, Matthew (born May 1991) and Mason (born January 1994). She resides in Weatherford, TX, on 20 acres that is part of the renowned Tailspin Airpark. Nancy began pursuing her private pilot license in 1990 but soon found out she was pregnant with her first son. Being a first-time mother, this delayed her completion of the necessary requirements for a private pilot certificate. Still determined, on Jan. 12, 1993, flying a Cessna 172, her dream of being a licensed pilot was fulfilled. Nancy has owned several airplanes, including a Piper Cherokee 140, two Beechcraft Bonanzas and a Cessna 310. When time permits, Nancy enjoys flying, hunting, traveling, arts and crafts. Nancy is proud to be one of the founding and original members of the Brazos River 99s chartered in November 1995.

MAYCAY BEELER, born June 17, 1955, in Washington, DC. Joined the 99s in 1984. She learned to fly in 1984 for a job assignment on the television show "PM Magazine." As the show's co-host/feature producer, she earned her license and shared the adventure with her viewers on TV.

MayCay went on to become a commercial pilot, flight instructor, and charter pilot for Piedmont Aviation. Her many national TV features included interviewing such aviation greats as Chuck Yeager, Dick Rutan, Jeana Yeager and Bill Kershner. MayCay even went on to marry the test pilot of Questair - an experimental kit aircraft company - the topic of yet another feature she produced.

The holder of five NAA world aviation records, MayCay is an avid promoter of flying. She served as the chairman of the Kitty Hawk 99s. At seven months pregnant, she earned her CFI ticket. Today, MayCay and husband Rich Gritter live with their 4-year-old son, R.J., in North Carolina.

Her mentor is 99 Evelyn Bryan Johnson, the woman with the most flight time in the world! She enrolled her in the 99s upon completion of her private pilot check ride. She was her FAA examiner! She is 80 years young now.

SUSAN BEGG, received her PPL in 1970 at Ottawa, Ontario. She rented Cessnas and Grummans for a few years, then decided ownership was the way to go. Her first airplane was a 1945 Aercoupe. It was after the purchase of a Cherokee 140 that she realized the pleasures of cross-country flying. The temptation to upgrade led to a partnership in a Mooney.

Marriage and having a child reduced her flying considerably, so she sold her partnership. Having full-time employment as project manager of Computer Systems cuts into flying, but without it there would be no money to fly. After a second child, and needing to fly again, she found an older but affordable Mooney which she currently enjoys. Home base is Rockcliffe, Ontario.

Susan's husband Alan does not fly but is very supportive, and has ventured to southern climates in the Mooney. Susan is an active member of the 99s and has held all positions in the Eastern Ontario Chapter (EOC). At present she is EOC vice chairman and Poker Run chairman.

MAYETTA WIEDEMAN BEHRINGER, learned to fly in 1945, and joined the 99s in 1947 in Wisconsin. Moved to California in 1949 in her Cessna 120, married Bill, a Navy pilot, and moved frequently, joining numerous 99 chapters. Flew an F9F at MACH 1.1 after taking flight physiology and survival training. As squadron commander of a Civil Air Patrol Cadet Squadron, she taught seven cadets to fly. A veteran pilot and flight instructor with commercial, single and multi-engine, CFI, CFII, ATP and 6200+ hours, she is also a veteran racer, and was inducted into the Forest of Friendship by Santa Clara Valley 99s. She is the mother of four (two of whom fly) grandmother of eight, and is the current editor of the chapter newsletter, *The Windsock.*

ALETA R. BELCHER, born March 19, 1951, Lubbock, TX. Joined the 99s in 1989. Flight training includes ground school at E-Systems Flying Club Greenville, TX; flight school at Love Aircraft, Greenville, TX, instructor Dennis Baer. She has flown C-150 #N74274, Piper Tripacer 150 and Cherokee 180-C. She has an AGI and teaches private pilot ground school at Commerce Muni. She and her husband Jim have three children: Tanya Inman, 21; Gabrielle Belcher, 16; and James Belcher, 15. They also have a grandson, Kenneth Inman, 13.

She and her husband own PA 7137D a Piper Tripacer that was #5000 off the line at Lockhaven, PA, in 1957. Their PA 28-180C (#8082W) is named *Wind Maiden,* their Tripacer is known affectionately as *Esmerelda.* They consider their airplanes personal friends.

Aleta has given an aviation weather forum called "Clouds—Reading the Sky" at Sun-n-Fun and Oshkosh and local EAA chapters, as well as her own 99 chapter, the Wildflowers.

She and her husband, Jim, produce aviation-related videos and training materials as a part-time, home-based career.

ELAINE LOCKHART BROWN BELL, born Aug. 5, 1928. She is a real estate agent, Capurro-Clark and Associate Realty (specializing horse-property). She and her husband, Robert E., have three children: Stanley Jr., Stephen and Allison. She joined the 99s about 1964, the Reno Chapter, 1964 (originally called Fallon Chapter). She has held the office of chairman, race chairman, etc.

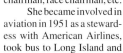

She became involved in aviation in 1951 as a stewardess with American Airlines, took bus to Long Island and started flying in a Piper J-3 (flying around the Statue of Liberty, sitting on three cushions).

Currently holds commercial rating and flies for pleasure. She owns a Cessna 182.

She holds a BA (University of Maine - Phi Kappa Phi, National Honor Society) with studies toward master's at Tufts College. Her professional history includes school teacher, flight attendant (three years American Airlines) and real estate agent. She is the mother of three and enjoys flying, skiing, golf, swimming, dressage (horseback riding).

Her membership in other associations, boards, etc.: president, Folded Wings International (former flight attendants); past president, Kiwi Assoc. (former American Airlines Flight Attendants); past president, Washoe County Lawyers' Wives; past chairman, Reno Area Chapter 99s; past chairman, Reno ARCA, California Dressage Society (Horsemanship); past county chairman, 4H, Horsemanship; member board of directors: California and US Dressage Organizations.

CLARICE BELLINO, soloed in 1972, and joined the 99s Inc. in 1973. As a charter member of the Palisades-North Jersey Chapter, her organizational skills led her to become governor of the New York - New Jersey Section. She flew the 1975 and 1976 Powder Puff Derbies, and the 1977 AWTAR, Inc. commemorative flight in her Piper Aztec N14136, as well as the Garden State 300.

Clarice is a registered nurse with a BS summa cum laude in anthropology. She is a member of alumnae association and hospital corporations working for fund-raising, future planning, and local health fairs.

Clarice is a member of Aircraft Owners and Pilots Assoc. Additionally, she is a member of Zonta International and is immediate past president of her Essex Country Club. She has been an Amelia Earhart public speaker for years.

Clarice is married to Dr. Joseph Bellino, an orthopedic surgeon, who is also a pilot. Eldest son, Christopher, is a captain with Northwest Airlines; son Dr. Michael is an emergency room traumatology physician, and daughter, Claudia is a chef/food service manager. Grandson Alexander is the joy of her life.

SARAH K. BENHAM, born Dec. 15, 1962, in Honolulu, HI, always dreamed of becoming a pilot someday, has achieved her private and is now working on her instrument rating. She hopes someday to fly in wilderness areas, and maybe the airlines in the future. She wants to thank all the pilots at Hawaiian Airlines, whom, while she worked there as a flight crew scheduler for nearly six years, graciously let her jump seat on flights all over the Pacific, and shared their experiences with her. She wants to thank her own High Country 99s and Judy, for continued support in attaining her flying goals. And a special "thank you" to her 12-year-old son, Christopher, for being the greatest, and most patient fellow, while mom is on her quest!

Flight training includes private pilot/Colorado Northwestern Community College; working on instrument. She has flown C-152, C-172, PA-Super Cub, T-Craft and DC-8-61 Sim.

RUTH BENEDICT, (then Ruth Taksel) became interested in flying when her Senior Girl Scout troop decided to become Wing Scouts in 1957. Knowing little about flying, Ruth sought help from the STL 99s, Civil Air Patrol and St. Louis Aero Club. After her Scouts graduated, Ruth remained involved in aviation, becoming editor of the *Aero Club* magazine.

Ruth worked in the STL Flying Club office in return for flying lessons, got her private license in July 1962, joined STL 99s, and earned commercial and instrument ratings by 1967. She trained in a Piper Colt, later flew Comanches, her own Cessna 210, and Aero Commander singles, for which she held the STL dealership from 1967 to 1970. Ruth flew as copilot to Valera Johnson in Powder Puff Derbies in 1965, 1967 and 1977 and the two of them flew in many other races, Bahamas Flying Treasure Hunts and lots of other events.

In 1970 Ruth moved to the Washington, DC area to become associate editor of *FAA Aviation News,* transferring to the Maryland Chapter of 99s. In 1973 she married Marshall Benedict, also with FAA and a former WWII "Hump Pilot." The Benedicts have flown on countless trips in their Pipers - a Cherokee 235, an Archer, and most recently, a Dakota.

In 1978 the two retired from FAA, moved to Arizona, and bought *Carefree Enterprise Magazine.* They sold that business in 1991 and are now retired and living in Scottsdale. Ruth, presently a member of Arizona Sundance 99s, has four children, including daughter Connie Arnold, also a 99, and eight grandchildren.

CARRIE BENNETT, born on Aug. 20, 1974, in Etobicoke. At the age of 14 in Abbotsford, BC, as she listened to the roar of jet engines above her head, she looked up at her Dad and said, "I want to be a pilot."

That dream became a reality in August 1991, when she became a glider pilot and in 1994, an instructor. She now has more than 246 hours on SGS-233A Gliders, C-172s, C-170s and the Cherokee 140. Now, at the age of 20, she is within months of receiving her commercial pilot's license.

She has completed first year aviation in BC and upon

return to second year, she will be doing her multi-engine rating, float endorsement and instrument rating.

Her advice to any aspiring young pilot is: Your dreams will only take you as far as you allow them to; determination and persistence are the keys to success.

SUSAN KIRSTEIN BENNETT, born in Loudon, TN. Joined the 99s in 1992. Susan has loved flying since her first flight on an airliner at the age of 3. After completing diplomatic assignments in Panama and Bangladesh from 1985-90, she finally learned to fly when she was transferred back to Washington, DC, in 1990. Since then, Susan, who now works in the State Department's Aviation Programs and Policy office, has earned a private certificate and instrument rating, obtained advanced and instrument ground instructor certificates, and logged more than 300 hours in the Cessna 150, 152, 172, 172RG and 182. She is currently working on a commercial certificate and plans to become a flight instructor.

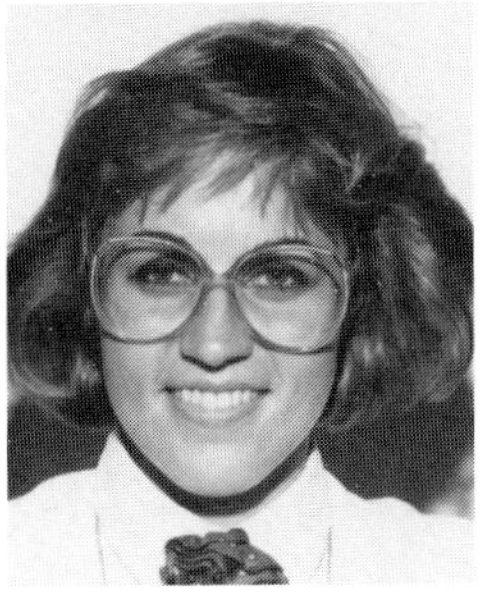

Susan got involved in the 99s as a student pilot in early 1992. She served as newsletter editor and treasurer of the Washington, DC Chapter from 1993-94, and she started a newsletter for the Old Dominion Chapter when she joined in July 1994. Other activities include Toastmasters International and the Women's Transportation Seminar.

A 1984 graduate of the University of North Carolina at Chapel Hill, Susan is married to Stephen Bennett, a foreign student advisor and enthusiastic passenger. They have two (non-flying) cats.

VIVIAN C. BENNETT, introduced to flying by husband, Joe, a retired Army officer (Artillery) who loved flying but due to heart problems could not qualify for a pilot's license. She was more or less "pushed" into flying. In 1968 became a member of 99s, sponsored by Mary Jane Norris. Acquired private pilot license Aug. 9, 1969.

Has logged 1,300+ hours, mostly in Cessna aircraft – trained in 152, owned two 172 Cessnas. Joe and she owned a charter service in Odessa, TX – owned a Cherokee Six which she flew occasionally.

The most exciting experience was flying to the space shot (Joe was with NASA after he retired from Army), then on to Freeport-Bahamas over one hour of nothing but water! She is a widow with one son and one grandson.

FRAN BERA, born Dec. 7, 1924. Started flying at 16. She became a free-fall parachutist, ferried surplus aircraft after WWII, holds an airline transport pilot license, rated in single and multi-engine land aircraft, single engine sea, helicopter, and hot air balloon. She is a rated flight instructor for airplanes, instruments and rotorcraft. She was a Federal Aviation Agency pilot examiner for private, commercial, multi-engine and instrument for 25 years, licensing more than 3,000 pilots.

She has more than 25,000 hours, has been chief pilot for aviation firms, a charter pilot, flight operations manager, operated a flight training school and aircraft sales business and was an experimental test pilot.

Fran became a member of the 99s in 1948, and is a member of the Whirley Girls, has served on WACOA for three years. She set a record as seven-time winner of the All Woman Transcontinental Air Race, and placed second five times. She has placed in many other races, too. She holds the world altitude record for Class C-1-d.

In 1975 her name was written into the Congressional Record in "A Salute to Women in Aerospace." In 1978 her name was placed in Memory Lane at the International Forest of Friendship in Atchison, KS. She was named 1980 "Woman Pilot of the Year" by the Silver Wings Fraternity.

Miss Bera is retired and lives in San Diego. In 1993 she flew her Piper Cherokee 235 to Siberia for the fun of it.

CLARICE ISABEL SIDDALL BERGEMANN, born in Watford, Ontario, Canada. She moved to New York City, then to Cleveland, OH, and finally settling in Alliance, OH, where she has lived most of her life.

She graduated from Boston University with a degree in health and physical education and taught briefly in the Alliance Public Schools and Mount Union College.

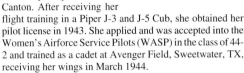

At the start of WWII, she worked in her father's office as a nurse assistant and first started flying at the old McKinley Airport south of Canton. After receiving her flight training in a Piper J-3 and J-5 Cub, she obtained her pilot license in 1943. She applied and was accepted into the Women's Airforce Service Pilots (WASP) in the class of 44-2 and trained as a cadet at Avenger Field, Sweetwater, TX, receiving her wings in March 1944.

She was assigned to B-26 school at Dodge City, KS and then transferred to Victoria, TX, to two targets for P-47 gunnery practice. She was then sent to Waco, where she was a test pilot. She returned home and instructed flying for a living, married and had four children.

She re-attained her pilot's license in 1978, obtained an instrument rating in December 1993. She is a member of the local Taylorcraft Flying Club and has served as their secretary for 11 years. She is active in the 99s having served as secretary and various committee chairman.

She still flies and has more than 1,500 hours of flying time. She was inducted into the Amelia Earhart International Forest of Friendship in Atchison, KS, in June 1993. Awarded "Pilot of the Year" in October 1990 by the Taylorcraft Flying Club and "Pilot of the Year" in October 1994 by the Lake Erie Chapter of the 99s.

ANNA "BOO" BERGMAN, Mission Bay Chapter - Southwest Section. The 99s have played a significant, dynamic role in her life since she joined the San Diego Chapter earning her ASEL in 1957. She has flown Pipers and Cessnas. She is a charter member both in El Cajon Valley and Mission Bay chapters, holding various offices and committee chairs.

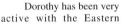

She's enjoyed Powder Puff Derby racing and committees; Pacific Air Race racing and committees; flying internationally with the Angel Derby; Pines to Pines; flying with the Flying Samaritans who fly medical teams and supplies to Baja, CA; flying to Alaska; attending 99s International Conventions; Southwest Section committees.

She is in the Forest of Friendship, thanks to the Mission Bay Chapter. The chapter also presented her with a plaque for 10 years as editor, publisher of the chapter newspaper, the *Mission Bay Omni.*

The many friends she's made and the new friends she's making have made 99s truly a special organization.

LYNN VANDEGRIFF BERNARD, born March 12, 1960. Started flying in 1988 and received her private license (SEL) in 1990. She married her first flight instructor in 1995, Marcel Louis Bernard, who became chief pilot of their local airstrip, Freeway Airport in Mitchellville, MD. Together, they have traveled the east coast from their home in Maryland to Maine and down to the Keys several times in a Mooney M20C. In 1994, Lynn flew her cousin, Bill Murphy, who then worked for the FAA, in his Comanche 250 from Victorville, CA, to Oklahoma City. In 1995, Lynn and Marcel received their sea plane rating in a Piper Cub at Jack Brown's school in Winter Haven, FL.

Lynn is part owner of Swamp Fox Communications, which produced two aviation videos. In 1992, she produced "Basic Aircraft Sheet Metal Techniques and Tools," which is an instructional video for kit plane home-builders. In 1994, Lynn produced "Tales From The Cockpit," which is a video made especially for children to introduce them to the world of flying, careers in aviation (military and civilian), and organizations for kids (EAA Young Eagles and Opportunity Skyway in Maryland). The video also includes a history segment (with an historic photo of Amelia Earhart), the EAA and Sun 'N' Fun airshows, and aerobatics from Patty Wagstaff, Sean Tucker, and the local flying circus in Bealeton, VA.

Lynn and Marcel are currently expecting their first child in April 1996. Lynn is active in her local Potomac Chapter 99s and is pursuing her instrument rating.

ESTHER BERNER, Indiana Chapter. Started flying in 1942, was able to talk her way into the WASP training program starting January 1943. After graduating she was based at Wilmington, DE, with ATC Ferrying Division.

After the war, she got a job flying a staggered-wing Beech for a radio president. Gave that up after a year to become a wife and a Hoosier.

Joined the 99s in 1955 ... served as treasurer, Indiana Chapter. Served on FAIR race board, flew most of the FAIR races, helped with airmarking a number of years.

In 1969 she was appointed by their governor to head the Aeronautics Commission. Retired from state government June 1977 after 16 years.

DOROTHY BERTHELET, learned to fly in 1989. Owner of a Cessna 182, she has logged many happy hours flying C-FBNC around North America. Within a year of becoming a pilot, Dorothy was busy helping organize a "Caribbean '91 Tour," which culminated with nine Canadian airplanes flying around the Caribbean together for two weeks. In 1994, she and her husband Bob flew the Alaska highway, ending up above the Arctic Circle. This year, she's headed for the Atlantic Provinces.

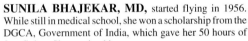

Dorothy has been very active with the Eastern Ontario Chapter, and is currently the chairman. She has organized and participated in Flying Companion Seminars, and numerous other chapter activities. Dorothy also is the East Canada Section reporter. In 1994, Dorothy was awarded the Ottawa Flying Club's (OFC) Presidential Trophy for the most valuable contribution made by a member. She has served on boards for numerous aviation organizations e.g. CASARA Ottawa and the OFC, and is the OFC's newsletter editor.

SUNILA BHAJEKAR, MD, started flying in 1956. While still in medical school, she won a scholarship from the DGCA, Government of India, which gave her 50 hours of free flying. She received her flying training on Tiger Moths, Sentinel L-5 & Luscombe Silvaire and got her private pilot's A-license in 1957.

Eighteen years later, taking time off from medical practice, home and family, she went through ground school and flying school again to renew her P.P.L.

Sunila is member of the Indian Women Pilots' Assoc. and was successively treasurer, secretary and vice president. She is also a charter member of the Indian Section, the 99 Inc., and was membership chairman and secretary for a few years. She took active part in organizing the World Aviation Education and Safety Congress, one held in Delhi and the other in Bombay.

A consultant anesthesiologist, Sunila stays with her surgeon husband Ashok in Thane in Bombay. She has two sons, both engineers and a three-year-old grandson.

MARGARET OLIVE FARRANT BIEDRON, born Jan. 29, 1924, in Frankfort, KS. Joined the 99s in 1987. Received her flight training at Stick and Rudder Flying Club, Inc. - Waukegan Regional Airport. She has flown Cessna 152 and 172. She and her husband Henry have three sons: John (Jackie), 46; Henry Jr., 41; and Christopher (Millie), 37. They also have two grandchildren, Nichole, 16, and Peter, 14.

Her educational background includes a BS in home

economics and inst. management, from Kansas State University, Manhattan, KS, a dietetic internship from Allegheny General Hospital, PGH, PA. She has been a member of the American Dietetic Assoc. since 1945, a registered dietitian since 1968, worked for Veterans' Affairs Medical Center for 35 years. Retired. Member of the Aux. Plains Chapter of 99s since 1987 and Women in Aviation. Flies with her husband who is also a pilot locally and in Midwest. Longest trip in Cessna 172 was to Bahamas Island. Fly annually in Illi-Nines Air Derby. Family member of EAA and AOPA.

LORETTA JEAN MCKINNON BIGHAM, born in St. Louis, MO, on Jan. 20, 1941. She was five when her family moved to Orange County, CA. She has an older brother and a younger sister.

Her brother and she attended "The Charles A. Lindbergh Elementary School," which was about a mile west of the Fullerton Airport. If any airplanes flew over when she was on the playground, she'd have to wave at them, of course! "After church on Sundays, Dad would take us to the airport to watch the airplanes take off and land. It was fascinating! This is when my dream of flying began."

She married a young man from church that she had known for several years. They had two children, Michael and Rebecca. A few years later, they decided to dissolve the marriage. At present, she's a proud grandmother of nine: two girls and seven boys.

Her brother usually calls on her birthday, as he did in 1981. This time he persuaded her to take a "demonstration" flight. The following week she did, and as the airplane left the ground she was sure she could see the old Chevy, with Dad, Mom and Lin waving to her. It was awesome! It was then that she realized, it was time to begin, to live her dream. She immediately started ground school; she received her pilot's license in 1982. In 1984 she received her instrument rating. She flies fixed-wing, single-engine land aircraft.

She's been a member of the 99s International Organization of Women Pilots, and the Fullerton Chapter since 1982. She's held the office of chapter treasurer, participated in events such as: the Shirts/Skirts Air Race, the Kachina Doll Air Rally, chapter fund-raiser airplane rides, airmarkings, EAA Eagle Flight. The Pacific Coast Intercollegiate Flying Assoc. held the Safety and Flight Evaluation Conference for the 1993 regional competition at the Hemet-Ryan Airport. She lives in this community, so this event was especially exciting for her. Fullerton Chapter provided the Friday evening barbecue and judges for various events.

Currently, she's a volunteer for the Temecula Police and the Riverside County Sheriff's Southwest Station and the Riverside Sheriff's Aviation Unit.

RANDA MOHAMMED BINLADEN, MD, Arabian Section. Licensed in Panama City, FL, in 1978. From a family of pilots, her brother Salem encouraged her to fulfill her dream. While she was in medical school in Egypt, he surprised her with an ultralight. Anxious to get a feel for it, she began taxiing around "like a baby." Before she knew it, she was airborne - what a thrill! Fortunately, cutting the power brought her safely to earth. While on her first cross-country in Florida, she experienced cockpit smoke and made an emergency landing in a farm. Luckily she escaped the upside-down aircraft unharmed. The 99s in Saudi Arabia learned that she was the first known Saudi woman pilot. Becoming a charter member of the Arabian Section, she attended the charter presentation dinner, Dhahran, Oct. 11, 1989.

As a pathologist, she practiced at King Faisal's Medical Center, Riyadh. In 1991, she received a master's in immunology, London. Now she lives in Jeddah with her husband Ameer and three children.

CAROLE WALLER-BINNS, born May 20, 1940, Detroit, MI. Joined the 99s in 1990. She has flown Taylorcraft, Glider, Cessna 152 and present plane Mooney M20C. She is employed with Walled Lake Consolidated Schools as office manager. She has three children: Victoria Waller, 32; Korrine Manners, 30; and Rebecca Waller, 21; and three step-children: David, Laurie and Karen Binns. She also has three grandchildren: John Harold, Steven and Nickolas. She is married to pilot and mechanical engineer, John.

Started her flight training in 1989. Her initial training was done in Cessna 150s and 152s. They owned a 1948 Taylorcraft at that time and she put a lot of taildragger time in that plane. She finished her flight training, after getting a good scare while doing stalls – put the plane in a spiral dive a couple of times, in 1992. They purchased a Mooney M20C in 1991 and she received her high-performance rating in that plane. Her husband and she planned a flying trip to Alaska in the summer of 1994, however, they were weathered out in BC Fort St. John, so they rented a car and drove the rest of the way. They flew Mount McKinley with a commercial group. It was a wonderful trip. They plan to try again after they retire from the working world. In the future, they would like to rebuild an old Stinson, or build a kit plane.

RENE BIRCH, deceased 1979, Maryland Chapter 1969, was an exceptionally active member of the Maryland Chapter during the 10 years she was with us, serving in many capacities. She was instrumental in acquiring the Chapter's tax-exempt status. Her primary interest and major contribution lay in air age education. She organized the speaker's bureau, set up week-long ground school mini-mester courses at the YWCA and two all-girl schools. Her work in aviation education was recognized in that she received the GAMA Award for the chapter the Safety Award for the Middle-East (now Mid-Atlantic) Section, and presented her Air Age Education program at a session of the Leadership Workshop in Oklahoma City in 1978. She and her 49 1/2 Phil designed and built the chapter's original three section display board and designed our chapter stationery. Rene had earned a commercial certificate with multi-engine rating. *Submitted by Doris Jacobson.*

DAREEN "DEE" BIRCHMORE, born in Meadow Lake, Saskatchewan, graduated as a registered nurse in Saskatchewan. Flew for Air Canada as a stewardess before getting married and having two children. Later returned to university and has worked as a psychotherapist for several years.

Received her flight training at Buttonville Airport. She has float rating, night rating and working on an instrument rating. She has flown Cessna 172, 182 and 206 Piper Archer, Dakota. Dee is an active member of the first Canadian Chapter of the 99s, COPA and the Canadian Seaplane Pilots Assoc..

DOROTHY BIRDSONG, born June 6, 1919, Mississippi. She joined the 99s in 1965. She soloed in Ohio in 1958 and received her private license in 1962, after four bad winters prevented her from flying. She and husband Charles joined International Flying Farmers in 1960 and the family moved to Florida in 1963, where she joined Florida Flying Farmers and was made Queen 1963-64. International Convention was held in Miami and she was host Queen that year. She also served as director of District Two which included five states in the Southeast Section. She was a member of Bonanza Society, AOPA and is still a member of Grasshoppers and life member of 99s, Inc.

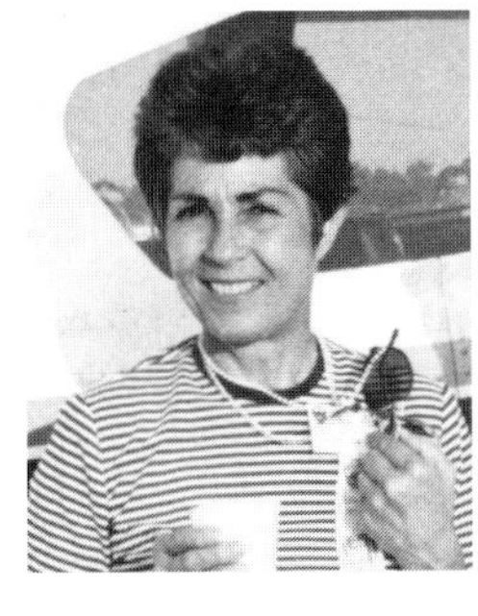

In 1965, eight girls from the area formed a new Suncoast Chapter. These charter members bid for the Powder Puff Derby terminus from Spokane to Clearwater 1966 of which she was co-chairman. Bad weather delayed the start two days, thus delaying all terminus activities except for a rushed Awards Banquet. She later served as chapter chairman two years, secretary-treasurer of Southeast Section two years and other offices. In the early 70s she served on Women's Advisory Committee on Aviation (WACOA) in Washington for three years. With cooperation of USF, the first aviation accredited class was held and MacDill AFB flew the class of 26 to Redstone Arsenal in Alabama on a VIP field trip to see training of astronauts.

She is a commercial pilot with instrument rating, twin engine land, single engine sea plane ratings and has logged 6,000 hours, accumulated flying 25 major air races, also three Alabama Petticoat races, three Bahamas Treasure Hunts, two Deltona races and other small ones such as Poker Runs and state-sponsored treasure hunts. She has flown to Managua, Nicaragua, three times in International Air Race or Angel Derbies. Her chapter honored her by planting a tree in the Forest of Friendship in her name.

During a landing on a sod strip on her first country solo flight she hit a pot-hole on the runway, while shooting landings and had to land without a nose wheel.

Her husband Charles is also a pilot. They have three children: Charles, Jr. (also pilot); Roger (deceased); and Jane. The have eight grandchildren and two great-grandchildren.

CHERYL A. BISHOP, born Oct. 11, 1952. She joined the 99s in 1978, Reno Area Chapter. She has held the following offices: chairman, vice chairman and secretary. She began flying in 1976 and owns a Luscombe.

Her hobbies include travel, gardening, flying, water and snow skiing. She is a member of the Reno Women, Reno Philharmonic Guild, and the Junior League of Reno.

EDNA HINES BISHOP, WASP 1943-44. Lifetime member of 99s and current member of "The Big Muddy" of Jackson, MS. Member of ILR (Institute for Learning in Retirement) Painting is a big interest now.

Soloed at Bishop, CA, in 1941. Trained for the WASP in Houston and Sweetwater, TX. Further training in pursuits, instrument training and officer's training in Orlando, FL. Stationed in Long Beach, CA, Air Transport Command and Liberty Field at Hinesville, GA. Flew a number of aircraft, including B-17, C-47, P-47, AT-6, L-5, and SB-2C.

Degrees from University of Southern Mississippi and University of Alabama. Retired from non-public schools in New Orleans as a psychometrist. Served as short-term missionary in Honduras, Belize, Chili, Nigeria and New Zealand.

Married to a former pilot and they have two sons, two daughters and six grandchildren.

JANICE "JAN" ELAINE BISHOP, born May 9, 1952. Employed with Washoe County School District. She and her husband, Herb have one child, Molly. She joined the 99s Reno Chapter in 1978. She currently holds the office of education chair and has held the office of vice-chairman in the past. She became involved in aviation in 1963. Started lessons at 11 years old, 1966, her dad and, eventually, her husband. She received her certificate Jan. 1, 1978. Her dad is the FBO and airport manager of Nervino Airport, Beckwourth, CA. She was raised near the airport and lived on the airport until 10 years old. She met her husband after returning from college as he was purchasing a Citabria which was ground-looped at Beckwourth. Her husband has completed a full career in plumbing and engineering and is now pursuing his hobby as a second career (as a flight instructor and aircraft mechanic) and is in the process of taking over Nervino Aero Service (as her dad eases into retirement after almost 50 years - Beckwourth Airport was renamed after her dad, Frank Nervino). They are starting a second generation of children growing up around aviation at the Nervino Airport.

She holds a rating of ASEL and currently flies for pleasure. She owns a Beech Bonanza. Her education background includes master of education and 18 years as a elementary school teacher (grades 1, 2 and 4). She is a member of Representative Council, Washoe County Teachers Assoc. and National Reading Assoc..

LISE M. BJERRE, born May 23, 1969, in Montreal, Canada. Joined the 99s in 1994. Flight training included Glider (1985), Private fixed wing and night rating (1986),

Glider Flight instructor (1988). Planes flown: Cessna 150, 152, 172; Piper Cherokee; Chipmunk/Gliders: Schweizer 2-22 and 2-33. Her educational background includes BS in biology (honors, Concordia University, 1993). Currently: MS in epidemiology (expected completion: June 1995, McGill University). Has been accepted to McGill's MD-Ph.D. program, commencing in August 1995. Extracurricular activities: cross-country skiing (member, McGill Nordic Ski Racing Club), hiking, cycling, tennis, piano, singing, handicrafts, literary and scientific reading.

ROSELLA BJORNSON, born July 13, 1947, in Lethbridge, Alberta. Flight training included private pilot license 1964, commercial 1968, and air-transport 1972. Her father owned his own aircraft while she was growing up so she started flying at a very young age. She took her first lesson on her 17th birthday at the Lethbridge Flying Club. She attended the University of Calgary for four years in a BS program from 1965-69. While attending university, she completed the commercial rating and instructors rating. From 1970-73, she worked as a flight instructor at the Winnipeg Flying Club. On April 1, 1973, she was hired by Transair to fly as a First Officer on the Fokker 28. At this time she was the first woman to be hired as a First Officer flying a jet for a scheduled airlines in North America.

She joined the 99s as a member of the Montana Chapter in 1965, since there was not a chapter in Canada. She helped to organize a chapter in Alberta in 1968 and when she moved to Winnipeg she assisted in forming a chapter there in 1971. She has held all the offices of the chapters and West Canada Section.

Now she is a captain on the Boeing 737 for Canadian Airlines International, based in Edmonton, Alberta. She is married to fellow pilot with Canadian Airlines, Bill Pratt, he is a First Officer on the 737 and they occasionally fly together. They have two children, Ken, 15, and Valerie, 11, who have a keen interest in aviation. They live on an airpark east of Edmonton with her Cessna 170B in a hangar in the backyard.

VIKI KAY ANDERSON BLACK, born Aug. 19, 1945, in Cincinnati, OH. Her husband, Carl V. Black, Jr., died Nov. 16, 1972, in a car wreck. They had children Robert Carl, 30, and Paul David, 25. They also had grandchildren Kaiyla Matise, 5, and Joscelyn Courtney, 1. Flight training included private, single engine, land - July 14, 1978. She owns a Cessna 172 (N80247). Flight experiences include 1978 joined 99s. In 1978 joined EAA, secretary 1983, 1984, 1985. Bicentennial of Air and Space Flight, flew pine seedling from 99s International Forest of Friendship to Cornelia Fort Airpark, Nashville, TN. On July 10, 1983, flew in airshow to celebrate opening of Nashville's Riverfront Park, attendance of more than 40,000 citizens. In 1984 and 1985 flew as a volunteer in American Red Cross Aerial Bloodline Program. On April 16-18, 1984, served as a timer at Nashville in the Americana Grand Prix Air Race. In 1983, 1984, 1985, flew in airshows presented by Nashville Chapter 162 of the EAA. In July 1986, flew in the Tennessee Homecoming Air Tour, landing at 47 airports in the state of Tennessee in one weekend.

Her hobbies include scuba diving since 1986, earning open water, advanced open water, and rescue diver certifications and logging more than 150 dives.

She received her BS in 1976, Vanderbilt University; MSSW in 1979, University of Tennessee; and Ed.D., completed 33 hours, Tennessee State University. She is employed as a social worker/counselor: rehab center working with traumatic brain injury clients and hospice working with terminally ill patients and their families.

JANIS L. KEOWN BLACKBURN, as a child she had two dreams – to fly in the Powder Puff Derby and to fly for Eastern Air Lines. "I've done both!" No one in her family flew, her father had died when she was five, she just loved airplanes!

She soloed on March 9, 1968, two days after her 20th birthday. She's made three parachute jumps, has gone hang gliding and has her commercial glider rating. Flying career: CFI at Deep Run Aviation, flew BN2, PA31, and N24A for Princeton Airways, went on to be a "freight dog" for Summit Airlines in the CV580, B727 S/O and F/O at Sun Country Airlines, joined Eastern Airlines on March 28, 1985 as an A300 S/O. Airbus told Eastern that she was the first female flight crew member in the world on the aircraft. She is now a B727 pilot with KIWI International Airlines. She ran the Newark strike center during the EAL strike and was the first female elected to the Master Executive Council for ALPA at Eastern.

For 10 years she instructed at the NJ Civil Air Patrol solo school at Lakehurst NAS. One of her greatest thrills was being there when her daughter soloed at 16 as a cadet at the school. One of the scariest moments was soloing her boyfriend there.

She flew in the 1976 Powder Puff Derby in her Mooney, M20E. She was so excited while taxing out to see her daughter, 6 at the time, with friends along the fence waving. She was also at the terminus, having returned from California via airliner. She has also flown the Garden State 300, Empire State 300, and Capitol Proficiency. She joined the Pike's Peak 99 Chapter in 1970 and transferred to the Garden State Chapter in 1972. She has been treasurer and chaired several committees as well as having been a founder of and chaired the First, 3rd, and 10th Garden State 300 races and co-chaired the 22nd. Member: 99s, ALPA, ISA+21, AOPA, MAPA, CAP, NJEAC, FAA Safety Counselor. 12,000 hours. Now divorced.

JUNE F. BLACKBURN, received her student pilot license in 1975. Joined the Greater Seattle Chapter of 99s in 1977. She earned her instrument rating in 1983 and flew Cessna 172s for 15 years.

She served as Amelia Earhart Chapter chairman in 1982-83. She was treasurer of Greater Seattle Chapter in 1990-91. She also served on several committees, including the Flying Companion Seminars.

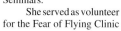

She served as volunteer for the Fear of Flying Clinic since 1980 and is now serving as director of clinic.

JOY PARKER BLACKWOOD, proudly received her private pilot license in 1982 after training at Buttonville Airport in Markham, Ontario. Joy's love of "warbirds" enticed her to learn to fly. Is a lifetime member of The Confederate Airforce and a squadron member of the B26 Martin Marauder.

She was the first "66" of the First Canadian Chapter and turned those numbers around in 1982. She was vice chairman and chairman of the 99s first Canadian Chapter, 1986-90. Governor of the East Canada Section 1994-96.

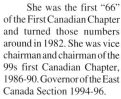

As a member of the International Long Range Planning Committee, Joy's goal is to see that the 99s purpose meets the needs of women pilots world-wide long into the 21st century.

Born June 4, 1955, in Toronto and spent her teenage years in Montreal. Joy and husband Billy Ray own event marketing/show production companies Bankten Communication Services Ltd. and Thoron Productions Ltd. in Markham, and Parwood Management Inc. in Dallas, TX.

KATHRYN JEAN BLAKE, born July 1, 1922, Eaton, CO. She began flying in 1943 at Reno Skyranch, NV, in a Taylorcraft. She continued to fly as a student/copilot until 1972, earned private license 1972 in her Cessna 182. Planes flown: Aeronca, Cessna 150 and Piper 140 and 235 as well as that 182. Family flights from California to midwestern states during many years. She assisted her husband as FBO of a small airport for a few years following WWII. Her husband, flight instructor more recently a corporate pilot.

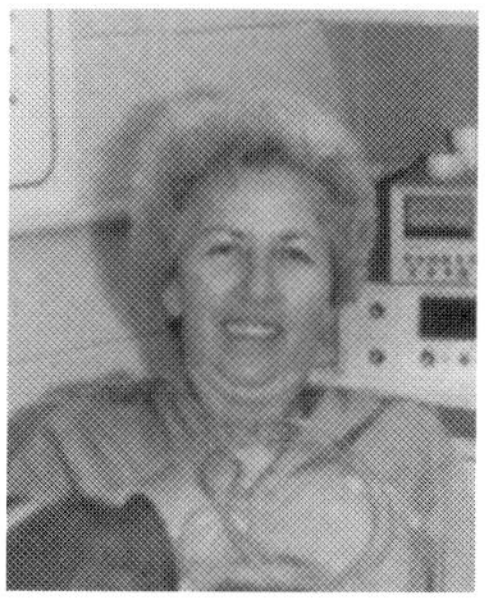

She is married to Francis F. and has three children: Claudia Stillwell, 58; Pamela Ann Frye, 50; and Bobbi, 44. They have the following grandchildren: Carrie Anthenien, 30; Michael Cordich, 28; John Cordich, 26; and great-grandchildren: Cody Anthenien, 7; Katelyn, 5; Dawson, 3; Andee, 21 months.

PIC experience includes flights to South West Section meetings. Flight to Lemore NAB as charter for a retired commander for a friend's promotion ceremony. March 1981, Blake became one of nine women, ages 55 to 65, for a series of tests to discover how they could withstand the physiological stress of weightlessness. She spent 24 days in the Ames Human Research Facility at Mount View, CA, for a sequence of events including nine days of controlled observation and tests, 10 days of bed rest and five days of recovery and post-bed rest tests of the cardiovascular system. Among the tests to establish medical criteria for space shuttle participation, Blake was spun on the big Ames centrifuge at 3 Gs to simulate re-entry conditions. At age 65 she joined 30 members on a climb to the top of Amelia Earhart Pike in the High Sierra above the John Muir Trail, three days with back packs.

MABEL ANESI BLAKELY, life member of 99s. Her first flying experience was in 1945 in San Fernando Valley, when she answered a newspaper ad: "Learn to fly with Pacific Pilots." She was employed as a watchmaker in Los Angeles, and had weekends free. Thirty minutes of instruction a week, in a 65 hp Taylorcraft was an exhilarating experience; her first solo flight was in an Aeronca Champ – 75 hp!

She moved back to Wyoming in 1946, married, in 1947, and "settled down," until her husband, a building contractor, needed a plane for his business, and she had the opportunity to start flying again – after 16 years! She attained her private license in their Cherokee 160 – proudest day of her life! AOPA held a 180 Course in Lander that summer, and she met Velda Benn, CFI from Washington, DC, Chapter, who told her about the 99s. She joined South Dakota Chapter, just two months after their charter. Then, with the help of Nikki Weaver whom she met at an AOPA 360 Course, helped organize Wyoming Chapter, and became its first chairman. Their Charter was presented by Donna Myers, int. vice president, on April 23, 1965. Being a member of the 99s has been a rewarding experience; this great organization keeps her flying with a purpose, now in her 1979 Piper Dakota.

HELEN BLANCHARD, born July 4, 1933, Davenport, IA. She joined the 99s in 1974. She has an instrument rating with a commercial certificate and owns a Cessna Cardinal 177 with her pilot husband, William.

They have five children: David, 42; Gary, 41; Melinda, 36; Debbie, 29; and Rusty, 28. They also have grandchildren: Miranda, 16; Jarod, 10; Karianne, 7; Christian, 4; and Matthew, 2. She particularly enjoys cross-country flights to visit family.

JO ANN HAMMOND BLAND, born in Neosho, MO. Learned to fly after many years in Tulsa, OK, in 1991.

Even though Jo Ann is the only member of her family who has a pilot's license, she owes her budding romance with airplanes as a little girl watching her older brother JD build wood and tissue paper model airplanes as his hobby.

She flies for pleasure and mostly in a Cessna 172.

Member of the 99s, Inc., treasurer Tulsa Chapter, and

Aircraft Owners and Pilots Assoc. Also many real estate and educational organizations. Bland's educational background includes BS from Pittsburg, KS, State and MA from Northeastern State in Oklahoma. She is a retired teacher and is now a broker associate in real estate in Tulsa.

She has three sons: David, 35; Paul, 33; Mark, 30; and two daughters-in-law, Patty and Molly. She also has a granddaughter Lauren who is 2 years. Her husband C. Wayne, is not a pilot but has encouraged and supported her in fulfilling her dream.

GLORIA JEAN BLANK, born Jan. 29, 1935, in Sweetwater, TX. She joined the 99s in 1982. Flight training includes private (working on instrument rating). She has flown Cessna 150, 152, 172, 177RG, 182; and Cherokee 140.

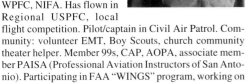

Her husband Chet Blank, 49 1/2 died in March 1994. She has four children and four step-children; and six grandchildren and six step-grandchildren.

Special 99s story: all chapter offices, section advisor; activities: USPFT, WPFC, NIFA. Has flown in Regional USPFC, local flight competition. Pilot/captain in Civil Air Patrol. Community: volunteer EMT, Boy Scouts, church community theater helper. Member 99s, CAP, AOPA, associate member PAISA (Professional Aviation Instructors of San Antonio). Participating in FAA "WINGS" program, working on number eight.

Worked as secretary, bookkeeper, taxes; assisted her husband who was a real estate appraiser.

PHYLLIS J. BLANTON, born Sept. 5, 1953, on her parents' farm outside Wakita, OK. She learned to fly in 1981, with her husband David as her instructor. Phyllis joined the 99s in 1982 and served in many positions on the board of the Kansas Chapter, including two terms as chairman. She organized and participated in many aviation education talks and exhibits for local aviation events and schools, and has been a registered Girl Scout leader for many years.

She has owned and flown several aircraft, including a Cessna 150 and 172, Aeronca Sedan, and a Mooney. She and her husband are active in the Experimental Aircraft Assoc. and own two custom-built prototype aircraft which they enjoy exhibiting at various events.

Phyllis enjoys volunteering at her kids' schools, teaching Sunday School and being an avid seamstress and quilter. She has enjoyed giving many people their first airplane ride and has participated as a volunteer pilot in the EAA's Young Eagle program. Phyllis and David have four children: Karen, Jeffrey, Kari and Ashley.

LORRIE BLECH, born in Eglin AFB, FL. Joined the 99s in 1971.

Received her flight training at Santa Monica, Van Nuys.

She has flown Cessnas, Pipers, Bonanzas, Barons, Starduster TOO, EAA Biplane, Cubs on Ford Trimotor Floats.

Employed as a flight instructor.

ANNA "ANN" AFFLICK BLEDSOE, at 14, after an airplane ride, she knew she would eventually fly! In 1949 obtained ASEL, joined San Diego Chapter; 1,200 hours were attained in C140, C170, Mooney Mite; as a charter member of Mission Bay she has been chairman, treasurer; instrumental in seeing eight honored in the Forest of Friendship; designed and constructed a distinctive vest honoring 99s which was a chapter fund-raiser; became an honoree in the FOF.

In 1949 Powder Puff Derby (2,800 miles) as copilot. Met future husband (an airline pilot) when he repaired an oil leak in the Mooney Mite during the 1950 PPD. The 99s met while traveling throughout the world have become life-long friends.

Still time for hobbies beyond travel and specialized cooking with recipes picked up from each country visited and sewing extensively custom designs.

LUCILE MARY BLEDSOE, C711, born Sept. 11, 1921, in Duluth, MN, resides with her husband Henry at Wray, CO. Manager of her families' multi-state cattle ranching operation, aircraft owner, pilot and instrument flight instructor, she devotes much time to supporting aviation education in the schools and FAA aviation safety programs in northeastern Colorado. In 1991 she received from International 99s, Inc. a plaque in recognition of her dedication to aerospace education. She uses her business aircraft to inspire students from local schools, participates in air races (and wins) (Mile High Air Race, Jackpot Air Race) but her greatest pleasure is teaching others the coveted pleasure of being an airplane pilot. She has logged better then 6,000 hours, and almost a hundred students have received their certificates under her instruction. She owns a B36TC Beech Bonanza, a Cessna 172 and T210. She is an FAA Counselor, Colorado Pilots Assoc. Area Representative, 99, Colorado Chapter, First Women initiated into the Blue Max Society Colorado Aeronautical Board past secretary, National Assoc. of Flight Instructors, Aviation Historical Societies, (Colorado and Kansas), Flying Farmers, AOPA Wray Aviation Authority Chairman and on October 1994 she was inducted into the Colorado Aviation Hall of Fame at the USAF Academy in Colorado Springs, CO.

She has owned B36TC-11BC, CT210-61BC, C172-5461K; and flown: Mooney, Viking, Glider Areo Commander, Pipers Marlow, C54KC, and Grumman.

She and her husband Henry A., have two children, Sandra Ann Bledsoe Bowman, 49, and Robert Edward Bledsoe, 48. They also have grandchildren: Ann Elizabeth Bledsoe Soehner, 20; James Jason Bowman, 20; Grant Henry Bledsoe, 18; and Matthew Bowman, 18.

PEGGE M. BLINCO, born Jan. 21, 1931, in Seattle, WA. Encouragement to be a 99 was half the thrill of learning to fly and obtaining her private license. She has had wonderful memories and associations with other 99s in her 24 years of membership.

She has flown the Palms to Pines Air Race, local Petticoat Derbies, poker runs, medical supplies for the DRF and helped organize and fly blood for the local Red Cross – all major highlights along with her three years as Northwest Section governor.

Only in the 99s could she have had the opportunity for hands-on experience in everything from a Breezy to the Goodyear Blimp!

Along with all of this, she has always had the full support of her husband, Stan – also a pilot. They have two beautiful daughters and two wonderful granddaughters.

LINDA BLODGETT, never in her wildest dreams did she ever imagine herself flying alone in an airplane, much less loving every minute of it. Fifteen years ago this would have been a terrifying nightmare, amazing what changes occur when the insidious sky virus attacks.

She obtained her private license in 1982 in order to overcome her fear of flying with her husband. They purchased a Cherokee 150 and a few years later she had an instrument ticket, she replaced the 150 with a PA28-180 (and also the husband).

She has managed to support this delightful habit of hers by teaching elementary school, which really means that she has time in the summer to fly.

She has been a member of all Ohms for the past 13 years and enjoys 99 Section and International meetings.

She and her copilot realized a dream last summer when they flew the Air Race Classic for the first time. She values the experiences and friendships she has acquired through the 99s.

SHARON GAYLIN BLODINGER, born in Baltimore, MD. She moved to Charlottesville, VA, in 1965 and now considers herself a "Native Virginian." She soloed in 1985, earned her private pilot certificate in 1986 and joined the 99s that same year. She flies the Cessna 152 in which she trained and has logged some 450 hours. She is an active member of the Virginia Chapter, having served four terms as secretary and currently serving as treasurer. Blodinger is proud to be a 99 and enjoys the camaraderie she shares with fellow pilots.

Blodinger's other interests include scuba diving and orchid culture. She is a member of the American Orchid Society and the Charlottesville Orchid Society.

She is married to Robert Blodinger, who is also a pilot and an attorney for whom she works as a secretary and paralegal. She has three children, John, Nancy and Harvey.

THELMA NADINE HARRIS BLUHM, born in Kansas City, MO, and grew up in Southeastern Kansas. She enrolled in civil pilot training program at Chanute, KS, Junior College in 1941; received her private pilot's license in May 1941. She obtained ground instructor's ratings in meteorology and navigation and taught CPT students and Naval aviation cadets.

She became a 99s member (Kansas Chapter) in 1942. She was accepted for WASP training June 1943, graduating class 43-W-7. After WASP's, she attended Kansas State College in 1945; moved to Ohio and obtained a flight instructor's rating.

Taught ground school at Brooklyn Airport, Cleveland, OH, where she met and married Norman Bluhm. She has eight children (one girl, seven boys); three oldest boys are pilots.

Worked as architectural drafter 1973 to 1989. Renewed pilot's license 1989 and rejoined 99s (Lake Erie Chapter) 1990. Appointed to Forest of Friendship 1992. Also belongs to Confederate Air Force, Experimental Aircraft Assoc. and restoration group Military Aviation Preservation Society.

MADELEINE DUPONT BOCK, born April 29, 1943, Germany (Stettin). Joined the 99s in 1990. Flight training includes single engine land, instrument rates.

She has flown PA-38, PA-28-140-180, 8KCAB-180CS, C-152-172, PT-40, PA28RT-201. Her total flight time is 780 hours.

She married Johann Bock Bach in 1972. They have three children: Andre, 35; Remo, 33; and Nicole, 18. They also have a grandchild, Joyce, who is 9.

LORRAINE BOEHLER, a North Dakota Chapter member since May 1974. She started flying when in high school in 1947, and received her private license in August 1949 at Washburn, ND. After marriage to Herbert in June 1951, the flying ceased and the next 22 years were spent raising a family of two boys and four girls. Now there are 12 grandchildren and one great-grandchild.

She resumed flying in 1973 at Bismarck and is a weekend pleasure pilot. She worked as a terminal manager of a freight company until retirement four years ago. Now her time is spent crocheting, sewing, and resting at the cabin by the river. She is a LTC in the Civil Air Patrol, but does very little flying any more.

FRAN BOHRER, a 54-year-old dream of flying came true for Granny Franny March 29, 1995, when she received her license. At 4, the flying mailman took her for a ride in his bi-plane at her grandparents' farm. Her mother said, "Just taxi, just taxi, — she won't know the difference." I said, "UP!" And away they went! She fell in love with "being a bird." In college, she "traveled the world" in a Link trainer for two years. Her dream went on hold for family, three children and seven grandchildren. One of her greatest thrills was when the wheels left the ground on her first solo and she was singing at the top of her lungs, "Somewhere over the rainbow bluebirds fly – now so can I!!" The road to her wings included radio failure, Florida thunderstorms, being "momentarily lost," watching a beautiful sunset as she landed at a controlled airport with a "dead" alternator (no flaps, radio, lights, etc.)

You are never to old to make dreams come true – Birds Fly – so can you!

JUDITH ANNE "JUDY" BOLKEMA, Chair, North Jersey Chapter 99s. She is a resident of New Jersey, earned her private pilot license in 1989, her instrument rating in 1993 and has logged 1,000+ hours. Planes flown include Citabra, Super Cub and Cessna 152, 172, 182 and 195B. In 1991 she purchased N75529, a Cessna 172, for recreation and pleasure.

Judy joined the North Jersey Chapter 99s in 1990 and has served two year terms as secretary and vice chair and is currently chapter chair. She chaired the 94 Student Pilot Forum and co-chaired the 1992 and 1993 Pennies a Pound for the chapter, placed ninth twice in the Garden State 300 and completed the Air Race Classic in 1993 and 1994. Trips to EAA in Oshkosh in 1993 and 1994 and flights from New Jersey to Florida, California and Montana have been special experiences in N75529.

Other memberships include EAA, AOPA, Lincoln Park Pilots, Ventura Publishers Users Group, Air Race Classic, Cessna Pilots and the New York City Art Students League.

Judy holds a BA in art education, MA in educational psychology, teacher of art, teacher of handicapped, and learning disability teacher consultant certification. She has taught art, learning disabled and emotionally disturbed classes and is currently writing technical documentation for a software developer, JST Services, Inc.

Widowed in 1985, Judy's two sons, two daughter-in-laws, two grandsons and John Tokar, a pilot and the 49 1/2 in her life, all support her flying adventures.

JOANNE GAIL BOLTON, born on Sept. 3, 1939, in Oneida, NY. Joined the 99s in 1987. Completed flight training at Kamp Airport, Durhamville, NY, with private license Sept. 11, 1987.

She has flown Cessna 150, 152, 172, Beech 23 Musketeer, Piper Cherokee 162, 180. Former co-owner Cherokee 180. She is divorced and has three children: James (born 1958, died 1992); Robert (born 1960); and Patricia (born 1961). Grandchildren: Megan, 14; Nicolette, 13; Robert, 11; Aaron, 9; Meredith, 8; and step-grandchildren: Krystal, 10; and William, 8.

Special story regarding flight experience: since she thinks all of her flying has been special, this is not an easy question. However, highlights of unforgettable flights are: copiloting a Beech 23 from Durhamville, NY, to Coeur d'Alene, ID, left three special thoughts; the view over the Thousand Island area of St. Lawrence River; flying over the mosaic of wheat, green and plowed fields in Northern Montana; and flying left seat over the Rocky Mountains. All were breathtaking and etched in her memory.

Flying left seat in a Cherokee 180 from Orlando to Key West, FL, the colors of water and islands over the Keys were truly outstanding.

Renting a Cessna 172 at Bar Harbor, ME, and flying around Pennobscot Bay was certainly a post-card view in every direction.

An early morning flight, solo from Burlington, VT, to Durhamville, NY, flying down Lake Champlain, was spectacular, with the mist still on the mountains on either side of the valley and sun shining above.

Active since 1987 in both Central New York Chapter and New York - New Jersey Section activities. Employed in office administration positions. Enjoys, besides family, camping and outdoor activities, cooking and of course - travel.

JUNE BONESTEEL, Phoenix 99s, got a BA in animal husbandry, and then switched to flying. Beginning in 1963 she soon was a CFI and charter pilot for several Phoenix area FBOs, starting an exceptional career in aviation. By 1974 she owned and operated June's Aviation, Deer Valley, AZ, had gotten her A&P and began flying helicopters. From 1975-79 she worked for an FBO doing charter, maintenance and instruction, getting the first female IA inspector authorization. She worked for the city of Phoenix as air support mechanic, police helicopters from 1979-85. At the same time, she instructed part-time and in 1981 became a pilot examiner. She added multi engine pilot examiner in 1982, flight instructor examiner in 1985, and airline transport ratings pilot examiner in 1986.

June was named Arizona CFI of the Year in 1978 and 1987, and National CFI of the Year in 1987. She stays perpetually busy giving advanced pilot instruction, pilot exams, seminars, writing technical aviation books, and teaching at aviation industry seminars. She is currently working on a seven book series on oral pilot test prep. June can't fit in much time for chapter activities, but they all know that when they need something only she with her unique talents can do, she will be there for them. She recently was a presenter at the Southwest Regional Expo 1995 in Phoenix, addressing the 99s Women in Aerospace seminar on flying careers.

DEBORA JEAN BONNARD and her mother, Dolores, started ground school together at Cosmopolitan, based at Republic Airport in 1981. They were flying Cessna 152 and 172. In September 1982 she decided to accelerate her flight training at Burntside-Ott Aviation, at Opa Locka Airport, Miami, FL, where she received her private single and multi-engine, commercial single and multi-engine, instrument rating, advanced ground instructors certificate and CFI-ASME. She returned to New York in 1983, where she worked for Nassau Flyers, Inc. as an instructor pilot. She achieved a Gold Seal by 1985 and added on her CFI-I; November 1988 she earned an ATP certificate.

In June 1993 she worked at a Girl Scout jamboree put on by the Greater New York Girl Scout Council. She spoke to the girls on the joys of flying. She also spoke to a group of Scouts in Suffolk County on aviation as a career. Because of her work at Nassau Flyers, she has become a FAA Safety Counselor.

She learned about the 99s as a teenager. In the spring of 1984, she finally hooked up with the Long Island Chapter by flying a Poker Run with her parents. She and her mother went to the June meeting together and they both signed up that night. By the fall of 1985, she was elected recording secretary and served for the next two years. She also became the 66/Student Pilot Committee chairman for the governor for the New York, New Jersey Section. She served as treasurer from 1988 to 1991, and on the board of directors. She has worked on all the chapter's flying events including USPFT, Poker Runs, Pennies-a-Pound, Ida Van Smith Club, Flight Day for Handicapped Cub Scouts, and fly outs. She put together an Air Bear program at Seaford Harbor Elementary School in 1993.

She flew herself and Jeandar, her dog, out to Lansing, MI, to visit her younger sister. In the fall of 1987, she worked for a few months for Tambrands as First Officer on their Beech King Air. This experience lead to ferrying a King Air from Republic Airport down to Sao Paulo, Brazil, in the late winter of 1988. That same year in December, her parents and she took 105VB on a long vogue to Angle Falls in Venezuela. This adventure lead to another ferry flight, in a C-210 from San Antonio, TX, to Boa Vista, in the Brazilian Amazon. After all of these big trips, her little hops up to Boston or down to Washington, don't seem like the big deals they use to be. She does still get a kick out of flying to Block Island, or East Hampton for lunch, and flying around Manhattan is still a thrill.

DOLORES LIEGEY BONNARD, born Oct. 1, 1927, in Pennsylvania, was always fascinated with airplanes. Received a private pilot's certificate in 1984 at Republic Airport in Farmingdale, NY. Got a seaplane rating that same year in East Hadden, CT, and also joined the 99s.

Has worked on projects like the USPFT; Poker runs, Pennies-a-Pound; Aviation Experiences for Children and right seat seminars.

Does most of her flying with husband, Vicente, an ATP Pilot, in their Cessna 182. Several trips across the USA, one of the most thrilling included criss-crossing the Grand Canyon. Has done island hopping in the Caribbean, gone to the Angel Falls in Venezuela and ferried a C-182 from Arlington, TX, to Rio de Janeiro, Brazil. Acrobatic experience has been brief so far but very exciting at the Brazilian Air Force Academy in their fighter pilot trainer the *Tucano*.

She has flown Cessnas 152, 172 and 182. She and her husband have four children: Vicente, 42; Debora, 40; Mark, 38; and Patricia, 37.

RACHEL BONZON, joined the 99s in 1960. Her flight training includes commercial, multi-engine, instrument, ground instructor. Planes flown Cessna 140, 150, 172, 182, 310.

She has two children and three grandchildren. She has participated in the following races: Powder Puff Derby, Angel Derby, Pacific Air Race, Palms to Pines, many local and regional races.

BARBARA BOOT, born Sacramento, CA. Joined the 99s in 1980. Flight training includes private third class since 1974, approximately 800 hours. She has flown C-172, Decathlon and Pitts.

She and her husband Bill Boot have three children: Patricia Fountain, Carol Cecoone and Jim Boot. They also have nine grandchildren.

She has flown in six Palms to Pines. She has been Corwing Airport Comm. Chairman since 1979 and started the Corning Antique Fly-In Airshow in 1992.

SUSAN BOPP, has lived in New Jersey for more than 40 years. She and husband, Gene, have been involved in business aviation for 15 years at Teterboro. Their charter company Air Charters, Inc. (formerly Air Fleet International) are responsible for nearly all of the organ transplant retrievals done in the tri-state area, as well as flying many well-known entertainers, sports figures and politicians.

Susan learned to fly after they added a flight school, Air Fleet Training Systems, four years ago. She has logged 300 hours and is currently working on instrument/commercial ratings, eventually hoping to do some of the flying for their charter operation. Memberships include AOPA, NATA, NBAA.

Other interests are tennis, roller-blading and roller-skating, biking and most sports.

Her husband, father and oldest son are professional jet pilots. Her youngest son and brother are currently private pilots. Another son is an engineer involved in the medical laser industry. Her three sons are Gene, 29; Jim, 26; and Ryan, 23.

Susan was born on July 16, 1944, in Blytheville, AR. She has been a member of the 99s for two years. She has flown the following planes: C-172 (Bonanza's A-36, F-33), Baron B-58, and Learjet (with assistance!)

MARGARET MARY "PEGGY" BOREK, joined the 99s in 1955. Holds a commercial SEL rating. Planes flown include: Cub, Aeronca, Taylorcraft, TriPacer, Ercoupe, Cherokee, Cessna 172 and 175, Twin Cessna 150. She learned to fly at Ernie Buehl's grass field in Somerton, PA. Tried gliders and helicopters and open cockpit. Active in 99s as chairman and governor of Eastern Pennsylvania and Middle East. Active with AWTAR from 1955 to 1965 in impound inspection and operations. Entered Dominican Sisters in September 1965. She has been active in retreat houses in Miami, FL, and Albuquerque, NM. Interests: ice-skating, hiking, camping, reading, sewing, gourmet cooking and swimming.

CARO BAYLEY BOSCA, born March 29, 1922, Springfield, OH. Joined the 99s in 1941. She served with the WASP May 1943 to November 1944, Class 43-W-7. Flight training included captain in 1941, WASPs, instructors rating. She has flown most civilian planes, SBD, SB-2C, AT-11, AT-9, B-25, B-26, and P-47. She instructed at Sunny South Airport, Miami, FL, and Jess Bristow's Air Shows. She and her husband Orsino Hugo Bosca have three children: D'Orsi Hugo, Caro-Gray, and Marcy E. Their grandchildren are Orson Bayley Humphrey, Caro-Mia H., and Elena Lee H.

She ended up her ten years of flying in a Pitts Special (third one Curtis ever built) 125. Entered aerobatic competition 49-50-51 in clipwing cub, Bradley Special and Pitts-Special in 1951. She was women's Aerobatic Champ in 1951. Also in 1951 took Piper Super Cub to 30,203 feet for international record for class II Aircraft. Official record. Received Bleriot Medal and Mademoiselle Women in Aviation Award for 1951.

NATALIE BOSSIO, a charter member of the Santa Clara Valley 99s (charted in 1954) who has been flying since 1947. She has served as chapter secretary and has held committee chairmanships for the chapter as well as the Southwest Section. Natalie has a commercial ASEL license, and also a lighter-than-air-balloon license. Natalie has logged more than 400 hours. She was a training officer with the Civil Air Patrol for many years, where she had the opportunity to fly a T33 at Hamilton AFB. Natalie worked for the Civil Aeronautics Administration's General Aviation District Office at Palo Alto Airport from 1952 through 1962, when she took an administrative position at NASA Ames Research Center. She retired in 1982 and lives in Santa Clara, CA.

RUBY L. WILLIAMS-BOSTIC, had a love of aviation since early childhood, but due to family commitment and other obligations she could only dream of flying. Her dreams became a reality July 30, 1978, when she obtained her private pilot license at Republic Airport. Ruby then invested in a single-engine airplane. Her greatest conquest was flying a single engine airplane from New York to Trinidad in 1985.

She has a BS and a MS in social work. She was employed for 30 years in social work until retirement December 1992 as Borough Director of Family Services in Queens.

Ruby devotes much of her time giving speeches, lectures and career seminars on behalf of aviation. She belongs to several aeronautical organizations – NAI - Black Pilots of New York; Civil Air Patrol; The 99s, Inc.; (International Organization of Women Pilots); OBAP; and The Tuskegee Airmen.

OLIVIA GRAYE BOTTUM, born in 1927 in Montgomery, AL. There were two Air Force bases there and she was quite accustomed to military trainers overhead. She never dreamed that one day she would fly in that same sky. After becoming an RN she joined the Navy Nurse Corps during the Korean War and eventually was stationed in Cuba, where she met and married Bill, a "Sea Bee." They flew to and from their Miami wedding in Navy bombers. Both left the Navy after the war and became the proud parents of two daughters, Olivia Lynn and Carolyn. After the girls became teenagers, she had more opportunity for other interests, the primary one of which turned out to be flying. She obtained her private license in 1969 and her instrument and commercial ratings in 1971. She has owned a Skyhawk, a Piper 180, and now owns a Piper Dakota, flying these with Bill as copilot on business and personal trips. Olivia has been very involved in supporting her local airport, serving on the Airport Advisory Board and being one of the organizers of a local pilots' association. She has been a 99 for almost 25 years, and was one of the organizers of an Ann Arbor Chapter. Currently she is a member of North Central Section. She has more than 3,000 hours and still finds flying a wonderful challenge, a spiritual experience, and an opportunity for self-actualization. One of her greatest joys is the beautiful trip from Ann Arbor, MI, to Bedford, MA, to visit 2-year-old grandson, Ian, and his parents, Carolyn and Peter.

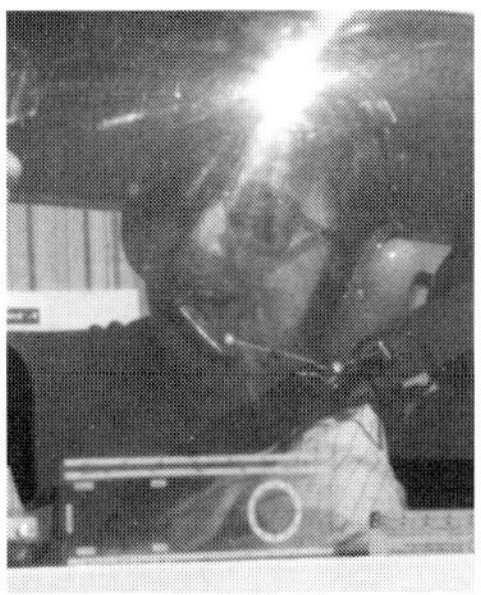

DR. LAUREEN NELSON-BOUTET, born Jan. 5, 1954, in Victoria, B.C., and is presently residing in Pickering, Ontario with her husband Marc. Laureen obtained her private pilot's license and night rating in 1990. She presently flies a Cessna 172 and a Piper Warrior.

Laureen has been a member of the 99s since March 1992. In 1993 she won the 99s Annual First Canadian Poker Run and flew to Switzerland via Air Canada for 10 days of downhill skiing.

Laureen holds a BS in agriculture and is a doctor of veterinary medicine (D.V.M.), both from the University of Guelph, Ontario. She specializes in small animal medicine and surgery, is a veterinary consultant to companies in the animal health business and works with her husband in commercial equipment leasing.

Laureen often speaks to students on non-traditional careers, plus enjoys sports in her spare time.

GLADYS-MARIE BOWDITCH, Alberta Chapter, 99s, has been flying since childhood, as her father is a private pilot and the family has owned several aircraft. In 1983 she joined the Namao Flying Club and obtained her private pilot license and glider license. After obtaining her glider license she joined the Air Cadet League as a pilot. She then obtained her commercial rating and qualified as a tow pilot on the Bellanca Scout, towing new glider pilots and their passengers to altitude. She completed her instructor rating in 1991, and instructed at the Namao Flying Club until May 1994. She is presently instructing at Great International Flight Training on a Citabria 7ECA.

Gladys-Marie joined the Alberta Chapter in 1986 and has participated in hosting the All Canada Section Meeting, presenting a Flying Companion Seminar, and participates annually in the poker run and airmarking activities. She is presently the secretary/treasurer of the Alberta Chapter.

Gladys-Marie was born May 30, 1957, in Oxbow, Saskatchewan. Her flight training included Namao Flying Club; Namao Soaring Club; Edmonton Flying Club; Great International Flight Training. She has flown the following planes: C-150, C-172, C-177, Mooney, Scout, Citabria, Schweitzer 2-33, 2-22.

GERTRUDE BOX, Ed.D., Ph.D., born Dec. 30, 1921, in Farmingdale, NY. She joined the 99s in 1994. Holds a private pilot license. She has flown Cessna 150, 152 Cherokee 140, and Warrior II 161.

First interest, 1931, $5 open cockpit, 15 minutes over Jones Beach, grandmother, parents and herself. Some Sunday afternoons 1937-41 watching fiancés friend fly own Piper Cub, usually hitting trees at end of runway. Saturdays friend repaired damage for Sunday's flight. Married in 1941, Feversky employee, promised they'd fly, jump when marriage. He never did go in a plane until 1978. Divorced in 1966. Her father died in 1970. Now could afford flying lessons with son and daughter. "No man to stop us now!"

She has eight children: Trudy, 54; Terry, 53; Ted, 52; Twinkle Star, 50; Tom, 47; Tim, 43; Tiara, 41; and Todd, 40. Grandchildren: Susan, 35; Shirley, 34; Beverly, 33; Bobbie, 32; Jonna, 31; Mike, 29; Theresa, 27; Bonnie, 26; Vathum, 25; Jake, 23; Joe, 21; Ted, 20; Garreth, 17; Jeffrey, 17; Keith, 12; Aubrey, 8; Tyler, 6. Great-grandchildren: Jennifer, 18; Jean, 14; Susie, 12; Tristan, 9; Vincent, 7; Elayna, 7; Jeremy, 7; Victoria, 6; and Justin, 4.

Her hobbies include bicycling, billiards, bowling, cabin boating, canoeing, dancing (ballroom, belly, line), dressmaker/designer, driving (NYS Certified Driver Ed Teacher, Alaska to Australia), flying, kayaking, knife-throwing/bullwhip, motorcycle (NYS Certified Motorcycle Ed Teacher, Yamaha 1978/Harley Davidson Sportster), musician, needlework, rifle-marksmanship, rowing, scuba diving, skating (ice/roller), surfing, swimming champ-master's, water-skiing, wet biking, writer (non-fiction).

JOAN MARIE BOYD, born March 18, 1932, Freeport, IL. Flight training includes airplane SEL and instrument ratings. She has flown 7AC (Aeronca Champ), Stinson and Cessna 182 Skylane. She and her husband John Kirtley Boyd have three children: Geoff, 25; Laura, 31; and Cathy, 27. They also have two grandchildren, Sarah, 7 and Zachary, 3. She started flying in 1958, took time out to be a mother, got back to flying in 1972, pilot's license in 1973. Joined the 99s in 1974. Earned her instrument ticket in 1976 and has now logged 1,000 hours. Her husband John, is also a pilot and retired retail store manager. Joan is a retired elementary teacher. She has participated in Flight Without Fear clinic and many Flying Companion Seminars. Member of 99s, AOPA, and Flying Farmers. Also co-chairman at annual EAA Fly-In, Oshkosh, WI. "Flying extends your horizons!"

MADALINE "LINDY" BOYES, scarcely ever called by her given name of Madaline – learned to fly while attending the University of California at Berkeley just after WWII. A private license was earned flying from the Oakland Airport and subsequently commercial and flight instructor ratings were added.

She graduated from U.C. with a BA in general curriculum and pursued a writing career, starting with the Oakland Tribune, for which she wrote a weekly column, "Aviation Roundup."

In 1957 she was called to New York to become associate editor of *Skyways Magazine*, then the official publication of the National Business Aircraft Assoc..

The Paris Air Show drew her to France in 1959 with fellow members of the International Aviation Writers Assoc. and she left *Skyways*. Flying opportunities with both French and Italian light aircraft manufacturers provided material for articles for *Cross Country News* (published by 99 Tony Page in Fort Worth).

Back home in California Lindy began freelance writing. She wrote for Tab Books, *Pilot's Weather Guide*, and co-authored a novel with the late David McCallister, *Sabres Over Brandywine*.

In 1963, Lindy joined the Peace Corps and spent two years in Brazil where she met 99 Anesia Pinheiro Machado.

Returning to the US in 1965, Lindy settled in Honolulu, HI, and founded the Aloha Chapter of the 99s.

Now retired from the Hawaii Visitors Bureau, she will relocate to San Francisco's East Bay Area.

Lindy's flying career includes six Powder Puff Derbies and two Hayward-Tucson Air Races. She soloed a Hiller 12-B helicopter and obtained a glider rating, both for stories for *Flying Magazine*. As a member of the Aviation/Space Writers Assoc., she logged pilot time in the giant B-26 (six engines) and flew with the Navy's Blue Angels and the Air Force's Thunderbirds.

ELAINE BRADBURY, born in Ontario, Canada, in 1962. Her interest in flying started watching her father build and fly model airplanes.

In 1980 Elaine attended Seneca College in Toronto, graduating in 1983 with a commercial rating and multi-IFR. Her aviation career began at Burlington Airpark where she instructed and worked in administration. In 1985 she moved to Central Airways at the Toronto Island Airport working in management positions and multi IFR instruction. Elaine flew with City Express of Toronto Island Airport in 1988 on the Dash 8. In 1989 she moved to Air Canada as a Second Officer on the B727. Currently she is training as a First Officer on the DC-9.

Throughout her career Elaine has been involved with the 99s in Ontario, Manitoba and Alberta.

She is married to Jim Waters, who holds a commercial pilot license. They have a son, William born in 1994.

MARY ANN BRADFORD, born Feb. 20, 1939, in Los Angeles, CA. She joined the 99s in 1989. Flight training included soloing at Long Beach, CA; Cross Country at Marlboro, MA; and pilot's license at Livermore, CA, in 1989. She has flown Cessna 150 and 152 (owns a Cessna 150). She is a registered nurse and worked as a pediatric nurse for many years, presently an allergy nurse at Kaiser Permanente, Pleasanton, CA. She has been married to William L. Bradford, Jr. for 35 years, he is a pilot and electrical engineer. They have three children: Lisa Bradford Carbelt, 31, accountant; William L. Bradford III, 26, professional pilot; and Karen Bradford, 25, bank customer service and student.

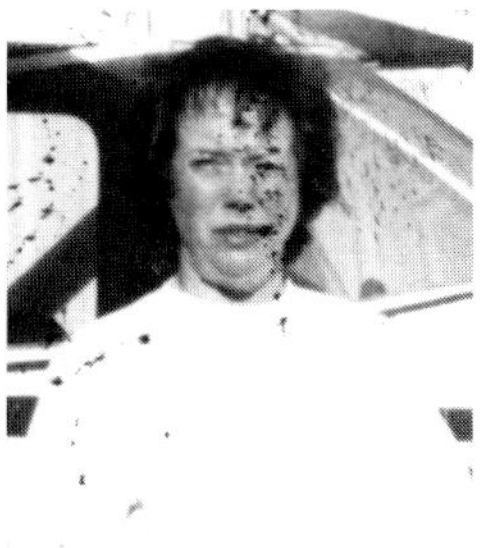

Her flight training was interrupted on several occasions by moves across the US due to her husband's employment. Although this was very frustrating, she was spurned on by the love of flying and the desire to finally become a 99. Upon arriving at Livermore Airport she discovered that it did not have a 99 chapter. That soon changed and along with several other women pilots they formed the Livermore Valley Chapter in 1990. She had the privilege of becoming its charter chapter chairman.

BRUNI BRADLEY, Phoenix 99s, was always a hard-working aviatrix, getting her private license in 1955, commercial in 1956, and instructor's in 1957. She now holds ATP, ASMEL, ASES and CFI ASMEL, backed up by 15,000 flight hours, and is an FAA Flight Examiner and an Air Safety Counselor. She not only examines pilots and instructs both flight and ground school locally, but often is called on for special jobs, like going to Colorado Springs in 1989 on an AOPA assignment to teach FBI agents the art of mountain flying and refreshing their instrument skills. She is scheduled to return for a long weekend soon giving an instrument refresher. Burni was named Arizona Flight Instructor of the Year in 1991.

Working as a Pan-American Airline stewardess, Bruni had to keep mental track of 16 foreign currencies and exchange rates. She married a Pan-Am captain. Fond memories are of flying their Cessna 210 over the Atlantic to Europe in 1970, visiting Africa (Morocco), and being the only non-Europeans in the European Air Rally in 1973. This was from Rome through Corsica, southern France, Luxembourg, Germany, Holland, England, and back to France. Bruni won the navigation prize for the Beacon Hill (London) to Paris route.

Since joining the Phoenix 99s in the 1960s, Bruni has contributed countless hours and her valuable knowledge on projects, especially the Kachina Air Rally and fund-raising air lifts. She also works hard in AOPA, CAP and other aviation related activities to promote aviation safety and success. Bruni probably explained herself best, saying "All I do adds up to furthering aviation, be it instructing or analyzing and improving present skills."

Her daughter, Brunhilde Jr., went to Annapolis and now, after 12 years in the Navy, is a lieutenant commander working as administrative officer of a naval squadron at Whidberg Island, north of Seattle.

EVELYN BRAESE, born July 3, 1927. Married Bill in 1945. They have three children and seven grandchildren. Managed Dyersburg, TN, Airport 1962-76. Solo J-3. Private pilot in 1964 and commercial in 1966. Joined the Cape Girardeau Chapter 99s in 1964 and held many official positions. Nominated and won 1967 Amelia Earhart Scholarship Award, used in obtaining her instrument rating. Accumulating more than 8,000 total hours. Qualified in many single- and multi-engine aircraft. Flying air taxi, charter and instruction. Also holds CFI and advanced ground instruction ratings. Organized and held two Powder Puff and one Angel Derby Stops at Dyersburg. Contributions to aviation include column in local paper for five years. Many guest speeches at local club dinners. She served many years as FAA accident prevention specialist. In 1976, qualified as FAA flight service specialist, earning many outstanding service awards. Will retire from the FSS in 1996 after promoting aviation for 33 years.

SISSI BRAINERD, born April 16, 1965, in Lorain, OH. Flight training was with Flight Safety International. She has flown the following planes: 152, 172, PA24, PA40, PA30 and PA. Employed as a flight instructor. When she joined the 99s she was looking for female aviation comradeship. What she found was great people that offered a wealth of knowledge and incredible support for her to reach her aviation goals. She owes many thanks to this great organization.

DOREEN BRANCH, joining the Potomac Chapter in 1989 was a really positive step for her, so that she could be with women who fly. When she was young, her mother would take her family to places where they could see airplanes, and Doreen had a college friend who took her for a ride. They flying bug bit her on about her third or fourth flight, and she thought, "This is for me!" She found out what kind of appetite she had though, for as a money struggling student, she often found the decision had to do with eating or flying. So she would go off to her mother's to eat, and spend her money on flying! She now has her private pilot's license and has taken on the secretary's position with Potomac Chapter. She flies the CE-150 and 172 and the Beech Sundowner. "It still is for me! Thanks, Mom."

LORI ANN BRAND, born in San Francisco, CA, and grew up dreaming of flying in Sacramento.

She started flying in 1977 while a student at San Jose State University. While in school she earned her private pilot license, instrument rating, and seaplane rating. She joined the 99s and became a member of the Santa Clara Valley Chapter.

Following graduation, she returned to Sacramento and earned her commercial and flight instructor certificates. She was an active member of the Sacramento Valley Chapter and won an American Flyers/99s scholarship in 1983. She used the scholarship to obtain her instrument instructor rating. Lori worked her way up the aviation career ladder. She instructed for more than 2,220 hours, flew as a Grand Canyon tour pilot and held a variety of other flying jobs.

In 1989, she became an employee of Embry-Riddle Aeronautical University in Prescott, AZ. She instructed and was a flight standards check airman.

In 1990, she was hired by the Federal Aviation Administration as an aviation safety inspector at the Portland, OR, Flight Standards District Office.

In 1991, Lori finally achieved a long-cherished goal. She earned her glider rating and CFI-G. She and her boyfriend, Jay, own a Let L-33 glider. They fly it every chance they get. Lori has competed in competition. She is a member of the Women Soaring Pilots Assoc. and the Soaring Society of America.

She currently lives and works in the Portland suburb of Hillsboro with her two cats.

ELIZABETH BAILEY BRANDSON, born July 20, 1924, in Deadwood, SD. Joined the 99s in 1987. Began training at the age of 50. She has flown the following planes: Cessna 150, 152, 172, 182; Hawk XP; and Piper Warrior.

She is married to Robert Ellis Brandson and has three stepchildren, three step-grandchildren, and two step-great-grandchildren. Her flight experience has been mainly short trips for lunch or overnight. Longer flights terminated in South Dakota and Oregon. Her husband, Bob who is also a pilot, and she fly together for fun.

LYNN BREDIN, First Canadian Chapter, loved flying and always wanted to be a pilot. As a child she spent a lot of time in the cockpit as her mother was a flight stewardess for American Airlines.

She began flying in 1989 at Toronto Airways, Buttonville Airport, and worked at the dispatch counter. Lynn got her wings in May 1991 and is now finishing up her commercial rating.

Lynn's first major cross-country was in a rental C-172 from Oshawa Airport to Myrtle Beach, SC. It was fun and very challenging. Over the Alleghenies, the transponder overheated, the VOR and ADF were unreliable. Lynn thought of Amelia. She didn't have the navigational aids they have now and it didn't stop her from doing what she wanted. Due to weather, an unscheduled stop was made at the New London Airport and Drag Strip in West Virginia! Lynn found the American aviation community went out of their way to make her feel welcome. They even rolled out the red carpet.

The return flight was just as challenging. After take-off from Myrtle Beach, a suction pump failure resulted in dead reckoning all the way home to Toronto.

In 1989 Lynn joined the 99s and has helped on several committees. She has flown in five consecutive Poker Runs. From 1993 to 1995 she has been responsible for organizing FCC's participation in the annual Toronto Aviation and Aircraft Show at the Toronto Pearson International Airport.

Lynn's ambition is to finish her training and apply for work in the airline industry as a pilot. Amelia Earhart has always been her inspiration and she would also like to fly around the world in an International Air Rally.

JACQUELINE BREEDEN BOYD, member of Nebraska Chapter since 1975. She teaches first grade and also teaches aerospace education classes to gifted fifth- and sixth-grade students. She is the advisor to their Aviation Explorer Post. She is one of three Nebraska teachers to have her certificate endorsed to teach aerospace education grades K-12. She has been chapter Air Age Education chairman for the past two years.

In May she entered the race for Grand Island's Hall County Airport Authority Board. She made it through the primary, and now has the general election to look forward to.

She is currently a member of Golden Triangle.

HARRIET BREGMAN, born Dec. 6, 1945, in Brooklyn, NY. She flies a 1958 Piper Comanche 250 and is building a Van's RV-6A with Kenneth Haefner, her significant other. Harriet gained her rating in June 1970 and has logged about 1,000 hours in 25 years and holds a private pilot's certificate with instrument rating, in single engine land and sea. In the last five years, she and Ken have flown to the International 99 Conventions in Las Vegas, NV; Portland, OR; Kansas City, KS; and Norfolk, VA; and has flown to various locations in Florida.

Harriet has been very active in the 99s since she joined in 1975. She has been vice-chairman and chairman of the New York Capital District Chapter on and off since the late 1970s. She has held various positions in the New York-New

Jersey Section, including secretary, vice-governor, governor and treasurer.

Harriet works for the New York State Department of Transportation in Albany, NY, and is currently working in their aviation division. She is a member of the Empire State Aerosciences Museum in Schenectady and a long time member of the Airplane Owners and Pilot's Assoc. (AOPA).

DORIS RENNINGER BRELL, although aviation has been part of nearly all her life, it was not until age 38 that she became a pilot. Marrying Henry Renninger in 1938 brought aviation into her life and Henry's illness turned her into a pilot. When he declared, "You're going to learn to fly" – she did!

Joining the 99s in 1957 opened a wonderful "Pandora's box," with the strong friendships, the camaraderie, and their many far-flung aviation activities. The responsibilities as a Section and International officer helped prepare her for a once-in-a-lifetime position as general manager of the prestigious Wings Club in New York City. There she met and worked with many of her aviation heroes i.e., Neil Armstrong, Ann Morrow Lindbergh, Jimmy Doolittle, Barry Goldwater. In 1992, the Club presented her honorary membership in "The Golden Eagles."

Her fondest aviation memory goes back to 1963, when she soloed a Bell 47G3B helicopter in two and a half days and seven hours at the AWA Convention in Dallas, TX. That same year, she became the first woman licensed helicopter pilot in New York state, and Jean Ross Howard-Phelan sent her Whirly Girl disk #59. Another proud first came in 1986, when she was elected a trustee to the College of Aeronautics, their first woman board member.

They lost Henry in 1980 but aviation brought their long-time friend Carl Brell back into her life. Together, they have four children, four grandchildren and two great-grandchildren.

HELOISE C. BRESLEY, born Sept. 19, 1925, farm near Cotesfield, NE. Joined the 99s in 1979. Flight training at Ord-Charles Zangger, instructor. She has flown the Cherokee 140 and Cherokee 180.

On April 6, 1947, married Dean. They have two children, Sheryl, deceased, and Mark, age 46. They also have grandchildren, Nicole, 17 and Jason, 15.

She used the plane a lot in her business as a farm/ranch Realtor and appraiser. A buyer can get a good look at the land from the air (impossible to hide the bad spots, such as a sand hill's blowout). She had the edge over land-bound appraisers too, when looking at subject properties or comparable sales.

BARBARA C. BREWER, a 1977 late start in flying at 55 years from Marlborough, MA, gave her 13 wonderful flying years with her friends and camera.

Joined the 99s, bought a Cessna 150 and flew all over New England. Two years later they retired to Cape Cod. Showing photos of scenic abstractions of Cape Cod shoals, patterns, cranberry bogs, and homes in galleries and camera contests gained news recognition, requests in various fields such as seal and bird research, conservation and Audubon volunteer, plus participation in supplying Brewster with photos for four covers of town reports. Requests for prints, postcards, slide shows, calendars and churches plus a new golf course led to mini-aerial photo business and setting up own darkroom. Meetings with 99s, air markings, plus convention in Alaska were memorable, but increased arthritis, plus back fractures led to plane sale, so photos are now used in her computer letters and on on-line.

LAURA BREWER, Santa Clara Valley Chapter, started flying in 1986 at Stockton, CA, and has commercial, CFI, CFII, and MEI ratings. She is a graduate of Embry-Riddle, where she was the chairman of the Daytona Beach Chapter and flew with the Embry-Riddle flight team. Laura currently works as a flight instructor at San Jose Airport with Trade Winds Aviation. She has logged more than 1,700 hours and is working toward her ATP.

Laura is also interested in martial arts. Laura joined the 99s in 1989 with the Sacramento chapter, has served as chairman of the Embry-Riddle Daytona Beach chapter and is currently a member of the Santa Clara Valley Chapter.

LILLIAN (MIKE MCGUIRE) BREWER, Phoenix 99s, grew up determined to fly from age 6, when a barnstorming pilot climbed out of the cockpit after a thrilling exhibition, pulled off helmet and goggles and was a girl! Doing test and repair of B-24 electrical systems in preflight at the Willow Run bomber plant gave Lillian the money to solo in 1944, and get her private ASEL in 1945 at Ann Arbor, MI. She married a pilot who got weathered-in in a blizzard, hence the old family saying, "It was a cold day when I met you!"

She liked the aviation field a lot, but didn't aspire for higher ratings, and after several years working for an FBO returned to the medical career for which she had trained, being lured by more money. She flew as much as time permitted, enjoyed raising two sons (no pilots), and riding many charter miles in relation to her work in health care administration.

She has always supported 99s, and helped in many local level activities, first in 1951 in the Michigan Chapter, then in 1961 in the Northern Arizona, and since 1987 in Phoenix. Since retiring, she went to Africa as a medical missionary and has done quite a bit of other foreign travel.

SUE BREWER, native Georgian, has lived on East and West coasts, Kentucky, and is currently living in North Carolina.

Brewer has been employed by public libraries for the past 20 years. Her educational background includes marketing/retailing degree and BA in business administration from Campbell University.

Brewer began flying in 1976, earning her private certificate on Friday, Oct. 13, 1978.

Brewer flies Cessna 172s, Cherokee 140s and proudly owns a Cessna 150 named *AME* (Amelia Mary Earhart), with 800 hours flight time.

Brewer joined the 99s, Inc. in 1989. She is a member of Cape Fear Aero Assoc., AOPA, EAA, and she attends the Forest of Friendship, EAA Oshkosh, Dayton Airshow and local flying events.

Brewer is currently in instrument flight training, married and she has a daughter and two sons.

KATHERINE A. MENGES-BRICK, armed with her BS from Boston University and an MA in psychology from New York University, she "entered the future" in September 1941 by obtaining her pilot's license in an Aeronca Chief at Teterboro, NJ, joining the New York-New Jersey 99s after 13 years of teaching.

December 7, 1941, Pearl Harbor, Civil Air Patrol was created; all their planes were moved, little more West each time, away from the coast. As intelligence officer in the first Bendix Squadron, she participated in military training missions.

With more than 200 hours in her Chief and Frank's Fokker SuperUniversal, Jacqueline Cochran recruited her for the WASP (Women Airforce Service Pilots). Graduated advanced ferry command pilot, and flew 1943 and 1944. Logged 1,016 military flying hours; flew high and low tow target missions, searchlights, smoke laying, tracking, chaff and radio control missions, some ferrying.

Hold SMEL, instrument and instructor ratings with more than 6,000 hours. Flown in every state Canada, Cuba; ferried to Alaska, Hawaii and Australia. Currently a member of UFO (United Flying Octogenarians) those who after 80 years hold a current BFR.

Past International president 99s; International secretary; executive boards and first budget chairman; governor and secretary New York - New Jersey Section. Charter member of Coyote Country Chapter, CA. Past VP National Silver Wings secretary P-47 Pilots Assoc.; secretary National Pilots Assoc.; member AFA (Air Force Assoc.); member AWA (Aviation Space Writers Assoc.); Altrusa; past president Fallbrook Garden Club.

Edited 99s *Thirty Sky Blue Years, Twenty - Transcontinental Sky Trails, Powder Puff Derby - The Record, Powder Puff Derby - Update.*

Powder Puff Derby (AWTAR, Inc. - All-Woman Transcontinental Air Race) board of directors 25 years; 13 as chairman/executive director.

Won Nolde Derby flying Vega BT-13, New York to Miami 1948. Flew numerous other races placing second, fourth, eight and eleventh in US, Canada and Mexico.

Served on FAA's Women's Advisory Committee 1968-71; appointed FAA Safety Counselor 1972-76; institutor/co-chairman "Colt for Kim and Women of Korea"; institutor/chairman Votive Model of AE's Vega to hang in Protestant Chapel at JFK International Airport.

Recipient: FAI (Federation Aeronautique International) Paul Tissandier Diploma 1973; AE Medals 1949, 60, 76; NJ Aviation Hall of Fame 1978; Sargent College Humanity Award 1967; Boston University Alumni Assoc. Award for Distinguished Service to the Community 1988/89; OX5 Pioneer Woman's Award 1981; OX5 Hall of Fame 1983; Silver Wings Paul Fromhagen Award 1993; WIAA Lady-Hay Drummond Hay Trophy, 1948.

Appears in *Who's Who in Aviation and Aeronautics,* 1983; *The World Who's Who of Women,* 1984; *Forgotten Wings,* 1992; *Lady Birds II,* 1993. Also in Memory Lane "International Forest of Friendship" Atchison, KS.

JANET ZAPH BRIGGS, Bay Cities Chapter 99s Charter Member, private license #19823. Janet learned to fly in 1930 at the Stanford Flying Club, Stanford University, where she also earned an AB in engineering April 1, 1931. She left to further her studies in Vienna, Austria.

in 1953 she was an employee of a steel company with headquarters in New York. As a metallurgist, she traveled extensively for the company.

Some planes Janet flew were the OX5 Waco, Kinner Fleet, OX5 Travelair and the Kinner Lincoln.

While on a business trip to Japan, she died in Tokyo, Jan. 24, 1974.

NELL BRIGHT, born June 20, 1921. Joined the 99s in 1943. Served in the Women's Airforce Service Pilots - WASP. She has flown the following planes: PT-19, BT-13, AT-6, AT-17, B-25, B-26, P-47, A-24, A-25, AT-11, AT-7. Her husband's name is Weldon and their children are Scott Jennings, 46; Margo Jennings Thurman, 44; Diane Bright, 45; Kim Bright, 43.

After receiving a BA from West Texas University at age 19, started private flying in Amarillo, TX. Qualified for WASP training in January 1943 with private license plus 75 hours.

Entered the seventh class for training at Sweetwater TX. On Graduation day, 20 members of her class learned that they had been chosen to attend B-25 school at Mather Field, Sacramento, CA, the first time any woman had trained on the B-25. There she earned her first pilot rating on the B-25 and instrument rating. She was then sent to the Sixth Tow Target Squadron at Biggs Field, El Paso, TX, where she flew nine different types of aircraft including the B-25, B-26, P-47, A-24, A-25, AT-7, AT-11.

"We would roll out big target sleeves and the boys would shoot live bullets at them. Sometimes they'd miss and sometimes they'd hit. There was never a dull moment," Bright said. "Sometimes we flew three to four missions a day and all in different aircraft. We enjoyed every minute of it because we loved flying."

Bright has been a stockbroker for many years. She lived in Phoenix before moving to Sedona in 1984 and was one of the first women stockbrokers in that area. She is presently with Fox & Company Investments of Scottsdale. She is married to a jazz pianist and they have four children. Bright is also active on the Sedona Jazz on the Rocks board of directors.

Bright attended the 50th Anniversary of her organization which was held last October in Washington, DC.

HARRIET BRIN, soloed a Taylorcraft in 1961 in Ogden, UT, and obtained her private license in a Cessna 150 in 1967. She later copiloted a Piper PA-20 and a Mooney Mark 21. She was a copilot in the 1969 Kachina Doll Race. She served as chapter chairman for Santa Fe Chapter 1973-76 and Monterey Bay Chapter 1978-80, and has served as airmarking chairman and as a presenter at Flying Companion Seminars.

Brin has a BS in business administration from Fort Hays, KS, State University. Her community activities include tax-aid assistance to the elderly, volunteer cataloging at the public library, and computer input for a religious education program.

She is a member of the 99s, Inc., Hollister Airmen's Assoc., and Experimental Aircraft Assoc..

Her husband, William, is also a pilot. They built a VariEze in 1978 which they still fly. The couple have three daughters, one of which obtained her private pilot license, and four grandchildren.

SHEILA "PAT" BRINNAND, born Aug. 17, 1935. She and her spouse have four daughters, ages 21, 29, 32, and 35. She joined the 99s Reno Chapter in August 1992. She began flying in 1988, it started as a Christmas gift from her husband and became something she had to do. She holds a private pilot license with single engine, land rating. She and her husband arrived in the US (Oregon) from London, England, in 1964, with two young children. They moved to Scottsdale, AZ, and after nine years moved up to Reno, NV, where they have firmly planted roots, becoming US citizens in 1992.

Her hobbies have run the gambit of hiking, climbing, skiing and outdoor pursuits in general. Now they are lucky to be fit and healthy enough to become "travelers."

JOAN BROCKETT, born Jan. 8, 1932, Derby, CT. Joined the 99s in 1982. Her flight training includes Niagara Falls (IAG) NY and Fort Eustis Flying Club (FAF)Felkar-Field, Newport News, VA. She has flown Cessna 150, 152, 172, BL-10 (7GCBC), PA-19 and AA-5.

Joan got bitten by the flying bug in 1975, while returning to Niagara Falls in a Piper twin from a ski trip to Vermont. It was a crystal clear night with lights twinkling from the cities on both sides of the black Lake Ontario. So at age 43 she began flying, joined the AOPA and the 99s, and has held the offices of secretary, treasurer and chairman for the Hampton Roads Chapter as well as historian for the Mid Atlantic Section.

A native of Connecticut she married her college sweetheart, Wilfrid "Bill" 41 years ago. A retired electrical engineer and non pilot, he helps to keep his wife on course and at altitude. Joan and Bill have two grown sons, Gunnar R. Brockett, 33 and Keith R. Brockett, 31.

Substitute teaching keeps her in flying money. An active member of her church choir, she also enjoys snow and water skiing, aerobics, sewing and crafts, and traveling.

LAUREL NANETTE BRONSON, born Jan. 28, 1958. She grew up in Davenport, IA. Received her BS in pharmacy from the University of Iowa, 1982. Moved to Indianapolis, IN, 1987. Employed as a pharmacist at Wishard Memorial Hospital. Received her private pilot certificate on Aug. 7, 1993, at age 35. Became a 99 in 1993. She likes being a 99 because it helps her connect with her sisters in aviation. She was inspired to become a pilot rather abruptly after accomplishing a large goal (purchase of a home). She was in search of a new goal that was exciting, challenging and fun.

Flying is by far the most personally rewarding activity she has ever pursued. However, when she must stay on the ground, she enjoys being a violinist with the Philharmonic Orchestra of Indianapolis. Last, but not least, she dearly loves her three cats.

BARBARA S. BROTHERTON, born in Oakpark, IL, on March 18, 1926. Joined the 99s in 1963. Flight training includes SEL instrument rating. Learned to fly in a Stinson Voyager, and has also flown Cessna 150, 170, 172, Mooney, all Piper single, and Bellanca. Has been married for 47 years to Thomas (WWII B-29 pilot). They have two sons, Tom II and Timothy (each received pilot's license at age 18). They also have six grandchildren, three boys and three girls, ages 12 to 7.

She has a BS in home economics/nutrition and a Ed.M. Flown lots of proficiency races (AWNEAR and Michigan Small Race) and eight major cross country races i.e. Angel Derby and AWTAR. Still flying. Past treasurer and governor of East Canada 99 Section. Has been officer of these: Greater New York (treasurer); First Canadian (chairman, vice chair and treasurer); Iowa (treasurer and chairman); Coyote Country (treasurer and vice chair).

ANDREA BROWN, joined the 99s in 1986, just four months after receiving her private pilot's license. She transferred from the Alameda County Chapter to the Santa Clara Valley Chapter in 1995. She has participated in the Hayward-Las Vegas Air Race every year since she became a pilot. She works on the Annual Hayward Air Fair Committee as aircraft registration chairman. She and her husband Steve, who is also a pilot, own a Cessna C-172 and a Citabria, and recently purchased an FBO at Livermore airport. She is currently the Amelia Earhart Scholarship chairman for the Southwest Section. Andrea enjoys her job as a technical support engineer for a software company. When she is not flying, she is dancing, fishing or camping.

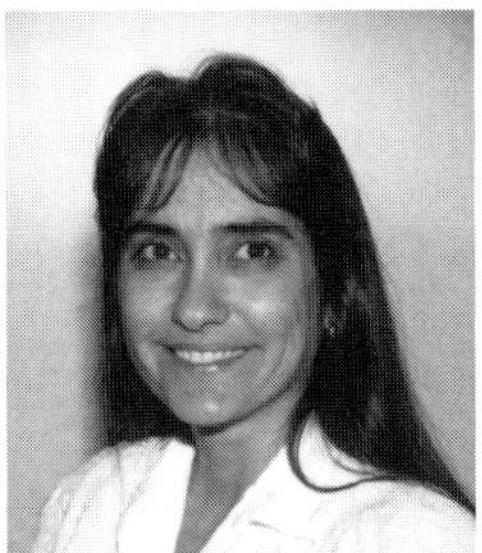

BETTY J. OVERMAN BROWN, born in 1923 in tiny Atalissa, IA, and grew up in Detroit, MI. She applied along with 25,000 others, for entrance in the Women Airforce Service Pilot training program and, in 1944, was one of a grand total of 1,072 graduate WASPs to receive her Silver Wings. Flight training was in the Stearman PT17, the BT13, and the AT6. She then flew AT6's daily, towing targets for cadet aerial gunnery practice until, regrettably, investment in this womanpower was dissolved in December 1944.

She is married to Ron, who is also a pilot, and has children Kathy, Rodney and Tyler, and grandchildren Sean, Patrick and Kevin. Her community activities have included service in Girl Scouts, on a regional planning board and on the Maine State Critical Areas Board. She has an avid interest in canoeing, kayaking, swimming and the out-of-doors.

Since retirement, they fly their PA18 Cub and Stinson Voyager to aviation gatherings the year around. She has about 1,000 flight hours, and single-engine land and sea ratings.

She is a member of the Women Airforce Service Pilots, the Women Military Aviators, the 99s, Women in Aviation, Aircraft Owners and Pilots Assoc., Silver Wings, and the National Women's Hall of Fame.

CAROLYN OLIVIA BROWN, born and raised in Phoenix, AZ, moving to California in 1983. It was here she became interested in flying. Started taking lessons in February 1988 and received her license on Sept. 1, 1988. She currently has more than 525 hours and is pursuing her instrument rating. She normally flies Grummans, an AA5 and an AA1C.

She has been the chapter chairman of the Long Beach Chapter for two years and is very active in chapter activities. She flew the Palms to Pines two years and participates in six to eight flying events every year.

She is also a member of the American Yankee Assoc. (AYA), AOPA and the EAA.

She has three children: Cari, 35; Christianna, 30; and Stephanie, 26. and also three grandchildren: Bradley, 13; Zachary, 5; and Breanna, 5.

HEATHER BROWN, born on June 29, 1967, in Peterborough, NH. She's a 1990 graduate of Embry-Riddle Aeronautical University. In addition to holding a BS in aeronautical science, she is a professional pilot with more than 4,000 flight hours. Licenses and ratings include: commercial pilot, single and multi-engine, flight instructor and instrument flight instructor. She has flown most factory-built single engine aircraft and is also an experienced experimental aircraft pilot. Additionally, Brown has flown the King Air turbo-prop aircraft, the CASA-212 turbo-prop transport, and the Fouga Magister jet trainer. She also regularly flies a long EZ experimental aircraft. Heather is currently chief flight instructor for New Garden Aviation, Toughkenamon, PA.

Organizational memberships include AOPA and the 99s, Inc. She previously served as secretary of the Delaware. Chapter of the 99s and is currently chairman-elect of the chapter.

JANICE A. BROWN, is an airline transport pilot, flight instructor, and an elementary school teacher. After receiving her degree from California State Polytechnic University, she moved to Bakersfield with her husband, architect Nash Brown. Janice then began a teaching career with Bakersfield city schools. During this time she fulfilled a childhood dream of learning to fly, receiving her pilot's license in 1975. She now holds an airline transport pilot certificate with instructor, instructor instrument, multi-engine and glider ratings. She has worked as a Part 135 charter pilot and is presently an Aviation Safety Counselor for the Federal Aviation Administration.

The brightest event of her flying career occurred in 1980 when she became the test pilot for the first solar powered aircraft, the Gossamer Penguin and the Solar Challenger. These became the first piloted aircraft to be powered only by the sun's energy. The Solar Challenger obtained altitudes of 14,300 feet and remained aloft for more than seven hours. Because of her involvement with this project, she received the International Harmon Award from President Reagan in 1982. This trophy is awarded for outstanding achievement in aviation.

Other memorable events in her career relate to the all women's transcontinental air race, the Air Race Classic. She entered this exciting event in 1987, 1988 and 1989 (winning fifth, third, and eleventh place honors.)

Presently, Janice is employed as a classroom teacher with Bakersfield city schools and also runs her own aviation company.

Aviation is a wonderful, exciting and challenging part of her life. Janice truly loves flying!

JEWELL BAILEY BROWN, born July 9, 1926, in Charleston, SC. Joined the 99s the first time in 1946, and 1988 the second time. Flight training included flight instructor in early years now commercial certificate with instrument rating. Planes flown are too numerous to mention. She was employed as a flight instructor at age 18, still flying but not for commercial purposes. She and her husband, William Brown, have two children, George, 37, and Mark, 25.

Started flying age 14; soloed 16th birthday 1943, youngest commercial pilot in state at age 18, flight instructor same year; took Super Cub 90 hp to 26,875 feet for altitude record Charleston, SC, 1949, broke previous record. Tried again in 1950 in 115 hp but missed by 300 feet.

KRYSTINE M. BROWN, born July 15, 1953, USA. Joined the 99s in 1985. Holds a private pilot's license. She has flown the following planes: Cessna 152, 172; Cherokee 6, Lance, Piper Arrow, Archer, Citation Lear Jet. She and her husband have one child, Jenifer Baum, age 7. The most memorable flight she did was demonstrating a 1995 Citation Lear Jet with a female copilot as her instructor, filled with all male pilots as riders. Her landing was the second best and gave her the thrill of her life.

LUCY BROWN, Bay Cities 99s Chapter Charter Member, Lucy learned to fly at the Stanford Flying Club, Stanford University, CA, where she was majoring in economics.

She was killed in a plane crash on Alameda Airport, June 3, 1933. She was not the pilot of the plane, but a passenger.

MARGARET L. BROWN, born Sept. 27, 1922, New York, NY. Joined the 99s in 1981. Military service included USNR (WR) and WAVES March 5, 1943 to Feb. 5, 1946. Flight training included private and SEL. She has flown the following planes: C-150, C-152, C-172, C-182, PA 28/161. She and her deceased husband, James T. Brown, Jr., had one child, Louise V. Brown, 36.

When she was very young, before the days of parkways, she was taking pictures with her Brownie box camera of bi-planes at a small grass strip and her favorite comic strip was "Buck Rogers." When she was 20 she joined the WAVES and was fortunate to be stationed at an NAS with carrier based planes, but the women were not trained to be pilots. After her husband passed away and her daughter was over 21 and away at college, she decided to try something she had always wanted to do and received her private certificate in March 1980 in Connecticut. She doesn't fly as often as she would like, but enjoys going to the 99s International Conventions.

MARION SCHORR BETZLER BROWN, learned to fly in her late teens on the CPT program while a student at USL in Lafayette, LA.

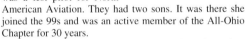

After two years teaching in a New Orleans High School, she joined the second class of WASP training in Houston, TX. After graduation at Sweetwater she was assigned to the 3rd Ferrying Group of ATC at Romulus AFB, Detroit, MI.

She later married Charles Betzler and moved to Columbus, OH, where he was a test pilot for North American Aviation. They had two sons. It was there she joined the 99s and was an active member of the All-Ohio Chapter for 30 years.

She is now married to JD Brown, also a long-time professional pilot. They live in Hammond, LA, where Marion keeps busy as a FAA Designated Pilot Examiner.

MARY LOU BROWN, Phoenix 99s, did not become a pilot because it was "something she always wanted to do," but become one she did, in 1965. Her primary reason for learning to fly was to augment her career qualifications as the administrator of a research program in the US Geological Survey, handling logistics for a remote sensing program. Once she was in the pilot's seat, she stayed there through multi, commercial, instrument, seaplane and helicopter ratings, adding the duties of project pilot to those of administration.

For the USGS she arranged for acquisition of sophisticated surplus military aircraft and sensing equipment for the collection of data on the nation's water resources. The aircraft included a US Navy T-34B, an Air Force Lockheed T-33 jet trainer, an Army Grumman OV-1B Mohawk, Twin Beeches, DeHavilland Beavers, Cessnas 180 and 310, a Bell UH-1F (Huey) helicopter, and a Sikorsky H-19 helicopter. She flew the aircraft on data gathering missions through out North America, including Alaska and Panama. She worked extensively in Arctic Alaska, including a year at Point Barrow, the northernmost village on the continent.

At the request of NASA, Mary Lou piloted the T-33, underlying the track of the orbiting Apollo 7 spacecraft across southern Arizona, photographing the earth from an altitude of 32,000 feet. She photographed the length of the Grand Canyon in infrared color from 35,000 feet to commemorate the centennial of John Wesley Powell's exploration of the Canyon.

Mary Lou, the proud mother of two and grandmother of five, lives in Tempe, AZ. Retired from the USGS, she is a principal and the administrator of a civil engineering firm doing studies in scientific hydrology. She is a member of the 99s, Whirly Girls, and AOPA, and owns a Cessna 182.

SANDRA A. BROWN, born May 20, 1941, in Rochester, NY. Flight training included private pilot's license and working on instrument rating.

Planes flown: Cessna 150, 152, 172; PA-28 Cherokee; and Aeronca Champ.

She is employed as lead switchboard at Genesee Hospital. She and her husband, Glennard, have two children, Erik, 31, and Melissa Hargreaves, 28. They also have a granddaughter, Kealy Marie, 2. She sews her clothes to help finance her flying.

KATHLEEN FAGALA BROWNE, Phoenix 99s, has a deep interest in aviation that has not only resulted in her becoming a licensed pilot – she also is a senior engineering programmer with Honeywell (formerly Sperry), and writes and tests software for flight management computers. She got her private ASEL in Phoenix, AZ, in 1984.

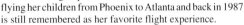

Aviation has been a family affair with Kathleen, who was inspired to be a pilot because her father flew for his business. As the single mother of two sons and a daughter since 1979, she finds that borrowing a plane and flying her children from Phoenix to Atlanta and back in 1987 is still remembered as her favorite flight experience.

Kathleen has been a 99 since 1985, and contributed to chapter success in many ways - she's held most offices, including two terms as chapter chairman, and is presently treasurer, as well as working on such special activities as the Kachina Air Rally, Scavenger Hunt-Poker Rally, Flying Companion Seminars, and a membership drive.

SANDI PIERCE BROWNE, born on Sept. 10, 1944, in Fort Worth, TX. Joined the 99s in 1970. Began flying in August 1966 at Tim's Airpark in Austin, TX. Got her private license in December of that year. Flew solo and formation aerobatics routines and was wingrider for The Flying Pierces Airshow for 12 years. She and her husband Stan Browne have four children: Chandelle, 27; Will, 30; Steven, 28; and Michael, 24. Their grandchildren are Meagan, 4 and Jacob, 1.

She has had more than six forced landings in her flying career. One time she landed her highly-modified clipped-wing Tcraft on a highway south of Savannah, GA. Her 5-year-old daughter Chandelle was following with her parents in their car. After Browne landed on the road, she parked the airplane over to the side under the trees. About one hour later her parents drove down the road. Her daughter pointed to the TCraft - "Look, there's my Mommy." Her mom and dad were horrified but her daughter thought it normal.

GALE BROWNLEE, learned to fly in 1963 at the Kingston, NY, Airport. Her first flight as a student gave her the desire to make aviation a new career. She had a commercial rating within a year of her first lesson and her instructor's rating soon after. Her first instructing job was in Danbury, CT, a short air commute in her Cessna Skyhawk. When a job opened at Kingston, she instructed for some time and became the first woman charter pilot at that airport. Gale founded the Hudson Valley Chapter of the 99s and in 1968 was the first recipient of the Doris Mullin Scholarship Award. Her Whirly Girl number is 141. Gale was instrumental in building the Benedictine Hospital Heliport which is named for her grandfather, the late Dr. Mortimer B. Downer. She raced in the Angel Derby International Air race in 1965. One of her most exciting aviation events was as co-captain of a ferry flight of a Cessna 414 from Poughkeepsie, NY, to Nairobi, Kenya, via the North Atlantic. She became an activist when the local utility proposed 2,600 feet stacks in the airport traffic pattern at Kingston, and thanks to a huge response to an aeronautical study (largely by 99s) and two years work, had them declared a hazard to aviation. The plant was never built.

Her early career was as a model in New York City, followed by design career and public relations. She now works a real estate broker in Woodstock, NY, and still flies for pleasure, does aerial photos for real estate and environmental causes. She also is the official town photographer. She has one daughter Pixie.

She is a member of 99s, Aircraft Owner and Pilots Assoc. and Whirley Girls (#141). She has 13,000 total hours. Her ratings include: single multi-engine land, commercial, instructor airplanes and instruments, multi-engine instructor, helicopter commercial.

MIRIAM ELAINE BRUGH, Bay Cities Chapter. The happiest day of her life was May 26, 1946, when she received her private license at Erwin-Newman Airport in Houston, TX. It was a funny, friendly little grass field and they learned to pull back on the wheel when they heard the weeds on the belly of the plane. The little red TCraft was a thing of beauty and so was life. Five of them learned to fly at the same time and all joined the Texas Chapter in 1946. Four of them have had many happy hours of flying and only wish Ruth Newsome could have shared in the great times, but it was not to be as her life was short.

She has served all chapter offices and on various committees. Her previous accomplishment is in the area of Airmarking, including the SOP for the Southwest Section.

BARBARA J. BRUSSEAU, born Sept. 26, 1929, Peoria, IL. Joined the 99s in 1964. Flight training included Marshall County Airport, Lacon, IL, C-75. A public school teacher for 38 years. She has flown mostly single engine: Piper, Cessna and Beech. Loved model airplanes as a child. Fulfilled a life long ambition and learned to fly in 1963. Added instrument, commercial, CFI ACFII in 1971. Has accumulated nearly 10,000 hours. A gold seal flight instructor with more than 7,000 hours of instruction. Enjoyed flying Illi-Nines Air Derby and other small races.

BETSY BRYANT, soloed in May 1993, instrument December 1995, commercial and instructor December 1994. Major aviation, business administration — ERAU and Aviation Salem Minor. Used to go to "Hollywood Burbank airport" to visit her grandfather who worked for Lockheed and watch all the planes. That was her first memory of flying aviation.

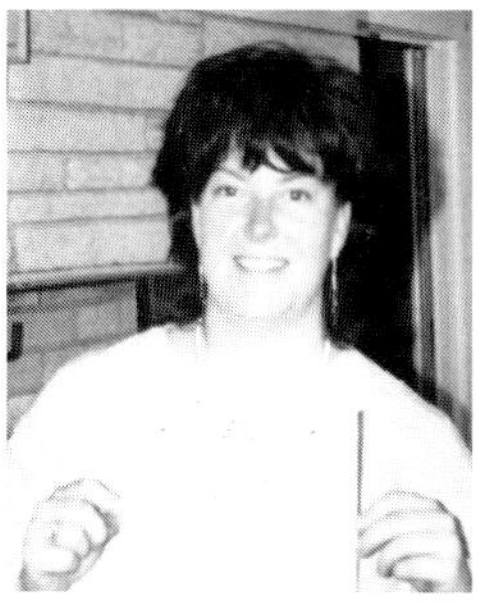

She used to fly back and forth on PSA from Burbank to San Francisco to visit her grandmother in San Francisco always had fun.

She hopes to start out with a regional carrier and from there who knows, she likes props. She would like to get her A&P when she's got enough money saved up, that would be a challenge. Worked for Mesa Airlines for one and a half years.

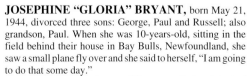

She flies a 1974 Beed Sierra which ferried back from Illinois with a friend, everyone should do a long cross country, what a learning experience airport/airplane people truly are the nicest folks around. She loved it!

JOSEPHINE "GLORIA" BRYANT, born May 21, 1944, divorced three sons: George, Paul and Russell; also grandson, Paul. When she was 10-years-old, sitting in the field behind their house in Bay Bulls, Newfoundland, she saw a small plane fly over and she said to herself, "I am going to do that some day."

When she was 13 years old her mom died and she had to take care of her eight brothers for two years. When her

father remarried she decided to go live in New York with an aunt. So at 15-years-old, she got on a plane and flew to New York and loved it. At 18 she got married and raised three sons. When her last son Russell went to kindergarten, she said that it's, time for her and Jan. 19, 1976, she went to D'Angelene Aviation in Islip Airport for her first lesson; on April 20, 1976, she soloed and Aug. 31, 1976, with 53 hours, she received her pilot's license and fulfilled one of her dreams.

She has met many wonderful people through aviation and flown many places with them.

There is no feeling greater than being at the controls going where you want, when you can, "weather permitting" for her that is. Because she is a VFR Pilot, maybe someday she will be on IFR Pilot, another dream.

MARJORIE E. BRYANT, born June 17, 1926, in Concord, NH. She learned to fly in 1965 in Pennsylvania. She joined the 99s in 1966.

In 1968 Marge and husband Jack started Bryant Aviation, Inc. an FBO at Pennridge Airport, Perkasie, PA. Marge was chief pilot and chief flight instructor. In 1982 Marge was a designated FAA pilot examiner.

During their 20 years operating as FBO they were awarded the Airport of The Year from the International Flying Farmers in 1971. In 1973, Marge was awarded the Flight Instructor of the Year from the ABE-FSDO and in 1981 she was awarded the Governor's Award from the Aviation Council of Pennsylvania. In 1992 Jack and Marge were inducted into the Forest of Friendship in Atchison, KS.

Marge has logged more than 25,000 hours and participated in two Powder Puff Derbies, 1971 and 1976. She has been an active member of the Pennridge Pilots Assoc., the Aviation Council of Pennsylvania and AOPA. In 1969 she cofounded the Upper Bucks County Aviation Technical School at the Quakertown Airport, PA. She is on the advisory committee of the school.

Jack and Marge retired from their FBO in 1986. In 1994 when she lost her husband to cancer she developed type I diabetes and can no longer fly solo. She is hoping the FAA study on diabetes will open the air to those with this disease.

MARY SNELL BRYANT, born in Mexico, MO, in 1949. Mary holds a BA from Northwestern University and an MBA from the University of Illinois. Mary is also a certified public accountant and a certified financial planner. Mary started flying in 1973 but worked for Fortune 200 companies and international management consulting firms in the financial and strategic planning areas for many years before pursuing flying seriously as a career. In 1985, Mary resumed her flight training and now holds an ATP as well as CFII/MEI ratings. Mary was an active racer in the late 1980s and won the Great Southern Air Race in 1988. In 1988, Mary joined Piper Aircraft and held a variety of positions including instructor in the Piper Training Center and, later, in the sales area, as Eastern Region sales director. In 1991 Mary was one of two individuals who founded Attitudes International, Inc., an organization principally dedicated to providing flight training in Piper aircraft. At this time, Attitudes is "Piper Aircraft Corporations' exclusively approved training school" and provides training to purchasers of new Piper aircraft as well as pre-owned aircraft. Attitudes specializes in training in Malibus, Mirages, and Aerostars, but also provides training in a variety of other aircraft as well.

Mary has been a member of the 99s, Inc. since 1986 and is also a member of many financial/accounting professional organizations as well as aviation-related organizations.

Mary resides in St. Petersburg, FL, with her husband Timothy C. Bryant.

MARCIA J. BUCHYNS, BC Coast Chapter, born in a suburb of Toronto and learned to fly through the Air Cadets Program. She moved west when she was 16 and got her glider pilot's license then her private license at age 18. She stayed with the Cadets Program for 11 years, becoming a glider instructor at age 20. She has been a tow pilot for the past seven years. Marcia has more than 1,000 glider flights and 200 hours of glider flying.

At 23 Marcia graduated from Simon Fraser University in business administration and also got her commercial rating. She worked in the accounting department at a local airport and when the opportunity arose applied for and got the job as airport manager. During her time as manager she had the chance to ferry an aircraft to Florida. In 1987, Marcia flew to the Reno Air races where she got to meet Chuck Yeager. Marcia joined the 99s in 1988 and in 1990 got a job with Transport Canada.

In 1992 Marcia became an airport duty manager at Vancouver International Airport, where she is today. She is recently engaged, a homeowner and raising eight pups. Marcia has about 1,000 hours total time.

MARY HARWOOD BUCKWALTER, born in Washington, DC, and now resides in Leesburg, VA, with her husband, Len Buckwalter. She had one son, Stephen and two stepchildren, Valerie and David.

She started flight training in Bridgeport, CT, in March 1981, and soloed June 1981 at New Haven Airport. She completed her FAA private check ride in December 1981 and immediately joined the Connecticut Chapter of 99s. She served every position in the chapter from program chairman to chapter chairman during the next seven years. She earned her instrument rating in 1988.

She and her husband founded *Avionics Magazine* in 1981, Mary is currently advertising sales representative for Eastern North America for the magazine's new owners, Philips Publishing. The Buckwalters also founded and operated a technical book and software publishing company serving the avionics industry - Avionics Communications, Inc. located at Leesburg Airport.

For seven years they owned a Cherokee 180, which they flew to Oshkosh and to the Bahamas often. They now fly a TB-20 Trinidad, out of Leesburg, VA, nicknamed *Trinny*. They fly the Trinidad extensively for business and pleasure. Their home is a half-mile from the airport, "in the pattern" of JYO.

CHANDA SAWANT BUDHABHATTI, India Section, at the age of 17 she wanted to fly and did! Breaking tradition's mold of India and became the third woman commercial pilot of India. In 1965, became a 99; 1976, charter and present governor of India Section; founder/president Indian Women Pilot's Assoc.; flew six air races receiving two trophies.

On International 99s Board 1984-88; she was the first non-US candidate for presidency of 99s, Inc.; pioneered/chaired major events that enhanced 99s' worldwide image; two "World Aviation Education and Safety Congress" in India – the largest 99 – sponsored activity outside the US and first of its kind in India.

Chanda's efforts to promote aviation amongst women, especially in India, is evident – since 1980, Indian Airlines has 20 women pilots, including three captains; in 1994, Air Force enrolled seven women as pilot officers.

She has received Awards of Merit and Appreciation Certificates, for her outstanding leadership and dedication in aerospace education from the World Aerospace Education Organization, the 99s, Inc., Lions Club, Zonta Club International, Rotary Club and Kutch Industrial Group. She was the first Asian woman to receive "Chuck Yeager Aerospace Education Award."

Chanda's dream – India Air and Space Museum and solo flight around the world.

BARBARA BUEHLER, has been eating her words since 1973, when she said to her student pilot husband and his instructor while pointing to the Cherokee 140, "You'll never get me in one of those things!" More than 2,500 hours later, and armed with a commercial rating and an instrument rating, she is still flying her Cherokee Six (47463) out of Teterboro, NJ, and loving every minute.

Born in New York, Barbara earned her BA with a major in music, married Edgar L. Buehler, a dentist, and moved to New Jersey to raise their two fine sons (her proudest accomplishment). She spent happy years planepooling her boys to college and back and forth to Cape Cod.

Retired from her business, Warp One, Inc., Barbara is still office manager for her husband's dental practice and busy with local community organizations.

She is still singing and playing the piano and is enjoying her grandchildren.

SANDRA KAY BUFKA, ratings include single engine land-instrument. She has 500 hours of flight time. Started flying in December 1987. Joined the 99s in January 1990. She is married and has two adult children. Owns and flies a 1993 PA28-236 a Piper Dakota. Her husband has been flying for 35 years and he owns a 1965 Beechcraft Baron and a 1983 Piper Aerostat. She does all the com/nav radios when they fly together.

MARIA S. BULLER, born in Syracuse, NY. Joined the 99s in 1971. Flight training includes ASEL Private (close to taking instrument check ride May 1995). She has flown Piper Tri-Pacer, Cherokee 140, Cherokee Warrior, Cessna 150 and 172. She and her husband, Dale F. Buller, have one child, Stephanie, 20.

Marcia's husband, Dale, introduced her to flying and it is an interest they have shared throughout their 25 years of marriage. They have made many wonderful trips and have been active in the Central New York Pilots Assoc., Syracuse Flying Club and EAA Chapter 486. Additionally, Marcia has been active in the Central New York Chapter of 99s.

MIRIAM PARSONS BURCHAM, learned to fly along with her 16-year-old son, got her license Aug. 1, 1969. As a 99, served as airmarking chairman, assisted with sectionals, served as registration chairman at FAA seminars; and publicity and public relations chairman of the start of the famous "Last PPD" (Sacramento). Made many decorations for tables and aviation displays.

Spoke on aviation for several school career days. Taken elementary school field day children on their first flight. She was design chairman of the "Silver Wings Museum" at Mather AFB.

She was born Dec. 21, 1917, in Oakland, CA. Joined the 99s in 1968. She has flown the following planes: Cessna 140, 150, 152 and 172. She worked as copilot for an aerial photographer for 15 years. She is a widow with two children, Arthur, 43, and Charles, 37. She also has the following grandchildren: Jeffrey, 14; Christyn, 12; Jennifer, 10; Daniel, 7; Roxanne, 5; Ellen, 3; and Eva, 1. Served as publicity and public relations chairman the start of the last (29th) Powder Puff Derby.

ROBYN JOY SINDELAR BURNS, born April 15, 1960, in Cleveland, OH, moved to St. Petersburg, FL, in 1973, and in 1974 was introduced to the Civil Air Patrol, where she realized it was possible to fulfill a lifelong dream – to learn to fly. It was not until 1988, however, that she made that dream come true.

She took on the challenge of learning to fly out of Albert Whited Airport, a small airfield surrounded by airspace complications that included the adjacent McDill AFB to the east with its heavy, fast, low-flying F-16 aircraft coming in from the south, the nearby Tampa International Airport to the northeast, the St. Petersburg/Clearwater International Airport to the northwest and recreational parasailers along the beach to the west.

Within 55.5 hours, she received her license, and has flown C-152m PA-28-181, AC112A, and has flown right seat in aircraft such as Grumman, experimental Long EZ, Porterfield Lp65 and in a bi-wing Breezy. Robyn hopes someday to fly in a Corsair and a Harrier.

Robyn has traveled throughout Europe and intends to continue trips abroad and plans to become fluent in at least one foreign language. She has been a 99 member since 1989 and has taught children in the Air Bear program and is also a published author in *Women In Aviation*, a publication. She plans to become more active in the Angel Flight program.

Her educational background is varied and includes a BS in chemistry. Currently she has a career in the environmental testing field. She intends to pursue her entrepreneurial desires to become self-employed in the near future.

GLADYS DAWSON BUROKER, born March 16, 1914, Bellingham, WA. Planes flown, most Tail-dragger manufactured 1920 and later. Worked as flight instructor and operation of their flight school. She and her husband, Herbert, have three children: Sally Simundson, 54; Linda Melhoff, 49; and Kelly Buroker, 42. She also has three granddaughters and three grandsons, ages 5 to 51.

Gladys learned to fly in an OX5 Waco 10, after five hours instruction Sept. 24, 1932. She barnstormed, parachuted at several cities in Washington and Idaho, and brought her own Travel Air 2000 in 1935.

Gladys instructed aviation courses at the all-male St. Martin's College in Lacey, WA. When Pearl Harbor was bombed, she trained hundreds of Army and Navy students for War Training Service.

She taught her three children to fly and exactly 50 years from her solo date, she soloed her grandson Mike Burris in Sept. 24, 1982. Buroker is rated to fly instrument, single engine land and sea, multi-engine, glider and hot-air balloon.

Buroker has been recognized for her aviation contributions by the Boeing Museum of Flight, OX5 Aviation Pioneers, the Federal Aviation Administration, the state of Idaho and others. She has been flying for 62 years and has enjoyed many as a member of the 99s.

MARY LEE BURNS, BC Coast Chapter, born April 30, 1950, in Vancouver, WA. She lived in the Northwest while growing up, ending in Vancouver, BC. After a year at university, she went to the Yukon for a summer job and ended up staying for eight years. During that time, she had a variety of jobs and also built two log cabins and lived in the bush with no electricity or running water for three years.

She returned to city living in 1976 and got a job with Indian Affairs as a local government advisor. In this capacity she traveled by small plane to isolated reserves along the Northwestern BC coast. The flying was often uncertain with lots of waiting for the weather to clear so she could be picked up. During this time she developed a fear of flying.

In 1982, she decided to learn to fly to overcome her fear. With hard work and an understanding instructor, she got her private license in 1983. Since then she has learned to love flying and flew to the 99s Convention in Anchorage in 1984. She is now self-employed as an interior decorator.

CATHY BURROW, born Catherine Patricia Burrow in Edinburgh, Scotland, June 21, 1941. Her father was Scottish and her mother, Irish. Emigrated with family to South Australia in 1954. Graduated from Woodville High, University of Adelaide and University of Papua New Guinea: BA (honors), diploma physical education, diploma sec. education.

Flying licenses: Australian ATPL, British ATPL, Papua New Guinea Senior Commercial and Marshall Islands ATP.

Captained Australian Women's Parachute team at World Championships in Leutkirch, Germany, in 1964. Papua New Guinea Accuracy Champion in 1971 and British Women's Parachute Champion in 1973.

Married Keith Burrow 1965. One daughter, Sam, born 1970.

First solo, Aero Club in Port Moresby, Papua, New Guinea, in 1966. First paid employment as an instructor and then an air taxi pilot from 1972 to 1976 in Essex, England.

Returned to an independent Papua, New Guinea, in 1977 and flew there commercially for three years - over 2,000 hours with Douglas Airways and Talair - single pilot ops.

Also flew Coast Watch in far north Western Australia and as a pilot for the Queensland Police Air Wing. Ferried a turbo commander from Portland, ME, to Darwin for Amann Aviation in 1977.

Command on the British Aerospace 748 and the Nomad with the Airline of the Marshall Islands from 1981 to 1984.

Flew Viscounts (British Air Ferries) with her husband of 30 years throughout Europe on night freight from 1988 to 1992. Keith has more than 13,000 hours. Her total hours are more than 9,000.

Semi-retired and currently own their own Aztec.

DIANA L. BURTON, a member of the Tulsa Chapter and formerly a member of the Oklahoma Chapter, has been flying since 1982. She met her husband, John, also a pilot, on the day she soloed, and married him exactly two years later in the tower of Wiley Post Airport. They own a Comanche 250. She joined the 99s as a 66 in 1982, and was the Oklahoma Chapter's first 66 to become a 99. While a member of the Oklahoma Chapter, she designed and manufactured a line of 99 jewelry for the chapter to sell. She also served as the chapter's treasurer. Now a member of the Tulsa chapter, she serves as membership chairman. She participates in the Oklahoma Chapter's Okie Derby Race and placed first in 1990.

Diana's interests include traveling and the design of jewelry. She and John are in the process of renovating a beautiful home overlooking the city of Tulsa. She has worked for American Airlines for the past two and a half years as an international tariff pricer.

NANCY BAILEY BUSHKO, born Feb. 6, 1931, Buffalo, NY. Joined the 99s in 1970. She has flown the following planes: Cessna 120, 140, 195; Fairchild PT-26, AT-6; Cub PA-17; Aeronca Seabee; Stinson 170-172; Swift; Navion; Culver; Piper; Tri-Pacer.

She learned to fly and received her private and commercial ratings at Stephens College, Columbia, MO, in 1949-51. Received her associate degree in 1951 from Stephens College. Summer of 1951, she was a "Line-Girl" at Oak Bluffs Airport, Martha's Vineyard, MA. Received her CFI and worked the following summer at Oak Bluffs Airport flying charter and instructor. While at Stephens College she flew in two intercollegiate air shows in Norman, OK. She soloed the Brantley and Bell Helicopters in the 1960s and taught her son to fly in 1980.

She is a proud member of the 99s, Inc. One of the first 99s she met back in 1950 was Nancy Harkness Love. She summered at Martha's Vineyard and flew her Bonanza to the Island Airport where Bushko worked. She was a lovely person and told her many stories about the (WAFS) she headed, and owning her own AT-6.

She is a former Brownie leader, charter member of the Brooklyn Historical Society, PA, and Lackawanna Historical Society. AOPA member, Cherokee Pilots Assoc. Has owned her own (5027-W) Cherokee 160, since 1961. She has logged more than 3,000 hours, charter instructor and pleasure. Member of the Silver Wings Fraternity.

Married to John Bushko, retired captain Pan American World Airways, and retired, VP SJ Bailey and Sons Inc. They have two sons, John Jr., 38; Todd, 36; and a daughter, Anna, 35.

JEAN R. BUSTOS, born May 7, 1935, in Los Angeles. Joined the 99s in 1988. Served in the USAF 1953-56, flight service. Held her private pilot license since 1985. Flies C-172.

She has a daughter, Lisa, 37-years-old, who loves to ride. Jean is the daughter of WWI pilot, has a flying family. She has flown to Washington State three times to visit sister, usually fly some locally, visit Lake Havasu often.

ELIZABETH ANNE BUZZELL, Bay Cities Chapter. Youngest female private pilot in Maine (pictured here in 1960 at 17 years old); commercial ASEL.

BA, University of Washington, Seattle; attended Hastings Law School, San Francisco.

An owner of Sutter's Mill Antiques, Coloma, CA, and fine arts appraiser throughout California. Married to Charley F. Petersen, president Petersen & Associates, general contractors, Sacramento, CA. Owns a ranch in Lotus, CA.

MAGGIE BYRNE, born September 19, in Mexico City. Joined the 99s in 1994. Started flying in September 1992. Flight training includes CFI and has flown Cessna 152, 172, 172RG. Employed as a flight instructor.

She and her husband, Bob, have three children: Michael, 14; David, 11; and Lisa, 5 weeks.

HELEN E. PARMENTER "BETTY" BYTWERK, in 1943 when she was 16 years old, she took her first flying lesson in a Taylorcraft. She soloed in six hours.

The next two years she worked at the airport, earning the weekly pay of two hours' flying time. She went to school mornings, worked afternoons, and spent weekends from sunrise to sunset at the airport. It was a wonderful life!

On her 18th birthday she got her commercial certificate (by-passing a private license). A week later she received her instructor rating.

In 1945 she helped start the "Winged Spartans," the Michigan State University flying club. To earn college money she ferried new Aeronca Champions from Dayton, OH, to Detroit.

She is married, has three grown children, three grandchildren, and has been a 99 for more than 20 years.

At the age of 68 she still flies for the fun of it!

MICHELEE M. CABOT, born June 11, 1938, Ft. Sam Houston, San Antonio, TX Air Force brat.

First flight: 1946 at eight years, over Luzon P.I., with Dad, early USAF test pilot Gen. Wm. M. Morgan, in biplane PT-15—can remember like it was this morning. Picked up languages while based abroad. Three years college: Pine Manor College and Boston University.

Began flying and got private license 1967, subsequent instrument and commercial ratings as well as comm. gliders (1980). Logged time in nearly every model SEL aircraft; L-19 tow pilot for Greater Boston Soaring Club. Joined 99s in early 70s. Current governor of New England 99s (1994-1996). Held chapter and section level offices; chaired Aviation Safety and Education committees both levels, plus chapter and section newsletter editor. Resurrected and chaired the NEAR in 1991. Flew Air Race Classic 1991.

Currently 2,800 hour pilot, co-owner T210. Member EAA, Aero Club of New England and NAA. Married Harold Cabot 1971. Sons: Andrew C. Cabot (born 1971-died 1991) and (first marriage) Gilbert T. Stair (born 1960) William Morgan Stair (born 1965); extended family Christine, Marshal and Adele Cabot.

Besides flying and managing and FBO for a few years, life has been a tapestry of volunteerism in the judicial court system, Junior League, choral singing (international tours), docenting and serving on local museums boards, as well as career stints in radio talk shows and newspaper restaurant reviewing. In all of it, flying has been the one constant joy and challenge.

PATRICIA "PAT" L. CAIN, born Aug. 30, 1927 in Philadelphia, PA.

Pat learned to fly at San Jose Airport in 1965 and joined the Santa Clara Valley Chapter in 1967. She has logged about 170 hours, having flown mostly with her husband before their divorce and now flies only occasionally. Has flown Cessna 150, 172, 182; Skylane.

Pat has her MBA in finance and works in real estate. She has a son, Richard F. Cain R., 40, and two grandchildren and currently lives in San Jose, CA.

JANET CALDWELL was born April 18, 1941, in Lincoln, NE.

Received private license in April 1976. She joined the 99s in 1977, and has flown a Piper Cherokee, Warrior, Lance Cessna 172 and 210s.

Enjoys taking trips with her family and shares piloting with her husband Carl. They have two children, Karen, 32 and Charles, 30; and three grandchildren. Karen enjoys being able to go places they would not be able to get to if they did not fly. She has participated in many flying activities. Also a member of AOPA, MCA and USPA.

MARGARET CALLAWAY will always remember a summer day in 1927 at Mexico City airport when she saw her first airplane. It circled over a crowd of thousands. It was none other than the "Spirit of St. Louis," piloted by Charles Lindbergh. "Oh boy, next to God, he is it!" she thought, and this little girl knew her destiny from then on. Margaret has been living on "cloud nine" ever since she earned her pilot license in 1945. Margaret not only liked to fly fast, she preferred aerobatics.

Margaret joined the 99s in 1947 and became a life member in 1978. She has logged 14,000 hours, almost 6,000 as flight instructor. She has trained many pilots, some now flying commuter and major airlines. She became a pioneer in the air race circle, helped organize two Baja Air Races and displays a large number of trophies in her home. She has been active in Flying Samaritans and the Forest of Friendship, but mostly she promoted the International spirit of the 99s in places like Cuba, Mexico, all of Central America and India.

Margaret married Lt. Richard Callaway in 1938. They have two daughters, Sandra and Margo. As far back as both girls can remember, they sat on pillows behind the controls of airplanes with their mom and even flew races with her, sometimes as copilot, other times as competitors. Margo at 17 and Sandra at 19 entered history as the youngest team in the 1959 Transcontinental Air Race from Lawrence, MA, to Spokane, WA. They placed third!

One of her major contributions to the 99s is on the International level. In 1977 she traveled to India and Nepal with other 99s. While there she met Saudamini Deshmukh (Minoo) and invited her to stay in her home in San Pedro and gave her dual instruction. Other 99s provided a Cessna 150 trainer and ground school; and Minoo's only expense was for fuel. A very grateful Minoo left seven months later for India with all her ratings except the ATP. Minoo became captain for India Airlines, type-rated for 737s and Airbus'.

BONNIE JUNE CALWELL, born May 28, 1954, in Newmarket, Ontario. Bonnie found her niche in life in 1976 after a close female friend introduced her to the world of aviation. She built her career around this field by working as a ticket/reservation agent, chief flight attendant, aircraft sales person, charter coordinating manager and in 1989 established her own company in the charter business in Toronto. At this time she also achieved her multi-IFR rating.

She is a 99 member of the First Canadian Chapter and was past chairwoman, aerospace education.

Presently she and her husband Peter Lubig (who is a regional commuter pilot) own and operate an executive charter operation in Calgary, Alberta. Calwell has a PPC on a Piper Navajo and is presently training as first officer on a King Air 90.

She is the proud mom of one son, one stepson and recently one daughter.

COLEEN ANNE CAMPBELL, born in Halifax, Nova Scotia in 1954, when it seemed that men had all the fun. During an aerobatics show, the pilot emerged revealing tresses of blond hair, a vivid memory even today. While dating in high school, Dave (her husband) and Coleen would go to the airport to watch airplanes. Today, 25 years later, they still go to the airport; but they fly airplanes instead.

In 1984, Dave got his license and in 1989 bought into an Arrow partnership, with plans to travel. Cautiously Coleen approached the prospect of learning to handle the airplane. Five years later, almost 1,000 hours, IFR rated, commercial soon to be completed and plans for CFI and CFII, she has passed Dave in training and experience. This achievement, she owes mainly to his continuous support and long hours of working with her on VOR concept and how to avoid S turns on an ILS.

Dave and Coleen own a 1975 Piper Arrow with another gentleman. They have flown coast to coast twice, Mexico and all over the Southwest. She works full-time as a hospital laboratory technologist. A member of the 99s for five years, she has participated in numerous projects and committees and was chosen San Gabriel Valley Chapter Pilot of the Year of 1991. At 40 it's exciting to say, "I fly airplanes."

DOROTHY B. CAMPBELL, spent her early years in Brooklyn, NY, where her family enjoyed Sunday afternoons at Floyd Bennett Field watching notable pilots like Wiley Post fly. As a teenager in Maine she was a member of the Civil Air Patrol but she never flew a plane until she was a grandmother.

Thanks to the Long Island Chapter's 66 program and 99 instructors, she earned her license at age 55 and her IFR rating two years later. She has a partnership with several 99s in a Cessna Skyhawk, which they frequently fly to Florida or Maine with convention trips as far as Oklahoma City.

Her most memorable flight was piloting a Piper Archer from Nelson on New Zealand's South Island (across gusty, spectacular Cook Strait) to Wairarapa, Masterton, on the North Island.

She recently retired after 45 years in nursing and health education and lives on Long Island with her favorite navigator, husband Harold. They have two sons, two daughters, two grandsons and a future 99 granddaughter Casey. Their favorite day trip is to Block Island for lobster lunch.

GRACE CAMPBELL, learned to fly at San Jose Airport in 1972 and her flight instructor was her husband. She joined the Santa Clara Valley Chapter in 1974, encouraged by Evelyn Lundstrom. Grace has logged 750 hours and has her instrument rating.

For many years, Grace flew her Cherokee 140 from Hayward Airport, but she recently moved to Las Cruces, NM, to be near her brother. She is active there, working with the chamber of commerce to bring more business to the local airport. Grace is currently a student at New Mexico State University, studying her life-long loves of geology and paleontology. She has two children. This summer, Grace is planning a flight to Nova Scotia, with a stop en route to Detroit to attend her 50th high school reunion.

JOANNE A. CAMPBELL, the first female pilot and only natural American citizen in her family. Encouraged by her spouse Bill and owning their own Cessna 172, she earned her license in 1995. The Campbell's are members of the International Flying Farmers and enjoy their fellowship. At the age of 4, she first recognized her ambition to become a pilot while helping her brother drop parachutes from the attic window. Today, she hopes to be a good role model for her grandchildren. Joining the North Jersey Chapter in 1992 has been both an encouraging and heartwarming experience. Joanne thanks you, the members of the 99s, for fostering and nurturing this lifelong dream.

SUE CAMPBELL, a life resident of Ohio started flying in 1988 at the age of 49. From solo to private ticket took some self-motivation and courage. You see, Sue's mentor, and friend, who planted the thrill of flight, died suddenly in 1989, shortly before she would have taken her last few lessons and checkride. Sue's instructors' encouragement kept her going. Then in June of 1990, another tragedy struck again. Sue's instructor was killed along with another student pilot in a plane crash only hours before she was to have had a lesson. Self-motivation, determination and the love of flying took over and Sue received her private ticket in 1991.

Sue learned of and joined the International 99s in 1992. She has served as ways and means chairman, Lake Erie Chapter, from 1993-1994. She is now a charter member of the Women With Wings Chapter, chartered in July of 1994, serving as alternate secretary and library chairman.

Sue works as a medical assistant and has two daughters and two grandchildren.

JANICE CANNELL, Winnipeg, Manitoba, began flying, 1979, in Cherokees, obtaining her private license in 1980. She then began flying Cessnas and in 1982 started aerobatics. In 1984 Janice joined the 99s; received her chapter's scholarship towards aerobatics and has been secretary and chairperson. In 1985 she competed in the novice category of Aerobatics Canada flying the Aerobat and passed. In 1986 she went into partnership purchasing a Decathlon; competed in the Basic category in 1987 flying her Decathlon and passed. In 1987 she attended the International Aerobatics Club judge's school and assisted at some competitions, then in 1988 obtained her commercial.

Janice has been a member of COPA and is working with them, representing the 99s, holding a Flying Companion Seminar to run jointly with the Barnstormers Air Show in 1995. She was secretary for the local Founding Executive of CASAR; has 80 skydives and 235 flights hang-gliding.

CAROL CANSDALE, born July 25, 1953. Her love of flying was fostered at an early age by her father, the late Robert Cansdale, who was an airline pilot. He started a logbook for her at age 9 and in 1969 Carol soloed at the age of 15.

A 99 since 1972, Carol graduated from the University of Washington in 1975. She worked as a flight attendant for Hughes Airwest from 1976-1979, while at the same time acquiring ratings and students. In November 1979 Carol started work as a pilot for Cascade Airways, a commuter airline in the Pacific Northwest. Employment at Cascade paved the way for a

subsequent job at Republic Airlines in April 1983. A merger in 1986 brought Northwest Airlines and Republic together as one; and today Carol is a captain flying DC-9 aircraft, the same aircraft her dad retired on. It's true when they say, "Everything comes full circle!"

Carol owns a 1946 J-3 Cub and supports numerous animal welfare organizations. In addition, she is a violinist with the Linden Hills Chamber Orchestra in Minneapolis, MN.

TERESE A. CAPUTO, born on July 15, 1959 in New York. She had flight training at MID Island Air Service and earned her commercial and instructor's licenses CFI and Multi. She has flown the Beech 99, King Air, Cessna 150-180s, Aztecs and Lear 31.

Caputo is employed at the Metro Air in Ronkonkoma, NY, a flight instructor and charter pilot.

While she was on the flight team for Dowling College in 1981-1982 she got to fly with Cliff Robertson. During the ground breaking for Dowling's new facility she got to meet Neil Armstrong in 1993.

MARY ELLEN CARLIN, learned to fly in 1988 at Palo Alto Airport in California, and joined the Santa Clara Valley 99s in 1992. She has her CFII and MEI ratings and is working on her ATP now, having logged 1,600 hours. She owns a Seneca-3 and an Archer and instructs at West Valley Flying Club at Palo Alto. She has flown the Palms to Pines race twice.

She has three sons and her non-aviation interests include biking, piano, golf and skiing.

LAVERNE BILLINGSLEY CARLTON, born in Texas. She is a widower and has two children, Sandra Teitsworth, 48 and M. Wayne Ezell, 47; and three grandchildren: Susan Teitsworth, 20, Ryan Ezell, 15 and Scott Ezell, 13. She retired from Chevron after 30 years.

She received her flight training at Taft, CA, in 1970 and flew planes such as: Cessna 172, 150 and 152; the Piper Cherokee 180; and the Cessna 172XP. She joined the 99s in 1971.

She and her husband purchased a used airplane in March 1975 and they flew several times with their 18 month old granddaughter between her folk's summer home and theirs that year. They had the plane annualed in October to find wing damage that was covered up. Grandpa like to have had a heart attack. He's gone now, but her granddaughter is a real joy.

AVA CARMICHAEL, a Texas native and a California resident, began flight training in 1967 along with her physician-husband. With the same enthusiasm that earned her a BA at Texas Tech University and an MA in theater at the University of Iowa, SEL, MEL and commercial tickets followed. She flew three AWTAR competitions with her flying pal, Wanda Cummings, and these memories evoke many hilarious conversations years after the fact.

Her source of great pride in the 99s is her authorship, with her husband of, *From White Knuckles to Cockpit Cook*, a book for "rightseat" flyers. Originally designed to comfort and instruct frightened passengers in light airplanes, it originated with a series of lectures to wives with pilot husbands. This led to the Flying Companion Seminars, and ultimately to the book.

Her logbook reflects trips extending from Merida, Yucatan, to Calgary, Alberta; from Key West to Santa Catalina Island, and hundreds of points in between.

MARGUERITE CARMINE, born Aug. 17, 1922, in San Francisco, CA. She started taking flying lessons and had flight lessons as a private SEL SES instrument rated. She has flown a Citabria, Piper Cessna 150, Piper Arrow, Piper Turbo Arrow. She joined the 99s in 1972.

She is a widow and has two children Dean Carmine, 48 and Janet Salinas, 46. She has one grandchild Franz Carmine, 19.

Carmine flew to Baja, CA, and western and midwestern states and Canada (west coast). She flew aerobatics in Citabrias out of Oakland, CA, float plane. She has held office of chapter chairman.

DEBORAH ANN KAEDER-CARPENTER, born in 1959, in St. Paul, MN. She learned to fly in 1981 and joined the Minnesota 99s shortly thereafter in 1982. Participation in the 99s helped Deb achieve her aviation career goal, as she was awarded an Amelia Earhart Scholarship in 1985 for a turbojet flight engineer rating.

Deb began her aviation career as a flight instructor, moved to Part 135 multi-engine charter pilot, then she landed a position with a commuter airline flying the Beech 99 and Metroliner. In 1988 it marked the beginning of Deb's career with United Airlines. Domiciled at Chicago O'Hare, she began as a 727 second officer. then moved to 737-300 first officer. She now flies the 767 and 757 for United Airlines and is now a 737-300 captain.

Deb remains active in general aviation, as she is an FAA-appointed accident prevention counselor. She and her husband, Steve Carpenter are demo pilots for Wipline Floats in the C-206 and Caravan. They owned a M6 Maule for several years and currently own a Piper PA12 on floats and a C-180 on amphibious floats. Her and her husband now reside in Inver Grove Heights, MN.

REVA HORNE CARPENTER, born Aug. 25, 1938, in Anson County, NC. She and her husband, Rex, started taking flying lessons after purchasing a 1964 Cessna 172E. They soloed on the same day and later got their licenses on the same day, March 31, 1984. Reva participates in the Young Eagles Program and has given several young people their first flight. She was a timer for the Air Race Classic at the Asheville, NC, stop in 1993.

She is a proud member of the Carolinas Chapter of the 99s and served as chairman from 1987 to 1992. She is also a member of AOPA and EAA.

She lived in Africa from 1962 to 1968 while her husband was working there with Voice of America.

She was a Cub Scout leader for six years and is a member of the First Baptist Church of Marshville, NC.

She has three sons, five grandchildren and expecting the sixth grandchild soon.

RIKAKO SUGIURA CARPENTER, born in Saitama, Japan, on June 17, 1932.

She is married to Royal E. Carpenter and has no children.

In 1969 she moved to New York City because of her job and she lived there for 13 years. She took flying lessons at Brookhaven, Zahns and Flushing Airports and got her certificate in 1979.

The most exciting and unforgettable moment of flying was her first solo, in her Grumman AA1B. Trying to descend for landing, the airplane did not sink at all with normal power setting that she used when flying with her big instructor. Suddenly, she realized that he had to reduce power more than before and made a nice landing. She was so happy.

Carpenter joined the 99s Long Island Chapter and transferred to Member at Large upon returning to Japan. In 1992 she established the Far East Section and became the first governor of the section. She works at Yokota Air Base Aero Club and flies as often as she can.

CAROLYN C. CARPP, a teacher and opera singer for most of her adult life, Carolyn learned to fly in 1978. She soon joined the 99s and held the offices of Western Washington Chapter, chairman Northwest Section governor and International director. Always involved in aerospace education, Carolyn was advisor to her school's Young Astronaut Club for eight years.

She won the Western Region Brewer Award twice, the Northwest Section Achievement Award in 1990 and 1993 and Western Washington Chapter's Pilot of the Year Award in 1994. Other awards include: 1994 Kirkland Rotary Outstanding Educator Award; 1990 Lake Washington School District Creative Integration of Technology, Math and Science Award; Washington Aerospace Education Inspiration Award; and the Wesley Crum CAP Award.

A mother to Chip and Curtis and a grandmother to Austin, Amanda and Alyssa, Carolyn enjoys flying her Cessna Cardinal out of Boeing Field in Seattle.

JOAN CARROLL, started flying in 1962 and after receiving her private license she decided her true interest in aviation was in acting as navigator when she and her husband took their two children on many cross-country flights. She joined the Shreveport Chapter of 99s in 1964 and proceeded to hold every office, serving as an enthusiastic social chairman for 10 years straight. The entire chapter was grief stricken when cancer took her from them in 1992.

Joan received her BS degree from University of Illinois and proceeded to receive a second degree from Centenary where she majored in education. She was an excellent bridge player, held offices in many organizations, loved hosting dinner parties and was an avid golf player. Serving many years as a Girl Scout leader and enthusiastic room mother, she exerted a major influence on the youth of the community.

KARLA CARROLL, born on April 3, 1958 in Kingsville, TX. Karla learned to fly in 1988 at Providence, RI. She has logged 600 hours, earning an instrument and commercial rating. She participated in the 1993 Air Race Classic. She is currently flying aerobatics and working on her multi-engine and certified flight instructor ratings.

Carroll has a BS in nursing, specializing in critical and emergency nursing. She is a member of many nursing organizations, and a firm believer in continuing education for nurses.

Carroll enjoys working with her hands as well has her head. She sews, makes porcelain dolls, works with stained glass, cross stitches and quilts. She also enjoys walking and swimming.

Carroll belongs to the 99s Inc., Aircraft Owners and Pilots Assoc., Experimental Aircraft Assoc., International Aerobatics Club and Rhode Island Pilots Assoc.

She has one son James who is a senior in high school.

LINDA CARROLL, learned to fly in 1991. She flies a Cessna 120. Her educational background includes a BA in business administration. She holds a state of Alabama real estate broker's license.

Carroll is a proud member of the 99s Inc., Aircraft Owners and Pilots Assoc. and the International Cessna 120-140 Assoc.

She is currently vice-chairman of the Alabama Chapter 99s.

She has one son.

WANDA "GWEN" W. CARROLL, born on Sept. 2, 1916. They had an air ambulance service. She flew the patients as she was also a registered nurse. Decided she should learn to fly. After a year of lessons, she received her license at age 52 in 1969.

Flying was a great experience, one of her very special achievements besides her family. Both her sons have their pilot license. She and her husband owned Watsonville Aviation. She also accompanied another pilot who was a registered commercial pilot and they flew prisoners from Santa Cruz County to their destination.

She has served as a council person on their city council and is presently serving on the Pajaro Valley Water Management Agency. She is a past president of Watsonville Community Hospital Service League, Watsonville Women's Club and Sowptimist International.

BETTY KIRBO CARTER, born July 22, 1937 at Mt. Vernon, TX. She is married to William F. Carter and they have no children.

The year she celebrated her 40th birthday, it occurred to her that her life was probably about half over; therefore it was time for her to do some of the things she deserved to do, like start wearing some of her good jewelry, use linens she had put away for a special occasion and complete her MA.

About a year after making the decision to do what she deserved to do, her husband and she were watching a TV documentary about Amelia Earhart's life. Her non-pilot husband turned to her and casually ask if she would like to learn to fly. Her immediate reply was, "Yes." Her husband wasted no time in gathering literature on flying from all the flying schools at Meacham Airport in Fort Worth, TX. After reading all the brochures, she began flying lessons in October 1978. When she talked with family, friends and co-workers about learning to fly, everyone was supportive. A work schedule was made out at the office, where she worked as a medical technologist at USAF Regional Hospital, Carswell AFB, Fort Worth, to accommodate her flying lessons.

In July 1979, she received her single-engine private pilot license. In September 1979, she enrolled in ground school for instrument rating. Near the end of the course, the instructor approached her and asked if she would like to visit a 99 chapter meeting. He had been contacted by a local 99 chapter in the D/FW metroplex who asked him if he had any female students in his class. She accepted the invitation. Not knowing what to expect from this organization, she was pleasantly surprised and greatly relieved to find a group of confident women who had families and careers unrelated to flying but shared one central theme, the fun of flying.

While enjoying the flying activities in the 99s, she continued her academic studies toward her advanced college degree, receiving a MA in management from Central Michigan University in December 1980, the same year she became a 99 member of the Golden Triangle Chapter, located in the Dallas/Fort Worth metroplex. Membership in the 99s has provided many happy moments while participating in flying activities like airmarkings, serving as judge for flying competitions and sponsoring safety seminars.

LUCRETIA CARTER, resident of Soldotna, AK, since 1966, born Oct. 12, 1932, in Omaha, NE. She taught elementary school (29 years) in Oklahoma, New Mexico, South Dakota and Alaska.

Carter's educational background includes: schools attended in Nebraska and North Dakota; graduated, Omaha North High, 1950; BS degree (elementary education) 1957, Bethany Nazarene College (now Southern Nazarene University) in Oklahoma; graduate studies in New Mexico, Colorado, South Dakota, Alaska; MEd degree (aerospace education) 1982, MTSU, Tennessee; Aerospace Education seminars: Alaska, California, Florida, Georgia, Tennessee, Colorado, Nevada and DC; aerospace ambassador; participant in NASA's Teacher-in-Space program (1985-1986).

Having become a student pilot in 1974, Carter finally completed private pilot certification on Aug. 4, 1984 while participating in 99s Northwest-Sectional Convention at Homer, AK. Carter participated in the Great Southern Air Race and in Alaska, has flown her C-172 west to Cold Bay (Alaska Peninsula) and to Kotzebue, (north of Arctic Circle).

She's a member of 99s Cook Inlet Chapter, AOPA, Civil Air Patrol (has served as observer with CAP Search Missions) and was a charter member of Nazarene Aviation Fellowship. She's also an avid photographer, traveler and cat lover, and serves on Soldotna's Main Street Steering Committee but considers church activities her greatest interest and highest priority.

SUSAN M. CARTER, born April 12, 1939, in Helena, MT, was raised on the west coast. She currently resides in Los Angeles and Idaho and is a member of the Idaho 99s.

She had long wanted to fly. Suzy took up soaring when her children were small, only to put it on the shelf until her children were grown.

She got her SEL license in June of 1994. Her flight training consisted of mountain flying, flying the Southern California airspace and flying in Hawaii where her daughter, Anne, now lives.

Some favorite flying experiences include: flying in the Swiss Alps; flying to Cody, WY, and Montana; flying the Idaho back-country with her son, Rob; and flying to Death Valley last spring.

Suzy is proud owner of 182 HR-A Cessna 182, and will soon begin working on her instrument rating. For her becoming a pilot in her 50s was a real challenge and has added a wonderful perspective to her zest for adventure.

KAY CASE, born Sept. 26, 1949, in Shell Lake, WI. She has a BS in medical technology and is active in research.

In 1989, she and husband Don, took introductory flights in La Crosse, WI, to explore the world of flying.

Kay was hooked and soloed after 11 hours of instruction. Earning her pilot license in May 1990, she has flown Cessnas 152, 172, and Piper Warrior. In July 1991, they purchased a Piper Cherokee 180. Flying across state to Washington Island and ever-changing weather patterns in the Great Lakes area, created the need for flying instruments. She received her instrument rating in July 1992.

Don and Kay's goal is to spend summers at Washington Island and winters in Pensacola Beach, FL, flying back and forth. Jay is a member of the Wisconsin Chapter of the 99s, La Crosse Area Flyers and AOPA.

She has two children, Matthew, 20 and Lindsey, 16.

MARTHA CASE received her first flight lesson as a gift from her future husband on their third date in December 1983. After she used the free lesson, she signed-up for ground school and her flying affair began.

She completed two ground school lessons and was so excited about flying that her future husband asked if she would mind if he signed up for the class, too. As their relationship grew, so did their interest in flying. They started a small flying club and purchased their first airplane, a Cherokee 140.

They were married four months later. She obtained her private license in November 1985, an instrument rating in August 1988 and a commercial rating in October 1994.

They purchased a Mooney 201 in 1991 and have continued to fly faithfully. She became a charter member of the Woman With Wings Chapter in 1994 and has enjoyed the friendship and knowledge of her flying friends ever since. Flying is a challenge to her on a daily basis. Just when you think you know it-it humbles you!

CHERYL CASILLAS, born Aug. 15, 1959, in Fairbanks, AK. Parents are USAF retired TSgt. Gary Kolb and Eleanor has a brother Tracy and sister Diana Todd. She is married to Macario Casillas Jr. and has two sons, Jimmy Kolb (Oct. 29, 1974) and David (May 24, 1985).

Living and working near Randolph AFB she became more intrigued every day as she watched the airplanes perform their maneuvers. She took her first lesson C-150 in February 1993. Cheryl's instructor, David Sheppard, made learning to fly very enjoyable. After many hurried trips out to Stinson Airport, San Antonio, between her school bus routes, her proud moment finally came when she earned her private pilot license on Oct. 31, 1993, in a Piper Cherokee 140.

The following year Cheryl completed her instrument and commercial written tests. Her quest for more information about her new world of flying prompted her to join several pilot organizations. There she met two 99s and was asked to attend one of their meetings, which she did, becoming a member in March 1995. Cheryl took her instrument training at Wright Flyers, San Antonio International, with another 99 as her instructor, Faye Makarsky. On May 6, 1995, she successfully completed her instrument checkride (C-172) with Chief Flight Instructor Fred Barney, also of Wright Flyers. Cheryl is currently working on obtaining her commercial rating. Her ultimate goal is to travel the airways of the sky as a full-time profession. She wants to thank all who have helped and encouraged her flying endeavors.

RAE CAWDELL, born in 1907 and died in 1991. Though she never drove a car, she earned a private pilot license at age 45, giving "fear of heights" as her reason for wanting to learn to fly.

Two months later, she joined the 99s, becoming a life member. She served as her chapter's news reporter and publicity chairman and attended many conventions.

She was co-founder of the Fairladies Annual Indiana Rally, winning several trophies in this event and in the Michigan Small Race. She also flew in the Illinois Air Derby, the Angel Derby and Powder Puff Derbies.

Cawdell was a longtime member of the prestigious Indianapolis Aero Club, twice serving as secretary and twice receiving their Nicholas Trophy as most deserving woman pilot.

Cawdell held a degree in drama and taught in Indiana. She received awards for her contribution to community theater and remained active in productions around Indianapolis.

Her creative writing talents were put to use in her employment in the advertising, newspaper and radio fields. She was also an executive secretary at Lake Central Airlines, which later became Allegheny Airlines and then USAir.

LOIS CHALMERS graduated from Albright Art School and Buffalo Art Institute, NY, in 1940 and went to Miami, where she was employed by the Intercontinental Aircraft Corp. on the production line, building Corsairs. She was also a "ham" operator, license number W4JGG (Jumpin Gravel Gertie).

In September 1942 she enlisted in the Navy, attended boot camp and radio school at the University of Wisconsin in Madison and assigned to the Jacksonville NAS where she spent the remainder of the war. After being discharged in New York City, she worked for the Civil Aviation Administration in the communications section using Morse code, radio telephone and teletypewriter.

After the CAA stint, Lois learned to fly in the Miami area. By 1964 she had acquired her private, commercial and flight instructor ratings for both airplane and instruments. She began instructing and rose to chief instructor at Tursair, Inc., an FAA 141 approved school. A year later she received her airline transport pilot rating for airplane, single and multi-engine land and sea.

FAA appointed her designated pilot examiner for all the above ratings in 1984, working out of the Miami GADO. She was also accident prevention counselor, Miami GADO/FSDO, performing countless hours of safety counseling during briefings, at club meetings, safety seminars, etc. She conducted uncounted numbers of flight checks in categories mentioned above without incident or accident.

Beside her membership in the 99s, she is a member of the Florida Air Pilots Assoc., Florida Aero Club, Miami Aero Club, Florida Women's Pilot Assoc. and is historian number 2 in OX-5. Lois served as an officer at various times in these groups and participated in the Angel Derby.

At age 60, she received her DC-3 type rating on her ATP certificate, which many considered to be quite a feat for a 110-pound lady. After more than 9,000 hours instructing, she is now taking things easy, doing quite a bit of traveling and has again taken up her first love, painting.

Lois is a life member of the Florida Goldcoast Chapter.

PAT CHAN, a native Californian, began skydiving at the Antioch Sport Parachuting Center in 1975 and received the Star Crest Recipient award in 1978. She became a jumpmaster and earned a C license. As a California Parachute Club member, Pat performed at the 1990 Livermore Air Show.

She learned to fly at Antioch in 1979; later earned the instrument and advanced ground instructor ratings and commercial pilot license.

Her 99 membership in the Bay Cities Chapter began in 1984 and included chapter offices. Activities included participating in Right Seat Seminars and airmarking airports.

Participating in the 99s aviation educational exchanges, she traveled to China in 1987; Russia, republic of Georgia; and Finland in 1988. She was a delegate to the 1994 World Aviation Safety and Education Congress in India.

Pat is serving as International and Southwest Section oral historian.

She has a BFA in graphic design from California College of Arts and Crafts and is working as a computer design artist specializing in presentation graphics. Future plans include multimedia and flight instructor training.

ANN L. CHANDLER, born March 11, 1922, in Bridgeport, CT. After a few lessons in the 3-Cub she bought her own airplane (Tri-Pacer) PA-22. She then found out she needed a better flying plane; bought a Citabria-7ECA, which gave her her own commercial rating in 1967. She was able to do airwork and all the wonderful maneuvers which she enjoyed. Ann then took lessons from a WWII pilot in Middlebury, VT (who was then a crop-duster). He taught her to really fly her Citabria-7ECA and to enjoy aerobatic flying. This was a real challenge for her but it gave her the world of enjoyment. Ann also enjoyed flying her husband George's Smith Mini DSA-1 bi-plane.

Ann's hobbies include: FCC licensed amateur radio operator, W10AK, hunting, organic gardening, golfing, painting and being jack-of-all-trades.

LENORA EATON CHANEY, deceased 1990, flew out of a bean patch in Maryland. Obtained her private pilot's certificate in 1944, added 1,600 hours to her log book in the following aircraft: J-3, J-4, J-5, Super Cub, Ercoupe, Aeronca Champ and Chief, PT-17; Stearman 220; Cessnas 120, 140, 150, 172; Taylorcraft; Luscombe 65, 85 and 90 HP; and Stinson 150. She was an aerospace draftsman for Westinghouse, a charter member of the Maryland Chapter, 1963 and a life member of International, 1985. She was also a member of the Maryland Flying Farmers and AOPA. She enjoyed organizing treasure hunts and mystery fly-ins, scavenger hunts and spot-landing contests for the chapter. She flew a solo trip from Maryland to Susanville, CA, in a 65-HP Luscombe (hope she had a soft cushion!)

BARBARA CHAPMAN, commercial, instrument rated pilot, realized a lifelong dream when she learned to fly in a two-place American Yankee in 1971. Besides past ownership of a clipped wing J-3 Cub and Cessna Skyhawk, she has most enjoyed the years since 1974 piloting a Grumman Traveler.

Barbara has been a member of the Florida Goldcoast 99s since the late 70s, forming many close and lasting friendships, including those with several WASPS which the 99s are privileged to have in that chapter. She also belongs to AOPA, American Yankee Assoc. and Florida Grasshoppers.

Her husband, Rollie, a retired real estate appraiser, shares her love of flying. He is a CFI and also enjoys building small race planes and sailboats. They have one daughter, Lynda, an RN, associated with hospice in the Columbus, OH, area, married to a professor in the dental school at Ohio State.

CAROLYN CHARD, Phoenix 99s, soloed at Sky Harbor Airport in 1973 and got her private ASEL in 1974. She joined the Phoenix Chapter 99s in 1975. Carolyn served two terms as chapter chairman and felt very pleased to be in this role when the Sundance Chapter was chartered. She also chaired the Kachina Air Rally in 1983, was airmarking chairman 1977-78, and served on many other committees, as well as, the sectional nominating committee. She flew in several Kachina Air Rally's and treasures winning a "Tail End Turtle."

Carolyn has 1,620 hours, many accrued flying her Cessna to provide outlying Arizona communities with on-site physical therapy. Weather didn't always cooperate and she especially remembers having to go back to Safford when she ran into "cotton balls" at San Carlos. A "white knuckle" recollection is of somehow getting into Durango, CO and tied down in a severe thunderstorm. She taught at several Fly Without Fear clinics.

Carolyn has a grown daughter living in New York, and is working in home health care. She spent a year and half stint in Sierra Leone with the Peace Corps recently.

ANDREA CHAY, born Nov. 2, 1953 in Lakewood, OH. She is married to James "Jim" Chay, who is a commuter airline pilot E120 captain for Brazilia.

Andrea received flight training at Cleveland Hopkins Airport, Cleveland, OH, on 5-K flights. She has flown C-152, C-172, AA5B Tiger, AA1B and C177RG. She joined the 99s in 1981 and has been employed as a FAA flight service station specialist since 1987. She is a member of the Professional Women Controllers. She and her husband have owned a 1973 AA1-B but now own a 1973 C177RG Cardinal.

They were on a cross-country flight from Cleveland to Houston. They rented a T-C210, turbo-Centurion with dual everything: dual horizon gyros, vacuum pumps, alternators and directional gyros, etc. They figured they had it all covered. About 30 minutes out of Cleveland, they noticed failure of first one system, then another, then another, until they were back to needle, ball and airspeed. Good thing the weather was severe clear! To add insult to injury, they had such strong head winds that they may as well have been in a C-172 for all the ground speed they were able to maintain. So, even with dual everything, it can all fail. Moral to the story: know your basics and don't become too dependent on all the fancy gizmos!

MARY BOVEE-CHESNUT, realized a dream held from the age of 8 when she received her private certificate at the age of 39. Born in Sheridan, WY, she grew up in Oregon. Married to Ray Bovee, a former B-17 pilot and a pilot for United Airlines, the family made homes over the years in several states.

Pilot training began for Mary in 1970 when she and son Michael, age 16, took off into the Illinois skies. Mike is currently a captain for American Airlines. Mary's youngest daughter, Robin Ruhwedel, is also a licensed and accomplished pilot. Two daughters, Debra Lanius and Pamela Swanson, chose to navigate on the ground. Eight grandchildren enjoy flying with Grandma.

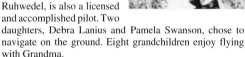

Widowed by a highway accident in 1979 and wanting to keep active in flying, Mary joined the 99s, co-founding the Cameron Park Chapter. Always involved with aviation, she held all chapter offices, was elected to three offices in the Southwest Section and held several committee positions. She served on local aviation boards, including a term as director of Cameron Park Airport District. Very active in Sacramento Valley Pilots Assoc., Mary has served 12 years on their board, two terms as president. In SVPA she met Al, also a pilot, whom she married in 1990.

Racing the Palms to Pines Air Race holds great memories for Mary. She and her co-pilot placed in the top 10 several times, once coming in third. She plans to fly that race again sometime with daughter Robin in her beloved *Charlie,* a 1978 Cardinal Classic. Another thrilling adventure was making a sky dive – that first step was mighty big!

PATRICIA V. "PAT" CHESTER, by profession a licensed counselor, Chester also holds a private and commercial pilot license. She is qualified to teach ground school. She is an advisor to the Civil Air Patrol and a member of both the Experimental Aircraft Assoc. and a lifetime member of the 99s. When the diminutive plane in her garage is finished (she and her husband are building it together) she'll fly it.

And when she flies the small craft, it'll be the 12th airplane that she has checked out to operate.

"Saying that I like to fly really isn't enough," she admits. Her approach is more metaphysical. "You really have to know how to fly to enjoy that different perspective that comes when you first lift off the runway," she says. "It's really a great privilege to get up off the ground."

KATHERINE SUI FUN CHEUNG, was born in Canton, China, and as a young girl practicing on a neighbor's piano, dreamed of studying to become a concert pianist. After immigrating to the US in 1921, she graduated from the Los Angeles Conservatory of Music.

While learning to drive on a dirt road adjacent to an airport, her fascination with "flying machines" took hold and in 1932, she became the first Chinese woman to earn a US pilot license. Aerobatics, "blind flying," numerous competitions and Katherine flew the needed hours for a commercial rating in 1935. She became a 99 in 1934.

Her accomplishments were recognized by the Forest of Friendship in Atchison, KS, in 1992, and in China. They are displayed at the World Aviation History Gallery in Beijing. She is cited in the *History of the Chinese in America, Chinese Women in America* and was recently included in a literacy program magazine designed for third grade school children.

Katherine has two children, two grandchildren and four great-grandsons.

CATHERINE JOAN CHICHESTER, born in Hartford, CT, joined the 99s in 1973. Her flight training is Put, instrument, commercial, CFI, CFII (SEL and MEL), ATP. She has flown the C-150, C-172, C-177RG, BE-76, C-206, PA-28-140. She flew as first officer on the MU-2 and CE-550 and has flown many others owned by flight schools or students.

Pilot employment consisted of flight instructor at Interstate Aviation, Plainville, CT; Copters of New England, Westerly, RI; traffic net pilot in Hartford, CT; first officer on MO-2, CE-550 for Barnes Group at Bradley Airport, Winsor Locks, CT.

She is married to Lyle F. Chichester (also a pilot) and they have four children: Catherine Wedmore, captain at American Airlines; Linda Caballero; Leah Wedmore/Banate; and Christine Black. They have three grandchildren: Tania Caballero, Martin Caballero and Zack Black.

She has flown their C-150 from Connecticut to California and Mexico; their C-172 to Alaska out onto the North Slope; their C-177RG to Baja, Mexico City, Grand Cayman, Jamaica, Trinidad, into Brazil to Boa Vista, Manaus and down the Amazon to Belem, returning through Cayenne Guiana, Tobago, the Windward and Leeward Islands, Virgin Islands, Puerto Rico and the Bahamas back to Connecticut; the C-172 and C-177RG have also taken them to the Bahamas and Newfoundland; the Duchess, C-206 and PA-140 have taken them across the US, down the Keys, Grand Cayman and to AEE and AOPA annual meetings.

They are both retired now (Lyle as a college professor and Chichester as a full-time flight instructor and corporation pilot) and they hope to continue flying for a long time.

BEVERLY "B.J." CHRISTENSEN, dreamed of flying since she was a little girl. In January 1992, she walked into her local airport in Greenville, MI, and said, "I want to learn to fly!" Her husband, Tom, had given her a ground school gift certificate for Christmas, and their two sons were in college. The day after her private pilot checkride, she started instrument training. Flying had become her passion and she was at the airport everyday after school. She joined the 99s Lake Michigan Chapter to meet other women pilots.

By June 1994 she resigned her position as high school business teacher to pursue a career as a professional pilot. Finishing CFI and CFII at AMR Combs, Grand Rapids brought more decisions to be made.

Flight Safety International, Vero Beach, FL, is where B.J. earned her multi-engine commercial instrument and multi-engine instructor ratings by May 1995.

B.J. plans to fly corporate jets internationally. To prepare herself for this challenge she is working on a fourth degree, this one in international business and she speaks Spanish and French and travels extensively. Many people dream big dreams, yet B.J. is actually achieving her dreams, one step at a time.

DORENE M. CHRISTENSEN, born Feb. 3, 1931, in North Dakota. She received her flight training SEL, instrument rating. All training was at Martin Aviation, Orange County, Airport, Santa Ana, CA. She has flown Piper-Cherokee 235, Cherokee-6, Comanche 260, Cessna 182 and 172. She is a graduate of Lomalinda University School of Nursing.

She is a widow and has three children: Janice M. Boswell, Dan Christensen and David Christensen. She has two grandchildren, Alicia M. Boswell, age 12 and Morgan Boswell, age 7.

She flew in Powder Puff Derby three times 1971 and placed seventh with her sister as co-pilot (both first time racers); 1975; 1976 with her daughter, Janice as co-pilot. She is a member of AOPA and EAA.

ELIZABETH DORRANCE CHRISTIAN, a native Californian, learned to fly in 1985. A VFR pilot, she has flown in 16 states, including Alaska. In the fall of 1986, Beth joined the Bay Cities 99s, where she has held many offices including chapter chairman. She served on the board of directors of Alameda Aero Club, is a member of AOPA and EAA. Her most memorable flight was one to Oshkosh in 1994 with Nancy Stock.

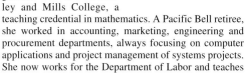

Education includes a BS in business from UCLA, graduate work at UC Berkeley and Mills College, a teaching credential in mathematics. A Pacific Bell retiree, she worked in accounting, marketing, engineering and procurement departments, always focusing on computer applications and project management of systems projects. She now works for the Department of Labor and teaches and tutors mathematics.

Mother of two wonderful sons, Earle and Clifford, she enjoyed five years as a Cub Scout den mother. Current community service focuses on visiting classrooms to talk about aviation and developing a Women in Aviation exhibit for the Western Aerospace Museum.

GAYLE E. CHRISTMAS, born Feb. 28, 1951 in Michigan, was raised in California. She is married to John Kleeman, an MD and a pilot. They have two children, Kellianne age 8 and Scott age 7 months.

She learned to fly in Washington, DC and joined the 99s in 1980. Gayle founded Executive Flyers, Inc., an FBO in Sacramento, CA, while still in her 20s. Has flown Cessnas all over the USA and Mexico. Charter member of Crater Lake Flyers Chapter of 99s.

MARY CHRISTOFFERSON, was born in Idaho and still has the exhilarating memory of that cool winter afternoon in 1979 when her instructor removed himself from the little Cessna 150 and told her it was time to solo. His instructions were to do three touch-and-go landings. Her heart skipped a beat but as she completed these landings successfully, there was a "war whoop" in that little cockpit that should have been heard all over Idaho. She had earned her "wings" and nothing could stop her now. Mary went on to fly larger aircraft and to get a sea plane rating and has been very active in the Idaho Chapter 99s. She has served as chairman and treasurer as well as on numerous committees.

Mary is happily married to her 49 1/2, Chris, and has two sons and three perfect grandchildren. Her advice to all who want to experience this thrill of flying is, "Listen to your heart . . . and never give up."

MARTHA CHRISTY, a Louisiana transplant and 99 lifetime member, was licensed as a private pilot in 1961. In 1976, Martha and husband Ray flew their twin Comanche to Europe. An account of the flight was chronicled in the *99 News Magazine*, December 1976. For added fun, they flew the same plane to Canada and Central America. Martha served in various offices, including chairman of the SHV Chapter and has enjoyed the fellowship of the group since joining in 1961.

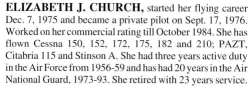

Besides the piloting experience, this member was in the field of education, retiring as principal for the Bossier Parish school system. Teaching experiences started with the couple's two daughters and continues with their three grandchildren, one who is well on his way to becoming a pilot.

The love of aviation lives on, and long live the 99s.

ELIZABETH J. CHURCH, started her flying career Dec. 7, 1975 and became a private pilot on Sept. 17, 1976. Worked on her commercial rating till October 1984. She has flown Cessna 150, 152, 172, 175, 182 and 210; PAZT, Citabria 115 and Stinson A. She had three years active duty in the Air Force from 1956-59 and has had 20 years in the Air National Guard, 1973-93. She retired with 23 years service.

She is married to Augustus "Bill" D. Church. They have five children: Charles Wayne D., 33; Jeannette L. Hughes, 32; Kenneth W., 31; Richard D., 29; and Elizabeth V., 17. They have five grandchildren: Daniel Hughes, 7; David Hughes, 4; Michael Paul Hughes, 3 months; Christopher Church, 6; Jonathan Church, 3; and one on the way.

Has been a member of AOPA, CAP and 99s since 1976. Made her one and only skydive in August 1976. Came away with two sprained ankles and a lot of aches and pains.

The first time she ever flew was when her dad took her to Bowman Field (it was only a cow pasture then) and Dick Molley took them for a ride over Louisville and out over their farm in Indiana. She fell in love with flying that day and always dreamed of flying. It's one of the reasons she joined the Air Force, but at that time they wouldn't let women fly. Elizabeth's husband talked her into trying it out and she ended up using part of her GI Bill going for her commercial rating. She had to give it up because of her working hours. Her last child, Elizabeth, had a lot of flying experience before she was born. Her instructor was always afraid she was gong to have her while doing turns and spins. Elizabeth was so big in front with her child, he had to add extensions to the seat belt, but she loved every minute of it.

PATRICIA CHURCH, born Oct. 28, 1939 in Stockbridge, MA. She is married to David Church and is the mother of sons, Michael and Brian.

Church, a part-time flight instructor, has logged more than 1,800 hours and has been an active member of the Bakersfield Chapter of the 99s since 1974.

She developed and taught a nationally recognized high school aviation program, conducted aerospace workshops for teachers and served in 1990 as an aviation and aerospace education ambassador to the Soviet Union and Poland. She has received numerous awards for her work and was named the National FAA Champion of aviation education for secondary schools in 1992.

In 1992, Church left her teaching position in California to become curriculum coordinator for Starbase, an aviation oriented National Guard program for elementary school children. In 1994 she began her current position in Pensacola, FL as director of Starbase-Atlantis for the US Navy.

JAN CHURCHILL, joined the Easter Pennsylvania Chapter in 1975 and helped with the terminus of the 1976 Powder Puff Derby in which she also raced. In 1976 she was a charter member of the Delaware Chapter and later served as chapter chairman. She has also raced in the Air Race Classic, Garden State 300 and the New England Air Race. Formerly was tax coordinator for the Middle East Section.

Pilot experience: ATP, CFII/ME, DC-3 type. Employed as airline pilot and corporate pilot, as well as, part-time flight instructor. MA from the University of Virginia, paralegal certificate from Widener University, graduate of USAF Squadron Officer School, aviation writer and photographer, articles on military aircraft and helicopters in numerous publications.

She has more than 10,000 hours including time logged in AT-6, U6-A, O-1A (L-19), O-2B, C-47, C141, C5A, F-16 and CH-47. From 1980-85 helped to restore and flew as co-captain on R4D that was 1983 Grand Champion Warbird at Oshkosh (also Best Transport, 1982). Owned and restored to original military configuration O-1 (L-19). Presently flying O-2B.

Author of *On Wings to War* about the women who flew for the US Air Force in WWII. Founder and president of the Delaware Aviation Memorial Foundation; first project is a life-size statue of a WASP erected at the former New Castle Army Air Base (now Wilmington Airport).

President for four years of EAA Chapter 240, Wilmington. Director for the EAA East Coast Fly-in in 1993. Regional editor for *EAA Warbirds Magazine*. Vice-president of EAA Warbird Squadron 8 in 1995 and EAA Warbird 2252 and Eagle Squadron 727.

Aircraft commander for the US Coast Guard Aux. since 1988, doing search and rescue, patrols and transporting admirals. For many years FAA accident prevention counselor and FAA aviation safety counselor.

Member of the Wings Club, the OX5 Aviation Pioneers, the Aviation/Space Writers Assoc., the American Aviation Historical Society, the American Society of Aviation Artist, the International Flying Farmers, Aviation Chapter of the Realtors Land Institute and the 8th Air Force Historical Society.

Currently owns an 0-2 which she flies regularly in air shows. Lectures on women pilots in WWII. Recently spoke at the Smithsonian Air and Space Museum, the National Archives and the University of Delaware WWII Lecture Series.

Interests: tennis, needlepoint, real estate sales, Reno Air Races, licensed A.K.C. dog show judge and author of *The New Labrador Retriever.* (Macmillan Co.)

BARBARA CLARK, was almost 40, when she decided to fly. It was a dream – and an imperative – which matured into sometimes terrifying but always exhilarating journey. Now, she considers flying to be her greatest accomplishment, though the inherent challenge is renewed with each flight. With 400 PIC hours, she rents a Cessna 172 from the FBO which provided her ground, flight and instrument training.

Memorable flights include the tense ones, where "close calls" taught valuable lessons; and the exhilarating ones where pride and joy surpass the capacity of mere words to describe. The Sacramento Valley 99s have been instrumental in keeping her aloft both literally and figuratively since she became a pilot in 1991. It has been her honor to serve at its 1993-95 chapter chairman. "My thanks to all its members, past and present." The contribution the 99s has made to her life is exceeded only by flight itself.

JULIE E. CLARK, was born to fly. Her first solo flight at the age of 19 launched her career in the sky. A pilot for more than 25 years and a captain for Northwest Airlines, Julie has logged more than 20,000 accident free hours in the air and is checked out in more than 65 types of aircraft. Julie is an award-winning pilot of the American Aerobatics, Inc., MOPAR T-34 Mentor for Chrysler Corp. Parts. For the past 15 years, Julie has performed before millions and has earned many aviation awards and honors including the Woman Pilot

of the Year Award from the Southwest Section of the 99s. Clark was recognized by the FAA in Washington, DC with the Certificate of Appreciation and in 1993 was inducted into the Forest of Friendship at Amelia Earhart's residence in Atchison, KS. To her credit, Julie has been featured in numerous books, articles and national television aviation specials and documentaries.

KATHERINE P. "KATHY" HAMBY CLARK,

became interested in flying after she had been a flight attendant for Braniff International for eight years. She told her husband Mike and father-in-law, Frank Clark, that she would like to learn how to fly. Mike knew of a good place where his brother had learned and took her to Mercury Flight Services at Hooks Airport in far northwest Houston. Carmon Neal and Mike Butcher taught her how to fly in a 180 hp Beechcraft Sundown, #66710 and she found actually flying the airplane, instead of walking up and  down the aisles, very exciting. She still remembers her three landing solo cross-country as one of the most exciting things she has every done!

She received her license in July 1991 and joined the 99s in October 1991. Later she learned to fly a complex airplane, a Beech Sierra, from Cheryl Olivier. She has flown Mike and their son, Chris, all over Texas and into Arkansas. Presently, she is the vice-chairman of the Houston 99s, where she helps as a program director and assists the editor in publishing their newsletter. She is also the secretary of Houston Area Pilots for Christ, another group she enjoys very much, and works part-time for her husband at Production Tool. It is fun to soar like an eagle above the earth! She is working on her IFR as time permits.

CAROLYN ELIZABETH CLARKE,

a native of Massachusetts, learned to fly in 1974, while residing in Salt Lake City, UT. Her flight time of more than 5,000 hours includes the Powder Puff Derby, Air Race Classic and Pacific Air Race as well as many hours of giving flight instruction. In 1981 she was the recipient of an Amelia Earhart Memorial Scholarship, which enabled her to get her multi-engine instructor rating and ATP. Since then she has added the single-engine seaplane and helicopter ratings.

Carolyn is a member of the Whirly Girls and the 99s, Inc. She has served as governor of the 99s Southwest Section, and chairman of the International Nominating and the Award of Merit committees.

Carolyn is married to Dana H. Clark, MD They have three daughters and one grandson. In addition to aviation, her interests include: designing and creating knitwear, needlework and figure skating.

CLARA THARPE CLAXTON,

born Oct. 31, 1921 in Ottawa, OH. She is a widow with one child, Peggy Ann Tharpe, 45. She has one grandchild, Aaron Tharpe, 17. She became acquainted with aviation while serving with the Army Air Force from 1942-45. Although not flying at that time, she developed a deep interest in flying. It was not until 1959 that she learned to fly through the Wright-Patterson AFB Aero Club. She flew in five national and International races, served as chairman, vice-chairman and secretary of the All-Ohio Chapter of 99s and was  awarded Women Pilot of the Year award in 1969. She worked and chaired numerous race stops and flew small and proficiency races. She has flown Cherokees, Cessnas, Mooneys, T-34s and Twin Apaches.

She is no longer able to fly due to physical limitations, but still thoroughly enjoys flying with friends and just being around and talking and reminiscing about their flying days.

FRANKIE BEA CLEMENS,

born Sept. 14, 1932 in San Diego, CA. She is married to Raymond A. Clemens. They have four children: Gloria, age 44; Carol, age 41; Russell, age 40; and Andrew, age 30. They have two grandchildren, Christopher, age 8 and Lindsey, age 5.

She joined the 99s in 1979 after attaining a private pilot license. She has flown a C-150, 152, 172, 182; P-150, 151; and P-Tomahawk.

She has flown four Pacific Air Races, one Palms to Pines and has been chapter chairman, vice-chairman and secretary. She was on Pacific Air Race Board eight years and has attended three International meeting and many sectional meetings. She has a plaque in the Forest of Friendship.

WINOMA A. "WINNIE" CLEMENTS,

born Jan. 19, 1932 in Little Rock, AR. She has been married for 40 years to Robert F. Clements and they have two daughters, Ann T. Clemence, age 39 and Jean E. Traviss, age 37. They have three grandchildren: Roy R. Traviss, age 15; Crystal M. Clemence, age 12; and Robert K. Clemence, age 9.

She became an RN at St. John's School of Nursing in May 1953 and has worked at hospital and nursing homes from 1953-80. She worked at Reno, NV VAMC from 1981-89 and  Audie Murphy VAMC from 1989-91.

Winnie loved to fly with her husband Robert, but soon she had the desire to be able to fly herself. She expressed this desire to him and for Christmas that year he gave her a gift of block flying time. They had just moved from Wichita, KS, to Atlanta, GA, and looked together to find a field nearby where she would enjoy flying. She learned to fly at Peach Tree DeKalb Airport, soloed and cross-countried. One cross-country she can remember was PDK to LaGrange to Anniston, AL. In LaGrange three Caribou airplanes were practicing touch-and-goes. They looked so big to her, as she was in a Piper Colt, but land she must to get her log book signed. So she tried her radio on unicom to contact them or LaGrange radio, neither would answer. But she noticed they had plenty of time in between them that she could follow one down and get safely landed. She did this without a problem. Before she received her pilot license, they moved back to Wichita. There she went back to Red Cross blood bank nursing. They traveled in Kansas, Oklahoma and Texas. She wanted again to return to flying. Her husband was working at Cessna, Aircraft Co. The Flying Club there gave a pilot's refresher course. She took the course and did so well, the instructor suggested she go on for her license. She soloed again and practiced much with dual cross-countries and solo cross-countries to receive her license on Nov. 2, 1972.

Her first flight was to take a school teacher friend out for lunch. She loved the trip and started flying lessons too. She and her husband flew to Hutchinson, KS, for dinner many evenings and practiced their touch-and-goes. They went on to land together in 42 out of 50 states. She enjoys the 99s of San Antonio as they are busy with fly-ins, runway markings, Bear Care in schools, safety rules, scholarships, etc.

NANCY CLEVELAND,

began flying in 1990 and now has her instrument and commercial ratings with about 450 hours. She joined the Santa Clara Valley 99s in 1991 and has served as corresponding secretary.

Nancy is a realtor who flies out of the Flying Country Club at Reid-Hillview. Her other interests include traveling, gardening and reading.

BARBARA CLEVER,

born Oct. 24, 1947 in Richmond, CA. She currently resides in Kona, HI, with her husband and four sons.

At a young age she was interested in airplanes but never thought at the time she would ever be able to fly one. The interest would resurface from time to time though, growing up in California. The real urge to fly was to come to fruition much later in life at the age of 43 when Barbara set up a flight lesson on Aug. 14, 1991 at her home airport in Kona in a Cessna 172. She soloed three months later. After much effort at overcoming obstacles, which included at one point no airplane to train in, Barbara was able to once again resume her training. She was helped and encouraged by a very special flight instructor and was also privileged to find and purchase her own Cessna 172 to train in. On April 11,  1994, she passed her checkride in her own airplane. It was a special day.

In October 1994, Barbara joined the Aloha Chapter of the 99s after being invited to attend one of their meetings. She has since had occasion to be involved in several 99s activities including a Compass Rose Fly-in at the Molokai Airport which involved working with other 99s who came from other islands to help.

Currently, Barbara is working on her commercial and multi-engine ratings. Her goal is to continue flying and someday fly commercially.

NANCY CLINTON,

born and raised in Pennsylvania. Nancy learned to fly in 1986 at Zamperini Field, Torrance Municipal Airport and then went on to earn her instructor ratings and airframe and powerplant ratings which enables her to maintain her Gulfstream American Tiger.

Nancy was honored by the 99s in 1991 as a recipient of an Amelia Earhart Memorial Scholarship for multi-engine rating. She joined the Long Beach Chapter in 1986 and she has been chairman  of membership, fly-ins, Flying Companion Seminar and Annual Poker Run. Nancy was vice-chairman and ultimately became chairman of her chapter from 1991-93.

She flew in the 1991 and 1992 Palms to Pines Races, winning Best Third Leg with her race partner, Ann McNeeley, in a Cessna 172 in 1992. The Long Beach Chapter named her Woman Pilot of the Year in 1988-89 and 1990-91. She has logged more than 1,700 hours and is currently working on her MEI and ATP ratings.

Nancy is very civic-minded and has worked hard to bring about positive changes at Zamperini Field. She is the air fair director for the annual Torrance Air Fair, and she has been instrumental in bringing school children, Girl Scouts and underprivileged children to the airport for the day. She is also working with the local EAA Young Eagles program when she is not flying with her clients. Nancy is a proud member of the 99s, AOPA, NAFI, Southern California Aero Club, Del Amo Flyers, AYA Grumman Pilots Assoc. and National Aeronautical Assoc. In 1992 she was named as an accident prevention counselor for the Western Pacific Region, and as such, works with the Long Beach FSDO to present informative FAA seminars. Nancy's current goal is to make females of all ages aware of the thrill of flight.

YVONNE C. COATES,

has been a member of the Alberta Chapter for 10 years and through this association has made lifelong friends. Her flying didn't start until she was in her late 30s because she was terrified of getting on an airplane. She received her pilot license on her 40th birthday and to this day she has 280 hours of very enjoyable flying.

She and her husband own a Piper Cherokee 180, which she now uses for CASARA, their local air search and rescue. She has been with them for a number  of years flying on exercises and a few actual searches.

In 1990 they flew from Calgary down through Montana, Wyoming, Nebraska, Kansas and Oklahoma to Texas and back. Great trip, would like to do a few more. Most of their flying is done in Alberta and Saskatchewan and once in a while over the Rocky Mountains to the Okanagan and Vancouver.

COGGINS FAMILY, Minnie Wade, Minnie Wade Coggins, Christie Coggins Dover and Carla Coggins Gilroy, all of Alabama, represent three generations of membership in the 99s.

Minnie Wade began the family tradition when she joined the organization in 1955, just a few months after the Alabama Chapter was established. She learned to fly from her flight instructor-husband, Charles Wade, at the Clanton, AL, Airport he managed until his death in 1978. Wade soloed in 1940 and soon after earned her private pilot certificate. A member of the 99s for more that 50 years, she served in every officer position in her state chapter and was elected governor of the Southwest Section in 1961. She frequented section meetings and national conventions with her Alabama Chapter friends. She and the group often would fly to her Florida beach house for a few days of bridge playing and relaxation.

Before her death in February 1991, Wade was named a life member of the organization. After her death, the Alabama Chapter established the Minnie Wade Memorial Scholarship in her memory. The scholarship is awarded annually to a female pilot and is used to further her aviation education.

Wade flew in numerous air races and Poker Runs. She hosted 99 chapter meetings at Gragg-Wade Field, the Clanton airport named in honor of her husband.

Wade had six children who became pilots under their father's instruction. In flying circles, the family was known as "the flying Wades," a title that remains appropriate today. Currently, nearly 30 descendants of Wade have soloed on their 16th birthdays. Three of the descendants include Coggins, Dover and Gilroy.

Minnie Coggins, 52, was the fourth child of Wade to take to the skies. She soloed on her 16th birthday, Sept. 3, 1957. The next year on her birthday, she earned her private pilot certificate. In 1960, at age 18, Coggins entered the Powder Puff Derby with fellow 99 and friend, Marie Carastro. From 1961-65, Coggins completed her commercial, CFE and ground instructor certificates, as well as her multi-engine, instrument and CFII ratings. She was the first women in Alabama to earn the certified instrument flight instructor rating. Having logged more that 3,000 flying hours, Coggins has instructed an estimated 200 students. Those students include her husband, Charles Coggins, and two daughters, Christie and Carla.

Coggins was appointed as an FAA accident prevention counselor in 1978. The Alabama 99s named her their Outstanding Alabama 99 in 1991. Earlier in her chapter membership, she was given a silver tray in recognition of her contributions to aviation.

Today, Coggins teaches mathematics at Northview High School in Dothan, AL, where she began an aviation career club for students interested in aviation. She has taken several of her students on their first airplane rides and encouraged them to become aviators. Coggins looks forward to watching her grandchildren: Crimsynn, 5; Courtney, 3; and Forrest, 1; solo in Clanton on their 16th birthdays.

Christie Dover, 29, soloed her parent's Cessna 172 on her 16th birthday, Feb. 24, 1982, in Clanton, where her instructor mother had soloed 25 years earlier. Dover then flew home to Dothan and got her driver's license later that afternoon. At 17, she was awarded a private pilot certificate and became the third generation of Flying Wades to join the 99s.

Dover attended the University of Alabama, when she was a majorette in the Million Dollar Band. She received a BS in education in 1990. She currently teaches third grade at Englewood Elementary School in Tuscaloosa, AL. Like her mother, Dover uses her profession to promote aviation. She is a member of the aerospace committee at Englewood, which sponsors an aerospace week at the school. They have made a model airplane and space shuttle which are on display at Englewood. Dover is married to Jim Dover and they have three children: Crimsynn, Courtney and Forrest. Since Dover took her first airplane ride at just 2 months old, she plans to get her children into the skies early, as well. Crimsynn took flight at 3 weeks old and Courtney at 2 months. Of course, since Forrest is male, he is not allowed to get ahead of the future 99s. He is still waiting for his first flight.

Christie's sister, Carla Gilroy, 26, took her first airplane ride when she was just 3 weeks old in 1968. Her mother and father were the pilots. She must have been hooked from then on, because she logged her first dual flight time at age 8. Gilroy soloed on her 16th birthday, April 12, 1984, at the Clanton airport where her grandfather, grandmother, mother, father and sister had all soloed. She received her student pilot certificate three days before her driver's license, since her birthday fell on a federal holiday. A year later, she followed the family tradition again by earning her private pilot certificate. Still in high school at the time, Gilroy enjoyed taking her friends for rides around their hometown of Dothan – although her friends never told their parents about the flights until they were back on the ground.

Carla received a BA in communication from the University of Alabama in 1990. While at the university, she was a majorette with the Million Dollar Band and a member of Chi Omega sorority. Gilroy also met her husband, Mike, at Alabama. She took him for a ride in her parent's Cessna 172 and, needless to say, Mike is now a pilot. Currently, Gilroy is living in Birmingham, AL, and studying for a doctorate in pharmacy, which she will complete in May 1996. The Alabama Chapter of 99s awarded her the first Minnie Wade Memorial Scholarship in 1994. She is using the scholarship to train for an instrument rating.

ROSCILLE COLBURN, born June 23, 1941 became involved in aviation in 1961 with United Airline as a flight attendant. She joined the Reno Chapter of 99s in 1978. She flies a Cherokee 180 which she and her husband own.

She graduated from University of Nevada in 1988 with a BSN degree and is now employed with Home Health Services of Nevada. She is also a member of the Oasis Hospice. Served on the Ad Hoc Committee, has served as vice-chairman, then chairman. Now she is a volunteer and coordinator.

She is married to Earl Colburn and they have two children, Diann Spencer, 31 and Steven, 30. They reside in Fallon, NV.

DANA L. COLES, born Nov. 6, 1963 in Encino, CA. She is married to Jerry Coles. Coles holds a BBA in finance from Central State University, Edmond, OK.

She joined the 66s in 1992 then joined the Oklahoma Chapter of 99s in 1993. She obtained her private pilot license in 1993. Dana participated in the Wings program, airmarking, flight companion seminars, Okie Derby. She has flown a Cessna 172, 150 and Piper Warrior and has accumulated more than 130 hours of flying time. She was elected in 1995 as treasurer in the Oklahoma Chapter 99s. She was elected in 1994 as chairman for the Ways and Means Committee and the Okie Derby decorations committee chairman and Okie Derby Hospitality Room Volunteer. She is a member of Aircraft Owners and Pilot Assoc.

JESS ANN COLLIER, of Scottish heritage, was raised on a ranch named "Glencoe" near Sturgis, SD. Her love affair with flying began in the 1930s when airports were few and far between. Planes would fly overheard, circling the ranch house, waggling their wings at four small children below, waving hello with dish towels. This little curly-haired Scot asked with disappointment, "Oh, why didn't they land, Daddy?" And one day they did!

That particular wintry day, her mother looked out the kitchen window and exclaimed, "Why, Rod, there's a plane in the field!" Four excited youngsters scrambled for the car, and as it approached the plane they could see it was on skis. They went back and got the pilot two cans of tractor gas and when he tried to pay, her father said, "Oh, no – you have given these kids the thrill of a lifetime!"

The picture and the desire never faded. Some 50 years later, this same girl bought four calves; they grew up and had four calves and she sold them to pay for her private pilot license in 1982. Jess is currently co-owner of one of the most successful accelerated flight training centers in the Southwest, North-Aire, Inc., at Prescott, AZ.

She never won an air race nor flew across an ocean – the only history she ever made was in her heart – she's just a pleasure pilot, but she's a 99. Dreams do come true!

PATRICIA COLLIER, born March 31, 1932, in Killbuck, OH. She has loved everything about flying all her life. She graduated from Grace Downs Airline and Modeling school in New York City and earned her private license in August 1962, followed by obtaining her commercial, instrument and multi-engine rating.

She has one daughter, two stepdaughters and six grandchildren. She has flown a total of 14 air races – Powder Puff, Small race and IWAR. Married to Myron Collier, a professional pilot and author. She acted as his copilot on many trips for the steel company for which he flew for 34 years. She has been a buyer for a department store, substitute secretary for local school system and active in Scouting and Woman's Club. Adopted and raised a grandson, Christopher, who is in the Navy studying to be an air traffic controller.

BARBARA JEAN COLLINS, has a BA degree from Queen's University, Canada. She qualified both as a Canadian and then international cytotechnologist following training at Ottawa Civic Hospital.

Barb became a private pilot in 1975 at Rockcliffe Flying Club, Ottawa. In 1979 she qualified for her commercial rating at Ottawa Flying Club. By 1980 she became a Class III flying instructor, upgraded to Class II in 1980, then received a Class I instrument rating in 1982, and her seaplane endorsement in 1986. Barbara taught flying at Ottawa Flying Club for four years. Three-quarters of her nearly 3,000 hours are from instructing.

For about 15 years, Barb has been a proud member of the Eastern Ontario Chapter of the 99s. The enthusiasm and encouragement of both her 99s friends and especially her husband, Paddy, have enabled Barbara to fulfill her lifelong dream of flying. What a splendid surprise it has been that the reality of flying surpasses the dream!

ELIZABETH COLLINS, born Nov. 17, 1917, worked in the Philadelphia Navy Yard. She was a crossing guard and worked at John Wannamaker's store in Philadelphia, also Schrafft's and Echelon Airfield and Collins Airfield. She has flown J3, PA22, Tri-Pacers, Cessna 140s, 170s and Cherokees. On one of the Amelia Earhart flights, the door of her 140 fell off while landing at Dover Air Force Base.

She has two daughters, Bonnie Lombardi and Luci Roban. She has one stepson, Philip J. Collins and one stepdaughter, JoAnn Sisco, and six grandchildren. After she married Harold Roy Collins (17 years ago), they attended Flying Farmer meetings which were always on the same day as the 99 meetings – so you won't know her – but she is proud to be a 99.

JEAN COLLINS is a charter member of the Santa Clara Valley 99s (charted 1954) after joining the Bay Cities Chapter in 1951. She learned to fly in Coeur d'Alene, ID, in 1945, but had to wait to 1947 (when she turned 17) to get her license. She got her commercial rating in 1951 and flew for a living in the early 1950s when she worked for a Navion dealer at Palo Alto Airport, performing maintenance pickup and delivery. Jean flew the AWTAR in 1949 and 1977.

Jean's interests include: art, raising birds, quilting, silk screening, upholstering furniture, skin diving and water skiing. She and her husband, Dick, are active in the International Swift Assoc., as they have a 1947 Swift they rebuilt which they have flown all across the country (including Alaska) and Canada. She and Dick also built their house where they in live in Portola Valley, CA.

MARY ROSS COLLIS, originally of Phoenix, AZ, learned to fly with "the big boys" at McChord AFB, AK, in 1986, learning that the expression "hold for wake turbulence" has special meaning when applied to C-141s and AWACS. Collis believes that it was magical using Mount McKinley as a reference point for 360 degree turns and flying above whales playing in the bay near Anchorage.

Her most memorable flying experience occurred in Alaska during a cross-country flight. Collis, flying at 3,000 feet, AGL, spotted a bald eagle flying surprisingly near to her aircraft's left wing tip. She clocked flying with the eagle for two minutes.

Collis has flown Cessna 152, 172 and the Piper Archer and Warrior. In Alaska, she flew a radio station's traffic report a few times.

An educator, Collis, earned her doctorate degree from Nova Southeastern University in 1996. She is married to Jerry Collis.

LOIS DIANE COLLUM, born on Nov. 11, 1963, in Bakersfield, CA. She was 20 years old when she got her pilot license. She was inspired by a special friend of her mother's that was involved with airplanes in the military. She received her pilot training from Janice Brown, a 99 also, who flew the Solar Challenger. Since getting her single-engine land pilot license, she has flown a Cessna 150 and 172. Collum joined the 99s in 1984. She got married to a 99s son, John, and they now own a Cessna 172.

ROSEMARY COLMAN joined the 99s in May 1974. She was vice-governor, Australian Section 1977-78; co-chairman International Convention, 1978; Hon secretary, treasurer and chapter reporter, 1974-76.

Rosemary started flying as a diversion from her high school music teaching, but her interests merged, resulting in a diverse career. Aviation-related achievements: 1991 – president of the Australian Institute of Navigation; October 1991 – chaired a session at the International Institutes of Navigation Congress, Cairo; 1990 – invited to address the Australasian Trade Fair on Training for the Future; 1990-91 – initiated and lectured in courses in flight instruction methods at the University of New South Wales; this was a first in the eastern states of Australia; 1994 – inducted into Rotary International; 1980 – elected Freeman of the Guild of Air Pilots and Air Navigators; 1990 – elected General Committee member, Australian Branch; 1976-77 – state president of the Australian Women Pilots Assoc.; 1977-79 – elected a director of the Royal Aero Club of New South Wales; 1978-92 elected councilor of the Australian Institute of Navigation.

Flying record: first solo June 1969, with Canberra Aero Club; unrestricted license 1970; night NMC 1975, with the Royal Aero Club of New South Wales, Bankstown, Sydney; Class III instrument rating March 1977; multi-engine rating, 1976. Rosemary won the Sir Donald Anderson Trophy in 1972 for the most meritorious progress by a woman pilot towards achievement of professional qualifications. She and the late 99 Margaret Kentley won the RACNSW Air Race in 1974.

BONITA G. COLONY, born Oct. 11, 1933 in Platteville, WI. She received her flight instruction at McNairy Field in Salem, OR. She has flown the Grumman Cheetah, Skyhawk 172 and Beechcraft Debonaire Her first three hours of flight instruction, in 1990, were a gift from Elaine, her daughter, who is a pilot. It gave her the opportunity to pursue a life-long dream of flying. She earned her license on June 7, 1991. She is a retired educator with a MA in education from the University of Alaska. She is also a private educational consultant training teachers in multi-sensory methods of teaching dyslexic students. She has trained teachers in the US and in Australia and will be training in Manila, Philippines.

Colony was married to Lee (deceased) and had three children: Elaine Crawley, also a pilot and 99s member; Wayne; and Gail Steege. She has one grandchild, Miranda Steege, age 6.

PATRICIA ANNE COMPTON, born May 30, 1956 in Endicott, NY. As a youngster in her hometown of Endicott, she would watch the airplanes take off and land at Broome County Airport. She realized her dream to fly in 1984 at the age of 27. While working as a stenographer in Miami, Pat simultaneously completed her BA at Barry University and earned her commercial pilot certificate, with instrument and multi-engine ratings. Support from the 99s came when she received the Florida Goldcoast Chapter Griner Scholarship. Pat also holds commercial glider and seaplane certificates.

In 1989, she married her flying mentor, Burt Compton. An accomplished glider pilot, he served as an inspiration for Pat's favorite flying – soaring. Together they share their love of sailplanes, classic aircraft and aviation history.

She was elected chairman of the Florida Goldcoast Chapter for two terms (1991-93). Pat was a recipient of a 99s Amelia Earhart Scholarship in 1995.

MARGARET "MIKE" CONLIN, a native of Kansas, earned a private license in 1967. She has co-owned with her husband, Mike, a Cessna 175, 172 and a Cessna Skylane, N22Mike. She holds a commercial license with an instrument rating.

Conlin joined the Michigan 99s in 1967 and flew in their small race on three occasions. She has held vice-chairman, secretary and treasurer offices with the Dallas Redbird Chapter and served as aerospace education chairman for South Central Section.

Conlin's early career began as an airline radio operator at LAX. She currently works with her husband in their aviation business consulting company. They have enjoyed extensive travel in the US and abroad. She has two daughters, one grandson and three granddaughters.

LINDA CONNELL learned to fly at Palo Alto Airport in 1982. She has her private ASEL license with about 170 hours logged.

Linda works for the Aviation Safety Office at NASA and is very interested in rotorcraft safety and emergency medical services. Toward that end, she is interested in getting her rotorcraft license. Linda joined the SCV 99s in 1992 and has been active in airmarking and promoting women in government aviation. She currently is flying a C-172 at Palo Alto and has two sons, one of whom is also a pilot.

AMY C. CONNER, born June 12, 1917 in Honolulu, HI. She married Jack Conner in 1941. They have three children: Kathryn, 45; George, 42; and Douglas, 40. She has four grandchildren: Alex, 7; Marianne, 7; Valerie, 5; and Laura, 6.

She received flight training and got private and instrument rating with more than 2,100 hours of flight time. She has flown C-150, Aerobat, Tri Pacer, C-172, 205 and 182. She joined the 99s in 1993. Decided to become a pilot when her husband, a WWII pilot, rebuilt a Tri Pacer. They flew it around the country, changed to a C-205. They flew that to Alaska, Guadalajara, coast to coast many times and various routes. She did the flying. Her husband lost his medical insurance after open heart surgery.

GAYLE CONNERS, born March 1, 1962 in Boston, MA. She received her commercial helicopter and fixed wing ratings, and IFR helicopter and glider. She has flown a Bell 47, Bell 206, Hughes 500, AS-350, B-19, B-23, C-150, C-172, Scout, L-19, S2-33. Pilot employment consists of a helicopter pilot and tow pilot.

She is married to Stephen Ryan and they have one child, Sean Ryan, 11 years old. She has been employed as an aircraft accident investigator for TSB of Canada. Conners was part of the Royal Canadian Air Cadets for 10 years. Flew helicopters in Northern Quebec, taught ground school and towed gliders. She also has a BAC in psychology. Loves to travel.

AUDREY MELENYK COOK always wanted to learn to fly since she was a young girl. Her desire led her to an MA in aerospace engineering and a pilot license. After spending many years as an engineer with jet engine companies, and flight instructing on the side, she decided to pursue a career in flying. She now flies a corporate charter King Air 200 and has ambitions to fly jets. Audrey is active in the 99s and the FAA Aviation Safety Program as a safety counselor. She enjoys setting up safety seminars and giving tours and talks to young children about aviation. She holds an ATP, single and multi-engine and a CFII single and multi-engine, and seaplane. Audrey is married to a professional pilot and has two children whom she hopes will someday fly also.

ANN L. COOPER, author of *Rising Above It* and *On The Wing*, co-authored *Tuskegee's Heroes* with her husband. A CFIAI and aviation writer, Ann published more than 600 magazine articles. She edits *Aero Brush* for the American Society of Aviation Artists and *IWASM Quarterly* for the International Women's Air and Space Museum.

Married to Charles Cooper, major general USAF (retired), former commander of the New York Air National Guard and executive director, Bell Communications Research, Ann and Charlie live in Beavercreek, OH. Ann has three grown children and is step-mother to four. She and Charlie enjoy seven grandchildren – soon an eighth — and their Cessna.

Ann is working on biographies of Nancy Hopkins Tier and three-time national aerobatic champion Patty Wagstaff. Anticipating three more equally outstanding books: the stories of Emily Warner, Dot Swain Lewis and artist Gil Cohen, Ann said, "Committing these remarkable stories to history is important – and an honor!"

NADINE MARCELLE COOPER, born Feb. 25, 1927 in Ponteix, Sask. Canada. She learned in 1967 at Regina Flying Club and received a private license with night end. Joined the 99s in 1971 when Saskatchewan formed a chapter. Ruby McDonald West, Canada governor, was their mentor, followed by Eleanor Bailey. Soon she was chairman of a brand new chapter. She served every level and every capacity of the chapter, also of the West Canada Section. In 1986, her own chapter collaborated against her daughter and her until she resigned and became a section member only. The Saskatchewan Chapter was dissolved in 1990, and the people who caused the problem are no longer members.

Memorable experiences: flying to Philadelphia for International convention in 1976 the year she was West Canada governor. The year of the Legionnaires disease and they were booked into the Benjamin Franklin Hotel, where the disease originated! They had their two children with them. Their daughter was 14 then, later took her pilot license at age 18. Traveling with their Arrow 200 to Whitehorse (Yukon) and Dawson City with friend flying a Cessna 172. A great trip. Flying with their Cherokee 180 to Vancouver and returning via Washington State, Montana, etc. Flying to Stratford, Ontario for a weekend 1,500 miles each way.

She is married to William A. "Tony" and they have four children: Zora Tanner, 40; David Cooper, 38; Melody Jackson, 33; and Brian Cooper, 30. They also have eight grandchildren: Marnie, 15; Jocelyn, 13; Laura, 7; Eric, 8; Nicole, 4; Kylee, 5 months; Tracy 8; and Reis, 5.

CONSTANCE COPELAND, received her private pilot license Dec. 14, 1964, for airplane single-engine land.

Constance joined the All-Ohio Chapter of the 99s, and when the Scioto Valley Chapter was formed she joined as a charter member. She has held the office of chapter treasurer.

Connie and her husband Ray are owners of a Cessna 340, which they use for extensive travel.

MARILYN FRANCIS HALL COPELAND, born Dec. 26, 1931, in Rich Hall, MO. She joined the 99s in 1962. She graduated salutatorian with a BS in education from Central Missouri State University and taught high school for four years. Flight experience included SEL with instrument rating, 3,800 hours to date in all 48 states, Mexico, Canada and the Caribbean. Twelve transcontinental air races, including 10 Powder Puff Derby Races (1976 flew solo in Piper Lance), Air Race Classic, Angel Derby and Shangri-La Grand Prix. John, her husband, was her avid supporter in races and all 99 activities. She married John Wheatley Copeland on June 13, 1953. He was her first flight instructor and flying booster for 40 years before his death from cancer in 1994.

In 1976, Marilyn and John purchased historic Rawdon Airfield in Wichita, which was closed and abandoned. They opened with FAA, built new runway, taxiway, buried power lines, constructed a new hangar and administration facilities and operated it as a private airport open to the public until Raytheon Beechcraft purchased in 1982 and renamed it Beech North. Marilyn was airport manager and John was facilities supervisor and president of Copeland Aviation. Both were aircraft dealers for Rockwell 114 and Grumman American Aircraft.

Copeland's 99 activities included Kansas Chapter offices and chairmanships including chapter chairman. Has created several musical skits at local, section and International levels with flying themes. Served South Central Section in offices and chairmanships, including governor. Co-chairman for two section meetings. Awarded Jimmy Kolp Award for outstanding work in section.

She was 99s International Headquarters chairman several years, including fundraising chairman for original 1975 International HQ Building and the 1988 two-story International HQ Building. Served on research committee for 99s 501-C-3 application. Organized first International Careers Seminar and original Career Bank at HQ. International board member and 99s president, 1982-84. Attended Sally K. Ride launch as VIP guest of NASA. Presently serving six-year term as Amelia Earhart Birthplace Museum board of trustees chairman for restoration. Due to her organizational efforts and support of 99s and Atchison, a $75,000 Kansas Heritage Trust Grant was approved.

International board member and speaker at two 99s World Aviation Education Safety Congress meetings in India. Board member of World Aviation Education Organization.

Kansas Aviation Museum activities included founding board member; founded Wichita Wright Brothers Celebrations, chairing several events. Has served on board of trustees since 1976 and served as president, assisting in securing 501-C-3. Awarded Kansas Aviation Honor Award of Kansas Aviation Hall of Fame in 1987. Honored in International Forest of Friendship.

Marilyn was active in dental auxiliary for 40 years. Served in offices and chairmanships, including president of Wichita District Dental Auxiliary, Kansas State Dental Auxiliary and president of the Auxiliary to the American Dental Assoc. Her national campaign re: baby bottle nursing decay has benefited many families. She appeared briefly on "Good Morning America" for both dental auxiliary and again for 99s. She was office manager in Dr. John W. Copeland's pediatric dental office nearly 40 years.

Present family members include daughter, Dr. Jo Hansen, an orthodontist; sons, David, airline-rated pilot and International marketing director of Caravan for Cessna Aircraft Company in Central and South America; Marc Hansen, Kelli Copeland and three grandsons who really like airplanes.

JACLIN CORDES, joined the 99s in 1992 shortly after learning to fly at San Jose Airport. She has her private license and flies with the Flying Twenties flying club. She works as a public relations officer for the San Jose Police Department and has helped the Santa Clara Valley with publicity for many of their activities.

SUSAN CORNELL, born Nov. 26, 1960, at San Luis Obispo, CA. She received her commercial MEL, instrument rating, CFI and CFII. Cornell is employed at Sierra Academy of Aeronautics. She has flown C-152, 172, 182, 310; BE-76, BE-18, Taylorcraft and Stearman.

She is a member of the 99s and Aircraft Owners and Pilots Assoc. She also attended Embry-Riddle Aeronautical University.

MARGARET COSBY, was born a New Yorker but is a Southerner by choice. Her first airplane lessons were in a WWII surplus PT-19 at Zahn's Airport, Amityville, NY. About the time she was to solo, someone totaled the aircraft. That, career, life-style and many moves put her flying on hold.

It was in Louisiana, when her son became a pilot and CFII-MEII before his 19th birthday, that he signed her off and she received her ASEL at age 55. She has flown Pipers, Taylorcraft, T-41, Cessnas and owns 199IQ – a C-177RG.

As soon as she located the 99s in 1980, Margaret joined the San Antonio Chapter and has held every office and most chairs. She plays a significant role in chapter flying activities and has judged USPFT Regionals. She served on the South Central Section advisory board as well as her community aviation advisory board.

Margaret has always been an office holder in civic and political organizations in the seven states in which she has resided. She is a precinct chairman and election judge. Her vocation was public relations but she now is a substitute and GED teacher.

Margaret is a major in the Civil Air Patrol and is qualified in all areas of their flying program. For three years, she was honored as the Texas Wing Public Affairs Officer of the Year. Although she finds her work with the CAP rewarding, her heart is always with the 99s.

CAROL COSNER, born June 4, 1948, in Barnesville, OH. She learned to fly at Big Beaver Airport, Troy, MI and earned a CFIAI. She has flown a Cessna 152, 172, 182 and a Cutlass. She joined the 99s in 1985. Cosner has been employed part-time for Mahan Flying Services at Oakland Troy Airport, Troy, MI. She was an Amelia Earhart Scholarship winner in 1988.

Her 99 activities include treasurer for two years, secretary and newsletter editor for two years. She has been involved with Air Bear presentations, Buckeye Air Rallies, Career Day presentations and Michigan Small Races.

She is married to Robert, also a pilot. He is a retired Ford employee. She has two children, Jeffrey, 21 years, and Adam, 13 years, Landefeld; and three stepchildren.

BETTY COSTA, born Oct. 14, 1931, in Los Angeles, CA. She earned a private/ASEL on May 11, 1992 and Wings I and II. She has flown a Cherokee 180. Costa decided to get a license after her husband, who had flown since he was a teenager, finally obtained his private/ASEL in December 1989. She joined the 99s in 1992.

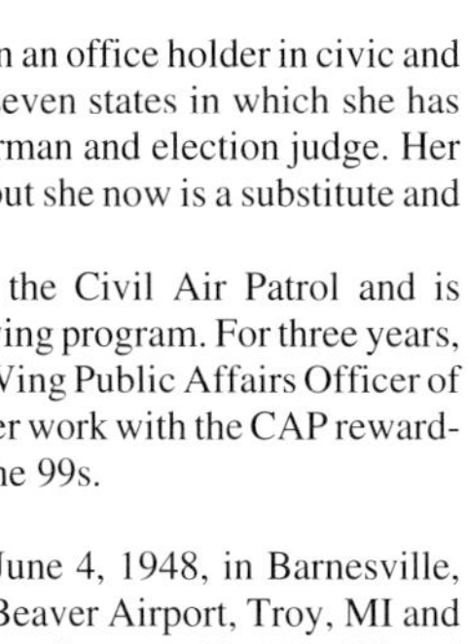

Costa is married to Lawrence and they have three daughters: Marie, who is a 66; Beverly and Sharon. She belongs to the High Country Chapter of Western Colorado.

LISA ANNE COTHAM learned to fly basically "because it was there." Her father bought a 1947 Luscombe, later swapping it for a Cessna 150. While attending Nicholls State University, she enrolled in ground school, organized the NSU NIFA Chapter and earned her private certificate in 1978.

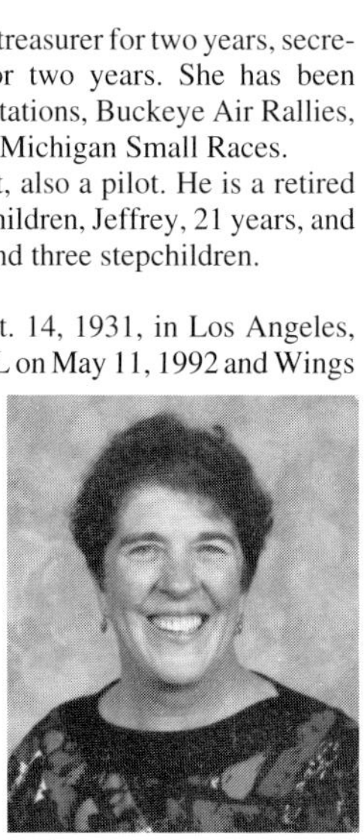

After a six year hiatus, Lisa, a new BFR in hand, attended a meeting of the South Louisiana Chapter of the 99s. Consumption is the only way to describe her involvement. A charter member of the Aviation Assoc. of Louisiana, she served as district director before being elected treasurer. She worked with the US National Hot Air Balloon Championships in 1989 and 1990, training new observers in 1990. She was chairman of South Louisiana and New Orleans Chapters and is on the 1995-1997 Southeast Section nominating committee.

In 1991, Lisa returned to flight training, going from private to CFI in four and a half months. She was a dispatcher for a commuter airline and freelance instructor before becoming administrative manager of the New Orleans Navy Flying Club. She is a mission pilot and check pilot in Civil Air Patrol and an FAA aviation safety counselor.

Lisa was a 1992 recipient of an Amelia Earhart Scholarship for a multi-engine instructor rating.

In 1994, a dream job materialized when Lisa was selected as curator of the Wedell-Williams Memorial Aviation Museum of Louisiana. Now employed by the Louisiana State Museum, Lisa has been promoted to branch director. Her job is to develop exhibits that spotlight the accomplishments of Louisiana's people and industry.

GENEVIEVE VATTEROTT COUGHLIN, born Nov. 4, 1953, in St. Louis, MO, received a BS in transportation, tourism and travel from Parks Aeronautical Institute in St. Louis in 1976. She received a private pilot license in 1976 and an instrument rating in 1980. She has flown C-150s, B-A33, Commander 112As, Grumman AA-1A Yankee, AA-5 Grumman Trainer; G.A. Tiger, AA-5 Cheetah, PA28-14D, PA28-151, Citabria, C-172, PA38-112, PA-32R-300/A, PA28-180, M20J, ATL-610J, PA24-260, M20J, PA-28-161. Coughlin joined the 99s in 1979.

Coughlin is married to James and they have four children: Courtney 13, Bridgette 11, Caitlin 7 and James 4. Employment includes: United Airlines in international sales since 1994; owner of a gift service and party planning-festival business since 1989; Flying Tigers Airlines 1978-85; Logistics-traffic agent; Capitol International Airways in charter sales 1976-77; Spirit of St. Louis Airport, Missouri, Hangar West general sales, front desk, flight demo's and sales.

She has early flying memories of flying up and above the bluffs flanking the Missouri and Mississippi Rivers and watching eagles and hawks airlifted on drafts, catching the setting sunset several times as it dipped into the horizon and knowing when to go back at the legal gas reserve.

Air race history: WOW Derby-Keokus Race, August 1980, youngest pilot; 99 chapter Air Meet, Chicago, August 1980, first place spot landing, third place in proficiency race; 99 Whiskey Run Air Race, October 1980, just fun!; 99 Air Race, May 1981, placed 25th out of 60; 99 Sectional Air Meet, October 1981, placed "second," which means the winner – first plans the next meet!; 99 Air Derby, May 1983, just fun and tornadoes; 99 Air Derby, May 1988, ditto!; Air Bears, 1988 until 1994, worked in the Chicago area schools in primary and preschool instruction.

KIM B. COUGHLIN, born Sept. 16, 1957, in New Haven, CT, joined the 99s in 1991. She has flown Cessna 150s and 172s. Flying runs in her family. Her two cousins flew F-16s; one pilots for Northwest and other flies for the National Guard. Her father, a commercial pilot who taught fighters in WWII, gave her the flying gift in 1983. Coughlin's father and she own and fly a Cessna 172 which they use to "tool around" and explore New England airports. Her husband, Tom, and she enjoys it to island hop. With her private rating, she has about 500 hours. A lifetime Connecticut resident she works in human resources for a Hartford-based insurance company. She has an undergraduate degree from Fairfield University and earned a MA the year she soloed. Her goal it to follow in her father's footsteps and pass this gift on to the next generation.

CAROL COUNTISS, born Sept. 6, 1945, in Grand Rapids, MN, began her training in a Piper Tomahawk. Since then, she has flown exclusively in a Grumman Tiger and Travel Air. She is strictly a pilot for the pleasure of it. Countiss and her husband, Chad, have two children, Deborah an anchorperson for ABC local news, and Spencer, a physician in Grants Pass, OR. They have one grandchild William who is 3 years old.

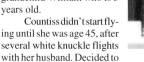

Countiss didn't start flying until she was age 45, after several white knuckle flights with her husband. Decided to take a "pinch hitter," that clinched it. She was hooked. She is a member of Scioto Valley of the 99s. Has logged 450

hours to date. Works for American Horse Shows, Inc. She also enjoys skiing.

ELLEN COUSSENS, born March 26, 1939, in Sydney, MT, intended to just to take a couple of flying lessons to conquer her fear of flying. Six months later she has a private license, a Grumman Tiger and joined the 99s in 1989. Elle loves the camaraderie of 99s and especially enjoys stories of women who had the courage to fly before she did. She has served in all chapter offices, including chairman. She frequently talks to community groups and schools, always encouraging others to fly.

Elle became interested in air racing, believing that "friendly" competition creates striving for excellence. In 1994 she entered her first Palms to Pines and placed third with race partner Susan Larson.

Her husband pilots an Aerostar, and every year they fly to EAA/ Oshkosh together. She is a member of AOPA, EAA and American Yankee Assoc.

The Coussens live and fly out of Rosamond Skypark, CA. Elle has three children and three stepchildren. They work together in nearby Lancaster as physician and registered nurse team in private practice of family medicine.

EVELYN SEDIVY COWING, Nebraska Chapter, active in aviation since 1960, holds a commercial rating with instrument, multi-engine, flight instructor and basic ground instructor ratings. She is former chairman of the Nebraska Chapter, served as South Central Sectional aerospace education chairman, received the 1970 AE Memorial Scholarship and participated in three Powder Puff derbies. She co-founded the Nebraska Assoc. of Aerospace Educators and is currently serving her third years as president.

Evelyn holds a Ed.M. in aerospace education and has been actively involved in numerous aerospace teacher workshops and teaching aviation ground school.

BARBARA C. COX was born in Pittsburgh, PA, on Oct. 3, 1940, married in 1958 to Norman, had four children and then decided to fulfill her childhood dreams, after several inspiring flights in her brothers' Cessna 172, especially after he sent her a logbook, with the first entry being her flight consisting of a take-off, flight, then the landing! She enrolled in the flight program at Community College of Beaver County the summer of 1991, first solo was Feb. 3, 1992. After passing the private pilot check ride on June 16, 1992, she joined the Condor Aeroclub, met other women pilots, then joined the 99s! Since that time, she has flown her training plane, a Cessna 152, then on to Cessna 172, then trained in a low-wing Warrior. The most rewarding experience is all her flight time in a 1936 Beechcraft C17B Staggerwing, which recently was consummated by winning three awards at the Buckeye Air Rally, June 9-11, 1995, for Oldest Aircraft, Best Female Team and Best-Scoring 99 Team! (Christine St. Onge is owner/pilot and Barbara Cox, co-pilot/ navigator)

She also had an hour flight in a T6! Not bad for just four years of flying. She is actively involved in the Wings Program, currently working for phase 5. "Wonder what the next 50 years will bring?" Cox has four children: Kimberly Young, Nov. 3, 1959; Mark Alan Cox, Feb. 26, 1961; Lisa G. Cross, April 1, 1962; and Brian J. Cox, July 26, 1972. She has seven grandchildren: Louis Young, 8; Chelsy Young, 7; William Cox, 15; Andrew Cox, 7; Alex Lizzi, 12; David Lizzi, 9; and Madalyn N. Cross, 6 months.

DEBORAH L. COX, born Aug. 19, 1949 in Passaic, NJ; ASEL, owns a Piper Cherokee 140 and flies out Greenwood Lake Airport, NJ. Debbie is a self-employed private duty nurse with a BSN degree from Dominican College. While in college, Debbie was awarded the Outstanding Athlete of America award. Staying active in sports, she trained for Olympic luge, has been on the national ski patrol and has many trophies for Hobie Cat sailboat racing.

Debbie wanted to fly since age 8 when she went up with her Uncle Walker; she accomplished her dream in 1987 when she obtained her pilot license. In June 1994, she and her nephew, Eric, also a pilot, flew to Montana for a family reunion. Cox is also a member of AOPA and International Women's Air and Space Museum. In June 1995, Debbie flew fly her plane in the Air Race Classic.

DIANE COZZI got her pilot license in 1971 and has been a dedicated 99 for 25 years, currently serving on the board of directors as International treasurer. She has also served as Chicago Area Chapter chairman, North Central Section governor, and on numerous committees at all levels, being particularly active in aerospace education and membership.

Diane is instrument rated, a licensed ground school instructor, and "hands-on" amateur A&P. She is a major in Civil Air Patrol and former squadron commander, director of aerospace education and public affairs officer for Illinois Wing. In addition, she is a director for the World Aerospace Education Organization headquartered in Cairo, Egypt.

She has attended virtually all International and section conventions since becoming a member and enjoys meeting 99s the world over. (Diane is pictured with her Cessna 172, 5411-*Honey*.)

JOAN COOMER CRAFT, began flying at an early age with her father and fell in love with airplanes. She still loves the smell of the old cockpit and the sound of those radial engines. Her dream to become a licensed pilot was not realized until later on in life after her family was raised and time and money allowed. Born July 14, 1937, some in family speculate the "lost" Spirit of Amelia Earhart found its way to Lexington, KY, and was reborn in her. She has flown C-172s, Cutless R.G., DC-3 (right seat) and Beech 18 (right seat).

Educational background: registered nurse, presently working in psychology. Active in community and church activities including teaching, leadership and committee work to preserve and protect their heritage, land and history.

Married to Gerald A. Craft, and has two children and one grandson. Her son, Daniel, is a police officer and captain Army Aviation Reserves, pilot OH-58 Scout helicopter. Daughter, Lisa Ann Miller, works in human resource development. Her husband, also a captain, flew AH-64 Apache helicopters during Army aviation career. Grandson, Travis Miller, age 5, is already an aviation enthusiast. Craft is presently a member of All Ohio Chapter, 99s and AOPA. Hobbies include: flying (anything, anytime), painting and sculpture.

JOY CRAIGHEAD, born July 1, 1938, in Campbellford, Ontario, Canada, joined the 99s in 1993. In the early 70s, Joy volunteered with a Canadian non-governmental organization (CUSO) as a health educator in Zambia, Africa. Given her eight years of skydiving in Canada and no sport parachuting in Zambia, Joy transferred her love of the sky to flying.

With her Zambian pilot license, she explored parts of Africa which were, for the most part, inaccessible by road. As well, Joy flew with a team in a single-engine Cherokee from Lusaka, Zambia, to Nairobi, Kenya, to enter the East Africa Air Safari – a unique opportunity to look up at the peak of Mount Kilimanjaro, and to look down at the plains filled with animals. The two days of flying, guided by instructions received in envelopes at the take-off threshold, sent them to the sparsely populated interior of Kenya, along the beautiful coastline highly populated with vacationers, and finally, to the idyllic island of Lamu.

On her return to Canada, Joy obtained her Canadian private license and instrument rating, and as a recreation pilot, she continues to explore parts of Canada and the US. She is presently raising funds for the purchase of an aircraft for use by a hospital situated in the foothills of the Himalayas in India. The next chapter in Joy's flying life may well be with an Indian license.

KAYE CRAIG received her private license in 1984, and became a 99 the following year. Her primary flight instruction was at Tucson's Ryan Field, in a 1952 Tri-Pacer. She was taught to fly by longtime 99 Mary Mercker. Craig earned her instrument rating in 1990, in the Cherokee 180 she and her husband, Kayl, now own. She also holds advanced and instrument ground instructor ratings.

She has served eight years as secretary of the Tucson Chapter. As aerospace education chairman, Craig organized a program for public schools in which chapter members made presentations on flying to students in grades three to eight. The Tucson Chapter received a recognition award from the FAA for this program.

A teacher for 24 years, she holds a BA from the University of Florida and a MA from the University of Arizona. She is a member of AOPA and Arizona Pilots Assoc. She enjoys writing, and being Nana to three step-daughters and their families.

MARDO CRANE helped organize the first All Woman Transcontinental Air Race (AWTAR) in 1947, and she was the chairman of the event until 1952. Mardo began flying in 1933 at Mines Field in Los Angeles (now LAX), got her license in 1940 and was a WASP during WWII. Her first instructor was Amelia Earhart.

She holds a commercial certificate with single and multi-engine ratings. She became a 99 in 1945 and was chairman of the Los Angeles Chapter, was editor of *The 99 News* and was an early member of AOPA. Mardo authored *Fly Down of the WASP* and *Ladies: Rev Up Your Engines,* a novel about the Powder Puff Derby. Mardo was the editor of the *Aviation News Beacon* from 1945-50. She was a member of the Aviation and Space Writers Assoc., the Author's Guild and Women in Communication. In 1979, Mardo was awarded the Barnstormer Trophy of the Antelope Valley Aero Museum.

PAMELA HOFFMAN CRANE, a native of St. Louis, MO, began flying lessons in 1960 at Weiss Airport near St. Louis, in a Piper PA-11 and soloed that year. Her lessons were intermittent for several years but following her marriage to Col. Carl J. Crane, USAF, retired in 1965 and a move to San Antonio, TX, Pam resumed flying in a Piper Super-Cruiser and later a Cessna Skyhawk which were owned by the couple.

Pam completed her formal instruction and received her license (SEL) at San Antonio International Airport in 1967. She had attended 99 meetings in St. Louis and joined the San Antonio Chapter in 1969. Most of Pam's flying after 1965 was for the purpose of assisting her husband in the development and demonstration of equipment for improvement of instrument flying safety. Carl Crane was a pioneer of instrument flying and continued to work in this field until this death in 1982.

In addition to the aircraft mentioned, Pam has flown the Luscombe Silvaire, Piper Colt, Piper Tri-Pacer, Piper Cherokee, Cessna 150, Cessna 182, Beech T-34, Grumman Traveler and Geronimo. A highlight of her flying career was an opportunity to make a flight and log stick time in the right seat of American Airlines Tri-Motor Ford in 1968.

Pam was employed by Department of Defense for 26 years as a cartographer and computer specialist and received

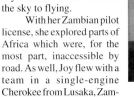

the USAF Decoration for Exceptional Civilian Service in 1961. Pam has five stepchildren and 21 grandchildren (including three pilots) and 28 great-grandchildren.

PHYLLIS FLEET NELSON CRARY, born April 24, 1909, in Montesano, WA, is a charter member, holding license # 8097, dated Sept. 1, 1929. Her first flight was in August 1927 with Barney Oldfield at Brook Field, San Antonio, TX. She learned to fly at Consolidated Airport and was taught to "fly the pants off the ship – don't let it fly you" and " remember – you can't be too careful." Her father, Maj. R.H. Fleet, established Consolidated Aircraft Corp. and designed and built the open cockpit Fleet Aircraft. Naturally, this is the plane that she "flew the pants off of." Since Consolidated Field was so small, she concentrated on becoming proficient in landing by "slipping in" which paid off by saving her life in a "dead stick" forced landing into a very tiny field. Crary served as a pilot for Gordon Mounce, who held the outside loop record. She did all of her flying in a Kinner Fleet and once flew from Los Angeles to Tacoma, WA, in four days, stopping frequently to promote the plane. One of the highlights of her life was meeting Amelia Earhart at her father's home. She also enjoyed doing stunts over the Niagara River. She studied engineering at Cornell University, was secretary of the Flying Club, and a member of the Society of Engineers. Crary's license was endorsed from 1929-33.

She married Gerald Carter Crary in 1932 and the Great Depression forced to her to stop flying. She had two stepsons and one son: Gerald C. Crary Jr. MD (deceased), William E. Crary, Ph.D., 63 and Fleet Nelson, 53. They have six grandchildren: Carter Crary, 44; Barret Crary, 41; Laura Steel, 33; Gray Crary, 31; Elizabeth Crary Hawkes, 25; and Grant Crary, 23. She also has three great-grandsons.

NANCY CRASE, Phoenix 99s, earned a private license in 1968, getting started by flying in the family 1946 Aeronca Chief. After a harrowing IAR in 1971, she added an instrument rating and a vow to avoid Mexico.

Although educated in microbiology at the University of Illinois, she and her husband Cliff launched a publication on wheelchair sports and recreation in 1975, and for the next 15 years guided the publication into being an international *Sports Illustrated* for wheelchair sports. After leaving permanently the day-to-day editorial affairs, she now happily continues to act as photographer and consulting editor for the publication while raising a daughter who arrived in 1990.

Although not currently active as a pilot, nor in 99 activities (she was Phoenix Chapter chairman 1976-77), flying will hopefully reenter the schedule with a course in aerobatics now that "Number One Daughter" has started kindergarten.

KATHLEEN MARY CRAVER, age 42, was born on Oshkosh, WI, and now resides in Waco, TX. She has a 22-year-old daughter, Heather Lisa, now attending the University of Wisconsin at Oshkosh.

Craver started flying training Sept. 14, 1994, soloed Nov. 3, 1994. She completed her FAA private pilot checkride Jan. 28, 1995. She joined the local 99 Chapter Feb. 7, 1995. She currently is a certified airframe and powerplant technician, having graduated from Texas State Technical College with an associate's degree in 1992.

Craver's plan is to continue with her flying training at Texas State Technical College and to obtain her instrument, commercial and certified flight instructor certifications. Then, she hopes to instruct students at Texas State Technical College in Waco.

CINDY CRAWFORD, born June 9, 1955, in Chillicothe, OH, was working as a nurse for a doctor who occasionally did flight physicals. She got asked to go flying by a handsome, eager pilot, Pete K. Crawford. Pete let her fly the plane and it was love at first flight. It was a short time later on a flying trip to Bowling Green he proposed marriage, on the ramp.

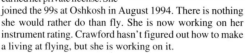

Crawford loved flying with him; he is instrument and commercial rated, but she knew she would learn to fly too. On July 14, 1994, she earned her private license. She joined the 99s at Oshkosh in August 1994. There is nothing she would rather do than fly. She is now working on her instrument rating. Crawford hasn't figured out how to make a living at flying, but she is working on it.

Of course she said yes to Pete's proposal, and he has recently purchased a restoration project that is neatly stored in their garage, a PT-26. Crawford hopes to be of some help to him and realizes this was probably part of the reason he was looking for a co-pilot for a wife, for life. She and her husband have one daughter, Chastidy A. Hall, 24 years old. Expecting first grandchild in February.

NICOLE CRAWFORD, started flying at age 11 with her father in his Stinson. She got her ASEL private with Dee Blum at San Carlos Airport and joined the SCV 99s in 1992. She has logged more than 100 hours and her current goal is to get her instrument rating. Nicole is an information security officer with National Semiconductor in Milpitas and lives in San Mateo, CA. Outside of flying, her interests include reading and art.

SHARON CRAWFORD, a native Californian, graduated from Berkeley in 1966 with a teaching credential, a husband and two infants. Flying came along as a hobby in 1976 and was a cherished supplement to teaching and homemaking. In 1989, after 27 years of marriage and 18 years of teaching, the opportunity to move into professional piloting came about and she took a job with United Airlines.

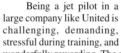

Being a jet pilot in a large company like United is challenging, demanding, stressful during training, and wonderfully rewarding. The scenery from the "front office" is truly beautiful – most of the time. Sharon's husband, Donald, is finally becoming accustomed to being supportive during training and to being alone during trip time. Daughter, Carolyn, a veterinarian in Northern California, says she tells everyone what her mom does. Son, Neil, a Ph.D. student in bioengineering at Arizona State University in Phoenix, says he's proud of his mom for making a career change.

MARY RAWLINSON CREASON, born Nov. 20, 1924, in Greenwood, DE, married William Creason and had four children: Kennard, 45, dentist; Yvonne, 43, dental hygienist; Stephen, 40, musician and music teacher; and Paul, 37, dentist. She has seven grandchildren: Jamina, 10; Nicholas, 7; Sonya, 4 ; Kevin Creason, 9; Sally Creason, 5; Eric Balison; and Katelyn Balison, 3.

She was accepted into the WASP program but it was disbanded prior to class in June 1944. Her sister, Mabel Rawlinson, departed for Kalamazoo, MI for Houston, TX, for WASP class W-3. She had given Creason her share in an Aeronca CP N22108. Creason had flown as a passenger with her. She never knew Creason was close to "barfing" most of the time as she chased cows doing her pylon 8s. Since she was one-eighth an owner of the airplane, she decided to learn to fly. She took lessons from CFI Eloise Smith, also in the One-O-Ate Club. On her second solo, she leveled off several feet too high and dropped in – broke the landing gear and ground looped, damaging a wing tip. Creason duly reported the incident/accident to sister Mabel. She wrote back that "it was nothing," once she bounced so hard she knocked out two silver (teeth) fillings!

Education experience includes a BS from Western Michigan University, Kalamazoo, MI, 1944 and apprenticeship and employment at the Kalamazoo Public Library.

Creason has more than 10,000 flight hours and 5,000 teaching hours. She has an airline transport pilot license and CFI and AEI and ground base and instrument rating. She has flown all single-engine and many light twins, SIC King Air. She is self-employed as the owner FBO The Aero Technology Company, flight instruction and dealership for the AG5B Tiger. Retired in January 1989 from Michigan Department of Transportation's Bureau of Aeronautics. She was inducted into the Michigan Aviation Hall of Fame on Oct. 14, 1995.

CAROLE CREVANI, a resident of New Jersey, earned her pilot license in 1994 and has logged more than 500 hours. She purchased N521CC, the last Cessna 172 built in the 80s. Carole is a member of the North Jersey Chapter of 99s and has participated in Poker Runs and Pennies-a-Pound. Memberships include AOPA and Lincoln Park Pilots. Cross-country experiences include flights from Texas, Kentucky and Florida as well as the opportunity to pilot a small plane in Africa.

Carol and her husband, Stephen, own a construction company and Crevani Farms, a standardbred breeding facility and training center. Carole holds an owner-driver-trainer license and has driven standardbred races, competed in 100-mile competitive trail rides with her horses, rides fourth level dressage and roller blades for exercise. Their family includes two sons, two daughters, their husbands and wives, and seven grandchildren.

IRIS CUMMINGS CRITCHELL, flying since April 1939, earned the private pilot certificate April 1940 at University of Southern California on the first Civil Pilot Training Program.

Completed the CPTP advanced aerobatics course and graduated from USC in 1941 in physical science. Taught on CPTP at Brackett Field, La Verne, CA, and for Stockton Junior College, Carson City Navy program until December 1942. Joined the WAFS through the WFTD at Houston, Class 43-W-2. From December 1942-44 flew in the WAFS, later known as WASP, ferrying airplanes within the US for Long Beach 6th Ferrying Group, ATC. Flew more than 18 types of military aircraft as pilot-in-command including P-51, P-47, P-40, C-47, B-25, A-20, P-38 and P-61. From 1946-48 she developed curricula, taught classes and instrument flight in BT-13s at USC College of Aeronautics.

Critchell became a 99 and charter member of the Long Beach Chapter in 1951. She was chapter vice-chairman and chairman over four years. Flew in 15 Powder Puff Derbies. With Alice Roberts, won in 1957 AWTAR and Michigan Small Race. Served on the AWTAR board seven years and the AE Scholarship committee six years.

From 1961-90 developed a special aeronautics curricula for Harvey Mudd College, directed and taught aeronautics program and served as chief flight instructor for 28 years. Presently, manages the aeronautical library at Mudd and flies professionally as FAA pilot examiner. She has four grandchildren.

RETHA MCCULLOCH CRITTENDEN, a charter member, was a grade school arithmetic teacher who took a ride in an airplane in 1928, and like most of us, just had to learn to fly. She paid $20 per hour for instructions and the flying school paid her $20 per hour for the publicity of having a woman student. She found she could then afford flying lessons. Retha claimed only two records in flying: one being the first woman pilot in Texas and the second being the first pilot to have a baby.

PATRICIA S. CROCKER, summer 1962, she was in a wheelchair but dreamed of walking again plus being a pilot. She concentrated on passing the flying medical, realizing her flying dream in 1963. Crocker soon joined the 99s, who set a teenager lifelong role-model standards. Marital breakup and raising her infant daughter necessitated relinquishing

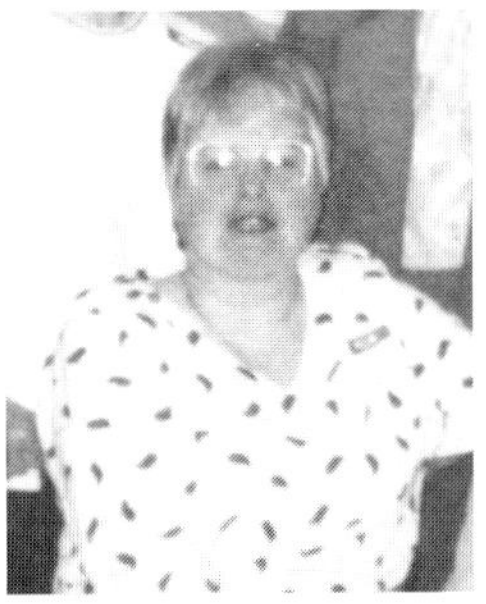

her flying dreams. She remarried; and also rediscovered flying joy in the 1980s via ultralights. Reinstated by fellow charter chapter member, Hilda Devereux, who recruited her for First Canadian Chapter before they formed Maple Leaf Chapter in 1969. She has held almost every chapter office, plus publicity/ public relations, national aviation magazine byline, section posts. Medically grounded by health conditions and muscular dystrophy, her flying dream now translates into chapter and section level aerospace education, scholarship, motivational and/or speaking specialties focus. Her occupations and interests include flying, professional writing, medical secretary, computerized hospital operating room department and the 99s.

KATHY GARDELLA CROFT, learned to fly in 1989 at Palo Alto Airport. She joined the SCV 99s that same year, with 175 hours. She now has logged 1,400 hours, and has her CFI, CFII and MEI ratings. Kathy was the recipient of the Marion Barnick Memorial Scholarship in 1991 for her commercial rating, and the AE Scholarship in 1993 for her MEI. Kathy has been active in airmarking and racing, and has flown the Palms to Pines and the Pacific Air Race. Kathy is a dietitian who works for the Marriott Corp. and is a flight instructor at Palo Alto Flying Club. Her goal is to fly for the airlines.

SELMA CRONAN, born May 6, 1913, in Ferndale, NY, has held a pilot license since June 1943. She has a commercial flight rating and all ground instructor ratings and is the mother of twin boys born in 1943, Thomas, father of two, and Jeff, who died in 1963.

She has two books published by the school division of Random House, magazine articles, a TV film and many studies done on behalf of commercial aviation. She became a 99 in June 1944 and has been active member since. Originally in the New York/ New Jersey Section, Sue served two years as treasurer and as governor, and four years on the nominating committee; served the national organization to read and approve the minutes of the 1955 convention, and as chairman of the budget committee. Since transferring to Florida in 1986, she has been nominating chairman for two terms.

Cronan flew many long-and short-course races, including three Powder Puff Derbies, four N. E. Air Derbies, the Michigan Small, International, five Bahamas Treasure Hunts and local contests. As a member of the Civil Air Patrol for 14 years, she was NY director of women's affairs, organized the first all-women's tactical squadron and the first girl Cadet training squadron.

In addition to the 99s, she has been a member of the National Aeronautical Assoc., National Pilots Assoc., AOPA, Aviation Writers Assoc., Women's International Aeronautical Assoc., Wings For Peace in Africa and an honorary in the Korean Women's Assoc. of Aeronautics.

Awards include a service citation from the city of New York (1942), the Amelia Earhart Medal (1960), the Lady Hay Drummon-Hay Trophy (1967) and many air race cups.

Cronan has been active in the Gulf Stream Chapter of the 99s, the Gold Coast Women's Vets, part-time writing, as a volunteer at a local hospital and a docent at the Moricami Museum and as a lecturer.

LENA M. CRONK, born April 25, 1934, in San Antonio, TX, has a private pilot license and has 400 hours flying time. She has flown the C-150, C-152 and C-172.

Cronk has flown in some air races but mostly flies for pleasure. She is a member of San Antonio Chapter. Held office as treasurer and presently vice-chairman and airmarking chairman. Besides working in 99s she is also a staff nurse in the operating room and directs a choir at church. She also belongs to AOPA.

She is a widow and has two sons, Valentine Nicholas, age 35 and Christopher Aaron, age 28. She has three grandchildren: Tara Lynn Cronk, age 15; Tamara Jean, age 11; and Taileah Ariel, age 6.

JANET CROSBY, a private pilot since December 1973, in Albuquerque, NM, has more than 2,900 hours. She holds commercial and CFI certificates and has an ASEL, AMEL and instrument ratings. Crosby has flown all single Cessnas, a Bellanca Scout, Beech 35, Stinson 108, Maule M4 and M5, DHC-1 Chipmonk, Pitts S-1S and S-2A and S-1T, Stampe, Citabria, Decathlon, Viking, Mooney, Cougar, Zlin, PA18 and owned about 20 planes. She says, "There is nothing like the feel of your first Lomcovak!" Crosby joined the Lake Tahoe Chapter of 99s in 1990. Crosby's flying activities include: aerobatics pilot at airshows, national competitions: Fond du Lac, Denison, TX, etc.; IAC regionals, Truckee Tahoe Airshow, International Balloon Fiesta, Albuquerque airshow, Monte Vista, CO, Midland Airshow, member of US Aerobatics Team 1990.

Crosby is a member of the NAA, IAC, EAA, ACA, 99s (five years). She received Lake Tahoe 99s "Woman Pilot of the Year" in 1990.

She is married to Mike, a pilot and former Coast Guard HU16 pilot; they currently have six cats. Crosby enjoys hiking, backpacking, horseback riding, cross-country skiing, fishing and anything outdoors. In August 1992, Janet and Mike completed their four-year project of building a monoplane, *Zoomy* (N2OOMY), from the ground up for Janet to use in aerobatics competition. Janet want to regain her position on the US aerobatics team in the near future!

MARJY CROWL, Phoenix 99s, was a charter member and initial vice-chairman of the Phoenix Chapter in 1949, but this was by no means her first effort in advancing opportunities for women in aviation.

Marjy was drawn into association with other women pilots by her dedicated work in recruiting and training Wing Scouts for many years. She and her husband operated a charter, crop-dusting and flight school in the Phoenix area. Marjy not only crop-dusted, charted and instructed in fixed wing aircraft, but got commercial and instructor's ratings in helicopters as well. She was also a slurry bomber pilot, and in 1961, spent three months fighting fires in Montana. Also, she completed the fourth annual All-Women Transcontinental Air Race from San Diego to Greenville, SC, in 1950.

A 1953 *Glamour* magazine featured Marjy and other 99s in an article commemorating the 50th anniversary of powered flight. Marjy and her husband and business partner, Cliff, are pictured with one of the crop-dusters they operated. She is in the open cockpit, complete with helmet and goggles, with him standing beside her. It explains what they do, going on to say, "This month marks aviation's' first 50 years. On Dec. 17, 1903, Wilbur and Orville Wright made their first flight at Kitty Hawk and opened the air age. Today there are exciting jobs for women in this man's field."

In the early years after the new chapter was formed in Arizona, Marjy flew many miles, including a trip to New York to attend 99s meetings and establish ties with other 99s all over the country. Marjy has continued to share her enthusiasm for aviation through the years, serving as chapter chairman 1950-51, and contributing her many skills on all functions, races, airmarking, whatever she could do that developed opportunities for women pilots, furthered aviation safety, etc.

Marjy was given life membership by the Phoenix Chapter in 1978, and was again honored in 1989 by having the chapter place a marker in her name in the Forest of Friendship (Atchison, KS). She is also a member of Whirly Girls.

STEPHANIE PENKIVICH CRUZ, the oldest of three girls, was born into aviation at Ellsworth AFB, Rapid City, SD. "I was the only SAC kid, mother was a quadriplegic and SAC personnel are barely home. Mom gave Dad the ultimatum, her or the Air Force. They compromised and Dad left SAC for the Air Defense Command and Duluth, MN, where we spent the next dozen years." Steph didn't take up flying though until much later. "Life forced me to take a look, my only regret is that I didn't start flying sooner."

Stephanie started lessons in 1989. By September 1992,

she had her CFE and her own airplane, a 1977 Piper Warrior. She joined the 99s in 1993 and achieved "Pilot of the Year" for 1993. She volunteered to assume the duties of recording secretary and newsletter editor when those positions unexpectedly became open. She was elected to the position in 1994 and still remains the newsletter editor and producer along with her faithful friends and volunteers.

"In addition to being an Air Force brat and later a Navy wife, I decided to find out how the other side lived and joined the Navy Reserves in 1987." Both Cruz and the Reserves survived, but neither will be quite the same. "Between the active Navy and Reserves, I've seen quite a bit of this planet, but I like the view best from at least 5,000 feet in my own plane with me at the controls."

Cruz is married (22 years on September 1995) to Virgil Cruz and they have a 16-year-old daughter, Theresa. She works for Wenk Aviation Insurance Agencies as a producer and flies and instructs on weekends and spare time. She averages about 150 hours per year and hopes to break 1,000 hours this year!

BETTY CULL, Indiana Chapter, earned her private pilot license in June 1941 on Civilian Pilot Training Program at Indiana University, Bloomington, IN, under Roscoe Turner School of Aviation.

Jefferson Proving Ground 1942, worked as PBX operator on firing line and then as only woman aerial observer (civilian) with Air Force at proving ground where they tested bombs and flares.

In 1944, received commercial, flight instructor and horsepower ratings at Embry-Riddle School of Aviation, Coral Gables, FL.

Then became first woman flight instructor in Indianapolis, IN, at Sky Harbor Airport. Later, instructed at Greensburg, IN. She has flown the Fair Ladies Annual Indiana Air Race seven times, placing fifth in 1961, fourth in 1968, first in 1969, third in 1970 and second in 1975.

Cull received her 99 25-year pin on Nov. 25, 1970. She owns a Piper Cherokee 140 and holds a commercial rating. Cull is a free-lance reporter/photographer based in North Vernon.

THELMA CULL, learned to fly in 1964-65 at Sacramento Municipal Airport after a driving trip to Alaska made her realize she was missing a lot of beautiful scenery.

Since then she has made three trips to view that scenery. Thelma and her husband, Neil, are native Californians who are now retired and ready to see the Western Hemisphere from their Cessna U-206.

Thelma has a BA and MA from California State University, Sacramento and taught elementary school for 37 1/2 years.

She has participated in the Powder Puff Derby, Air Race Classic, Pacific Air Race and Palms to Pines since joining 99s in October 1965. She held offices in the Sacramento Valley Chapter and Southwest Section, serving as governor 1980-82. Had the privilege to co-chair the 1976 PPD – the last official and largest race. Holds a private certificate with about 999 hours.

BARBARA ANN CULLERE, born Feb. 24, 1959, in Teaneck, NJ, trained with Leroy Heidrick in a PA-28-140. She has flown a Cessna 172, J-3 Cub and Cherokee 140. She is married to John R. Cullere, also a pilot, and has one stepson, Jason Allen Cullere and a daughter, Alyssa Jeanne Cullere. As a student pilot, she learned that she was pregnant with her daughter. She did all her cross-countries pregnant and took her check ride when she was eight months pregnant.

Cullere is a teacher. Joined the 99s in 1994. She also owns a 1941 Aeronca Chief which is being restored and she is assisting her husband in restoring his 1945 J-3 Cub.

MARA KRISBERGS CULP, born in Riga, Latvia. Her formative years were spent in Sweden and Argentina before immigrating to the US in 1957.

Mara started her flying lessons in a Meyers 200 at Hawthorne, CA, in 1962. She rapidly progressed through commercial, multi-engine, airline transport pilot, instrument flight instructor and ground instructor ratings. She was active in the Orange County Chapter of the 99s and air race organizations.

In 1965, Mara was employed by Elpac, an electronics company in Newport Beach, CA, to fly a Cessna 182 mostly to military airfields with some recreational visits to Mexico. For the next 10 years, Mara's flying career was diversified and adventurous; it included flying for John Wayne during filming of the movie "Cowboys"; Newport Beach Volkswagen dealer, Chick Iverson, employed Mara to fly to numerous auto races throughout the US; Mara also "covered" horse racing circuit, flying horse owners, veterinarians and jockeys to Ruidoso, NM, and other race/training sites.

By 1971, Mara was type-rated in a Lear Jet 24/25. During the next four years she made eight Atlantic crossings in a red Lear with home bases in Europe and California while flying for Volkswagen/Porsche magnate, John von Newman. Passengers included people famous in the entertainment business.

By 1974, it was time for a career change: Four years of college, a BS degree in dental hygiene from the University of Iowa and 15 years employment as a dental hygienist with Saudi Aramco Oil Company in Saudi Arabia. Mara developed the hygiene services from six to an international group of 42 dental hygienists. Although women can not drive automobiles or fly airplanes in Saudi Arabia, Mara together with a dozen other frustrated women pilots, formed the Arabian Section of the 99s.

Mara looks back at her air racing days of 1966-76, which included pylon racing at the National Air Race circuit in a stock Meyers/Aero Commander 200, winning the Powder Puff Derby in 1969 and the International Air Race in 1970. Twenty-plus other races were enjoyed in that time period. In 1996, Mara will retire from her dental hygiene position overseas and hopes to once again become actively involved in the air race/ air show circuit. Home base will be an RV on the move.

ELSIE E. CULVER (HANSEN), born near Pablo, MT, on Aug. 2, 1917. She married early and at 19 had a daughter and at 22, a son. She moved to Missoula, MT, in 1949, went to work in the office of Johnson Flying Service, Inc. In 1951, she began flying lessons in a Piper PA-11 and J3, earned a private pilot license that year. After several years of flying in a Piper Tri-Pacer, Cessna 140, 182 Skylane and Piper Comanche, she earned her commercial pilot license in 1961.

She was one of the original members of the Montana 99s where she served as secretary, vice-chairman and state chairman in 1964. Elsie was race chairman for the race in September 1965, from Missoula, to Kalispell to Cut Bank, to Havre and the terminus in Great Falls. Elsie received the cup for best time between Cut Bank and Havre.

In 1970, she and husband, Bob, moved to Anchorage, AK; in 1971 they moved to Juneau, where her husband worked for the FAA in flight standards as flight inspector. Elsie worked for the US Forest Service. Ten of the 13 years in Juneau, they lived on a 60 foot cabin cruiser they had brought up through the Inland Passage.

After retirement in 1984, they moved to Montana to Flathead Lake. Bob passed away there in 1985. Elsie still lives there and is very active in bowling several times a week plus tournaments, and in the winter, downhill skiing at Big Mountain, near Whitefish. She also takes care of her small cherry orchard plus gardens. In addition to two children, Hazel Foley-Jones, 58 and Dr. Derald G. Smith, 55; she has three grandchildren: Kevin Foley, 35, Maureen Foley Garcia, 33 and Pat Foley, 29 and three great-grandchildren.

DIANAH CUMMINGS, a resident of Oklahoma, began learning to fly in 1990. She has flown Cessnas 150, 152, and 172, Piper Cherokees 140 and 180. A member of the 99s, Inc., Aircraft Owners and Pilots Assoc., Green Country Aviators and Angel Flights, Inc., she is participating in the FAA Pilot Proficiency Award Program where she has earned the Phase I Wings. She is married to Richard Cummings, vice-president of manufacturing at Liberty Glass Co., Sapulpa, OK. She has one daughter, Kimberly; one son, Stephen, one stepson, Rick; and three beautiful grandchildren, Brittany, 2 1/2 years old, Ryan, 3 and Morgan, 6 months.

She is currently serving as news reporter for the Tulsa Chapter of the 99s, Inc., editor of the chapter's newsletter and is airmarkings chairman.

DEBORAH CUNNINGHAM, got her license in 1979 at Reid-Hillview airport and joined the SCV 99s the same year. She is a CFI, CFII, MEI, ATP and has logged about 3,400 hours to date. Debby has two children and her husband, Bruce has his own auto repair shop. Bruce and Debbie ride motorcycles, fly airplanes and drive vintage Mustangs. Debby has been chapter chairman and chapter coordinator for student pilots for many years. She has flown the Pacific Air Race, the Palms to Pines and the Salinas His and Hers. Debby was chapter Pilot of the Year in 1988, won the chapter Service Award in 1985 and won an Amelia Earhart Scholarship in 1988 for her CFII. She is currently working on getting her AS degree in aviation technology and wants to fly for the commuters. Her interests are family, Bible study, gardening and skiing.

JUANITA CURLEY, received her license in 1990 and flies a Piper Archer 181. She is chairman of the Michigan Chapter, a member of the National Aeronautics Assoc., Aircraft Owners and Pilots Assoc., Michigan Aviation Assoc., Michigan Pilots and Owners Assoc. and a charter member of the Michigan Aviation Hall of Fame. She and her 49 1/2 set a national record in 1993 by doing a touch-n-go at all 128 public paved airports in Michigan. First as navigator, then as pilot, she placed many times in the top four in the Michigan Small Race, last year receiving the rotating trophy for the Best Score-Michigan Lady Pilot.

She has a degree in criminal justice, is active in the community, was in Girl Scouts for 30 years, and is an avid genealogist. Juanita is currently working on her instrument rating.

Married to her flight instructor, Pat, recently retired from Ford Motor Co. They have three daughters and three grandchildren.

DIANA M. CURTIS, born Feb. 16, 1956 in San Antonio, TX. She resided there until her marriage in 1978.

She received her pilot license in 1990 and has logged more than 300 hours in Cessnas and Grumman Americans. She is currently working on her instrument rating.

Curtis is very active in the Fort Eustis Flying Activity and has participated in the two gymkhanas between Fort Eustis and Langley Aero Clubs. She placed first in her class and category in these competitions. She also enjoys working with the local EAA Chapters and their promotion of the Young Eagle program.

Curtis is a proud member of the 99s, Inc., Aircraft Owners and Pilots Assoc. and the Fort Eustis Flying Activity. She is currently secretary for the Mid-Atlantic Section of the 99s and is also vice-chairman for the Hampton Roads Chapter.

She is married to Charles T. Curtis Jr., who is also a pilot and a computer agency analyst for the Virginia State Health Department. Both she and her husband now reside in Newport News, VA.

ALICE A. CUTRONA, born March 17, 1926, in New Jersey, joined the US Cadet Nurse Corps from 1943-46. She is a retired nurse educator with a BSN and M.Ed. She is married to Fred Hartmann, a retired high school teacher. She has two children, Jerry M. and daughter, Valerie A. Fisher. She has two grandchildren, Nicole Lea Cutrona, 16 and Antonio J. Cutrona, 14.

She had primary training in New Jersey and instrument in Florida. She has flown a Cessna 150 and 172.

Flying has brought many happy hours: going to new places, meeting with other women who are of similar interests, being challenged to maintain and sharpen skills and the fact that she has accomplished what few others have. Cutrona has enjoyed working for the 99s in various chapter offices and committees, section committees, helping with FAA Safety meetings, speaking to various groups about the 99s, participating in several air races and, best of all, having fun while doing these things.

DIANA DADE, born Jan. 20, 1937, Brooklyn, NY. Joined the 99s in 1980. Flight training includes glider/sailplane (trained in Schweizer 2-22). She has flown the following planes: Schweizer 2-22, 2-33, 1-26, 1-34, 1-36, ASTIR Twin, GROB 103ACRO, Schweizer K6 and KA8B (owns it). She is married to William R. Dade.

She is employed as a registered dental hygienist and is a graduate of Fairleigh Dickinson University.

Other than her many Garden State Chapter activities, she is also involved in both Central Jersey Soaring Club and Freedom's Wings, International, a group dedicated to encouraging disabled people to fly specially-equipped sailplanes.

One of her favorite flights was with fellow G.S. Chapter member Alice Hammond's 49 1/2, John. (He was the type of pilot who could fly a barn door.)

She's organized, as well as flown in, several sailplane competitions. Her 49 1/2, Bill, who is also a pilot, is just as involved as she is in their various aviation pursuits.

ELIZABETH "LIZ" DAFFIN, a flight attendant for British Airways, who learned to fly in 1988 in Texas. She possesses commercial, CFI, CFII, MEI ratings, and is currently chief pilot for the Palo Alto Flying Club. Liz is a part owner of a M20K. When she joined the Santa Clara Valley Chapter in 1990, she had logged 1,350 hours. She also enjoys hiking, camping and cycling.

LISA BUSWELL DAHL, young owner of Buswell Aviation at the Salem, OR, McNary Field, is one of the most highly-respected pilots and instructors on the field.

Dahl was born in 1955, grew up at her dad's FBO in Lakeview, OR, and learned to fly as a child. She soloed at 16, got her private license at 17 and a commercial rating at 18. Thousands of hours and a long string of ratings later, she flies a King Air for Entek Corporations, gives check rides for the FAA, and operates her busy FBO with a fleet of 10 aircraft, including seven passenger turbo-prop planes, and six instructors and charter pilots. She serves on several state and local aviation boards.

Lisa and her husband, John Dahl, met when she began giving him instrument instruction. They married in 1981. Their daughter, Kara Jeanne, born in December 1983, took her first airplane flight at the ripe old age of one week, and can already keep a plane straight and level.

Dahl's pet airplane is an old Luscombe. On the ground, her favorite pastime is horseback riding with Kara.

Dahl is a charter member of Oregon Pines Chapter.

CECILIA M. DALZELL, born March 30, 1957, in Huntington, NY. Lived most of her life in the Danbury, CT, area. She is married to Carl Dalzell, an electrical and computer engineer.

Cecilia got her private pilot certificate in 1993. She has flown a little over 200 hours, mostly in Cessna 152s. Her longest airplane trip so far is from Danbury, to Presque Isle, ME, and her highest is to 11,000 foot MSC. She flies for the beauty and freedom of it, and loves to watch the changing seasons.

Cecilia has a BS in biology from Fairfield University and will have a master of library science and an MS in instructional technology from Southern Connecticut State University by January 1996. She has worked as a veterinary assistant, photomask production technician, process engineer and technical trainer. On her own time, she paints watercolors and volunteers in her community and church.

VIOLET DANIEL, born Dec. 25, 1927, Detroit, MI. Joined the 99s in 1986. Received her flight training at Detroit Metro. She has flown 150, Warrior, 172, Archer, Bonanza, Stinson, Bell Helicopter. She and her husband, George Daniel, have five children: Geoffrey, 30; Linda Stueves, 32; Diane Holt, 36; Patricia Birch, 38; and Nancy Daniel, 40. Their grandchildren: Tara Birch, 9; Eric Birch, 6; Kevin Birch, 4; George Holt, 3; and Griffin Holt, 1 1/2. Once children were out of house, on Mother's Day husband gave student log book and said, "Come fly with me." In their Bonanza, they've flown as far west as Catalina Island and as far east as Treasure Key Bahamas. Every flight is different still learning! Most enjoyable 99 activity is Annual Pinch Hitter at Ann Arbor, MI.

VAL DARLING, Nebraska Chapter, received her private license in 1973 and has been an active member since that time. She and her husband Don share piloting their Cessna 172. They have three sons and two granddaughters.

Teaching in the Aurora elementary schools is her professional career. Other than teaching and flying, she enjoys golf, bridge and sewing. At the present time she is preparing to take the Nebraska state realtors exam.

She has been treasurer of their chapter as well as assisting on various committees.

In July 1975 she spent a month in Germany. While there, she met several exciting and interesting 99s. On one occasion they flew to the North Sea. Their hospitality left nothing to be desired.

She sincerely enjoys flying and encourages her students and everyone to share the thrill.

KIM DARST, stays busy running her business, K.D. Helicopters, which she operates out of her home in Blairstown, NJ, functioning as the chief flight instructor at Trinca Airport in Green Township, NJ, and working as a flight engineer on 727s for KIWI International Airlines. She currently holds all seven CFI ratings, making her a member of a very small but elite group of instructors in the world.

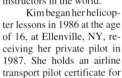

Kim began her helicopter lessons in 1986 at the age of 16, at Ellenville, NY, receiving her private pilot in 1987. She holds an airline transport pilot certificate for airplane single engine land; airplane multi-engine land and the Douglas DC-3; a commercial certificate for airplane single engine sea, rotorcraft helicopter, rotorcraft gyroplane, instrument helicopter, glider; ground instructor for advanced and instrument. In addition she holds both an A&P and I.A. rating. Her future plans include working on her balloon rating and multi-engine sea. Despite all this she is also a designated examiner for helicopters in the Allentown, PA, FSDO. Her current fleet of aircraft includes a Bell 47 G-2, Piper J3 Cub, Cessna 172 and her favorite, a Cessna 195 fondly nicknamed Clyde.

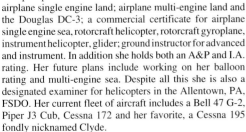

Despite all of her commercial airline involvement and fixed wing time, her first love has always been and remains helicopters. She has been a member of the Whirly Girls Organization since receiving her helicopter certificate and is known as #577.

HELENE DAEMEN-DARVEAU, born Jan. 31, 1951. Pincourt, Quebec. Education and training: 1989, Flight Surgeon Course, Canadian Armed Forces, Toronto, Ontario; 1985, certificate of special competence in emergency medicine Canadian College of Family Physicians, Toronto; 1992, acting director, Occupational Health Services, East Air Canada Centre, Dorval, P.Q.; 1991-92, Medical Consultant, Air Canada Dorval; 1991 - Civil Aviation Medical Examiner Transports Canada; 1989-91, Aviation Medical Officer, Dorval Airport, Dorval.

Numerous articles in publications on diving and flying. Professional interest: Medication and flying; flying after diving; human factors and stressors in aviation; physiology of decompression sickness and clinical applications; inner ear barotrauma. Other professional activities: 1988 - examiner, certification in Emergency Medicine Canadian College of Family Physicians.

Hobbies: Aviation: private pilot license with 200 hours and member of the 99s (Women pilots association). Diving: Active diver with 350 hours, completed training in night and wreck diving as well as underwater photography. Also enjoys reading.

MARGARET "PEG" DAVIDSON, born Feb. 26, 1928, in New Haven, CT. Joined the 99s in 1963. Flight training includes private pilot license, commercial and instrument ratings. She has flown Cessna 205 for 32 years, Cessna 172 and Cessna 310 occasionally. Employed with D.S. Davidson Co., Inc., family nuclear instrument design and manufacturer. She and her husband, Donald S. Davidson (an electrical engineer) have four children: Russell, 44; Carol, 43; James, 41; and Richard, 40. Their grandchildren: Owen, 13; Kevin, 11; Adam, 10; Andrew, 8; and Robert, 6.

In 1963 her husband bought a Cessna 205 and wanted her to learn to fly it. He didn't have a license either, but they both took lessons. In August 1963 she got her private license, in 1964 commercial and in 1965 her instrument rating. She has flown for business and pleasure. Flying two Powder Puff Derbies, International Air Race (Angel Derby), proficiency and efficiency air races.

In 1963 she helped organize the Connecticut Chapter of the 99s and became the first chairman. Since then has served the chapter on most committees. As newsletter chairman, she named it *Triple Charlie,* which stands for Connecticut Cockpit Chatter and has written *Triple Charlie* for more than 10 years. Many years were spent reviewing and distributing the Ruth Nichols memorabilia which would have been destroyed except for their chapter concern and dedication to the project.

Served as governor of New England Section of the 99s and held many offices and active on committees. In 1970 served as chairman of the International Convention at Breton Woods, NH. Served as All Woman New England Air Race (AWNEAR) chairman and/or race board for five years and flew many Round Robin Races, winning three races.

She was awarded the Section Merit Award. Served on the Powder Puff Derby board of directors for seven years (five as secretary) flying to many terminus locations and most board meetings. Timed and scored and helped plan many air races.

Graduate of the University of Connecticut with a BS in nursing. Membership: 32 years with the 99s and Aircraft Owners and Pilots Assoc.

URSULA MALLUVIUS DAVIDSON, Florida Goldcoast Chapter, spearheaded the building of the Aviation Institute at Broward Community College near Fort Lauderdale.

A designated pilot examiner, Gold Seal flight instructor, ATP and commercial glider pilot, Ursula has more than 5,000 hours flight time. She learned to fly in the Bahamas in 1973. Winning an AE scholarship in 1977 gave her the career boost she needed. She shows her appreciation by creating new scholarships, serving as Southeast Section AE scholarship chairman and helping others further their careers in aviation.

A native of Cincinnati, OH, Ursula has a BA from Ohio State and a master's from American University. Her son, Steven, is an attorney in Washington, DC.

Ursula has held all the offices in her chapter and has coached NIFA flight teams, judged SAFECONs and served on the NIFA Council. She is active in the University Aviation Assoc., serving as a trustee. She was the first woman elected president of the 200-member Greater Miami Aviation Assoc. As chairman of the scholarship committee, she helped raise the endowment to over $500,000.

DONNA DAVIS, learned to fly at Van Nuys Airport in 1973. She joined the 99s in 1974 and has been a member of the Santa Clara Valley Chapter since 1984. Donna has her commercial and IFR ASEL and her ASES ratings, with about 850 hours logged. She has flown the Palms to Pines, the Midwest Valley Air Derby and the Mini Air Derby and was San Fernando Pilot of the Year in 1979. She has been active in airmarking and has flown charity flights. Her husband Michael is also a pilot. Her other interests include golf and sports.

DOURELLE JAY DAVIS, learning to fly was a way to prove something to herself. The first time she left the pattern on her own, she leveled off, trimmed the plane and yelled at the top of her voice, "I did it ... I am free." Oh, what a feeling. When she pilots a plane now, she feels the same freedom as she did then. Being in control of a plane reminds her that she alone is responsible for everything.

Her husband, Charles, is a professional pilot, and they share the joy of viewing their wonderful earth from high above. He brought her roses when she passed the check ride just in time to be a charter member of the Aux Plaines Chapter (April 1978). The 99s give her the opportunity to share adventures, make good and lasting friendships and "sell" general aviation.

They have five children and one grandchild.

EDNA AUSMUS DAVIS, born Sept. 3, 1909, in Speedwell, TN. At the age of 6 she told herself she would fly an airplane. At the age of 55, on June 29, 1965, she got her private pilot license and joined the Tennessee Chapter of 99s. Davis has served as vice chairperson, chairperson, news reporter, award and nominating committee. She bought a Cessna Skyhawk 172 and enjoyed every minute she had time to fly. She worked as a timer for both Powder Puff and Air Classic Races. Davis taught school 29 years, children from first grade through the twelfth grade.

FRANCESCA S. DAVIS, Caribbean Section 99s, born in Waterbury, CT, April 2, 1930. Her interest in flying began at age 12. During high school, she became a member of the Waterbury Squadron Civil Air, soloed at age 16, in an Aeronca. After graduating from high school and Boston Aviation Training School, she was hired by Trans World Airlines as clearance agent. On reaching age 21, transferred to flight hostess, flying domestic and international routes until marriage in 1956 to TWA Captain R.W. Davis.

She formed the first Girl Air Cadets in Newport Beach,

CA. After their move to the Bahamas with daughters Maria and Jaime, she began flying lessons, receiving her SEL pilot's certificate December 1971, instrument rating March 1975.

She's flown four PPDs and five Angel Derbies, placing first in International division once. Activity in the Caribbean Section 99s since 1977 includes secretary, vice governor and governor, presently vice-governor.

Other volunteer work is with Women's Clubs in Indiana during summer months, in Grand Bahamas in winter months and with the Hugh O'Brian Youth Foundation Leadership Seminar program as organizer, chairman and presently, corporate president of the Bahamas Seminar.

JANET SISSON DAVIS,

born Feb. 8, 1926, in Wilmington, NC. She learned to fly in Quincy, IL, where she soloed in 1942 at 16 years of age. At 18, she passed her FAA private pilot check-ride in Champaign, IL, while attending the University of Illinois. Currently, she holds a commercial certificate and instrument rating. She has four children, 11 grandchildren, and one great-grandchild. She joined the Garden State Chapter in 1984, transferring to the Kitty Hawk Chapter in 1991 after retirement from her position as associate dean for administration at the Rutgers Graduate School of Management.

Janet owns a Cessna 182, hangered at Dare County Regional Airport, Manteo, NC. She and her husband, Dick (deceased in 1995), have flown to 47 of the lower 48, Alaska, Newfoundland and Nova Scotia. She flies search and rescue missions for the Outer Banks Flotilla, USCGA.

Janet lives in Southern Shores on the Outer Banks of North Carolina, a few minutes from where it all began on Kill Devil Hill. Her aunt, Jeannette Lempke Sovereign, was International president of the 99s in the mid 1940s. Her daughter, Carol Yunker, holds a private pilot certificate and is a 99. Janet holds a BA from the University of Illinois and an MBA from the Rutgers Graduate School of Management.

Janet has flown the following planes: Aeronca Champ, Piper Cub, Interstate Cadet, Cessna 150, Cessna 172, Cessna 177, Cessna 182, Cessna 206, and Mooney 206.

LUANA DAVIS,

born Dec. 2, 1942, in Alameda, CA. Joined the 99s in 1972. A military service was not applicable since the opportunity did not exist. She would have been honored and thrilled to have had been able to attend the Air Force Academy and continued with a flight career in the Air Force. She has been employed as a pilot with Federal Express since 1979.

Began learning to fly at age 11 from father who was a pilot and captain at Pan American World Airways. Father's name is Gerry Mahan. She has flown approximately 38 makes and models of aircraft and has accumulated more than 11,000 hours of flight time. Her certifications include: airline transport pilot; airplane single and multi-engine land; airplane single engine sea; flight instructor airplane single and multi-engine-instrument airplane; ground instructor-advanced-instrument; flight engineer-turbojet powered. Type ratings: Lear Jet; DA-20 (Falcon 20); B-727; DC-10; MD-11.

She is the first female captain in the world of the MD-11.

Competed in the last Powder Puff Derby; Angel Derby in 1977 with a sponsorship from Dobbs Inc.; Pacific Air Race-1977; Air Race Classic-1977-third place with a sponsorship from the Big Boy Restaurant chain and awarded the annual Beechcraft Award at the conclusion of the Air Race Classic; First place 1977 Palms to Pines Air Race with the sponsorship from Big Boy Restaurants.

NANCY DAVIS,

born June 2, 1944, in Brooklyn, NY, and was married Oct. 13, 1962, to William R. Davis. Her flying experiences started on a sunny Sunday drive past Long Island Republic Airport. Upon seeing a small aircraft on approach she said to her husband Bill, "Now that our two boys are in high school, that is what I would like to do." Nancy enrolled in flight school and completed her training in January 1978. Nancy participated in her first 99s poker run in the spring of 1978 and was introduced to the 99s and became a life member in Long Island Chapter. She was correspondence secretary, uniform committee chairman and historian, and currently holds a license for single-engine land and sea and 10 hours in lighter-than-air free balloon.

She enrolled in New York's Farmingdale University and received her degree in aerospace technology in 1985, graduating Phi Theta Kappa and registered in the National Dean's List honor society registry. A highlight of her love for flying came in 1981 when she was a hostess at the 50th anniversary of Floyd Bennett Field in Brooklyn. She was companion for the day to Muriel Earhart, the sister of the 99s founder Amelia Earhart. The guest list included names from the early days at Floyd Bennett Field: Paul Rizzo, founder of Barren Island Airport, and astronaut Fred Hayes of Apollo 13. The images of Howard Hughes, Wily Post and Amelia Earhart taking history-making flights from this airport inspired many to follow them and Nancy is proud to be a small part of it. Bill and Nancy will retire to Florida in 1997 and will live close to an aviation theme park now under construction.

Davis has flown the Piper 140, 151, 181, J-3 on floats; Cessna 150, 172 on floats; 65K cubic feet Raven free hot air balloon.

Her children are William R. Jr., 32, and Steven M., 29.

HELEN S. DAVISON,

lifetime Pennsylvania resident. Learned to fly in 1968 and joined 99s same year. Active commercial pilot, instrument rated. Organized Greater Pittsburgh Chapter and charter member. Current chairman and served seven previous non-consecutive terms. Section officer and two terms as governor of Mid Atlantic section.

Participated in many air races: Powder Puff Derby, Air Race Classic, proficiency races. Won the Air Race Classic and placed in the top 10 three times.

Flies a Piper Lance regularly and has raced a 182 Cessna and 182 RG. Also raced a Beech Musketeer.

Earned a BS in home economics and English. Taught senior high school for many years. Co-sponsor of Aviation Explorer troop. Belongs to several pilot organizations. Married to Ellison L. Davison, a mechanical engineer, also a pilot. They have two sons and two grandsons.

DONNA M. DEAKEN,

First Canadian Chapter, East Canada Section, the 99s Inc. Donna is a recreational pilot with a scientist's career, expanding from forensic scientist and expert witness to manager, environmental health and safety.

As a child she wanted to "drive" an airplane: The only thing her father said a girl could not do. In 1977 she received her private pilot license as a surprise gift to the man she wished to, but didn't marry, then went on to earn an instrument rating.

In 1978 she joined the 99s for the educational seminars and flew Operation Skywatch missions when possible. The years 1980-81 saw her chairing the chapter air education seminars, which were awarded the Federation Aeronautique International Aerospace Education Award. During 1982 and 1983 as chapter chairman, she planted the tree from the Forest of Friendship with the lieutenant governor and introduced to the chapter Ski Day, Bunch for Lunch, APT day, and led the sponsorship of a member in the Canadian Aerobatics Competitions.

As governor of the East Canada Section during 1985 and 1986, she initiated the Canadian 99s local and national Science Fair Awards in aviation to encourage careers in aviation and aerospace. From 1987 to 1989 she was a member of the board of directors, Canadian Owners and Pilots Assoc.

SARAH M. DEAL,

First Lieutenant, in 1992, graduated from Kent State University's aerospace technology program and was commissioned in the US Marines. After basic school, Deal was sent to Memphis to Air Traffic Controller School. Upon completion, she had the distinct privilege of being the first woman Marine to be selected to attend flight school. In 1994, Deal reported to Pensacola NAS to begin her training. Although, she already possesses her CFII and MEI, Deal still had to learn the "military" way. Her training included survival, the "swim,"

and then, basics and instruments in the T34C, turbo Beechcraft. Upon completion of fixed-wing Deal moved in to the Bell Jet Ranger T57 for basic and instruments in the T57C. Deal expects to earn her wings in March or April 1995. She is a charter member of the Women With Wings Chapter.

BETTY EILENE DEBUAN,

Indiana Chapter, North Central Section, a Terre Haute, IN, native, married with five children, two daughters and three sons. Her husband, Curt, and two sons are also pilots. She has 12 grandchildren and three great-grandchildren. She works with her husband in the family funeral business.

Betty stared flying in 1965, and joined the 99s in the fall of 1966. She has held all chapter offices and served on many committees. She flies a 1968 Cessna Skylane, has flown to many section and International meetings, including Vancouver and Alaska.

Her memberships consist of the Wabash Valley Pilots, Aircraft Owners and Pilots Assoc., Experimental Aircraft Assoc., and Indianapolis Aero Club. She serves on the Indiana State University aero space advisory board and is a first lieutenant mission pilot and squadron finance officer with the Civil Air Patrol.

FIORENZA DEBERNARDI,

the only Italian woman airline pilot entitled to qualification as captain. Her father was one of the most renowned aviators of his time, so aviation has always been Fiorenza's life.

She began to fly with her father and during several years made tourism and air races. Then she acquired her commercial license and in 1967, she was hired by Aeralpi and became the first airline woman pilot in Italy (one of the first four in the world). She worked as airline pilot 18 years with Twin Otter, YAK 40, DC8. For the YAK, a Russian trijet, she took all training courses in Moscow, and flew all over the world. With the YAK she flew also for Olympic Airways in Greece.

Also glider and glacier pilot, approximately 7,000 hours/flying. Very proud because there is an ISA scholarship in her name and she is mentioned in the Forest of Friendship in Atchison.

LORNA DEBLICQUY,

born in Blyth, Ontario, Nov. 30, 1931, began flying in 1946. With ALTP fixed wing, commercial helicopter and private glider licenses, she has enjoyed flying for almost half a century.

Lorna has worked instructing on wheels, skis and floats; as a high Arctic Islands pilot on a tundra-tired Beaver; as a New Zealand instructor and charter pilot; and has flown famine relief in Ethiopia. She was the first Canadian female aviation inspector.

Among her numerous

awards in recognition of her contribution to the advancement of women in aviation, are the governor general's Persons Award, the Order of Ontario, and the Trans-Canada McKee Trophy, Canada's oldest and most prestigious aviation award.

First Canadian Chapter chairman 1952, a proud holder of the 99s Award of Merit, grateful winner of a 1970 A.E. Scholarship, and a participant in several AWTAR's, she feels it a privilege to have been a 99 for most of her flying career.

Joined the 99s in 1950. Military service included RCAF (Reserve) 1951-53, Communication Op. Received flight training mostly in Ottawa (Ottawa Flying Club). She has flown light singles and twins, twin offer, DC-3, Bell-47 and various sailplanes. Formerly employed with various clubs and schools in Canada and New Zealand; Atlas Aviation, Resolute Bay NWT etc., Transport Canada (DETE, still). She is divorced and has one child, Elaine, 29. "All flying experience is special! Aren't we lucky?"

MARY ELIZABETH DECANTER, born June 24, 1924, in Rhinelander, WI. Joined the 99s in May 1979. She has flown J-3, Ercoupe, Warrior II, Cessna 172. DeCanter has two children, Gina Marie Carter, 44, and Todd A. Cirilli, 41. She also has two grandchildren, Tara Meredith, 21 and Todd Meredith, 18.

A flight in a Stearman as a birthday gift in 1935 started her love of flying. In 1944 during nurses training, United Airlines told her she couldn't be a stewardess because she was too short, wore glasses and as a nurse, was needed on the war front.

At home in 1946, before she could drive, she learned to fly a J-3 and an Ercoupe.

After marriage, two children and a move to Oregon, she returned to flying.

She was instrumental in forming the Oregon Pines Chapter after a Palms to Pines Race, the race terminus, then at Independence, OR.

Flying highlights: Palms to Pines Air Race 1978, 1979, 1980; Chuck Yeager, (convention speaker), handed her the 1979-80 Oregon Pilots Assoc. Outstanding Pilot Award; and flying the Oregon Trail, 1991 and 1993.

She has served in the 99s as chapter chairman twice; Section A.E. Scholarship chairman; Section Forest of Friendship; state secretary of OPA; chapter president, secretary and treasurer of Polk County Pilots; president of the Aviation Breakfast Club of Portland; a member of the N.W. Ercoupe Club, and the Flying Farmers. She has been active in the children's activities of the Oregon Air Fair.

BARBARA L. DEEDS, born Feb. 22, 1939, in Thurston, OH. She learned to fly in Ventura, CA, in 1970, and went on to obtain her commercial pilot license at Santa Paula Airport, CA. A charter member of the Santa Paula 99s since 1973, attended six Southwestern Section meetings, held various offices in the chapter, flew two PARs and was on the committee to open Camarillo Airport for general aviation. While living in California, she participated in physiological training at the Pacific Missile Test Center, Point Magu.

Living on a farm in Ohio with a 2,000 foot grass landing strip, a 172 Cessna and three other pilots in the family, one of which is husband, Chuck, UAL pilot.

A member of the Scioto Valley Chapter, she flew in two Buckeye Air Rallies and learned to fly aerobatics. Barb and Chuck have one daughter, two sons and three grandchildren.

RUTH DEERMAN, born and raised in El Paso, TX. She learned to fly in 1944. She is a commercial pilot with flight instructor and instrument instructor ratings, advanced ground school and instrument ground school instructor. Helicopter pilot (Whirly Girl #78), past board member of Whirly Girls, her suggestion of "Collective Pitch" was chosen for the Whirly Girl publication. Past International president of 99s, founded the 66s, Winner of AWTAR (Powder Puff Derby) in 1954. In 1983 was the first woman inducted into the El Paso Aviation Hall of Fame.

Honored with memorabilia at War Eagles Museum in Santa Teresa, NM, where her plane, "Cotton Clipper Cutie," she won the AWTAR in, hangs from the ceiling. Honored with a granite plaque in the International Forest of Friendship in Atchison, KS. Presented the Jimmie Kelp Award in 1975. Texas Flying Farmer Queen in 1955. Past vice president of the El Paso Aviation Assoc. FAA Accident Prevention Counselor. Member Instrument Pilots Assoc. AOPA and NPA.

She has always taken an active part in civic, church and community affairs. Some of her achievements include 10 year board member of the American Cancer Society, where she drove cancer patients for treatment, past president Providence Memorial Hospital Auxiliary, past Queen Daughters of the Nile, past Worthy Matron of Eastern Star ... served on committees for Women's Department of the Chamber of Commerce, past treasurer of the Woman's Club of El Paso, former board member of the YWCA, active member of First Baptist Church and past president of the WMU, member PEO. She had articles published in *Flying* magazine in 1954 and 1956.

She joined Mary Kay Cosmetics in 1969 and is an independent senior sales director of the company, having won awards every year for recruiting and sales.

DEBI DEHAVEN, has been racing stock single engine aircraft since 1976. She enjoys competing in proficiency races and poker runs as well as in the Air Race Classic, and annual all-women transcontinental speed race. Her more than 2,300 hours have led her to many new air strips in many different states. Debi is rated to fly commercial, multi-engine land and instrument.

Her love of flying has involved her in many areas of volunteer work in general aviation. Debi has served as the chapter chairman of the 99s and has volunteered on committees to promote FAA safety seminars, airshows and proficiency racing. When the Red Cross called, she volunteered and flew emergency relief missions.

In order to insure an ample supply of future pilots, she has participated in the Adventurer program, which takes "young future pilots" for their first flight and gives them a glimpse of general aviation. Debi is a sales/marketing executive as well as a professional model.

LINDA DELL'OLMO, proud owner of a Piper Cherokee 180 based at Pocono Mountains Municipal Airport, PA, started her flight training after much "encouragement" from friends from whom she continually begged rides.

She joined the Keystone Chapter as a 66 in October 1993, soloed on New Year's Day (in a Tomahawk), and obtained her license October 1994. Linda most enjoys the focus of the Keystone Chapter (education). She has participated in many 99 activities including instructing boy scouts at their aviation merit badge camp out weekends.

Linda is a CPA, and her educational background includes a BS in accounting from George Washington University and MBA concentrating in finance and information systems design from Rutgers University. She is also active in karate, trap shooting and scuba. She is a member of the 99s, Inc., Aircraft Owners and Pilots Assoc., and Cherokee Owners Assoc.

BARBARA REIBEL DELONG, born Dec. 10, 1933, developed a love for flying when her father took her for a ride in a Ford Tri-motor at the age of 8. Her middle years were devoted to husband, daughter and three sons. When the youngest went to West Point, her husband encouraged her to pursue flying. Her goal was to solo by her 50th birthday and she made it a day early. After her husband's early death, she added instrument and commercial ratings. Leaving a position as assistant treasurer in the local school district to became secretary at an FBO was a major decision, but put her closer to airplanes. She owns a C-172 and has flown in several all-Ohio Buckeye Air Rallies and won the spot landing in 1994. In addition to a loving family, she enjoys quilting, sewing, reading and sharing her enthusiasm for flying.

LINDA LEE DENETT, born on July 6, 1950, in Biddeford, ME. She married Paul Denett when she was 18 and they had two children, Scott and Michael.

She met Linda in 1978 at Leesburg, VA, Airport. She already had her private pilot license and was working for Century Aviation as a dispatcher.

Linda and she became best friends. As a student pilot, Linda helped her with the bookwork and as her career progressed she was one of her biggest supporters. They flew together as much as possible, ferrying broken airplanes or flying anything they could as much as possible.

Linda ran the flight school for Dulles Aviation in Manassas, VA, from 1980-83. About the same time, she joined the Potomac Chapter of the 99s. She was chairman or vice chairman for several years and had a great talent for raising money. She put lots of energy into fund-raisers and organizing sectionals.

Linda died on Oct. 12, 1994, from Lymphoma. She was 44. She was loved by everyone in the Potomac Chapter, and she is greatly missed. *Submitted by Robin Hosenball Kidder.*

EDITH LITCHFIELD DENNY, First Canadian Chapter, became a 99 in 1960. She learned to fly because she was scared of flying. By 1964 she and her husband Wally both held Class I multi-instrument ratings (sea and land). Edith logged many hours at the controls of the early Goodyear blimps.

Edith re-activated the Canadian Section in 1966 and served as governor for two years. She traveled a great deal overseas and was elected chairman of the 99s International nominating committee in 1969, International executive board in 1970, 1971 and International coordinator through the 70s.

Contributing enormously to the 99s organization, Edith established sections in Iceland, Finland, Kenya (East Africa), Zimbabwe (Flame Lily) and Colombia, S.A. (Orchid). She also sponsored members at large in Sweden, Equador, Japan and East Africa. She has served as a director of COPA (Canadian Owners and Pilot's Assoc.) is a member of AOPA, NAA, Wingfoot, Lighter-than-Air Society and the Australian Airwomen's Assoc.

Edith and Wally Denny are now members of the prestigious UFO Organization (the United Flying Octogenarians). They have flown all over Canada, from coast to coast, Southern America, the Andes and the Caribbean Islands in their magic carpet, a twin Piper Aztec. They reluctantly sold their airplane a year ago.

Presently retired, living in Litchfield Park, AZ, Edith has 10 grandchildren, nine great-grandchildren and still travels extensively all over the world with Wally.

SHARON DENSMORE, born July 26, 1943, in Fordyce, AR. Joined the 99s in 1991. Flight training included: private 1991; instrument 1992; commercial 1993; AAS degree in aviation/professional pilot 1993. She has flown lots of planes: SEL; favorite: toss-up between Citabria for aerobatic training in 1991 and V-tail Bonanza for commercial training. Used aviation for Oklahoma Aerospace Academy, directed av/aerospace program that she takes into schools statewide. She has three children: Laura D. Jergensen, 24; Valerie Densmore, 21; and Sarah Densmore, 16.

"From chasing the light-house lapping waves of Maine's rocky coastline, to flying the sparkling Pacific waters out of Maui to the other islands, to skimming the mountain ridges to come in for a taxi-way landing in Angel Fire and a dirt-bottle neck canyon runway in Idaho, to being engulfed by and chased down by one of Oklahoma's sudden T-storms, to savoring the footless halls of the slick night air above Oklahoma, flying has exploded my life with new people and great vistas! Is a transatlantic flight in my future?"

CAROLE B. DEPUE, born Westchester, PA. Soloed Sept. 9, 1959, at Santa Monica Airport, CA. Received flight instructor rating 1965 and instructed for Claire Walters Flight Academy for three years, then moved to Las Vegas, NV; began flight and ground instructing at Nevada Aviation, which was to become ScenicAirlines. Flew customers to the Grand Canyon in 1968 and 1969, then moved to Pocatello, ID. Las Vegas 1975 - owned Grumman American dealership and flight school. Flown in eight Powder Puff Derbies, four AWTARS, four Pacific Air Races, two Henry Oyhee races, three Palms to Pines and others. Held all chapter offices and SW Section treasurer. Won Amelia Earhart Scholarship 1978, second place in Citabria Aerobatic Contest and did two air shows. Invited by Duane Cole to participate in national finals in Reno. Ratings held: commercial pilot, instrument, CFI airplane and gliders, ground instructor. Member 99s, AOPA, NAA.

She has flown the following planes: Cessna 150, 172, 180, 170, 182, 210; Piper 140, 160; Comanche Cub-Grumman Amer; Tiger; Cheetah; Citabria 150 hp; Bonanza H and A36, T-34; T-craft; Aeronca all small a/c.

She married Ben W. DePue in 1968, they have two children, Genevieve Golden, 25, and Beniah DePue, 24. They also have grandchildren: Robert Golden, 6; Victoria Golden, 3; Alex Golden, one month; and Honey DePue, one month.

SAUDAMINI DESHMUKH, born March 9, 1952, airline pilot with Indian Airlines since 1980. Presently deputy general manager. The following are among her many achievements: Captain Fokker Friendship F-27, August 1985. First female check pilot in Asia F-27, March 1987. Captain All-Women-Crew Flight F-27 Nov. 29, 1985. First such flight in world on any aircraft, among IATA member airlines. Captain Boeing 737: Aug. 1, 1988. First female jet captain in India. Captain All-Women-Crew Flight Boeing 737: Sept. 16, 1989. First such flight on jets in Asia. Captain Airbus A-320, May 30, 1994. First A-320 female captain in Asia. Captain of all-Women-Crew Flight A-320: June 9, 1995. Probably first such flight in world on A-320.

Her awards include: 1977, Achievement Award, the 99s Inc., San Francisco, USA. 1988, Certificate of Appreciation, the 99s Inc., Shangri - LA - USA. Airlines with women pilots were honored. Saudamini represented Indian Airlines. 1991, the Forest of Friendship Award, Atchison, USA. Felicitations from many organizations.

SUZE DETOMBE, born Jan. 22, 1959, in Tamworth, England, moved to Scarborough, Ontario, in 1963. As a teenager she became one of the first girls in Canada enrolled in the Royal Canadian Air Cadets and served with 631 Squadron. She received her private pilot license at the Brampton Flying Club in March 1993, at which time she became a member of the 99s First Canadian Chapter. Suze produced the 94/95 Chapter membership directory and served on the chapter's nominating committee. She is on the organizing committee for the third Canadian Women in Aviation Conference and is a member of COPA. She has a night rating and did a mountain flying familiarization course at the Calgary Alberta Flying Club. In March/April 1995 she is flying to the Bahamas with three other 99s and is pursuing further licenses and ratings. She is keenly interested in aviation safety and aviation psychology.

AMY DEVRIES, earned her private license on her 17th birthday, Dec. 12, 1992, in Sulphur Springs, TX. She immediately became involved with the Wildflower Chapter after being told about the organization by her flight instructor and after meeting other 99s while she was a student pilot.

She is currently a student at Southwestern University in Georgetown, TX. She flies a Piper Warrior whenever free time allows her to. She is an active volunteer with Mount Vernon Airport fly markets and also a member of AOPA. Recreational flying allows her to be active in other hobbies such as hiking, biking and locating historical sites.

HILDA DEVEREUX, born London, Ontario, June 19, 1921, received her pilot license at age 43 and in the next 15 years attained commercial rating, IR.II. Completed instructors course, 1970 flew Angel Derby. Has held most chapter and section offices, including governor in 1972 when International Convention was held in Toronto. Served on International committees. Owned two beautiful airplanes, CES 182, CF-JGT and Commander 100, CF-UXX which took her on the Angel Derby to Nassau. Today she spends her time between her apartment in London Ontario, and her condo in Largo, FL, and enjoys the friendships of Maple Leaf Chapter 99s, members of Florida Grasshoppers, COPA and Silver Wings Club.

In 1990 her chapter honored her by sponsoring her induction into the Forest of Friendship in Atchison, KS. Her favorite pastimes are hangar, flying with the Bunch for Lunch on Wednesdays and attending International Conventions.

NANCY DEVEZE, began her love of aviation when she first saw Amelia Earhart in news reels on her family's first television. She and her brother often told their parents they were going off to church, but instead they would watch planes take off and land at O'Hare Airport.

June 9, 1962, she was 17 years old, still in high school and did not legally drive a car, but she had her first flight lesson in her dad's Ercoupe. March 1993, with exactly 40 hours, she earned her private pilot certificate. Her dream was to have a career in aviation or be an artist, but instead she married a pilot who was living her dream, and she went on to law enforcement, which was his secret dream. Her creative side is now just a hobby.

By 1966 her husband was in the Army. She continued flying a Cessna 140, a Cessna 150 and a Champion Citabria in Germany where her husband was stationed as a flight instructor.

She and her husband own their own aircraft, a Cherokee 140 with a 160 engine. They have taken trips that last close to a month at a time. From northern Illinois (her home base), she has flown their Cherokee as far south as Key West, FL, west to Montana and Wyoming, north to Canada, and east to Vermont. One year when they were planning their trip to Texas, they could not find a cat-sitter. Three and a half weeks they smuggled an 11 lb. cat into the best of hotels. Can you imagine their 11 lb. calico cat being walked on a leash through the Alamo? She has photographed her low-altitude photo flights around Mount Rushmore, the Statue of Liberty and Niagara Falls. She has landed her aircraft in a total of 32 states and Canada.

Highlighting her love of aviation, she recently, (1994) had a "dog fight," with a Boeing Stearman she flew versus a roll of Charmin 300 toilet paper. Another great experience was to fly WWII B-17G. She hopes to purchase a Cherokee 235 and fly "North to Alaska."

JOAN DEWITT, learned to fly at Reid-Hillview Airport in San Jose, CA, in 1977. She has logged over 700 hours and has her instrument rating and a commercial license. Joan joined the Santa Clara Valley 99s in 1981 and enjoyed flying activities in her Piper Warrior.

She and her pilot husband have two children and two grandchildren. Joan works as an office manager for a software manufacturing company, and enjoys bridge and hiking.

SUZANNE MALONEY DEWULF, born Green Bay, WI, USA. Holds a private pilot license and has flown the Cessna 150, 172 and 182. She and her husband, Marcel DeWulf, have one child, Denise Monique DeWulf Martin, 29. They also have grandchildren, Alexis M. Martin, 2, and Quinn P. Martin, 6 months. She is a charter member and first chairman of Aux Plaines Chapter (North Carolina Section). Earlier joined the 99s with Chicago Area Chapter. Held safety chairman at section level.

DOROTHY KATHLEEN DICKERHOFF, born in Clovis, NM, in 1933. At the age of 9 or 10, she went to a birthday party dressed to represent what she wanted to be when she grew-up - a pilot! At 16, she was a cadet member of the CAP and took her first flying lessons, which were interrupted by a move and then college. She worked for TWA as a hostess on the Martins and the beautiful Constellations, watching in the cockpit as often as possible. Then came marriage to George and three children, Bonnie, Randal and Robert. Finally in 1968 at the age of 35, she became a pilot, and in 1971, joined the 99s. She taught ground school at the airport and at a nearby community college. After obtaining her commercial certificate, she began flying charter for several small FBOs and doing freelance pilot service. During this time, she obtained an instrument rating and flew in the 1976 Powder Puff Derby. Later she was office manager, bookkeeper and assistant chief pilot for the local FBO. She got her multi-engine rating, and after receiving an Amelia Earhart Scholarship, an instructor's rating, and instructed for the same FBO. For a while, she and a friend bought and sold airplanes. Her daughter received her license at the age of 17. Aviation has been and continues to be a big part of her life. She tells her six grandchildren she feels very blessed to be a part of it.

RUBY KATHRINE SMITH DICKERSON, born on Oct. 5, 1936. Her flying career began in 1970 when husband John's flying lessons interested her, and she began training in Raymond, MS. She received her private certificate in April 1971, and followed it up with the commercial certificate and instrument rating. After transferring from Jackson to Birmingham, AL, in 1977, she obtained the multi-engine rating, CFI and CFII.

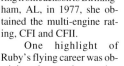

One highlight of Ruby's flying career was obtaining the seaplane rating during the International Convention in Anchorage, AK, in 1984.

Ruby and John both took early retirement and moved to Natchez, MS, in 1990. They now live only two miles from the Natchez-Adams County Airport, where their Skyhawk is hangared. Ruby is still an active CFII, and a member of Alabama Chapter 99s.

MARILYN DICKSON, First Canadian Chapter, in 1987 she attended the 99s International Convention with her friend Jean Franklin, simply because she wanted an excuse to go to Vancouver. Once there she went to a few sessions and met such fascinating women pilots she was hooked. What struck her most was the number of women who learned to fly in their 40s and 50s. It wasn't to late for her.

Immediately upon returning home she found an instructor, 99 Adele Fogle, and began taking lessons. A few days before her private flight test, she and Jean bought C-FEVJ, a C-150 with which they have had wonderful adventures. In 1989, she earned her commercial license and has logged many hours as a Skywatch pilot.

Jean and she have both returned to university and currently are "between planes." She looks forward to earning an instructor's rating next year after she completes her doctoral dissertation.

ANNA JO DIESER, born Hickory Flat, MS, April 17, 1941. John, her husband of 33 years, is a building contractor. Their son John, age 30, is an accountant.

She has wanted to fly since she was 4 years old. She became interested in airplanes, especially the B-17, when her parents talked about the planes during WWII.

She took flying lessons in 1967 at Salinas, CA, Air Service. She soloed a Cessna 150 in September of that year.

She received her private license and joined the Monterey Bay Chapter of the 99s in August 1968. By 1971, she had obtained her commercial, instrument and multi-engine ratings. She received the Monterey Bay Chapter Pilot of The Year trophy in 1971.

Dieser has flown the Cessna 150, 172, 177, 182, 337, 310, 401, 414 and 421. She has also flown the B model Bonanza. She received her multi-engine rating in a Cessna 310B. Her dream came true when in 1994, she was able to fly the B17 G model belonging to the EAA B-17 Historical Society.

Her highlights have been: Flying new planes to California from the Cessna factory in Wichita, participating in the Pacific Air Race and the 1976 Powder Puff Derby, owning her own Cessna 182, and flying the EAAs B17. Currently she is a member of the Jackson Gold Dust Chapter of the 99s. She is proud to say that she has been a 99 for 27 years.

ANNA MARIE DIETRICH, (March 12, 1909 - Nov. 27, 1994). Anna 50 plus, learned to fly at Brookhaven Airport, Long Island, NY, and made three parachute jumps as a member of the Long Island Skydivers. As a private pilot, she flew her Aerocoupe whenever time permitted. In 1962, she joined the 99s as a charter member of the Long Island Chapter. She participated in many of the chapter and section flying and fund raising activities. She flew two New England All Women Air Races and in 1972, with Ellie McCullough, came in third in her Piper Cherokee 140. Although she did not hold any offices, Anna loved flying and offered the use of her airplane for all 99 sponsored events.

Anna retired to Florida in 1974 and later moved to California to be near her daughter, Betty. Anna passed away on Nov. 27, 1994. May she be flying to her heart's content.

JAN DIETRICH, Bay Cities Chapter, soloed at 16, joining the 99s in 1947. That year, with less than 100 hours each, she and her twin, Marion, won a speed handicap race. With less than 200 hours each, they placed second in the AWTAR against 87 planes.

Jan was president of the University of California Flying Club flight operations, an FAA FI, was one of 13 women to pass the preliminary tests for astronaut, and ferried a plane to Germany in 1965.

In 1960 she was one of the first 10 US women to obtain an airline transport pilot license, with multi-engine, flight instructor, and seaplane ratings. Jan then went into corporation flying – Convair's, D-C 7s, and in 1968 was the first woman in the US to receive an airline transport pilot license in a 4-engine jet. In 1967 she was the first to receive the Flight Safety Award from the NBAA with over 12,000 hours.

She is currently on special assignment to NASA and received a certificate of commendation from the administrator.

SANDRA RUTH DIETZ, born Oct. 3, 1958, in Batesville, IN. Joined the 99s in 1994. Private training, Greensburg, IN, licensed in January 1990. She has flown Cessna 150 and 172. She and her husband, Dennis have one child, Elizabeth Louise, born Oct. 10, 1990. Dietz has wanted to fly since she saw "Peter Pan" in the early 1960s, in fact, it upset her so that she hasn't watched "Peter Pan" since! She is a secretary for Batesville Casket Co., and lives in Yetamora, IN. She is also a member of AOPA.

ELIZABETH CROWLEY DINAN, learned to fly at Grand Central Airport, Glendale, CA, age 17. Has been a San Fernando Valley 99 since 1958. Holds commercial, CFII, seaplane, multi-engine and ATP and instructs in San Luis Obispo, CA.

She was 1972 Chapter and Section Woman Pilot of the Year for her work with Wing Scouts and awarded the 1977 Amelia Earhart Memorial Scholarship to obtain multi-engine rating. Has flown in several All Women's Transcontinental Air Races, All Women's International Air Races, Air Race Classic, All Women's Baja California Air Races, Palms to Pines and various local air races.

Climbed Amelia Earhart Peak in 1987 to commemorate AE's 50 year disappearance. Loves vintage and WWII aircraft and belongs to OX-5 Aviation Pioneers as historian, Silver Wings, Vintage Airplane Assoc., AOPA and board member of San Louis Obispo Pilots Assoc. and Accident Prevention Counselor.

Has traveled in Europe, Africa, Asia and South America. Lives in San Luis Obispo, CA, with her husband Don, dog Ted and three cats.

DELORES ELAINE DITTON, born April 6, 1934, Bedford, IN, started learning to fly in 1979, following a childhood dream. Owes success in getting license (August 1981) to Judy Graham (99 deceased) who introduced her to 99s. Started with Indiana Chapter, then she encouraged Judy to start a local chapter which became the Three Rivers Chapter in which Ditton held various offices.

Soon purchased a PA28-180C. Obtained an instrument rating in March 1991. Educational background as registered nurse with BS degree in health arts.

Member 99s, Inc., AOPA, Cherokee Pilots Assoc., Greater Fort Wayne Aviation Museum.

"Being a 99 has helped me in all aspects of life by meeting so many multi-talented, inspiring women spanning several generations of accomplishments."

She has flown Cessna 152 and 172; Piper Tomahawk; Cherokees 140, 160 and 180. She and her husband Louis have two children, Cynthia, 34, and Ryan, 31. They also have grandchildren: Benjamin, 2, and Stephen, one month.

LORI LOUISE HARALDSON DIVERSEY, born April 20, 1960. Growing up, mom Nancy, a Chicago Area 99, wanted her to join her in her love of flying - I said, "I'd jump first." Her chance came, as a freshman, at Illinois State University. After parachuting a couple of times, she decided it was time to stay in the plane. She began instruction October 1980 and earned her private license May 1, 1981.

The next fall she transferred to Arizona State University, Tempe. Pursuing a BS in business administration, she also took advantage of the aviation program offered, joined Alpha Eta Rho (Aviation) fraternity and became a member of the Phoenix Chapter 99s. She participated in various NIFA activities and was a pilot in the 1982 Kachina Doll Air Derby. She worked briefly at Chandler, AZ, Airport.

As copilot for dad Wayne, in several Illi-Nines Air Derbies, she received a first place trophy, proficiency category, in the 1982 race.

She has flown the Cessna 150, 152, 172, a Tomahawk Decathlon and Mooney 201. Simulator time in an ATC-610K at ASU and a T-37 at USAFB, Chandler, signed by Captain David Berkland.

After graduation, May 1983, she began work at the Chicago Mercantile Exchange. She became a Chicago Area 99. While working toward the instrument and commercial ratings, she took her most memorable flight - under the hood from Chicago to Tampa and back.

In 1984 she became a commodity trader and met her future husband, John. She was able to give him his first ride in a small aircraft. Got married March 1987. She is at home, Park Ridge, IL, with their children: John III, 7; Kyra, 5; and Ryan, 9 months.

Her community and family activities include helping with the 99s project "Air Bear" aviation presentations. Was the 1993 vice president of the Park Ridge Newcomers Club. Currently she's a "room mom" and "computer mom" for son's class, and it doesn't stop there.

She hopes to raise her children to the wonders of aviation and to the discoveries to be made.

GWEN JAKSICK DIXON, born May 16, 1939, Reno, NV. Flight training includes single engine, land, sea, multi-engine instrument; type rating, Cessna Citation 501 SP. She has flown Super Cub, Aeronca, C-182, T-210, Cessna Citation 501SP. Gwen is a widow with three children: Stan, Wendy and Todd Jaksick; and a grandchild, Lexey Jaksick, 7 years old.

She learned to fly in Carson City, NV, in an Aeronca 0. Her instructor was Joe Williams, who's motto was "Learning To Fly The Hard Way."

She completed her private training in September 1974 and went on to get her instrument and multi-engine ratings.

Gwen joined the local chapter of 99s in 1975. Her most recent accomplishment has been her type rating in a Cessna Citation 501 SP.

In 1975, Gwen flew to Fairbanks, AK, and received her seaplane rating. She has been flying for 21 years and has enjoyed her many years as a member of the 99s.

This September was Gwen's 27th year as a volunteer for the Reno Air Races. She started as a home pylon judge for the races and now works in the Race HQ administering flight credentials.

Gwen is a native Nevadan.

JOANN HOWEY DOBBINS, born Oct. 13, 1935. Joined the 99s in 1993. Flight training includes private pilot and stewardess (Capitol Airlines) 1958-61. She has flown Cessna 150 and 152; 120 Taildragger; Bonanza E-35. She and her spouse, Robert, have four children: Deborah Ann, 34; Robert Jr., 33; Brett, 30; and Heather, 27. They also have grandchildren: Samantha, 7; Ashley Brighton, 5; Devin Frazer, 5; Robert Dobbins III, 7; and Anika Dobbins, 5. She received her private pilot certificate on Dec. 22, 1993, "greatest thrill of my life."

RUTH SCHILL DOBRESCU, Phoenix 99s, is a life member of the 99s. She got her private ASEL in 1965, and became a charter member of the Long Island Chapter, having shared her husband's interest in aviation enough to learn to fly when he did. He's a retired TWA airline pilot with many years of experience. Their two girls have given them six grandchildren. Their shared record of volunteer achievement in both aviation and civic betterment is awesome.

In the 99s, Ruth has held many chapter offices (including chairman) and section offices (including governor). She was on the International board of directors six years and chaired many committees at the International level. She worked atop the Empire State Building as a timer at the 1969 London (to New York) Daily Mail Great TransAtlantic Air Race. She flew in two air races. The Dobrescus headed a national drive to get a commemorative stamp issued on behalf of Long Island's Cradle of Aviation Museum honoring the 50th anniversary of Lindbergh's solo across the Atlantic. Ruth continued with limited issues in honor of other aviation feats, including one for a Russian crew which had flown 13,300 miles in 1929 to land at Curtiss Field, and been greeted by Charles Lindbergh.

This lady, who believed in taking on anything that really needed it, had started caring for foundling babies in big hospitals as a teenager "because I felt I was making a real difference in someone else's life." She followed this rationale through years of tireless work in successful civic beautification, earning lots of appreciative press coverage. But of all the recognition Ruth has had, she is especially proud of the award she was given in 1977 at Long Island University along with another recipient, astronaut Walter Schirra, for their contribution to aviation.

MARGARET DODSON, born Aug. 31, 1946, in Harvard, IL. Joined the 99s in 1993. Flight training includes private pilot training at Dacy Airport, Harvard Island, commercial pilot training ASEL and MASEL at Mount Olive, NC. She has flown Cessna 150 and V-Tail Bonanza. Dodson is self-employed farmer and housewife.

She and her husband Robert have two children, Debbie, age 26, and Robin Marie, age 15. They also have a granddaughter, Shelby Marie Etes, age four months. Dodson acquired her private pilot license in March 1987. Acquired commercial rating in April 1994, and is presently working on becoming a CFI.

DEBRA TOMLINSON DOLAN, born April 16, 1955, in Manchester, CT. Joined the 99s in 1988. Flight training, 1981-88, primary and inst. training at Windham, CT; commercial and multi-engine training at Groton - Norm London Airport, 1995. She has flown the following planes: Cessna 150 and 152; Cessna Cardinal; Cherokee 140; Piper Cherokee Arrow; and Piper Seneca. Her education includes AS, Mfg. Eng.; member of Phi Theta Kappa. She is married to Col. William Robert Dolan, who has ATP C550 C500 BE1900 BE300. She is past chapter woman in Connecticut. She flew the last civilian airplane into Rentschler Field, East Hartford, CT, Dec. 12, 1994.

GIBBY DOMBROSKIE, started flying in 1991 and completed her private pilot in 1993. In 1993 she joined the 99s and the All-Ohio Chapter. At an early age she would fly with her father, who later was also her flight instructor.

She was involved in the six-state, six-stop, flight of the five-handled ceremonial shovel for the observance of Earth Day 1995. Being part of the first leg of this historic event was a great honor.

Other organizations to which she belongs are: AOPA, EAA, International Women's Air and Space Museum, Flying Angels, Inc., and Dayton Pilots Club.

Gibby is also lucky in the fact that she has a husband who shares in her enthusiasm in flying and is also a pilot himself.

NELDA MCDOWELL DONAHUE, a resident of Garfield, AR, was born Nov. 16, 1936, in Shreveport, LA. She earned her private pilot certificate in Arizona in 1984; most of her hours have been logged in a Piper Cherokee 235.

While a member of Phoenix Chapter of the 99s, she held several offices and participated in airmarkings, air lifts, and Flying Companion Seminars. She is currently a member of Arkansas Chapter and serves as vice chairman and newsletter editor.

A retired public administrator, Donahue is a quilter and March of Dimes Birth Defects Foundation volunteer. She also serves Alpha Xi Delta, a Panhellenic fraternity, as regional financial advisor working directly with seven collegiate chapters of the organization in Arkansas, Louisiana and Texas.

Donahue is also a member of Aircraft Owners and Pilots Assoc., International Institute of Municipal Clerks, and Arizona Municipal Clerks Assoc.

Donahue's background includes an AA in business from Central Arizona College. She is a certified municipal clerk, certified professional secretary, and received the Quill Award from IIMC.

She is married to Jerry Donahue, also a pilot and a retired fire chief. She has two daughters, two sons, two stepchildren and eight grandchildren.

PATRICIA LAMBERT DONALDSON, Phoenix 99s, first aviation experience was her father's taking the family to Teterboro when she was small. She got a commercial glider license at Elmira, and then soloed a fixed-wing at Pai-Waukee. Getting her private ASEL in Phoenix in 1959, she joined the 99s at once.

Pat held several offices in the Phoenix Chapter and was chairman in 1963. She served as section secretary and then governor in 1967. She recalls as governor flying all over the section presenting charters - it seems 99s were really growing then. She flew many races, sometimes with her mom as copilot. She flew solo in the 1960 Powder Puff, winning an award for the last leg.

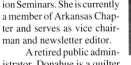

Her funniest memory was flying back to New Jersey, picking up her mother, and visiting all the family, offering rides at each stop. All the female relatives went up but she couldn't get a single male kin to go. On the return trip there were generator problems and rain. She went to the weatherman but she and her mother were ignored in their dresses and heels. The FAA was more cooperative, and they made it to Phoenix safely. Pat hasn't flown for over 15 years, but she still has a spot in her heart for aviation and the 99s.

ACHSA B. PEACOCK DONNELS, started to fly in 1923 at age 19 in Fresno, CA. She has no idea what impelled her to learn, but was fascinated with airplanes and engines. The airplanes at that time were JN4D's, called Jennys, and her teacher was Loxla Thornton. Thornton had no arms; his left arm was off below the elbow and had a hook on it that was tied to the stick. His right arm was off below the shoulder, and he used the shoulder to activate the throttle. Those years, people who flew just enjoyed the wonderful clear, clean air: No smog, no regulations.

She had objections from her family when they found out, and her flying was curtailed for awhile. A year later, she and her boyfriend flew his Jenny and as time went on, he became very chauvinistic. She was able to work around that. She received her private license in March 1928 (#3289) and bought a Waco 10 airplane. Her limited commercial license came along in August of that year and she received her transport license in 1931. She taught flying during all of those years. She and her husband and their partner had flight schools in Bakersfield, Salinas and Fresno, and an airline to the old Grand Central Airport in Glendale, which ran twice a day. They also had an airline from Fresno to Alameda and San Francisco. Later on, they had a contract with the Mexican government to carry mail from Mexico City to Los Angeles.

In November 1929, she became a charter member of their new 99s Organization of Women Pilots. These wonderful girls have been one of the joys of her life. She tries not to miss any sections and internationals.

In July 1990 she was pleased to received the Katherine B. Wright Memorial Award which had previously been presented to Anne Lindburgh and Moya Lear.

TERRI MARTIN DONNER, born in 1960 in Louisville, KY. At age 2 she moved to Madison, WI, where she grew up and took flying lessons at age 17 at Morey Airport. She flew pilot in command in the Air Race Classic at age 19, while a meteorology major at the University of Wisconsin. Terri completed her flight instructor rating in 1980 and was hired by the Wisconsin Bureau of Aeronautics as the state airport inspector. In 1981, she was awarded the Amelia Earhart Memorial Scholarship from the 99s for her multi-engine rating.

In 1985 she was hired by Stevens Aviation in Louisville, as a flight instructor, and worked her way through aircraft sales and charter. She got her airline transport rating in 1986 while six months pregnant with her second son.

In 1989 Terri was hired by United Parcel Service as a flight engineer on the Boeing 727. In 1991 she upgraded to First Officer.

Terri married Ray Donner and they have three children, Nicholas, Charles and Amelia. Ray has a private fixed-wing license and a lighter-than-air rating. They own their own hot-air balloon and the family spends evenings and weekends flying all over the country.

Terri has been very active in the 99s since age 18, serving as chairman of both the Wisconsin and Kentucky chapters. She is also a member of ISA plus 21, the International Society of Women Airline Pilots. Terri was part of the first all-female B727 flight crew at UPS.

FLORENCE C. DOOLEY, born Jan. 28, 1930, in Troy, NY. As a small child she always wanted to touch the clouds and made that possible by earning her pilot license in 1978. She now has logged over 300 hours of flight time. She has three children, two sons and a daughter, and three granddaughters. She has worked for the New York Senate as an administrative assistant for 19 years. She also worked as education coordinator at a local hospital for six years before that.

Her flying adventures started with a ride in a Piper Cub on floats. She has flown all around the Northeast in that Cub, both on floats and on land. Other planes she has flown include Piper Warriors and Cessnas 172, 150 and 152. Flights have taken her to New Orleans, Florida and the Bahamas. She has also flown in poker runs and other flying contests here in the Northeast. She has some soaring and ballooning experience and one lesson in hang gliding.

She is a second lieutenant in the New York Civil Air Patrol as a pilot and observer. While in Anchorage, AK, for a 99s International Conference, she visited the Anchorage CAP group and ended up flying 10 hours with them, some of the time in a Beaver. This proved to be a memorable experience.

Dooley's other interest include quilting, painting, flower arrangement, skiing and playing the piano. She also is a master gardener with the Rensselaer County Cornell Cooperative Extension.

She has three children: Bernard H., 45; Jeff, 43; and Kimberly, 39; and grandchildren: Laura, 9; Rachel, 5; and Emily Alida, 3.

FAY DOORNEWEERD, born in St. Cloud, MN, on June 17, 1943. She and her husband, Dennis, reside in Port Richey, FL, where Dennis works at the VA Outpatient Clinic.

After their children graduated from high school, Dennis pursued a long-time dream of learning to fly. Because of his "age" he encouraged Fay to get her license "in case she had to get the plane down." In 1992, Dennis, Fay and their son David all became licensed pilots. In addition, their daughter, Susan, has soloed.

In 1993 Dennis and Fay bought a Cessna Skyhawk II. They have flown up the eastern coast of the US visiting Susan, David and his wife, Christina, and friends.

Fay, a homemaker, is a member of the 99s and the Florida Suncoast Chapter of the 99s and is actively involved in her church and Bible Study Fellowship International.

TRINA DORSEY, a fourth-generation Californian, was born in Los Angeles. She learned to fly in 1993 at the Van Nuys Airport, the busiest general aviation airport in the nation. Her interest in flying began while traveling throughout the Far East and Asia with her family when she was very young. On occasion, she was allowed to sit right-seat in various aircraft and talk to controllers. This left an impression on her that would later lead her to pursue becoming a pilot. Trina flies Cessna 172s and is a member of the 99s, Inc., Aircraft Owners and Pilots Assoc., Angel Flight and the Flying Samaritans, where she also volunteers as a translator and medical/dental assistant to provide health care to remote parts of Baja, CA. Trina's educational background includes a bachelors' degree in biological science. Other community activities include American Red Cross volunteer in the advanced first aid team and disaster services, Los Angels Unified School District's Adopt-A-School program, and a lector at her local church.

DENISE A. DOSTOLER, born Jan. 22, 1957, in Worcester, MA. Flight training includes private, instrument, commercial, advanced ground instructor. She has flown PA28-161, 181, and Grumman AA-5B. Her flight experience began in 1977 when she skydived 3,000 feet out of an airplane, but she got so scared that she then decided that her dreams of flight would continue, but would remain inside of an airplane.

Denise joined the 99s in 1985 shortly after she obtained her private pilot license. Since then, she has obtained her instrument rating, commercial and advanced ground instructor's licenses, and logged more than 1,000 hours in Piper Warriors, Archers, and Grumman Tigers.

Denise has been a very active member of the 99s, serving in positions of Connecticut Chapter membership chairman, secretary, vice-chairman and chairman, as well as New England Section secretary. She is presently the NES vice-governor and membership chairman.

Denise is an avid island-hopper, where she spends most of her flights flying out to Block Island, Martha's Vineyard and Nantucket. Another favorite flight of hers is the "Hudson River Route" where she frequently takes passengers down the Hudson at 500 feet (or less) AYL for a real tour of New York City and the Statue of Liberty.

She is a senior structural engineer, part-time art student, and has an expertise in drawing airplanes.

MELODY DOUGHERTY, born Sept. 17, 1956, in Cape Girardeau, MO. She received her private pilot certificate in April 1992, and her multi-engine rating in April 1995. She flies a 1974 Piper Cherokee 140 and a 1963 Piper Comanche 250. She received her multi-engine rating in a Cessna 310. She has logged over 380 hours and is currently working on her instrument rating.

Melody is a proud member of the 99s, as well as the Aircraft Owners and Pilots Assoc. She has served as chapter secretary and is currently vice chairman of the Dallas Chapter of the 99s. She has participated in the chapter's annual poker run as well as numerous other chapter-sponsored flying activities. She has also participated in the EAA's Young Eagles program.

Melody graduated from Southeast Missouri State University in Cape Girardeau, July 1978, with a BS in business education. She currently is an asset management analyst for Invesco Realty Advisors in Dallas, TX.

In addition to flying and her activities with the 99s her hobbies include dancing, snow skiing, hiking, camping and reading.

PAULINE LOUISE DESJARDIN DOUGLASS, born April 6, 1945, in Worcester, MA. Joined the 99s in 1992. Soloed in San Salvador, El Salvador. "As soon as I returned to the US, I told my dad, let's go flying." They went out to the airport, and only then did she tell him she was taking him flying and she was sitting in the left seat. Many years have passed, but in her mind she can see the smile on his face, and how proud he was after they taxied in, and secured the Piper Colt. For a whole year she never let on in her letters home, she had been taking

flying lessons. Private and commercial, Hollywood, FL; instrument, Ft. Lauderdale, FL; multi-engine, gliders, Opalocka, Tamiami, FL; sea plane, Winter Haven, FL. She has flown the following planes: Cessna 150, Piper Colt, Tri-Pacer, Mooney MK21, Cessna 172, Piper J-3s, PA-23, SGU 2-22, Blanik L-3. More than 20 years of aviation experience encompassing positions as corporate, charter and airplane flight attendant. Eight years of supervisory experience including positions as a chief flight attendant and flight attendant instructor.

Her recreational and professional travel experience throughout Central and South America, the US, Caribbean, Europe, the Middle East, Asia, Africa and Saudi Arabia.

Most of her adult life, she has made her living as a flight attendant. Her deceased husband, Don, was an airline pilot. Her step-son Kurt has a commercial airplane and helicopter rating. He's building time, hoping to transfer to police helicopter patrol. Her nephew David has 23,000 hours towing banners for Aerial Sign Co. in Hollywood, FL.

IONA FAITH "FAYE" DOUTHITT, born May 23, 1922, in Long Beech, CA. She became interested in flying in 1939, when her boyfriend (later husband) took her up in his Ford V-8 powered Arrow Sport. They married in 1941, son born 1942 (commercial multi-SEL), a daughter in 1945, (her son has CFI, multi and instrument) received her private in 1942 in Piper J-3 at flying school in Wellton, AZ.

Husband Bob has CFI, SEL and A&P. Bought their first Navion 1951, still own and fly it along with the Piper and Arrow Sport. Flew five transcontinental Air Cruises, Philadelphia to Palm Springs, won in 1958. Entered 1958 PPD and won the first All Woman's Fun Race, San Diego, CA, to Fallon, NV, 1965.

Joined the American Navion Society in 1959 and in February 1995, they were made life members.

After 50 odd years of flying, they are both quite active in their flying clubs. Many fun experiences, few scary ones, but wouldn't trade them for anything.

BILLIE M. DOWNING, Eastern New England Chapter, New England Section, obtained private pilot license in October 1967 and became an active 99 in March 1968. Has held almost all chapter and section offices and was governor from 1982-84. Helped with two successful International Conventions held in this section; the first at Bretton Woods, NH, and the second in Boston, MA. Married to Stuart who is also a pilot, they fly for the sheer joy of it and for travel. Has participated in many local air rallies and had the joy of being the first place copilot in three of

them! Has flown to every state except Hawaii; Canada and the Bahamas. Flying around Alaska was very exciting. They have just completed the most exciting and that was 4,100 miles around the outback of Australia. Now retired from 30 years with the Federal Government, the last years as a data management specialist buying electronics for the Air Force. Flying is their chief occupation. Fly Cessna 172.

EDNA M. DRAGOO, one summer, after the last day of school, she had her first flight lesson at age 55. She flew every day the weather allowed. Suddenly, it was Labor Day and time to go back to the classroom, and she hadn't soloed! Taking stock of things, she decided to change instructors. It was not long after that that she soloed, completed her cross-country work and went for a check ride a month later. What a thrill it was and still is!

After building some time, she started instrument training and obtained that rating, sharing the training experience with a friend and finding that it was a help to them both. They went to ground school together, completed the written exam and the check ride.

A few years later, she and another friend flew together, and she persuaded her to keep right on with accomplishing more. So, on to commercial maneuvers and the commercial license; then the CFI and CFII.

Since their 99s Section awards a scholarship each year, she applied for it and received it, putting the funds toward her CFI training. It is one of the many ways their chapter and section members help one another, and she is very appreciative of the help and support that they gave her.

It's a thrill to fly! Retired from teaching? Yes … but not from flying; not from life!

LINDA MAE DRAPER, pilot flight instructor, Hivert, Inc., owner/operator, Soar Minden. Joined the 99s in 1980, Reno Chapter. She has held the AE Scholarship Chair. She became involved in aviation in 1967 - solo glider, a friend had bought her a gift certificate for glider time. Current ratings: ATP; Comm. ASEL and sea; AMEL and sea; Inst.; Glider; CFIa/CFII/CFIG. She flies an Experimental Jungster II, for both pleasure and business. Background information: University of Missouri, sociology, art, Spanish studies, various special pilot training classes.

Holder of several Nevada state soaring records. Glider and power aerobatics, flying experimental aircraft and homebuilding, sailing, skiing, aerobics, mountain trekking (Nepal), Oshkosh, Warbird Crewmember. She is a member of the Carson Valley Historical Society Board of Trustees and Genoa Courthouse Museum, Docent.

SHEILA DRAYSTER, flying was a dream come true when she went up in a Cessna 172 on a job assignment in 1985. She started her lessons the following week and got her license in 1986 and immediately joined the 99s. She flew the 25th Pacific Air Race and navigated the 25th Palms to Pines. She has flown many poker runs, Lil' Ole Airplane Rallys, Jim Longs and other races. She loves going to the Southwest Section meetings and visiting with other women pilots whom she has met over the years. She is married to Mike Drayster and they have three daughters and six grand-

children. They just opened a country store in Las Vegas.

MARIANA "MARDI" DREBING, Renaissance Chapter, Michigan, born Aug. 4, 1933, Sacketts Harbor, NY. Joined the 99s in 1978. She has flown the following planes: C-150, C-172, C-177, Rockwell 112TC and Rockwell 114. She and her late husband Carl Drebing had three children: Ric Shepard, 37; Anne McGlone, 35; and H. Peter Shepard, 33. They also have grandchildren: John, 8; Sarah, 7; Marc, 1; McGlone/Ahren Drebing 14; and Martin Stith, 2.

In January 1976 she started to accomplish a childhood dream of flying with the firm encouragement of her late husband, Carl, who had been a Navy pilot in WWII. She became a 99 in July 1978. They shared many enjoyable trips with the 99s, Carl being an avid 49 1/2. He also encouraged her participation in the 99s to become a chapter officer, her current duties including being treasurer and 501 (c) 3 chairman. She is now approaching the 1,000 hour milestone and hopes to continue flying in her Rockwell 114.

ELYSA J. DRILLETTE, LCDR, has been a pilot for the US Navy since 1982. The youngest of seven children, the former Elysa Jean McDermott was born and raised in Massachusetts. She attended the University of Michigan on a Navy ROTC scholarship, graduating from the engineering college, with honors, in 1982.

Upon graduation E.J. was commissioned an ensign and selected for Naval flight training. She completed training in Pensacola, FL, and in November 1982, reported to Corpus Christi, TX, for primary flight training in the Navy's now retired T-28 Trojan aircraft.

Ensign McDermott was one of the first women to be selected for jets and completed the jet training syllabus, previously closed to women. She completed aerial combat training, bombing and strafing and, ultimately, aircraft carrier landings in both the T-2 Buckeye and the A-4 Skyhawk. Later, LTJG McDermott became the first woman pilot to fly the EA-7L at VAQ-34 in Point Mugu, CA. The EA-7L was a single-engine, light attack jet modified to fly an electronic warfare mission.

In May 1990 LT Drillette resigned her active duty commission and accepted a reserve commission. Upon relocating to Dallas, TX, she began flying the DC-9 with VR-59, at NAS Dallas. Six months later she was recalled to active duty in support of Operation Desert Storm, where she transported troops from Germany to northern Saudi Arabia. She was on the targeted airfield during the first Scud missile attack against Dhahran, Saudi Arabia, and watched as the Patriot missiles destroyed the Scuds.

LCDR Drillette is still a Selected Reservist and flies the DC-9 for VR-59 as an Aircraft Commander and NATOPS instructor pilot. She holds a commercial instrument rating, ATP and DC-9 type rating. Drillette is married to Tom Drillette, a former Navy F-14 pilot and current pilot for Northwest Airlines. They have three daughters, Casey,

Jamie and Lindsay and reside in Granbury, TX, where she is a charter member of the Brazos River Chapter of the 99s.

M. IRENE DRIZOS, born Dec. 5, 1926, in Nashville, AR, and raised in Dallas, TX. Married a Yankee serviceman and eventually wound up in Middletown, NY, in 1956. In 1988, having raised and dispersed her seven children, she retired from her accounting/computer operating job to care for her terminally-ill husband whom she buried after 43 years of marriage. With the time and means now at her disposal, she decided to pursue a lifelong dream of flying. After trying balloons, helicopters, gliders, and powered fixed-wing, she chose airplanes and received her private pilot certificate from Quade's Flight School in Orange County, NY, in 1992 at age 65. She joined the IWPA as a 66 during her flight training and has continued as a 99 thereafter. Other fun things she does besides flying, are writing poetry and visiting family and friends from coast to coast, gathering data for her memories.

Her seven children are: Karen Stewart, 49; Donna James, 48; Gary Drizos, 46; Dina Sage, 44; James Drizos, 42; Clark Drizos, 37; and Alex Drizos, 34. Her grandchildren are: Nicole Drizos, 24; Joshua Stewart, 23; Justin Drizos, 22; Kelli Dodd, 21; Stephen Drizos, 21; Jessica Drizos, 20; Peter Drizos, 19. Great-grandchildren: Brooke Carron, 4 and Caleb Stewart, 4 months.

Has flown Piper Cherokees 140, 180 and Warrior II.

LINDA LANGRILL DRUSKINS, Midland, MI, has been flying since 1979 and flight instructing since 1987. She joined the 99s Michigan Chapter in 1980, has held all offices and currently serves on the Mary vonMach Scholarship Committee. She is involved in the Experimental Aviation Association's Young Eagle program, and has spoken to career days and school, social, aviation and church groups about careers in aviation.

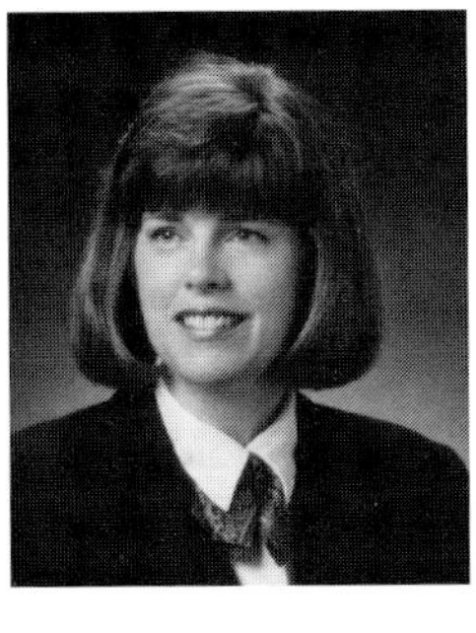

Druskins graduated from Central Michigan University with BA and MA degrees in English and worked at the Dow Chemical Company until 1988, when she left to pursue aviation full-time. She has flown corporate charter and overnight air freight routes and holds an ATP and a Citation I & II type rating.

Druskins is a mother of twins, David and Christina, born Nov. 25, 1993. She is married to Craig Druskins and has three step-daughters and two grandchildren.

JEANETTE A. DUDEK, it was 1944 and she wanted to do something fantastically wonderful and what could be more spectacular than piloting an airplane? Completing ground school, she trained at Cleveland, OH, Hopkins Field and earned her ticket in August 1946 – 727737. Being the first ticketed member in the flying club at Euclid Avenue Airport, she took their Taylorcraft N43770 to Florida with the Gulf Air Tour in January 1947 – no radio, no equipment – all VFR. Landing was a matter of circling the field and waiting for the green light. After marrying in 1948, flying became secondary. Eventually her husband became interested and they purchased Skyhawk II, N78755. She re-soloed and renewed her proficiency while he took lessons, earning his ticket in 1974.

Together "we danced the skies on laughter-silvered wings .. and chased the shouting wind along and flung our eager craft through footless halls of air ..." Flying was their greatest togetherness, until one day in 1977, Ted flew alone to New Horizons ... "and touched the face of God."

Flying added a new dimension to her life, and she will always be a pilot, sans wings notwithstanding.

She now flies the countryside in her fantastic sports car, and she also enjoys stopping to watch a plane fly by and imagine herself at the controls.

BERTIE DUFFY, began her flying career in 1975 as a student pilot at Burbank Airport. She learned to fly in a Cherokee trainer and after earning her license began flying a Comanche which she and her ex-husband owned. She determined that there must be airplanes which would give her then thrill of the barnstorming days, which is the feeling she wanted.

She flew with a friend in a Stearman and her world was forever changed. They decided to attempt a restoration project and found a Stearman advertised in the paper, which was delivered to the hangar at Van Nuys in boxes. That was the beginning of nights and weekends for two years, spent with copies of blueprints from the Smithsonian putting tab A into slot B, and eventually cranking up and taxiing out for a maiden flight.

Bertie sells flights in the airplane from Whiteman Airpark in the San Fernando Valley. Her passengers are a medley of people who learned to fly in WWII, today's military pilots, airline pilots, and those people who simply wish to experience flying with the wind in your face, the roar of the round engine and the hum of flying wires.

Bertie is active in the 99s, International Organization of Women Pilots. Valley women pilots founded the local chapter, based at Van Nuys Airport, in 1952. Among its 90 members are four former WASPS.

She teaches a ground school for private pilot students at Kennedy Adult School which is preparation for the pilot license issued by the Federal Aviation Administration.

JUNE JOAN DUGGER, born on Jan. 28, 1929, in Haskell County, TX. She is a grandmother of six who took flying lessons in her 50s and joined the 99s in 1985 in Corpus Christi, TX. She and her husband Wayne own a 172 Cessna and she is the senior pilot by one hour. Neither daughter Linda or son Wayne Allan are pilots. Joan was Tip of Texas Chapter, in Corpus Christi's chairperson for several years and was chairperson of the Air Race Classic of 1993. Her husband was co-chairman. She flew in the 1994 Air Race Classic as a passenger crew of four which was the first time four crew members had entered the race. Joan is a retired school teacher and teaches calligraphy at Del Mar College and is an active member of her church. Deciding to fly late in life has been one of her most challenging accomplishments.

CES BRAV ROSE DUMAS, born Sept. 27, 1913, resident of San Francisco for 76 years. Served military WASP in 1978. Joined the 99s in 1943. Flight training include: both Kinner and Warner bi-wing fleets (Mills Field) now S.F. Airport; also Avenger Field, Sweetwater, TX. Planes flown: Piper Tri-Pacer, Beach Bonanza, PT-19. Also all metal Luscombe, bi-wing Waco (Military Hatch). Her husband is Alvis "Jack." She has one child, Robert "Bob" Rose, son of late husband, Attorney Adrian W. Rose. Her first solo, S.F. Airport, bi-wing fleet June 9, 1939. Flew in several air shows (1939-40) (introduced all metal Luscombe) (Redding, Corning, Reno, etc.) (P-19 while in training WASP 1943) copiloted "Spirit of San Francisco" AWTAR Powder Puff Derby also copiloted Spirit of San Francisco AWTAR, Powder Puff Derby 1957-58.

CHRISTINE WOODWARD DUNCAN, born March 3, 1953, Grand Rapids, MI. Joined the 99s in 1994. Flight training includes private Jan. 26, 1992 and instrument rating Sept. 30, 1993. Purchased her own 1980 Piper Archer in 1993, and has also flown Cessna 152 and 172. She is married to James Duncan and has one son and two stepdaughters: David Woodward, 14; Erica Duncan, 20; and Becky Duncan, 14.

Her father has flown since the early 40s and owns a 1975 Cessna 172. Her brother has flown since 1971 and owns a 1979 Piper Archer. Even after years of flying with them, she decided to overcome a reoccurring dream she had of crashing in a plane (with her at the controls) and learn to fly. After her solo, she never had that dream again! She got her private license in six moths and her instrument rating in six months, while she had a full-time job of running an FM radio station. She bought her own plane two years after she started lessons because she's totally addicted to flight! She flies for fun and relaxation but plans on getting her commercial and CFII in the near future.

Her career is in broadcasting. She currently oversee three radio stations (two FMs and one AM) in Indianapolis for Emmis Broadcasting as the senior vice president/general manager. Funny, she can talk on an airplane radio, but is scared to death to talk on her own radio stations!

MARION ELIZABETH DUNLAP, learned to fly in a J3 Piper Cub in 1947. Owned a Piper Tripacer, 180 Cherokee, 180 Comanche a T-34, Twin Comanche. Has a commercial rating. Joined Civil Air Patrol in 1961 and the 99s in 1963. Is a life member of the Silver Wings and an active FAA safety counselor. Owned a private airport in 1952. In 1965 built and operated a commercial airport in Pennsylvania, Bellefonte Skypark which was a Piper flight center. Been active in air mail celebrations in Pennsylvania. Has five daughters; three soloed and one went on to get her private rating.

Is a past chapter chairman of the 99s, and a past Group Commander of the CAP.

DOROTHY A. "TAYA" DUNN, a need for some new and challenging activity that would be both a diversion and a complement to her work as a teacher, that's what led her to begin flight training in Citabrias at the Oakland Airport in 1969. She continued her training, after a long interruption, on St. Thomas Island, the Virgin Islands, and the result was a rather spotty cross-county (and over-water) training experience.

Although her currency comes and goes, she always enjoys flying again no matter how long she may have been away. In 1976 she joined the Bay Cities Chapter of the 99s.

The future? Well, someday she does hope to operate a non-tower airport in the foothills of the Sierra Nevada Mountains.

SUSAN DUNN, learned to fly at Palo Alto Airport in 1991 and has logged 280 hours. She flies a Piper 180, and joined the Santa Clara Valley 99s in 1992. Her husband is also a pilot, and they have three children. She owns her own business, a cruise travel agency in Palo Alto.

LOUBELLE M. DURAND, Phoenix 99s, whose family had always been deeply involved with model aircraft and had their own plane which her husband and son flew, was inspired to get her own license after attending a Pinch Hitter course at Oshkosh. She learned to fly in Phoenix, getting her private ASEL in 1977. She flew a Cessna 150 on a cross country and was very impressed at the attention and kindnesses she received when she stopped, until she realized her plane was emblazoned in large letters "SKY NEWS KOY" (a Phoenix radio station).

1978 was the year LouBelle became a 99, and she is still very active since retiring from teaching. Her daughter, Mary Ross Collis, is also a Phoenix Chapter 99, and her son was a member and officer of the local EAA many years. LouBelle has held several chapter offices and worked on 16 Kachina Air Rallys, as well as many poker runs, airlifts, scholarship committees, airmarking, sales, etc., and she still finds time to do quite a bit of traveling.

LAURA J. DWYER, born March 28, 1955. Joined the Tucson Chapter in September 1993. From August 1991 to December 1992, Davis Monthan AFB; private pilot, airplane single engine land, December 1992. Current occupation, hearing officer; former active duty Air Force, administration officer. Incentive rides: F-4 Phantom with her then-boyfriend, now-husband as PIC; F-15 Eagle, BFM 1V1 fighter training mission; restored fleet biplane. Co-owned a Starduster Too biplane.

Born in Massillon, OH, to Bill and Peg Matthews. Two sisters, Patti and Nancy, are married with two kids each; respectively, husband Tom, sons Sam and Ben; husband Brian, son Joshua and daughter Brianna.

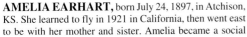

She is currently flying Cessna T-41C's out of the Davis Monthan AFB Aero Club and will soon be working on her checkout for the Beech T-34 (military trainer, single engine, complex). Her husband Bob is a former Marine aviator and Top Gun graduate before he interservice transferred to the Air Force to be an F-4 instructor pilot and aggressor pilot in Europe. He is now retired and working for flight safety as a Learjet ground and simulator instructor.

AMELIA EARHART, born July 24, 1897, in Atchison, KS. She learned to fly in 1921 in California, then went east to be with her mother and sister. Amelia became a social worker in Boston, at which time she was invited to make an Atlantic crossing as a passenger.

The transatlantic flight whetted her appetite for repeating the flight alone, which she accomplished four years later. Meantime, Amelia competed in the Transcontinental All Women's Air Derby in 1929 from Santa Monica to Cleveland and was among the 14 who finished eight days later. Shortly after the Derby (labeled the "Powder Puff Derby" by Will Rogers), the 99s was formed and Amelia became the organizations first president.

Amelia undertook many speed and distance flights and demonstrated a new autogyro around the country. She encouraged women to fly and wrote and spoke regularly on aviation. Her husband, G.P. Putnam, supported her activities enthusiastically.

The story of her fatal 'round-the-world attempt in 1937 is well known and its outcome raises questions even today. Amelia's continued importance to the 99 organization is exemplified through the AE Scholarship program, the airmail stamp in her name and the International Forest of Friendship near her birthplace. Amelia remains today the personification of "woman pilot."

LOIS M. EATON, Phoenix 99s, got her Private ASEL at Robertson Airfield, Plainville, CT, in 1989. She is a true aviation enthusiast, but also has many other interests. She grew up in Butte, MT, got a teaching degree, married and raised two children. After teaching 20 years in Connecticut she went into real estate sales in Arizona and joined the Phoenix 99s in 1989.

Lois has many happy memories of rain and a rainbow on her solo flight, and says she was very grateful later, when she flew into an ice storm, to be able to turn back and find an airport for a safe landing. Her hobbies are art and travel, and she and her family consider it a great experience that they had a refugee family live with them for almost a year. Currently, she teaches children about the desert, and tutors homeless children. She has served as chapter treasurer, and worked on many 99 activities.

RITA EAVES, Oklahoma Chapter, started flying in 1949 in an Aeronca Chief, obtained her pilot certificate in 1951 and joined the 99s in 1952. She held all chapter offices (several times) and was PPD stop chairman twice. At age 9, at the dedication of Will Rogers Airport, she watched aerobatics and at the completion, saw a woman step out of one of the planes. From that day on, Rita knew she wanted to fly. The inspiring lady was 99s own Dorothy Morgan, who years later sponsored her into the 99s.

After Rita and Leonard were married she convinced him they should learn to fly so they purchased an Aeronca Chief. Her first instructor was certain women should never fly. The second instructor soloed her in six hours.

Leonard and Rita became active in the EAA and built a plane called "Cougar." After flying the Cougar several years Leonard designed and he and Rita built the *Skeeter* which was completed in 1966. At this printing they are still flying *Skeeter*.

Rita has been active in the 99s and EAA having attended (and worked) 36 consecutive International EAA Conventions flying a home-built all except four years. Rita retired from S.W. Bell and took up golfing. She enjoys training her dog, traveling, photography as well as golf and flying.

She has flown the following planes: Aeronca Chief, Cessna 140, Cessna 150, Stinson Voyager 150, Piper J3, J3 with floats, PA28-180, Fairchild 22, homebuilt aircraft - *Eaves Cougar*, *Eaves Skeeter,* Pierce Arrow.

NANCY K. EBERT, born Jan. 7, 1962, in Heron Lake, MN. Joined the 99s in 1991. Flight training includes ATP: multi-engine - BE1900 type rating. She has flown BE1900C, BE99C, PA24, BE76, C182, C172, C152, C172RG.

A Minnesota native, she first started taking flying lessons in 1987 after receiving them as a Christmas gift from her husband, Phil. While working at the Minnesota Pollution Control Agency as a computer operator, she acquired her rating and then became a full-time flight instructor. She currently flies as a Beechcraft 1900C Captain for a commuter airline, GP Express Airlines.

As a member of the Minnesota 99s, she has participated in many activities including: airmarking, Red Cross shuttle flights and daffodil deliveries. She is also a member of the 99s, Inc., Aircraft Owners and Pilots Assoc., Comanche Owner's Society and Experimental Aircraft Assoc.

Nancy is married to Phil, a mechanical contractor and 49 1/2, who is working on his private pilot rating. Together they enjoy flights in their Piper Comanche and are building a Kitfox IV.

AILEEN EGAN, Governor of the British Section, learned to fly in 1982 in Fort Worth, TX, and joined the 99s shortly after this. She did her IMC rating in France and her twin rating in England.

She is committed to the idea of encouraging women to fly and to ensuring that those whose dream is a career in aviation have the opportunity to do. This has taken her to all corners of the world to participate in events for women pilots.

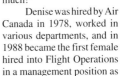

She is chairman of the British Women Pilots Assoc. and is a member of the British Parliamentary Aerospace Group. She is a founding member of the European Women Pilots and a member of the World Aerospace Education Organization. She is currently setting up the first Women Pilots' Museum in Europe. Her aim is to make this a living museum.

A management consultant, she has a master's degree in business administration from Brunel University. Her report on airport and airspace congestion - "Europe-The Overcrowded Sky" has been used throughout the aviation industry worldwide.

Aileen plans to continue to develop her flying skills and to become a successful Rally Pilot.

YVONNE MAY EGGE (LANTZ), born Feb. 16, 1944, in Lebanon, PA, and now resides in Addison, TX. She took her first airplane ride in a PA-12 (Piper Super Cruiser) at Kutztown Airport in Kutztown, PA, at the age of 10 years old. From that time on, she was hooked on flying.

At the age of 15 she met her future husband. Airplanes were the first topic of conversation and have been ever since.

In June 1964 she took her first flying lesson at Hershey Air Park in Hershey, PA. Lack of funds prevented her from continuing her flying activities until October 1975, when she and her husband bought their first airplane, a 1969 Cherokee 140B. Her husband renewed his flight instructor certificate for the sole purpose of teaching her to fly. In the quest for her private pilot certificate many exciting things happened but three things stand out. One of the most memorable things, of course, was the day she soloed at Manassas, VA. The first solo was as exhilarating as her first airplane ride. Another memorable moment occurred when she flew the required long cross country. One of the legs included a stop at Hershey, PA, where her formal flight instruction had begun many years before. This particular stop was all the more special because her proud father met her at the airport to see her fly for the first time, not as a passenger, but as a real pilot! The third thing, of course, was finally getting her private license on May 6, 1977. The FAA examiner was country western singer and song writer, Everett M. Noel. He even gave her an autographed copy of his record, *Bring it on Home to me Darling,* to add a lovely touch to an already wonderful day!

In July 1978 Yvonne and Ken bought a beautiful all-original 1962 Piper Comanche that came off the assembly line a month before they graduated from high school. They have enjoyed taking many long and short trips as time would allow.

Since becoming a member of the Dallas Chapter of the 99s, she has really enjoyed the camaraderie of these wonderful women flyers.

DENISE EGGLESTONE, First Canadian Chapter, East Canada Section, began flying in 1976 after a friend insisted that flying lessons "really didn't cost that much."

After a $10 familiarization flight, there was no turning back. By 1980 she had her private, commercial, multi-engine, instrument and instructor rating and, yes, she says, "It really did cost that much!"

Denise was hired by Air Canada in 1978, worked in various departments, and in 1988 became the first female hired into Flight Operations in a management position as a B727 flight operations instructor. In 1992 Air Canada sold off its entire fleet of B727s. Her job became redundant and she was laid off.

Denise enjoyed a short instructing career in the early 80s and was also right seat on a Navajo Chieftain flying charters and cargo.

She joined the 99s in March 1988 and became Montreal Chapter chairperson. Transferring to Toronto in 1991, Denise joined the First Canadian Chapter, becoming secretary of the East Canada Section. She is currently vice-governor.

Denise was awarded an Amelia Earhart Memorial Career Scholarship in 1993. One of 19 winners of the scholarship that year, she used the monies to regain her instructors rating.

In 1994, Denise was awarded the ISA+21, (the International Society of Women Airline Pilots) Canadian Citizen Scholarship and she used those monies to regain her multi-engine instrument rating.

Weather permitting, Denise flies weekly for the Ontario Ministry of the Environment with environment enforcement officers on board photographing illegal dumping of waste material by industries.

Hired by Peninsulair in 1994 as a flight instructor, Denise continued to pursue her shelved dream of becoming an airline pilot. In 1995 Denise received "the" unexpected phone call and without hesitation immediately rejoined the ranks of Air Canada Flight Operations.

Denise co-chaired the 1995 Canadian Women in Aviation conference, operates a small business, "Just Plane Crazy" with a fellow 99, is a Girl Guide Aviation Badge examiner and a judge for local science fairs and she loves to travel!

Denise's husband Jim is an Air Canada A320 First Officer and their two children, Lauren, 4, and Ian, 3, are budding aviation enthusiasts.

SUE A. EHRLANDER, who dreamed of flying from the age of 10, made it happen in Saint John N.B. in 1973. So intense was her love that she started work at the Saint John Flying Club two weeks into her training. This led to her meeting her future husband, Ralph, then a charter pilot. Sue won the 1973 CFI Trophy for the highest marks on her private pilot written exam and flight test.

Sue joined the Eastern Ontario Chapter in 1979 when she moved to their area with her husband and daughter, Sabrina. She served as treasurer, vice-chairman, and chairman of the chapter between 1980-85. Sue transferred to the First Canadian Chapter in 1985 after her move to Toronto, and to the Maple Leaf Chapter in 1989.

She has been very active in aerospace education at the chapter level. She served as East Canada Section A.E. Scholarship Chairman from 1984-90. In the East Canada Section, Sue served as treasurer, vice-governor and governor from 1986-92. He is very proud to be a charter governor of the International Council of Governors. She also served as the first chairman of the International Grievance Committee.

Sue, was a member of the 1993 and 1995 organizing committees of the Canadian Women in Aviation Conference. She served as the local chairman for the 1995 International Convention in Halifax N.S. Sue is a proud member of the 99s, Inc., Canadian Owners and Pilots Assoc., United Empire Loyalist, St. John Ambulance Brant County Branch, and the Brantford Flying Club.

LORRAINE F. ELAM, Phoenix 99s, got her private license in 1992 at Deer Valley Airport (in a Phoenix suburb) and joined the 99s in 1995. She has served on the nominating committee, and as corresponding secretary.

Lorraine considers her most exciting aviation experiences as surviving an accident caused on take-off by a wind sheer, and doing a spin in a Decathlon. In addition to flying, she is married and has nine children, is a ham radio operator, and worked at Kerley Industries for 30 years, most recently as vice president of Human Resources. She is still going to college.

Hobbies are flying, reading, helping the aged, and with her husband, rebuilding a 1938 Stinson SR10E Reliant.

JEAN T. ELLINGSON, born Dec. 20, 1929, in Albert Lea, MN. Flight training includes commercial rating, her husband Lem taught her to fly; 1,541 hours total time. She has flown 172, 182, 150, Citabria. She and her husband Lem Ellingson, have two daughters, Pam, 43, and Kim, 41. Their grandchildren are Ted, 13; Dan, 12; and Mike, 8.

Jean learned to fly in 1969 and 1970, when her husband Lem earned his civilian ratings after having been a WWII Navy jet fighter pilot and instructor. She joined the 99s in 1971 and has held most chapter offices and attended many sectional and International meetings. She flew the Powder Puff Derby in 1976 as copilot with Kathryn Hach in a 182. She used to fly to professional meetings and to transport her children back and forth to college. Her present flying usually involves flying co-pilot with her husband on trips.

Jean's educational background includes a BS and MS. She taught business education and English at the high school level for 32 years. She is presently involved in several volunteer endeavors.

MARY G. ELLIS, joined the 99s in 1994. She grew up in a family that traveled at least every weekend to a cottage (no matter where they lived, they moved every three or four years) and made other long distance trips. Then she married an Air Force pilot and traveled some more. When they finally stayed in one place long enough, she found a ground school course, started her flight lessons, bought a C-150 and is still traveling and having the time of her life. "It's to enjoy, you only go around once and the sky is still blue."

LOTFIA EL-NADI, the first Arab woman to throw off the veil and learn to fly. She took flying lessons in 1933, soloing at the age of 25 in a Gypsy Moth out of Cairo, Egypt. She received her private pilot license (Egyptian pilot license number 34) that same year.

It was considered a dishonor for Egyptian women of a certain class to work, but Lotfia decided that she would not tolerate the subjugation and would try to make something of her life. The first and only flying school in that part of the world had opened just one year earlier when Lotfia decided she had to fly. She began working for the school as a secretary and took all of her wages out in flying time. The school wanted publicity about her flying and required that she inform her parents of her activities. Many news articles spoke of her accomplishment and noted that she brought honor to her country.

She flew five years before medical problems resulting from a fall ended her flying as a pilot. However, during those years she participated in numerous air races on the African continent in addition to taking aerobatic instruction. To obtain free flying Lotfia was known to hide in airplanes that she knew were going for test flights. After they were airborne she would pop her head up and talk the pilot into letting her fly. This permitted her to obtain experience beyond what she could pay for out of her salary. After her active flying days ended, Lotfia worked as a secretary at the Aero Club of Egypt so that she could continue to be in touch with aviation.

Lotfia El-Nadi has received many honors for her accomplishments. She was honored again in 1989 as Egyptian aviation's first lady during the celebration of the 54th anniversary of civil aviation in Egypt. At that time she received the coveted Order of Merit, the highest award of the Organization of Aerospace Education. She is a member of the Arabian Section 99s. Lotfia El-Nadi's life has certainly reflected her belief that "You must never regret anything – look for the future."

ANN ELSBACH, learned to fly with Amelia Reid in 1973 at Reid-Hillview Airport in San Jose, CA. She has her commercial ASEL, CFI airplane, and commercial glider ratings and has logged more than 10,000 hours, most in tailwheels. Ann has been the manager of the West Valley Flying Club at Palo Alto Airport (and now San Carlos) for more than 10 years, where she also instructs. She is a partner in a 1954 C180 and a 1979 F19 Taylorcraft. As a new pilot (only 100 hours), Ann flew an L-2 Taylorcraft from Kentucky to Reid-Hillview, and has since logged time in many aircraft, her favorite begin the Kreider Reisner 21. Ann is a member of the EAA, AOPA and other local airport associations. She has a daughter in San Diego.

MINA C. ELSCHNER, born in the Netherlands, joined the 99s in 1960. Flight training includes private SEL - commercial rating and 2,200 hours flight time. She has flown the following planes: Cubs, Aeronca Chief, Aeronca Champ, Ercoupe, Cessna 150-172, and Piper Cherokee. Her husband Howard is deceased. Elschner has two daughters, Barbara and Beverly (mature young adults) and four grandchildren and two great-grandchildren, ages 3 1/2 and 2.

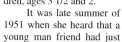

It was late summer of 1951 when she heard that a young man friend had just purchased an airplane for $500! On her way home from work that very evening she stopped in at a small sod field and asked "the man there" (FBO) if he could buy her a plane for $500 also. The following Saturday she was the very proud owner of an Aeronca Chief side-by-side, a 65 hp cream-colored vehicle that she couldn't fly ... N 21309. On Sunday Mr. Landers of the Ramapo Valley Airpark in Mahwah, NJ, gave her an orientation ride in the first of her planes. That field later became the site on the Ford assembly plant.

That was her beginning of a 44 year ecstatic and exhilarating love affair with the world of aviation. Joining the 99s organization in the summer of 1960 has kept her interest and activities keen in all phases. She's certain that the 99s have kept her active on the front burner of aviation activity these many years ... to this very day. Life-long friendships are made and a warm camaraderie is the order of the day, within the 99 circle.

CAROL A. EMMENS, North Jersey Chapter, fate intervened - that's how she explains why she learned to fly. She planned to take an adult school course in antiques, but ended up in an aviation ground school course instead. And what changes it brought to her life. She gained confidence and successfully started and built her audiovisual and computer business, Presentation Media, Inc., into a multi-million dollar company.

After graduation from college, she taught, earned her MLS degree, became a reference librarian, and a freelance writer. Among her books are: *An Album of Television, An Album of the Sixties, Knights, Stunts and Stunt People,* and *The Abortion Controversy.*

She now owns a Piper Cherokee 235; she's flown from New Jersey to Minnesota and up and down the East Coast. She hopes to fly in all 50 states and if dreams do come true, around the world.

She is also the proud mother of a son named Scott.

IRENE P. ENGARD, someone said she couldn't do it, so she received her license at Orange County Airport, June 27, 1978. She joined the Orange County Chapter Aug. 21, 1978.

She moved and lived in Nevada from October 1978 until November 1987. She was a member of the Reno Area Chapter from August 1978 until June 1995.

She served Reno as chapter secretary/treasure, treasurer, newsletter editor, Forest of Friendship and APT chair and airmarking co-chair.

She started attending section meetings in the fall of 1980 and has missed only five. At the section level, she has served as treasurer, Forest of Friendship and Aerospace chair and is currently the editor of *The Southwesterly.*

She owns a tax preparation business and is a representative for Primerica Financial Services. She holds an ASEL rating and has flown 14 different aircraft. She is married to Alan Engard, computer whiz, and they have two dogs, Tasha and Shara.

KARA ENGLISH, became so discouraged while taking flight instruction that she was going to quit. Then, she met the 99s who encouraged her to obtain a new instructor. Training took on a whole new meaning. Kara received her pilot license in the summer of 1993 and immediately became a member of the 99s. Kara has done most of her flight training in Ohio where she was born and raised. Currently, she is pursuing a degree in airport management at Embry-Riddle Aeronautical University located in Daytona Beach, FL. Kara is also a charter member of the Women With Wings Chapter, American Assoc. of Airport Executives, AOPA and Women in Aviation.

ELIZABETH ENGSTROM, has belonged to the Santa Barbara, Phoenix, Long Beach, and Tucson chapters. Flying experiences have been the Powder Puff Derby, Pacific Air Race, Kachina Races, Shirts and Skirts, Shamrock Derby, Tucson Treasure Hunts, and other events; also the Goodyear blimp, Columbia, for 2.8 hours, and a seaplane.

She is a volunteer for Girl Scout and school aviation programs, pilot for Wright Flight, Aviation Days, and a mentor to high school aviation students. Betty has served her chapters as membership and scrapbook chairman, set up Flying Companion Seminars, and conducted an aviation radio program.

She dreamed of being a pilot since 1929, when she was a passenger on the Ford Tri-motor, which had carried Mrs. Lindbergh to Mexico City.

A retired teacher and the mother of a daughter and a son, a professional pilot, and grandmother of two, she still flies the plane she bought 23 years ago.

MARTY L. ENNISS, born April 10, 1952, grew up with a WWII Air Force pilot, Lt. Col. Byroads and loved hearing his stories of the war. In 1970 Marty married a man stationed at Moffett Field, a Navy base where she had attended many air shows with her father. Early in her marriage she told her husband of her dream. She wanted to learn to fly and surprise her father by taking him flying. This dream took over 20 years to come true.

Marty learned to fly in Temecula, CA, at French Valley Airport (F70). On Dec. 28, 1994, she took her check ride and obtained her private pilot certificate. On December 29, she surprised, really surprised! her father and took him flying. During her training she logged over 90 hours, discovered she loves flying, hates being lost, won't fly without a GPS and probably will never fly to Big Bear alone. She joined the 99s and values their encouragement and support. Marty flies with the support and encouragement of her husband, Gary and two children, Jennie and Jon.

Marty flies a Cessna 150 and a 172. She has also flown an Ercoupe, a Tomahawk and enjoys playing in a Citabria.

BEVERLY MAHONEY-EPSTEIN, San Gabriel Valley Chapter. She has a commercial CFI Instrument SEL. A fear of flying started her on her flying career in 1971. After she received her private license she still was uncomfortable with flying until she entered the Palms to Pines race, based in California, in 1972. She had 105 hours and fell in love with flying. After that race she and Peggy Marsh won the Low Time Race Team in the 1973 PAR. In 1974 Peggy and she placed eighth in the Angel Derby, also winning Low Time Race Team. Racing in the Powder Puff Derby, she and Marge Robbins placed sixteenth in 1976 and then she and her daughter won the Palms to Pines after placing eighth in the race the year before. In 1977, Beverly and Katie Moskow placed third in the Palms to Pines. Her Dream Machine, a 1969 PA-250 Piper Comanche, helped her reach for the stars and even touch a few.

She has used her CFI to teach flying at Chino, Cable, Fullerton and Camarillo Airports and taught at Chaffee College with the college flying club.

She joined the 99s in 1973 and has held many positions in the San Gabriel 99s and has gained a great respect for women pilots and their flying machines.

LYNNE ERBAUGH, Women With Wings, born Sept. 2, 1948, in Ravenna, OH, has been a lifetime Ohio resident. Currently flying a Cessna 172, she trained and soloed in a Cessna 152 at Kent State University Airport. Lynne received her private certificate December 1991, and went on to receive her IFR September 1994. Lynne has participated in the Buckeye Air Rally and has done aerial photography for the Ohio Department of Agriculture. Lynne is a proud charter member of the Women With Wings Chapter and attended the International Convention in Norfolk, 1994.

Lynne's educational background includes a BA in liberal arts from Denison University, study abroad in Florence, Italy and a secretarial certificate from Katherine Gibbs Secretarial School in Boston, MA.

Lynne is married to Marty Erbaugh, an entrepreneur in Hudson, OH. They have three daughters: Carolyn, Darby and Lindsay.

LOIS A. ERICKSON, an instrument rated private pilot, still flies her Piper Cherokee 180C (N99SW) which she has owned since 1968. Lois previously owned a Cessna 140, and has time in gliders, helicopter and other complex aircraft.

Lois has competed in AWTAR, Pacific Air Race and Palms to Pines air races. She has served as a competition judge for regional NIFA events, timer for the first Air Race Classic and several Palms to Pines air races.

She joined the Wisconsin Chapter, North Central Section in 1970, later transferred to Las Vegas Valley Chapter, Southwest Section and currently belongs to the MT. Shasta Chapter, Redding CA.

Lois has served as chapter secretary, vice chairman and chairman. She has served on the section level as Aerospace Education and Forest Of Friendship chairman. In recent years Lois has been treasurer, secretary, vice governor, and governor of the Southwest Section, spokesperson for the International Council of Governors, past International secretary and currently is International vice president (1994-96).

A professional nursing administrator consultant, Lois is an RN with a BS in nursing, and MBA in health services administration. She has been honored in *Who's Who in American Nursing* and Sterling *Who's Who Executive Directory*. Her husband Donald is also a pilot. Lois and Don are co-owners of D&L Quality Feed and Pets in Corning, CA.

KIM ERNST, born May 27, 1954, in Inglewood, CA. When she was 7, her parents jumped at the chance to take their whole family flying on this drizzly, Sunday morning when there was advertised "Pennies-a-Pound plane Rides." This ride made a lasting impression.

Her next flying opportunity was in college when she enrolled in the flight attendant program and was invited to join the flying team. They competed in the Pacific Coast Intercollegiate Flying Assoc. and the National Intercollegiate Flying Assoc. She competed in events such as the message drop and aircraft identification.

Once she graduated from college, her flying adventures came to a screeching halt, until one sunny, summer, Saturday morning in 1978, she drove to the Fullerton Airport and enrolled in her first flying lesson. She received her pilot certificate in 1979 and joined the Fullerton Chapter of 99s in 1980. She immediately became active, holding offices of chapter and vice chairman, treasurer, fly-in, air races, public relations, Forest of Friendship, APT, membership, nominating aerospace education and was newsletter editor for four years. Section level office was Forest of Friendship chairman. She has also attended several section and International conventions.

Planes flown: Cessna 150, 152, 172, Beechcraft Sundowner, Grumman Cheetah, and Mooney 201. Her flying experiences have led to other adventures, such as: Skydiving, gliding, parasailing and traveling abroad. She hopes there will be more to come.

LISA EROTAS, born in South Bend, IN, in 1961 and now makes her home in Phoenix, AZ. She has a 7-year-old son, Alexander, who shares her interest in flying and accompanies her as often as possible.

Although her interest in aviation began at age 2, when she visited her first airport and exclaimed, "Yookie all the Yights!" She didn't take her first flying lesson until July 3, 1992, and soloed on July 25, 1992. She has completed her private SEL, private MEL, commercial SEL, instrument ratings and is currently preparing for her commercial MEL, which she will complete in October 1995. She has trained mainly in Cessna single/multi-engine aircraft, Piper Aztec, Beechcraft Bonanza and King Air 200. Lisa's plan is to continue gaining multi-engine experience, and learn as much about the aviation business as possible. Her ultimate goal is to fly for a major carrier and then purchase an FBO for charter/lease operations. Once achieved, she will help someone else to learn as she was helped.

Lisa has been a member of the 99s since September 1994.

DOROTHY ESTEP, an ex-WASP, who began her flying career in 1943. After graduating from Berkeley, and teaching high school for one year in Fresno, she read an article about the WASPs, quit her job and decided she wanted to enlist. She learned to fly in Yerington, NV, as there was no flying in California in 1943 and then joined the WASP as a member of the 44-9 class in Sweetwater, TX. Dottie instructed students from Sequoia Union High School from Palo Alto Airport for the State Department of Education, Pre-Flight Orientation Program. In 1972 she was listed as having over 950 hours on her commercial flight instructor ASEL license and was interested in gardening and piano, flying once a month for pleasure. She joined the Bay Cities Chapter of the 99s in 1945, transferring to Santa Clara Valley in October 1954.

JENNIE JANE ESTERBROOK, born Aug. 5, 1963, Frankfurt, Germany. Joined the 99s in 1981. Her father was her inspiration to start flying. He would tell her stories of when he was 20 years old back in 1934, when he flew the mail in a Stinson. So in 1977, at age 13, she began flying a C-150 at Buckley ANG in Colorado and loved it! Being a woman she was discouraged by male flight instructors to fly and told she was going through a phase - she proved them all wrong!

In August 1981, she received her private pilot license in a C-150; March 1985 her instrument in a C-172; May 1985 her commercial in a T-34; and in May 1990 her multi-engine in a BE-76 Dutchess. She received a professional pilot bachelor's degree from Metropolitan State College in 1985, even though she was told by the department chairman she didn't need a degree - she should get married and forget flying as a career.

She stared flying professionally a King Air E-90 in 1990, as a copilot for Public Service Company of Colorado, a utility company and is currently still there.

When she is not flying, she enjoys volunteering in the 99s Flight Without Fear Program. This program is not only beneficial to their students, but also to her and the passengers she is in contact with everyday.

"I love to fly - I love the people - I love the scenery - I love the challenge!"

BARBARA EVANS, obtained pilot license in 1948 in a J3 Cub and joined the 99s in 1950. Served as governor of New York - New Jersey Section 1953-55. Served as chairman of the PPD Terminus and 99 Convention in 1955 in Springfield, MA. Elected to International 99 board 1955-57, International treasurer 1957-59 and International secretary 1959-61.

Executive vice-chairman/treasurer of All-Woman Transcontinental Air Race (PPD) 1961-77. Flew many route surveys, edited the official race program for 13 years, co-edited the 1974 Powder Puff Derby Album and the 1983 PPD History book. Flew four PPDs, the AWNEAR (winning copilot) and the IAR.

In 1959 co-edited *Thirty Sky Blue Years,* a history of the 99s, for the 30th anniversary. Served six years as trustee on the Amelia Earhart Memorial Scholarship Fund Board. For many years was Alice Hammond's assistant on the issuance and selling of the Official First Day Covers for the benefit of AEMSF. Upon Alice's death in 1993 has continued this annual project.

Has attended all but a few of the last 45 International 99 conventions plus a lot of section meetings.

Since 1990 has served as trustee on the International Women's Air and Space Museum in Dayton, OH.

Having been a 99 for 45 years and knowing members from all geographic areas, she feels she is qualified and capable of serving the members well on the nominating committee.

BRONETA DAVIS EVANS, born in Story, OK, July 31, 1907, soloed an OX5 Eagle-Rock in 1928, went on to fly a Coffman monoplane, Curtiss Pusher, Taylorcraft, Aeronca Chief, Cub Coupe, J3 Cub, Cub Cruiser. Many of those planes were utilized by Broneta and her late husband, Tod Davis, in their ranching operations.

Broneta earned her commercial rating in 1942. During WWII, she was one of three women commissioned by the US Army to fly Civil Air Patrol planes.

Her crowning year in aviation was in 1983, when she was selected the International Flying Farmers' "Woman of the Year" and was inducted into the Oklahoma Aviation Hall of Fame.

Broneta was the only woman among 38 charter members of Flying Farmers organized in Stillwater OK, in 1944 elected first secretary of that organization which became the International Flying Farmers. She authored a history *Flying Farmer Organization: First in Oklahoma.*

She joined the International 99s in 1945, held most offices in the Oklahoma Chapter, regionally and nationally, including International president from 1957-59. Broneta flew in three transcontinental Powder Puff Derby Air Races, flew with a group of 99s from Key West to Cuba and with Flying Farmers to Central America.

She owned a total of 16 airplanes including a Stinson 165, Cessna 170 and her last plane a Piper Arrow which she affectionately called a great "Old Ladies Airplane." Oklahoma lost an aviation historic treasure when she passed away, Dec. 31, 1994.

ELAINE KAY EVANS, GDAC 99s, a Michigander, has lived in the Birmingham and White Lake areas most of her life. In 1981 she acquired her private license in a Piper Archer. She has flown a Cherokee Six, Decathlon and Warriors. Currently she flies and proudly owns her "very own" Piper Warrior II.

With a BA degree, English and speech majors, she has taught in secondary schools, worked in two family-owned contracting businesses and has sold computer systems. Today she is venturing into technical writing.

Over the years she has flown throughout the US with her husband, Richard (a pilot) and her two children. The newest family member, grandson Luke, is currently the most enthusiast aviator.

As a past chairwoman of GDAC 99s, she makes sure there is room in her schedule to swim three times a week at a local fitness center. Elaine feels health is a priority if one wants to continue flying.

JUNE EVANS, received her private pilot license on March 2, 1990, in Fremont, NE. Two weeks later she began a six month leave of absence from her job as a staff pharmacist to attend flight school at American Flyers in West Chicago, IL, where she achieved her commercial through certified flight instructor, instrument ratings. Upon completion she returned to Omaha, NE, where she became a member of the Skyhawk Flying Club and instructed part-time. She also joined the International Woman Pilots Assoc. and was elected the Omaha Chapter chairman for 1994/95. Also June joined the Civil Air Patrol (99th Pursuit Composite Squadron) and accepted the position of flight/safety officer and is a cadet orientation, instructor and check-pilot for the squadron.

Currently June is training for her multi-engine rating in a Cessna 310 and is preparing for the Women's Air Class race in June 1995 with fellow pilot and friend Jeanne Peters. She is also applying for an air charter certificate and is hoping the business will move toward flight instruction and air ambulance. Her ultimate goal and dream is to fly or work for NASA.

June is working as a surgical pharmacist at a local Omaha hospital. She enjoys spending time with her family and is taking ice and in-line skating lessons. Also, she is a certified scuba diver and enjoys biking, traveling and reading.

MARIJANE EVANS was teaching school in Virginia in 1967 when her husband, an Air Force pilot, persuaded her to take flying lessons "just through solo" so she could understand his love of flying.

New military assignments for her husband led to new flying experiences for Marijane as she went from private pilot at the Langley AFB, VA, aero club to commercial pilot at Eglin AFB, FL, and added her instrument rating at Hickam-Wheeler Aero Club in Hawaii. Her favorite plane at the aero club at Norton AFB, CA, was a Beech T-34, which gave her experience with retractable gear. It also beat freeway commuting to Pearblossom, where she got her glider rating at the Great Western Soaring School, and to Santa Paula for aerobatics lessons at Michael Dewey's.

Marijane combined her interests in flying, writing and photography to embark on a new career as a freelance writer. She flew aircraft ranging from a restored 1936 Ryan STA to the USAF T-38 jet.

After getting a multi-engine sea rating in 1978, Marijane went to work for Antilles Air Boats as First Officer flying Grumman Mallards.

Marijane maintained her affiliation with the ALOHA Chapter in Hawaii, since there were no 99s in the Virgin Islands or Puerto Rico. She received The Amelia Earhart Memorial Scholarship and their own chapter award, which she received in 1979.

In 1980 Marijane was hired by American Inter-Island as a First Officer on their Convair 440s in the Virgin Islands. She married another AII Convair pilot, Vince Sipple and they now live in Ceiba, Puerto Rico.

SALLY HOLT EVARTS, born Sept. 30, 1951, in Dickinson, ND. Joined 99s in 1994. Flight training includes 123 hours. She has flown Cessna 172 and Citabria 7GCBC.

Evarts has two children, Austin, 11, and Ian, 9. She's had her license for one and half years and is the current chairperson for the High Sierra Chapter of 99s. She's been awarded a scholarship for backcountry flying training in Idaho. That is her desire – to become a great backcountry pilot!

SHIRLEY S. EVERETT, born Oct. 9, 1929, in Powell, WY. Joined the 99s in 1977, Wyoming Chapter (held all chapter offices). Her flight training includes PSEL rating December 1974. She has flown and co-owned with 49 1/2: Piper Tri-Pacer, two Cessna Skylanes, and two Cessna RGs. She was a federal government secretary for more than 40 years. Her husband Richard died in November 1985. She has two children: Marsha and Rick, and also three granddaughters and two grandsons.

Enjoyed providing assistance to the Air Race Classic when the route included Wyoming locations as stop chairman and timer. Has attended EAA Fly-In Oshkosh, WI, several times. Special interests include stained glass.

MARGARET J. EWERT, "Peggy" learned to fly in 1977 at Reid-Hillview airport in San Jose and joined the SCV99s immediately. She holds a private license with an instrument rating and belongs to the EAA, AOPA, Santa Clara Airmen's Assoc., and Angel Flight. Peggy and her husband, John, fly their Bonanza A36 (N6174N) out of San Jose just about everywhere and have been to Alaska several times. Peggy is the current vice-governor of the Southwest Section and has been the treasurer of the SCV chapter for years, as well as section treasurer. She has been a very active 99, participating in fly-ins, airmarking, seminars, and has flown the Palm to Pines Air Race. Peggy has won the chapter service award three times and has won the chapter Pilot of the Year award. She has logged 2,000 hours. She and John have four children, one of whom is also a pilot. Her non-aviation interests include sewing, beading and computers, as well as her full-time business with John operating Ewert's Photo shops in Santa Clara.

NANCY J. EZELL, president of J.B. Ezell Construction Co., Inc. DeVel Development, Inc. Pinnacle Land, L.C. and vice president, Pinnacle Development, Inc. At 55 years of age Nancy took a pinch hitter course in her Piper Super Cub tail dragger that just seemed to continue until she received a pilot license.

Two incidents happened that called for her persistence during this time. Right before soloing, her husband decided to have the plane painted causing a two month delay. Then he took the plane to Alaska for two weeks right before the check ride. All this time he was cheer-leading her on to completion within the year.

Despite the delay, she reports this learning process was a wonderful confidence builder. She was thrilled about becoming eligible to be a member of the prestigious 99s Women's Pilot Organization.

Since she recently became a widow, she especially enjoys the activities and fellowship of this pilot group. With less than 100 hours, her professional flying activities consist of photographing land development from the air!

NANCY M. "PAT" FAIRBANKS, born July 5, 1925, in California, KY. Joined the 99s in 1964. Flight training included private pilot 1948, now ATP, SMEL, SES, glider, helicopter, CFI-AIMH. She has flown general aviation planes, Piper J3 to King Air and has 8,000 hours of flight time. Fairbanks was employed as CFI at Cardinal Air Training 1965-92. Now CFI at Cardinal Helicopter Training. She has been married to Charles Don Fairbanks for 50 years. They have three children: Donna, Ruth and Carol. Carol is also a CFI. They also have grandchildren: David, Shannon, Chris and Lisa; and one great-granddaughter, Devon.

Don taught her to fly in 1948. Received her private pilot certificate on July 24, Amelia's birthday. Took time out to raise three daughters. Began flying again in 1962. Earned her commercial and CFI in 1965 and has been instructing since that time. Joined the All-Ohio Chapter of the 99s in 1964. Held all chapter offices including chairman in 1975 and 1976. Won the All-Ohio Achievement Award in 1967 and 1982. She was the recipient of an A.E. Scholarship in 1975 to acquire her helicopter rating. She has placed in the top 10 in Powder Puff Derby, the Angel Derby, Air Race Classic and many small races. She and Don own a Mooney 201, and a Robinson R-22 helicopter. Memberships include 99s, Whirly Girls (#256), Silver Wings, AOPA, Mooney Aircraft Pilots Assoc., Daughter of America and Eastern Star. She enjoys renewing friendships from PPD days, conventions, Forest of Friendship. They are very active in Silver Wings and Ohio Chapter is hosting the Convention in September 1995. She hopes to see more eligible members of the 99s become members of this organization.

MARYLOU FALCO, born Albany, NY, Oct. 19, 1948. Joined the 99s in 1984. Member of Civil Air Patrol since 1985. Started training in August 1983, licensed March 1984, instrument rating 1987. She has flown the following planes: Cessna 152, 172, 172RG, 182, 182RG; Beech A-36, A-23; Piper PA28-161, 181, 236. She worked as a pilot shop consultant for Richmor Aviation.

During college (mid 60s) and high school, worked at the local FAA-GADO office as a student aide. Later as an adult responsible professional medical person, took up flying lessons for fun. Total hours about 800. Along with flying, Marylou is an accomplished trophy-winning competition pistol and rifle shooter.

CHARLENE FALKENBERG, "Char" joined the Indiana Chapter in 1963. Transferred to Chicago Area and later became charter member of Indiana dunes and Illiana Cardinal Chapters. She was instrumental in forming both chapters. Served at all offices and committees on chapter level-treasurer and 501(c) 3 for North Central Section. Started, published and served as editor of North Central Newsletter, *Waypoint* from

1980. At International level she served as director, secretary, vice-president and chairman of memberships and flying activities committees. In 1984 she became a trustee of the Amelia Earhart Memorial Scholarship Fund and is still serving as chairman and permanent trustee. Holds commercial with instrument and multi-engine ratings and AGI and IGI certificates. Has taught ground school since 1964 and runs her own school. She is a FAA written test examiner and accident prevention counselor.

She and pilot hubby, Walter, owned Mooney 9199V for 21 years in which she participated in many Powder Puff Derbies, Air Race Classic, Angel Derby and many state races.

CONSTANCE FARMER, born Feb. 15, 1947, in Dupont, LA. She moved to California in 1971 and started her flight training at Hawthorne Airport in 1978. Learning to fly was a major achievement in her life. After earning her private pilot license, another 300 hours was required to overcome a basic fear of flying.

She has owned a Cessna 172, 182, Citabria 7KCAB, Piper Arrow and Miller Twin Comanche. At 1,300 hours with multi-engine, instrument, commercial ratings, she is currently working on a type rating in a T39 (North American Rockwell Saberliner).

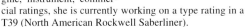

Connie is a member of the Long Beach Chapter of the 99s, Inc., Flying Samaritans, Airborne Relief Medicine and Experimental Aircraft Assoc. Her flying activities have included: Back to Basics and Shirts and Skirts Air Races; helped build a school for handicapped children in San Felipe, Baja, Mexico, transporting medical personnel to clinics in Baja and mainland Mexico; helped organize Flying Companion Seminars and flies children for the EAA Young Eagles program. She has flown from California to the Bahamas in a C182, toured Alaska in a twin Comanche and explored the Grand Canyon in a Citabria.

She is a system consultant for a major defense contractor working on air traffic control programs for Canada and Saudi Arabia. Merrill, Lori and Jean are her three wonderful children.

EVELYN V. FARNHAM, born Feb. 7, 1928, life-long resident of Iowa. Learned to fly in 1979 and owns a Piper Cherokee 180 along with husband, Paul, who is also a pilot. Joined the 99s in 1979. She has flown the Cessna 150 and 172, and Cherokee 180. Holds a RN degree from Iowa Methodist School of Nursing and attended Iowa State University. Worked in Ceilandia, Brazil teaching hygiene, first-aid etc.

Active in United Methodist Women holding offices on the local, district and state levels; served on national committees; was certified lay speaker in the United Methodist Church for over 25 years (most recently the director of Lay Speakers School). Does public speaking for various clubs and organizations. Reads the newspapers on radio (live) for the visually impaired and print handicapped. Member of the 99s, AOPA, Flying Farmers, Greene Aero Club.

Married to Paul Farnham, an agriculturist. They have three sons and one daughter; seven granddaughters and five grandsons. Their children are as follows: Dean, 40; Judy, 39; Dale, 38; and Doug, 36. Their grandchildren are: Trent, 19; Caleb, 15; Tyler, 15; Candice, 13; Leilani, 12; Dan, 11; Malia, 10; Christina, 10; Ashley, 7; Alex, 6; Rosalind, 4; and Pohai, 3.

In 1987 she traveled to Saipan where she visited the jail where Amelia Earhart (alledgedly) was held as well as the hospital and other sites … also visited with the Saipanese … and came away with the strong feeling, "Yes, she was there."

JOYCE MARIE FARRELL, born April 25, 1929, Los Angeles, CA. Joined the 99s in 1987. Started training in March 1977, soloed July 5, 1977, FAA check ride May 19, 1978. She has flown Cessna 150, 152, 172, and the Piper Cherokee.

Her children are Michael Farrell, age 39, and Steven Farrell, age 38. She also has grandchildren: Jesse, 8; Janna, 14; Matt, 18; Sean, 14; and Daniel, 16.

She has flown to Anchorage, AK, in a C-150 again to Fairbanks, AK, in a C-172.

She has visited many airports in the western US as well as Baja, CA, in the C-150. Participated in the Palms to Pines race, 1992 and 1995.

MARJORIE E. FAUTH, young Marjorie Hook traded bookkeeping for flying lessons at Mills Field, now San Francisco International Airport. She started in 1931 in a 40 hp low-wing Salmson-Klemm and got her license in a fleet biplane. She joined the Bay Cities Chapter of 99s when it was but a month old and has remained active in chapter affairs for 46 years.

Her description of the early years of flying: "Aviation today and personal flying prior to WWII is like day and night. We had no radio, only rudimentary instruments and the only navigational aids were a few lines of beacons which covered only the main air routes and were practically invisible by day. Rita Hart and I flew her Fairchild 24 to New York for the 99s 20th birthday and anniversary convention in 1949. It was a completely unregimented flight. We came and went when we felt like it, flew the course and altitude of our choice, and with great exuberance and joy indulged in cloud-busting and hedge-hopping whenever conditions and the mood was right for it." Marjorie flew aerobatics at country air shows, covered the various widely scattered cow-towns of Nevada as a traveling payroll auditor, served numerous years as permanent trustee of the Amelia Earhart Scholarship Fund, became International 99 treasurer and treasured the friendships of 99s around the world.

BARBARA LOUELLEN FEADER, earned private pilot license and joined 99s at age 48. Added commercial ticket and instrument rating at fifty. Flew to Kalispiel, MT, and took mountain flying course then flew to Couer de Alene, ID, for International Convention as delegate from Maryland Chapter, thence to British Columbia. Attended International Convention in Canberra, Australia, as delegate for Delaware Chapter. After convention, flew safari covering about half of Australia, including outback.

Second Lt. Civil Air Patrol eight years. Flew Angel Flights two years; blood for Red Cross one year. Senior Pilot Coast Guard Auxiliary five years. Five One Charlie now belongs to daughter-in-law, Lana Taft. She buys and restores old houses for profit and fun. Writes books and short stories, yet to be published, but still maintains her second class medical and flies locally. Her next adventure; flying safari in South Africa if gods be favorable. Will be 76 in November 1995.

CHARLENE FEE, born Sept. 4, 1929, Pennington Gap, VA. Joined the 99s in 1973. She holds a private, commercial, instrument and is a certified flight instructor. She has flown fixed wing-Piper Cherokees and Cessnas. She instructed at Painesville Flying Service, Painesville, OH.

She and her husband Denver have two children, Michael, 43, and Sharon, 34. They also have grandchildren: Daniel, 19; Jennifer, 16; Jill Ann, 14; and Jeffrey, 11.

In 1969 she had her first flying lesson and from that day on she was hooked on flying! She drove a school bus to finance her lessons and earned her ratings and instructed for many years for the sheer love of flying.

LOIS FEIGENBAUM, Cape Girardeau Area Chapter since 1963; International president (1976-78); flies her Baron N99LF; honored by Southern Illinois, Inc., only woman to receive Area Appreciation Award; Commendation from FAA Administrator; 1972 appointed to Women's Advisory Committee on Aviation by President Nixon, serving as co-chairman in 1975; honorary chairman of Bishop Wright Annual Air Awards; board of electors, International Aerospace Hall of Fame; board of electors, Notable American Women; accident prevention counselor since 1972; first woman to receive AOPA's Laurence P. Sharples Award for outstanding contribution to general aviation; appointed Midwest regional representative for AOPA; appointed national chairman United States Precision Flight Team; listed in Jane's *Who's Who in Aviation Aerospace*; vice-chairman World Aviation, Education and Safety Congress and special advisor to Prime Minister Rajiv Gandhi; asked by President Reagan's Transition Team to accept FAA assistant deputy administrator for airports. She declined, it meant separation from family.

During her presidency, the Powder Puff held its last race, and she turned the direction of the organization to aerospace education. The 99s History Book was conceived and started; "For the Fun of It," a film on the history of the 99s, was produced; first 99s travel tour of India was made; Aviation Aerospace Education Weekend Seminar was held to teach 99s teachers how to incorporate aerospace into the classroom; 99s started displaying at aviation trade shows with a well-planned booth.

She has flown in five AWTARS and five Air Race Classics, and has lectured, spoken, conducted seminars, and done public relation work for General Aviation. As vice-president, she made a tour around the world, meeting with members in Italy, Korea, Japan; and, she charted the Indian Section.

EDYTHE GORDON FEIN, born in Leominster, MA, a few miles down the road from Fitchburg Airport, where she first soloed on Oct. 8, 1944 in a Taylorcraft. She managed to accumulate 20 hours of flying time that year including takeoffs and landings on skis.

After a 20 year lapse, she went back to flying in the 1960s. Since then, she has logged about 900 hours in the Piper Arrow, Turbo-Arrow, Piper Warrior, Citabria, J3 and glider. Commercial and instrument ratings were earned in the 1970s. She flew medical equipment for her husband's company, but most flying has been recreational, including the Garden State 300 Air Races.

Besides being a long-time member of the 99s, she also belongs to the EAA, the Silver Wings Assoc., and a newly formed Aviation Club of Senior Citizens in Whiting, NJ.

Most memorable flights: first solo, of course; flying her son and his cello to Toronto, Canada, the customs inspector was amazed, had never checked a cello through general aviation before; and flying to Florida with Miran Schwartz, a young Israeli 99 with only 20 hours of flying time.

LYNDA SUE FELDSTEIN, a resident of New Jersey, received her license on March 16, 1995, flying 100 hours in a Cessna 172 out of Teterboro. She became a member of the North Jersey 99s, as a student pilot. Also a member of AOPA, she enjoys flying for recreation and pleasure.

She is a CPA with an MBA and mother of four wonderful children. Her husband George is the president of a leading electronics manufacturing company.

Lynda has been taking lessons for only six months. The first time she was in a small plane was with her husband during his lesson. She flew with him when he earned his license and decided she should learn to land. She found that she loved flying and decided to become a pilot. A year ago she would never have believed that she would ever get into a Cessna 172; especially to fly one.

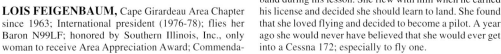

93

DENE CHABOT FENCE, joined the 99s in 1950, and served in the USAF. She has flown C-150, C-182, C-172, and PA28-180. She has one child, Lorena Fence. She learned to fly in 1974, bought her first and only airplane, referred to as *Ms. Cherokee,* before finishing her private pilot license. C.F. has logged more than 1,700 hours, and participated in many air races since 1985 and finished in the top 10, 13 out of 17 entries through 1994: ARC, Palms to Pines, PAR, Shamrock Air Derby, Valley Air Derby, Mile High. Past chairman of Fresno Chapter 99s. Educational background: AA American River College; BS California Institute of Technology, University of San Francisco; MS University of San Francisco. Practicing engineer for 20 years, now owns and operates Vitamin Villa Health Food Store in Carson City, NV. Entered into *Who's Who in the West* in 1987 for scientific and academic achievements. Currently a member of the Southwest Section AAS, AOPA, Air Race Classic.

JEANNE FENIMORE, a San Fernando Valley, CA, resident for four decades watched daily air traffic at Burbank Airport as a child. Through association with her parents' small aviation business, Aircraft Sparkplug Service, her ambition to become a pilot was fostered. She received her pilot license in 1971, and during the next 24 years added instrument and multi-engine ratings. She has logged more than 2,000 hours of primarily cross-country flying.

Jeanne is an avid air racer who participates in many local and national events. She flew her first of three Powder Puff Derby in 1973, and earned eighth place in 1976. She flew three Angel Derbies, winning the 1979 race which traced the first PPD route of 1929. She won the Palms to Pines race in 1974. Fenimore finished in the top 10 in almost every Pacific Air Race held from 1972 through its end in 1990, and won in 1979 and 1989.

Mrs. Fenimore has been an active 99 since 1973. She has held all chapter level offices, and served on almost all committees; many of them for more than one term. Throughout its 20 year history, the chapter sponsored Jim Hicklin Memorial Air Race found Jeanne a committee member or, for six years, the co-chairperson. She was chairperson of the Chapter's Valley Air Derby from 1979 through 1983. She received the San Fernando Valley Chapter Woman Pilot of the Year Award and the Chapter Service Award.

Her dedication to aviation extends to membership in AOPA, Cessna Pilots Assoc., and National Council of Women in Aviation. She also serves as a member of the board of the Van Nuys Airport Assoc. In addition, Jeanne is a continual volunteer worker for the local YMCA.

Wife of William Fenimore, an aerospace test engineer, she has a son, Christopher, four stepchildren, and six grandchildren. In her spare time, she can be found working on her 1967 Cessna 182, or enjoying her hobby; flying.

DONNA M. GRIMES FENSKE, born Dec. 20, 1940, in Sigounvey, IA. Joined the 99s in 1975. Received her flight training at Strand School of Aviation, Kalispell, MT. She has flown Cherokee 140 - 180, Maule, and Scout. She is married to John Fenske, who is a flight-ground instructor, commercial, instrument and multi-engine. She has four children: Kevin Smith (wife Lynne); Craig (wife Diane); Dina Marion (husband Lou); and Kristen Brown (husband Jim). She also has one grandchild, Shannon Smith, age 4.

Fenske took an AOPA Pinch Hitter course, overcoming a flying fear, and proceeded immediately to becoming a licensed pilot in Kalispell, MT, soloing in a Cherokee 140 and receiving a private pilot certificate in 1974.

After moving to Alaska in 1974 she and her husband bought a Maule which they both flew, beachcombing and exploring throughout the States ever since. Donna is also float rated. She joined the 99s in 1975 and is a member of Cook Inlet Chapter.

In 1978 Donna, husband John and four children moved to Homer. Since that time she has worked as a public health nurse for the state of Alaska. This involves considerable flying to remote villages providing community health services. She received her BSN from Loretta Heights College in Denver, CO, in 1962; MA from fielding in 1978, Santa Barbara, CA; and a MPH from Loma Linda University, Loma Linda, CA, in 1985.

When Donna retires, she and her flight instructor husband, plan to devote much time to flying around Alaska in their Maule.

NOHEMA FERNANDEZ, born in Cuba, emigrating to the US in 1961. She obtained bachelor of music, master of music and doctor of musical arts degrees from DePaul University, Northwestern University and Stanford University, respectively. She has developed an International career as a classical concert pianist, performing worldwide and making solo and chamber music recordings.

In 1991 an invitation to be flown to a concert engagement in a single-engine airplane stirred up a passion for flying. She started flight training soon after that, obtaining private and commercial certificates, an instrument rating, and, most recently, a CFI-airplane certificate. Based in Tucson, AZ, Fernandez is active with the Tucson 99s since 1993 and received one of the chapter's scholarships for the CFI training.

Now, she frequently pilots her Cessna 182 to concert appearances. Recently, she has flown herself from Tucson to concerts in Arizona, New Mexico, Texas, California, Nebraska, and Mexico. Fernandez is a professor of music at the University of Arizona, and frequently introduces her students to the joys of flying.

SYLVIA FERRARE, always enjoyed living life to the fullest. Flying has proven to be an experience in freedom for Sylvia and that is why she obtained her private pilot license in 1993. She currently flies a Piper Archer but hopes to own an Areospatiale Trinidad soon. Sylvia's other flying related activities include hang gliding, paragliding, ultralighting, hot air ballooning, and the somewhat flying related sport of bungee jumping. Sylvia has also enjoyed the unique experience of flying in an open cockpit biplane. When she isn't flying, Sylvia enjoys participating in practical shooting competitions and is an active member of several practical shooting organizations. For relaxation Sylvia likes to paint, sculpt, draw and play the piano with her three Persian cats by her side. Learning to ride a Harley motorcycle and participating in cross country air races with her mother are among the goals Sylvia has set for the future.

LYLA FERREL, born April 12. She is an executive assistant for Bonanza Casino. She has four adult children, two sons and two daughters.

Joined the 99s Reno Chapter in 1977. She has held the following offices: chairman, three terms; vice chairman; membership chairman; secretary, two and a half terms.

Ferrel became involved in 1969, "Just could not resist." She has a private pilot license with glider rating and owns a Cessna 152 II.

LINDA LEE FETSCH, born July 1, 1942, in Salem, OR. Her flight training includes private SEL, September 1984; instrument, August 1986; commercial, February 1987; MEL, March 1987; CFI, May 1987; CFII, May 1988; MEL, March 1990; ROTO private, April 1992. She is the proud owner of '77 Archer flown to 39 states including Alaska. As an instructor and charter pilot, she flies various Cessna, Piper and Beech aircraft. She is employed by Buswell Aviation, Salem, OR, as flight instructor and charter pilot. Fetsch and her husband Jerry have two children, Randy, 31, and Karen, 28. They also have grandchildren: Danielle, 8; Matthew, 6; Andrew, 7; and Kyle 5.

In the spring of 1984, she and her husband visited their local airport to have luncheon at the restaurant for the first time. Watching the activity on the field, she suggested that he learn to fly so they could fly for pleasure and for business. He said he was not interested in learning to fly, but if she wanted to, go ahead and learn to fly. Two hours later they left the airport having paid down a deposit on lessons. Two days later she had her first lesson which was her first experience in a small aircraft. By September 1984 she had her private pilot license and truly was "Bit by the Flying Bug." She flew as much as possible, continued to gain experience and ratings. She knew by 1987 she wanted to share her love of flying with others and began to teach. She and her husband purchased a 1977 Archer in 1985. They have traveled to 38 states so far including Alaska. She intends to fly into and visit 49 states in all, plus all providence of Canada in her Archer before her flying days are ended. Two years ago, she decided to learn to fly a helicopter. That is the most challenging and rewarding flying rating she has accomplished to date.

JOYCE FINDLAY, for more than 25 years, she followed her Air Force husband from base to base listening to "war" stories and longing to "punch holes in the clouds." The closest she got to fulfilling her dream was a high speed taxi in an F104, quite impressive in and of itself, but still a far cry from the real thing.

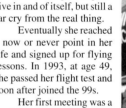

Eventually she reached a now or never point in her life and signed up for flying lessons. In 1993, at age 49, she passed her flight test and soon after joined the 99s.

Her first meeting was a section event. She loved it! It was so exhilarating listening to similar sorts of flying tales, but this time it was women sharing their experiences.

Her dreams are still unfolding and her adventures continue. Over the past year, she has logged time in a C-180 Float Plane, a BH47 Helicopter and a T33 jet!

NITA FINEMAN, knew she wanted to fly when she was 7 years old. At age 15, she joined the Civil Air Patrol. Her squadron was based at historic Douglas Field, Orchard Place, which later became O'Hare International Airport. Her first flight was in a Ford Trimotor on a sightseeing trip over her home town Chicago for her 16th birthday!

She married her high school sweetheart Sidney Fineman in 1946. With his encouragement and support, she soloed in a J3 Cub in August 1948 and received her private pilot license two months later. She earned her instrument rating in 1961 in her Piper Tripacer, *Small Fry,* which she cherished and flew for 20 years.

After her husband's passing in 1973, Nita continued to work full-time as a receptionist and part-time as a travel agent. In 1989 she moved to San Diego, CA. She travels extensively and loves to listen to the music of Glenn Miller.

Nita completed her autobiography in 1993 entitled *Flight to Life.* She is a life member of the Chicago Area Chapter 99s and is proud to have served as chapter chairman as well as governor of the North Central Section.

She has two son and two grandsons for whom she is grateful and very proud.

BARBARA FIORAVANTI, born Feb. 10, 1944, Worcester, MA, USA. Joined the 99s in 1984. Received her private pilot license in 1982; instrument rating in 1988; certified ground instructor and FAA safety counselor. She has flown Piper, Tomahawk, Warrior; C-150/2; Beech Sierra; Piper Arrow. Pilot employment included FBO Rensselaer (county) Airport, 1984-91 (Poestenkill, NY).

She and her husband Eugene have two daughters, Wendy, 26, and Deanna, 24. Fioravanti especially enjoys introducing "landlubbers" to flight, seeing faces turn from fear to smiles. Many happy times as FBO at small private airport, hangar-flying, help-

ing green students become fine pilots, assisting A&P/AT with aircraft maintenance. Pride of running a quality operation.

AMY FISCHER, born Nov. 29, 1955, in Franklin, NJ, and raised in Sparta. Joined the 99s in 1993. Her first hours flying went unlogged. While project specialist on Navy P-3s, she flew Hawaii to California. She learned to fly at Patrick AFB Aeroclub in 1986. Her career required extensive traveling and she would get checked out in nearby flying clubs when time permitted. This gave her experience flying around the Oregon Cascades, float plane piloting in Anchorage and an international flight to Dawson City, Yukon. In 1992 she took a month off and flew cross country in a Grumman Tiger; San Diego indirectly to Stuart, FL, and return, all VFR. She resides in Cocoa Beach and is an investment broker at Merrill Lynch.

She has flown the following planes: Piper, PA-28, Cessna Centurion and Aeronca.

EDITH COPELAND FISCHER, got "hooked" on flying after completing the AOPA Pinchhitter course at Williamsburg-Jamestown Airport (JGG). She completed her pilot training through the encouragement of her husband, John Fischer, a flying optometrist, and received her private pilot license in April 1976 in a Cherokee 180C.

Edith joined the 99s in October 1976 and has been very active in the organization on both chapter and section levels. She has served as chapter chairman (three times including current term), vice chairman, and secretary; also, secretary for the Mid-Atlantic Section.

Edith is retired from Newport News Public Schools, Virginia, as a teacher certification specialist. Sadly, her husband died of cancer on March 15, 1994. They had no children.

MARGARET FISCHER, Ph.D., J.D., born in New Jersey. She has flown the following planes: Rotorcraft; Bell and Robinson helicopters; Beechcraft Duchess; Cessna 152; Piper Tomahawk.

She is employed as personnel officer for Civil Air Patrol.

"No derby I've entered, no prizes I've won, flying for me, is just wonderful fun."

LINDA E. FISH, a friend gave her her first small airplane ride. They discussed the expense and time involved in learning to fly. He recommended an instructor whom she met two years later while on jury duty. This person offered to take her on an introductory flight and the next day became her instructor and flying mentor.

She started her flying lessons in October 1987 at Post Mills, VT, and received her private license in July 1988. Shortly afterwards, she joined the Northern New England Chapter of the 99s and met a group of women who inspired her to continue with her flight training.

In May 1990 she purchased a 1967 Piper Cherokee 140 which continues to provide her with the utmost pleasure every time she flies it. She has shared her love of flying with family members and many friends. They understand her preoccupation with flying once they experience the beauty and thrill of flight.

In January 1991 she obtained her instrument rating and over the next few years enjoyed flying all over the New England area. She found that she was consumed with flying and wanted to become an instructor.

She won the New England 99s Scholarship and received her commercial rating in July 1994. She finally reached her goal by attaining her CFI in July 1995. This represents a new beginning in her flying career which she looks forward to as a rewarding challenge.

CAROLYN FISHER, born in New Orleans, raised in Vicksburg, MS, attended college in Virginia, served with the Peace Corps in both the Philippines and Bangkok, taught college classes in Europe and served the Air Force in Florida, Mississippi, Alabama, Alaska, Illinois, Virginia and California. She began flying soon after joining the Air Force as an air traffic control officer in 1974. She has logged more than 1,300 hours and holds ASEL, AMEL, CFI and CFI-I ratings. Due to her Air Force career, she has been a member of numerous chapters of the 99s and has helped charter two: the Florida Panhandle and the Sedona Red Rockettes. After the completion of her Air Force career in 1994, Carolyn settled among the glorious red rock of Sedona, AZ, where she is active with tutoring activities, 99 activities and the beginning of another career.

She is married to Miles "Gary" Fisher, a pilot, formerly an Air Force officer and currently a practicing CPA in partnership with another 99. They own a 1976 Cessna Cardinal RG.

DIANE TRIBBLE FISHER, says flying "leaving all the cars of the world behind." At age 9, Diane was introduced to flying by an older cousin who loved model airplanes. She earned her private license in 1968 at age 20; joined the 99s in 1969.

From 1976 to the present Diane - more than 4,500 hours - has earned her ATP, CFI, AI, ANP, FE, SEL, MEL and all ground ratings. She has owned and operated her own FBO, been chief pilot and charter pilot for local FBOs, and for over five years flew as flight engineer for the airlines on national and international flights. Currently she instructs at Long Island-MacArthur Airport.

Diane has a BA in education and master's degree in aerospace technology. Diane's most recent 99 position was as the LI chapter chair from 1993 to 1995. She has led many committees and received numerous appreciation and club awards. Her most memorable 99 experiences include flying the last Powder Puff Derby, the Angel Derby and the Garden State 300.

Her airline stories include such thrillers as losing pressure in a Boeing 727 on a flight from Nassau in the Bahamas to Chicago; having to dump 10,000 lbs. of fuel into the Gulf of Mexico when the Boeing 727 had to land right after take-off due to a faulty transponder; while ferrying federal prisoners in chains, being handed a gun and told to "shoot to kill" should the door to the cockpit open; being told to make herself "look like a man" if she chose to leave the aircraft in Cairo, because of that country's poor regard of women fliers.

As flight instructor she's experienced engine failure, faulty gear mechanisms and complete electrical failure but still loves the job.

Diane is surrounded by airplane people. She is married to William Fisher, a retired airline engineer. Her brother, sister-in-law and stepson work for the airlines. Except for Bill she introduced her family to aviation. "I started them," she proudly says.

PAMELA S. FISHMAN, private pilot, secretary, 99s Oklahoma Chapter. A long time ago in Ohio, proud parents recorded "hi" and "airplane" as their daughter's first words. Pamela navigated her way through Capitol Hill, the Smithsonian, law and nuclear medicine.

On course, but doubting her instruments, she began lessons at Godfrey Field. Her instructor, told her, "if you don't quit waiting for days when the wind is seven knots or less, you'll never get to fly." Six months into her lessons Pamela and her hus-

band moved to Oklahoma "where the wind comes rushing down the Plains."

Pamela and her husband David, a judge, enjoy life on their ranch with an airstrip. They ride horses and raise Bouviers. Pamela is working on her instrument rating and participates in the Okie Derby, Summer Space Camp and enjoys teaching children about flying. Pamela is secretary of the 99s Oklahoma chapter. Pamela is a certified firearms instructor, teaching basic pistol, personal protection and home firearms safety.

Now, Pamela prefers not to fly when the winds are calm.

JEAN FLAKKER, born Oct. 24, 1962. Joined the 99s in 1986. Her flight training includes commercial, ASEL and instrument ratings. She is employed as a structural engineer. Flakker has flown Beech Sundowner, C-152, C-172, Piper Arrow and Cub. "I found my freedom when I learned to fly."

VIRGINIA FLANARY, as a third generation Californian and 1930s child, loved "up" places. Tree branches, roof ridges and water towers were hangouts for reading, dreaming, viewing or planning sways, swings or leaps. When "grounded" she compensated by looking up, at tree tops, mountains, birds, clouds, rainbows and airplanes. As a pilot in 1964 she preferred aerial views and destinations over speed, precision and fame. She joined the Orange County 99s to hold offices, serve committees and write newsletters, and they steered her toward higher ratings and

proficiency to make her 1973 Pilot of the Year. Favorite flown airplanes are Cessnas 150s to the 337; Luscomb, TF8 to the Aero-Commander 500 – high winged viewing ones. She's discovered parachute silk more efficient than umbrellas, and balloon baskets more exciting than bridge rails. Supplementary interests: medical, geography, travel, arts, antiquities, design, words, intellectual challenge, humor, positive thinking, humanity and still – high up places.

RUTH S. FLEISHER, Florida Goldcoast Chapter, started flying and instructing ground school for Civilian Pilot Training program in Rochester, NY, in 1940. Served with the Coastal Air Patrol on Long Island, NY, during war for a year.

Graduated from WASP program, was assigned to testing and instructing staying until disbanded in December 1944.

Worked as a flight and ground instructor, and Air Traffic Controller (Tower) in Philadelphia, PA. Joined the USAF Reserves and worked as a controller and Flight Facilities Officer at bases in US and RAF Station in England. Retired as a major, USAF Reserves. Flew as charter pilot and instructor, and instructed for several USAF Aero Clubs.

Has flown in several air races, including the AWTAR and acted as timer and chief judge for the IAR race.

A 99 since 1942 she has served in various chapter and section offices and worked on many committees.

A member of the WASP, the Woman's Military Aviators, IWAS Museum, Glenn Curtiss Museum, American Aviation Historical Society and the AOPA.

DELRYN R. FLEMING, born March 20, 1945, in San Angelo, TX. Joined the 99s in 1994. Completed her private pilot license in January 1994. Married to Skip Kilmer, private pilot, and owns a 1961 Piper Cherokee 160.

Fleming learned to fly, at first, simply for reasons of survival. If her husband Skip Kilmer, was going to fly his newly purchased Piper Cherokee 160, she wanted to be part of the decision making process. After receiving her license in January 1994, she now sits left seat one way and right seat the other. With

not many hours to her credit, she highly reveres the power of flight.

A native Texas born in San Angelo, Fleming has a BA from Southwestern University, Georgetown, TX, and a MA from East Texas State University in Commerce, both in English. She teaches composition both traditionally and on computers, at Brookhaven College in Dallas, TX. In addition to the Dallas Chapter of 99s and AOPA, she is a member of the American Assoc. of University Women and the Methodist Church. She has one adult son attending graduate school, Jeffrey Cookston, born May 2, 1970.

PAMELA FLEMING, joined the Arkansas Chapter in April 1966. Flew 1969 Powder Puff Derby. Transferred to Chicago Area Chapter in 1969. Flew 1971 PPD, also flew about 20 proficiency races. Transferred to Indiana Dunes Chapter about 1978. She has been FAA accident prevention counselor since 1975. CFI from 1966 to 1994 and airport manager at La Porte Airport, 1974-75.

IRENE W. FLEWELLEN, born 1922, Chattanooga, TN. Earned pilot license in 1954, joined the 99s, flew 8th Annual All Woman Transcontinental Air Race same year. Flew 1971, 1977, PPD and served in various capacities for AWTAR, Angel Derby and ARC. Has commercial and SES ratings with more than 1,500 hours total time.

In 1963, bought land and developed Dallas Bay Skypark (1AO) near Chickamauga Lake in Chattanooga. Still active as owner/partner in the business and recognized as one of the few female airport owner/managers in Tennessee. A Cessna Pilot Center since 1968 and has flown mostly Cessnas.

Devotes most time promoting aviation and 99s. In 1974 served as SE section treasurer, rest of time-officer for Tennessee Chapter. Lifetime member 99s, received Golden Merit Award, Glad Hand Award, and inducted into International Forest of Friendship, Atchison, KS, sponsored by Tennessee Chapter. The first female pilot invited to join Chattanooga Flyers Club, oldest flying club in America.

She has four children: Jo F. Cantara, 53; John G., 51; David W., 44; and Beryl F. Hartwig, 38. She also has six granddaughters and six grandsons.

CLAIRE FLORIO, She learned to fly in 1983 at San Jose, CA, and joined the Santa Clara Valley 99s in 1993. She is an aircraft mechanic who enjoys hiking, skiing and gardening. Claire has worked for San Jose International and United Airlines. She won the chapter's Marion Barnick Scholarship award in 1987 and has served as chapter student pilot program chairman.

JEAN F. FLOWER, born April 12, 1936. She and her husband Wesley A. Flower have two children, Kimberly, age 31, and Kristina, age 28. Flower has two loves, flying and painting. Through flying she's able to see the world from a different perspective than that of the average person. With her paint brush, she tries to capture these images on canvas that are seen from a pilot's point of view. She's fascinated by the sky forms and land shapes that are constantly shifting with the changing seasons.

Trying to capture that ultimate feeling and thrill a pilot feels or experiences in the air is her goal in painting a picture.

She learned to fly in 1967 and currently is part owner of a Cessna 172. She lives outside of Plainwell, MI, and is a member of the Lake Michigan Chapter of the 99s. She currently is a artist member of the American Society of Aviation Artists and has taught art classes at Western Michigan University and the Kalamazoo Institute of Arts.

ANITA LOUISE FLOYD, born Feb. 3, 1930, Oklahoma City, OK. Joined the 99s in 1968. She is currently serving as vice-chairman of the Oklahoma Aeronautics Commission. Learned to fly in Ada, OK. Flown Mooney single engine and gliders. She is a widow and has three living children: Eve Hawley, Jana Dile, Paula Kay Floyd; and one son Ray, deceased. Grandchildren: Russell Floyd, Amber Floyd Crawford, Thomas Hawley, David Hawley, Jason Hawley and Leah Dile.

She first became "hooked" on flying as a teenager during WWII. Lived in Arkansas in a small town. The local airport was grass strip outside of town, owned by father of boy she was dating. He got his license at 16 and she flew with him after that. An older brother was an Air Force flight instructor who occasionally came home on leave and would take her up and do aerobatics. Her flying stalled when she went to college. It took her 16 years to get back in the air. Both her husband and her son also became pilots.

DOROTHY FLYNN, Sacramento Valley Chapter, learned to fly in a C-150 rather late in life; age 50, to be precise. Other commitments, responsibilities and six children prevented her from considering it earlier. She is an instrument rated commercial pilot, hold a CFI certificate, with about 1,500 hours and the proud owner of a Piper Arrow.

She has served as chairman, vice-chairman and secretary of their local chapter, been honored as chapter Woman Pilot of the Year twice and received their chapter's Service Award.

She flew for the Sheriff's Aero Squadron, Angel Flight, EAA's Young Eagles, belongs to International Flying Farmers, serves on the California Aeronautics Aviation Advisory Committee and works with local aerospace education programs. She has flown the Air Race Classic and Palms to Pines.

She loves the freedom and challenges of flight; always trying to improve, but rarely quite satisfied. Climbing up through the clouds into the sunshine is truly euphoric.

HELEN FOEGER, private pilot license May 5, 1972, and has 960.6 hours (1994). Private certificate and ASEL rating. Joined the 99s in 1974. She is a Lake Tahoe Chapter charter member of 1975. She is now retired. Previous occupations included: ski instructor at Yosemite and owned Luggi's Sport Shop at Ski Incline for 22 years. She has flown C-172 (owned N12503); Marge Schwed and she bought *Tiger Libby* in 1973. Foeger is a widow with two sons, Chris and Jimi.

She helped at ARC in Sacramento and Palm Springs, attended 99s International Convention in San Francisco, Reno Air Races, Watsonville Airshow, 99s section meetings at La Jolla and Apple Valley. Her nonflying actives are skiing, sailing and golf.

Foeger is a member of IV Chamber of Commerce, IV Golf Club, Truckee Tahoe Plane Talkers, 99s (21 years).

She picked up her plane at Truckee after having repairs on the engine (carburetor was sent out to Sacramento). She took off on runway 19 but couldn't gain more than 350 feet altitude. She called in and traffic held off as she limped in and headed for maintenance. Airport cars came out to meet her with red lights blinking. She got out and said she was fine and with that her knees buckled and she was on the ground. The part was sent back to Sacramento again!

ADELE FOGLE, an experienced flight instructor with a senior commercial pilot's license with 4,000 hours. As president and owner of Aviation International (Canada) Inc., a pilot training and charter service company in Guelph, Ontario, she has introduced many young people to the world of flight. The school regularly hosts charitable events. Her company also is involved in air cadet training programs. A regular competitor in air rallies, she has criss-crossed North America from Alaska to the Bahamas. She was a member of the only all-women team to participate in the 1986 Israel balloon race on the occasion of that country's 40th Anniversary. Adele is past chairman of the Toronto Chapter of the 99s.

Adele has two grandchildren. Co-captain in following: recent Around-the-World Air Race 1994; New York-Paris Race 1985; Air Race Classic 1989, 90; Great Southern Air Race 1992, 93; and co-director, pilot, Air-O-Sols - a flying team to measure air pollution.

BEVERLY WARD FOGLE, started flying in 1972, and earned ground instructor ratings to help pay the flying bills. She had the distinction of signing herself off to take the commercial pilot written test! Upon earning the commercial license and flight instructor (instrument) rating, she taught for North West Flight Services at Pearson Airpark in Vancouver, WA, for three years. She served on the Airport Advisory Committee for Pearson for eight years and in 1995 joined the board of the Pearson Air Museum. During the 80s and 90s she continued to instruct, giving many flight reviews and shepherding several students through training for the instrument rating.

Beverly flies her Cessna 172 (which she has improved with a 180hp engine and long range fuel tanks) widely throughout North America including Alaska. In 1995 she joined others from the Alaska Airman's Assoc. in Friendship Flight 1995, flying across the Bering Sea to Russia. She would like to fly to Europe, and thinks a route through Russia would make more sense for a 172 than trying to cross the Atlantic.

Professionally, Beverly is a certified financial planner and owner of Cambridge Financial Management Corp., an investment advisory firm based in Vancouver. She has clients throughout the western United States and reaches many of them by air. She earned her master's degree from Kansas State University and has two grown children living in Seattle.

MARY F. FOLEY, "Bunny" learned to fly in 1957, and holds commercial and instrument ratings. Has flown: Piper Cub; Aeronca 7AC; Cessna 120, 150, 152, 170, 172; Beech Musketeer and Beech T-34; Navion and Stinson Voyager. Joined 99s in 1959 and is a life member. Also member of Silver Wings, Aerospace Medical Assoc., Aerospace Physiology Society, Space Medicine Branch of Aerospace Medical Assoc., International Women's Air and Space Museum, AOPA, National Space Society.

Worked in Aviation Medicine Research Laboratory at the Ohio State University doing research in the altitude chamber, in flight studies, underwater and respiratory physiology.

Member of United States Air Force Reserve for 30 years, retiring at rank of colonel. Last 20 years of service was spent in research at the Aerospace Medical Research Laboratories at Wright-Patterson AFB, OH, working in zero-gravity, human factors, hot environments and acceleration.

Presented papers at numerous national and International meetings; 14 journal publications with several in preparation at present time.

JANE TAYLOR FOLSOM, born Oct. 7, 1928. Joined the 99s in 1984. Private instruction by John Leach at Waterbury Oxcford Airport, CT. She has flown the Cherokee 140. She and her husband Earl have four children: Lawrence E., 43; Virginia Folsom Rave, 41; Jill Schmidt, 38; and Charlotte Clark, 34. They also have grandchildren: Corrine Folsom, 17; Patrick Folsom, 15; Shelly Folsom, 13; Brian Schmidt, 9; Steven Schmidt, 7; Jeffrey Schmidt, 3; Kate Rave, 9; Helen Rave, 3; and Jennie Clark, 5.

Folsom had a long time association to people who fly, (two brothers and husband). Soloed in 1977 and got her license after first grandchild was born in 1978. Has made cockpit covers for husband and son and others. Some are fullcovers. They've flown across the country and to Florida several times.

LAURA LYNNE FONSECA, born Oct. 29, 1944, in Queens, NY. Holds a ASEL and has flown the following planes: Grumman Trainer; Traveler; Cherokee 140, 180, 181; and Cessna 150, 172. Fonseca was employed as an air ambulance nurse for two years in 1980 and 1981. She has two children, Jennifer Anne Fanders, 24, and Christopher Andrew Fonseca, 19.

Fonseca is a member of the Eastern Pennsylvania Chapter. She earned her license in a Grumman Trainer on Sept. 9, 1975. Joined the 99s in November 1977. She is an RN currently working part-time to support her flying habit. She also works part-time as a travel agent to satisfy her urge to travel.

All of her flying is done for pleasure and membership in the 99s gives her places to go and an excuse to fly there. She attends every chapter meeting that her work schedule permits. Lunch with the Bunch is her favorite flying activity and she attends every one for which the weather is VFR. She is a past treasurer of her chapter and also past air marking chairman. She also participated in many Pennies-a-Pound and helped her chapter with the Terminus for the Air Race Classic at Phila North East (PNE). She is currently flying a Cherokee 180 and hopes to continue for many more years.

JUDIE FORBES, Ph.D., born Sept. 27, 1942, in Fullerton, CA. Joined the 99s in 1993. She served in the US Navy as air control tower operator from 1960 to 1961. She holds a private pilot license and has been flying since 1980. Forbes has flown the following planes: Piper, Tomahawk, Cherokee, Arrow and Dakota.

Forbes is married to Ralph Hawk and has three children: Laurel Schader, 34; James Resha, 32; and John Resha, 28. She also has grandchildren: Jessica Resha, 6; Kirsten Resha, 4; and Jason Resha, two days old.

LYNNE DONIVAN FORBES, born Oct. 8, 1941, in Philadelphia, PA. Joined the 99s in 1986. Her father, Clarence Donivan, was a sharpshooter in the US Marines during WWII. Flight training includes Pennridge Aviation, Perkasie, PA; instructors: Margaret Bryant, FAA examiner and James Soti and Bill Thomas, aerobatic instructors, Venus, FL. She has flown Pitts 52B, Stearman, C-150, C-152, AC112. Employed as a corporate pilot for Cynosure Corporation, Second in Command, AC112. She is married to James A. Forbes and has the following children: Constance, 30; Daphne, 27; Alexandra, 28; stepchildren: Karen, 40; Joan, 38; Frank, 36; Susan, 32. She also has grandchildren: Cory, 9; Christopher, 8; Nichole, 7; and step-grandchildren: Holly, 12; Sunny, 7; Levi, 3.

Her flight experience has been a metaphor for her continuing education, learning to "complete the task at hand" through "the art of flying" has made all the difference to her self and others.

PATRICIA ANNE FORBES, born April 30, 1939, in New York, NY. Joined the 99s in 1968, Started flying in January 1968, currently holds following ratings: commercial, instrument, certified flight instructor, all ground instructor ratings. She has flown the following planes: C-150, C-152, C-172, C-182, C-172RG, C-206, C-210, and Comanche 250 and 260. She is married to David M. and has two children, Sheryl, 30; and Karyn Rosol, 28.

In 1970 charter member of Golden West Chapter served as treasure and chairman. A 1972 chairman of Powder Puff Derby Start-San Carols, CA. From 1978-80, member of SW Section Nominating Committee, 1980-82, treasurer of SW Section; 1982-84, secretary of SW Section; 1984-86, governor of SW Section; 1988-90, International treasurer of 99s; 1990-92, International vice president of 99s; 1993 to present, vice-president of National Council for Women in Aviation and Aerospace. Currently field representative for EAA's Young Eagle Program, FAA Accident Prevention Counselor, and director of finance for 1996 World Precision Competition. Member of AOPA, EAA, NCWA NAFI, SCPA, TCPA Silver Wings. Has flown numerous air races including 1970, 71, 72, 75, 76 Powder Puff Derbies placing in the top 10 in three. Won the 1979 recreation of the first all Women's Air Race from Santa Monica to Cleveland. Flew numerous Pacific Air races winning twice.

Currently owns a C-172, C-210, a fast build Lancair 360 kit, and a 1929 Davis Dk-1 which is a restoration project.

ESTHER BRUNEN FORDHAM, born in Fort Pierce, FL, her first flight was with her brother at age 13. How exciting it was going to the airport at Vero Beach, FL, and watch DC-10s land.

The Greater Miami Air Show in 1951 was represented by two 99s Blanch Noyes and Arlene Davis. Esther was runner-up in a beauty contest for Miss De-icer staged by the airshow and "Smilin-Jack" comic strip. They were photographed together but did not know each other.

She married a pilot, Ace Fordham, of Goldsboro, NC. Flying as a family with four children in a 182 and 210 for safety, she gets her license in 1964. Her four children learned to fly – three become license pilots.

A charted member of Kitty Hawk Chapter of 99s - Petticoat Pilots, Goldsboro - Wayne Pilots Assoc., an accredited flower show judge, avid gardener, a Junior Garden Club Leader and four grandchildren.

LAURIE ERMENTROUT FORTE, born June 8, 1959, Abington, PA. Joined the 99s in 1985. Received her flight training for private from Marge Bryant (also a 99) on Dec. 10, 1985. Has flown the Cessna 152 and 172, and Beechcraft T34A Mentor. Married Michael (also a pilot) in 1985. They have three Dachshunds dogs. In 1986 purchased the Cessna 152 that she and her husband both soloed in. In 1992 bought a Cessna 172 and in 1993 bought a Beechcraft T34A, and sold the 152.

As a member of the Pennridge Pilots Assoc. she has been the editor of the monthly newsletter since February 1993. She also runs the Aviation Awareness Days her association sponsors as well as the Young Eagles programs.

One of her most memorable times as a 99 was June 1995. She was fortunate enough be to able to participate in the first all women's airshow at the Quakertown Air Festival, Quakertown, PA. The four featured performers were Julie Clark, Susan Dacy, Joann Osterud and Patty Wagstaff. One of the performers who was supposed to be there was Jan Jones, however, she was killed in a crash in Ohio a month before. Jan was on her way to Quakertown to do some pre-show publicity. Jan was honored at the opening of the show each day with a Missing Man formation performed by four T-34 pilots, herself being one of them. Nancy Sarver flew the lead, Forte flew the #2 position, Julie Clark flew #3 and Terry Deardan flew #4.

She spends a lot of hours practicing formation flying with other T-34 pilots and finds it very challenging as well as enjoyable. Unfortunately there are not a lot of female T-34 pilots in their immediate area so it was especially exciting to be able to have the four of them honor Jan at the Quakertown All Women's Airshow.

Forte and her husband practice formation flying together, most of their vacations are based around flying events (i.e. Oshkosh, Reno and Phoenix Air Races; and Sun-n-Fun).

Memberships: International Council of Air Shows, Inc.; Warbirds of America; Experimental Aircraft Assoc.; T-34 Assoc.; 99s Inc., International Women Pilots Assoc.; Aircraft Owner and Pilots Assoc. Has been a private pilot for 10 years.

MIRIAM "WINKIE" FORTUNE, born on April 4, 1930, in Bowling Green, VA. She received a BS degree in 1951 then worked for several years as a hospital dietitian before joining the Air Force as a first lieutenant in 1959. She is a retired lieutenant colonel.

Miriam learned to fly in Minot, ND, while assigned there as an USAF officer. A friend who was a nurse introduced her to flying. Her reluctance to fly vanished after her first solo flight.

She has had many gratifying experiences since joining the 99s in 1977. She has served in all capacities; committee memberships to chapter chairperson. Her most memorable venture as a 99s was flying right seat on a flight from Panama City, FL, to Haiti. It was incredible flying over the breath taking jeweled waters of Cutlas Bay and the surrounding area. This flight measured 1267 nautical miles and the aviators spent 11 days and 10 nights in splendor on the islands. Miriam cherishes the camaraderie of pilots who love to fly.

JONNIE FOX, born in New Orleans, LA, where she learned to fly at the age of 33. Having joined the 99s immediately as a 66er, Jonnie went on to become chapter chairman of the New Orleans 99s from 1989-91. Accomplishments under her chairmanship included a "Welcome Home Troops" barbecue/fund-raiser for Desert Storm Squadrons, Aerospace Education school trips, airfield and aircraft tours for deaf children, and active participation in the Blue Angels yearly airshows.

Jonnie is also a first lieutenant in the Civil Air Patrol having actively participated in scores of SAREX's at WING level.

"I am still amazed at times when I reflect on Nov. 11, 1987, when my CFI stepped out of the C-152 at Lakefront Airport in New Orleans and said, "It's all yours!"

She couldn't believe the day had come that she would be operating this heavy piece of equipment all alone. She was 34 years old.

Her father was a CFI at the time, but was adamant about not teaching her. He was there for her solo though - the proud father of his "chip off the old fuselage." Mom would never come. She despised the fact that her husband of 23 years and now one of her "little girls" were both engaging in what she thought to be a ridiculous and risky pastime. It wasn't until she began calling the house and asking to speak to Dad more often than her, that she became suspicious.

Jonnie now resides in San Diego, is a professional singer, member of the Mission Bay 99s and leaves them with her favorite prayer: "Glory be to Thee my God, this night for all the blessings of Your light. Keep me, Oh keep me, King of Kings, Beneath thine own Almighty Wings."

KATHY FOX, born in Montreal, Quebec and remembers having a passion for air and space from the time she was old enough to know what an airplane was.

Although she took her first ride in a plane at the age of 13, she decided to put her dreams of flying on hold while she earned her BS from McGill University. During that period, she took up skydiving and at one point had over 300 takeoffs, but no landings (in a plane, that is). Kathy went on to become the first female, youngest ever and longest serving president of the Canadian Sport Parachuting Assoc.

After graduation Kathy became an Air Traffic Controller in 1974 with Transport Canada and worked at various towers in the province of Quebec, culminating her operational career at the Montreal Area Control Centre in 1992. During the same period, she earned her MBA and learned to fly obtaining her private, commercial, instructor and multi-IFR ratings. She currently holds an Airline Transport rating with over 3,500 hours and serves as a part-time flight instructor and designated flight test examiner at the Rockcliffe Flying Club near Ottawa.

She continues to work at Transport Canada Headquarters in Ottawa as superintendent of ATC Rules and Procedures.

JEANETTE FOWLER, born March 22, 1939, in Los Angeles, CA. Joined the 99s in 1977, held various offices. Flight training includes commercial, SEL, MEL, instrument, some aerobatics and some helicopter. Planes flown: Citabria, Beech Sport, Sundowner, Sierra, Musketeer, F33 bonanza, A36 Bonanza, B36 Bonanza, V35 Bonanza, V35A Bonanza, V35B Bonanza, Beech 76 Duchess, E55 Baron, B58 Baron, C90 King Air, Sweitzer Gliders, HU 269C helicopter.

She owns a Beechcraft A36 Bonanza, N2048B, which she uses in her export business. She has been married to Tom Fowler for 38 years. Their children are Rick Fowler, Stacey Salas and Chris Fowler. She has flown the following Air Races: Air Race Classic, Pacific Air Races and Salinas His and Hers. Her accomplishments include: World and National Speed Records held for C1C Aircraft (registered records); original board member and served 15 years as board member for AirLifeLine; over 180 hours for AirLifeLine with medical supplies, doctors, transplant teams, people in need of medical treatment in United States and Mexico; many flights for Direct Relief in Santa Barbara; Woman Pilot of the Year Sacramento Valley 99s.

At present she owns Sunburst Ltd. an export company that has sales worldwide. She started it herself in 1976. Worked for Red Cross. Member of California Melon Research Board 12 years, Cub Scout leader, member of Aircraft Owners and Pilots Assoc., Air Race Classic, Western Bonanza Society and World Bonanza Society.

She has logged 2,050 hours to date.

Memorable experiences include: seeing patients being able to see for the first time in many years after cataract surgery in Mexico. Many older people had never seen their own grandchildren.

Hair raising experiences: Engine failure in a twin right at take off in the Sierra Nevada Mountains. Flying against oncoming traffic, in a descent, at Van Nuys, CA, Airport (with their permission) at the end of a World Speed Attempt.

CAROL M. FOY, curiosity and the desire to preserve oneself lead Carol to learn to fly in 1990 after the purchase of a Mooney by her pilot husband, Bob. After two lessons she was hooked, and currently holds a commercial/multi-engine certificate. The early support of the Austin Chapter 99s, of which she is currently chairperson, was instrumental in her success.

When not working as a landscape architect for Austin Parks, Carol works as delivery pilot. She's gathered a variety of experiences and some great hangar tales flying strange airplanes cross-country. She thinks flying the Air Race Classic is great experience too. Spreading enthusiasm for aviation, earning her instructors rating and logging more multi-engine time are Carol's future goals.

Flying her Mooney with the fold up bikes stashed inside to new places is great! Carol marvels at the self confidence and joy this skill of flying has brought to her life.

JUDITH "JUDY" FOY, spent her childhood in the southwest, and moved to Wisconsin in 1962. Her interest in aviation began in 1961, when she worked for an airmotive distributor in Hawaii. However, it was 1986 before she had the opportunity to rekindle that interest. She earned her private pilot license and became a member of the 99s that year. She and her husband, Ron, own a Cherokee 235 which they have flown to Alaska, Florida, Nova Scotia, California and many points between. She has more than 700 hours of flying to her credit.

Foy holds a bachelor's degree in business administration and has worked for her local school system for 23 years. She has held offices in many professional, civic and church organizations.

She has served the Wisconsin Chapter as newsletter editor, director, airmarking chair, secretary and other committee chairs.

As well as belonging to the Cherokee Pilots Assoc. and AOPA, she is involved with the Young Eagles Program of the EAA. Additionally, she is a volunteer pilot for AirLifeLine – flying patients for non-emergency medical treatment.

KAROLYN FRAINE, Phoenix 99s, graduated USAF pilot training and received her commercial multi-engine rating in March 1989. She got her ATP multi in February 1994 and her ATP single in May 1994. She joined the Phoenix 99s Chapter in September 1994. From January 1995 to the present she has been a 727 Second Officer with United Airlines. Before that she had spent three months of 1994 as DC9 First Officer, American International Airways, and from May 1988 to May 1993 had been in the USAF as C12F (BE200) instructor pilot, C23A (Shorts 330) Aircraft Commander, and T-38 - T-37 pilot training.

Her husband, Bob, is a USAF F16 pilot, and her hometown was Fort Lauderdale, FL. She graduated from Smith College, MA (1987), BA math. Her interests are flying, traveling, painting, singing, hiking, running, rollerblading and swimming.

MEARL M. FRAME, born July 24, 1932, Collinsville, IL. She learned to fly at Lambert Field in St. Louis, MO, in 1965. She joined St. Louis Chapter in 1966. Received multi-engine license in 1967. Received IFR in 1983. Flew right seat in the 60s in North American Sabreliner with husband Bill.

She has been involved in many 99 activities from pinch hitter to safety seminars. She was honored at the Forest of Friendship at Atchison in 1991 by Renaissance Chapter in Michigan. She has been a member of St. Louis, Michigan, Greater Detroit, Renaissance and Tucson Chapters. She is currently vice chairman of the Amelia Earhart Memorial Scholarship trustees.

She is married to William G. Frame, a retired corporate pilot. They have traveled extensively in corporate planes in the USA, Europe, Mexico and the Islands. They have three children and five granddaughters.

They have personally owned a Cessna Cardinal, N#10139, a Cessna 180 N#2755C, and presently own and fly a 182RG, N#2744C.

Flew the Berlin Corridor from Dusseldorf. Crossed into E. Berlin at Checkpoint Charlie, very grim city. Very happy to return to West Berlin: stood at the Brandenberg Gate and watched the Russian soldiers with their submachine guns and cried. They did not expect the wall to come down in their lifetime and were so thrilled when it did.

ELISABETH ANNE FRANCE, soloed when she was 16 and received her private pilot license when she was 17. She will be continuing her flight training at Ohio University College of Engineering and Technology in Athens, OH, this fall, proceeding toward a BS degree in Airway Science.

She currently flies a Cessna 152 and 172 out of Oakland-Pontiac Airport in Michigan.

NAN FRANCE, born in 1937, but didn't learn to fly until 1970 when her Air Force pilot husband bought her flying lessons for a birthday present. His career kept them on the move (28 houses in 31 years) so her flying was done in a wide variety of places. In four years in Germany, she managed an aero club and organized the only International general aviation fly-in ever held on an American Air Base. She earned a CFI in 1976 and CFII in 1978, teaching flying for about 10 years.

She is a captain in Civil Air Patrol, a member of 99s Inc. and a member of Aircraft Owners and Pilots Assoc. She is married to John France who has retired from the Air force and teaches ground training for American Airlines. They have three children and six grandchildren.

LAUREN FRANCIOSI, born on June 8, 1954, in Evansville, IN. Lauren's aviation exposure began in 1977 as a flight attendant. She met her husband while working for the airline. He gave her a first ride in a single-engine and she was hooked. Lauren obtained her private in 1995 after some 10 plus years of job relocations and new instructors. She is currently working on her instrument.

Lauren has been involve with AirLifeLine, Civil Air Patrol and is a current member of the 99s, Airline Owners and Pilots Assoc., and Cessna Pilots Assoc.

She is happily married to Carlo, who is also a pilot as well as a flight engineer on the 747. They have three dogs ... Mayday, Curtiss-Jenny, and Tailwind; two cats, General Doolittle and Lindbergh; one airplane, a Cessna 172, N6798H.

APRIL FRANKE, born April 17, 1949, in Adrian, MI, is a third generation pilot, giving her lots of time in the air. She began flying lessons at age 16, but with too many other distractions she put off flying lessons until age 40 when she seriously pursued her private license.

She owns a Cessna 150 and at this writing is restoring a 1957 Cessna 172 with her father and husband.

April and her husband of 28 years, Rusty, have two daughters, Tonya and Nicole, both now in college.

Since moving to Bradenton, FL, in 1983, April and her family have built a private sod airstrip at their home and play host to several ultralight and experimental planes along with their own.

April is a proud member of the 99s since 1993 and AOPA since 1989.

COLLEEN JOYE SKINNER FRANKE, born Nov. 3, 1932, in Dallas, TX, earned her private pilot license on Easter Sunday 1994, at age 61. She flies a Piper Tomahawk and a Piper Cherokee 160, both planes owned by her fiancée, Ray Roark. She and Ray fly together almost every weekend near their home in Olivia, TX, flying out of Calhoun County Airport at Port Lavaca and Jackson County Airport at Edna. They also have flown cross-countries in Texas and Oklahoma. Her favorite flights are along the Texas Gulf Coast where they land at Matagorda Peninsula to enjoy the Gulf of Mexico in its natural beauty.

On the day of her first solo, she took her first aerobatics flight with US Aerobatic Team member, Debbie Rihn. And, later, had her check ride with Debbie Rihn, her biggest thrill in her short flying career.

Franke has two sons, and six grandchildren, all but one of whom have flown with her. Her 82 year old mother claims to be her first "civilian (non-pilot) passenger." Her sister also enjoys flying with her when they're together.

She is official court reporter for the Calhoun County Court at Law and has been a court reporter for 20 years. She hopes to retire in 1996 and have time for some real cross-countries.

Franke is a member of 99s, Inc., Aircraft Owners and Pilots Assoc., Experimental Aircraft Assoc., National Council for Women in Aviation/Aerospace.

ELLEN A. FRANKLIN, USAF, Captain, her mother, Romaine J. Ausman, got her interested in flying. She would take her flying, during her lessons and practice flight. As she graduated high school, her interest included flying. She went to Colorado State University and was playing the drums for an AFROTC parade when four F-4 Phantoms made a low pass over the parade. WOW! She promptly visited the AFROTC recruiter who began to tell her women can't, this and that. To prove him wrong and show the world, she signed up.

During her AFROTC career she began flying lessons with her aviation mentor Jean S. West, her cousin. Got her private pilot license and instrument, later commercial. She had some bad grades, so she couldn't be an Air Force pilot, but she could be a navigator. Commissioned on July 18, 1988 from Arizona State University as a ZLT. Next its on to Navigator School at Mather AFB, CA, graduating on Aug. 29, 1989, the only woman in her class. Her first assignment was to Davis-Monthan AFB, AZ, where she navigated the EC-130 Compass Call. After completing initial training in the C-130 and EC-130, six months later she was on her way to Desert Shield. She came home after four months just as the war started, and begged to go back. She went in time to fly in combat three times. Wow, what an experience! Came home and later got an assignment to Little Rock AFB, AR, where she is today.

Recently completed Navigator Instructor School and also got a regular commission in the Air Force. Her goals are to make the Air Force a career and hopefully command a flying squadron someday. "Bless all who slip the surly bonds at flight who have come before me and those who are still to come."

BETTY SKELTON FRANKMAN, soloed at age 12, officially at 16, June 1942. She instructed WWII Veterans on the GI Bill in Tampa, FL, and became internationally famous as an aerobatic pilot, winning the International Feminine Aerobatic Championship in 1948-49-50 at the Miami Air Maneuvers. Presenting exhibitions at Cleveland National Air Races in 1948-49, she also flew AT-6 in the Halle Trophy Race. She established light plane altitude records and attempted a P-51 world speed record, ending unofficially due to engine explosion.

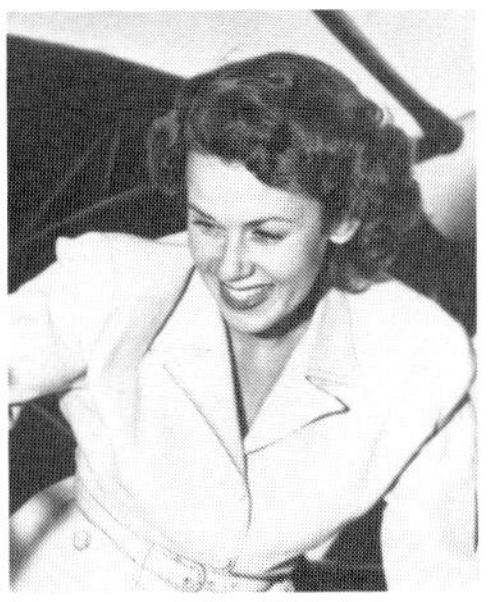

Betty's plane, N22E, the second Pitts Special ever built, serial #2, she named *Little Stinker.* Representing the US in England and Ireland, she took the tiny plane onboard the Queen Mary to Europe. In 1985, the craft became a part of the collection of the National Air and Space Museum of the Smithsonian Institution in Washington.

In 1959, Betty was the first woman to undergo NASA physical and psychological tests given the original seven astronauts, resulting in *Look* magazine cover story. She also established automobile history with transcontinental records across US and South America, and world land speed record for women on Bonneville Salt Flats with jet car top speed of 315 mph.

Betty was the first woman inducted into the EAA International Aerobatic Hall of Fame and the NASCAR International Motor Sports Hall of Fame, as well as four other halls of fame.

A native Floridian, Betty and her television director husband, Donald Frankman, reside across the lake from Cypress Gardens in Florida. She originally joined the 99s around 1944.

ETTA SUE MURPHY FRANTZ, born in El Dorado, AR, in January 1930. Grew up on a dairy farm and had never had a plane ride until she was in her late 20s. In the fall of 1959 Joe Frantz invited her to attend a Louisiana State University football game. As it happened, they went to Baton Rouge, LA, from El Dorado, AR, to the game in his Cessna 140. It must have been her second time to be in this little plane. On their return trip Joe showed her how to keep the plane straight and level and on course. He went to sleep for about an hour and left her flying. That was it! She wanted to learn to fly. They were married in May 1960 and he soloed her that summer; however, she got pregnant and was unable to finish getting her license. Seventeen years and three babies later in January 1977, she became a private pilot. Joe bought her a new 1979 Cessna 172. In 1979 they went to John Brown Seaplane Base in Florida and got their seaplane ratings.

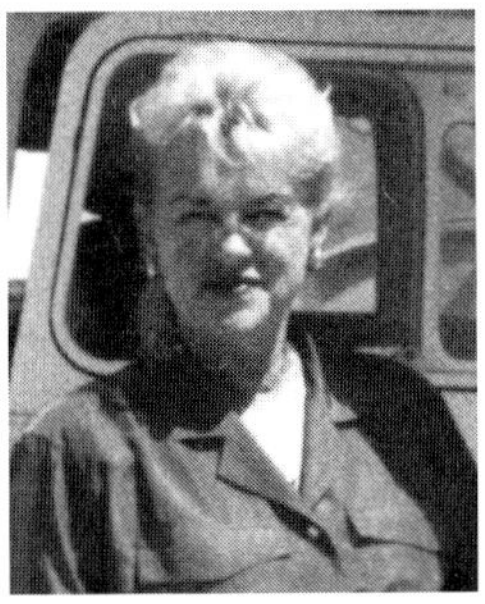

Two of their three children are private pilots. Now number three just decided she wants to become a pilot also. She now has more than 600 safe flying hours. Their family will continue this exciting hobby because their four grandchildren love to fly.

She joined Houston Chapter 99s in 1982 and continues to be active as membership chairman this year. Flying has put her in a special category and she is thankful it has been a very important part of her life.

RUTH ELLEN ELLIS FRANTZ, born Jan. 20, 1930, in Kansas City, MO, moved to the Chicago area in 1958 where she and her husband Robert Waldo Frantz, also a pilot, raised four children: Margaret, Robert, Shirley and Randall. Ruth received her private pilot license in 1972 and became a 99 in 1973. She is also a member of the Northeast Pilots Assoc., Illinois Pilots Assoc. and United States Pilots Assoc. Ruth and Bob have enjoyed more than 20 years of flying together in many air derbies, as well as trips as far away as Point Barrow, Alaska and Aruba in their Cherokee Arrow. During that time Ruth has accumulated more than 1,000 hours as PIC, mostly in the Arrow. She has been employed in the analytical laboratory of the Quaker Oats Company's Research and Development facility for 23 years and is the proud grandmother of Matt, Ted and Chris.

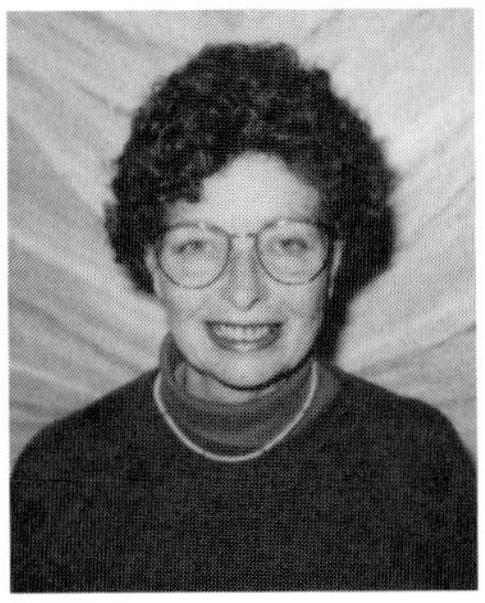

CATHY FRASER, started flying in 1979 at the age of 17. Shortly after completing her private license in Montreal, she became a 99. Over the next three years, she earned her commercial, multi-engine, instrument and instructor ratings and began teaching flying at the Rockcliffe Airport in Ottawa, Ontario. There, for a brief time, the staff was comprised entirely of women! After moving to Toronto in 1985, Cathy continued to teach flying and became a flight test examiner. In 1986, an Amelia Earhart scholarship permitted Cathy to get a type endorsement on a Piper Navajo which, two days later, secured her a job flying cargo. From there, she became a charter pilot flying King Airs and later a corporate pilot until 1988 when she was hired by Air Canada as a B-727 Second Officer. Furloughed since 1993, Cathy recently received her call back notice and will be training as an A320 First Officer as of June 1995. She has been an active member of the 99s, holding many positions at the chapter and section level, a Skywatch pilot and past ECS Governor (1992-94), currently chairman of the Montreal Chapter and member of the International Award of Merit committee. Married to Dr. Dave Williams, a physician and astronaut; they have a son, Evan, and will be relocating to Houston, March 1995.

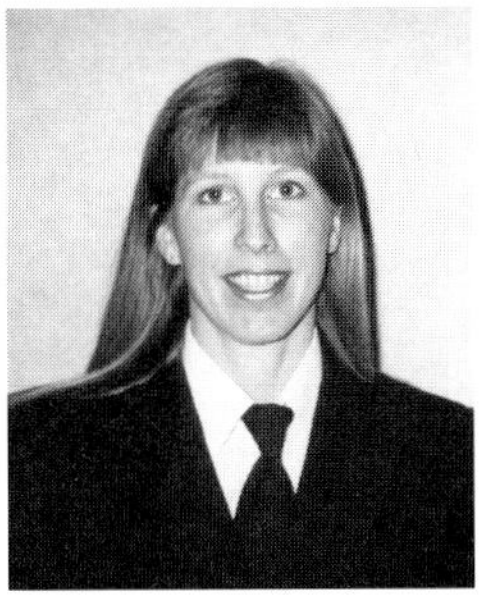

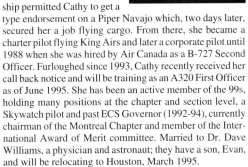

MARVEL L. "HUTCH" FREDERICK, Pittsford, NY, licensed private pilot in 1967. Joined Finger Lakes Chapter of 99s, 1974.

Spouse, James Frederick, attorney and instrument rated private pilot. Flew family Cessna 172 throughout the United States and Canada. Made quilted wall hanging for the 60th anniversary of the 99s honoring Amelia Earhart, Louise Thaden, Sally Ride and the charter 99s. The wall hanging was raffled at the 1989 International Convention in New York City and raised money to send a 99 to space camp in Huntsville, AL.

RUTH MAY FREDERICKS, learned to fly in 1988 at San Carlos, CA, and joined the SCV99s in 1991 with her commercial, multi-engine and instrument ratings and about 450 hours. Since then, Ruth has earned her CFI, CFII, MEI and ATP, as well as a 747 flight engineers rating, and has logged 2,400 hours.

For the last year, she has been working for Air Midwest USAir Express flying Beech 1900s, but now is back in the SF Bay area waiting on a job from either American Eagle or Skywest. Ruth won Santa Clara Valley Pilot of the Year award in 1994. Ruth is active in aerospace engineering and is a member of ALPA. Her goal is to be captain on an airline.

GAYLE GORMAN FREEMAN, born Feb. 14, 1955, in Mansfield, OH. She is currently president of Manairco, Inc., an airport and heliport lighting equipment company based in Mansfield. She is also chief executive officer of Mapco, which manufacture and repairs aircraft galleys. Gayle is the first woman to hold both of these positions.

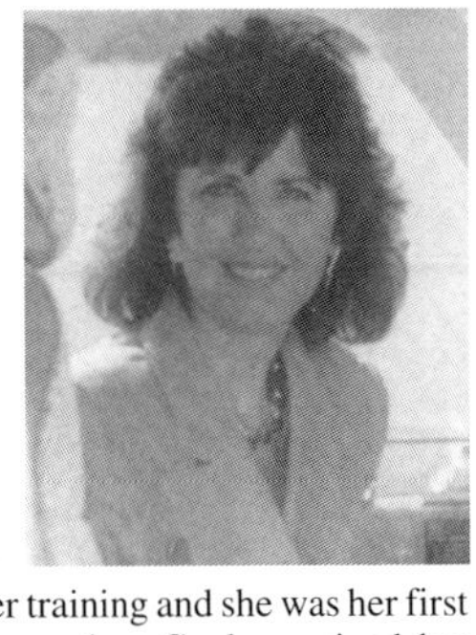

Gayle started learning to fly at the age of 15 and is a helicopter and glider pilot. Gayle's mother, father and husband are pilots, also. When Gayle was only 7 her mother received her helicopter training and she was her first passenger. Twenty years later, when Gayle received her helicopter rating, her mother was her first passenger.

Gayle is currently a member of the 99s; International Organization of Women Pilots and Whirly Girls member #293 of which she has served as secretary and vice president.

Gayle is co-chair of the Ohio State University Capital Campaign; on the board of directors for the North Central Technical College Foundation; the New Beginnings Alcoholic Recovery House for Women; and the Illuminating Engineering Society of America, serving as chairman.

Gayle and her husband Dan currently reside in Wooster, OH, and have three children: Curtis, Stewart and Elyse, and also one German Shepherd.

One of the more enjoyable things Gayle does is visit schools in her helicopter and talk with students about careers in aviation. "Most young people are still surprised to see a woman fly the helicopter." She adds, "When my husband and I fly into an airport, and I have been (flying) it in the left seat, people still ask him questions about the airplane!"

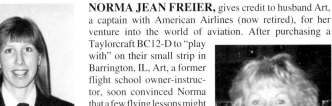

NORMA JEAN FREIER, gives credit to husband Art, a captain with American Airlines (now retired), for her venture into the world of aviation. After purchasing a Taylorcraft BC12-D to "play with" on their small strip in Barrington, IL, Art, a former flight school owner-instructor, soon convinced Norma that a few flying lessons might be fun. The solo flight in the Taylorcraft proved to be the inspiration for Norma to continue in a Tri-Pacer to gain a private license. After entering the Illi-Nines Air Derby in 1969 and returning home with a trophy, Norma was hooked. She has continued to compete in most of the small races in the Midwest and now owns and flies a Cessna 182.

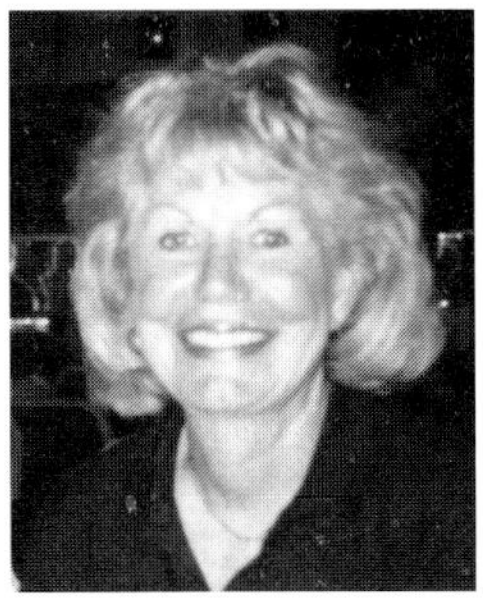

Active in the Chicago Area Chapter, Norma has served on the board and most committees; she was a board member of the Illi-Nines Air Derby for 15 years, serving as chairman for three years; and was liaison for International and EAA for eight years when she co-chaired the Friendship Tent at Oshkosh.

SHERI K. FREY, born and raised in Baton Rouge, LA, where her love for flying began at the age of 16. Her high school sweetheart came from a family of pilots and he had a Cessna 150 which they flew in often. The father had a Learjet; she will never forget the first time he took them up.

She later became a buyer for a boutique and spent a lot of time flying to and from the Dallas Market. After moving to Connecticut in 1986 she became involved in real estate investments and lost her itch to fly.

She left Connecticut in 1989 driving to Atlanta, GA, to return to the buying/market industry. During the trip the engine seized in the car which left her stranded. She was hired by USAir two days later as the BWI Chief Pilots secretary. Ten months later she realized she was on the wrong side of the desk.

She was accepted at Embry Riddle Aeronautical University and resigned from USAir. She graduated April 1993 with a BS in aviation business administration. She obtained her private pilot license and instrument rating, later obtaining a commercial and multi-engine rating in 1994.

Presently she is flying charters and building flight time, recently took the flight engineer written, and she is now seeking student loans, financial aid and grants to be able to attend AVTAR to further her career possibilities.

CONNIE FRICKER, born March 4, 1908, London, England. Joined the 99s in 1972. She has flown the following planes: Cherokee, Diplomat, Seneca, Apache, Aztec, Cougar, Tobago, Cessna 172, Cessna 210, Reims Rocket, Fuji, Jodel, Beagle (Aerobatics) Tiger Moth, Auster, Rallye, Arrow, Float Plane, Tampico, Airship Fuji.

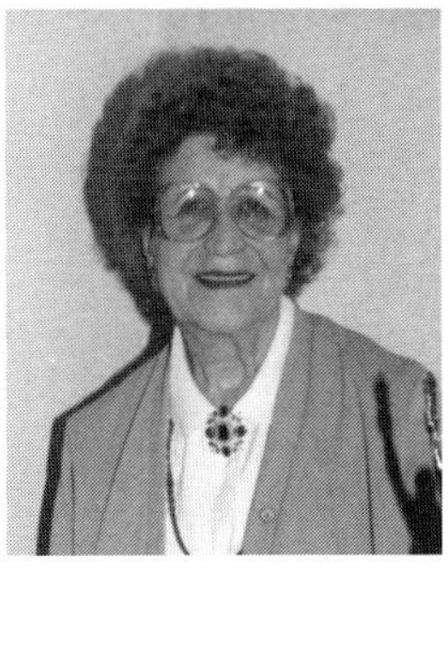

Her husband Alan died in 1987. She has two children, Mary, 52, and Paul, 48. She also has grandchildren: Jago, 24; Emily, 23; Giles, 22; Victoria, 21; and Charlotte, 11. Participated in 25 International rallies and received 29 awards.

Retired from government service. Served on the committee of the BWPA for 10 years and for part of that time was vice-chairman and responsible for foreign touring. She was vice-governor and hospitality chairman of the 99s (The British Section of the International Women Pilots Assoc.). Commenced flying in February 1968. She has flown more than 1,060 hours as of December 1981.

Mrs. Fricker was almost 60 years of age when she decided to learn to fly. She obtained her PPL and went on to gain her IMC, night and twin ratings. She had only 30 hours P.1 when on July 5, 1969, she made her first long distance flight to Malta. From then on her consuming interest became foreign touring and competing in air rallies. She distinguished herself in the field and collected 24 awards overall. The major ones are as follows: four times winner of the GEC Overseas Services Ltd. Trophy (Isle of Man Rally). Twice winner of the Brazendale Cup (Isle of Man Rally). Winner long distance flight in one day, achieved 579NM and Concours D'Elegance (Isle of Man Rally); Runner up Ladies Trophy (Shamrock Rally). Won the Mid Med Bank Trophy in the 1978 Malta Rally. Won the Arbuthnot Trophy in the 1978

Malta Rally. Timed Section Winner in the 1978 Malta Rally. Sixth overall in the 1978 Malta Rally. Presented with the Biggin Hill Cup for Rally Achievements. Presented with the Alan Cobham Trophy for Air Rally Achievements. Presented with AOPA Trophy for Air Rally Achievements. Presented with MONA Aero Club Trophy for Air Rally Achievements. Plaque in Forest of Friendship, June 1993. Award Guild of Air Pilots and Navigators, London, November 1993. BWPA Achievement Award, November 1994.

BLOSSOM RUTH FRIEDMAN, received her private pilot certificate in September 1971 and became a member of the 99s in June 1973. She has logged over 300 hours flying Cessnas and Pipers for fun and recreation. She helped to found and was a charter member of the North Jersey Chapter. Served as membership chair, vice chair, chair, nominating committee chair and historian of the chapter and has worked in many positions at the New York-New Jersey Section over the past 22 years.

Blossom is retired from an advertising agency in New York City where she worked for 23 years. A love of planes and flying is a life long passion and Blossom never fails to participate in and help with flying activities whenever the opportunities arise.

DOROTHY FRIEDMANN, born Feb. 8, 1946, in Santa Monica, CA. Joined the 99s in 1989. Flight training included Gunnell Aviation, Santa Monica Airport, private and instrument; and Fantasy Haven Airport, Tehachepi, CA, glider rating. She has flown the Cessna 152 and 172, and Schwettzer 233 and 136.

She is married to Wolfgang A. Friedmann. Participated in Pylon Sudao, Reno National Air Race Championships and Phoenix 500 Air Race. First solo, she is told the windsock jutted straight out as soon as she took off and therefore, the extreme shear at the approach end was unexpected. After a perfect approach, her craft was thrown into a sharp right bank and all she remembers seeing was the tower filling the windscreen. They say the split "S" was beautiful. All she knows is she landed mid-field center line. If only providence stays around to help her with other deeds!

HARRIET FULLER, a lifetime resident of Massachusetts started to fly in 1966, at which time she was already a grandmother. She joined the 99s in 1968 and served in all offices of the Eastern New England Chapter and all offices of the New England Section. She served on the International Board and was a trustee of the Amelia Earhart Birthplace Museum. She has attended every International Convention since 1969.

She has a commercial license, single engine land and sea with an instrument rating. She has logged more than 1,700 hours and flown in many local air races plus one Air Race Classic. She flew fire patrol for the state. Always appreciates a new challenge, so soloed in a helicopter just to see if she could do it!

Married to Howard Fuller, who is also a plot, she has two daughters and one son, Howard, Jr., who was the first pilot in the family. She has four grandchildren.

CARLYN D. FULLINGTON, born in Lake Charles, LA, and learned to fly in 1990. Fullington attended Sowela Technical Institute and graduated in 1994 with an associate degree in Avionics (Aviation Electronic). Fullington owns and flies a 1959 Cessna 172. She is a proud member of the Lake Charles 99 Chapter and the Southeastern Museum Conference.

She has served 20 years in the Christian Science Society in Lake Charles. She is married to Benjamin K. Fullington who is also a pilot and an A&P instructor at Sowela Tech. Fullington has six children and six grandchildren.

In 1994 Fullington was instrumental in organizing the Chennault Airpark Aviation Museum. She worked to obtain the 501 (c) (3) status and charter. She has contacted politicians and members of the aviation community to develop interest.

Fullington has aspirations of founding a museum which will house aircraft as well as an archive of aviation memorabilia for southwest Louisiana.

Special story regarding flight experience: Fullington's first cross country flight became exciting when the weather closed in and she lost track of her position. She maintained 800 feet and watched for towers. Lake Charles Approach control was contacted and told about the situation. They sent two A-10s to locate her. Fullington was very comforted to see them circle in front of her. She made her way to the airport and landed safely.

DOROTHY JOHANNA FULTON, Oct. 14, 1918 to February 1985. She was the first of three daughters born to Johanna C. Jensen-Fulton and Harry L. Fulton, originally from Ridgefield Park, NJ.

At age 15, she sold her two-wheeler bicycle for $3 and took her first flying lesson. Teaneck High School was the first high school in the nation to offer a complete two-year aviation course. Dottie signed up for the course at age 16.

At 17 she made her solo flight and at 18 received her private pilot license. At age 18 1/2 Dottie made the headlines of the New York Post and New York Journal. Flying the high school's Aeronca out of Teterboro Airport, Dottie encountered engine problems over Teaneck. Upon hearing the engine "rev" she immediately cut the switch knowing the crankshaft was about to shear off. Her quick reactions allowed the prop to pass the plane after it sheared off, then plummet to the ground. Although 2,000 feet in the air, five miles from the airport, and with the air speed of 90 MPH, Dottie glided the disabled plane back to the airport making a perfect landing with no injury to herself or damage to the aircraft. She later recovered the prop in a back yard in Teaneck.

Dottie attended New York University, where attained her ground and flight instructor's license. Post-graduate work included teaching at Teterboro, Teaneck High School and Bergen Junior College in Teaneck. She owned her own aircraft and taught flying out of Princeton, Williamstown and Morristown, NJ.

In 1942 Dottie signed up for the service and was one of the original 28 WAFS later known as the WASPs. After serving two years in the Air Transport Command - WWII, Dorothy received an honorable discharge from the US Air Force. She was the recipient of three medals: WWII Victory Medal, American Campaign Medal and WWII Honorable Service Lapel Button. After leaving the command, she ran flying schools and taught in New Jersey and New York. Dottie was featured in *Look* magazine article on the WAFS.

Dorothy was given a military service at her burial in 1985.

IONA ELEANOR FUNK, Greater Seattle Chapter, and a life member of the 99s, Inc. She received the Northwest Section Achievement Award in 1971, and currently is chapter historian.

Iona received her private pilot license in 1950 at Vancouver, BC. Her first solo flight was in an Aeronca Champion. Another favorite is the DH 82C Tiger Moth – "I always enjoyed flying Dear Old Tiger" also Stinson, Cessnas and Beechcraft Sport and Sundowner. She has a seaplane rating - Cessna 172. Her aviation experience includes
several years as stewardess for Alaska Airlines. Other interests include raising thoroughbred horses, travel and gardening - especially roses.

WALLY FUNK, born February 1939, in Taos, NM. Joined the 99s in 1958. Received her flight training at Stevens College, PUT; Oklahoma State University, CA, CFI -glider, seaplane, MEL, AI, IGI, ATP. She has flown all Pipers, Cessnas, Beech, Mooney, Stearman, Waco, AT-6 and DC-3. Funk was employed as chief pilot for three schools, first FAA woman inspector, first NTSB woman investigator.

Graduating first in her flight training class of 24 at Stevens College. Wally went on to Oklahoma State University, where she captured top honors as The Outstanding Female Pilot, The Flying Aggie Top Pilot and by taking home the Alfred Alder Memorial Trophy two years in a row. Stevens College recognized her aviation accomplishments in 1964 and named her the youngest woman in the college's history to receive their distinguished Alumni Achievement Award.

She landed a job as the first woman flight instructor at Fort Sill, OK, at age 20. She taught US Army officers, soloing more than 400 service men and sending a total of 500 on to private, commercial and instrument ratings. Wally and 24 female pilots were secretly chosen to undergo preliminary astronaut testing. The women endured the same grueling tests that the Mercury astronauts had previously completed. However, the mission was scrubbed almost as mysteriously as it had begun in 1963. "Naturally I was crushed," recalls Wally. "I'd have given my life for the space program." In the years that followed, she embarked on a three-year tour as a Good Will Flying Ambassador, covering 50 countries and 80,000 miles throughout Europe, Africa, the Mid-East and Russia. Back home, she flew everything from balloons to gliders to DC-3s, and competed in many air races, including the Pacific Air Race where she won first place in 1975.

As the 58th woman to earn an ATP, she applied and was granted an interview by one major carrier, only to be told by the personnel director that she was head and shoulders above the male applicants. The fact that they didn't have a women's restroom in the flight department prevented them from hiring her.

She became the first female FAA Inspector in 1971 and was the first female in the FAA's System Airworthiness Analysis Program in 1973. The NTSB in 1974 made her the first female Air Safety Investigator. Today, with 13,600 hours of flying, Wally has dedicated herself to educating pilots on how to put safety and common sense into flying, maintaining a steady course on the lecture circuit with seminars like "How to Fly and Stay Alive."

RABIA FUTEHALLY, born in 1935. Got PPL in 1962 when her first daughter was 9 months old. Has flown over Western and Southern India extensively, for pleasure and business with husband, father and brothers. In 1965 became member of 99s through the Sponsorship of El Cajon Valley Chapter. Charter Member of Indian Women Pilot's Assoc. which was formed with five members in 1969. Founder member of India Section of 99s in 1975. Junior Honorary Secretary World Aviation Education and Safety Congress held in New Delhi in 1976
and again in 1994 in Bombay. Has three daughters who fly, one of whom also sails. Hobbies: Indian Classical Dancing, has performed extensively; reading, cooking, child development.

NORMA LAMKIN FUTTERMAN, a commercial pilot with ASMEL, SES, instrument, glider and BGI ratings, has accumulated 3,500 plus hours since learning to fly a Citabria in Santa Monica, CA, in 1967. Since then she has flown many different aircraft and has owned a C-150 and a C-310. Last year she traded her A36 in on a 58-P Baron. She has especially enjoyed cross-country racing and has flown many races, including AWTAR, AWIAR, Air Race Classic (presently is on the ARC board of directors), Palms to Pines, PAR, etc. She joined the Los Angeles 99s in 1970 and has held many chapter offices, including chairman.

Born in Arlington, KY, Norma majored in Art at Murray State University before moving to California. She and husband, Charles (recently deceased), have two children

(Arthur and Anne) and three grandchildren (Spencer, Christopher and Amy).

In the past she was involved with many Beverly Hills community groups, the library, Cub Scouts, PTA, Little League, etc. Next to flying and golf, her primary interest is music and she is still very active in various music endeavors supporting such organizations as the LA Philharmonic.

MARY "CORKY" CRONIN GAFFNEY, born July 9, in Pittsburgh, PA. Joined the 99s in 1976. Holds a private pilot license with instrument rating. She has flown the Cessna 150, 172, 182; Cherokee 180; and gliders.

Her husband is Jack Gaffney. She has four children: Dr. Timothy Cronin, Patrick Cronin, Kathleen Medina and Brian Cronin. She also has grandchildren: Rory and Chelsea Cronin; Kelly and John Cronin; Patrick and Michael Medina. Flying has always been exhilarating for her.

MARY W. GARDANIER, born Feb. 12, 1914, in Newberg Township, MI. Joined the 99s in 1946. Flight training includes single engine, land and sea, commercial ratings. She has flown the following planes: Curtiss Robin Challenger, Cub, Aeronca, Stinson 501C, Sea Bee and Cessna.

Her late husband was Gerald S. Gardanier. During the war she flew search and rescue missions. Flew passengers to raise war bond money. Given charge of tallying in international voting ballots 1951 and 1952. Been treasurer of her chapter most of its existence. Gave pep talk to grade school children.

GEORGE-ANN WALTERS GARMS, born Oct. 31, 1924, in Springfield, MO. Joined the 99s in 1961. Flight training includes private pilot license, Concord, CA and instrument rating. She has flown Aeronca Champ, C-172, C-182, and PA28.

Learned to fly at a small airport in Berkeley, CA, in 1960. This met with strong disapproval from her family who considered it irresponsible for a young mother. Her husband Walter was heartily in favor as he badly wanted to fly himself. Flying turned into a lifetime of adventure and pleasure for the whole family. With her family or alone she has flown their C182 throughout North and South America, across the North Atlantic to spend a summer touring Scandinavia. More recently, George-Ann and fellow 99 Mary Lee McCune flew from California to Nome, Alaska, across Siberia and back.

George-Ann is active in the San Joaquin Valley Chapter of the 99s. She is the 1995 president of the Mount Diablo Pilots Assoc. and occasionally transports patients for AirLifeLine. She participates in the Flying Companion Seminar and gives aviation related programs for schools and civic groups.

Her late husband was Walter I. Garms, Jr.; and she has two children, Margaret Garms, 43, and Walter I. Garms, III, 42. Also has one grandchild, Joseph Turner, 13.

PATRICIA GARNER, born Aug. 3, 1938, in Augusta, GA, and grew up in Hephzibah, GA. She is married to Edward Garner and they had three children: two sons, Tinley, 36, and William, 33, and one daughter Glynn, 30.

While living in New Jersey in 1973 she began her pilot training at a small country airport with no 99s and no other women pilots. She had young children and most of her friends were not at all interested in her flying adventures. Edward's encouragement and support were very important at that time.

Patricia and Edward moved to Maryland and she joined the 99s in 1979. In the Potomac Chapter, she has served as chairman and other positions with special emphasis on aerospace education and airmarking.

She has been a docent at the Smithsonian Institution National Air and Space Museum since 1986, where she can share the fascinating legacy of flight with visitors from all over the world. She also gives tours to school groups hoping to inspire future contributions to aviation and space exploration. Another hobby is calligraphy, which is very much like flying, although on a much smaller scale. The enclosed picture is of Patricia and Edward standing next to the *Spirit of St. Louis*.

Garner has flown the Piper, Cessna, Beech and Mooney.

SYLVIA GARRATT, First Canadian Chapter member since 1977, when she joined as a mature private pilot. Although busy raising four children, she regrets not starting earlier. Husband Phil flew from his teens – originally owned a Stinson Station Wagon, then a Cessna 180 and in turn two Cessna 185s, on which she took all her flight training.

Now grounded, their eyes constantly turn skyward with many fond memories: some business flights but mostly pleasure trips covering many parts of Canada (especially the north, the Maritimes and Newfoundland); much of the US plus twice island hopping in the Bahamas. With Canada's many beautiful lakes and rivers, float flying remained her first love.

She's just retiring from ten years on the executive board of Mission Air Transportation Network, which is since its inception. Akin to Corporate Angel Network, they board patients with varied medical problems (sometimes has flown over 300 per month) and it's been a very rewarding endeavor. Additionally she is involved with the Lake of Bays Heritage Foundation, and the Toronto Chapter of the Parkinson Foundation of Canada.

She plans to remain associated with the 99s and is very proud of what so many young — and not so young – women are accomplishing in the aviation field. Her present excitement is that a 14-year-old granddaughter recently began flight training.

"Here's to CAVU and many happy landings!"

TERRI GARRISON, throughout the 1970s, fulfilled her childhood dream of skydiving (over 500 times), traveling North America and competing in "10 man" speed star competitions with her future husband (skydiver and jump pilot) Don Shackleton.

After achieving her BA in mathematics and chartered accountant designation, an introductory aerobatic flight in a Decathlon led to her private pilot license in 1978.

Two year later, Terri and Don purchased a "super" DeHavilland Chipmunk (modified by Jean-Paul Hunevult) which has provided them with many hours of aviation fun from aerobatics to cross-country trips from the Toronto area to a variety of destinations including Vancouver and Oshkosh.

Terri has enjoyed her membership in numerous positions with the 99s First Canadian Chapter including vice-chairman, treasurer and perhaps most enjoyable as chairman of the chapter's 1994 Poker Run event.

Fulfilling another major dream in January 1995, Terri gave birth to future 99, Justine!

ROBIN GARTMAN, a most independent person, began her aviation career upon receiving her driver's license. With a way to the local airport, she was airborne! With an interest in the mechanical aspect of aviation, Robin breezed through autoshop and began working as an A&P mechanic's assistant. Robin earned her wings while attending college. She is a private pilot with an instrument rating. Robin has been an active 99 since 1986, holding many offices within the Palomar Chapter. She has been instrumental in making things happen to get the community involved with aviation in the San Diego area. Robin plans to blend her two passions (biology and aviation) together to create the ultimate job. In her free time, Robin works with the local EAA Chapter 14 to bring other facets of aviation into the public eye. Robin and her husband plan to build and fly as many airplanes as possible.

KATHRYN G. GASKER, All-Ohio Chapter, 99s, born in Cleveland, OH, she graduated in 1938, from Case-Western Reserve University. In 1940, she was accepted into the Civilian Pilot Training Program, the lone female in a class of 10, flying before working hours and sandwiching in ground school. Although rough going at first, she received her license 77720-41, first in her class, in February 1941, the year she became a 99. In 1943, recently married to one of her flight instructors, an AAF officer, she declined WASP training. In 1946-47, she was chairman of the All-Ohio Chapter and of the International Convention. Restoration of airmarking was priority. She also did volunteer PR for the National Air Races and for UNESCO.

From CAP activities came an interest in disaster-relief, furthered by Red Cross employment, then 16 years with the Cuyahoga County Medical Society, earning commendations from the Surgeon General and FEMA and citations in *Who's Who of American Women* and the *Dictionary of International Biography*. As research associate at the American Medical Assoc., she prepared and presented the fundamentals for Emergency Medial Technician certification and registry to ambulance and rescue groups country-wide. Plagued with the physical condition which clipped her wings in 1960, she joined VA's Department of Medicine and Surgery in 1971, from which she retired in 1987.

Happy retirement has included volunteer activities at the Air and Space Museum, regional air shows, airplane watching, and many community activities.

MARLA GASKILL, obtained her private pilot license in November 1994, and immediately joined the Women With Wings Chapter. Marla's husband Manny D'ostroph is also a private pilot. The two plan on earning their instrument ratings together, flying a 172 from New Philadelphia, OH.

Marla, an Ohio State Trooper, realizes that women in a man's world must go the step further to prove they can do it! Marla is also a licensed practical nurse and someday hopes to complete her degree in criminal justice. Marla is a member of AOPA and the LPN Nursing Assoc.

MARY FRENCH GATIPON, born in National City, CA, lived in New Jersey, Georgia, and moved to Baton Rouge, LA, at 19. Love of aviation began as Georgia home was near glide path of Columbus airport. Dad, Gene, was a California/Alaska pilot and Grandfather Horace held aviation ratings. Flying since 1970, licensed SEL 1974 in a Cherokee 180, loved to fly with husband, Bill Gatipon (New Horizons 1987) in his Mooney Mark 21 2962L.

An active member of South Louisiana 99s since 1975, has served as chapter chair, vice chair and treasurer. Is a real estate broker, professional photographer, LPN, administrator of Baton Rouge Aircraft Pilots Assoc. (BRAPA) and Louisiana Legislative lobbyist on behalf of aviation interests.

A graduate of LSU with BS, has one son, Bill Benton, III; three step-daughters: Suzanne, Ann, Lynn; and five grandchildren: Emily, Michael William, Alexandra, Kurt and Victoria.

MARLAETTE "MOLLY" GENTRY, Phoenix 99s, was married to Ed, who had been in the Air Force and the Reserves but hadn't flown for several years. He decided to become an active pilot again. Molly flew with him, and then she took a pinch hitter course. At 58, she decided to get her license too and succeeded in getting a private, ASEL in 1985. With working full-time as an executive secretary it was a struggle, but she made it!

As well as being deeply involved in 99 and 49 1/2 activities, she and Ed fly for Flights for Life and the Civil Air Patrol. Some of the middle of the night flights to deliver blood or plasma to the outlying areas of Arizona have been exciting, especially when there was no moon. Molly has held several chapter offices, including Kachina Air Rally co-chair, and is currently chapter secretary.

VIOLA GENTRY, born in 1900 in Gentry, NC, soloed in 1925, and has kept her license active ever since. She was first interested in flying when she saw Lt. Omer Lacklear in 1919, landing his Jenny on top of the St. Francis Hotel in San Francisco. Robert Fowler, the first to make a transcontinental flight before WWI, was one of her instructors at Curtis Field, Mineola, Long Island, NY, where she flew a Curtis Jenny, and was also helped by Floyd Bennett.

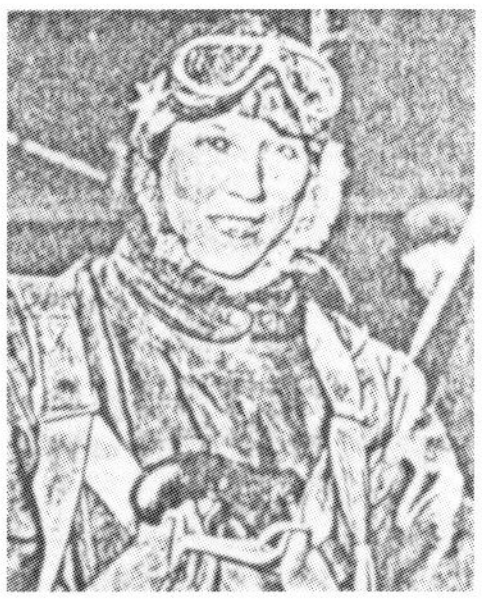

In 1926 she joined the FIA (Federation International of France), the only available license then. She set the first women's solo record, around Curtis Field for most of a day on Dec. 20, 1928. Another attempt ended in tragedy when her copilot, Jack Ashcraft of the Gates Flying Circus, was killed and she was in the hospital for 22 months.

Still in casts, and accompanied by a nurse and a newspaper reporter, she attended the charter meeting of 99s in 1929. She won a prize at the Annette Gipson races for women – the booby prize, the Royal Jackass Flying Trophy! She redeemed herself, however, when she set a 10-day refueling record in 1934. She also is among the distinguished few, with Charles Lindbergh, to be given the Birdman trophy for courage. Joined Long Island Chapter in 1965.

In 1967 she was appointed to the Humanities Research Center University of Texas, as consultant to the History of Aviation Collection. And in 1972, as navigator for Ruth Johnston, she flew in her third PPD, at age 72.

PAULINE GENUNG, deceased, started flying in 1964 and became a 99 in 1965, joining the Indiana Chapter. She served on various committees and became the chapter chairman (1971-73). She served as International 99s PR chairman. Her most memorable moments in the air were shared with *Jennifer,* the 1946 Globe Swift she loved so much. *Jennifer* was awarded Best of Show in the Swift category during the 1975 EAA Convention at Oshkosh, WI.

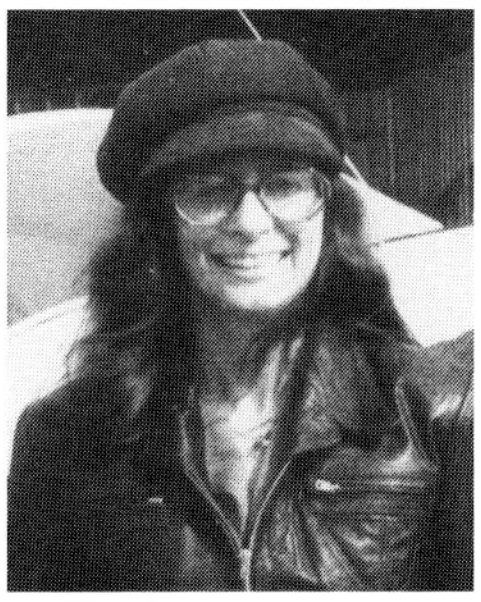

Pauline initiated the start of a newsletter called *Pireps from Pauline* which is continuing today under the name of *News and Views.*

She flew in the last Powder Puff Derby in 1976 as pilot of a Cardinal 177 RG. Not winning but finishing the race was a sense of accomplishment.

Pauline had the opportunity to fly many aircraft in her capacity as aircraft sales representative for Sky Harbor, Inc. in Indianapolis, IN. She went on to New Horizons in 1980 due to an unfortunate airplane accident while serving in this capacity. She held ratings of ASMEL, with instrument and CFI-A.

SONJA GERFEN, one of her earliest memories was the concerned look on her mother's face when she became her dad's first private pilot passenger. She was offered flying lessons as a favor to her dad, but before finishing, married and raised her children.

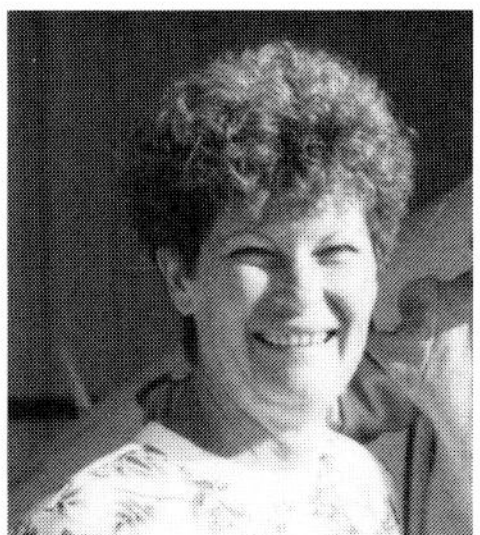

Sonja is a retired elementary teacher, now actively involved in a family owned trucking business as well as her husband's civil engineering firm. She has a BA from Cal Poly and a MS from Pepperdine University.

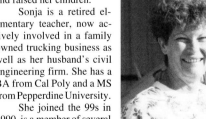

She joined the 99s in 1990, is a member of several local flying clubs, AOPA, and a director of the Santa Maria Museum of Flight. She enjoys being a grandma, playing golf and flying her Cherokee Warrior.

The Palms to Pines Air Race in 1993 and 1994, rate high on her "most exciting events" list. Her significant others, are husband, Herb, three children and two most wonderful grandchildren.

MARGARET C. GERHARDT, born July 9, 1914, in Seattle, WA. Joined the 99s in 1941. Flight training includes CPTP 1940. She is a widow with three children: Catherine, 46; Lois, 43; and Gordon, 42. Also has grandchildren: Krisine, 16; Alvin, 13; Stewart, 7; Ethan 4 (all Rectors).

Her mother believed that eventually there would be a airplane in every garage. When the CPT Program was started, she suggested Gerhardt apply. She got her private license in September 1940, and joined the 99s later that year. Was married in 1942, and after John went overseas, went to Sweetwater as a trainee in Class 44-WWI. When he was discharged in 1946, they were busy with running a business and raising their children. Still did some flying. Was copilot in the 1955 AWTAR and assistant route surveyor in 1958. Has been timer for about five AWTARs and one Palms to Pines.

After Pearl Harbor all private planes were grounded in the 200 mile area from the coast. She had a Kinney Sport Wing hangared in Alameda. It had the prop removed and was stored in Mountain View. After much paper work, she received permission to fly it to Reno, where she hoped to sell it. On Feb. 20, she and Helen Kelton flew it out of Mountain View. Around Sacramento they were buzzed by two P40s RON in Auburn. Later learned in NOTAHs that Auburn was not safe to land, too wet and soggy. The Kinner was open cock pit and it was a cold trip in February, but her friend did sell the plane for her.

DARLA JEAN GERLACH, born in Jefferson City, MO, and currently lives in Anchorage, AK. She is married and does not have any children (yet). She became a student pilot July 27, 1978, during her junior year of high school; soloed Oct. 24, 1978, and received her private pilot license Aug. 27, 1979.

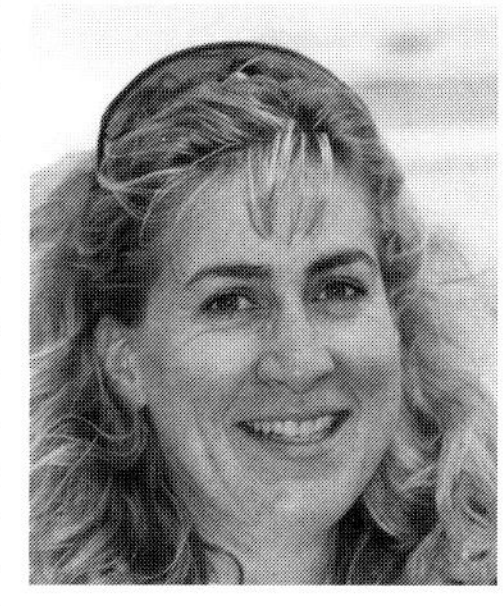

Darla attended Parks College of St. Louis University, a predominantly male school, and graduated in December 1982. She double-majored and received a BS in meteorology and a BS in aviation management. While attending Parks College, she continued flight training and received a commercial pilot certificate, Dec. 3, 1982. She is rated to fly instrument, single-engine and multi-engine land.

She joined the 99s in 1983 and has served as chapter vice-president, secretary and the newsletter editor, for the Greater Kansas City Chapter. She received a MA in business administration, management, from Golden Gate University in June 1989.

Darla currently works for the FAA as an air traffic controller in Alaska. In February 1993, she participated in a Russian/American Air Traffic Controller Exchange Program. She lived for two weeks in the home of a Russian air traffic interpreter in the Russian Far East city of Anadyr. While in Russia, she was given a lifetime opportunity to fly an Aeroflot AN-24 (a twin-engine, turboprop aircraft) from Providenya to Anadyr, Russia, with the plane full of Russian passengers!

She is an active member of the Professional Women Controllers (PWC), 99s, National Council for Women in Aviation/Aerospace (NCWA), and the American Meteorological Society (AMS). Darla is a certified Emergency Medical Technician (EMT-I) in the state of Alaska.

She has flown the following planes: C-150, C-152, C-172, C-182, C-310, KC-135, F-4, TU-154, and TU-64.

JEANNE GERRITSON, learned to fly at San Carlos Airport in 1988 and currently has logged about 300 hours She joined the Santa Clara Valley 99s in 1994 and is interested in air racing and airmarking. To celebrate her 50th birthday, Jeanne and her two children took their first skydiving jumps. Her current goal is to run a marathon. She has two grandchildren and lives in Half Moon Bay, CA.

MARGUERETTE GERRY, Bay Cities Chapter 99s Charter Member, Private Pilot #9853. Rita learned to fly in a Curtiss JN4, at Redwood City Airport in 1928. She flew before one had to be licensed and when complying with the new regulation, accompanied by her unlicensed instructor, he said, "Ladies First," and she therefore had a lower license number than her instructor.

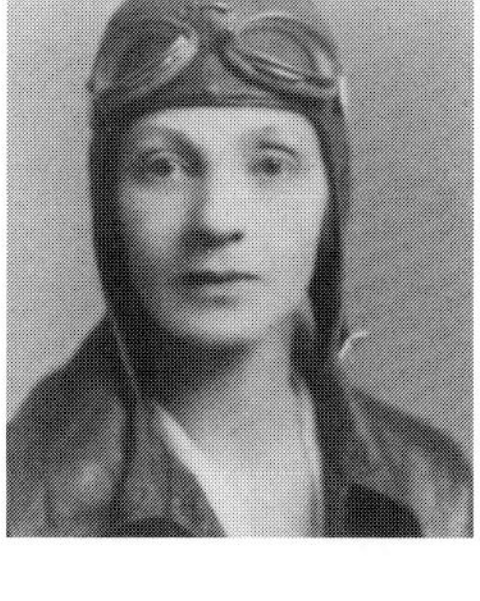

She claimed to be an original (charter) member of the 99s but could not prove she sent in her $4.00 as a fire on her tomato farm destroyed her records. There may have been a mix-up as one of the original 99s was named Margaret Perry.

Rita was a very active pilot and encouraging to all. She flew the early planes; International, Eaglerock, Standard. In 1942, she married Al Hart, and they flew a Cadet. They also ran a Boatel on the San Rafael waterway. Rita's favorite saying was "We flew when it was fun." She died in 1972.

LOUISE POIRIER-GIACOMO, Montreal Chapter (East Canada Section), since her birth on May 2, 1957, in Lachine, Quebec, most of her dreams have come true. After completing her education, she worked at a construction company for a few years. She enjoyed it so much that she decided to become an entrepreneur and start her own business renting heavy machinery. Today, she is happily married and has two beautiful children, Marie-France, 12, and Carol, 10. They also have two dogs and a bird.

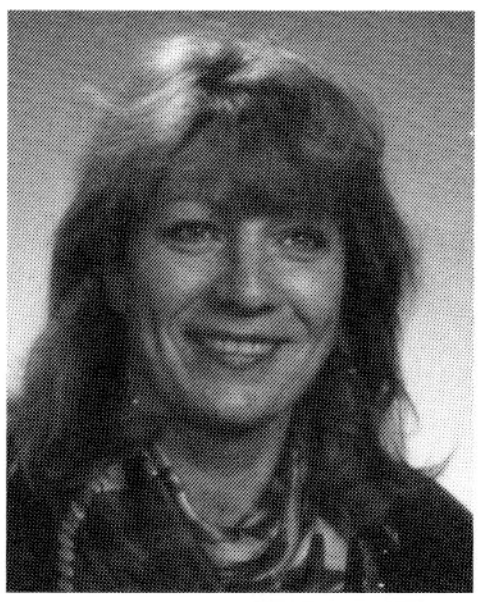

When she has a few moments alone, she enjoys painting, reading (mostly books about psychology and aviation), and a variety of sports such as downhill skiing, golf, sailing, swimming and tennis. Since discovering it in 1991, her passion is flying.

She has obtained licenses for private, night rating and sea-planes. She is presently studying for her commercial rating and she would also like to get her helicopter's license. She has experienced flying a Pelican, Cessna 152, 172 and Seaplane 182. Her most wonderful experience to date is to have flown over the Grand Canyon, NV, in a Cessna 172 and her next challenge is to fly to Banff, Vancouver by helicopter. Her memberships include the Canadian Owners and Pilots Assoc. (COPA), the Seaplane Pilots Assoc. and she is especially proud to be a member of the 99s, International Women Pilots. Being a member of the 99s, it gives her the chance to meet some very extraordinary people who have brought her great support and enduring friendships.

ETHEL GIBSON, born Feb. 10, 1915, in Pittsburg, PA. Joined the 99s in 1965. She is a RN and widow of Walter "Hoot" Gibson, an enthusiastic pilot; mother of John, grandmother of Heather, age 23, and John Jr., age 20. She holds a private ASEL and instrument ratings.

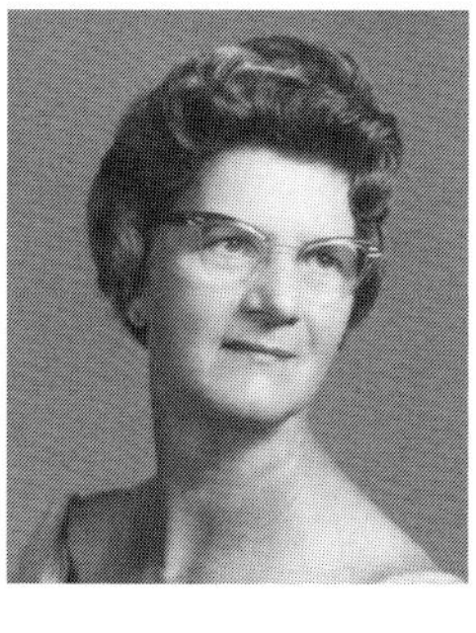

She is a charter member of the Florida Grasshoppers, a charter member of Suncoast Chapter of 99s, a life member of International 99s, the Florida Flying Farmers, International Flying Farmers also many nursing organizations, licensed in Florida and Pennsylvania.

She served as chairman, vice-chairman, secretary and headed most of the committees of the chapter, served in official capacity at two Powder Puff Race Terminus, and raced in many Powder Puff Races, an Angel Derby and an Air Classic.

Ethel and "Hoot" owned a Cessna Skylane 182 and flew all over the USA and Canada.

Retired now, she still supports her chapter and flying friends.

LOUANNE GIBSON, born in Los Angeles, CA, in 1957, and raised in Orange, CA. She grew up playing all the sports she could. LouAnne attended college on an Athletic Scholarship playing on the Women's Golf Team. She graduated with a BA in journalism.

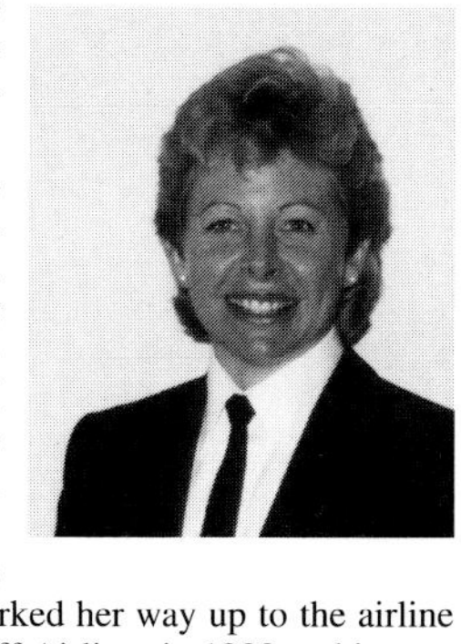

LouAnne worked as a California Highway Patrol Officer until she discovered flying. She soloed in eight hours and received her private, CA, instrument and CFI in eight months, with great support from the Long Beach 99s. As a flight instructor, 135 charter, bank check pilot and commuter pilot, she worked her way up to the airline industry. She flew for Braniff Airlines in 1988 and is now

presently flying for American Airlines and loves it. Playing golf is still her favorite pastime.

She has flown the following planes: Cessnas, Beeches, Pipers, 727, 737, MD-80. First officer and member of Long Beach Chapter 99s.

M. RITA GIBSON, life member 99s, purchased a 40 hp Taylorcub in order to learn to fly. Her first lesson was her first airplane flight. Soloed February 1944. Private license in March 1945. joined the 99s in Phoenix 1951; served as secretary. Transferred to Long Beach Chapter in May 1952; held all chapter offices. Worked in all phases of chapter's five starts and two finishes of PPD. Copilot in 1957 PPD; pilot in 1961. Inspection chairman and pilot in Palms to Pines Air Races.

Transferred to Washington, DC, Chapter 1961; to New York-New Jersey Section 1963. Charter member of Long Island Chapter, 1965; chairman 1966-67. Rejoined Long Beach Chapter in July 1967.

Husband, Paul was an FAA engineer – flight test pilot. Six children including four pilots; one a Navy test pilot and astronaut with five shuttle missions and married to one of first six women astronauts, who has flown three missions; 10 grandchildren.

BEVERLY A. GIFFIN, born and lived most of her life in Ohio. Since 1987 she and her husband have resided in Colorado. She learned to fly in 1969 and is presently piloting a Thorp T-18 which she helped her husband build. They are currently building an RV-4 in the basement of their home. She has been a member of 99s holding various offices in Scioto Valley and Pike's Peak Chapters. Bev is also very active in EAA, locally and internationally, having attended every EAA convention in Oshkosh since 1972 and is currently one of the chairmen of the event. Bev is an RN, graduating from St. Luke's Hospital in Cleveland.

She was born Feb. 23, 1926, in Bucyrus, OH. Joined the 99s in 1969 or 1970. Flight training includes private pilot ASEL. She has flown Cherokee 140, Cessna Skyhawk, Beech B-23 Musketeer, Thorp T-18. Married to Walter (a retired engineering prof and pilot) and has two children, Steven C., 30, and Rebecca Guddendorf, 28. Her grandchildren are: Mathew Giffin, 8; Zachery Guddendorf, 4; and Justin Guddendorf, 7 weeks.

JACKIE GILES, First Canadian Chapter, learned to fly in 1975. Her interest in aviation came about when her husband Walter, was involved with the Ministry of Natural Resources and his position necessitated an enormous amount of flying on business. They appreciated the advantages and convenience of air travel.

Jackie taught biochemistry part-time at the University of Toronto for 30 years. No one in the family, not even the children, knew that she was taking flying lessons in 1975. She graduated from ground school with a 99 percent mark on the written exams and received a trophy. Her son remarked that it should be more important to be able to fly the plane! However, her mother, who had never flown anywhere, even commercially, was ready and anxious to fly with her – a real vote of confidence.

Although Jackie's husband didn't have a pilot's license, he enjoyed flying tremendously. Many long distance flights were made with a group of flyers from the Oshawa Flying Club in various rental aircraft – New Orleans; Brownsville, TX; Las Vegas; and the Bahamas, were some of the interesting destinations.

In 1977 Jackie added a float endorsement and a night rating to her license. She joined the 99s in the early eighties and served as membership chairperson for three years. Jackie also flew several missions for Operation Skywatch.

A partnership in a Cherokee Six enabled them to make more long distance flights – across Canada west to Vancouver – the return trip took them north through Prince George and Dawson Creek. Several flights east to the Maritimes, Quebec, also south to Florida. They covered a lot of territory.

Following the sale of the Cherokee Six, Jackie bought shares in a Cessna 172 at Buttonville Airport. She is a staunch member of First Canadian Chapter, keeps current flying and plans to attend the 99s International Convention being held in Halifax in 1995.

URSULA GILGULIN, born Aug. 2, 1945, in Zurich Switzerland. Flight training includes all in Colorado Mountains, ATP, CFIAI, SEL and MEL, DPE, written test E, NWS observer, airport manager, accident prevention counselor. She has flown most single engine under 200 hp and has 6,500 hours. Her aviation employment includes: a FBO owner, air taxi owner, flight school owner, scenic air tour owner, world famous mountain flight training program, and operate contract station for national weather service. Married Robert in 1965.

BFA and MFA in fine art and art history from Pratt Institute in Brooklyn. Long struggled to keep FBO (North America's highest 9927 feet) Leadville Airport, Inc. afloat while teaching art and aviation ground schools at Colorado Mountain College at night. Active in community economic development planning. Ursula's art work is represented by two Colorado galleries and is included in several collections.

PAULINE GILKISON, interest in aviation started at an early age. It was when she was lifted up over the crowd of people watching the parade for famous Charles Lindbergh in Chicago. Suddenly his car stopped in front of them long enough for "eye contact" and her dream to fly began.

Born on a farm in Harvard, IL, and educated in Chicago where she later attended the University of Illinois, hospital volunteer work lead her to join their flying club. She joined the 99s Chicago Area Chapter in 1973 and was assigned to work with collegiate aviation.

Working with the National Intercollegiate Flying Association's Executive Director, (then) Harold S. Wood and their knowledgeable 99s all led her to help raise funds for NIFA through the 99s. She remembers her 99 aviation writer/author Page Shamburger who helped her enjoy NIFA. The 99s, Inc. a long time supporter of collegiate aviation, continue to raise funds for the NIFA and NIFA Foundation. Judging at SAFECONs (Safe Flight Evaluation Conference) was good volunteer work for them. "Thanks to husband John for buying the Cessna 172 Skyhawk and working with her and NIFA. Mother of Barbara Mays, Patricia Montone, Karen Gilkison, Dawn Gilkison, your support was appreciated." She was greatly honored when the Alpha Eta Rho Award for 1992 was presented to her at the University Aviation Assoc. awards banquet.

BETTY HUYLER GILLIES, a charter member, received her private license on May 6, 1929, and her transport license on June 16, 1930. From then until 1939, she was active in aircraft sales and flight instruction as well as charter flying. She was the International president from 1939-41. During WWII, she joined the Women's Auxiliary Ferry Service and became a commander. She has flown in the AWTAR and served as chairman of the AWTAR board. She is still active in the 99s.

First solo flight Dec. 23, 1928, OX-5 Travelair, Roosevelt Field, Long Island. Flew for Curtiss Wright, Waco Sales, Grumman Aircraft, Country Club Flying Service, Hicksville. Married B. Allison Gillies 1931 – three children.

During WWII, was a WAFS/WASP in Ferrying Division of the Air Transport Command, New Castle Army AB, Wilmington, DE, from September 1942 to WASP deactivation December 1944. Rated 5P; ferried single, multi-engine, pursuit aircraft within USA-Canada.

After competing four years in 99's All-Woman Transcontinental Air Race, was race chairman nine years, 1953-62. 99s charter member; president 1939-41.

Appointed by President Johnson in 1964 to first FAA Women's Advisory Committee. Awarded a Paul Tissandier Diploma by Federation Aeronautique Internationale in 1977; received NAA's Elder Statesman of Aviation Award in 1982.

GLEN JUDITH GILLIES, a California resident, comes from an aviation oriented family. Granddaughter of a Charter 99, Glen earned her rotary and fixed wing licenses in 1986. She joined the 99s at that time, and has been active with the Palomar Chapter for several years. Glen participated in the Air Race Classic in 1990, attended the World Aviation Education and Safety Congress in Bombay, India in 1994, and attends section and International meetings of the 99s whenever possible. Another favorite aviation activity is the Forest of Friendship gathering in Atchison, KS. Currently Glen teaches at a San Diego massage school, while maintaining a small private practice.

A former cross-country truck driver, Glen enjoys seeing the world from the perspective of a pilot. Glen is a member of the AOPA and ALEA. She is currently studying Hindi, and is active on a volunteer basis in the local HIV+ Community.

CHRISTINA GILLILAND, born Sept. 9, 1966, Montreal, Canada. Joined the 99s in 1990. Began flying at 15 in 1981, in Toronto Canada, soloed on 16th birthday. She has flown Cessnas, Piper Seminoles, 50360 (Shorts-Turboprop). First officer with American Eagle flying a shorts 360. Married to Brent Gilliland, pilot-American Eagle.

Since the age of 3 when her uncle took her for a flight in his aircraft, Christina always knew she wanted to be a pilot. As a teenager, at 15 she worked as a flight dispatcher in exchange for flight hours. She soloed at 16 before being able to drive a car! After graduating from the University of Toronto with a degree in international relations, Christina moved to Miami, FL, where she earned a second degree in aeronautical science while completing her flight training.

Christina shared her love of flying with others as a CFI, CFII and CFMEI for four years. She also founded her own company; Precision ATP with her husband Brent Gilliland specializing in ATP training.

Christina is a proud member and former chairman for the 99s in Miami.

She now flies for American Eagle as a Shorts 360 First Officer and aspires to fly for a major airline in the future.

VIRGINIA NORDMANN GILREATH, born June 13, 1930, Charlotte, NC. Flight training included commercial rating, single, multi-engine and instrument ratings. She has flown all single and twin Piper Cherokee aircraft; 140-180 Arrow 200, Lance 300, Seneca.

She had three engine out landings in a corn field, in Nashville, TN; 29 Palms, CA, in the desert; and in a federal prison, near Lakeland, FL. Landed safely all three times! Keeping her cool, having excellent training, knowing her aircraft, "saved my life."

Her flying career began in November 1968 in her home town of Charlotte, NC. Has logged more than 8,000 flying hours. A charter member of the 99 Blue Ridge Chapter, Southeast Section in 1974. Served as vice president, secretary, chapter reporter. On the Southeast Section 99 Nominating Committee, Southeast Scrapbook chairman.

Flown as pilot in command in her Cherokee Arrow 200 aircraft in the Powder Puff Derby 1973 -1975 - 1976;

the Angel Derby as copilot navigator 1978 in a Cessna 172. Member of the Air Race Classic, worked impound at Terminus in 1976 and 1985.

Employed by Excel Telecommunications Inc. Married J. Edwin Gilreath on June 29, 1951. He has been a pilot since 1968, and founder of Gill Manufacturing Co. in Charlotte, NC, for 38 years. They have one child, Debra Ann Gilreath Morgan, 43, and grandchildren: Tony Cannon, 23; Carrie Cannon, 20; Adrienne Morgan, 15; and Amanda Morgan, 11; and one great-granddaughter. Now live at Ocean Isle Beach, NC. Soon to build a retreat home in Big Sky, MT.

JEANNE GIVEN, commercial rating, CFI-I and ground school instructor. Flew three PPD races, 1968, 1969, and 1971, two of them with her sister Pat Davis. Flew the Sky Lady Derby and two IARs, 1973 and 1976, with her daughter, Jeanne Willerth, a licensed pilot. Has been chairman, vice-chairman, secretary and treasurer of the Nebraska Chapter of 99s.

PATRICIA GLADNEY, member of Santa Clara Valley Chapter. Pat learned to fly at 16 while in New Jersey in 1934. Following graduation, she worked as secretary for Aero Insurance Underwriters in New York. They inspected airplanes around the country before insuring them. She also flew to meet various small plane manufacturers to fly their new models and report the plane's performance and comfort. In 1936 she took the exam, newly required by the CAA, to teach, and began more than 50 years of flight instructing. In 1941 she moved to California to instruct in the Civilian Pilot Training Program. She joined the WASPs, stationed at Williams Field, AZ. Pat joined the 99s in 1935, was the first Amelia Earhart Scholarship winner in 1941 for her instrument rating. She flew 24 AWTARs and several Palms to Pine races. Pat logged more than 20,000 hours, held ASMEL, commercial and instructor. She died in August 1993.

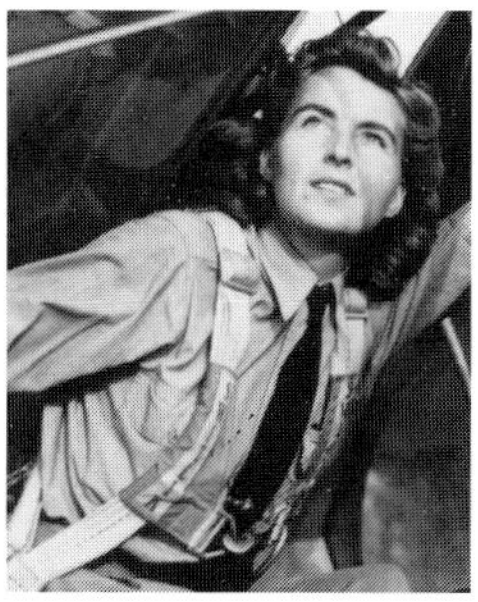

MARY GLASSMAN, born in August 1944, Toledo, OH, relocated to southern California in 1963. Although interested for years in learning to fly, it wasn't until 1984 that she devoted the time and expense to obtaining her license. She still remembers sitting in the left seat with tears in her eyes as the flight examiner shook her hand and told her she was now "a pilot." Ten years and six grandchildren later, she is proud to be chairman of the San Fernando Valley Chapter of the 99s for a second year.

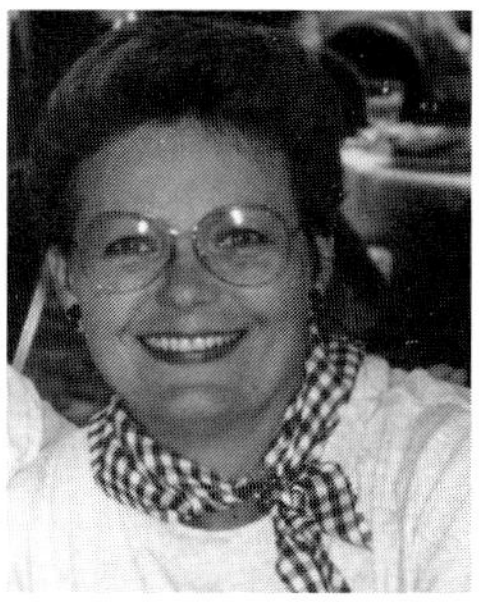

"One of my most memorable flights was in 1993. I flew from Van Nuys, CA, to Jackson Hole, WY, for a family reunion. That was the first time I, as PIC, flew beyond the borders of California or Nevada flying over five states: California, Nevada, Utah, Wyoming, Idaho."

Introduced to air racing in 1992 by fellow 99 Paula Sandling, Mary has participated in three Palms to Pines Air Races. Flew with veteran air racer Mary Rawlings in the 1994 Air Race Classic.

Mary Glassman lives in the San Fernando Valley north of Los Angeles. Her two daughters and their families are also in the San Fernando Valley as is her job of 27 years with a local grocery chain.

She has two children, Michelle Maryan, 32, and Lisa Perry, 30; also grandchildren: Lyrissa, 10; David, 10; Christopher, 6; Lydnsey, 4; Jenna, 4; and Nicolas, 3.

PAULINE GLASSON, her life has been flying 365 days a year-soloing in 1934 in a bath tub Aeronca with homemade floats, too ignorant to be scared. Wasn't successful at all in Sky Writing, tried banner towing, crop dusting (no spraying), aerial ambulance then pipeline patrolling ... finally decided on aerial photography, with husband Claude as photographer and flight instructing. Organized Wing Scouts #1 Chapter (high school girls) taught several to fly. Flew *Little Toot*, built by George Meyer, on its first flight, a dream of a airplane, several out there today.

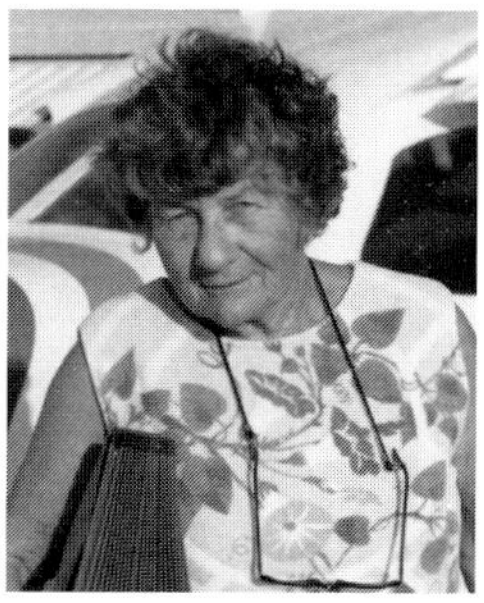

Planned, painted and repainted 706 Air Markers (Corpus Christi 99s) on roofs in towns/cities covering all remote areas.

Flew 24 of the 29 All Woman Transcontinental Air Races (Powder Puff Derby), several Angel derbies, International Air Race and thus far 18 Air Race Classics ... never number one.

"Of course, being a 99 has meant much to me!"

ANN GRIFFIN GLESZER, born June 21, 1981, in Columbia, CT. Joined the 99s in 1944. Military service included WASP. Flight training included private instruction, Civil Air Patrol, Avenger Field - WASP. She has flown the following planes: Piper J-3, PT-17, BT-17, AT-6, AT-11, various Piper, Cessna models, Kaman Helicopter K-125 and K-190. Gleszer has been employed as a flight instructor and flight test pilot. She is married to Kenneth M. and has two children, Douglas D., 44, and Glenn T., 42; and also grandchild, Douglas D. Griffin, II, 21.

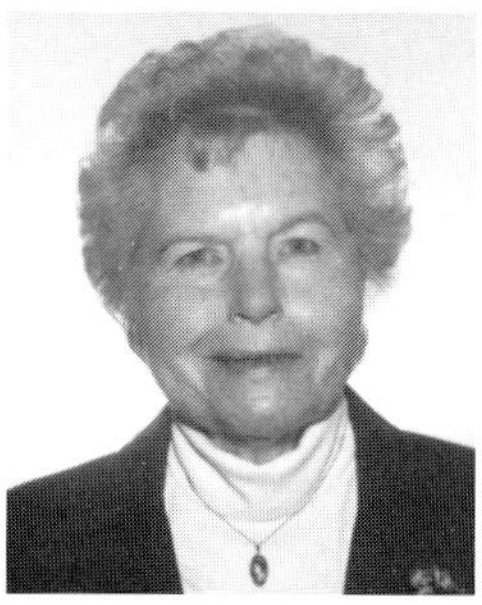

Flight experience, test pilot, AAF, Test Pilot Charles Kaman Helicopter development of servo-roto blade, before CAA certification of K-125, K-190 which was featured in *Life Magazine* issue 1948. Now major helicopter producer of well-known UH-2 used for rescue duty for Navy. K-Max Heavy Lift latest development. Before joining WASP she worked as air traffic controller in Experimental Test Department of East Hartford, CT, (United Technologies) Pratt & Whitney Aircraft, along with Ed Granger and eight or nine test pilots, developing the engines that were necessary in the WWII effort. Many notables, general, admirals, Charles Lindbergh came in to consult with this most important project.

She first joined the 99s in 1947, but her husband Thomas was assigned to a five year business project in Europe in 1950; therefore, she dropped 99 membership when she went abroad.

TRAUDE GOMEZ, born July 3, 1931, in Berlin, Germany. She has flown Cessna, Piper, Beechcraft B55 and Group Glider.

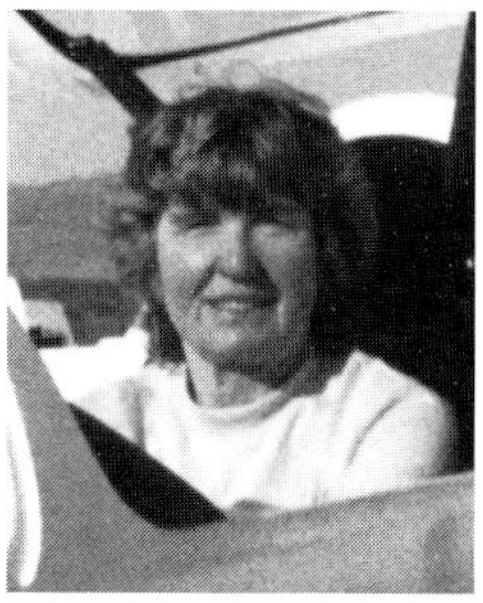

Private pilot 1.16.1974. Member of the San Gabriel Valley Chapter 99s. Has been flying high with the 99s and her family ever since. They enjoyed airmarkings, fly ins and 99s conventions together. Vice chairman 1978-79, chairman 1984-85, and news reporter. The chapter voted her pilot of the year in 1979. She served as the chapter's institutional representative for the Aviation Explorers in 1978-79, with flights to the Colorado Springs Air Force Academy, San Diego aircraft carrier and Air Force bases. Aviation project leader for the Monterey Park Equestrian 4-H Club from 1975-78.

She enjoys gliding with her son Miguel, traveling with her journalist daughter extensively throughout Europe and Central America, and teaching aquatics with her daughter Gabriela and aqua aerobics for the YMCA. She teaches health and safety classes for the Red Cross. Her grandsons Skyler and Willi share her enthusiasm for flying and swimming. She has great memories of visiting 99s in Australia and New Zealand, of their flights over the Great Barrier Reef and the New Zealand Alps. She now lives with two aerobatics pilots; Goldy Blue and Amber, two beautiful Macaws. They both love to fly to her.

LOIS BROYLES GOODRICH, as a young girl her heroine was Amelia Earhart. To follow her dreams, she began flying lessons fall of 1948, before having her family. Renewing her dreams in 1965, she earned her private pilot license ASEL. Joining CAP as captain, flying search and rescue missions in Piper Cub. Earning commercial with instrument rating 1975.

Joined the Michigan Chapter 99s in 1968. Held offices of program/newsletter chairman two years, chapter chairman 1975-77. Was Michigan Chapter representative at International Forest of Friendship dedication in Atchison, KS, July 1976. Revisited in 1986 with Honoree Leah Higgins. Enjoyed attending many International and section conventions, meeting wonderful 99s, is now happy life member.

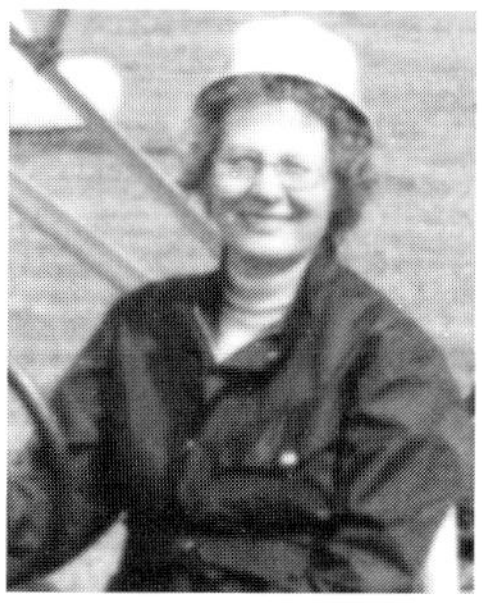

With her beloved 49 1/2, commercial pilot "Marty" Broyles has owned a 1947 Aeronica Sedan, 1958 Skimmer, a 1960 Piper Comanche N 99LB. Flew many Michigan Small Races, plus many other Proficiency races with Marty as copilot. Flew Arkansas Air Derby 1974, third place 1975 Best Female Pilot trophy awarded by Arkansas 99s. Also PPD 1975 and 1976, Angel Derby 1975 and 1977. Marty and she flew several Bahamas Flying Treasure Hunts placing third in 1972. Before Marty's death in 1977 with several members of EAA Chapter 13, Detroit, MI, they built a Breezy #N61DC.

Married Charles J. Goodrich, a P-47 fighter pilot in 1979. Flew with him in his home-built Cougar and Navion until his death in 1985. Was staff anesthetist at Crittenton Hospital, Rochester, MI, until retirement in 1991.

BARBARA A. GOODWIN, born Dec. 26, 1934, in Fort Wayne, IN. Her favorite activity when she was four years old was to visit Smith Field Airport in Fort Wayne with her parents. She did not know then, that in 1984, five years after her husband, Bob, earned his private pilot license, she would take her check ride for her private pilot license in November and join the Lake Michigan Chapter of the International 99s in December. All of her flight training was done in and around Kalamazoo, MI. She received her instrument rating in 1987 and her commercial rating in 1989. The commercial rating was possible to acquire because she was awarded an Amelia Earhart Scholarship from the International 99s in 1988. With help from several Lake Michigan Chapter members who already had their CFI licenses, Barbara started her CFI career in 1992. She acquired her ground school teaching license in 1985 and teaches private pilot ground school through public school Adult Ed.

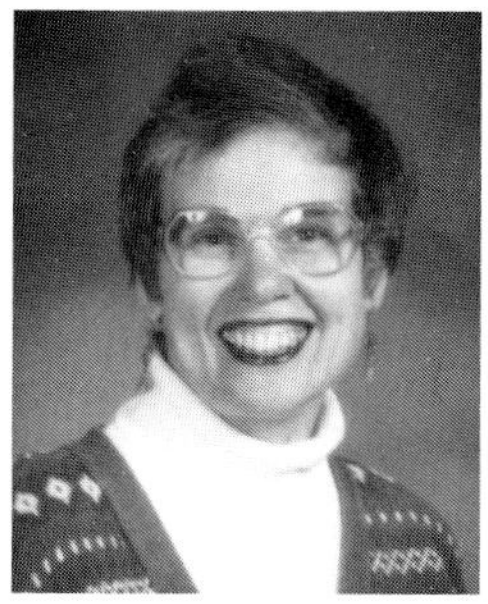

Barbara owns a Cessna Cardinal with 49 1/2, Bob, and has flown a Cessna 152, a Cessna 182, a Piper Archer, Piper Arrow and a Grumman Tiger. In 1988 and again in 1990, Barbara flew in the Air Race Classic cross country race as copilot with Mary Creason. In 1990 Barbara and Mary placed fourth out of 40 race airplanes.

Barbara was chapter chairman for the Lake Michigan Chapter of the International 99s from 1991-95. Barbara has earned four sets of FAA Wings and is a Young Eagle Flight Leader for the EAA Young Eagle's program. She was president of the Kalamazoo/Battle Creek Aviatrix organization for two years.

Barbara teaches sixth grade science and math, is married and has three grown children: David, Katherine and Steven. She also has two grandsons, Ryan and Adam.

MARTY GOPPERT, after spending many hours in the right seat with her airline captain husband Tex, she decided she had better know how to handle the airplane if anything happened to him. As an RN, she knew one should think of these things! Her military service began in 1966 in the Army Nurse Corps, and she served for three years. In 1979 she started flying in earnest and obtained the private pilot license; it seemed like so much fun and such a challenge, that she kept right on going. She built time as a flight instructor. She volunteered on an air ambulance service, thanks to her nursing background; and when there was no patient on board, she flew legs to build multi-engine time. That same year (1979) she joined the Potomac Chapter

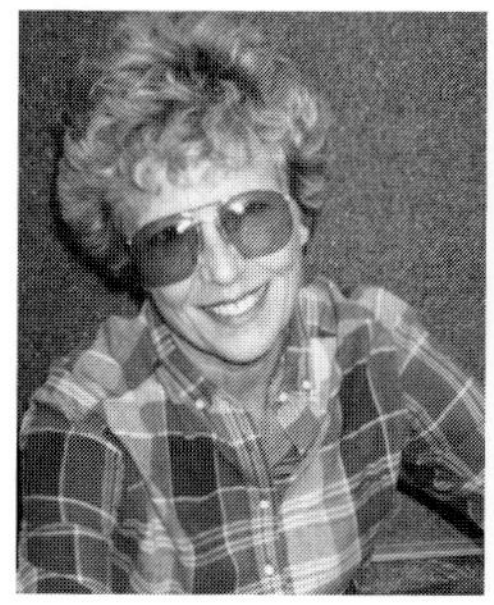

99s. Eventually she served as Aerospace Education Committee chair, and then chairman of the chapter. At the same time, exciting things were going on with her career, as she completed the CFII and ATP and was hired to fly Jetstream 31s for one of the USAir Commuters. She did this for two years, and what an experience it was! But it doesn't compare to what she does now with the Flying Circus in Warrenton, VA. She flies a 1941 Stearman every week in the summer. The pilots whom she flies with are spirited and fun-loving performers. She has kept up with nursing as well, for it was her nursing background that got her into aviation.

MARGE GORMAN, commercial pilot ASMEL, instrument, glider, helicopter. Flies V-35 Bonanza, Beech Duke and Staggerwing.

She has 52 active flying years with more than 5,000 hours.

MARY J. GOWANS, born March 10, 1926, in St. Louis, MO. Joined the 99s in 1993. Flight training includes Free Spirit Aviation, Livingston County, Howell, MI. Has flown C-150, 152, 172; Tomahawk and Waco. Learned to pilot an airplane when retirement, after over 20 years of teaching high school and community college, left time to do things on week nights and weekends. On Jan. 8, 1993, she completed the FAA private pilot check ride. After earning her private license, she has started working on an instrument rating and finished the written test for instrument and commercial rating. Some of her most exciting days were in Hawaii flying a Waco over the Pacific Ocean and looking back at the beautiful Hawaiian shoreline and flying a Cessna 150 over a glacier in Alaska. She just intends to continue flying and trying different single engine planes and a glider now and then.

JANNA GOWTHROP, previously know as Janna Imlay or Shea, began flying in 1974. The daughter of a Northwest Airline's pilot, she first flew in a Piper Vagabond at age 5. Employed as an aviation safety inspector/safety program manager with the FAA, her mentor is Emily Warner of Frontier Airline's fame. Her aviation background is varied, with experience as a CFII, charter pilot, tour pilot and corporate pilot. She is ATP type rated in the Citation III, Falcon 10 and Gulfstream IV.

Janna holds a BS in aerospace science/management earned in 1977 at Central Washington University.

Gowthrop joined the Rainier Chapter of the 99s in 1974, moved to the Phoenix Chapter in 1978, and became a charter member of the Arizona Sundance Chapter where she maintains her membership, although living in Colorado. She has participated in many races, including the Powder Puff Derby, Air Race Classic, Shamrock Air Derby, Palms to Pines, Pacific Air Race and Mile High Air Derby. She held the position of race chair for the Shamrock Air Derby its first four years of operation.

Janna is a founding board member of the Fantasy of Flight Aviation Education Program.

Married to Rick Gowthrop, who is also a pilot, Janna has two children and two stepchildren.

JUDY OVERMYER GRAHAM, earned her private license in June 1968 and her instrument rating in 1976. She flew the last Power Puff Derby (fulfilling a childhood dream) in 1976 as copilot of TAR 190 and received the Low Time Team award. She flew the IAR in 1977 as PIC for the first time in any race, placed ninth and won the Rookie and Low Time Team award. She participated in numerous Indiana FAIR Ladies' Proficiency Rallies. Judy joined the 99s in 1972 and was a member of the Indiana Chapter 99s before helping establish the Three Rivers Chapter 99s. She held offices in both chapters and as airmarking chairman for the North Central Section. Judy received an AE Scholarship for her multi-engine instructor rating. Judy shared her love of flying with everyone, including her husband and three sons. She was a flight instructor, did air photography work, flew cargo, and was a charter pilot at DeKalb County, IN, airport. For a year, she was a first officer for Britt Airlines. Flying for Judy put the world "in slow motion. When you're up there, you see the big picture. It puts things down there in perspective." Judy was born Jan. 8, 1941, and grew up in the Leo, IN, area. Judy was fearless in the sky, but ironically met a tragic death in 1986 in a car accident just three miles from home. Judy many times said, "Some people think flying is more dangerous than driving, but it's just not so."

LISA GRAHAM, born Feb. 27, 1970, Georgetown, Ontario, Canada. Flight training included Brampton Flying Club, and Sault College Aviation Program. She has flown the following planes: C-152, C-172, PA 28, and PA 31. Her pilot employment included flight instructor, Brampton Flying Club and flight instructor and charter pilot, Skyways.

Graham loves traveling and after an introductory flight at the Brampton Flying Club with her father and sister, she knew she wanted to become a pilot. She immediately started private pilot training and changed her high school courses to prepare her for an aviation career.

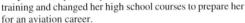

She attended Sault College's Aviation Technology program and graduated in April 1992. During this three year course she obtained her commercial rating and multi-engine instrument rating.

Returning to Brampton Lisa became a member of the First Canadian Chapter. She volunteers for Skywatch, where 99s fly ministry of the environment investigators for aerial photography. The photographs are used to prosecute environmental offenders.

She completed her instructor rating and started working as a dispatcher and then a flight instructor for the Brampton Flying Club. She now works for Skyways as a captain on a Piper Navajo Chieftain doing passenger and cargo charters. She also instructs for the company and has accumulated more than 1,500 hours.

Lisa recently received her airline transport license and would like to work for a major airline.

CYNTHIA GRANT, flying life began when she obtained her private pilot license in 1990. Six months later she earned her tailwheel endorsement in an Aeronca 7AC Champ and shortly thereafter she bought a 1946 Luscombe 8E. Cynthia believes in continuing training. She has taken an aerobatic course and participates in the FAA's Wings Program.

Cynthia loves to take long cross-country trips. She particularly likes the challenge of flying without radio navigation aids, using pilotage and dead reckoning. Her first two long trips, to Leadville, CO, and to San Francisco, CA, were done strictly the "old-fashioned" way. She has also flown to Dayton, OH, to Carlsbad, NM, and to various destinations in Texas.

Cynthia's other favorite flying activity is sharing the excitement of aviation. She participates in the EAA's Young Eagles program, and she has given 40 Eagle flights to young people so far. She has also introduced many adults to the joy of flying a small plane.

Cynthia is an advocate of pure VFR flight and of flying classic airplanes as an inexpensive, safe, and fun way to get more people into aviation. She feels very fortunate to be able to see the country from the air, and loves to share that good fortune with others.

Having grown up in San Antonio, TX, Cynthia has lived in Austin for the past 28 years. She works as a computer programmer to earn money for flying and other pursuits (traveling and outdoor activities such as caving, hiking and birding). In addition to the 99s, she is a member of EAA, AOPA and the Luscombe Assoc.

FRAN GRANT, began her flying lessons in 1939 at Mills Field at the Morrison Brothers FBO. She obtained her private license through the Competitive Civilian Pilot Training Program at Oakland Airport in the summer of 1940. Her lessons in the CPT program started July 17 she soloed July 25, and got her private ticket Sept. 15, 1940. Her first passenger was her new husband, Norm.

During the summer of 1940, Fran found out about the 99s from three members of the Bay Cities Chapter. Fran became a member in 1941 and has been a lifetime member since 1967.

Since 1941 Fran has logged more than 1,000 hours of flying time. Most of her flying was before 1957, when an accident curtailed her flying regularly. Her first cross-country out of California was to the 99s International Convention in Albuquerque. Other favorite cross-country flights include a trip to the Cleveland Air Races in 1946 and a flight to the 1949 International Convention in New York. Her flying career was made up of many trips to 99 conventions, section meetings and starts of the Powder Puff Derbies. Fran served 20 years on the "start" committee.

Fran is a member of California Aerospace Education Assoc. (past president of the Northern Section), the California Aviation Council, the Civil Air Patrol (education officer), the American Society of Aerospace Education, the County of San Mateo Aviation Committee, the Western Aviation Museum Trustee (currently a member of the board of directors), the Aircraft Owners and Pilot's Assoc. and Zonta International (past president of the Redwood City Club).

Grant cofounded the Fear of Flying Clinics with SCV member Jeanne McElhatton in 1976.

She received the 1979 Service Award for serving as the National Congress of Aerospace Education Workshop director, and also in 1979, the California Aerospace Education Paul Mantz Memorial Award, for five years outstanding service in the Aerospace field. In 1980 she was a co-recipient of the Frank Brewer Memorial Award for outstanding achievement in Aerospace Education. In 1983 she and Jeanne received the Amelia Earhart Silver Medallion for the FOFC, and in 1992, the SCV chapter sponsored Fran into the Forest of Friendship.

Fran is also the mother of Bruce and Judith. Fran and Norm (a retired electronics engineer with the US Navy) now have four grandchildren: Daniel, Elaine, Darby and Kelsie; and two great-grandchildren, Kelsie and Emily. Fran's career has included teaching, clerking, office management and medical insurance specialist. In 1976 she became the administrator for the Fear of Flying Clinic. She has a BS in physical education, mathematics and science from San Jose State University and is an accredited teacher for general elementary, junior high and special secondary for PE and math. Fran earned her certificate of medical administration from the University of California, Extension Division.

SHIRLEY CLARK GRANT, Montreal Chapter, East Canada Region, born Oct. 28, 1935, in Lincolnshire, England. Joined the 99s in 1981. Flight training included fixed wing commercial, aerobatic, multi-engine, and IFR. She has flown: C-152, C-172, 7ECA, 8KCAB, CAP 10, AA1B, PAZT, BE76, BE55. She is married to Daniel Grant and has four children and four grandchildren. She is the business manager of a private school in Montreal. Owner of a Grumman AA1B, based at home grass strip, St. Mathias Grant (SX5).

VEDA GRANT, learned to fly in 1932 when she was 21 years old. She flew a WACO and her brother was her first passenger.

105

She was the first woman pilot in Maine. Her license, issued by the CAA, is #20400 and she still uses the same number on her automobile plate. She joined the 99s when two other women pilots came to Colby College in the 1930s. One of her prize possessions is the sheet music for the 99s song which she displays on the piano.

In January 1942 she signed up when the CAP was started. She was a lieutenant and served for many years.

At 81 years of age she has many memories of the years that she has been flying out of Augusta Airport. She met Wiley Post when he arrived in the Winnie Mae. Amelia Earhart's visit is also a cherished memory.

An avid cat lover all her life, the accompanying picture of her shows Tailspin Tommy, who loved to fly and went with her on many flights.

GINGER A. GRAVES, received her private pilot license in April 1992 through Flight International, at Newport News/Williamsburg International Airport, Newport News, Virginia and joined the 99s immediately thereafter. She is a member of the Hampton Roads Chapter even though she now lives in Daphne, AL. Ginger is a pharmacist with Wal-Mart.

KATHLEEN "KATHY" GRAY, born Dec. 7, 1925. She is a retired RN, BSN. Her husband is Worlin "Wug" and she has children: Leslie, James, Allen, David and Doug. Joined the Reno Chapter 99s in 1969. Gray currently holds the APT chair. She began flying as a stewardess for Pan Am, 1948-56; began flight training in 1968, licensed in 1969. Holds the ASEL rating and owns a Cessna 206, Turbo. Received her BS in nursing in 1948 from the University of Southern California.

During her employment with Pan Am she flew internationally, marrying a Pan Am pilot and raised their family in Reno. Became a pediatric nurse, then a school nurse, retiring in 1990. Her hobbies include skiing, golf, tennis, flying and reading – she loves to travel. Memberships include: World Wings International (former PAA flight attendants); Folded Wings (former Flight attendants); American Association of University Women; Lakeridge Women's Golf Club, Rosewood Women's Golf Club, Pan Am Retirees Assoc.

MARJORIE M. GRAY, Long Island Chapter, New York - New Jersey Section. The turning point in Marjorie's life occurred on July 4, 1937, when she took her first flight at Westhampton Airport, Long Island, NY. Exactly one year later, to the day, she soloed and soon after joined the 99s. She obtained her commercial, instructors and seaplane ratings. In 1942 Jackie Cochran telegrammed Marj to join the Air Transport Auxiliary in England but Marj declined as she was presently at La Guardia Airport as air traffic controller trainee. After being contacted again by Jackie Cochran, many joined the first class for military training at Ellington Field, Houston, TX, (later in August 1943 called the WASP). She then earned her instrument and multi-engine ratings. Marj opened up an FBO at Teterboro Airport in New Jersey in 1946 to 1947. She also served as governor of the New York - New Jersey Section in 1946. Marj joined *Flying* magazine as associate editor in 1953-60.

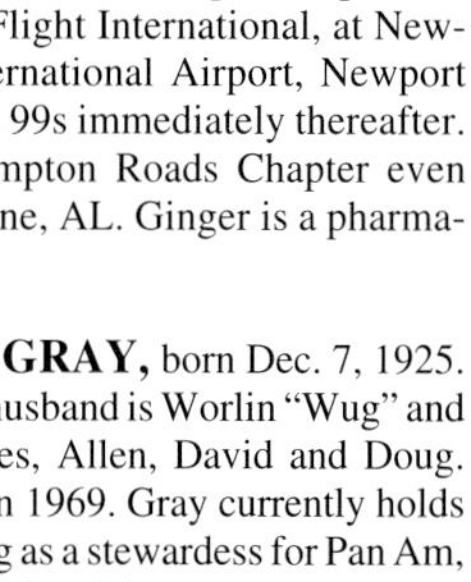

In 1956 she was honored with the Lady Hay-Drummond Hay Award for Outstanding Achievement in Aviation. After moving to Long Island in 1965 she worked as a technical writer and retired from Grumman Aerospace. She often reflects on her exciting experiences in aviation. Marj retired from the Air Force Reserve as a lieutenant colonel after 22 years of service and received the Meritorious Service medal.

RITA GRAY, born May 13, 1931, in Shippenville, PA. Her fascination with flying began in grade school while she watched WWII Aviation Cadets practice over her father's farm.

She learned to fly at the age of 19 at Parker D. Cramer Airport, Clarion, PA, soloing the Piper J3 on snow skis after seven hours and 20 minutes dual.

Flying was interrupted by her marriage to Dave Gray a former Air Force pilot and mechanical engineer. She raised eight children, one of whom is a corporate pilot. During this time she spent many years as a Girl Scout leader and Cub Scout den mother. She has also been very active as a lay minister in her church, especially for the shut-ins and hospitalized of her parish.

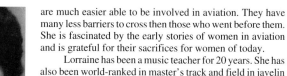

In addition to a love of flying she had a desire to parachute. She realized this dream at the age of 58 by completing a tandem jump from 11,500 feet. The experience was such she plans to do more.

She plans to reactivate her flying by joining her pilot husband in sailplanes and general sport aviation.

Gray children's names and ages: David, 37; Brenda, 35; Tom, 33; Jim, 31; Yvonne, 27; Joe, 25; Barb and Bev, 23. Also grandchildren: Nick, 12; T.J., 11; Courtney, 9; Brittaney, 7; Kaleigh, 6 months; Brach, 3; Brooke, 1; and Kasey, one month.

JANET C. GREEN, Phoenix 99s, always wanted to fly and after retiring from active duty with the USAF as lieutenant colonel, medical technician (or, rated as biomedical laboratory officer). She got her private license at Action Aviation, Glendale, AZ, in 1986 and immediately joined the 99s. She was corresponding secretary when her reserve status won her a spot in Desert Storm.

Meantime, she and her husband (retired lieutenant colonel USAF), became part owners of Action Aviation where Jan was manager from 1986 until they sold out in February 1992. In 1989 they tried to fly their Warrior to Michigan. Stuck four days in a stationary front at Omaha, they had to turn back, and got into trouble with wind-shear in Texas and two stuck valves in New Mexico, finally making it home.

Jan always remembers taking a BFR with Phoenix 99 Bruni Bradley when the engine quit. To Bruni's "That's fine – let's go" Jan reported it wasn't her idea – they restarted it, got home safe but scared. Jan is now director of aerospace education, Arizona Wing of the CAP, and is a charter member of WIMSA (Women in Military Service Assoc.). Her household consists of a husband, four cats, two dogs and a bunch of fish.

JANET F. GREEN, born Dec. 19, 1925, in New Orleans, LA. Joined the 99s in 1961. Flight training includes private, commercial, multi, instrument, basic instructor, air. She has flown: Aero Commander 520, 560F, 500B; Cessna 150, 172, 182, 187RG, 210; Mooney M20F 310-401. She did some charter and instruction long ago. Green is married to Donald and has four children: Gayden, 42; Travis, 41; Colin, 40; and Courtney, 38. Her grandchildren are: Shannon, 13; Taryn, 6; Eammon, 7; Kyle, 5; Owen, 3; Wylie, 5; and Matthew, 4. Enjoyed Powder Puff and Angel derbies and flying all over US, Canada, Mexico and Central America, Bahamas and Cayman Island, Puerto Rico. Soloed daughter and two sons. Two are pilots. Has served as past International president 99s.

LORRAINE GREEN, born May 13, 1952, in Evanston, IL. Joined 99s in January 1990. Her training includes her private pilot certificate in single engine land and sea planes. Green has also done some aerobatic training as well as glider training. She has flown primarily in Cessna 172s and 152s. She's also had the opportunity to fly in a Beech Staggerwing, Pitts S-2B, and Grob Twin Glider.

She is married to Woody Green and they currently have no children.

Has always wanted to fly. Her mother was a pilot, in fact she flew and had her certificate before she had her driver's license. Her father was a cattle rancher and the airplane was a necessary tool to help "survey" the crops. She remembers her first airplane ride. Her grandfather took her up in his Piper Supercruiser. She sat in the back totally awed. The small ranching community over which they flew looked like toy land, so small and unreal. Green was eight at the time. Little did she know she would be the proud owner of a plane some 30 years later! She finally decided to learn to fly when she turned 39. She considers herself lucky to have had the opportunity to be able to fly. Green's story is not spectacular or flashy, but is becoming more and more common. Women are much easier able to be involved in aviation. They have many less barriers to cross then those who went before them. She is fascinated by the early stories of women in aviation and is grateful for their sacrifices for women of today.

Lorraine has been a music teacher for 20 years. She has also been world-ranked in master's track and field in javelin and race walking. Her competition has taken her all over the world including Moscow, Russia and Melbourne, Australia.

RUTH G. GREEN, joined the New York Capitol District Chapter at the formation in October 1970. She is married and has two children and two grandchildren. Purchased C-172 on April 15, 1966, for commuting to business in Maine, first flight ever on April 24, 1966, private pilot license on June 17, 1967. Purchased more power C-182 in June 1967. For more speed to get from business in Maine, New York, and Florida, purchased C Skymaster (337) on July 6, 1972. Obtained multi-engine with CLT in October 1972, instrument in June 1976, commercial pilot on June 26, 1985, basic ground instructor on March 7, 1983. She has 1,000 hours of flying time.

Held offices, chaired several committees. Worked on many air races and section meetings and 1979 International Convention. Member of Grasshoppers, AOPA, and Silver Wings. Started a flight school in 1973.

MARCIA ELLEN GREENHAM, commercial, instrument, multi-engine, CFII ASMEL. First flight lesson was in November 1968; received private license in August 1969; commercial/instrument 1974; multi-engine 1975; CFI and AGI, 1980; instrument instructor, 1981; multi-engine instructor, 1984. Worked as full-time instructor 1980-83; flew air-taxi 1983-86; flew as corporate pilot in Beechcraft Duke 1986-94.

Member of All-Ohio Chapter and North Central Section. Served as chapter secretary, vice-chairman, chairman and on numerous committees. Was NCS board member, treasurer and currently serving as historian. Currently has 6,200 hours plus and still actively flying. Has been co-owner of C-310, C-172, and since 1978 has been single owner of Mooney.

Registered nurse; married to Glenn Greenham, two children, five granddaughters.

BEVERLEE K. GREENHILL, born Oct. 15, 1935, in Chicago, IL. Joined the 99s in 1976. Started flying in the early 1970s and has more than 2,400 hours in single and multi land and sea. Flown many air races with her husband Chuck, and daughter Stacey who are both pilots. Has flown an assortment of airplanes – started with a Cessna 150, 152, 172, 172RG, 182, 182RG, Turbo 210, Skymaster; Aerostar 601P, 700; Lake Amphibian 200EP; Grumman Goose; Widgeon; Albatross Twin Seabee; Great Lakes; Citabria; Piper single engine; Queen Air; AT6; Baron 55; Mooney; Skipper; Turbo Commander; Seminole; (not in that order) and she flew her own 1989 F33A Bonanza.

She belongs to the Chicago Area Chapter 99s and loves it, AOPA, American Bonanza Society, Seaplane Pilots Assoc. and many others.

She is married to Charles "Chuck" and has three children: Stacey, Tammy (spouse Cal), and Steve (spouse Marcie). Her grandchildren are Paul Jason, 2 1/2; Jennifer Faye, 3 weeks; and another expected in May 1995. Her husband is rebuilding a Hawker Sea Fury and owns and flies a P-51 Mustang, an Albatross, a Grumman Goose and a Grumman Widgeon which he rebuilt.

Special flight experience: "We were flying at about 2,400 feet IFR light chop over a layer and severe weather under us. The pilots (male) below kept yelling about the severe turbulence – ATC asked about our turbulence – I answered "light chop." The male pilots yelled over the radio that I didn't know what I was talking about. ATC asked if I heard what they said, I repeated it again, "light chop." The male pilots below started to yell nasty things at me. ATC asked me to answer them, I keyed the mike and said, "When your bra straps snap, that's severe turbulence." Everything was quiet after that.

LUANA LYNNE GREER, Nee: Milligan, born May 5, 1941, in McComb, OH. Joined the San Diego Chapter in 1968. Flight training includes SEL, Gillespie Field, Santee, CA. She has flown C-150, PA28-140, and PA28-180R.

Married to Robert and has one child, Robyn L. Greer, 14, born Sept. 30, 1980. She was charter chairperson of the Inland California Chapter 99s in 1974. She's currently active in El Cajon Valley 99s serving as treasurer. She flew in 1976 Powder Puff Derby and she and her husband have flown across US several times.

HELEN GREINKE, born June 10, 1916, in Chicago. Joined the 99s in 1945. Flight training includes March 31, 1939, first training lesson; 1941 commercial, flight instructor ASEL. She has flown the following planes: Piper J3, J4; Aeronca Chief; T-Craft 40, 50, 55, 65 hp L-2; Cessna 120, 140, 150, 172, 175, 180, 185; Piper Cherokee 180-235 hp; Kinner Bird (:05 dual tri-motor Ford!). Her pilot employment consisted of flight instructor Bloomington, IL, and TAFT TX Centralia, IL. Participated in the 1946 All-American Air Maneuvers Miami, fourth place closed course race; 1947, 1st All-Woman Air Show, Tampa, FL, second place closed course race; 1948 2nd AWTAR Palm Springs Cato Miami, FL, third place; 1949 1st International Race Montreal Canada to Miami, second place; 1950 2nd International Race Montreal to West Palm Beach; 1951 AWTAR; AWTAR 1953 through 1959; 1959 Dallas Doll Derby; AWTAR 1960 through 1962; 1965 through 1967.

ELIZABETH "BETTY" GREPLY, First Canadian Chapter. Betty learned to fly in 1956 at the Gilles Flying Service, now Buttonville Airport, north of Toronto. At that time, almost 40 years ago, the airport was a grass strip. She got her license in a Aeronca Champ (on skis in the winter) with no radio, a real accomplishment in those days.

Betty got a night endorsement in 1964, kept current and encouraged her friend Edie Luther to get her license in 1969. They did a lot of pleasure flying together and enjoyed attending fly-in breakfasts at various airports in Ontario. Betty flew in several Poker Runs to support the chapter's only fund raising project and joined the 99s in the 70s.

The Guelph Air Rally, sponsored by FCC, was a highlight for Betty and Edie (they didn't come first, but didn't come last either!) They were a well-known team around the circuit in the early years of Operation Skywatch and contributed many hours of volunteer flying.

Betty has attended two conventions and is a long-time member of First Canadian Chapter.

THON GRIFFITH, born April 25, 1923, San Francisco, CA. A lifetime resident of California, began flying in 1961, joined the Orange County Chapter when it formed in 1962 and held most chapter offices. She served on the International board of directors for 10 years and as International president 1978-80.

She flew a Beech P58 Baron as a corporate pilot for 15 years. Over a period of years Thon and her late husband George flew throughout the US, Canada, Alaska, through the West Indies into South America and made numerous flights to Mexico and Central America.

From 1967 to 1972 she was the Powder Puff Derby Route Director/West Coast Representative. She few the IAR, three AWTARs and many West Coast races. She served on the Aeronautics Advisory Committee for Cypress College and is a past director of the National Pilots Assoc. (NPA).

Thon was chairman and later treasurer of the Amelia Earhart Memorial Scholarship Board 1982-91 and in 1992 was awarded the 99s Award of Merit.

An accomplishment not to be overlooked is to have reared four terrific children. The fun is enjoying teenage grandchildren. Her four children: Bailey Abbott, 47; Sheldon Abbott, 45; Stancie Foss, 42; and Morgan Abbott, 39. Her grandchildren are Ashley, 17; Taylor, 15; Carrie, 13; (all are her daughter's children, Foss).

ANNE A. GRIMM, born Aug. 24, 1926, in Colfax, IL. Joined the 99s in 1962. Flight training includes private pilot ASEL. She has flown Cessna 150, 172 and 175. Grimm is divorced and has three children: Karin, 47; Allen, 46; and Eric, 28. She has two grandchildren, Rhonda, 26; Michael, 22; and one great-grandchild, Nathanial, 2. Her flight activities include: 1959 Dallas Doll Derby; 1960-61 AWTAR "Navigator"; and 1962 and 1965 AWTAR copilot.

WENDY GRIMM, born on Jan. 25, 1961, in Pittsburgh, PA. She learned to fly in 1984 in Morgantown, WV. While a college student at West Virginia University she worked her way up to general manager and charter sales manager at a local fixed base operation. She migrated to the Newport News, VA, area to work as a dispatcher while finishing her professional pilot certificates and ratings. Wendy worked her way up on the flight line from flight instructor to chief flight instructor of a FAA Part 141 Flight School. Much of Wendy's time is spent in aviation safety programs and activities as an FAA accident prevention counselor. Wendy served as chapter chairman of the Hampton Roads 99s, from 1992-94. Becoming an Amelia Earhart Memorial Scholarship recipient in 1993 allowed her to realize her goal of becoming an airline transport pilot. Wendy continues to live the dream of teaching others her relentless love of flying.

Her first student was a physically challenged paraplegic. He received his private pilot certificate in a 1947 Ercoupe. All the controls were designed to be interconnected with the airplane yoke. He taught her the true meaning of dedication and perseverance. It was one of her most challenging and rewarding experiences as a flight instructor.

MARY GRONEWALD, a native and lifelong resident of California, learned to fly in 1982 at Riseto, CA. She flew copilot in the 1985 Pacific Air Race. She has served as chairman and treasurer of the Inland California 99s. She coordinated the chapter activities for the 1989 60th Anniversary Reflight of the Women's Air Derby.

Gronewald has a BA in chemistry from Immaculate Heart College and an MSW from San Diego State College. She retired as associate director of the San Bernardino County Department of Public Social Services. She has served as member and officer in many professional organizations. Her community activities include the United Way and YWCA. Gronewald is a member of the 99s Inc. and AOPA.

She is married to Jack Gronewald, who is also retired and a pilot. She has a step-son, step-daughter and three grandchildren.

NANCY PATRICIA "RED" GUERNSEY, born Oct. 12, 1955. Newark, NJ, became the birthplace of brown-eyed, red-haired Guernsey. Her flying lessons began at "Safair" in Teterboro Airport, while studying for her BE at Manhattan College (being the second woman graduate with a degree in mechanical engineering). She completed her flight training at Republic Airport in Farmingdale, Long Island, while simultaneously studying towards her MS in nuclear engineering at Polytechnic University in Brooklyn (again, being the only woman in the major!) and working in Long Island's then-booming aerospace industry. During her years as a systems engineer at Grumman Aerospace she did design work on the Navy's F-14, C-2, EA-6B and several futuristic aircraft – including NASA's space shuttle. When employed by Harris Corporation/GSSD, she did electronic support systems work for the US/Canadian versions of the F/A-18. "Red" is presently employed as a project engineer for NYC's DOT, Division of Bridges/Roadways. She is a published writer and member of Mensa. Her mother, Dorothy, lives in Irvington-on-Hudson, NY.

MURIEL EVELYN GUGGOLZ, known to her friends as Meg, was born Feb. 28, 1904, in New York City. She was a 1926 graduate of Cornell University and held a nursing degree from Johns Hopkins University.

A registered nurse in the Army Nurse Corps, she served in the South Pacific, Germany, Japan and Korea. Meg had enlisted during WWII and was recalled in 1952 to serve in Korea. She retired from the Army in 1964 at the rank of major.

Throughout most of her 20 years of military service, she worked as a nurse. But between 1945 and 1952 she settled in Santa Fe, NM, because she said it was the most beautiful place she had ever seen. During that time she opened a shop and used her GI benefits to learn to fly. For a time prior to her retirement she was an air traffic controller for the US Army.

Meg joined the 99s in 1957 and was an enthusiastic life-time member. She made several cross country trips with members of the group and around the time of her 80th birthday, she and her older sister journeyed to a 99s convention in Alaska.

Licensed as a private pilot for single-engine aircraft, over the years she flew a Cessna 150, 172 and 182.

In addition to her interest in aviation, she was an avid weaver. And as a young woman she was an enthusiastic athlete competing as a fencer in the 1932 Olympics in Los Angeles.Meg, who never married, spent the last 10 years of her life with her sister Marjorie in Pennsylvania. She died Feb. 9, 1995, at the age of 90.

MARIE ELLEN GUY, born May 31, 1951, in New York City, NY. She learned to fly in 1983. After earning a dispatcher's license she gradually moved westward; working for both domestic and international airlines.

Marie is a proud member of the 99s, the Aircraft Owners and Pilots Assoc., the Piper Owners Assoc. and is on the board of directors for the San Juan Pilots Assoc. in Washington state. She is active in the community and is the treasurer of the Animal Protection Society in the San Juan Islands.

She is married to professional pilot Robert E. Guy, Jr. They own an aircraft leasing and aerial photography business and enjoy flying their Piper together.

She has flown C-150, C-172, C-180, PA28, PA32.

SANDRA M. HAAG, Phoenix 99s, got her private ASEL in 1972 at Sky Harbor Airport, Phoenix, AZ. She was born in California and has been in Arizona 30 years. After a BA from the University of California at Long Beach she became an elementary school teacher, and now is a business owner/manager. Sandi was married in 1964 to Lucien C. "Luke" Haag. He is a criminalist, and also an instrument rated pilot. Sandi and Luke have flown the length and breadth of Arizona starting up their Forensic Science Services business, and to many states throughout the contiguous United States to attend professional meetings. Not only their two sons, now grown, but also their dogs have been good flyers and ridden many miles with them.

Sandi, who is currently chapter corresponding secretary, has been chapter treasurer, Cloud 99 editor, and Phoenix 99s Memorial Scholarship chairman. She works tirelessly on all chapter activities, but her greatest contribution has been as route chairman of the annual Kachina Air Rally for many years, being helped by local pilot and chapter friend Dave Ellis. They are all very saddened by the sudden loss of Dave to new horizons when he was in a recent crash, and they will all miss him greatly at this year's race. In 1980 Sandi not only worked on the route with Dave, but was overall chairman of the Rally.

Sandi's many other interests are astronomy, piano, gardening, wildlife, bicycling, and home projects. Neither of her sons has taken time to become rated a pilot yet, but she says they made excellent right seat partners and can no doubt get licenses when they want to.

JOSEPHINE B. HADFIELD, born March 11, 1917, Twin Falls, ID. Joined the 99s in 1983. She was 59 years old when she earned her private pilot certificate. She was 11 years old, when she had her first flight in an Alexander Eaglecock, a biplane, with an Hispano-Suizi, 150 hp engine.

She graduated from Salt Lake General Hospital with a RN in 1939.

Her husband Dale a physician, started to fly in Elko, NV, in 1947 and became a board member of Flying Physicians and Civil Aviation Medical Assoc. Even though her love for flying was there she stayed grounded helping to raise five children: Virginia Le Pore, 60; Sue Terhaar, 50; Dale B. Hadfield, 48; Scott R. Hadfield, 46; and Jody Born, 40. She also has grandchildren: Brett Linbo-Terhaar, 27; Tor Linbo, 24; Jason Terhaar, 25; Todd Terhaar, 28; Michael Le Pore, 23; Scott-David Hadfield, 20; Dale S. Hadfield, 17; and great-granddaughter, Adreianna Linbo-Terhaar, 7.

When she was 53 years old she started lessons and at the same time, a son, Scott, flying a P3 for the Navy and at present a captain for Alaska Airlines flying an MD-80. Her younger daughter a former 99, obtained her private, commercial and instrument ratings.

She has been a member of the 99s since 1983, with three years as chairperson of Mt. Tacoma Chapter, Tacoma, WA. She is and has been a member of OX-5 Aviation Pioneers, International Flying Nurses Assoc., Medical Auxiliary, Scouts, Camp Fire, and 4-H. She is a 33 year member of Summit Presbyterian Church, Bremerton, WA. She has checked out in six different aircraft, the last a Piper Turbo Arrow.

LOIS BROOKS HAILEY, born Jan. 18, 1995, Reno, NV. Graduate University of Nevada 1936, BA in Spanish and music. Taught school in Minden, NV, until 1943. Started flying in 1939, and bought a Taylorcraft in partnership with brother and friend. Soloed in 1939, private license in 1940, and commercial rating in 1941, instructor rating 1945, examiner in 1945.

Joined the WASP in January 1943 with class 43-3 and was assigned to Tow Target work along with most of her class. First they were assigned to Ferry bases, then they were all called to Washington, DC, where they met with Jacqueline Cochrane in the offices of Gen. Hap Arnold in the Pentagon. They were told by Gen. Arnold that they had been assigned to a special new and experimental (for women) duty. They were not to talk about their work.

Her assignments were to Camp Davis, NC, then to Camp Stewart, GA, where they would be the first group of women to learn to fly radio controlled targets. These would be the PQ 8A and PQ 14. Her last assignment was to Biggs Field, El Paso, TX, where they flew Lockheed B34 and 26, Curtiss Helldiver, Douglas A24, Beechcraft AT11 and AT7, and C45. They sprayed tear gas on troops in the desert, flew for searchlight tracking at night, towed sleeve targets and flew the Radio-controlled targets, the PQ3A and PQ14.

After the WASP she taught flying in El Paso on the GI Bill of Rights. When that phased out she went back to public schools teaching band and orchestra in elementary through high school. Joined 99s in 1945 as charter member and first chairman of El Paso Chapter.

ISABELLE MCCRAE HALE, Mission Bay Chapter, Southwest Section, by 1940 she had her ASEL and had helped form South California Flyers. Civil Air Patrol beckoned; she was member #501 and flew training missions in Alaska.

When the WASP was formed she trained in Sweetwater, TX; ferried; flew target planes for two years; flying B26, AT6, BT13, BT15, UC78; then was a flight nurse till the end of the war.

Joined the Los Angeles Chapter in 1940; became a South Dakota 99; charter member in El Cajon Valley, and Mission Bay. Active, holding all offices, and was governor of the Southwest Section.

Participated in 13 Powder Puff Derbies, four Pacific Air Races; Palms to Pines. She owns a 260 Comanche and has over 3,500 hours; a CFI, instrument rated.

The Indian Women Pilots Association was formed through her help and she is an honorary life member of the Indian Women Pilots Assoc. Her doctor son owns a AT-6.

AULEEN KATHERYN HALL, born in Oneida, NY. Schools included Syracuse, NY, University, Los Angeles Lumbleau School of Real Estate, and American Airlines Stewardess School. Flew as a stewardess for American Airlines where she met and married AA Captain Al K. Hall, Jr. Al, now deceased, was manager of Flying Instruction on the Boeing 707 and 747 at the American Academy of Flight in Fort Worth, TX, and an avid and enthusiastic 49 1/2. Many a time he made the AA simulators available to Fort Worth, Dallas and Golden Triangle 99 Chapters.

Al was her sole flight instructor and she received her SEL in 1966 at the Ventura County Airport in Oxnard, CA. Living in Pacific Palisades, CA, at the time, she attended 99 meetings of the San Fernando Valley and Los Angeles Chapters. Ground school was at the Santa Monica, CA, Airport. She finally joined the San Fernando Valley Chapter because in 1968 the Powder Puff Derby started at the Van Nuys, California Airport. As a result, she was the publicity chairman for the start of the 1968 PPD. She has served as chairman of the Fort Worth 99s, Public Relations chairman and 99 news reporter for the FTW 99s. Also was a member of the International 99 Credentials Committee and served as publicity chairman of the South Central Section of the 99s. When she moved to the Washington, DC, area she transferred to the Washington, DC, Chapter of 99s.

Her hobbies are travel, golf and bridge. Besides the 99s, she belongs to the Kiwis, organization of former and now, on-line AA Flight Attendants, is president of her community's Private Golf Club for both men and women members, belongs to the Order of Eastern Stars Fraternal Organization and the USGA Golf Associates.

Currently she is a travel consultant of American Airlines and is employed as a travel coordinator for the community in which she lives. She selects the trips, prepares the ads, sells and escorts groups of their residents both locally, long distance and on cruises.

While living in the San Francisco area, she was assistant editor of the *National Golfer* magazine.

HARRIETT HALL, Alabama Chapter 99s, joined the Mississippi Chapter 99s in 1974. Her younger brother, Lee Monroe, was a CFI at Auburn University and he taught her to fly in his Cessna 150. She has two sons, Joe Hall, III, and Michael Hall. Ruby Dickerson, an Alabama 99s, taught Joe to fly and she started teaching Michael to fly.

Hall was about 5 years old when she had her first plane ride. She sat in her dad's lap. Her uncle thought she was getting sick when she asked him to open the window. She just wanted to see better.

As a teenager she joined a Wing Girl Scout Troop where she decided she wanted to be a pilot. In her Wing Scout Manual, Amelia Earhart said "We women have a rough, rocky road. Each accomplishment, no matter how small, is important. The more women who fly, the more become pilots, the quicker we will be recognized as an important factor in aviation." She has a twin brother, Harry Monroe, who was allowed to do the things she wanted to do, but she had to beg, plead and even sneak around to do things like learn to fly. Hall managed to travel the rough, rocky road Amelia Earhart described.

She is proud to be a 99. She has never regretted the time, money or struggle to become a pilot and she will always be thankful for women like Amelia Earhart and organizations like the 99s. The 99s offered her an opportunity for wonderful friends, great fun and flying.

Several years ago, she had the opportunity to fly the Goodyear Blimp, *America.* When she got in the pilots seat and the pilot went to the back of the gondola, some of the ladies gasped. The pilot said "Oh, she is a good pilot. I saw her land in the crosswind earlier."

JULIA "JUDY" CORBETT HALL, Perry, GA. "Would you like to learn to fly?" This second most important question her husband, Jerry, asked her stared a lifetime of adventure for their young family. During the summer of 1970, they both started flying lessons which eventually led to commercial ratings, multi-engine and instrument ratings for both and a single engine seaplane rating for her. She joined the Georgia Chapter of the 99s (later to become the Deep South Chapter), and the adventure began.

Conventions, section and chapter meetings were a reason to fly to exotic, new places with new flying friends. They explored places they might otherwise only have dreamed about. They met people with their heads in the clouds, literally. They took their two daughters and their son with them whenever possible.

Working her way through all chapter offices and then Southeast Section offices up through governor, she was then ready to serve as International director and treasurer. A lot of work and fun.

Air Racing was another way to go places and meet interesting people, so she flew the 1976 AWTAR (Powder Puff Derby), two Angel derbies and two Air Race Classics as well as numerous smaller races, treasure hunts, scavenger hunts and other flying events.

Although not actively flying now because of a medical problem, the adventure continues, and her hope is to be able to share this adventure with their four granddaughters and one grandson. It started with a question, and Yes was the right answer!

SUE HALPAIN, born March 5, 1940, Texas. Joined the 99s in 1974. Flight training included private single engine land, 900 hours, owns and flies Beech Bonanza A-36. Has flown Cessna 150 and 172, Cherokee Six, Bonanza S, Bonanza A-36.

She is an independent music teacher, adjudicator, clinician and pianist, residing in Bethany, OK. Learned to fly in 1973. Sue and her husband Bill who is an electrical engineer, own a Beechcraft Bonanza A-36 which they have flown over most of the US, Canada and the Bahamas. They enjoy a unique system of traveling self-contained in the A-36 taking along a take-apart motorcycle for ground transportation, and sleeping in the back of the airplane, camping style. She has one child, Philip Halpain, 32.

The Halpains are avid supporters of the 99s and the flying community. They won first place in the Okie Derby Proficiency Air Race and participated in a 99 sponsored parachute jump. Sue has held all Oklahoma Chapter offices and committees and is a recent past South Central Section governor. "The opportunity of flying has offered so many golden experiences and friendships!"

JUANITA HALSTEAD, a charter member of Alabama Chapter 99s, has been flying since 1952. Her husband, C.F. Halstead, wanted her "to learn enough to land his plane." She learned enough to earn her private, single and multi-engine and instrument and to log more than 7,000 air hours in 40 years. The Halsteads rarely drove their car for long trips, preferring to fly their Piper Twin Comanche which they park in a hangar next to their private runway beside their Montgomery home.

Juanita flew two good friends, former governors George and Lurleen Wallace, all over the state.

She flew in many air races including the Powder Puff Derby and the Air Race Classics. She enjoyed the competition. Juanita flew two Lob-lolly Pine seedlings to represent the state of Alabama, to be planted in the Forest of Friendship. Years later she was thrilled when Alabama Chapter 99s honored her in the Forest of Friendship.

Juanita was appointed to the FAA Women's Advisory Committee on Aviation.

Juanita passed away Feb. 27, 1995.

MARIE HAMANN and ARDYTH WILLIAMS, are a unique mother daughter 99s team. In their case it was daughter "Ardy" who was first to become a pilot, going on to distinguish herself as a professional flight instructor and FAA air traffic manager. Marie first soloed at age 46, got her private pilot certificate, and instrument rating and is currently working to become a CFI. It was, in fact, instructor Ardy who gave mother her last check ride before Marie took her private flight exam.

Marie Hamann

Ardyth Williams

Marie grew up in a flying family. Her uncle and aunt were magicians and used airplanes for transportation and stunts. Their stage names were Mardoni and Louise. Mardoni owned many different planes: a Piper Cub, with a rabbit coming out of a hat, painted on the tail, and a Fairchild, Stagger-wing Beach, to name a few. He was a CFI and A&P and taught Navy Cadets during WWII. Mardoni is written up in Ripley's *Strange As It Seems* for some of his stunts and his first parachute jump while handcuffed and shackled. Marie helped cover and dope fabric, watched engine overhauls and planned to solo at 16, but regrets she didn't do it then!

Ardyth Williams born July 11, 1956, Aurora, IL, learned to fly in 1976, while a student at Purdue University and earned ratings, CFI AIM ATP and ground instructor advanced and instrument. She graduated from Purdue's School of Aviation Technology and stayed on as chief flight instructor at Purdue University Airport. She also flew copilot for Indiana Airways and Skystream Commuter Airlines.

Ardy became an air traffic controller at Indianapolis ARTCC and later area supervisor and area manager. She went to Washington, DC, as an FAA specialist, and also traveled the US lecturing at safety seminars. She next advanced to assistant air traffic manager at New York ARTCC and then manager operations branch eastern regional office. She is currently in Washington working for upper management in FAA Headquarters. In her spare time she teaches flying and is a CAP Captain. She has logged more than 3,300 hours in 27 different airplanes.

Marie Hamann, member of the Chicago Area 99s from 1979 to 1990, taught Air Bear classes, flew Illi-Nines Air Races, worked Flying Companion Seminars, etc. Marie is currently a member of the San Diego Chapter. She teaches aviation English to foreign students at Skyline Flight Academy in Ramona, CA. Her husband, Tom is a retired Chicago ARTCC supervisor and 49 1/2. Marie and Tom are both instrument rated and have flown their Cherokee 180 from coast to coast, around the periphery of the US, and into Canada.

Ardy and Marie love flying and are proud to be 99s.

JUANITA A. HAMBLIN, born June 27, 1931, in Wausaukee, WI. Joined the 99s in 1988. Flight training includes private pilot license. She has flown Cessna 172. Married to Gene and has four children: Debby, 43; Becky, 41; Joel, 38; and Rachel, 35. Grandchildren: Melissa, 23; Matt, 21; Christina, 21; Laura, 19; Kelly, 18; Kristy, 18; Christopher, 17; Ken, 17; Josh, 16; Mike, 13; and Noah, 7. Also has two great-grandchildren, Megan, 4, and Mike, 2.

Her first airplane ride was from a neighbor's cow pasture when she was around 11 years old. A friend of their neighbors was giving rides to their children and very kindly gave her a ride also. She loved it! About six years later she looked up into the Wisconsin clear blue sky in June and watched a plane flying by. She said to herself "Someday I'm going to fly up there." When they married her 49 1/2 bought an Interstate airplane with the intention of learning to fly it. However, it was destroyed in one of Illinois strong wind storms. The following years were spent making a living and raising children.

In 1982 her husband again purchased an airplane with the intention of learning to fly it. This time it was a 172 Cessna and he did indeed learn to fly it. They were living in Illinois and they flew to Wisconsin to see family, to Minnesota to see friends, downstate Illinois to see his family, and then all the way to Las Vegas for a convention he was attending. She learned some about navigating and returned from that trip determined to earn her private pilot license. In 10 months she obtained her license and they flew their 172 to Alaska that year. It was a marvelous adventure and they learned much more about aviation on that trip. They also learned that there are wonderful people at all airports. It was a thrill to be landing at airports where the WAAFs had landed in the planes they were ferrying to Alaska to help the Russians beat the Nazis in the 1940s. They both flew and she had the privilege of landing and taking off next to the ocean in Nome, AK. What spectacular views of ocean, mountains, wilderness and forest fires. What wonderful people there are in that newest state. There are 99s who have flown farther and seen more but there are no 99s who enjoyed their trips more than she did. A dream did come true for a young girl in Wisconsin.

LINDA L. HAMER, achieved her pilot certificate in 1976 as a result of an unusual gift of five hours of paid dual instruction. Her first aviation organization was the 99s. Hamer participates at the chapter level in aviation education and air racing with the 99s, Inc.

She became involved with aerobatics and aerobatic competition. Aerobatics opened up a whole new realization of flight for her. Linda is the current president of the International Aerobatic Club, a Division of the Experimental Aircraft Assoc. She is the first woman to ever hold that position. Hamer is on the board of the Experimental Aircraft Assoc., US Aerobatic Foundation and the Exec. Committee of the National Aeronautic Assoc. In Illinois she chairs the IL Aviation Hall of Fame and has served on the IL Aviation Forum. She works with FAA and the IL Div. of Aeronautics to produce and participate in safety and education forums. Hamer is a charter member of Women In Aviation, International.

Hamer started her own aviation flight training school, Aerial Enterprises, Inc. in Peru, IL, and is rated as a CFIAI, MEI and ASES. Flight time is 3,000 plus hours. Her school offers basic through advanced pilot training and aerobatic training. Hamer has been focused in the field of safety and education throughout her aviation career.

Married to Gerald Hamer, she resides in rural Peru. Their four children are grown and on their own. She and Gerald farm and operate an aviation refurbishing business. They are now setting up an airpark to give others an opportunity to "live with their airplane" as they have done for many years. Aviation has been a part of Hamer's lifestyle for more than 30 years.

NADINE ROSE HAMILTON, born July 2, 1927, in Chattanooga, TN. She is married to Tim and has two children: Paulette Pearson, 44, and Carolyn Rose, 51. Also has grandchildren: Alabama Folsom, 25; Christy Jackson, 24; Chuck Prickett, 28; and Will Pearson, 15.

She is the first woman to receive an associates in applied science in the flight technology course offered at Wallace State Community College. She received her private pilot license on Nov. 13, is, now, a member of the Aircraft Owners and Pilots Assoc. and has received an invitation to join the 99s. Mrs. Rose is qualified to fly as a private and commercial pilot, single and multi-engine aircraft and with instruments.

Served as 99s chairman 1986 and 1987; Outstanding Pilot for Alabama 1986 and 1987, received trophy. Formed Civil Air Patrol Squadron in Cullman, AL. She has 300 flying hours and received her pilot certificate on Nov. 13, 1980.

ELLEN VANDEVENTER HAMLETT, grew up in Grapevine, TX, and lives at Northwest Regional Airport (the sixth largest private airport in the United States), Roanoke, TX. She dreamed of flying throughout her years of schooling, and after graduation from Texas Wesleyan University, Fort Worth, TX, took her first paycheck and went to the closest airport: Grand Prairie Municipal. Ellen joined the Golden Triangle Chapter in April 1971 and has enjoyed many terrific years of airmarking, flying and camaraderie with her chapter.

Aviation brought her husband, Gerry, and her together in 1978 when she and several friends decided to be checked out in a Cessna 172 at his flight school. Ellen logged her multi-engine hours under the watchful eye of her instructor/husband and took her check ride when she was five months pregnant in 1980. She has two sons, Geremy, 15, and Collin, 12, who are both eager to fly when they are old enough.

The 99s have always been a strong foundation for Ellen and she treasures her friends in the Golden Triangle Chapter, South Central Section.

STACY HAMM, Phoenix 99, born Nov. 14, 1949, in Downey, CA, the second of six children. Recognizing a desert rat when they bore one, her parents moved to Arizona in 1950. She grew up right along with Phoenix as it evolved from a small cattle and cotton town to become a major metropolis. Sweating out her student hours in the peak of the desert sun, she obtained her pilot certificate in September 1977. Three years later, inspired by a poster in the Falcon's Roost Restaurant, she joined the Phoenix 99s. Immediately the chairman put her to work creating Phoenix's first Flying Companions Seminar. That success prompted her to head up Fly Without Fear programs, Girl Scout Aviation Badge and Interest Projects, and eventually serve in chapter offices.

During her term as chairman, Phoenix earned the First Place International Safety Education Award which they accepted at convention in Baltimore, MD. Hazel Jones took a liking to her when they met at convention in Anchorage, AK, and appointed Hamm her International membership chairman. She served under three presidents and with the Each One Reach One membership campaign, they raised their membership from 5,887 in 1983 to more than 6,500 members in 1988.

Beginning with her first section meeting in Sacramento, she was a vocal contributor to section business meetings, encouraging training programs for chapter chairmen. She served as treasurer, vice-governor and governor (1992-94) of the Southwest Section. Her most notable accomplishments were the deliveries of a few rousing speeches and convincing them somehow to spend most of their hard earned money on scholarships. During these years she also served the Arizona Cactus Pine Girl Scout Council as a troop leader, event coordinator and neighborhood chairman and volunteered for the American Red Cross and ASU's Public Television Station. She served on the Arizona Aviation Safety Advisory group, the only group of its kind in the country. They represent all segments of the aviation/aerospace industry, assisting their Flight Standards Office in their safety and awards programs. But it is through the 99s that she has realized friendship, education and service to others as important lifetime achievements.

ALICE HAMMOND, after graduating from the University of Michigan in 1927, Alice Hirshman attended the Curtiss-Wright School of Aviation, and received her private pilot license in 1931 in a 38 hp Curtiss-Wright Junior. She then earned her commercial, instrument and seaplane ratings. In 1933 she won the first closed course race for women in Michigan. In 1936 Alice married John S. Hammond II and they had three children.

In 1939 Alice proposed the Amelia Earhart Scholarship for Women. Alice was the AEMSF Historian and a permanent AEMSF Trustee.

At least one of the AE scholarships awarded each year is entirely funded solely due to the efforts of Alice Hammond. Each year she diligently finds another aviation anniversary to recognize. She then proceeds to create an exciting new edition of a first day cover of the 1963 Amelia Earhart Airmail Stamp.

In 1941 Alice was called into civilian service to start and command the first all-women's flying squadron of the Civil Air Patrol. This squadron remained the largest of its kind during WWII. She was promoted to the Great Lakes Region staff of the CAP as coordinator of women and later became executive officer of the region. She remained in that position as a lieutenant colonel until 1961, when she was transferred to Philadelphia, PA. There she performed search and rescue missions and gave orientation flights to new CAP cadets. Later she served the CAP as a consultant to the Illinois Wing.

Alice was selected as a charter member of the FAA Women's Advisory Committee on Aviation, 1964-68, under President Lyndon B. Johnson.

Alice served as the president of the 99s from 1951-53 and was the first president to compete in a Powder Puff Derby. Alice served for 60 dedicated years, holding nearly every possible position, including being the first editor of the 99s newsletter. She died in 1993.

JEAN FRANKLIN HANCHER, First Canadian Chapter, after learning to fly in northern Ontario in 1976 and joining the Winnipeg Chapter of the 99s, she changed careers and lifestyles and moved to Toronto. Although she joined the First Canadian Chapter immediately, flying took a back seat to a busy career. Her interest in flying was rejuvenated when her friend Marilyn Dickson got her private license and together they bought a C-150 in 1988. She also started law school, which made it a memorable year.

Since then they have had fun flights throughout southern Ontario and major trips to Oshkosh and Kitty Hawk. Landing at First Flight Field was a tribute to the Wright brothers and a personal spiritual pilgrimage with her flying partner.

Now as she completes her articling year with the Department of Justice and looks forward to being called to the Ontario bar, she looks forward to returning to her first love - flying.

MARCIA LYNN HANE, born March 21, 1947, in Cape Girardeau, MO. She is married to William Lee and has eight children: Leah, MeLanie, Pat, Bill, Mike, Greg, C.L. and P.J. Her grandchildren are: Samantha, Nicholas, Mitchell, Chris, Ben, Kylie, Jen, Madison, Cody, Dillon, Courtnee, David, Jason, Bill, Dan, Tom, Katrina, Tricia, Laura, Dave and two on the way.

She is a graduate of Stephens College in Columbia, MO. Her previous occupations include medical director for acute care facility; consulting director at Menard State Penal Institution (maximum security) and consultant in skilled care facility. Hane's nonflying activities include: interior design, gardening (especially roses), reading, sailing and travel.

Hane has more than 1,000 flight hours and holds a private pilot license with ASEL rating. She joined the Lake Tahoe Chapter 99s in August 1990. She has flown the following planes: Beech A36TC (own N1122B), Glasair RGIII (own N86MH), North American P51D Mustang (own N151X), North American T6 (own N4269P), North American P51D, dual control (own/syndicated N4151D) Beech 19A, Glasair RGII (owned N86MH), C-172, Piper Warrior II, SNJ, Boeing B-17G, C-140. Since 1984 has participated in over 500 airshows/events all over the world, the farthest being in Wanaka, New Zealand.

She is building hours in their T-6 and will soon be flying Ho! Hun! — a P51D Mustang. Since Bill is PIC on B17G *Sentimental Journey* owned by Arizona CAF wing, she would like to pass their flight program and fly right seat to his left seat position (just once!). After she gets through the T-6 program this will be possible.

Hane is a member of ICAS, CAF, VFW, Stephens College Alumni, AMRA, My Sisters Guild, and 99s.

EVANGELIA HANLON, finds the skies friendly. Soloed October 1952, licensed in 1953. Favorite planes: Cessna C-140, C-171 and C-182. She has been a charter member 99s, 1954, California Southwest Section. Participated in the 1958 Powder Puff Derby, San Diego to Charleston, SC, sixteenth place; named Personality and Woman Pilot of the Year. She is also a member of Coast Guard, Bush pilot.

In 1960, flying from Palmilla, Baja, CA, flew into a severe dust storm and had to make a 180 degree turn onto a nearby airstrip in the village of El Rosario in the Tropic of Cancer. When returning home her group of three women pilots went on radio and television asking for aid and medical supplies for the people who lived there. She also wrote a short story called "Dust In My Eyes."

She is one of the originators of the Flying Samaritans, and played Santa for 50 children, while in a cart pulled by a burro, and filled with toys and gifts.

The greatest joys in her life are her children, grandchildren and great-grandchildren.

PATRICIA "TRISH" J. HANNA, is a retired teacher with a BA in English and MAT in English, speech and drama. Flight training includes private pilot license in June 1984; instrument rating, March 1992; Wings Program: Phases 1, 11, 111. Aviation Assoc.: International Women Pilots, 99s, chapter chairman, Oregon Pines Education, air/space; Northwest Section offices: Section News Reporter, Nominating committee, currently; Oregon pilots: Salem Chapter 1983-present, chairman two times, vice chair two times, OPA state award 1988, Salem Airport Days, "Fly with Y" chairman two years, worked on this project several years, 1984-92; Seminar, Willamette University, math and science careers for teen girls; Oregon Air Fair: assisted in children's area, 1992-93. Her special interests include art, theater, needlework and reading. She directed *Oliver* for Salem Boys Choir in October 1994.

CORAL BLOOM-HANSEN, born Feb. 7, 1916, in Alvin, IL. Joined the 99s in 1947, Bay Cities, then Charter Sacramento, now Aloha. Flight training includes 1943 CFI in Ohio and Illinois. She has flown mostly single and multi civilian planes. Served 22 years as department chair Aero at Diablo Valley Col in Concord, CA.

Hansen has two children, Roger Bloom, 44, captain, Northwestern Airlines, and Dan Berryman, 23, new private pilot. After retiring from college teaching moved to Honolulu. Persuaded to come out of retirement to be chief pilot of 141 school at Hawaii Country Club of Air. Retired again 1985. Has more than 23,000 hours. Now flying just for fun!

EDNA L. HANSEN, a lifetime resident of Ohio, was born in Marietta and learned to fly at Carl Keller Field in Port Clinton in 1979. She flies a Cessna 172 XP II which is co-owned by three others. Her main flying destinations, besides the usual rides to and for family and friends, are to sites of meetings of the National School Orchestra Assoc. of which she is executive administrator.

A graduate of Capital University, she was Orchestra director of the Port Clinton City Schools for 26 years. In retirement she escorts student groups for Noteworthy Tours of Sandusky, plays bass in the Firelands Symphony and does substitute teaching.

Edna belongs to the All-Ohio 99s, AOPA, St. John Lutheran Church, Delta Kappa Gamma and other music and education organizations.

She is married to James Hansen and has a son who teaches USAF Navigation in California, a daughter and four grandchildren.

MARY HANSEN, born July 28, 1953, resides in Hay Springs, NE. Married to David and has two children, Christine Hansen Chasek, 25, and Jon David Hansen, 21. Also has one grandchild, Ashley Elizabeth Chasek, born June 15, 1994. Education: Mid Plains Vocational Tech College, 1973, dental roentgenology, L&D Aero Service, 1986, private pilot. September 2, 1987, will always be another delightful date when Mary became owner of a 1974 Cessna 172.

Mary worked for Dr. A.B. De Castro as a dental lab technician until 1981 when she joined her husband in business. They purchased Hay Springs Lumber Company in January 1980, which they still own and operate today.

Mary has served on several Hay Springs City Boards and Organizations in Nebraska and she was elected to the Hay Springs City Council in 1988 for a four year term and reelected to a second term in which she served as president of the city council. Mary was elected mayor of Hay Springs in November 1994.

She has flown Cessnas 150, 172, 182, Piper Arrow, and tried gliders and air balloons.

MARJORIE HANSON, born Aug. 7, 1919, in New Castle, PA. She learned to fly in California and received her SEL on Sept. 25, 1974, while a resident of Orange County, CA. She flies a Beechcraft Bonanza, S35, N14FM. She has participated in air races and flew the Powder Puff Derby Commemorative Flight in July 1977.

Hanson joined the Orange County 99s in September 1974. She served as an officer in that chapter from 1982 to 1985 and was chairman in 1985/86. Since moving permanently to Prescott, AZ, in 1987 she has served on the Prescott Airport Committee from 1988 to 1992, the Special Prescott Airport Taskforce in 1992 and the Prescott Chamber of Commerce Aviation Committee from 1992 to the present (1995). She is currently the secretary/treasurer for the Yavapai Chapter of the 99s.

Marjorie received her undergraduate education at Mount Holyoke College in South Hadley, MA. Her principal graduate training was at Fullerton University in California. She is now retired from an administrative position in Orange County, CA.

Her husband, S. Fred Hanson, is a pilot and they enjoy a coast to coast flying life. They share three children and nine grandchildren.

NANCY LOUISE PEELOR HARALDSON, took a ride in a small plane with her family and loved it. Several weeks later she saw a sign that said, "Sally's Flying School" and she said to her husband, "If Sally can do it – I can do it!" He encouraged her and earned his license after her.

Earned her license Feb. 3, 1973, and joined the Chicago Area Chapter of the 99s. Never did she dream when she took her first fun flight in 1970 at Starved Rock State Park, IL, that her life was to change and take a new direction and that she was going to gain such a wonderful extended family along with a love of flying.

She has enjoyed being an active pilot and member, chairing and working on many events, projects and programs. Working at airshows, airport tours to Boy and Girl Scouts, five years her chapter's International news reporter, chairing two chapter air-meets, three years chapter Ways and Means chairperson, presentations to the Zontas and EAA meetings, helping at flying companion classes at Safety Seminars, presentations at career day workshops in schools and at a teacher workshop, airmarking which is painting numbers on runways.

She holds ASEL, instrument and commercial ratings and has owned a Cherokee 6/300 and since 1981 a Mooney 201. Also, member of AOPA.

Her husband Wayne and children, Lori (John) Diversey, Kelly (John) Cossum, and Kevin Haraldson. Grandchildren: John, Kyra and Ryan Diversey and Taylor and Zachary Cossum.

Her whole family has enjoyed and benefited from the 99s and they became a flying family.

She loves working with children and being a role model (especially for girls) while giving aviation presentations in schools. Her favorite is the "Air Bear Airlines" presentation for kindergarten, first and second grades. Phase 1 is a classroom re-enactment of a flight to Disney World complete with a flight simulator, props and roles for everyone. For five years she has chaired this program and has given it to hundreds of classes and several thousand students over the last eight years. Phase 11 is a trip to the airport plus a walk around and sitting inside a small airplane.

"The 99s have given me extra reasons to fly and keep proficient. Flying to International and Section Conventions, 22 years of speed and proficiency races, Poker Runs, chapter Fly-ins, and Fly-In campouts at airports. I've flown two Grand Prix (2,200 miles) cross country speed races. Another thrill was being part of the 99s opening fly-bys at the Chicago DuPage Air Show in 1993-94 and hopefully 1995!"

ETHELYNE A. HARBY, born May 23, 1927, in Florida. She was 62 in 1991, when she took her first flying lesson. She bought a Cessna Skyline 182 that same year and got her private license. She then completed her instrument training about one year later at Herlong Field, Jacksonville, FL, and has logged more than 500 hours.

Her husband of 47 years, George, does not fly but does enjoy flying with her. She has four children: Ginna, 44; Susan, 42; Hugh, 37; and Brooks, 36; and nine grandchildren. Her oldest grandson is taking lessons in her plane now.

She flew all over the states of Florida and Georgia and flies to her summer house in North Carolina. She has flown to the Bahamas.

She joined the 99s in 1993 when their First Coast Chapter was formed. She is also a member of Aircraft Owner and Pilot's Assoc.

She first had a desire to learn to fly when she was 16 and little did she know that finally at 62 she would fulfill that dream. The thrill and excitement of flying her own plane is unexplainable, only a pilot would know.

ZOAN HARCLERODE, Phoenix 99s, holds an ATPC certificate (AMEL) and a commercial (ASEL). Also, she has her CFI certificate with instrument and multi-engine instructing privileges, as well as advanced ground instructor and instrument instructor certificates. Zoan was appointed an FAA accident prevention counselor in 1992.

She left a successful career in the hospitality industry to become a full time aviation instructor in 1989, and along the way has married an aviation medical examiner and had two little girls, both of whom had an enviable amount of flight time before they were born.

Zoan graduated from Phillips University in 1981 and began flying in 1987. She has more than 2,800 hours, raced in the 1990 Air Race Classic, as well as competing in various proficiency rallies and aerobatic events, also teaching aerobatics. She has been one of the most active members of the Phoenix chapter sine 1987, and has worked very effectively as Flying Companions chairman.

SHERRY BRUNSON HARDIN, born Jan. 13, 1941, Haynesville, LA. Joined the 99s in 1990. Received her flight training in Shreveport and has flown Piper Warrior.

She married David Leon Hardin in 1961; he is also a pilot. They have two children, Jo Ann Hardin Mack, 32, and Amy Hardin Heflin, 26; and grandchild, Madison Taylor Mack, 2.

JOYCE AUTRY HARDING, born July 5, 1937, in North Carolina. Joined the 99s in 1958. Flight training: 1956 Stephens College, Columbia, MO, fixed wing 1964 glider. She has flown the following planes: Cessna 120, 140, 150, 170, 172, 182, 195, Piper Tri-Pacer, Archer, Stearman, Piper Cub, Aeronca Champ, Pitts and several gliders. Her husband Ray is deceased. She has two children, Jeff, 33, and Greg, 31. Also has grandchildren: Jessilynn "J.J." 2 1/2, and Michael, 14 months.

Received her private pilot license at age 19, commercial at age 20. participated in two NIFAs, several Poker Runs and local races. Worked on many Powder Puff Derbies on both coasts. Has flown all over US, Caribbean and parts of Mexico and Austria. Harding has more than 2,000 hours of flight time. Member of 99s, NZ Air Women's Assoc., Museum of Flight Seattle, and Zell AM Zee (Austria) Soaring Club.

ELLEN HARDWICK, born Oct. 4, 1948, in Johnson City, TN. She learned to fly in 1986, in Knoxville. She earned BS and MS in education at the University of Tennessee and taught elementary and middle-school math for 17 years. In 1992 she became an Aerospace Education Specialist for NASA, working at the Goddard Space Flight Center, in Greenbelt, MD. In her work she travels in 11 states, doing teacher workshops using NASA's mission to demonstrate the integrated applications of science, mathematics and technology. Her professional highlights include facilitating a telephone downlink from the Space Shuttle to a classroom in the Bronx, appearing as a presenter on a live interactive video-conference for teachers and flying in "microgravity" on the KC-135. Ellen is a Captain in the Civil Air Patrol and has served as the External Aerospace Education officer for the Tennessee Wing and the National Capitol Wing.

KAREN HARKER, Southwest Section, Phoenix Chapter 99s, (maiden name Sommerstedt) was born in Anaheim, CA, on Feb. 5, 1963. She learned to fly in Hawaii, and soloed in Hilo - General Lyman Field, at age 20. Karen would fly while on "shore leave" from the cruise ship SS *Constitution,* which she worked aboard.

In 1984 she became employed with Ozark Airlines as a flight attendant and thereafter with Western Airlines (which is now Delta). Karen completed flight training in Salt Lake City, UT, where she became a certified flight instructor, CFII and MEI. She met her husband, David who is also a pilot, in Salt Lake City.

Karen has logged time in 35 types of aircraft from gliders to seaplanes to the Learjet. Karen currently works for Delta Airlines as a flight attendant, based in Los Angeles, CA. She also flies part-time for Cutter Aviation, Phoenix, AZ. Karen has been a member of the 99s for 10 years.

VIRGINIA HARMER, born in 1948, Orange County, CA. Joined the 99s in 1991. Flight training includes private pilot license and instrument rating. She has flown Warriors, Archers and now owns and flies a 1983 Archer II. She married Dennis in 1968; he is also a private pilot since 1993. They have two children, Kerry, 25, married to James, both love to fly with them; and Mike, 21, will fly if he can ride right seat. As a member of 99s and AOPA, she sees aviation education as the most important tool in expanding their aviation community. She loves to share her aviation experiences with children and adults with whom she comes into contact. She helps instruct boys and girls groups in attaining their aviation badges. She and her husband enjoy flying cross-country and entering pilotage competitions. They also love cruising around southern California. She also enjoys helping with Flying Companion Seminars, Air Fairs, Pacific Coast Intercollegiate Flying Assoc. (PCIFA) among other aviation activities.

BARBARA L. HARPER, born Nov. 25, in San Diego, CA. Joined the 99s in 1970. She served contractual NAFI-Air Force from 1975 to 1984, T-34 and T-41. Received her flight training at Petaluma Sky Ranch, CA, and Southwest Skyways Torrance, CA. She has flown the following planes: Cessna Series, Piper Series, Beechcraft most, DC-4 and C-119L. Employed with Continental Airlines B-727 Flight Engineer. Married to Dale and has two children, Tora, 20, and Darci, 30; and grandchild, Emilie, 1.

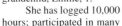

She has logged 10,000 hours; participated in many 99 Air Races. Former member of the board of directors Pima Air Museum, Tucson; former chairman Aviation Hall of Fame; current member and former commander of Pima County Sheriff Aero Squadron. Assistant stand/eval Civil Air Patrol. Member Tucson Airport Authority; past treasurer World Aviation, Education and Safety Congress, 1986. Executive board 1994 World Aviation, Education and Safety Congress. Previous member Arizona-Mexico Commission. Educational background: BA and graduate studies archaeology, University of Arizona, Professional Member Arizona Archaeological and Historical Society, Wyoming Archaeological Society. Consultant Junior Achievement and FAA accident prevention counselor.

SUE HARPER, Phoenix 99s, got married to pediatrician MD, pilot, Bill in 1953. Three children later, she soled at Saguaro Aviation at Sky Harbor Airport, Phoenix. She got her private ASEL in July 1965, and totally astonished a CFI by getting checked out in a 172 and having her last kid that October. In her spare time that year she joined the 99s, got her commercial, instrument and multi and acquired part interest in a Turbo 210.

As chairman of the Phoenix chapter in the 1971-72 era, Sue was a member of the Phoenix City Council, Aviation Advisory Board and the Arizona Pilots Assoc. She was the main pilot flying the Episcopal bishop around to reduce his travel time to all his outlying parishes.

She enjoyed taking trips with the family. A favorite memory is of taking the whole gang to the east coast, and getting to be PIC the whole way, with little or no "help" from Bill.

Two of her tribe went on to be pilots. The youngest, George, doesn't fly but is in the Navy working on avionics, while Kendall, the other escapee, is married to an aerospace engineer. She has one grandchild, and three on the way, which gives her a chance to go visiting to San Diego, Tulsa, Falls Church and Elgin, IL.

Another memory – FBOs at Sky Harbor cringing when she walked through the door because they knew she would hit them up for something for 99s. She recalls with great love that they always came through for her. She treasures the friends she's made in 99s. There are still several she keeps in touch with, and says "friends are always valuable."

SUSAN K. HARPER, began her first flying lessons when she was 19 years old from her airline pilot father. Her brother is a pilot as well, plus her mother and sister also took flying lessons. Susan didn't take flying seriously until after her divorce, and received her license in 1989. She joined 99s in Colorado and two years later transferred to Idaho where she changed her focus from an airline flying career to private ownership and all the activities it presents such as camping, long cross country flights, and the camaraderie enjoyed by flying with other aircraft owners. She flies a 1947 C-140 which she is extremely proud to own. Susan has a bachelor's degree from the University of Southern Mississippi and supports her flying as a human resource specialist for the Federal Government.

SYLVIA I. HARPER, Received private pilot license in Bryan, TX, in 1949.

Received commercial rating, multi-engine land and instruments at Aviation Training Enterprises, Inc. in Chicago, IL, in 1973.

Flies a Beechcraft Bonanza N7209E F33A with 300 hp conversion.

Joined the 99s about 1968. Belongs to the Colorado Chapter.

ELLEN MARIE NOBLES-HARRIS, and her husband Jerry Harris started flying in 1989 after a month long trip to Alaska. Both got their private certificates and a Cessna 172 in 1991. They both have their instrument ratings and Ellen is now working on her commercial and hopes to get her flight instructor certificate. Their main flying activities include flying some place for lunch.

Ellen joined the Delaware Chapter in 1991 and immediately became chapter treasurer and eventually their airmarking chair. In 1994 she became the International airmarking chair and the following year, Mid-Atlantic Section treasurer. She works with Anne Eriksen to make needlepoint airplanes for their chapter to sell.

She was born in Scarsdale, NY, and lived in Massachusetts, Kansas and Mississippi while she was growing up. She grew up sailing and participating in Girl Scout activities. She graduated with a BE in chemical engineering from Vanderbilt University and is a licensed professional engineer. She currently works for Mobil Oil as a project engineer and was the first women engineer in any of Mobil's US refineries in 1971. She has been active in the American Institute of Chemical Engineers, local political activities and served on the local Municipal Utilities Authority.

Ellen has a brother and sister and one step-daughter.

LAURIE SMET-HARRIS, born and raised in Kenosha, WI. She learned to fly in 1987. After obtaining her CFIAI she worked at Waukegan Aero Gateway Technical College and presently at Kenosha Aero. Smet-Harris is currently building time flying freight in a BE-18 and flying a C-340 corporate.

Smet-Harris' educational background include a degree in aeronautical science with Presidential Honor's and Dean's List. She is also an advanced ground instructor.

Smet-Harris joined the 99s in 1988. In 1993 she joined Illi-nines Air Derby board of directors representing the Aux Plaines Chapter. She is the NIFA chairman for her chapter. She also belongs to NAFA, AOPA and EAA. Smet-Harris participates in pinch-hitter seminars and career days.

She is married to David Harris who is a captain for United Airlines. They have a 2-year-old daughter Ellyn Rhiannon.

MARY H. KOKO-HARRIS, wife of John and mother of Kate, Charles and Heather, flies out of Wiscasset, ME. A white knuckle flier in both commercial and private planes, Koko took her first lesson in January 1992 and received her private pilot certificate in August 1992. It was a real victory for her as in the ten years of flying with her husband the only control she had mastered was cabin heat. Koko credits her success to the patience of her instructor, Perry Neilson. Koko lives in Newcastle, ME, where she and her husband own and operate Country Farm Furniture Store, a retail furniture business. Flying is a wonderful way to visit the family who are scattered in Vermont, New York and New Hampshire. Koko is presently working on an instrument license. The membership and scholarship chairman of the Katahdin Wings Chapter of the 99s, Koko is trying to figure a way to fly more and work less.

RITA M. HARROLD, private pilot, SEL (1983), instrument (1991), with more than 500 hours, Rita flies Cessna 172s and a Piper Warrior out of Mustang Aviation, Essex County Airport, Caldwell, NJ. A member of the 99s since 1983, she is past secretary of the North Jersey Chapter.

After a busy business week in New York or traveling around the country as director of Educational and Technical Development for the Illuminating Engineering Society of North America, Rita finds flying, despite its challenges, offers relaxation, and a sense of "getting away from it all." It helps put a new perspective on life at 3,000 or so feet.

"Thanks are owed to a friend, who, almost 20 years ago, let me share the skies from the right seat of a 1940s Aeronca Chief and find out what a wonderful activity this could be."

In 1995 she hopes to start working on a helicopter rating.

CAROLYN ZAPATA-HARSHBARGER, Golden West, 99s, Inc., born Sept. 13, 1930, San Francisco, CA. Joined the 99s in 1973. She has lived all her life in the San Francisco Bay area. Her flying career began in 1972 when she received her private pilot certificate. Since then Carolyn has acquired more than 11,500 hours flight time, serving as chief pilot/chief flight instructor at San Carlos Airport, CA.

Having chaired all committees and held all offices in the Golden West Chapter, she also found time to fly the Powder Puff Derby and Hayward-Las Vegas Air Races. Carolyn and her husband, Orville, own a C-172 and a C-310.

She has five daughters, 13 grandchildren and one great-granddaughter. In addition to raising her family, Carolyn has been an FAA Safety Counselor, taught Aeronautics at the College of San Mateo, was active in Civil Air Patrol and the Girl Scouts, serves on the Hayward-Las Vegas Air Race committee and as council member of Grace Lutheran Church, Hayward.

She has flown the following planes: Cessnas 152, 172, 177, 206, 210, 310; Piper PA 28-150; Grumman AA5A, B, BE 33, 35, T-34.

VIVIAN HARSHBARGER, Bay Cities Chapter, decided to learn to fly because she was afraid of commercial airplanes. She had never been in a small plane and didn't know anyone who had. That was in 1969 on her 47th birthday. Her daughter pointed out that she couldn't do this because she was a grandmother, and her employer could only see her imminent demise. She now has commercial, CFI, and instrument written passed.

She joined 99s in 1970, and entered her first air race in 1971. Placing 18th caused immediate addiction, and she's since entered about 14 races. Her most memorable experience with 99s was as chairman of 1977 International Convention which took four years to plan. It was an exciting, enlightening, interesting undertaking.

EDWEENA DODGE HARTLEY, spent 22 years living in Anchorage, AK. She flew throughout Alaska with husband Atlee R. Dodge, who was killed while flying a Beaver airplane. She acquired a new fear of flying. After marrying another pilot, Bill Hartley, and moving to Seattle, WA, she decided to take flying lessons to alleviate her fear of flying. She got her fixed wing rating in 1986 and went on to get her helicopter rating in 1988. She currently flies a Bell Jet Ranger helicopter. She is #622 of the Whirly-Girls and is currently serving as secretary of the executive board. She and Bill reside in Auburn, WA, on a private air ranch. They spend their winters on a private air ranch in southern California. They have three children and two grandchildren. In addition to her love of flying, Edweena enjoys motorhoming, calligraphy and rubber stamps and antiques.

MARIAN BURKE HARTLEY, born Jan. 1, 1936, in Atlanta, GA. Employment as secretary included a merchant marine firm, customs brokerage, and FAA. In 1957 she married Dan Hartley. They moved to Seattle, WA, in 1961; their family consists of four children and six grandchildren.

Marian's community activities have included Camp Fire leader, Sunday School teacher, and MOPS teaching leader. She earned her pilot certificate in 1987, is a member of AOPA, and joined Greater Seattle Chapter of 99s in 1989, where she has served as secretary and newsletter editor, and also enjoys airmarkings, Young Eagles flights, and helping with Flying Companion Seminars.

Marian and 49 1/2 Dan currently own a Cherokee 6, and enjoy flying to visit relatives and friends across the country, as well as flying in the local area.

MAY E. HASKELL, retired Civil Service/Fallon Airmotive. She is married to Richard and has children: Mike, Liz, Loretta and Tom. Haskell's employment history includes social security service representative and veterinary technician. She joined the 99s in 1972 and the Reno Chapter in 1973. She has also been a member of the following chapters: Monterey Bay, Lake Tahoe and Placer Gold and held the offices of secretary, chairman and various committees. She became involved in aviation in 1944 – her father was a pilot and she has always been around airports. Haskell holds a private pilot license and owns a 1964 C-172 and 1947 Aeronca Champion. Fly-In Chairman LVPA, most of her family is aviation oriented and that is her main hobby. Memberships: Lahontan Valley Pilots Assoc., Truckee Tahoe Plane Talkers, Silver Wings and Kowina Anna.

MARDELL MAIN HASKINS, oldest of identical twin girls was born June 24, 1938, in Lindsay, CA.

From age 15 Haskins dreamed of racing airplanes. An avid racer with more than 1,000 hours, she has participated in ARC, Great Southern, Palms to Pines, Pacific, Shamrock, Fullerton, Rialto, Back to Basic, Great Pumpkin, Denver and Valley Derby. Mardell's dreams were more than realized when navigating with Esther Grupenhagen they placed first in 1990, 1991 ARC and second in 1992. She also has over 20 other first or top 10 finishes.

A graduate of Virginia Intermont College, Bristol, VA, and CalPoly, Pomona, CA, she has a BS in public administration and political science. Community activities include semi-pro softball and other sports, Scouting, church organist, musician in ensembles, symphonic and dance orchestras.

A heavy equipment operator in southern California, Haskins occasionally flew to job locations or escorted students to son's Wilderness Survival School in remote Utah mountains. A 30-year construction career ended when she was severely injured November 1992 while oiling on a Drill Rig. Currently in therapy from seventh surgery, with more planned in near future.

Memberships include Women in Construction, AOPA, ARC, Florida Race Pilots Assoc., and Moapa Valley Art Guild with numerous awards in art, sewing, sports and only achievement award ever given by Inland 99s. Currently Scrapbook chairman for Southwest Section, Inland chapter and ARC.

After husband Fred's retirement from construction they moved to their cabin at Lake Meade, NV. They have eight children, 25 grandchildren and one great-grandchild.

MARY HASLIP, competed in the first Women's Air Race in 1992. Having been taught pylon racing by her husband Jimmy Haslip, she became an active racer with her first race in 1930. In 10 days at the 1931 Cleveland Air Races she successfully flew six high Performance airplanes despite the fact that she was unfamiliar with them. In 1932 Mary broke the women's speed record in a Wedell-Williams (Harry Williams hated the notion of women flying his airplane and she had only 15 minutes to get familiar with the plane). This record stood for seven years.

Mary was a test pilot for Spartan Aircraft Co., the American Eagle and Buhl Aircraft Company. In 1982 Mary was the first women inducted into the Oklahoma Air Space Hall of Fame at Kirkpatric Center Oklahoma City. Her husband was also inducted at the same ceremonies held on Dec. 17, Kitty Hawk day. In 1985 Mary was guest speaker at a reception at the National Air and Space Museum, sponsored by Bendix Aerospace Company. She was honored as an Outstanding Pioneer Woman Pilot by the OX5 Aviation Pioneers. Mary is a member of the Los Angeles 99s.

CECILE HATFIELD, became a member of the 99s in 1964 and has held office in the Gold Coast Chapter, Florida. She is a licensed pilot and ground school instructor. Cecile served as president of the Florida Women Pilot's Assoc., and also as chairperson for the "Angel Derby," all Woman International Air Race. After organizing and conducting air races for many years, Cecile went to Law School in 1972. In 1991 she opened her own aviation law practice in Miami, FL. Cecile served as associate general counsel for Piper Aircraft Corporation 1987-88, and was with the US Department of Justice for nine years as an aviation trial lawyer.

Cecile graduated from the University of Florida, where she was elected to the Hall of Fame and Mortar Board.

Received her law degree from the University of Miami in 1975. She served as chairman of the Aviation and Space Law Committee of the American Bar Assoc.; is a member, director, Florida State chairman and contributing editor of the Lawyer-Pilots Bar Assoc. She has lectured and written articles on aviation litigation for the American Bar Assoc., Lawyer Pilots Bar Journal, SMU's Journal of Air Law and Commerce, Embry Riddle Aeronautical University, and Lloyds of London Aviation Symposium. In 1986 ERAU chose Cecile to serve as chair of its Annual Aviation Law/Insurance Symposium. She continued to serve in that position.

Cecile is a member of the Royal Aeronautical Society, London, England, and was an honoree in 1993 at the 17th annual celebration of the International Forest of Friendship, Atchison, KS, for her contributions to general aviation. In 1994 she was chosen to be the general counsel of the 99s, Inc., International Women Pilots.

DOROTHY J. HAUPT, took her first flight Oct. 2, 1959, in a Piper J-3 and soloed three months later on Jan. 2, 1960. Then, she was hooked. Received her airman certificate flying a Tri-pacer 125. In April 1960 they purchased a Piper Tri-pacer 160 and was flying Piper J-9 and PA 12. In November she took part in her first air race with copilot Mickey Clark in her Tri-pacer. March 1961 they purchased a Cessna 175 and during the next three years she entered many air races, including the Sky-Lady Derby, St. Louis Air Races and the Michigan Small Race. They raced all over the country and came in anywhere from first place to "tail-end tony."

With their base at Walston Aviation in Alton, IL, she had the opportunity to ferry many planes from the Cessna factory and other designations and was able to experience many types of aircraft.

She's always enjoyed meeting people from various walks of life and is proud of her participation in the 99s, Inc., where she's been a member for 35 years.

KAREN HAUSTEEN, Phoenix 99s who got her private ASEL in Phoenix in 1989, was introduced to 99s by her flight instructor, Nancy Rogers, and joined Nancy in being a member. She has served two terms as treasurer, but can't fit as much flying time as she wants into her very busy schedule right now, and really values participating in 99 functions to stay in touch.

Karen is the happily divorced mother of three teenage sons and she says they're "terrific," all taller than she, and not interested much in aviation – yet! She is an RN working in the specialty of hand surgery, and as an RNFA in the op room. She loves to see the surprise on faces when people learn she is a pilot because she sees it encourages them to think they can do something that exciting too.

BARBARA HAVENS, Santa Clara Valley Chapter, learned to fly at Palo Alto Airport in 1991 and has logged 300 hours. She is a member of the Sundance Flying Club at Palo Alto. Barbara joined the Santa Clara Valley 99s in 1991 and has been active in airmarking. She is the planning manager at Watkins-Johnson, a semiconductor company in Scotts Valley. Barbara is active with the Navy Reserves.

LUCY THELMA HAVICE, born March 25, 1925, in McAllen, TX. She started flying lessons in late 1968 when living in Ann Arbor, MI. Her first solo flight was in a Cherokee 140 on March 17, 1969, (St. Patrick's Day). The next day she met Bernice Steadman (99s President 1969-70) who introduced her to the 99s, making her a 66 ("upside down 99").

After moving to Bartlett, IL, she received her private pilot certificate with an ASEL rating on Feb. 21, 1971, in a Cherokee 140. Helen Sailer sponsored her to join the 99s and the Chicago Area Chapter. Later, she moved to Louisville, KY. She checked out in a Cessna 150 there, and moved on to Carmel, IN; Costa Mesa, CA (Orange County Chapter); Albuquerque, NM; and now resides in Austin, TX.

Thelma has served as chapter vice-chairman, secretary and membership chairman. She has attended several section meetings and International Conventions.

She is married to Jim Havice and they have three sons: Robert, Michael and Philip; and six grandchildren: Kelly, Michelle, Scott, Pandora, Adam, and Elizabeth.

Thelma's hobbies are golf, oil painting, bowling and collecting silver souvenir spoons.

Thelma loves the 99s and supports all its principles and endeavors with great interest and enthusiasm.

MARIAN CYRIL HAVILAND, born Feb. 13, 1951, in Brooklyn, NY. She grew up in Anchorage, AK, the second daughter of an electrical engineer who built radio-controlled model aircraft; with the help of her mother, a chemist, musician and teacher, together they took the family to numerous airshows, hoping he could learn to fly and perhaps someday rebuild an antique airplane. Although he died before realizing his dream, Marian moved to California and after studying architecture and art, she decided to explore aviation.

After earning her private pilot certificate in 1988, Marian went on to become an instrument rated pilot, and has also received her commercial pilot certificate in both single engine and multi-engine aircraft. Marian is an advanced ground instructor as well and is currently working on becoming a flight instructor.

While being employed as a legal secretary to be able to afford to continue flying, Marian is currently nearing the completion of her degree in professional aeronautics through Embry-Riddle Aeronautical University's Moffett Center campus center with a minor in Aviation Business Administration. Marian is currently a member of the 99s, Inc., has assisted with simulator training for private, commercial and instrument students at Foothill College in Los Altos, is a member of AOPA (Aircraft Owners and Pilots Assoc.) and has attended meetings of Bay Area Black Pilots Assoc. Marian is a participant in the FAA's "Wings" Program (Phase VII in Progress) and she has most recently been working on a new group of paintings and graphics involving airplanes and aviation for upcoming shows.

KATHERINE "KIT" HAWKINS, born Waco, TX. Raised in Kettering, OH, and lives in New York, NY. Received her BA in psychology, anthropology and sociology, from University of Texas in Austin. She is the owner/producer of Harmony Productions (children's film production company, including animation); and owner/baker, Sweet Cakes (chocolate desserts). Executive board member ASIFA East (International professional animation society) and director/young mensans coordinator/scholarship chair, Greater New York Mensa.

Feeling as if she has been reincarnated without wings, she always wanted to fly. She began flying while attending UT at a small, grass-strip airport outside of Austin, paying for lessons by washing airplanes and baking for restaurants. Completing her license in Dayton, OH, (birthplace of aviation) in 1981, she continues to fly at Moraine (OH) Airport and the historic Stormville (NY) Airport. She loves to share the sky with her family and friends and encourages all women to taste the freedom of flight.

RUTH HAWKS, All Ohio Chapter, born Dec. 22, 1941. She joined the 99s in 1991. Learned to fly because of her husband's interest in flying. They own a Cherokee 140 and have their own grass strip at their residence.

LOIS C. HAWLEY, born Aug. 28, 1924, in Indianapolis, IN. She started lessons in 1943 but was unable to continue lessons until 23 years later, obtaining her license on Aug. 24, 1967, and an instrument rating May 1975. She was co-owner in a 1960 Cessna 175 and then later co-owner of a 1964 Cessna 182 Skylane. Joining the 99s, Inc. in October 1968, she has been active in her Indiana Chapter, was elected chairman for the years 1994-96 and received the North Central Section Governor's Service Award in 1994. She has logged 982 hours. She was copilot in the 1976 AWTAR and in many FAIR (Fairladies Annual Indiana Race) races, serving on the FAIR board and becoming chairman in 1978. Lois co-chaired the North Central Section meeting held in Indianapolis in 1989. She has attended nearly all the section meetings and International 99s Conventions. One of the most memorable was flying her Cessna 182 with her friend and co-owner to Vancouver, BC, in 1987. Lois is a member of AOPA, PIA and a charter member of Cessna Skylane Society.

MIRIAM P. HAWORTH, private, SEL, joined the Oklahoma Chapter in 1965, two months after receiving her license. Transferring to the Nebraska Chapter in 1966, activities have included cofounder and editor for the newsletter *Chatter Frequency*. Chairman of the 1968 Section Committee; public relations chairman; co-chairman Nebraska Air Race; head timer and assistant of operations AWTAR; chapter chairman, two years; initiated the Nebraska Flying Poker Game and "football rides" for pilots. She sponsored the University of Nebraska Red Barons, active in NIFA. She was copilot in the Nebraska Race, Skylady Derby, and Illinines Race. Aviation awards included second place Chapter Achievement Award, two years and Governor's Chapter Achievement Award. Mimi is the mother of three children, holds a Ph.D., and is a psychologist for the Nebraska Department of Correctional Services.

SARA PAYNE HAYDEN, born Aug. 29, 1919, in Granite Falls, NC. Secretary/bookkeeper in Charlotte, NC, when WWII started. One night she saw a newsreel showing women ferrying airplanes, and decided that was what she wanted to do. She borrowed $200 to learn to fly and get the 35 hours required to apply for WASP training. The wings were earned with Class 44-W-10, "The Lost Last Class."

After the WASPs were disbanded in December 1944, Sara got her commercial rating and flight instructor rating and continued flying. She joined the Carolina's Chapter of the 99s in 1946, the US Air Force Reserve in 1949, and was called to active duty in 1951 as a WAC/WAF recruiting officer. She married Dr. Frank Hayden in 1953 and left the Air Force.

Sara has been an active member of the New England Section and Eastern New England Chapter of the 99s, as well as earlier in North Carolina. She has held several offices, served as legislative chair, and flew in Transcontinental Air Race and several derbies. She received the New England Section Honor Award in 1994.

Sara is a past vice president of the WASP organization, a founder of the Women Military Aviators, past president, and remains their clerk of the corporation.

She is past commander of the Methuen Women's Post 417, American Legion, life member, post adjutant and Boys State chairman.

Sara continues to speak to various groups to promote women in aviation, both civilian and military, and is the Veterans Affairs person from the WASPs and area women veterans.

The 99s were the source of the women pilots available for military flying and training. After the WASPs disbanded, they were not well organized as a group until 20 years later in 1964. Many of them belonged to the 99s; and at the convention in Cincinnati in 1964, WASP Marty Wyall requested permission to call the WASPs together. Some 80 of them were there, the beginning of the round-up. The 99s are number one in their contributions to women in aviation.

BERNICE "BEE" FALK HAYDU, born Dec. 15, 1920, Bradley Beach, NJ. Learned flying, Martins Creek, PA, 1943. Joined WASP February 1944. Active duty Pecos AAF, Pecos, TX, as utility, engineering test pilot. After WASP, obtained flight instructor, ground school ratings. Ferrying and Cessna Dealer under own business "Garden State Airways." First year sold 20 aircraft. In 1947 joined veterans group operating flight school and instructed in Iselin, NJ Airport closed because Garden State Parkway construction eliminated one runway. Attempted "show biz," taping sample radio programs under name "Wings Over NJ" Unable to locate sponsor. Became buyer of aircraft and parts for Indamer Corp., NYC, offices for India company. In 1951 married Joseph Haydu, WWII flight instructor. Children: Joseph, 42; Steve, 41; and Diana Potter, 40. Also grandchildren: Katheryn, 7; Kristen, 5 1/2; and Sara Marie, 6 months.

They owned and flew various aircraft throughout their marriage, many from WWII. She has participated in two Powder Puff Derbies, and one Angel Derby. As of 1995 still flying. An organizer and historian of Women Military Aviators Inc. (current women military aviators). WASP uniform on display in Smithsonian.

She has flown the following planes: Taylorcraft, Piper J3, Waco RNF, Piper J5, Stearman PT17, AT6, BT13, AT17, UC78, Fairchild PT19, PT23, Aeronca Champion, Ercoupe, Stinson, Cessna 120 and 140, Fairchild PT26, Cessna 172 and 310, Beech D18, Meyers 200, Cherokee 140, Comanche 260, Beech Baron, Beech Mentor T34, Cessna 210L, Beech Debonaire Bonanza F33A, YAK 52.

TOMMY HAYES, born March 3, 1926, in Ft. Worth, TX. Obtained her pilot license in 1974 and flew Cessna 182s and her own Piper Comanche, logging more than 325 hours, until, widowed, she had to sell her plane.

She has been a member of the 99s since 1974 and served as membership nominating, vice-chairman, alternate delegate 1978 Australia International Convention, and Pacific Air Race Chairman 1982. She is a member of the San Diego Aero-Space Museum and participated in the Powder Puff Derby 1977 Commemorative Flight.

Education: San Diego Evening College, AA in English; San Diego State College, BS in computer management; National University, MBA and JD (law degree); and continuing studies in seminary classes.

Community activities include: Girl Scouts (40 years) leader, neighborhood chairman, volunteer adult trainer, board member, member finance committee; Kaiser Hospice Volunteer, earned Star Award 1993; San Diego Blood Bank, Three Gallon Club; Hospital "On-Call Clergy; and American Business Woman's Assoc., named Woman of the Year 1977.

She has two daughters and two grandsons, one by each daughter: Emily Ann Hayes Ward, 49; Gerrit James, 19; and Lou Edra Hayes Woodruff, 47; Cord Lee, 28. After 27 years, she retired from a banking career to work as office manager of an engineering firm before returning to her law studies. Now working as court services clerk in the San Diego Superior Courts, she is a church elder, sings in choir, plays in the Bell Choir and is a member of the Mustard Seed Players.

LINDA CAMPBELL HAYNES, born June 22, 1951, in Detroit, MI. Joined the 99s in January 1988. Inspired by her father's love of flying and airplanes, decided to learn to fly. She received her pilot license in May 1987. She has logged most of her hours in 152 Cessna's but has had the opportunity to fly a DC3 and a NASA T-34 trainer. Linda, after learning of the 99s during Oshkosh 1987, joined the Michigan Chapter in January 1988. She has held several offices since that time, including Scrap Book Organizer (which took a first at North Central Section) and secretary for two terms. She has also served on the Michigan Small Race Committee.

Linda's educational background includes an elementary education degree from Western Michigan University and a MA in library science from the University of Michigan. She is currently employed by the Troy School District as a media specialist.

Other outside interests include volunteering for Alpha Phi Sorority, activities at the First United Methodist Church, Bible Study Fellowship, golfing, scuba diving, skiing, reading, antiquating and travel. She has recently traveled to Belize, Costa Rica, Mexico, the British Virgin Islands, Europe, the British Isles, and throughout the US and Canada.

Linda is married to John Milo Haynes who is also a pilot and a financial manager for Volkswagen of America.

WYN HAYWARD, was able to pursue her life long ambition of flying when she received her private license in 1972 in a Cessna 150. She progressed to a Cessna 172 and 182. In 1974 she obtained her commercial rating in a Beechcraft Debonaire.

Wyn has been a member of the Tucson 99s since 1972 and has been APT for 20 years. She has held the position of vice chairman for two years and various other chairmanships, including co-chairman of the 1986 Southwest Spring Sectional.

She was first introduced to racing by driving ski boats in the National Speed Boat and Water Ski Assoc. Sanction Races. She has carried this love for racing over into flying. Wyn has flown one All Women's Trans-Continental Air Race (Powder Puff Derby), three Pacific Air Races, His & Hers Air Race and Tucson Treasure Hunt. She was awarded the Barry Goldwater trophy for first place in the Kachina Doll Air Race. She took first place in the Shirts & Skirts Air Race and has flown in the Shamrock Air Derby, receiving one second place and two third places. Wyn developed a desire for aerobatic flying, learning basic aerobatics in a Citabria. Her more serious aerobatic flying was done in a Christen Eagle, where she received her aerobatic patches with stars in Basic & Sportsman Competition. She participated in the annual Eagle Exchange fly-in at Laughlin, NV, being the only woman piloting her own Eagle.

She is an active member of the IAC, EAA, Southern Arizona Aerobatic Club, Phoenix Aerobatic Club, Eagle Exchange and World Beechcraft Society.

Wyn and her husband, Jason, take turns flying their Beechcraft Bonanza A-36 and their two Christen Eagles.

BILLIE LOUISE HEAD, 99s Houston Chapter, born Feb. 4, 1920, in Bay City, TX. Joined the 99s in 1986. Served in Women's Army Corps during WWII, attached Army Air Corps, retired T/Sgt. from AF Res. 1970. Flight training includes private license SEL. Started in 1984 at age 64, check ride in 1986 at age of 66. She has flown C-152, C-172, Piper 140, and Citabria T-6.

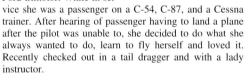

Loved planes all of her life, flew first time in the 20s in an open plane with barnstormers. Flew in other planes as a passenger including a Ford trimotor. While in service she was a passenger on a C-54, C-87, and a Cessna trainer. After hearing of passenger having to land a plane after the pilot was unable to, she decided to do what she always wanted to do, learn to fly herself and loved it. Recently checked out in a tail dragger and with a lady instructor.

HEATHER HEAPS, realized a life-long dream to fly shortly after she moved from Canada to San Luis Obispo, CA, in 1988. On June 28, 1993, her 43rd birthday, Heather attained her private pilot license. She is currently rated in the Cherokee Warrior and Archer plus the Cessna 152 and 172.

Heather is an active member of the Santa Maria Valley Chapter. She is currently the chapter secretary, chairs various committees and looks forward to becoming the chairman soon. Heather is also a member of AOPA, the Los Padres Pilots Assoc., Cal Coast Flyers and is a board member of the San Louis Obispo Pilot's Assoc.

Heather has always found flying to be exciting and finds new tales to tell from each flight. She rarely misses flying to the SW Section meetings and recently experienced quite an adventure on her first Palms to Pines air race in August 1994.

LENORA ASH HEATHMAN, Kansas 99s, born West Virginia, has lived most of her life in Wichita, KS. Encouraged by her husband, Jack, to learn to land their airplane, she began flying training April 28, 1954, soloed in a TriPacer May 26 and passed the private pilot license check ride June 24, 1954.

She acted as family co-pilot in their Beech Bonanza until their two sons, Kim and Michael, and daughter Sherry, became enthusiastic about flying. Each one learned to fly and soloed on his or her 16th birthday. Daughter, Sherry, has become an aerospace engineer.

In December 1975 Lenora followed her husband's example by earning an instrument rating and in June 1976, the commercial rating. She and her husband have enjoyed many years of flying on business and pleasure trips.

She has been a member of the Kansas Chapter of the 99s since 1975 serving as chapter chairman in 1982-83, and continues to enjoy belonging to the 99s.

FRANCES MARIE HEAVERLO, Greater Seattle Chapter, started flying in fall of 1971. Joined the 99s in July 1973. Served as scrapbook chairman, news reporter, nominating committees, Air Age Education chairman. With WWII Navy Pilot husband owned a Piper Cherokee 180. Made numerous flights to Mexico, the western half of the US, with fellow 99, Kay Sterns and husband, to the Dominican Republic for the 500th Anniversary of Discovery of the New World by Columbus. Additionally, a jaunt through Guatemala, Costa Rica and the Yucatan.

Presently, president of Adworks Co., Inc. A past president of the Women's University Club of Seattle. Was active in Alpha Delta Pi Sorority serving in the various advisor capacities with the active chapter. Has been active in Children's Hospital Guild and on the Seattle Opera Guild Board.

Born and raised in Seattle, attended the University of Washington. Was a ballet dancer and teacher and still is modeling and doing TV commercials.

The true joy of belonging to the 99s comes from the association of the many friends, one is privileged to share experiences with. One such experience was at their chapter Blakely Island Fund Raiser, which was a basket social, at which the baskets were auctioned to the highest bidder and the winning bidder won the pleasure of lunching with the basket's preparer. Her basket was last to be auctioned (her husband being the auctioneer). One of the members wanted to buy her basket for him unbeknownst to either of them, and because of his special diet, Heaverlo was as determined as she was, to be the winner. They both kept raising the bid until lunch was about $105, at which point the truth came out and a good laugh was had by all.

NANCY HECKSEL, born Dec. 12, 1934. Her love for aviation is a family affair, her three brothers were career US Air Force.

She learned to fly in Florida in 1967. She joined the Florida Space Port Chapter and became a 99 in 1968, a life member since 1984. A move to Michigan in 1969 prompted transfer to the Michigan Chapter. She is a charter member of Greater Detroit Area Chapter. Active in chapter activities, she has held all elective offices at chapter level. She has been chairman for the GDAC Pinch Hitter, hospitality chairman and debriefer. Nancy has been International North Central 49 1/2 chairman since 1983.

She has flown in numerous air races in the Midwest, has been chief timer for AWTAR, timer for Air Race Classic, judged at numerous NIFA SAFECONs, and chief teller at several International Conventions.

A charter and life member of MAPA, she helped organize fly-ins and attended MAPA conventions for the last 20 years. Member of Silver Wings.

Her community has benefited by her work with Girl and Boy Scouts as a leader and advisor for aviation activities. She also participated in the Fly High for Dystrophy, International Book Project and numerous FAA Safety Seminars.

Her husband Warren is also a pilot. Daughter Linda and grandson Aaron are willing passengers in their Mooney (N99NH). Son, Phillip is a SEL, instrument rated pilot.

JESSICA WIMMERS HEDGES, 99 member since 1963, life member North Central Section, Scioto Valley Chapter. In 1960 solo in Aruba, Netherlands Antilles. In 1963, American pilot certificates in Phoenix, AZ. From 1965-80, chief pilot and charter pilot at Sky Harbor Airport, Phoenix, AZ. Served 15 years FAA pilot examiner until leaving for Columbus, OH. In 1965 ferried Cessna 150 from Wichita, KS, to Lima, Peru. In 1966 ferried Piper Aztec from Miami, FL, to Johannesburg, South Africa. From 1966-67, charter pilot and instructor at Grand Central Airport, Johannesburg. In 1975 ferried multi-engine Sky-Van (Short Brothers Northern Ireland) from London, England to New Delhi, India. From 1969 to present, AOPA Air Safety Foundation Flight Instructor at 158 weekend flight clinics. Raced in three Powder Puff Derbies.

Certificates: Netherlands Antilles, private A South African, commercial, instructor, MEL. American: commercial, instructor, instrument, CFII, ATP, MEL, helicopter, seaplane, ground school instructor, and Gold Seal instructor.

ALICE I. HEGEDUS, born Jan. 7, 1928, in Perth Amboy, NJ. Joined the 99s in 1979. Flight training includes private, instrument and SEL. Has flown the Piper Saratoga, Lance, Warrior, Cessna Cardinal, 152, 172. Married to Ernest J. Hegedus and has two children, Susan Vander Haar of Muskegon, MI, and Judith Lee of Colonia, NJ. Grandchildren: Ernest Lee, 16; William Lee, 12; Sarah Vander Haar, 11; and Melissa Vander Haar, 9. She had a grandson when she got her license.

The 99s have given her reason to use and improve her flying skills. She has given over 200 first-time flyers rides during their Pennies-a-Pound flights. She has flown in the Garden State 300 Proficiency Contests, participated in Safety Seminars and Poker Runs. She has flown her Piper Saratoga to International Conventions from New Jersey to Vancouver, BC; Portland, OR; Shangri-La, OK; and Orlando, FL.

In the chapter and section meetings she has held offices and made many good friends.

ROBERTA ANN HEGY, born in Milwaukee, WI, on Dec. 14, 1949. Her interest in flying began at an early age. Her father flew when he was younger and would take her to work at Mitchell Field in Milwaukee where he was employed at the local National Guard base. They had a flying club and she could fly with some of the guys who were instructors. She would spend many Saturdays at Mitchell Field just to watch the planes come and go. But it wasn't until she was 38 that she really got the chance to become a pilot. In 1988 she asked her husband Richard if she could realize a lifelong dream of flying. He agreed and scheduled her first lesson in over 20 years. After 10 hours of lessons, they purchased their Cherokee N44132. She soloed in it three hours later and took her check ride in February 1989.

They now spend many days flying from their home at Air Troy Estates where she and Richard with their Cherokee and their Macaw, Oliver, live happily ever after.

BETTY SNOW HEISE, born in O'Fallon, IL, Sept. 17, 1946. Soloed 1965 in Lincoln, NE; earned private license in 1970, taught by Shreveport 99 Helen Hewitt. Joined 99s and served as Abilene, Texas, Chapter chairman.

Owned 1949 Piper Clipper PA16. Flew 1976 Powder Puff Derby with mother Evelyn Snow, as mother-daughter team.

Married 1966 to Air Force pilot James Heise, has two children: Brent, 24; and Chris, 22; and one grandchild, Kiaya.

NORMA JANELLE HELLMANN, born Jan. 21, 1949, Honolulu, HI. Earned private pilot license, SEL, 1975 in Nashville, TN. Glider rating 1979 and basic ground instructor 1985 in Rapid City, SC, repairman A/C Builder 1994 in Billings, MT. Member of South Dakota Civil Air Patrol 1976 to 1984.

Earned Certificate of Recognition for Life Saving by organizing and flying blood airlift from Rapid City, SD. Participant of CAPs International Air Cadet Exchange as guest of Swedish Air Force 1979, flew a Saab 105. As CAP squadron commander got to fly a B-52 as guest of USAF in 1981.

Bought and restored a 1947 Cessna 120 in 1985. Built own home-built Corben Jr. Ace aircraft from 1990 to 1993 and test flew it. Member of North Dakota Chapter 99s. Former South Dakota Chapter 99s chairman. FAA accident prevention counselor for Rapid City, SD, FSDO from 1985 to 1990. Member EAA, AOPA, and NAA.

MICHELLE BIGHAM HELMS, second career is her first love. Michelle received her BA in communication management from Mississippi State University in 1987, and her MA in journalism from the University of Georgia in 1989. That was the year she began working as a TV reporter in her hometown of Greenville, MS, and took her first flying lesson. After that lesson the plans changed! A career as a pilot became the new goal. With the love and support of her family; Mom, Mary Evelyn; sister, Niki; and commercially licensed, instrument and multi-engine rated father, Gene, she began working on her new career.

She earned her private license in August 1990, her instrument rating in December 1990, and her commercial and multi-engine in March 1991 all with the help of a very special American Airlines pilot/flight instructor, Brent Helms, who five years later became her 49 1/2.

Michelle flew fiber optics patrol and worked as a production test and corporate pilot for American General Aircraft Corporation flying AGAC Tigers, an Aztec, and a Grumman Cougar. She also worked as co-captain on an MU-2 for Brent Transportation of Greenville until 1993 when she was hired by Atlantic Southeast Airlines.

She currently serves as first officer on the ATR-72 base in Dallas-Fort Worth. She has her airline transport pilot license and just recently earned her B-737 type rating.

BECKY HEMPEL, is a native Texan and Austinite. She flies a Piper Archer, Warrior and Bonanza A36. Becky has 200 hours and holds ASEL and instrument ratings, and participates in the Air Race Classic. She is also a homebuilder and has an RV-6 in progress, along with her A&P father and flight instructor. She has also participated in the Expanding Your Horizons "She's the Pilot" program for young girls.

Becky earned a BS in the medial sciences from St. Louis University, and is a physician assistant-certified. She participated in many national and state PA associations.

Becky is a member of the 99s, Inc., AOPA, EAA, and Air Race Classic. She is currently membership chairperson of the Austin Chapter 99s.

Her husband Bobby Krejci, although not a pilot himself, has taken the AOPA Pinch Hitter Course. He enjoys flying with Becky and supports her flying related endeavors.

GWEN HEMS, First Canadian Chapter, had her first plane ride in 1948. The airport was Barker Field in Toronto and the plane was a Tiger Moth. Five years later she married the pilot of the Tiger Moth. (Incidentally, he had to sell the plane in order to finance the marriage!) Gwen started flying lessons at the Brampton Flying Club in 1971. She graduated in 1972 being the only female in a class of 16.

Gwen and her husband Les have owned several aircraft. Gwen built up most of her hours in their Cessna 150. She also has a float endorsement and flew their PA11 a few years ago. In her early flying days, Gwen was a passenger in a J3 Cub that developed serious engine trouble. The forced landing resulted in the Cub turning over in a hay field. However, it didn't discourage her and she went on to log more than 1,300 hours of pleasure flying in Canada and the US.

A project, dear to her heart, was building a Peter Bower's "Fly Baby" – a single seat, low wing airplane. After working on it for about 10 years, with much help from her husband and friends, it was sold. At that point it was approximately three-fourth's completed. She is happy to say that the plane was eventually finished and is now flying.

Gwen's family, her husband, two children and their respective spouses, are all licensed pilots, the men flying professionally. This leaves "Hamish" her Golden Retriever, the only member of the family who is not a pilot. Who knows, Gwen says – someday he might qualify as a "Flying Companion."

DEBORAH RUTH MARSDEN HENDERSON, born Jan. 30, 1952, in Tokyo, Japan, to missionary parents, a Phi Kappa Phi and 1975 cum laude graduate of Southwest Missouri State University, worked for original Braniff International Airlines. Married Larry Henderson at Rizal City, Philippines and accompanied him on construction projects in the remote Guajira Peninsula of Colombia, they returned to the US in 1986. Dealer in Colombian emeralds from 1984 until beginning a second baccalaureate in 1988.

Took first flying lesson on eighth wedding anniversary June 28, 1990, at CP Aviation in Santa Paula, CA – instantly electrical engineering was out and flying was in. She flies B-727s for Express One International, Dallas, TX, and is assistant chief instructor at The Aviator, Addison, TX; was previously a Metroliner III First Officer with Lone Star Airlines, Ft. Worth, TX, and flight instructor at Bourland Academy, also in Fort Worth.

Debby has flown over 35 different models of airplanes, gliders and helicopters and her accomplishments include SA-227 and ASMELS ratings on her ATP; Citation type with commercial privileges; Gold Seal, ASME, instrument, glider and ground instructor ratings and a flight engineer's certificate. Some of her goals are, typing the 727, completing helicopter training, becoming a designated pilot examiner (give 49 1/2 his checkrides!) and buying or building an airplane.

Member and former secretary of Fort Worth 99s, attended 1992 national convention. Participates in aerobatic competitions and is a member of NAFI, IAC, EAA and AOPA. When not flying enjoys water skiing, tae kwon do and playing the piano. Debby and Larry live in Palo Pinto County, TX, with their six cats and three-legged dog BoBo.

DOROTHA ELROD HENDRICKS, (deceased Nov. 29, 1993) was born in Warren County, KY, on Jan. 26, 1915. She obtained her license on Oct. 15, 1941, at Muncie Indiana Airport. Passing all WASPs tests but never serving, she was active in the Civil Air Patrol, Bond Rallies and Blood Bank during WWII. Joining the 99s, Inc. in December 1941, she was instrumental in reactivating the Indiana Chapter, being elected chairman in 1945. She held all chapter offices and was the historian until her death. Dorotha served two terms as secretary-treasurer (1946-48) of the North Central Section and was International scrapbook chairman 1990-92.

Dorotha was the first to receive the Governor's Award in honor and recognition of the dedicated service given to the North Central Section. She was also the first Indiana member to become a life member of the 99s, Inc. and was honored by her chapter with a 50 year pin in 1991. In 1979 she researched and wrote the 40th anniversary history of the Indiana Chapter, which was founded at Purdue University. Dorotha was a member of Goodyear Blimp Club, Killbuck Retirees, Pendleton Historical Museum, Silver Wings and the Indianapolis Aero Club. In 1990 she was placed on Memory Lane in the International Forest of Friendship.

VONNE ANNE HENINGER, born Jan. 4, 1932, in Oakland, CA. Joined the 99s in 1982. Flight training included North Field, Oakland; Spirit of St. Louis, St. Louis, MO; and Van Nuys, CA. She has flown Cessna 172, and Piper 140, 160, 180. Her five children are: Tom, 43; Shelley LaCour, 41; Heidi Young, 39; Kurt, 37; and Lisa Anderson, 35. She also has grandchildren: Jenilee Heninger, 13; Tom Heninger, 4; Christian LaCour, 11; and Courtney LaCour, 9.

One of the most memorable flights she made was to Guaymos, Mexico with a group from a flying club she belonged to out of Scottsdale, AZ. On the return flight they ran into weather and were blown off course and suddenly noticed the big "lake" they saw below was actually the Gulf of California. In nearly IFR conditions they made their way to Tucson, Intl. and landed; grateful to be on terra firma.

TOOKIE HENSLEY, born and raised in southern California, Jan. 8, 1935, in Brawley, CA. Learned to fly 1973 and has logged 5,000 hours. Participated in numerous air races: Powder Puff Derby, Angel Derby, Shirts and Skirts, Palms to Pines, Fresno 400, Pacific Air Race, Riverside Air Race, Rialto Air Rally, The Great 150 Air Race, Shamrock Air Derby. She has raced and been on the board of director for the Air Race Classic. Has set up tower and airport tours and scholarships for young people. Charter member of Inland California Chapter and helping form Rio Colorado Chapter, held all offices at chapter level.

Served on the airport land use and county airport commission. Loves to water ski, play tennis, read and be mother to Pamela, 41; Daniel, 39; and grandmother to five grandchildren.

Received her private, commercial and instrument 1973-74-75, then the Amelia Earhart Scholarship 1985 was able to get her CFI in 1987. In 1991 she got her CFII. In 1993 started her own flying school at Eagle Airpark on the Colorado River just south of Laughlin, NV. In 1994 got her multi-engine. She has four airplanes and has been teaching basic instrument classes at Mojave Community College for the last seven years.

She belongs to AOPA, NFIA, Air Race Classic and 99s.

Married to Don Hensley for 42 years who is also a pilot.

GAYL HENZE, currently the governor of the Mid-Atlantic Section, started flying in 1978, with the encouragement of her husband, Herbert, at Philadelphia International Airport. After getting her private, her first passengers were her three children. It was while attending the Reading Air Show in 1980 that she met Kate Macario, who introduced her to the Eastern Pennsylvania Chapter of the 99s. She started on an airmarking committee and served on all chapter offices. She earned her instrument and seaplane rating, her commercial certificate, and in 1984 was hired as pilot for Penn Fishing Tackle. While vice-governor of the section, Gayl worked to further the goals of the International organization by chairing a nominations and elections committee, and working on the bylaws and standing rules committees. The experience of being a 99 and a governor has been greatly enhanced by the association with fascinating people in North America and around the world.

WRENN R. HERMAN, born Dec. 30, 1962, in Norfolk, VA. Joined the 99s in 1991. She has flown the following planes: PA28 140, PA28 160, PA28 180, F90, B-200, B-100, Kingairs, C-150/152, C-172, C-172RG, Be-77, 7KCAB, C-207, MV21.

She became a flight attendant for Piedmont Airlines in 1987. In 1988 she decided she would prefer to be flying the airplane. She began taking lessons and received her private license in 1889.

After receiving her private license she successfully completed her instrument, commercial, flight instructor and instrument instructor ratings. She worked on a rating every vacation she had from her flight attendant job.

While working on her private license she met her husband Kevin. Kevin is a pilot at DHL Worldwide Express. He has been a positive influence and supporter of her flying. They have a son, Connor, born Nov. 7, 1994. Presently she is instructing at Lincoln City Airport in Lincoln, NC. Her goal is to fly for a scheduled air carrier.

MARY B. HERMANN, born Nov. 28, 1946, in Glendale, CA. She learned to fly in Glenwood Springs, CO, in 1987. Received her license July 25, 1987, and her instrument rating in September 1989. She has flown the following planes: Cessna 150/150, 172, 182; Piper Turbo Arrow; Cherokee 180; Archer and Warrior. As a student pilot flying from Glenwood Springs, CO, to Rifle, CO, at 0630 in May 1987, the sun was just beginning to rise on the mountains and as she looked down on the highway and all the traffic, she realized that she was no longer a "ground pounder." She was free up there above it all. The air was very smooth that morning and she felt very peaceful and totally free. "Now I know how a bird must feel," I said to myself. To this day she cannot stand to see a bird in a cage. She does not want to deny them the exquisite feeling of freedom at sunrise in the high mountains.

She still relives the feeling of freedom that she had that morning every time she flies. She now lives at an airport in the Colorado mountains and even though she doesn't have much time to fly she still has the memory of that beautiful morning in 1987 to keep her going.

She has met many nice people since she became a pilot and she hopes to continue flying at least another 40 years. She has always loved airplanes and she will continue to love them long after she is too old to handle the controls herself.

ELLEN HERRING earned her pilot license in 1973 after flying with private pilot husband Chuck and deciding he shouldn't have all the fun! She joined the 99s in 1974, and across the years has belonged to the Central New York, Indiana Dunes, Shreveport, and Chicago Area Chapters. Along the way she has flown in, or over, 41 of the 50 states, and has flown in air rallies in Ohio, Indiana, Illinois, Iowa, and Arkansas, winning the Illi-Nines twice. The Herrings own a Cessna 172, which was recently upgraded with a 180 hp engine. Does it climb!

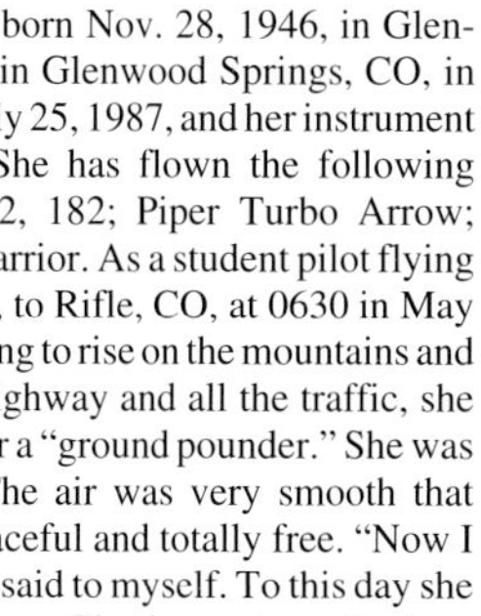

She and Chuck have two sons, Tom, who also has his pilot license and Steve, who prefers rafting and skiing. When not flying, she is a business systems analyst for a finance company.

Ellen is currently the Amelia Earhart Scholarship chairman for the North Central Section.

DOROTHY MEANS HERRINGTON, born Jan. 17, 1920, in Moberly, MO. Her father was a pilot and one of the pioneers in aviation in Missouri. When she was 7 years old, father's instructor arranged for her to have her first airplane ride. That was truly the most exciting day of her life. It was then that she decided her first real goal in life was to someday be a pilot. She graduated from high school in Fayette, MO, and attended Central Methodist College before transferring to the University of Missouri in Columbia. Within a few days she was told that the US Army ROTC program at the university had been selected to train officers for the US Army Air Corps and permission had been granted for one woman to be included in the program. Her father was on active duty and he immediately sent his written approval, but her mother refused to sign the permit. She knew that within a few years she would become the wife of a young officer graduating from USMA in May 1942 and he also had hopes of becoming a pilot. She said that one pilot in a family was enough. Most of her friends were in the group of ROTC cadets who later joined General Chenault's Flying Tiger Squadron.

Her first husband, Lt. Frank D. Waddell, was killed in a B-26 crash at McDill Field shortly before their son, Frank D. Waddell, Jr. was born. Her second husband, Maj. William P. Oliver, Jr. USMC, and she had a son Thomas Means Oliver, who at the age of six months. Within a few months they divorced.

She became a US Civil Service employee in St. Louis. She began graduate school at the University of Missouri and became a first grade teacher in the Fayette Public Schools. She married Lt. Kenneth F. Herrington, Jr. in February 1950.

In June 1961 she received her BS in education with life certificate to teach and in August 1961, she received a MA in guidance and counseling with life certificate as a guidance counselor. She was selected for graduate school for post-masters study in counseling psychology at the University of Texas, Austin, and volunteered to spend one year helping the Job Corps Program at Texas Educational Foundation in San Marcos. She was employed as a school psychologist with the Milwaukee Public Schools.

The Veterans Administration counselor advised that she was eligible for the GI Bill for education and this included flight training after obtaining a pilot license. In November 1974, she earned the license by completing successful flights in Cessna 150 and 172. It was then that she became a very proud member of the 99s Assoc.

She has six children and 13 grandchildren. She continue to take FAA medical examinations each year in the hope that very soon she will be financially able to fly in a Cessna 182.

SANDRA HERRON, born in Peoria, IL. Currently Aspen 99s chairman started flying in 1973. On her first solo cross country flight, she ran in the back door of Bellanca Aircraft Corp. in Alexandria, MN, and asked for a quick tour to see the planes, while her little Cessna 150 was being refueled a few yards away. Within a month, she was a certified pilot with her own Bellanca Super Viking. Now an instrument rated pilot she is working as a real estate broker and developer in the beautiful Colorado Rocky Mountain ski resort of Aspen, where she has lived a sporting life for 26 years.

Having attended the University of Miami in Florida, worked as a flight attendant for Eastern Airlines, had a stint in modeling and as a professional musician playing electronic keyboards. She was vice president of Chicago Career College, started and got state approval for the first licensed travel agency school in Illinois, and went on to work for Playboy Clubs, International in Chicago as International Bunny Mother during the height of the Playboy Club scene.

Tiring of the cosmopolitan life in the big city, she moved to Aspen in 1968 where she discovered the joys of skiing, snowboarding, fly fishing, kayaking, horseback riding, hiking, dirt bikes, mountain bikes, tennis, ice skating, and above all - flying. As a member of the local EAA, ICAS, and Aspen Pilot's Assoc., she can hangar talk with the best of them. Reno Air Races are high on her list of travel destinations, along with the Albuquerque Hot Air Balloon Festival, Oshkosh, and the Paris air show. Having lived in France, working for Club Med for eight years, fluent in French, as having worked in Club Med villages, worldwide, has added to the appreciation of freedom of travel, the outdoors, and diversified cultures. "And I'm not finished yet!" Her favorite part of town is always the airport, the gateway to freedom, challenge, and always a new source of wonder and delight. "Especially the Aspen airport, where the world is coming to us, and what they don't bring in, we wing our way out to go find it for ourselves."

DEBORAH HERZOG, born April 16, 1960. She is a business owner, Pacific Research. Married to Gregory Herzog and has children, Eric, 14, and Casey, 12. Joined the Reno Chapter 99s in November 1994. Herzog became involved in aviation in 1993, encouraged by husband and friends. She earned her private pilot license and is currently studying for her instrument rating. Herzog owns a Piper Lance PA32/A. She is an amateur radio operator registered with the FCC technician class rating. Also involved with the Mono County (races) Emergency Services, Search and Rescue.

CHRIS HETTENBACH, born on Jan. 30, 1954, in Milwaukee, WI. She has also lived in Colorado, Washington state and currently resides in Texas. Chris soloed on Aug. 27, 1988, in her brother's Cessna 172. She received her private license in 1989. Since that time, she has logged more than 2,000 hours of flight time and holds commercial, instrument, ASEL/AMEL and CFII ratings. In 1993 and 1995, Chris was a recipient of the Amelia Earhart Memorial Scholarship which she used to obtain her CFII and ATP/ Citation Type ratings. Chris is a flight instructor in the Dallas area and has her own business at Addison Airport. She is now teaching students in the same C-172 in which she did her first solo. Chris has been flying as copilot in a Cessna 421 and flies a Cessna 206 for a local aerial photographer. Chris' goal is to fly corporate aircraft or for a major airline. Chris has been an active member of the Dallas 99s since 1989 and has held several offices including vice chairman (1993-94) and chairman (1994-95).

Chris also has a degree in veterinary technology and has been working in the veterinary field for the past 17 years. She still does relief work for several local clinics when she is not flying. Chris is an avid woodworker and, in her spare time, she does woodworking and makes stained glass windows and glass airplanes.

BELLE HETZEL, was a school teacher who flew because she just couldn't help herself. Born in Avoca, IA, in 1889, she taught history for 27 years in Omaha high schools, and taught in other schools in Council Bluff, IA, and in Colfax, WA, for 13 years before that. In 1952 she retired early so that she could fly full time instead of weekends and holidays.

Belle learned to fly at age 40 at Boeing School in Oakland, CA. In 1937 she was largely responsible for organizing the Missouri Valley Chapter, starting with five members, all officers. By 1947 she had held many section and International offices, and was elected International president. She was never a racer or adventurer, just an individual who found something mystical about flying. She started one of the first high school aeronautics classes in the country as soon as she herself learned to fly in 1930. In 1948 she used an inheritance to buy herself a Cessna 140, and traveled constantly from then on.

In 1958 she sold her plane to take a round-the-world air trip, the first of four round-the-world trips. During the war she had organized and commanded a woman's squadron of the CAP, and also taught ground school and Link Trainer classes, as well as continuing her high school classes in history and aeronautics. Now she wanted to fly, and so she did. In 1961, age 71, she flew alone to attend her 1911 50th anniversary class at the University of Michigan, undoubtedly singing all the way, her usual habit.

Four years later, while in Berlin she suffered a severe stroke and spent many months in bed. But she fought back – she intended to fly again. She died in November 1971, age 82.

HELEN S. HEWITT, a resident of Shreveport, LA, since 1952, learned to fly at Downtown Airport in 1959. Since then she has acquired her private license, commercial, instrument, multi-engine, flight instructor, flight instructor instrument, airline transport pilot and all ground instructor ratings. She taught all ground school courses at Shreveport Aviation for 15 years before her retirement and also organized and conducted several Pinchhitter courses. She chartered the Shreveport 99s in 1960 and has held all offices in the chapter. At the time of her retirement, Helen had logged 3,000 hours.

Helen received her BA from Allegheny College and her MA from Louisiana State University. She is married to Forrest A. Hewitt, a petroleum geologist who is also a commercial and instrument rated pilot. They have four sons all who learned to fly from Helen and they are justly proud of six grandchildren and one great-granddaughter. She is an active member of St. Marks Episcopal Church, DAR, several community clubs and regularly plays golf at Northwood Country Club.

DOLLY HICKLIN, learned to fly at Van Nuys Airport in 1970. She has participated in numerous air races including the Palms-to-Pines, Pacific Air Race, Valley Air Derby and the Salinas Great Pumpkin Classic.

For 20 years (1974-93), she co-chaired The Jim Hicklin Memorial Air Rallye, an all-men's, cross-country, stock aircraft race held in memory of her late husband. Sponsored by the San Fernando Valley Chapter of the 99s, more than $25,000 in proceeds from the Rallye was awarded to young adults seeking careers in aviation.

A member of the 99s since 1985, Dolly was recipient of the Rookie of the Year Award and in 1991, the Trixie Ann Schubert Award for service. She has been with the Los Angeles Department of Airports, Public Relations Department at Van Nuys Airport in a part-time position since 1975 and is a full-time legal secretary.

Dolly has a son, Jim, and a granddaughter, Heidi.

BETTY HICKS, her accomplishments are very numerous and span flying, education, golf, and cooking. She began flying in 1958 to get to her golf matches. She is an ATP rated CFII MEI and authored the "Ground School Workbook" for private, commercial, and instrument students. Her writings (over 300 publications) include an article for *Flying Magazine* and the scripts for the "Invitation to Fly" video series.

She taught ground school at Foothill College for many years, and was the program coordinator when she retired in 1993. Betty won the National Aeronautic Assoc.'s Certificate of Honor in 1963 for her service in aviation education and was appointed to the nation's "Women's Advisory Committee on Aviation." Her fights have taken her to Mexico and Canada and she flew the Powder Puff Derby in 1960, 1963 and 1969.

Betty has been a processional golfer who was selected Woman Athlete of the Year in 1941 and was the first president of the Women's Professional Golfer's Assoc. Betty is also a gourmet cook and published a cookbook in 1986.

CLARE BROOKS HIGGINS, B.C. Coast Chapter, since learning to fly in Toronto, Ontario, in 1974, her Cessna 140 and then 172 has carried her over some of the most beautiful landscapes in the world – Southern Canada, North, South, East and West in the US, the Bahamas and Mexico.

Recent trips have been into the interior of B.C., and down the Pacific Coast, and over the Grand Canyon with her husband Paul and son Kellan.

What she loves most about flying is exploring new areas, camping at airports. The future? "Northern B.C. and beyond, I hope ..."

MARY BROCK HIGGS, born March 11, 1943, in Mt. Airy, NC. Joined the 99s in 1984. Her flight training includes private pilot, single engine land and sea, multiengine land ratings. She has flown the Cessna 172 and 337.

She and her husband Derek have two children, Elizabeth Higgs, 24, and Catherine Higgs Maingot, 27.

Member of Gov. 99s Caribbean Section and Bahamas Air Sea Rescue Assoc. Participated in the Great Southern Air Race in 1986. Higgs is a certified scuba diver, underwater photographer, HAM radio licensed operator. She has lived in the Bahamas since 1964. Her husband is also a pilot.

DELLE HIGHTOWER, born April 4, 1929, in Searcy, AR. Joined the 99s in 1965. Has flown Cessna 150, 152, 172, 182, 210, 206; Piper Cherokee and Arrow; Beechcraft Bonanza; Mooney M20C, 201 and F Model Mooney.

Her husband's name is Gene. Hightower took her first airplane ride in March 1964, having been tricked by her husband into doing so, and she was bitten by the flying bug.

As fate would permit she soloed July 26, 1964. Many despairing and traumatic moments followed before she obtained her private pilot license on April 17, 1965. Her flight instructor was killed in an airplane crash shortly after they had completed a lesson. Also killed in that accident was a good friend who was also a fellow student pilot along with two other men whom she knew. Perhaps that accident has made a more safety conscious pilot of her. Since then many years and lots of fun places have passed beneath her wings.

KATHLEEN A. "KAY" HILBRANDT, born in Long Island, NY, Jan. 26, 1924. She moved to New Jersey two years later. Kay started flying in 1941 at Staten Island Airport, NY, and in 1942 moved to Pennsylvania where she worked and flew at Martins Creek Airport.

In 1943 she worked as a mechanic and plane captain at Eastern Aircraft Division in Trenton, NJ, preflighting TBM Avenger Bombers.

In May 1944 Kay entered the WASP (Women Airforce Service Pilots WWII). After gradation she was stationed at Eagle Pass Air Force Base, Eagle Pass, TX.

Kay returned to New Jersey after the WASP disbanded, got her instructor rating and instructed at Lambrose and Mellor Howard Seaplane Bases in Ridgefield Park, NJ, from 1945 to 1948. In 1948 Kay joined the Bendix Corporation at Teterboro, NJ, and worked there until she retired in 1981.

Kay flew copilot with Selma Cronan (99 member) in 1960 in Selma's Cessna 172, in the Powder Puff Derby.

Kay joined the 99s in 1945 and held the offices of treasurer, secretary, vice governor and governor from 1954 to 1962, of the New York-New Jersey Section.

Kay is currently a member of the WASP, WIMSA (Women in Military Service of America), AAFHA (Army Air Forces Historical Assoc.) and the Aviation Hall of Fame of New Jersey.

JOYCE HILCHIE, realized her dream of flying in her 50th. Joyce on obtaining her private license in 1981, joined the Colorado Chapter of 99s. The 99s encouraged Joyce to obtain her ratings.

In 1994 Joyce added sea plane to her instrument, commercial and CFI. While a member on the Colorado Chapter she participated in many programs and helped organize the chapter's Companion Flyer Course, Air

Bear, Chapter Scholarship program and bi annual trips to the decompression chamber.

Joyce loves flying air races and organizing safety seminars; however, she spends some of her time acting as company pilot for Douglas W. Hilchie Inc.

After moving to Arizona Joyce became a charter member of the Yavapai Chapter of 99s, and at present serves as their chairman.

Look for Joyce in her beautiful white, blue and black Comanche which she has flown since 1985.

GERALDINE M. HILL, born Nov. 15, 1910. Joined the 99s in November 1939, but left San Francisco for a year and didn't renew until 42. Flight training, small aircraft at Mills Field, S.F., Oakland Municipal, and S.F. Bay Airdrome. Private license November 1939.

Most small single engine planes available in the late 1930s and 1940s. Entered WASP training December 1942 and served until December 1944.

During the 1950s, she flew seven All Women Transcontinental Air Races. Flying was a wonderful experience.

SHERRIE HILL, Women With Wings Chapter of 99s. When her husband, David, and she were married in 1988, as a wedding gift, he presented her with a homemade gift certificate for flying lessons. She was thrilled, but she must not have been ready because she kept putting it off. After flying right seat for another five years, she decided it was time to at least learn how to land in case of an emergency. David suggested that she sign up for one of the 99s Flying Companion Seminars. She walked away from that seminar thinking "If they can do it, so can I." So, she started  her lessons thinking that she would just solo and stop there. On a gorgeous, sunny day, July 1 she made her first solo flight! She doesn't think that she ever came down from the clouds that day. After that, there was no stopping her. She knew she had to go all the way! Nine long and often times frustrating months later, she got her wings and joined the Women With Wings Chapter of 99s.

JEANNE B. HILLIS, a resident of The Dalles, OR, in the Columbia River Scenic Area flies her C-182 from the city airport in Dallesport, WA. She and her husband used their planes for business, transportation and pleasure, including 29 trips to various sites in Mexico. She is chairman of the Columbia Gorge 99s and chaired the Safety Seminar for six years. She belongs to the local EAA Chapter and serves on the city Airport Commission.

Her current professional life is in free-lance art and lectures; is part of an air-  conditioning-heating business. Her education includes a BA in fine arts and education qualifications. She has served on local and state committees in the arts; was named Woman of the Year in 1967, recipient of Soroptomist award in 1989 and AAUW Honor in 1990.

PEGGY HINE, private certificate, ASEL, glider rating. Joined the 99s in 1986, Lake Tahoe Chapter.

MARGARETHA ROSE KLOEPPLE HINMAN, also known as Marga was born in Freeburg, MO, one of eight children. Marga joined the US Coast Guard at the age of 20 and was one of their first female members known as the SPARS. During her tour of duty as a projectionist she was transferred to San Francisco, CA, where she met and married her husband a glider pilot with the US Army Air Force.

Marga became interested in learning to fly while attending Civil Air Patrol meetings in Marin County, CA, and while flying with her husband. Marga earned her private pilot license in June 1969 in a Piper Colt and has the distinction of being the first woman in Marin County, CA, to have earned her CAP, Mission Pilots rating in the Squadrons T-34 Beechcraft on loan from nearby Hamilton AFB. Marga has been a proud member of the 99s since 1969 and has flown several aircraft including a restored Stearman bi-plane and a helicopter.

Marga has one daughter, a pilot in training for her flight instructor rating and three sons and six grandchildren. Marga says that learning to fly has given her a wonderful sense of freedom and personal accomplishment.

KRISTINE HINTERBERG, began flying ultralights in 1988. Realizing the limitations of ultralight flying, she received her private license in February 1989 at West Bend, WI. She holds an instrument rating, and flies mostly for recreation and travel. She and her husband Jack, also a private pilot, spend many enjoyable hours in their Piper Cherokee Archer.

Kris joined the 99s and the Wisconsin Chapter in 1989 and has served as the chapter secretary, treasurer, and newsletter editor. Elected Chapter chairman in  1995, she will serve in this position for two years.

A former teacher, Kris is the general business manager of Hinterberg Design, a manufacturer of quilting frames and other wood products. She teaches fitness classes at the YMCA, and enjoys playing tennis and sailing. She has three children Anne, David and Paul Hinterberg.

JEANETTE BALDERSON HINTON, born May 26, 1936, Virginia. Joined the 99s in 1984. Received her flight training at Martin State and Baltimore Airpark. She flew PA28-181 Cherokee Archer. Hinton learned to fly in 1983. Hinton is proud member of the Maryland Chapter 99s, Inc. A winner of the spot landing contest three consecutive years. Also member of AOPA.

She is married to Carroll Hinton also a pilot and they own and operate a trucking company She has one son Michael, two daughters, Katrine and Diane. Grandson Joshua, granddaughters: Cherie, Erica and Brittany.

ILSE E. HIPFEL, born April 18, 1953, in Los Angeles, CA. A native Californian, her dream to fly began in 1970 when her father bought an excursion with a chartered airline to Europe. Her dream of flying came true when Ilse and her best friend Maureen Oster engaged in duel flying lessons. Through many hours of exhausting studies and solo flying they soon acquired their pilot license in 1988. Ilse prefers to fly a Piper, Dakota but will venture out in a Cessna 182. Ilse has accumulated several hours of flight time, many of which  were spent in air races and local air rallies. However, she is known best for her annual organization of the chapter's successful Poker Run.

The 99s have enjoyed her contribution of enthusiasm when holding various chapter officers positions along with her appointment as Southwest Section Membership chairman. Also a member of AOPA and selected as one of the few safety counselors for the FAA.

BEVERLEY A. HIRZEL, born May 24, 1946, Indiana, PA. Joined the All Ohio 99s in 1976. In 1977 she was co-chairman and publicity chairman for the First Air Race Classic, Toledo Terminus. Meeting so many experienced women pilots was, for her, inspiring and enlightening. Many firm friendships came out of that event. Next was the Buckeye BAR Rallies that she co-chaired or was the treasurer. For those that flew the BAR at Toledo Metcalf she is sure you remember! Airmarking has also been an added experience at several airports in Ohio. Flying activities are the best way to learn about 99s, make friends and encourage new pilots.

She and her husband Bill still fly their M20G when time permits, and she continues to encourage young women to fly and seek adventure.

She also keeps busy with her own bookkeeping and tax prep business; property manager for local rentals; and always many county civic volunteer boards.

JANET ANNE RUSSELL HITT, started flying in 1946, and joined the 99s in 1948. Janet is an ATP rated FAA examiner, a flight instructor, ground instructor and has more than 5,000 hours. She currently works for GADA at San Jose.

Janet and her husband Bill have two children. Bill is a retired naval aviator. Janet was a charter member of the Aloha Chapter, and is now a member of the Santa Clara Valley Chapter. She has worked as a charter pilot and company pilot for various companies over the years in the US as well as the West Indies and Puerto Rico, and has served as chief flight instructor at several clubs. She is also an FAA aviation safety inspector. Janet was named SCV Aviatrix of the Year in 1974 and Flight Instructor of the Year that same year by the San Jose GADO. Janet participated in National Intercollegiate Air Shows between 1947 and 1949, and in the 99s All Women's Air Maneuvers in 1950, where she took first place in several categories. Janet is a member of the International Society of Air Safety Investigators and the Aero Club of Northern California.

JEAN F. HIXSON, (1922-84) began her flying career with an airplane ride at age 9 in Hoopeston, IL, and ended as a colonel in the Women's Airforce Reserves.

Jean began flying at 16, got her license at 18, joined the WASP and flew B-25s then became a flight instructor in Akron, OH. She held an instrument and instructor rating for single and multi-engine aircrafts.

She received a BS and MA in education from Akron University and taught for 31 years. She received awards in the field of education, aviation and journalism.

Jean served three years on the Citizen Advisory Board on Aviation at Washington, DC, conducted air education workshops in Akron and the University of Wyoming and was one of two reserve officers to be the first women to tour NATO bases in Europe.

In 1957 she was the fourth woman to fly through the sound barrier in an F-94C jet fighter. Then in 1958 was given the opportunity to pilot the F-102. In 1961 she was one of 13 women selected for training, but the program canceled.

Jean was active in many aviation organizations, held a life membership in the 99s, was the recipient of the 1948 Amelia Earhart Scholarship Award and flew in seven Powder Puff Derbies. In her last year Col. Hixson was awarded the Meritorious Service Medal by the US Air Force for her work in the Air Force Aerospace Medical Research Laboratory at Wright-Patterson AFB, Dayton, OH.

BARBARA HOBSON, a native of Minnesota, soloed in Boulder, CO, in 1976. Shortly thereafter she received her private license flying a Cessna 172.

In 1978 Barb joined the Colorado Chapter of 99s. She has held various positions within the chapter, including chapter chairman. At this writing Barb is the local chapter historian.

Barb recently retired from a 33 year teaching career. She will now have time to fly, play golf and ski during the week. She may even get a few new ratings.

MARIE HOEFER, born March 6, 1935, in San Diego, CA. She soled Feb. 15, 1964, in a Piper Cherokee 140 at Long Beach, CA, received her license Nov. 30, 1964, and in 1974 her commercial rating. She had always been interested in learning to fly and flying in general. After graduating from college she worked as a stewardess for United Air Lines where she met her husband, Rick, a pilot. They were married in 1959 and by 1962 had their three boys: Steven, Mark (twins) and Dan. Ten years later they got their girl Teri. Steve is now a DC-10 copilot for United.

Joan Merriam Smith had contacted her, after her flight around the world, about helping organize pilot's wives into taking pinch hitter sources. But it was Barbara London that told Rick about the 99s and who eventually sponsored her into the 99s. One of the best things she ever did was join the 99s. She has always belonged to the Long Beach Chapter and held every office at one time or another. She has attended numerous section meetings and conventions and is very proud and honored to have her name in the International Forest of Friendship.

In 1971 she flew her first air race, Palms to Pines and went on to fly seven more. Last year she flew it with Teri! She has also flown three Powder Puff Derbies, seven Pacific Air Races, three Shirts and Skirts and four Air Race Classics. Won the Pacific Air Race in 1984 and placed in many others. Needless to say, she is hooked on racing.

Rick and she travel extensively and has met 99s in many other countries. Last October she had her first glider flight in Innsbruck, Austria with Gudrun Henle's (99) husband, Wolfgang, a cosmonaut trainee.

For the past 16 years they have owned a Cessna 172 and flown it all over the US. It is on lease back at the El Toro Aero Club on the El Toro Marine Air Station.

She belongs to AOPA, Silver Wings and the San Diego Aero Space Museum but it is the 99s that have given her life long friends. She's proud to say both of their car license plates refer to the 99s. They read; I'm a 99 and LGB 99s.

SHIRLEY JEAN DINGMAN HOERLE, her first lesson, April 15, 1967, at Highland Airport near Utica, NY. Soloed June 11, 1967. Private pilot ASEL July 10, 1968. Commercial pilot rating, Aug. 18, 1970. Instrument rating, Nov. 8, 1969. Multi-engine rating, July 6, 1978. Copilot Powder Puff Derby, July 5-8, 1971, 25th Annual All-Woman Transcontinental Air Race.

She has made several long distance cross-country trips from her home in Central New York State to Florida; Las Vegas, NV; and Freeport, Bahamas (on their honeymoon). One of the founding charter members of Central New York Chapter of the 99s. First Chapter Chairman, Central New York Chapter.

Unfortunately she hasn't piloted a plane since 1980. The basic reason is because of the demands of being the mother of young children and later the mother of very busy teenage children. Additionally, her husband retired in 1985, selling the business and, shortly thereafter, their airplane.

MARION HOF, developed an interest in flying while writing a biography on Amelia Earhart. She felt that in learning how to fly, one could better understand the life of a pilot.

Hof was born in 1950 in Germany. She was educated in Ireland, Italy, Switzerland and Germany and graduated with a degree of the Colleges of Music in Wurzburg and Darmstadt.

Actively engaged in sports, riding, Judo, fencing and Tae Kwon-Do, she held functions as trainer for Tae Kwon-Do and is the author of a book on Budo-Sports (Martial Arts).

Hof's curiosity and interest were awakened for Amelia Earhart while reading a newspaper clipping. Pursuing more information about Earhart's life and disappearance, she realized that very little was known in Europe about the famous American aviator. As a result of two years of research, she has published her own book of Amelia Earhart.

She now manages her own large farm business in south England. All her spare time goes into trying to gain more experience and skill and enjoying the beauty and excitement of flying.

Currently, Hof holds a powered glider license, a license for ultra-light aircraft, and a PPL. Amelia Earhart commemorative flight to Burry Port in 1991, together with Dr. Angelika Machinek. Inducted into International Forest of Friendship, Atchison, 1992.

HAZEL HOHN, born October 13. She is a freelance writer and has four children. Joined the 99s in the early 1950s and the Reno Chapter in 1962, originally this chapter was known as the Fallon Chapter. Has previously been a member of chapters in New Jersey and New York. Hohn has held the offices of secretary and newsletter reporter. She became involved in aviation in 1937, met Amelia Earhart and began taking flying lessons at Newark Airport, NJ, in 1940. Before that, for a year, she had been taking unofficial lessons with a pilot who did not have an instructor rating. She holds a commercial, seaplane, instrument and multi-engine rating (military).

Hohn has a AA degree and is a senior at University of Nevada, Reno. (Air Force cadet training in 1943 and a graduate WASP). She is a professional writer of aviation as well as children's books and poetry. Taught writing at Western Nevada Community and Old College.

Her memberships include: Combat Pilots Assoc., Military Women Aviators, Air Force Assoc., WASP WWII, Aviation-Space Writers Assoc., Experimental Aircraft Assoc., Liberator Club, Alliance for Animals, Marauder B-26 Club Aerospace Education Assoc.

ELISE HOIT, soloed on a grass airstrip in 1958 while a student at Auburn University, AL. Since then she has earned her commercial, multi-engine, and instrument ratings. She flew her Cessna Cardinal in the last Powder Puff Derby. A 99 since 1968, she is a charter member of Rainier Chapter 99s and has held several offices. She is also a member of the Puget Sound Antique Aircraft Club, and Aircraft Owner and Pilots Assoc.

Elise is a graduate of California State University at Sacramento. She is an elementary school librarian, and has organized a Young Aviator's Club at the school where she teaches.

Elise enjoys the yesteryear flying atmosphere of the scenic grass airstrip she lives on, which is home to many antique and classic aircraft. She owns a Piper Pacer, and has owned and flown many different types of aircraft.

Husband, Edward Hoit, is a training pilot with the Boeing Company. Elise has two sons, two step-children, and five grandchildren. In the photo is the oldest Tiger Moth belonging to the Tiger Club in England.

HELEN CROCKETT HOLBIRD, has lived in Oklahoma City most of her life except for seven years in California. Helen learned to fly because of the encouragement from her husband Roy. He loved the freedom flying gives. Helen learned to fly in their first aircraft, a Cessna 150. After receiving her certificate in 1979 Roy was her first passenger. They then bought a Cessna 172.

Helen had read articles about the 99s but didn't know how to connect with them. A pilot friend gave her name to the Oklahoma Chapter membership chairman and she became a 99 after her first meeting.

In addition to presiding as Oklahoma Chapter chairman of the 99s, Helen is secretary of Oklahoma Pilots Assoc., Personnel officer of Civil Air Patrol, and belongs to AOPA.

As a 99 Helen gives presentations to school and/or civic groups about the purpose of the 99s.

She is a widow and has two children, Rory Matthew Lewellen, 40, and Carole Ann Whitlow, 37. Grandchildren, Elizabeth Anne Whitlow, 12, and Robert Earl Whitlow, 10.

KATHERINE ELIZABETH HOLCOMBE, born April 18, 1936, Wallowa, OR. She has sons, Kevin, 35, and Brent, 34. Grandchildren: Lauren, 7; Nicole, 5; and Nicholas, 2. Work for Kaiser (health care); Musician (Oboe/English Horn); make/sell gemstone jewelry, details airplanes to support flying habit.

At 9 years of age saved pennies for first flight in a Cub hangared on a grass field by her home; still remembers the thrill! But, in her 30s a second C-182 night flight over San Francisco Bay left her terrified after the pilot throttled back to prove he could glide to nearby airports. At age 45 during her third small plane flight on a crystal clear night, was asked to take the controls and responded, "I can't fly, I'm a woman, I'm too old!" The pilot Mark Baird, proved her wrong. Became partners in his 1976 Cessna 150, quickly renamed "Spirited Lady," and earned her private pilot Aug. 14, 1984 (only "claim to fame" is "acing" private written). Has logged 498 hours, member of 99s, AOPA, and Aircraft Pilots of Bay Area. Through long labors of love, their bird is a "Cream Puff" (custom paint/upholstery, overhauled engine IFR avionics) and they have flown her to New Orleans via Grand Canyon, Arizona, New Mexico, Texas. Also flown partner's C-180 tail dragger while he photographed Grand Tetons, Glacier, Yellowstone, Rushmore, etc. Spent time in C-172s but C-150 fits her size/pocket book best. Starting IFR rating. Will fly as long as $$/health allow (mother is 92). "For me, flying is not a luxury but a necessity of heart; to lift aloft is pure joy!"

NANCY O'NEIL HOLDEN, learned to fly at the Windsor (Ontario, Canada) Flying Club and received her private pilot license on Oct. 20, 1984, at age 41, winning the Sheila Scott Award for women's excellence. The following year she and her husband Ross bought C-GMEM, a green and white Cessna 172 that has been her companion now for 10 years.

In 1985 she earned her night rating and in 1988 her commercial rating. While mainly flying in Ontario, often with son Scott, she and Ross have also made flights to Florida and the east coast.

Nancy says that joining the 99s is one of the best things she ever did. A member of Maple Leaf Chapter for 10 years, she has served as chairman, vice-chair, Poker Run chairman and news reporter. At the section level, she has been reporter for the *99s News*.

Meeting fellow 99s at chapter and section functions and at the four International Conventions she has attended has been a highlight of her life. Flying to her first International Convention at Tulsa, OK, with Karin Williamson and Grace Morfitt in Karin's 172 was a wonderful experience never to be forgotten. Nancy enjoys taking first time fliers for a flight to share in their joy and each year takes the winners of the 99 science fair award for their prize winning flight. She also enjoys horseback riding and made two parachute jumps in 1991, one to celebrate her 48th birthday.

WENDY L. HOLFORTY, grew up in Rochester, MI. She graduated from Michigan State University with a BA in criminal justice and became the first female patrol officer in East Lansing, MI. Wendy had always known she would fly, but didn't think seriously about it until a fellow officer took her for a ride and "showed" her how to fly the airplane.

While a police officer, Wendy earned her private and commercial ratings, instrument and multi-engine ratings, and instructor and instrument instructor certificates. During her first solo, the pilot that took off ahead of her flew all the way around the pattern lining up on the opposite end of the active runway. Both airplanes were on short final at the same time and when it became apparent that the other pilot really planned to land on the wrong end of the runway, she executed a go around. The other pilot landed on the grass next to the runway as if he truly expected her to land also.

Wendy served on the Aviation/Aerospace Education Council of Michigan, was the director of Aerospace Education for Michigan Wing, Civil Air Patrol, was selected as the Outstanding Young Woman of America from Michigan, and has been listed in Who's Who of Rising Young Americans.

After 10 years as a police officer, Wendy left police work to serve as an adjunct professor in the Department of Aviation and Flight Technology at Lansing Community College. Wendy went back to school and earned a BS in aircraft engineering and an MS in mechanical engineering from Western Michigan University. Wendy is pursing a Ph.D. in aeronautics and astronautics at Stanford University. She also studies Russian with the hope of participating in international space research and sings in the Stanford Russian Chamber Choir.

LU HOLLANDER, a communications professional, is a "for the fun of it" private pilot who has held her license since 1970. She has more than 1,000 hours in Cessna 172s and 182s, a Piper Dakota and an Aero Commander 112TC and 30 minutes in a United Airlines Boeing 747 simulator. Her favorite flight occurred when she took her 90-year-old grandfather up for his only flight to view his lifetime farm home near Old Glory, TX.

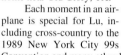

Each moment in an airplane is special for Lu, including cross-country to the 1989 New York City 99s Convention and many other long-distance flights.

Lu has been an active member of the Oklahoma Chapter since joining the 99s in 1979, holding most chapter offices. She edited both chapter and section newsletters at various times and, at the International level, served as editor of *The 99 News* from 1980-86.

In 1988 she was elected to the International board of directors and has since served as secretary (1990-92) and International president (1992-94). Her 8-year stint on the International board will conclude in 1996.

Other accomplishments in the organization include design, editing and production of the coffee-table size *History of The 99s, Inc.,* produced for the organization's 50th anniversary in 1979, and *Sixty and Counting,* the 60th anniversary supplement to the first volume.

In addition, Lu is the first woman to be named to the board of trustees of the Oklahoma Air and Space Museum in Oklahoma City and currently serves as its secretary.

ROBIN H. HOLLEY, born Sept. 14, 1943, in Fairbanks, AK. Flight training includes: single engine, land and sea at Fairbanks, AK. She has flown the Cherokee 140, Cessna 150/150 on floats, Cessna 185 on floats.

Married to Curtis. Spends half a year in Chandler, AZ, in hangar house at Stellan Air Park. Spends summer in Fairbanks, AK, using Cessna 185 on floats to reach remote cabin and loves fishing. Husband is pilot for United Airlines. She worked as flight attendant on DC3 and C46 in Alaska, 1975-76.

CLAIRE M. HOLMBLAD, born July 7, 1940, in Worcester, MA. Flight training included Auburn/Lewiston Airport, Auburn, ME, private pilot license. She has flown the Cessna 152 and Piper Cherokee 180.

Married to Roy V. Holmblad and has two children, David, 32, and Duane, 27. She also has two grandchildren, Anna Holmblad, 3, and Seth Holmblad, 1. She never thought about flying. It started when she gave her husband a Christmas gift of two flying lessons. He then gave her a birthday present of the same. She soloed March 21, 1992, and they both got their private pilot licenses on July 28, 1992. They bought their Cherokee 180 in January 1993 and have been having fun. She cannot imagine her life without flying!

HELENE BRADFORD HOLTON, Iowa Chapter, North Central Section, entered the magical world of flying in 1941. She owned a J3 Piper Cub. She received her private license in 1942, her commercial rating in 1960, and her instrument rating in 1987. Through the years Helene owned different single engine planes with three fellow pilots in Chariton, IA.

Helene is a life member of the 99s, Inc. She has served in all offices of the Iowa Chapter. She was a charter member of the Iowa Air Age Education Council. During the summer of 1957 she taught an air age education workshop to a group of elementary teachers at Drake University, Des Moines, Iowa. As a follow-up during the following year, Mrs. Holton flew her 1948 Stinson Station Wagon to various towns in Iowa to visit those teachers and their students. Helene also brought the enchantment of the Air Age to her own classroom. Many of Mrs. Holton's former students say she inspired them to choose a career in aviation.

Probably Helene's most memorable flight was when she helped ferry a Piper Pacer 135 to Anchorage, Territory of Alaska in 1958 (before Alaska became a state).

Since retiring from 35 plus years of teaching school, Holton now owns a Cessna 150 which she has flown to many places in the United States including 99 meetings. She has one sister, Doris Mann; one son, Daniel; three grandchildren, four great-grandchildren, two ex-husbands, and one very precious black cat.

JILLIAN HONISETT, born Oct. 28, 1934, in Hampshire, England. Received her flight training at the Sherwood Flying Club and Leicesterhire Aero Club at Nottingham and Leicester, England.

She has flown the following planes: Piper Cherokee; Cessna 150, 172, 152; De Havilland Chipmunk; Piper Tri-pacer; Rollason Condor; Robin 2100; and Piper Warrior.

She has held a private pilot license for 25 years and been a member of the British Women Pilots Assoc. for the majority of that time. This has enabled her to be involved in numerous aspects of flying at home and overseas, meeting hundreds of like-minded people in all walks of life. A very enriching experience.

MISAKI HONJO, joined the 99s in 1991. Helicopter training at Texas and Oklahoma. She has flown the helicopter 269C, Huse 300, R-22 Robinson 22, and Jet Ranger.

She was employed as copilot at Japan Flying Service. She flies AS365N (French made Aerospatiale) as copilot for Japan Flying Service, which is Toyota Automobile Group, 40 to 50 hours monthly. This flight is to take VIP to various places. Therefore, not only flying technique but also it requires how to take care VIP. She experienced taking president of Ford Co.

She enjoys working as copilot and cabin assistant meeting many wonderful people and learn many things through her job.

PAULA G. H. HOOK, born on Dec. 5, 1944, in Bedford, IN, but grew up near Indianapolis. After earning home economics education degrees from Ball State University and St. Francis College, she moved to Ohio in 1989 to pursue a Ph.D. in adult education/rural sociology at the Ohio State University in Columbus.

While serving as county staff with the Purdue Cooperative Extension System in Auburn, IN, Paula earned her private pilot license in 1976, joining the 99s later that year. In addition to several local committee leadership roles, Paula served two and a half years as Indiana Chapter chairman and became a charter member of the Three Rivers Chapter. While in Indiana, she also flew as pilot or copilot in several local air races.

As a graduate student, she now maintains NC Section membership while working part-time as an academic advisor with University College at OSU.

KATHERINE M. HOOPER, born Nov. 19, 1957, in Belleville, NJ. Joined the 99s in 1994. Flight training includes private pilot license. She has flown the 152, Siai Marchetti SF260, T-6, Super Decathlon, and T-34. Hooper has a son Raymond Lees, 17. She wanted to fly all of her life and finally had an opportunity to do so in 1991; she got her license in July 1994. She loves aerobatics and has done a lot in the Marchetti SF260 and was checked out in Super Decathlon last fall. She is now planing on going for her commercial rating, and of course her IFR, multi-engine ratings. She will continue to work on aerobatics and she hopes to get her own Warbird (which she loves) someday.

Hooper simply loves to fly, she loves the smell of aviation fuel, she loves being around airplanes, but mostly in them, looping and rolling.

JILL ANN FAULKNER HOPENMULLER, born Sept. 10, 1946, in New York, USA. Joined the 99s in 1980. Flight training includes single engine land, instrument, seaplane, rotary. She has flown Cessna 150, Cessna 172, Aeronca 15AC (floats), Cessna 140, Bell 206, Mooney 201, Robinson R22.

Jill is an instrument rated pilot. She flies Cessna 172s. Jill also likes to fly floatplanes and helicopters. As a youngster Jill never envisioned that she, a girl, would fly airplanes. Her father was an instructor in the US Navy during WWII. Her maternal grandmother retired after 25 years as an aircraft mechanic for American Airlines at Kennedy Airport.

Jill started flying in September 1975 after a sightseeing flight with her husband Steve, in Bar Harbor, ME. In 1976 flying their Cessna 150, Jill soloed in February and earned her license in October, before Steve earned his license. Jill added her instrument rating in November 1979, six weeks before she gave birth to her son, Alex.

Joining the 99s in 1980 Jill is a Long Island Chapter member and a past chapter chairman. She is the current governor of the New York - New Jersey Section.

LINDA J. HORN, Colorado Chapter, has always been interested in flying and enjoyed commercial flights and flights with friends, but never considered doing it herself until her ex-husband decided he was going to learn to fly in 1972. Well, if he was gong to learn, she was at least going to be able to get the thing on the ground and to know where they were. So, she decided to take ground school with him. Turns out, it was lots of fun and she decided to go ahead and get her license. The husband didn't last, but the love of flying did, and also she got the airplane.

She has been a member of the Colorado Chapter since 1975 and has always been active in many of the chapter activities. She has been chapter treasurer since the early 80s and it seems to be a lifetime job.

BARBARA HORNBECK, first chapter chairman and charter member of Idaho Chapter 99s. Born March 6, 1926, in Kansas City, KS, spent her early years on a cattle ranch in Nebraska. In 1940 she moved to Boise, ID, and after several years as a legal secretary, she married Marvin Hornbeck. Marvin was a pilot and operated a dude ranch in the primitive area of Idaho back country. It soon became evident that Barbara would have to learn to fly. This was the only access to the ranch other than horseback. She took lessons from a famed Back country pilot, Bill Woods, and Barbara and Marvin raced to see who would become the first commercial pilot in the family. Of course, Barbara won.

In the "early fifties" the Hornbecks bought land on the middle fork of the Salmon River and developed Pistol Creek Ranch. They flew many different types of aircraft from a J3 Cub to a C-206. They hauled everything imaginable, from ranch guests to gasoline, food and supplies, and occasionally a Shetland Pony for their daughters, Pat and Jackie. Pat states she thought it was a great privilege to ride in a car. All she knew was airplanes and horses. She never saw anything by car until she was grown. The Guest Ranch became a watering hole for Hollywood celebrities and visiting politicians. It was the site of many 99 flying activities as well. It was a wonderful life and they loved it. Barbara stopped flying in 1965, after her husband was killed in a plane crash. She died in 1971, not aware of the great contribution she and her husband had made to the science of Back country Flying. The 99s owe her a debt of gratitude for starting and nurturing the local chapter and leaving a legacy of flying and fun in Idaho.

HELEN HORSEY, born in the Blue Ridge Mountain Range of Maryland. The romance of flying captured her heart when she was 12. Resolved then she was going to be pilot, each day for 34 years she scanned and savored the glories of the sky.

It seemed like eons waiting for opportunity, but with a passion for nature and the beauties of earth and sky, she continued to dream and learn, reading books on "How to Fly" and other aviation literature. She became a licensed clinical social worker and obtained a MA from Howard University in Washington, DC; the profession kept her busy, but not too busy to marry and give birth to two sons.

Finally in Baltimore, MD, the doors of a Beechcraft Sport parted and at age 46, Helen climbed into the left seat for her first lesson. Soon she had her private pilot certificate and was off on jaunts around the countryside in the Beech Sundowner. Later she took a job in Palmdale, CA, and flew the high desert for a while, transistioning to the Piper Warrior.

Helen now flies Archers out of Brackett Airport in Pomona and lives in nearby Brea, CA. Engaged in study for her written instrument test, Helen, at age 63, looks forward to an IFR rating and more time to fly upon retirement. She says, "I've always journeyed latitudinally and longitudinally, but now I travel in the third dimension." Her brother, Bobby, a dentist in Pasadena comments, "I have two sisters. One – you can't get her into a plane. The other – you can't keep her out!"

CAROLYN FAY HORTON, born Nov. 23, 1936, Bear Creek (Search City) AR. Joined the 99s in 1983. Private pilot license and SEL rating. Divorced and has five children: Carol Peerson, 38; Michael Borick, 32; Joanna Borick 25; Monica Borick, 24; and Verna Marie Borick, 21. Also has grandchildren: Tiffany Wytaske, 17; Richard Dean Wytaske, 15; Matthew Martinez, 6; Layla Ahngel Teles, 3; and baby Stover expected in April 1996.

While working in the field of her parents Snowball, Arkansas Ozark farm, she dreamed of someday flying. June 1982 at Claire Walters Academy, Santa Monica, had her first lesson with instructor David Davis AKA "Rip" Bell who in July 1980 Detroit murdered his wife. Eight years later was caught and now serving life sentence. March 1983 at age 47, she was a private pilot and 99. June 1983 bought a 1966 Cessna 172 and still owns N4441L. Active member of Palms 99 and participates in air races. Founder of Aviator/Aviatrix Single Pilot Assoc.

Joined Angel Flight 1992 volunteering to fly for non-profit medical group. October 1993 the celebrated 2500 mission of Angel Flight flew with cancer patient Kathy Ebner from San Jose to SMO.

April 1993 joined Civil Air Patrol Squadron 51 SMO. Average is two years for flying ratings but by September 1993 had her rating as CAP transport pilot, mission pilot standard, high mountain qualified search pilot, cadet orientation pilot and CAP radio operator. Her daughter Joanna Borick became a member of CAP ground crew.

Serving as public affairs officer for Squadron 51, January 1994 the big earthquake hit there. Camera and video in hand and working with others in Van Nuys American Red Cross she filmed an emergency air lift of Los Angeles County Sheriff helicopter. This video footage was used by TV news stations and was requested by CAP National Headquarters. June 1994 received Public Affairs Officer Award from Group One in California Wing CAP. Dreaming of opening Harrison, AR, Airshow September 1996.

MARILYN HORVATH, Indiana Dunes Chapter, North Central Section. Received her private pilot license in January 1990. Served as chapter secretary 1993-94 and 1994-95; and chapter vice chairman 1995-96.

NORIKO HOSOYA, born Feb. 10, 1961, in Shizuoka, Japan. Joined the 99s in 1989. Flight training included Miho Airport, Chofu Airport and North Las Vegas Airport. She has flown FA20, PA28, C-150, C-152, and C-172.

Hosoya is married to Shozo and has one child, Saeko, 3. Born in Shizuoka where a small airport "MIHO" is near by, she convinced her parents to support her to take flight lessons. At age of 18, she moved to Yokohama to be close to the college Tsurumi-Daigaku, so she flew from Chofu airport at this time. But she was so busy studying and did not have time to fly. After graduating the college, she went to Las Vegas for flying lessons at North Las Vegas airport; there she got the license. Now, she works as a dentist, housewife and mother, not active flying but dreaming someday if she could fly the plane to take care of her patients. This dream might be impossible in a small country like Japan.

MAURINE HOUCK, joined the Santa Clara Valley Chapter in 1982. She has been a very active member, serving on several committees and offering her home for chapter events. She and her husband Warren have flown their Bonanza A36 to Alaska. Maurine is a serious poker player.

MARY D. HOUGH, born Sept. 7, 1946, in Chicago, IL. Born, raised and still lives in Chicago. Started flying in 1970, received her private and joined the 99s in 1971. Holds private instrument rating and flies a Beech Debonair. Planes flown: C-150, C-172, C-182; Piper: Pacer, Cherokee, Warrior, Aero; Beech: Bonanza, Debonair; Tiger; BO 209. Participates in proficiency meets and x-country air races: the Angel Derby and Powder Puff Derby (fifth place in 1974). Chairman and board member of the Illi-Nines Air Derby. Served as chapter treasurer, VP and the current chairman.

Studied math at Loyola University (three years) and photography at the American Academy of Art (two years). Currently, an IOM member of the Chicago Mercantile Exchange and trades commodities.

Has two children, Jane, 27, and Jason, 26. Credits the 99s for the opportunity to make life long friendships and improve flying skills through aviation activities.

ROBIN ROBBINS HOUSE, born Nov. 4, 1960, in Mesa, AZ, she was raised in Louisiana where she cherished two childhood dreams (one to ride horses the other to fly).

She married Ike House in 1982, and after their third child was born, he encouraged her to adopt a hobby. When choosing between horses and planes, her husband said, "go fly, it's cheaper." Following Ground School then lessons with Floyd Lindsey, she soloed on her 28th birthday (wondering if she'd reach 29) and received her private pilot license May 22, 1990. She has flown 150, 152 and 172 Cessnas. She finds flying challenging and awesome. A favorite experience was in riding in a B-17 after an air show. She is a physical therapist and her husband is a dentist. Their children are: Dwight, 10; Reuben, 9; and Brittany, 7. She is also getting to pursue her other dream on "Major" (a thoroughbred) who loves to jump (fly) over fences.

BETH G. HOWAR, fell in love with flying the first time she flew in a small plane – the back seat of a C205. She vowed to learn to fly when her youngest child entered kindergarten – and she did, in 1973. She became a 99 as soon as she qualified, earned a bunch more ratings, flew a Powder Puff Derby and a Palms to Pines Race, and then got serious.

After apprenticing for a few year as a CFI and Chief CFI, she opened her own flight school, called Above All Aviation so that they would be first in the yellow pages, in 1984 and has been enjoying full-time airporting ever since. She operates 13 airplanes and has several flight instructors including her son. She often participates in school career days and enjoys encouraging young women to choose goals that excite them with the aim of finding a career that is a source of pleasure and satisfaction.

PHYLLIS HOWARD, a proud Okie pilot from Oklahoma, learned to fly in the Oklahoma wind in 1965 with the FAA Flying Club. She added to the private pilot certificate, a commercial rating, instrument rating, multi-engine rating, flight instructor certificate with instrument and ground instructor.

For Phyl, aviation achievements, highlights and experiences include being instrumental in organizing the first annual Okie Derby, a proficiency air race, in 1978. The 17th annual Okie Derby remains the main resource for the Oklahoma Chapter to fund aviation scholarships. Phyl's main interest in aviation is air racing. She particular enjoys competing with best friend 99, Jan Perry.

Phyl is employed by the Federal Aviation Administration as special assistant to the director, AVN. She is associated in some airshow circles with the famous flying exhibit, the FAA DC-3. Phyl was the first woman FAA DC-3 program manager. She considers the 99s, especially the Oklahoma Chapter, as an extended family.

Granny Phyl shares the joy of flying with her 15-year-old granddaughter, Cheree Hall, student pilot, in a Cessna 172, N1922V. Husband Bob is favorite air racing partner.

EVELYN GREENBLATT HOWREN, born July 28, 1917, in Atlanta, GA. A native Atlantan, was bitten by the flying bug when she was 18. She soled in a Piper J3 Cub at Candler Field, Atlanta, GA, in early 1941. By December of that year she earned her private rating, joined the Civil Air Patrol and helped organize the first all-woman CAP squadron. She remained active in CAP for many years and earned a command pilot rating.

In June 1942 she was one of three pilots named to the first class of eight women Air Traffic Controller Trainees, later becoming a member of the first class of 30 Women's Air Service Pilots, Ellington Field, Houston in April 1943. At disbanding of the WASP in December 1944, Howren test-flew and ferried 20 or more types of planes for the military. By the end of WWII she had logged 3,000 hours flight time, with instrument, instructor's and twin-engine ratings and had flown more than three dozen civilian craft.

Howren became a flight instructor at Candler Field. In 1947 she established Flightways, Inc., a full service, fixed-base operation that was sold to Lockheed in 1968.

Evelyn Howren was one of very few women, active in both management and flight operations of any fixed-based operation in Georgia. A member of the 99s in Georgia and an organizer of the Atlanta Women's Aero Club, she flew in the All Women's Trans-continental Air Race in 1951 and was appointed a captain in the United States Air Force Reserve that year. She was secretary-treasurer of the Georgia Aviation Trades Assoc. from 1950 to 1965.

SUE HRINDAK, a native of Dyer, IN, earned her license in June 1994, at the age of 55. Although relatively new to the sport of flying, Sue comes from a interesting aviation background. Her father, Steve Beville, was an Air Force Captain during WWII and became well-known as pilot and co-owner of "The Galloping Ghost," a P-51 Mustang with a winning record during the days of Cleveland Air Races. Now a retired corporate pilot, he continues to fly. Her brother, Steve, is a pilot for a major airline.

Sue, a bookkeeper, has two children and two grandchildren. She and husband Mike, now retired from NCR Corp., are yearly visitors to Oshkosh and the Reno Air Races. Sue loves flying, and is proud to be a member of that elite group of women pilots, the 99s.

JOAN HRUBEC, is a native Ohioan who has been a member of the All-Ohio Chapter for 45 years. Her aviation interests started to materialize during the WWII era in the form of model airplane building and competitions. She served as a CAP cadet during high school. Stephens College in Columbia, MO, allowed her to pursue flying and in 1949 she became a licensed pilot.

Worked 32 years in the specialized machine tool industry in production and inventory control management. Purchased two airplanes, flew numerous races and officiated at many flying events. Has served the 99s on chapter, section and International level.

Since 1985 has been the administrator for the International Women's Air and Space Museum in Centerville, OH, and serves as a trustee on the board. Memberships include: Zonta Club of Cleveland, Silver Wings, AOPA and Women in Aviation, International.

In 1957 received the "Ohio Woman Pilot of the Year" award of the All-Ohio Chapter, was winner of the 1967 Indiana FAIR race, and was copilot of the first place team of the 1955 International Air Race.

LYNNE KASTEL HSIA, born Los Angeles, CA. Joined the 99s in 1980. Flight training included 1977 private pilot license; 1983 instrument rating. Has flown the Piper Cherokee. Employment: city planner, budget analyst, MA in city planning, University of Southern California.

Married to Yuchuek (city planner) and has one child, Rebecca, age 7. Chairman, Monterey Bay Chapter 99s, 1984-86. Member of Southwest Section Board 99s, 1986-present. Participated in and organized air races, flying companion seminars, county air tours, and pinch hitter courses. Member 99s, California Pilots Assoc., and Aircraft Owners and Pilots Assoc.

RUTH CLIFFORD HUBERT, born in Lakeland, FL, March 1971. First solo in January 1940; obtained private license April 1941; commercial rating November 1942; and flight instructor rating February 1943. She first joined the Florida chapter of 99s in 1941. She joined the Civil Air Patrol in 1942 and was active in this organization until she joined the Women Airforce Service Pilots (WASP) in 1944. Was test pilot for US Army Air Corps at Macon, GA. After the WASP disbanded she was chairman of the Florida Chapter of 99s (1946-47) during which time the Florida Chapter planned and staged the First All Woman Air Show in Tampa, FL. The First Transcontinental Air Race for Women from Palm Springs, CA, to Tampa, FL, (later known as the Powder Puff Derby) was held in conjunction with this Air Show. In 1977 she was honorary terminus chairman for the 30th and last, Powder Puff Derby. Has been active in Search and Rescue Operations for Civil Air Patrol for 20 years and holds rank of lieutenant colonel.

She is still flying at the age of 78 with more than 2,200 hours. Married Peter Hubert in 1946. Two children, Clifford Hubert and Linda Hubert. Two grandchildren, Patrick Benson and Samantha Hubert.

CHARMA HUDDY, first obtained her glider rating in 1975 and her private pilot single engine land in 1986.

She is a member of the Scioto Valley Chapter of the 99s. She has held chapter offices of secretary, vice-chairman and chairman. She and her husband John are active in sport's car clubs.

DOROTHY "DOTTIE" HUGHES, born July 5, 1930, in Houston, TX. Flight training included ASEL. She has flown Cessna 150, Cherokee 140, and Cessna 175. She has been married to Bob for 44 years and has two children, Richard, 41, and Monica, 39. Also has grandchildren: Travis, 13; Trisha, 11; Regan, 10; John 9 1/2; and Robyn, 6 1/2.

Flying an airplane was one of those things that only the rich and intelligent ones could do. It was just unattainable, let alone fly on an airliner for her. Her first flight on an airliner came when she was 37 and the unbelievable sight below has never left her. Her husband Bob has had a long, deep passion for flying and was able to fulfill his dream of getting a license after retiring from the Navy, so she got to fly in small planes. He so wanted her to fly – however, the landings were where her uncertainties laid. Then one time he asked her to learn so she could land in case something happened to him while they were flying. Needless to say – it changed her life when he worked with her on landing (now one of her confidences in flying).

Since joining Golden Triangle Chapter of 99s in 1974, God has placed her in the presence of very knowledgeable and fun women who have a common bond – Flying! Through the years she has served as chapter chairman, secretary, treasurer (presently serving again, and again …), scrapbook, and money-making chairman.

Her other activist include bowling, golf, sewing, handwork (cross-stitch), ceramics, cake decorating, volunteering at one of local hospitals in Arlington, TX, and enjoying five grandchildren.

Non-flying people are always amazed when they learn she's a pilot and say, "You fly a real plane?"

KELLI HUGHES, was born Feb. 21, 1958, in St. Joseph, MO. She learned to fly at Southern Illinois University in Carbondale, IL, in the late 70s. She was to achieve collegiate fame in 1982 at the National Intercollegiate Flying Assoc. National Airmeet held in Bakersfield, CA. Becoming the first woman ever to earn the distinction of All Around Top Pilot since the inception of the sport in 1928.

She would go on to fly a trans-Atlantic flight in 1983 from Reykjavik, Iceland, to Fort Worth, TX, in very hostile flight conditions and later restore a T-6G Texan which would win best T-6 at Oshkosh in 1993.

She was president of the Fort Worth Chapter of the 99s in 1986-87 and again in 1993-94 during which time her chapter hosted the South Central Spring Section.

Some of her community activities include chief navigation judge for the NIFA and sponsorship of a young boy at Happy Hills Boys Ranch.

She is currently employed with American Airlines as a Super 80 pilot with more than 12,000 hours.

MARY MASON HULL, learned to fly in 1965 at Williamsport Lycoming County Airport in Pennsylvania. She has also lived in Vermont, Costa Rica and Texas.

Mary flew copilot in the 1967 Angel Derby from Montreal, to Miami, with Pilot Helen Brass Porter Sheffer and the last Powder Puff Derby in 1976 from Sacramento to Wilmington, with Pilot Joyce Conklin Williamson. She owned 1947 Navion 8865H and flew it to 1971 International 99s Convention in Wichita, KS, with her 2-year-old son. She was delegate to 99s Conventions in Coeur d'Alene, ID, Washington, DC, and served as Central Pennsylvania Chapter chairman.

Hull's educational background is a BS in home economics from Mansfield University, PA. She worked as an electric utility home economist, teacher, small business owner, antique dealer and real estate salesperson. She says home economics taught her how to read the instructions!

Some of her community activities include: Girl Scout leader, church council member, chamber of commerce member, past president of Pennsylvania Radio Common carriers, past secretary of Pensionados and Rentista Assoc. of Costa Rica.

Mary is married to Carl R. Johnson, a non-pilot and retired engineer. She has two sons, Eric and Jason Hull, two stepdaughters, one stepson and six grandchildren.

JUDITH HUMPHRIES, born in Chicago, IL, on April 9, 1949. She has spent roughly one-third of her life in each of three locations: Illinois, California and Colorado, and makes her home in Grand Junction, CO. Judy joined the High Country Chapter of 99s three months after the chapter was chartered in November 1990. She has served as membership chair and chapter chairman.

Judy obtained her private certificate in August 1990. She owns a Cherokee 235 which she flies for business, pleasure and 99s activities. A real estate appraiser by profession, she has been in private practice for 15 years. In addition to several professional associations, Judy also belongs to the Cherokee Pilots Assoc. and is the Grand Junction Area Representative for the Colorado Pilots Assoc.

JEAN HUNT, was 63 when she got her pilot certificate, in 1991, in Woodland, CA. The death of her husband motivated her to learn to fly.

Her father flew. He was a Quiet Birdman. She flew a BE-23, Sundowner.

Her instructor was Robert Griffith while Aaron Schiff and Jon Stephani assisted.

RACHEL SNEAD HUNT, born Dec. 24, 1922, in Knoxville, TN. Being one of the lucky ones to have been around when Charles Lindbergh made his flight, at the age of five years flying was to be part of her life.

Her birth year made it possible for her to serve her country in the United States Marines. She went to Lakehurst, NJ, to the parachute rigger school and later to Parris Island, SC, to pack, repair and service parachutes for F4U's and SNJ's during WWII.

After the war she learned to fly under the GI Bill at the University of Alabama in Tuscaloosa. The first solo was in J-Cub. (September 1946). After their daughters were in college she went back and earned an instrument rating with husband Albert, has flown all over the US, also Eastern and Western Canada and a few trips to the Bahamas. They owned a Cessna 172. While in the service, her greatest disappointment as a buck private about to have her first pass and very first flight home, PFC arrived from another air field and she had to give up her flight. She did get up to the rate of sergeant before discharge.

FRANCES HURITZ, there first flying lesson June 20, 1972. Soloed October 1972. Bought 1963 Cessna 150. Acquired pilot license October 1973/sold Cessna/bought 1969 Cherokee N7864N. Reprinted 1983 white with red/blue stripes; installed newest avionics. Flew Brantly B2B helicopter five years. Instrument rating 1985. 675 hours. Joined Chicago 99s 1975. (Born Feb. 13, 1927.) Entered banking 1944. Celebrated 50 years banking July 7, 1994. Senior vice president/director, First National Bank of Lincolnwood.

Graduate, School of Banking, University of Wisconsin 1983; Illinois licensed real estate broker 1960; Internal Revenue Service Licensed

Enrolled Agent 1970; certified professional secretary 1971. Chairman/board member, Lincolnwood Chamber of Commerce, 1983-92; appointed commissioner 1994 on Lincolnwood Economic Development Commission.

Hobbies: flying, travel, opera, ballet, theater, golf, playing piano and accordion. Visited 64 countries and seven continents. Met 99s in Australia, New Zealand, Kenya, Finland, Israel, Hawaii, Alaska, Chile and a Seattle 99 in Antarctica.

MARI HURLEY, Imperial So-Lo Chapter of the 99s, Inc. Born Oct. 27, 1929, in Riverside, CA. She moved to El Centro after her marriage to Cliff Hurley, and taught homemaking in the schools for 37 years.

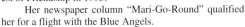

She earned her private pilot license in 1970. Now holds commercial, instrument, MEL, and SES ratings.

She owns a twin engine Piper Geronimo, and has logged almost 3,000 hours.

She was the first chairman of the Imperial So-Lo Chapter of the 99s. The chapter sponsored her in the Forest of Friendship in 1995.

Her newspaper column "Mari-Go-Round" qualified her for a flight with the Blue Angels.

Mari is a charter member of the Baja Bush Pilots, and the Dawn Patrol, and a director in the California Pilots Assoc. She also belongs to AOPA, San Diego Aerospace Museum, and Emergency Volunteer Air Corp.

She received the Certificate of Merit for outstanding efforts as a teacher in California Aerospace Education Assoc. from the FAA.

CHERYL "CHERRI" HUSSAN, from Norwalk, OH, obtained her pilot license in 1985 and instrument rating in 1987, Cherri has owned a Cessna 150, 172, Smith Mini-Plane, and currently a Luscombe 8A. She was the first student in an experimental prototype "Buddy-Baby." An active 99 member since 1986 of All-Ohio Chapter, she won their 1990 Achievement Award.

Cherri is pictured here with the rotating trophy which is the All-Ohio Chapter's Achievement Award. Each annual winners' name (since 1956) is engraved on the trophy's brass plaques. In 1992 a new trophy was purchased and this one is now in Centerville, OH, at the Women's Air and Space Museum on display. Cherri has logged more than 750 hours, has flown and co-chaired Buckeye Air Rallies, hosted and helped with many 99 events, and won the Aviation Activities Award for her chapter. Memberships: AOPA, Luscombe Assoc., NRA, OH21, St. Anthony's Catholic Church, Huron County Airport Authority secretary, Republican Women's Club Member, and treasurer Wheeling Yacht Club. Monroeville High graduate, attended Firelands and Associated Schools in Florida.

Managed a travel agency. Married Ben Hussan – a multi-engine instrument commercial pilot. Office manager for his steel company (USS) and environmental partnership (AIC) and started Cher-Air. They enjoy aerobatic flying, hunting, including bird hunting with Cherri's Brittany – Ozzie. She has two stepdaughters, four grandchildren.

BARBARA M. HUSTON, born April 25, 1932, in Burlington, IA. Joined the 99s in 1991. She has flown Cessna 152 and 182.

Married to Gordon and has three children: Roberta, 39; Rebecca, 37; and Michael, 32. She received her pilot license in 1983 and does most of her flying in her C-182. They belong to EAA, IFF, Venice Cloud Busters, and Lincoln Park Pilots Assoc. Her husband is a retired EAL Captain and they enjoy seeing "The birds eye view" they get flying their plane.

CHRISTINA K. HUTSON, born Nov. 23, 1930, Melrose, MA. Joined the 99s in 1964. Flight training included private, instrument and commercial SEL. She has flown Cessna 150, 172, 182, and T-206; and Piper 28-181. She is an RN and owns a business, Alpha Medical Resources, Inc. (rents medical equipment to motion picture and TV industry). Hutson is a widow and has three children: Paul R., Christina A., and Mark T. Grandchildren: JoAnn, 13; Danae, 10; Christina, 7; and Matthew Jerome, 1 1/2. She enjoys golf, skiing and scuba diving. Has enjoyed pleasure flying for 30 years. Flew in local air races (DAR). Currently owns a PA28-181. Loves flying and the 99 organization.

JOYCE ILVES, has lived in California 48 years and in Orange County over 45 years. She has two children a girl and boy and four grandchildren, one is a boy.

At the end of 1978 her husband who had taken lessons before asked her if she would like to learn to fly. It sounded exciting but had doubts that she could do it as they know women don't fly. Her first lesson was in a C-152 in Chino, CA. After that she was hooked. After she had soloed they had to discontinue their lessons due to financial problems, but she knew she would continue

later. In 1984 they decided to continue their flying lessons, this time out of Fullerton, CA, in a C-172 so one of them could sit in the back and observe. After she soloed they bought their first plane, a 1969 C-172. She took her check ride on Oct. 4, 1984. It felt like a dream. In spring 1985 the Fullerton 99s had a membership social in the airport lobby. It was wonderful to be able to talk to other women with the same interests. She became a 99 and went to her first International Convention that July.

At the end of 1985 they bought a C-205 and in 1989 a Comanche C-260. This, in turn, was traded in on a Seneca II in 1994. She received her twin rating in Las Vegas, NV, June 25, 1994. She passed her check ride but her fellow student (a male) didn't. "My husband, Henry, was so proud of me as was I."

She is self-employed and has owned a blueprint and drafting supply store for over 15 years. It is a great feeling to be not only a woman pilot but a business owner also.

She is continuing her flying and hopes at least one of her grandchildren will follow in her footsteps. They really enjoy flying with her. "Thank you to the 99s for being there for us."

LAURA INGALLS, a 99 member from 1930 to 1937, was called in *Clipped Wings*, published in 1942, one of the most distinguished pilots America has ever produced. In 1934 she was presented the Harmon Trophy as the world's outstanding woman flier. Two years later she came in second in the Bendix race, with a $2,500 prize, the winners being Louise Thaden and Blanche Noyes.

A tiny woman, she had every advantage when young, studied in Paris and Vienna, was an accomplished linguist. She graduated from nursing school at 16; was a talented musician; went on a theatrical circuit as a Spanish dancer. But planes lured her to the airports again and again; she walked miles just to look at them. Later on, her varied talents included great mechanical aptitude, and she cared for her own plane.

In 1928 she learned to fly a Roosevelt Field, soloing in 13 hours. Two years later she completed another course at the Universal School of Aviation in St. Louis, the first American woman to graduate from government approved school, with a Limited Commercial license.

From 1930 to 1939 her spectacular rise was meteoric. Muskogee, OK, promised her a dollar for every loop she completed at their field above the record-breaking 344 she had flown in St. Louis. They thought they'd have to shoot her down: she did 980. They same year she did 714 barrel rolls in a single flight. In 1934, she was the first woman to circle South America or fly over the Andes; in 1935, she broke records for non-stop east-west, and west-east transcontinental flights.

Then in 1939 she dropped anti-war pamphlets over the White House, was grounded, then imprisoned, and dropped out of sight after her release in 1943. She died in 1961.

KRISTINE "KRIS" GANDL IRVIN, born July 24, 1958, in Kirkwood, MO, and grew up in Little Rock, AR. She received a degree in accounting from the University of Arkansas in 1980 and is currently a CPA in Little Rock and a member of several professional organizations.

Kris received her private pilot license in 1988 and shortly thereafter purchased a 1958 Champion 7EC. She currently owns a Cessna 172 and part interest in an H35 Bonanza. She also flies a 1946 Piper Cub and has some time in a rare 1972 McCulloch J-2 gyroplane. Kris plans to get her rotor-craft-gyroplane rating in this aircraft. The Cub and the gyro are owned by her finance, Ron Herron who is a pilot and an A&P mechanic and IA.

Kris is a member of the Arkansas Chapter of the 99s, Inc. and has served as chapter chairman. She is also a member of several other aviation organizations including the American Bonanza Society, Aircraft Owners and Pilots Assoc., Experimental Aircraft Assoc., U.S. Pilots Assoc. and Arkansas Pilots Assoc.

ANITA ISRAEL, born in Calgary, Canada, Nov. 14, 1945. First exposure to Women in Flying was with a woman missionary who would land in their field on the farm in Northern Canada. She would sing songs about her flying. That began her love of flight. After living in the remote areas of Northern Canada, where the only way in or out was by small plane, it was still a desire. However her father felt women didn't need to fly, so she let the desire die for 30 years. After being married to two pilots and owning their own plane Cessna 180, she decided it was her turn in the sun. Three days before her 45th birthday on Nov. 11, 1990, she became a private pilot. It was by far her greatest accomplishment, and a desire fulfilled. Her son Gordon Trzinski's belief that women can and should follow their dream, has been her greatest encouragement. Both her mother and father are proud of their daughter who can fly.

She has been a proud member of the 99s, Inc. since 1987, first as a 66 and now as a 99. She now serves as vice chair in the Western Washington Chapter 99s.

ALDINE VON ISSER, born May 29, 1933, joined the 99s in 1968. Learned to fly Tucson 1967. Holds a commercial ASEL and is professor – UA. She used to fly commercially for hotel in Punta Chivato in Baja, CA. Her husband, now deceased, and two sons were all pilots.

BEVERLY REICH IVY, born June 14, 1963, Long Island, NY. Joined the 99s in 1994. She has flown single engine Cessnas, BE55/76/90, DC-3. She is married to Jack Ivy. Her father is a retired DC-10 AA Captain and her best friend's mother, Sandra Simmons, was first female pilot hired at Braniff in 1974. She remembers Sandra being active in the 99s and participating in Powder Puff Derby. Ivy became a flight attendant for SWA in 1986 and it was in 1991 got her private license. From there she met her husband who is SWA f/o and F-16 pilot, and flight instructed until her first job flying freight in DC-3s for a company based out of San Juan, Puerto Rico. Currently, Ivy is secretary for the Dallas Chapter.

MARLENE JACOB, born Feb. 11, 1971, in Kitchener, Ontario, Canada. Marlene learned to fly in 1990 at 18 and recently got her night rating. She flies Cessna 150/52, 172, Katana DV20 and Piper R-200. Currently she is working at the Rockcliffe Flying Club as she is training for her commercial pilot license.

Jacob's educational background includes a BA in French and German from Carleton University in Ottawa. She was the founder of Carleton University's Women's Rugby Club in 1991. As well, she has participated in Student Work Abroad Programs in Britain and Germany.

She is a volunteer for CASARA, Ottawa's search and rescue team, as well as being a member of the 99s and Canadian Owners and Pilots Assoc.

Marlene looks forward to one day flying for a major Canadian airline.

RUTH JACOBS, born March 23, 1943, Bartow, FL. Joined the 99s in 1973. Flight training includes private pilot license and SEL. She has flown T-craft, Ercoupe, C-182, C-170, and Helio Couner.

Jacobs is married to Jim and has children, Dwayne Kersten (deceased) and Sherman Kersten, 37. She also has grandchildren Kalista Kersten, 8; Christine Kersten, 6; Jason Kersten, 17; and Keisha Kersten, 12. Learned to fly in Schoolcraft, MI, in 1969. Jacobs has flown from Alaska to Russia and Mongolia. She helped organize several flights to Russia and Mongolia; founder and first chairman of Cook Inlet 99s. Hobbies include making quilts for Shriners Children's Hospitals.

DORIS JACOBSON, charter member Maryland Chapter 1963. Born in Sheboygan, WI. She took her first flight lesson on May 5, 1962, soloed June 7, 1962, completed her FAA private pilot check-ride on Dec. 11, 1962, at Sheboygan County Airport in a Cessna 120.

She and her CFII husband, John, moved to Bel Air, MD, in January 1963, where she attended Washington DC 99 meetings until the Maryland Chapter was formed. Doris has served in all officer positions at chapter and section level including governor of the Middle-East (now Mid-Atlantic) Section 1977-79, as well as membership chairman for many years. She is active in Air Bear presentations, speaker's bureau, ground school mini-mester courses at local schools. She maintained a scrap book for the chapter over the years and compiled the history of the first 10 years of the Maryland Chapter and gave history reports at the 20th, 25th and 30th anniversary celebrations of the chapter. She will always remember her cross country trip to Kenosha, WI, when the neighbors came over to check her out to see what sort of person flies to a girlfriend's house for a cup of coffee!

RUTH EVELYN MOVCHAN JACQUOT, born June 18, 1920, in Kongsberg, ND. Joined the 99s in 1947, at the Alameda Naval Air Station. She worked at the Navy base in 1943, building the Folding Flap for the 503-C and becoming better known as "Rosie the Riviter." She went to Stead Army AB in Reno; the war ending and so was the WASP so she returned to Alameda NAS with 90 hours of flying time and a broken heart.

Jacquot came in second in her first air race and has won four trophies for racing. She returned to school to study engines. After graduation she was hired as crew chief on the Flight Line. They had many distinguished guests, even from the Pentagon. No matter what she was doing she had to run up an airplane for them. This is where she met her husband, Lt. Lyle Jacquot, in 1949. Within five months they were married Nov. 18, 1949, at the Naval Air Station Chapel.

They moved to Seattle, WA, Sandpoint. Their baby boy was born Oct. 17, 1952, Curtis Dean. Five months later she entered the Powder Puff Derby. From Lawrence, MA, to Long Beach, of the 54 planes entered, they came in fourteenth. They were living at Phoenix, AZ. Candice Lynn was born at Berkeley, CA, Nov. 13, 1954, now living at Maui, HI. Lyle was a produce buyer for Safeway Stores.

Jacquot became a member of 99s San Francisco Bay Cities Chapter in 1947. She holds a private pilot license and single engine land. Listed in *Who's Who of American Women* 1979-80. Memberships included Lodi Garden Club, Century Assembly Church, and Women's Ministries.

KATHY JAFFE, lover of antique aircraft, she owns a 1948 Stinson and has piloted such exotics as a Tiger Moth, Stearman, Mooney Cadet, Super Cub, Champ, Spartan Executive, C-195, RV-6, Waco, AT-6 and the coveted P-51. In England she went wing-walking on a 450 Stearman as a member of the famous Cadbury Cruchie Flying Circus team. Kathy also loves flying competition aerobatics in a Pitts S2A.

She is a director in North Jersey's Chapter of the 99s and was secretary. She's vice president of the Mid-Atlantic Pilots Assoc., secretary of the Morristown Airport Pilots Assoc., and a representative for the Mid-Atlantic Aviation Coalition, also an active member of International Aerobatic Club Chapter #52, EAA #238 and belongs to AOPA. She does FAA Taildragger seminars, and calligraphy for FAA achievement certificates.

Kathy's husband, Michael Jaffe's, a materials scientist. They have two great sons, Sean and Josh.

GRETCHEN L. JAHN, born Nov. 17, 1951, in Summit, NJ. Joined the 99s in 1985. Flight training includes 750 hours, commercial/instrument ratings and SEL. She owns a Cessna 182RG.

Married to Karl Sutterfield. Raced in Air Race Classic, Mile High Air Derby, Jackpot Air Classic and Shamrock Derby. Active in Colorado Chapter of 99s, Colorado Pilots Assoc., member of AOPA, Cessna Pilots Assoc., Aviation Technical Education Advisory Committee for DRCOG (Denver Regional Council of Governments). Jahn has a BA and MA in experimental psychology. She has been owner of a computer consulting company; currently engineering manager of a software development firm.

TERESA D. JAMES, born raised and educated in Pittsburgh, PA. Her interest in flying was heightened when she met the good looking pilot with the silver airplane. In 1933 she soloed a OX-5 Travelair in four hours 20 minutes at Wilkinsburg, PA, and the following year, she obtained her private license #31249. To earn money, she performed acrobatics throughout the East Coast featuring her 26 1/2 turn-spin. In 1938 she obtained her commercial license the same year she flew the Air Mail in Wilkinsburg, PA. She was the first woman to obtain primary instructor rating at Buffalo Aeronautical, NY, 1940. Earned her secondary rating in 1941 to teach inverted flying at Max Rappapart School of Aviation Roosevelt Field in Long Island.

September 1942, received telegram from General "Hap" Arnold. She was ninth to report for flight check at New Castle Army Air Base for formation of WAFS (Women's Auxiliary Ferrying Squadron). First WAF to fly coast to coast. Served 27 months. She flew 54 types of civilian and military aircraft, including P-51 Mustang and P-47 Thunderbolt. Her husband, George Martin, married in 1942, was shot down in a B-17 over Europe and listed MIA. She learned the facts of his death 40 years later in France.

In 1950 she was commissioned a major in USAF Reserve and served in several commands including Alaska, where she gained a "taste" of "bush flying" on skis and floats. She retired after 27 years of duty with several commendations. Her WAF uniform was donated to Smithsonian Air and Space Museum in Washington, DC. Received OX-5 Outstanding "Pioneer Women's Award" for promotion of aviation.

In 1993 she was inducted in OX-5 Hall of Fame. In 1980 featured in the PBS video, "Silver Wings and Santiago Blue." In 1992 CBS "Top Flight," a video documentary with Gen. Jimmy Dolittle and WWII Aces. Member of International Forest of Friendship. Participated in Air Race from Montreal to Florida. In 1992 book released by Jan Churchill, *On Wings to War – Teresa James, Aviator.*

MYRA JAMISON, received her private license (single engine land) in May 1966.

She became a member of the All-Ohio Chapter of the 99s and later became a charter member of the Scioto Valley Chapter. She has held various chapter offices and is the scrapbook chairman. Jamison has worked with Senior Girl Scouts in an aviation program.

BETH ANN JANSSEN, born Tonawanda, NY, April 30, 1962. Flight training included civilian, Hobby Airport, Houston, TX.

She has flown the following planes: Pitts Special (S1-T), ATR-72, Embraer 110-120, Cessna Citation, Westwind, Sabreliner, Stearman, Taylorcraft, and various other little, funky tail-draggers.

Her pilot employment: Atlantic Southeast Airlines, ATR-72 First Officer, DFW Airport. Joined the 99s in 1990. She was lucky enough to grow up around aviation, being the daughter of American Airlines pilot. She earned her private pilot certificate in Kerrville, TX, and then flew in the Houston area through flight instructor. After flying for a Part 135 charter company in Houston, she was hired by Atlantic Southeast Airlines (Delta Connection). She is currently an ATR-72 First Officer based at Dallas-Fort Worth International Airport. For fun, she likes nothing more than to jump into her Pitts Special S1T to practice for the competitions and air shows in which she participates.

ASTRI ERICA LARSEN JARRETT, born in Risor, Norway, came to the USA as a student to improve her English. She met a medical student, Joe Jarrett, in her typing class. She had only a year's visitor's visa, but a distant cousin, US Senator Henrik Shipstead of Minnesota found a clause in immigration law permitting a permanent visa for an immigrant whose family could vouch for him or her.

Joe and Astri were married after his graduation from the University of Pennsylvania School of Medicine and his internship. Astri received her RN degree from Garfield Memorial Hospital, later earning her BA at West Virginia Institute of Technology, serving as substitute publicist for the college. She received her MA from West Virginia University.

They had four children: Sandra, Scott, Joe Jr. and Jennifer. Astri earned her private pilot license with commercial rating, and at age 80 still assists from the right front seat of their Skylane. She has been a member of AOPA for 20 years, and was one of the first women pilots in the USA to be appointed as an accident prevention counselor, serving in that capacity for more than 20 years. She is past president of the New River Flying Club and was secretary of that club for eight years. She is past president of the Fayette Study Club. She served as president of the Fayette County Girl Scout Council, involved in the American Red Cross Bloodmobile in Fayette County for many years and taught nurses' aides classes. She is past president of the Fayette County and West Virginia State Medical Alliances, receiving the Outstanding Member Award from the latter organization in 1995. She is an ordained deacon in First Presbyterian Church of Oak Hill, where she has sung in the choir since 1942, served as president of the Women of the Church and as Sunday school teacher. The Jarretts have crisscrossed the US, and have made several trips to Alaska, Mexico and Canada. In 1975 they flew a rented Cessna 172 around southern Norway. Over the years they have also flown a Beechcraft Baron, a V-tail Bonanza and a Bellanca.

BARBARA KIBBEE JAYNE, born Albany, NY, July 25, 1914. Joined the 99s in 1938 or 1939, sponsored by Betty Gillies. February 1938 graduated Ryan School of Aeronautics, San Diego, CA, commercial, instrument rated pilot. Ryan check pilot until May. June 1938 private flying Troy Airport, NY. Got instructor's license June 1939, started teaching Glens Falls Airport, NY. November 1939 was rerated for instructor's rating by CAA inspector sent from Washington, DC, for rating pilots for the Civilian Pilot Training Program. December 4, 1939 started instructing on CPTP program for Rensselear Polytechnic Institute, Troy, NY, first woman flight instructor on the US CPTP.

June 1941 hired by the Aviation Country Club, Hicksville, Long Island, NY, for flight instructing and Grumman Aircraft Engineering Corp. courier flights. Added multi-engine land and sea ratings.

July 1943 Teddy Kenyon, Lib Hooker and Kibbee

checked out in Grumman Aircraft's F6F Military Fighter Aircraft becoming full fledged factory test pilots on their various military aircraft and quietly pursued this occupation until the end of WWII.

In 1946 she ran Annapolis Airport, Annapolis, MD, including GI Program and FBO operation. Added student examiner rating.

Moved to California 1950s. Own Cessna 175 business and pleasure. 1970s and "Fly-Yourself Safaris" with Betty Gillies – Australia – South Africa etc. 1980s in Forest of Friendship.

"Nope – throughout it all never "dinged" anything – and, of course a great deal due to our wonderful mechanics. Wonderful life!"

MARION P. JAYNE, is the leading cross-country air race pilot in the world. She has won more races then any other pilot; and to top her racing career in 1992 and 1994, she flew her Piper Twin Comanche in the around the world air races. In 1992 she flew with her daughter Nancy Palozola, placing second. In 1994 her copilot was her daughter, Patricia Keefer and they placed first.

Her flying career started in 1964 when her husband, George took up flying for business. She learned to fly at the Elgin Airport, Elgin, IL, in a Piper Colt. Her first air race was the Illi-Nines Air Derby. She now holds ratings of ASMEL, Air Transport Pilot and multi-engine, instrument flight instructor. In over 30 years of flying she has flown over 30 different types of general aviation aircraft.

She is a mother of four and a grandmother of seven. After her husband died in 1970 she became a business woman. She built a six court indoor tennis facility with a health club. Invested in 11 hangars at the Landings Condominium Airport, Huntley, IL, and created the leading aviation gift catalog, Tailwinds Catalog of the Skies.

Marion promotes general aviation as a motivational speaker and has organized and acted as race director in two successful air races, the Grand Prix Air Race and the US Air Race and Rally.

A past member of the Chicago Area 99 Chapter and is now a member of the Dallas Chapter.

ALISON JEFFERY, BG Coast Chapter, West Canada Section, was raised in Ladner, a small town in the Fraser Valley. In 1953 she married her husband, Jim, and moved to a dairy farm east of Mission where they spent the next many years working together on their farm and raising five children.

Although she drove all kinds of farm machinery she was always quite happy to keep both feet firmly on the ground. However, her husband always wanted to learn to fly, so for his 40th birthday she gave him an introductory flight lesson. He loved it and went on to earn his private license and joined the Mission Flying Club. She mustered enough courage to go up with him twice, but was terrified both times. Finally in 1980 she decided to take a few lessons, just enough to get over being afraid. After about six sessions it became a real challenge and with the support of Jim and a very special, patient flight instructor she went on to earn her private pilot license in March 1982. She joined the 99s in May of the same year, and has been a member ever since.

Always feeling the need to upgrade she decided to try for a commercial license, reaching that goal in March 1984. In May 1986 she received her IFR rating. To this point all her flying had been in their little PA28, GHEL. In July 1987, with 600 hours of flying time in HEL, they sold her and purchased a M20F, GJAI. Her total time is 960 hours. They have made several wonderful trips, the Baja, twice to Alaska, Texas, Winnipeg and many, many shorter trips.

She and her husband have had a great deal of pleasure out of flying, and she's really glad she conquered her fear and joined him in the sky.

LYNN MARIE DAVIS JEFFERY, born in Flint, MI, on April 19, 1958. Lynn realized her love for flying after her husband began taking lessons and signed her up for a Pinch-Hitter course in 1986. Lynn earned her private pilot (ASEL) in 1987 and instrument in 1989. In 1990 Lynn and copilot, Bob Epke, won the Michigan SMALL Race (Back to Basics Precision Cross Country). In 1991 they placed second in the SMALL Race by a margin of one second. She has served the Michigan Chapter as vice chairman, treasurer, and as chairman of 501c3, 66, AE Brunch, membership and the Michigan Small Race (92, 93, 94). Lynn has participated in airmarkings, poker runs and flown children with cancer to summer camp. Although her busy schedule of career and family has only allowed an accumulation of 300+ hours, she has flown Cessnas' 120, 150, 152, 172; Pipers' 140, 160, 180; an Aeronca Champ, and the Texan AT-6. Besides her involvement with the 99s, Lynn participates with Michigan Aviation Assoc. and is a voting member of the Michigan Aviation Hall of Fame.

Lynn works in hospital laboratory management having primary responsibility for satellite STAT labs, education, safety, and quality improvement. She holds a BA in chemistry from Hope College, a MSA in administration from Central Michigan University and is an ASCP registered medical technologist. Lynn and husband, Joe, live next to Dalton Airport in Flushing, MI, with their three children: Joseph Scott, 12; Emily Jeanne, 10; and Olivia Beatrice, 5. In addition, Lynn serves her community as a Kiwanian, on a college advisory board and on a women's health advisory board.

LORRAINE MAE JENCIK, born July 27, 1938, Norwich, CT. Joined the 99s in 1968. Flight training with Flight Safety, Inc.

She has flown various single and multi-engine piston; type ratings in: G-159, LI329, DA10, LR55, CL600. Employed with Xerox Corp., Westchester County Airport, White Plains, NY.

Previous employment night freight; certified flight instructor; charter; ATP; 1968 PPD finished 10th; 16,000 hours flight time.

BARBARA JENISON, of Paris, IL, born Jan. 14, 1909, in Chicago, IL. She learned to fly during WWII at Hyde Field near Washington, DC, while her husband was a naval officer on carriers in the Pacific. She joined the 99s in Washington, and upon returning to Illinois in 1954 became an active member and chapter president of the Central Illinois Chapter. She also was an active member of the Civil Air Patrol, and for a number of years was in charge of women cadet activities for the Great Lakes Region with the rank of lieutenant colonel.

She promoted private aviation interests through newspaper and magazine articles, and served on the Illinois Aviation Education Advisory Committee during the administration of three governors, and as a member of the FFA women's advisory committee under two presidents. She participated in AWTAR competition until the final air race, placing as high as third and in international women's air races in the 1950s. She held SEL, multi-engine and instrument ratings, flying both private and military aircraft, including with the Navy's "Blue Angels." In 1979 she gave up her pilot license but continued to enjoy flying "on the right."

LESLIE TUCKER JENISON, began to fulfill a lifelong dream of learning to fly in the fall of 1991. Her idea was to begin flying lessons and surprise her husband with his own flight training as a holiday gift. Well, after sneaking out to the airport, as well as locking herself in the bathroom to read her ground-school instruction, she decided to move the date of his surprise forward to their wedding anniversary in November.

Since that time, she has earned her private pilot certificate, multi-engine and instrument ratings. She has approximately 550 hours, most of which is in C-172, a C-120 tail-dragger and a C-421. She also has several training hours in an R-22 helicopter and an L-29 single-engine Russian trainer jet. She and her husband would flip a coin to determine who would be PIC. Since they were able to locate and restore the 1947 Cessna 120, they are now both able to fly airplane on short trips (giving their three daughters more elbow room).

By profession, she is a registered nurse specializing in women's healthcare. She has worked in obstetrics, an outpatient ambulatory care women's center and teaches community health classes for girls and women.

Jenison joined the 99s before the ink on her temporary private pilot license was dry and really enjoys the companionship of this enthusiastic group of women! Their NE KS is a very active flying group. The friendships that her 99s membership has brought her is wonderful, and she enjoys seeing other 99s at various flying events. She finds their accomplishments to be a great source of inspiration to continue upgrading her own skills as a pilot.

BETTINA JENKINS, born Oct. 7, 1967, in Toronto, Ontario, Canada. Joined the 99s in May 1991. Presently working on multi-instrument. She is a commercial pilot with a multi and float rating. She has flown twin Comanche, Piper Seminole, T-34A, 180 on floats, 172RG etc. Employed with "Skywatch" Environment Canada.

She has been employed with Ministry of Environment, Toronto, Ontario, since 1990; aerial surveillance, flying all over northern and southern Ontario, looking for illegal dump sights, chemical spills and other environmental hazards. Her other employment included Globespan Hawaiian Holidays, Vancouver, BC; Marlin Travel Agency, Toronto, Ontario; Branksome Hall School for Girls, Toronto, Ontario; Raven Ski Club, Collingwood, Ontario. Flight training at Brampton Flying Club: commercial pilot license, multi-engine endorsement (PIC 35 hours, total 55), float endorsement, certified on 13 different airplanes, presently working on a multi-instrument rating; total time: 505 hours. Jenkins has volunteered with the University of Toronto Broadcasting Radio; Brampton Newsletter; and 99s (Women's Pilot Assoc.). Her special achievements include advanced scuba diving license, Round Robin Tennis Team, Nissan Auto Racing Courses, Spenard David Racing Course, and synchronized swimming. Jenkins many travels have included: India, Alaska, Mexico, Caribbean, Europe, South America, Scandinavia, Canada, and United States.

DOROTHY RING JENKINS, earned private license in 1935. Joined 99s. Received limited commercial and then transport license in 1936. Performed in local all-girl airshows in Chicago acrobatics and racing. Carried air mail from Chicago to Moline, IL, on Air Mail Day in 1937.

Was CPT instructor at Sky Harbor Airport, Northbrook, IL. Moved to Phoenix, AZ, with husband Charles just before the beginning of the war. Joined Tucson Chapter in 1954, has held all chapter offices, flew in 14 Penny-a-Pound events given by that chapter. Did instructing and some charter.

Flew AWTAR 1958, 60, 61. Is now, proudly, a life member of the 99s, courtesy of her chapter.

JEANETTE J. JENKINS, born Oct. 6, 1917, in New Philadelphia, OH, where she grew up and went to school. She graduated from high school and attended the local branch of Kent University part-time. When the local Chamber of Commerce sponsored a CPT (Civilian Pilot Training) program, she enrolled and earned her private pilot license in 1941. She joined the 99s and shortly thereafter began to work in the office at the local airport. During WWII years 1943 and 1944, she served as a WASP until they were deactivated in December 1944.

She returned to work at the airport and worked there for 10 years until she went to work for the Timken Company. Until she retired, she served 14 years as a member of the local Airport Commission. She is now a life member of 99s.

PATRICIA E. JENKINS, Idaho Chapter 99, Whirly Girl #316, raised in Olympia and schooled in Seattle, WA, a quantum leap took Pat from the city to life on a remote cattle ranch in southeastern Oregon when she married.

Her visionary husband desired an airplane to cover the 100,000 acre ranch, and within five years they were both flying. In 1970 they bought their first airplane, an old C-182, and traded beef for their initial flight instruction. Being sixty miles from the nearest town, an airstrip and hangar on the ranch were easily justified. Pat found aviation to be the perfect cure for her remote and sometimes lonely lifestyle, and within a few years earned a commercial certificate with instrument privileges and was flying a C-185. She joined the Idaho 99s in 1972, gratefully embracing the female companionships. In 1980 a Hughes helicopter was added to the stable and Pat's commercial certificate soon included a rotary rating. The yellow Hughes 300C was named *Woodstock* and became a cowpony, chuckwagon and pickup for the ranch. At this writing, Pat has over 5,000 flight hours, mostly helicopter, and is still herding cattle with *Woodstock.*

MARY C. JENSEN, born Jan. 8, 1921, and raised in Oklahoma City, OK, which of course is a long way from the seashore. When she took a week vacation to Alaska in 1948, she learned about seafood. She was a single parent and sole support of four children when she decided to learn to fly. So the lessons were slow and long time between, but she finally got her license in 1964.

She found a side by side two-person Taylorcraft on floats and bought it. So she earned a ASES license, too. She learned where the best fishing rivers and lakes were and the best places to get fish, clams, crabs, abalone and mussels.

She taught her children to fish and used to ferry her youngest son and some of his 12- to 13-year-old buddies into a remote lake and leave them for a week at a time to fish and camp out. Her son is now an airline pilot.

She sold the Taylorcraft and bought a used Cessna 180 on floats and wheels and skis. She joined the Civil Air Patrol and trained in search and rescue; she retired as a lieutenant colonel. Jensen really enjoyed the 30 years she donated to the CAP.

She had to stop flying when she had open heart surgery, but she is a lifetime member of the 99s and hopes to have her memorial in the Forest of Friendship.

She loved flying and hopes more and more ladies go into aviation related fields.

MARY H. GOODRICH JENSON, charter member 99s, born in Hartford, CT, Nov. 6, 1907, now resides in Wethersfield, CT. A student pilot in 1927, flying out of Brainard Field, Hartford; instructor, the late Percival H. Spencer, aeronautical pioneer noted aircraft designer and builder. Soloed in an OX-5 Waco in 1929; became first federally licensed woman pilot in Connecticut.

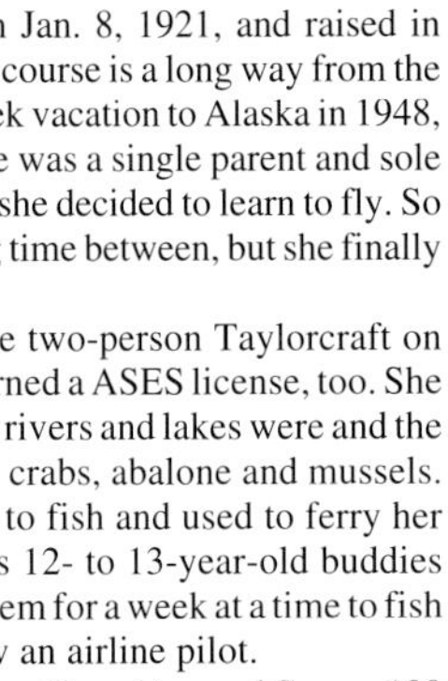

US Department of Commerce, Aeronautics Branch, private pilot license No. 9410, 1929. Federation Aeronautic International, 1930 sporting license, No. 263. Certificate No. 7495, signed by Orville Wright, chairman. NAA Contest Committee.

Flew her own KR21, a Fairchild built experimental biplane, in many sporting events, air meets and exhibitions. Attended organization meeting of 99s at Curtiss Field, Valley Stream, NY, November 1929.

Aviation Editor of the *Hartford Courant,* 1928-33, writing a daily column covering developing industry – Pratt & Whitney, Sikorsky in state, air mail, passenger transport, as well as pilot activity. Career opportunities in New York, Virginia, Texas and California limited pilot progress. Retired in 1942 as story research editor for Walt Disney Productions, returning to Connecticut with husband, Carl Daniels Jenson, who soon left for WWII service – later served in Korean Conflict. Son William, Arlington, VA, a grandson and granddaughter, Carl and Kristin; daughter Ann Jenson White, two grandsons, Daniels and Dean. Lt. Col. Jenson, AFR, died in 1993.

NANCY JENSEN, has been a private pilot since 1976 and joined the 99s in 1977. She lives on a private airstrip southeast of Seattle, from where she flies her Cessna 180.

Nancy is past chairman of the Greater Seattle Chapter of the 99s and governor-elect of the Northwest Section. She is also an active member of six other flying associations. Nancy has flown in several Petticoat Derbies and as copilot in the Palms to Pines Race. She enjoys airplane camping in Idaho, Alaska and Canada with her husband, enjoys her two grown children and adores her two young granddaughters.

Nancy has a BS in education and is a busy volunteer for the Flying Companion Seminar, Fear of Flying Clinic, EAA Young Eagles and other aviation education activities.

GENE NORA JESSEN, an Evanston, IL, native, she worked her way through Oklahoma University teaching flying in Aeronca Champs. She was one of 13 women pilots participating successfully in the Mercury Astronaut research program in 1961. In 1962 she became a sales demo pilot for Beech Aircraft Corp. in Wichita flying the entire line in a 48-state territory. She and Bob Jessen were married and later settled in Boise, ID, starting a Beech dealership where Gene Nora operated a Pt. 141 flight school.

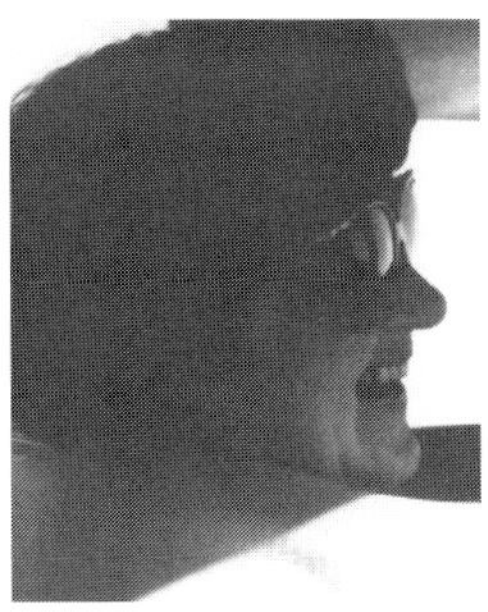

Gene Nora has participated widely in aviation: serving on the FAA's Women's Advisory Committee, aviation columnist for the Western Flyer and Idaho Statesman, FAA Safety Counselor, board member of the Idaho Aviation Hall of Fame, 10 years on the Boise Airport Commission and also Girl Scout and YWCA boards. Special honors have included Idaho Statesman Distinguished Citizen, INAC Achievement Award and YWCA Pioneer for the Future Award.

99s activities have included virtually all committees and offices on the chapter, section and International level including chairman of the 99s Resource Center/Museum and International president. Aviation credentials include commercial, ASMEL, ASES, IFR, Gold Seal CFI A&I, all ground. She currently flies a Bonanza.

The family business is FBO Boise Air Service operated by Bob with Gene Nora writing aircraft insurance for Sedgwick James. However, their greatest achievement has been their two beautiful children Taylor and Briana and now Briana's husband Tom LeClaire.

HELEN E. JESSUP, born Dec. 17, 1991, Oklahoma. Joined the 99s in 1979. Served in the Air Force Auxiliary (CAP). Flight training included FBO - Norm Seward. She has flown the Tri-Pacer, Aeronca, Cherokee, 172, Bonanza. She is married to H. Allen and has three children: Jacqueline Grantham, 48; Shelia Webb, 41; and Mark Jessup, 39. Also has grandchildren: Sara, Sabrina and Loran Grantham, Travis Webb, and Casandra Jessup.

Helen, organizer and first chairwoman of the Waco-Centex 99s Chapter, was born on a small farm in Copan, OK, where she was very active in 4-H and won a state meat judging and a trip to Chicago. She earned a BSHE from Oklahoma A&M (now Oklahoma State University).

Helen began flying after a pinch hitter course in the family aircraft (Tri-Pacer). She got her private pilot license in 1963, while being a mother of two girls and a boy, a Sunday School teacher, and a Camp Fire Girls director. Her husband was an electrical engineer from Oklahoma A&M. He was employed by Rockwell in Waco, TX, and active in Civil Air Patrol. Helen joined the CAP and trained to be a mission search pilot. She was the pilot in command on many search missions, participating in locating seven downed aircraft. At 70 years of age, she retired from the CAP with the rank of lieutenant colonel.

DELORES GRACE JEWETT "DODIE," born April 1, 1936, in Warren, OH. Saved money for her husband's birthday gift for lessons for private pilot license. When asked what she would do if flying with children and something happened to him, prompted her first lesson in January 1968. Received private pilot license in 1972, has commercial license, instrument rating, multi-engine rating, basic ground instructor, owns a Piper Cherokee 140, logged over 1,700 hours.

Joined 99s in 1973. Charter member Lake Erie Chapter, held all offices at chapter and section levels. Member first Council of Governors 1990; Lake Erie Chapter Pilot of the Year in 1978 and 1988; Chapters Achievement Award 1985; A.E. Scholarship winner; participated in several proficiency races; in 1980, with son Bruce, won the Lake Erie Air Derby Proficiency Race during Cleveland National Air Show; past aviation explorer advisor, Zonta, YAF, EAA, BPW, NIFA judge; facilitated Girl Scout troop in earning their aviation badge. FAA aviation safety counselor, member AOPA, ECOPA, CAP, Cherokee Pilots Assoc., Silver Wings Fraternity, Paul Bunyan Clan, ROAL.

Received certificate of recognition from the FAA in 1988.

She and daughter Jennifer Syme were first mother-daughter to be inducted together at the same time into the International Forest of Friendship in 1994.

Joined the US Coast Guard Auxiliary 1979; presently, USCG Auxiliary District Staff Officer Air Operations 9ER; district liaison to US CAP; one of six auxiliarists selected by NAVCO for Auxiliary Aviation Quality Action Team to US Coast Guard Headquarters in Washington, DC.

Received US Coast Guard Merit and Service awards including Award of Administrative Merit for Air Operations in 1994; Coast Guard Unit Commendation Ribbon, and the most Prestigious Secretarial Award, DOT Gold Award Ribbon.

Received the Earl W. Murphy Award in 1983, 1988, 1994, for Excellence in Air Operations within Seventh Division of 9ER, USCG Auxiliary.

Married in 1960 to Harlan, an engineer; son, Bruce, a pilot with United Airlines; daughter, Jennifer Syme, healthcare representative for Pfizer Pharmaceutical and a 99.

During children's growing up years, was little league team statistician, girls' softball coach, assistant coach synchronized swim team and swim team; swim team parent for 12 years. Worked in Geneva School System eight years as an educational aide and district attendance officer; employed 10 years as CSR for Huntington National Bank in Kent, OH. Enjoys flying her Cherokee 140, fishing, boating, water skiing, photography and traveling United Air Lines, especially, when her son is at the controls of the B-727.

SHANNON L. JIPSEN, born Feb. 4, 1966, in Kansas City, MO, and reared in Lee's Summit, MO. She attended Central Missouri State University and graduated with a BSBA in marketing (1987) and a MS in aviation safety (1988).

She began flying in 1986 – attained her private, instrument, commercial, CFI, ME, and CFII by November 1988. As a graduate assistant she taught flying, private pilot ground and technical report writing from September 1987 to December 1988. Then Shannon began teaching at Executive Beechcraft (MKC) and attained her MEII rating. In 1990 she earned her ATP and flew Part 135 Charter in various King Airs. July 1991, she was hired by United Parcel Service as a B-727 flight engineer. July 1994, she was asked to be a B-727 check airman for F/Es. From August 1994 to January 1995 she taught IOE (Initial operating experience) to new hires. March 1995, she upgraded as a B-727 first officer.

Shannon's involvement with the Greater Kansas City 99s began in 1988 when she finished her MA and moved back to Kansas City. She taught some of the Flying Companion Seminars the GKCC sponsored for a few years; she was the safety education contributor to the GKCC newsletter; she co-organized and presented the GKCC 50th Anniversary program in 1990; she gave Air Bear presentations; she was the 66 chairman for the chapter for several years, and helped with many other activities including section and International conventions. Shannon worked on the Wright Day Dinner committee one year and asked one of her special friends, Mr. Robert Gilliland, the first SR-71 pilot to come speak. She also was elected chairman of the GKCC in 1991; however, had to resign when hired by UPS as she was based in Louisville, KY. Shannon remained involved with the Kansas City group until February 1995 when she moved her membership to the Kentucky Blue Grass Chapter and began working with those members to organize the 1996 NCS spring section meeting and utilize her talents with the Kentucky Chapter. Shannon also was elected to the North Central Section 1994-96 nominating committee.

CHARLOTTE BLOECHER JOHNSON, began flying in 1981 at Lompoc, CA, as a pleasant pastime to share with her husband Don. Since that time she has accumulated 400 flight hours as a private pilot in their Piper Archer. The Archer has proven to be a fun, convenient way to visit friends, enjoy a faraway weekend getaway, or a way to experience that special flying feeling one gets from that fascinating perspective from "up there." Charlotte graduated from California Polytechnic State University with a degree in mathematics and is employed with an Air Force contractor at Vandenberg Air Force Base where, among many duties, she oversees the complex computer systems utilized to predict aviation weather. Charlotte is also a charter member of the Santa Maria Valley 99s. Her most memorable flight was an aerobatic lesson in a Decathlon at Santa Paula Airport, a birthday gift from her husband. Flying upside down was a complete thrill!

ELAINE NOLAN JOHNSON, learned to fly at Taunton Municipal Airport in East Taunton, MA. She joined the Eastern New England 99s in 1988. She is a nurse practitioner whose specialty is emergency medicine. She has worked as a paramedic and a flight nurse on a fixed-wing air ambulance in Massachusetts. Now living in Alaska, she works in a rural clinic in 'the Bush,' as well as doing Medevac flights. She has recently started to fly a friend's Taildragger, a Cessna 180, in Bush, AK. She is also a member of the Aircraft Owners and Pilots Assoc., the National Flight Nurses Assoc., and the International Flying Nurses Assoc.

ELINOR JOHNSON, in 1960 the youngest of their four children started school and she learned to fly at Highland Park Airport, near the corner of Coit and what is now LBJ Freeway, Dallas, TX.

After getting commercial, instructor, and ground instructor ratings, she taught ground school which paid for instrument and instrument instructor ratings.

She and her husband owned a Bonanza which they flew all over the states, the Bahamas, Mexico, and Guatemala.

Pat Jetton and she flew in several Powder Puffs, coming in seventh in the last race. They flew Angel Derbies, coming in second, last, and disqualified. They flew in almost all of the first Air Race Classics, winning in 1980.

She's had a great time sightseeing and racing, and has met some wonderful people in racing, aviation, and in the 99s.

EVELYN BRYAN JOHNSON, born Nov. 4, 1909, in Corbin, KY. Joined the 99s in 1947, has held all Tennessee Chapter and Southeast Section offices. Began flight training in Oct. 1, 1944, Knoxville, TN. She has flown all SEL and MEL below 6,000 lbs. also SES. Employed with Morristown Flying Service, Morristown, TN, 1947 to present time. She is a widow and has one child, Morgan Johnson, Jr. Also has grandchildren: Mike, 24; Chad, 21; and great-grandchild, Christopher, 1. Johnson flew Powder Puff Derby five times; International Race one time; flight instructor since May 15, 1947; pilot examiner since 1952. Flying time to date 53,850 hours. She has been an airport manager in Morristown, TN, since 1953, and owned and operated Morristown Flying Service 1949 to June 1982.

Honors received: Carnegie Medal (bronze), 1959; National Flight Instructor of the Year 1979; Outstanding Alumnus Tennessee Wesleyan College, 1981; Service to Mankind Award Morristown Sertoma Club, 1980; FAA Safety Counselor of the Year, 1984; Mississippi Valley FSDO, Kitty Hawk Award, FAA, 1991; Stuart G. Potter Aviation Education Award, 1993; Aviation Distributors and Manufacture Assoc. 1993; Elder Statesman of Aviation, 1993; National Aviation Assoc.; Women in Aviation Pioneers, Hall of Fame, 1994; Award of Merit, 99s in 1994.

Member of Women's Advisory Committee on Aviation 1970-73, board of director National Assoc. of Flight Instructors for eight years. Member of board of directors Silver Wings Fraternity. Appointed by governor of Tennessee 1983 as member of the Tennessee Aeronautics Commission, re-appointed two more times, present term goes to June 1996. Her biography written, by George Prince, in a book entitled *Mama Bird,* 1993. Johnson was honored in the Forest of Friendship in 1991.

LISA NICOLE JOHNSON, born in Oakland, CA, and raised in the city of Alameda remembers growing up under the departure flight path of Oakland Airport's North Field. Was always very intrigued with airplanes and flying. On Sundays, her father would take her and her sister to the Old Oakland North Field passenger terminal to watch the airplanes. In 1979 she started flight training and on Oct. 31, 1979, received her private certificate. She moved on to earn her instrument rating and has more than 1,500 hours of flight time.

Lisa owns and maintains a Cessna Hawk XP which she purchased in 1982. She has flown this airplane all over the western US. Lisa is a proud member of the East Bay Chapter of the 99s, Inc., Aircraft Owners and Pilots Assoc. and is a past president of the Aircraft Pilots of the Bay Area, Inc.

MARION BECKER MARRIOTT-JOHNSON, born Feb. 25, 1924, in Nebraska. Joined the 99s in 1967. Flight training includes private, instrument and commercial ratings. She has flown Cessna 150, 205 and T210. Married to Edward and has three children: Linda Marriott Wallis, 49; Sandra Marriott Carr, 47; and Robert Andrew Marriott, 41. She also has eight grandchildren, ages 28 through 6.

Flying – what an exhilarating broadening experience! A wonderful way to visit new places and meet many interesting people. Since her first take-off over 30 years ago, the world opened up with flights to Alaska, Bahamas, Canada and all over the continental US. During that time, she saw many advancements in navigation and communication.

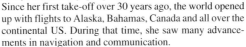

In addition, the undeveloped short, sod strips were changed to asphalt and concrete. Her participation as a member of the San Gabriel Valley Chapter has included begin chapter chairman for three terms, vice-chairman for one term and chapter treasurer for two terms. The highlight of flying has been the wonderful fellowship of the 99s, with many becoming life-long friends.

MARJORIE JOHNSON, a retired teacher who started flying in 1990 at Palo Alto. She and her husband, Frank, recently got their instrument ratings in their Mooney 20F, and just purchased a Mooney 20K. Her interests include aerospace education and she has published 80 mathematical papers. Marjorie and her husband are active in the Mooney Pilots Assoc. Marjorie has logged more than 400 hours and also flies with the Palo Alto Flying Club. Marjorie and Frank have flown their Mooney across the country and back, and also took a flying trip through Australia.

NANCY RUTH JOHNSON, born March 7, 1917, in Beckley, WV. The first child of George Allen and Lina Pearl Redden Johnson was born March 7, 1917, in Beckley, WV, where her father was teaching. The length of the term and salaries for teachers were uncertain. Mr. Johnson took a job as trainman on the Chesapeake and Ohio Railroad, and the family moved to Quinnimont on the New River. When Nancy was ready, there was no high school in Quinnimont, but the C&O gave passes to employees an their family. A school pass enabled her to commute daily to Woodrow Wilson High School where she graduated with highest honors in 1933. Another school pass let her commute to Montgomery to New River State College for a year of college. Career choices for girls were limited. Nancy could be a nurse or a teacher. She chose teaching and earned a Standard Normal teaching certificate from Marshall College, Huntington, WV. She then taught in three one room mountain schools. The family moved back to Beckley where Nancy enrolled in CPT, Civilian Pilot Training given by Beckley College and Mount Hope Airport.

Nancy had traveled many miles by train to begin an education and by foot to the one room mountain schools. Like Thoreau, most of her travels had been to and from Quinnimont. After her first flight lesson, she knew the way to go was to fly. After she earned a private pilot license, Nancy heard about an organization for women pilots, founded by Amelia Earhart. Now with a license, she could be a 99. She also heard about a program for women: WAFS, which required two hundred hours. She flew every weekend to build flying hours.

After the Japanese attack and the great need for pilots, the WASP program was begun. A private license was enough to qualify for entrance. Nancy applied, was accepted and reported to Avenger Field, Sweetwater, TX, April 23, 1943, class 43-W-6. After graduation from 318th. AAFFTD, Nancy was assigned to ATC Second Ferrying Group, New Castle Army Air Base, Wilmington, DE, where she ferried various aircraft, many PT19s from Hagerstown, MD, to training bases. Sealed confidential orders meant a delivery to Canada.

In March 1944 Nancy went temporary duty to Dallas, TX, checked out in the DC3, the C47 and the C49. From Dallas she reported to Fifth Ferrying Group, Fairfax Field, Kansas City, to move B-25s. The WASP were civilian pilots, under military discipline, flying military planes. They had been promised militarization when they were sworn in, but remained on officer status. From Kansas City Nancy attended AAFSAT, Army Air Forces School of Applied Tactics. She also went back to Sweetwater for advanced instrument training. When the bill to militarize the WASP came before the all male congress it was defeated. Males were saying that women were taking men's flying jobs. Rumors flew about the WASP program. Nancy was transferred to Training Command, Santa Ana, CA, then to Victorville, CA. Here she was assigned to engineering and flew testing missions in the AT-11 that an Army sergeant could adjust and repair the AFCE Automatic Flight Control Equipment.

Here the rumors became real. No more training for women. No more flying for women. The program was to be deactivated Dec. 20, 1944.

The base gave Nancy MATS to Cincinnati, then she took a bus to Beckley. She knew she needed a college degree so she enrolled again in Marshall College, Huntington, WV. Flying jobs for women did not exist in 1944. "No one would fly with a woman" they were told. Nancy was one of the lucky ones. She got a job at the Huntington Airport giving flight instruction, sightseeing rides over town and charter trips flying Aeronicas. The work at the airport continued until she earned an AB in 1946. Although she was quite busy at the airport, the owner was a pilot, as was his son and his son's friend. There wasn't enough work for four pilots, so Nancy didn't wait to be told, she went back to Beckley hoping to find another flying job.

The Air Force Reserve gave commissions to the WASP after the program closed. Nancy was promoted to captain, but a disastrous accident caused her to be transferred to retired reserve.

The WASP were invited to the Soviet Union in 1990 to meet the Soviet Women Pilots called "Night Witches" by the Germans. Bob Hope was in Moscow at the time to tape a program. The WASP were invited to hear him and also

invited into the Embassy Residence to meet Bob and Mrs. Hope. An embassy wife snapped a picture as Nancy was privileged to talk with Bob Hope. The other girl is Mary Lou Neale, a member of the first class of WASP and who came to West Virginia to organize the first 99 Chapter which Nancy joined.

Enter Nancy's home today and you will see above the doorway to the living room a propeller with one blade roughened. On reporting for work one Monday morning the propeller was standing in a corner of the office. "I'd like to tie it around your neck," the airport manager said. "I told you not to solo that guy. You did and he caused me to get a prop. It's your fault. I want you to put it where you can see it every day. Look at it and say, "I don't know everything." Nancy had a student who was quite heavy. By the time they got altitude, a stall or two, a spin and the lesson was over. The manager said, "Give him all the dual he wants, but don't solo him." Nancy thought that unfair and soloed the student, which gave the manager an excuse for a mistake. She well knows she doesn't know everything, but she does know flying is more fun than not flying.

SHIRLEY CUNAN JOHNSON, born May 3, 1928, in Bakersfield, CA. Joined the Bakersfield Chapter 99s in 1982. Flight training single engine -land (VFR). She has flown the C-152, C-172, C-23, PA28-161. Married to G. Ted Johnson, and has eight children: Jane C. Wade, 48; Linda G. Kuntz, 43; Stevan Giumarra, 42; Nancy G. Scanlan, 41; Giovanna G. Escasany, 38; Gary Giumarra, 36; Peter Johnson, 26; Kylie Johnson, 25. She has 10 grandchildren: Tieada Wade, 24; Bryana Kuntz, 8; Orlando Escasany, 7; Karl Wade, 22; Sean Scanlan, 6; and Alexander Giumarra, 1. She also has one great-grandchild, Brecca, 1 year.

Johnson's husband is also a pilot (IFR) and they own a V35B (Beech-Bonanza). They are members of AOPA, American Bonanza Society and World Beechcraft. Flew to Oshkosh 1992 and Bahamas in 1994, moved to Sun Valley, ID, three years ago.

Her special recollections from her time with the Bakersfield, CA, Chapter 99s: Getting to know a group of very special, wonderful women. Fly-ins. Air Markings. Co-chairing/hosting the 1984 Fall Southwestern Section in BFL. Co-chairing/hosting the 1987 Air Race Classic start in BFL. Attending other Southwest Sections. Helping out at air shows. Helping with Safety Seminars and sponsoring Flying Companion Seminars. Hosting Fuel Stop in Bakersfield, CA, for Hayward/Las Vegas Air Race 1989.

VALERA G. JOHNSON, learned to fly in 1962 for safety reasons after she learned her husband, Belmont (known as "Dinger"), a former WWII pilot, planned to get back into flying. But once in the air, flying became her life. She got her private certificate in 1962, commercial in 1963, instrument and instructor rating in 1964 and a commercial helicopter certificate in 1975. In her 23 years of flying she averaged more than an hour per day of flight time in the family aircraft – which included a Cessna 182, Bonanzas and a Bell 47 G4 Helicopter – instructing and flying charter. Val and Dinger got their initial flight instruction at Vichy, MO, before spearheading a successful drive to get an airport for their hometown of St. Clair, MO.

Val flew in 10 Powder Puff Derbies, four Angel Derbies, the Bahamas Flying Treasure Hunt, a flight to Alaska and countless other events. A spark plug in the STL 99 Chapter, Val served in many capacities including chairman. For years she chaired a very successful annual "STL Chapter 99 Rummage Sale" in St. Clair, gathering sale items folks left for her at various airports and hosting volunteer 99 members overnight at her home. She also served as North Central Section governor.

Sadly, on April 15, 1985, Valera Johnson, flying alone in a brand-new Bonanza, died in a power-on crash near her home airport. Friends felt she must have lost consciousness before the crash. Oddly, one year later, her husband Dinger, also alone, died in a crash.

ELLA E. G. JOHNSTON, born Aug. 6, 1933, to Bennett and Marcia Ellingson, in Lawrenceburg, TN. Ella now resides in Nashville, TN. Ella has a son, David M. Gilmore, who is an air traffic controller. Flight training began on March 9, 1970, and soloed May 4, 1970. She received her pilot license on Dec. 15, 1970.

In the early 70s she owned a fashion school and used her flying skills to travel to location for her fashion shows.

One special story – while flying at 3,000 feet she encountered a mass formation of migrating Monarch Butterflies. This was truly a beautiful moment in flight.

At present Ella is employed by a large cooperation. She is the flight coordinator in the aviation department and works directly with the pilots in scheduling all flights for the corporate departments. Her flying skills are a great asset in this position.

LINDA THOM JOHNSTON, born Dec. 2, 1953, San Gabriel, CA. Joined the 99s in 1975. Learned to fly in 1975. She has flown Cessna 150, 172, 182, 182RG; Piper 140, 180; and Archer. She has a child, Sandy Johnston Falcey, 29, and grandchildren, Michael, 5, and Daniel, 3.

Linda has logged more than 750 hours of flight time, has competed in a number of air races, including the Powder Puff Derby, Pacific Air Race, Palms to Pines as part of a "mother-daughter team." She is a member of 99s and has been actively involved in chapter activities over the years. This includes chairman of the Jim Hicklin All Mens Air Race out of Bullhead City, AZ, (a 99 Chapter Air Race).

Besides college, Linda has been a real estate broker, managed a ComputerLand and started her own computer business in Los Angeles.

Linda now resides in Portland, OR, and is the owner of a leading computer consulting company in the Pacific Northwest with over 30 employees.

KAREN KAY JOHNSTONE, born Aug. 31, 1971, Englewood, CO. Joined the 99s in 1991. Flight training includes commercial, multi, instrument, CFI and MEI. She has flown the following planes: Katana (DA20/DV20), C-152, C-172, C-182, CT210, BE18, Seminole. Her first lesson was at age 14 in a Cessna 152 at Centennial Airport in Englewood, CO. Her flight instructor was also a woman and was very supportive of her beginning a career in aviation. This first experience was so positive and encouraging that she's never had to look back or hesitate about becoming a woman pilot.

ELLY WALRAVEN JONES, Bay Cities Chapter member since 1963. First female ever hired by Pan American World Airways as an aircraft mechanic (during WWII). Also worked in that capacity for US Army Air Corps and US Navy. Finally decided to learn to fly in 1961 along with husband, Roy, and son, Dale. Earned commercial pilot license in 1967. Present occupation: licensed vocational nurse and housewife. Though no longer able to actively pilot, enjoy traveling by land, sea or air sharing the good life of semi-retirement and its amenities.

HAZEL HENSON MCKENDRICK JONES, quit Texas Women's University to learn to fly and join the WASPs, but the WASPs were discontinued just as she completed her 35 hours. Hazel went to work for FAA, instead. She won many awards and honors from FAA, Department of Transportation, and AOPA. She won an Amelia Earhart Scholarship from the 99s. Active in 99s, NIFA, USPFT, Powder Puff Derby and Whirly Girls, Hazel belonged to the Dallas Redbird Chapter. Hazel was International president of the

99s 1984-86. She retired from FAA in 1973. She owned Jones Testing Service and gave written pilot exams.

Hazel eventually received a MA from TWU. She raised one son, Mike, after her husband died; married Roys Jones in 1971. Hazel died Dec. 10, 1990, followed by Roys soon after he had accepted an award for Hazel.

JOAN BARBARA JONES, a resident of Bucks County, PA, learned to fly in 1968. She joined the 99s in 1970 and has held various offices at the chapter and section level. She was a winner of the Amelia Earhart Scholarship and flew in the Angel Derby.

Joan has a BS in education as well as an MA and is a member of several educational as well as aviation organizations. She has been an aviation professor at Mercer County College for 10 years and serves as advisor for the college flight team. As their assistant chief instructor Joan conducts the Part 141 Flight checks and has more than 3,000 flight hours. She also teaches an aviation camp for youth at the college during the summer.

Joan is a single parent with two teenage daughters, Kristina and Lauren.

MARY CATHERINE AKA KATE CLAYTON-JONES, born in Tonbridge Wells, England, on Christmas Day in 1964. The daughter of a RAF 111 pilot turned crop duster and a circus related mother, she was raised in Jamaica until she was 10, and then in England. She began her pilot training at Somerset Air Services in New Jersey in 1989 after obtaining a degree in geology from the University of Colorado. Her only ambition was to become a private pilot. Although she had always wanted to learn to fly, she had been led to believe that one had to have 20/20 eye sight to become a commercial pilot, an asset she was only fortunate enough to have with glasses. After her private license she realized that an instrument rating in the north-east was a must, so she completed that early in 1991. Kate wanted to fly, so she passed a first class medical, with a waiver, and in August of 1991 passed her multi-engine commercial instrument check-ride, in Northampton, MA. She then started training for her instructor rating.

Kate has been a flight instructor for three and a half years now; and although she is starting her own 135 cargo and charter operation out of Orange, MA, she will still continue to teach. She says "you learn to fly when you have taught someone else. To turn that fledgling student, that summoned the nerve to admit that they wanted to try something new, into a pilot who realized a dream; that is her reward."

With more than 2,500 hours, the operation of a cargo/charter/instruction business, a passion for aerobatics, a member of WNE 99s, AOPA and several other pilot organizations; Kate has seriously embarked on a career in aviation.

PAMELA E. JONES, born Nov. 7, 1959, Oklahoma City, OK. Joined the 99s in 1983, Oklahoma Chapter membership chair, treasurer. Flight training includes commercial rating and ASEL.

She has flown C-152, C-172, 1948 V-Tail Bonanza, C-182, Baron, Starduster, Shrike Commander, C-150, Aero Commander, C-421, Bonanza E225, Piper Warrior, C-414, BEJ35, Piper 180, PA24-260, PA28-51, Grumman.

Married to Kevin LeGrande. Flew the 1990 Air Race Classic and received the "Mary Pearson Award" for best first time racer.

PATRICIA L. JONES, September 1984, made a turn off the interstate that would change her life forever. One hour after making that turn, she had signed up for flying lessons at Hanscom AFB, MA. Her instruction continued through a very cold and snowy winter with her first solo at nine hours

and her private pilot certificate on April 27, 1985. Four days later she moved back to her hometown of Melbourne, FL.

She did not fly again for 10 months until she checked out in a C-152 at Melbourne Airport. Over the next six years she had limited flying time. In April of 1990, she became a 99 (Florida SpacePort Chapter) and the proud owner of Stryker, her yellow Labrador Retriever.

She began judging at National Intercollegiate Flying Assoc. meets and going to section meetings with Bonnie Gann (who gave her her first introduction to twins, B55 Baron N99ZG). In 1992 she was elected chapter chairman of the Florida SpacePort Chapter and was a member of the Southeast Section nominating committee. She has also been on the board of directors for the Florida Race Pilots Assoc. which runs the Great Southern Air Race.

In February 1994 she moved to Greenville, SC, to work toward her advanced flight ratings. She now has more than 250 hours and will soon complete her instrument and commercial ratings.

In March 1995 with many more flight hours and higher level of confidence, she returned to Melbourne, FL, where she now works as a senior computer systems analyst in support of the Eastern Test Range at Patrick AFB.

Her copilot, Stryker, loves to fly (log book hours are adding up).

Jones deepest wish is to complete her instructor ratings so she may pass along her love of flying to others.

LEANNE TENNANT JOPSON, earned her private certificate Dec. 18, 1988. Her husband was responsible for getting her started. She decided that if she was going to fly around the country with this guy she wanted to know more about it.

She learned to fly in their Stinson 108-1 Voyager and had lots of fun flying to Baja, Alaska and lots of other places close to home. They own an N3N (open cockpit bi-plane WWII Navy Trainer). Her most memorable flight was flying it from Concord, NH, to Boise, ID, in June 1994.

She joined the 99s in March 1990. She was treasurer in 1992-93 and is currently chairman of the Idaho Chapter. She loves all the flying activities organized by the 99s at all levels and is proud of her scenic route award she won with Anita Lewis in the Palms to Pines Air Race in 1993.

ELIZABETH JORDAN, Dallas Redbird Chapter, earned her private pilot license in 1968 at Weiss Airport in St. Louis, flying a Cessna 150 until she and her pilot husband Richard, purchased a Tri-Pacer. Later she checked out in a Citabria. In 1975 the Tri-Pacer was replaced with a Piper Cherokee 180, which they still own and fly. Her logged hours total almost 1,000 hours and includes an instrument rating. She and her husband moved from Dallas to Boerne, TX, in early 1995 and fly from Boerne Stage Airport.

Elizabeth joined the Greater St. Louis Chapter of the 99s in 1973, transferring to the Dallas Redbird Chapter in 1983. She served as chairman of each chapter and on many committees.

She has a paralegal associate's degree, a BA in business administration, retired from real estate work, and is enjoying flying, watercolor painting, their two children and three grandsons.

CAROL JORGENSON, learned to fly at San Carlos Airport in 1991 and she joined the Santa Clara Valley 99s that year. Carol owns her own company, where she acts as a manufacturing representative for premium incentives. She owns a Baron, which she flies out of San Carlos Airport. She has logged 750 hours, including a flight to Florida for International, and has her AMEL, ASEL, commercial and instrument ratings. Her interests include gardening, cooking and outdoor sports.

KATHRINE "TRINE" JORGENSEN, born June 3, 1966, in Oslo, Norway. Currently residing in Lakewood, CO. Her father traveled a lot and sometimes she, her mother and brother would go along. The trip to Disney World was exciting; Canada was on the agenda; Banff and Vancouver were beautiful. Boating in Seattle came in later years.

As a youngster, this shy little girl grew up to be a great skier, even made the Junior Olympics a couple of times. As she grew she invested her time in teaching other young people how to race, in Vail, CO. After graduating Lakewood High School in 1984, this young woman earned her private pilot license and then left to attend a year's worth of school in Stranda, Norway. Upon being flight team captain, an Alpha Eta Rho member, a commercial pilot rating in hand and graduating from Metro State College, it was time to hit "The Real World."

She attended fly-ins to Santa Fe for the ladies shopping weekends. She attained a regular job at her flying club. She earned more ratings and flight time. Calhan, CO, became a yearly event, flying into a dirt strip for breakfast. And the National Intercollegiate Flying Assoc. became a yearly vacation to judge the National SAFECON.

The Colorado State Patrol was looking for a "Few Good Men." She completed the patrol's motor school and became the first female State Trooper on the motorcycles for Colorado. She has been on the local news and on Real Stories of the Highway Patrol, most recently in an article in *American Iron*.

NANETTE JOZWIAK, born on May 26, 1952, in Galt, Ontario, Canada. Her real encouragement to fly came from her father. She became hooked on flying after taking her first lesson at the London Flying Club in 1977. She was awarded Pilot of the Year for highest marks in attaining her private pilot license. To build flying time she borrowed a Cessna 140 and flew across Canada, through the Rockies and back through the USA along the Grand Canyon. She entered the Governor General's Cup air rally, in Weyburn Saskatchewan, where she placed third overall and was given an award by the 99s for being top female scorer. Her first job was at the Waterloo Wellington Flying Club as a flying instructor. She then accepted a position with Voyageur Airways and was appointed chief flying instructor, as well as flying medevacs, charters and scheduled flights. Her next job was with City Express as a captain on a ST-27. Her total flying time was now 6,000 hours. She was hired by a major carrier in 1986; she is second officer on the L-1011 with Air Canada.

She has a MA in kinesiology from the University of Western Ontario and Ontario Scholar. She was Athlete of the Year for three consecutive years in high school, and was a member of the national rowing team for Canada. Nanette volunteers her time to speak at schools during career days, offering encouragement to young girls who may want to choose flying as their career. Nanette is a member of the 99s First Canadian Chapter as well as a member of ISA. Nanette is a very grateful for the support her parents and sister have given her during her many endeavors.

MARGARET JUHASZ, private pilot since 1974, instrument rating in 1984. She has been a 99 since October 1974. Retired from 27 years service at Samuel Moore & Company last 17 years as advertising administrator. Checked out on Skyhawk, Cardinal RG, Cutlass RG, and Archer. Learned to fly as an alternative to driving for 12 hours each way to her sister's place in Eagle River, WI. Most of her flying has been to that area, as well as LaCrosse, WI, where her niece and her three daughters live.

JERRY ANNE JURENKA, a native Texan, has been flying since 1984, logging more than 1,300 hours in her Cessna 182. She flew in two Air Race Classics, receiving the Mary Pearson Award in 1993 with copilot Jody McCarrell. Jerry Anne has been an aviation competition judge for NIFA and USPFT since 1985 and in 1993 was appointed an International judge for world competition by the FAI. She serves on the steering committee to host the world competition in Texas in 1996.

Jurenka is past state president of the Texas Aviation and Space Education Forum, state president of the Texas Pilots Assoc. and a director of the US Pilots Assoc. She served on Governor Ann Richard's Commission for Women, heading a task force on women's history. She is past chair of the Longview Commission for Women, current president of Women in Longview and serves on the Gregg County Airport Economic Development Board. She is an alumna of Leadership Texas and member of Leadership America Class of 95.

A member of the Texas Dogwood Chapter since joining the 99s in 1984, she has continually been a delegate to section and International meetings. She participated in the Aviation Education Exchange Tour to the Soviet Union in 88, was elected to the International nominating committee 1990-92, served on the Amelia Earhart Birthplace Museum Board of Administrators 1993-94, and was awarded the South Central Section Outstanding Service Award in 1993. She enjoys presenting to regional civic and aviation groups programs on the disappearance of Amelia Earhart, one of her greatest heroines. A special thrill was to be in Florida this year with the Mercury 13 for the midnight blast-off of the Discovery carrying the first female, shuttle pilot, Eileen Collins.

Jerry Anne and her husband, Ron, have three grown, married children and currently two grandchildren. They are both active in the Methodist church and live in Longview, TX.

CLAIRE JUSTAD, received her pilot license in February 1952, starting in a J-3 Cub. Eight formed the Idaho Chapter on Oct. 15, 1954. She was the treasurer. There are four of them still living! In 1956, she checked out in their Beech C Model Bonanza. The day before Laura Conner (her copilot) and she was to leave for Palo Alto, CA, to enter her first Powder Puff Derby. The Bonanza blew a piston at 200 feet as she took off from the Boise airport. Bob, her husband, quickly bought a G Model Bonanza and they took off the next day. She and Laura again flew the Powder Puff in 1959. She chaired several Idaho derbies in the sixties and enjoyed them immensely.

They had four children and both she and Bob took them on many, many trips throughout the US. Mostly, they spent much time in the Coeur d' Alene area in north Idaho. Their only son, John, is the only one to take up flying. John was murdered in Boise on Dec. 20, 1990. Their loss of John also took a heavy toll on Bob's health. He deteriorated rapidly and he passed away 15 months later on March 23, 1992. In 1984, they sold their last (thirteenth) plane, a Beech Baron because neither of them could pass their flight medicals. She received her multi-engine rating in 1963.

Flying was a big and enjoyable part of their lives. The companionship and memories of their fellow pilots will always be dear to her.

NELL ELIZABETH JUSTICE, presently a second officer on the Boeing 727 for United Parcel Service. Previously, she served as a captain and first officer from 1987 to 1994 with WestAir/United Express, a regional carrier headquartered in Fresno, CA. She flew the BAe 146, EMB-120, and EMB-110 at WestAir. She has been in aviation since 1977, soloing on Oct. 31 of that year.

She learned to fly in the aviation program at Big Bend Community College in Moses Lake, WA. She has wanted to fly for as long as she can remember.

Nell has experience as

a flight instructor, corporate pilot, and airline pilot. She has a BS in aeronautics and astronautics and three years of experience working as a civilian flight test engineer at Edwards AFB. She holds an airline transport pilot certificate with single and multi-engine land and Boeing 737, Embraer 120 and Embraer 110 ratings. She is also a flight instructor with single engine, multi-engine and instrument ratings, and a ground instructor with advanced and instrument ratings. She has more than 7,000 hours of flight time which includes 1,100 hours as an instructor, 5,200 hours of multi-engine time, and 2,000 hours of jet time. She is a charter member of the Antelope Valley 99s.

KAREN M. KAHN, born May 1, 1949, in Portland, OR. Joined the 99s in 1970. Flight training Gnoss Field, Novato, CA, Sierra Academy, Oakland, CA. She has been actively involved in aviation for the past 27 years. She holds all ratings through ATP, including an MD-80 type rating and was the first woman to be type-rated in a Lockheed JetStar.

Her other ratings include CFII, MEI, flight engineer: turbojet, seaplane, helicopter (including helicopter IFR) and glider. In airline service she's flown the Boeing 727 and DC-10 on domestic and international routes and is currently flying as captain of the MD-80 based in Houston.

Prior to starting her airline career in 1977, she instructed at a large flight school in northern California and operated her own weekend ground school teaching private, commercial and instrument courses.

After speaking at numerous career workshops and assisting many pilots accomplish their flying goals, she now provides professional advice and personal resource management through her firm, Aviation Career Counseling.

In addition to her airline flying and counseling service, Karen's an FAA accident prevention counselor, past president of ISA +21 (the International Society of Women Airline Pilots), former airport commissioner in her hometown of Santa Barbara, writes a bi-monthly column for the *99 News* and is a frequent speaker on the Pilot's Audio Update, a monthly aviation cassette series.

On her days off, Karen enjoys walking and biking, as well as non-scheduled flying with her husband (captain for a regional airline) in their Beech Baron.

MARILYN KAMP and her husband Bernie built Kamp Airport, which is a general aviation airport. They bought the land in 1975 and opened for business two years later. They have a 10 aircraft fleet and run charter, flight school, aircraft maintenance, avionics, sell fuel, pilot supplies, and buy and sell aircraft. She even lives at the end of runway 28, so you can say she lives and breaths aviation.

Kamp Airport is home base for the Central New York Chapter of 99s. She has been a chapter chairman for CNY two times, she also sits on the New York - New Jersey Section Board and is International Operation Skywatch chairman.

Operation Skywatch is a volunteer environmental aerial surveillance program that she helped start in the US.

Because she has run a flight school for 15 years, she has had the opportunity to influence many young people into the field of aviation.

She has four daughters and three granddaughters.

Through the 99s, Operation Skywatch and being in the aviation business has given her the opportunity to travel and meet other interesting people through out the world. Five years ago she traveled to Russia with a group of 99s. This was by invitation of the Soviet Government. They met Russian pilots and met with first woman in space, Valentina Treskinof. In January 1994, she met with Israeli 99s in Tel Aviv, Israel.

RONNIE KAMPS, got interested in flying four years ago. Her husband received his private pilot license and she was flying with him. He would let her take the yoke, she fell in love.

There was a spouse course – it teaches you how to land in case of emergency. The day she went for a test flight, she took the plane off and landed. When she got back to the FBO she decided to go all the way. She received her private pilot license Nov. 19, 1994. She intends to get her IFR rating, and after that the sky is the limit! She joined the 99s in February 1995 to meet women with the same interests and goals in aviation safety and education.

JEANETTE KAPUS, Wisconsin 99s, early in 1941 Milwaukee Area Women pilots organized an Easter Wisconsin 99 Chapter. Prior to 1941, Melba Beard formed a Central Wisconsin Chapter when she lived in Madison. However, when she left, it became inactive and the two groups merged. Hosts of the 1973 International Convention in Milwaukee. Jeanette Kapus joined the 99s in 1945, know as the "Spin Queen." Jean spun a J-3 Cub 64 turns from 9000 feet and recovered with her altimeter indicating 1000 feet AGL, breaking the world's record of 50 plus spins.

Over the years she has participated in: Magic Carpet Goodwill flights for the handicapped. WWII, had six members who joined the WAFS (later WASPS), one WAC, one WAVES, one Wisconsin Army Nurse was first Wisconsin woman to get her pilot license on GI Bill of Rights and WI First Woman airport operator (Kenosha) taught Army and Navy cadets to fly. In the 1970s, Ann Roethke's African Flying Safari. Delivered Daffodils for the American Cancer Society. Flying Companion Seminars. Airmarking Projects. Sponsored Annual EAA 99s Dinner during the EAA Convention. Members participated in Illi-Nines Race, winning the 1989 Rookie Pilot trophy in the speed trophy. Members that are pilots fly Midwest Express Airlines, United Airlines and one FAA Air Traffic Controller.

DEBORAH CHISHOLM KARAS, received her license in 1966. The first airplane flight was in a Cub at 13 years old and first glider flight in Regensburg, Germany, while working as a nurse aide in a German hospital. Karas has logged time in Cessna 172 and 182; Aeronica Champ and DC-3 and holds a SEL, commercial rating. She has participated in chapter air races, activities and ferried a Cessna along the Alaskan Highway to Fairbanks, AK.

Karas' educational background includes a BA in German Literature, accounting degree as well as a RN working in long term care and nurse aide educator. Community activities include older adult quality care advocacy and senior issues. She is a member of various nursing associations as well as a proud member of the 99s, Inc. enjoying the wonderful friendships.

She is married and has four boys who also support her interest in aviation.

ANNE HARRIS KATZ, enjoyed flying since childhood but didn't earn her pilot license until her middle years. Then she entered the aviator's life "full throttle."

Early experiences as the pioneering and adventurous member of her family – known for always "pushing the envelope"—ultimately led her to the challenge of flight. She began her adventures on the water, when as a youngster she sailed in summer and iceboated in winter. Lessons learned about wind and weather then have come in handy when "sailing" through the sky. As a college student she was one of two dozen selected from the USA for NASA's Space Biology Institute, an aerospace studies program. This experience led to a career in science. Educated through the doctorate as a biologist, she established herself in academia, first as professor then academic dean. But always there was her desire to break earth's bonds.

In 1991 she got her chance, earning a private pilot license. She joined the 99s and Civil Air Patrol. She continues to draw on her background as an educator in all her new vocational and avocational roles – focusing now on aviation and aerospace education. With her husband, and fellow pilot, she developed a publication about aviation museums and historic aircraft collections. They share a cockpit in single-engine Cessnas and Pipers – flying together on CAP missions, honing piloting skills on cross-country trips or on the lookout for flights in historic birds at air shows and aviation museums.

ROCHELLE B. "SHELLY" KATZ, born Aug. 10, 1939, in Philadelphia, PA. Joined the 99s in 1975. Flight training includes PVT 1973; COMM 1976; instrument 1980; and advanced ground instructor 1979. She has flown Piper singles. Employed as a flight school manager from 1978 to 1989. Designated written test examiner from 1982 to present.

Married to Alvin and has three children: Michael, 34; Jeffrey, 32; and Robin, 28. She also has grandchildren: Jonathan, 7; Matthew, 4; Joshua, 3; and Jordan, 3 months. As a commercial pilot she enjoyed doing sight-seeing rides and introducing non-flyers to aviation and to a woman pilot.

DEBI KATZEN, born Dec. 2, 1949, in Washington, DC. Joined the 99s in 1990. Her flight training included private pilot license and ASEL. Katzen has flown Cessna 150/150 Taildragger, 172, PA28, Tri-Pacer, Cherokee Six. She has two children, Lisa, 26, and Shawn, 22. Married and raised a family before learning to fly. She is president of Designs Unlimited Architectural Graphics (sign manufacturer). She learned to fly in a J-3 Cub, and hardly knew what radios and electrical equipment were until late in her training. Still frequently flies with only a compass to navigate. She's logged over 500 hours in her 150 and several other airplanes. As chairman of the Potomac Chapter 1994-96, she organized several airmarkings and participated in local air shows. She's been a member of AOPA since 1989, and flown her Taildragger from Detroit to Florida and many points in between.

RUTH MYCOLE KAZMARK, born Sept. 11, 1942, Elmira, NY. Joined the 99s in 1973. Flight training included commercial glider, single engine land and sea multi-engine. She has flown the following planes: Cessna 150, 152, 172, 182, 210; and various gliders and home -builts; Piper Cub; Mooney; Apache; Luscona; Ercoupe; Piper Seminole. She is part owner of FBO Hollywood Aviation.

Kazmark has one child, Michaelene; and grandchildren, Blake and Max. Member of 99s and Grasshoppers, Aircraft Owners and Pilots Assoc. Started flight training in Elmira, NY, finished in Ft. Lauderdale, FL. Glider training Schweitzer Air in Horseheads, NY.

HAZEL SNYDER MILES KEFFER, earned her license (number 203181) on Aug. 23, 1945, at the Detroit City Airport. She joined the 99s in Michigan in 1945. She moved to California and joined the Bay Cities Chapter in 1948. Most of Hazel's flying experience is in Luscombe's, although she also has flown Tri-Pacer's and Aeronca's. A VFR pilot, she has done everything from lazy 8s to snap rolls.

Hazel has one son, Fred Miles III, a structural engineer and one granddaughter, Sadie, age 10. An avid volunteer worker, she has spent decades working for a wide variety of community volunteer assignments which include Cub Scouts, Red Cross, Oakland Museum, and YWCA board of electors.

An active Republican, Hazel has been a delegate to the national convention, and inspector on the election board and an active member of the Women Voters and Women's Republican Club.

Her husband, Cdr. George Keffer, USN, Retired, holds pilot license #4753.

NICHOLE KEGEL, born Aug. 11, 1974, Tacoma, WA. Joined the 99s in 1993. Flight training includes private and instrument through local FBO (total time 300 hours). She has flown C-150 and their own Cessna 210 (bought in 1994). She owns aerial photography business with parents and fiancee, Rick Vander Ley. Started flying in February 1992, one year and 100 hours later, they (herself and Rick, CFI) hit three Canadian geese at night at 3,000 feet over Hillsboro, OR (airport). Lost their windscreen in a rented C-172. They now have the Canadian goose head mounted on a propeller, hanging over their fireplace.

HELEN J. KEIDEL, a late bloomer, receiving her pilot license at the age of 52. Her interest actually dated back to her childhood and after graduation, she inquired into becoming an airline stewardess. However, you had to be a registered nurse ... so that idea was out.

Marriage and the raising of three children became her primary interest until she went back to work after they had all graduated. One woman at work was just learning how to fly, and it rekindled her interest. Could an older woman learn to fly? Yes, and it became the thrill of her life.

Helen joined the Lake Erie Chapter as a charter member in 1974 and has served both as elected officers and committee chairmen over the past years. She has participated in airmarking, local air races, safety seminars and other activities of the chapter and still claims that flying is the biggest thrill in her life.

CAROL KEINATH, born Nov. 16, 1944, in Coxsackie, NY. Her love for flying took off at the age of seven when she accompanied her parents on a transatlantic flight to Germany. Although she was very young, she knew that someday flying would be a part of her being. At 22 a friend took her for a ride and encouraged her to meet other female pilots. Carol began taking lessons in a Piper Cub at a grass field in Freehold, NY, but after an instructor did not keep a few appointments she decided to quit. However, the bug was in her system and she decided to fly again at other airports Athens, NY; North Adams, MA; where she soled after 12 hours. Continuing training in NYS, she received her pilot license from William Kobelt at Kobelt Field on Aug. 10, 1974.

She had some memorable experiences during her student time: two engine failures and living to talk about them, a close encounter at 2,500 feet with a Bonanza flying 10 feet above and meeting a female line person for the first time. The line person greeted Carol with enthusiasm and exuberance as Carol was the first female pilot she had land on her airstrip.

Carol joined the 99s to converse with those who had a similar love for flying as well as becoming intellectually challenged and working on various projects. She has held the office of secretary and worked on various chapter projects.

CHARLOTTE KELLEY, soloed in 1945 at Leicester Airport, MA. Holds the following ratings with her private pilot license: seaplane, commercial, instrument, multi-engine, helicopter/rotorcraft and balloon (all acquired by 1955). Appointed First Woman State Aeronautics Commissioner in the US for Massachusetts 1956-66. A member of the First Federal Aviation Advisory Committee 1964-68. Reappointed 1964-68, to the project of Women to Provide Assistance and Advice to the Federal Aviation Administration; reappointed in 1972-75. Appointed to the first CACOA (Citizen's Advisory Committee for Aviation in the USA). Served as director of the National Aviation Education Council for five years.

First woman to pilot a US Jet (T-33), a Massachusetts National Guard plane with General Charles Sweeney, commanding officer. Trained with the US Navy to Pilot the World's largest Blimp in the world, at the Naval AB, South Weymouth, MA, 1955 (ZPG-2). Selected to be the First Woman Civilian to train and break the sound barrier. Trained at Mitchell AFB, Long Island, NY. First woman civilian to go through the altitude chamber (receiving a military certificate for same) and practiced parachute drops. Broke the sound barrier by diving an F-94 Star Fire from high altitude at Bedford AFB, Lawrence, MA, in 1957.

First woman, helicopter-rated pilot to fly the Lockheed-Hughes "Rigid Rotor" experimental helicopter at the factory, located in Culver City, CA. This was the prototype for the Cheyenne, used by the military.

First woman to hold office in Aero Club of New England, founded in 1909. Secretary for two years, and vice president for two years. (1948-52)

Member of US Helicopter World Championship Team and selected one of six to represent the US of America in world competition. No men were on the team, which trained at the Bell Helicopter factory in Dallas, TX. Official pictures of the team are in the Smithsonian Institute and the San Diego Aerospace Museum. The US placed sixth in the world. In 1978 Vitebsk, Russia, enroute they were trailed by Russian "MIGS" buzzing by the plane. They did land at a Russian Military Base. The US placed second in the world. In 1981 Pietrikov Tribulanski, Poland – the team consisted of all military men, who placed first.

In 1984 Castle Ashby, Great Britain, performed the duties of chief judge for the USA. The first woman judge for any country, for Rotorcraft/Helicopters. Kelley met HRH Prince Andrew and Miss Sarah Furgason, who later obtained both her fixed and rotary wing licenses. US placed first.

Entered the International Air Race in 1954, Hamilton, Ontario, Canada to Verdero Beach, Cuba. Finished race in fifteenth place, flew on to Havana for a reception by the then President Batista. Founded the All Woman New England Air Race. Founded the Kachina Air Race, Phoenix, AZ. Was a pilot and/or copilot in many of the above races and did win and place in many of them.

Also Angel Derby in 1969, New Orleans to Managua, Nicaragua; Pacific Air Race, 1969-70; All Woman's' Transcontinental Air Race; San Diego, Atlantic City; Calgary, Canada, to New Orleans, 1971; Van Nuys, CA, to Savannah, GA; Illi-Nines, Small Races, etc.

Memberships: AOPA, NAA, FAI, CAP, HAI, HAA, International Order of Characters, etc. Member of the 99s since 1946, current life member. Governor of the New England Section: held all offices and assisted the section to form smaller chapters for smaller groups. International APT chairman (for proficiency). Whirly-Girls: member since 1955 and awarded #21 in world (15th commercial rating), first secretary, 1971-73, and second president 1973-75. Helicopter Club of America, one of two founders.

In 1961 honored by President John F. Kennedy in the Rose Garden as one of the 13 famous Woman Helicopter Pilots. In 1967, honored by President and Mrs. Lyndon Johnson at the White House and the Rose Garden with Beautification Award for the Phoenix Airport at Sky Harbor. 1994, elected president and Whirly-Girl Scholarship, Inc. Associate in Arts; bachelor of science; master of education teacher – vocational subjects – New York University, etc.

JANE ZIEBER KELLEY, May 30, 1935 - April 20, 1978, became a private pilot in 1968, about the time the Aloha Chapter of the 99s was organized, and she quickly became one of its most supportive members, taking special pride in her work as APT chairman and in aero-space education projects. In 1971 Jane and another Aloha 99, Beth Oliver, teamed up to fly in the 25th Powder Puff Derby.

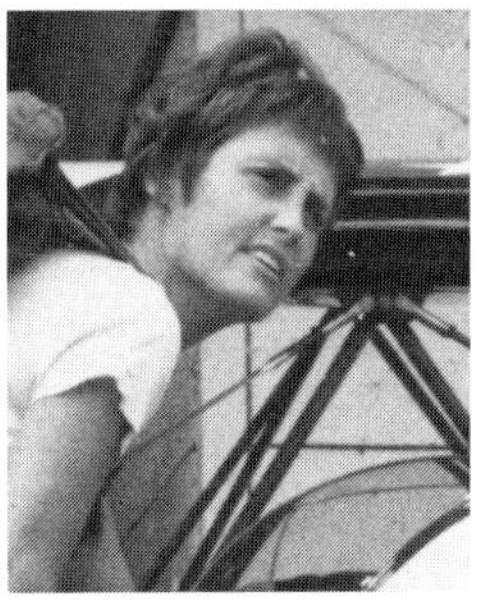

Quick to praise others and encourage them in their endeavors, Jane was modest about her own accomplishments. She held commercial SEL and MEL, CFI, instrument and instrument flight instructor, multi-engine sea, and ATP ratings, plus the distinction of being the first jet-rated woman pilot in the Hawaiian Islands.

Her greatest joy, though, was flying aerobatics. Jane worked as an aerobatics instructor for Polynesian Sport Aviation, flew in local air shows and practiced diligently in her own single place Pitts Special, N21JK. Jane belonged to the Aerobatic Club of America and to the International Aerobatic Club, and took every opportunity to travel to the mainland to participate in aerobatic competitions, both as a contestant and as a judge.

While judging an International Aerobatic Club competition near Tucson, AZ, Jane became ill and was rushed to the Tucson hospital, where she died on April 20, 1978.

Born and raised in northern California, Jane was a graduate of Stanford University, where she met her husband, Richard. They had five lovely children: Kathy, Chuck, Linda, Bitsy and Colleen.

At the time of Jane's death, the family requested that in lieu of flowers, donations be made to the Amelia Earhart Memorial Scholarship Fund. Contributions from Jane's many friends totaled over $13,000, enough for the 99s to add a full scholarship each year, and a fitting tribute to an outstanding 99.

JANE KELLOGG, born Oct. 28, 1939, Pontotoc County, Ada, OK. Joined the 99s in 1987. She regrets having never lived close to a chapter to be active, however, she loves the publications she receives. Her flight training included private pilot license and instrument rating. She has flown Cessna 150, Tomahawk, Grumman Tiger, 172, Cherokee Archer, Cherokee Warrior. Married to Don and has five children (no pilots, but one spouse is a first officer for American Airlines). She also has nine grandchildren who all love to fly.

She had always said, "Someday I'll be a pilot and drive to the airport in my red sports car." Well, dreams have a way of getting sidelined. Having children, getting a college degree herself, then teaching school and raising children kept her from her dream. A divorce caused her to start dating and the guy that finally captured her heart owned a Cessna and had been flying several years.

She thought it was finally her time to get that pilot's certificate and started taking lessons. But, with three children in college and only one steady job, the dream had to stay in the background. In fact, they even had to sell the plane to keep them going.

Obviously, the story ends happily. Professional success beyond their wildest dreams came a few years later to both herself and her husband, and the money finally wasn't an issue any more. They started shopping for a plane in August 1987 and her earlier instructor had one to sell.

A part of the sales price was that he "threw in" her lessons to complete her certificate with a catch! She had three months to do it in as he was moving to Chicago with his wife – a United Airlines Captain! She flew Friday afternoon and evening, Saturday all day and many Sundays. But, at 47 she did it!

Sadly, the Warrior that was purchased was destroyed in a hangar fire – "we all cried!" "Today we fly a beautiful Cherokee Archer but still miss our Warrior. Oh yes, the red sports car? Well, a few years go, my husband surprised me with a red Fiero to use as an airport car – some airport car!!"

CYNTHIA ATKINSON KELLY, trained for her private pilot certificate while a senior in high school. One week after graduation she became licensed. That was the beginning of her career in aviation which has led to her most recent accomplishment, a type rating in the Cessna Citation.

After earning her private pilot license in 1983, Cynthia went on to study aerospace science at Metropolitan State College in Denver, CO, later transferring to Embry-Riddle Aeronautical University. She earned her flight instructor certificate in 1986 and began flight instructing. After logging more than 700 hours she left her home in Denver for a flight instructor and charter position in Texas.

Flying single-engine charter helped Cynthia to excel as a pilot. She recalls many memorable flights in and out of busy commercial airports but most interesting were the flights into remote dirt and grass strips. Picking up multi-engine time whenever possible and logging more than 1,200 hours, she got her first big break into corporate jet flying in August 1987. Cynthia accepted a position with Phoenix Air in Georgia and began flying right seat in a Learjet. For the next two years she traveled domestic and international airways, transporting passengers, freight, and flying precision profiles for military training exercises.

Cynthia's career has included flying for a commuter air carrier; but her true love is corporate aviation, to which she returned as a free-lance pilot. She hopes to one day own an aircraft management business with her husband.

Born May 8, 1965, in Denver, CO. Joined the 99s in 1993. Kelly's flight training includes commercial, instrument, ATP, CFII, MEII, CE-500 type. She has flown the following planes: CE-150, 152, 172, 182, 177RG, 210, 310; PA28, Navajo; Cheyenne; Seneca; EMB120; LR23, 24, 36; Citation; Diamond. She and her husband Terry have one six month old child, Samantha.

MARY KELLY, joined the Hampton Roads 99s Chapter when she became a pilot in 1979. She is a multi-engine commercial pilot and CFII with 3,850 logged hours, 2,450 of those as an instructor. She has owned her own flight school and served as manager of three Oklahoma airports: Altus Municipal, Hatbox Field in Muskogee and presently, Tenkiller Airpark.

Kelly received an Amelia Earhart Scholarship in 1985. She was inducted into the Forest of Friendship in 1987. She became an accredited airport manager (AAE) through American Assoc. of Airport Executives in 1990. She serves on the board of directors of the National Biplane Assoc. and is past president of the Oklahoma Airport Operators Assoc. She is feature writer for the *Oklahoma Aviator* newspaper.

She as pilot and her husband as navigator earned awards from the NAA/FAI for National and World Record Flights: in a Cessna 172RG from Harbour Grace, Newfoundland to Londonderry, Ireland commemorating the 1932 solo flight of Amelia Earhart; in a Cherokee 180 from Oklahoma where Will Rogers was born to Barrow, AK, where he and Wiley Post were killed, and in a Cessna 414 for completing a 60th anniversary flight around the world commemorating the 1931 flight of Post-Gatty.

Mary's husband, Joe Cunningham, is also a pilot, ground instructor and public relations consultant. She has one daughter.

PATRICIA A. WEIR KELLY, joined 99s in March 1967, after getting her license in October 1966 in New Jersey. Her first 19 hours of instruction were in a J3 Cub. She has seaplane, multi-engine ratings, instrument and commercial licenses. She owns a Piper Archer and her husband is building a Glasair.

She has held the following chapter offices: chairman, treasurer, Amelia Earhart Scholarship, nominating committee, registration and treasurer for Wis-Sky Run, registration chairman for International Convention - Milwaukee, 1973, registration chairman for North Central Section meeting - 1980, and co-chairman EAA, 99s tent 1994. She flew the 30th Anniversary Commemorative Flight, Powder Puff Derby, July 1977, and flew an Air Race Classic.

Besides her love of flying, she enjoys knitting, needlepoint, sewing, photography, hiking, skiing and gardening.

She has two children, acquired seven more after marrying her present husband Paul, also a pilot, and together they have 12 grandchildren.

GERALDINE JOANN PALLISTER KEMICHICK, Greater Detroit Area Chapter of 99s. Yearning to fly was born with the sun glinting off the silver wings of a DC-3 in flight. Desire became reality half a lifetime later in 1984 when she earned a private pilot license in a Cessna 172.

In between desire and fulfillment came a bout with polio, college, marriage, raising three daughters and two sons, learning to scuba dive, church involvement, more education and working as a library technician.

Jerry is a member of the 99s and the Aircraft Owners and Pilots Assoc. She has held various offices and committee chairs in the Greater Detroit Area Chapter where she was instrumental in implementing the Air Bear Program for the chapter.

Husband Robert, a retired automotive design leader, is also a pilot. They have three grandsons and have been married for 44 years.

CYNTHIA GARDNER KEMPER, born in Bedford Hills, NY, on April 18, 1919. Always wanted to learn to fly. For five years during WWII served as an Aircraft Communicator with CAA (now FAA) at Charleston, WV, and Elmira, NY, airports. She was trained in the third class in which women were admitted. They were also required to take and transmit weather.

She and her husband, Charles, learned to fly in 1959 at a small grass field near Naval Air Test Center Patuxent River, MD, in an Aeronca Champ. Charles was sent by Sikorsky Aircraft to the Test Center to work with Navy pilots.

She joined 99s in 1963 as a member of Connecticut Chapter, in which she held several offices. During this time she became friends with Nancy Hopkins Tier and has been enriched by knowing her and learning of her early aviation experiences. Teddy Kenyon had a wealth of memories to share. She had the opportunity to interview Clarence Chamberlin several times before he was a featured speaker at 99s Annual Convention at Bretton Woods, NH, in 1970.

They have a son and daughter-in-law living in Weston, CT. Kemper and her husband now live in Asheville, NC. She is a member of Carolinas Chapter. They have joined Western North Carolina Air Museum located at Hendersonville, NC, airport. Many WWII planes here!

MADELINE B. KENNEDY, born Oct. 25, 1943, Gaffney, SC. Joined the 99s in 1974. Flight training includes 20 hours dual to qualify for private pilot license. She has flown Cessna 150 and 172.

Married to Marcus and has a son M. David Kennedy, Jr.

The photo is of her in the cockpit of restored military plane, which was on display at an airshow sponsored by the Warbirds in Spartanburg, SC, in 1987.

SUSAN J. KENNEDY, Treasurer Southwest Section 1994-96, born in San Diego. During the war she would watch the P-38s practice dog fighting over their canyons and the novice paratroopers fill the sky with white "chutes" She never dreamed of flying. Then one day the opportunity arose for her to go flying on her lunch hour. She had been in a big airplane once. They arrived at Hawthorne Airport and walked up to a Cessna 152. Her heart was pounding with excitement. They took off, she saw LAX, then the ocean, then the Palos Verdes Peninsula. She was so excited that after a perfect landing all she could do was expound on the thrills of the last hour. She knew she wanted to fly.

One of her husband's acquaintances was a member of the Civil Air Patrol and owned a Piper Archer. He took both of them up and let her husband operate the controls. Carl was hooked again. They did start saving their money for those much-wanted flying lessons, five or six times a day. She convinced her husband that they should talk to someone at Torrance Airport about flying lessons. They had a meeting with the chief pilot at Peninsula Aviation. Suddenly, they were the proud owners of a 1980 Warrior.

She received her private pilot license in February 1982 and became a 99 that month. Then came the Pacific Air Race. With less than 500 hours between the two of them and not knowing anything about air races, Jacquie Sprague and she entered the PAR. In 1982 they decided to purchase a new 1983 Piper Turbo Arrow. They started instruction for the instrument rating which they both now have. They share left seat time, but somehow Carl has managed to acquire about 200 hours more than she. They have been to Baltimore, Vancouver, and Shangri-La for Convention and Oshkosh twice and most of the Southwest Section meetings.

When they moved to Sacramento, their next door neighbor reintroduced her husband to pheasant hunting and skeet shooting. She is learning how to shoot a shotgun, but only on the clay pigeons. She enjoys watching the dog work a field, point a bird and hold that point. They acquired a fully trained German Shorthaired Pointer. His name is Bandit and he is a joy. Misty and she have been in training. They have been in shows and hunt tests.

LISA KENNY, born in Richmond, BC, on April 28, 1995, and raised in Abbotsford, BC. Within three months after graduating from high school she earned her private pilot license and was accepted into the UCFV Aviation Diploma Program.

Lisa's first flying job was with Timberline Air (Chilliwack, BC) where she flew a King Air 100 and C-177. To date, she has worked in flight operations for Coastal Pacific Aviation (Abbotsford), Wilderness Airlines (Vancouver, BC) and Air Tindi (Yellowknife, NWT).

Lisa joined the 99s, Inc. at the Women in Aviation Conference in Orlando, FL, and is a member of the BC Coast Chapter. She enjoys doing the publicity for the chapters events. Lisa is a new member of the BC Aviation Council.

Her father, before retiring, was a DC-10 captain for Canadian Airlines International. It was a beautiful day on the BC coast when she flew one of her flights on the King Air 100. On the return leg of her trip, Vancouver center radioed her captain and she said that Canadian 3 wanted to talk with us on a separate frequency. They tuned in the frequency and it was her dad and his crew outbound from Vancouver for Taipei. They compared their altitudes, VOR radials and DMEs and sure enough they passed by each other and each took photos. "I guess you could say that our careers did cross paths!"

LENORE KENSETT, born in St. Joseph, MO, earned her private license in 1959 in Chanute, KS. She has been an active 99 since 1974. She married James W. Kensett, an optometrist, also a pilot. They have a son, three daughters and six grandchildren. She was business manager of her husband's optometric practice in Chanute for 22 years. Kensett attended high schools in St. Joseph, Dallas and Fort Worth, TX, attended Park College (MO) and graduated with a BS in business administration from University of Kansas.

Genealogy, travel, flying, aircraft home-building and restoration, and church and community service are their continuing major interests, although over the years they counted tennis, skiing, and sailing dominant interests as well.

Her two most cherished accomplishments are flying in the 1976 Powder Puff Derby and as volunteer campaign director, successfully reaching their $2 million goal in the Depot Restoration Project in Chanute in 1992.

JEANNE "JAN" KENT, New York resident, earned private certificate in 1980, followed by instrument and commercial. She owns a C-172 based at HPN. Kent has participated in the Garden State 300 for last seven years and in an Air Race Classic; and has flown extensively around the USA.

Kent earned a BA at Cornell University and MS in special education at College of New Rochelle. She has been a teacher in various fields and local community activist.

Kent has been a member of North Jersey 99s since 1987. She has held all chapter offices and New York - New Jersey Section treasurer, director and membership and ways and means committee chairs. She is also member of AOPA, Westchester Aviation Assoc., Teterboro Hall of Fame of New Jersey, and New York State Aerospace Education Council, Inc.

She is married to Leonard P. Kent. They have three children and four grandsons, none pilots, yet.

ROBERTA A. "BOBBYE" KESTERSON, born Sept. 13, 1925, in Warrensburg, MO. Joined the 99s in 1964. Flight training includes private and SEL. She has flown 46 Luscombe.

Her three children are Scott Ellis, 40; Sheree Gail, 34; and Karla Deane, 27. Bobbye started trading bookkeeping for flying lessons at the Centralia, IL, Airport in 1957 where she met her husband, Deane. She received her license in July 1959, about the time they started teaching son, Scott, age 4, in their 46 Luscombe. She joined the 99s as soon as she got her license but dropped out to have Sheree Gail, then rejoined in 1964. They had another girl, Karla Deane, in 1968 and Deane was killed in their home-built in 1973. Bobbye sold the Luscombe

about a year later (no longer having a mechanic) and hasn't flown much since except with Scott. He soloed on his sixteenth birthday and still flies for a living. Bobbye has been a life member since 1984.

SHARON KAY KETCHUM, Mt. Diablo Chapter 99s, Concord, CA. Private pilot since October 1970, 475 flight hours, joined the 99s in December 1970. She has held the following offices: chairman, vice chairman, secretary, treasurer, membership chairman, news reporter and almost every committee. Her occupations have included secretary, currently homemaker, self-employed Ketchum Displays (Christmas displays and decorations). She is married to Gerald "Sam" Ketchum, Sr., also a pilot and has two children, Sharilyn, 25, and Gerald, Jr., 24. They also have their son-in-law, Dean, and their daughter-in-law, Susan and four grandchildren: Paul, 7; Mark, 6; Trevor, 3 months; and William, 1 month old.

They have made many cross-country trips in their Bonanza A36 to the mid-west, east coast, Bahamas, Mexico and Baja, CA. She flew in the PAR (Pacific Air Race) which was very exciting and lots of fun. Almost 25 years ago, she was attending her first 99s Christmas Party and their son, Gerald, Jr., decided it was time to be born. They left the Christmas party early along with a new friend, 99 Sis Breuner (also a nurse) and Dr. Riley. They headed for the Mt. Diablo Hospital. It was a very exciting Christmas party! Just after starting flying lessons, she found out she was expecting. She continued flying and got her license in October before Gerald, Jr. was born in December 1970. You can imagine the surprise on the faces of the controllers when she would call in "student pilot" and then get out of the plane with her big tummy. "Did I get respect, or what! Remember that was 25 years ago and not many lady pilots, especially pregnant ones!"

BETTY KIDD, born in Ada, OK. She received a BS in education from East Central State College in Ada and a MA in music education from North Texas State University. She was a public school music teacher for many years in Wichita Falls, TX.

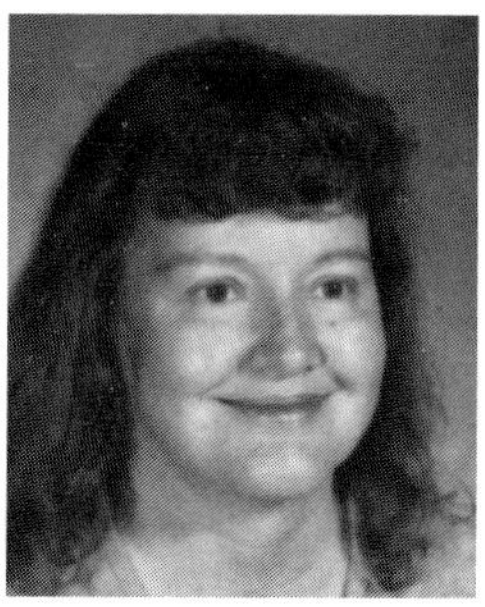

She earned her pilot license in 1978 and her instrument rating in 1980 and has logged almost 500 hours. She has flown 150s and 172s, Piper Warriors and Piper Archers. One of her most unusual flying experiences was flying a wedding cake to Tulsa, OK, for a good friend and ex-student who owns a bakery in Wichita Falls.

Betty joined the 99s in 1979 and has held many local offices. She has also attended many South Central Section conventions. She is a member of Aircraft Owners and Pilots Assoc.

She is also a member of Texas Retired Teachers Assoc., Beta Sigma Phi and Wichita General Hospital Auxiliary. Betty enjoys doing volunteer work for the hospital, reading, walking, theater, cross-word puzzles and being with her family.

ROBIN HOSENBALL KIDDER, born May 11, 1954, Cleveland, OH. Certificates include: ATP, typed in SF-340 and SA-227. Flight training, General Aviation at Leesburg, VA, and Washington Dulles. She has flown most single engine Cessnas and Pipers, Navajo, C-340, EMB-110, SA-227, SF-340, DC-9. Kidder has been employed with Dulles Aviation as instructor; Pegasus Aviation as pilot, Navajo; Resort Airlines as pilot, Navajo; ComAir Airlines, F.O. EMB-110, SF-340, captain SA-227, SF-340; and Midway Airlines, F.O. DC-9.

Started flying after completing college. She was working on a film about the Flying Circus Aerodrome in Bealton, VA, when she went for a ride in a Stearman (bi-plane). She thought that flying would beat working for a living, and she set out to make flying her career.

Kidder worked at night at the telephone company to pay for her flying lessons. She got her CFI, CFII, MEL, and MEI and then she started working for a small commuter flying Navajos out of Dulles. Although that airline went bankrupt, she got the experience to get a job with ComAir, a large regional airline. After flying as a scab captain at ComAir, she left for Midway Airlines, where she flew DC-9s as a first officer.

She is currently at home raising her two children, Lydia and Audrey. "Hopefully I will resume my flying career when my girls are a little older."

DR. MARTIZA R. KINCAID, born Oct. 14, 1948, in Cuba. Learned to fly in the Island of Guam, where she obtained her private pilot certificate. Following her interest in aerobatics, she took lessons at Hart Air in California and later at Kaimana Aviation in Hawaii. At Kaimana she met Jim A. Kincaid, aerobatic instructor, designated pilot examiner and airshow pilot, who she later married. Together Maritza and Jim opened a flight school in Ponca City, OK. Kaimana Aviation, Ponca City, opened its doors in January 1995, and it is a comprehensive PAR 141 flight training center, training pilots from private to ATP, including aerobatics and banner tow operations.

Maritza is the general manager for administration and Jim, the general manager for operations. Maritza continues to hold her 99 membership with the Aloha Chapter in Honolulu, HI.

Maritza continues her flight training and she is presently working on her instrument rating. She enjoys flying different types of aircraft and hopes to one day be able to fly the Pitts Special which they personally own. She presently flies other aircraft owned by the school, including Cessna 172, Piper Warrior and Citabria. She is also looking forward to one day flying the Travel Air which is the multi-engine trainer at Kaimana Aviation.

Maritza, who holds a doctorate in education, is also the publisher of Kaimana's official newsletter *The Diamond Flyer*, with a circulation of 300. She has been also published by *Woman Pilot*.

She has five children: Alex, 14; David, 11; Lillion, 8; Michael, 5; and Randy, 3.

LESLIE A. KING, born Aug. 13, 1948, in Schenectady, NY. First expressed an interest in flying at age 13 in a careers presentation. Guidance counselor's comment was "someday you'll realize you're a girl." Gave up the ambition until mid-1974, when discretionary income allowed for flying lessons at Farmingdale, NY, Republic Airport, weather permitting (about once every three weeks).

Laid off from airline sales October 1974 – decided to attend a full-time flight school – Spartan School of Aeronautics, Tulsa, OK. After all, didn't want to reach 40 wondering "what would have happened if…?" During and after flight school, worked as a ground/flight instructor for Spartan and briefly for Houston Piper, until a job was landed flying Twin Otters for Crown Airways, an Allegheny Commuter out of DuBois, PA, April 1977. Upgraded to captain in June 1978.

Hired by United Airlines December 1978, seventeenth woman pilot. Became Captain King once again in February 1991, United's fourth female captain.

Married to John H. Zimmerman, also a UAL captain. They were the first married couple to fly together as a cockpit crew for UAL. Goals yet to be achieved: to reenact but complete AE's final flight; to become Doctor Captain King.

MARTHA BENHAM KING, born Jan. 27, 1943, in Riverside, CA. Joined the 99s in 1991. Her ratings include ASEL and instrument. She has flown the Grumman and AA1. Married to Bill King and has one child, Shawn Benham, 30. Member of Civil Air Patrol Squadron 150.

DALE B. KINTOP, born Sept. 27, 1952. She is an accountant. Married to Jeff and has two children, Krista and Caitlin. Joined the Reno Chapter 99s in September 1990. She currently holds the office of treasurer. Became involved in aviation in 1990 – her business partner had a plane and rekindled her interest. Her father flew for the Air Force and owned a biplane. She currently flies for both pleasure and business. Kintop has a BA in education and business. Currently a partner in a small tax/accounting/bookkeeping office. She loves to ski, camp, fish and fly. Other memberships include vice president of Northern Nevada Enrolled Agents.

FAYE L. KIRK, born May 24, 1926, in Kentucky. Joined the Bay Cities Chapter in 1953 and then was a charter member of the Santa Clara Valley Chapter in 1954. She decided to get her pilot license after she and her husband, Bob, crashed a BT-13 in the bottom of the Ohio River. She was licensed in May 1953. She flew the 1970 Powder Puff Derby with Marion Barnick in her 1946 Ercoupe. She and her husband attended the International Convention in San Juan in 1974. Faye was the postmistress in Stockton, where she was periodically in trouble with the postal department for insisting on flying a Confederate flag. Two of her children are also pilots, and one is a 99 (the other is male). She now has grandchildren and enjoys fashion and flowers and lives in Stockton, CA.

Kirk has flown Cessna 150 and 172, Ercoupe, Cessna 182, Cessna 210, Cessna 411, and Twin Comanche. She is employed by Kirk's Lumber and Dinsmore Lodge. Her husband Bob is deceased, their children are: Claudia, 40; Karen, 49; and Jim, 46. Her grandchildren: James George Dilling, 5; Olivia K. Dilling, 2; Bob St. Denis, 26; Nicole St. Denis, 24; Matt St. Denis, 21; Jessicah Kirk, 26; Sarah Kirk, 21; Meggin Kirk, 24; and Ben Kirk, 18.

MARY LYMAN KIRK, earned her license in 1966. In 1968 she was responsible for organizing the Mt. Tahoma Chapter of the 99s. She held most of the offices for the chapter as well as offices in the Northwest Section. She was state vice president of the Washington Pilots Assoc., public information officer and administrative secretary of the Washington State Aeronautics Commission, member of the OX5 Aviation Pioneers, named in 1967 edition of *Who's Who in American Women* and was feature writer of 99s news for *Western Flyer.* In 1969 the Pierce County Chapter of the Washington Pilots Assoc. named Mary as their Pilot of the Year.

Mary was the mother of four, widowed from Lyman and married Robert E. Kirk, a strong supporter of Mary's aviation interest although not a pilot himself. Mary was a corner-stone of the Mt. Tahoma chapter, always encouraging, happy and supportive. She died Dec. 22, 1992. *Submitted by Francis Huff Postma.*

DEBI D. KIRSCH, born in Michigan, but fell in love with flying as a passenger in a small plane going over the Rocky Mountains. She first soloed on Friday the 13th, August 1982, an appropriate date, since the right main tire blew out on impact. She instinctively knew she was a born pilot when she was able to keep the plane on the runway without consciously having any idea of how she was doing it!

Despite this beginning, she went on to commercial, instrument and multi-engine ratings. Her second most memorable experience was flying over the glaciers and geysers of Iceland, and she plans many more adventures with her 6-year-old daughter who already has hopes of following in her mother's footsteps.

DEBORAH DAVEY KIRSCHNER, an Illinois resident, has been flying since 1985 but has been a pilot since 1992, thus qualifying her to teach for the Chicago Area Chapter's Flying Companion Course. Her flying sometimes takes a back seat to her career as a computer consultant.

She is married to John Kirschner, a pilot and her inspiration. They own a Cardinal 177RG. Their house literally has an airplane in every room, from paintings or sculptures to 1/4 and 1/3 scale remote control aircraft that they both build and fly.

Deb has flown a Piper J3 Cub (preferably in the summer!), done aerobatics in a Bellanca and has sky-dived. Deciding to stay in the plane from now on, she has assumed the role of treasurer for her 99 Chapter. She has coordinated and flown in the Fly-By Opening Ceremony at the DuPage Airshow each summer.

Deb's most memorable experience was her private checkride. The inspector told her she "flies like a man." This was meant to be a compliment as well as a joke. Her first medical had listed her sex as "M"!

BETTY KJELLBERG, Phoenix 99s, who rejoined the Phoenix 99s in 1994, had previously been in the chapter during the mid-80s and was legislative chairman. She learned to fly at Sawyer Aviation in Phoenix, getting her private ASEL in 1984 and her instrument in 1986. Betty has a MA of aviation management from Embry Riddle Aeronautical University and is currently in association management.

In 1993 she flew to Alaska and considers that her best aviation experience. She has served on the board of directors for the Arizona Pilots Assoc., as well as being their vice-president, news-letter editor, and chief of legislative affairs. Currently she is on the Scottsdale Airport Advisory Commission, the board of Scottsdale Leadership, and is a member of Soroptimists, International - Phoenix.

NANCY BUCK RICHTER KLEIHEGE, a native of Denver, CO, learned to fly in an Air Force flying club in Lubbock, TX. On her solo cross country she got weathered out, landed at the small airport of Knox City, tied down her Cessna 150 and spent the night with a nice airport family. Two years later, they were glad to see her show up with her fellow Abilene 99s to airmark that runway.

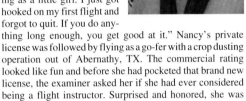

When asked about her flying career, Nancy says, "No, I never dreamed of flying as a little girl. I just got hooked on my first flight and forgot to quit. If you do anything long enough, you get good at it." Nancy's private license was followed by flying as a go-fer with a crop dusting operation out of Abernathy, TX. The commercial rating looked like fun and before she had pocketed that brand new license, the examiner asked her if she had ever considered being a flight instructor. Surprised and honored, she was inspired to obtain her CFI, instrument, CFII and ATP.

After a stint flying charter, Nancy realized she preferred working with people and the airplane. After instructing for FBOs, she started her own successful instrument instruction business. She also flies with the 99s, Civil Air Patrol and Wings for Wildlife.

In 1988 she married Mike Kleihege, her favorite instrument student. "Some people will do anything to get an instructor when they want one." Now Mike and their two daughters, Michelle and Elise, have their very own.

MATHILDE MAE KLEMENTS, born McDonald, PA. Joined the 99s Los Angeles Chapter in 1959. Flight training was at Hawthorne Airport, Hawthorne, CA. She has flown Ercoupe, Cessnas 140, 150, 172, 182, Navion, Piper Tri-Pacer 22. Her first introduction to flying was during a vacation in Pennsylvania, after she had moved to Los Angles, CA. She soloed in a Ercoupe. Years later she decided to pursue her dream of becoming a pilot. Two years later she was proud to say she received her private pilot certificate. Shortly thereafter she joined the Los Angles Chapter of the 99s.

During her active years as a member she served as secretary and treasurer in the chapter. She attended many of the Southwest Section Conventions, also attended two International Conventions including the convention in Australia.

She participated in local air races (PAR) and many local fly-ins. Her greatest flight was along the Caribbean Islands with stops in Haiti, Puerto Rico, St. Martinique, Venezuela, Surinam, S.A., flying over the Panama Canal, Mexico and the US. She's proud to be a 99, all her memories are pleasant ones.

EMILY PORTER KLINE, born Nov. 8, 1920, Michigan. Joined the 99s in 1992 and served in the Women Airforce Service Pilots 1943-44. Flight training included Sweetwater, TX, in service; Behan, IN, and Muskegon, MI, in a Piper Cub. She also flew PT-19, BT-13, AT-6 and UC-78. Her husband Charles is deceased. They had four children: Philip, 45; John, 43; Thomas, 38; and Caroline Kline Sassone, 41. Her grandchildren are Rebecca E. Kline, 14; Kimberly, 4; Christopher, 1; Daniel Sassone, 10; Charle H. Kline III, 13; John P. Kline, Jr., 9; Benjamin V. Kline, 5; Amber Kline, 19; Jason Kline, 22; and Elise Sassone, 6.

She learned to fly in Muskegon, MI, in 1942. In 1943 she joined the WASP, and trained in Sweetwater, TX. After graduation she was assigned to Greenville Army AB, Greenville, MS. There she flew as test pilot for production, line and maintenance. They flew BT-13s and VC-78s. Excitement happens when, you as test pilot, have to rescue planes that cadets have force-landed in cow pastures! She had OCS at Orlando, FL, and 30 hours of instruments. She was being re-assigned to B-24s when they were deactivated December 1944.

CHARLOTTE SCHRIER KLYN, born June 14, 1916, Indianola, IA. Joined the 99s in 1969. Flight training includes Ken Hoffman Flight School, Broomfield, CO, private to ATP. She has flown the following planes: Cherokees, Comanche, Cessna 172 and 182, Bonanza, Aztec, C-310 and B Baron.

Married to Andrew William Klyn and has two children, Charlene Klyn de Juarez, 50, and Vincent S. Klyn, 45. Their grandchildren are Javier Juarez; Enrique and Patricia Juarez; Lucas Klyn, Bailey Klyn, and Nathan Klyn.

She instructed for Hoffman Flight Center and Star Aviation, Denver from 1968-72. In 1972 she received the FAA Flight Instructor of the Year Award for the state of Colorado and subsequently the Rocky Mountain Region and was also designated a flight examiner by the FAA. From 1972 to 1982 she taught and supervised the flight training for Metropolitan State College, Denver, CO, Aerospace Science Department.

ALONA KNAAN, born Oct. 15, 1939, in Israel. Flight training included private pilot license, single-engine. She has flown Poker, Piper, Cessna and Shiroky. Knaan has two children, Ayiv, 29, and Istar, 36; and a grandchild, Ohav, 10. She is a widow.

Her way to flying started after the "six day war." Those days she worked as sound man for the CBS news in Israel, 1967.

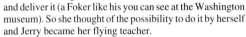

During the war their crew was stuck at the Suez Canal. A pilot named Jerry Renov had to land with his Foker in Sinai, get the film and deliver it (a Foker like his you can see at the Washington museum). So she thought of the possibility to do it by herself and Jerry became her flying teacher.

On "Yom Kippur" war 1973, she flew by herself from the Golan Heights to Tel-Aviv with the film to be the first in the satellite. Their story was sold to all the television stations around the world.

On 1974 she stopped working for CBS news, but she never stopped flying. She flies with her granddaughter, Ohav; she is 10-years old and no wars for her, just dreams about peace.

One day she got a call from Professor Zipora Altman. She was a pilot and she told her about the 99s. First she was a member at large, then a few friends got pilot licenses and they created the Israeli Section; unfortunately Zipora is no longer with them. Knaan would love to meet with the 99 pilots of the peaceful countries of the Middle-East.

JOAN E. KNAPP, Olympia, WA, earned her private and commercial in Wichita Falls, TX. After moving to San Francisco Bay, instrument, all ground and flight (airplane) instructor ratings, and ATP were achieved. She instructed at Pacific Piper, San Jose Airport, and taught at West Valley Community College. After she passed the Flight Engineer Written (B-727), she applied to United Airlines when they first opened the cockpit to women to be turned down based on age.

After learning in a C-150 (1972), she has flown various planes including Navajo Chieftain. In Olympia, she was chief flight instructor (Vagabond Aviation), first officer (Air Olympia), and accident prevention counselor (Northwest Mountain Region).

Joan has logged more than 3,000 hours, and participated in Powder Puff Derbies (1976-77) and Pacific Air Races (1977-78). She joined Santa Clara Valley in 1974, served in many capacities including chairman, was involved with aero space education and Flying Companion Seminars, and won the Service Award of the Year (1979). She was active in Mt. St. Helens and is now a member of Mt. Tahoma.

She also belongs to AOPA, Girl Scouts and Zonta. She is married to DeWayne Enyeart, an anesthesiologist and has two children, David and Elizabeth.

CAROL B. KNICKERBOCKER, born Nov. 24, 1965, in Baltimore, MD. She has flown the Cessna 152 and 172. She has been a private pilot and 99 since the spring of 1992. She has been interested in the field of aviation for most of her life. She took the private pilot ground school, which was being given at a local community college having only been in an airplane once. She decided when she passed the written test to learn to fly. She worked weekends at a local airport in addition to having a full-time job, not in the field of aviation, in exchange for flying hours. After almost two years she obtained her private pilot certificate.

Currently, she works for the Commonwealth of Pennsylvania, however she is trying to pursue a career in the field of aviation as an air traffic controller. She is also the secretary of her local chapter of the 99s, a member of AOPA, and flies as time and money permits.

LAURA LOUISE KNIPMEYER, born June 28, 1958. Joined the 99s in 1994. Flight training at Century Air, MMU from October 1993 until Aug. 19, 1994 (date of ASEL). Civilian employment, Sandoz Pharmaceuticals, 1986 to present. She has flown the PA28-140, 160, 181. Married to Ken Knipmeyer. Her parents are M. Joan Lindsey and Lawrence Clark.

Her neighbor took her up in his Cessna 140A a few times, about eight years ago. Acting on a lifetime dream, her husband studied on two different occasions for his pilot license, but stopped for two years following a busy period at work. During his training she learned about the navigation, and they imagined that she could help him with just this part in their flights together. They talked about getting a plane when they retired.

In September 1993 she was sent to the Center for Creative Leadership as part of her training for her job as a project manager. One of the people in her class was Whit Smith, a pilot from Corning, NY, who had flown himself to this intense, one-week course in Greensboro, NC. She spent a lot time with Whit on those first few days of the week and asked at great length about his training and his airplane. During their only free time on Wednesday, he took her out to the airport for a ride in his Cessna 172 to a neighboring uncontrolled field. They taxied on to the big Greensboro runway and as they began the take-off roll, she said, "I love this part." Whit made fun of this remark, but later all she could think of was what she had revealed. Her true feelings had come out and she decided that if she wanted to fly, she should take it directly into her own hands.

At the end of the week she returned home with much new awareness. On Saturday she told her husband she was going to start flying, and she made an appointment with the flight school. She recommended that they start shopping for an airplane right way, since the purchasing process could take as log as the training. She wasn't going to wait until retirement; she was going to start that day. And she did it!

Her first lesson was the next Tuesday. Her first flight was the following week. Her husband restarted his own training within the month. They purchased an Archer II the following May, in which they each received their ASEL certificates in August. Although they fly cross-country every weekend, neither of them can get enough.

PAM KNOLINSKI, born Feb. 16, 1958, in Richland Center, WI. At the age of 8, on her first Penny-a-Pound airplane ride, Pam decided she wanted to fly. Pam started her flight instruction April 1993 and joined the 99s in June 1993. Since becoming a third class private pilot May 12, 1994, she has logged time in Cessna 150 and 172, Archer and Arrow. Pam and her husband John, own and operate an automotive parts and accessories store in Wisconsin Rapids. She is also church treasurer and serves on the Wisconsin Rapids Chamber of Commerce board of directors. As Wisconsin Chapter 99s member, Pam enjoys chairing public relations, 99 news reporter, scrapbook and historian since 1994 and is currently chapter vice chairperson. She is a member of the Aircraft Owners and Pilots Assoc. and Mid-Wisconsin Flyers. Knolinski is married with two step-children, Mike, 23, and Amy, 21.

MARY JO KNOUFF, learned to fly in 1952 in Missoula, MT, soloing in an Ercoupe, before purchasing a Taildragger Taylorcraft to complete her private license. She joined Montana Chapter of 99s in 1954.

During her career, Mary Jo, a professional educator, worked in aviation education for state of Montana, Cessna Aircraft Company, and FAA in Washington, DC. Her flying experience includes most single engine and light twin aircraft. She holds commercial license with single, multi-engine, instrument and glider ratings.

She is married to George W. Knouff. She has a son, Mark R. Janey, and three grandchildren: Lauren, 16; Clay, 14; and Morgan, 6. She has two stepchildren: Jerry Knouff and Cathy Levinson; and four step-grandchildren: J.J. and Tessa Knouff, 18 and 16, respectively; and Alec and Scott Levinson, 12 and 10.

In 1985 Mary Jo was awarded the Frank G. Brewer Trophy for outstanding service to aviation education. The trophy is on permanent exhibit in the National Air and Space Museum.

PATSY KNOX, Arabian Section, licensed in Gaithersburg, MD, in 1966. Having been a dietitian in the US Public Health Service, she was able to use VA benefits to get a commercial with instrument, multi-engine, and flight instructor ratings.

Joined the 99s and Colorado Chapter in 1977. As membership chairman, chapter doubled its membership and won the Great Race trophy, 1980. Greatest flying thrill was pilot in her first air race, Angel Derby, 1978. Few Hughes Air Race Classic, 1979 and Mile High Air Derby, 1993. In 1989, organized Arabian Section in Dhahran, Saudi Arabia and served as governor until 1994. Lead civilian tours including 99s to Dhahran AFB for the Langley, VA. First tactical fighter wing in 1990 prior to Desert Storm. Sharing the joy of flying with 99s throughout the world highlights her fond 99 memories. After 12 years in Saudi Arabia, now lives in Broomfield, CO.

ETHEL C. KNUTH, first soloed in February 1955 at the Sky Harbor Airport in Indianapolis. She was 48.

She received a private license in 1955 and became a member of the 99s. She also had a commercial license.

In 1955, Ethel's Easter and Mother's Day gifts were flying lessons. After her son and husband received their private licenses, they were a flying family.

She was newsletter reporter, secretary, treasurer, vice chairman and chairman of the Indiana Chapter, later serving at section and International levels.

In 1963-64, she served as air education chairman for the North Central Section and was secretary from 1962-64.

In 1959, she was Indianapolis' Aero Club Woman Pilot of the Year.

Knuth was very active promoting aviation everywhere she went and proved to many that age was not an obstacle.

She participated in these races: 1960 pilot in the Powder Puff Derby flying her Cessna 170. The copilot was Sophia M. Payton, who said Knuth gave her the opportunity to fly in her first all woman's transcontinental air race. Knuth's sister, Thelma Dick, and her husband sponsored the team.

1961 pilot in the Powder Puff Derby. Delia Sanders, Indianapolis, was copilot.

1963 copilot in Indiana Air Race (she flew seven Indiana races).

1964 pilot in Michigan Small Race with copilot Tannie Schlundt.

1965 pilot in Powder Puff Derby with Schlundt as copilot. Her sponsor was Stokley Van Camp.

ALEXIS KOEHLER, a dedicated member of the 99s since 1976 and a pilot since 1970, is the proud owner of a Piper Archer. She has logged more than 1,500 hours to date, and is instrument and multi-engine rated. A member of the board of directors for the Florida Race Pilots' Assoc. who organize the Great Southern Air Race, Alexis has participated as a racer and/or board member since the race's inception in 1985. She has served in the officer positions as vice president, president and treasurer. For the 99s, Alexis has served at the chapter and section levels and at the International level, she has served as a director and as treasurer, and as a trustee of the Amelia Earhart Birthplace Museum in Atchison, KS. Alexis is also a member of AOPA, the Florida Aero Club, and a director for NCWA.

Having traveled extensively with the 99s around the world in 1978, the World Aviation and Education Congress in India, educational exchanges to China and Russia, Alexis is a born traveler. She plans to continue traveling as much as possible.

Retiring early as a manager from Southern Bell in 1991, Alexis owns her own cleaning service. Her educational background includes engineering from Georgia Tech and business administration from Florida Atlantic University.

Alexis has a son, who is a pilot; a grandson and a granddaughter due July 1, 1995.

ANN EMBRY KOENIG, a life long Dallas resident, first took the pinch hitter training in 1984. As her desire to fly increased, she earned her wings in 1988 and added an instrument rating in 1990.

With experience in single engine Cessna, and Piper aircraft, her favorite airplane was the Mooney 201 until the purchase last year of a Navion Rangemaster. She is currently thoroughly enjoying this airplane.

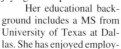

Her educational background includes a MS from University of Texas at Dallas. She has enjoyed employment in Engineering at Chance Vought and Texas Instruments. She currently teaches physics at Richardson High School. She was chosen Outstanding Physics Teacher in Texas 1986.

She is married to Hank Koenig who is also a pilot. She had a son and daughter, a step-son and step-daughter, and three grandchildren.

She is a proud member of the 99s, Inc., Aircraft Owners and Pilots Assoc., Texas Aerospace and Aviation Teachers, American Navion Society, and Southwest Navion Air Group, Assoc. of Texas Professional Educators, American Assoc. of Physics Teachers.

ANITA KOHFELD, joined the 99s in 1986 in the Santa Clara Valley Chapter. Anita is active with the Northern California Antique Aircraft Assoc. and its annual Watsonville Antiquers Fly-In and Air Show. She works at San Jose State University in the Aviation Department.

JIMMIE HUDSON KOLP, born in Copperas Cove, TX, and came to Electra, TX, as a child. When as a teenager she viewed her first plane in the air, she was determined to fly. She started her flying lessons in February 1928, in an OX5 Eagle Rock.

She received as a Christmas gift in 1929 a white single-engine Spartan and her own airport from her husband C.F. Kolp, independent oil producer. She owned and operated the Electra Airport for forty years.

Mrs. Kolp was the first woman in Wichita Falls, the second woman in Texas and the 37th woman in the United States to earn her pilot license. She joined the 99s in 1930, and in 1942 became the first woman to be commissioned in the Civil Air Patrol. She belonged to the Silver Wing Assoc., the Texas Aviation Advisory Council, the Texas Chapter of OX5, the Sportmen's Flying Club and the Texas Flier's Assoc. She also served on President Lyndon Johnson's First Women's Advisory Committee on Aviation in 1964.

In 1950 Mrs. Kolp received the 99 trophy for "Contributing the Most to Aviation." In 1968 she was awarded the Tiny Broadwick Trophy during the 13th Annual National Reunion of the OX5 Club. In 1979 she and Senator Barry Goldwater were honored as Aviators of the Year by Service City of Southwest Automotive.

Mrs. Kolp had well over 5,000 hours when she died in 1970. Although Jimmie Kolp received many awards and honors, she is best remembered as a gracious, lovely woman who unselfishly devoted her lifetime to aviation.

JEAN ELIZABETH KOPE, born in Reedley, CA, and now resides in Renton, WA. Jean learned to fly in a Cessna 150 at Martha Lake Airport in Washington state. She owns a Cessna 172 with her husband, and has approximately 1,000 flying hours. She holds a commercial pilot certificate with SEL, MEL and instrument ratings. She also has flown a number of solo flights in gliders.

Her most memorable flights have been cross country flights from Seattle, WA, to Louisville, KY, including an emergency landing on a road in Wyoming. She has also enjoyed flying to Texas, Oklahoma and California.

In 1979, Jean earned a BA in English, and in 1986 a BS in aeronautics and astronautics, both from the University of Washington.

Currently working as a flight test engineer at Boeing, Jean is responsible for testing navigation and communication equipment in flight and on the ground for all Boeing aircraft models for FAA certification. Recent testing included the Flight Management Computer Function on the 777 aircraft.

Her husband, David Denney, is a captain in the US Air Force, and is a C-141 pilot stationed at McChord AFB in Tacoma, WA. She has a 24-year-old daughter, Lori Franz, living in San Francisco, and a 2-year-old daughter, Shannon.

KATHRYN E. KOSHAN, born July 30, 1958, in Daytona Beach, FL. Having attended Duquesne University School of Pharmacy in Pittsburgh, PA, Kathy returned to Florida in 1980 to practice clinical pharmacy in Ormond Beach. That same year, Kathy married long-time "best friend" Will D. Weatherbee, whose interest in aviation led him to a degree from Embry Riddle Aeronautical University and a career in aviation. Kathy became interested in aviation through Will, and was encouraged to pursue a pilot license in 1987.

Moving to Orlando that year, Kathy became a Florida Spaceport Chapter 99. A "two-pilot-family" needed a plane, so in 1989, a Piper Cherokee PA-28 was purchased - N6228W. The 1964 model was lovingly restored and updated for IFR flight by Will (an A&P/I.A.) and Kathy. It has since been flown across the US several times and to the Bahamas frequently.

Kathy has held the offices of treasurer and vice chairman for the Spaceport Chapter, and was aerospace education chairman, airmarking chairman, and newsletter editor for the chapter. Other aviation affiliations: EAA, AOPA, NAA, FSAACA (FL Sport Aviation/Antique-Classic Assoc.), NCWA (National Council of Women in Aviation/Aerospace) a charter member and currently the Southeastern Regional Representative for NCWA. Kathy is employed as a clinical pharmacist for a 267-bed hospital in Orlando. Also a Florida licensed consultant pharmacist, Kathy is an active member in the central Florida Pharmacy Assoc. and the central Florida Society of Hospital Pharmacists.

ANGELA KOVACS, born and raised in the Bronx, NY, Angela started flying at the Teterboro School Aeronautics when Dave Van Dyke was its president. During that time, she became the first TSA woman to ferry a new Cessna aircraft back from Wichita, acquired the private through to CFI ratings and was awarded the Instructor of the Year for the Teterboro District. A licensed high school teacher, Angela initiated a team-teaching ground school and presently teaches ground and flight classes at Air Fleet. She has more than 5,000 total flight hours with over 3,000 hours at night – her favorite time to fly and teach. Many of her students did their first solo flights at night. As a FAA appointed safety counselor, she stresses night flying safety for all levels of training. Angela joined the 99s and soon became a charter member of the former Palisades, now North Jersey Chapter – a really special group of woman pilots!

CATHERINE KOVAR, a woman with the distinction of being the only member of the 99s living in Atchison, KS, the birthplace of Amelia Earhart, learned to fly in 1989. She flies a Cessna 172 and is currently studying for her commercial exam. She is a member of the Northeast Kansas Chapter 99s.

She was born in Atchison in 1943 and was raised on a tobacco farm in Weston, MO, just a few miles from the Missouri - Kansas border. She is an avid hunter, a member of the National Ski Patrol, a Winter Emergency Care Technician, and a member of the United States Parachute Assoc.

Kovar is a transportation officer for Atchison Engineering Company, operator of an underground defense storage facility. She has been employed with the facility for 33 years.

Kovar has one daughter, Carrie, who resides in Falls Church, VA.

ELMA PEARE KOZAK, born in Dublin, Ireland, Nov. 28, 1950, and emigrated to Canada in 1975.

The relatives with whom she stayed in Montreal were actively involved in flying. She was asked one day if she would like to "have a go" herself. That is how she got into flying.

She joined North Country Flying Club, Clinton County Airport, Plattsburgh, NY, where Jack Perkins was her instructor. She drove a long distance from Montreal to Plattsburgh after work to attend Dr. Perkin's Ground School at the local University and also to fly in her free time. In 1977 she obtained her FAA PPL at Plattsburgh, NY, and joined AOPA.

In 1978 she moved to Vancouver, British Columbia, and joined the Pacific Flying Club. Gretchen Matheson was her instructor there for her commercial license and she quickly introduced her to the 99s which was just getting off the ground at that time in BC. She also joined Canadian Aeronautics and Space Institute, British Columbia Aviation Council, Canadian Owners and Pilots Assoc. and volunteered for Abbotsford Airshow.

She subsequently obtained her Canadian PPL and night rating, commercial, multi/IFR, flying instructor's rating/instrument teaching endorsement. She worked in secretarial/administrative positions while advancing her flight training. In 1983 she was hired by Pacific Flying Club as part-time flying instructor.

In 1986 she started her own computer services business. Her husband, Alexander Kozak, was killed in 1989. He obtained his PPL shortly before his death.

KIMBERLY GILLETTE KRAESZIG, is willing to bet that she started her flying career at the youngest possible age. Her mother was expecting her while she was taking flying lessons. Her mom received two solo certificates, one for her and one for "Baby Kraeszig." She was forced to discontinue her lessons due to her pregnancy, but the love of "soaring over the highest height" was already firmly established-for both of them.

Her dad was a pilot in the Air Force at the time and also flew general aviation aircraft on the side, so she can't remember a time when flying wasn't in her life. When she was 16, she began taking lessons after school; working at a grocery store to finance them and begging for flying money at birthdays and Christmas. She earned her ticket the summer between her junior and senior years of high school and fully intended to attend the Air Force Academy and make it a career. However, her two-inch thick glasses prevented that, and she ended up pre-med at Baylor University.

She flew fairly frequently and was usually current. She met her husband, Brad, at the time an Air Force ROTC cadet and private pilot, during college, and he says he married her because her daddy had an airplane.

Her father is currently a first officer on an American Airlines MD 80 crew; and her mother, tired of always being in the right seat, recently went back and earned her license. She is also chairman of the Brazos River 99s, recently chartered in November 1995. Her brother is an aviation major at Louisiana Tech, and her husband is a weapons director on the AWACS aircraft.

She doesn't get to fly now as much as she wants to; medical school demands much of her time. She was born into it and she married into it. They plan to buy a plane as soon as possible, so that their children can begin to fly in utero, also!

LANA EDWARDS KRAESZIG, after flying right seat and back seat for 26 years she became a private pilot on Feb. 16, 1995. She flew an orange and white Cessna 150 affectionately called *The Pumpkin* and an M35 Bonanza with a very nervous husband copilot. Her husband Chuck flew in the Air Force and is an American Airlines pilot; daughter, Kimberly got her license at age 17; son Michael is an aviation major at Louisiana Tech University; and son-in-law, Brad is also a pilot. They live on an airstrip and conversation at their dinner table can get very boring if you do not fly.

She is activities director for the Classic Bonanza Assoc. and a member of EAA Chapter #983. She is also cofounder and first chairman of the newly organized Brazos River Chapter of the 99s. They were officially presented their charter in Granbury, TX, on Nov. 11, 1995, by South Central Section Governor Carole Sutton. She attended her first section meeting in Grand Junction, CO, in September and plans on begin an active life-long 99.

Her future goals are to leave her husband in the back yard as she takes off in the Bonanza and possibly go on to get her instrument rating.

PHYLLIS J. KRAMER, born June 21, 1928, in Norton County, KS. Joined the 99s in 1978, Colorado Chapter, South Central Section.

First soloed July 24, 1969 in a L-16, N4953V, tandem, stick, tailwheel, Cont 85 engine, at Sky Ranch Airport, Aurora, CO. Pinchhitter Course (AOPA): May 1972; PA24-250 Comanche (right seat); next pilot training solo, Aug. 3, 1976, PA28-140 Cherokee. First x-country solo, Sept. 18, 1976, PA28-140 (APA - Cheyenne - APA). Best learning training/x-country solo, Oct. 30, 1976, PA28-140 (APA - Yuma - LaJunta - full stop/return LaJunta to APA). Private pilot certificate checkride on Feb. 21, 1977. December 16, 1984 received high performance aircraft certification. September 1985 Mile High Air Derby, copilot PA24-260 Comanche with Cory Kroll.

On Nov. 5, 1985 she was re-certified tailwheel C-140, N5683C. August to September 1987, one of most interesting trips in C-140 N1806V was a trip with son Ben P. from BJC to Craig, CO. He took first and third legs; she flew the second and fourth legs. On Nov. 24, 1990, gave her 6-year-old grandson Benjamin C. his first ride with her as PIC.

On Dec. 21, 1991 she lost her wonderful friend and copilot, her husband Ben G. The shared so many hours in the air together, both in "his" airplane, *Waltzing Matilda* Comanche N7891P and in hers, the C-120 *The Broom.* They made many trips to Alaska in a Comanche. The loss is still very keen in her mind. They honored Ben G. for his many contributions to aviation with a memorial at the Forest of Friendship in Atchison, KS, in June 1992.

In May 1992 she became the proud owner of N5163L, a PA28-180D, a very nice 1968 Cherokee *Dancin' Bear.* Shadow, their constant companion, a part Australian Shepherd/Blue Heeler, thus became her companion.

She was able to accomplish a long X-country as PIC in August 1993. The 1993 INT Convention in Portland offered the opportunity. The first leg was to fly from her home in Western Colorado to BJC, another first, a flight across the mountains of Colorado by herself. Her friend, Clancey Maloney, came up from COS to join her.

Kramer has children: Benjamin P. Kramer, born Feb. 7, 1951; Robert M. Kramer, born Aug. 9, 1952; Leslie Kay Kramer Manley, born Oct. 19, 1958; and grandchildren: Benjamin C.A. Kramer, born Aug. 15, 1984, and Kathryn Marie Kramer, born June 24, 1987.

Joined the Colorado Chapter SC Section 99s, Inc. in 1978; chaired various committees; served as vice chairman. Instrumental in chartering High Country Chapter (Western Colorado), SC Section, in November 1990, charter member and chairman. Also member of South Central Section, serving as treasurer June 1, 1992-94, and secretary June 1, 1994-96. Member of Colorado Pilots Assoc., serving as vice president for two terms: November 1992-93 and November 1993-94. Served in the Civil Air Patrol as member of Emergency Services Squadron; appointed as Colorado Wing Radio Station Licensing Officer.

JUDY KRAS, began flying in 1991 after considering the idea for over 20 years. She started flight instruction at Freeway Airport, MD; continued at Quantico Marine Corps Base, VA; and received her private pilot license at Brackett Airport, CA. She and her husband, Hank, fly a Cessna 172, their 25th Anniversary gift to each other.

Born in Oak Park, IL, she spent her first 18 years in Berwyn. After graduating from the University of Illinois in Champaign, she lived with her husband in many places throughout the world while he served in the USAF. "Home" included Mississippi, England, Texas, Alaska, Colorado, northern California, West Germany and Virginia. They now live in southern California with their two teenagers daughters. Judy earned her MA in language and learning disabilities in 1976 and teaches special education. She is a member of the San Gabriel Valley Chapter of the 99s and the Aircraft Owners and Pilots Assoc.

ESTHER F. KRAUTH, born April 11, 1942, in Lax. A native Californian, soloed a Cessna 150 in 1971 at Torrance Airport, where her instructor threatened her "if you don't finish this course and get your license, I'm never going to teach another female." The 99s were a great opportunity for Esther to participate in two Baja and an Angel Derby air race. Esther instructed at John Wayne Orange County Airport in a variety of aircraft culminating her instructing career as a CE-500 instructor at Martin Aviation. In 1979 she was hired by Western Airlines as a GIB on a B-737. In 1981 Esther was hired by Air California to fly F/O on the B-737 progressing to captain in 1986; American Airlines acquired AirCal in 1987 and Esther then flew the B-767/757 as a first officer. She has also flown captain the MD-80 and is now a Boeing 757/767 captain.

Esther and her fiancé Jim Horn live in Southern California where Esther has been a member of the 99s, Orange County Chapter for more than 20 years. Esther has three children: Gary, Randy and Renee. At this time four grandchildren: Randy and his wife Eileen have produced Erin, 3, and Margaret, 1. Renee and her husband Craig; Dylan, 4, and Gabrielle, 1. Randy, her second son, also a pilot, graduated in 1984 from West Point and went on to earn his wings in helicopters. At this time he is vehicle manager for Rockwell on the Atlantic/Mir projects.

Esther is also an accomplished figure and landscape artist.

ALICE C. KRICK, born May 7, 1938, in Washington, DC. Flight training at Freeway Airport, Mitchellville, MD. She has flown PA28-180, PA32-260, and LA4-200.

Employed with realtor association, Caldwell Banker. Married to Benjamin and has four children: Alice

Dollins, 38; Naomi Musti, 37; Marchel Wilson, 35; and Benjamin III, 30. Krick also has 13 grandchildren at this writing. She began flight training in 1978, received instrument rating in 1981. While attending 99 Convention in Anchorage, AK, received sea plane rating. Her husband Benjamin is a mechanical contractor and a pilot.

EVELYN KROPP, studied music in Cologne, Germany, to become a concert pianist, but gave up career in favor of marriage to Dr. Peter Kropp, a psychiatrist. They have two sons, Ralph and Robert. Besides English speak German, French, Spanish (Latin). Flight training: 1967, private pilot license; 1968, instrument rating; 1969, commercial and multi-engine; 1971, CFII, AGI, IGI; 1976 commercial seaplane; 1978, ATP ME; 1979, ATP SE; 1985, CFI ME. More than 4,700 hours. Worked as CFI and charter pilot. Since 1977 FAA appointed safety counselor. Races flown: IAR in 1972, 1974, 1975, 1976, 1977, 1979, 1981. PPD 1973 and 1976. ARC 1980, GSAR 1985 and 1989. Also flown Michigan SMALL race, AWNEAR, Empire State 300, Garden State 300, FAIR, Illi-Nines Air Derby. Placed up to second place in those. Joined the 99s in 1968. Headed most committees on chapter and section level. Connecticut Chapter chairman 1973-75, New England Section Governor 1980-82. Served on International Nominating committee. In 1984 AE Scholarship winner (CFI ME). In 1991 Woman of the Year of the New England Section. In 1992 placed in Forest of Friendship in Atchison, KS. Newest hobby: 19-month-old granddaughter Andrea Elise who loves to fly with her.

JANICE R. KUECHENMEISTER, became a member in 1952 of the All-Ohio Chapter of the 99s, Inc. A past chairman, vice chairman and treasurer of the chapter. A past governor, vice governor and treasurer of the North Central Section. Also was chairman of the International Convention of the 99s held in Cincinnati, OH. She started flying in 1944 and got her spins, and then money became a little hard to come by. Eventually she got her SEL in 1951. She entered Ohio State University in 1952 and was one of the founders of the Buckeye Glider Club in Columbus, OH. The glider ticket was obtained in 1955 using winch tows.

She belongs to the Jet Flyers Inc., a flying club based in Hamilton, OH. The club originated at General Electric Company in Evendale, OH. The aircraft used in the many races she entered like the SMALL, FAIR, IAR, ILLINII, 99 conventions and sectional meetings, were all Jet Flyers aircraft and she is extremely grateful for their use. She was a copilot in the Powder Puff Derby once with Pat Fairbanks and worked at stops for the Powder Puff Derby and IAR as a timer.

AIMEE KUPRASH, first flight in a Cessna 150, N17170, Cottage Grove, OR; Aug. 2, 1977. First solo in same plane, same place, Feb. 11, 1978. Owned, operated and leased back a 1977 Cessna 150, N714PE, from 1978 to 1986. Received private pilot certificate, March 13, 1979; #576480337. Completed commercial rotorcraft-helicopter certificate April 17, 1980, at Albany, OR, in a Bell 47. Accumulated 20 hours of helicopter agricultural spray training in a Hughes 269 at Benson, MN, in May of 1980.

Joined Whirly-Girls in 1982 becoming member

#374. Attended Lane Community College in Eugene, OR, from 1980 to 1988; completing an associate of science degree in aviation maintenance in 1982 and an associate of applied science degree in flight technology in 1987. Obtained Airframe and Powerplant mechanic certificate on March 14, 1983. Became a member of the Willamette Valley Chapter of the 99s and served as treasurer from 1984-88.

Competed in the Oregon Petticoat Derby in 1982, 1983 and 1985. Acquired instrument airplane rating May 30, 1985. Obtained commercial airplane single engine land privileges June 2, 1986. Awarded the 99s Amelia Earhart Memorial Scholarship Award in 1986 for multi-engine training.

Participated in US Proficiency Flight Team Regional Competition and USPFT National Competition in 1986. Competed in the Northwest Petticoat Derby at Bremerton, WA, in 1986. Completed commercial multi-engine land rating July 8, 1987. Participated in Palms to Pines Air Race in August 1987. Received flight instructor certificate Aug. 14, 1987.

Awarded the 99s Northwest Sectional Achievement Award in 1987. Obtained instrument flight instructor rating June 30, 1988. Returned to Hawaii, where had resided as a child 1956-68; in July 1988 and joined the Aloha Chapter of 99s, serving as secretary from 1989-90 and treasurer 1990-92. Did flight instruction for Great Bend Air and flew Cargo Part 135 in Cessna 206 and Cessna Caravan 208 for Hutchinson Air from 1988.

Flew Cargo Part 135 as a first officer in a Shorts Skyvan for Inter-Island Air in 1990. Participated in The Apuepuelele Air Competitions in 1988, 1990, 1991 and 1993. Received the Aloha Chapter of 99s Scholarship Award multi-engine instructor rating in 1990. Flew in Beech 18 for Scenic Air Tour 1994-95 and for Polynesian Airways, 1995. Flew Skydivers in Cessna 182s for Skydive Hawaii, 1995. Flight instructor for Petrides Aviation, 1994 to present. Obtained FAA Pilot Proficiency Safety Award Level 6, 1995.

MADELINE KURRASCH, DDS, born June 19, 1941, in Minneapolis. Flight training at Santa Monica, CA, and has flown Grumman Tiger *Blue*. Learned how to navigate on an Air Race. (Rialto Air Rallye in 1986). Has flown Palms to Pines nine times, Air Race Classic, PAR, Mile Hi Air Derby, Shirts and Skirts and the Valley Air Derby.

Dr. Madeline Kurrasch is a prosthodontist, practicing in Torrance, CA. She received a BA and BS from the University of Minnesota and a DDS from UCLA. She is boarded in prosthodontics and is a member of many professional organizations, but is especially proud of being the first woman elected to the Pacific Coast Society of Prosthodontics.

Community service includes being on the advisory board for the Assistance League of San Pedro-Palos Verdes. She is a member of the San Fernando Valley 99s.

BARBARA LEE KURTZ, San Antonio Chapter, private ASEL. In 1968, joined the San Antonio Chapter with her mother, Vel Morgan. In 1971 Vel organized the Coastal Bend Chapter. Coastal Bend offices include chair of membership, aerospace, APT, and air and space museum committees; news reporter; vice chairman; registration co-chairman 1977 fall sectional. Air Races: pilot 1973 Baytown Kiwanis Proficiency Air Race; copilot 1973 Skylady Derby. In 1992 Coastal Bend disbanded. Barbara and Vel rejoined the San Antonio Chapter. Resides in Garland, TX. Family: Steven, husband; Stephanie Dowlen, daughter; Scott Dowlen, son-in-law. Division secretary, Eastfield College, Mesquite, TX.

EDITH PEARCE KUZENKO, born in southern coastal New Jersey. She grew up hearing the "dogfights" of Naval combat training out of NAS Pomona (ACY): Wildcats, Corsairs, Dauntless, Avengers. She was determined to learn to fly. She did in 1949 at Teterboro. Her instructor, Michael T. Kuzenko, had instructed Army aviation cadets and he put her through the same thorough training program. So disciplined, on her first solo at Teterboro, she refused to hurry on landing when tower radioed "chug along Cub, Arthur Godfrey behind you." He circled the field.

Edith and Michael were married in 1952. He is a veteran China-Burma-India "Hump" pilot and is an air safety inspector with the FAA. She is an active member of patriotic, genealogical and fraternal organizations and a hospital volunteer. Both are avid sailors. They own a Cherokee 140. Together they formed E&M Enterprises and have patents, copyrights and distribute aviational instructional material. She is a life member 99.

NANCY R. KYLE, earned her private pilot certificate, SEL, in 1990, and since then has gone on to obtain the commercial, CFI, and CFII certificates and instrument, MEL, and SES ratings. Nancy also holds advanced and instrument ground instructor ratings. She owns a C-172, N2609L, which is based at Pottstown Limerick Airport.

Nancy a lifelong resident of Pennsylvania, received a BS from Pennsylvania State University and an MBA from LaSalle University. She is responsible for investor and public relations for SunGard Data Systems. Professional activities include membership in the National Investor Relations Institute and the Association for Corporate Growth, where she is also the vice president, programs, for the Philadelphia Chapter.

Aviation activities include: an officer and board member of the Aero Club of Pennsylvania; a trustee of various aviation scholarship funds; and a member of the 99s, AOPA, EAA, Valley Forge Taildraggers, Pottstown Aircraft Owners and Pilots, Inc., and The Wings Club in New York City. Nancy is also an aviation safety counselor for the FAA and is the editor for FSDO 17s newsletter.

JAMIE MARIE MICALLEF LABADIE, a native Californian born 1964, began flying at the March AFB's Aero Club, earning her license in 1991. She has completed an emergency maneuvers and introduction to aerobatics course in a Pitts S2-B, logged most of her time in a Beechcraft Musketeer and flown 235 hours in a Cessna 120, 150, 172 (T-41), 185, Comanche and Bonanza. She also has had the wonderful opportunity to pilot a 1930s Consolidated Fleet biplane, a T-34 WWII trainer, a T-33 jet and a balloon.

Jamie has a BA from Dominican College and a MA from Thunderbird, both in business and another MA in aeronautical science from Embry-Riddle University. She is a cost accountant, member, vice chairman and newsletter editor for the Inland CA Chapter 99s, a member of the Corona Pilots Assoc. and AOPA and commissioner on both the Riverside County Aviation and Airport Land Use Commissions.

BEVERLY A. LABRIE, Idaho Chapter, born July 25, 1946, in Boise, ID. She started her flying in 1973 after loosing a bet with her husband. She was a happy passenger that had no intentions of flying herself. Upon the loss she was hand delivered to an instructor in a rented Citabria Scout. The day she soloed her husband bought back a 1946 Aeronca Champ, which just happened to be his first airplane also.

She moved up to a Piper Super Cub for several years, then her Champ was rebuilt with wing tanks and a larger engine. She still flies as an antique with no electrical, hand prop, a few gauges, no flaps, heel brakes and a stick which adds up to a lot of fun.

She has been an active member since 1977. LaBrie has held various chapter offices and is currently Northwest Section Airmarking chairman.

LaBrie has flown the Aeronca, Citabria, Super Cub, Cessna 182 and 170 and Stinson.

MADELINE LACARRUBBA, during the Vietnam War, was living in Macon, GA, near Warner Robbins AFB. B52s would fly low and slow over her home. She'd stare at their endless undercarriages, waking her curiosity of flight.

After her daughters left for school, Madeline left for the municipal airport where she met Mr. Arnold, the owner of Arnold Aviation. During the demo ride Mr. Arnold pointed to a smoke stack seven miles away and said: "Fly to it." Then he closed his eyes. Unsure if he was only sleeping, she nudged the rudder. He promptly woke up.

Madeline soloed, Dec. 7, 1967, at age 30. That historic day in US history ranks as one of the most exhilarating days of her life.

After earning her private license, Aug. 7, 1968, she moved to Long Island. The pastor of her church introduced her to the second pilot in the congregation, Marilyn Hibner. Madeline joined the Long Island Chapter in February 1972.

She headed many chapter committees, including AE Scholarship, Ways and Means and the banquet for 1979 Convention in Albany. During her tenure as chapter chair, 1977-79, she incorporated the Long Island Chapter and finalized its 501C3 with the IRS. Madeline actively participates at section meetings and conventions.

Madeline is also active in her community as a Girl Scout Leader, Sunday School teacher and school election board inspector. She works as a travel agent, which allows her to maintain her interests in art, photography and foreign languages. In addition to bi-coastal daughters and one grandson, various pets have always been a part of her life.

"Flying opened up a whole new interesting world to me. It's been a great adventure."

CATARI LACORAZZA, "Cat" joined the SCV99s in 1988 and got her license on her birthday, April 25 at San Carlos Airport. She has two children and runs a successful catering business. Cat is a native of Bogota, Columbia and has worked as an occupational therapist.

SUZANNE LAFONTAINE, born Dec. 24, 1960, in Montreal. Joined the Montreal, Canada 99 Chapter in 1994. Licenses included private pilot, glider pilot. She has a BA in mechanical engineering. Airworthiness engineer at aircraft manufacturer: Canadair.

She started her flying at the age of 16 after obtaining a Royal Canadian Air Cadet Gliding Scholarship followed by a Powered Flying Scholarship. Her keen interest in aviation led her to study the design of aircraft where she obtained her mechanical engineering degree. Her career then focused on the certification of aircraft, at Canadair. As an airworthiness engineer, her work called on her flying experience and knowledge of the regulations governing the design and testing of aircraft. This work fueled her interest to further develop her flying skills; she is presently pursuing the advancement of her ratings.

In the 99s she discovered a world where other fascinating women shared similar interests and fulfilling careers in aviation.

She hopes that someday her love of flying, traveling the world and speaking many languages will come together in the form of an adventurous flight odyssey around the world.

SALLY LAFORGE, received her private license in 1952 learning to fly at Santa Monica airport in an Aeronca Champ, and presently owns a third partnership in a Cessna 182. She now holds a commercial and an instrument rating.

Member of the Los Angeles Chapter since 1954. Held all chapter offices at least once. Flown five ATWARs, two Angel Derbies, Pacific Air Race and

Palms to Pines. She is a retired captain in the Civil Air Patrol and flew a number of search missions as search pilot.

Graduated from UC Berkeley with a degree in mechanical engineering and received a MA in engineering from UCLA. Worked for Hughes Helicopters for 36 years. While at Hughes, Sally was manager of the performance section and was responsible for providing the performance analysis of all the helicopters built by Hughes. Presently a consultant for RAND.

In addition to being a member of the 99s, Sally is a member of the American Helicopter Society, and AOPA.

FERN LAKE, born April 22, 1934, in Council, ID. Joined the 99s in 1962. Certificates and ratings: private, commercial, flight instructor, CFI instruments, commercial glider, multi-engine, instruments, ground instructor. She has flown various types and models during flight instructing. Co-owned Piper Clipper, Luscome, Aeronca, Citabria, J3, C-175, C-170, C-172. Pilot employment included Sandpoint Flying, Sandpoint, ID, and Coeur d'Alene Airways, Coeur d'Alene, ID. Married to August Ames Lake for 43 years

and has three children: Mike, 42; Leslie, 38; and Diane, 34. Grandchildren: Jessica, 11; Joshua, 8; Andy, 8; Lindsey, 6. Learned to fly in 1961 at Boise ID, and later moved to north Idaho. At different times was chairman of both the Idaho Chapter and Intermountain Chapter. Served on numerous 99s committees, including chairman of the 1975 International Convention at Coeur d'Alene, ID. Was chairman of an AWTAR stop at Boise and an Air Race Classic at Coeur d'Alene.

Served on the Coeur d'Alene, ID, Airport Board for 16 years. Was an Idaho Search and Rescue Mission coordinator and a FAA accident prevention counselor. Received a Flight Safety Award from the Spokane, WA, office of the Federal Aviation Administration for Outstanding Support of Flight Safety Program Activities in the General Aviation Community.

Flew most of the Idaho Air Derbies, winning one and chairman of one.

Has flown various types and models of aircraft plus co-owning the ones mentioned above. Other occupation is a registered nurse and school nurse. A high point of her career was soloing her two teenage daughters in the same day. Has received much moral support from her A&P instructor husband, also he is a commercial pilot and designated mechanic examiner for the FAA.

LINDA LAKE, married to Roger, school district administrator, private pilot; their children are Robert, 32, and David, 29. She is the owner/partner: of Business Options, a multidimensional financial management services company; Oxygen Plus, providing oxygen and respiratory therapy in the home setting; and Western Collections, a full service collection agency. Their various business lines employ 28 persons with branch offices in Grand Junction, CO.

Her desire to fly began in her early teens but she was unable to accomplish this goal until 1989. At that time, she earned her private pilot license, as did her husband, and purchased a Cessna 172. Currently she has more than 700 hours and holds ratings in ASEL and glider. She has been qualified to fly C-170s, C-182s and has 20 hours in an Experimental Long EZ. They use their plane for travel around the US, and she is able to use it for business in travel to her various clients. She is a charter member of the High Country 99s and has served as treasurer since its' inception. She is a member and qualified Mission Pilot of the CAP and has completed Wings Level IV and Mountain Flying courses.

GEORGIA LAMBERT, received her pilot license in 1966. Joined the 99s in 1967.

She has flown two Powder Puff Races, two Angel Races, many Palms Races, also Pacific Air Races, CA. Has done airmarking, officer of chapter. Put on flight instructors

revelation clinics. She attends two Southwestern Section meetings each year and has attended 12 conventions.

Lambert was honored that the Los Angles Chapter sponsored her in the Forest of Friendship 1984.

Flying, racing and 99s have changed her life. She is thankful for all her 99 friends.

EDITH RAYMOND LAMM, born Aug. 27, 1939, in Wilmington, NC, and lived near Rose Hill, NC, until she met her husband Richard Lamm and moved to Black Creek, NC, where she now lives.

She started flight training July 1978 and completed her FAA private pilot checkride on April 4, 1988. She joined the local 99 Chapter sometime in 1981. She completed her instrument checkride on Feb. 4, 1991, and at the present time is working on her commercial rating.

She and her husband had gone to Hickory, NC, to the Wings Weekend and were returning to Goldsboro, NC, when she had vacuum pump failure.

She and her Aunt Ella Mae were flying from Goldsboro, NC, to Wilmington, NC, in a B24R Sierra when she had total electrical failure.

She has flown Cessna 150, 152, 172, 172RG, 182, 210, 206; PA28-151; B24R; Sierra, and PA44. She has also been in several of the coastal tours from Monteo, NC, to Wilmington, NC. She has all of the Phase Ten Wings.

SHERRILL OTT LAMONT, born Sept. 9, 1946, in Eugene, OR. She is a life long resident of Oregon and learned to fly in 1994. Sheri took her flight instruction at Lane Community College in Eugene, where she is currently working on her commercial license and her instrument rating. She has also mastered her tail wheel endorsement and is currently flying a 1947 Piper Super Cruiser. She and her husband also have a Bonanza.

Sheri is 1995 co-president of the Willamette Valley Chapter of the 99s an has been very active in the chapter sense joining a year ago.

Sheri and her partner founded Lady Aviators, a company making quality flight apparel for lady pilots.

She is married to Bruce Lamont a pilot and local business man, and they are members of The Oregon Air & Space Museum, AOPA, T-34 Assoc., Cub Club, EAA, and Warbirds of America.

JANNA LAND, born Feb. 12, 1964, in Goldsboro, NC. Joined the 99s in 1994. Flight training, Part 61. She has flown Cessna 152, 172, 177, 182, 210, 421; P68C; PA23, 31, 24, 28. Employed with Capitol Aviation, Lincoln, NE.

Janna earned her wings in June 1983. She did the majority of her flight training with her father, who is a retired Air Force pilot. She has logged more than 1,500 hours by flight instructing, ferrying airplanes, flying skydivers and aerial photography.

While building flight time and experience, Janna attended Metropolitan State College in Denver, CO, where she received a BS in aviation management. She has been employed by United Airlines since 1986 in reservations and in customer service, but says her ultimate career goal is to be a pilot for United.

Janna currently lives in Lincoln, NE, where she is a member of the Nebraska 99s.

JANET KAHAUNANI BAL LANDFRIED, a native of Hawaii, has lived in California since 1964. She earned her private pilot license in 1979, and soon also earned her instrument rating and advanced ground instructor license.

As a teacher at Redlands High School she has taught history, government and economics; and taught aviation ground school in the evening. She earned a BA at northern Arizona University, as well as a MA in history from Bowling Green (Ohio) State University, and a MA in economics education from the University of Delaware.

In addition to membership in professional organizations, Janet is a member of the Inland California Chapter of 99s, AOPA, EAA, and the Cardinal Club. She and her husband, Jim, are also members of local aviation groups at her home airport in Redlands. Since a bout with cancer, Janet leaves most of the active flying to Jim.

Jim and Janet score air races in southern California, and have been the start timers and judges for the start of several Air Race Classics. They also raced together in the Aircraft Spruce and Specialty Great American Air Race in 1993. They recently purchased a lot at Sky Ranch in southern Utah and plan to build a retirement home there.

NANCY LANE, born July 7, 1937. Employed as a window clerk for USPS. Married to Richard and has two children, 37 and 35. Joined the 99s in 1980 and the Reno Chapter in 1981. She has also been a member of Lake Tahoe. Lane has held the office of scholarship chairman a couple of years for Reno Area. She became involved in aviation in 1970 because she was tired of being home while the family was at the airport. Her ratings include ground instructor and she owns a Cessna 172. Lane's other interests include golf, fish, and traveling. She is a member of LVPA, Lakontan Valley Pilots for 10 years.

ROSEMARY LANE, born March 16, 1927, in Indianapolis, IN. An active pilot since 1946. In 1948, at 21, she joined the Washington, DC, Chapter of the 99s. She owned a PT-19, held a commercial license and was a certified flight instructor.

In 1951 Lane became chief of Army Aircraft Requirements. She was key in delineating the future structure of Army aircraft capabilities. In 1954 flying a Cessna 195, she became one of the few female corporate pilots – at less than half the salary of her male predecessor, of course!

Later, as one of the first women professionals at the RAND/System Development Corporation, Lane designed computer-based war-games for US Air Defense Command training. Finally, for 10 years prior to retirement, she was USAF Space Division budget officer for Global Positioning System development, and chief of program control for Satellite Systems at Sunnyvale. Using her Cherokee Archer, Lane also owned and operated Fly-Inns, Ltd., an air taxi service in southern California. Now semi-retired and living in Cambria, CA, Lane is a volunteer with both the Flying Samaritans and the San Luis Obispo Hospice program.

Her racing experience includes: Cleveland-New York City (1950), Innsbruck Austria Air Rally (1973) Palm-to-Pines (1980), All-Woman Baja, CA Air Races (1979, 1982).

Flight training included commercial pilot, flight instructor ASEL, instrument rating with more than 2,000 flight hours. She has flown Piper J3, Taylorcraft, PT19, Fairchild 24, Cessna 150, 172, 182, 195, and Piper Archer.

LAURIE LANEL, a native Oregonian, born 1967, caught the aviation addiction at a young age and earned her private pilot certificate in April 1986 during her senior year in high school. She graduated from Redmond High School in 1986 as valedictorian and earned an AA in Liberal Arts from Central Oregon Community College in 1988.

She now holds her commercial, instrument, and multi-engine ratings and is flying as a first officer on an Aerospatial Corvette Jet.

Other aviation experience includes seven months volunteering at a T-34 restoration shop in Tennessee. She helped navigate on many cross country flights in T-34s and a T-28. She has been on the Redmond airshow committee and is a long standing member of the EAA, Warbirds of America, and AOPA. Laurie was the assistant manager for the Sunriver, Oregon Airport prior to her current job.

The photo was taken after a flight in the late Rick Brickart's T-33, *The Red Knight.* Laurie and Skip Holm were observing a training flight of a Saab J35 Draken owned by Chris Mullin Jr.

PATRICIA LANZI, born Aug. 24, 1933, Balto, MD. Joined the 99s in 1987. Has a private pilot license and is currently working for instrument rating. She has flown the Cessna 152 and 172, Grumman Tiger, Piper Warrior and PA28-180. Married to Joseph, and has five children: Michael, 38; Joseph, 37; Phillip, 36; Mary Beth, 33; John, 31. Her grandchildren are Anthony, 14; Shannon, 13; Joseph, 14; Debbie, 12; Nicholas, 10; Joshua, 5; and Jessica, 2.

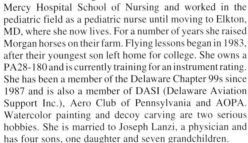

Pat Lanzi was born and educated in Balto, MD. She has a RN in nursing from Mercy Hospital School of Nursing and worked in the pediatric field as a pediatric nurse until moving to Elkton, MD, where she now lives. For a number of years she raised Morgan horses on their farm. Flying lessons began in 1983, after their youngest son left home for college. She owns a PA28-180 and is currently training for an instrument rating. She has been a member of the Delaware Chapter 99s since 1987 and is also a member of DASI (Delaware Aviation Support Inc.), Aero Club of Pennsylvania and AOPA. Watercolor painting and decoy carving are two serious hobbies. She is married to Joseph Lanzi, a physician and has four sons, one daughter and seven grandchildren.

SUSAN LYNNE FAWLEY LAPIS, born Sept. 21, 1946, Elizabeth, NJ. Joined the 99s in 1994. Flight training includes private pilot license and instrument rating. She flies Cessna 182.

Married to Dr. James L. Lapis and has two children, John, 18, and Beth, 16. She fell in love with general aviation airplanes in a Cessna float plane while in Alaska on a cruise.

Her husband's father was an early aviator who once flew with Amelia Earhart and flew for American Air until he retired.

SAUNDRA NIX LAPSLEY, learned to fly in the 1960s as a school girl in Woodward, OK. She joined the 99s in 1966. At Oklahoma State University she continued flight training, later working as flight instructor. After college she pursued a career in aviation advertising, writing, and teaching.

In 1975 she proposed that *99 News* magazine be brought to headquarters to be published by 99s there. She became the first of almost two decades of headquarters-based editors. Career moves with her husband, FAA inspector Tiner Lapsley, took Saundra to the Dallas area, Puerto Rico and Houston. Every new home brought opportunities to make more 99 friends.

Retiring with Tiner to the Rio Grande Valley of Texas in 1984, Saundra repeatedly served as RGV Chapter chairman. She organized a reunion for charter members of the 99s in 1985.

She and Tiner remain active in sport aviation and Christian lay ministry.

SUSAN LARSON, a native of San Jose, CA, owns a Cessna 182 which she has raced in numerous Air Race Classics and Palms to Pines. A pilot since 1978 and instrument rated in 1986, Larson has logged more than 1,300 hours. She joined the San Joaquin Valley Chapter in 1979 and has served in many capacities.

Training was accomplished in a Citabria 7ECA Taildragger at Reid-Hillview Airport. Her affection for the airport has kept her involved in political issues as they affect local airports. She has served as acting president of the Coalition for Responsible Airport Management & Policy and as chairman of its PAC. In 1993 she was awarded AOPA's Presidential Citation at Phil Boyer's 99th Town Meeting for her general aviation advocacy.

A graduate of UC Berkeley with a degree in business, she worked in San Francisco for three years acquiring her CPA certificate then joined her family's pallet manufacturing business in 1975. With her brother David, she co-owns and co-manages the business. She is a member of the board of the National Wood Pallet & Container Assoc. and is a past president of the Western Pallet Assoc.

Currently the governor of the Southwest Section, Larson credits 99s activities and valued friends with keeping her flying, and her Mom, Dad and pallets for making it all possible.

GUDRUN LASHBROOK, born June 18, 1948, in Herborn, Germany, and moved to the US in 1968 becoming a citizen in 1973. She was involved in real estate in Florida for several years as a broker/owner. Although she had long been interested in flying, it was not until 1981 that she took her first flight lesson. From that time her interest in flying was consuming and additional ratings quickly followed. Between 1981 and 1987 she was certified as a private pilot single-engine land, instrument airplane, commercial single and multi-engine land, single-engine sea, helicopter and glider. She also became certified as an instructor ground school, instrument, airplane multi and single engine and glider. With the additional ratings and experience she broadened her involvement in general aviation logging 7,000 hours to date while serving as the executive vice president of a fixed base operation, chief flight instructor of a 141 school with examining authority, chief pilot of a Part 135 operation and an FAA aviation safety counselor. Gudrun has been a strong supporter of the "Wings" program, organizing and conducting one of the more successful programs in central Florida. She is continuing her achievements in general aviation by preparing for her CFI (helicopter) checkride.

PEARL BRAGG LASKA, born April 29, 1909, in West Virginia. Soloed a Kinner Fleet at the Bluefield, VA, Airport in 1933. She was recipient of primary, secondary and instructor civilian pilot training courses. She instructed Navy students at the Kanawha Flying School and Army Air Force students in Pennsylvania. After four months as a WASP trainee, Pearl went to Alaska where she instructed, ferried aircraft and flew the bush. Her most memorable trip was in 1946 when she flew a 1939 Piper J4 from Asheville, NC, to Nome, AK. In 1955 she became a member of the 99s and continues flying now for fun.

She has flown Stearman BT13; Cub J3, J4, J5; and Cessna 140, 150, 172. Married to Lewis Laska first and then Ed Chamberlain. She has one child, Lewis, 48, and grandchild, Jenny, 10.

LINDA K. LAUER, born Oct. 7, 1963, Wichita, KS. Flight training includes private single engine and instrument ratings. She has flown Cessna 150, 172 and Whittman Tailwind, an experimental her father built. Married to David K. Lauer, MD for five years and has two children, Kyle, 3 1/2, and Landon, 18 months; she is expecting their third child in April.

Learned to fly in 1980 while in high school and financed her instruction by working part-time at a local department store. She learned to fly in her parents Cessna 150. She joined the 99s shortly after obtaining her license. Initially she was not very active because she was pursuing a degree in marketing at Wichita State University.

Lauer has been membership chairman and vice president of their local chapter. She's enjoyed flying in many local and regional air races and flight competitions including their Kansas Sunflower Rally.

One of her most memorable flights was while spending a summer in Oregon. She rented a plane and flew over Mount St. Helens which had recently erupted. The land around the smoldering crater left only the stripped trees lying in perfect order on the miles of desolate mountainside. It was breathtaking.

JAN ELAINE LAURO, born April 22, 1952, lived in Napoleon, OH, until marrying Dan Lauro. She learned to fly in 1987 in Bryan, OH. Earned her ground school instructor certificate in 1988.

Jan was the owner of a C-150. She joined the Three

Rivers Chapter in 1989 and is a member of the EAA.

She is a nurse for the Defiance County Health Department and a Hospice nurse.

She has two sons, Daniel and Kristopher, who enjoy flying and have plans to get their pilot's license. Her husband, Dan, is presently working on getting his private pilot license.

MARGARET E. LAVAKE, "Peggy's" involvement with aviation began in 1980 while on a trip to Jackson Hole, WY. Her father, a retired airline captain, offered a flying lesson at a local airport and she was hooked. Unlike most pilots, Peggy's training began in a Super Cub on floats. Her x-country flights were made without radios or navigational equipment. She received her private license on wheels and floats in 1983. In 1984, Peggy was involved in a near fatal plane crash as a passenger. She recovered from injuries and went on to obtain her instrument, commercial, multi-engine, AGI, CFI and CFII ratings. She has worked for Computer Flight, Nelson Flying Service, Mustang aviation and flown copilot for Syrec-Mee Aviation. Peggy has logged 1,400 hours of flight time. She holds a MA in viola performance, and works part-time as a respiratory care practitioner in a local hospital. She has been a member of the Northern Jersey Chapter 99s since 1987.

ROSALIND SUE LAVIN, (nee Wells), began her flying career on Dec. 3, 1971, at Bader Field, Atlantic City, NJ, in a Cessna 150. She has logged more than 3,500 hours and gained proficiency in a Cherokee 180, a Bonanza V35B, a Cherokee Arrow, and a Grumman Tiger all at Blue Bell, PA. "Roz" has participated in many air races such as the Garden State 300, Bahamas Islands Treasure Hunts, and Holly Runs. She has flown the East Coast from Maine to Key West extensively and has explored all the islands from Nantucket, to the Bahamas.

Roz's educational background includes a BA in English and a MA in medieval English literature from Temple University. She became a member of the National Historical Honor Society and the National English Honor Society while in college. She has acquired New Jersey and Pennsylvania Nursing Home and Personal Care Home Licensing Certifications from Stockton College, Pomona, NJ, and Philadelphia Community College, as well as New Jersey and Pennsylvania Real Estate Licenses. She is also a former teacher of English, history and music in the Philadelphia Public School Systems and classical pianist. She has been a member of the Opera Guild of the Philadelphia Opera Company, Mensa, and the Brandeis University Women's Literary Guild.

She is an active member of the Wings Field Pilot Assoc., the 99s, the National Aircraft Owner's and Pilot's Assoc., the American Yankee Assoc., the American Bonanza Society, a life member of Harley Davidson Owners Group, a founding member of the Executive Riders Limited, and a member of the Keystone Region Rolls Royce Assoc.

Roz has attained private pilot, commercial and instrument ratings while raising her two children, Scott Andrew Lavin and Stacy Beth Lavin. Roz also has dabbled in ballooning, bungee-jumping, gliders, parasailing, long-distance bicycling, water sports and motorcycling.

Roz is married to Robert Edwin Lavin, a private pilot with commercial and instrument ratings. In addition, Roz is the vice president/chief executive officer and the director of operations of a chain of personal care residential facilities that she jointly owns and operates with her husband of 27 years.

LINDA CHANDLER LAW, born April 22, 1947. Self-employed, currently under contract with the Legislative Counsel Bureau, state of Nevada, owner, Business Services of Carson City, bookkeeping and computer services. Married to David and has one child, Bretia. Joined the Reno Chapter 99s in 1990. Currently holds the office of chairman and has held the office of vice chairman, executive committee.

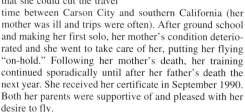

Completed ground school in 1988. She decided to get her pilot certificate so that she could cut the travel time between Carson City and southern California (her mother was ill and trips were often). After ground school and making her first solo, her mother's condition deteriorated and she went to take care of her, putting her flying "on-hold." Following her mother's death, her training continued sporadically until after her father's death the next year. She received her certificate in September 1990. Both her parents were supportive of and pleased with her desire to fly.

She current holds ASEL and flies every chance she gets for pleasure and business. Owns a Comanche 250. Law has an AA in business and finance, real estate and mobile home licenses and California private teaching credential for Heald Business Colleges; licensed Nevada real estate agent - inactive, National Assoc. of Female Executives. Her interest includes grandkids, travel, flying, softball, reading, solving puzzles and knitting. Other memberships include vice president (and charter board member), Boys & Girls Clubs of Western Nevada. Member: EAA and Comanche Society.

CODY LAWRENCE, joined the Santa Clara Valley 99 Chapter in 1992, after getting her license in 1979 at San Jose. She is currently a member of both the Palo Alto Flying Club and the Multi-Engine Flyers at San Jose. She has logged more than 475 hours and is interested in getting her instrument rating. She has been an SCV airmarker, and flew the 1983 Salinas His and Hers, placing second with D. Bennet in a Grumman Tiger. In the 1970s she worked for World Airways at Oakland, getting landing and uplift rights for worldwide charter aircraft.

She is an artist and technical writer.

MARGARET C. LAWSON, born Dec. 12, 1935, in Chicago, IL. Joined the 99s in 1967. Flight training included CFIA, I, M, ATP, and more than 10,000 hours (BS from USC). She has flown most singles and many light twins. Worked as designated pilot examiner for LAS FSDO. Married D. David Lawson and has two children, David and Monica Lawson Lavigne; and grandchild, Jacob Lawson Lavigne, 1 1/2 years old. Member of 99s San Gabriel Valley, AOPA, NAFI. She flew M20F from El Monte, CA, to Oslo, Norway and returned.

JUNE LEACH, born Nov. 13, 1937, in Pampa, TX. A special education teacher, resides in Orange, CA. She began flying after a friend, with his own plane and instructor's rating, convinced her that she could learn to fly. She took her first lesson in 1974. A career in flying was never intended, only for pleasure. California weather provides a lot of time for pleasure flying.

Shortly after receiving the certificate, she joined the 99s. She has been active at the chapter level, as well as the International level. Friendships made in the 99s have been cherished.

June was born in Texas, received a BS from West Texas State University and a MA from the University of Tulsa. She is married to Harold Leach, a non-pilot, but a wonderful copilot.

Other activities include being a member of Council for Exceptional Children, church, gardening, cooking and sewing.

BRIANA JESSEN LECLAIRE, native of Boise, ID, was taught to fly as a teenager by her mother. The family fixed-base operation made it natural for Briana to work at the airport, grow up in aviation, and tag along to 99 activities all over the country.

Briana's college years started at Willamette University in Salem, OR, then continued at the University of Idaho and included a study-abroad semester in England. After graduation Briana spent a year in law school and decided that wasn't the career direction for her.

She worked for the Jessen FBO partner Senator Steve Symms in Washington, DC, where she became enamored of politics and co-worker (and fellow Idahoan) Tom LeClaire. Briana then worked for the General Aviation Manufacturer's Assoc. (GAMA). She and Tom married in 1994 and reside in Moscow, ID, where Briana is employed by Horizon Air.

GWEN BJORNSON LEDBETTER, Phoenix 99s, got a solid start in an aviation career at a young age, getting a degree in aviation administration from University of North Dakota School of Aviation in 1982. In school she served as secretary, treasurer, and vice president of the Student Aviation Management Assoc., and was a member of the UND flying team. She joined the Phoenix 99s in 1990 and has served as membership chairman as well as participating in many activities.

Gwen, who has more than 2,200 hours flying time, now works as a dispatcher and flight instructor, and does some aviation consulting and technical writing. She holds a commercial ME/SE, CFI-I-M, and Aircraft Dispatch certificate, and has over five years experience as a professional dispatcher. Gwen's husband is a flight captain for a major airline, and they have two small sons.

SANDRA J. LEDER, Ph.D., learned to fly as a young teacher in 1969, which led her into aerospace education. In 1984 she was a "Teacher in Space" applicant. She received a grant from the 99s to attend space camp for teachers. In 1993 Dr. Leder was selected as an "Outstanding Young Astronaut Chapter Leader" in the US. She was recognized by the FAA for her contributions to aviation education.

Dr. Leder founded "A Week in Space," an aerospace education day camp. She has presented her idea for the camp at conventions of the National Science Teachers Assoc. She has taught workshops in aviation and space for teachers.

Working tirelessly to promote aviation in southwest Louisiana, Leder was the first woman appointed to the Lake Charles Regional Airport Authority. She is on the board of directors for the Louisiana National Airshow and the Chennault Airpark Aviation Museum.

CAROL JOAN LEE, born on Nov. 1, 1952, in Houston, TX. A native Houstonian, she attended the University of Houston on a scholarship and graduated with a BS in biology in 1975. Carol is still an active member of the U of H Alumni Organization.

Since her early years, she can remember running outside her house when she heard an airplane flying over and being fascinated with flying. On March 14, 1981, Carol obtained her private license through FAA examiner and fellow 99, Mary Able. In May 1981 she joined the Houston

Chapter of the 99s and has been a member ever since. Over the years, Carol has served in several capacities and is presently the Houston Chapter chairman. She also continues to support the monthly Federal Aviation Administration's Safety Seminars as an aviation safety counselor and is a member of the Aircraft Owners and Pilots Assoc.

Being intrigued with vintage aircraft, Carol also joined the Confederate Air Force in 1981, where she became the first woman to become a "colonel" in the West Houston Squadron of the West Texas Wing. Since joining the Squadron, she has served on numerous committees and has held several positions including recruiting/information officer, public information officer (PIO) and newsletter editor. In 1989 she received the Squadron's "Award of Excellence" and was honored as "Colonel of the Year" in 1992. For her meritorious service to the squadron, the West Texas Wing, and supporting CAF Headquarters through the Oral History Program and in public relations, Carol was also awarded the Confederate Air Force Letter of Commendation in 1992. During her tenure as Wing PIO, she received several awards for her Outstanding Work for the editing/publication of the Wing *Logbook* newsletter, and in 1993 received the Wing's Pulitzer Award.

Carol has been a part of the Wings Over Houston Airshow ever since its inception in 1984. On the Airshow Staff, she is presently volunteer coordinator, ground transportation coordinator for the performers, military and VIPs, and assists with the media. At the 1994 Airshow, she received an "Award of Excellence" at Ellington Field in Houston, TX. In a beautiful glass and oak case, Carol is the proud recipient of the first US flag to fly over the CAF's new International headquarters in Midland TX, and the Texas State capitol for a day.

The experiences and memories of the variety of aircraft she has had the opportunity to fly in over the years are priceless to Carol, as well as meeting some of the best airshow performers and legends in aviation history.

NELDA K. LEE, born Sept. 14, 1946, in Carrolton, AL. She is manager of F-15 flight simulation and flight test engineering at McDonnell Douglas Aerospace. Her career began 26 years ago after receiving an aerospace engineering (AE) degree at Auburn University. Nelda learned to fly fixed wing aircraft in 1969 and received her private license at Auburn University along with her AE. When Nelda was four years old, her dad was a student at Auburn University and he would take her to "piz-sics" (physics) class with him and also to the airport to fly with friends. So a love for engineering and flying began early in Nelda's life!

Nelda's engineering career continued to develop and so did her flying. She joined the 99s Greater St. Louis Chapter in 1971. Her flying advanced to a commercial license with instrument and multi-engine ratings. In 1977 Nelda was recipient of the tenth annual Whirly Girl Scholarship and obtained a helicopter rating. Nelda has served both the 99s and Whirly Girls in many offices and committees. She has flown in two Air Race Classics. In 1994 Nelda joined Women In Aviation International as a charter member and currently is on the board of directors. The ultimate and most memorable flying event however, occurred in April 1980, when Nelda flew the F-15 Eagle and logged the 1.5 hour experience. Absolutely outstanding!

Nelda is a member of the society of Flight Test Engineers, Auburn Alumni, YWCA, International Clown of America, US Tennis Assoc. Officials, and Baptist church. Her hobbies are tennis, golf, travel and photography.

PATRICIA LEE, First Canadian Chapter, took first airplane ride in a Tiger Moth at Baker Airfield in Toronto, 1944. Years later, 1967, after impromptu ride in a Citabria aerobatic, she was hooked and wanted to take flying lessons immediately. She didn't know how to break news to husband, Ray, and three children, until daughter, Karen Patricia, received her private license in 1970. Patricia followed shortly, receiving license in 1971; then son, Gordon, in 1976. All flew with Tom and the late Bob Wong, Central Airways, at Toronto Island Airport.

A 99 since 1974, she is member of COPA; past chairperson for annual Poker Run, 1976; past volunteer in a program called Operation Skywatch; past APT chairperson; volunteer since inception, 1986, for Mission Air Transportation Network; present FCC Communications chairperson; and also East Canada Section International Forest of Friendship chairperson.

Her hobbies include traveling, reading and being an avid "QiGong" follower.

VICKI BALFANZ LEE, born Oct. 20, 1944, in Monroe, LA. Joined the 99s in 1992 or 1993. Flight training included private pilot license, instrument rating and working on commercial. She has flown Cessna 150, 152, Piper Lance Turbo six place retractors. Married to Darrell and has four children: Tiffani Lee Mastronardi, 25; Brittani Lee, 22; Derek Lee, 18; and Austin Lee, 13. She was seven months pregnant with her fourth child when she landed in Baja, Mexico on uphill dusk, dirt strip, littered with cow pies.

MARJY LEGGETT, a lifelong resident of Washington state was born May 6, 1948. Inspired by her uncle, she had a lifelong dream to fly. A birthday gift from her husband, Bill, started her private pilot lessons in Pasco, WA, in 1984. She earned her instrument rating in 1987. She has flown a variety of Cessna, Beech and Piper aircraft and competed in the Palms to Pines and Air Race Classic.

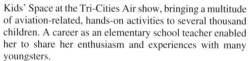

An active member of the Mid-Columbia 99s, she has held a variety of offices and helped with numerous activities. She coordinated the Kids' Space at the Tri-Cities Air show, bringing a multitude of aviation-related, hands-on activities to several thousand children. A career as an elementary school teacher enabled her to share her enthusiasm and experiences with many youngsters.

Leggett is grateful to her husband, Bill, also an educator, who although not a pilot himself, continues to support her flying activities.

ELIZABETH "BETTY" LEHMAN, began flight training July 1977, in a Cessna 150. She is a wife, and mother of three. Her husband was a flight engineer with Seaboard World Airlines, now, FedEx. She soloed in 13 hours and continued her training in the Cessna until February 1979, when they purchased a Grumman Traveler, N10GW, quite a change for Betty, since the low wing of the Grumman and the bubble windshield gave a much expanded view which took some getting used to! Continuing her training in both aircraft, a day after her 50th birthday, she passed her flight test and became a pilot! She was a member of the Garden State Chapter of the 99s. Moving with her husband to Virginia, Betty joined the Virginia Chapter of 99s and flies her Grumman out of Hummel Field in Topping, VA.

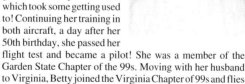

Born Aug. 8, 1929, in New York City, NY. Married to Gerhardt E. and has three children: William, 34; Howard, 32; and Thomas, 30. Also has grandchildren: Stephanie, 11; Megan, 10; Russell, 9; Katie, 8; Jarrett, 6; Brandon, 3; and Shelby, 2. Flew navigator with Nelda Ewald, Chicago area, in Garden State 300. Very, very windy that day. They lost one airport, briefly and came in twenty-eighth in a field of 30, but had a great time!

SHIRLEY LEHR, learned to fly in 1966 in Sacramento, CA, when her husband gave her flying lessons for her 30-something birthday. Her first passengers were her husband and two young sons.

She soon learned about and joined the Sacramento Valley Chapter of the 99s and got thoroughly involved in their activities. She went to the Powder Puff Derby start in Monterey and picked up the air-racing bug. She entered the 1971 race, which was its 25th anniversary, from Calgary, to Baton Rouge. There were 150 airplanes in this anniversary race and they were number 25.

The next year she decided to fly the race solo from San Carlos, CA, to Toms River, NJ. Her family met her in New Jersey for the festivities and they flew back to California together. In 1973 she tried the PPD again with another 99 as copilot. On the second day of the race after leaving Prescott, AZ, the prop-governor line broke, spreading oil all over the plane and the windshield, leaving the only vision out the side widow. The engine quit while they were over the town of Holbrook. They found a place to land on Route 66, which was under construction and only about a half mile from the airport. Neither of them were injured but that was the end of racing for them that year.

The start of the last PPD was in Sacramento with 200 airplanes entered. She has flown other races and participated in other flying contests.

She received and Sacramento Valley Woman Pilot of the Year award in 1973 and 1991. In 1980 she started to work for the FAA at the Sacramento FSDO and retired in 1994.

She has started her own aviation consulting business to assist air taxi companies and repair stations become FAA certified. She writes operation and training manuals, inspection procedure manuals and letters of compliance.

She owns a Mooney and has commercial, instrument and flight instructor ratings for single and multi-engine airplanes. Her husband, Ernie, flies with her. They have two sons and five grandchildren.

LOY ANNE COX-LEIBLIE, born Sept. 9, 1958, in Toronto, Canada. Soloed at Deland, FL, Dec. 11, 1982. She has flown the following airplanes: Cessna 150, 152, 172, 172RG, 210, 402; D95; DC6; BE100, 200; WWII24; DV20; and PA44. Married to Donald Eugene Leiblie.

Born in Toronto, grew up in Chicago began flying after graduating from Loyola University with a BS in biology and moving to Ormond Beach, FL. Moved to Miami and obtained all certifications and ratings. Has been a flight instructor since 1985 and is currently chief ground instructor at Miami-Dade Community College. Has served as chapter treasurer and currently secretary and safety education chairman. Attending ERAU, working on MA in aeronautical science, aviation and aerospace management. Has flown for commuter airline in Miami, currently logged more than 2,300 hours.

ANN LEININGER, graduated from Oklahoma University with a BS in mathematics and began her flight training in 1987. She currently holds an airline transport rating and her CFII/MEI ratings. To date, she has accumulated more than 2,500 hours.

She was hired by Oklahoma University as a flight instructor in January 1990 and was promoted to assistant chief flight instructor in 1991.

Ann has participated as an instructor for the Oklahoma Aerospace Academy and the NASA Space Grant Consortium. She is the coach/advisor for the University's NIFA team, Sooner Aviation Club, and Alpha Eta Rho.

Ann is the vice chairman for the Oklahoma Chapter. She participates on many committees including her chapter's Flying Companion Seminar and the Okie Derby. Ann frequently speaks to school children about aviation careers.

While aviation is Ann's love, she enjoys cross stitch and sewing. But one of her other true loves is playing with her cocker spaniel, Penelope.

FRANCES FERGUSON LEITCH LEISTIKOW, in the summer of 1927 left Kansas City to visit cousins in Ponca City, OK. There she met her future husband Al Leitch and stayed three years. Al taught her to fly his JN-4D and she soloed in April 1928.

When she was taking her flight test before the Department of Commerce inspector in September 1929, it began to rain heavily. The inspector asked Al, "Is there any way you can signal her down?" She completed the required maneuvers in the downpour and was approved for private license #8695. (Her daughter's is #2,172,188.)

After their marriage she flew a Curtiss Robin and kept her license renewed for four years. Her husband continued to fly and test flew during the War.

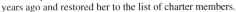

She had been off the rolls of 99s for many years when they found her a few years ago and restored her to the list of charter members.

PAT JUDGES LEMMIN, born March 18, 1940, in Toronto, Canada. Joined the 99s in 1972. Flight training included multi-engine, commercial and instrument ratings. She has flown Cessna 150, 172, and 310.

Married to Fred and has two children, Arlene, 32, and Charles, 27. Learned to fly in Toronto, Ontario, in 1970. Participated in Air Racing for many years in USA, Canada, Mexico and Bahamas. Raced solo in her own Cessna 310 aircraft C-FMSM for 15 years. Except for the first year, she always placed in the top ten. Raced in the Angel

Derby, Air Race Classic, Grand Prix, Great Southern Air Race and many smaller regional air races. Placed second in the 50th anniversary of the original All Women's Air Derby from Santa Monica to Cleveland in 1979. Moved to Belleair Beach, FL, in 1980.

ANDREA LENDE, born April 13, 1963. First soloed on April 10, 1988. Private certificate, July 11, 1988; instrument rating, July 2, 1990; commercial rating and flight instructor certificate on March 1, 1992. She was the recipient of the Amelia Earhart Scholarship which granted her the opportunity to add a multi-engine rating to her flight instructor certificate. The highlight of her aviation career was the day she passed the checkride and earned the right to instruct in a twin engine

aircraft. She could not have accomplished this dream if the 99s had not believed in her and granted her the funds to complete the training. "Thank you 99s!"

MARY ANN LENGYEL, born May 12, 1952, in Tampa, FL, joined the 99s in 1994. She learned to fly in 1992 after setting a goal to obtain her private license before age 40! Her instructor, an air traffic controller at Springfield Regional, provided insights and understanding of ATC which proved to be invaluable! She flew a Cessna 150 to earn her private. The next day, her husband bought her a Cherokee Arrow 180. One day while practicing approaches, they noticed the landing gear was not coming up. After several low approaches (200 foot ceil-

ing) it was confirmed the landing gear was three-fourths up. They were circling for more than two hours, receiving suggestions from her first instructor. The last suggestion was to pull up the back seat and cut the hydraulic line. After attempting this, she notice a rod to the gear extender was displaced. She manually forced the rod back into place and Bingo! Three green lights! Even though they had three hours of fuel left, her instructor didn't want to resume her lesson! Thanks to Jack Reynolds, in the tower and Chris Condon in the cockpit, a definite team effort. (*Readers Digest* version)

Some of her community activities include being involved in Project Graduation, A+ Schools Program, Arts as a Basic and many fund raising projects for her children's school.

She is a proud member of the 99s, Inc., Aircraft Owners and Pilots Assoc., involved in the Wings Program. She is married to Mike Lengyel who is a perfusionist in heart surgery. She is the vice president of a small cooperation that has produced two superior products, Chad, 18 years and Courtney 16 years. She is a company pilot for Auto USA in Springfield, MO. Her next goal is to fly in the Powder Puff Derby or the Air Race Classic. Hopefully, before she's 50!

NANCY LEE LEOTA, born Dec. 22, 1932, in La Grande, OR. Has four children: Linda, Zane, Noel and Starr; nine foster children; 17 grandchildren; and five great-grandchildren.

In September 1953 she learned in her first flight at 20 years old to control the plane. Surprise, the pilot said her efforts to keep the wings level, sitting in middle of back seat, her hands down or lifting up to keep weight shifted was not what was happening!

A student permit, bumming rides, instruction with any who were willing. Tail Draggers, Pipers, Taylorcraft, Stinsons, Interstate Cadet, and a whirlwind hour with a WWII flight pilot in his Swift DID! In 1959 flying Alaska bush, single parenthood got top billing. In 1979 flying again, Cessnas, Bellancas, joined 99s.

Nancy moved to Hawaii, shared rides with Aloha 99s and a nodding acquaintance with "fish spotting" and Apupulele.

"Life is what happens while we make our plans. I planned to be the modern Amelia, I became who I am, memories, a licensed pilot, a 99, flying my broom far more than any plane!"

MARIE CAMILLE LEPORE, born in 1920, in New York, NY. She has flown the Taylorcraft, Pipe Cub, Cessna 150 and 172, Beechcraft, Luscombe, Navion, Bonanza, etc. Married to Irving A. Lepore. Became a member of New England Section in 1945. Charter member ENE Chapter, 1962. Chaired and served on numerous committees including two International Conventions. Flew many air races. Initiated and organized a "Special Fund" for the advancement and upgrading of ENE pilots – named the Marie Lepore Fund in her honor.

As aerospace chairman for 99s and deputy for Massachusetts Wing Civil Air Patrol, promoted and organized in-service teacher workshops and participated in four national workshops.

Carter member and incorporator of Massachusetts Aviation Council. Appointed to Women's Aviation Advisory Committee FAA-NE. First woman appointed to commissioner of Worcester Airport. Appointed CO Lackland AFB National Special Encampment for CAP honor cadets. Awarded Woman of the Year in Aviation, USAF Assoc. Twice awarded Woman of Year Honor Award from Northeast 99s and special for promotion of aerospace education. Recipient of Achievement and Meritorious Service Awards from USAF Assoc. and Civil Air Patrol.

Achievement Award from Soroptimist International Assoc. for work with youth of Massachusetts in Aviation Education. Retired as lieutenant colonel from CAP. She pursued an art program. Her works are shown in museums and galleries in New England and Florida, also in Massachusetts State Capitol, Boston. Recipient First Place and Honor Awards from juried shows. Member Artist's Forum, Brevard Watercolor Society, Space Coast Art League.

ROSE LEPORE, born in Minnesota in 1941, raised in Iowa, and has been a Texan since 1964, currently residing in Houston. She flies in a Mooney M20J and although a relatively low time pilot, loves air racing, and has competed in several speed races, including the Air Race Classic, which she and her partner won in 1993.

Husband, Bill, a stockbroker, is also a pilot and very supportive of Rose's flying. Between them, the Lepore's have six children and 12 grandchildren, (one son, Todd, died in 1993 at 24).

Rose has been an active member of the 99s since 1982, chairing various chapter committees and serving as Houston Chapter chairman. She served on the International nominating committee and is currently aviation scholarship advisor for South Central Section.

Rose's vocation is Healthcare Materials management, her avocation – books, grandkids, and air racing.

KATHRYN MARY GAMBLE-LERCHNER, is presently completing her studies in law and expects to specialize in the area of aviation and aerospace law. She earned her commercial license at age 18 and continues to maintain her float endorsement and ultralight license. Her involvement in Canadian aviation has included: director of Aerobatics Canada, Contest Organizer and Judge at National and International Aerobatic competitions, Canadian Aerobatic Team - Ground Support, Aerospace Education Coordinator 99s,

regional director of the Recreational Aircraft Assoc. (RAA) and most recently vice president of the RAA. Kathryn, a lifetime member of the 99s, has been a member of the 99s since she was 15 and 1/2 when she joined as a 66. She learned to fly a tail dragger and still maintains her love of flying grass roots type aircraft. Kathryn's husband, Anton, is an Apiarist (Beekeeper) and her son Kristian (July 15, 1993) loves to fly, especially when his grandfather and mother take him with them.

Lifetime member of EAA, Aerobatics Canada, 99s, Recreational Aircraft Assoc.

CATHY LESHER, Women With Wings, flew with her husband Tom since they were married in 1976. She started to fly when she took her first lesson on April 27, 1992, achieving her private ticket less than a year later on Feb. 2, 1993, in a Beech Skipper. Cathy has time in a Piper Colt and is currently flying the family's 1968 Piper Comanche 260B.

Cathy is secretary/treasurer of D&K Supply & Equipment Inc., in East Canton, OH, along with the daily routine of raising two teenage boys, Jake and Jess. She is an active member of the

Church of God. Even though business and everyday life keep her busy, she is close to completing her instrument rating.

Cathy joined the 99s as a student and appreciated the support and encouragement to complete her ticket. She is a charter member of the Women With Wings Chapter.

JUDY A. LESTER, started flying in December 1979 with a solo course given to her for Christmas by her husband (49 1/2) Al. She received her private SEL in May 1982. They bought their Cherokee Warrior II in August 1982 and flew to Georgetown, Exuma in the Bahamas. Since then they have traveled Florida and the Bahamas all for pleasure. She also helped organize and flew in the Florida 400 Air Rally for three years.

She joined the Florida Suncoast Chapter 99s in 1984 and has held various chapter positions; newsletter editor, aerospace chairman, chapter secretary, vice chairman, chapter chairman 1993-95, and section by-laws chairman 1991-93.

She has always enjoyed attending and meeting other members at the International Conventions and the Southeast Section Meetings; and she has participated directly in the planning of the section meetings that their chapter has hosted.

She has been active in the preservation of their historic waterfront airport, Albert Whitted, where the first scheduled airline flight took place and she relishes in any activity that promotes flying.

She is presently employed by Interiorscape Company as office manager and certified operator in charge, and she is a corporate partner in a commercial grounds maintenance company.

She and her husband enjoy traveling and meeting people; so flying and being part of the 99s has given them the opportunity to do both with pride.

LOIS LETZRING, started flying in 1970 at San Jose, and got her private license there in 1972. She holds a commercial, instructor, MEI, CFII, ATP (multi) rating with more than 4,500 hours to date.

Lois has served as treasurer, *Windsock* editor, and various chairmanships since joining the SCV99s in 1972. Lois was awarded the 1983 SCV Pilot of the Year award and has participated in races, including the Powder Puff Derby in 1975. She had to fly the race solo, because her copilot was killed in a plane crash earlier in the year. She has also flown the PAR, the Palms to Pines and the Hayward to Las Vegas Race. Lois has been editor of the *Southwesterly,* the Southwest Section newsletter. She joined the 99s after reading Mardo Crane's book *Ladies Rev Up Your Engines,* although at first she thought the 99s were fictitious!

THERESA DAY LEVANDOSKI, took her first flying lesson at Van Nuys Airport, Van Nuys, CA, in the mid 70s. At this point she has approximately 1,500 hours or more.

She participated in two Air Race Classic's so far. Also flying in many other aviation sponsored flights. She has held many different positions in the Monterey Bay Chapter of the 99s, including chairman. She's worked on many chapter functions. She especially enjoyed flying the local politicians over the areas they govern. It was quite an experience to help in the forming of their local airport pilots association and being the first vice president. A very enlightening experience was serving on the local Mayor's Airport Study Committee to research the usefulness and or closure of their local airport. After much opposition, the facts prevailed and the airport is still open. Along with those previously mentioned, she belongs to numerous Aviation Organizations.

Through all of her studying, flying, and late night meetings, her two daughters Angela and JoAnn were and still are very supportive of her. They have often helped work on some of her events. She always appreciates them. She finds aviation has been very giving to her. It has allowed her to fly in many different types of aircraft (The Concord). She's met a variety of people. Flown to many different places and been able to fly in different countries (Ireland). But most of all, flying has given her many lasting friendships. She's especially fond of her friendship with her own C-150. Her 29 year partnership with AT&T as a systems engineer has helped in making much of the above possible.

IRENE LEVERTON, soloed in Chicago, in 1944. Joined the 99s in 1945. After CFIA rating in 1948 instructed on GI programs, flew as AG pilot three seasons. Received MEL&S and IRA, became first officer

DC3s taking military personnel throughout states and Alaska. Had own instrument school then flew air taxi on USFS fires for three corporations, ferried Pacific and Atlantic, organized, directed two 135 operations and one 141 school. Flew cargo, was captain on multi-engine commuter into San Francisco and finally night runs in an air ambulance.

Organized and directed first Illinois 99 Airmeets, Women's Pylon Racing Assoc. at Reno, was Women's National Champion 1964. Organized original Women Airline Transport Pilots Assoc. in 1968. Designated FAA pilot examiner 1981 and continues today. Hold ATP, MEL, CFIA SEL, MEL, and glider with TT 25,188 hours. In 1961 passed astronaut testing for NASA. Specialize in flight training as consultant. Own Aviation Resource Management in Prescott, AZ. Completed three years of university. Flew "Great Race," London England-Victoria, BC, eleventh place in 1971.

BARBARA LEWINSKI, born Feb. 27, 1949, in Detroit, MI. She took her first flying lesson with a Simmons pilot and then continued her training at the Marquette airport. After being instructed by three instructors she received her private pilot license on July 20, 1989. She was a 66 before and became a 99 in September 1989. During training she flew a Warrior and Archer. A few weeks after received her private license she was certified to fly a high performance plane, a Piper/Lance.

Barbara enjoys traveling with her husband, Gary (instrument rated pilot) to Las Vegas, Florida, Hilton Head, Colorado and many short trips to Las Vegas, Florida, Hilton Head, Colorado and many short trips in her area in their Piper/Saratoga. They share the flying responsibilities and take turns being pilot-in-command. They also run a bar/restaurant operation for 17 years now in the Marquette, Michigan area. She has a daughter, Meshel and she lives in Madison, WI, age 24.

Since her 99 Chapter is some 400 miles away, she seldom gets to attend meetings but she hostess the August meeting in her town and some of the ladies and husbands fly up for the weekend. She keeps in touch with the newsletter and 99s magazine.

AFTON REVEL LEWIS, Bay Cities Chapter 99s charter member, private license #22632. Afton learned to fly in 1930, at the San Francisco Municipal Airport, "Mills Field," San Bruno, CA. She and her

instructor made a forced landing on Twin Peaks, hurting no one, and with slight damage to the Kinner Plane. Another time, she was on final approach in a Fleet when she was run (?) into by another plane, which managed to land. Afton only got her plane to the dike. She had a cut on her head, and her plane was severely damaged. Soon, she was back in the skies.

She flew many different type of aircraft; the Travelair, Fleet, Kinner, Stinson, Waco and Aeronca.

Afton stopped flying in 1948. She found a new hobby - gem cutting and mounting, for which she earned many awards. She died Feb. 23, 1986.

ANITA LEWIS, Governor, Northwest Section 99s, Idaho Chapter, International chairman for the 99s "Twenty-First Century Fund." Married for 35 years to Vic, grown children, Maria, Leah and Chris and granddaughter Mallorie.

As a child she could spread her wings (arms) and fly. This was a recurring dream until she actually began to fly, in an airplane. She no longer dreams about flying because she has the wonderful privilege of being able to do it in real life. Since 1978 she has become commercial and instrument rated, received her certified flight instructor and become multi-engine rated. Entering the arena of aviation has opened up a new world to her. Her husband, Vic, and she have explored the Western United States by C-182. They have marveled at beautiful sunrises and "aahed" at magnificent sunsets. Flying over Yellowstone National Park on a clear pristine daybreak is a rare treasure few people will find.

Her most difficult flying assignment so far, has been her solo trip across the western US in a 1946 7AC Champ. She saw mountain passes, ridges, roads, wild-life, ranches, trading posts, geological features from volcanoes and ancient glaciers. She knows what a tarn is, but she saw very few mountain crests except from below looking up. It was the experience of a life-time. Not only for the flying, but for the fact that it was the first time in her life, that she had been alone for two weeks.

Her three children are grown and the 99s have become her avocation. She has held every job in 99s on the local level and is currently serving as governor of the Northwest Section. In 1990 she was given the daunting job of creating and developing a perpetual fund or foundation for the 99s. The "21st Century Fund," is dedicated to preserving and enhancing the 99 legacy. She has hopes of this fund becoming a "ten million dollar" extravaganza by the end of the century.

BONNIE LEWIS, earned her private pilot license on Nov. 13, 1976. In the past 19 years Bonnie and husband, Tom, have owned and flown two Cessna Skyhawks, N1626Y and N12122, and most recently a Cessna Skylane, N2814Y. Bonnie and Tom's two children, Nancy and Paul, each earned their private pilot license in N12122. Bonnie and Tom each earned their instrument rating in N12122.

Bonnie was a member of the Indiana Dunes Chapter of the 99s four years. They organized and participated in many flying activities, put on flying companion seminars, and introduced aviation to children in the public schools. Then Bonnie moved to the Minnesota Chapter for nine years where they enjoyed many flying activities: flew blood for the Red Cross, flew daffodils for the Cancer Society and hosted a NC Section Meeting. Now Bonnie has been in the Dallas Chapter for six years. In Dallas they enjoy flying activities, airmarking, pancake breakfasts, and most recently a trip to the Forest of Friendship to honor a member of the chapter who is a WWII WASP. Bonnie is now looking forward to becoming a charter member of a new 99s Chapter, the Brazos River Chapter, based on the Pecan Plantation where she and Tom will soon build a home on the airstrip. Bonnie has served in many chapter offices and committees including aerospace education, secretary and newsletter editor.

Bonnie and Tom have flown their airplanes to the Bahamas, Alaska, California, Oregon and Washington, DC. They have flown into the Oshkosh Airshow 18 years in a row, several times into Sun'n'Fun, and once to the Albuquerque Balloon Festival. Bonnie and Tom are now building an experimental RV6A aircraft in their garage. Flying is a big part of Bonnie's life.

CAROL J. LEYNER, born in August 1942, in Alamosa, CO. Joined the 99s in 1979. She has flown Grumman Tiger, Cheetah, Cessna 150 and 180, Stinson 108, and Citabria. Married to Robert and has two children, Beverly, 30, and George, 26.

Her desire to fly and her love of airplanes began at 14 when they moved across near the Floating Feather Airport outside of Boise, ID. In 1961 her college roommate took her for her first small airplane ride out of Boulder Airport. She learned to fly in 1978 when her family gave her lessons for her birthday after their first trip to Oshkosh by car. She and her husband went through flight training together and in 1979 with licenses only two months old, they flew to Oshkosh.

She joined the 99s in 1979 and was the first coordinator of the Flight Without Fear clinic. She has been chairman of the Colorado Chapter, telephone chairman, and chairman of the 1995 Mile High Air Derby. She was president of the Boulder Municipal Airport Users Group and was instrumental in saving open space around the airport. She serves on the Boulder Airport Advisory Board. She has held many offices in EAA Chapter 648 and Colorado Antique Airplane Assoc. She spent two years on the Colorado Air Fair Board and has been treasurer and now is president of the Rocky Mountain Sport Aviation Committee for the Rocky Mountain Regional Fly-In. She and her husband own a 1944 Howard DGA 15P, 1946 Stinson Voyager, and recently acquired a 1936 Lockheed 12A serial #5 (the smaller version of Amelia's Electra). She and her husband are working on their A&P license and plan to restore the Howard and Lockheed themselves after they retire.

JANET LEE LIBERTY, Western Washington Chapter of 99s, earned her private license at age 41 and joined the 99s in 1984. She purchased a 1952 Cessna 170B in 1985. In 1986 the engine failed but she

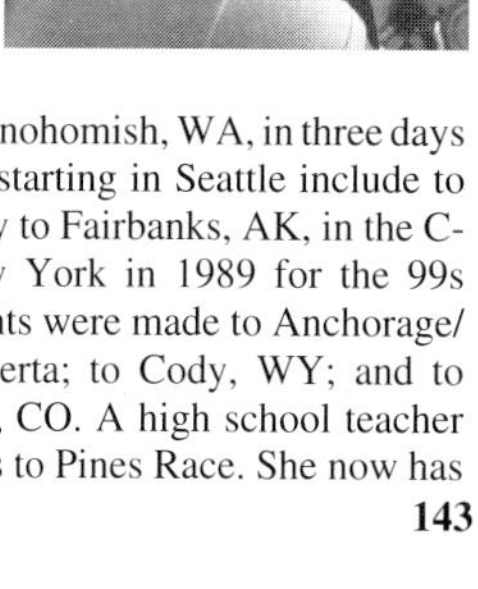

landed safety at a state emergency airport. Always very involved in 99s, Jan has served as chapter chairman, was honored Section Achievement Award winner for 1989 and chosen Pilot of the Year for Western Washington Chapter 99s in 1993. She holds a commercial certificate SEL and SES and is instrument rated. In 1990 Jan purchased a 1955 Cessna 180 in Florida and flew it home to Snohomish, WA, in three days (24 hours). Other round-trips starting in Seattle include to San Diego, the Alcan Highway to Fairbanks, AK, in the C-170. She was copilot to New York in 1989 for the 99s convention. In the C-180, flights were made to Anchorage/Homer, AK; to Calgary, Alberta; to Cody, WY; and to Pagosa Springs and Leadville, CO. A high school teacher with a MA, Jan flew the Palms to Pines Race. She now has

1,600 hours total time, and has landed at 221 different airports and seaplane bases in Washington state alone. "There is no end to the wonderful fun you can have as a pilot and member of the 99s."

BARBARA LICHTIGER, a resident of Coconut Grove, FL, learned to fly in 1991. Her educational background includes a BA in physiology from Hunter College in New York City. She worked as both a volunteer and honors science teacher for secondary schools in Dade County. In addition to flying, she loves SCUBA diving and traveling; she and her husband have journeyed to all of the seven continents.

Barbara is married to Dr. Monte Lichtiger, who decided to become a pilot after being his wife's first passenger. They own a Piper Turbo Arrow. They have two daughters and a son who live in Miami, Dallas, and Charlotte. Barbara is especially proud that Frances Sargent, one of their most illustrious 99s, has been her flight instructor.

GRACE LIENEMANN, became an active member of the Michigan Chapter 99s in December 1983. She's been serving tenure as assistant historian on several scrapbooks featuring the very beginnings of Michigan Chapter's growth and development. Journalism, free lance writing has always been her active outlet, writing guest columns for local newspapers and flying publications. Plus visiting elementary schools in her area and talking aviation safety. She is a member of the AOPA and Civil Aviation Counsel (CAC) of southeastern Michigan.

Her copilot is 49 1/2 Fred. They became partners some 40 years ago and have shared many adventures. Their first being road rallies, then four children and building a cottage. They shared flying lessons and soloed on the same day. Today they have eight grandchildren, each a potential pilot.

LEAH S. LIERSCH, born Kansas City, MO. Joined the 99s in 1954. She has flown the Luscomb, Cessna 140, 150, 172, 182; PA22, PA24. Frequent family trips to watch airplanes Kansas City, MO, flying in friend's biplane at 10 ensured love of aviation. World War II gave an opportunity to work as aircraft communicator with CAA, now FAA, resulting in a 38 year aviation career. A young woman air traffic controller recently said "Thanks for being a trail blazer." She proudly received her pilot license in 1954 and became a 99. Flew in two Pacific Air Races, was an official in all 25 PARs, an official and contributor to many AWTARs. As chairman of El Cajon Chapter they sponsored the Indian Section. Flew light aircraft; was treated to left seat of C-195 with experimental crosswind gear, DC4, DC6, handled controls of a T33 breaking the sound barrier. She loves aviation and continues to keep active through 99s, San Diego Aerospace Museum, Silver Wings.

ELEANOR A. LILIENTHAL, born April 3, 1928, in Ong, NE. Joined the 99s in 1973. She has flown the 150 and 152, Aerobat, Citabria, 172 and 182.

Homemaker and member of the Nebraska Chapter since 1973. Currently serving as chairman. Started flying on Feb. 6, 1962. Earned ASEL on May 8, 1963. Total of 800 hours. Completed Cessna 150 aerobatic course on July 3, 1970. Has assisted with Nebraska Chapter's Aviation Art Festival, International Aviation Art Contest, Flying Poker Games, Air Markings, AWTAR, Air Race Classic, and Pilot Safety Seminars. Eleanor has four children, nine grandchildren, one great-grandchild, and one stepson. She is vice present of a family owned farm, England Farms, Inc., Doniphan, NE. Her husband Ronald is the FBO, assistant airport manager, and flight instructor at the York Municipal Airport, York, NE. They have shared many happy hours in the air together. Her hobbies include gardening and refinishing old furniture.

AGNES SANBORN LILJEGREN, joined Phoenix 99s in May 1968; was promptly appointed airmarking chairman. Next four years were spent planning, organizing and painting names and numbers on airports throughout Arizona. Many thanks to sister 99s and friends making those projects such fun and success. Dale Liljegren was her project design engineer on all the above.

Fascinated watching airplanes land and take off at Phoenix Sky Harbor, she never dreamed she would be flying one. Then Dale introduced her to Luke AFB Aeroclub's ground school and she was hooked. She earned the honor to solo on Luke's 10,000 foot runway. Received her private in September 1967 and commercial in May 1971.

December 9, 1968 she experienced the first emergency landing with her future mother-in-law, Kate Liljegren as a passenger. She had logged 100 hours and on Oct. 3, 1971, with 296 hours she made a second emergency landing.

Christmas Day 1968 she and Dale were married. Dale was an excellent pilot and A&P mechanic. Her success was due to his support and encouragement.

In their 1960 F1A Aircoupe she won first place in the 1969 Kachina Doll Air Race, second place 1970 and first place in 1973. Took third place in 1971, flying a Cherokee 160.

In 1977 she flew the Aircoupe in the 30th Commemorative PPD from PSP, CA, to Tampa, FL, logging 41 hours and 26 landings upon returning home. Won first place spot landing at Monroeville, AL. Flight was very good except her radio navigation quit upon leaving home.

In 1981 won a second spot landing at Hesperia, CA, again placing both wheels right on the line.

During her flying career she logged 600 hours, she was in Civil Air Patrol Search and Rescue 11 years. In 1972 she and Dale joined the Arizona Flying Farmers. In 1983 she was crowned Queen to represent Arizona at the International Flying Farmer convention in Dayton, OH. That year she drove to Dayton and on to Detroit, MI, and Windsor Ontario, then to northern Michigan returning through Wisconsin, Iowa, Nebraska, Colorado, New Mexico to Arizona.

DOROTHY D. LIMBACH, born in Albany, NY. Graduated from Rutgers University in New Jersey, with a BA in political science and history. Graduate work at Western Reserve University in Cleveland, OH. First flight in 1940 from Cleveland to New York in United Airline DC3.

Her interest in flying was expanded greatly during the time she volunteered at the Cleveland USO. In 1945 she married a Navy man she met at the USO, and in 1946 they moved to California and settled near Santa Monica Airport. By 1952 they had three children and not much time to think further about learning to fly. However, with temptation staring her in the face daily as she drove to work through Santa Monica Airport, by the early 60s she had determined that she must learn to fly. With a friend at work (and unknown to both her husband and his wife) they bought a 1946 Luscombe and found a good instructor for $5 per hour. Unfortunately, before she had her first lesson she lost her husband to a heart attack, going ahead with lesson as planned turned out to be the best tonic for recovery she could have had!

In 1966 she joined the Los Angeles Chapter of 99s and has found her life greatly enriched by her association with these exceptional women.

As well as her three children has a wonderful stepdaughter, seven grandchildren and a great-granddaughter.

Hobbies are (in addition to aviation) hiking and camping, music, reading, spending time with family.

LINDA LEE LINDGREN, born April 25, 1940, in Alva, OK. Flight training includes private, instrument and commercial. She has flown (owns) Cessna 172, Piper Arrow and Turbo Commander. Married to Edward, pilot for Delta Airlines. She has three children: Nannette, 33; Carolyn, 30; and Jennifer, 25. Grandchildren are Ryan, 6; Meghan, 3; and Connor, 6 months. Received private license at age 45, instrument rating at age 50, commercial license at age 55. Working on multi-engine rating by end of 1995. Total hours - 600. BA in home economics, author, designer and TV demonstrator of craft and needlework designs.

DOROTHY LAFITTE LINDSEY, 399182, Shreveport Chapter 99s, South Central Section, born July 4, 1921, in Shreveport. Before learning to drive at age 10 she had learned to ride all the horses, including the gaited ones. She began to think there must be something else "to drive." When she was 13 a friend of the family called to invite them to an airshow in Bossier Parish. After being shown how to get in the cockpit and looking at the four or five instruments on the dashboard – she knew she could handle this little red plane. She knew that she was gong to learn to fly the red plane since all three of these aviators were now her friends and were teaching her how various parts of the plane worked.

On her 18th birthday she soloed, legally, at Porterfield, in a plane which had the same instruments as the Aeronca. In 1941 she began working for United Gas Pipeline Company as a legal secretary. Weather permitting – she and the three other girls flying from this strip were in the air. They had three Porterfields, two Taylorcrafts and one Piper Cub. These planes had no radio equipment, but were easy to check your "dead reckoning" with the railroad tracks.

In 1942 she had sufficient flying time to become a private pilot. During the war years a Civilian Pilot Training Program was initiated at the field and they assisted the young men with ground school, weather and what navigation they knew. It was not until 1958 when Floyd and she were married, that she began flying again. They have had several airplanes but she thinks their first one, a Vagabond, was her favorite because it reminded her of the red Aeronca.

MARIKAY LINDSTROM, watched the P38 circle overhead. It was 1943, the pilot was her father signaling to pick him up at Long Beach Airport. There was no phone at home. It wasn't long after he was killed in a plane crash.

That was the beginning and earliest memories of her exciting flying appetite. It took her until 1984, "growing up," being a single working mom to be in the right place and frame of mind to pursue flying.

Her husband Paul told her if she could finance her own plane, she could have one! After following up an ad in the paper, prevailing upon a good flying buddy who just happened to own part of a bank, it was possible to buy N33321, her pink, red and white Cherokee which she flies to this day and flew single pilot through Canada and Alaska. Marikay works at Big Bear Airport Unicom and flies charter.

Goal realized for the 57 years old grandma of two, and Woman of the Year of the California Optometric Assoc.

GAIL P. LINGO, born May 13, 1936, in Eaton, PA. Moved to New Jersey in 1954. Employed by Lingo, Inc., Camden, NJ. Married John E. Lingo, Jr. in 1965 who is now corporate officer of company. Has two sons, Douglas and Jeffrey.

Learned to fly in 1969. Husband was her flight instructor. Joined 99s in 1970. Owned 1949 Ryan Navion from 1970-87. Flew many times in Bahamas Flying Treasure Hunt which was sponsored by Bahamian Board of Tourism.

99 Activities: chapter chairman, treasurer, Penny-a-Pound, airmarking, and fund raising. Section Mid Atlantic: treasurer, committee for Powder Puff Derby Terminus' in Pennsylvania, New Jersey, and Delaware. In 1976 chairman of International Convention in Philadelphia, member of

AOPA and Navioneers. Other activities include Hospice Volunteer, choir member, local ski club officer and ski instructor.

Employment: accounting background, officer and bookkeeper for family business which will celebrate 100 years in business in 1997! Also enjoys being part-time travel consultant.

LAUREL LIPPERT, received private pilot license in April 1990, at Truckee, CA. Has 850 hours of flight time as of 1995. Her ratings include commercial (1993), ASEL, instrument (1993), AMEL (1995). Joined the 99s in April 1990. Freelance writer and editor. She has flown C-172, C-182, Grumman Tiger, C-182 RG, C-340, Seminole (PA44-180). Married to Tom (becoming a pilot). Has a sister who is a pilot in Quincy. Palms to Pines Air Race 1991, CAP mission pilot, recipient of the 1993 Lake Tahoe Chapter scholarship for a commercial certificate and the 1994 Amelia Earhart Memorial Scholarship for a multi-engine rating.

Non-flying activities include snow skiing, hiking, bicycling, tennis, searching for the world's best coffee shops. Member of Civil Air Patrol, Friends of the Library, Truckee Tahoe Plane Talkers, in-kind writing and photography for various non-profit organizations, 99s (five years).

In 1992 she and her husband flew to Minnesota with stops in Idaho, Montana and North Dakota to visit family and friends. For the first time, she saw her North Dakota birthplace, the valleys and small Midwest towns that she knew as a child all neatly laid out beneath her. She landed at an airport in Lisbon, ND, where her 92-year-old grandmother lives and on a grass strip at her cousin's farm. She smelled fresh-mown hay from an open cockpit and spotted her parents' cabin on one of the million lakes in Minnesota. That trip and its aerial perspective made the old familiar places look refreshingly different and new. Climbing out of the Grumman Tiger, she expects she might have looked good to them, too.

JERRIE P. LISK, born July 22, 1939, in Sampson County, NC. Joined the 99s in 1988. Flight training included private pilot license in 1988, and instrument rating in 1990. She has flown Cherokee Warrior, Archer, Arrow, Saratoga, and Cessna 172, Boeing 737 in flight simulator (a funny thing just once). Married Ronald and has two children, Bryan, 22 (born Oct. 3, 1972), and Brandon, 19 (born Oct. 13, 1975). She also has one grandchild, Sarah Elizabeth Neckyfarow, age 15 months. Longest trip made in her Cherokee Archer was from North Carolina to Grand Canyon and back. After becoming a pilot and buying an airplane, her husband was inspired to become a pilot as well as her oldest son; her youngest son is ready to began his training now. She sells real estate and occasionally flies to the beach to show property. However she uses her plane primarily for fun and family trips.

MIMI LITSCHE, Fingerlakes Chapter, private pilot, single engine land and sea, since 1977. Cessna 150 owner since 1979.

Occupation: clinical dietitian with the Department of Veteran's Affairs and assistant professor with Fingerlakes Community College, Canadaigua, NY. Flying puts the world in the right perspective, helps her reduce her stress and it's a "helluva" lot of fun!

MARY ANN LITTRELL, ATP airplane, single and multi-engine land. Ratings include CFI, CFII, airplane-single and multi-engine land, Learjet advanced ground instrument. Joined the Oregon Pines Chapter 99s in May 1980. She has held the office of secretary. Started flying lessons in Salem in the fall of 1978, when her youngest child was 4. She had a degree in physical education from San Jose State College in California and was looking for something to teach when all of her children were in school. She did all of her flight training in Salem and completed private, instrument, commercial CFI in the next two years. She taught in Salem and Independence for the next eight years. Her husband Earl is a professor of accounting at Willamette University in Salem. Has son Don, and daughters Mary and Helen. She also got CFII, multi and ATP ratings.

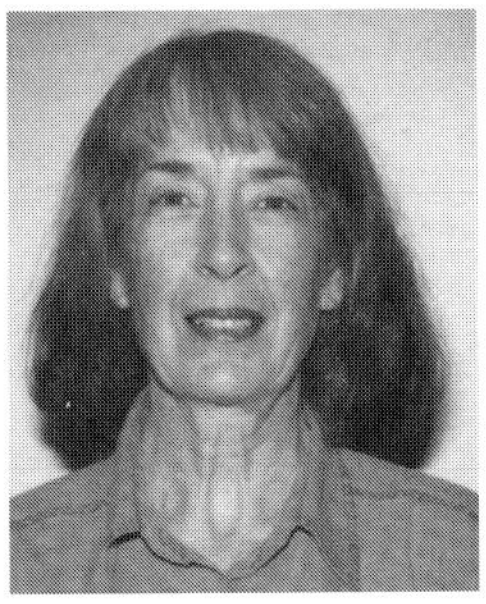

After teaching for a few years, Ann started flying charters and eventually was chief pilot at Buswell Aviation in Salem. She had to opportunity to fly copilot on a corporate jet and included that with her charter flying and flight instruction. In 1989 she was offered a position with an air ambulance/charter company. She moved to Medford and spent the next two and a half years flying for Pacific Flights. She flew a Cessna 340, a King Air 100 and copiloted a Lear 24 and 25. She earned her type rating in the Lears in 1991. In May 1992, Ann moved back to Salem. She went to work for Flightcraft in Portland and is flying King Air 90, 100 and 200 and copilots on King Air 300, on the Beech Starship and on the new Citation Jet. She was 40 years old when she soloed!

LINDA LITWIN, private pilot single engine land, with ratings of instrument and commercial. Owns an Arrow 111, based at Gaithersburg Airport, MD, with 650 hours and Wings V. She has been chairman of the DC Chapter for the past three years and is on the board of directors of Montgomery County Airport Assoc.

Graduated from the University of Kentucky in 1969. Has two sons, one graduated from Cornell University the other attends the University of Virginia; at present only have two dogs and one horse. Owns and is vice president of Orthodyne Lab, Inc. In her spare time she plays tennis, golf, skiing, trains and breeds horse, runs and weight trains.

MARGARET PIRZ LLAMIDO, lifetime Long Island, NY resident, first flew Cessnas in 1986, in an attempt to make "Mom's Taxi" more efficient by taking to the air and avoiding the Long Island Expressway when daughter, Elisa opted to attend New York Military Academy high school. Those weekend furloughs were murder on the family van, demanding four 120-mile trips on the LIE and Palisades Parkway, while Long Island MacArthur Airport was only minutes from home and Stewart International only minutes from the school. Since the NYMA years, the family pilot has beaten a sky trail between Long Island and SUNY Buffalo, and recently returned from a flight to the Bahamas. "My next goal is to fly to Alaska."

"Maggie Owl" became her nickname after husband, Felix, gave her a Cessna P210 'Riley Rocket' with call sign N600WL in 1992. Needing an on-line name for her new interest in computers, she found the name came in handy.

Maggie is an RN with a BS in religious studies from Molloy College and an MA in theology from Seminary of the Immaculate Conception. She's a member of the 99s, Civil Air Patrol, COPA, Cessna Pilots Assoc., and Cessna Owners Organizations and got her instrument rating in 1994.

She's married to Felix Llamido, MD, a general surgeon, with whom she also has two sons, Michael and Kevin.

MAYBELLE L. LOCKHART, BC Coast Chapter, began flying in 1990, many years after being raised in a Veteran's development in Lethbridge as the daughter of a WWII navigator-mechanic father who didn't feel women should fly.

She was encouraged by the enthusiasm of her son, Ken, a new pilot, despite her concerns about safety. She completed a copilot course in June 1990 at Boundary Bay. In December 1991 she completed her private pilot license. Her greatest thrill was receiving a certificate for this at the PFC graduation ceremony at the same time Ken obtained his commercial license. She learned of the 99s through a friend at another flying club and joined as a 66 in January 1991. May made her first major cross-country flight over the Rockies from Vancouver to Edmonton in July 1994 and became a 300 hour pilot. She is now working on her commercial. She remains active in her chapter working towards safe flying. She plans to fly herself to the convention in Halifax.

DORIS LOCKNESS, Mount Shasta Chapter, born Feb. 2, 1910, Bryant, PA. Doris began her flying career in the late 1930s in Wilmington, CA, and is still actively flying today at age 85, holding all commercial ratings in all categories of aircraft. World War II found Doris at Douglas Aircraft working as a liaison engineer on the C-47. In 1943 she joined the WASP, Women's Airforce Service Pilots and trained in Sweetwater, TX.

A long time member of the 99s, she has been a member of Long Beach, San Fernando Valley, Cameron Park, and the Mount Shasta Chapter. For several years Doris flew her war bird a L-5 Stinson Vultee named *Swamp Angel* to military and civilian airshows. In 1992 Doris flew the Palms to Pines Air Race. She received her FAA Pilot Proficiency Wings Phase X (10) from the FAA in September 1994 flying an R-22 helicopter at age 84.

In 1986 Doris was inducted into "Memory Lane" at the International Forest of Friendship. A life member of the OX5 Aviation Pioneers she received their Legion of Merit award in 1984, she was again honored with their Outstanding Women's Award in 1987 and again honored in 1989 by being inducted into the Hall of Fame. Doris was the fifty-fifth woman in the world to receive a helicopter rating in 1963 and became a Whirly-Girl. She is also the second female rotary wing pilot in the United States to obtain a commercial gyroplane rating in a 18A Air and Space Gyroplane. During the National Helicopter Championships held in Las Vegas, NV, in 1992 she served as a judge. She is a charter member of Helicopter Club of America. In 1991 the National Aeronautic Assoc. (NAA) presented her with their Certificate of Honor award recognizing her achievements in aviation.

Doris holds several offices in the following aviation organizations: 99s Inc.; OX5 Aviation Pioneers; Women's Airforce Service Pilots (WASP); International Women Helicopter Pilots (Whirly-Girls #55); Helicopter Club of America (HCA); International Pioneer Helicopter Pilots (Twirly-Birds); Women Military Aviators (WMA); Aero Club of Southern California; United Flying Octogenarians (UFO); International Women's Air and Space Museum (IWASM); and Federation Aeronautique International (FAI).

ILA FOX LOETSCHER, a charter member, was one of twin girls born Oct. 30, 1904, in Callender IA, to Dr. and Mrs. Charles Fox. She became Iowa's first licensed woman pilot, winning her wings on Sept. 1, 1929, license #7738. Amelia Earhart called to congratulate her when news of Ila's first solo flight made headlines. She told Ila of the organization licensed women pilots were planning and encouraged her to press on in the fabulous world of flying. Later she asked Ila to chair a tri-state membership drive. When the meeting date was set, Ila was able to report she had 100 percent recruited for her area – she was the only woman with a license and she was joining! From 1931 to 1933 Ila worked for the Alexander Eaglerock Sales Corporation in Royal Oak, Michigan, giving demonstration flights to prospective buyers of the bi-wing Eaglerock. A sign in Royal Oak said Fly to Estral Beach … Miss Fox, pilot.

On June 11, 1933, Ila and David Loetscher were married. When she and her husband lived in New York she was an active member of the 99s chapter. Ila recalls the wonderful fun of meeting, in person, Amelia Earhart; of welcoming a new member, Jackie Cochran. She remained a lifelong ardent supporter of women in aviation.

Years after her flying days were over, after her beloved 49 1/2 died and she lived alone on Padre Island, TX, she became the internationally famous Turtle Lady.

LILLIE A. LOHMAN, (Danek Normington), born Sept. 9, 1919. Began flying in 1971. In 1977-79 she was chairman of the Indiana Chapter of 99s. Her principal project was flying the legislators of Indiana. She has been interested in state legislation since. In 1976 she flew in the Powder Puff Derby and won the Indiana FAIR race the same year, placing third in the same race two years later. As copilot for her husband, she was in second place in the Illi-Nines race. She has also been secretary of the Indianapolis Aero Club. She holds a commercial license with instrument rating, single and multi-engine. Her career has been in music, teaching on both public school and college levels (Indiana

State University, Phillips University, Judson College, AL) and playing concerts in Alabama, Indiana and Texas.

DIANNE LOLLAR, born Dec. 14, 1955, in Jasper, AL. Joined the 99s in 1994. Flight training includes private pilot license May 14, 1994; aerobatic training, and now pursing instrument rating. She has flown Cessna 150, 172, 180, and Skybolt. Educational background, BS in foods and nutrition University of Alabama, MA in public health, University of Alabama at Birmingham. A registered dietitian she is a member of many processional organizations and a published author. Member of 99s, Inc., Aircraft Owners and Pilot Assoc.; historian of EAA Chapter 1061, Jasper, AL.

BARBARA ERICKSON LONDON, is a member of the Long Beach Chapter and has been since 1951. She started flying in 1939 in the Civilian Pilot Training Program as a sophomore at the University of Washington. She took her training on seaplanes at Lake Union in Seattle. She then took the secondary and cross country classes and received her commercial and flight instructor ratings. With this she started instructing in the CPT and WTS Program in 1941 and 1942. She was instructing in Walla Walla, WA, when she heard about the Army Air Corps recruiting women pilots for the Experimental Ferry Group, the Women's Auxiliary Ferrying Squadron.

She went to Wilmington, DE, in October 1942 and passed the physical and flight check and became one of the original 25 WAFS stationed at the second Ferry Group in Wilmington. In January 1943 she was transferred to Long Beach, CA, to the sixth Ferry Group where she was appointed squadron commander of the WAFS. During the time they served in Long Beach they had up to 80 women pilots in the Squadron. She was awarded the Air Medal by General Arnold in 1944 for Outstanding Service. This medal was meant to show the valuable job that all women pilots were doing for their country.

After the war she married Jack London, Jr., and had two daughters, Terry and Kristy. She served on the 99 executive board for nine years and was executive director of the Powder Puff Derby for 15 years. She either raced as a contestant or served as a board member for 20 years.

She went to work in sales for the Piper dealer in Long Beach in 1966, and later formed her own company for aircraft sales in 1970. After the death of her partner, her youngest daughter Kristy came to work with her and got all of her pilot ratings and became the company pilot. Terry was hired by Western Airlines in 1975 and now is a pilot for Delta since their merger with Delta and Western.

NANCY LOOMIS, started to fly in October 1991 when her significant other who is a pilot arranged an introductory flight lesson for her to see if she would like flying in small planes. She was quickly hooked and started taking lessons on her own and received her private pilot license in June 1994. During that time she flew Aeronca Champs, a J3 Cub, Cessna 150s and 152s and finished up in Piper Cherokee 140s. She is now the proud owner of a Cessna 150.

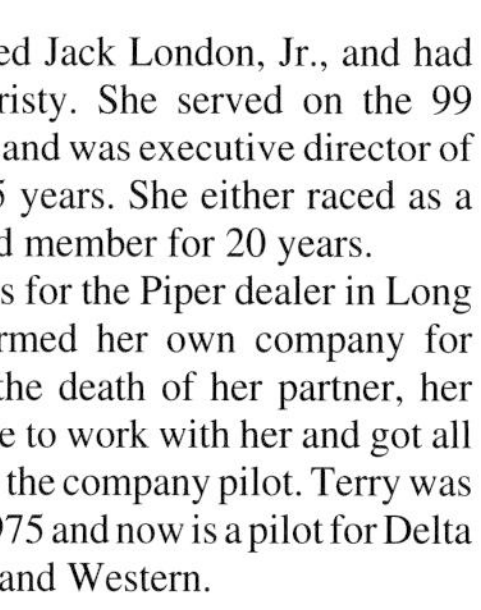

Nancy is a member of the Austin, TX, Chapter of the 99s, the Aircraft Owners and Pilots Assoc., and the Experimental Aircraft Assoc. She hopes to eventually build her own experimental aircraft, but is happy flying N3214X for now.

Nancy flies with her significant other, Matt Lawrence, and works as a certified nurse-midwife in Texas.

SUSAN LORICCHIO, of Jersey City, NJ, became intrigued with flying at age 8. While at her parents' summer home near the seashore, a "bright yellow airplane" would ritually fly low and rock its wings to entertain her. One lazy summer afternoon in 1981, she got her first prop plane ride – a 1929 Standard. There was no cure for this fever. Within three weeks her flying lessons were underway, and she obtained her private pilot certificate flying mostly Cessnas. Barnstorming in biplanes is still her first love, though she's even managed to log time in the Goodyear Blimp N1A!

Besides the 99s, Loricchio's aviation activities include: the Experimental Aircraft Assoc.'s Young Eagles Program (coordinator) and AOPA's Project Pilot (mentor). She's co-produced radio and cable television programs, and coordinated with the FAA and EAA, "hands-on" exhibits at the renowned Liberty Science Center.

In Jersey City, Loricchio served as historic preservation commissioner and is music director in her church. She is a graduate of the Hartt School of Music in Connecticut.

BETTY MCMILLEN LOUFEK, joined the Los Angeles Chapter in 1944. She raced in the 1948 Powder Puff Derby with twin sister Claire Walters and again with her in two Palms to Pines in the 1980s.

A 1951-56 PPD board member, she was publicity chairman, 1954-56. She developed a national 99 publicity network, making the PPD a nationally-recognized event. She created the official program format, still used by others. She persuaded the 1955 board to allow famous cartoonist-pilot Bill Mauldin to race so he could write the story for *Sports Illustrated.*

She flew small planes for pleasure, and a sailplane in which she established official US women's records in 1948-50. She was the first American woman to fly the standing wave.

Today a Palms Chapter member, she was Palms to Pines Air Race publicist for 11 years. She has two daughters, two sons, and four grandchildren. She holds a BA in anthropology and journalism. A world traveler, she writes articles illustrated with her own photographs.

RUTH LOVE, member of Lake Erie Chapter 99s since 1962. When she took her first airplane ride with the barnstormer Lindbergh, she didn't dream that her future would include flying. She was too young to flirt with the two pilots. She jumped up and down with joy to see a plane on the ground. Noone had $2 for a ride at Castleberry, AL. The pilots agreed to give her friend and her a ride for a batch of fudge candy. According to the book *WE*, the pilots were on their way to Mobile, then to Texas for flying lessons. At that time she didn't expect to get a pilot license and belong to the 99s.

Marry Merrill, who had a desire to fly and get to fly in a single-engine plane in all 50 states and 26 territories.

That her son Ronald would become an Air Force pilot and jet instructor. He would fly 13 trips in a C-141 to bring soldiers home from Vietnam. That he would be chosen to represent the US at Supreme Headquarters Allied Powers of Europe (SHAPE). That he would retired as colonel, after 26 years in the Air Force with 5,000 hours of flying, including the C5.

At the retirement dinner, as vice commander of Dover AFB, the room was filled with accomplished C5 pilots. She thought of her first ride with the open cockpit and the hair blowing in the wind. When the plane banked, it threw her against the side of the plane. She sat wondering "How did this come to pass?" Now, she's taking instructions to get wings of a Cherub. When she crosses over she would like to look like Raphael's painting of the Cherub on the postal Love stamp.

GAYLE CONKLIN LOWE, first soloed in 1983, and earned her private pilot certificate in 1984. On the day she took her private checkride she joined the 99s. Since that time she has actively participated in such air racing events as the Kentucky Air Derby and the 1993 Air Race Classic. She continued her flying aspirations and completed commercial and instrument ratings, ground instructor advanced instruments certificate. Currently she is working on her certified flight instructor certificate with funds she received from an Amelia Earhart Scholarship.

Gayle is employed as a regional quality nurse for the mid-west district ambulatory surgical division of Columbia/HCA Health Care. She is married to David M. Lowe who owns and operates Lowe Aviation at KY 80 Airport in Sacramento, KY. David is also president of the International Cessna 120/140 Assoc.

When not flying Gayle enjoys underwater photography, teaches scuba diving as a certified master scuba instructor-NAUI and is in the process of rebuilding a 1957 Cessna 150 (with her husband's assistance) for the purpose of flight instruction.

PATRICIA LOWERS, joined the SCV99s in 1989 as soon as she got her license, and has served as secretary, vice chairman and now chairman. Pat loves flying her Cessna 172 and attends many fly-ins. She is a skydiver, with over 400 sport parachute jumps, and has logged more than 700 hours. A single parent, her son is an accomplished horseman - having won many ribbons for show jumping. Pat is a computer engineer for Hewlett-Packard Company and is currently managing editor of a technical journal. Pat has received the Chapter Pilot of the Year and Service Award. She has flown the Palms to Pines Air Race five times, taking tenth place in 1993. Pat enjoys skiing, roller blading, tap dancing and softball.

ESTHER POWELL LOWRY, thirty years ago went on her first plane ride on a Sunday afternoon and had her first lesson on Monday morning. She soloed in a Piper J3 Cub. She earned her private, commercial, instrument and instructor ratings the first year of her flying. She has logged better than 6,000 hours.

Esther has raced and placed in the Powder Puff Derby, raced in several Angel Derby, winning in 1976. Placed several years in the Air Race Classic. Esther was one of the organizers and original directors of the Air Race Classic and has served as president for three years. At present she is chairman of the ways and means committee.

She is a member and has served as president of the Deep South 99s. Has held International chairmanship for the airmarking committee. Has been a safety counselor for seven years. Member of the Presidents Women Advisory Committee on aviation during Nixon Administration and a member of the Citizens Advisory Committee on Aviation during Carter's administration.

She owned and operated her own Holiday Inn, served on the board of directors of the International Assoc. of Holiday Inns and on most of the committees. She is one of the original organizers of a national bank and member of board of directors. She has consulted on hotels in Cayman Island, Jamaica, and Equidor. At the present time she is CEO of the timeshares of Sky Valley. Married to Carl Lowry. They have seven children, seven grandchildren and four great-grandchildren.

LINDA LEE LUCHS, born April 14, in Philadelphia, PA. Joined the 99s in 1980. Flight training includes Summit Airpark, North Philadelphia Airport, Oceana. She has flown the PA128-140; Cessna 150, 152, 177RG, 172; PA28-161; PA38-112; PA36; FRASCA; F-14A. Married Joseph H. Fisher on Oct. 25, 1975. She has one son, J. Lee Fisher, age 15.

Eight months pregnant when she obtained her ticket. She had been told by a male pilot that the first passenger took up should be a nonpilot, so of course she decided that her darling husband Joe would be her first passenger. As she waddled around the Cessna for the preflight inspection, her darling rested on the plane and practically tipped it over. Joe was not impressed that they were going flying in a "toy." After the outside preflight inspection, they were seated for the pre-start preflight inside the plane and since she had only flown with instructors who knew to move their knees, banged the yoke right into her darling's knees. The flight itself went fine.

Fifteen years have passed since that flight. Last summer she attended her first International Convention. She was able to go to Oceana Naval Air Station and while there "completed aerobatics with superb control of Aircraft" as recorded for her log book by the deck officer.

As proud as she is that she was able to become a pilot, she is honored to belong to the 99s.

SHIRLEY W. LUDINGTON, after husband Ramsey purchased a Cessna 172, decided she should have a few lessons. Well, she fell in love with flying and obtained her private license on March 23, 1973. She joined the Central New York Chapter of the 99s in June of 1974 at the first chapter meeting following the chapter's chartering.

An active member, she served the chapter on various committees and as chapter treasurer, vice chairman, and chairman. She also served New York - New Jersey Section as a board member, treasurer, vice governor and governor. She served on the first Council of Governors. Shirley earned her instrument rating after being 'weathered in' on a college information trip with her daughter. She chaired two section meetings and worked on the 1979 and 1989 International Conventions. The section entered her in the Forest of Friendship June 1992.

Shirley has not been able to be very active since 1991 when she was diagnosed with Multiple Myeloma - bone marrow cancer, but she still gets to several chapter meetings a year.

REBA J. LUDLOW, began her flying hobby late in life. After having a friend fly her over Mt. Hood in Portland, OR, in 1969, she was definitely bitten by the "Aviation Bug." However, it took another 16 years for her to begin training. Of all things, for a 40-year-old, she wanted to do aerobatics! Lessons began in St. Augustine, FL, in a Citabria. Never having been exposed to any facet of aviation, flying the front seat of the Citabria was exhilarating, but it was also expensive for a single mother. She persevered, however, acquiring her private pilot license in 1990. The "love" was only beginning. In 1994 she acquired her instrument rating along with a partnership in a Cessna 172. Undaunted, she is now a partner in a Cherokee 180 and working on her commercial rating. She can be found flying around the Florida and Carolina skies with her husband, Dick, scouting out great places to eat!

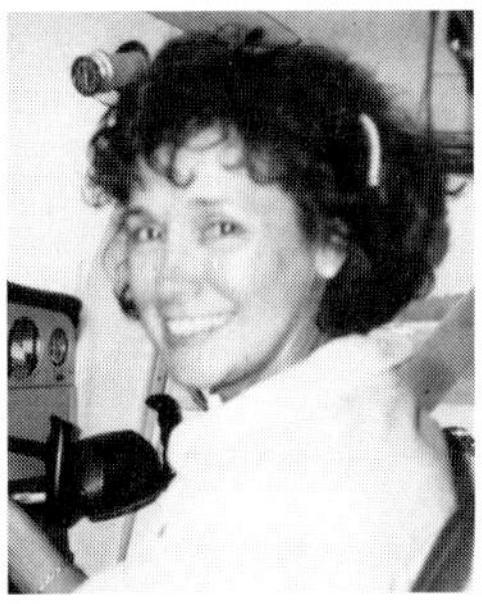

Born Oct. 14, 1944, joined the 99s in 1993. Flight training included Craig Field, Jacksonville, FL, North Florida Flight Center. She has flown Cessna and Piper.

FLORENCE MARIE LUENINGHOENER, life member of the Nebraska Chapter, earned her private pilot license in a J3 Cub at the Naha AB Flying Club on Okinawa in 1961, where she was on sabbatical, teaching dependent children. The island was so small that she had to hop a ride in the belly of a C-130 to Tokyo to take the written exam and came back to Lincoln with the restriction that she could carry passengers only within 40 miles of Naha because she couldn't fulfill the cross-country requirement there.

In 1963 Florence obtained her commercial license in order to fly in the Powder Puff Derby that year. The Bakersfield, CA – Atlantic City, NJ, flight fulfilled a dream held years before she was ever in an airplane. Her first flight was actually her first lesson.

Born in Avard, OK, June 1, 1916, Florence spent more than 45 years as a schoolteacher, mainly of science and aerospace. She is now widowed but has two sons, a grandson, stepson, stepdaughter and granddaughter.

EVELYN C. LUNDSTROM, born Sept. 3, 1918. Joined the 99s in June 1968. Worked as a civilian on instrument repair in Florida during the early 1940s. Started at Palo Alto Airport in September 1957 in a 65 HP Stinson, continued training and flew many types, including Pipers and Cessnas. She has flown Piper 140 through Comanche 260; Cessna 140 through 182. Rallye-Minerva, a French STOL aircraft. She always flew for the pleasure of it and for sharing with many magnificent women (and men) that she met through flight.

Her husband Oscar is deceased. She has two children, Signe A. Lundstrom, born Jan. 11, 1947, and Arlis L. Lundstrom, born Feb. 6, 1958. Signe earned her pilot license when she was 17. She also has grandchildren Eric C. Johnsen, 21; and Megan L. Lundstrom, 4.

She treasured the chance that she had to introduce flight to youngsters. She treasures the special memories of the caring sharing people that she met in her years of flying. Her niece, Laura Staudt is a 99, she's proud to say. She is a life member, a member-at-large, as she is no longer an active pilot. She served in many chapter offices and worked on committees for International Conventions; was treasurer for Convention of 1977 in San Francisco. Flew in two AWTARs, and other races, i.e., Pacific Air Race, Palms to Pines and the wonderful Fun Filled Fiesta for Frugal Flyers.

PAULA LUPINA, born on June 27, 1954, in Highland Park, MI, and now resides in Kalamazoo, MI. She literally jumped into aviation when in 1977 her first flight ended in a parachute jump. The jump was fun but she wanted to be "in the air" longer. In 1986 this lead to her becoming the first woman USUA registered ultralight owner/pilot in the state of Michigan. She flies her Phantom ultralight on long cross country trips and as a competitor in 1993, she finished fourteenth in the US, the only woman listed in the top 20.

Paula became acquainted with the Lake Michigan Chapter in December 1987. While not eligible for 99s membership she still became an active participant in chapter events including 66s chair from December 1990 to June 1993. In September 1991 she started her GA training at the urging of her non-pilot husband. After passing her checkride in June 1992, she became a 99s member. In August 1992 she became a partner in a Cessna 172.

Paula is a member in numerous aviation organizations as well as a member of the General Aviation Committee of the Michigan Aeronautics Commission from its January 1992 inception. She also is a fund raiser for a shelter for homeless families and an active member of a women's professional group.

EDIE LUTHER, First Canadian Chapter. Edie's first flight was in a Harvard in 1944 when she was in the RCAF (Royal Canadian Air Force) Women's Division. The pilot did aerobatics and Edie was airsick. Not a good beginning!

However, in 1968 Betty Greply took her flying again in a Piper Cherokee and she decided to learn to fly herself. Edie soloed later that year and got her wings in February 1969. She also became a grandmother for the first time.

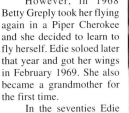

In the seventies Edie and Betty did a lot of pleasure flying, cross-country to different airports in Ontario. Edie added a night endorsement to her license.

Week-ends, on Sunday evenings, Edie was pilot for Gord Craig flying traffic reports in a Cessna 182. Edie logged 96 and a half hours on these assignments, a great confidence and time builder.

In 1978 and 1979 Edie and Betty teamed up together again to fly missions for Operation Skywatch – they were a well-known twosome and contributed a lot of their time to the beginning of Skywatch.

Edie has attended two conventions and flown in several Poker Runs. She has been and is a great supporter of First Canadian Chapter. This flying grandmother now has five grandchildren and works full-time in Toronto.

JOYCE RICHARDSON LYKSETT, born March 23, 1935, in Minnesota. Joined the 99s in 1984. Flight training at Mongham Field, Fort Worth, TX. She has flown Cessnas, Citation, and Bonanza. She has two children Catherine and Elizabeth; grandchildren, Patrick Ian, 4, and Oliver Maxwell Merriman, 6 months. Learned to fly after daughters graduated college. Edna Gardner Whyte at 81 years was first instructor. Learned to fly mid life to fulfill life long ambition. First airplane she saw as child was barn stormer. Decided them to learn to fly. "Jennie" wicker chairs for passenger. Has had aerobatic and helicopter experience.

JOAN LYNUM, BC Coast Chapter, West Canada Section. Her husband learned to fly in the 1970s, fulfilling a dream that had begun when he took instruction on Harvards in the Norwegian Airforce. Once he was licensed he wanted company on his trips. She was a "white-knuckled" passenger, but before a cross Canada trip with their three children in 1980, he convinced her to take some lessons. Despite having a 152 door fly open on her first lesson, she hung in reluctantly, until it all amazingly began to make sense.

She received her license in November 1981 after four canceled flight test and went on to do the required flying for her commercial, but went back to work and didn't have time to finish. She and her husband have done four trips to the Baja, AK, and all over the US and Canada. Along with her original flight instructor she's flown to New Orleans for Convention. She joined the 99s in August 1982 and has held all chapter offices. She is presently section governor. In her nonflying life she is a social worker, presently on a special project with the government of British Columbia and an enthusiastic grandmother.

MELINDA MITCHELL LYON, born Dec. 27, 1941, in Columbus, OH. Started flying lessons in Danbury, CT, and finished up at ATE in Santa Monica in June 1978. She immediately joined the San Fernando Valley Chapter of the 99s and has been a consistently active member, serving on several committees and participating in many events. Her major interest has been the scholarship programs and she has chaired both the Future Women Pilots Program and the SFV Chapter/JHMAR Career Scholarship Program. She has been on the board twice as chapter vice chairman and once as corresponding secretary.

Melinda has flown two of her own planes in many all women's air races. The longest race was the 1978 Air Race Classic from Las Vegas to Fort Destin, FL, in her Cherokee 140. The most fun was the Baja Air Race in 1982 when she upgraded to a Cessna 182.

Melinda and her 49 1/2, John, have owned several planes including a Cessna 172, a Mooney M20F, a PA28-140, a Cessna 182, a Meyers OTW (biplane) and are in the process of restoring their Meyers 200.

She holds a BA in sociology/anthropology and an MS in early childhood education. She has worked in the business community as well as in education. Melinda is currently working on her Ph.D. in Gerontology at UCLA and teaching her 11-year-old son, Carl to fly.

KATE MACARIO, in 1942 she couldn't wait to finish high school and get a job so she could start flying lessons. It took several years traveling a distance to airports, once a week, sometime, getting only a half hour in a J3, to finally obtain her license and buy an Aeronca Champ. In 1950 she married and subsequently had seven children, but continued being involved in aviation. As a family, they restored at least a dozen wrecked aircraft, flying them briefly, selling them, and starting all over again. They ran the Macario Airport a 3,500 turf field for five years, until the property was sold. She became a 99 in 1955 and for 40 years has remained an active member holding nearly every office, running Pennies-a-

Pound rides every year and Lunch with the Bunch Fly-Ins every month. Her fondest memories are her participation in the Powder Puff Derby Air Races. She worked at many Termini, as chairman and on impound and inspection, and more recently, at several Air Race Classics.

R. CANIVET BRICK MACARIO, first flights were "BB" (before birth). At 7 1/2 months her mother and she flew the Montreal to Miami AWIAR (All-Woman International Air Race) in a Ryan Navion, mother, Kay Brick, as pilot of course.

"PB" (post birth) at two weeks her father Frank and mother flew her in their Aeronca Chief from Teterboro, NJ, around the Statue of Liberty. Thereafter she was settled into the little piggy-back carrier of the Chief, surrounded by luggage, only her head showed.

She developed some navigation skills in their Fairchild 24. Since then she received her license and joined the Eastern Penn 99s Chapter. Some of her experiences include flying the F-25 Bonanza, Cessna 150 and 195, Lear 35, Piper PA31 P Navajo, Ryan ST 3KR and Boeing Stearman.

She is married to a corporate/instructor pilot, Michael; their daughter, Danielle, is also learning to fly as it must be in the genes. It should be as her four grandparents are pilots and both grandmothers are 99s (Kate and Kay).

SHIRLEY MACDOUGALL and her husband, Neil, started flying lessons at the Oshawa (Ontario) Flying Club on the same day, after reading Frank Kingston Smith's *Flights of Fancy*. Her lessons were interrupted when she became pregnant, because "You can't take a passenger on a student permit." She soloed a Tri-Champ and received her private license in 1965.

With her husband, she has flown across Canada from coast-to-coast and from Mexico to Alaska. As a team, they've taken part in the Bahamas Flying Treasure Hunt, the Wyoming Air Tour and many rallies. They came third in the prestigious Governor General's National Air Rally.

After flying with Civil Air Rescue in Canada, they joined the Brownsville, TX, squadron of the Civil Air Patrol in 1992. They have flown as pilot and observer on many special missions as well as on practice searches.

In 1994 they flew a CAP crew to Los Angeles in a Cessna 182RG, where they joined 2,000 other CAP volunteers doing earthquake relief. As squadron historian, she is trying to trace members who flew anti-submarine patrols from Brownsville in 1942-43.

Shirley joined the 99s in 1966. She was the first 99 appointed to the board of directors of the Canadian Owners' and Pilots Assoc., registration chairman of the International Convention in Toronto and, from 1980-82, chairman of the First Canadian Chapter in Toronto.

JOAN MACE, Professor Emerita retired first woman chairman of Aviation Department at Ohio University began flying in 1943. She was a CFIA-IME, ATP and FAA designated examiner.

She joined the WASP in 1944, however, they were disbanded before her class began. After receiving her CFI at OU, she flew for the GI Bill. There were 22 CFIs, with Joan being the only female.

In 1948, she married and had three sons. College was started in 1968. She graduated summa cum laude, along with her oldest son.

During her aviation career, she was: Top Female Pilot of the Year, Chief Check Pilot for the CAP, Second Place PPD and the Laursen Award from UAA. In 1988, she established the BA in aviation, brought Alpha Eta Rho and United Airlines Internship program to OU. In October 1992, Joan was given the OU Alumni Medal of Merit Award for outstanding accomplishments in aviation education.

Born June 7, 1924, in Columbus, OH. Flight training at Columbus, OH, and Ohio University, Athens, OH. She has flown Cubs; Interstate Cadet; Apache; Aztec; King Air; PT19; Aeronca; Cessna 152, 172, 210; Baron B-55; and Bonanza. She is a widow and has three children: Mark, 38; Pat, 37; and Mike, 36. She also has two grandchildren, Brian, 5, and Randy, 11. Joined the 99s in 1946.

DIANE TITTERINGTON-MACHADO, grew up to the sound of sonic booms in Dayton, OH. Her father was a wind tunnel engineer at WPAFB. She skipped high school once – to go to the Air Force Museum. As a registered dental assistant, she wired patient with braces for six years. Diane learned to fly in 1973 and has a commercial instrument license. Renting, ferrying and borrowing various aircraft, her log book is a Heinz of makes and models, from a J3 Cub to a BE90. Her flying includes ferrying aircraft from the factory, flying fire patrol and a number of air races. A radar qualified air traffic controller at Houston Center until 1981, she enjoyed familiarization rides which included being a passenger on a carrier landing and catapult takeoff, dozens of airline jump seat rides, flying a T38 and flying the Goodyear Blimp. She was the Professional Women Controllers Convention chairman in 1981.

Charter chairman for the Houston North 99s, Diane has held all chapter offices except secretary (she hates taking notes) and has also been a member of the Houston and Orange County chapters. She is a member of the EAA, AOPA, NATA, ICAS and IGAB and for two years was the chairman of Leadership and Ethics for the International Group of Agencies and Bureaus.

In 1988 she acquired The Aviation Speakers Bureau. Diane helps organize hundreds of safety seminars and places aviation speakers, celebrities and specialists at events in North America and Europe. She is an aviation marketing specialist and aviation editor.

Her husband, Rod Machado, is a CFII/ATP, a humorist, aviation educator and author.

CELESTE MACIVOR, began flight training after high school. Her first trip in an airliner encouraged her to apply for stewardess job. She flew for Colonial Airlines and later Eastern continuing flying lessons. Transferred to Miami, married husband Jim who later became Eastern pilot.

After children were grown she finally received her private license. Celeste competed all ratings including Commercial SMEL Multi and instrument as well as commercial glider rating.

Celeste has a business degree from Barry University, Miami and has been a member of Goldcoast Chapters since 1978.

Born in Honesdale, PA. Joined the 99s in 1978. She has flown the Piper Cub, PA11, Cessna 150 and 172; P28-140, P28, 151, P23-250, P28-180, DC-3, SG's 2-33A. Married to James and has four children: Sharon, James P., Michael J. (deceased), and Timothy. Grandchildren, Bridgette and Blake, 15 months and 3 1/2.

DR. BARBARA MACLEOD, Maya archaeologist and epigrapher, cave and jungle explorer, singer-songwriter, former fearful nonflier, now a born-again aviation zealot and infinitely happier human being, making up for lost time. She got her private certificate in July 1994, and jumped right into aerobatics, which she pursues with fervor, chutzpah, and any money she can squirrel away. She's been pushing the local long spin frontier, having done 52 consecutive rotations (an unofficial C152 world record) in an El Nino winter in Texas! She's now working on her tailwheel endorsement, and hopes to find a way into competition aerobatics in the near future. New 99, joined in 1995.

MYRA MACLURE, BC Coast Chapter, joined the 99s in 1981. She started flying 18 years ago when her kids finally "grew up." She has been interested in flying ever since the RCAF did a barnstorming session in her hometown of Swan River, Manitoba. The rides were $5, but she couldn't afford it – that was enough money to rent a house for a month. Myra's 50th birthday present from her husband, Lorne, was flying lessons.

Her flight training was done in Abbotsford, BC, and Edmonton, Alberta when Lorne was transferred. Myra completed her night rating and her commercial license.

Myra has traveled twice around the world and has lived in Pakistan. She is a strong supporter of the chapter's annual safety seminar.

When Myra flies herself, Lorne navigates using his experience as an RCAF Bombardier.

CYNTHIA S. MADSEN, born in Chicago. She has been married to Ralph for 28 years, certified dental technician (DCT). They have one child, Craig Madsen.

Earned private pilot license Easter Sunday, 1985, from Howell Airport, Crestwood, IL; got instrument rating May 1989. Used new instrument rating to fly to 99s International Convention in New York in 1989, did first hard approach of 300 and 2.

Joined Chicago Area Chapter 99s in May 1986. She was chairman of the chapter from 1992-94. Other chapter work was as secretary; way and means chairman for over three years; chairman of legislation, airmarking. Participated in many activities - organized an ice cream social at Howell Airport in New Lenox, IL, for hundreds; participated in the air demonstration marking the 50th anniversary of the Battle of Midway at Midway Airport in Chicago; participated in the 99s pre-airshow flyby at DuPage Airport in Illinois for three years.

She is now chairman of Flying Activities for the North Central Section.

She has won the Illi-Nines Air Derby twice, and in 1994 in the Keokuk, Iowa Air Derby predicted fuel consumption for the race exactly - zero variance.

She owns a Piper Cherokee 140 (*Tango*) 150 hp which she bought August 1985. She designed her paint scheme herself.

Her husband learned to fly in 1989, and has been (and still is) a great navigator and copilot in all air races. He has won awards for his activities and support as a 49 1/2. They flew Tango from Chicago to Las Vegas for the 99s International Convention, and as of this writing they're planning to fly her to Halifax, Nova Scotia.

She is active with the EAA Young Eagles program and has flown a number of children out of Meigs Field at the Chicago Lakefront.

NELL SELLERS MAGOUYRK, from early childhood was always fascinated with airplanes, especially Stearman biplanes, vintage "Warbirds" and gliders. May 1992, her dream finally came true at age 47. That same month, she proudly became a member of Dallas Redbird Chapter of 99s, private pilot ASEL having flown Cessnas 150, 152 and 172. Offices held: aerospace education chairman, membership chairman. Education: BS business administration/economics, paralegal certificate. Memberships: EAA, AOPA, TAAT, WINGS, AAUW.

Current activities: ILT CAP (testing officer, Cadet Aerospace Education Program), Frontiers of Flight Museum/Airshow, AAUW presenter Expanding Your Horizons in Science/Math, National Congress in Aviation/Aerospace Education Aerospace Program/Christmas Party Amelia Earhart Learning Center, designs/creates aviation jewelry (Plane Crazy Creations).

Favorite aviation experienced: EAA Oshkosh 1992 meeting Tuskegee Airmen, ride in 1931 Stinson Trimotor, Gee Bee R-2 Performance; Aerospace Symposium USAF Academy (Nephew: senior cadet); USAF Museum; NASA (Langley AFB).

She is married to Don and has a daughter Donna, 28.

SANDRA LYNN MAHONEY, born Feb. 11, 1958, in Upland, CA. Joined the 99s in 1978. She is a private pilot and has flown Comanche 250, Cessna Cardinal, Cessna 140.

She has one child, Darius Ryan Hutchison. She won the Palms to Pines Race with her mother, Bev Mahoney – Epstein in 1976 – competed the year before and placed eighth. Competed in the Angel Derby as a team with her mother in 1978.

PAMELA HARTER MAIRS, born March 11, 1961, in Akron, OH. Currently a Michigan resident, Pamela always had a fascination with aviation along with anxiety about actual flight. This was the precursor that led to flying lessons. She earned a private license July 22, 1994. Pamela is a psychologist with a MA in counseling psychology from Western Michigan University. She is a member of the 99s, Inc., the Lake Michigan Chapter, Aviatrix and the secretary to the local pilot's club in Three Rivers where she earned her rating.

SUE MALADY, lives in Sparta, NJ, with her two grown children. She completed her basic training at the Andover Flight Academy in Andover, NJ, in 1993. Shortly after receiving her private pilot she purchased and still owns her 1980 Cessna 172 Skyhawk and has recently added to her growing fleet with the purchase of a 1962 Cessna 210 Centurion.

Since receiving her certificate she has added to it the rating of multi-engine land and is currently working on her instrument and rotorcraft helicopter.

When she's not spending time at the airfields with her planes or flying off on some trip, she stays busy working full-time as the director of nursing at the Dover campus of the Northwest Covenant Medical Center. She hopes to own her own private airstrip one day.

JILL R. MALCOLM, born Nov. 10, 1965, in Baltimore, MD. Joined the 99s in 1990. Flight training included private pilot license. She has flown Cessna 152 and 172, and Piper Cub.

Married to John and has one child Peter Devon. She began flying in 1988 to pursue a dream. Flying is now an endless adventure through which she's enjoyed many hours aloft and met so many wonderful people. The 99s have been a support group and friends.

JOYCE MALKMES, Long Island Chapter, New York - New Jersey Section. At age 55, time for a change, learn to fly! The dreams ceased and became a reality after her first demo ride at Brookhaven Airport on Long Island. Two years later she received her SEL ticket in 1980. For the first year of training she kept it a secret from her husband and mother by telling them she was picking up plants for their flower shop and taking a course at a community college.

She was the first 66 in their Long Island Chapter to follow through and become a 99. She flew as navigator in several proficiency air races in Washington Capital district, in which they came in first place. She has participated in airmarking, Pennies-a-Pound, Corporate Angel Network, Ida Van Smith flying clubs (introducing young children to aviation), spot landing contests and poker runs. She was also president of Aerocats and the Grasshoppers.

Attending every International Convention since Boston, has attended northwest section meetings in Homer, AK; Bozeman, MT; and Fairbanks, AK. She has also flown to Oshkosh and Sun and Fun in Florida and attended the Forest of Friendship in Atchison, KS. A real thrill was attending the New Zealand Australian air rally in Christ Church, New Zealand several years ago.

As a member of the 99s she's had the opportunity to meet such people as Jessie Woods, the wing walker, Edna Gardner White and Amelia Earhart's sister. At age 70 she is still flying her Cessna 172, playing tennis and golf.

PAULINE L. MALLARY, Fairburn, GA, hitchhiked 10 miles to take her first flying lesson at Ross Field Airport, Benton Harbor, MI, in 1952. Four years later, after attaining her private license, she learned to drive a car. She has accumulated 4,300 plus flying hours, commercial, flight instructor, sea plane and instrument ratings and has owned four airplanes.

Her air racing career began in 1958. During the past 36 years she has competed in 63 air races all over the US, Canada, Bahamas and out islands, Mexico, and Central America to Nicaragua, placing consistently in the top 10. As pilot-in-command she won 10 air races, including the 1977 Angel Derby.

From 1960 until 1966 Pauline was employed at South Haven, MI, as a flight instructor, charter pilot and office manager for South Haven Flying Service. She also ferried pilots, parts and chemicals to dusting locations for Bob Mueller Crop Dusting Service.

In 1966 Pauline married Eastern Airlines pilot L. Peter Mallary and has resided in the Atlanta area since 1969. Daughter Tracy was born in 1970.

Active in the 99s, she became a charter member of the North Georgia Chapter, served as chairman in 1971 and 1972, and was the first woman appointed by the Atlanta GADO as a volunteer Accident Prevention Counselor for the state of Georgia, serving four years in this position.

In 1988 she became a Memory Lane Honoree in the International Forest of Friendship in Atchison, KS.

A member of the Silver Wings Fraternity since 1978, Pauline was presented the Veda Dyer Williams Memorial Award in 1989. She is presently serving on the board of directors for the Georgia Chapter of the Silver Wings Fraternity. Her other memberships include AOPA, Atlanta Flying Rebels, and all racing organizations.

During air racing season Pauline is a consultant for Patton Aviation Weather Service in Atlanta.

TAYA DUNN-MANGNALL, born Oct. 13, 1942, in Richmond, VA, and moved to northern California in 1946. Trained in Citabria 7ECAs at Oakland Municipal Airport and in Cessna 150s on St. Thomas in the US Virgin Islands. She flew to Puerto Rico to do her cross country training from St. Thomas.

Planes flown: Citabria 115s and 150s and Cessna 150s and 152s. Had the opportunity to fly a DC3 briefly in Caribbean Islands and a Constellation over Peru and Columbia. Licensed in 1971. Joined 99s in 1976.

In 1981 she married Steve Mangnall, who is currently working as a data processing business analyst. She taught elementary school for several years, dispatched at a flight school for a couple of years, and has worked as a data analyst for the past 22 years.

Most of all, she has enjoyed all the fantastic people she has met and the events that she has participated in since becoming a member of the Bay Cities Chapter of the 99s.

KRISTEN MANSEL, born July 18, 1973, in Glendale, CA. Joined the 99s in 1991. Flight training included PPL, instrument rating, commercial rating, multi-commercial and IFR. She has flown Cherokee, Warrior, Archer, Dakota, Saratoga and Beechcraft Duchess. She has been involved in aviation since the age of 14. Her father and brother were taking lessons and she was always the "tag along sister." Finally, on her 15th birthday, she was given a very special introductory flight. She immediately pursued her private pilot license, receiving this rating two days before high school graduation. Almost immediately, she checked out in a high performance airplane and flew with her family across the country.

She started her instrument rating. To help finish up she received a scholarship from the Sacramento Valley 99 Chapter. Being Buck Scholar allowed her to pursue her commercial single and multi-engine ratings. She is now working on her certified flight instructor rating.

She is involved in the Mt. Diablo Chapter of the 99s for which offices she held: Annual Proficiency Test chairwoman, secretary, and treasurer. She is also involved in Angel Flight. She is a student at San Jose State University where she anticipates a Spring 1996 graduation. She will graduate with a degree in aviation operations. She works as the marketing assistant at a local FBO.

MARGARET MANUEL, in 1941, her husband, Malcolm, an engineer, and she, a social worker, took a private pilot course at Minneapolis International Airport. They accumulated flight time in a flying club and the Civil Air Patrol. In 1946, they sold their home and moved to the Carleton College Airport at Stanton where they remained until 1991 as airport owners and operators. They had a GI bill flight school and ROTC program. In 1949, she became a charter member of Minnesota 99 Chapter and flew to meetings throughout Minnesota on wheels, floats or skis usually accompanied by her young son and daughter. They did airmarking, plane rides, wrote news stories and gave airport tours which became popular at Stanton for school children, parents and teachers. In 1950 she and her copilot Marilyn Henderson flew in a 99 International Air Race from Montreal to West Palm Beach. Weather delays enabled them to go sightseeing in New York City and Washington enroute. Another fun escapade was flying a seaplane to their chapter meeting on Lake Ida. A high wind blew the plane a few miles down the lake. They found it after dark by listening for metal floats banging against rocks. The plane was OK, but four of them got poison ivy. Over the years she has flown in light aircraft to almost all lower 48 states and bordering Canadian provinces, Cuba (before Castro), the Bahamas, Mexico and Baja Peninsula. She is still a 99 and her sons' career is also in aviation.

BETTY LOU MANWARING, raised in Beaver, PA, 25 miles northwest of Pittsburgh. A boyfriend, who had flown in WWII, took her for her first plane ride in 1948. He did an inside loop and a spin which were very exciting, but when he wanted to fly under the town bridge, she thought that might be a little too exciting. She didn't fly in a small plane again until her husband learned to fly in 1968 when they lived in Salt Lake City.

In 1973 they moved to El Paso, TX, where they bought their first plane, a 1958 Skylane. Two years later her husband returned to Salt Lake City to start an engineering company while she stayed behind to sell the house. Seeing the need to know how to land, she began taking lessons at Sunland Airport getting her license in 1976. Her husband was not aware of this until one week-end when he came to El Paso, she presented him with her solo certificate and her tailess shirt. This is the only time she has ever seen her husband speechless.

In 1976, she joined her husband in Salt Lake and shortly thereafter, they bought their second plane, a 1971 Skylane. A few years later, she joined the 99s in which she has held various offices. For her husband's company, she is an officer, the accountant, Girl Friday, and flies their present plane, a Cessna T-210, throughout the US.

LYNN ADKINS MARKERT, Women With Wings Chapter, grew up in Ohio and, after studying and working in Peru and Brazil for four years, she returned home where she spent her South American savings on flying lessons. She

learned to fly at Clever field in a Cessna 120 that had no electrical system; practice maneuvers usually were held over the strip mine moonscape south of New Philadelphia. She received her license in August 1959 and joined a C-140 flying club, flying for about a year. Then ... a very long break until last November when she began taking instruction in a C-172 at Akron-Canton (CAK) and studying to get current. She also joined the Women With Wings Chapter of the 99s.

During her non-flying years she married and had two children, Brooke Molina, 33, whose son, Sam, is two, and Nelson Guda, 31. She went back to school, ending up with BS and MS degrees in chemical and environmental engineering; worked in jobs ranging from textbook writer to environmental engineer; and was married a second time to George, a fellow chemical/environmental engineer who has a strong interest in flight and aircraft history. They both retired a year ago from the industrial environmental business and bought a trailer to better explore the environment of the outdoors. They plan to do some of the exploring by air.

She also plans to add other ratings to her current SEL, starting with instrument training, and to incorporate flying experiences into a novel she is writing.

JACQUELINE HITE MARSH, born Oct. 21, 1929, in Roanoke, VA. Joined the 99s in 1981. Flight training included ASEL and instrument rating. She has flown Cessna 152, 172 and 182, Cardinal RG, Turbo Piper Arrow, Piper Warrior, Champ, Luscombe. Married to James and has two children, Mark L. Marsh, 45, and Bonnie E. Pruett, 42. Her grandchildren: Marsh, 16; Erica, 15; Jennifer, 13; and Jonathan, 8.

Went to China with 99s, flew her Warrior from St. Augustine to Anchorage, AK, and back. Has MA in music education. Taught in public and private schools. Organist and/or in Lutheran church's throughout southeast. Active in ABWA, hospital auxiliary, toured Europe with north Florida Women's Chorale, active in local and section 99s.

MARGARET E. MARSHALL, PHD, Phoenix 99s, is a native Arizonan, born in Phoenix. She is a psychologist who got her private pilot license at Scottsdale in 1985 and joined the Phoenix 99s the same year. She added her instrument rating in 1987. She has flown quite a bit in Arizona and New Mexico, and really enjoyed a charity flight delivering Christmas toys to Mexico.

Her solo professional practice has kept her too busy to do as much in aviation as she would like, but she really enjoys the 99s and had her biggest flying thrills participating in Kachina Air Rallies. She and Bill Preece won second place and the spot landing contest in 1986. In 1987 they came in first. Barry Goldwater had donated the trophies and was there for the presentation. She also enjoys teaching in the Flying Without Fear classes.

Margaret has two grown children, a daughter who is a critical care nurse, and a son who works at Allied Signal in Phoenix after getting his A&P from Spartan in Tulsa, OK, and his BS in aerospace from Embry Riddle in Prescott, AZ. He planned to interest his mother in flying by taking her for a ride when he got his private, but went into the service too soon to get it done. She promised him to go out and take a trial ride - she loved it and got her private license.

TOODIE MARSHALL, private pilot in 1956 at Livermore, CA. Received glider, ASEL (1971). Joined the 99s in March 1994, Lake Tahoe Chapter. She is a realtor associate with Tahoe Resort Properties. Has been Pacific Bell service rep, secretary for Royal Research. She has flown gliders: Schliecher AS-K 13, Hummingbird motorglider, TG-3, 1-26, Carbon Hornet, MG-23SL, AS K-21, Grob 2 place, owns a Standard Cirrus - 13D; Powered: C-150, C-172, Turbo 206, C-180, J3 Cub, Super Cub, Scout, briefly owned a Christen Eagle.

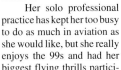

Married to Bob, who is also a glider and power pilot. Both her father and brother flew. Her dad designed the Hummingbird Motorglider in 1953. She has two daughters and two step-daughters.

Marshall teaches an ancient form of meditation called reiki.

Her memberships include Truckee Tahoe Plane Talkers, soaring Society of America, Airsailing Glider Port in Nevada and Pasco, Mercedes Benz Club of America and the 180-185 International Club, active in 4-H in Livermore, 99s (one year).

Towing gliders was thrilling and challenging. Since she was the only female, it was fun to compete with the men! And she could hold her own in fact, she could get back on the ground after a tow faster than some of the fellows and without abusing the airplane, plus she knew where the sink was. On top of that her dad was the chief tow pilot and abusing the Super Cub was a real no no! The critics, late in the day, would then sit next to the hangar holding large cards (with numbers on them) up in judgment whenever a tow pilot made a landing. Great fun, but you had to have a good sense of humor!

CAROLYN G. MARTELL, from Caldwell, ID, received her private pilot certificate in 1983 and completed her instrument rating in 1992 with a 1990 Amelia Earhart Memorial Scholarship Award. Most of her flight hours have been in a Cessna T206U.

Carolyn is a member of the Idaho Chapter of the 99s and has served as chairman and co-chairman. She has been the oral historian for the Northwest Section since 1985 and is a member of International Resource Center Oral Histories Committee.

Carolyn is president of Aviators Historical Foundation and is active in collecting oral histories of pioneer aviators. She is the president of Martell Farms, Inc. and is involved in agriculture associations.

Her participation in fund raising events has resulted in one of her designer quilts being donated for display at the 99s Headquarters building. Another of her commissioned designs is displayed on the Boise State University Campus.

Carolyn has two sons who are both pilots.

BETTY J. MARTIN, born Dec. 2, 1929, in the Southern Illinois countryside near Dongola, IL, where she graduated from a one room country school and the Dongola High School. She received a BS summa cum laude in nursing from Northern Illinois University in January 1974, after putting her husband through the School of Engineering at the University of Illinois. Her nursing career included hospital, public health and home care nursing from which she retired in December 1994.

Betty received her private pilot license in September 1983. She began flying lessons at the Elgin Airport. When Elgin closed she finished at DuPage which is near St. Charles, IL. In 1985 she joined the 99s and in 1986, the International Flying Nurses Assoc.

She is married to Jim Martin who is a commercial pilot and an electrical engineer.

DOROTHY ALICE MARTIN, born in Evanston, IL, and now resides in Lawrence, KS. She started flight training July 5, 1986, and soloed Oct. 9, 1986. She completed her private pilot check-ride May 27, 1987. She joined the Northeast Kansas 99s in 1989.

Her flying experience has been in the AA1A American Trainer, AA5B Grumman American Tiger, and BE 35 Beech Bonanza. She is a RN CNOR currently working in the local hospital operating room.

Her husband, Dick, has been a pilot since 1961. Daughter Cindy is a RN in an ICU, son Greg is a computer science major at Kansas University and son Glenn is an electrical engineer in the US Air Force.

EILEEN MARTIN, born in Brooklyn, NY, and grew up in Dalton, MA. An elementary teacher in Pittsfield for 28 years, she holds a BA from the Elms College, an MA in education leadership from North Adams State College and a CAGS in education psych. from American International College.

Her experience includes carpentry, dog-grooming, dog obedience instruction and years of work with 4-H youth organizations in the dairy goat department. A member of the American Dairy Goat Assoc., she has operated a dairy goat farm in Richmond, MA, with her partner, Barbara Finn, for the past 20 years.

After years of skywatching, she learned to fly at the Pittsfield Airport four years ago and joined the 99s a year later. Flying in a rented Cessna 172, she enjoys visiting and dining in the New England area.

GISELE ELISABETH MARTIN, President, G.E. Martin & Assoc., Inc., Certified Educational Institution, private adult coaching. Canadian private pilot license No. 163311 obtained at 48 years of age, on the seventh day of the seventh month, 1977. Member of the 99s, Inc., First Canadian Chapter, since the year 1979. Flown in private Cessna 182 from Ontario to British Columbia, California, and Florida in North America; to Costa Rica, Belize and Honduras in Central America; as well as to many other destinations in Canada and the USA.

JUDY ANN MASON, born Dec. 13, 1938, in Enid, OK. She has flown PA18, Cessna 140 and 170B, Smith Miniplane, Piper PA12 Super Cruiser. Married to David and has two children, Kathy Ann Cundieff, 29, and Sandy Ann Neumann, 26. Grandchildren: Katie, 9 1/2; Kimberly, 5; Clint, 3, Cundieff; Jennifer, 7; and Starla, 8 months, Neumann.

Her fascination for flying started at a very young age. In the mid to late 40s her parents took her younger sister and her to a flying circus; what a wonderful experience that was. She remembers the old bi-planes; of course they were newer then, with the wing walker performing his routine, and the bi-plane with a ladder hanging down under it's fuselage, and a man reaching for the ladder as the bi-plane would make a pass down the runway, climbing and performing on the ladder. The ribbon cutting, and the aerobatics. What a sight to see, and the start of her dream for her future course in life. Her parents would tell her of their seeing Amelia Earhart, Will Rogers and the Spirit of Saint Louis, and she would later in her life realize the significance and privilege they had experienced. If only she had lived during that era of history!

As a child growing up after the war, and not being endowed with the extra finances to take flying lessons, she would spend as much time out at the airports as she could, meeting and talking with pilots about their planes and hearing their stories and adventures. Most of the planes she came in contact with had no electrical starters, and some they even had to take a crank tool and put it in a slot in the left side of the engine cowling and turn it several times; then, the pilot would hit the starter and blue smoke would pour out; if the pilot was lucky, the engine would catch and the old radial engine was music to her ears. She still gets goosebumps today when she hears and sees a radial engine plane. Some day her dream is to own an old antique classic radial engine bi-plane.

In the mid to late 1950s her father was able to start taking flying lessons and got his student license, and soloed an Aeronca Champ. She was excited for him and knew then that some day her turn would come. Her dad's good friend, Jim, took them up in his Stinson Voyager: and this was her first time ever to be in a plane. Later her dad and she joined the Tulsa, OK, Civil Air Patrol. She was a second lieutenant and spent several years going to SAR-CAPS and encampment. After their flight observer time, one of their instructors would take her up in the CAP PA18 plane and give her actual flying lessons. He signed off into her log book 10 hours of flying instruction. By then she was fully convinced she would pursue this adventure.

It was now 1958, and after graduation from high school, and time to continue her education in business and secretarial school. Later on in 1959 she was hired by Sinclair Oil Company as a receptionist/secretary. No longer did she have the time to continue her flying, and would have to put aside for now any hope of finding the time. In 1960 she became very much interested in applying for an airline job as a stewardess; and so, started her new and exiting career. She was to fly for Continental Airlines for three and a half years and thoroughly enjoyed it. She quit the airlines in 1964, and married her long time friend and pilot Dave Mason. They met each other in 1958 at Harvey Young Airport in Tulsa, OK. In fact their first date was in a J3 Cub; she'll never forget the intentional spins he introduced her to!

Before Dave and she got married in November 1964, he was a flight and ground instructor and charter pilot. In May 1964 he hired on Delta Airlines as pilot/second officer. They now knew she would continue her flying, with Dave as her instructor. In September 1965 they bought into partnership in a Cessna 140 owned by a fellow pilot; the Cessna 140 was a fun plane to fly; she put many hours on it, and soloed it on Nov. 2, 1965. They decided to start a family and she continued to fly up to the week before their first daughter was born on July 20, 1966. Her flying slowed down at this time due largely to needing a bigger airplane and having their second daughter on Feb. 7, 1969. They now started looking for a four place airplane that was a tail dragger and decided on their current 1952 Cessna 170B 2698D; she learned to fly it and soloed it on Aug. 9, 1969. This aircraft is a great and forgiving airplane a pleasure to fly. She received her private ticket on May 26, 1972. Many fly-ins and vacations have been logged in this airplane with many exciting adventures.

In the Spring of 1973, on April 7, she fell in love with this darling little bi-plane called a Smith Miniplane. She looked like a butterfly with her orange, white and black spots on her. She was the sixth owner and second woman to fly her. She named her *Lady Bug*. They were to become one and fly to many fly-ins, where she was to receive several awards during the more than 300 hours they flew together. She won awards for spot landing, ladies choice award at the Texas AAA Fly-In in Denton, TX, and her most proud award, the EAA Headquarters Best Custom built for 1974 at the Tulsa/Tahlequah Fly-In on October 11, 1974. They had their picture in the International EAA magazine. She will never forget her love and devotion to this little bi-plane! She served her well, and was a true friend! She owned *Lady Bug* for over nine years.

As time moves on, and their oldest daughter Kathy soloed on her 16th birthday, they begin to realize her need for some stick time if she were to ever fly a bi-plane. She really wanted a two place bi-plane; but, that was not to be for now. She sold her bi-plane on Sept. 25, 1982, and in October 1982 bought her 1947 Piper PA12 Super Cruiser which Kathy did solo. They have taken the PA12 to many fly-ins and it is a joy to fly.

As she continues to fly to fly-ins and personal trips she looks forward to many more years of flying fun!

Joined the 99s International Women Pilots Organization and the Dallas Chapter 99s in July 1993. She is an associate life time member of EAA International; and a member of the Longview Piney Woods EAA Chapter 972. She is an associate member of the Antique Airplane Assoc. and a member of the Dallas Antique Chapter. She has been a member of the AOPA since 1972. She is a member of the International Cessna 170 Assoc.

WYNEMA "NEMA" MASONHALL, Oklahoma Chapter, learned to fly through the CPT Program at Frederick, OK. Joined the 99s in 1941 and became a WASP trainee in 1943. Was a school teacher and computer on a seismograph crew. Served as *99 News* letter editor from 1956-60 and continued to assist with the *99 News*. Held most of the chapter and section offices. Was SCS historian; also edited and published the SCS History Booklet in 1971. Attended all SCS meetings on a consecutive basis, including the Spring Section meeting in El Paso, TX, in May 1993. The city of El Paso honored her and gave her a Key to the City for her dedication and support of activities of 99s. After returning to her home in Minco, OK, Nema quietly passed away within a few days, on May 26, 1993. All South Central Section members miss her and doubt anyone will surpass her record of attendance, over 70 SCS!

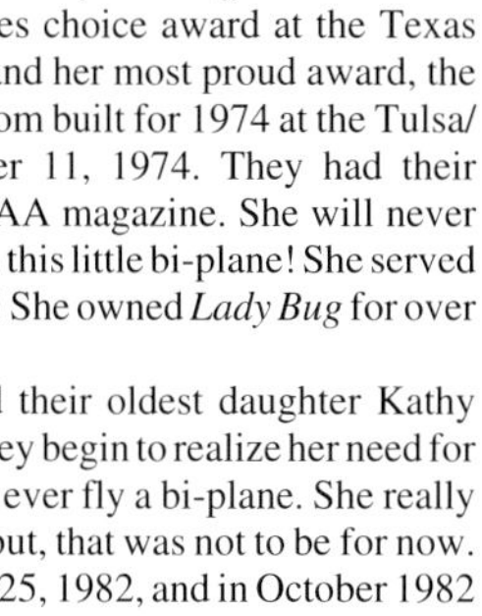

ANGELA MASSON, soloed in 1967 at Cloverfield, in Santa Monica, CA ... the same place where Amelia Earhart had soloed 46 years earlier. But how aviation had changed! When Amelia flew the Powder Puff Derby in 1929, she was among 14 who finished eight days later. By 1972, Angela flew the race in less than three days finishing in a field of almost 100. (That year, age 21, she was the youngest solo entrant, and Viola Gentry, at age 89, was the oldest solo flier!) Amelia was a charter member of the 99s, women who have earned the right to fly. Angela became a charter member of ISA+21, women who have earned the right to fly with the airlines.

Angela earned her BFA, MA MPA, and Ph.D. at the University of Southern California. There, before women were allowed to fly in the military, she wrote her famous dissertation on "Elements of Organizational Discrimination: The Air Force Response to Women as Military Pilots." This paper was presented before the Congressional Hearings which admitted women into the military academies. A copy of her paper can be found in the Congressional Library.

A pilot with American Airlines for almost 20 years, Angela has worked as flight engineer, copilot and captain on both domestic and international routes. She is rated on seaplanes and gliders, and has flown as copilot on the Boeing 727 and 707, and is typed-rated on the Boeing 747, 757, 767, DC-9, DC-10 and A-300. As a speaker for the FAA, her frequent lectures and articles on aviation continue to enthuse and excite kids of all ages.

JUDITH MASURA, born July 19, 1946. Joined the 99s in 1990. Received private pilot license in 1984. She has flown Cessna 150, 152, 172, 182; and PA22. Married to James and has three children: James, Jr., 27; Julie, 25; and Jeremy, 23.

Licensed at 37. She began flying at Gray Army Airfield, Fort Lewis, WA. She owns a Piper Tripacer which she helped restore.

In 1960 after joining the Civil Air Patrol, Judith dreamed of flying. Her first flight was in a Super Cub at Galt Airport, Ringwood, IL. She hung around Galt Airport whenever she could. Not until 1984 did she fulfill her dream of flying.

Judith is a CAP search and rescue observer. She joined the 99s in 1990 at Norma Koukils' invitation. This year, Judith chairs the Mt. Tahoma Chapter for the second time. She is an OX-5 historian.

Judith has advocated aviation since 14. She enjoys the solitude of flying into the air above the Puget Sound and escaping the hustle of life on the ground. Her future plans include advanced ratings and more pleasure flying.

ELIZABETH ANN MATARESE, began flight training on Oct. 31, 1970, in Sanford, ME. In February 1972, she obtained the private pilots certificate and in 1975 the instrument rating. In 1980 now a resident of the Metro Washington, DC area, she obtained a commercial pilot certificate and in 1986 the single- and multi-engine land instrument flight instructor certificate.

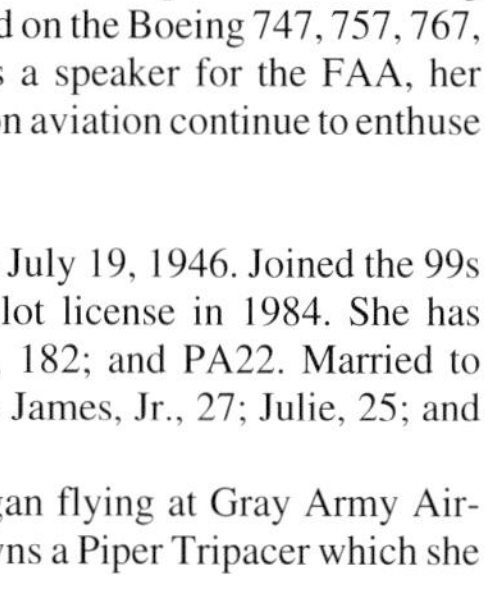

Born in Framingham, MA, Matarese attended schools in Brookline, graduated from Rutgers University in 1963, and moved to Maine, where she attended the University of New Hampshire and earned a MA degree in 1971. She instructed in English studies at both the secondary and college levels, 1961-74.

Matarese joined the State Aviation Administration in Maryland as the aerospace education specialist. She established a safety education program for the state's 9,000 pilots and edited the aviation newsletter, *Flight Plan,* and created the "Careers in Aerospace" program for students and developed aviation workshops for teachers.

In 1984 she became the first woman to serve as the director of the Maryland State Aviation Administration's General Aviation Office, and serves on the President's Advisory Board - University of Maryland, Eastern shore.

In April 1987 Matarese joined the Federal Aviation Administration (FAA) in the Eastern Region, where she was the first woman to hold both airport certification inspector credentials and the instrument flight instructor certificate. She is the only woman to have served as an FAA airport certification specialist. Matarese is now a technical program analyst in the Flight Safety Division. She was the first woman permanently appointed to this position and was the team leader for the *Single Engine Passenger Operations in IMC - Part 135.181* study.

Matarese conducts the FAA General Aviation Airports Program at the Hershey Airports Conference each year.

In February 1995, she was selected as one of the 10 Women in Aviation, at the annual award luncheon hosted by the National Aviation Club, of which she is a member and has been elected to the board of Governors for the past two terms. She is a charter member of the Maine Pilots Assoc. and the Mid-Atlantic Helicopter Assoc., a member of the Aircraft Owners and Pilots Assoc., the Experimental Aircraft Assoc., the International Aerobatic Club, and the 99s, Inc. in which she has held elected offices in two chapters as well as a section vice-governorship. She is currently the co-chair of the International Aerospace Education Committee for the 99s.

She is the mother of two sons. She has been elected corporate/executive secretary of the National Aeronautic Assoc.

SUE MATHEIS, Greater St. Louis Chapter. For her, a mother of five, flight began reluctantly at 40 – one hour later, she was completely skyhooked.

Earned her ASEL in 1965. Then followed a stimulating association with the St. Louis 99s, a commercial rating in 1968, 99 chairmanship in 1969-70, and an instrument rating in 1973.

Flying opened the world of radio to her: she did traffic reporting for KMOX-CBS Radio plus news, features and documentaries. In 1977 she was the first media-person to broadcast from NX211 – a replica of Lindbergh's Spirit of St. Louis.

Through a Whirly-Girl Scholarship, she received a commercial rotorcraft rating in 1975, making her Whirly Girl #194 and the first woman issued a Missouri helicopter license.

Born March 8, in New York , NY. Joined the 99s in 1965. Flight training included fixed wing, commercial/instrument ratings and rotorcraft commercial rating. Her favorite plane to fly is the Piper Comanche. She did charter work plus ferried new planes for Cessna and Piper. Her children are: Scott, 46; Terri, 41; Todd, 39; Michael, 37; and Leslie, 32. She also has six grandchildren; seventh due in August 1995.

CHERYL MATHER, born in May 1956, learned to fly in 1979 in Sacramento, CA. She flies a Cessna 172 with some of the 150 hours logged in an Aeronca Champ. In June 1983 she received her Airframe and Powerplant Mechanics certificate from Sacramento City College.

From 1985-94 she married Kevin Mather who is a mechanical engineer and technical sales rep., they backpacked the South Pacific, purchased a home, then traveled in South America with Isaac at 10 months old, he is now six years. Their other son, Joshua is four years old.

In August 1994, after 10 years of not getting into the left seat of an airplane, it was very exciting to once again be at the controls.

Cheryl is currently the secretary and aerospace chairman for the Sacramento Valley Chapter of the 99s.

DOLORES MATHEWS, became a 99 in 1975. She received her private certificate in 1973 and her instrument rating in 1976. She flies/owns a C-172, and with her husband Floyd has just moved to Rosamond Skypark, an aviation community, following the loss of their home in Woodland Hills, CA, as a result of the 1994 earthquake.

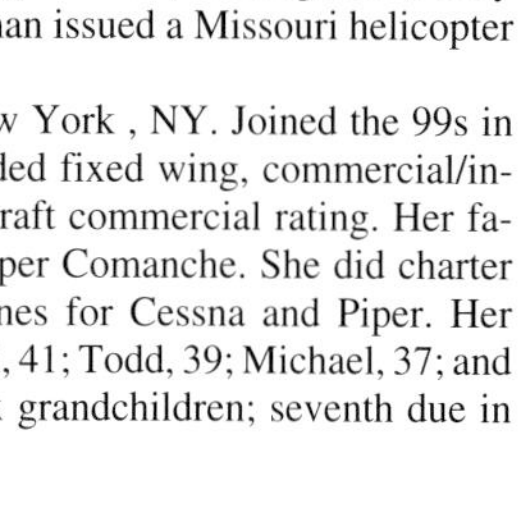

An outdoor enthusiast, Dolores skis and windsurfs. After a career as a secretary, she took early retirement, and then free lanced for several years. She and her husband divide their time between their Rosamond home and a vacation place on Lake Nacimiento.

LINDA MATHIAS, born in Norfolk, VA, began flying in 1971. In 1978, she earned her certified flight instructor certificate and won an Amelia Earhart Memorial Scholarship to finance her instrument instructor rating. During 23 years as a pilot, Linda raced the last AWTAR in 1976, participated in several aerobatic competitions and has served as a designated pilot examiner since 1988. In 1989 she married Joe Mathias, a retired Piedmont Airlines B-737 captain, and, in her spare time, Linda apprentices with Joe doing antique aircraft restoration. Their current project is a basket-case 1936 WACO YQC-6 custom cabin biplane which is about half done; their cross-country flying is done in Linda's 1956 Cessna 180. Linda holds a BS in secondary education from Old Dominion University and a MS in computer systems management from Naval Postgraduate School in Monterey, CA; she is employed by the United States Atlantic Command as a computer specialist.

ANN E. MATHIEU, born Aug. 10, 1930, in USA. Joined the 99s in 1973. Flight training included single engine, commercial, instrument, multi-engine ratings. She has flown Comanche 180, copilot Navajo, Aztec. Married to Ralph and has eight children: Carol, 42; JoAnn, 42; Lynn, 41; Matt, 39; Christine, 38; Betsy, 36; Tighe, 34; and Tracy, 32. Grandchildren: Mathieu, 18; Shane, 17; Kaylif, 8; Megan, 7; Kyson, 6; Avery, 5; Molly, 5; Kayla, 5; Nicole, 2; and Scotty, 3.

While flying around our beautiful country and the Caribbean, it became apparent she was crazy not to take a pinch-hitter course. With eight children at home, what would she do if anything happened to her husband in flight? Well, that pinch-hitter rolled on into a private, commercial instrument and multi-engine rating, subsequently a job at Silver Ranch Airpark in New Hampshire.

Once in the 70s, when landing at Quebec City Airport, she realized she became a part of a new breed. When the customs officer came out to the plane he asked if she flew this plane. He explained that he went off duty two hours earlier, but had stayed on to check in the first woman pilot he had experienced in his 30 years of duty!

"It's an honor to be counted among the 99s."

LINDA M. MATTINGLY's first airplane ride was in her brother's 1946 Funk at the age of 7. Years later she signed up for ground school and passed the written test before taking her first lesson in a Cherokee 140. As a 20 hour student pilot, she experienced an engine-out crash landing on take-off which inspired her very strong commitment to safety. She earned her private license and joined the 99s in September 1979, instrument rating in 1985, followed by a commercial license, advanced ground instructor, CFI and CFII. In 1987, she was one of 10 99s to win an Amelia Earhart Career Scholarship for a multi-engine rating, and received her medal in Van Couver, Canada at the International Convention.

The 99s have helped her develop many friendships over the years. As a member of the Indiana Dunes Chapter, she served as chapter chairman, secretary and chairman of many committees including safety, airmarking, ways and means, scrapbook and achievement awards. For two years she served as North Central Section board member, and recently became a charter member of Women In Aviation International. Flying ACS daffodils and participating in many air rallies and spot landing contests have been fun and educational experiences.

In 1986 she flew an 1,800 mile pipeline patrol covering all of northern Indiana. She's earned nine Pilot Proficiency Wings and is an accident prevention counselor.

She is currently an aviation insurance underwriter, earning a CPCU (Chartered Property Casualty Underwriter) designation consisting of 10 graduate insurance courses.

Her varied interests include gardening, stained glass, scuba diving, boating, belly dancing, furniture refinishing, sewing and crafts. She's designed and made many aviation and space art needlepoints.

VIRGINIA DARE MATTIZA, learned to fly in 1978, upon encouragement of her Scappoose, OR, high school classmate Evelyn Urban. She purchased a Mooney 201, obtained her instrument rating and instructor's certificate. Joining 99s, she served in all offices of the Austin and Heart of Texas chapters, and on South Central Section and International committees. She participated in the Air Race Classic and Palms to Pines air races with copilot Evelyn, and is a 1985 honoree in Forest of Friendship.

Mattiza taught in the US and in international schools in Japan and Korea. She founded the junior high for Gulf Oil in Ulsan, Korea.

She volunteers with her community newsletter and as school election judge.

A daughter and son live nearby in Austin, and a married son, also a pilot, lives in Atlanta. Mr. Mattiza is deceased.

She currently works for Sprint Relay Texas which facilitates telecommunications for the deaf in the US and worldwide.

LAURA BECHER MATTUCH, became interested in aviation during high school. The summer before going away to college, she completed her private pilot check-ride and became instrument rated. While earning her math degree at Doane College, she worked at the local airport and earned her commercial pilot license.

After graduation Laura joined the 99s and worked ground operations for two airlines. Since then, she married an air traffic controller who is also a pilot. Laura now works at Duncan Aviation in Lincoln, NE.

SUSAN MAULE, born Oct. 19, 1960, Jackson, MI. Flight training included general aviation and airline. She has flown 737, SA227, F28, Maule's, Piper's, Cessna's CIC.

Grew up in Michigan and took her first flying lesson, from her father Ray, at the age of 7. On her 16th birthday she soloed in 12 different types of planes, two of which were Maule's manufactured by her grandparents, June and B.D. On Susan's 17th birthday she received her private license, single engine land and sea.

In 1978, Susan moved to Georgia and attended Valdosta State College. After graduating with a social science degree she earned her ratings and moved to San Diego to work as a CFI. In 1985, she was hired as a first officer for Wings West Airlines. She was type rated on the SA-227 Metroliner and flew as captain until being hired by Piedmont Airlines (now USAir) in 1986 as a first officer on the Fokker F-28. She now flies the Boeing 737-300/400 out of Baltimore.

Susan has accumulated more than 10,000 hours of flight time and has flown more than 100 different types of airplanes. She has been a 99 since 1981 and has raced in the Air Race Classic and Pacific Air Race.

JANET JAN MAURITSON, born April 16, 1925, Muncie, IN. Joined the 99s in 1959. She has four children: David, 46; James, 44; Barbara, 43; and Nancy (deceased at age 34). Grandchildren: Chelsea, 18; Amy, 18; Julia, 16; Eric, 15; and Austin, 5.

Began flying in 1958 and joined Tulsa 99s in 1959. Jan has more than 11,500 hours in more than 65 makes and models of aircraft. She holds ATP, CFII, all ground instructor ratings, was FAA CIRE and FIE examiner and safety counselor. She has AS&MEL and ASES ratings.

Jan worked in flight training at Tulsair Beechcraft for 18 years and recently retired, after eight years, from Airport Operations at Tulsa International Airport.

Other flying activities include racing in the Skylady Derby, Powder Puff Derby and Air Race Classic. Jan has also made "Lifeguard" flights for Akdar Shrine 1979-83 and helped promote and provide transportation for participating Tulsa citizens to San Luis Potosi, Mexico in their mayor's Sister City Program 1980-84.

Jan was inducted into the Oklahoma Air and Space Hall of Fame in 1991, is executive director of the Oklahoma Airport Operators Assoc., board member of the National Biplane Assoc., recipient of 99s Jimmie Kolp award, USPA Member of the Year 1990, Oklahoma Aviator of the Year 1991 and numerous other awards.

KATHERINE MAXFIELD, learned to fly in 1990 at Reid-Hillview Airport. She owns a Bonanza A36/TC, which she uses to stay active with Angel Flight. Kathy is a self-employed marketing consultant who also enjoys hiking, gardening and piano.

EDYTHE SALO MAXIM, thought it might be a good idea if she learned how to land the Ercoupe. Airplanes subsequently owned were Taylorcraft, Stinson Voyager and Beech Musketeer.

Maxim acquired instrument rating and commercial in 1970 and 1971.

In 1952 joined the 99s where she held several positions, chairman of the All-Ohio Chapter, and served on various committees. She became a charter member of the Lake Erie Chapter in 1974. In the North Central Section, she was secretary, vice governor and historian. In 1958-60 she was president of the Cleveland Women's Chapter of the National Aeronautic Assoc.

She promoted the organization of the Kentucky Bluegrass Chapter 99s in 1966. She has been a member of the board of governors of the Western Reserve Aviation Hall of Fame in Cleveland.

She participated as pilot or copilot in numerous air races, including the Michigan Small and Indiana Fair races, in the latter taking a first place; several Angel Derbies, including the 1979 50th Anniversary Commemorative First Women's Air Derby from Santa Monica, CA, to Cleveland, Lake Erie Air Derbies, Grand Prix Air Races, Air Race Classics and Buckeye Air Rallies.

Maxim won several honors, including the All-Ohio Achievement Award in 1958 and 1966, and the Lake Erie Pilot-of-the-Year Award in 1975. She received the silver wings Woman-of-the-Year Louise Thaden Memorial Award in 1987 at the Dayton US Air Force Museum.

In June 1994, she became an honoree in the Forest of Friendship, which is in Atchison, KS.

JUDY MAXWELL, born March 5, 1963, in West Palm Beach, FL, with her parents owning a flight school during her childhood, you would think she would have gotten her license at 16, but she showed very little interest.

Then in 1993, with age 30 quickly approaching, she felt like her life was ho-hum. So she decided to change careers: from legal secretary to what could be more exciting, but aviation. She began her lessons immediately and bought a Cherokee Warrior (PA28-151). She is presently working on her CFI and CFII with hopes of opening a flight school.

Judy is presently a member of: the 99s Inc., as well as secretary of the Gulf Stream Chapter; and Aircraft Owners and Pilots Assoc.

Judy's husband, Tony Maxwell, is a plumbing contractor in Lake Worth, FL. Tony enjoys flying but won't get his license because he says, "Why should I, Judy is my private pilot."

DIAN WARD MAY, born April 29, 1946, Bladenboro, NC. Joined the 99s in 1972. Served as a medic 1966-70, Wiesbaden, Germany. Flight training includes commercial, instrument, ME, CFI and CFII, MEATP in General Aviation, AC-8, DC-10, B-747 Flight Engineer and B-737 and B-767 ATP, United Airlines. She has flown Cessna 150, 172 and 182, PA 140 and 180, Twin Comanche, Aero

Comm, Beech Travel Air B-737-757, 1977 to present. Employed as flight operations instructor and pilot United Airlines 1977 to present.

Married to James W. May and has one child, Heather Jean May, 11.

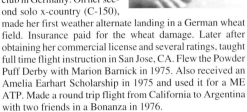

She took her first flying lessons in a German civilian club in Germany. On her second solo x-country (C-150), made her first weather alternate landing in a German wheat field. Insurance paid for the wheat damage. Later after obtaining her commercial license and several ratings, taught full time flight instruction in San Jose, CA. Flew the Powder Puff Derby with Marion Barnick in 1975. Also received an Amelia Earhart Scholarship in 1975 and used it for a ME ATP. Made a round trip flight from California to Argentina with two friends in a Bonanza in 1976.

Was hired in 1977 as a flight operations instructor for United Airlines and is currently flying as a B-757 first officer based in Denver.

JUDITH ANN MAY, born in St. Petersburg, FL, April 3, 1946, and was moved to West Virginia a month later. She recalls hearing the sound barrier being broken frequently and now realizes that it probably was Col. Chuck Yeager. His parents lived close by and he was known to fly over the area often. She feels that this was her call to the air, along with watching Penny fly with Sky King on television.

She has always loved aviation and learned to fly in 1990 along with husband Bill, a podiatrist, and sons William, a commercial pilot and James, working on an engineering degree. Judy has flown grandson, Dre' Maverick in their Cessna 182 as part of the Experimental Aircraft Assoc. Young Eagles program.

Currently she is working on an Aviation Technology degree from Marshall University. She is a member of the 99s, Inc., Aircraft Owners and Pilots Assoc., Experimental Aircraft Assoc., and Civil Air Patrol.

She has flown south to Florida and as far west as Billings, MT, and would love to race planes sometime in the future. Jackie Cochran is her favorite female pilot and she enjoys reading and collecting books about all pilots.

GRACE BIRGE MAYFIELD, was a pilot and parachutist of great skill and daring. She was born in Brurdett, NY, in 1913. At 13 while hanging around the local Watkins Glen, NY, grass strip, a local doctor offered her a ride in his JN4.

In 1939 she received her private pilot license. While learning to fly, she attended Asbury (Kentucky) College, where she earned an AB and the University of Kentucky, where she was awarded an MA in anatomy and physiology. Mayfield took a job as a PBX operator at the Cleveland (Ohio) Municipal Airport. She left the PBX job to work at Hornes Flying School as a flight instructor, bringing many flyers into the Civilian Pilot Training program.

During WWII, Grace served in the Women's Air Force Service Pilots (WASPs) "Ferry Command" beginning in December 1942. She was sent to Romulus, OH, to fly Aerocobras to Montana, where Russian pilots would then take them to Russia. As a WASP, she ferried P39s, P47s and P51s, picking up over a thousand hours.

In 1942 while ferrying a P39 to Montana, visibility was so poor that she couldn't see the runway. She told the tower that she was low on fuel. The ceiling was just 15 feet. She found the runway and came in so screaming hot that if there hadn't been six inches of water on the strip, she would have overshot it. As Grace taxied back up field, the engine sputtered and died. She had run out of gas.

After the war, she found a position as an air traffic controller in Denver, where she met and married George Mayfield, a fellow flight instructor. They taught all three of their children to fly. Grace became a science teacher and started an aeronautics program at Thomas Jefferson High School. Grace was inducted into the Colorado, Aviation Historical Society's Hall of Fame in 1983, making more than 65 jumps after the age of 50. She was an active Colorado 99 and past chairman of the chapter.

Grace passed away on Aug. 8, 1994.

ELLEN M. "PEGGY" MAYO, born Klamath Falls, OR. Joined the Wisconsin 99 Chapter in November 1964, shortly after earning her private license. She went on to earn her commercial license, then instructor with instrument instructor rating. She has held private pilot licenses in Canada and South Africa.

Peggy has also belonged to the Intermountain, Far West, and Florida Panhandle Chapters. She has held numerous offices in these chapters from membership chairman to chapter chairman. She flew in the last two Powder Puff Derbies and two Air Race Classics.

Peggy and her husband, Bob Mayo, are retired and are living in Florida on Steep Head Farm Airport which they spent a number of years building from a sand bed in order to fulfill a lifetime dream. They currently own a PA28-151.

She has flown PA28-151, Cessna 150, 182, and 177RG. Married to Robert L. Mayo and has two children, Robert A. Mayo and Sandra L. Niles. She also has four granddaughters and three grandsons.

CATHERINE E. MAYR, Wisconsin Chapter. Private pilot since 1987; instrument rated. Dated a pilot who introduced her to flying, and became engaged to him on the day she passed her private pilot checkride. How many husbands come complete with an airplane?!?

Currently North Central Section Governor; has served as chapter chairman, and NCS secretary. International 99s/EAA liaison and chairman of 99s exhibit at EAA-Oshkosh for the past four years. Member of International 99s Long Range Planning Committee.

Has been privileged to meet 99s from all over the world; 99s are some of her most treasured friends and are part of her most memorable experiences. Highlight was a trip to India in 1994 for the World Aerospace Education & Safety Congress. Enjoys proficiency air racing with fellow 99s, and long cross-country trips with her 49 1/2 Mike.

ROBERTA "ROBBIE" COLVARD MCBRIDE, begged her parents at age four to allow her to take a flight when a barnstormer came to town. They wouldn't agree but her love of flying started then and continued.

A lifelong resident of Texas, she sampled living in many, many Texas towns as well as in other states.

She married Robert McBride in 1955. He was a career officer in the Air Force and when they arrived back home to Texas in 1976, she and her youngest son, Chris, took flying lessons together, soloing the same day in a Luscombe 8F. All her three sons are pilots.

She has been an active member of the 99s, Inc., since 1981; vice chairman, secretary, scrapbook chairman in a chapter and Pilot of the Year; is a member of AOPA, ABS and WBS.

She and Bob presently live at Breakaway Airport, Cedar Park, TX, and fly a PA12 and an F33 Bonanza.

SIEGRID SIKORSKY-MCCALLUM, born in Cologne, Germany, and now resides in Belchertown, MA, with her husband Harold. Vacationing in Connecticut she obtained her private pilot certificate in 1972 and joined the 99s shortly thereafter. Multi-engine, instrument and flight instructor ratings followed during the next two years. She also obtained a German private, commercial and flight instructor rating in 1976.

Siegrid worked as ground and flight instructor

in Hangelar, Germany, and later at Sikorsky Memorial Airport in Bridgeport, CT. She owned a Cessna 172, a Piper Cub and currently a PA28-180. Siegrid flew numerous rallies and races in Europe and the US including the 1981 Paris-New York- Paris Race in a single engine PA28-300 and the last Powder Puff Derby in 1977. In 1980 accepting an invitation from Jerrie Cobb and Ruth Lummis she visited them in Mitu, Columbia, South America, to get a taste of jungle mission flying and to witness the incredible work both were doing. She enjoyed being an FAA accident prevention counselor for many years at Bridgeport Airport, organizing safety meetings and two very successful air rallies in 1978 and 1979.

The IOC, International Order of Characters, an international pilot organization, presented her the Pilot of the Year award in 1982. Presently Siegrid owns her own aviation consulting company and enjoys flying after more than 20 years for business and pleasure.

NORMA "JODY" MCCARRELL, Oklahoma Chapter 99. In 1970 she made her visit to the Golden Triangle Chapter of the 99s. She got involved in NIFA, 25 years ago and she has just been nominated for the NIFA council. She is one of the senior judges of NIFA as well as the International committee chair for the 99s in 1994-96. Has served in every position for SAFECONs including chief judge in 1994 at Parks College, St. Louis. Got involved with the 99 Air Age Education programs and attended several National Air Age Education Council meetings.

In 1974-75 was co-editor of the "Let's Go Flying" coloring book.

She helped coordinate the first Air Age Education Workshop in July 1978 at the University of Oklahoma. Also on the board of directors of the AWTAR, she resigned to race in the great Bicentennial Race of 1976; also flew the 1977 Commemorative Flight. Flew the Air Race Classic in 1993 and 1994 with Jerry Ann Jurenka. After having served as chapter chairman Golden Triangle Chapter, SCS Air Age Education chairman, and SCS nominating committee, she got involved in 1983 with the United States Precision Flight Team, serving as International chairman of that committee since 1986. Helped run as chief navigation judge the World Precision Flying Championship in 1985 in Kissimmee, FL. Was appointed an International judge of Precision Flying by the FAI in 1984. In 1987 was presented the Jimmie Kolp award by the South Central Section. During these years she completed her commercial, instrument, flight instructor, CFII and became an accident counselor. Also built up more than 15,000 hours as a corporate pilot flying twins and turbo-props. In 1992 put in a bid to run the second World Precision Flying Championship first time ever it was given to the same country twice; is currently serving as the competition director.

GLADYS MCCASLIN, born in Oklahoma on Aug. 4, 1929. She and her husband Ben moved to California in 1946. She learned to fly in 1959. With Ben's encouragement she got her license, flying from their field in Boron, CA.

In 1966, when their daughter Holly graduated from high school, they moved to Lexington, OK.

Their grandsons learned to fly in a Cessna 120, at McCaslin field. All are still active in flying.

Gladys has flown the following Cessnas 120, 170, 172, 180, 182 and 195. She is active in the International 99s and the Oklahoma Chapter, helping with the Okie Derby Air Race. She hopes to devote more time to the 99s, since she has retired from her hair salon and boutique. She has been a hairdresser for 45 years.

ELIZABETH GAY MCCAULEY, ASEL, Commercial. Soloed at Okmulgee, OK, Airport, December 1958, in a 1956 Cessna 172 under instruction of Cliff Rands. Purchased 1959 Cessna 172 with husband, who then took up flying. Commercial license, July 1962.

Joined Tulsa 99s in 1959; secretary and *99 News* reporter. Flew Sky Lady Derby, 1961, and co-chaired and flew Dallas Doll Derby, 1961. Flew charters, art tours and airport rides.

Transferred to Washington, DC, Chapter 1964; to Bay Cities in 1974.

Resumed flying after 12-year absence with H.J. McMurdo, Oakland (CA) Airport in 1975. Chairman, Hospitality Committee, International Convention, San Francisco, 1977.

FIONA ELIZABETH MCCHESNEY, born in Sussex, England, in 1965, and moved to California in 1984 to live with her brother Ron Smith. Upon his suggestion and encouragement Fiona started what she now hopes will be a long career in aviation. She is the first female pilot in her family (hopefully not the last) and the third commercial pilot following in the footsteps of her father and brother. She has logged more than 100 hours of multi-engine time and has had the good fortune to fly a diverse group of aircraft from C-414, Navajo, C-310, Lake Amphibian, Stearman, to a C-152. She hopes to reach her goal of flying as a first officer for a major airline within a few years.

As a newlywed Fiona spends time with her husband John, playing coed soccer, tennis, movie watching, reading and working on their home together.

Fiona is a dedicated member of the 99s, serving as both Bay Cities Chapter chairman for the last two years and as a member of the legislative committee for southwest section. "The 99s has a very important role in the aviation world, and as women pilots we must take the time to become involved."

ALANNA MCCLELLAN, received her private pilot license on Aug. 15, 1985, at Bakersfield, CA. Had more than 575 hours by 1994. Flight training includes private pilot license, ASEL and instrument. Joined the 99s in June 1987, Lake Tahoe Chapter. Employed as sharedraft manager for ATM. Previous occupations included physical education teacher, owner of Thrifty Car Rental. She has flown Piper Archer (owned N2903A), Trinidad, Grumman Tiger, C-172, Cherokee, Bonanza, and T-34. McClellan is single and has two sons (both CFIs), and one daughter (logged some hours). Attended 99s section meetings, Reno Air Races, Palms to Pines Air Race 1989 and 1991. Her non-flying activities are swimming, snow skiing, water skiing and traveling. Member of Truckee Tahoe Plane Talkers, 99s (eight years).

She feels that almost every flight has a story or incident to be told from the first solo at Buchanan Field when she bounced nicely three times and asked her instructor in the tower if that would pass for three touch-and-goes, to flying cross country to North Battlefield, Saskatchewan, and learning how to find airports with the NDB/ADF and the wonderful experiences: of flying the Palms to Pines (good and bad), listening to the accomplishments and dreams while interviewing candidates for their scholarship and selling hot dogs to make their dreams possible. She thinks these are the things Amelia Earhart had in mind for them as 99s!

BONNIE MCCLINTOCK, as registered nurse and paramedic, her nursing experience is in intensive care, while her paramedic experience includes a private ambulance service and a fire department. Currently she is an instructor for the Stark County Paramedic Program in Canton, OH.

She began flying with a fellow-paramedic who had his private license and owned a 1956 Cessna 172. She grew tired of being a passenger and got her license in a Cessna 152 at Martin Air Field, a grass strip, on Oct. 16, 1983. She occasionally used her paramedic skills on jump school students who made rough landings.

In N6802A, the 172, she gathered hours, both right and left seat, gaining invaluable experience during their VFR trips to Florida, Texas, New England and Niagara Falls. Her friend has since become an ATP for a charter company and remains a good friend and mentor.

She is now progressing on her instrument rating in a Cessna 150, based at Akron Canton Airport and has just completed Phase V of the Wings Program.

She occasionally blends her medical interest with aviation by helping with mock aircraft disasters and FAA safety seminars.

Her summer vacations are spent in Oshkosh, where she tent camped with a group who gathers in the same spot every year. She is pictured here with Patty Wagstaff at Oshkosh 1993. She loves the thrill, challenge and sense of accomplishment she gets from flying. Her goal is to continues advancing her ratings but more importantly just to have fun!

ERDINE I. "DEE" MCCOLLUM, born Feb. 22, 1926, in Indianapolis, IN. Served in the USAF. Flight training includes US Flying Service, Albert Whitted Airport, St. Petersburg. Has flown FL. J3 Cub, Cessnas, Piper's, Comanches, T-38, Commuters 114-116, etc. Married to Bruce and has three children: Bruce III, 45; Jack B., 43; and Richard L., 40. Her grandchildren are Moya, 21; Crieg, 17; Matthew, 6; and Kathyrine, 5.

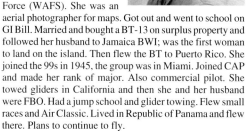

Started to fly in the 1940s but went into the Air Force (WAFS). She was an aerial photographer for maps. Got out and went to school on GI Bill. Married and bought a BT-13 on surplus property and followed her husband to Jamaica BWI; was the first woman to land on the island. Then flew the BT to Puerto Rico. She joined the 99s in 1945, the group was in Miami. Joined CAP and made her rank of major. Also commercial pilot. She towed gliders in California and then she and her husband were FBO. Had a jump school and glider towing. Flew small races and Air Classic. Lived in Republic of Panama and flew there. Plans to continue to fly.

CONSTANCE ANN MCCONNELL, member of the All-Ohio Chapter since 1978, was introduced to flying at an early age by her dad. Connie has served as secretary, vice chairman and chairman of her chapter. She has been chairman and co-chairman of the Buckeye Air Rally and news reporter/public relations chairman.

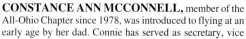

Connie has a BS in education and her CFII and multi-engine ratings. She is the owner of a polished Aluminum 1946 Cessna 140, and is the Ohio representative for the Cessna 120/140 Organization. Her husband, Jim, also enjoys flying and owns a 1941 Navy Stearman.

Connie spends time, when not flying, taking care of her Bassett Hounds, and her hobby is collecting Bassett figurines, prints, etc.

JEAN O. MCCONNELL, born Dec. 28, 1931, in Douglas, WY. Joined the 99s in 1972. Flight training includes private, commercial, instrument, CFI, CFII. Flew SEL - GEN Aviation. She has three children William, Sherry and Donald. Also has grandchildren: Jennifer, 17; Dana, 13; Christopher, 11; and Evan, 10.

Learned to fly in 1971, after a demonstration flight with former WASP Gene T. FitzPatrick, who was a flight instructor at Hawthorne, CA. She flew the Palms to Pine Air Race with 70 hours, subsequently flying in the Pacific Air Race, the Baja Air Race, Air Race Classic and several proficiency races. After buying a PA-28, she continued taking instruction, earning commercial, instrument, CFI and CFII.

She is a member of the Civil Air Patrol, the 99s and AOPA. She worked with the Flying Samaritans, flying medical personnel into Mexico for clinics and medical evacuations from the Baja Peninsula.

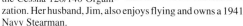

After retiring in 1994, she moved to Arizona, became a partner in a Cessna T210, and built a home at Montezuma Heights Airpark, a private airport, in Camp Verde. Still an active flight instructor, she looks forward to learning about the many experimental aircraft which abound in the area.

JUNE MCCORMACK, born June 30, 1928, in Oklahoma. Joined the 99s in 1973. Flight training included SEL SES private, will log 3,000 hours by end of 1995. She has flown 737JM (personal), C-172 (racer), C-206 (Jim's) to haul grandkids, 1946 Taylorcraft. Flies for pleasure, did commute to work from desert over 5,000 foot mountain to oceanside for 15 years. Married to Jim and has three children: Jan, Joe and Ken; pilots and auctioneers. She has 10 grandchildren from 3 to 23, all love to fly.

In 1994 winner of 25th Palms to Pines, won final twenty-fifth place of Pacific Air Race; placed top 10 in ARC, Baja and Mile Hi Derby – 30 races. San Diego Chapter chairman in 1975; charter member of Borrego Springs and chapter chairman 1985-90 and 1992. Attended World Aerospace Conference in India; donated time and money for youth training at Borrego Airport (SD County). Lives on private airport, Borrego Air Ranch since 1976, rides ATV to paint wild flowers in Anza Borrego Desert.

JILL S. MCCORMICK, member since 1941. Chapters (active) New York - New Jersey; Atlanta; Washington, DC; Boston and Indiana.

Aviation achievements – private license, sea-planes 1941; rode bicycle 30 miles to obtain land rating 1941; WASP (Women Airforce Service Pilots) March 1943 - December 1944; US Air Transport Command (civilian worker) in France and Germany, 1945-48. Ferried some liaison aircraft. EAL, TWA and FTL – link instructor and crew coordinator, 1948-54. Purdue University, School of Aviation Technology – 1955-76. Organized and taught the Professional Pilot Program – major subject, advanced instrument flying.

FAA ratings achieved – commercial and instrument; multi-engine; CFI, instrument and airplane, and ground instructor, instrument.

Air Races entered – PPD 1961, numerous Michigan SMALL Races, four in 1964, and organized and flew several FAIRs. Recipient of the Amelia Earhart Scholarship award, 1964. US Air Force Reserve – retired major.

OLIVE ALLEN MCCORMICK, born Aug. 26, 1902, Goderich, Ontario, Canada. Joined the 99s Indiana Chapter Oct. 3, 1947. Pilot status: ASEL – Active Single Engine Land. Has more than 2,064 hours. She has flown and owned Piper J3, Stinson Voyager, Piper Tri-Pacer, Cessna 195, Comanche, Aztec, Cessna 206.

In 1952 Muncie, in to Kenyon College in Gambier, OH, to visit her son, Allen. In 1955 ninth All Women's Transcontinental Air Race Powder Puff Derby Long Beach, CA, to Springfield, MA, cash award $50 for best score, finished twelfth with daughter, Audrey, as crew. In 1958 International Air Race for Women Welland Ontario, Canada to Grand Bahamas, finished sixth with daughter, Audrey. In 1960 Eastern Provinces of Canada: Quebec, New Brunswick, Prince Edward Island, Nova Scotia with son, Arch. In 1961 Western Provinces of Canada: Manitoba, Saskatchewan, Alberta, British Columbia, plus Yukon to Fairbanks, AK, with son, Arch. In 1964 Auckland, New Zealand: North and South Islands in rented Cessna 172 with husband, Cassius. In 1964 Sydney, Australia: Cookstown to Aukunim Aborigine Settlement in rented Cessna 205 with husband, Cassius. In 1968 Pretoria, South Africa: Windhock, West Coast, Cape Town, Rhodesia in rented Piper 235 with Cassius. In 1968 Versailles to Nice-Cannes, France in rented plane with Cassius. In 1979 Concorde to Paris, France with Cassius; Audrey and granddaughter, Olivia.

CHARLENE MCCULLOUGH, Tulsa Chapter, earned her private license in 1978 in a 1956 model Cessna 172, purchased at auction by husband, Ron. The airplane had been abandoned after sustaining rudder damage in a tornado. Ron set to work restoring the airplane inside and out, and soon he and Charlene began learning to fly it. Son, Mike, then 17, and daughter, Cheryl, became pilots as well.

An active 99 since 1979, Charlene has served in various chapter capacities including chairman. She has flown a Cessna 182, Mooney, J3 Cub and Stearman N2S3, all of which have been "part of the family" from time to time. A

former elementary teacher, it was not uncommon for her to leave school at 3:30, make the short drive to the airport, get the plane ready, take a half-hour flight, tie up, drive home and have dinner on by 6:00.

Although IFR rated, her airplane of choice is the Stearman. Daughter, Karen, is an eager passenger, as are other family members and friends. Charlene, Ron and Mike are active in local antique airplane organizations and are currently restoring a 1943 Stearman PT-17.

MARY LEE MCCUNE, born Nov. 12, 1938, in Los Angeles. Joined the 99s in 1989. Learned to fly in 1981-82 in Concord, CA. She has flown Cessna 172 and 182; owns a 1982 C-182. Married to Michael and has two daughters.

Began taking flying lessons in 1981 because she was convinced that her husband would soon get his license and expect she and their two daughters to fly with him. She soon got "hooked" though he dropped flying before his solo. The first big family trip was to the tip of Baja, CA, in 1982. On that trip, Mary Lee learned two valuable lessons: not to count on flying tips from the "man" sitting in the right seat (where her male instructor had always sat); and not to rely on guide books for advice about unexplored airports.

Fourteen years, an instrument rating and 1,800 hours later, flying has become Mary Lee's highest priority. She has flown her family and friends on numerous long trips to Canada, Alaska, across the US, and several times to Mexico.

In June 1992, Mary Lee and fellow 99, George Ann Garms, flew Mary Lee's Cessna 182 over the Bering Sea, across Russia and into Mongolia for a spectacular, three-week flight. Returned by way of Alaska. Total flight and return to Concord, CA, five and a half weeks (116 hours of flight). Probably the smallest/slowest foreign/private plane to fly across Siberia and Mongolia. Flight time in Russia and Mongolia, three weeks (65 hours). She and George Ann are still studying the Russian language so another Russian adventure may be in the future.

Until Mary Lee joined the 99s in 1989, her only pilot friend had been her former instructor. Belonging to an organization full of women who fly continues to be a great source of pleasure to Mary Lee. Those Tuesdays are sacred, when Mary Lee flies off to meet her 99s friends for their lunch meeting, always at a different airport.

NANCY MINOR MCCURRY, born Jan. 15, 1958. Joined the 99s in 1990. Flight training includes private pilot ASELS. She has flown Aeronca Champ, J3 Cub, Cessna 140 and 172. Employed by AT&T Bell Laboratories (1979 to present). Married to James D. McCurry.

MARGO MCCUTCHEON, Toronto, Canada, a former nurse and mother of four made her dream come true in 1976 when she obtained her private pilot license. To learn how to really fly, she obtained a commercial license with instructor and instrument ratings. After a few hundred hours of instructing she flew for a private cooperation.

She also joined the 99s to meet women pilots who knew how to have fun. A member of the First Canadian Chapter, she has served as chairman, has frequently flown in poker runs, air rallies and long distant flights across Canada and US to and from conventions.

In 1983 Margo and copilot Heather Sifton flew to the Forest of Friendship, Achison, KS, to a tree seedling's "fly away" ceremony. They were the first in a relay team of pilots to distribute seedlings to Toronto, Montreal, Halifax, and St. John's Newfoundland. In June 1985, flying Baron C-GOVQ, Margo and teammates, Adele Fogle and Daphne Schiff the only all-women team, competed in the air race from New York to Paris and brought home a sizable memento, the women's cup. Margo and partner, Anna Pangrazzi, had a taste of "speed racing, the most exciting form of competition" flying the Air Race Classic 1990.

As a volunteer member of a team of women pilots, in a program called Operation Skywatch, Margo has been very active in spotting and recording pollution from the air, in cooperation with the Ministry of the Environment in Ontario. Her ultimate dream for Operation Skywatch was achieved in 1994 when the award winning film ""Angels of the Sky" was aired on TV and distributed to other countries.

VIRGINIA "GINNY" MCDANIEL, born in Tampico, Mexico, on Oct. 3, 1923. She obtained her pilot license in 1980 at 57 years of age. Her son was her instructor and that was an experience for both of them. She likes to state that she put all the gray hair in his head. Ginny owns the Grumman Cheetah that she flies. She has three children: John Luther (whose father was a jet pilot) and Sherry and Rick McDaniel (whose father was a bomber pilot in WWII).

Ginny has a BS and a M.Ed. and is a licensed professional counselor. After retiring as a school counselor she began a new career as program manager with a mental health program. She has been actively involved in the 99s since joining as flying activities chairman and is now the SA 99s chapter chairman.

MARY J. MCDONALD, born Oct. 20, 1947, in New London, CT. Flight training included Waterford Flight School/ ATE Training/Flight Safety Inc. She has flown G-II, G-III, DA-10, DA-50, BE-200, Baron, numerous SE Pipers. She is a self-employed air charter business/ Cessna corporate pilot - American Can Co., and Primerica. Married Angus McDonald and has three children: Nicole Suisman, 21; Jarred Suisman, 19; and Tony Suisman, 18.

McDonald began her flying in 1968 with a $5 introductory lesson and fell in love with aviation. After obtaining her private, commercial, instrument, and flight instructor's certificates in the first year, she then began to flight instruct and do charter out of the Waterford Airport, Waterford, CT. In 1970 with a Cherokee 180 and a Beechcraft Baron, she opened a charter business, Fishers Island Airways/Air Leasco, Inc. operating out of Fishers Island, NY, and then out of Groton, CT. In 1978 her corporate career started when she was hired as a copilot for American Can Company, flying out of Westchester County Airport, NY. After begin type rated in the Falcon 50 and Falcon 10, she retired as a captain in 1985. She has one daughter, two sons, three stepchildren and three grandchildren.

JEANNE MCELHATTON, began her flying career just out of high school in 1946 who joined the SCV 99s in 1956. She has flown many PPDs, International, Pacific and Palms to Pines Air Races. She was active in developing direct relief flights and flew an emergency mission for earthquake victims to Managua, Nicaragua. She also flew to Valparaiso, Chile in her Cherokee 235. Jeanne was cofounder of the renowned Fear of Flying clinic. She is host/ instructor of the award winning "Invitation to Fly" video ground school program. She is an Aviation Safety Analyst for the NASA-ASRS program and contributes to Call Back. She is single, multi and instrument rated and is a CFI with more than 6,000 hours. Jeanne is an accomplished teacher and speaker for many 99 programs. She is mother to three children and has taken them on many trips across the continent and into Canada and Mexico. She currently co-owns a C-182 in which she enjoys using for traveling.

COLLEEN MCGRADY, born July 10, 1954, in Eagle Mills, upstate New York. An introductory flight Sept. 11, 1991, in N5404B enticed her to pursue her private pilot certificate which she "earned" on April 7, 1993. Flying took a great amount of courage for Colleen as her 31-year-old brother, Donald McGrady, was killed while flying his helicopter on Aug. 30, 1987. One week after her first solo, her flight instructor suffered second degree burns in an accident involving an experimental aircraft on it's maiden flight. Colleen found the courage to complete her second solo on April Fool's 1992.

Colleen is employed by the FAA as an aviation technician. She has flown Cessna 152 and 172, a Piper PA-11 on "floats" and is currently pursuing a commercial/instrument rating. Colleen is a member of the 99s, AOPA, actively participates in FAA seminars and will earn Phase III Pilot Proficiency Wings in April 1995.

Colleen has a 22-year-old daughter, 21-year-old son, and 7-year-old thoroughbred horse which she has raised and trained.

EMMA L. MCGUIRE, born Feb. 24, 1914, in Cleveland, OH. Joined the 99s in 1960. Flight training included ASMEL, instrument, CFI flight instructor, Commercial, sea plane, glider, and aerobatics. She has flown Cessna 150, 172, 182, 210, etc. Piper Apache, and Citabria. Employed by Claire Walters Flight Academy, Santa Monica, in 1960s (1968-82). She is the widow of Ralph McGuire and has three children: Michael McGuire, 57; Mareva Cramer, 60; Marlene Mejia (deceased). McGuire also has 13 grandchildren and 14 great-grandchildren.

Her introduction to flying was in 1960 at age 46. She was married with three children, who were married, with families. While working in the family grocery store she became acquainted with Jan Vawter. She was learning to fly and talked about her lessons. One day she stopped by to ask her to go to the airport. A short flight in a Cessna 150 was all she needed. Her husband, Ralph, bought her a 172.

She took lessons from Claire Walters, and became a private pilot. In 1961, with her private license, Jan and she entered the Power Puff Derby and she flew all the following: PPD, 1962 through 1977. She also flew the ARC, Angel Derby, Palms to Pines, Pacific Air Races and other small races.

She continually worked on new ratings obtaining her commercial, instrument, CFI, multi-engine, sea plane, glider and aerobatics (owning two Citabrias). She was a participant in the eleventh Coast Guard District. Her best pleasures were flying races and teaching students.

GRACE MCGUIRE, is a member of the Long Island Chapter of the 99s and has had a long love affair with flying. She began flying in 1970 at Red Bank Airport, NJ, and holds commercial and instructor licenses with multi-engine land and sea ratings and instrument rating. She also has solo helicopter time. She was the first female to graduate from the North American Institute of Aviation, NJ, with instrument and instructor ratings. She taught flying at Preston Airport, NJ. She was thrilled to be checked out in an F-15 jet fighter at Langley AFB in Virginia.

She chaired the 1994 Garden State 300 Air Race at Toms River, NJ, and has participated in numerous air races, such as the King's Cup air race from London to Londonderry.

She holds the rank of captain in the Civic Air Patrol, Pineland Squadron. She has also served as 99 Airmarking chairman for New Jersey. She was the first woman to set foot on Howland and Baker Islands, South Pacific. She was the 1993 honoree at the International Forest of friendship.

Amelia Earhart's life has always been an inspiration. At some future time she hopes to duplicate and complete Amelia's 1937 around-the-world flight using the same type Lockheed Electra L-10E aircraft which she is restoring at Lakehurst Naval Air Warfare Center, NJ.

JANIE MCINTIRE, like many other women pilots, she too was one of those kids that was always looking skyward at the airplanes. Her cousin took her for her first airplane ride, her father has his pilot license and as they say, she was hooked!

In November 1979 she started taking her flying lessons. In July 1980 she was finally a "real pilot," she thought, when she received her license! She can remember her first solo as if it were yesterday. Shortly after receiving her license she joined International and Scioto Valley Chapter 99s. She has been active in the chapter holding various chapter and committee offices.

She has been employed as a grants administrator by the Ohio Department of Transportation, Division of Aviation since 1985. She was fortunate in that her flying "hobby" was able to turn into a career.

BARBARA MCINTOSH, is a lifetime Ohio resident. At the age of 44 she began her quest to become a pilot and received her certificate in June 1992. She flies a Cessna 150 and a Cessna 172.

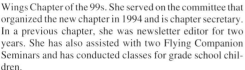

Barbara works for a large bank supply company as a service dispatcher. Her hobby is being a beauty consultant for Mary Kay Cosmetics.

Barbara is a member of the Aircraft Owners and Pilots Assoc. and a charter member of Women With Wings Chapter of the 99s. She served on the committee that organized the new chapter in 1994 and is chapter secretary. In a previous chapter, she was newsletter editor for two years. She has also assisted with two Flying Companion Seminars and has conducted classes for grade school children.

Barbara is married to Tom McIntosh, a non-pilot. He teaches secondary level truck and diesel mechanics at a career center. They have one son and a daughter-in-law.

MARY CATHARINE "JARY" JOHNSON MCKAY, first aviation experience was in the glider club at the University of Michigan, where she learned to fly a Franklin glider, and later flew it at the international glider meet at Elmira, NY.

At Metropolitan Airport in Van Nuys, Jay learned to fly a Piper Cub. During WWII, when Jacqueline Cochran started the Women's Flying Training Detachment at Houston, TX, Jary was accepted in the second class to be trained. There she flew primary, basic and advanced military trainers.

After graduation from training, Jary was assigned to the Air Transport Command base at Romulus, MI, as a Women's Air Force pilot (WAF). (This designation was shortly changed to Women's Airforce Service Pilot, or WASP). There she started out by ferrying liaison planes such as the Stinson L-5 and the Fairchild PT-26. She was sent to Palm Springs for fighter (pursuit) training, and from then on ferried P-47s, P51s, P-40s, P-39s, and P-63s from the factories to training bases or overseas departure points.

After marrying and rearing a family of three daughters, Jary took flying lessons again and activated her commercial pilot license. She also earned an instrument rating, which she keeps current. She now belongs to the Air Spacers Flying Club at Santa Monica Airport and flies a Cessna 172.

Jary belongs to several aviation organizations. As a member of Angel Flight she transports patients from their homes to medical clinics and coordinates missions flown by other pilots. Her other memberships are in the Confederate Air Force, the 99s, the Aviation Breakfast Club, Silver Wings and the Women Airforce Service Pilots, WWII.

Jary has four grandchildren and two great-grandchildren.

FELICITY BENNETT MCKENDRY, born in 1929, in Spencerville, Ontario. As a young farm girl she built model airplanes. With savings from her teacher's salary, Felicity became a private pilot in 1951 at the Kingston Flying Club. In 1952 she competed in the prestigious Webster Trophy zone trails and joined the 99s.

Offered a job, Felicity became the first female commercial graduate at KFC and then one of seven female instructors in Canada. More than 2,300 hours were logged on 20 types including wheels, skis and floats. Her husband Spence's Air Traffic Control career dictated moving finally to Ottawa. She was one of the first female chief flying instructors in Canada. While raising two children, Felicity became a designated flight test examiner at Rockcliffe Flying Club in 1984. Retiring to Kingston she let her license lapse in 1992.

Highlights of 4,500 hours and three air races. In 1955 with Dorothy Rungeling was the first non US AWTAR team. In 1982 judged the Canada wide Governor General's Shield Competition. In 1986 received Rockcliffe Flying Club's life membership. In 1993 inducted into the Forest of Friendship by her Eastern Ontario Chapter.

Volunteer activities: United Church of Canada, Youth Camps, the 99s, Alcohol and Drug Concerns, Meals on Wheels, Parkinson Foundation. She enjoys her family and three grandchildren.

ELIZABETH ANNE BUZZELL-MCKENZIE, ASA; Bay Cities Chapter, born Jan. 2, 1943, Biddeford, ME; daughter of Dr. Benjamin R. Buzzell, Belfast; and Dorothea Perkins Towne, Kennebunkport; raised in these towns as well as Cape Elizabeth, ME. Great-grandfather, Capt. Fordyce C. Perkins, Kennebunkport, ME, and brother of Governor George C. Perkins, California (1883). Grandfather, Hon. Hodgdon C. Buzzell, Belfast, president of Maine Senate, 1925; Uncle, Hon. Hillard C. Buzzell, Superior Court Judge, Belfast, ME. Learned to fly (1959-61), being the youngest female pilot in the state of Maine, at the former Port of Maine Airport, South Portland, ME (a 1,200 foot dirt strip). What a change to fly out of Boeing Field, Seattle, WA, when came to the World's Fair in 1962. Graduated University of Washington, BA, 1964. Attended Hastings College of Law, San Francisco; later, manager of a Piper Dealership in Oakland, CA. Commercial ASEL. (Always preferred by Cessna 172!)

Always interested in history, objects and genealogy, became an accredited senior appraiser with the American Society of Appraisers in 1981, of antiques, decorative arts and residential contents with assignments throughout the country for private clients, insurance companies, moving companies and attorneys.

Married to Donald Stewart McKenzie, a native of Seattle, retired from Chemcentral. Divide time among homes in Seattle and Whidbey Island, WA, and a ranch in Lotus, CA, as well as traveling throughout the country and Europe to keep-up on the antique and fine arts market. Many of her happiest hours were in the air in a Cessna 172 and being proud to be a 99.

JAN MCKENZIE, born July 2, 1951, in Fort Worth, TX. Joined the 99s in 1973. She has flown Stinson 108-3, J4 Cub, J3 Cub, Cherokee, C-172, Mooney, Bonanza, Maule, and Decathlon. She has one child, April Clemen, 24.

One of the reasons they fly is because they can get to more places faster. Quick transportation for business or pleasure gives them a big jump on earth-bound mortals. That's why she has a Mooney.

But they also get to see those places from a perspective that few others share. The beauty of the Alaskan wilderness is just breathtaking from the air, so great that she has toured it twice. She has enjoyed the sight of a Caribbean Island creeping over the horizon into the sea below. Only pilots and artists can really appreciate things like this.

But flying lets her do more than feel the wonders of place and perspective. It is really a time-machine as well. She learned to fly in a classic Stinson 108-3. The taildragger era really was a golden age. What a privilege to experience first hand just a little bit of what it was like in those days! She loved touching the aviation past so much that she bought a 1939 J4 Cub and flew around the country.

One experience from the past she could have done without was an engine failure in her Stinson at 1,000 feet – not all that uncommon in pre-war airplanes. She had to land it in a soft field. But she was safe and so was her daughter.

What will be her next time-travel adventure? She wants to know what Louise Thaden might have felt. Her next goal is to buy a Beech Staggerwing.

JEAN WINGFIELD MCLAUGHLIN, born Oct. 25, 1925, in Kentucky. Began flying in 1960 in self-defense – she just wanted to know how to land in case something happened to the pilot. Since then she has taught many, many copilots her own pinch-hitter course, flying them from the right seat – some of them male, some female, one 14-year-old who was the oldest of three children who flew with the father.

Now, 18,000 plus hours, she is still teaching primary, advanced, instrument and multi-engine. Many of her students have gone on to become professionals in aviation which makes her very proud. Just recently, she was inducted into the Illinois Aviation Hall of Fame and in 1992, she was recognized by the National Aeronautical Assoc. with a certificate of honor for her work in grass-roots teaching. She has taught in everything from antiques to jets and seaplane through helicopters.

She raised three children and one son followed her into aviation where he now flies Lears and Challengers as a corporate pilot. Her husband and her former flight instructor is co-owner of a small business with Jean where he gives checkrides as a designated examiner and they both teach. His logged hours run into the 30,000s.

"We will continue as long as our health permits and it seems that spending so much time in the air has made us extremely healthy, especially when we see some of our friends who are withering on the vine!" quoted from Jean.

SISTER MARY LORETTA MCLAUGHLIN, born in Brooklyn, NY. Before school opened Sister Mary Loretta was working at the mother house of her order, the Sisters of St. Joseph, in Brentwood, Long Island, where retired and sick nuns live. Recently she was among a group of nuns taking a minicourse in auto mechanics, given by another nun, a certified auto mechanic. She holds a BS from St. John's University, and did graduate work in science and science education at three other universities. Between teaching assignments she was a medical librarian.

Her passion for planes began with Lindy's epic transatlantic flight. While in high school she hoarded her nickels and dimes, meant for lunch and carfare, saved $3.50 to pay for her first flight

"I didn't tell my parents when I went to Floyd Bennett Field in Brooklyn. It was an open cockpit biplane, the Blue Bat."

The pioneer pilots of the 30s were her heroes. She wept when Wiley Post and Will Rogers crashed at Point Barrow, AK. She remembered the exact date Aug. 15, 1935. She also was brokenhearted when Amelia Earhart disappeared.

"It was July 3, 1937. I stayed up listening to the radio until it went off at two in the morning."

Sister Mary Loretta's memberships included 99s, IWASM, AOPA and library organizations.

RENATE MCLAUGHLIN, born in Germany in 1941. After undergraduate study in Germany, she emigrated to the USA in 1963. In 1968 she earned a Ph.D. in mathematics from the University of Michigan. Since 1975 she has been professor of mathematics at the University of Michigan – Flint.

In 1964 she married James C. McLaughlin. They have two sons, Frank and Kevin. She and James have jointly owned three airplanes: C-150, C-172RG and now a C-172.

Renate and James took flying lessons together and earned their private pilot certificates in 1980. Renate joined the Michigan Chapter of the 99s immediately and in short order became newsletter editor and then chair of the chapter. She has been particularly interested in the Forest of Friendship project of the 99s.

Her most memorable experience with the 99s occurred when, during the International Convention in Anchorage, she sold her banquet ticket to help finance a seaplane rating out of Lake Hood.

CATHERINE MCMAHON, born Dec. 5, 1948, in Hoosick Falls, NY. She was graduated from New York State University at Albany in 1970 and received a MA in business from San Francisco State University in 1989. While working on her MA, Catherine met a young woman pilot who convinced her that learning to fly was not an accomplishment reserved only for military personnel and the very wealthy. With the help and guidance of the 99s, Catherine received her private pilot certificate on Sept. 21, 1991, in Oakland, CA.

Catherine is a proud member of the 99s and the Aircraft Owners and Pilots Assoc. She is also an instructor and air observer for the United States Coast Guard Auxiliary Flotilla 11N-02-03.

JESSICA MCMILLAN, born in Hartford, CT, on March 29, 1952, and moved to her present location of Boulder, CO, in 1971. She became a graphic artist after graduating from college in 1977. Her interest in aviation was sparked through her husband, Bruce, a pilot and airplane owner. She received her license and became an active member of the Colorado Chapter in 1984. Jessica is currently the chapter secretary, news reporter, and co-chairs the companion flyer course. She is past PR chairwoman and has worked to produce the Mile High Air Derby where she and partner Gretchen Jahn placed third in 1994. Jessica flies a Bonanza, has more than 600 hours, is IFR and commercial rated working towards her CFI.

Jessica has more than 20 years experience as a ski patroller. She is a member of the Arapaho Basin Volunteer Ski Patrol where she was the 1993-94 patrol director. She is an EMT, CPR instructor, winter emergency care instructor and an avalanche instructor.

MARIE ELIZABETH MCMILLAN, Las Vegas Chapter 99s, 6,500 hours. Federation Aeronautique International Sporting License No. 3. Holds 656 World & National Aviation Records for Speed, Time to Climb and Closed Course in Class C-1-c Group I. Speed records to all Caribbean Islands and all major cities in Mexico. International media recognition was given to a single trip in 1983 when more than 300 World and 300 National aviation records were achieved. That same trip was also the focus of a US nationally syndicated television show "Hour Magazine."

Born Exeter, CA, Aug. 1, 1926. Soloed at Thunderbird Field in Las Vegas, NV, on Memorial Day 1961. FFA Commercial certificate, single and multi-engine land, seaplane, instrument and glider ratings. Certified flight instructor. BA received in 1994 from University of Nevada Las Vegas in anthropology, archaeology and ethnic studies. After earning commercial pilot license, ferried aircraft from factory to Las Vegas for John Seymour who was Howard Hughes personal pilot and executive for Las Vegas Aircraft Sales and Service for Summa Corporation.

Represented the US as a delegate to the Federation Aeronautique (FAI) World General Conferences in Czechoslovakia 1984, India 1985, Bulgaria 1989 and Hungary 1990. Supervised FAI National Record Flight for Indian Airlines from Katmandu, Nepal to New Delhi, India 1985. Received FAI "Paul Tissandier World Diplome" from Prime Minister of India, Rajiv Gandhi in 1985; National Aeronautique Assoc. (NAA) "Elder Statesman of Aviation" award, 1991; Rolex watch used for timing records entered into the Rolex Watch Museum in New York City, 1993; 99 International Forest of Friendship, 1994; and Soroptomist International "Woman of Distinction Award," 1995. Numerous other recognitions by the media and professional associations.

Held all Las Vegas 99 Chapter offices. Woman Pilot of the Year 1973 and 1976. Southwest Section Pilot of the Year 1976. Two Amelia Earhart Medals. *Who's Who of American Women* 1979 through 1985. Foremost Women of Twentieth Century 1985. Memberships have included: Civil Air Patrol, Nevada Safety Council Aviation Advisory Committee, AOPA, NAA, FAI, SSA, AWTAR, AWIAR.

Flew races: Powder Puff Derby, Angel Derby, Fresno 400, Hayward Air Race, and Tucson Treasure Hunt.

Married to Dr. James McMillan. Children: Jack Daly, Jeffrey, Jarmilla, James (Michelle and Jacqueline deceased). Two grandchildren: Brian, Monique. One great-grandchild: Travis.

Volunteer flights for: Wings for Direct Relief, LIGA of Nevada – Flying Doctors and Nurses of Mercy.

EDITH GENEVA MCNAMEE, born in Knoxville, TN, April 30, 1925, as a child, moved to Bakersfield, CA, with her parents.

In 1944 she married Wayne McNamee and they became parents to Judy, Nancy and David and grandparents to Timothy, Annie, Terran, Nathan and Christy. She is a member of First Baptist Church, American Legion Auxiliary and Kern County Probation Department Auxiliary. In addition to flying she enjoys reading, cooking, planning trips and gardening.

Interest in flying began in 1978 after husband Wayne became a licensed pilot and their corporation purchased a Cessna Cardinal. He wanted her to learn how to land the plane in case of an emergency, so she took a few lessons liked it and continued on, taking the Aviation Ground School class at Bakersfield College, finding a flying instructor and received her private pilot license in July 1981.

On a flying trip to Macinack Island, MI, in August 1981 she joined the 99s, Inc. during a stopover at the EAA Fly-In at Oshkosh, WI. She is an active member in the Bakersfield Chapter and has earned her Phase IX wings in the FAA Safety Program, her auto license reads "A BFL 99."

Visiting children and grandchildren in Oregon and Washington and using the "Bird" for vacations in "faraway places" made for wonderful flying experiences every year.

HEATHER MICHELLE MCNEAL, learned to fly in Salem, OH. She took her first couple flying lessons in the summer of 1989 before she left to go to England for a year as an exchange student. Upon return to the US she continued flight training and received her private pilot license on April 22, 1991. She became a member of the 99s in 1992.

While working on her instrument rating at Youngstown Municipal Airport, Heather was also attending college at Youngstown State University. In September 1992 she began working for an ambulance service in Youngstown, OH, and graduated with an associate degree in paramedicine in June 1993. On Sept. 20, 1993, she received her instrument rating.

Heather decided to continue her flight training at Emery Aviation College in Colorado Springs, CO. During one of her mountain flights she landed at Leadville CO. This is North America's highest elevation airport at 9,927 feet. She also completed a United Airlines weekend Training Orientation Program and logged an hour of flight time in the Boeing 737 simulator. Heather graduated from Emery with an associate degree in Aviation Technology in July 1994. She completed her commercial and flight instructor certifications for airplane single engine and multi-engine land, including instrument and ground instructor ratings.

Heather plans to graduate from Youngstown State University in June 1996 with a bachelor's degree in allied health and she hopes to continue flying for many years. She is engaged to James W. Drake. They plan to be married on June 8, 1996.

MARY JANE MCNEIL, member of Long Beach, CA, Chapter. She soloed May 1976, licensed November 1976; became a 99 in September 1977.

Mary Jane owned a Cessna 172 for 13 years and accumulated 1,000 hours and flew the western area of the North American continent.

She flew many charity missions to reservations and orphanages in Arizona and Mexico with the Hawthorne Disaster Wing and the Rotary Club.

She flew many air races, including the Palms to Pines, Henry Ohey, All Woman Baja Air Races and was a timer at Guaymas, Mexico for the Angel Derby, on the way to Acapulco, Mexico.

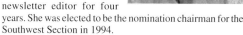

She was honored by the LGB Chapter by placing her name in the Forest of Friendship, Atchison, KS, in 1989.

Chapter activities include: Fund raising by organizing and hosting the AGT Flight Instructor Revalidation Clinic for 10 years; chapter newsletter editor for four years. She was elected to be the nomination chairman for the Southwest Section in 1994.

She was employed in the aerospace engineering industry for 35 years.

KATHY MCPHERSON, born Dec. 28, 1947. Employed as operator/AT&T and flight instructor/general aviation services. She has one child, Heidi. Joined the Reno Chapter 99s in 1990. She currently holds the office of secretary and has held uniform chairman. McPherson became involved in aviation in 1988 with a scholarship from AT&T. Her ratings include instrument and CFI (SEL); commercial, flying for business and pleasure. Graduated UNR in 1992; NV real estate license; lifeguard (Reno); National Ski Patrol. Her other interests include astronomy, religion and philosophy; skiing, swimming, animals. Member of National Assoc. of Flight Instructors and Friends of Washoe County Library.

ELAINE HARRISON MCREYNOLDS, a fourth generation Texan, started flying in 1978. She flew mostly Grummans until she bought a Piper Archer. She has an ASEL license and 203 hours. Elaine joined the Dallas Redbird Chapter in 1981, and has served as chairman, vice-chairman, treasurer, historian, and scrapbook chairman. Hobbies include photography and genealogy.

Active in the community, she is presently on the board of the Institute of Management Accountants, and will be treasurer of Nancy Horton Davis Chapter of the DAR next year. A certified public accountant, she belongs to AICPA, TSCPA and Dallas Chapter of CPAs. She holds BBA and MBA degrees from the University of North Texas. Elaine and husband Robert have a son and daughter and seven grandchildren.

JEANNE L. "BECKY" MCSHEEHY, born Dec. 28, 1917, in New York. Joined the 99s in 1940. Learned to fly in 1940 in a Piper Cub J3 through the 35 hour primary course of the CPTP (Civilian Pilot Training Program) while working in Washington, DC. She also went through the secondary course of the CPTP, 40 hours in a Fairchild PT-19 in 1941. She then joined a small flying club with a Luscombe 8B and built up her hours to 200.

Becky came out of WWII with four sets of uniforms: Civil Air Patrol, WAAC (Women's Army Auxiliary Corps), WASP (Women Airforce Service Pilots) and American Red Cross with overseas service in Italy and Germany.

During her year's service in the WASP she logged about 500 hours (for a total of about 700) on the Stearman PT-17, Vultee BT-13 and North American AT-6. She was assigned to Mariana Army Air Field, AZ, as an engineering test pilot.

Becky became a member of the 99s in 1940, was a charter member in 1946 of the San Diego Chapter, and currently is in the Mission Bay Chapter. She also is a member of Silver Wings, the San Diego Aerospace Museum and Women Military Aviators. Unfortunately, the last entry in her log book is dated 1947. However, she put in 22 years in the US Air Force Reserve and retired as a lieutenant colonel.

MARGARET MEAD's
first interest in flying is documented by this photograph from 1940! Twenty-one years later she earned her private license in Southern California. By 1967 she had a Airline Transport Certificate and had logged several thousand hours of air taxi and flight instruction.

Since airline jobs were not open to women, Margaret opted for an aircraft sales career and became the first female district sales manager in the business jet industry. She held that position with Learjet from 1972 until forming her own business jet brokerage and consultation firm in 1983.

Air racing was her obsession until the early 1970s. She chalked up several first place finishes including the Powder Puff Derby in 1968 and 1970, and the Angel Derby in 1972. She also piloted a Piper Aztec in the transatlantic "Great London to Victoria Air Race" in 1971.

Now from Santa Fe, NM, Margaret flies a turbo charged Pipe Twin Comanche. She travels the Rocky Mountain states, Pacific northwest and California coast for business and pleasure. In the summer you can find her fly rod and waders along side her brief case in the airplane's baggage compartment!

JACKLYNN "LYNN" HAWKE MEADOWS,
born Dec. 6, 1945, in Colorado Springs, CO. She learned to fly at NAS Moffett Field in 1978. She currently owns a Cessna 172, logging more than 500 hours. Joining the Santa Clara Valley 99s in 1981, she moved her membership to Lake Tahoe 99s in 1987. She has held all chapter offices including chairman twice.

Meadows taught elementary school for 17 years for the Sunnyvale, CA, school district. Since her move to Truckee, CA, she has been a typesetter and graphic production manager for several local newspapers.

Meadows has been active in the 99s not only locally but as chairman of the 1991 fall section meeting at Lake Tahoe, as circulation manager of the *Southwesterly* newsletter since 1993, and by chairing the 1995 Air Race Classic start from Reno, NV. She raced in the Palms to Pines Air Race in 1989 and has twice been given the Lake Tahoe Chapter service award.

She is married to Tom Meadows who is also a pilot and local CFI at the Truckee-Tahoe Airport.

BERNADETTE MEAGHER,
born July 15, 1950, in Milwaukee, WI.

Has a private pilot license and flies Cessna 152 and 172.

Meagher has one dog, Breeze (Whippet) who is 14 years old.

She began private training in Santa Monica, CA, finished in Milwaukee, WI (Timmerman Field).

SHAUNA MEGILL,
born Sept. 11, 1947, in Yellowknife, NWT, Canada. Joined the 99s in 1979-80. Flight training included private pilot license; single-engine, fixed wing airplanes; endorsed for night flying. She has flown Cessna 150, 152, 172; Beech Sport, Sundowner; Grumman Yankee, Cheetah, Tiger; Piper Cherokee. Megill is a conference interpreter by profession, working in an aviation organization (the International Civil Aviation Organization).

NONA W. MEINEN,
Phoenix 99s, has a glamorous aviation history because of the mostly-Hawaiian-setting, but it really is unique. Her husband has been a major airline pilot for over 37 years, and her son has also become one. She got her private ASEL in 1981 to ride as safety pilot with the son, who was working on his instrument and commercial ratings. She loved flying, and the 99s, which she says her instructor told her she had to join.

All this was in Honolulu, where she had much flying experience before recently moving to Phoenix. Examples – a scary dead stick landing (clogged carburetor) into the airport with her 85-year-old mother as a passenger and a scarier forced landing, from over water, into a sugar cane field. That time she had the good fortune of great updrafts, and the presence of a 22,000 hour pilot. A sheared valve wrecked the engine, but the aircraft was OK.

Nona and her husband motor home and fly, and are in the process of buying a Beech Sundowner with aerobatic capability. She served two years as chapter chairman and three as treasurer of the Aloha Chapter. She has flown the Apuepuelele (annual proficiency race) in Honolulu many times, winning it twice.

KATHY MELBY,
born and raised in Dallas TX, and is now residing in Oregon. She was the first in her family to earn her private pilot license, much to the amusement of her three brothers. She completed her checkride with Jane Phillips, who is also a 99, in September 1994. She joined the 99s in November 1994. She hopes someday to obtain her instrument, commercial, CFI, and multi-engine ratings. With these ratings she would like to teach flying to at-risk adolescent females.

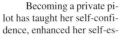

Becoming a private pilot has taught her self-confidence, enhanced her self-esteem and thrilled her adventurous spirit. She would like to pass this feeling on to other women. On one of her first trips after receiving her license, she was coming in for a landing at Eugene Mahlon-Sweet airport in a Cessna 152. She had a very strong left cross-wind. When she touched down her left wheel refused to touch the ground. She tried everything her limited experience could imagine to not flip over on the only active runway in front of the observation area! Finally, the wing just dropped. She taxied to her parking area. She sat for few minutes and went over the landing in her mind. It finally dawned on her that she could have aborted the landing. She has no problem aborting a landing now! When she calculated the cross-wind component, it was near the maximum manufacturers recommended cross-wind component of 14 knots. What a lesson learned!

LAURA MELBY,
interest in flying began while watching her father-in-law fly a Rans Ultralight almost on a daily basis. Laura began her flying lessons in a Cessna 152 and now flies the family Cessna 182. She received her private pilot license on Sept. 8, 1994, just 40 days before her son Dalton William was born, and became a member of the Women With Wings Chapter on Sept. 10, 1994. She is married to Duane Melby, an engineering assistant, who is also working toward getting his private pilot ticket. Laura is employed as a court stenographer for the Portage County Courthouse.

ROBBIN MELCHIORRE,
born on March 26, 1950, in Joplin, MO. In 1978 she learned to fly at Wagon Wheel Airfield in Lambertville, MI. She purchased her first airplane, a 1977 Cessna 172 the same year and used it to obtain her instrument rating in 1979. Since that time she has logged more than 900 hours, most of them in Cessna 172s. Robbin joined the All-Ohio Chapter of 99s in 1979, and transferred to the Potomac Chapter in 1980. In 1993 Robbin met and married her

husband, Joseph Melchiorre, at Eagle Crest Aerodrome in Lewes, DE, a residential airport community where they both live. She navigates when they fly his 1964 V-Tail Bonanza.

Their wedding reception was held in the hangar and they flew away with Robbin in the left seat of the Cessna in her wedding dress.

She has a daughter, Stacy, and a granddaughter, Madeleine, six step-children and three grandchildren. She flies for pleasure and in her business selling total hip and knee replacements to orthopedic surgeons.

AILEEN SAUNDERS MELLOTT,
earned her SEL rating in 1956, logged more than 4,000 hours; includes twin Bonanza, a Baron; four hours of T-33 time.

The only lady on San Diego County Airport Commission; director at SD Aerospace Museum. Air races: Powder Puff Derbies, first in 1959 and 1960, second in 1961; International Air Races, second in 1960 and first in 1962. Selected as Pilot of the Year, Citizen of the Year, (1961); Woman of the Year (1968). Helped organize first Baja Air Race from Long Beach to La Paz. A forced landing at El Rosario,

Baja, CA, led to the formation of the Flying Samaritans to fly-in medical teams and supplies monthly for the natives.

Most memorable is flying above the Arctic Circle, around Alaska; and the forced landing over the Locondonia Jungle, spending two days with the Indians. An active charter member of the Mission Bay Chapter.

She has a son, lovely daughter-in-law; and two grandchildren.

MARY A. MERCKER,
born June 29, 1932. Joined the 99s in 1957, Tucson.

She has held the following offices: chairman, treasurer, aerospace education, and scholarship. Learned to fly at Phoenix Sky Harbor in 1949-50.

Her ratings include private 1950, commercial in 1958, multi-engine in 1985, CFI in 1978, CFII in 1980, CFII, Pres. Alpha Air, secretary treasure Manasco. Brought home Aviation Underwriters trophy for Phoenix College in 1950. Taught ground school privately through 1950s, for Pima College in 1980s.

Now President Pima Aviation Assoc. (ground schools). Current activities include director, APA, treasure 99s member, AOPA, NAG.

BARBARA MEREDITH,
BC Coast Chapter, both parents were pilots so it should have been no surprise to her when they asked if she wanted to take lessons when she was 16. Taking lessons at that point meant getting to the airport at 6:30 a.m. and after waking up her instructor she'd do the walkaround until he was ready. Her license wasn't issued until she turned 17 and shortly thereafter she joined the 99s as an Alberta Chapter member as there were no BC Chapters at that time.

Over the years, Barbara has held almost all the officers position both at chapter and at section level. She was co-treasurer and hospitality chairman for the 1987 Convention in Vancouver. Barbara also developed and was first editor of the section newsletter, Tracking Outbound.

She has allowed her license to lapse because she has neither the time or resources to fly, but she continues to be active in the 99s.

She has her own business, providing contract administration and accounting services, and a one year old daughter, Meredith.

EUGNIE M. MERRELL
no bio submitted

THYRA MERRILL,
Bay Cities Chapter 99s Charter Member. The widow of the late Russell Hyde Merrill. She

and her husband both had commercial and transport licenses, and they pioneered an aviation center in Alaska. Russell was killed in a plane crash in 1929, and Thyra moved to Palo Alto with their two small sons.

Thyra was a charter member of Bay Cities Chapter in March 1932, but left the Bay Area in August 1933 to return to Alaska, where she planned to establish herself in aviation.

SYLVIA M. MERRITT, born Aug. 13, 1924, Pittsfied, MA. Joined the 99s in March 1975. Earned private license in 1974 at Turner Field in PA. She has flown Piper Cherokee 140 and 180, and Cessna 150 and 195. Married to A.H. "Bill" Merritt, 1922-90. They had four children: William, 51; Candace, 47; Lynn, 38; and Jim, 35. Grandchildren are Laressa, 20; Christie, 19; Sean, 17; Leah, 17; Drew, 5; and Jeffrey, 3.

In 1973 Sylvia's husband, Bill, decided "its now or never" and purchased a Cessna 195. He had not flown since his Army Air Corps years when he flew the B26 and A26 during WWII. For Christmas Bill gave Sylvia one Cessna and one Piper introductory flight and in July 1974 she had her private license.

Youngest son, Jim, soloed gliders at age 14 and got his glider license on his 16th birthday, his power license on his 17th birthday and is now commercial and instrument rated and a certified ground instructor.

Daughter, Lynn, earned her private pilot license in 1976 while studying engineering at Georgia Tech.

In March 1975 Sylvia joined the 99s and has enjoyed their support and fellowship ever since. She has attended every International Convention except one since becoming a 99 and has made many wonderful friends all over the world.

Sylvia has also been a member of AOPA since 1981.

TERESA MERTZ, a Northwest Airlines flight attendant for the past 10 years, learned to fly at Pampano Air Park in 1994 and hopes to continue and achieve the instrument, commercial and multi-engine licenses.

Terri was inspired by her grandfather who served in WWI in the Army Air Corps and encouraged by her mom and three sisters.

The most thrilling flying experience was in a Navy Air Corps Stearman Trainer doing aerobatics. Also, some helicopter experience in a Robinson 22 flying over the east coast beaches of southern Florida.

Educational background includes a BS in education from Plattsburg State University with a minor in Spanish. She has been married for five years to Dean Mertz who flies C-6s for the New York Air National Guard and is employed as a copilot on the LIO-11 with a commercial airline.

A former ice skating instructor and an avid rollerblading enthusiast. Terri is a member of the Teamsters Airline Division and Aircraft Owner and Pilots Assoc.

In the future, she hopes to use her flying ability to aid the medically and physically challenged victims of national disasters and would like to volunteer time towards the preservation of animals and wildlife. Eventually, she hopes to work her way from the back of the aircraft to the front of the aircraft on a full-time bases.

JULIE G. MESSERRLY, born May 13, 1966, in Stamford, CT. Joined the 99s in 1988. She is single and has no children. Flight training includes Institute of Technology (private through multi-engine commercial instrument) Melbourne, FL; American Flyers (certified flight instructor), Fort Lauderdale, FL; certified multi-engine instrument flight instructor in Groton, CT. She has flown Cessna 152, 172, 172RG, 182RG; Piper Warrior, Archer, Arrow, Seminole, Aztec, Seneca; Beechcraft Baron, Learjet 31, Beechcraft 1900C.

Employed as operations coordinator at Westchester County Airport; flight instruction at Danbury, CT; copilot Learjet 31 private corporation; currently copilot 1900C for a private corporation.

She was flight instructing, and finally had the opportunity, time and money to treat herself to a pleasure flight, and, it was on her birthday. It was one of the most enjoyable flights she has ever experienced! She flew over mountains, noticed wonderful cloud formations, snapped a few photos, landed in a "new" airport, and noticed small features and interesting objects she hadn't observed before. It brought back the joy and special feeling of flying, much like that first initial flight lesson.

Flying back from the Bahamas to New York at 41,000 feet in the Learjet, there was a tremendous, beautiful, bright glow from the east. She couldn't figure out the source of the glow. A few minutes passed, and her puzzle was solved; it was the moon! She had never seen the moon so bright or large before. What a vision! What a privilege to be flying at such an altitude and witness such a great sight!

PATSY LEE MEYER, born May 7, 1951, in Lincoln, NE. Joined the 99s in 1987. Flight training included private license with instrument rating. She has flown Cessna 152 and 172, Piper Arrow, and Saratoga.

Married to Steve Meyer and has one child, Melanie, 18. Owning an aircraft has been an enjoyable experience. The ability to go when and where you want outweighs some of the monetary anxiety! They have flown to a couple of section meetings which is always fun!

SHARON KAY MEYER, member of the Nebraska Chapter. Has been news reporter, vice-chairman, chapter chairman. Worked on various other committees and projects. Chapter Achievement Award winner, 1976. Private license, 200 hours. Received BS and MA in elementary education and specialist in educational administration from the University of Nebraska-Lincoln. Former elementary teacher (taught in Nebraska and for the Department of Defense Schools in Okinawa and Germany). Enjoyed taking students for airplane rides. Currently employed as consultant in elementary education, Nebraska State Department of Education, Lincoln.

RUTH SCHUSTER MIELE, born May 26, 1930, in Youngstown, OH. She flies Cessna and Piper aircraft. The owner of Cessna Pilot Center, FAA 141/135 certified charter. Married to Anthony and has two children, Lisa, 35, and Mark, 33.

Her father's interest in aviation resulted in her first ride at age 7. Prior to that age, her "China Clipper" toy airplane, spurred her interest. When her children began school, her flight training began. She received her license, and while teaching elementary school she and her husband began a flight training school, with her husband as instructor and she as the manager, their little school grew into an FAA Part 141 school with a Part 135 Charter.

After her husband's death, due to cancer, she now finds herself carrying on with the business. She has remarried, continues to fly, with new plans for AmAir Inc.

As a charter member of the Lake Erie Chapter of the 99s, she finds her aviation experiences have enriched her life immensely.

What a thrill it is to be able to enjoy the first solo, first solo cross country, and finally a pilot license with the many students they've had in the last 25 years of their business.

CAROL MIGHT, born in Ohio in 1961 and began flying in 1979 at Kent State University. She has a commercial instrument multi-engine land rating and is a certified flight instructor. She received her BS in aerospace technology while flight instructing for the University.

In 1984 she was hired by the FAA as an air traffic controller at Cleveland Air Route Traffic Control Center. She became an air traffic instructor and was active in educating pilots in air traffic procedures. She received an achievement award for her work in aviation as well as her volunteer work with the community as a crisis counselor for the Cleveland Rape Crisis Center.

She currently works at the Mike Monroney Aeronautical Center in Oklahoma City as an international air traffic control instructor while studying Spanish at the University of Central Oklahoma. She is a member of AOPA, Professional Women Controllers and the 99s.

JOYCE MILLARD, a lifetime Michigan resident. She began flying in 1991. Millard is a proud member of the 99s, Inc., Aircraft Owners and Pilots Assoc. and Experimental Aircraft Assoc.

She owns a Piper Cherokee PA28-150 and has also logged time in a Piper Warrior.

Some of her community activities have included CPR instructor, St. Vincent Food Assistance Program and domestic violence counselor. She is married to Ken Millard, who is also a pilot and an electrical engineer. She has three daughters.

Millard and her husband are currently building an Adventure Air Amphibian and are planning many land and water adventures in the near future.

ALMA J. MILLER, born in Millersburg, OH, and has been a resident of Ohio all her life. Her husband Eli owns and operates a general contracting business for which she performs the secretarial duties. She is also involved with the Winesburg Fire Department as a volunteer EMT.

She has always enjoyed the great outdoors. She loves to travel, fish, work in her gardens, read, and fly airplanes!

Eli started flying when he was 43 years old. After her first ride with him she was totally hooked! Eli has an instrument rating. She obtained her private license when she was 45 years old and plans to get her instrument rating this year.

In 1992 they bought a Piper Dakota and had it completely refurbished a year ago. It is such a joy to fly that she can hardly quit! Their future plans include a trip to Alaska in 1996 with their own airplane.

She became a member of the 99s in 1994, and is a charter member of "Women With Wings."

DONNA MILLER, learning to fly in South Korea was an interesting experience. It was a 10 hour trip commute to the only aero club in the country. Navigation was critical. The sectional charts very clearly state, "Aircraft infringing upon non-free Flying Territory may be fired upon without warning." Like looking into a rear view mirror, communist countries are closer than they appear.

Donna then moved to Frankfurt, Germany. A brisk 95 mph commute on the autobahn to the flying club was a refreshing change and viewing castles from above made for wonderful Saturday afternoons.

From Germany, Donna returned to Colorado. She became involved in a project to help catalog Captain Jeppesen's memorabilia for display in the Jeppesen Terminal at Denver International Airport, and learned aviation history first hand from a true pioneer. Donna is the chairman of Flight Without Fear and a historian member of the OX-5 Aviation Pioneers.

DORIS E. MILLER, Garden State Chapter, life member of 99s. Started flying in 1965. Earned commercial license with instrument rating. Joined 99s in March 1967 (Greater New York Chapter) and became charter member of Hudson Valley Chapter in 1968. Chapter chairman for two years. Served as New York - New Jersey Section treasurer, vice-governor and governor. Co-chairman of 1979 99s International Convention. Also served as registration chairman for that convention. Moved to New Jersey in 1978 and transferred membership to Garden State Chapter. Served as chapter chairman. One of her greatest joys was owning a Piper PA28-180. She flew for pleasure and her favorite activity was air racing. Flew two AWTARs and four ATWARs. Entered several AWNEAR and Garden State 300 contests. While a member of the Hudson Valley Chapter founded the Empire State 300 Proficiency Contest and was awarded the Amelia Earhart Medal by that chapter.

Presently active as much as possible with Garden State Chapter. Also, member of advisory commission to the Flight Technology Department of Mercer County Community College and trustee of International Women's Air and Space Museum.

LEE LEGER-MILLER, born in Louisiana. Received her pilot license in 1966, and also has commercial, instrument and multi-engine ratings. Logged more than 5,000 hours. She is the current owner of a British Beagle B206S and has flown Aircoupe, Beagle, Beechcraft Baron, Dutchess, Travelair, B-26 Martin Marauder, Cessnas 172, 177, 182, 206, 210; North American T-6, and SNJ, Piper Apache, Seminole, and Piper PA34.

Cooperate and pleasure flying, US, Mexico, Canada, and Europe. Ferried Piper PA34 transatlantic FLL/ France.

Memberships include Florida Gulfstream Chapter of the 99s, past chairman, vice chairman, and treasurer. Confederate Air Force, operations officer B-26 Martin Marauder SQDN., copilot and sponsor. Life member Air, Sea Rescue, Bahamas. Also member of Ocean Reef Club; Chub Cay Club, Bahamas; Bath Club; National Organizations for Women; Navy League, French Riviera; Chevalier Du Tastevin; Chevalier De Meduce; France-ETATS UNIS; Alliance Francaise; and Aircraft Owner and Pilots Assoc. She is a certified scuba, Coast Guard licensed ocean operator, motor vessel, USA Eastern Seaboard, Atlantic Ocean, Gulf of Mexico 100 miles off shore and Bahamas.

Married to Adason and has three children: Paula A. Gonzalez, Medina F. Fritz and Gerald D. Fritz, Jr.; and stepchildren: Joan, Margaret, Julie, Susan and Tom Miller.

MARILYNN L. MILLER, member of Scioto Valley Chapter, North Central Section. Marilynn is married to husband Robert, mother of two married sons, Ron and Tom and grandmother of four grandchildren.

Started flying in 1959, received her private pilot license in 1961 and promptly joined the 99s. Now holds a commercial license, single and multi-engine land, single and multi-engine sea, glider rating and instrument rating. Marilynn spent some time towing sailplanes.

She has held offices in All-Ohio Chapter and Scioto Valley Chapter and at section level was treasurer, vice-governor and governor. She has flown one Powder Puff, four Angel Derbies, four Air Race Classics, four Michigan SMALL races, three each Illi-Nines, Arkansas and Kentucky Air Derbies, and two each Indiana Fair Races and Lake Erie Air Derbies.

NANCY A. MILLER, born May 20, 1943, in Topeka, KS. Her love of flying began while working as an airline hostess for Braniff Airlines. She learned to fly in Emporia, KS, in 1970 and later that year moved to Texas where she joined the Dallas Chapter of the 99s. Nancy moved to Memphis, TN, and served as chairman of the Memphis Chapter. She lived in Michigan for 10 years and was active in the Michigan Chapter. She returned to Memphis and then to Moody, AL, where she joined the Alabama Chapter and served two terms as chairman. Besides her full-time career as a postal inspector, Nancy raises llamas and serves on the board of directors of the Southern States Llama Assoc. She is an active Big Sister and is also on the board of directors of Big Brothers/Big Sisters of Greater Birmingham.

TESS MILLER, her pilot license is only nine months old and is still considered a new recruit in the First Canadian Chapter. At this stage in her flying career she doesn't have a lot of experience in which to draw upon and therefore would like to dedicate her insert to her dear friend and chapter president, Bev Bakti. Bev took her on several flights before she ever began her pilot's license and her excitement and enthusiasm for flying was contagious. She soon began ground school and flying lessons at the Oshawa Flying Club with lots of guidance and support from Bev. She was there when she wrote the exam, her first solo and when she finally took the big test. "Bev, thank-you kindly for everything. You are a great person an excellent leader and my best friend. Thanks a million!"

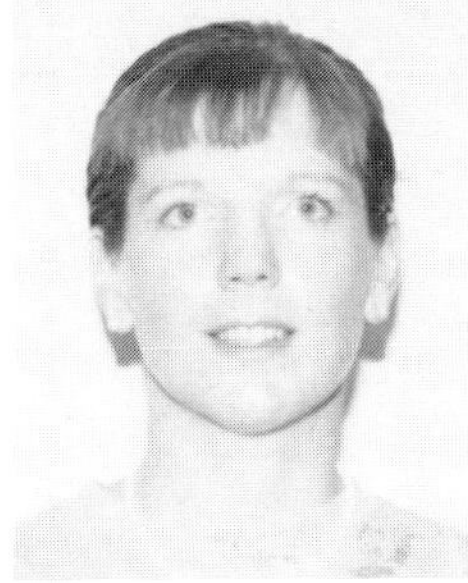

JUNE MILLS, MB, ChB, FRCP (glas) FRCP(C), DCH, Dip. Av. Med. Born in Becontree, Essex, England, graduated at the University of Birmingham U.K. Traveled and worked through Europe, India, Australia, Burma, then settled in Saskatchewan for 19 years before moving to British Columbia, Canada. Licensed in 1970 and became a 99 straight away; total flying hours now 2,500; commercial, with night, block airspace and instrument ratings. Granted diploma in aviation medicine, after six months training at Farnborough in 1976. Has since written articles, given lectures and counseled problem pilots.

Member of the Saskatchewan 99s since 1970, and was chairperson for some years; also member of the Saskatchewan and International Flying Farmers. Now member of the Canadian Rockies, presently the newsletter programmer and vice governor of the West Canada Section. Was previously chapter chairperson as well. Governor's Member of the Year Award 1989.

Member of the Aerospace Medical Assoc., Canadian Society of Aerospace Medicine; previous president of the Saskatchewan University Women's Club, programme chairperson for the Penticton Club (B.C.). Some "Fear of Flying" counseling; CASARA (Canadian Air Search & Rescue Assoc.). Hiking and photography.

RANELL MINEAR, of Akron, OH, got her private pilot certificate on March 8, 1985. She always wanted to fly, even when she was in high school. She went to ground school in the early 1970s, but had to cancel her flying plans when her mother passed away and she had to work for the family business.

In the mid 1980s, she had the chance to begin flying again and found that flying was time saving in out-of-state jobs in her awning/tent rental business. Her personal flying is limited to visiting friends in West Virginia and in-laws in Tennessee.

Ranell is a member of the Women With Wings Chapter of the 99s. She is married to Wayne and they own a Piper 235.

DORIS MINTER, born April 18, 1927, in Pittsburgh, PA. Joined the 99s in 1961. Flight training includes ASEL, commercial, instrument, A&P mechanic. She has flown Aeronca, Cessnas, Bonanza, three hours in helicopters and aerobatic planes. She volunteered surveillance Santa Monica Police Department for two years. Married to Rex Minter and has three children: Christie Thobe, 46; Laurie McGinnity, 38; and Thomas Minter, M.D., 37. Her grandchildren are as follows: Nani McGinnity, 20; Nick Thobe, 15; Matt Thobe, 9; Jamie McGinnity, 9; Jason McGinnity, 6; Danny Minter, 5; Billy Minter, 4; and Sarah McGinnity, 3. Currently a member of San Fernando Valley 99s. She was chapter chairman of Santa Monica Bay Chapter.

Doris took her first flying lesson in 1948 at Bay Meadows Airport near San Francisco, CA. Everything was great until her instructor, an Air Corps veteran, pulled the throttle back to demonstrate that the plane would not fall out of the sky if the engine stopped. She was frightened because she thought he wanted to spark. Doris didn't get back to flying until 1960 when she met Claire Walters.

Minter's confidence was restored by Claire's habit of painting her nails while her new student practiced landings. Minter now holds a commercial license with an instrument rating and has accumulated 2,000 hours. She is also a licensed A&P mechanic.

Doris has participated in many air races including 12 transcontinental races and several Palms to Pine and Pacific Air Races.

Doris has been chair of the Santa Monica Airport Commission and a volunteer surveillance pilot for the Santa Monica Police Department.

Doris and her husband, Rex, have flown their Bonanza throughout the US, Mexico and Belize. She also flew to the 99 convention in Anchorage, in 1984.

MADGE RUTHERFORD MINTON, born March 22, 1920, in Greensburg, IN. Joined the 99s in 1940 when she was a student pilot in the Civilian Pilots Training Program at Butler University. She earned her private pilot license in May 1940, and was the only woman in a class of 23. In August she was denied admittance to the advanced CPTP course which included aerobatics in heavier planes because she was female. That same day, she wrote a letter to Eleanor Roosevelt suggesting that women be used as ferry pilots when male pilots were needed for combat. On Aug. 16, 1940, Madge received a letter from Jack Gram, chief of the CAA Private Flying Development Division, that read, "Mrs. Roosevelt has requested that we make a direct reply to your letter of August 8 ... as no discrimination will be made against pilots of your sex." After graduation, Madge applied to the Royal Air Force to fly in the Air Transport Auxiliary but was rejected because she did not have enough hours. However, in February 1943 she received a telegram from Jacqueline Cochran recruiting her for the WASP. Shortly afterward, she reported for flight training at Avenger Field, Sweetwater, TX. After graduation, she was stationed with the Ferrying Division of the Air Transport Command at Long Beach, CA. In October 1943 she married Sherman Minton then a medical officer in the US Navy. As a WASP, Madge ferried trainers, twin-engine transports, four single-engine pursuit planes (the P-39, P-40, P-47, and P-51) and flew B-17s in the Army Airline Service. Madge is currently serving as a regional vice president for the P-47 Thunderbolt Pilots Assoc.

Madge resigned from the WASP in August 1944. Subsequently she has been mother to three daughters. In the late 1940s, Madge and other Indianapolis WASP and 99s members sponsored a Wing Troop for Indianapolis Girl Scouts. From 1983 to 1988 Madge served as an airshow staff member of the Indiana Wing of the Confederate Airforce. CAF regulations and time forbade her to more than just observe the warbirds she had once flown. Madge turned to scuba diving as a substitute for flying. She and Sherman scubaed as a team to catch sea snakes whose venom they collected to use in a medically oriented research project at Indiana University School of Medicine.

SUSAN "SUE" MIRABEL, lifetime resident of New York where she resides with husband Jim, six cats and a dog. Sue obtained her private pilot license in 1980 and is currently a proud co-owner of a Cessna 172 and 2448E. Since joining the Long Island Chapter of the 99s in 1981, she has held the positions of chairman, vice-chairman and secretary. Other organizational memberships include AOPA, Antique Airplane Club of Greater New York and the Aerocats Flying Cub (recently disbanded).

Sue has participated in the Garden State 300 Air Race, and Nutmeg Air Rally as well as in Poker Runs and other 99s activities. Some of her most memorable flights have been to conventions in St. Louis, New Orleans and a return from Orlando. No matter where she goes, however, Sue believes it's the camaraderie of other 99s that makes each trip special.

JUDY MITCHELL, learned to fly in Tulsa, OK, while she was working the road as an Oklahoma State Trooper. She was fortunate enough to be transferred to the Aircraft Division and was the first female pilot for the patrol. Every day was interesting and exciting flying for the patrol, working traffic, manhunts, searching for missing people and downed aircraft, drug interdictions and transportation flights in a Cessna Skylane and sometimes copilot in the Governor's King Air.

She received her multi-engine, CFI, and CFII ratings, and then retired from OHP to fly a Cheyenne for Panhandle Telecommunications Company based at Guymon, OK, with her husband, who is their chief pilot. She loves aviation because it offers endless challenges, it is a confidence builder, and because she loves to travel.

MARCIA E. MITCHELL, born in Glendive, MT, March 22, 1953. She learned to fly in a 1956 Cessna 172 and soloed on April 19, 1980. Marcia obtained her pilot licensee on March 28, 1987, and joined the 99s immediately after becoming eligible. She is a past president of the Glendive Hangar of the Montana Pilots Assoc. During her term in office, she designed and produced a patch for the hangar featuring a smiling airplane and the club name sewn in a red, white, and blue design.

Marcia and her husband, Keith, own and operate a construction company, Pine Street Inc. This business specializes in work for the State, Federal and Utilities Company as well as a general store selling more than 200 items ranging from monumental granite carpet, furniture, jewelry to small engines and many things in between.

Marcia has three stepsons, Kelly and his wife Cheryl, Kevin and his wife Dineen, Justin and his wife Michelle also two granddaughters De Lona, age 4, and Danielle, 19 months.

OLGA M. MRACEK MITCHELL, hooked on flying at age 12 her mother paid for her first flight at Barker Airfield in Toronto. She joined the University of Toronto Flying Club in her senior year and flew a Cessna 140 and Fleet Canuck at Toronto Island Airport.

Flying was put on a back burner for nearly 20 years while Olga was at graduate school in Toronto, looking after two babies, and working at Bell Labs in New Jersey. At Bell Labs, Olga met Dot Kirby, an accomplished pilot and 99 who flew her own Comanche. With this inspiration, Olga earned her private certificate at Colt's Neck and Somerset Hills in 1973 and joined the Blue Sky Flying Club at Solberg Airport to work on instrument and multi-engine ratings and commercial certificate.

In 1991 Olga retired from 28 years in telecommunications research to devote more time to sport aerobatics, competing in several regional contests and at the IAC Championship at Fond du Lac. In 1995 she earned CFI and CFII certificates.

Olga was born in Montreal, Canada, and grew up in Toronto, graduating from University of Toronto with a Ph.D. in physics. In 1963 she moved to Summit, NJ, with her husband Peter and two daughters, Janice and Susan. Olga and Peter now have two granddaughters, Emma (age 3) and Madeleine (age 1), daughters of Janice and Scott O'Shea.

Olga is a member of the Garden State and Northern New Jersey Chapters of the 99s. She is also a member of EAA, IAC and AOPA.

PAMELA ANN MITCHELL, born May 6, 1955, at Otis AFB, MA. Graduated from Colorado State University in 1975 with a BFA and went to work for United Airlines as a flight attendant. When she realized that the guys "up front" were having much more fun than she was, pouring coffee in the back of the plane, she decided to learn to fly. She soloed in 1977, got her instrument and commercial ratings, and promptly ran out of money!

In financial desperation, she started Deliverance, Unltd., a worldwide aircraft delivery company. Her first trans-Atlantic ferry was a Cessna 210 from Wichita to Dusseldorf, Germany, and her last was a Cessna 150, solo, from Memphis to Salzburg, Austria. In between other jobs she has always gone back to ferry work and has had some really fascinating trips, "crossing the pond" almost 40 times and stopping in places as diverse as Narsarssuaq, Greenland; Vagar in the Faroe Islands; Crete; Luxor, Egypt; Calcutta; Kota Kinabalu; and throughout some beautiful spots in the Far East.

The end of 1980 brought a job as production test pilot for Cessna Aircraft Company's Wallace Division, testing multi-engine pistons, turbo-props, and finally the Citation business jet. When Cessna had cutbacks in 1982, they offered her a position as the National Spokeswoman for Cessna's Learn to Fly program. In that capacity, she flew a C-172RG all over the country doing television, radio, and print interviews and "first flights" for reporters and cameramen. She thinks the prevailing impression was supposed to be "if she can learn to fly, anyone can!"

After the airline industry got back on its feet, she went to work finally in the front of an airliner! Republic Airlines hired her first as a B-727 Flight Engineer, then as a copilot on the Convair 580. Just short of the end of her probationary year, the airline down-sized and furloughed a group of about 100 pilots, including her, since she was at the very bottom of the seniority list.

She promptly went to the other side of the Minneapolis/St. Paul International Airport and went to work for the competition, Northwest Airlines.

She had found her niche in life! After one year at Northwest, she upgraded to Boeing 747 Flight Engineer and started flying international flights. She got a copilot bid in 1989 and in 1990 went on to Northwest's (and the industry's) newest aircraft, the Boeing 747-400. She now flies the state-of-the-art-long-range jumbo all over the Orient to the major Pacific Rim countries.

She has been involved in the International Society of Women Airline Pilots for about 10 years now and is on the executive board of the Society, called ISA+21. She is also the delegate from ISA+21 to the IFALPA Conference (International Federation of Airline Pilot Assoc.) for her fourth year.

Her hobbies include golf, tennis, snow skiing, and playing the piano, and oddly enough, travel! Her goals are to see and experience as much of this incredible world as possible before she dies, and each time she crosses off an item on her "To Do and See" list, she just adds two or three more to the bottom!

PATTY L. MITCHELL, born April 17, 1943, in Wichita. She wanted to be an airline pilot since the fifth grade when her father gave her a flight in a TWA Super Constellation. Graduated from high school in Derby, KS, and went to college in Abilene, TX, after getting turned down by the Air Force for flight training. Graduated from Abilene Christian University in 1967. BS in psychology. Had two children while in college. Went to work in social work in Houston and taught school.

She finally started flying in 1974 in Houston, TX. Obtained her private license in three months while substitute teaching. Motto was, "Teach a day, fly an hour."

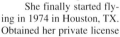

Started working on commercial and instrument. Went after the CFI and a job teaching flying at Houston Hobby Airport. Every three months she got another license including all ground instructors, CFII, then multi-engine, multi-engine instructor.

Finally got the ATP and flew Part 135 charters, instructed, and got a corporate job flying a Beechcraft Duke. Was the first successful woman pilot for Metro Airlines in Houston.

Married Robert Mitchell and moved to Stevensville, MT, and flew Twin Otters for the Forest Service under contract to Empire Airways, was their first DeHaviland Twin Otter woman pilot.

In 1982 she went to work for the state of Montana, Aeronautics Division as the director of Aviation Safety. In 1984 with another 99 and her husband built a hangar and opened an FBO in Bozeman. Flew the Citation as captain and started applying for a job as an airline pilot with many airlines and also started graduate school in counseling and human development. She finally got hired by United Airlines, Dec. 19, 1988. Finished graduate school classes three days earlier.

Based in Chicago on the 727 as second officer (flight engineer) and then went on the flying as engineer on the DC10 based in Los Angles. She flew 11 day trips to the Far East. Now she has upgraded to first officer on the 737 based in Denver. Plan to take a bid on the DC-10 or 747 as first officer. She lives on a cattle ranch in southeast Montana with her husband, and an assortment of dogs, cats, birds, horses, mules, and cows.

SUSIE MITCHELL, Oklahoma Chapter, born in Evansville, IN, came to Oklahoma at the age of 10 with her parents, two sisters and two brothers.

While working for the General Electric Company as an electronic technician, Susie met her husband, Phil, an electrical engineer who has been flying since his college days.

Susie learned to fly in 1983 and joined the Oklahoma Chapter 99s in 1984. She has held every chapter office and chaired most chapter activities including the annual "Okie Derby" a proficiency air race. She has not missed a section meeting or an annual 99 convention since joining the 99s.

Susie and Phil own a plastic injection molding company in Yukon, OK. They fly for both business and pleasure. They own a Beech 58 Baron and a Cessna 172XP. Between them they hold memberships in the 99s, QBs, AOPA, EAA, American Bonanza Society, Cessna Pilots Assoc., Oklahoma Pilots Assoc., Oklahoma City Aviation Club, Scottsdale Pilots Assoc., Seaplane Pilots Assoc. and several business aviation organizations.

After retirement, they plan to reside in Scottsdale, AZ, where they already own business property on the Scottsdale Airport.

LISA ANN MIXON, been a member since 1989, holds a private pilot license and has 175 hours. Lisa began flying at 16, before she learned to drive a car! She had an introductory flight in a LearJet with Russ Christener and stunt flying training with Edna G. Whyte.

Lisa is secretary for the Dallas Redbirds and is a member of AOPA, and Frontiers of Flight Museum. She also is treasurer of Junior Group Marianne Scruggs Garden Club and recording secretary and treasurer for Equest. Lisa is committee chairman for: Weekend to Wipe Out Cancer, Susan G. Komen Luncheon, Symphony 10K Race, Marianne Scruggs Garden Club Flower Show. She also belongs to Bryan's Friend's, Junior Group of Dallas Symphony, Texas Auxiliary of the National Kidney Foundation, Friends of Timberlawn Psychiatric Auxiliary, Women's Council of Dallas Arboretum and Botanical Garden.

Lisa is an original member of Reba McEntire's Fan Club, Texas Highway Patrol and Sheriff's Committees, National Assoc. for Female Executives for Mortgage Loan Closing, Juvenile Diabetes Foundation, CHANCE, Women's Auxiliary Children's Medical Center, Ronald McDonald House Friend.

Lisa has a BA from the University of Texas at Dallas in psychology and is fluent in Spanish. Honors include *Who's Who of American Students*, Rotary Club Student of the Month, and Dallas Cowboy Cheerleader Finalist, 1985. She is a former bank teller/mortgage loan closer.

ANN MOFFAT, born March 23, 1946, in Aspen, CO. Member since December 1991. She has a private pilot license with glider rating with 1,200 hours of flight time.

She has flown Super Cub, Cessna 120 and 195, Blanik, Schweitzer, and Cirrus. Self-employed entrepreneur and teacher.

Married to Andy. Presently she and Andy are restoring an abandoned WWII Airport. She has flown coast to coast, including Alaska and Hawaii, Mexico and Belize. She's experienced fire in the cockpit, non-working landing gear, lightning strikes, severe up and down drafts, turbulence, flying backwards in headwinds and sick passengers, but she will never give up the freedom it provides. The Lord is their Shephard, the Angels our Guides, If you stay out of Bad weather you shall not need them as much!

MADELEINE MONACO, started to fly in 1978. She is active in politics on the local level as an Alderman and airport commissioner from Prospect Heights – the city that is co-owner of Palwaukee Airport near Chicago. Former chapter chairman and former International legislation chair, she is currently North Central Section secretary.

Madeleine has a degree in applied behavioral sciences from National Lewis University; is assistant business manager of a Labor Union in Chicago; and co-founded Palwaukee Airport Pilots Assoc., an aviation action organization honored by the FAA and AOPA. Aviation education has been a long time activity for Madeleine, and she is a member of AOPA, NAFI and the Chicago Flight Instructors Assoc. She was honored with the THEO Award by the Chicago Flight Instructors and was runner up for Flight Instructor of the Year for FSDO III in 1993.

She has owned several airplanes and currently flies a Piper Arrow. Her ratings include CFII, MEI, and she is seaplane rated with more than 2,200 hours of flight time. She enjoys proficiency races and loves to travel. Madeleine has two grown children, Michelle and John.

HAZEL SHIRLEY MONROE,
born Sept. 1, 1924, in New York. Joined the 99s in 1969. She has flown Cessna 150, Cherokee 150, Arrow, Air Coupe and Piper J3.

She is a retired primary teacher. Soloed in a Piper J3 Aug. 23, 1947, acquired a private license, 1966 in a Cessna 150 and acquired 30 jumps as a parachutist from 1963-66. Joined the Iowa 99 Chapter in 1969, flying a Cherokee Arrow. Moved to North Carolina and joined the Kitty Hawk 99 Chapter acquiring a Cherokee 140. Became chapter chairman 1978-79 and again in 1985-89.

She served on the airport commission in her community, was elected as the Southeast Section Amelia Earhart Scholarship for eight years and continues to serve scholarship chairman of the Kitty Hawk Chapter.

After retiring from teaching, backpacking long distance trails and traveling became her hobby which led to chairmanships and committee work in aviation and hiking associations.

She was married to Dr. H.B. Monroe, retired Air Force major, Community College president, now deceased. They have two sons and one daughter: Gary, Terry and Susan. Her grandchildren are April, Daniel, and Scarlet.

KAREN MONTEITH,
born June 30, 1953, in Kankakee, IL. Joined the 99s in 1986. Flight training included private 1980, Seaplane (SES) 1986, AGI/IGI/Instrument 1995. She has flown Cessna 120, 150, 152, 170, 172, 180; PA 28; J3; 8KCAB; Clipped Wing Taylorcraft. Started flying in 1979 at Capitol Airport (SPI), Springfield, IL. Received private certificate in August 1980. Began working for the FAA at Quad City Airport, Moline, IL, in November 1982. Obtained single-engine sea rating at Quad City Seaplane Service October 1986.

Joined 99s as member of Quad City Area Chapter in November 1986, serving as chairman from 1988-90. While in Quad Cities, she also began timing the Illi-Nines Air Derby, an annual event she looks forward to each year. Transferred to Milwaukee Mitchell in 1990 and became a member of the Wisconsin Chapter. In 1994 received an AE Scholarship for instrument rating which she completed in January 1995. Hopes to begin commercial training soon.

MARILYN JO "BRITT" MOODY,
born in Clinton, IA, she learned to fly in 1979 in Seattle, WA. She has more than 1,100 hours, her instrument rating and Phase V of the WINGS program. She joined the 99s in the early 1980s and has held the office of treasurer and chairman of both the Mount St. Helens and Western Washington Chapters.

She is currently the treasurer of the Northwest Section. She received that Sections Achievement Award in 1992 and her local chapters' "99 of the Year" award the same year. She was also featured on the cover of the 1994 January/February issue of the *99s News*.

She owns a Beech Bonanza and has flown from Seattle to 34 of the 50 states and to the Bahamas. She has flown one Petticoat Derby and the "Palms to Pines" Air Race three times.

In 1993, her article, written about a cross country flight from Seattle to New York with copilot Jan Liberty was published in the American Bonanza Society's "Fabulous Flights."

She is also the treasurer of the "Fear of Flying Clinic" and has been a staff member since 1984.

She and husband, Jack live on Whidbey Island in the waters of the Puget Sound. They retired early and divide their time between flying the Bonanza and cruising the San Juan Islands in their 30-foot Bayliner. She is captain while in the air and he is captain while aboard ship.

CATHERINE DWINNELL-MOORE,
changed careers in 1987 from geology to flying. Since then she has flown skydivers in California, flight instructed out of their nations highest airport in Leadville, CO, and spent eight months based in Kennedy, NY, all in order to achieve her aviation goals.

Upon completion of her private pilot checkride in 1987, she received an honorable mention for an excellent checkride in the California *FAA Aviation News*. Her excellence in aviation extends to her current flying, as a first officer for American Eagle flying ATR-72's out of Dallas/Fort Worth and Chicago O'Hare.

Her husband Tom, is an active 49 1/2, who joined with Catherine in October 1991. (High Country Chapter). They own and fly a 1956 C-180, based in Buena Vista, CO. Although Tom is not a professional career pilot, he shares her enthusiasm and helps to keep her current on general aviation news items.

Her aviation goal is to fly large jets for a major airline, and have enough time off for her and Tom to travel the country in their Cessna 180, as well as, participate in more 99s fly-in activities.

DONNA MOORE, RN,
Women With Wings Chapter, working in quality management, has a degree in industrial management and MBA. Before earning her private ticket, she turned to Tae Kwon Do for an outlet, earning a third degree black belt. Now for relaxation Donna takes to the skies in her Warrior with her non-flying, accountant spouse, Allan. Earning her license May 1991, she joined the local 99s Chapter and became active at the chapter and section levels, chairing committees and holding offices. Donna is a charter member of Women With Wings.

Being an Ohio native and realizing the importance of the instrument rating, she earned her IFR and has a goal of CFI.

Donna is active in Angel Planes, Lifeline Pilots, Young Eagles, and East Central Ohio Pilots Assoc., promoting general aviation and providing transportation for those less fortunate. She has participated in local air races, aerospace education, aviation safety seminars, and flying companion seminars. The sky's the limit!

EVELYN MOORE,
learned to fly in 1984 after her husband got his pilot's license and asked her if she wanted to learn to fly also. Learning to fly at Burke Lakefront in Cleveland, OH, was a real challenge, since it is located beside Lake Erie. The flying part was fun and finding the airports were challenging. Moore only planned on learning enough to handle the plane in case of an emergency, but after the instructor let her handle the controls on the first flight, she was hooked. Moore is a proud member of the Lake Erie Chapter, AOPA and Civil Air Patrol.

Moore works as a food service supervisor at Fairfax Elementary in Mentor. Moore has been a Scout leader and a Big Sister.

She is married to James Moore who is a flight instructor and an electrical engineer. They have three sons, four grandsons, and one granddaughter.

HELEN L. MOORE,
member of the Nebraska Chapter, 99s died at her home on Monday, Feb. 20, 1995. She was a lifetime member of the 99s. Her pride and joy was the 1946 Piper Super Cruiser she purchased at the factory and owned until her death. She was the secretary of the Flying Conestogas Flying Club for 16 years and an active member of the Gage County Historical Society. In 1993 she arranged to donate the land for a permanent site marker where the Oregon Trail entered Nebraska. Whenever 99s had questions of a historical nature concerning flying, she usually had the original newspaper clipping commemorating the event.

KAYE COMBS MOORE,
Kentucky dental hygienist, soloed an Aeronca Champ on her Aug. 2, 1958, birthday at Bowman Field, Louisville. In 1960 Kaye was a dental hygienist in Sarasota and joined Florida Aero Club, Grasshoppers, AOPA, Civil Air Patrol, and 1965 joined the Suncoast Chapter of 99s. Attended first section meeting in Ashville, NC.

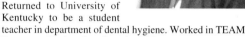

In 1965 married military pilot, R. Dan Baker and became member of San Antonio Chapter then Orange County, CA, Chapter 99s. Returned to University of Kentucky to be a student teacher in department of dental hygiene. Worked in TEAM Program with dental students in 1973.

In 1973 married Dr. H.M. Bohannan, Dean, University of Kentucky College of Dentistry. Became chairman of Kentucky Blue Grass Chapter 99s. Butler Toothbrush Company was Kaye's sponsor for 1977 Powder Puff Derby Commemorative Flight.

From 1974-76 demographer for Kentucky's Council on Higher Education. Part-time dental hygiene clinical instructor, University of Louisville.

In 1976 at the request of Fay Gillis Wells, Kaye obtained a flag of Kentucky from Governor Carroll for the International Forest of Friendship. In 1978 Kaye was an honoree in Memory Lane and enjoys the annual celebrations in Atchison, KS.

In 1987 Kaye, Sheri, Coin Marshall, and nine others flew three airplanes from Lexington to Anchorage and were welcomed by the Alaska 99s!

In 1994 Kaye Combs Moore was PIC of Cessna 172 "4X-CGG" over Jerusalem, Tel Aviv, Sea of Galilee, Golan Heights and Massada. This "First Pilots Tour of Israel" was fantastic especially meeting the Israeli 99s.

Moore is a charter member and has served on the board of the Aviation Museum of Kentucky, Blue Grass Airport, Lexington.

SUSAN D. MOORE,
born March 30, 1939, in New Jersey. Joined the 99s in 1993. Flight training included private pilot instruction. She flew PA28-180 and Cessna 172. Married Frank and has four children: Alice Moore, 27; Michelle Cardoza, 25; Ted Moore, 24; and Dave Cardoza, 22. She started flying in July 1992, as a result of real encouragement from her spouse and from a woman pilot whom she happened to meet almost by chance. She continues to be her mentor as she makes progress. She got her ticket in September 1993.

It was last April that she first flew her husband away for a weekend trip to the Pennsylvania Dutch country. After that they were really hooked on the excitement and adventure of flying trips. She has flown 100 hours in 94, mostly cross country trips. She has just started her instrument instruction and hopes to complete it soon.

BONNIE M. MOREHEAD,
born Aug. 19, 1956, in Cleveland, OH. Joined the 99s in 1995. Flight training included private and instrument; Skypark 154, Wadsworth, OH. She flew Cherokee 140, Piper Lance PA32, Cessna 152 and 172, Mooney 201. Married to James B. and has four children: Melodie, Whitney, Wesley and Jeffery.

After taking her private pilot checkride in Lorain, OH, she joined Women with Wings in March of 1995, during the Women in Aviation conference in St. Lewis. Bonnie plans to continue with her ratings and aviation education, hoping one day to transfer her professional energies towards aviation. She is married to Jim a professional pilot (UAL Cap 767) and they have four wonderful children. When she is not in a cockpit or with her family she consults small businesses on financial strategies. Bonnie sits on several boards for private enterprises of these, is a manufacturing firm, a development corporation and a non-profit lending institution.

Feel free to e-mail her at: INTERNET: N3959X@AOL.COM.

Spending most of my life quite conventionally,
Looking up at the sky and dreaming ...
I soloed on July 11, 1994
This marked the day I looked down, smiling;
The Earth ...
Something much larger than I thought,
Names and places on maps signifying life
In ways I hadn't ever thought of or imagined.

GRACE MORFITT,

born May 3, 1953, in Toronto and learned to fly in London, Ontario 1989, while studying for her MA in education. Grace remained in the London area after graduation to do work in educational research, and married fellow pilot Ron Riley, an AME and owner of Ron Riley Aero in Grand Bend, Ontario. Grace was able to pursue a life long interest in flying and mechanics, completing the coursework for her AME in 1992.

Grace has been an active member of the 99s since she was a "66" and has been Poker Run coordinator for five years. She is currently serving as Maple Leaf Chapter chair, and was instrumental in the development of educational programs such as the Theory of Flight presentation and assisting members of the local Girl Guides to achieve their aviation badges.

Current interests include compiling a history of women in aviation maintenance in Canada.

VELDA M. "VEL" MORGAN,

discovered airplanes at age 9 when she saw an autogiro land on a golf course. In 1943, eight years later, she soloed a 65 hp Taylorcraft. College, marriage, and children, took up the next 23 years, and in 1966, Vel returned to flying. After earning a private license, she joined the San Antonio Chapter in 1968, and has busily enjoyed being a 99.

She has held all chapter offices, served as South Central Section nominating and airmarking chairs, and as International airmarking chair. Vel has "raced," "timed," and "airmarked." Together with husband, Speck, she serves as an aviation prevention counselor. She belongs to: AOPA, EAA, Order of the Eastern Star, and Texas League of Vocational Nurses.

Vel, an LVN, works as a hospital pharmacy IV technician. Her second job, a part-time paramedical examiner, supports Baby Blue, her Cessna 150F.

Favorite 99: activity: airmarking; memory: Jimmie Kolp Memorial Award Recipient; and 99: daughter Barbara L. Kurtz.

BECKY HOLZ MORGENTHAL,

born Aug. 5, 1947, in Altadena, CA. Joined the 99s in 1992. She has flown C-172 and Superhawk. Married to Roger M. She is a graduate of Wilson College, Chambersburg, PA, and she now resides in Carlisle, PA, where she is the executive director of the Cumberland County Bar Assoc. She earned her private pilot license on Nov. 9, 1992, and joined the Central Pennsylvania Chapter of the 99s shortly after that. She is currently the secretary of that chapter.

Although she did not become interested in flying until later than many women, aviation is now very important to her, and membership in the 99s has enabled her to meet many people and make new friends. In 1995, she had the opportunity to attend the International Convention in Halifax, Nova Scotia.

Becky and her husband own a Cessna 172 "Superhawk," and she hopes to attain her instrument rating in the near future.

CAROL CRAIG MORRIS,

born Aug. 26, 1937, in Acworth, GA. She received her pilot's license in Dallas, TX, on her 30th birthday. She has owned a Cessna 206 and Cherokee 180 and logged time in a H model Bonanza, Twin Comanche, helicopter, and hot air balloons. She has participated in several air races in the South Central Section. A two time chairman of the Dallas Chapter, and vice chairman of the Fort Worth Chapter, she has also chaired many committees. She was a Texas Pilots Assoc. board member and organizer, Cub Scout Leader, Circle Ten Council board member, north Texas Hot Air Ballooning Assoc. member, and a member of the 99s since September 1967.

In addition to a real estate broker's license and insurance licenses, she founded Holly Outdoor, Inc. which celebrated its 20th anniversary in 1995. As owner she is actively involved in advertising organizations and also in her community. She has three sons, one daughter and two grandchildren.

MARILUPE "MARY-LOU" RODRIGUEZ MORRIS,

born in Chihuahua, Mexico. One of nine children of Marina Avila and Francisco Rodriguez.

At 11 years old, in this same city, she organized school breakfasts for the poor children of School #56. Later she represented her school on the basketball team and became captain. In 1970 she immigrated to the United States with the hope of being followed by her family, a dream that was attained a year later.

Mary Lou began her education in high school, continued in a community college and continued past that as well. After nearly 24 years of study and hard work, and after giving life to her three children, John C. Morris, Jr., Nicole Marie and Walito, being a single mother, Mary Lou enrolled in the School of Aviation. In 1994 she became a diplomatic aviator in Southern California. This same year, this woman opened new doors of opportunity for the first time for women, was elevated to president of the board of the Professional Resource Network (PRN), Ontario Chapter, California.

Morris, with three professional titles, is part of the Women's Pilot Association, Chapter International 99s, Inc. Without a doubt, some (her licenses as pilot and her knowledge) accredit Mary Lou to a position without limits within this organization. In the universe of knowledge, Mary Lou says the more she learns the less she knows. Each time she discovers something new, she realized that there is an immense amount of learning in the world.

God has given her life in order to undertake a career in whatever direction she chooses.

GLADYS MORRISON,

began her pilot career in central California in 1945 and by 1947 was a partner in a FBO based in Concord, CA. She flew in air shows in a modified Piper Cub and by 1951 was the first woman licensed by the state of California as a crop-duster. Crop-dusting was then added on to the FBO operations with several bases.

The next 14 years were spent as a charter pilot, ground instructor, crop duster and FBO management. By 1966 she was writing aviation training manuals and instructing in an approved flight engineers' school. She also instructed in AOPA Flight Clinics from 1966 to 1984.

Head of Aviation Department, Yavapai College, 1968-78. Then in 1982, Gladys was named Arizona, Western Region and National Flight Instructor of the Year.

Named in 1993-94, *Who's Who of American Women* and 1994-95, *Who's Who World Wide*. She is currently president of North-Aire, Inc., an international flight school.

PAM MORRISON,

35 years ago her indoctrination into general aviation began in Southern California. Her future husband had been a pilot for some time and he got her hooked on the joy of flying. Since, then, they have owned four different airplanes. Over the years she has learned to navigate and read charts from the right seat, but she felt that it was important for her to learn to fly as they now make several trips across the US each year.

She has two children, Tamara Ann and John. Tamara Ann presented them with their first grandchild, Matthew.

With the help of her husband, Ed, the 99s Lake Erie and the newly formed Women With Wings Chapter, she received her private pilot license in 1994 and hopes to be checked out in the Mooney this coming spring. Now she wishes she had done this 35 years ago.

PATRICIA A. MORRISON,

born Dec. 5, 1949, in Pensacola, FL. Joined the 99s in 1993. Pat is a lifetime Florida resident, learned to fly in 1992 with the encouragement from her husband. She flies a Cessna 172 and has logged 180 hours, all of which were with Eglin AFB Aero Club. She has had two years college, majoring in physical education (secondary education). She was a certified aerobics instructor for 12 years and worked retail 20 years. She is currently employed by First National Bank & Trust. Married to Bill Morrison for 25 years and they have a son, Rob, who graduated from the University of West Florida (history major).

ELAINE STREHLOW MORROW,

of Delano, MN, the first thing she did when she earned her pilot license was to send a copy of it to her brothers – who had said when she was growing up that she couldn't be a pilot because she was a girl! Elaine received her license in 1986 in time to participate in the annual daffodil flights for the American Cancer Society. Since then Elaine and her husband, Glen have purchased a Cessna Cardinal RG and a hangar at Maple Lake, MN, so they can continue participating in Minnesota Chapter 99s, North Central

Section 99s, EAA, Young Eagles and general aviation activities. Elaine earned a bachelor in business degree from the University of Minnesota, Carlson School of Management, night school, and then went skydiving to celebrate. Elaine works full-time as a technical computer consultant for Analysts International Corporation, has one daughter and three grandchildren (two of whom she's grooming for future 99s).

BETTY H. MOSELEY,

a Kentuckian by adoption, lives in Lexington, hails from her home state of Utah. Her husband, Kent, a WWII Air Force pilot was responsible for her learning to fly. Her favorite flying is over the blue grass countryside in her 1969 red, white and blue single engine Piper Cherokee *Smitten Kitten*. Loving every facet of flying, she has logged more than 2,000 hours sharing cockpit time with her husband.

After soloing, she decided she must fly the famed Powder Puff Derby. Upon contacting the race board, she was informed to fly the race she must have an advanced rating to fly as pilot in command. She continued on to upgrade her flying skills toward that goal.

In 1970, she represented the city of Lexington, KY,

flying the *Smitten Kitten* in the Indiana Fairladies Air Race with her husband as copilot. The race was great fun and gave her a taste of racing competition.

In 1971, she was Kentucky's first representative in over 10 years to fly the Powder Puff Derby in the *Smitten Kitten.* On the next to the last leg of the race, she heard a fellow racers "mayday" call. She helped her to land safely, forfeiting any chances of having a respectable score or winning. Safety for her fellow racer was all that mattered. Because of that act, she received many awards. In December 1971, the Federal Aviation awarded her the second highest award given a civilian – a beautiful inscribed silver medallion and citation. She has actively promoted safety in aviation ever since, serving as a flight safety counselor with the FAA, participating in numerous flight safety seminars, speaking to grade and high school and college level students concerning the field of aviation.

In 1972, she served as chairman for Kentucky's First Aviation Week. In October 1972, she became a feminine first to fly with the Kentucky National Guard in their RF-101 photo reconnaissance jet on an actual low level mission over Kentucky and Indiana.

In March 1973, the national publication of the International Women Pilots Organization, *The 99 News,* presented the Outstanding News Award to Betty as a woman pilot in the US. The cover of the magazine bore her picture climbing into the National Guard jet just before the flight. Betty, a professional model for many years, modeling for Coca Cola and Chevrolet, never dreamed she would grace the cover of a magazine in a Supersonic Jet in flight overalls and gear!

Betty also was one of 32 women appointed to serve on the Women's Advisory Committee in Washington to the FAA. She served as a member of that committee for five years.

She is presently serving on the bard of trustees of the National Aviation Hall of Fame in Dayton, OH.

She is a proud member of the Kentucky Bluegrass 99s. At 70-years young, she is still actively flying the *Smitten Kitten.* Betty attended Brigham Young University and Indiana University. She still teams up with her dentist husband as his office manager and bookkeeper.

Her hobbies are flying for sheer pleasure on a beautiful day and of course one could never leave out the joy of personally taking care of the *Smitten Kitten* keeping her sparkling for 26 years! She also enjoys tennis, walking, the world of interior design and entertaining. Their family consists of two kitty cats, "Airport" and "Lil Sweetie."

GRACE ANNE MOSHIER, born Jan. 31, 1950, in Potsdam, NY. Her ratings include SEL. Has flown Cessna 150 and 172. Married to Victor Lee and has one child, Daniel Lee, 23-years-old.

Became seriously interested in flying after attending a "Pinch Hitter" course in November 1992. This only whetted her appetite for flying, so she continued in private pilot training earning her license in April 1993. She was awarded "Student of the Year" honors in a Part 141 school in the face of keen competition.

She became interested in the goals of the 99s as a student, becoming a member in 1993. Her interest in aviation has continued as evidenced by serving as secretary and later as president of a local pilot's association. Her enthusiasm for flying has been recognized by her appointment to the County Aviation Advisory Board where she is instrumental in the planning and development of the two county airports.

Future goals include attaining an instrument rating and a flight instructor rating where her interest in both teaching and flying will be put to good use in imparting her enthusiasm for flying to others.

MAUREEN "MO" MOTOLA, born March 9, 1941, Ventura, CA. Joined the 99s in 1974. Flight training includes private, commercial and instrument ratings. She has flown the Piper Warrior, Piper Arrow, Grumman Tiger, and C-182. Married to Danny Motola. She is a native Californian, learned to fly in Santa Monica with Claire Walters in 1974. Became a charter member of the Palms Chapter and later joined the Orange County and Lake Tahoe Chapters. Maureen logged most of her 1,200 hours in Pipers and participated in the Palms to Pines Air Race, Air Race Classic, Angel Derby, Baja Air Race and Salinas Great Pumpkin Classic. Earned her commercial and instrument ratings along with 49 1/2 husband Danny, who has enjoyed the same zest for adventure and shared in all of her flying experiences. Currently both reside in Truckee, CA, and sell real estate in the area, occasionally giving aerial tours for a real perspective. Both also enjoy hiking, canoeing, skiing and snow shoeing.

REBECCA ANN MOUHOT, born March 21, 1944, in Houston, TX. Joined the 99s in 1972. Flight training SEL, multi-engine, instrument ratings. She has flown American Yankee, Cessna 152, Grumman, Cheetah Piper Warrior, Piper Arrow, Beechcraft T-34 Mentor, Beechcraft Debonair, Piper Seneca II, Piper Cherokee.

Born in Houston, TX, 1944 was graduated from Sam Houston State in 1966 with a BS in journalism and teacher's certificate in secondary education. She flew as a flight attendant for eastern airlines before earning her private pilot license in 1968 at Stone Mountain, GA. After moving to Fort Lauderdale, FL, she became a member of the Florida Gulf Stream Chapter 99s. In the summer of 19??, Rebecca as copilot/navigator and her friend Gene Merrill flew the 19?? Powder Puff Derby Commemorative Air Race an 1,800 mile route originating in California, terminating in Ohio. Before moving to Sarasota, FL, in 1983 Rebecca earned her multi-engine and instrument ratings.

She is the widow of Emile W. Mouhot. In Sarasota, Rebecca enjoys her three sons: Lee, 26, Will, 12; and Joe, 24. She is a realtor for Mount Vernon Realty, is a member of the Florida Suncoast 99s, Valiant Air Command, Executive Women's Golf Assoc. and Silver Liners.

PENNY E. MOYNIHAN, San Gabriel Valley Chapter. Friends had invited her and her husband to join them in their Mooney 231 for a four-day weekend in Mexico. On the third day, they caught up with their flying club's "fly-in" in Guymas. She was immediately hooked by the romanticism of flight. She decided then and there, "If these people can fly, so can I!" The next thing she knew she was taking flying lessons. Her husband had signed them up for "team training" the week after their return. They continued their team training through their private pilot certificate and onto their instrument rating in 1987. And then their own flying adventures began. Their 1967 Mooney M20F has taken them north to Alaska above the Arctic Circle, south to the tip of Baja, CA, and beyond to Puerto Vallarta, west throughout California and the western states, and east to the Eastern Seaboard and New England.

Her most memorable flying adventure to date is the three weeks they flew to Alaska. They learned to fly "Alaskan VFR" as the locals call it, the inevitable result of the omnipresent low ceilings but generally good visibility that lures pilots into flying so low that the wildlife is in easy view below. They flew hundreds of miles up the Yukon River and over the vast tundra dotted with thousands of ponds and lakes as far as the eye can see, passing over a part of the world where they know humans have never trod.

She has been a member of the 99s since 1991 and has served as events chairman for two years, vice chairman for one year, and is presently in her second year as chapter chairman.

MARYLOU MUELLER, born Dec. 7, 1943, in Springfield, IL. Joined the 99s in 1979. Flight training includes Chicago Area Chapter 99s, commercial, ASMEL, and instrument rating. Married to Ted Mueller and has one child Joel Davis, 27, an attorney in Chicago. Mueller learned to fly at Elgin Airport, Elgin, IL. Her instrument and commercial and multi-training at DuPage Airport, West Chicago. She currently owns a Piper Arrow (T-Tail) and has flown several Illi-Nines Air Derbies and the Great Southern Air Race, 1995. Mueller is also a member of Lifeline Pilots and AOPA.

DONNA ARENDELL MULLINS, a native Floridian, Donna received her flight training in 1993 from her husband Fred. "Learning to fly was one of the greatest and most challenging experiences of her life, and being able to share the joy with her husband made it even more special."

Donna started her training in a multi-engine airplane and although a relatively low hour pilot, one-fourth of her training has been in multi-engine airplanes.

Donna flies to the Bahamas Islands every chance she gets and tries to visit a different island each time. She has flown to Bimini, Rock Sound, Exuma, Marsh Harbor, Treasure Key, Staniel Key and North Eleuthera.

The summer of 1994, she and her family flew from Tampa, FL, to Sedona, AZ, in their Cessna 310.

She has one daughter, Erin, age 12, and is a member of the Suncoast Chapter of the 99s in Tampa, FL.

MARI MURAYAMA, born on April 27, 1957, in New Brunswick, NJ. She got her private certificate while attending the University of California at Berkeley. After graduating with a degree in landscape architecture in 1979, she moved to Colorado.

There, she found a job as a camera operator with an aerial survey company. After a few years running the camera, she decided that she would rather be flying the airplane. She got her instrument ratings and commercial certificate in the early 1980s and moved to the front seat. She flew for aerial survey companies in St. Louis, MO, and Boise, ID, before finding a job back in Colorado in 1986. That year she also joined the Colorado Chapter of the 99s.

Currently living in Elizabeth, CO, she flies a Cessna 206 and a Cessna 320 for Rocky Mountain Aerial Surveys, a company based at Denver's Centennial Airport.

JEANNINE MURPHY, earned her private pilot license on April 2, 1994. She flies Cessna 152s and 172s and learned aerobatics in a Citabria. Born on Dec. 16, 1965, in Colorado Springs, CO, she has a BA in history and philosophy from St. Mary's College of California. Currently she is researching aspects of Jewish and Hindu mystical traditions for a master's thesis. March 13, 1994, was one of the most blissful days of her life before earning her license. For her long cross-country, she flew from Hayward to Red Bluff to Lincoln and back. Even though the radio failed on her way to Red Bluff, this flight was her best performance to date, and was celebrated by friends with champagne as soon as she stepped out of the plane back at Hayward. Jeannine received encouragement and many classic flying books from her brother, Michael, a retired fighter pilot and an Air Force Reserve squadron commander. She will always remember the friendship of Welmoed DeBly, her instructor and Bernadette Hayward, the chairperson of the Golden West Chapter of the 99s.

LINDA RICE MURPHY, born Oct. 11, 1941, in Kunkle, OH. A native of northwest Ohio and resident of Indiana since 1978 learned to fly in 1980. While a student pilot she was attending an air show at Smith Field in Fort Wayne when she was invited to join a group of female pilots at their 99s meeting. What a joy, as she had not yet met any women pilots. Soon afterward, she received her private pilot license and became a member of the Indiana Chapter. She served as secretary of the chapter in addition to participating in many activities including compass rose and runway paintings, flying poker runs and several air rallies.

Special highlights were being asked to copilot with Margaret Ringenberg, a veteran air racer and later winning a rookie award herself, in addition to the many great friendships made.

She was very instrumental in starting the Three Rivers Chapter and is a charter member. She served as PR chairman for several years winning many North Central Section PR awards.

In addition to being a proud member of the 99s, Inc., she is a member of the greatest service organization, Fort Wayne Central Lions Club, and serves on its board of directors.

She is married to Hobart S. "Bo" Murphy, a mother of three, stepmother of two and grandmother of seven including a little "Amelia."

Linda had a long financial career and for the past 10 years has served as a travel consultant. She holds a master cruise counselor certification.

MICHEAL MURPHY, born in Nova Scotia, Canada. She started her pilot training in Nova Scotia and completed it in Alberta in 1982. She grew up in a family of 12, none of whom had any interest or involvement in flying. Her interest in flying came when as a child she was told that she couldn't fly airplanes. When asked why, the reply was, "because you are a girl!" She wanted to prove this person wrong and fell in love with flying.

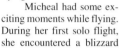

Micheal had some exciting moments while flying. During her first solo flight, she encountered a blizzard which was next to whiteout. With no visual reference she contacted the tower of her last departure. After about 15 to 20 minutes ATC had picked her up on radar and directed her to safety. The storm grounded all aircraft for sometime and Micheal made it home two days later.

Micheal joined the 99s in 1988 to encourage and support women pilots, and to be among other women who were interested in doing the same.

Micheal has a skydiving certificate to her credit. She has coached judo and in 1979 placed first in the National Judo Championships. Micheal has taught ballroom and modern dance. She enjoys experiencing different cultures and has traveled extensively around the world.

Micheal is married to Randy Gay and has two girls; Charley and Darby whom were born in 1993 and 1994 respectively.

She hopes to pass the flying bug on to her two daughters and add two more to the 99s.

PATRICIA "PATTY" LAMBERT MURRAY, a native Californian, learned to fly in 1966 out of Lindbergh Field in San Diego, earning a private pilot license in 1967. Aviation took a back seat for several years working on and receiving a BA in history at CSULB, and raising two wonderful children, Linda and Michael. Michael is now a professional pilot.

She has logged more than 1,600 hours and her certificate reads: "Airline Transport Pilot; Airplane multi-engine land; Commercial privileges; Airplane single-engine land; Private privileges; Airplane single-engine sea; Lighter than air free Balloon (limited to hot air balloons with airborne heater); Glider Aero Tow." She was awarded the Orange County Chapter of 99s Pilot of the Year for 1994-95.

Her introduction to and her love of flying she owes to Peter, her husband of 29 years who eagerly gives up the left seat to a woman! He is a captain with American Airlines flying the Boeing 757-767.

Joining the Orange County 99s was one of her best decisions. It is a terrific organization and an ongoing incentive to reach for higher goals.

BARBARA MURREN, learned to fly in 1980 at Reid-Hillview Airport and got her license and joined the SCV99s in 1981. She has her commercial and instrument ratings, with almost 1,500 hours, and has flown the Classic, the Pacific Air Race and the Palms to Pines air races. She has been very active in the chapter, serving as chairman 1989-91, won WPOY for their chapter and SW Section in 1985, and has won the SCV Service Award twice, in 1986 and 1992. She has been active in airmarking, acts on the board of directors for the Northern California Aero Club and San Jose Airport Safety Committee. Barb works as a real estate agent in Santa Clara County with her husband, John.

PATRICIA MUSSER, born in Pennsylvania, a 30-year resident of upstate New York, and now residing in Maine, learned to fly in 1979. She and her husband Henry, also a pilot, restored a 1946 Aeronca Champ which they enjoyed flying on skis. They now fly a Cessna 172.

Musser is a graduate of Pennsylvania State University and taught high school English for 33 years. Now she owns and operates a travel agency.

Her community activities include hospital board of directors, historical society, and theater groups.

She is proud to be a member of the Katahdin Wings Chapter of the 99s, a member of AOPA, and Cessna Pilots Assoc.

While teaching she created an aviation video and presented workshops as part of career education.

She has two daughters and a German shepherd who loves to fly.

CHRISTINA F. MYERS, born Oct. 25, 1913, in San Diego, CA. Joined the 99s in 1980. She has flown Cessna 142 and 180. Employed as a registered nurse. Married to Earl and has two children, Joe C. Myers, born Nov. 7, 1935, and Geraldine, born Aug. 8, 1938. Her grandchildren are Terry Welsh, 33; Eric Welsh, 30; Britt Welsh, 28; Anna K. Myers, 30; and Daniel Myers, 17. Since she's gotten her private pilot and also her husband has his, they have enjoyed flying from Chicago to El Centro and they even got to fly to Lima, Peru several times. She belongs to So-Lo 99s, Imperial.

DIANE MYERS, born May 29, 1945, in Philadelphia. She has flown Cessna 150, 172, 182, 310, 402, 421; LR 35, 55, B727, B737, DC-10, etc. Employed with Continental Airlines and married to Ralph Myers.

Today the students are from Guatemala; tomorrow, from Indonesia; and next week from Pakistan. This past year she has been teaching B737 Simulator for Continental's Contract Training Department. When training slows down, she flies B737 Maintenance Test Flights. And sometimes she flies "line trips" – as DC-10 first officer – to London in the summer or Honolulu in the winter.

She started flying in 1973; her husband Ralph was her instrument and ATP instructor (before marriage). He owned a seaplane then they did furtive "splash and dashes" in the California lakes before SPA approval.

They still fly general aviation airplanes together and sometimes he lets her land (hopefully not at DC-10 flare height). She joined the Orange County 99s, in 1974, awed by Shirley Tanner's "over 1,00 hours." In their chapter's FCS she tries to explain why the airplane flies!

Home is San Clemente, leisure activities include rollerblading (four marathons), biking and skiing.

VELMA F. MYNSTER, Greater Kansas City Chapter, their children had flown the nest and her husband Tom arranged for her first flight lesson. She loved it, and so their first Stinson. Received her license in 1974 and joined the 99s in 1976. Tom restored a second Stinson in 1978. Having held most offices in their chapter, she had the honor of serving as chairman in 1982-83.

There have been many wonderful experiences with the 99s – the fellowship as they planned and worked on projects and the joy of flying together, which they did quite often. What fun they had!

The years have flown,
Our family has grown.
The Stinson has been sold,
And our flying put on hold.
My 99 friends are still wonderful, our grandchildren a joy;
And my sweetheart of over 50 years has a new toy.
He is building a Starduster to grace the sky.
And then, how exciting, once more I will fly!

DIANA J.W. NAPOLI, born Jan. 3, 1946, in Hampshire, England, joined the 99s in 1990. She received flight training in May 1982 checkride 052690. She has flown Cessnas 152, 172 and 182.

Napoli started with a pinch-hitter course in 1982 in a Cessna 152 on a hot and humid, bumpy July day in south Florida. Fixated on the instruments, resulting in a nauseating afternoon, and not much inclination to continue! She kept up with the training on and off just to learn how to land in case something happened to her husband while flying. Napoli met up with Fran Sargent, flight instructor and friend extraordinaire, who instilled a love of flying and respect for safe procedure, and resulted in her semi-permanent 66 status, becoming a coveted 99. What a privilege to be a pilot and a 99.

She is married to Dominick J. Napoli Jr.

MARCIA A. NELLANS, Three Rivers Chapter, North Central Section. She received her license in August 1979 and joined the 99s in June 1980. By 1995 she had accumulated more than 250 hours. She flew to Anchorage, AK, with Wisconsin 99 Eva Parks in August 1984. Nellans flew the Great Southern Air Race placing fifth with Indiana 99 Margaret Ringenberg in 1986.

Nellans has served four years as chapter chairman of Three Rivers Chapter. She has served as Three Rivers Chairman of Daffodil Days Delivery for the American Cancer Society for eight years. She got her SEL private license with Warriors and 172 primary aircraft flown from Warsaw Airport, IN. Flew last two Indiana F.A.I.R. races in 1980-81 and other local proficiency races. Nellans has given several first flights including one young man who is now a professional pilot.

KIRSTEN NELSON, has been flying since 1990 and joined our chapter in 1991 with 175 hours logged. Since becoming co-owner of the last Charger, N235H, that same year, she has logged another 300 hours, including several trips from home in California to the East Coast. She belongs to the Civil Air Patrol and the Reid-Hillview Airport Assoc. She owns Clear Ink, a computer and communications consulting business and enjoys flying, reading and swimming. Kirsten holds multiple chairmanships and is a very active member of our chapter. Cherokee 235H is based at Reid-Hillview Airport.

BARBARA NERROTH, born in Waukegan, IL, in 1952. She was raised in Antioch, IL. Both her parents were aviators sparking her initial interest in aviation. After many years at the school of hard knocks, she received her private pilot license in 1990 at Waukegan Aero from Laurie Smith-Harris. Barbara worked on her instrument rating and finally received this prestigious rating on Dec. 30, 1993. While this training was taking place, she also was working on her master's degree in long-term health administration. In December 1994, she completed her master's degree. She currently works as surgical coordinator for Dr. Walter Fried in Gurnee, IL. Barbara flies a Piper Arrow from Rainbow Airport in Franklin, WI. She has been a 99 member since 1989.

JANE NETTLEBLAD, started in the CPTP program at the University of Colorado flying a Cub JC3 in the primary program. She went on through the secondary course, flying a Waco UPF through the aerobatics program. Nettleblad was enrolled in the instructor program, and the day they went out to the airport to start this course a telegram came from the government saying that women were not allowed in any more flight programs. Needless to say, she was very disappointed. Later on, she received a telegram from Nancy Love telling her she was eligible for the WASPS but since she was married with a baby, she didn't feel like leaving her family. She had some hours in a Travel Air J5 and purchased a Kinner Sportster which she dearly loved.

Nettleblad took a back seat to her husband who became a captain for Frontier Airlines. Capt. Nettleblad was the first to fly with Emily Howell on her initial flight with Frontier Airlines, the first woman pilot to fly for any airlines in the US.

She co-sponsored the Colorado Chapter of 99s with Donna Myers in September 1941. There were seven members present at the meeting in her home in Denver. She has been Colorado chairman, secretary, as well as, International membership chairman.

Nettleblad has three children: two sons, Bob and Bill Nettleblad and one daughter, Christine Welch; nine grandchildren and seven great-grandchildren. Her son, Bill, has a private pilot license. His daughter, Tracy, 16 years old, has soloed and is working on her private license. She is in CAP and hopes to enter the Air Force Academy.

DANA E. TAYLOR-NEU, born Oct. 30, 1966 in Portland, Or, grew up in northern California and began flying at age 15. At 16, Dana obtained her private pilot license and within a few years, flew the Palms to Pines Air Race with Jane LaMar. After college, marriage and moves from California to Florida to Michigan, she remains a recreational pilot. She is married to John Neu.

GOLDA NEUMAN, became interested in flying and was selected for the Future Women Pilots in 1989. She obtained her private license in 1991, and immediately joined the 99s. In 1993 she was selected Rookie Pilot of the Year. Golda has participated in many SFV Chapter events, both before and after becoming a member. She has helped with the Air Expo at Van Nuys Airport, Poker Flights, FIRC, FCS, FWP programs, socials and section planning. She has served for two years as Chapter treasurer and is currently vice-chairman.

Golda is involved in church and various other activities, including election polling, as well as enjoying quilting, needlework and tole painting. She is married to John Neuman (also a pilot and a retired aerospace engineer) and has two married sons and four grandchildren.

LINDA H. NEUMANN, born Oct. 21, 1944, in Aurora, IL, joined the 99s in 1986. After getting her private pilot license and instrument rating (ASEL), she got her commercial rating at Wisconsin Aviation, Inc., in Watertown, WI. She has flown a Piper Warrior, Archer and Arrow; a Cessna 152, 172, Hawk XP, 182, 182 RG Turbo and Beech Duchess.

Both she and her husband, Paul, are pharmacists. They are also both private pilots. They own a 1977 Hawk XP-II.

Neumann is a member of EAA, Cessna Owners Organization, AOPA and International Fellowship of Flying Rotarians. She has held every office in the Wisconsin Chapter of 99s with eight years of service.

NANCY J. NEUMANN, born and raised in Huntington, Long Island and Sag Harbor, Long Island and New York. With two older brothers and two younger sisters, she was somewhat of a tomboy dressed in lace. Her oldest brother, Drexel, presently a Delta pilot, inspired her to fly.

After graduating from Pace University in White Plains, NY, she worked for Amtrak Railroad for five years. In 1987, she was sworn in as a Suffolk County police officer and joined their flying club.

In May 1989, she became a licensed pilot and continues to punch holes in the sky. She joined the 99s in 1989 and has been an active member since. She was the chapter treasurer from 1993-95 and is presently the vice-chairperson.

MELANIE NEUMEIER, born Sept. 2, 1956, in Michigan, married Rudolph Neumeier and has two children, Gretchen age 22 and Angela 16. She has flight training in flight safety at Vero Beach with type ratings of B300/BE3B and B-737. She has flown C-152s, 172s, C-210, PA-38, PA-24, PA2T, PARO, C-441, B-200, BE-300, BE-38, WW24, C-650, C-550. She is employed by United Express BAEATP. In February 1992 she was the recipient of American Flyers Judith Resnik Memorial Scholarship.

LOIS NEVILLE, became involved with aviation in 1942 when she belonged to a high school aero club (wanted to be air stewardess). Built a model airplane for a contest and won first prize (her first airplane-right in a three-passenger Stinson at Chamberlain Air Field in Minneapolis, MN). Her first flying lessons were at Palo Alto, CA, in 1947, but nurse's salary couldn't support flying so she quit until 1961-62, when her former husband who was a new pilot wanted her to be a co-pilot. She joined the Reno Chapter in 1963-64 and has been chairman and secretary. Her current rating is private license.

Neville worked as an RN from 1947-54, raised four boys and after 21 years at home with them, returned to work as an RN for 18 years. She retired on July 31, 1992. She is married to Bob Neville.

Neville has fond memories of flying their Cessna 182 with ex-husband, Dave Williams to Alaska, the Arctic Circle, Fairbanks, Deanli, Talkeetna, Anchorage, etc. and as far south as Honduras for a dive trip (scuba dived at Quanaja). She had many wonderful trips to Baja, CA, too. She loves to ski, hike, swim and travel.

KATHLEEN NEWHOUSE, joined the 99s in 1993. She currently serves as the chairman of the Renaissance Chapter in Michigan. She is a CFII and a co-owner of an airplane business called Vintage Air Tours. She and her husband, Russ, spend their summer weekends flying passengers in their 1941 Stearman and 1929 Travel Air. Kathleen also offers instruction for tail wheel sign-offs in an Aeronca Champ. Although she is quite busy, she still finds time to promote aviation in schools and community activities.

LORRAINE NEWHOUSE, born Jan. 6, 1928, in Miami, FL. Went to school in Phoenix and Tempe. Married a pilot in WWII and moved to Tucson in 1950. She joined the Tucson Chapter in 1952, the Phoenix Chapter for one year in 1973 and back to Tucson in 1974. She has had held all offices at least once. She was chairman of SW Convention, Treasure Hunts, five Pennies-a-Pound, air tours. She learned to fly at Tempe Airport, Arizona in 1948 and soloed private in 1949 at Tucson International Airport with a commercial in 1958 and SEL.

She worked for Brunswig Pharmaceutical Co. in the office and in telephone sales for the Mt. St. Telephone Co. Back in the late 40s she worked for Ruthie Rheinhold Sky Harbor, Phoenix; Tempe Airport and Freeway Airport, Tucson.

Newhouse has been in nine Powder Puff Derbies from 1957-60, 1971-77. She experienced two engine failures both at Prescott, AZ. Her second passenger was her mother. Working at Tempe Airport, Tempe, AZ, in 1948, she received $25 a week and one hour flying time. She was the only girl flying at that airport, also the only girl to register for aeronautical class in high school in Tempe in the late 40s. Has flown medical supplies to doctors, TV camera crews and sick people. Was in Civil Air Patrol (first woman pilot in the 501 Squadron). Remembers flying her antique airplanes around the country.

Her present husband has been a pilot for American Airlines for 38 years. They fly their antique airplanes around the US to Fly Ins. She has owned the following airplanes: Cessna 140, Stinson 150, Navion 260, Bonanza C, B and V models. Newhouse and her husband, Raymond, now have a CK Bird, C-3 Aeronca that his father bought new in 1931, Beech Bonanza B model. She has four stepsons. Two fly for airlines and all of their children and grandchildren have the love of flying.

She is a member of the 99s Antique Airplane Assoc., EAA and the Assistance League of Tucson. She also enjoys watercolors and oil painting.

D. ANN NEWCOMBE, Phoenix 99s, born in Pasadena, CA, on March 23, 1927. When she was 5, her father kept taking her out to Burbank and Glendale, CA, airports to watch the stunt pilots, wing walkers and movie folk ply their trades. She was fascinated! She hadn't yet flown, but at college tried for an aviation degree. They had no idea of what she had in mind, but set out a course of study that included trucking and foreign languages. The trucking really didn't sit well, and she had trouble with English which made the whole thing come up a cropper. Anyone for business?

She got her first ride in an airplane shortly after that, making a trip from Monrovia to National City and back in a Luscombe 65. Loved it! Spent some time sailing the coastal waters, got married, had a son and flying got a little lost. Came to Phoenix, met some pilot types and bummed rides until her son joined the Marines and 13 days later she was in ground school. Getting to son's graduation from boot camp and taking the FAA written was a traveling nightmare, but once that was over the flying and fun began.

Ann joined the 99s in 1977 and has enjoyed being active and holding many offices. Somehow, the over-riding job has been writing, editing and/or distribution of the newsletter. This has been unending. She has flown a couple of Kachina Air Rallies, three Tucson Treasure Hunts, has airmarked airports all over the state and enjoyed the activities, whatever they were. Her favorite is Flying Companions, which, to everyone's dismay, she refers to as "the chicken wives." So far she can chalk 18 years of fun and friendship up to the 99s.

JUANITA NEWELL, Phoenix 99s, in 1933, handsome young Elgin Newell arrived in a Travel Air 2000. She gingerly climbed into the open cockpit. They married in 1935, bought an OX5 Robin, and Elgin gave his bride an occasional flying lesson, but, she says, they always came back not speaking. Elgin was to be out of town for two weeks, so Juanita enrolled in a ground school course, got her private ASEL in about 30 days.

Thus began great aviation careers for both Newells, working together when they could. Those early arguments were way behind them, and Juanita wants her many 99 friends to know that it was Elgin's ever-patient encouragement that got her ratings for her – after the private, she says she had "matured" and never wanted another instructor. He taught her to fly a Fleet, very cautiously because of its tendency to flat spin.

After Pearl Harbor, they moved to Phoenix and she spent all her time at Sky Harbor getting her commercial and instructor's ratings. She often instructed or chartered from dawn to dusk during the war, even taught aerobatics in a Stearman. She tried having her own flight school but went back to work for an FBO, and then got her first executive pilot job in 1947 flying a Bonanza for a heavy equipment company.

She considers 1949 very memorable because she joined the 99s. The Phoenix Chapter was chartered that year with her as the first chairman. She was made a life member in 1978, and she was honored by the Phoenix Chapter with a marker in the Forest of Friendship in 1989. When her husband retired as chief pilot for the state of Arizona, the Newells moved to a ranch below the Sierra Nevadas in California, where they train horses and show German short-haired pointers in field trials.

ALICE MARIE "BUNNY" NEWMAN, RN, born April 9, 1939, in Fairview Park, OH, joined the 99s in 1983. She received flight training and got her airplane single engine land certificate. She has flown a Cessna 172 and 182, Beechcraft Sundowner, Trinidad and Tobago. On her first flight after obtaining her pilot certificate, from Van Nuys to Palm Springs (with her husband as a passenger) she lost all electrical. They were safely landed at Palm Spring after getting a green light by using lost communications procedures. She has served as member or chairman of various committees.

She is married to Dr. Byron Newman and they have one daughter, Judi Walden, age 34, who is married to Mark Walden. She has one grandchild, Sara Nicole, age 1.

ELIZABETH "BETTY" PETTITT NICHOLAS, remembers when she was younger, she was privileged to have Lindbergh smile at her and her brothers and sisters as he came out the driveway of a neighbor. Her sister was working for an airline that flew to Montreal and she got them a weekend pass. When the plane ran up its engines before take off, Nicholas had the feeling of a bird being restrained.

So when she read about the WASP program, she quit her job as secretary to the president of a music foundation and applied. After training, she served for three months as maintenance test pilot on the AT-6 before her deactivation. She was one of those invited back to the same field in Alabama as link instructors and being the only instructor who knew some Spanish, taught Link to the entire Mexican student officer group. On returning home to New Jersey, she joined the 99s in 1946.

She remained in flying for 41 years, doing a stint as a ferry pilot on Cessnas and Aeroncas for her friend, Bee Haydu; was a co-pilot/hostess on a small airline between Newark, NJ, and Atlantic City. She then accepted a job as skywriter in Indiana, using her beloved AT-6 for three years.

She finally joined the Aeronautics Commission of Indiana as pilot and editor for their publication until her marriage in 1953. During that time, she served as lieutenant colonel with the Civil Air Patrol in charge of operations.

JUDITH NICHOLLS, born Jan. 1, 1956, in Hamilton, Ontario, Canada and grew in Niagara Falls, Ontario. She started flying in 1991 and received her commercial license in 1993 at the Ottawa Flying Club.

Judy graduated from nursing in 1985. She has studied theater at Brock University, electronics at Algonquin College, art at the Ottawa School of Art and completed the CNE program at Lantec.

She has worked as a Byward market vendor, as a nurse, an electronic technician, and is about to embark on a career as an airborne system specialist (A.S.S.) operating electronic survey equipment on board aircraft at remote world-wide locations.

Her hobbies include flying, playing piano, artistic photography and computers.

Judy volunteers as a pilot/navigator with Ontario Civil Aeronautical Search and Rescue Assoc. She joined the 99s in 1993 and is currently membership chairman for the Eastern Ontario Chapter. She is a proud 99 and is grateful for all the support and encouragement from her fellow members.

LAUREN TRENT NICHOLSON, spent lots of time around airplanes growing up, while her parents worked for airlines including Piedmont, Mid-Pacific and Air Hawaii. At the age of 15, she jumped at the opportunity when her father's friend offered to give her a flight lesson in his J-3 Cub in North Carolina. The short flight encouraged her to return the following summer to learn to solo. That summer she also began researching her Aunt Mary Webb Nicholson, a pioneer woman pilot. During high school, Lauren's interest in flying grew when she joined the Aloha Airlines Explorer Program. Infected by the flying bug, Lauren enrolled in the flight training program at Purdue University. Immediately after earning her private license there, she joined the 99s. She also became president of Purdue's Women in Aviation Organization and began flying in the Air Race Classic in 1995 for Purdue.

When finished with her training at Purdue, Lauren hopes to return home to Hawaii, build flying time, earn her A&P and fly commercially there. She is also interested in missionary aviation, aviation education and aviation writing.

MARY WEBB NICHOLSON did parachute jumps and office work in exchange for flying lessons in 1928. Signed off by the Wrights, she became the first licensed woman pilot in the Carolinas, and joined the 99s as a charter member. She continued flying by barnstorming and flight instructing. Among her students was her brother, Frank, who later flew in WWII and then became chief pilot for Piedmont Airlines.

Mary held several office positions for the 99s, including governor of the New York-New Jersey and Southeast Sections, and board member for the Amelia Earhart Scholarship.

Jacqueline Cochran hired Mary to be her secretary, a position she filled for five years. Mary helped Jackie to organize 25 American women pilots to ferry planes during WWII for the English Air Transport Auxiliary, which freed the men to fly in combat. Mary joined the last group of women who went to England, determined to put her flying skill to good use, despite rigorous training and England's bitter winters. While there, she ferried many planes including Spitfires and Hurricanes, and was promoted to second officer. Mary was ferrying a Miles Master when an engine failure forced her to make an emergency landing. She was killed when she struck a small building at the end of a field.

Mary's love for flying is well remembered by her family and shared by her brother and two of his children, who are on their way to flying for the airlines.

JANICE GERBER NIELESKY, born Jan. 2, 1944, in New York, NY, received her pilot license in 1978 in Fullerton, CA, and joined the 99s in 1979. She has flown C-150, C-172, Cherokee 160 and Bonanza A35. She has participated in several air races and has been active in chapter affairs since joining. Jan has been particularly involved in NIFA and PCIFA serving in almost all judging capacities including chief judge of PCIFA in 1994.

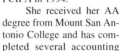

She received her AA degree from Mount San Antonio College and has completed several accounting courses. She has worked as a bookkeeper for the past 18 years.

Jane is also very involved in youth ministry and works with high school age young men and women. She is currently enrolled in a national certificate program in youth ministry.

She is married to Edward Nielesky, a pilot and aircraft mechanic. She has one son, six stepchildren and 10 grandchildren.

MERRILYN MARTIN NOBLE, got her private license June 1, 1990, at South Lake Tahoe, CA, and also has her ASEL rating. She joined the Lake Tahoe Chapter of 99s in April 1991. She has flown a Piper Dakota and owns a N9090X. Noble has flown in Reno Air Races and the Truckee Tahoe Airshow. She is also a member of the AORN, Alto Alpina Bike Club, volunteered for "Death Ride" the world's toughest triathlon.

Noble is an RN at Tahoe Fracture Clinic and also worked in OR at Barton Memorial Hospital. She is married to Norman, a pilot, and has one son and one daughter. She enjoys biking, hiking, cross-country skiing and running.

JOHANNE NOLL, born Aug. 31, 1927, in Denmark, took up flying when her youngest took off to college. She joined the 99s in 1983. She has flown the Cessna 150, 152, and 172; Piper Warrior and Archer II.

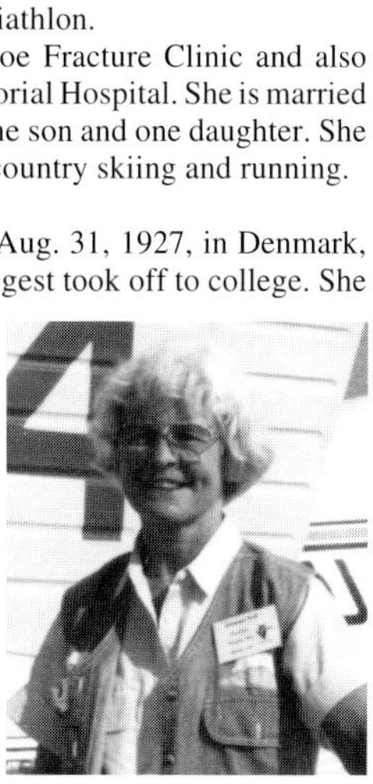

She is married to Hans. They have four children: Kirsten, 46; Lucas, 44; Katharine, 36; and Elizabeth, 33. They have two grandchildren, Isabel Nicole, 1 year and Jessekai, 21.

At a safety meeting she met her neighbor Antonia Reinhard and they have flown together since. They participated in Illinois Air Derbies, flown Poker Runs, and moved to Seattle together to continue to fly together.

DOROTHY AILEEN NORKUS, born Feb. 15, 1969 in Southfield, MI. At the age of 10, she moved to San Diego, CA. A lifetime of interest in aviation is what led to flying lessons and aviation as her career path. After many months she finally got her private license in May 1992, and as luck would have it, the maiden flight was cut short by an engine failure!

Dorothy graduated from San Diego Mesa College in June 1994 with an associate's degree in aerospace. She has also worked for Southwest Airlines since 1989 as a ground service agent/training coordinator and was the Agent of the Year in 1994. In May of the same year she received an instrument rating and is currently training for a commercial license flying the Cessna 172-RG, as well as C-172 and C-152. She is a volunteer for the AOPA Air Safety Foundation, AOPA member, San Diego Chapter of 99s treasurer and secretary as well as a charter member of Women in Aviation International.

NANCY NORRIS, a lifetime New Jersey resident, started training in a 152 at Lincoln Park Airport. Her initial interest in flying was due to fear of small planes. A friend a fellow pilot bought her first log book and intro flight. From that first flight she began training and ultimately bought a 182 in 1989. Nancy enjoys a career as a licensed health administrator and owns a nursing home in West Orange, NJ. She is a member of Health Professionals, served on the NJ Dept. of Health regulator board for licensed facilities and has served on several committees for Dept. on Aging in New Jersey. As a member of the 99s Assoc., she has participated in the Poker Run and Pennies-a-Pound, and is currently a member of AOPA.

BARBARA K. NORTHROP, became a private pilot on Aug. 28, 1978 with a ASEL rating. She has logged 932 hours and has flown C-182 (own N8354S), C-152, 172; Citabria, C-210, 180; J-3 Piper Cub, C-340. She joined the Lake Tahoe Chapter in August 1979.

Flying activities include: business flying, orientation rides for school kids, attended Oshkosh in 1989, Watsonville Airshow, Reno Air Races, Travis Airshow, Beckwourth Fly Ins, Shuttle Landing at Edwards AFB and Alaska air tour.

Northrop is married to Lowel (a pilot). They have three daughters (all pilots). Northrop has been a board member of TTAD for 14 years. She is president of Northbilt, Inc. General Engineering Contractors but was an assistant to an oral surgeon, EMT-1 with Alpine Meadows Fire Department and Firefighters for six years. She is involved in St. Nicholas Episcopal Church activities, Tahoe Women's Services, Truckee Tahoe Plane Talkers, and has been in the 99s for 16 years. She enjoys Alpine skiing, hiking, local homeowners association, St. Nicholas altar guild.

Best trip ever: from TRK to Arctic Circle via Whitehorse, to Fairbanks, around Mount McKinley, to Soldotna to Iliamna to Girdwood to Anchorage area to Delta to Northway, Whitehorse, Skagway, Juneau, Ketchikan, Bellingham, TRK in 10 days – over 6,000 miles, 22 hours tach time, all without incident, just two women pilots, her daughter, Pat, and herself.

KAREN NORTHROP, got her private pilot license on Nov. 12, 1979 in Truckee, CA. She also has ASEL and ASES (1981) ratings. She joined the Lake Tahoe Chapter of 99s in October 1989. She has flown C-150, 182, 180 and 172; PA 185 (seaplane), PA23-180, C-185RG (amphibious), and J-3 Cub.

Flying activities include: Mountain Flying Seminar at Quincy, Truckee Tahoe Airshow, Oshkosh EAA 1989, Beckwourth (Nervino) Fly-ins, photographic flights for research on Lake Tahoe, parachute static line jump, Flying Samaritans trips to Mexico.

Northrop works as a physical therapist. She belongs to the Therapeutic Riding for Exceptional Children (volunteer), Tahoe Handicapped Ski School (volunteer), American Physical Therapy Assoc. (member) and International PNF Assoc. (member).

She enjoyed flying to Oshkosh EAA in 1989 with her mother and was one of the best flying events. She has also enjoyed combining her physical therapy skills with her love for flying via trips to San Quentin, Mexico to work with amputees, etc., with the Flying Samaritans. Flying seaplanes at Lake Tahoe and learning from one of the best instructors around how to fly by the "seat of her pants" has been invaluable. Learning to sky dive was challenging. She will never forget the family trip to the Bahamas in their C-182, December 1979, or landing on the beach at Bahia de Los Gonzaga, Baja, Mexico.

Northrop is single but her whole family, parents and two sisters are all pilots.

KIMBERLEI A. NORTHROP, got her private pilot license on Sept. 21, 1982 at Truckee, CA. She has a ATP certificate with an ASEL, AMEL, instrument, B-707 type ratings. She joined the Lake Tahoe Chapter of 99s in October 1989. Northrop is a US Air Force pilot and has flown the C-152, 172, 182; T-41, T-37B, T-38 and KC-135A, Q,R,T. She has 2,484 hours flying time. She belongs to the USAF Academy Assoc. of Graduates, Women Military Pilots Assoc. and received Lake Tahoe Chapter "Woman Pilot of the Year" in 1994.

Flying for the Air Force is one on-going story! From T-37 aerobatics, to flying fingertip formation in the T-38, to Desert Storm sorties in the KC-135, or an early morning sortie flown during a horrible exercise which turned into one of the most spectacular flights she has ever flown (sunrise over Glacier National Park, fresh snow on the mountains which were bathed in rosy sunlight with a beautiful sunrise in the east and a full moon setting in the west). Or Desert Storm with F-4s in tight on the wing during refueling with another wonderful sunrise (turning those gray-blue F-4s pink!) with the moon setting off of their wing. Returning home to Michigan in one of its winter blizzards with blowing snow reducing the visibility to minimums, crosswinds close to their aircraft limits, and looking for the trees that defined the runaway for them. Watching spectacular Northern lights for hours while crossing the ocean (nice of someone to provide some in-flight entertainment for us). Or flying the J-3 Cub with Al Richardson – almost as memorable as her high school graduation flight in a float plane into the lakes in Desolation Wilderness or finding the spring waterfalls in 54S.

PAT NORTHROP, got her private pilot license in August 1970 at Stillwater, OK (OSU as a Civil Air Patrol cadet) and an ASEL and instrument rating in 1990. She joined the Lake Tahoe Chapter of 99s in March 1982. Northrop has also been involved with Truckee Tahoe Plane Talkers. She has flown the C-150, 172 and 182.

Northrop is a program coordinator since 1986 of UNR Elderhostel, ski instructor/supervisor at Squaw Valley since 1977 and director of Women's Ski Program. She also holds a California state teaching credential (1975-90) as a substitute teacher. She enjoys skiing, mountain biking, catamaran sailing and rollerblading.

She went to Oshkosh in 1989, Watsonville in 1986, Truckee Tahoe Airshows 1970 to 1995. CAP (Tahoe-Truckee Sqdn. Commander since 1991, member since 1967. Nevada wing director of CAP cadets programs 1985-1990.

For the first 10 days in June 1995, she and her mother, Barbara Northrop, and N8354S (C-182) through Canada into Alaska via the Alaska Highway to Whitehorse, then down the Yukon River to Eagle, Arctic Circle Hot Springs and Fairbanks. Their one and a half hour flight around Mount Denali (wx CAVU) complete with views of climbers, helicopter landing areas and this mountain peak that towered another 8,000 feet over them, was most breathtaking! Staying over in a log cabin with friends at Soldotna allowed them side trips to Homer and Iliamna through Lake Clark Pass. A return flight through Northway, Whitehorse, over White Pass into Skagway, down the coast with stops in Juneau, Ketchikan, Bellingham, WA, and finally at Truckee made this a wonderful aviation adventure!

BLANCHE NOYES, a charter member, was a young actress who relinquished the theatre when she learned to fly in 1928. She was one of the first women transport pilots. In 1929, Blanche tried her wings at competitive racing and placed fourth in the National Air Races. Several years later, in 1936, she was co-winner, with Louise Thaden, in the Bendix Air Race. It was the first year the race permitted women contestants. Blanche was the first International treasurer and a past International president. She was chief of airmarking staff with the FAA for 35 years. Blanche was honored by virtually every aviation organization in the world.

PATRICIA HELEN NOYES a.k.a. **PATRICIA PRENTISS,** was born in Albany, NY, on Jan. 17, 1947. At an early age she moved to Southern California. After her graduation from high school, Pat went to work for Autonetics, a division of North American Rockwell. She was assigned a temporary position working with the Air Force on parts inventory for the FB-111 aircraft. This was Pat's first exposure to aviation and instilled the initial desire and motivation to become a pilot. Becoming a young mother delayed that goal until the age of 30 when she received her private license. Two years later Pat was teaching flying and loving every minute of it. She has flown all Beechcraft single-engine series and all Cessna single-engine series. From 1979 to 1991, Pat excelled in several careers, as an aviator and as a general contractor in Orange County. She has two children, Dasna, age 22 and Byron, age 28.

At present, Pat is flying for a private corporation on a Citation III. She holds a private, commercial, instrument, CFI and ATP with type ratings in CE-500 and CE-650 aircraft. Pat has been very active in the 99s, Inc., which she joined in 1980. She has been Orange County Chapter chairman for two terms, and also Southwest Section secretary for two years. Pat chaired, formulated, organized and continually upgraded 10 very successful Flying Companion Seminars. Pat has always been available to help and encourage others and has been involved in numerous activities which promote women in aviation. Pat was the recipient of the Orange County Chapter Pilot of the Year award twice and also received the Southwest Section Pilot of the Year award in 1991.

One of Pat's proudest moments occurred in South Bend, IN, when she was pre-flighting the Citation III. A young line service lady who was learning to fly, approached with a shy smile and stated, "You're my inspiration!" Little did she know that her comment was an inspiration to Pat as well.

YAE NOZOKI, was born in Yamagata, Japan. When she was a teenager, she convinced her parents to spend the money that they were planning to spend for her marriage on flying lessons. Nozoki started taking flying lessons at the First Aviation School in 1937. Before WWII, she worked as a stewardess. After WWII, she worked for the aviation department of the Traffic Bureau. During this period she got a commercial pilot license. She established the Japanese Women in Aeronautics and dedicated her life to aviation. Because of this Nozoki received the "Ranju-Hosho Award" from the Japanese government. She is now president of the Japanese Women in Aeronautics and a life member of the 99s.

JOYCE C. NUNNERY, her first flight in any airplane was in a Piper Pacer on her honeymoon in 1952. On her 40th wedding anniversary, she passed the private pilot checkride in a Cessna 150. The impetus for learning to fly was the completion by her husband of a bright red experimental RV-6 taildragger. The initial goal of pinch-hitter training soon got out of hand and led to the pilot license. Flying activities currently range from frequent short hops out and back for lunch, to cross-country destinations on both east and west coasts. Nunnery is a member of the 99s, Florida Panhandle Chapter; Aircraft Owners and Pilots Assoc. and EAA Chapter 108.

Primarily a homemaker, Nunnery has a BS and has taught in Florida schools. Her avocation and continuing education has been anthropology and archaeology; she is a member of the Florida Anthropological Society and has been both docent at local museum and county museum teacher. Interests include water sports and sailing, having sailed with her family on a four year trip around the world aboard a 38-foot yawl.

She is married to Cliff Nunnery, a retired Air Force pilot. They have one son.

DELORES HELEN NUSBAUM, born in Milwaukee, WI, and moved to northern Indiana in 1966. She now resides with her husband, Bob, in Union Mills, IN. She is a retired director of nursing and an avid gardener. In 1984, at age 56 she passed her private pilot checkride and in 1993 her instrument checkride. Joined the 99s, Inc. and the Indiana Dunes Chapter of 99s in August 1984. Served as chapter chairman from 1993-95 and is presently the North Central Section Scrapbook chairman. Yearly at the Oshkosh EAA, in conjunction with another 99, Phyllis Webb, she conducts a Flying Companion Seminar. In 1991, at a Safety Seminar hosted by her chapter she met her present husband Bob. They were married in February 1993 and share their mutual love of flying.

ADRIANNE ARMOR O'BRIEN, was born Jan. 26, 1942 in Long Beach, CA. Her grandfather, Robert J. Armor, was a pilot and member of The Early Birds. Adrianne always had a fascination with aircraft, but never thought about obtaining a pilot license until she met pilot and aviation enthusiast Patrick O'Brien. He managed a flight school and arranged several lessons for her. A single mom, she vowed to learn to fly when her children became teenagers. By then they had married, Pat was a flight instructor himself, and they decided to buy a Cessna 172. Pat taught Adrianne and two sons to fly. Adrianne worked to get her instrument rating and after that flew just "for the fun" of it. Active in the Orange Chapter of the 99s, Adrianne continues to fly to 99 section meetings and various vacation destinations with her husband. She has a son, daughter, three step-children and three grandchildren.

BLANCHE MCEOWEN O'BRIEN, born Nov. 13, 1916 on the H.E. McEowen farm where in 1928 Lansdowne Airport was established honoring US Navy Cdr. Zachary Lansdowne, commander of the ill-fated airship, *Shenandoah,* who was a native of Greenville, OH. (This was a very early fixed base operation before Port Columbus or Dayton Municipal airports in Ohio.) Many early military pilots from McCook and later old Wright Field landed at this field, other early pilots landing here included Dewey and Blanche Noyes, Wiley Post, Gladys O'Donnell and Walter Beech. Brother John McEowen managed the field in those early days.

As co-owner with her husband, John O'Brien of Greenville Flying Service they had a CPTP contract before WWII. Blanche soloed a Piper J-3 from Lansdowne Airport Nov. 26, 1939, got her private license # 177538 just four days before Pearl Harbor, and moved with her husband and daughter, Peggy, and son, Jack to Ocala, FL where John was with Greenville Aviation School training Army primary flight cadets. Blanche was a member of CAP at Daytona Beach and Greenville, OH, 1942-48.

Returning to Lansdowne Airport after WWII she was a member of the Greenville Girl Scout Council and had a Wing Scout Troop which did airmarking, and was state chairman of the All-Ohio Chapter of 99s during the 50th Anniversary of Powered Flight in 1953 and a member of Gen. James Doolittle's National Committee to observe this anniversary. During that year Blanche along with Blanche Noyes was pictured in Ken Bodecker's (Lindbergh's mechanic) *Bodey's Book of Aviation Personalities.*

Along with flying service duties she served as president of the Greenville Business and Professional Women's Club in 1957 and for many years was a member of the Darke County Republican Executive Committee.

The O'Briens moved to Delaware, OH, in 1963 for John to give the flight training to ROTC students for Ohio Wesleyan, Otterbein, Denison and Kenyon Universities living there for 23 years before retiring to Ocala, FL, in 1987. She is a member of the Ohio Wing OX-5 Aviation Pioneers, Ohio Silver Wings and the International Women's Air and Space Museum near Dayton, OH.

In Florida, Blanche is a life member of the Suncoast Chapter of 99s, a member of Florida Aviation and Historical Society and Ocala Chapter of Florida Aero Club.

The O'Briens have been married over 62 years and are still active in aviation circles.

MARY O'BRIEN, was the first woman examiner of airmen with the Department of Transport and Communications. Mary's career in aviation began in 1964 in the NSW town of Walgett, where she was born, when she was taken for a flight in a Tiger Moth and decided to become a pilot. By 1967 she had worked at various jobs to fund her flying training and now had a commercial pilot license. Mary became an aviation theory instructor at the College of Civil Aviation in Sydney.

In 1970 Mary was offered a job in Singapore. She found herself ferrying aircraft around the world, dropping payrolls into isolated mining camps in Asian jungles, transporting freight and passengers and searching for ships. She obtained an airline transport pilot certificate with an instructor and instrument rating and flew extensively in the Americas from Alaska to Guatemala.

In 1980 she brought the first Grumman Cougar light twin aircraft to arrive in Australia, crossing the Atlantic, Europe, the Middle East and India on the way. Mary became chief flying instructor in Tamworth until joining Wards Express. She was the first woman to captain a Lear jet in Australia.

She worked in the department of Aviation in 1985 where she wrote aviation theory exams for the various pilot licenses. In 1987 she moved from the theoretical examination area in Canberra to the operational area in Sydney. She is the only female flying operations inspector in Australia. In her role as the district flight operations manager with the Civil Aviation Authority she manages a team. She is also involved in the Civil Aviation Authority's accident prevention and pilot education programs.

Mary, who holds an Australian airline transport pilot license, has flown over 30 different types of aircraft. The recipient of the Nancy Bird Trophy, she is past national president of the Australian Women Pilots Assoc. and the governor of the Australian section of the 99s.

PAMELA O'BRIEN, received her private pilot certificate on her 40th birthday and has been a member of the Sacramento Valley Chapter of 99s since 1990. She started flying lessons jointly with her husband, Dale, and when he had to quit because of health reasons, she couldn't stop flying. She received her instrument rating in 1991 and has more than 700 hours. She learned to fly in a Piper Archer II, owned a Piper Dakota for five years and is currently the proud owner of a 1983 Bonanza F-33A. Besides the 99s, she is also a member of Aircraft Owners & Pilot Assoc. and the American Bonanza Society and has participated in Flying Companion Seminars, airmarking and publishing the chapter newsletter, *Pre-Flite.* She won the Woman Pilot of the Year award from her chapter in 1994.

MARY LYNN O'DONNELL, began her love affair with aviation as a skydiver in 1972. While recuperating from a skiing injury she missed "being in the air," so she went to the local airport for an airplane ride – which turned into a lesson that turned into a license that turned into a career. After that first lesson in 1974, she got a job with Piper Aircraft Corp. where she flew everything they built. Lynn left her systems analyst job at Piper to make a career out of flying.

First job: Ferrying airplanes to Europe, 52 solo transatlantic crossings. Then PT-135 charter and commuters. In 1984, first PT-121 job as a B-727 FE flying cargo, then Eastern Airlines, then Pan American World Airways, then United Airlines.

Now a B-767 first officer, Lynn still skydives and flies small airplanes. She and husband, Robert Hadow, own a C-172. She is a member of 99s, AOPA and International Society of Women Airline Pilots.

ELEANOR "ELLIE" MCCULLOUGH ODORICO, learned to fly as a member of the Republic Aviation Corporation Flying Club, Farmingdale, L.I., NY. Soloed June 1946, private license November 1946, later commercial – instrument, private glider, 16 registered parachute jumps. Accrued more than 4,200 PIC hours, more than 2,500 air race connected hours from 1963 until present time. Flew numerous Powder Puff and Angel Derby races, Air Race Classics, New England All Women Air Races, Garden State 300, Michigan Small, Great Southern Florida 400 and many others, winning some.

Ferried aircraft from Wichita, KS and Kerrville, TX to Long Island, NY. Served seven years as an FAA safety counselor. Executive secretary to Alexander Kartveli, 18 of 25 years employed at Republic. Worked with Dr. Theodore Theodorsen and John Stack after they left NASA. Employed six years in Grumman Marketing Department. Conducted four marketing surveys for Mitsubishi re MU-2 aircraft through IRM International. Employed 16 years as executive assistant to city manager of Vero Beach, FL.

Joined the 99s in 1962. Was governor New York/New Jersey Section; founded and first chairman of Long Island Chapter; chairman of Greater New York and Florida Gulf Stream Chapters; vice-governor Southeast Section; International historian, nominating committee, credentials committee, chairman of contest committee. Served on many section and chapter committees.

Memberships: life member of 99s, AOPA, Silver Wings, Grasshoppers, Ladybugs, past member of International Order of Characters, L.I. Early Flyers, Long Island Aviation Council, ZONTA, L.I. Skydivers.

Ellie owns and flies *Schatzy* a 1967 Piper Cherokee 140. *Schatzy* has birthday parties, gets greeting cards and get well cards when she goes in for her annuals. Ellie and husband, Lou, spend many hours giving rides.

ELLEN L. PARKER O'HARA, started flying in 1963 as a result of returning to the flatlands of Chicago from living in Switzerland. Enjoying the beauty of clouds from a mountain top was not possible in the "before Sears Tower" years! So Ellen called a flying club that was seeking a new instructor and asked it they wanted any new members – the pause before the answer seemed to say, "Good grief, it's a girl!" (Later that thought was confirmed.) The following Saturday, Ellen met the seven male club members who asked the new instructor how she did on the introductory flight: "She'll get her license before any of you do," said the instructor, much to the embarrassment of the female fledgling. And she did, too, after 52 hours.

Ellen earned her commercial/instrument ratings and has taught a ground school for over 10 years at the high school where she teaches business subjects. Her 600 hours of flying time include a few in helicopter and 10 in aerobatics (great fun!). She has been a 99 in Chicago Area Chapter since January 1966, and has been chairman and held various positions.

SHANNON COLLEEN O'HARA, an Arizona native, began flying in 1990 at Falcon Field in Mesa, AZ. Since then she has been a very active pilot, becoming a flight instructor, instrument instructor and multi-engine instructor in 1994. Shannon has logged more than 500 hours – just the beginning of what she hopes to be a promising career in aviation.

Shannon graduated from Arizona State University with two BS degrees in aviation and aviation management. While at the university, she also became involved with the ASU Flight Team, part of the National Intercollegiate Flying Assoc. She currently enjoys her position on the team's coaching staff. Shannon also spends a great deal of time working with the Arizona Sundance 99s, participating in airlifts and air races.

Shannon grew up across the US and around the world but will always call Arizona her home. She looks forward to the excitement and challenge of a future in aviation.

SHIRLEY T. OHL, born May 19, 1934 in Cleveland, OH. Ohl joined the US Marine Corps in 1954 and began her flying career when she received her private pilot license in June 1979 at Norfolk, MA airport. She also has a SEL rating. She has flown a C-172. Learning to fly at age 44 and joining the 99s were thrills of a lifetime for her.

She married Charles B. Ohl and had one son Charles B. Ohl Jr. They have two grandchildren, Ashley B. Ohl, age 12 and Charles B. Ohl, III, age 8.

PATRICIA ANN OHLSSON, born July 26, 1939, in Springfield, MA, feels that she is the proudest 99 there is, but she does know that there are many more that feel the way she does. Her first general aviation aircraft ride was in a 1969 Mooney to Danbury, CT, September 1972.

Two years later, September 1974, her ex-husband took her to Islip Long Island Airport and they went for a ride in a Grumman Traveler. When they got back on the ground, the instructor and Pete were talking and the instructor said, "When would you like to start lessons?" Pete replied, "Oh, it's not for me, it's for my wife." Ohlsson's exact reply was, "What are you are crazy! I can't learn that," but she knew for another $5, she could go for another airplane ride. Her next ride was in October 1974 and she was hooked.

Ohlsson's career mushroomed from there. Ronni Minnig introduced her to the Long Island Chapter of 99s, March 17, 1976. For 10 years she never missed a meeting or a section meeting and flew to 15 conventions. She has served as their hostess, secretary, treasurer, vice-chairman and chairman. Ohlsson served in many other chairmanships and activities within the chapter and New York/New Jersey Section. Flow Air Races in New York, New Jersey, New England and Washington, DC, capturing many first prizes.

On Feb. 9, 1979, she purchased her 1976 Grumman Tiger N74359, and formed Busy Bee Airways. (I use to be a beekeeper.) Together they have flown some 2,000 hours to date.

Then with the encouragement of Doris Abbate and her Long Island Chapter members, she applied for and was awarded the Amelia Earhart Scholarship 1983, still the proudest day of her life. Ohlsson currently holds a commercial rating in SEL, MEL and gliders. Instrument rating, ground and flight instructor.

She and her husband, Lenny, work as a team in the field of aviation. Lenny was her instructor for her multi-engine and glider. He owns Spruce Creek Fly-in Realty, Daytona Beach, FL, "Live Where You Can Fly." Together they sell property. The best of both worlds, sell property, where airplanes taxi past your office, flight instruct and ground instruct and do some charter work too. When they got married, Nov. 14, 1991, their wedding cake was the Spruce Creek Fly-In runway 5/23, with lights, planes departing and landing.

Forbes Magazine, Feb. 14, 1994, wrote a short article in which Lenny and she were featured, called, "Living on the Tarmac."

In 1993-94, chapter chairman Spaceport Chapter of 99s in Florida, her wings are still with Long Island and has transferred back, although still active with the Spaceport Chapter. "If the road gets bumpy, let's take to the skies."

JEANNE KELLEY OHNEMUS, born Feb. 1, 1926, in Newton, MA, joined the 99s in 1975. She earned her private pilot license and commercial and instrument ratings at Hanscom Field in Bedford, MA, in 1973. She earned a multi-engine rating at Haverhill with Howard Dutton as instructor in 1976 and presently taking a water rating in Metheun, MA.

She is married to Clifford Ohnemus and they have five children: Jan, Cindy (a private pilot), Kim, Cricket and Kelley. She has six grandchildren: Tim, Dianne, Samantha, Gregory, Stevie and Will.

In the 1970s and 80s, it was awesome flying over spectacular glaciers and mountains. There were many trips to all the islands and in the Bahamas in a Comanche 250. Best trip was in their Comanche 260B in 1989 from Windsock Village, NH, to Bellingham, WA, to Juneau, Ketchikan, Sitka, Talkeetna, Yakutat and Fairbanks, AK.

CLAIRE M. OJALA, received her private license in June 1963, in a Cherokee 160, she immediately checked out in a Beechcraft Bonanza, joined the Michigan Chapter and flew the Angel Derby with Donna Blake Sentas in the same month. What an adventure for two private pilots. The experience was only topped by flying the Powder Puff Derby in 1965. Also flew many small races. Other aircraft she checked out in and flew were a Piper Cherokee 180, Piper Arrow 200, Cherokee 6, Cessna 140, 172, 182 and 210. Held several offices and many committees at chapter level. After the birth of their son, Galen, Ojala curtailed flying and became involved in other ways.

The most fulfilling, was developing and implementing the Mary von Mach Scholarship for the chapter. This scholarship has enabled many young ladies to experience flight through solo. Many have gone on to earn their license and other ratings to become successful in their chosen fields.

Ojala attended 21 International conventions and numerous section meetings. Most of the convention experiences, she shared with her husband, Keith, and their son. They enjoyed the travel, meeting so many interesting 99s and families and sharing experiences. Thanks to Keith, a super pilot in the USAF flying the B-36, for introducing her to aviation and encouraging her to learn, even when she would become very ill. It has given her 33 years of adventure and pleasure.

JOYCE OLDAKER, a charter member of the Women With Wings Chapter, believes that life begins after marriage. Joyce married Bob, a fireman, in 1954, raised her family, then pursued a career in nursing, climbing the educational ladder of RN, BSN to MSN. Joyce currently is nurse manager of an OB department. When Bob started flying, Joyce decided to also take to the skies.

She earned her private license in 1993 and is currently working on her instrument rating in the family Cherokee 6. Being able to fly has made the family closer, with one son in New York and another in Connecticut.

MARION ROSE OLMSTED, born May 24, 1923 in St. Louis, MO. A desire for flight began as a youngster with a ride in a Ford Tri-motor and furthered with a ride in the Curtiss-Robin St. Louis Number 1 endurance plane with Forest O'Brien, St. Louis Lambert Field. The Civil Air Patrol was joined at it's conception in 1940 with participation in an all woman squadron.

During WWII from January 1943 until August 1946, she was a flight operations person at the 4810th AAFBU at the National Guard Hangar at Lambert Field. In conjunction with her CAP and flight operations work, flight training was started in August 1943 at St. Louis Flying Services Kratz Field, the purpose being the qualification for the WASP. Olmstead was accepted and scheduled to join the August 1944 class, however the WASP training program was canceled with the July 1944 group.

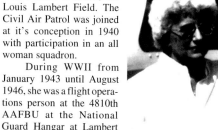

Her private pilot SEL license, #115794, was obtained in October 1944 at Lakeside Airport, Granite City, IL. Flight after that time was continued at Parks Bi-State Airport in East St. Louis, IL, and Cedar Rapids, IA. In 1957 she began FAA air traffic controller training at the St. Louis ARTCC. Women were not readily accepted at that time and she was subsequently transferred to the Cedar Rapids flight service station in October 1958. She remained there as an air traffic specialist until her retirement with 26 years of service to the aviation public in May 1984. It was a long interesting aviation career in Iowa.

DARLENE OLSON, earned her private pilot license in 1970, joined the 99s in 1972 and became a member of AOPA in 1974.

During her 25 years in aviation, she has flown the plains of the Midwest, the beaches of Mexico, the congestion of California and the mountains of Montana. Her most memorable flight was ferrying a Bonanza V-tail from Maryland to California.

Darlene does extensive volunteer work with the young people and schools. This has afforded her an excellent opportunity to bring the 99s educational program, Air Bears, into schools in western Montana. Her success with Air Bears is her most rewarding role as a Montana 99.

An independent travel agent, she is married to Jerry Olson who also is a pilot and a CPA. They live with their son, Matthew, on a ranch in Montana.

RUTH M. OLSON, Phoenix 99s, met and married a pilot when she was in college, and is happy that she has had the opportunity of flying in many types of aircraft – from open cockpit to the nicest of modern planes. Ruth became an elementary school teacher when she got out of college and had five children herself. She had to wait until she and her husband got them raised for her to actually begin flying lessons, which she did in 1970, getting her private ASEL in Phoenix in 1971. She also joined the Phoenix Chapter of 99s in 1971, and has served in many ways, mainly enjoying serving on rules committees and assisting with Fly Without Fear workshops.

In 1994 Ruth and other Phoenix Chapter members, Mary Lou Brown and Ruby Sheldon, flew to Alaska in Mary Lou's Cessna 182. Ruth enjoyed it greatly as beautiful country "and certainly our last frontier." She is amused that "three women flying in the wilderness" made quite an impression on a few of the natives. Ruth, now retired from teaching, says she spends her time enjoying herself.

BARBARA MARIE-PIPER OMSTEAD, Capt. USAF, of New Hampton, NH, was born June 23, 1964 to MSgt Frank F. Piper and Leona A. Piper and has two sisters and four brothers. She joined Civil Air Patrol in August 1978 and was introduced to flying through the Cadet Orientation Flight program. She earned a flight scholarship through a high school aeronautics class and soloed at the Cessna Pilot Center at Laconia Airport, NH.

On Aug. 28, 1987, she graduated from the University of New Hampshire with a BS in mathematics and a ROTC commission as a second lieutenant in the USAF. She continued flying during technical training at Tyndall AFB, FL, and earned her private pilot license while on temporary duty at Luke AFB, AZ. She was assigned to the 601 Tactical Control Squadron in then West Germany July 1988-April 1991. She enjoyed the crowded skies (and actual IFR conditions) of Europe. Flying was put on hold when Barbara spent August to October 1990 in the "Sandbox" of Saudi Arabia as a specialized intelligence officer for Operation Desert Shield. Her warrior experience was again called upon January to March 1991 for Operation Desert Storm, when she was sent to Turkey as the commander of a deployed combat crew. In May 1991, Barbara was transferred to Tinker AFB, OK, to fly as an aircrew member on the E-3 Airborne Warning and Control System (AWACS) aircraft. She earned a promotion to captain and became a primary instructor and a standardization and evaluation flight examiner. She joined the 99s in Oklahoma in December 1992.

On Nov. 6, 1993 she married Capt. David A Omstead, her best friend and fellow war veteran. Barbara has logged more than 1,500 hours in Piper, Cessna, experimental, F-15 and E-3 aircraft. She is pursuing a master of aeronautical science degree and flies for pleasure.

GENEVIEVE KRIMM ORANGE, worked at Hermitage Airport, Richmond, VA, in spare time and weekends in exchange for flying lessons. First lesson was Dec. 6, 1941. Parents gave consent on the condition that she would always fly wearing a dress. (So men at airport would never forget she was a lady.) Obtained instructor rating at 20 years old in 1943. She was the youngest flight instructor and first female instructor in the Richmond Tidewater area.

Organized the Richmond Women's Flying Club. Became second lieutenant in Civil Air Patrol as transportation officer. In 1955 became a co-pilot on Lockheed Lodestar with Alex Lowery until 1955.

Stopped flying in 1955 due to other commitments. Active in Virginia Aviation Museum and on board of directors. Joined Virginia Chapter of 99s in 1995.

PATRICIA ORCUTT, born March 5, 1926, became involved in aviation in 1965. She learned to fly to get from one golf tournament to another. She got so involved in flying that she blew her handicap and didn't have any tournaments to go to!

Orcutt first joined the 99s in 1968. She has been a member of the Nebraska and Aloha Chapters and is currently a member of the Reno Chapter. She was treasurer for four years and chairman for two years. She was also a newsletter editor and fundraising chairman for 15 years.

Orcutt has a ATP, single and multi-land rating. She also has a private privilege glider rating. She currently flies for pleasure in an Aztec PA-23-250 *Fat Albert,* which she owns.

Orcutt is married to Ken Orcutt. She is a retired FAA operations inspector. Orcutt's first flying job was as a charter pilot for Dillingham Marine; flew tours of the State of Hawaii (Part 135); taught instrument procedures to Army helicopter pilots (UH-1H).

LEE O'RILEY, obtained an aviation diploma from the University College of the Fraser Valley in conjunction with Coastal Pacific Aviation Ltd., in Abbotsford, B.C. She received the highest achievement award for a first year student and after graduation instructed part-time for the school. At the same time, she flew part-time as a King Air co-pilot for Timberline Air Ltd., Chilliwack, B.C. At 21, she found a full-time King Air job working for Awood Air Ltd. in Thunder Bay, Ontario. After one year of flying medevacs and gold miners, she left the minus 40 degree Celsius weather for spring back on the west coast.

She's flown a King Air ever since, mostly into logging camp strips and coastal towns. She's looking forward to a hopeful career with Air B.C. and Air Canada. In 1991, Lee won the Jack Sanderson Shield of the Webster Memorial Trophy competition for placing runner-up as best amateur pilot in Canada, an event sponsored by Air Canada, flight operations.

Lee joined the B.C. Coast Chapter of the 99s after being invited to speak at a meeting about the Webster Competition.

Currently, she is working towards a BS in professional aeronautics as an independent study student. She lives on her houseboat under the approach for Runway 26 at Vancouver International Airport.

MARILYN ORLOFF, was SCV Chapter chairman in 1981-83, after joining the chapter in 1974.

She was a mathematics instructor at West Valley College and a speaker for many Flying Companion seminars.

She currently lives at Pine Mountain Lake, where she uses her C-152 to commute for her job with United Airlines at San Francisco where she flies B-757 and B-767.

A. LEE ORR, married Buford B. Orr, Nov. 11, 1950 (they will celebrate their 45th anniversary this year 1995). Mr. Orr was in the aviation business. After three daughters, Lee became a lifetime veteran of aviation, obtaining her private license through multi-engine, flight instructor and instrument ratings.

Lee is a business woman professionally engaged in general aviation. As owner of an FBO (fixed base operator) she serves as chief pilot for a 30-year-old FAA 141 approved flight school. (The first woman to serve as chief pilot in this 30-year-old school.) Lee serves as airport manager, Shelby Airport, NC. She is an FAA examiner, FAA written test examiner, FAA aviation safety counselor for both North and South Carolina.

Lee is the recipient of the Spartanburg Junior Women's Club's First Achiever Award and the Amelia Earhart Scholarship Award.

Lee has recommended thousands of students for their license, many who are professional pilots. Currently, she is teaching an 81-year-young man and a 74-year-young woman to fly. Lee has taught two sons-in-law to fly and is currently teaching her youngest student ever, her 3-year-old grandson who has been flying from the left seat since he was 6 months old.

Lee joined the 99s, Inc., in May 1971. She has served all offices on the chapter and section level. Lee was elected to the International board of directors for a two year term and was appointed to serve an additional year by the current board members. Lee is presently serving as International secretary of the 99s.

ALLEGRA A. OSBORNE, born Jan. 22, 1933 in Rockville, CT. Over the past 50 years Allegra has dedicated her life to aviation. She loves nothing more that to talk about flying! As an FAA designated pilot examiner she spends much of her time flying with and examining pilots all over the region. Allegra is an active FAA safety counselor who recently received her phase 12 Wings. She holds ratings in the following categories: ATP, CFI A and I, SMEL and ground school instructor.

Her forte is instrument instruction. Over the years she has developed an accelerated course utilizing a computer based simulator program, which enables her students to learn quicker, simpler and with a much safer approach.

Allegra is constantly on the go as is exemplified by her many interests , ranging from white water rafting, skiing, biking, karate and hiking. Allegra resides in Massachusetts with her husband of 40 years, Phillip and their German shepherd, Nicke.

She is quick to add that if it is not just to survive but to thrive, every pilot must become involved in making things happen. She feels that all pilots must set a positive example for the non-flying community. "Be excited about flying, talk positively-be safe, stay legal-have fun. Take someone along the next time you fly."

The next time you visit Lawrence, watch for a gold, brown and white Cessna Cardinal. You'll likely see Allegra Osborne at the controls – still having the time of her life. As an FAA aviation safety counselor and dedicated flight instructor, she welcomes fielding aviation safety and education questions.

ANNA M. OSBORN, born Nov. 13, 1934 in Plaquemine, LA, joined the 99s in 1978. She is a charter member of the Aux Plaines Chapter, North Central section. She has a private license and has flown Cessna 150, 152, 172, 172RG, 182; and the J-3 Cub.

She is married to John, also a pilot, and they have one child, Barbara Mohs, age 31. They have one grandchild, Daniel Gray Mohs, 1 years old. She was also first woman president of the Stick and Rudder Flying Club in Waukegan, IL.

Best flying has been flying her 1944 J-3 Cub to Oshkosh and back. For 11 years they have camped under her wings. She served as EAA Antique Classic Division as co-chairman, manpower at the convention.

SUSAN "SUE" H. OSBORNE, born Dec. 7, 1941 in Oakland, CA, was hooked, having grown up in an aviation family, her standard reply to, "Do you have a license?" was, "I'm a good passenger." In 1987 she had a career change, she started managing the Aspen Flying Club. Figuring she couldn't sell people on learning to fly if she didn't know how to fly, she started flying lessons, got her private license, then multi-engine rating. Now more than 500 hours later, she is working on the instrument rating. She is having the time of her life.

Osborne joined the Colorado Chapter in 1994, but had participated over the years with the Mile High Air Derby. She is the safety chairman for the 1995 Mile High Air Derby. She is fortunate that she is involved with aviation in a unique position, both in training and recreational aspects of the industry. She and her husband, Norm, are the proud owners of several aircraft, so they have a good choice when they want to go on trips. Norm is instrument multi-engine rated and is part owner Aspen Flying Club, the club she manages. They enjoy flying out to California to see their son, Steven, age 31, and daughter, Elizabeth, and especially her granddaughter, Kassandra, 7.

Osborne is a second generation 99, her mother Thelma Drew was a very active member of the Sacramento Valley Chapter of 99s when she was alive. She would be very pleased to know she is enjoying the 99s. One of her goals is to fly the Air Race Classic with her sister (she is a private pilot too) in memory of their mother, as she flew two of the Powder Puff Derbies.

LYNN O'SHAUGHNESSY, went for her first small airplane ride in 1980. Two weeks later, she bought N6086A a 1956 Cessna 172 and began taking lessons. At first, she only wanted to learn to fly for fun, but she soon realized that she had a burning desire to pass along that fun to others by teaching. By 1983, she had earned her instrument, commercial and CFI. Her CFII came in 1985 with the help of the Amelia Earhart Scholarship.

During this time, Lynn became the advisor of Aviation Explorer Post 555, became active in the 99s, and was appointed an accident prevention counselor. She also assisted in refurbishing and updating N6086A, under the guidance of a mechanic. She flew in several Michigan Small Races and helped organize many flyouts and activities at her local airport.

In 1986 Lynn realized a dream as she started her own flight school, Free Spirit Aviation. By 1987 she quit her computer job and dove into her new career full time as she began construction on an 84 by 100 hangar with offices at the Livingston County Airport. For seven years she operated a first class 141 flight school and maintenance shop.

In 1990 Lynn met and married Bob O'Shaughnessy. She eventually sold the hangar and business to spend more time with her husband, but still owns N6086A and is an active 99, CFI, safety counselor and pilot examiner.

DIANE OSTER, born Aug. 20, 1948, grew up in San Diego. She joined the Tucson Chapter of 99s in 1989 after earning her private license in 1987 in Tucson. She earned her instrument rating in December 1994 and is working on her commercial rating. She will follow with the multi, seaplane, and ultimate goal is the DC-3 type rating. The only other member her family that flies was her uncle who flew a P-38 during WWII. He was a general in the Air Force. She has flown PT-22 for five years and has sign-off in an AT-6. She has flown a lot of antique aircraft and has taken aerobatics lessons.

Oster is the owner/manager of Precision Machine Shop manufacturing components for aerospace, medical and computer industry.

MAUREEN OSTER, born Aug. 5, 1948, in Pennsylvania, a busy county executive joined the flying community and the 99s in 1988. Encouraged by her colleague and friend, Ilse Hipfel, Maureen found the skies a welcome relief from the hectic business world and the art of flying an exhilarating challenge. She can be found in her favorite Dakota en route to her quiet get-a-ways in Northern California and with her long-standing flying buddy, Ilse, at Jim Long and Palms to Pines air races. A supporter of the flying community and the San Gabriel Chapter of the 99s, she looks forward to the days she can spend in the skies and in support of the 99s organizational events, a commercial pilot rating and a volunteer pilot for the Flying Samaritans, these are dreams and goals that this woman intends to pursue: To be one of the special few is one of her greatest loves.

JOANN OSTERUD, joined the Lake Tahoe Chapter of 99s in 1995. She received private pilot license in 1967 and has ASEL, AMEL, instrument and helicopter ratings. She has accumulated more than 12,000 hours flight time and has ATP and FE certificates. Osterud has flown many, many types of airplanes such as Hiper-Bipe, Ultimate 10-300S, P-51 Mustang, Hawker Sea Fury and Boeing 727.

Osterud was the first female airline pilot with Alaska Airlines in 1975. She holds the world record for consecutive outside loops (208 on July 13, 1989) and the world record for inverted flight distance (658 miles in 4.5 hours on July 24, 1991). She has been airshow performing for the past 26 years, among other things. She is first female to compete in the unlimited division at the Reno Air Races in 1987. She is currently a co-pilot for United Airlines.

MARY OSWALD, learned to fly in 1979 – not because it was a lifetime dream but because she was a poor passenger in small planes. Achieving her license put fun into flying!

Mary has been an active Alberta 99 since 1979 and has helped organize Air Rallies, Poker Runs, airmarking, safety seminars, Flying Companion seminars and Alberta's 75 Anniversary Fly-Out. She served as governor of West Canada Section. She was the first woman to sit on the board of directors of the Alberta Aviation Council and was the first woman to be elected their president, 1992-93. She is a member of COPA.

Mary's career is in education, with an M.Ed. from the University of Alberta. She has taught at all levels and worked as a teacher consultant for many years.

She is married to Gordon who is also a pilot. They have three grown children and five grandchildren.

DR. S. TONI OTIKER, her desire to fly began early in life, but without financial backing it stayed a dream. Her family was in an economic slump. After becoming a registered nurse and obtaining other degrees, time and finances were scarce.

It was not until she married Col. Harold Otiker, completed her Ph.D. in health sciences/behavior sciences, moved to a farm in Peru, IN, that she began to realize her childhood dream – some decades later. She finished ground school, soloed and obtained her private pilot license at Grissom AFB Aero Club and promptly joined the 99s. Flying still was limited because she was into animal health and participating in Angus cattle shows.

Otiker is now more involved with the 99s, International, sectional, Illiana Cardinals Chapter meetings and the International Forest of Friendship. Harold died October 1992 and was honored at the Forest of Friendship in 1993.

BERTHA "BEBE" C. OWEN, born June 22, 1934 in Bay City, MI, joined the 99s in 1976. She received Part 91 flight training from a local instructor and her husband,

Verwayne "Curley" L. Owen. She has flown the C-140, 150, 170 and 172. Owen and her husband have one daughter, Pamela M. Owen who is also a 99!

Her most memorable flights were those to Oshkosh. One month after receiving her license she flew from Maryland to Oshkosh and again flew to Oshkosh in formation with two Luscombes several years later in a challenging flight due to weather conditions.

Owen is the proud owner of a polished, 1948 Cessna 140 and works with the International Cessna 120/140 Assoc. as their parts coordinator and state representative coordinator. Her husband and she recently retired to Aero Acres, an airpark in Florida.

PAMELA M. OWEN, born May 23, 1959 in Saginaw, MI, joined the 99s in 1985. She received her Part 91 flight training from her father and a local flight instructor. She has flown the C-140,150, 170, 172; PA-23, PA-44, EMB-120, ATR-42 and ATR-72. She is presently employed by American Eagle – ATR 42 and ATR 72. She was employed by United Express, Atlantic Coast Airlines and Brasilias in 1992.

One memorable flying experience was recent relief/ humanitarian flight to Anguilla and St. Kitts in the Leeward Islands after Hurricane Luis toured the Caribbean. As an American Eagle flight officer, she was given the opportunity to fly an ATR 42 loaded with water, food and general supplies to these devastated islands. It was depressing to view the destruction caused by Luis, especially to the beautiful island of St. Marten. As they unloaded, the smiles on the needy faces were most rewarding. It is a real experience to combine the joy of flying with the satisfaction of aiding our friends in need.

Thanks to the 99s, Owen's airline career is progressing. Her goal to follow in her father's footsteps, a flight officer for a major airline, hopefully is in the future soon!

MARTHA OZBUN, born March 25, 1924, joined the Indiana Chapter of 99s in 1952, after receiving her private pilot license. She started flying a Piper Super Cruiser at Reese Flying Service in Muncie, IN. At that time she flew WACO UPF7 and Stinson.

Ozbun and her husband, Virgil, were members of the Civil Air Patrol in the 1950s. Virgil was also a pilot and retired after 42 years as a mechanical engineer. Ozbun is retired after 38 years in the beauty salon business as a licensed master's beauty colorist. They have a son, Gregory, 34, and daughter, Janet, 37 and six grandchildren (one set of twin girls).

Her flying has been for pleasure which has taken them over the greater part of the states.

SUSAN PAL, born Jan. 24, 1936, emigrated from London, England to Canada in 1952 and obtained an AA degree in fine and commercial art. She also attained rank of able seaman in the Canadian Naval Reserve.

Susan met her husband in Vancouver, Canada. They were married in Los Angeles and later moved to Seattle, WA, where they started their company, producing documentary, entertainment and corporate films.

Susan took her first flying lesson on St. Patrick's Day in 1975 and flew a C-152 solo from Munro, MI to Seattle a year later. She quickly

obtained her SES rating and commercial/instrument rating and in 1990 completed a BA in psychology at the University of Washington. More recently she passed the flight engineer's written exams. She has flown the Cessna Cutlass, Aeronca, Beechcraft RG and C-177B. She is currently a corporate pilot for Pal Productions, Inc.

The flying club she joined on obtaining her private license was owned by a crusty pilot named Virginia Hubbard. She sold Pal her first aircraft, N714TD, one of the first C-152s off the line and the "picture" plane for the Cessna brochures and Jeppson IFR videotapes. As she was leaving for Michigan to fly *Four Tango Delta* back to Seattle solo, Virginia said to just keep flying west and stay low enough to read the street signs!

Chairman of the Greater Seattle Chapter, 1980-82, she has held positions as chapter secretary, vice-chairman, sectional co-chair and is currently chapter safety chairman and registration coordinator for the annual Flying Companion Seminar. She has also participated in several Petticoat Derbies, and created and taught an aviation badge program for Girl Scouts. She has logged more than 1,200 hours and flies a 1974 Cardinal Classic. Her hobbies include photography, boating, residential design, camping, hiking and landscape painting. Susan and her husband have two grown children, Paul G. Pal, 29 and Anna F. Pal-Doak, 27, and divide their time between Seattle, Friday Harbor, WA, and Vancouver.

VALERIE G. PALAZZOLO, learned to fly in 1977 at the age of 19. She spent many years at Detroit City Airport acquiring her instrument, commercial, multi-engine and flight instructor ratings. After accumulating many hours from flight instructing, Valerie flew as a charter pilot for a freight company and then as a corporate pilot. In 1987 Valerie was hired at American Eagle Airlines as a first officer on an ATR-42 aircraft. She checked out as a captain for the airline on both the ATR-42 and ATR-72. In 1991 Valerie accepted a position as an aviation safety inspector with the Federal Aviation Administration. She is type rated in a DC-9 aircraft and presently works at the Detroit Flight Standards district office as a air carrier principal operations inspector. Valerie stays current in general aviation activities along with being very active as a speaker for elementary, high school and college aviation career programs. She is married to Russell Palazzolo and has two sons, Alexander and Zachary.

MARY PANCZYSZYN, one of her earliest memories was a visit to their local airport when she found only one single airplane parked on the field – a faded yellow Piper J-3 Cub. It wasn't until after she married that she began flying and low and behold – completed all flight training and received private pilot license in none other than a bright yellow Piper J-3 Cub.

In 1965 she joined the Chicago Area Chapter of the 99s and served as corresponding secretary, ways and means, nominating, public relations, co-chairman 99s Exhibit in Friendship tent at the EAA convention for nine years and Illi-Nines Air Derby Race board for the past 25 years. On Jan. 14, 1984, Panczyszyn had the privilege as chapter chairman to preside at the 50th Anniversary Commemoration of the Chicago Area 99s.

Flown many races as pilot and co-pilot. Received the Chicago Area Chapter Service Award in 1980 and North Central Section Governor's Award in 1986. Graduate of St. Francis Hospital School of Nursing, Evanston, IL and at present charge nurse of the pre-surgical holding area in the department of surgery.

ANNA M. PANGRAZZI, born Jan. 6, 1959 in Detroit, MI, learned to fly in 1977, the year she graduated from high school. She went on to get her commercial and instrument rating in 1980.

In 1982 Anna married John Kucher and moved to Toronto, Ontario. There she attended York University and received a BA in economics. After graduation she started working for a firm selling airplanes. It seemed a perfect way

to combine business with flying! In 1988 she incorporated her own business, Apex Aircraft Sales Ltd., and since then she has consistently sold approximately $ 2,000,000 worth of airplanes a year!

Anna joined the 99s in 1984 and has held many positions including chairwoman of the very busy First Canadian Chapter. She has also been the coordinator of Operation Skywatch, a joint project between her chapter and Ontario Ministry of the Environment in aerial surveillance and photography for the investigations and enforcement branch.

In 1990 Anna flew in the Air Race Classic with her flying buddy Margo McCutcheon in Margo's Beech Baron. She participated in the making of the Operation Skywatch documentary called "Angels in the Sky." She has flown to three International conventions, Vancouver in 1987; New York City in 1990; and Norfolk in 1994.

As part of her job selling airplanes, Anna has the opportunity to fly many different kinds of planes. Her favorite airplane though is the Mooney 201, so she concentrates on selling that model so she can get to fly them as much as possible!

Anna is proud to be a member of the 99s, Inc. and has met many flying buddies through her chapter. She looks forward to years of camaraderie and hopefully some more air racing!

JUDY PANTAGES, joined the SCV 99s in 1991 after getting her license in 1989 at Palo Alto Airport. She has her private ASEL and logged more than 500 hours. She belongs to the Palo Alto Airport Assoc., AOPA, and the West Valley Flying Club, where she served on the board for several years. She is an active supporter of General Aviation, coordinated Airport Day at Palo Alto Airport in 1993. Her favorite flights have been over Hawaii where she takes her vacations. Judy is the technical director of respiratory medicine at El Camino Hospital and likes to read, write and hike.

SYLVIA L. PAOLI, a commercial pilot with instrument rating, has logged more than 3,000 hours since 1968. She was general counsel for the 99s, Inc. for 14 years and actively continues her support at local and section levels.

Her background includes a BA from Colorado College and a MM from New England Conservatory of Music, and over 20 years as a concert pianist and recording artist, on the list of International Baldwin Artists. Switching careers, she obtained her JD from Western State University College of Law and has been practicing law since 1977.

She is a long-term member of AOPA, past board member of California Pilots Assoc., Chairman of Fullerton Airport Commission and one of the founders of NCWA, for non-pilot women of aviation.

Paoli is actively involved in her church and in other community activities. She is married to lawyer, Peter. They have four children, an "adopted" fifth child, and eight grandchildren.

SHEILA RAE PAPAYANS, born Sept. 6, 1941 in Los Angeles, CA, joined the 99s in 1985 after receiving private, instrument and commercial rating. She has flown C-172 Trinidad, C-152 and Bonanza A-36. She is married to Serge and has three children: Jennifer, 33, Michael, 30 and Michelle, 12. She has three grandchildren: Michael, 6, Jessica, 5 and Nicole, 2.

Soon after getting instrument rating flew coastal route to Alaska and back. Goals are to fly across Atlantic to Europe and to become a CFII. She has her husband, Serge, and two passengers, George Peterson and Walt Hackman on Flying Samaritan trip. Sheila acts as pilot then as interpreter and triage on trips to Panta Prieta and Bahia de Los Angeles in Baja, CA. She never gets to rest on these trips, because she can wear so many hats.

GEORGIA ELAINE PAPPAS, learned to fly in 1966 at Hanscom Field, Bedford, MA and has had a career in aviation ever since. She currently holds a commercial pilot license with instrument and multi-engine ratings and has been a member of the 99s since 1967.

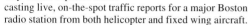

Georgia is currently manager of terminal services at Boston-Logan Airport. She has been at Logan for over 20 years and was the first female operations shift manager.

Georgia also spent several years as Boston's Skyway Patrol reporter broadcasting live, on-the-spot traffic reports for a major Boston radio station from both helicopter and fixed wing aircraft.

In her early days of flying, Georgia was an air race enthusiast. She raced in numerous air races, placing first in the New England Air Derby, second in three consecutive All Woman's New England Air Races and completed several transcontinental Powder Puff Derbies and intercontinental Angel Derbies. She also enjoys aerobatics and competed in one aerobatic competition.

In addition to being a member of the 99s, Georgia has been elected director of the Aero Club of New England for over 20 years. Georgia also received an appointment to serve on the Presidential Women's Advisory Committee on aviation. She attended Northeastern University and majored in business administration.

BARBARA A. HARRIS-PARA, born Aug. 9, 1945 in Providence, RI, began flying as a lark. She needed something in her life to jazz it up a bit. She saw that the community college was offering a course for ground school, she took the course. Para did have a fear of heights, which took a few hours to overcome, but eventually she soloed, and started on her cross-countries. In a little over 16 months she managed to get her ticket, and then her husband began his training; they soon bought their first plane a B-19 Sport. After their first long cross-country to Detroit, they bought a 200 Arrow, which they still have. She has flown Pipers, Beechcraft and Cessna. Para has accumulated 1,200 hours flight time with six hours in multi-engine airplane. She managed to have an engine failure, but got in safely. The other ratings came one a year over the next few years. Para joined the Garden State 99s in September 1990.

A slight problem occurred shortly after getting her CFI, she was diagnosed with breast cancer, and had to do six months of chemotherapy. Of course, that meant no solo flying, and she had to give most of her students to other instructors for that period. Para proceeded one to her CFII two years later. She won the Alice Hammond Scholarship from her chapter of the 99s; Alice passed away before she could show her the ticket. Para's next goal will be her multi-engine rating. She would like to become a flight engineer. She is now very involved with the 99s, as she is the chapter chairman, this past year. She hosted with their chapter, the fall section meeting.

Para is an industrial technology teacher for the seventh and eighth grades in Somers Point, NJ. She is married to Fred Para.

DOLORES "DEE" PARK, born in Maples, MN on June 12, 1934, is now living in Eatonton, GA. She has five sons: Charles, 40; Robert, 38; Michael, 34; Christopher, 30; and James, 27 and six grandchildren: Brian, 19; Becky, 17; Andrew, 7; Stephen, 4; Jonathan, 4; and Samuel, 1 month.

Dee started her flight training in 1983. At that time her husband to be, a pilot, suggested she take the "pinch hitters" course since they were flying together more and more. She didn't stop at that and went on to get her private pilot license becoming another flying grandmother. She joined the 99s in 1984.

Retirement has settled Dee and her husband, James C. Park Jr., in the sunny south in Georgia. Their Cessna Skyhawk 172 provides wonderful fun transportation back to the cold north to visit family and friends and great pleasure trips through the south.

Home base is Milledgeville, GA (MLJ), a small, neat, well-run airport with a very active FBO, a fun place.

ALMA B. PARKER, born on July 5, 1919 in Lake Panasoffkee, FL, and attended Stetson College. Alma was married in 1940 and moved to Easton Shore, MD. She has two children, Charles L., 50 and Jackie Vroman, 47. She has two grandchildren, Stephanie Vroman, 27 and Jimmy Vroman, 25. Her husband went into the counter intelligence service in WWII. She started flying at Beville Square near Webster, FL, a grass field where they often had to chase the cows off before they could fly. She obtained her private license in 1947 and her commercial in 1956.

A helicopter offering rides at Pinellas Park caught her eye one day. She fell in love with the helicopter and made up her mind that one day she would obtain a rotorcraft rating. While visiting her daughter in Seattle, WA, in 1976 and 1977 Alma took helicopter training at Queen City Helicopters on Boeing Field. She completed her training and obtained her rotorcraft rating in Clearwater in 1979.

In 1978 she took glider training at Venture Soaring, Clearwater Executive Airport in a Blanik glider, and obtained her rating. While attending a 99s International meeting in Alaska, she took seaplane instruction and obtained her rating at Brown's Seaplane Base in Winter Haven.

Alma joined the 99s, International Organization of Licensed Women Pilots in 1948 and has been participating in fly-ins, poker runs, etc. around the state, and attending many International conventions. In 1954 Alma joined the Pinellas Senior Squadron of Civil Air Patrol, advancing to the rank of major. She has worked in personnel, recruiting and emergency services, flown many missions as a rated pilot and observer, Sundown Patrol, and participated in flight clinics and other flying activities. She has flown the Stinson, Aeronca, Cessna 172, Luscombe, Grumman, Stearman, L-16A Taylor Craft, Ercoupe, Piper Cub, Cessna 210 and Piper Apache. She will do her Phase 9 as soon as she can.

She has attended National Congress of Aerospace Education seven times and National Board meetings on a number of occasions. Alma is a graduate of both Region Staff College and National Staff College. She is a member of the Suncoast Hunter's Assoc. in Tampa.

SARA PARMENTER, born in 1924, had her first flight at age 7 in an open cockpit bi-plane and was hooked. She began flying in 1944 while with the CAP, planning to join the WASPS; but when they disbanded, she went into nurse's training. After graduation, joined the Air Force Nurse Corps and resumed flying, getting her license in 1961. She stayed in the Air Force Reserves for 22 years, flying her Cessna 140 to Air Force bases for duty. She has flown and owned several different taildraggers, including a Stinson Voyager and for 23 years, has owned and still flies her second 140. She is working on an instrument rating, Phase VIII wings and plans aerobatic training. Sara has been married 45 years to a retired Army orthopedic surgeon and lives on a farm, raising peacocks and Pekinese. Nursing specialty is obstetrics and hobbies are flying and gardening.

CAROLYN PARMER, born Sept. 26, 1936, in Frankton, IN, joined the 99s in 1993. She always had an interest in aviation but never considered the possibility of becoming a pilot. Then on Sept. 25, 1989, her husband arranged a bi-plane ride as a birthday surprise, and life was never quite the same after that. On June 27, 1990 she took a demo flight and on Dec. 20, 1990 she soloed. She passed her private pilot checkride on Oct. 10, 1991. She has flown the Piper Warrior, Piper Archer and Cessna 172.

Parmer is married to William Parmer and has four children: Don, 39; Patricia, 35; Pamela, 32; and David, 27. She has six grandchildren: Mike, 15; Charlie, 13; Danny, 12; Becky, 10; Michelle, 8; and Katy, 7.

YVONNE "PAT" PATEMAN, was a ferry pilot and an engineering test pilot with the Women Air Force Service Pilot (WASP) from March 1943 to December 1944. Between WWII and the Korean War, she delivered civilian aircraft from factories to airfields throughout the country. Pateman also served as assistant manager at Monrovia Airport, California and as chief flight instructor at Culver City Airport, CA.

She was recalled to active duty in 1951 at Hamilton Air Force Base, CA, where she served as an intelligence officer. Pateman's overseas assignments were to the Philippines during the Korean War, Japan in the late 50s and to Vietnam where she was chief of the Warning Division for 7th Air Force during 1969-70. Her last assignment prior to retirement in 1971 was as chief of the China Air and Missile Section of the Defense Intelligence Agency.

Pateman was active as a flight instructor in military flying clubs both in the states and overseas. She has been published in *Minerva,* a quarterly report on women and the military, and in *Aviation Quarterly.* In 1990 she produced a 23 minute video, "We Were WASP." It covers the history of the WASP from 1942 until 1944. Pateman has also completed a novel, *Tomorrow Comes Last.* She is currently working on her non-fiction book titled, *Women Military Test Pilots, Yesterday and Today.*

She learned to fly at Wurtzboro, New York in 1942. She received her commercial license from the CAA following service in the WASP, her instructor rating in May 1950 at Culver City, CA and added instrument rating in February 1976, at Byrd Field in Richmond, VA. Pateman has logged more than 5,000 hours.

MARILYN PATIERNO, a lifetime New Jersey resident, earned her private pilot license in August of 1990. She flies a Beech Sierra B24R, which is based in Lincoln Park (N07)

Marilyn's educational background includes a BS in business management from the Madison Campus of Fairleigh Dickinson University. She is self-employed.

Her lifestyle is a combination of business, church, family and flying. She is involved with the Jacksonville Chapel, the Lake Valhalla Civic Assoc., the Caldwell Pilot's Assoc. and the North Jersey Chapter of the 99s. She initiated the Thursday Night Dinner Get-Together at the Landing Strip Restaurant at Lincoln Park Airport. Currently, she is treasurer of her local 99 Chapter and thoroughly enjoys the recognition she receives as she flies around the country wearing her 99 jacket.

She is married to Dick Patierno, who is also a pilot and a project developer. They have a son Michael, a daughter Susan and three grandchildren: Michael, Jennifer and Allison. Marilyn is known to sing the praises of the GPS and looks forward to her first trip to Oshkosh in 1995.

PAMELA PATRICK, age 45, received her private pilot license in August 1992. She received her high performance rating in June 1994. Her father was a pilot from 1965 to 1975 and an aircraft salesman for Cessna. He quit flying due to loss of medical. She received a bachelor's degree in medical technology from DePaul University and became a registered veterinary technician through Purdue University.

Patrick is the past secretary of the Indiana Rotorcraft Assoc. She started building a public use airport in April 1994, which was completed February 1995. It offers commercial maintenance and other services pending. Patrick presently flies Cessnas and Bonanzas and is working on airframe and powerplant mechanic's license with only a few months remaining.

Patrick is married with one 10-year-old daughter. She is a farmer and manages a veterinary practice located next to "Pam's Place" airport. She keeps really busy with work and Farm Bureau (county president) and the Indiana Agricultural Institute.

MARCIA PATTEN, got her ratings at Aris Helicopters at San Jose Airport, getting her helicopter rating in 1983 and then her ASEL in 1984. She now has her commercial CFI in helicopters. She worked at Aris Helicopter for many years, logging more than 7,000 hours, mostly in Bell JetRangers and the Hughes 300. She joined the SCV 99s in 1986 and has served on the Marion Barnick Scholarship committee twice. She currently is working for Membrane Technology, a chemical engineering firm, where she writes patents and does technical editing. She also works part-time for the ASRS, editing the *Callback* newsletter. Marcia is a member of the Whirley Girls and lives in Palo Alto, CA.

BETTY PATTERSON, is a computer programmer who has been flying since 1990 and has been with the SCV 99s since 1991. Betty has her instrument rating for ASEL and AMEL, is working on her commercial rating and currently has logged more than 600 hours. She received the SCV Pilot of the Year award in 1993. Betty belongs to the Northern California Antique Airplane Assoc. and works on the West Coast Fly-In and Airshow held annually in Watsonville. She is also involved in the California Antique Aircraft Museum and is a member of AOPA and EAA. She owns a Warrior, N40042, which she keeps at Frazier Lake Airpark. Betty also enjoys sewing.

CAMILLE MINOR PATTERSON, became a member in November 1989 and joined the Texas Dogwood Chapter, serving on several committees and as chairman for two years. Moved to Fort Worth, TX, in February 1993 and joined the Fort Worth Chapter, serving as membership chairman. Participated in EAA's Young Eagles and FAA's ACE and Wings Program. Earned awards such as Try Again Harder in spot-landing contest and first in navigation. Flying was a dream and a very hard earned goal due to family and health obstacles, but persistence (hardheadedness) won. Bought Skyhawk II while working on private pilot license and still working on how to make graceful landings. Met husband, Jim, through aviation while helping pilot/A&P hubby build a KR-2S. One son loves flying; one son hates flying; but foster daughter loves to fly from the back seat. Motto: "Keep Flying!"

KENNIE RUTH PATTERSON, was born in Oregon in 1940. Her desire to fly began in the third grade when she would go out to the airport with a neighbor boy, who would wash and fuel the planes. Patterson had always considered flying to be an unattainable goal; but her second husband, who was a pilot, and her two daughters, Kirsten and Elisa, encouraged her to get her license. They then bought *Mariah*, their 1959 Square-Tail Skylane. She was a delightful toy when they lived in California, as there were numerous opportunities, places and the 99s friends to share flying and playing. That Santa Clara group of ladies proved to be an exciting and wonderful introduction to the 99s.

Patterson has approximately 950 hours and holds ratings of ASEL, private and instrument. She is currently a member of High Country Chapter of 99s, Flying Aces and AOPA. She has completed wings and level V.

She and her companion, John Nelson, live in Grand Junction, OR. They are both retired teachers and each has two daughters, ages 26 through 28.

PUD PATTERSON, began flying in June 1980. By 1983 she had earned a commercial license, instrument rating, multi-engine rating, flight instructor ratings (CFI, CFII) and instrument and advanced ground instructor ratings. Formerly a professional photographer and journalist, she combined these interests with flying and opened Friday's Aerial Photos and Flight School in 1984 at Rusk County Airport in Henderson, TX. In 1987 she was named a FAA Gold Seal flight instructor. After many hours of instructing and flying as a fire patrol pilot for the Texas Forest Service, Patterson earned an airline transport pilot's license in 1988. In 1993 she was appointed a FAA accident prevention counselor. A member of the Texas Dogwood Chapter of the 99s since 1981, she is married to attorney, J.R. Patterson Jr., also a pilot. They have one son, Ben.

SHIRLEY A. PHILLIPS PATTERSON, a Reno Area Chapter 99s member, is a retired USPS and retired Keno supervisor. She joined the 66s in 1983 and the Reno Chapter 99s in June 1984. She became involved in aviation in 1983. She had taken a ride with a friend and bought Ground Study Home Study Course the same day. Two weeks later started flight instruction. Worked for CPA preparing tax returns in exchange for use of his aircraft during student pilot training (four hours work for one hour flying time). She has private license and flies strictly for pleasure.

She was treasurer of the 99s from 1984-86; bylaws chairman 1986-88; air race coordinator 1985-87; legislation chairman 1984-86; poster sales 1987-92; newsletter editor 1988-90; (appointed by council of governors to Special International Bylaws and Standing Rules Committee 1990-91; wrote student pilot program 1989; and designed scholarship and recognition certificates and 25th Anniversary pin 1989.

Patterson is a charter member of Air and Space Museum, Smithsonian; NAG member; associate member of the Smithsonian; member International Assoc. of Flying Turtles since 1990 and member AARP.

She is an artist and calligrapher for Ducks Unlimited and Nevada Waterfowl Assoc. She attended UNR Business School and Truckee Meadows Community College. She was a Reno Little Theater set painter 1984; performer 1985 and assistant director and stage manager 1986-88; co-found Nevada Air and Military Museum, director 1986-88; adjutant/secretary 1987 and commander 1987-88 of Washoe County Sheriff's Air Squadron, search and rescue coordinator and dispatcher 1987-88. Currently co-founder of charitable and educational organization, comprised of International Forest of Friendship honorees and others interested in aviation and aerospace. She has one son, Robert L. Nichols II, two daughters, Julia A. Nichols Setzer and Patricia L. Nichols Cole, and three grandchildren, Christina and Johnna Nichols and Zeke Cole.

ELLA MAY PATTISON, is a retired elementary school teacher. Her husband, Earl, is a retired aeronautical engineer. Their family consists of a grown daughter and a son and their spouses.

In 1973 at age 47, she took a pinch-hitter's course when her husband resumed flying after many years. She got hooked on flying and earned a single-engine land license and instrument rating, flying Grumman-American TR-2 Trainers, Travelers and Tigers in a flying club. Patterson took many of her third, fourth and fifth grade students for rides until she retired in 1988. She is active in Long Beach Chapter 99s. She has served as newsletter chairperson, secretary treasurer; helped with airmarking, air fairs, seminars, Poker Runs and other chapter projects. She collects aluminum cans for scholarship money for pilots. In March 1994, she completed an emergency maneuvers flight training course in a Robin and found that she loved doing aerobatics!

JANICE W. PAUL, born Sept. 29, 1926 in New Berlin, NY, began flight training in 1975 at age 48. Received her private pilot license in 1977. She has logged more than 1,100 hours. Paul received about 75 hours instrument instruction. She has flown the C-150, C-172, PA28-180 and PA28-181.

She was married to Roy D. Paul (deceased) a senior chemical engineer. She has two children, Dr. Charles Paul, 38 and Christine Paul, 40. She has two grandchildren, Alice Paul, 5 and Carla Paul, 3 and a half. Paul has a BA degree from Syracuse University.

Furthest flight to Oshkosh Airshow. Based at Brainard Field, Hartford, CT. She co-owns a beautiful PA28-181 Archer II, Angel Wings Partnership. She joined the 99s in 1991 and has been membership chairman CT Chapter 99s.

SUE PAUL, born July 4, 1945, in Austin, TX, took her first flight at 14 years old in 1960 in a WWII PT-19 with her future husband, John (17 years old). She has flown Cessna 172, 182; Warrior, Archer, currently owns a 150 hp Citabria.

Paul and her husband are the founders of the Warhawk Air Museum, Caldwell, ID, a WWII Air Museum. They privately own two Curtiss P-40s and a North American P51B Mustang. She has three children: Cassy Lindsey, 29; Sharon, 27 and John Curtiss, 25.

JO E. PAYNE, born in San Antonio, TX on Feb. 14, 1937; is married to husband, William B. Payne, attorney; one son, R. Michael Payne. Jo began flying in 1973 and earned private license, ground instructor and instrument rating.

In 1973 she was secretary of Carolinas Chapter. From 1974-76 she was half owner NC FBO. She was employed by Beech dealer in North Carolina as a special representative – flight training, charters sales 1976. In 1976 the family moved to Austin, TX and in 1977 finding the Austin Chapter disbanded, she reorganized it and became chairman. She was FAA accident prevention safety counselor.

She and Austin Chapter member, Diane Hadley conducted "Introduction to Aviation" activities in West Lake Elementary showing the students basic flying operations and the kids made and flew their own hot air balloons. Also worked with Young Civil Air Patrol students. She conducted workshops for teachers and 99s showing them how to interest students in flying. She was active in South Central Section and served as aerospace education chairman.

In 1985 she earned her bachelor's in science education with teaching license from the University of Texas in Austin; 1985 taught physical science at Seguin High School and 1986-89 at New Braunfels High School.

Jo's love of aerospace education and safety was paramount in her teaching career and work with young people. Jo died of a stroke June 8, 1989.

JOAN TANNER PAYNTER, Bakersfield Chapter, born in Los Angeles, CA, on May 3, 1929, and moved to Bakersfield, CA, when she married William Paynter in 1951. She learned to fly in Bakersfield in 1968 and joined the 99s immediately. Ratings include commercial, instrument, single and multi-engine land. She has flown as an air taxi pilot since 1974, has operated Payntaero Air Charter since 1980 and has logged more than 5,000 hours. She has flown the Cessna 150, C-172, Cardinal, Beechcraft, Musketeer, Debonair and F-33A Bonanza.

Joan has owned three Beechcraft, the most recent an F33A Bonanza. She and her sister, Shirley Tanner, have teamed to fly several Powder Puff Derbies, placing in three, including a narrow second place finish in the last 1976 race. She placed in the 30th Commemorative Flight, an Air Race Classic, Palms to Pines race and raced in four Pacific Air Races.

Offices held in the 99s include chapter chairman twice, Southwest Section secretary and nominating chairman and International nominating chairman. Aviation memberships include 99s, Inc., Aircraft Owners and Pilots Assoc., National Council for Women in Aviation and California Pilots Assoc.

Most unusual charter passenger . . . an injured dog whose master had to get him immediately to a special veterinary surgeon in Sacramento, CA. Time was vital and fortunately he survived because of the quick emergency flight.

Community recognition of air racing endeavors include an award from AAUW and a Commendation Resolution from the Kern County board of supervisors. Civic and community activities include American Cancer Society Kern Unit board of directors, auxiliary to Children's Home Society and Memorial Hospital Foundation. Joan graduated from the University of Southern California. She and her husband have two children, a son, Kevin, 38 and a daughter, Laura, 35, and two grandchildren, Michele, 9 and Benjamin, 1.

SOPHIA M. PETERS PAYTON, took her first lesson at Norton Field, Columbus, OH, in 1944 and 51 years later is still flying. Joined the 99s in 1946; private 1945; commercial glider in 1946; commercial and instrument rating, made aviation her career. She was in aircraft sales, charter, ARTC simulator operator, airshows (aerobatics in her glider).

Ask her what her true love is, and she will say two things: "my late husband, Neal, retired USAir pilot and then air racing!" Her first air race in 1947 with *Twin-Irene*, she placed second. She has flown approximately 74 races, 21 transcontinental races and International races plus regional races. She was in the top 10-55 times; AWTAR she placed second, fifth, seventh, eighth; IAR she placed second, third and sixth; two time winner Michigan Air Race; three time winner Indiana Race; Air Race Classic, 1978, first, fifth; flown 66 different makes and models airplanes and gliders. She was co-chairman of first Indiana Race; first chairman Greater Pittsburgh Chapter. Payton devoted many hours on 99 committees.

MARGARET JULIET MCCLAIN PEAKE, born Nov. 18, 1916 in Pittsburgh, PA, learned to fly a private airplane and got her pilot license in 1961. She has flown J3 Cub, Cessna 150 and Cessna 182. She has flown the Powder Puff Derby in 1978 from Palm Springs, CA, to Tampa.

She is married to Ervin C. and they have two children, Richard J., 52 and Margaret Jean Eastin, 47. Peake flies for pleasure only. Her educational background includes: faculty grad school of social work Carnegie Mellon. Field work instructor and faculty of grad school social work, Rutgers, the state university; field work instructor with Trenton State College, taught course to the state police.

MARTY PEARCE, Phoenix 99s, one of Phoenix's outstanding aviation professionals, also had a 20 year career as a US Navy nurse, ranking as a lieutenant commander by discharge in 1975. She not only has contributed in many ways to the success of the Phoenix 99s, but has used her skills to teach and train in many areas of aviation. She has an EDD from ASU and taught as a tenured professor in the ASU School of Aeronautical Technology.

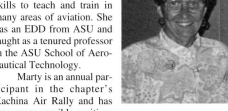

Marty is an annual participant in the chapter's Kachina Air Rally and has won every possible position: first, second, third, fourth and fifth. She flew the last Powder Puff Derby and placed second, flying solo in the 1986 Air Race Classic. She won third in the 1995 Air Race Classic with fellow Phoenix 99, Shirley Rogers. She has served the 99s as chapter chairman, and spent many hours supporting NIFA, being FCIFA sectional chairman and instructing in the CAP.

Her flight ratings are: COM-ASEL, AMEL, ASES, AMES, IA, glider, CFI – ASE, AME, IA and glider. She belonged to the Arizona Aviation Safety Advisory Group, the Arizona Educational Council, and now is an FAA accident prevention counselor. She is in AOPA, EAA, SSA and NAFI. Marty has encouraged many of her ASU students to join the 99s and/or apply for 99 scholarships.

JEAN HANMER PEARSON, received her private pilot license in 1941. Currently holds a commercial with instrument rating and flies own Cessna 172. She was operation officer in all-women CAP squadron before graduating from WASP training in class 43W3. She later joined the US Navy, commissioned an ensign and assigned to Air Training Division as deputy chief of Naval Operations for Air in Washington, DC. She remained in Navy on active and reserve duty as an aviation officer and is now lieutenant commander, USNR (Ret.).

Pearson was aerospace and science writer for the *Detroit Free Press* and science writer for the *Detroit News* for over 30 years. She won the Aviation/Space Writers Strebig Memorial Award for excellence in aviation writing for a series on "Man in Space." Also named "National Headliner" award winner by Women-in-Communications, Inc., for a series on Antarctica. She accompanied the first four women scientists to do scientific research on Antarctica, and with them, flew to the South Pole where they were the first women to land at the geographic South Pole. She also flew over the North Pole with the Japanese imperial family in a Scandinavian Airlines inaugural flight over the pole.

Pearson was a member of the first FAA WACOA and served as chairman and currently as a permanent trustee of the 99s Amelia Earhart Memorial Scholarship Fund. She flew in 10 All Women Transcontinental Air Races. She is past president of the National Assoc. of Science Writers, a board member of the Michigan Aviation Hall of Fame, second vice-president of the Michigan Chapter of Circumnavigators Club (members have circled the globe), is a member of the American Aerospace Medical Assoc., American Institute of Aeronautics and Astronautics, Silver Wings, American Medical Writers Assoc. and numerous other aviation, science and writers associations.

Pearson received a BA and a M.Ed. from Wayne State University, MPH degree from the University of Michigan and honorary doctor of humane letters from Wayne State University. She is married to Morton C. Pearson, an attorney and also a licensed pilot.

DEBORAH ROSE PEEL, born in Toronto, Ontario had always dreamed of flying but never thought she could ever become a pilot. At the age of 35 she finally decided to pursue the challenge of flight. She started her flight training in March 1994 at the Toronto Buttonville Airport. On June 4, 1994 she did, alone and unassisted, take a Cessna 150 into orbit and furthermore did land (on the same runway!) without loss of blood or confidence, thereby completing her first solo flight. As a result of great patience from her flight instructor, Greg Frazer, as well as her husband, Douglas, she successfully completed her flight training and received her private pilot license on Nov. 17, 1994. In recognition of her excellent academic marks and flying ability she was awarded the Award of Excellence from the First Canadian Chapter of the 99s. Deborah is still flying, has completed her night rating and plans to continue her flight training to obtain instrument and commercial certifications.

CAROLE PENDLETON, born Oct. 26, 1937, in Cleveland, OH, joined the 99s in 1956. She was a corporate secretary/pilot for Penco Tool Inc. She has a commercial and instrument rating. She has flown Cessna 172 and Cherokee 180. She is married to Lee Pendleton and she has two children, Jennifer (J.J) a second year medical student, and Cindii, a college student in Columbus, OH.

In the late 1970s her husband bought her a partnership in a Cherokee 180. She was extremely fortunate to have Jon Speer as a partner. Between 1977-85 they flew many cross-country air races. Due to Jon's navigational genius and timing efficiency they won trophies in many of the races they entered.

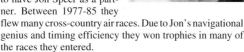

One memorable cross-country race they were able to take along her two young daughters. Thanks to Jon they won the race. None of them will ever forget that special event.

PHYLLIS M. GODDARD PENFIELD, born on the Island of St. Helena, Southern Mid-Atlantic, in 1897, earned her pilot license on March 15, 1929 after only three hours and 45 minutes of flight instruction. She was one of the original charter members of the 99s.

She was educated in England and Canada and became an American citizen in 1931. Together with her first husband, Norman A. Goddard, she operated a small commercial airport between Imperial and El Centro, CA. They established the Palo Alto School of Aviation on the Stanford University campus in 1927. They were also involved in the Dole Air Race. Her husband entered the race and was unhurt in the attempted take off crash. In January 1930, her husband was killed in a glider crash.

She continued to run the aviation school until selling it. She then established the Stanford Flying Club, which she managed until marrying Thomas Petts Penfield, a pilot for the Hanford School of Aviation in Santa Maria. She continued to fly until 1935. She died in Santa Maria in August 1984, at the age of 87.

ISABEL L. PEPPLER, born Feb. 26, 1934 in Hanover, Ontario, Canada, learned to fly in 1966 after flying right hand seat with her husband since the beginning of their marriage. She is married to Bill Peppler who is the general manager of the Canadian Aircraft Owners and Pilots Assoc. Isabel is the owner of and publisher of *From The Ground Up,* a ground school manual that is widely used in pilot training in Canada. Peppler has flown the Cessna 150, Cessna 172 and Bonanza.

She has held the office of governor of the East Canada Section of the 99s and has been chairman and treasurer of the Eastern Ontario Chapter, and served on the organizing committee when the Powder Puff Derby started in Ottawa, Canada.

She has one daughter, Carla, 34 and two sons, Rand, 31 and Graeme, 29. Aviation has been an integral interest of the Pepplers and the whole family has enjoyed many flying trips throughout Canada and into the United States.

LYNN MARIE PERRINE, born Sept. 7, 1949, in Cleveland, OH. A long time Miami resident, Lynn earned her private pilot license in 1989 flying a Cessna 172. She joined the 99s shortly thereafter. She transitioned to a Piper Arrow 180 hp for her instrument rating in 1992 and attained her commercial rating in 1993. She has logged 400 hours.

Lynn was encouraged to fly by her husband, Don. She has a 16-year-old daughter named Corinne, two stepdaughters, a stepson, and four grandchildren – three of whom are triplet boys!

A registered nurse, Lynn is self-employed in the field of corporate health promotion/wellness programs. In 1992 she earned a BS in professional management with a business specialty. In 1994 her work took her to Russia and Poland as part of a preventative medicine delegation from the US. She also did some consulting last year with businesses in San Paulo and Rio De Janeiro, Brazil.

Lynn is a 23 year member of the Rosicrucian Order AMORC and a three year member of BOTA (Builders of the Adysum). In addition to a love for flying, Lynn enjoys travel, reading, studying foreign languages and cultures, interior decorating, step-aerobics, walking, swimming, biking, ballroom dancing, and scuba diving and snorkeling in the beautiful waters of South Florida with her daughter, Corinne.

LORI PERRY, was in her late 30s in 1985 when she began taking flying lessons. Obtaining her private on Jimmy Doolittle's birthday in 1987 was a major step in overcoming a lifelong fear of flying. She and pilot husband, Al, own a 1976 Beechcraft Sundowner based at CAK in Ohio. Also instrument rated, she enjoys the many friends she has made through aviation.

An elementary art teacher, Perry is a former commercial artist. She holds a BS in design from the University of Cincinnati, and both a BS in education and a MS in educational psychology from the University of Akron. She is also a licensed realtor.

As a member of the Lake Erie Chapter, she served as chairman in 1990-92, and earned the Pilot of the Year award for 1992. Other memberships include AOPA and East Central Ohio Pilots, where she has made presentations on one-tank trips.

PAULINE "PAULIE" M. PERRY, born Sept. 7, 1927 at Mahaska, KS. Had first plane ride at age 8 with a barnstormer who flew her over the family farm. She is an RN from Lincoln General School of Nursing, Public Health Certificate from the University of Minnesota.

She got her private license in 1958, commercial instrument, commercial glider ratings also. Pauline and late pilot husband, John, re-covered her Aeronca Chief and built her experimental BD-4 999PP, both 99 white and blue. Timer for four Powder Puff Derbies, hostess for one. Served as Nebraska Chapter treasurer, corresponding secretary. She has been a member of the 99s since 1965. Panhandle Petticoat Pilots, EAA, AOPA, board of directors Scottsbluff C of C Air Transportation committee. She is currently a member of the Colorado Chapter.

She has three children: Susan Perry Grasmick, RN, BSN; Bill a private pilot; and Mary Lee Atenhan, songster and trackster. She has five grandchildren: Randall J. Perry, 14; Reid B.G. Perry, 11; Roark D. Perry, 8; Ashley R. Atenhan, 11; and Ryan S. Atenhan, 7. She works at Sew and Sew Fabric Shop. Hobbies: flying Citabria, sewing, swimming, biking, tennis and golf.

IRENE PETERS, her twin, Sophia M. Peters Payton and she learned to fly at Norton Field, Columbus, OH. Joined all Ohio Chapter in 1946. Pilot Twin Queens on the all Ohio Air Tour. In 1949 they were Columbus Air Race Hostess at the National Air Races in 1947. She and Sophia flew their first air race, placed second in a J-3 Cub in 1947. After they flew their first air race together in 1947, they said they would do it together again in 50 years.

Peters later married Vic and decided to become an executive housewife. Stayed home to raise her family: two boys, Victor and Steve and daughter, Janet. Her hobby then became her children and their activities. She has four grandchildren: Steve, Erica, Katrina and Christopher.

JEANNE M. PETERS, born May 9, 1940, in Omaha, NE, has had a love of airplanes since childhood and began flying in May 1974. She received her private pilot license two and a half months later. Jeanne has SE/ME commercial, instrument rating, CFI and basic ground instructor, with more than 2,100 hours in her log book. She received flight training at Sky Harbor, Omaha; Hangar One and ProFlite in Omaha. She is a check pilot for the Civil Air Patrol. From 1977-81 Jeanne owned Alpha Flying Services, a Part 135 operation out of Omaha, NE and from 1980-86 owned and operated Aviation Insurance Services, and with American Aviation Underwriters, helped to write an insurance policy for the 99s who owned aircraft.

Jeanne Peters has participated in three air races: the 1976 Powder Puff Derby and both Baja Air Races. She plans to race with the Omaha Area 99s Chairman, June Evans, CFII and pharmacist in the 1995 Air Race Classic.

Jeanne currently teaches behavioral disordered students at Marrs Junior High in Omaha. She has a master's degree in special education and plans to attend law school after retiring from teaching. She is married to Charles A. Peters Sr., has three children, four stepchildren and numerous grandchildren.

JO ELLEN PETERS, is the proud recipient of an Amelia Earhart Memorial Scholarship in 1995 for the Airline Transport Pilot rating which she completed July 15, 1995.

Jo Ellen lives in Janesville, MN. She is married to Matthew Peters, police chief in Janesville; her son, Brian Wood, 24, recently completed a five year tour in the Marine Corps and is now attending School of Mines, Rapid City, SD; and her daughter, Kim Wood, 20, is attending Mankato State University, MN; both majoring in electrical engineering.

Jo Ellen earned her private pilot license in 1984 in Custer, SD. Since then she has obtained the instrument, commercial and flight instructor ratings in both single-engine and multi-engine, instrument instructor and ATP. She has an AS degree from Mankato State University.

She has been a full-time employee in Nicollet Co. since 1989 and a part-time flight instructor in Mankato since 1989 and a part-time flight instructor in Mankato since 1988.

She joined the 99s in 1991 and is a member of the Minnesota Chapter. She is serving as public relations chair for 1995, and is involved with the Skywatch program for southern Minnesota.

Jo Ellen is a captain in the Civil Air Patrol, Mankato Squadron. As the squadron's check pilot, she is responsible for ensuring all pilots are proficient and competent. She has been a volunteer instructor for the Minnesota Wing Solo Encampment for two years and logistics coordinator for the encampment in 1995. She has flown on three ELT missions and one downed aircraft mission. Her goal is an airline flight officer.

DOROTHY A. PETERSON, born Aug. 27, 1914, became involved in aviation in 1944. Peterson and her husband had started a model plane club in Harvey, IL. They decided to fly the real thing. She and her husband, Herbert, received their private pilot license in 1945. They flew Porterfield, Cub, Luscombe and Aeronca. She joined the Reno Chapter of 99s in 1990.

Those days, they did spins, stalls, 720s, spot landings, pylon, eights and more. They had to wear a parachute to do the spins. At one large airshow they came in one-two in the spot landing contest. Their first plane was a Taylorcraft ($250, so you know how long ago that was!). The last plane was a Cessna 182.

Peterson is a retired registered nurse and enjoys hunting, fishing, reading and contesting. She is also a member of the hospital auxiliary. She has one son, Norman N. Peterson and one grandson, Derek Peterson, age 4 and a half.

JAN PETERSON, learned to fly in 1981 after graduating from college as a dental hygienist. Having grown up in a family with a test pilot for a father, it was inevitable. Ratings include: IFR, commercial, CFI and seaplane. Planes flown: Cessna 152, 210, Piper Lance, Taylorcraft and Decathlon.

Peterson has enjoyed flying to Mexico with LIGA to perform dentistry. Flying to Alaska with her dog, Aileron, to obtain a seaplane rating in Anchorage, was a real highlight, as well as, flying aerobatics.

She is a chapter chairman of the Orange County Chapter of 99s. It was been a rewarding experience having met so many wonderful, dynamic women. Moved to Boise, ID in 1994. Married to Tim Peterson who is also a pilot and A&P. Currently she is a member of the Idaho Aviation Assoc. and is an advisor to the Explorer Scouts.

LAUREL RAE PETERSON, a career in aviation has been Laurie's high priority since childhood. Laurie started flying at the age of 16. By age 25, Laurie had earned her ASEL and AMEL commercial, instrument, CIF-I-ME, and ATP. Thus far, her career includes more than 4,800 hours of flying charter, instructing, dropping skydivers, flying traffic watch and flying as a first officer in a Lear 35 and Lear 24. Currently she has a commuter flying job as a first officer in a BE1900 with Mesa Airlines. She also works part-time as a first officer in a Westwind for a Tucson-based company and as

an instructor at Flight Safety International (Learjet) in Tucson.

Laurie has devoted many years of service to the 99s, Civil Air Patrol and the Pima County Sheriff Aero Squadron. In 1995 Laurie graduated from Embry-Riddle Aeronautical University with honors, having earned a BS in professional aeronautics. She was elected to *Who's Who Among Students in American Universities and Colleges*. With a greatly appreciated scholarship from the 99s, donated by United Airlines, she obtained her flight engineer rating in 1994. The Tucson 99s have been a major source of support and inspiration for Laurie, who plans to remain a member for life.

LINDA G. PETERSON, Fort Worth Chapter, Texas, born March 3, 1946, in Lueders, TX. She started flying in Midland, TX and got her private pilot license in March 1973. Peterson has an instrument rating and multi-engine rating. As soon as she got her license, she became a member of the High Sky Chapter, Midland, in 1973, then was a charter member and officer of the Space City Chapter, Houston. Moves have made it possible for her to belong to the Lubbock Chapter, Lubbock, TX, and now the Fort Worth Chapter.

Peterson is a BOMA certified real property administrator. Her husband, Clyde, is a flight instructor and musician. They have three daughters: a mechanical engineer, a music teacher and a pharmacist. She will retire from her position with the US Government in December 1996, after 21 years. She flies for pleasure and is looking forward to doing more of it after retirement.

SANDRA PETERSON, was born at Travis AFB in California. Her family soon moved to Las Vegas, NV, where she grew up. It was a dry lake bed in the Nevada desert where she first got the flying bug. Her father let her attempt to fly his Piper Cub. Almost 20 years later, in 1990, she decided to reclaim her love of flying and get her pilot license.

In 1993 Sandy flew commercially to Australia and earned her Australian pilot and ultralight licenses. Other aviation highlights include two flying trips to the EAA convention in Oshkosh, WI.

Sandy's educational background include a BS in medical technology. She is a member of the 99s, Inc., AOPA, EAA and the International Cessna 170 Assoc. Her father, Andrew Peterson, is a retired captain from Northwest Airlines. Her brother, Drew, is a captain for Alaska Airlines. Sandy owns and flies a 1948 Cessna 170.

PATRICIA PETOSKY, learned to fly in 1965 in Midland, TX. In just nine months, she obtained her commercial, CFI and her first job as a flight instructor. She obtained instrument instructor, all ground instructor ratings and an airline transport pilot rating by 1969. After working as a flight instructor, charter pilot and corporate pilot in various locations in Texas, New Mexico and Colorado, she went to work for the FAA as a flight service specialist. For a time, Patricia was aviation inspector for flight standards, and eventually ended up an ATCS in the terminal option.

Patricia joined the 99s in 1965 and has been a member of the various chapters. She has flown almost every single and light twin general aviation aircraft and currently owns a C-182. After a 22 and a half year career with the FAA, during which she was the first woman to manage a terminal facility and a radar facility in the Southwest Region, she retired in 1995. She and husband Bill Darby, who is also a pilot and retired FAA ATCS, plan on putting their C-182 to good use by traveling extensively.

MARIA IRENE PETRITSIS, born April 25, 1961 in Watertown, NY. As a child, flying between homes in New York and Alabama sparked an interest in aviation.

After receiving her bachelor and master's in music education, she moved to south Florida. She taught music part-time and worked full-time as a receptionist at Pompano Air Center where she learned to fly in 1986. In 1987 she joined the 99s, Inc. and served as secretary and vice-chairman of the Gulf Stream Chapter. In 1993 she became chapter chairman. As chairman she organized the chapters first annual Open House, Air Bear presentations and participated in the EAA's first International Young Eagle Day.

Currently she is a director of music at Floranada Elementary School in Fort Lauderdale, a National School of Excellence. In 1995 she earned recognition as Floranada's "Teacher of the Year."

SUZANNE PETTIGREW, is the French Canadian daughter of an air traffic controller and pilot who owned up to seven aircraft at one time. She stared flying on Piper Cubs on wheels, skis and floats as well as Aeronca Champion, Taylorcraft and C-120. She received her commercial twin IFR in 1983 at age 20 from the Chicoutimi Aviation College. She went on working at an all English maintenance shop to learn the language and better her career opportunities. She instructed for a few years in different parts of the Province before she was hired as the first female pilot at AIR Inuit. She flew Twin Otters and Hawker Siddeley 748 in Northern Quebec landing on ice strips and fishing camps dirt strips.

She joined Air Canada in 1989 where she flew as a second officer on B 727. When the recession forced the airline to lay off, she went on flying DASH-8 for Air Nova, the Maritimes connector to Air Canada in Halifax. She was recalled at Air Canada in May 1995 to fly the DC-9 out of Montreal. This will be the twelfth time she moves cities for her flying career. She was introduced to the 99s when she joined Air Canada and has been a member of the Montreal Chapter since then.

SHARON BECK-PFEIFFER, a lifetime Delawarean, began flying in 1986 when she obtained a job at the local airport, Summit Aviation. Shari has held various positions at the airport and is currently their contracts administrator. Shari received her BA in economics from the University of Delaware in 1990, an associates from Delaware Technical and Community College in aviation management in 1994, and is pursuing her MBA from the University of Delaware.

Shari served as Delaware Chapter vice-chairman from 1992-94, Mid-Atlantic section news reporter from 1993-94 and chapter chairman from 1994-present.

Shari has participated in numerous airmarkings, flying companion seminars and Pennies-a-Pound. She has tended to the Delaware 99s booth at local airshows and given presentations on the 99s to various aviation groups. Shari is married to Charles Pfeiffer, who is an airframe and powerplant mechanic.

SHARON ROSE GROOME PFEIFFER, born Jan. 3, 1952, earned private pilot license on Aug. 6, 1987. She joined the Indiana Chapter in January 1988. She has held offices of secretary and treasurer of the Indiana Chapter of 99s. Pfeiffer usually flies the Cessna 172.

She has one son, Michael, age 12. Pfeiffer has had rheumatoid arthritis since she was a child so there are some physical things she can't do. So, having the ability to fly a plane is very precious to her.

BETTY HAAS PFISTER, born on July 23, 1921 in Great Neck, NY, but has lived in Aspen, CO, for the past 40 years. She learned to fly while a student at Bennington College. In 1943 she entered the WASP program, and flew military aircraft for nearly two years. After the WASP disbanded, she ferried war surplus aircraft, and also flew for several non-scheduled airlines as a co-pilot in DC-3 aircraft. She has also flown the Taylorcraft, Cub, PT-19, BT-13, AT-6, B-17, B-24, and P-39.

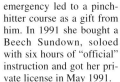

Betty also worked as a flight instructor in California, and was an assistant engineer in the automatic pilot laboratory of Bendix Aviation. In 1954 Betty married Art Pfister and moved to Aspen, CO. Betty and Art raised three daughters: Suzanne Kelso, 40; Nancy Pfister, 38; and Christina Smith, 35 and now have three grandchildren: Juliana Pfister, 9; Daniel Kelso, 1; and Chase Smith, 1. Unfortunately none of them have shown any interest in aviation!

Betty served on the Women's Advisory Committee on Aviation, and she has been a member of the 99s for many years. She founded a chapter of the 99s in Aspen. She is a Whirley Girl, and flew as a pilot and later served as an International judge at several of the World Helicopter Championships. Betty was involved in getting the FAA to install a control tower in Aspen, even though the airport did not meet normal FAA criteria. She also planned and supervised construction of the Aspen Valley Hospital heliport, the first hospital heliport in Colorado.

Betty founded the Pitkin County Air Rescue Group, and served as its president for 23 years. This organization of the best pilots in Aspen, takes responsibility from the Civil Air Patrol of searching for overdue or downed aircraft within a 50 miles radius of Aspen. They provide quick response by pilots who are familiar with the local area. They have been instrumental in saving the lives of almost 30 people. Betty was inducted into the Colorado Aviation Hall of Fame in 1984. In 1992 she received the NAA Katherine Wright Award and in 1994 she received the NAA Elder Statesman of Aviation award.

Probably her most frightening trip was one in a hydrogen balloon across the Alps in 1977. They ran into some very bad thunderstorms and had to make an emergency landing in some very high trees. Getting themselves and the balloon down to the ground was interesting to say the least!

CAROL A. PHELPS, a commercial pilot ASMEL-CFII, joined the Connecticut Chapter in 1974. An Air Force veteran, she earned her rating thorough the GI Bill. She received the A.E. Scholarship in 1981 to complete her CFII. Carol has been public relations chairman for the New England Section and chapter chairman.

Carol lives in Waterford, CO, with her husband, Tom, also a pilot. Their son Brian lives with his spouse in New Hampshire. She is a photojournalist with the *Norwich Bulletin*, where she's also written about her flying experiences. Her most exciting assignment was documenting a trip aboard a Flying Fortress. Stick time on the B-17 was an unexpected bonus.

Carol discovered flying when an assignment on General Aviation brought her to Waterford Airport. Laurie Reeves, a 99s member, offered a skyride and impressed Carol with her finesse in handling an aircraft. She enrolled in flight school two weeks later, with Reeves instructing her through CFII. When not teaching, she flew charters for Yankee Airways.

These days, she looks to combine photography with flying, recording landscapes from the air and writing aviation articles. She and Tom pilot their Cherokee 140 on aerial photo missions.

ARDITH PHILLIPS, of Albuquerque, NM, was 27 before seeing the inside of an airplane. Her first flight was a photo shoot. The photographer got sick. Eight years later, when her husband, a Vietnam veteran pilot, decided to get back into flying, her fear of not being able to handle an emergency led to a pinch-hitter course as a gift from him. In 1991 she bought a Beech Sundown, soloed with six hours of "official" instruction and got her private license in May 1991.

She and Phil own a Cessna L-19 Bird Dog, a Cessna 182 Skylane and a Pitts S1C and head the International Bird Dog Assoc., a worldwide organization, with 500 members, dedicated to preservations of the "littlest warbird," sponsoring and maintaining a museum in Albuquerque.

She is a member of the Lake Tahoe 99s, AOPA, EAA, Warbirds of America and is past president of the IAC Chapter 47. She has been involved in the Oshkosh 1990-94; Truckee Tahoe Airshow; airshow fly-ins with Bird Dog (Best Warbird Award at Casa Grande, AZ, 1992) (Kirtland AFB open house) (Cannon AFB open house); and local IAC officer. She owns and operates a real estate brokerage, Shield Properties, Inc. and Properly Staged Interior Decorating. She and her husband, Phil have three children and two grandsons.

ELIZABETH A. "BETTY" PHILLIPS, born March 1, 1933 in Brooklyn, NY, has three children: Christine A., 36; Yorke C., 32; and Roger C., 30 and three grandchildren: Joshua Williams, 13; Michael Williams, 8; and Trevor Phillips, 2.

After flying with hubby, Yorke, 20 years (Ch. Eng., commercial pilot license, IFR, with 1,870 hours); north to Calgary, south to St. Eustacius, east to St. Pierre et Miquelon and west to Colorado and in Hawaii (keeping her feet away from pedals!), started flight school in October 1991 at age 59, at Ellington, CT and retired from teaching kindergarten after 23 years (private license April 1992, IFR rating January 1994).

As a personal goal, on Jan. 19, 1995 she landed their C-172, based at Westover CEF, at Boston's Logan Airport, 48th and last unrestricted airport landing in Massachusetts. Also has landed at the seven Rhode Island airports and is close to all Connecticut airport landings. Besides the pure flying education, also gained has been meeting many interesting people and learning airport histories. Having also an FCC advanced class amateur radio license has made Morse code idents easy as well as contacting their children and HAM friends that they are arriving imminently

Phillips joined the 99s in 1994 after enjoying a Poker Race, and they both belong to AOPA and the Cessna Pilot's Assoc. Safe flying has been their goal. She has completed the third year of the FAA Wings Program.

ELIZABETH T.D. PHILLIPS, grew up wanting to fly. She was in the Air Force for 20 years.

Then her husband had a heart attack and he was retired from flying. Their six children were grown or almost. She flew, got her instrument rating and logged more than 600 hours. Then Phillips had a stroke; all good things must come to an end. But her association with the 99s doesn't and she is proud of it. "Keep 'em flying!"

SYLVETTE E. PHILLIPS, has been told that she's "plane crazy" and she tends to agree. She started flying while her husband, John, was taking lessons. He thought that she should take what he called a "crash course" which his instructor was quick to change to a "right seat course." She did, but did it secretively for 10 hours. Phillips presented her husband, John, her log book for Christmas in 1979. It was not easy to keep a secret, considering they both flew the same aircraft and had the same instructor. Boy, was everyone glad when Christmas was past. She soloed on her 14th wedding anniversary in September and went on to get her license in December on her son's 13th birthday.

Phillips has started collecting airplane Christmas ornaments, pins, earrings, watches, etc. and has them all over the house. She decorates two wall trees with aircraft ornaments each year and almost always has some kind of airplane jewelry on. She was one of the charter members for their Livermore Valley Chapter of the 99s in April of 1980. Although non-current at present, she enjoys being a part of this great organization.

GLENDA PHILPOTT, born Nov. 19, 1937 in Sydney, Australia, joined the 99s in the early 1970s. She started learning to fly in October 1971, while teaching high school students. Taught school in N.S.W schools for 12 to 13 years. She gained her private license, instrument rating single, then twin, commercial license and instructor rating.

Philpott purchased a single-engine Beech Musketeer and obtained a private license on that aircraft and also instrument rating. Purchased an Aztec aircraft in San Jose in January 1976 and flew the aircraft back with Jim Hazelton to Sydney – over five days. San Jose to Honolulu, to Tarawa, to Vila to Sydney. Gained commercial license and instrument rating on that aircraft.

Became involved with the Fear of Flying Clinic in 1977 when 99s conference in Canberra, Australia – and have been involved ever since, only missing two nights in 16 years of operation. Has attended 99s conferences in Canberra, Australia; Boston (1981), Alaska, Baltimore, St. Louis, Honolulu, New York, Orlando. Has visited the Forest of Friendship on several occasions, and has a stone there under the Australian tree. Her special friends in 99s are Willy Gardner, Gavilan Chapter; and Janet King from Detroit Area Chapter. Many other 99s have been very kind to her on visits to the US.

Philpott has retired from teaching, still runs the fear of flying course (Fearless Flyers Inc.) has completed her real estate license course and is involved in Strata Management. Her license plate is "FLY-99S" She is also a member of the Australian Women Pilot's Assoc.

She is married to Albert Andrew and has two children, Mija, 19 (Korean) and Tenni, 17.

SUSAN CLARK PHILPOT, was born on April 19, 1939 in McNairy County, Tennessee. Grew up in Dearborn, MI. She has been a rental agent for Avis Rent-A-Car at DTW for 27 years.

In 1976 she met Frank Philpot, a commercial artist and pilot. They were married in 1979. After many flying vacations she decided she should know how to handle their Cardinal in case of emergency, and in 1982 got her pilot license flying out of DTW. She has flown Cessna 150, Skyhawk 172 and Cardinal 177.

Working full-time and flying in her spare time she attended her first 99s meeting in 1983, and joined the Greater Detroit Area Chapter that same year. Her chapter activities include: participation as membership chairman, activities chairman and hospitality chairman, through section meetings and annual Pinch Hitter courses. Other involvement's include treasure hunts and Compass Rose paintings.

Susan has a son, Mark Butera, and a daughter, Leah Butera, three stepchildren: Dave, Michelle and Jennifer Philpot; and three grandchildren: Chelsea Butera, 4 and Morgan Butera, 2; Anneliese Wilson, 4, live in California. Her two little granddaughters, that live in Michigan, love to fly and accompany Susan and Frank on many outings.

AILEEN PICKERING, born in San Pedro, CA, had her first plane ride at age 16. By 1949 between college semesters, she learned to fly while working as a ranch hand at Pancho Barnes Fly-in Dude Ranch in the Mojave Desert.

She got her license in April 1950, joined the Los Angeles Chapter of 99s and flew her first Powder Puff Derby in June. She placed third.

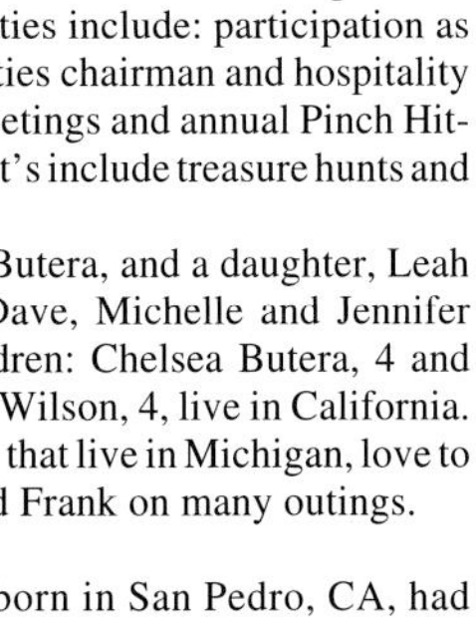

Pickering obtained the charter for the Long Beach Chapter in 1952, becoming their first chairman. She returned to the LA Chapter the following year. In 1980 she was given that chapter's Pilot of the Year award.

Pickering ferried airplanes to help finance college and graduated from Cal-State Long Beach in 1953 with a BA degree in political science. During this decade she got her seaplane and commercial ratings.

She was a realtor from 1960 to 1992, and also flew in 50 air races including: Powder Puff Derbies, ARCs, Baja, Grand Prix, PAR and Palms-to-Pines.

In 1990 she and 99 Marion Jayne flew a 1950 Bonanza in the World Vintage Air Rally from London to Brisbane, Australia. Crossing the Atlantic via the Azores and Portugal, they were joined in London by 99 Sammy McKay as a passenger on this unique flight. Their flight ended short of the terminus-in Alice Springs with engine failure while on base leg. With three hours of fuel still on board, they crash landed, but walked away unscathed from another good landing.

Pickering has one daughter, two grandsons and one great-granddaughter.

JEAN ALLEN PICKERING, born July 10, 1933, in Parkersburg, WV, started to take lessons after her husband had received his license and they had purchased their first airplane, a Cherokee 235. She then soloed and earned her private pilot license. She has also flown all Pipers, Taylorcraft, Decathlon and single-engine Cessnas.

In 1975 Pickering joined the 99s and worked on the last Powder Puff Derby. The West Virginia Mountaineer Chapter was formed and she became a charter member, going on to hold all the offices in the organization at the local level and governor of the Middle East Section for a term.

In 1976 with Parkersburg as the race stop for the PPD, they entered the race in the slowest of the slow and had the most fun on a budget that could be imagined. "Racing" in a Decathlon is not quite the way it sounds, but they finished successfully and had a great experience. The race route was from Sacramento, CA, to Wilmington, DE.

She has gone on to earn the commercial, flight instructor instrument, BGI, seaplane single-engine and multi-engine, commercial and instrument. Her multi-engine was earned with the assistance of a 99s scholarship in 1986.

Among the highlights of Pickering's 5,000 plus hour career of flying are the many people she has met in aviation and the wonderful country that allows them to view it from above. Also she counts the opportunity to teach her three sons: Chip, 38; Mike, 33 and David, 27; to fly . . . two earned their licenses and her daughter, Bev, 36, married an airline captain. She and her husband, Charlie, continue to fly and the most recent addition to their hangar is a 1941 training glider. Hopefully she will still be around to show some of her seven grandchildren: Aubrey, 13; Neil, 10; Justin, 10; Aron, 8; Sean, 5; Allen, 6 and Victoria, 1; the wonders that abound when you earn your wings.

Pickering is a successful sign painter for 25 years. Other hobbies are quilting and flying new planes.

CAROLYN M. PILAAR, in the summer of 1968, while a student at Western Michigan University, one of the guys took her flying in his Stinson. She changed her major to aviation and earned her flight ratings along with an airframe and powerplant mechanic rating. Pilaar competed in National Intercollegiate Flying Assoc. and won the 99s Women's Achievement Award in 1970,

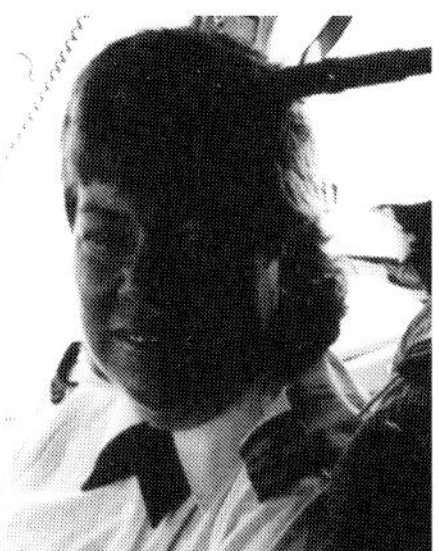

She then moved to Greenville, SC, where she flew a Beech 18 as chief pilot (Part 135). In 1974 Pilaar opened Carolyn's Flight Academy. She also instructed at Greenville Technical College and in 1976 was awarded South Carolina Flight Instructor of the Year.

Pilaar has raced the Powder Puff Derby, Angel Derby and Air Race Classic with several top 10 finishes. In 1977 she began competing for the United States Precision Flight Team and has been a team member ever since 1979. Pilaar was Top US Pilot in 1985 and 1995 and at the World Precision Flying Championship competition in 1990 and 1992, she was named Top Woman Pilot. She is currently a chief judge for the National Intercollegiate Flying Assoc. and FAI International judge.

In her more than 25,000 hours, she has flown for Air Virginia (American Eagle); flew SA227 Metroliners; based in Berlin, Germany for Pan American World Airways; flew a B727; and for DHL World Airways on a DC8.

Pilaar holds both a single and multi-engine airline transport pilot, certified flight instructor single and multi-engine with instruments, commercial glider, seaplane and rotorcraft-helicopter, flight engineer-turbojet B727 and DC8, ground instructor, basic, advanced and instrument and airframe and powerplant mechanic with inspection authorization.

Carolyn's Flight Academy is the oldest flight school in South Carolina. They also offer charter services under their Part 135 certificate.

MARGARET "MARGE" PINCIOTTI, born October 1930 in Riverside, NJ, has been interested in aviation for several years. The desire to become a pilot firmed in 1975 when Dr. Jim Bardsley started to take the Pinciottis on short trips, trim his Navion and allow one of them to "fly." That was the introduction to a new dimension, responsibility and language! Their spare time turned from golf and crafts to student lessons in their *Pussy Cat* (a gentle 1962 Cessna Skyhawk they were lucky to find).

The next year was a learning experience with happy memories of taxing in the winter mud, student air work, April solo, spring cross-countries, hangar tall stories. Marge attended 99 meetings as a solo student. Karin Minauro nominated her to the Eastern Pennsylvania Chapter at the March 1977 meeting – just after Marge and Dick got their ASELs. The intervening years have given them time to log trips to Virginia, New Hampshire and Key West – with all possible side trips. Besides being a vice-chairman of the chapter in the mid 1980s, Marge has helped the Eastern PA to update their scrapbook, toted the "Country Store" of sundry items for several years. Now, Marge participates in all possible chapter events and scholarship fund raisers. Whenever a seat in someone's plane is available, Marge is happy to even the "weight and balance." Selling the *Pussycat* was a deep loss for Dick and Marge.

MARY C. PINKNEY, transplanted from the Midwest and the University of Michigan to California, Mary met and married a former Air Corps pilot and aviation became a lifestyle.

Civil Air Patrol was the start. Pinkney became her squadron public information officer and a qualified search observer. She also served as squadron cadet commander and guided cadets to the USAF Academy and the USAF; and served in numerous cadet summer encampments and exchange programs. She was honored in 1957 as CAP Member of the Year, California wing.

A business woman since 1947, she was also the owner of "Gravel Mary's," a concrete products service, that became the name of her Cessna 182, flown in numerous country-wide and local air races.

Active in community affairs, Mary worked on air shows at the local Torrance Airport and was a frequent speaker for civic organizations.

A 99 since 1959, Mary has served all Long Beach Chapter offices and numerous committee chairmanships; attended International conventions and section meetings; and traveled as a delegate to the World Aviation Education and Safety Congress in India.

Her travels worldwide offered many opportunities to make friendships of International 99s and to serve as hostess during their visits to the USA.

MICKY PINKSTON, of Missouri has logged more than 2,600 hours since receiving her license in 1976, and current ratings include SELS, AGI and CFII.

Pinkston is past trustee for the AE birthplace. A Greater Kansas City Chapter member since 1978, she's a two-term past president, chaired numerous regular and special committees and received the Blue Maxine Award.

Pinkston is the 1995-96 president of Missouri Pilots Assoc. – recipient of its Aviation/Blackford Awards. She actively supports FAA programs; chaired FAA/MPA Wings program since 1989; organized and conducted numerous FAA safety seminars; and serves as accident prevention counselor.

In addition to her Piper addiction, Pinkston has enjoyed racing modified cars, water skiing, horseback riding and golf. She owned businesses and office and advertising manager and accountant before 1985 retirement from Allied-Signal.

She is married to LtCol Floyd Pinkston (USAF Ret.) and has two pilot sons, three stepchildren and 12 grandchildren.

SARA PIPER, born Nov. 29, 1931, became involved in aviation in 1978. She was interested in gliders.

Piper joined the Reno Chapter of 99s in 1982 after obtaining her private pilot with glider rating.

She is a retired Federal Government employee.

She is also a member of NavFe and AARP.

ELIZABETH "BETTY" JEAN PIPPEN, born on Nov. 23, 1935, in Washington, DC, and living in Upper Marlboro, MD, with her husband Jack and four children: Nita Jean Sady, Sandra Lockrow, Shari Robinson and Jack Pippin Jr., soloed in 1970 at Professional Flight Service in Friendly, MD.

After the dream of learning to fly was realized, she went to earn her private, commercial, instrument and instructor ratings. Her son, Jack, 2 years old at that time, traveled many hours in the back of a Cessna 172 while mom worked on her instrument rating.

Betty worked as a flight instructor at Professional Flight Service. During this time, she also worked with the Department of Commerce, Census Bureau, as the budget analyst for the 1980 and 1990 decennial censuses.

While living in the Washington, DC, area she was a member of the Washington, DC Chapter. Since moving to DeBary, FL, she is a member of the Florida Spaceport Chapter and is an active full-time realtor with Trophy Real Estate in Volusia County, FL. Pippen has four grandchildren: Paul D. Sady, Fallon L. Robinson, Casey C. Robinson and Brandon C. Pippen.

DIANE PIRMAN, learned to fly in 1983, holding single and multi-engine land, commercial and instrument instructor ratings. She married Mark Pirman in 1965 and they have two sons, Eric and Bradley.

Diane and Mark, also an instructor, currently fly a 1967 Twin Comanche and Cessna 172. Past aircraft include a Cessna 210 and 170s, Comanche 250s and a 1946 Aeronca Champ.

Diane belonged to San Luis Obispo Chapter serving as secretary (1984-86) and vice-chairman (1986-87). A charter member of the Santa Maria Valley Chapter, she served as its first chairman (1988-90). She was the Southwest Section's Woman Pilot of the Year for 1993 and was elected Southwest Section Secretary for 1994-96.

In 1974 Mark and Diane established DiMark International, distributors of security hardware. Diane is current active manager. In the past they owned and operated retail lock shops in Santa Monica and Santa Maria.

M.S. "TERRY" PITT, born Aug. 6, 1929 in New York City, learned to fly in 1990, at age 61. Her husband, then 64, also learned to fly. She has earned her SEL in a Piper Cherokee. Pitt joined the 99s in 1990. She is also a member of the Civil Air Patrol Aerospace Education Program and of the Connecticut Aerospace Education Council.

She is married to Harold T. Pitt, attorney, and has six children. She also has 17 grandchildren, ranging in ages from 20 years old to 1-month-old.

Learning to fly was one of the greatest stretches she has made in her life. Working with and socializing with people in aviation continues to be a delight. Now, retired last year from advertising, and having earned a second master's degree (first in education then in writing), she is writing novels about flying and working to support aerospace education modules in school curriculum.

BONNIE PLOTKIN, born March 31, 1965, in Buena Park, CA, joined the 99s in 1994. She has 115 hours of flight time based out of Long Beach, CA, Airport. She has flown Robinson R22, Enstrom and Hughes 300 helicopters.

In 1992 she and her husband flew over the LA riots on a photo flight. At one point they were circling a hot spot with four other helicopters, two of which were going in the opposite direction, 200 feet above us with cameramen hanging out the side of the 'copter.

She is married to Dorry Plotkin and has three children: stepson, Sean, 17; daughter, Kristy, 6; and son, Morgan, 4.

DOROTHY JANE PLUMMER, born June 23, 1937 in Passaic, NJ. From a young age she wanted to fly. Anytime she was able, she would watch the planes take-off and land at Tetersboro Airport and in the summer at Provincetown, MA, Airport. After her two children, Jenny and Frank, got into their teens she finally had the time to learn to fly. She spent her time at Somerset Airport flying Piper 140 and 180 in 1977, finally being able to the join the others in the Delaware Chapter of 99s. The many wonderful times she has had flying were well worth the effort. Plummer now lives on a sailboat called *Plum* most of the year, but she still enjoys flying.

Plummer is married to Robert Cavallaro, MD.

Special memory was a wonderful VIP tour via helicopter over Washington, DC, where only special permission to a lucky few got to go. It was a chance of a lifetime with CAP.

DANA ANNETTE PODEWELTZ, born April 18, 1944, in Merrill, WI, learned to fly in 1969 in a J5 Cub; her instructor was John Hatz, who designed the Hatz bi-plane. She didn't have a lot of time for flying, having three small children and a full-time job. Then they started a heating and cooling business, which took up a lot of her time. She did fly along with her husband, Wayne, and they also belonged to EAA and local chapters. They made a private airstrip by their home, which is on the Green Bay Sectional Aeronautical Chart.

In 1970 Podeweltz and her husband restored a 1939 Aeronca Chief. She has a 1940 Taylorcraft, which she plans to restore. They have a 1965 Cherokee 140 and are building a Hatz bi-plane.

In 1992 she decided that she would finish what she had started years ago and get her private license. She passed the written test and looked up an instructor, soloed and passed her check-ride June 29, 1993. On Oct. 31, 1993 she broke her ankle in three places when she slipped off the icy wing of their Cherokee.

Podeweltz and her husband have three children: Jeffrey, Todd and Tyler Anne and five grandchildren: Ryan, Heather, Jennifer, Darrin and Kassandra.

BERTHA "URBAN SIEGL" PODWYS, born Nov. 28, 1922, on a farm in Berwick, PA. There she learned at age 6 that the "big bird" high in the sky was an airplane that had people in it. After being taken to Berwick Airport by her brothers and watching the air show, she thought someday she would learn to fly.

At age 16, due to loss of her father and home, they moved to New York. She spent weekends at the LaGuardia Airport watching the planes come and go. During WWII, she worked as a mechanic's helper for Consolidated Aircraft in Bloomsburg, PA, where Navy airmen were trained to solo in UPFs. Later she worked at Grumman Aerospace. She couldn't afford to fly, but at least she could work at the aircraft manufacturing plant.

She has four children: Joan R. Rudolph, 51; Godfrey "Fred" H. Siegl Jr., 47; Jean R. Holloway, 45; and Robin J. Dombroski, 43. She has seven grandchildren: Deborah A. Rudolph, 30; Dawn M. Libardi, 28; Jeffrey S. Holloway, 20; Andrew S. Dombroski, 13; Brynn A. Dombroski, 11; Freddie H. Siegl III, 10; and Jessica Morgan Siegl, 8. She has one great-grandchild, Jacquelyn M. Libardi, 1, and one born in June.

In 1976 they retired to Florida due to her husband's heart condition. After the loss of her first husband in 1981, she met her second husband in 1986. He flew in a flying club in Detroit, MI, in his middle 20s, but hadn't flown in 40 years. They had lunch at Cedar Key, where an airplane sat in a yard nearby, which surprised her, since there wasn't an airport in sight. This led to conservation about flying. When he learned that she had been interested in flying all her life, he suggested she learn to fly. She had her first lesson the following week. She then studied ground school and soloed Feb. 18, 1990. They both got their license in their retirement years. Her husband, Stanley, on June 20 and she on June 23, 1990. Both of them have had the pleasure of flying ever since. The drone of an airplane engine overhead still turns her head up. It's never too late to learn to fly, that is, if you can pass your physical!

MARYLOU POHL, born Nov. 23, 1949, in Philadelphia, PA, has been in the Colorado Chapter of 99s for five years. She has a single-engine rating and has flown the Cessna 152, 172 and Mooney 252. She has been newsletter editor for the Colorado Chapter's *Crosswind Chatter* since March 1991. She received an award from South Central Section for Outstanding Reporting for 99s news magazine in September 1991.

She is still amazed in today's "enlightened" generation that some men are still incredulous when learning a female is a licensed pilot. She is married to Raymond F. Pohl.

KATERI L. POLEN, born April 8, 1969, in Elmhurst, IL, has always loved airplanes, ever since she saw the Blue Angels as a young child. Her original goal was to enter the Navy, but she was unable to meet the physical requirements. Since then she has graduated from Embry-Riddle and become an avid aerobatic pilot and competitor. She hopes to represent the US in World Competition and perform in airshows.

She has a commercial, SEL, instrument ratings with more than 260 hours flight time. She has flown the C-172, PA-28, CAP 10B, Pitts S2B, Mooney 20J, Arrow and PA-12. She is a customer service manager at FBO. She is married to Michael Polen.

LINDA LOUISE GERING POLEN, born March 15, 1948, in Atchison, KS, same as Amelia Earhart, began her aviation career when her pilot husband, Terry L. Polen, entered the aviation world in 1983. Her beginnings were reading the private pilot manual to fill unoccupied time. Once they purchased their own aircraft, flying lessons began from the safety aspect. She had her flight training at Ponter County Airport, Valparaiso, IN, and Key West, FL.

She took the challenge beyond being able to land their Piper Cherokee 140. Owning their own aircraft offered Polen the frequent availability to continue, and she secured her private pilot license. They upgraded to a complex Arrow PA28R 200 and she pursued the instrument rating. She has also had one lesson in aerobatics. Every time she has flown it has given her an exhilarating feeling of experiencing the "other world" – especially the one with the clouds beneath your wings.

LAURA POMERLEAU, learned to fly in 1991 at San Carlos Airport in California. She joined the Santa Clara Valley 99s in 1992.

Laura has worked at the flight desk at Golden Gate Aviation and has flown for traffic watch.

Laura has a degree in aviation and has her CFI and CFII ratings. Her other interests include skiing and reading.

MARILOU SOANES POORE, born Jan. 22, 1930, in Buffalo, NY, decided at age 57 to earn her pilot license as her husband is a pilot. She took flight training from Deanna Robertson (a 99) at Riverside Airport, Tulsa, OK. She has flown a Cessna 150, Cessna Skyhawk XP and 727 simulator.

Becoming a 99 was a very proud moment. Since then, they have enjoyed flying their Hawk XP from Tulsa to Colorado, New York and Florida. They will soon be flying their newly acquired 1946 Aeronca Champ. She was married in 1950 to William H. Poore, also a pilot, and they have two children, Robert, 43 and Jeanne Argomaniz, 40. They have two grandchildren: Ryan Argomaniz, 14 and Jonathan Argomaniz, 14.

KITTY MAYNARD POPE, Phoenix 99s, who began flying at Raleigh, NC in 1971, got her private ASEL in 1977, and by 1981 held instrument, commercial and single and multi-engine instructor credentials. In 1975 she married a veteran USS Independence (Navy) air traffic controller and they both began college one year later. They finished together in aeronautical technology.

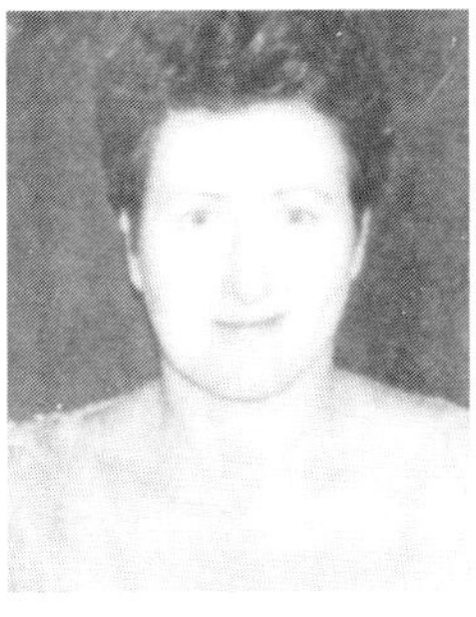

Kitty says she gave up as a flight instructor pretty quickly because most people didn't want to be as good as she wanted to teach them, and she learned patience after she became a parent much later. She and her husband had careers in aerospace, with Charlie leaving and starting his own business two years ago. She has worked for Allied Signal 15 years as an engineer on installation and flight testing of jet engines, both in fixed wing craft and helicopters, here and abroad.

When Kitty was getting her ratings, she took all kinds of flight jobs to build up her time, including going with a flight instructor on charter trips and surprising him by not minding the cadaver runs. One of her unique flight experiences was a job flying as test pilot on an Aerocommander 1000 for an experimental 1,000 hour endurance test, using three crews to alternate and flying all over the Southwest.

She joined the Phoenix 99s in 1979, has held several offices including chapter chairman, and worked on countless projects. Her favorite projects include Fly Without Fear, Flying Companions, Airlifts, Kachina Air Rally, NIFA, Airmarkings, Fly-ins and organizing the chapter's 40th Anniversary party.

The Popes have a daughter, Christina, born in 1985. Also, Kitty is an active member of the American Helicopter Society, the United Way and the Parent Teacher Organization.

SHELLY POPE, began flying as soon as she finished graduate school in Arizona and moved to Mountain View, CA in 1991. She got her license in 1993 at Palo Alto, and joined the SCV 99S that year.

Shelly worked at Johnson Controls World at NASA Ames as a space scientist and recently moved to San Diego.

DOTTIE PORTS, in 1943 married Bud, a commercial pilot. Although she shared his flying experiences with him, she did not start formally flying until spring of 1959 as her husband's student. Received private license later the same year. In 1960 she was a contestant in Sky Lady Derby race and received her instrument rating the same year. Charter member and helped form the Shreveport Chapter, of which she was first vice-president and chairman the following year, plus many chapter and sectional appointments and committees. Recently member of Silver Wings Club.

In 1969, received commercial pilot license; 1970 was the first contestant to fly Citabria in Powder Puff Derby race, contestant again in 1975, Cessna Cardinal.

Presently operating Dottie Ports Realty, founded in 1970, and utilizes her airplane in her business.

She is active in church work. Happiness is being a 99.

JANIE POSTLETHWAITE, Golden West Chapter, San Carlos, CA. ASEL and sea, instrument rated, advanced ground instructor. Active in 99s since 1972, she received her private while five months pregnant. Both of her sons and her husband all became pilots with her help. They still own their, once new, 1977 182 Skylane 21DF.

For years, was highly active in ELT searching and training seminars – traveling throughout the US, and Canada. Co-founder of the Happy Flyers, an International organization of Hams and pilots (my amateur radio license is WB60DQ). Was co-author/producer of search and rescue multimedia production on radio direction findings techniques for ELTs. Was active in CAP as pilot and communicator both in Florida and California.

One of the few fully recovered bulbar polio victims and mother of five, Postlethwaite is presently a registered electrologist. Divers interests and capabilities include: licensed scuba diver, licensed beautician, apiarist, former Scout Leader, (still Aviation and Bee Merit badge counselor), photographer, former roller skate dancer, water skier, writer, La Leche League counselor, graduate of Police Officers Training (POST), etc. Participated in 99s air races and other competitive events. Was chapter Woman Pilot of the Year in 1976; Southwest Section Woman Pilot of the Year, 1978. Proud mother of 1995 Naval Academy graduate, who received Laureate award as Aerospace Leader of Tomorrow (from *Aviation Weekly* and *Space Technology* magazine), presented at the Smithsonian Air and Space Museum. The fact that he was an instructor pilot for the same plane used for primary training at Pensacola, while still a midshipman, is probably one of the highlights of her flying memories.

FRAN HUFF POSTMA, member of the Mount Tahoma Chapter since 1985, was born in Pontiac, IL, and moved to the Tacoma area in 1957. She received her BS and MBA and is presently teaching at Pierce College after having taught many years at the secondary level. She has been chapter treasurer, secretary, vice-chairman and chairman and also enjoys the friendship, fun and sisterly support while attending sectional and International meetings.

Fran has her private license and instrument rated and flies mainly a 172 and Cardinal RG. She has taken glider training and working on a commercial rating. Fran's husband and son are also pilots. The seed for flying was planted in Fran's late teens when a friend's father took her for a ride and turned over the controls to her. That seed of interest was kept dormant for 20 years until she was able to take her first lesson. Fran's interest in flying was an attraction to her present husband, James. A flying date was one of their first.

Fran and Lisa Asplin, developed an aviation program entitled "Women in Flight" which has been presented to numerous women's organizations as well as the Museum of Flight in Seattle to educate and encourage women to aviation. Fran has also taught aviation in summer programs for fourth – eighth-grade students. She is a member of AOPA, EAA, ZONTA, Delta Kappa Gamma, a city council member, been involved in many community activities, programs and continuously supports aviation in many ways,

ILOVENE POTTER, born May 19, 1918 in Tacoma, WA, earned her license in 1941 flying an Aeronca C-3 the Taylorcraft, Cub, Waco and the Porterfield. Resumed her flying career in 1960, earned her private, commercial, instrument, multi-engine, seaplane, helicopter, instructor and advanced ground instructor ratings.

Ilovene was the first woman in the state of Washington to receive a helicopter rating. She was captain of the first US helicopter team to compete in a world championship in 1973. The year following, became a judge and observer for the next championship held in Vitebsk, Russia. (The US took second place.) After teaching her husband to fly the chopper, they became the first husband-wife team in this state.

In 1984 she became a certified Federation Aeronautique International judge. She served two years on WACOA, women's advisory committee. Every office in the 99s, chapter and section, has been served by her, as well as, two years on the International Board of Directors and the International Nominating Committee as its chairman. Treasurer and chairman of the nominating committee for the Whirly Girls two years each, became their president in 1984.

She has competed in nine Powder Puff Derbies, one Angel Derby, seven Palms to Pines races, one Lipstick Derby and six Pacific Air Races. She has received the AOPA Pilot of the Year Award, the NW Section 99s Achievement Award and Past Presidents Achievement Award and her induction into Memory Lane in the Forest of Friendship. Today, she remains an active vital force working on the board of directors of the Museum of Flight. She is a widow and has three children: Ted, Gregory and David. She also has seven grandchildren.

LINDA SUSAN POTTER, born Nov. 21, 1949, in Santa Monica, CA, joined the 99s in 1989 after receiving her private pilot license. She has flown the Cessna Cardinal 177TP (red, white and gold). She is a school teacher and married to Thomas C. Potter. They have two children, Erin Michelle, 12 and Lisa Nicole, 10.

In 1994 Potter's family flew cross-country following parts of the Oregon and Santa Fe Trails and touring historic cities of Philadelphia, Williamsburg, Washington, DC, Mystic Seaport, Gettysburg, Boston, Kansas City, etc.

PEGGY VINING POTTER, private pilot license #17138, dated Aug. 23, 1930. Joined the 99s Oct. 12, 1930 and served as vice-governor, Southwest Section. Was a member of the Aeronautical Committee of the California State Chamber of Commerce. Participated in most California air meets 1930-38. Made 998 flights. Planes flown: American Eagle (OX-5), Waco (Sieman), Barling (LeBlond), Fledgling (Challenger), Glider (10133), Stinson (Lycoming), Moth (Gypsy), Travelair, (J-65), Fleet (Kinner), Ford Tri-Motor (J-6), Monocoupe (Lambert), Robin (OX-5), Bird, Emsco (Challenger), Curtis Jr., Taylorcraft, Piper Cub, Great Lakes (Continental), Fairchild, Aeronca, Stearman, Arrow Sport (Ford).

Married in 1939 to Charles M. Potter, Army photographic and transport pilot. Soon after their marriage, they went to Saudi Arabia, where Mr. Potter was drilling superintendent for Standard Oil Co.. Later they lived in Bogota, Columbia for many years.

BROOKS M. POWELL, earned her private license, SEL, commercial and instrument rating. She joined the 99s and became an editor for *Runway Lights* the NE Kansas 99s newsletter. She has been on the Forest of Friendship committee, co-chairman for the NE Kansas 99s Pinch Hitter course, 1991 and past chairman of 99s.

She has also been a member of Junior League of Topeka, Inc., past chairman of Jr. League Sustainers, Shawnee County Medical Auxiliary, lieutenant in Civil Air Patrol, member of Kansas State Aviation Committee, member of SCS Governor's Advisory Board, Historic Perspective, Mulvane Art Center women's board, Historic Topeka Inc., St. David's Episcopal Church Altar Guild and allocations committee of United Way.

Education: Women's College, University of North Carolina, two years; Washburn University, Topeka, KS, bachelor of fine arts; and University of Kansas, School of Interior Design, bachelor of fine arts. She had experience working under Elizabeth Phelps (clothing design), 1948-50; legal secretary, 1950-52; instructor in summer art school 1962-79; artist/painter; has participated in numerous shows; paintings in private collections; free-lance interior design, landscape design 1978-81; and receptionist/bookkeeping for husband 1991-93.

She is married to Benson M. Powell II MD and has eight children and nine grandchildren.

BETTY L. PRAKKEN, learned to fly in 1960 in a Super Cub. A Penny-a-Pound ride when she was 12 was such a thrill, she knew flying was a must.

Since first joining the 99s in 1960, she has held all chapter and many section offices and committee positions and presently serves as a member of the International Nominating Committee and is also a director of the Northwest Section.

She has participated in the Palms to Pines air race as well many local contests, and was co-chairman of the 1993 International Convention. In 1994 she was presented with the Northwest Achievement Award.

Betty resides in Canby, OR on an airpark with her husband, Gordon, who is retired from the US Treasury Department. They have four children and six grandchildren.

JOANNE PRATER, received her private pilot license, single-engine land on Aug. 31 1972. She is a member of the Scioto Valley Chapter of the 99s which she joined in 1981. She has served as treasurer for the chapter and is currently the secretary. JoAnne is a registered nurse serving in the operating room for same day surgery patients.

BEVERLY FRANZ PRICE, a lifetime Michigan resident, learned to fly in 1964. She is a flight instructor, ASMEL & S and is a FAA designated pilot examiner. Price has logged more than 7,000 hours, and is active in the FAA Accident Prevention Program, and the Pilot Proficiency Award Program. She was director of flight training at Air-Flite at Tri-City Airport and has trained many pilots.

Price had received many awards: Flight Instructor of the Year, Grand Rapids FSDO; Accident Prevention Counselor of the Year for the FAA Great Lakes Region. She is active in the CAP, 99s, MI State General Aviation Committee, AOPA and local aero clubs. She participated in many races including the Powder Puff Derby, MI Small Race, Ill-Nines and Indiana Fair Ladies.

Price is married to Douglas Price and has a daughter, Cindy Bewick and a son, Steve Price.

KATHERINE PRICE, a Canadian 99, residing near Toronto, Ontario learned to fly in 1983 at age 15 and obtained her license in 1985 while in high school. In 1989, she received her commercial license and became a volunteer pilot for Operation Skywatch, a program established by the Ontario Ministry of the Environment and the First Canadian Chapter of 99s to monitor environmental concerns.

After graduating from the University of Waterloo in 1991 with an honors BS in biology, Price obtained her multi and instrument ratings. In 1992 she was awarded an Amelia Earhart Memorial Scholarship for an instructor rating, completing this rating in 1993.

Price has served as chapter secretary, past co-chairwoman of Skywatch and organized many 99 events. She is currently employed in plant biotechnology research, completing her MS degree and pursuing a flight instructing position. Katherine wishes to thank the 99s for 10 years of support, friendship and fun.

SUANN PRIGMORE, Mt. Shasta 99s, became a private pilot on April 5, 1985 and immediately joined the Mt. Shasta 99s. She earned her instrument rating and commercial certificate in 1987. She has taken aerobatics lessons and earned her Phase VI Wings. She owns and flies a Grumman Tiger.

As of February 1995, Prigmore has logged 1,200 hours.

Prigmore is a member of the American Yankee Assoc., Redding Area Pilots Assoc. (RAPA) and AOPA. She taught several Flying Companion Seminars, worked at air shows and coordinated fly-ins and other chapter events. She has been chapter treasurer and recording secretary.

She is an advisor to her local Aviation Explorer Post and has been an aviation presenter at AAUW's Math and Science Conference. She has raced in the Palms to Pines Air Race seven times – six times finishing in the top 10, once eleventh and even won it in 1991.

Prigmore believes her favorite flying adventure was her flight to Alaska as PIC in June 1994. This was a flight for just the girls. Another fun project of hers was the making of the video, *Ditto Flight,* which they sent to the Rush Limbaugh TV Show. Unfortunately, (or fortunately) he quit showing home videos the week they sent theirs in.

The Echo Tango Program, emergency training, is a program dear to her heart. Therefore, with the help and sponsorship of the Mt. Shasta 99s, they put into action the Echo Tango Program. The purpose is to increase the chances of successfully performing emergency procedures in the event it should ever become necessary. The requirement is a minimum of one hour airwork, with a competent flight instructor. The goal is to make emergency procedures second nature, strengthen self-confidence and foster calmness and clear-thinking whenever required. She hopes Echo Tango will eventually be adopted by all 99s Chapters.

VIRGINIA M. PROCTOR, born May 6, 1919, Prescott, AR, has been a member of Memphis Chapter since 1965. She received flight training and received her private license and commercial, instructor, instrument (SE) and multi-engine ratings. She has flown the Luscombe, Cessna 172, 150; Mooney Mark 21, Executive, Cherokee 6, Cherokee Lance, Comanche 180 and Piper Apache.

Proctor has served as chairman and also held all Southeast Section offices, including governor from 1975-77. Flew one AWTAR as co-pilot and one Angel Derby as pilot. Was the first woman appointed to Arkansas Aeronautics Commission and served 14 years. Chairman of the commission one year. Also on Wynne Airport Commission several years. She was also an FAA accident prevention counselor several years.

Proctor was married to Everett (deceased) and has three sons: Douglas, Richard and Kenneth. She also has six grandchildren: Tracie, Leslie, Mary Claire, Kelly, Kyle and Stephen. They range in ages from 28 to 14.

ADRIENNE VOLLMER PROKOP, became a licensed pilot in 1989. She flies a Cessna 182 and is a member of the 99s, Inc., Northeast Kansas Chapter and the Aircraft Owners and Pilots Assoc.

Prokop attended Beloit College and holds a BS degree from Northwestern University. She has been active in ophthalmic administrator and nursing organizations. For 10 years served as president and CEO of the first free standing ASC in Kansas devoted exclusively to eye surgery.

Community activities include participation in the Junior League, Medical Alliance, and served as founding chair and president of the Topeka Assoc. for the Gifted.

Married to Bradford S. Prokop, MD, an ophthalmic surgeon and pilot; they have two daughters, Janet Prokop Pregler, MD, Linda Prokop Woofter and two sons, Edward C. Prokop and Alan S. Prokop. Interests include travel, swimming, golf, theater and music.

MARIAN BANKS PROPHETT, Mission Bay Chapter, learned to fly in 1955 and joined the 99s in 1956. She served as secretary and chairman of the San Diego Chapter. Treasurer and vice-governor of the Southwest Section. She has ATP, SEL, MEL, CIFI and instrument rating with more than 3,000 hours. Prophett was on the AWTAR board for 16 years and held offices of secretary, vice-chairman and chairman. She has raced in 17 Powder Puff Derbies and placed first, second and third. Raced in 10 PARS and won in 1968 and 1985. Flew in seven Angel Derbies, two Palms to Pines and won in 1971, five Air Race Classic and won in 1983. Served three years on Women's Advisory Commission on Aviation, more than 25 on the San Diego Aerospace Museum's Board and is on Silver Wings National Board of Directors. She helped start the San Diego Chapter of Silver Wings and served as president two years. She is in the International Forest of Friendship.

ANA CAMBEROS PROVINCE, born in Tiajuana, Mexico. She was attending the University of Baja, CA, when she met, and married Dr. Fred Province. She continued her education in the United States, and graduated Sigma Cum Laude, with a Phi Beta Kappa Key. She earned her pilot license in her 1946 Ercoupe, and was a guest of the Ercoupe designer, Fred Weick, when he received the prized Guggenheim Aerospace Medal. She is the director of a weekly television program titled "Aviation Theater," and has filmed from hot air balloons, ultra-lights, blimps, gliders, tow-planes, antiques, home builts, experimentals and helicopters. She organized and teaches aerospace merit badge classes to Girl Scouts. Ana had the honor of appearing on the cover of the July 1994 issue of the *99s News,* magazine.

JOYCE PRYOR, born July 25, 1941, in Orlando, FL, received her pilot license in 1992. She flies a Cessna 172-L and is closing in on logging 300 hours. She has finished Phase I of the Wings program and has taken a 10 hour aerobatics course.

A lifetime Southerner, she currently lives in Alabama and worked for 20 years as an interior designer. She is married to Peter Pryor Jr., an electrical engineer and also a pilot. They have a son and a daughter and three grandchildren.

Joyce is a member of the 99s, Inc., Aircraft Owners and Pilots Assoc. and Cessna Pilots Assoc. She is currently membership chairman for the Alabama Chapter of the 99s, Inc.

LULU MAE STEGEMAN PURDY, born June 30, 1929, in Deuel County, NB. She and her husband, Charles, were owners of Cessna 182 N9263G from 1971 to 1990. Together they have flown around the US touching the four corners, sharing 1,431 hours.

Lu was recruited as a Minnesota Chapter 66 in September 1982. Her solo work was done in Washington. She obtained her license July 16, 1984 at Tacoma Narrows Airport. She joined the Minnesota Chapter in September, transferring to Colorado Chapter a year later. There she has served as director, membership and telephone chairpersons.

Lu was a Camp Fire leader; has a BA degree from Metropolitan State University, Minnesota; qualified to administer MBTI. Daughters: Linda Lu (1950) lives in Minneapolis, MN, a ballet dancer and waitress with twin daughters, Ashley and Larissa. Rebecca Ann (1953) lives in Tacoma, WA, a pilot and CPA with sons, Mario and Marco.

MARTHA M. PURDY, Nebraska Chapter since 1964. Husband, son and Purdy logged 1,700 hours in Cessna 172 flying coast to coast, Bahamas, Mexico and Canada. Flying days about over because of an eye problem, but it was a great experience and her 500 hours as PIC was a thrill. Two International conventions, seven or more section meetings. Several chapter offices and committees. Timed three AWTARs and two Nebraska Air Races. Retired teacher. Three children and four grandchildren. Enjoys painting in oil and watercolor, bridge, gardening and golf.

MARY JEAN PYATT, the first woman senior air safety investigator and investigator-in-charge of the National Transportation Safety Board's (NTSB) famous Go-Team for major aircraft accidents around the world, which is headquartered in Washington, DC.

Jean's career in aviation began as a flight instructor at Santa Monica, CA, airport and continued as a Federal Aviation Administration (FAA) air traffic control specialist, FAA air safety investigator in Washington, DC and FAA principal operations inspector at the Los Angeles, CA, flight standards district office (FSDO) and NTSB air safety investigator in the Los Angeles field office. She holds an ATP certificate, CFIAI for single and multi-engine airplanes, GIAI and seaplane and glider ratings.

Jean co-owned a Cessna 182 with 99 Marilyn Twitchell. They flew N299PT in numerous air races, including the Powder Puff Derby, Angel Derby and Palms to Pines. Jean's 1977 "All American Glamour Kitty," Sissy Earhart, has logged several hundred hours, including air races and domestic as well as international airline flights and is still a frequent flyer at 19. Sissy won her fame dressed as barnstorming pilot of the 1920s and was featured in the November 1977 *99s News.*

Jean was born Oct. 25, 1928 in North Carolina. She was educated at the University of North Carolina, Greensboro; Wellesley College; UC Berkeley and UCLA. Her first career was in modern dance. She was a university professor, co-authored *A Pocket Guide of Dance Activities,* conducted master classes and performed at concerts and arts festivals throughout California and on educational television.

MARY PYNE, born Sept. 27, 1924 in Saskatchewan, joined the 99s in 1977, after taking flight training at Bar X Aviation, Medicine Hat, Alberta and Mitchinson's Flying Service, Saskatoon, Saskatchewan, Canada. She has flown a Cessna 150, Cessna 152, Cherokee Warrior, Cherokee 140 and Beech Sundowner.

She has been featured in *International Flying Farmer* about 1986. She owns and flew a Cessna 150 CG-VIC to health centers in Northern Saskatchewan until 1989. Western Section Member of the Year 1993. Flew with Dr. June Mills to Alaska in 1984 in June's Beech Sundowner, to attend 99s International Convention. She is a member of Saskatchewan and International Flying Farmers.

Pyne is a widow and has three stepchildren: Caron, Donna and Jule. She has three step-grandchildren: Julia, 14; Caley, 12; and Desmond, 11. She has a BS in nursing, 1962. She is a private pilot but used her plane to fly to nursing assignments.

SATU KIMBERLEY PYYSALO, born Aug. 25, 1970, in Finland, has lived in Arizona for two years. She joined the 99s in 1993. She received flight training at North American Institute of Aviation, Las Cruces, NM; Aero Mech., Scottsdale, AZ and Arizona Flight School, Prescott, AZ. She has flown the Cessna 150, 152, 172, 172RG, 182; DC-6B and Beech TravelAir. She holds a commercial, single and multi-engine land, and instrument rating. She also has a ground instructor – advanced and instrument rating. She has a total of 380 flight hours.

She belongs to the AOPA, Angel Planes, CAP and is in the National Ski Patrol. Pyysalo is a college student and is planning to attend Embry-Riddle Aeronautical University.

VIRGINIA RABUNG, Aux Plaines Chapter, joined the 99s in 1953, purchased her 1946 Cessna 140 currently based at Galt Airport, Illinois and flew the 1953 AWTAR. Two air races to Cuba (1955-56) and another to the Bahamas followed, all solo in her 140. Has also flown Piper Cub, Porterfield, Taylorcraft, Ercoupe, Cessna 182 and 310.

Participated in the 99s AOPA "Fly-It-Yourself-Safari" of South Africa, Southwest Africa, Swaziland, Mozambique and Inhaca Island (1968). Took part in 99 sponsored European tour, Forest of Friendship honoree.

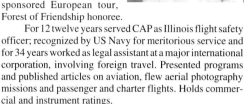

For 12 twelve years served CAP as Illinois flight safety officer; recognized by US Navy for meritorious service and for 34 years worked as legal assistant at a major international corporation, involving foreign travel. Presented programs and published articles on aviation, flew aerial photography missions and passenger and charter flights. Holds commercial and instrument ratings.

PEARLE BAIRD RAGSDALE, born in Running Water, TX (near Plainview), to Milton Joseph Baird and Lillie May McGee Baird. She graduated from Wayland Baptist College and Lippert's Business College, both in Plainview.

In 1940 she married Robert Ragsdale Jr. The following year they moved to Austin and together established Ragsdale Aviation, which they operated until 1984. In the spirit of the 99s dedication to the promotion of women in aviation, Pearle was proud of the contributions of Ragsdale Aviation toward furthering this effort.

Prior to WWII, few women were involved in aviation. In fact, Pearle's first encounter with aviation was in Lubbock. She had many friends who were the wives of pilots, but none of these women were actually pilots themselves. Determined that she wasn't going to be just a "pilot's wife," her husband taught her to fly; and she earned her pilot license in 1941. In addition to time as pilot-in-command, she logged hundreds of hours as co-pilot in various aircraft including multi-engine airplanes and seaplanes.

During WWII, Ragsdale Aviation trained a large number of women pilots, many of whom went on to join the WASP and a few who found careers in aviation. In the years that followed, Ragsdale Aviation sponsored many programs designed to encourage women to fly.

The Austin Chapter of the 99s was originally organized in 1964, and Pearle was a charter member. They had a very active group and received several awards for their achievements, but over the years the group broke up and the chapter eventually became inactive. In 1977 the chapter was reorganized and Pearle was involved in the rechartering effort. This chapter remains active today.

Pearle died peacefully in her home in Austin, TX, on Oct. 4, 1993. She is survived by her husband, Robert and many family members and friends who remember her fondly and miss her greatly.

SHARON RAJNUS, a native Oregonian, learned to fly in 1976 in a Cessna 120, then moved into the Stinson 108-3, Maule, Helio, C-172RG and others. Rajnus' aviation heritage began with family engineering history at Lockheed during WWII and continues today as an aviation artist with the American Society of Aviation Artists (ASAA). The unique perspective of flight has aided the artist in countless scenes and gives a flavor unobtainable by other means. Paintings, most often executed in oils or watercolor, are shown at exhibitions across the United States, are included in many collections and range in subject from early flying history to current times.

Educational background includes a BS at Oregon State University in science and art. Excursions through the back country and off-airport operations contribute to the scenes, and paintings often depict antique and historical aircraft including early flying boats and WWII amphibians. Rajnus includes landscape and skyscape as a consistent major element, considering these as giving the aircraft a "sense of place."

She is married to Donald Rajnus who is also a pilot, has two children, and paints from a studio in the country only four minutes from the nearest "grass strip."

JANE RALSTON, claims she was born with a desire to fly, which she did as soon as possible. The place was Bowman Field in Louisville, KY, the year 1944. A BS in chemistry supported this avocation.

On Bowman Field, Jane pioneered women flying professionally. Her first commercial job was "hopping" passengers on weekends and evenings. Eventually tiring of the chemistry lab, she accepted a position with an FBO and supported herself by instructing and charter work for seven years.

Jane flew a J-model Bonanza in the 1962 Powder Puff Derby for fourteenth place. A few years later she flew the race again as co-pilot.

In view of a more secure retirement, Jane worked her last 20 years for the federal government in Washington, DC, as a computer programmer analyst. Jane is a member of the 99s (1947), AOPA (1945) the EAA on the Aero Club of Louisville since 1944.

SIGRID RAMELLI, born in WWII in Germany, survived the air battles over her country in those days and always wondered what it would be like to be up there, above it all.

She came to the United States in 1960 on the *Queen Elizabeth* and remembers her first flight on a commuter airline, from New York to Florida. It was not until 1971 when she met her husband, Gary, private pilot. He flew her to Catalina Island (25 miles from the California coast) on their second date, not knowing that he pushed the magic button in her. Sigrid took flying lessons soon after they returned home. She earned her private license in 1972, soon to be followed by commercial and instrument ratings.

In 1982 Sigrid entered her first all women air race, organized by the 99s, leaving from Long Beach, CA to the destination of Cabo San Lucas, Mexico. She won first place and joined the 99s. Never having been approached for membership before, she found out that most girls assumed 'Sigrid' must be a man's name! Sigrid's second love, photography, has enabled her to contribute many photos for 99 publication.

Sigrid has flown for the 'Flying Samaritans' into Mexico, Direct Release International air transport and presently flies Angel Flight missions, moving patients to or from their treatment facilities. This pilot loves to fly in the spirit of the 99s: friendship, education, safety and the international fraternity of women aviators – and she does it all in their A-36 Bonanza. "Oh, Lord, it is good to be able to fly . . ."

NANCY RAMIREZ, joined the SCV 99s in 1989 after getting her ASEL that year. She started flying in the 1982 and flew gliders on and off before taking her first powered lesson in a Tomahawk in 1987 at South County. After taking ground school at Amelia Reid's in San Jose, she switched to taildraggers and received her private license in a Citabria.

She currently has about 100 hours in gliders and 200 in aircraft with engines. Her aviation education includes brief lessons in skydiving, bungee jumping, ultra-lights and an airship. She has even tried simulated skydiving in the Las Vegas aerodium.

Nancy's 99 contributions include airshow coordinator, airmarking, hospitality and PR efforts, and she was awarded the chapter Service Award in 1993. She works as a manufacturing support technician for Sun Microsystems.

Nancy likes to roller skate and ballroom dance and is raising two children on her own.

MARIE FASANO-RAMOS, born July 11, 1940, in New York joined the 99s in 1986 after earning a commercial pilot and multi-engine ratings. She has flown Cessna 172, Cherokee 180, Twin Commando and whatever else she could rent.

She has a master's degree in nursing and counseling/psychology. She has a holistic nursing practice. Her favorite flight was flying low over the blue waters to Baja, CA.

MAUREEN RAMSAY, born July 16, 1963, in Arcadia, CA, joined the 99s in 1988. She has flight safety recurrent for King Air 300, British Aerospace type-training for JetStream 31. She has airline transport pilot for airplane MEL, commercial pilot for airplane SEL, advanced ground instructor, flight engineer (written), radiotelephone operators permit and first class medical (no restrictions). She has flown the C-150, 172, 206, 310; PA-28; BE-200, 300; and BA-3100.

Ramsay has been employed as a regional airline pilot July 1994-present: Reno Air Express, San Jose, CA; captain, JetStream 31; December 91-July 1994: charter pilot for R.M. Aviation, Inc., San Jose, CA; captain, King Air 300; March 91-August 91: photography pilot for Pacific Aerial Surveys, Oakland, CA, captain C-206.

Her educational background includes a BS in aeronautics (May 1988). Her interests include skiing, volleyball and soccer.

NANCY JEAN RAND, earned her private pilot license in 1971 at Regina, Saskatchewan and joined Maple Leaf Chapter, East Canada Section that same year. She transferred to Alberta Chapter, West Canada Section on moving to Lethbridge, Alberta in 1973. Over the years, she has held all chapter and section offices culminating with serving as section governor 1991-93.

Recreational flying has given her many of her happiest hours – partaking of the beautiful Western Canadian scenery; mountains and prairies beneath her wings. Back on the ground, friendships have grown as they airmark windswept tarmacs, lead Flying Companion Seminars, organize Poker Runs, attend yearly International Conventions to meet with other women pilots from around the world.

Rand's husband of 32 years, Duncan and she have raised four children. She works as an RN in the surgical suite at their regional hospital. Recently she started a private practice specializing in therapeutic touch. To further her research skills, she has started going to the university toward a bachelor of nursing degree. She continues to treasure all the friends she has made since joining this great organization – the 99s.

MARGARET M. RAPPAPORT, MD, Ph.D., born Nov. 16, 1947, has 20 years of administrative and clinical experience in health services practice. She is a consultant to business, industry and governmental agencies with expertise in human resources training and media communications. She is the executive director of Reach New Heights, Inc. During the last four years, Margaret has lectured on aviation psychology and human factors to aviation audiences. She has been featured speaker at the SSA Convention, EAA Oshkosh, Women in Aviation Conference and the International Symposia on aviation psychology.

Margaret is air-minded. She is a power, seaplane and glider pilot. She is the current International membership chair of the 99s, International Organization of Women Pilots. She is director for Region 2 of the Soaring Society of America. Margaret is also a member of the Aircraft Owners and Pilots Assoc., the Philadelphia Glider Council, the Wing's Field Pilots Assoc. and the Chatham Airport Pilots Assoc.

She is married to Herbert Rappaport and has two children, Amanda, 22 years and Alex, 15 years.

SARAH RATLEY, born Aug. 30, 1933, in Kansas City, KS, joined the 99s in 1951. She has commercial ASEL, AMEL, instrument, rotorcraft-helicopter, glider, flight instructor and A and I ratings.

She is also a member of the AOPA, MPA, Whirley Girls, FLATS, Mercury 13. She has flown in the Powder Puff Derby, International Air Race, Skylady Derby, Michigan Small Race and many other races.

PAMELA S. RATLIFF, joined the 99s the same year she earned her private pilot license, 1990. She began her flight training in an Aeronca Champ; soloed a Cessna 152 and took her check-ride in January 1990.

In 1988 Pam bought into an airplane building project a good friend was involved in and became the third partner in the ownership and construction of a J-6 Karatoo. It is an "experimental" design; two place, side-by-side, taildragger. The project was completed in August 1989. Because of a freak accident, Pam and one of her partners crashed the Karatoo in July 1990. The plane was pretty broken up as was Pam's face and head! Pam and one of her original partners decided to rebuild their plane which flew again in October 1992. Pam is very active in Chapter 91 of the Experimental Aircraft Assoc. in Kansas City, MO, and currently serves as secretary, treasurer and newsletter editor.

Other planes she has flown include: Cessna 150 and 172 and has 15 minutes in the left seat of a B-17. Pam has worked on the FAA sponsored WINGS program and as of 1995 has earned Phase II. She is a lifelong resident of Missouri and hangars her "homebuilt" Karatoo at East Kansas City Airport in Grain Valley, MO.

MARY RAWLINGS, born Nov. 17, 1936, in Lansing, MI, raised in a sparsely populated farming area, came to California with her family as a teenager, and has since been a resident. Rawlings interest (mostly fear), in flying developed in the mid-1970s, while producing seminars throughout the state of California for the escrow industry. A short two hour familiarization flight opened up a whole new world. She soloed on Aug. 7, 1980; by October 1980, had purchased a Warrior II and on Nov. 27, 1980, accomplished a successful check-ride for her certificate; obtained her instrument rating in July 1984.

Rawlings discovered the 99s in November 1981 and air racing the next year. She has raced more than 70 races including seven Air Race Classics, all of the Mile High Air Derbies; trophied seven times (out of 12 races) in the Great Pumpkin Classic and many others. Rawlings has logged more than 2,500 hours of flying time, the last 1,000 hours in her Piper Dakota.

Rawlings in a member of the 99s, Inc., Aircraft Owners and Pilots Assoc., Air Race Classic, Inc., and many trade organizations. She is chairman of the Valley Air Derby, a cross-country speed race flown in October, sponsored by the San Fernando Valley Chapter. She lives in the San Fernando Valley, flies out of VNY Airport; has three grown children: Bonita Walker, 40; Mary Rios, 36; and Richard Rawlings, 34 and three grandchildren: Rhiannon Walker, 11; Samantha Walker, 9; and Allison Rios, 5.

DOROTHY "DOTTY" HARGRAVES RAY, started flying lessons on her 50th birthday in 1992, and got her private pilot license two years later. She can most frequently be found in the skies of the southern Mojave Desert near the birthplace of the U-2, the SR-71 and other Stealth aircraft.

She and her husband, Paul, own a Mooney M20A and an experimental taildragger called a Murphy Rebel, which they built themselves. She has flown the Cessna 150, 152, 172; Luscombe, Beech Musketeer, Mooney M20A. She received flight training at Hesperia Flight Academy in Hesperia, CA.

Dotty is a third generation native of Southern California. She was in the Air Force in the 60s, lived briefly in Scotland and studied commercial art in Los Angeles. She has an AS degree in computers and does database programming on a consulting basis. She has three children: Daniel, 31; Amy, 24; and Emily, 21. She and Paul publish a newsletter for builder of several kitplanes, and travel to many airshows each year as members of the press. They hope to continue this last activity well into their retirement years.

The perfect day: First she was buzzed by a B2 south of Edwards AFB (not in restricted airspace) then sang opera to herself over the intercom while she explored the beautiful desert from 2,000 feet AGL.

GEORGIANNE RAY, with an introductory flight coupon in hand, began her adventure in flying. In her excitement she got lost driving to her first lesson at Smartt Field, St. Charles County, MO! Her instructor, Jerry Gunkel, saw to it she flew many different aircraft from Cessna and Piper trainers to a Cessna Citabria (aerobatic), taildraggers, to her most exciting time as PIC in a Beech Baron. She earned her commercial license, SEL, MEL and instrument rating.

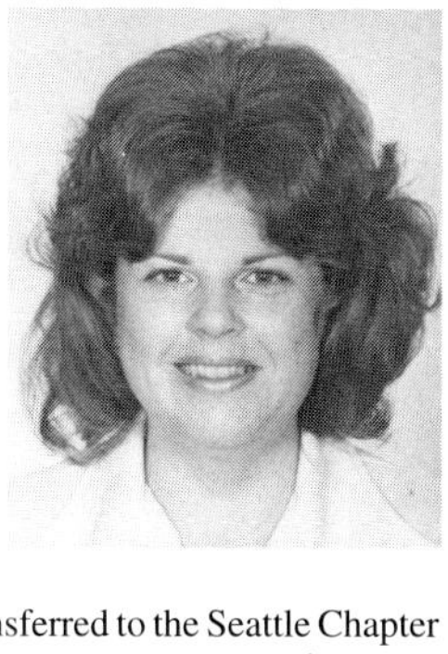

Ray joined the St. Louis 99s (1973) and served as Air Age education chairman. After she married Lee Ray and moved to Seattle, WA, she transferred to the Seattle Chapter and joined AOPA. With Lee's encouragement and support she used her knowledge of aviation to work on the noise advisory committee for Boeing Field and the sea-tac overflight committees.

Ray divides her time between photography, bowling, being a community activist and surfing the Internet.

HILDA RAY, born in Long Beach, CA, and grew up in southwestern Michigan. It was in Michigan where she was introduced to flying at the age of 15, and knew that one day, she too would fly. However, it took several years to accomplish that goal. In the interim were college, marriage and five children. Hilda took her first check-ride when her fifth child was 3 months old. Since then, she has taken two other rides, commercial and instrument. She is the first women in Walker County, AL, to accomplish these certificates.

She served for two years on the Walker County Airport board. Hilda has been an active member of the Alabama Chapter 99s, having held at one time or another all of the chapter offices, currently as treasurer. She has served on several Southeast Section committees as member or chairman.

Hilda has attended Southeast Section meetings and also a number of International conventions the first of which was the Vail, CO International Convention. She has enjoyed all of the meetings for the social aspects as well as the educational facets of the meetings.

Hilda has introduced several other women flyers to the 99s. She has given young people their first flying experiences, talked to school children about the joys of flying and shares her love of flying wherever she can.

JANICE R. RAYMOND, born in Kansas City, MO, and has resided in Overland Park, KS, on 99th Street for 21 years. Her husband encouraged her to get her private pilot license in 1982 when they purchased their Bonanza. Jan celebrated her 50th birthday year by attending ground school in 1986 and was issued her ASEL in 1987. She began flying a Piper Tomahawk, then a Warrior, a Beechcraft Sundowner and now a Bonanza. She is the only woman in her family to hold a private pilot license. She and husband have flown to many cities in the US and some islands in the Bahamas. She is currently working on being "signed-off" in their Bonanza.

Jan divides her time between family, teddy bear collecting, and leader in several different organizations. She has served as vice-chairman and chairman of the Northeast Kansas Chapter of the 99s. She considers her association with members in her chapter and the South Central Section to be some of the most rewarding of her many organizations. She is married to Robert Raymond who is also a pilot and an attorney.

MARY W. RAYMOND, born March 31, 1945, in New York City, NY, had her first flight in a small plane in July 1982 in a Grumman Traveler and began flight lessons in July 1983 at Harvey Field, Snohomish, WA, in Cessna 152s and

172s. Her husband, Bud, is also a pilot and most recently they owned a Grumman AA1B. She attended meetings of the Western Washington Chapter as a student pilot, and joined in March 1985, after getting her private license in February 1985.

She has held all offices of the chapter: chairman 1991-92; vice-chairman 1986-87; secretary 1990; and treasurer 1994-96; newsletter editor 1990 and 1993-94.

Raymond grew up in Pendleton, SC, and migrated to Seattle area in 1979, via the Southern California area (1966-1979). She graduated from the University of Washington with a BA in business administration (1983) and accounting (1995). She is married to Bud and has five stepchildren: Beth, Brett, Bruce, Beverly and Becky. She has five grandchildren: Greg, Whitney, Ashley, Andrew and Jessie.

LAURA B. READ, started flying in the spring of 1992, the day after she sold her company. Her instructor was Don Martel of Sanford Air and he let her land and take-off the first day out. She always wanted to fly and her Uncle Frank and Aunt Rose Andrew (99) were role models to her. They still fly their Cessna from Urbana, IL to the family farm in Palmyra, IL.

Read got her license in June and the next day made a flight over the ocean to Nantucket to deliver a plane. Three weeks later while returning from China, she rented a Cessna 172 in Hong Kong and flew at 500 feet through Victoria Harbor, out to the east border of China and back through the New Territories. Bruce was with her and they also had a guide.

After they returned home they made several trips to Maine, dropping their dog, Jack, off to the training camp in Rangely and going to a party in Bar Harbor.

In October, Read flew to St. Louis and back from Sanford, ME, using only piloting skills. She put in 27.8 hours of solo flying on that trip.

Later that month, she got her float plane rating from Folsom's Air Service, flying with Max, Bob and the late, great, Charlie Coe. It was there that she realized that she was pregnant.

Baby Helen put in numerous pre-natal flight hours, some of which involved getting caught in a sudden thunderstorm while returning from Martha's Vineyard.

After Helen's birth, Bruce and Laura flew to Quebec City, Canada in -15 degree weather and then to New Garden, PA, for a wedding. Last week, they flew with a falconer to search the forests for nests. Now Helen has a sister, Anne, and they all fit in a Piper Warrior.

SANDRA LINELLE REAGAN, born March 20, 1946, in Orlando, FL. She learned to fly (1991) after her son, Patrick, went in the Navy. She always wanted to fly but was told "girls can't fly." She proved them wrong. Sandra is now a commercial pilot, instrument rated, ASEL, AMEL and is an advanced, instrument ground instructor. She wrote, developed and teaches the curriculum for the aviation tech program at a local community college, where she is a student advisor. She has a BA in psychology and public relations.

Sandra was an Amelia Earhart Scholarship winner in 1994, and was outstanding member of the year for the Colorado Chapter (1994). She is the vice-chairman of the Colorado Chapter, co-chairman of the Flight Without Fear program and is a FAA aviation safety counselor. She enjoys participating in air racing and aerobatics and participates in the community by giving aviation career seminars at local middle and high schools, as well as helping Girl Scouts and Boy Scouts with their merit badges in aviation. Her son Patrick is married to Heather, and a grandbaby (girl!) is on the way. Sandra wants to be a role model for "girls who can fly," and has a dream of going into space.

NANETTE GAYLORD REATHER, born Sept. 5, 1942, in Akron, OH, received her private pilot license in April 1971. She currently holds an ATP, Gold Seal CFI certificate with SE, ME and IA ratings and GIA and I certificate. She has accumulated more than 5,000 hours. She has flown all Cessnas, Piper, Beech, Mooney and Grumman American singles and numerous twins.

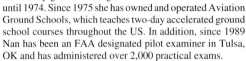

Nan's initial training was from Ross School of Aviation, Tulsa, OK, where she later worked as a ground instructor, flight instructor and finally general manager until 1974. Since 1975 she has owned and operated Aviation Ground Schools, which teaches two-day accelerated ground school courses throughout the US. In addition, since 1989 Nan has been an FAA designated pilot examiner in Tulsa, OK and has administered over 2,000 practical exams.

Nan has been a member of the Colorado Chapter where she held all offices through chairman, and is currently a member of the Tulsa Chapter of the 99s. Fun flying has included the 1976 Powder Puff Derby, and 1977 IAR, and she is the proud winner of the 1977 PPD commemorative flight with Pat Udall. She is currently working on an MS in aviation at Oklahoma State University. She is married to Timothy Reather.

MORGEN ANN GRESCHEL REEB, born Oct. 10, 1960, in an Army hospital in Frankfurt, Germany. She took her first flying lesson and soloed at the age of 16 and obtained her private pilot license right after her 17th birthday. She has flown the B-727, BA-3100, BE-02(190), IE-99, PA-31TI, PA-31-325, PA-60-600, PA-23-250, PA-44-180, C-421, C-303, AE-840, Stinson 6060B, T-34B, M-20C, TB-20, GAA-8B, BE23-160, B35C-33, 7ECA, PA-32R-301, PA-28RT-201T, PA-28-201, PA-38, PA-28-181, PA-28-180, PA-28-161, PA-28-151, PA-28-140B, PA-22-135, C-210, C-206, C-182RG, C-182, C-177B, PA-172RG, C-172, C-152, C-150.

Morgen attended George Mason University and earned a BS in business administration. After obtaining her instrument, commercial, multi-engine and CFI she went to work as a flight instructor and charter pilot at Leesburg, VA. She also flew parachute jumpers at Hartwood, VA. She obtained her CFII, MEI and ATP.

Morgen's airline career began with Colgan Airways flying Beech 99s and 1900s. After Colgan was bought by Presidential Airlines, she checked out as captain on the BA-3100. Morgen was hired by American Airlines as a flight engineer on the Boeing 727. After two years, she checked out as a co-pilot on the Boeing 727. She is married to Frank B. Reeb who is also a pilot.

FLORA BELLE (SMITH) REECE, born Oct. 21, 1924, at Sayre, OK, took her first flight which was also her first lesson on February 1943 at Will Rogers Field. Her first solo was at Nuckols Field, Oklahoma City. She flew before driving. Lloyd Catlin taught her to fly and to drive.

Reece worked as girl Friday at Nuckols and chartered in Waco and Fleet. She did some instructing on commercial which was allowed at that time. Reece joined WASP (Women's Airforce Service Pilots) November 1943 to December 1944. WASP training she flew the PT-19, 17; AT-6, BT-13 her class jumped from PT-17 to AT-6s. She was assigned after graduation to Foster Field, TX, where most of her time was in the AT-6. Flew co-pilot on C-60. Sent to Harlingen AFB, TX, where they towed targets in B-26s, then to Enid, OK, for deactivation.

In an AT-6 at Avenger Field, Reece called the tower and asked permission to go above the clouds to practice. The clouds closed in and when she found a hole to come through the clouds she was lost. She landed safely in a field and people came and built a bon fire and brought food. They waited with her until the base truck picked her up. It was raining and continued for days. Reece was embarrassed carrying her parachute to flight line. She became Cross County Katy to Mr. Turner, her instructor. He allowed her to stew until she finally asked was going to happen. He said, "You did great, they couldn't complain."

She married Ralph Reece on Aug. 1, 1945 and she had three children: Connie Kay Reece (Fox), born 1946; Cherryl Susanne Reece, born 1948, died 1964; and Russel Alan Reece, born 1949. Her grandchildren are: John Fox, born February 1968; Jim Fox, born November 1969; Susan Marie Reece, born February 1973; and Pamela Reece, born August 1974. She now flies a Cessna 150 and never remembers not desiring to fly and trying to find a way to get into the sky. She loves it.

LISA REECE, from Georgetown, ME, learned to fly a Bellanca Citabria shortly after she earned her private pilot license. She has since logged more than 400 hours in tailwheel aircraft. She has flow extensively throughout the United States, Canada and Mexico.

Lisa combines her love of flying with that of the great outdoors. Flying and backcountry camping in Baja, Mexico has been her greatest adventure. Throttled-back, flying 500 feet over the aqua shores of the Sea of Cortez, the wing "kissing" the rugged edges of the desolate desert mountains, is the type of flying Lisa loves to do. When not on one of her flying vacations Lisa can be found flying someone over the beautiful coast of Maine in her Cessna 185.

Lisa is a member of the Seaplane Pilots Assoc., AOPA, the Katahdin Wings Chapter of the 99s, an EAA member and a participating pilot in the Young Eagles program. For several years she worked at a small FBO. Currently, she works as a graphic designer and free-lance writer.

DANA ANN REED, born Sept. 3, 1953, in Boulder, CO, joined the 99s in 1993. Earned her private pilot license and instrument rating (single-engine land). She has flown the C-172, C-182, T-34 Mentor and Piper Cherokee.

Reed is married to Bill Rathfon and she has two children, Amy Caven, 20 and Ryan Parker, 18. She is employed as an electrical engineer. She was given aerobatics training in T-24 for a Christmas present by her husband one year. Family thought he was nuts! Got it all down on video tape.

KAREN REED, born Oct. 12, 1951, Cleveland, OH, she graduated from Brecksville High School in 1969 and earned a BS in marketing, University of Northern Colorado 1973. One of the first female medical sales representatives, she won numerous sales awards. In 1982 she earned a masters of business administration, Baldwin Wallace College and married Bill Reed, gerontologist. Lauren was born in 1984 and son, Taylor 1987.

Reed earned a private pilot certificate, 1985; instrument rating, 1988; commercial certificate, 1990; advanced ground instructor, 1990; and certified flight instructor, 1992. She organizes and teaches private pilot ground school and has logged 1,500 hours. They own a Cessna 182P and have taken numerous trips throughout the US, Canada and Caribbean with their children. She was 1990 Pilot of the Year, Lake Erie Chapter and is a charter member of the Women with Wings Chapter.

DEBORAH J. REEDER, born April 18, 1963, in Cleveland, OH, soloed 1979 got her private pilot license in 1983 and instrument rating in 1985. She joined the 99s in 1994. Reeder has flown the Cessna 152, 172, 206, Maule M5/235. She is an ophthalmologist and is married to Fred Reeder.

LAURIE SPENCE REEVES, first flew in February 1968, at PBI, completing CFI A and I in just two years. She began work at Waterford, CT, in July 1970. While there, she earned the MEL rating, became assistant chief flight instructor and was FAR 135 director of operations and check airman. As a Connecticut 99, she held the offices of treasurer and chapter chairman, headed committees for airmarking, membership, New England Air Rally and was a New England section nominee for an AE Scholarship. Laurie has acted as an FAA accident prevention counselor and was a member of the first FAA New England Women's Advisory Committee. She flew for Pilgrim Airlines in 1978 until being hired by United Airlines where she now flies as co-pilot with type ratings on B747-400 and B-757 and B-767 accumulating more than 11,000 career hours. Her husband Ed, an IA, owns and operates Reeves Air Inc., in Westerly, RI.

ELAINE REGIER, born in Oklahoma City, OK, joined the 99s in 1988. She received flight training in 1987-88 in Chickasha, OK, and has flown mostly Cessna 150 and 172. Regier used aviation and aerospace education in her pre-first grade class of 6 and 7 year olds to encourage the students to work hard and say, "No," to drugs and alcohol. One activity involves a trip to the local, unattended airport. Each child has a chance to sit in the plane, look at the instruments, see how the controls work and ask questions. Then the students help with preflight and she does a takeoff and landing for them. One little boy found it hard to believe that his teacher flies airplanes. The first thing he said to her after landing was, "You really flew!" He thought she had been teasing them all year!

JANE M. REGO, born Dec. 6, 1952, Fall River, MA, soloed Aug. 3, 1992; received her certification in ASEL since May 1993; currently working on and half way to instrument rating; has 270 hours and has completed Phase I and II of FAA Wings. She has flown the Cessna 150, 152 and 172 and is presently owner of a Piper Cherokee Warrior 150.

Getting her private certification was her life's dream. Rego wished she had started long ago. She flies for fun and adventure. "It's a quiet place up there, a great place to get away. I look forward to my long trips and flight training." Rego is currently employed at a local hospital, working as a staff clinical laboratory technologist. In her spare time she is actively involved with Girl Scouts, family and of course flying as much as she can. She resides in Assonet, MA, with two dogs, two cats and bird and fish.

DIANE ELIZABETH JONES REICHEL, born in Jamestown, NY, in 1938. She has been interested in aviation all her life, but it wasn't until recently that circumstances permitted pursuit of this interest. After raising her family, she went back to college and received her BS degree in accounting from Wright State University in 1990, graduating with honors. She then completed her private pilot training in her Cherokee PA-28 180, receiving her license in November 1991.

Since joining the 99s in 1992, she has enjoyed attending All Ohio meetings and special activities such as airmarking, helping with the Buckeye Air Rally, flying Young Eagles, North Central Section meetings and looks forward to many future 99s activities. Her husband, Roland, who recently retired from Wright State University, also has his private pilot license. They belong to EAA and AOPA, and enjoy their yearly pilgrimage to Oshkosh.

Reichel's oldest son, Dennis, was a flight engineer on an Air Force C-141. Their two younger sons, Greg and Lee, were both Cadet Commanders in the Civil Air Patrol prior to college. They now have two daughters-in-law, Wendy and Pam, and grandsons, Dean and Andy.

She recently purchased a 1947 Navion A which is being restored and looks forward to flying it. Other interests include sailing, bicycle touring, music, gardening and photography.

AMELIA REID, holds just about every rating and is a regular performer at airshows. She started flying in Nebraska in 1944. A first flight with barnstormer, Evelyn Sharpe inspired her to get her license. Amelia worked towards her rating with a Navy department secretarial job and working as a "gas boy" at the Congressional Field. In 1946 she moved to California, where she took a computer job at Ames Research Center. After losing her job due to pregnancy, she decided to go into business for herself. She began instructing in 1960 with her husband, Bobby Reid, whose family owned the Reid-Hillview Airport. She bought her first plane, a Taylorcraft L-2. When the airport sold to the county, she took complete charge of the business, which then had five aircraft. Amelia now has more than 20,000 hours and continues to teach aerobatics and taildraggers.

JOANNE WIXON REID, started flying with a family friend in the late 30s at the age of 11. Marrying at an early age, she had four children which did not allow time or money for flying. At the age of 50, and the family grown, she decided it was time to get back into aviation earning her SEL and SES ratings and joining the 99s at that time. She and her husband have owned two Fairchild 24s, Stinson 108, Cessna Cardinal 177B, 1946 Piper Cub, Cessna 185 on amphibious floats and a Waco VKS-7, the last production cabin Waco.

John H. Reid, her husband of 50 years, has been a pilot since 1944 and is a retired fire chief. They have four children, six grandchildren and two great-grandchildren.

M. JOANN REINDL, born Jan. 30, 1935, in Leavenworth, KS, joined the 99s in 1964. She has a commercial license and Gold Seal flight instructor, SMEL instrument and sail plane rating with 16,000 hours flight time. She has flown Cessnas, Pipers, Beechcraft, single and ME land and sailplanes.

She is employed with Executive Beechcraft in Kansas City, MO. Kansas Senator Bob Dole recently met with Mary, who was in Washington, DC to be honored as Flight Instructor of the Year, by the Federal Aviation Administration (FAA) and the general aviation community.

Joanne was the nationwide pick of a panel of judges from FAA flight operations staff and the general aviation community. She has been serving as a flight instructor since 1968 leading pilot clubs in safety and educational efforts. Even after a serious automobile accident in January of 1974, which led many doctors to be doubtful of her flying future, she continued with selfless determination to be a leader in the pilot community. While still hospitalized in June of 1974, she was awarded a Certificate of Recognition by the Central Region of the FAA for her support and participation in the accident prevention program. Joanne's determination was demonstrated further when on the last of day of December 1974, doctors who had told her that she would never walk again, were surprised to hear that she was going flying that very day.

ANTONIA (TONI) WEHMANN REINHARD, born Oct. 4, 1938, in New York City. During her formative years she expressed an interest in the women who flew in the military service. It was after she moved to the Midwest in 1960, married, raised five children: Christopher, Timothy, Matthew, Jacqueline and Geoffrey, and divorced that she earned her private license in 1983 at Palwaukee Airport, Wheeling, IL. She was the first female student of CFI Merry Schroeder to earn her license.

Toni was an active member of Aux Plaines Chapter from 1983 to 1993, serving as secretary, newsletter editor, chapter and section aviation activities chairman. Toni has raced in and served as chairman of the 24th annual Illi-Nines Air Derby in 1993. She volunteered at EAA, Oshkosh, participated in poker runs, written letters in support of general aviation, and designed and completed an aviation quilt embellished by chapter members. She is working towards her instrument rating and is a regular participant in the FAA Wings program.

Toni is a graduate of Earlham College, Richmond, Indiana and Harrington Institute of Interior Design, Chicago. After designing model homes for several years, Toni has her own custom pillow quilts business. Community service included teaching first aid, CPR and swimming and being an election judge.

In 1993 Toni transferred her membership to the Western Washington Chapter when her flying partner, Johanne Noll, was retiring to Seattle. "When Johanne suggested I move also, so that we could continue to fly, my initial reaction was, I couldn't possibly do that! But once the flying bug bites there's no telling what one might do . . .I moved a month later!" Toni joined Wings Aloft Flying Club at Boeing Field and the Museum of Flight and together they continue their love of flying in the mountains.

JUDITH "JUDY" B. REINHART, born Lebanon, TN, on Aug. 15, 1941; moved to Fort Worth, TX, June 1946 and had her first flight with pilot father in a Stearman, August 1946.

Married George William Peter Reinhart, June 14, 1969; moved to Austin, TX, August 1971. Raced sailboats of all sized on Lake Travis 1971-78. Pete reintroduced her to flying March 1978, with glider rides. Received private pilot (SEL) license, July 1980 and immediately joined the Austin Chapter of 99s. Earned glider rating June 1981. Flew Cessna 140, Cessna 152, Tri Pacer and Mooney.

Active member of Austin Chapter served as secretary, and chairman 1984-86. Active in South Central Section, attended first Section Meeting, Fall 1980. Served as treasurer, 1986-88; secretary, 1988-90; vice-governor, 1990-92; and governor, 1992-94.

Received bachelor degree in education from Lindenwood College in 1964; master of library and information science from University of Texas, Austin, 1982.

Judy taught school and was a school librarian for 31 years; retired May 1995. She loves to help others, is involved in many organizations and will begin a new career following retirement.

RUTH REINHOLD, Phoenix 99s, who went to new horizons in 1985, was famous for her friendliness, devotion to aviation, her accomplishments, and her modesty about them. For many years she was Barry Goldwater's executive pilot, including his campaign for president. From an office job in 1933 at the airport which would become Phoenix Sky Harbor, she entered a career of many years as a professional pilot and an aviation manager. She recalled her early role as nurturing, and said they all bonded. She flew an OX5, a B-23 bomber and most private aviation planes, as a commercial pilot, flight instructor and race pilot.

Ruth published a book in 1982, *Sky Pioneering: A History of Aviation in Arizona,* and in 1986 was honored by entry in the Forest of Friendship by Phoenix and Arizona Sundance Chapters. Much of the growth of aviation and the 99s in Arizona can be traced to Ruth's constant public relations effort supporting them on radio, TV and in writing, but chiefly by flying to lots of 99 and other aeronautical meetings. She also devoted many hours to chapter projects and to inspiring other women by the obvious fun she had.

SHIRLEY RENDER, born April 1, 1943, in Winnipeg, Manitoba. She learned to fly in 1973 (private pilot license) and currently flies a 1948 90 hp Luscombe. A former social worker and teacher, Shirley switched careers in the 1970s and became an aviation historian. She has both a BA and MA from the University of Manitoba.

In 1983 she was the first Canadian to be awarded the Amelia Earhart scholarship for research. This grant helped to cover her expenses when she wrote *No Place for A Lady, The Story of Canadian Women Pilots, 1928-1992,* the first book ever written on Canada's women pilots.

She is the former president of the Western Canada

Aviation Museum, and still remains the editor of the *Aviation Review,* the museum's 32-page magazine.

In 1990 Shirley was elected a member of the Legislative Assembly (MLA) of Manitoba and was appointed legislative assistant to the Premier. She maintains her currency in flying and still acts as an aviation consultant to many groups. She was a guest speaker at the 1994 International Women in Aviation Conference in St. Louis. Her book is being considered for TV production.

She has received numerous academic awards for her master's thesis on commercial aviation as well as receiving awards for her work in promoting Canadian aviation heritage Prix Manitoba Award (1990) and the Canada 125 Medal (1994) to name two. Shirley is married to Doug Render, a professional engineer and private pilot, and has two children.

CAROL L. RENNEISEN, born Nov. 1, 1956, in Winona, MN, joined the 99s in 1988. She was in the USAF and AF Reserve (1980-87) as an aircraft maintenance and transportation officer 926th Fighter Group (major) 1987-present. She soloed in 1978 in Lubbock, TX; with additional flight training in Arizona, Louisiana and Connecticut (helicopter). She has flown anything with wings she could get her hands on.

She has been in the Civil Air Patrol since 1985 and is currently Louisiana Wing flight operations officer, check-pilot, mission pilot. Flew missions for disaster relief efforts after Hurricane Andrew in Louisiana and flooding in Missouri. She has a BS in animal science, from Texas Tech University and an MA in aviation management from Embry-Riddle Aeronautical University.

She is a member of AOPA, Airborne Law Enforcement Assoc., AF Reserve aircraft maintenance officer for 12 years.

JOYCE S. REVELLE, born Nov. 18, 1927, in Peoria, IL, learned to fly at 16 in 1944, license in 1945. She married in 1946 and no more flying in early marriage and while raising children. She started to fly again in 1976. Upgraded license in 1984 with instrument rating. She has flown J3 Cub, Cessna 150, 172, 152; Grumman Tiger, Mooney, Beech A36 Bonanza.

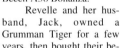

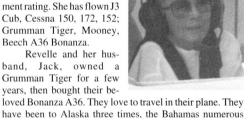

Revelle and her husband, Jack, owned a Grumman Tiger for a few years, then bought their beloved Bonanza A36. They love to travel in their plane. They have been to Alaska three times, the Bahamas numerous times, Cozemel, Belize, the Cayman Islands three times, the Turks and Caicos and out west to ski many times. She has also flown to many of the conventions. In fact when Revelle and Jack go any place in the US they always fly the Bonanza. It is the only way to go. Revelle has two children, Barbara Jo Revelle Sweetman, 49 and John Charles. Revelle, 48. She has three grandchildren: Katie Ann Sweetman, 15; John R. Revelle, 14; and Joanna Revelle, 6.

DOROTHY "DODIE" RIACH nee GLASS, earned her private license in 1991 at the age of 65 so that she could stop worrying about her older son who became a pilot three years earlier. (She has stopped worrying.) For the first time in her life, she learned how to study and refers to her ASEL as a genuine graduate degree on top of the one in science from McGill University. Recently retired from social work, Dodie enjoys her career in the sky along with her son and her Aero Lunch Bunch (four lady pilots). She handles publicity for the North Jersey 99s, is a member of the Mid-Atlantic Pilots Assoc. and flies 172s out of Morristown airport. Once a professional singer (soloist with the Buffalo Philharmonic), she is currently singing in two choral groups, studying piano and floral water color, and playing bridge. Above all, she considers her family of three children and three granddaughters to be her greatest achievement.

MARY JANE RICE, soloed in 1938 in a J-2 Cub from a farm field near Hector, MN, where there was no airport. She joined the original chapter of Minnesota 99s after her private certificate in 1940, then became a charter member of the reorganized Minnesota chapter in 1949. She received her commercial certificate in 1944 flying a Taylorcraft. Since 1945 she has lived the Minnesota aviation scene as the active secretary-treasurer of Willmar Air Service, Inc., the FBO she and her husband founded and operated in Willmar, MN, a Mooney sales and service center since 1956. Her 1,000 plus hours spanned mostly older taildraggers, PTs, BTs, Mooneys, culminating when she and her husband, John L. Rice, were inducted into the 1994 Minnesota Aviation Hall of Fame.

CHRISTINE ANSLOW RICHARD, born April 5, 1951, joined the Tucson Chapter in 1986. She started October 1985 and got her license April 1986 at Tucson International Airport; her SEL April 1986. She has worked with the Flying Club she belongs to organizing a Fun Day event spot landing flour bombing contest. Has been on Tucson 99 Chapter's Treasure Hunt committee, four years now on the Clue committee. You really get to know the territory looking for good clues and its fun to fly with other chapter members. In December 1993 in a 1946 Taylorcraft, she went on a long cross-country with chapter member, Ellen Ausman Franklin from Tucson, AZ, to Little Rock, AR. She has been vice-chairman, chairman, treasurer and hunt committee chairman.

Richard has been married for 12 years to best friend and fellow pilot, Blaise Richard. Blaise was the person who inspired her to learn to fly even though she had never gone up with him. His knowledge of weather and map reading had always impressed her when she asked him about it; he said his pilot training gave him the knowledge to do these things.

Richard and her husband both fly. He has been working three-and-a-half years on his A&P rating and will soon take his tests. In the future she may get involved with her A&P rating but for now enjoys VFR with club members and 99s.

LORRAINE RICHARD, born Oct. 23, 1933, in Revere, MA, started flight training in July 1953 and got her private license in November 1953, commercial November 1956, limited flight instructor, July 1958, multi-engine August 1961, CFI-A August 1968, instrument November 1978, Part 135 certificate aero commander 560 in August 1977 all in Nashua, NH. She has flown over three dozen different type ASEL and MEL with more than 5,000 hours. She has been a free-lance flight instructor for 20 years.

Owned a 1951 Mooney Mite for 23 years (1963-1986). Flew it to EAA Oshkosh three times in the 1970s (won best Mooney Mite award in 1972). Appointed FAA accident prevention counselor in 1979 till present (1995). As a 99, she was vice-chairman of Northern NE Chapter in 1978. Currently, she is chairman of this chapter along with chairing this chapter's AE Scholarship Program. Has been AE Scholarship Program chairman for New England Section for six years (1988-94). Coordinator of the last All Women New England Air Race held in Manchester, NH, in 1975. Was chairman of the 1979 New England Air Race flown out of Manchester, NH.

She received the New England Section Aviation Honor Award in 1986 for the above activities along with being co-chairman of her chapter's Girl Scout Career Day held from 1984 to 1993 in Manchester and Lebanon, NH.

LINDA STEPHENS RICHARDS, born Sept. 18, 1959, joined the 99s in 1990 after receiving flight training at Twinkletown Airport, Walls, MS. She has flown Cessna 172 (her husband and she own N61780).

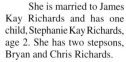

She is married to James Kay Richards and has one child, Stephanie Kay Richards, age 2. She has two stepsons, Bryan and Chris Richards.

Favorite places to fly: Sugar Loaf Key in the Florida Keys and Gaston's in Lakeview, AR.

JOSEPHINE IVETICH RICHARDSON, at age 80 of rural Decatur, has been inducted into a flying hall of fame.

Richardson a pilot for 54 years and has operated airports in the Decatur area for decades, was present when her name was added to Memory Lane in the International Forest of Friendship. She was born on March 15, 1915 in Chicago. Her brother, James, opened his airstrip (in Decatur), which Josephine helped run. She soloed on Aug. 2, 1940 in a J-3 Cub after only seven hours and 20 minutes of dual (flight) with her instructor, John Wright, a relative of the Wright brothers.

She now holds a commercial pilot certificate, with airplane, single-engine and helicopter ratings. In 1955, she bought 20 acres just outside Decatur, IN, and built her grass strip. She married Wes Richardson in 1959. He was a strong supporter of her flying, as well as helping in her reflexology and massage therapy business. He helped her buy her 1959 Cessna 172 in 1960, which she still owns. He died in 1994.

Richardson participated in a Powder Puff Derby and won the Indiana FAIR Race. She also participated in commemorative air mail flights. She is a life member of the 99s, the OX5 Club and a charter member of the Three Rivers 99s (of Fort Wayne). She received four patents; wrote a book about the first pilot in Adams County, IN, and painted pictures. Richards on is a carpenter, stone mason and massage therapist. She continues to manage, maintain and operate Decatur Hi-Way Airport.

HELEN RICHEY, (1909-1947) was a tomboy who ran away from home at age 12 but was promptly brought home again. Her father, a school superintendent in McKeesport, PA, bought her an open cockpit bi-plane, a Bird, in 1931. She graduated from ground school in 1929, and in the 20s and 30s was in many national air shows, at her best at aerobatics. But she early decided not to be a stunt flier. She earned transport and limited commercial licenses and became a pilot's pilot, at the stick with such passengers as Amelia Earhart and Clarence Chamberlain.

In 1933 with Frances Marsalis, she won a world's endurance refueling record for women, nine days, 21 hours and 50 minutes. She also held women's altitude, speed and endurance records. In 1935 she became the first woman to be hired by an airline as co-pilot, flying passengers, mail and freight for Pennsylvania Air Lines between Detroit and Washington, DC. Less than a year later she was forced out by bitter objections of the male pilots.

In 1936 and 1937 she worked with three other early members of 99s in a government airmarking job. She was the first woman to get an instructor's license, in 1940 from the CAA; she taught in Philadelphia, Boston and Los Angeles. She was the only woman assigned to train military pilots at the Pittsburgh Butler airport. In 1942-43, Helen became commandant of the American Wing of the British Air Transport Auxiliary, and back in the states in 1943-44, worked with the WASPs. After the war, she was despondent over not being able to find a job in aviation and took her own life.

LAURA JANE QUILLIN RICHTER, born Jan. 16, 1934, in St. Louis, MO, began flying in 1969 in St. Louis, MO, and became a 99 in 1970. She has flown the Cherokee 140, Cessna 150 and Apache. After doing the Powder Puff Derby stop in St. Louis in 1971, she moved to San Antonio, TX, and transferred membership to the San Antonio Chapter. She has held all chapter offices and chairman for four years, plus committee chairmanships, and has been South Central Section public relations and nomination committee chairman.

Addressed national business women's groups on air age education and attended National Congress on Aerospace Education. Helped with the Angel Derby Race, completed physiological training at Brooks Air Force Base, sponsored local and regional proficiency flight contests, judged US Proficiency Flight Contests and was a candidate for the NASA Teacher in Space Program.

As a teacher, Richter presented aviation programs for Career Week, sponsored an aviation club which visited the

Confederate Air Force Airshow and is actively involved with Air Bear presentations at the elementary school. She has six children: Gail, 39; Mark, 37; Scott, 35; Eugene, 33; Laura, 27; and Paul, 25. She has nine grandchildren: Jennifer, 17; Aaron, 13; Jessica, 11; Mark, 12; Chris, 12; Alisha, 10; Kevin, 8; Leanora, 3; and Ryan, 1.

ELYNORE RICKABAUGH, born on a farm in South Dakota, she did not become acquainted with airplanes until after they moved to their California home. She played the piano in the Philharmonic Band in the fifth grade and the drums in high school. After a career of photographic modeling, married and then they bought their first airplane.

Flying lessons, a Christmas gift from her husband after their twin daughters were grown and now a grandmother, started her adventure of flying in 1969. She joined the 99s and has held chapter offices of treasurer, membership and chairman.

Rickabaugh has flown in many air races and attended many Southwest Sections and Internationals.

Besides being an aircraft broker, Rickabaugh and her husband owned and operated an aircraft cabinetry interior business which she continued operating after his death. Flying taildraggers to V-Tail Bonanza and *Dimples,* her Cessna 172, is an integral part of her life and she is grateful to experience the joys of flying.

LAURA RICKS, born April 19, 1936, in Santa Monica, CA, joined the 99s in 1987, after she had received flight training at Van Nuys Airport in 1986 in a C-172. She has flown 152 (spins), 172, 172RG and T-210.

She has been married to Jack Ricks for 40 years and has three children: two sons (deceased) and one daughter, Danielle. She has three grandchildren: two in Australia, Cody and Savannah; one in Tarzana, Alexa.

She and her husband have flown their Cessna T-120 from the Caribbean to the Arctic Circle and back again. They have rented an A-210 in Australia and flown the perimeter of Queensland and the Great Barrier Reef. She has flown the Palms to Pines three times. The Salinas His and Hers, the Tucson Treasure Hunt and the Valley Air Derby (as chase).

Ricks has a BA in psychology from California State University Northridge and an MA in human factors engineering from the same school, which she earned while raising two children. She was employed by Kaiser Permanent Medical Care Program for 17 years as a management engineer. Now retired, she hopes to fly more often. As a member of the San Francisco Valley 99s she has been membership chairman, airmarking chairman, special events chairman, and assisted others with numerous fund raising events. She has also been awarded the Trixie Anne Schubert Memorial Service Award in 1992 (second place) and the Woman Pilot of the Year 1994. (San Francisco Valley Chapter).

MARGARET C. RIDDLE, holds an MA from the University of North Carolina at Chapel Hill. She taught history at the college and university level for 10 years, worked as director of the Governor's Policy Office for the state of North Carolina, and only began to fly in 1985.

She possesses the airline transport pilot certificate with the multi-engine rating and a commercial certificate with single-engine and multi-engine ratings. She is a Gold Seal certified flight instructor, multi-engine flight instructor, instrument flight instructor and holds the advanced ground and instrument ground instructor certificates. Riddle currently serves as a federal aviation administration safety program counselor and as chief flight instructor for the Part 141 program at First in Flight Aviation, Franklin County Airport, Louisberg, NC.

Margaret joined the 99s in 1993 and currently serves as chairman of the Kitty Hawk Chapter in North Carolina. Her husband, John, teaches at NC State University and they have a daughter, Erika.

SONDRA RIDGEWAY, a charter member of the Northeast Kansas Chapter since March 1967, began flying in a Cessna 172 in 1963 at her husband's insistence "for safety's sake" when they were both in the airplane. With 58 hours and several months later she was a private pilot. Air racing consumed them during the 1960s and 1970s. Their vacation time was spent participating in weekend proficiency races.

During the early 1970s, she upgraded her skills to commercial and instrument and changed airplanes to a Bellanca Super Viking. She worked full-time on the refueling crew at Philip Billard Airport and Jayhawk Aviation as a CFI while finishing CFII.

In 1975 a dream came true when Ridgeway flew the Powder Puff Derby air race from Riverside, CA, to Boyne Mountain, MI, in a Cessna 172.

In 1976 they purchased a Piper Apache. She added multi-engine and multi-engine instructors to her credentials and their Part 135 charter company, Blue Sky of Topeka opened for business in April. They are currently flying a Beechcraft Bonanza A36.

In the ensuing years she has acquired her ATP, flown Piper Navajos for a commuter airline in Arkansas and a short stint with the FAA at the Little Rock FSDO.

EUDORA ALETHA RIEMERS, born June 9, 1941, in Gary, IN, joined the 99s in 1990. She had her first plane ride when she joined the Civil Air Patrol at 15 in a Stinson. She later joined the Air Force and took hops.

Her ex-husband was a pilot and they owned a Tri-Pacer and a BD-5 (never finished). Despite flying observer and doing search and rescue with Civil Air Patrol for over 21 years, she never met a woman pilot until she was over 30 years old. She never thought an ordinary woman could fly. She now goes to schools for aerospace education to do career days and teaches workshops through the 99s.

She continues to do and teach search and rescue, is a Mentor pilot through AOPA, and flies Young Eagles through EAA. One of her greatest joys has been going up alone and knowing that she is a part of the great space adventure.

MARY SCHULTE RING, born Nov. 27, 1913, in Wallace, WV, is a West Virginia hillbilly who has had a thrilling life, mostly because of aviation. Learned to fly in 1943 at Sky Harbor School of Aviation. Received a commercial license with all instructors ratings. She has flown the 40 hp S-1 Piper Club, Luscombe, Taylorcraft, Cessna, Tri-Pacer, Piper Clipper, PT-19 Fairchild, PT-22 Ryan, PT-23 Steerman, PT-26 Fairchild, BT-13 Vultee, AT-6 No. American and Pitts S-2.

She instructed at Sky Harbor in Phoenix and flew search and rescue in the southwest. Owned a B1-13 which she and her husband and her 9-month-old son, she flew from San Diego to Cleveland air races in 1948. Flew a Steerman over San Diego with a neon sign wired to rack on the lower wing for advertising. Flew in several AWTAR races. Was chairman of SD Chapter of 99s in 1950 when the race started in South Dakota and ended in Miami. Ring is a personal pilot for Main Aircraft.

Forced landing: throttle cable broke on final approach. Veered to left and bounced the plane in a vacant parking lot and jumped the fence . . . a thriller. Hitchhiked with the Air Force from west coast to east and back. Was allowed to sit in pilot's seat in B24, B25, B17 and others.

She and her husband, Roger, started an aircraft manufacturing company where they made parts for all planes beginning with the DC-3 to the 757. This was a 25 year project. Now it is 1995. Ring is 81 and flying this Pitts S-2, soaring like an eagle. She has one child, Roger Ward Ring (deceased) and two grandchildren, Jason Ward Ring, 24 and Roger Royce Ring, 14.

MARGARET J. RINGENBERG, born June 17, 1921, took her first airplane ride at the age of 7 from a farmer's field in rural Indiana. She was called to be a WASP in 1943, became a flight instructor in 1945, and has flown as a commercial pilot ever since. She has been racing since 1957 in various races including the Powder Puff Derby, Classic Air Race, Grand Prix, Great Southern, Denver Mile High, Illi-Nine Air Race, Kentucky Air Derby, Indiana FAIR, Michigan Small Race and others. She has won numerous times and has more than 150 trophies. From the time

she soloed in 1941 until she completed the 'Round the World Air Races in 1994, at the age of 72, she has spent 40,000 hours in the air. She is married to banker, Morris Ringenberg. They have two children, Marsha and Michael, and five grandchildren: Jon Wright, 19; Joe Wright, 17; Joshua, 15; Jairus, 14; and Joala, 11. She has been active in Girl Scouting, church activities, museum work, and has been a proud member of the 99s since 1945. She is currently flying, speaking about her experiences, and cooperating in a book that is being written about her.

She has been inducted into the Forest of Friendship, Sagamore of Wabash (governor of Indiana highest award), Air Race Classic, WASP, Women Military Aviators, Inc., Greater Fort Wayne Aviation Museum (board member), EAA Chapter II, the Women's Memorial (charter member), Women in Aviation International, Silver Wings (gold) and worked with Girl Scouts many years.

LYNN D. RINGER, born June 22, 1939, in Fargo, ND, learned to fly just before her 50th birthday. That birthday was particularly meaningful as she had survived a serious ovarian cancer earlier. Celebrating with a new SEL rating and a climb up the Grand Teton in Wyoming made for a significant life passage. She has flown the Cessna 150, 152, 172 and the Piper Warriors.

Her background as a medical technologist and her experience with cancer led her into cancer patient work with hospitals and the American Cancer Society. She is the co-founder of an international patient service program of the ACS known as CanSurmount.

Lynn and her husband, Al, are active hikers and skiers. She is also a violinist with the Boulder Philharmonic Orchestra.

Lynn has participated as Aerospace chairman for the Colorado Chapter of 99s and particularly enjoys working with the Air Bear program. She hopes to soon obtain her instrument and commercial ratings and continue flying "just for the joy of it."

LYNN RIPPELMEYER, her love of flying has taken her from flight attendant to captain from a J-3 Cub on floats to a B-747. Flying began as a hobby, but turned into a career when Lynn became one of the first women in the US to be hired by a major airline in the 1970s.

Lynn has flown for Air Illinois, TWA, Seaboard World, Flying Tigers, People Express and Continental Airlines. Her aircraft have included a DHC-6, B-727, B-737 and B-747. She was type-rated and flew captain on the three Boeing aircraft. Lynn was part of the first all-female airline crew (1977), the first woman to fly the Jumbo Jet B-747 (1980), one of the first women in the US to earn her captain's stripes (1982) and the first woman to captain the B-747 across the Atlantic (1984).

A member of the 99s, she has been honored with placement in the Forest of Friendship. Lynn lives in California with her husband and four children.

DORIS MCPHERSON RITCHEY, born Dec. 6, 1928, in Pueblo, CO. Her "first flight" was for her 16th birthday. She graduated from Stephens College, Columbia, MO, with her commercial instructor rating and "Pilot of the Year 1948." Back home in Pueblo, Doris instructed a women's flying club.

After marrying Frank Ritchey in 1949, they lived in Denver, Wichita, Yuma and settled in San Diego in 1960. Their family includes: Cheryl Black, Wycliffe Bible Translator Missionary; Keith Ritchey, lawyer and Glen Ritchey, contractor. Their nine grandchildren's ages are 3 through 17.

Doris graduated from San Diego State University and began teaching in 1967. Honors include Cajon Valley School District "Teacher of the Year" in 1980 and 1989 for creating aerospace lessons for students and teacher workshops. She was International 99s Aerospace educator of 1988 and installed in the Forest of Friendship in 1990.

Doris owned a Stinson Voyager and a Cessna 182, and has flown 29 different aircraft including a Navy TV-2 (T-33) jet. She raced the Powder Puff Derby 1975 and 1976, Palms to Pines 1973 and 1978, and raced or timed Pacific Air Race from 1971-1990.

Major Ritchey has flown for Civil Air Patrol since 1952 in Colorado, Kansas, Arizona and California. She presents aerospace workshops at CAP National and Regional conferences.

Although retired after 27 years classroom teaching, she continues teaching summers College for Kids, and Young Astronauts at Reuben Fleet Space Theater and airport field trips for students. After 50 years Doris Ritchey still loves to fly.

JANE JENNINGS ROACH, earned her private pilot license on Oct. 4, 1979. She is a charter member of the Oregon Pines Chapter of 99s. Served as chairman, treasurer and membership chairman. Wrote monthly article, "The Left Bank," for *Val Airscripts.*

She has flown the Palms to Pines Air Race, Best Time First Time in Race, 1981; Best Score Team under 500 Flying Hours, 1982. Fourth of July Parade with the 99s and plane on float. Oregon Pines Chapter members presented seedlings from Forest of Friendship to Oregon governor, Victor Atiyeh, August 1983. Flew own plane (alternating with husband) to International Convention, Anchorage, AK, August 1984 and to numerous Northwest Section meetings. Flew in International Fly Over to commemorate 50th anniversary of Amelia Earhart's loss, July 2, 1987. Flew VHLBS, Cessna 172, March 1990; flew VHUEQ Tiger Moth, (the plane had been in the Ansett Air Race, 1961) in Australia, March 1990. Flew biennial Oregon Air Tour, summer 1988. Flew Russian mountain climber to Pacific Ocean during era of *peristroika*, Sept. 13, 1990. Benefit flights and introductory flights include Oregon Museum of Science and Industry benefit, Aug. 23, 1988; Salem Airport Day benefit flights for YWCA in 1983-88, 1990-92. Introductory flights for sixth graders for five years; for Chemeketa Community College students in English as a second language class, May 1994.

Flown to three National Cardinal Club Conventions in Reno, NV; Bend, OR; and Boyne Mountain, MI.

Past and present memberships: Oregon Pilots Assoc.; secretary, Polk County Chapter; secretary/treasurer, Experimental Aircraft Assoc. chapter; member, AOPA; member, AWPA.

KAY MARIE ROAM, born in Minneapolis, MN, on Nov. 12, 1938, to Lawrence and Marie Doherty. One brother, Dennis. Graduated from the University of Minnesota, June 1960, with a major in chemistry. Married Gary Roam on Sept. 17, 1960. Three sons: Mike, 1961; Karl, 1962; and Dan, 1964. Divorced in 1985. Lived in several locations throughout the Western US.

Learned to fly in 1978 in Billings, MT. In 1981 became a specialist at the Billing Flight Service Station, and earned the instrument rating. Transferred to Oklahoma City in 1985 to be an instructor at the FAA Academy. Became very active in Civil Air Patrol and 99s and took postgraduate classes in psychology at Oklahoma City University. Earned commercial license and CFI and CFII ratings.

Transferred to Prescott, AZ FSS in 1989 and purchased 1969 C-172 *Red Roamer.* Involved in planning for first Women in Aviation conference and has attended all subsequent conferences, speaking at several. Started traveling about to make her third trip to Moscow, Russia to visit son, Dan, who owns a business there and his French wife, Isabel. Went to France for their wedding. Karl is married to Frankie Malamud. They have one son, Dan, and have master's degrees in geography from UC Berkeley and are working on their doctorates. Mike is computer teacher and math department chair at St. Ann's School in New York City. Have also toured Turkey and traveled extensively in the US.

ALICE ROBERTS, born May 14, 1918, in Bisbee, AZ, joined the 99s in 1953. Received a commercial license with instrument rating. Started flying in 1953 at her husband's insistence. Joined the 99s in 1953 and immediately entered the AWTAR. Flew 14 Powder Puff Derbies, one Angel Derby and numerous small races. Held all chapter, section and International offices (except Secy.) and was 99s president 1965-67. Served on the AE Scholarship Fund as permanent trustee. Retired with more than 3,000 hours. Winner of AWTAR in 1957. She has flown the Piper Tri-Pacer, Piper Cherokee and Beach Bonanza.

Roberts is married to Charles and has two children, Charles Roberts and Barbara Roberts Pine. She has five grandchildren and seven great-grandchildren

CAROL ROBERTS, born in Alhambra, CA, started flying in 1989 and joined the 99s in December 1992.

She has a private pilot license, and high performance airplane certification.

She has flown the C-150, C-172, C-182, Warrior and Archer.

She is married to Eugene Roberts and she has two children, Jennifer and Rebecca.

DIANE C. ROBERTS, born on Sept. 24, 1956, in Laconia, NH, joined the 99s in 1986 after learning to fly in Southwest Colorado at Cortez Airport, Cortez, CO.

Roberts moved west to attend college in the Rocky Mountains where there is lots of snow. While at school she went on a Penny-a-Pound ride, and she was hooked, the rest is history. She has traveled with the construction industry, and has enjoyed flying at many a small airport. She got a BS in 1978 at Fort Lewis College, Durango, CO, and is now a training coordinator with Granite Construction Co. in Tucson, AZ.

SHIRLEY BROWNFIELD ROBERTS, a native of Coshocton, OH, was first in a small plane in June 1967 and immediately became addicted to flying. She began flight training on Thanksgiving weekend that year at the privately-owned Strongsville Airport (now gone) near Cleveland, OH. First solo came on a brilliantly sunbathed 3 degree F. day, Jan. 5, 1968. She received her private pilot certificate, Aug. 25, 1968, and became a member of the 99s All Ohio Chapter that November. She obtained an instrument rating on July 2, 1970.

Shirley is a civil engineering graduate of Ohio Northern University and a registered professional engineer in many states. During her nearly 15 year FAA airports division career, she and her two sons, Louis and Matthew Szabo, moved to Texas where in mid-1973 she purchased the first of three Beechcraft Bonanzas she has owned. She was a member of the 99s Houston Chapter for two years and has been a member of the Golden Triangle Chapter since moving to her present home in Colleyville, TX, in 1975. Over the years, Shirley has used her aircraft for business and pleasure and often says, "It's a pleasure to do business by flying!"

She considered her ability to fly and own an aircraft an asset to her firm, Aviation Alliance, Inc., which performs airport planning, design, construction management and other airport and airspace consultancy. Her focus is on faith, family, flying, friends, fitness and fun – which includes her work.

STEPHENIE RANDOLPH ROBERTS, born Jan. 21, 1953, in Hobbs, NM. Stephenie attended public schools in Hobbs and graduated from Texas Tech University in 1975 with a BS in special education. She also attended the Judevine School for Infantile Autism and has done postgraduate studies at the University of Texas at Tyler.

She is married to Randall L. Roberts, an attorney and has resided in Tyler, TX, for the last 20 years. Stephenie taught school in the Tyler Independent School District before beginning her flight training. Upon receiving her private pilot certificate, Stephenie went on to obtain her instrument rating, commercial rating, instrument instructor, multi-engine and multi-engine instructor rating.

In 1989 Governor William Clements appointed Stephenie to serve on the Texas Department of Aviation Board. Presently Stephenie is serving on the Tyler Airport Advisory Board and is past chairman of the aviation committee of Tyler Area Chamber of Commerce. Stephenie is also a director for the 1996 World Precision Flight Competition. She is a member and officer of the Texas Dogwood Chapter 99s, National Assoc. for Flight Instructors, AOPA, Texas Assoc. of General Aviation, National Assoc. of Women in Aviation, charter member of the National Council for Women in Aviation/Aerospace, member of the Tyler Airport master plan committee, owner and operator of Flights of Fancy (an aviation jewelry company) and part-time airshow announcer at various airshows in Texas.

In addition to Stephenie's involvement in the field of aviation, she has been active in community and charitable events in east Texas for a number of years. She presently serves on the board of directors of the Tyler Area Chamber of Commerce, is a past board member of the Junior League of Tyler, Inc., the Texas Rose Festival, Eisenhower International Golf Classic, past president and public relations director of many different organizations in years past and past participant of Leadership Texas. Stephenie was recently awarded the coveted 1994 Gertrude Windsor Award for Outstanding Volunteer Service by the Junior League of Tyler, Inc.

She is also serving in an advisory capacity with her flying partner Dan Shieldes, co-owner of Jazz Air, a new regional start-up jet carrier based at Fort Worth Meacham Airport.

ALICE ROBERTSON, learned to fly in 1979 at Palo Alto, CA. Although she always wanted to fly, Alice gives credit to Janet Hitt for actually getting Alice out and flying. She now has her instrument rating, ASEL and has logged 750 hours. She has flown her Bonanza A36 to Mexico and throughout the Southwest. Alice joined the Santa Clara Valley 99s in 1979 and has been active in airmarking and aerospace education. She owns an aviation insurance company and her goals include getting her commercial and working in aerial photography.

DORIS B. ROBERTSON, born March 25, 1928, in Detroit, MI, joined the 99s in 1970. She has participated in the Pacific Air Race. In 1969 she was the fortunate recipient of a scholarship towards a private pilot license sponsored by the Los Angeles Chapter of 99s and received her license in January 1970. Has flown a Cessna 150 and 172.

LORENE (WRAY) ROBERTSON, born April 19, 1915, in Camargo, OK, with a desire to fly from age 7. Entered nursing training after high school, traveled with a dance band as a vocalist for one year. Soloed in 1945 in one week at Redding, CA. Became a private pilot and joined the 99s in 1946 and has remained a loyal member since joining.

Was a member of Civil Air Patrol for many years, is a licensed parachute technician and rigger. Her career was as a civilian, working for the Navy and Naval Air Stations in San Diego and Alameda, CA, as a quality control specialist in aircraft, instrument, parachutes and avionics maintenance, retiring after 30 years of service. She currently has a commercial pilot certificate with instrument and flight instructor ratings with more than 3,000 hours. She wrote the regulations for and started the Hayward-Las Vegas Air Race in the 60s which is still a popular race. Has participated as a pilot in five Powder Puff Derbies, five Hayward-Las Vegas races, and as a co-pilot in two Air Race Classics, six Palm to Pines races and numerous local flying competitions. Is a member of AOPA, 99s Inc., National Assoc. of Flight Instructors and National Aeronautic Assoc. Married Gerald F. Robertson who is also a pilot and a licensed aircraft mechanic. Has owned a Waco Taper Wing, Aeronca Chief, Cessna 120, Stearman, Waco UIC, Cessna 180, Cherokee 180 and a Cessna Cardinal. Has never been without an airplane since 1945. She and her husband are currently flying the Cardinal and live in Dayton Nevada Airpark. They recently moved from Cameron Air Park where they lived for 17 years. Wray was instrumental in starting the Cameron Park Chapter of the 99s and was their first chapter chairman.

BETH A. ROBINSON, received private pilot license, Oct. 14, 1994, began her dream to be a pilot at age 8 but knew no pilots, had no access to an airport, and her desire to fly was not taken seriously. Thus, her dream went underground. Then came college, marriage, three sons, employment, sons off to college and then back to school for her Ph.D. Life seemed to be charting an earthbound path. Then fate stepped in. Robinson met a woman her age who was a new pilot, and the dream was born again. Her husband, as always, was supportive and life as they had known it began a new journey. At age 48, forty years after the dream was born, the dream came true. Life will never be the same . . . now to share it with others, to open up the "high" life to more women.

DOROTHY CALLAHAN ROBINSON, born in Blue Ridge, VA, began flying in 1982, in Bakersfield, CA. Joining the 99s and logging many hours in a Cessna 182, she moved to an A36 Bonanza. After two years and needing new challenges, she bought her own "taildraggers" – an Aeronca Champ and a Cessna 170 which she flies today.

During four years as chapter chairman she hosted a Southwest Section Meeting and Air Race Classic Start. Next came oral history co-chairman for International and Southwest Section 99s – positions she still holds.

Dorothy is an active FAA safety counselor and Region II vice-president of California Pilots Assoc. Among numerous memberships in aviation organizations are AOPA, NCWA, National Aeronca Assoc., International Cessna 170 Assoc. and Minter Field Air Museum.

Educational background includes a BA, Roanoke College; MA, Vanderbilt University; postgraduate work, UCSB and CSUB. Professional organizations: NEA, CTA, Phi Delta Kappa; non-professional/community: SPCA, Audubon Society. Married in 1967 to Frederic W. Robinson, but widowed in 1979, Dorothy has two stepchildren, seven grandchildren and several great-grandchildren.

DUANA M. ROBINSON, Mission Bay Chapter – Southwest Section, currently flies 737s as first officer and will be training for 757/767 domestic and international flights. Her father's airline experience as captain 31 years, kindled her love of flying. She soloed July 1973 in a C-150; was a charter member and chairperson in Huston North 99s; transferred to Mission Bay Chapter.

Robinson was the youngest woman first officer on a DC9 in 1978. She received a BS in aeronautical studies and was a CFI and commuter pilot at age 20. Rating in Citation CE-500; first officer in MD-80.

Vacation flying has been flying in Africa, flying a C-402 with her father, who is now with Mission Aviation Fellowship in Kenya. Areas have included Mali, Banako, Timbuktu, Niger River and west border of Mauri. Activities include ISA+21 (Women Airline Pilots), Newcomer's Club and singing bass with the Sweet Adelines who have won medal status in international competition. She also enjoys reading, tennis, handicrafts and community volunteer work.

RUTH ROCKCASTLE, was born on Nov. 6, 1936, in Uxbridge, MA. She completed nurses training at Jackson Memorial Hospital in Florida, a BS in healthcare and administration and a master's in healthcare administration, Webster University.

Ruth learned to fly in 1981 at Glennview Naval Air Station and was her husband's first student. She joined the 99s Chicago Area Chapter in 1982 and immediately became involved in the leadership of the chapter. She served as secretary and chairman and became involved in the Air Bear program during her chairmanship and from 1986-1991, during which time assisted in presenting the program to over 5,000 children.

Ruth is director of emergency services at Leesburg Regional Medical Center in Leesburg, FL, and lives in Oxford, FL, with her pilot and flight instructor husband, "Rock." She has one son, two stepdaughters, one grandson and a granddaughter.

PATRICIA GLAAB ROCKWELL, born to fly! Flushing Airport, NY, 1953 in J-3 Cub; soloed within two months, 1954 PPL, 1955 SES. Marriage and three sons then a commercial 1974. IFR 1977; CFII 1981; CFI 1982; various local air races and contests. Published writer of short stories and aviation newsletters LI Chapter news reporter (nine years) for *The 99 News.*

Current member of the Long Island 99s, Silver Wings Fraternity, Aviation Space Writers Assoc., Antique Airplane Club, Early Flyers, EAA, AOPA, canoe and ski clubs. Former member of Seaplane Pilots Assoc. and Taildraggers Assoc.

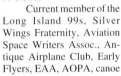

99 Chapter: Served on most committees from hostess to chairman. Most proud of forming Annual Flight Club Day to fly children in Ida Van Smith Flight Clubs. Also, reinstated 66 Program with new SOPs and format, creating a successful, long running student program. Both programs are still in effect today, years later.

Section: Collected information and original documents for New York/New Jersey Section history. International: Won Amelia Earhart Memorial Career Scholarship in 1980 to help earn CFII.

Personal feeling of accomplishment: Helped young woman to overcome her fear of flying. She eventually used this help to fly cross-country (commercial) to attend the birth of the baby girl she would later adopt, after again traveling in a commercial airplane.

ANN RODEWALD, born Feb. 20, 1950, in Fairbanks, AK, her love of flying started soon after she was born. Growing up, she frequently flew with her Air Force career pilot father in a wide variety of planes from certified to experimentals from grass strips to airfields that are now closed. She finished ground school and passed the written exam during high school while learning to fly a Mooney. Many years later, while living in California, she came back to flying when lessons came as a Christmas gift. She earned her private pilot license in 1990 and her commercial in 1994. She is very active in the 99s and has participated in Right Seat Seminars.

Ann has a master's degree in textiles combined with an eclectic work background and diverse knowledge of a wide range of subjects. She currently works with computers doing research and project management and is rebuilding an antique airplane.

Ann has flown the following airplanes: Cessna 150, 152, 172, 172RG, 182, Citabria, Funk, Globe Swift, Helio Courier 395, Mooney 20D; Piper Cherokee, Warrior, Archer, Arrow and Comanche; Breezy, Defiant, Lancair 235, Long Eze and VeriEze.

ROSEMARY L. RODEWALD, her love of flying started in 1943 during WWII when she became a riveter for Northwest Airlines in St. Paul, MN. When the war ended, she married her high school sweetheart, Warren A. "Rodie" Rodewald, a US Air Force career pilot. Together, they invested their life savings and bought a PT-26 in which Rosemary started flying lessons. Years later, she continued lessons in a Mooney 20D and a Breezy. Built by Rodie, the Breezy is similar to a 1912 Curtis Pusher.

Shortly thereafter, the Rodewalds moved to Hawaii where Rosemary continued her flying in a Cessna 172, a rebuilt Helio Courier 395, an experimental VeriEze and Long Eze, then earned her private pilot license in 1982 in a Cessna 150. After getting her certificate, she and Rodie began building a Rutan experimental Defiant, a four place, twin-engine, fiberglass airplane. It took three years working full-time to complete. First flight was Aug. 3, 1987. Currently they are building a Questair Venture. This project once again puts Rosemary on the operative end of the rivet gun.

In addition to a deep love of flying and building experimental aircraft, Rosemary, is a psychologist, an author, gourmet cook and quilt maker. She and Rodie have three children: Jill, Ann and Thomas and three grandchildren: Kaylan, Peter and Dana. Daughter Ann is an avid 99, flying enthusiast and holds a commercial certificate.

NANCY RODGERS, got her license in 1973 and joined the SCV 99s in 1974. She loves to race, having flown the AWTAR, the Air Race Classic, the PAR (14 times), the Palms to Pines (14 times), the Salinas His-and-Hers (six times), the Fresno 400 (four times) and the Grand Prix. Nancy was one of the instigators of the chapter's Hollister Air Meet and won the Chapter Service Award in 1979 and 1984. She served as chapter secretary, nominating committee chairman, section nominating committee, and is an active airmarker. She keeps the SCV 99 trophy case at the San Jose Jet Center. Nancy has worked for Optima Publications, checking out airports for the *Pilot's Guide to California Airports.* Nancy's interests outside flying include keeping track of her five daughters, (which have each raced with her over the years), driving her 1954 MG and tennis and bridge tournaments. Nancy has logged more than 1,100 hours.

ROBERTA "BOBBI" BURNS ROE, grew up in a home full of stories told by her mother about barnstorming in the 20s and pushing the envelope on a Jenny. These occasional glimpses into her mother's past sparked a desire in Bobbi that took until the 80s to satisfy. As a second generation female flyer, Bobbi holds a special zest for flying unmatched by many of her contemporaries. Her passion guided her to obtain her private license in 1984 and instrument rating in 1985. Since then, she has never stopped searching for new and more challenging aircraft to fly including: seaplanes, twin-engine, gliders, hot air balloons and aerobatics in a homebuilt, as well as, a reporter along for a ride on the Goodyear Blimp.

In the early 90s, an idea was born when another female flyer asked Bobbie what the best magazine was for a female pilot to read if they wanted more information about female role models in aviation? With this question, and five years of full-time work, Bobbi now owns and publishes *Woman Pilot,* a national aviation magazine devoted to promoting and sharing the spirit of flying with women around the country.

Bobbie has been active in the 99s for approximately 10 years as a member in numerous chapters including Iowa, Finger Lakes, NY, Houston, Greater Seattle and Chicago chapters. She is currently on the executive board of directors for Women in Aviation International, as well as, a member of several other aviation associations.

Bobbie has been married to Dave, a distinguished civil engineer, for almost 35 years and has three grown children and one grandchild. Her daughter Susan is employed by the American Medical Assoc. and is a mother to Christopher.

Her son David is an attorney in Michigan, and her son James is an electrical engineer with Kodak in New York.

NANCY ROGERS, made her first flight in a light aircraft in 1967 when she took up the sport of parachuting. After competing in the World Parachuting Championships in Yugoslavia, she turned in her parachute and in 1977 earned her pilot license. Nancy holds an ATP and flight instructor rating. She flies a T-182RG, a Cessna 152 and has logged more than 3,600 hours. Nancy serves her community as a volunteer pilot for Flights for Life and the Flying Samaritans. She is a five time winner of the 99s Kachina (proficiency) Air Rally.

Educated with a BS in business management and a MS in aeronautical science, Nancy has been employed for the past 29 years with Allied-Signal Aerospace Co. as a human resources professional. Aviation related memberships include the 99s, the Arizona Pilot's Assoc. and the National Council for Women in Aviation/Aerospace.

RITA ROGERS, born in Toronto, Canada, had first flight at age 10. Attended Queens University; learned to fly at Toronto Flying Club; soloed in Tiger Moth; flew Canucks, Aeroncas and several other types of aircraft. Continued in California for several years and lived in North Canadian Bush for a few years. Opened flying service at Toronto International Airport, Polaris Aviation, in the 70s with one aircraft, had three more in two years, then bought Markham Airport. Ran a flying school and operated the airport for 10 years. Had 23 aircraft, eight instructors and a staff of 23. Had over 150 tie-downs; was chairman of first Canadian Chapter of 99s in 1983-84; flew in several air races including the Angel Derby, the Antique Air Race from Vancouver, BC, to Toronto, Ontario; flew to Alaska with two other 99s; has flown with the 99s in Arizona, California and Florida.

Married to the Honorable W.F.B. Rogers (retired justice), lives in Canada for six months each year. Has two sons, Stephen, a scientist in California and Mark, an entrepreneur in Toronto, Canada.

SHIRLEY J. ROGERS, Phoenix 99s, was born on the east coast and moved to Phoenix after being married a few years. All went well until her husband was injured and was home all the time. To keep him occupied she started taking him out to airports, as he was really into types and names of aircraft. One thing led to another, and one day he said, "Why don't you learn to fly?" Why not! She got her license in 1972 at Turf Airport, and they bought a Piper Super Cruiser taildragger. They enjoyed this for a while, but decided to get a traveling airplane and invested in a 1959 straight tail C-175. She worked five years to pay for it, and then had time to go flying.

She joined the 99s in 1972, new ticket in hand. She became membership chairman and then, everlasting – flying activities chairman. Phoenix Chapter had many wonderful activities. She started the annual Poker Run, did a couple of progressive dinners and brunches and generally kept the chapter on their wings.

Her personal flying included a trip to Maryland that only got as far as Oklahoma, followed later by three completed trips. The first of these convinced her that an instrument rating would make her much more comfortable in the stuff they call air outside Arizona, so she aced that. She has flown eight Kachina Air Rallies and three ARCs, placing third in the 1995 Air Race Classic along with fellow Phoenix 99 Marty Pearce. She came in fifth in the 1995 Sundance Derby. She has wanted to race for a long time and is now really getting into it!

In spite of having flown them everywhere she hasn't been able to interest her daughter or two grandsons in flying. There is a new great-grandbaby she has her eye on – maybe she can attract this generation.

PAT ROLLINS, born Aug. 7, 1954, learned to fly in Phoenix, AZ, in 1994-95. She received her ASEL on July 23, 1995 and joined the 99s. Her first check-ride will be forever etched in her mind. She has had 10 years in the aviation business with Republic Airlines, American West Airlines and Morris Airlines. She married on Aug. 15, 1995 and she and her husband went through flight training together. She is currently enrolled at the University of Phoenix to receive a BS in business management.

TANNA ROMBERG, Phoenix 99s, born Feb. 21, 1953, got her private license Jan. 3, 1989 at Phoenix Deer Valley Airport, AZ. She joined the 99s shortly thereafter and has served two years as membership chairman.

Tanna really wanted her son to learn to fly, but as he was colorblind this was not an option for the airlines. So, burdened, with this money she has saved for flying lessons, and no student, she chose to use it herself. She is working on her instrument rating and is up to Phase IV of the Wings program.

Tanna missed many of the outings as she worked in the Deer Valley Pilot Shop for the last three years, and this required her presence on weekends, an airport's busiest time. But she has gone to Oshkosh five or six times and knows many people who recognize her and feels right at home. She reports that she belongs to many other groups, flying and non-flying, but feels the friend she has made in 99s are truly special.

BEVERLY ROMERO, born March 19, 1942, in Blairsville, PA. A graduate of Indiana University, she is currently the director of laboratory services at Redding Medical Center, Redding, CA. Bev has a son, Esteban, 23 and a daughter, Carlita, 20.

Beverly received her private pilot license in 1982, her instrument rating in 1986. She owns a Piper Archer II and joined the 99s in 1982. Bev has flown in the Palms to Pines Air Race and the Air Race Classic. Bev created the nine point Compass Rose in 1986, was SW Section membership chairperson for four years, SW Section and International airmarking chairperson from 1986-94, 1988 SW Section Woman Pilot of the Year, chairperson of the 1988 Spring Section Meeting in Redding.

Beverly is an AirLifeLine pilot, a member of the Flying Sheriff's Posse, AOPA, chairperson of the Mount Shasta Chapter, 1985-90. Beverly is also a Soroptomist.

MARY JANE "JANIE" RORICK, age 37, with more than 3,500 flight hours, freight pilot New Mexico Flying Service, Albuquerque, NM. She has ASEL, AMEL, commercial and instrument ratings; CFI, CFII and an AS in aviation flight at Vincennes University, 1990.

It seems pretty dry when stated like, that but flying has been anything but dry for Rorick. She always wanted to fly; her earliest memories include some desire to fly. She started in an airplane on loan from a friend, a 150 named *Vicki,* 383VU. After several years as a private pilot, she went back to school, giving up a career as a dietitian to become a commercial pilot/flight instructor. Rorick bought her own "baby," a 1941 Stinson 10A named *Sam,* NC31537. Flying became more than a hobby, more that a job, it has become a way of life. She flies for her job, she flies for fun and she even dreams about flying. It is her life.

Rorick also scuba dives, skydives and enjoys cross-stitch and bowling. So, flying is the magic glue that holds her life together; it shows her the interconnectedness of her life on this fragile planet. Best wishes and blue skies to all of you out there who share this passion. We are friends, though we have never met.

ANN ROSENBERRY, soloed at 16 and received her private at 17, all under the watchful eye of her instructor father. Her father, Jim Fulkerson, was formerly an FBO at Gillette, WY. Her mother, too, was a pilot. A highlight of Ann's early flying was helping fly a new Cherokee 180 home from the factory at Vero Beach, FL.

Ann has been a member of the Nebraska 99s since 1975 and has been involved in airmarking and the aerospace education poster contests. Today her flying is a hobby with husband Del an enthusiastic passenger. Ann has about 300 hours and a multi-engine rating.

DOROTHY "DOLLY" ROSS, born Dec. 23, 1930, in Michigan, obtained her flight training at Hillsdale, MI, in 1965. She has flown the Cessna 172. She has flown in seven Michigan Small Races and six Indian Fair Races. Won several awards. She also served as scholarship chairman for Michigan Chapter.

Her education includes a BS and teaching certificate from Western Michigan University. She is a member of Sigma Sigma Sigma Sorority.

She is married to George A. Ross, a retired pharmacist, and now a realtor in Naples, FL. She has two children, Betsy Ross, 39 and Ginger Landers, 37. She has three grandchildren: Ross Landers, 10; Ruth Landers, 8; and Gordon Landers, 6.

PHYLLIS SANDARS ROSS, has loved aviation ever since she can remember. First airplane ride at age 17. First lesson in 1951 and joined PanAm as a flight attendant in 1951, and management in 1958 (10 years total). She got a private license in 1968 and joined the Reno Chapter of 99s. Joined Hughes Airwest in 1972 as base manager flight attendant, regional director, south with Republic Airlines (now Northwest) (11 years total).

She earned a BA from the University of Washington. She also enjoys art, custom-painted tiles and oil painting, flying travel, politics, sewing and hiking. She is also a member of the Folded Wings (organization of past and present flight attendants in Reno) and World Wings (former PanAm flight attendants).

Ross is married to DeWitt and has one child, Kristi Drake.

POLLY ROSS, born in Chicago, IL, on July 28, 1952. Her father was a WWII and Korean War fighter pilot and her grandfather flew in WWI. Her father, Bill Ross, started restoring Warbirds in the early 60s and helped found Warbirds of America. He was well known for his Spitfire and P-38, among others. Polly had flying in her blood!

She got her private license in 1972 at Pal-Waukee Airport in Illinois and earned her instrument rating in 1992 in Aspen and Grand Junction, CO.

Polly graduated from Mount Holyoke College in 1974 and has lived in Aspen since 1977. She is a member of Pitkin County Air Rescue, and has helped put on the Aspen Air Show for 11 years. She joined the 99s when Betty Pfister, a former WASP and a 99 since the 1940s, started the Aspen Chapter in 1982. She finally bought a plane in 1988; a 1979 Piper Turbo Dakota. She has since flown it to New England, California, Chicago and all over the West. Her most memorable and exciting adventure was a three-week trip to Alaska with a friend in 1993, flying all the way from Aspen in her Dakota.

Her other flying passion is hang gliding, which she has been doing since 1979. She earned her advanced rating in 1982 and has flown her hang gliders all over the US. Her photo appeared (from a wing tip-mounted camera at 15,000 feet over Telluride, CO) in the 1992 National Hang Gliding calendar.

WADAD "POOTCHIE" ROTZINGER, began flying in 1974 with the Civil Air Patrol. She completed her private pilot certification in 1975 at Wiley Post Airport and joined the Oklahoma Chapter 99s in 1976. In 1977 she received her instrument rating and in 1978 her commercial rating and ground school ratings, both private and advanced. Poochie taught ground school at Central State University, where she was in the safety department from 1978 until her early retirement in 1989.

Poochie's husband, Ed, was her first passenger and has encouraged her flying. She is very active in many flying activities: the 99s, Oklahoma Pilots Assoc. and the Civil Air Patrol. She is the flying activities chairman for the South Central Section of 99s, an accident prevention counselor with the FAA, AOPA member and has completed Wings 11.

Poochie and Ed are proud parents of a daughter, Carol and two wonderful granddaughters, Regan and Cara.

PATRICIA L. ROWE, earned her private license in August 1970, and joined the Santa Barbara Chapter 99s the same year. She served as their chapter chairman for two years. From 1974 to 1988 she was a member of the Santa Clara Valley 99s. During this time she chaired numerous committees and received their Amelia Earhart Service Award in 1978 and the Amelia Earhart Pilot of the Year in 1981.

Rowe has owned her Comanche 260 for 19 years and earned her instrument and commercial rating flying it. Air racing has been a love and a challenging part of her flying; and she has flown the Powder Puff Derby, Pacific Air Races and Palms to Pines Air Races. She is a charter member of the Santa Maria Valley Chapter 99s which she is presently a member. In 1994 Rowe was honored by her chapter to be inducted into the International Forest of Friendship. She has been an active member since 1979 with the International Comanche Society and was elected to serve as the first women president of the International Comanche Society from 1987-88.

Learning to fly has been a wonderful experience and it definitely has changed her life. One of her greatest enjoyments has been the opportunity of sharing flying with other women pilots she has met over these past 25 years.

JUDITH M. ROWLAND, has been a pilot since 1982. She received her instruction from a jet corporate pilot at Elyria City Airport in Elyria, OH, in a C-152. After soloing, she finished her training in her Piper Colt.

They moved up from the Colt into a Piper Cherokee 180 and presently have a Cherokee Arrow II. She and her husband, Bruce, enjoy vacationing with their Arrow and attend the chapter flying activities when ever they can. They also make their annual trip to the EAA Convention in Oshkosh.

In 1989 she joined the Lake Erie Chapter 99s. Unfortunately she was also diagnosed as a diabetic and has become ineligible to obtain the needed medical certificate to be PIC.

Prior to flying, Rowland advised 4-H horsemanship for 10 years and showed her quarter horses in the Central Ohio Circuit. She still owns and enjoys caring for her horse. Since they live in a rural community she can have her horse at home.

She works full-time as a buyer of components used in manufacturing at York International, Elyria Ohio plant. Other interest include golf, bicycle riding and skiing.

FLORENCE E. ROY, an interest in the female airplane mechanic that appeared in the old "Steve Canyon" comic strip was the beginning of Florence Roy's "romance with flying." A dream that began with a 9-year-old girl taking her first airplane ride at the old Marcy Airport grew into a reality of becoming a member of the International Organization of Licensed Women Pilots founded by Amelia Earhart.

In the mid-1970s, she followed her husband's pursuit by joining him in taking flying lessons at the AMA Airport in Hamilton, NY. It was there she realized the excitement of her first solo flight. Obtaining her pilot license was no easy matter due to the lack of any formal instruction program in this area. Florence had to search out her own instructors in order to fulfill all the requirements necessary in the areas of navigation, radio and maneuvers. In 1976 she received her pilot license after successfully completing the required FAA written and oral tests as well as the flight check-ride. Her required long cross-country flight followed the Utica, Wilkes Barre, Batavia route.

The Roys (which include husband, Roger, and two sons, David and Daniel) have owned two planes, a Piper Tri-Pacer and a Piper Cherokee 140. However, over the years, Florence estimates that she has flown in more than 50 different types of aircraft.

Florence enjoys the freedom of flying over to Glens Falls, NY, for dinner or down to Rhinebeck to catch an airshow or out to Champaign, IL, to visit relatives. Added to this is the continuing pleasure of flight over the beautiful Mohawk Valley. But the convenience attached to this hobby is not her primary pleasure – believe it or not, it is the joy and delight she feels when performing the graceful rolls, climbs and turns of aerobatics. Florence truly has a "romance with flying."

RUTH NEVADA MARSHALL RUECKERT, started her aviation education at San Francisco's Polytechnic High, by learning to build airplanes. She started flying July 1929, at Mills Field, San Bruno, in an open-cockpit OX-5. Her flight school gave her a reduced rate. They said if she could learn to fly, the men certainly could. A working girl, Ruth took a street car and then walked to Mills Field in the mornings to fly before going to work. She earned her license Nov. 2, 1929, and her limited commercial on July 6, 1931.

She joined the 99s in April 1930. Ruth helped charter many other 99 chapters. Ruth organized and became first chairman of the Bay Cities Chapter. Her passion for record keeping made her the perfect historian for the National 99s, the section and the chapter. She also served as the 99s national treasurer. She competed in the Powder Puff Derby. She also served her country as a WASP. Through the years, she compiled 25 99 history books which were displayed at the Smithsonian.

Ruth married a pilot, John Fredrick Rueckert, in 1933 and had one daughter, Marsha Ann. Ruth last piloted a Cessna in 1973. She continued to be an active 99 and was an inspiration to all. She died in 1993.

JANE RUEHLE, first flew in Colorado in Pete Bartoe's DeHavilland Tiger Moth. Walt, Jane's husband, a captain with Frontier Airlines, said, "Hang on here, Hon," pointing to the fuselage tubing. The flight was excellent; she had to pry her fingers loose when it ended.

Jane learned to fly in a Piper J-3, became a private pilot in 1975, earned an instrument rating in 1984 and after 1,084 hours is still learning, enjoying a J-3 and a Varga Kachina. In 1977 she flew with Jane Stevens in Stevens' BC-65 Taylorcraft from Palm Springs, CA, to Tampa, FL, on the 30th anniversary of the Powder Puff Derby. She belongs to the Colorado Chapter of the 99s, Antique Airplane Assoc., AOPA and Colorado Pilots Assoc.

Originally from New York State, Jane moved to Colorado in 1965 as a postdoctoral research fellow in immunochemistry. She worked in medical research, and was the administrator of the cancer research program.

GERDA SOMMET RUHE, born in Germany, on Jan. 28, 1933, firstborn of a local dentist Fritz Sommer. She finished college with BA in modern languages in Germany. Went on to continued language studies and attended interpreter's college, 1953-55. She worked as flight attendant for Lufthansa Airlines from 1955-57. In 1958 she married Heinz Ruhe and emigrated to the US with him when he took over his father's wild animal import/export business. In 1961-63, they gave up the animal business, became traveling consultants to zoos for the installation of baby (petting) zoos and for designing open-area safari-type animal parks.

From 1963-65 Ruhe became a US citizen in Southern California, where they operated Jungleland amusement park in Thousand Oaks. In 1965 they bought a home in San Leandro and ran the Oakland Baby Zoo in Knowland Park. In 1974 Ruhe lost her husband through a heart attack and moved back to Germany for a while to recuperate from the loss.

In 1975 she returned home to San Leandro, CA. Started taking flying lessons as grief therapy. After getting a private pilot license continued for commercial/instrument and multi-engine ratings. Also got a ground instructor license.

From 1976-84 joined the Bay Cities 99s. Taught as ground instructor at Sierra Academy of Aeronautics in Oakland while doing her advanced flight training. Took other odd jobs to finance her flying lifestyle. In 1985-86 left teaching to attend college and obtained a paralegal degree. Started working for an attorney but quit to return to ground school teaching.

From 1987 to present, Ruhe is working as promotion/copywriter/administrator assistant for an artist studio and publisher. Freelancing as ground school instructor for adult education night classes and weekend flight seminars.

PAULA RUMBAUGH, private pilot, single-engine land, instrument rating. Paula fulfilled her lifelong dream to fly when, in her early 40s, she began primary flight training. She earned her private pilot license in June 1988, and her instrument rating in December 1990.

Paula joined the Scioto Valley Chapter 99s in July 1988. She served as secretary to the chapter from 1990-92, vice-governor from 1992-94, and is finishing the first year of a two year term as chairman. With fellow 99s Carol Countiss and Amy Yersavich, Paula owns a 1973 Grumman Traveler. To date she has flown in two Buckeye Air Rallies (with her husband, taking the Rookie Award in 1992 and second place in 1994) and three Air Race Classics as navigator.

DOROTHY WETHERALD RUNGELING, born May 12, 1911, in Hamilton, Ontario, Canada. She joined the 99s in 1951 after obtaining a private pilot license in 1949, commercial 1951, instructor rating in 1953, senior commercial in 1954 and airline transport in 1958. She has flown the Aeronca, Fleet Canuck, Piper Cub, Piper PA20, Stinson, Cessna 120, 140, 170, 172, 180, 182; Piper Apache, Piper Comanche, Twin Cessna, Luscombe, Silvaire, Bonanza G35 (her favorite plane), Cessna 310 and Bell G2 helicopter. She joined the 99s in 1951 and has been governor of the Canadian Section during the 1950s.

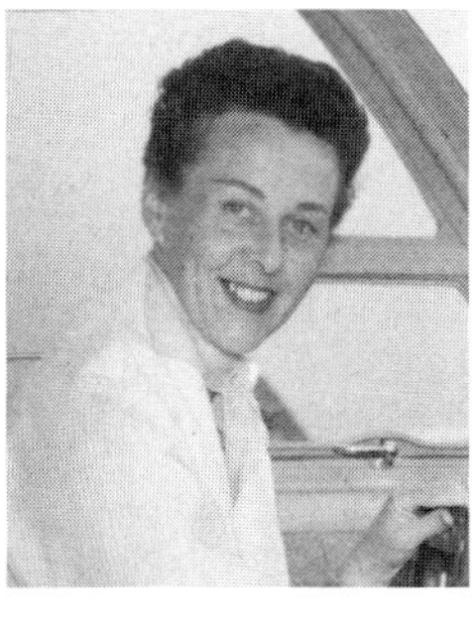

Rungeling is an instructor at Welland Airport. She is married to Charlie Rungeling and has one child, Barry, 52. She has three grandchildren: Charlie, 29; Barry, 28; and Albert 26.

She was the first Canadian woman to solo in a helicopter in 1958. She won the Governor General's Cup Air Race at Toronto Exhibition 1953; won Governor General's Cup Air Race in Toronto, 1956; was the only women in both these races. After many years of trying she succeeded in getting the city of Welland to airmark the city and shortly afterward, a lost pilot gave credit to this marking for saving his life. Flew many times in both the TAR and IAR and was the first Canadian woman to enter the TAR. Wrote aviation stories for *Air Facts*, New York City and several Canadian magazines as well as being an aviation editor for the *Welland Tribune,* a daily newspaper. This column twice won her first prize in a nationwide competition for aviation writers. Her biggest thrills came when planes started taking off at the start of an air race. These races were wonderful and they were a wonderful group of women who flew them. She had to give up flying because of health problems in 1969.

Last Flight: I've flown beneath the moon and stars/And felt the night's sweet kiss/But all the while I wondered/

What's beyond all this!/So when I break Earth's tethers/For that last fantastic flight/I'll soar above the stars and moon/ And look down at the night.

MARION BABE WEYANT RUTH, born Feb. 7, 1918, in Lansing, MI, started flight training in 1934 at Driggs Skylark and flew the E-2 Cub and J5 Waco. She finally soloed on Oct. 27, 1936. She has flown all Aeroncas, Luscombe, Pipers, Taylor Craft, Bonanza, Bellanca, Comanche 180, 260 and all single-engines. She joined the 99s in 1937.

Ruth was the last woman to be trained under CPTP, written in CAA book, *Putt-Putt Air Force;* one of five women trained by Link Aviation Devices in WWII as a Link trainer instrument instructor and flew Aeronca Champ out of two city blocks, She taught Air Force pilots in Link as a civilian instructor in the Air Transport Command. Flew in three Powder Puff Derbies and one International race.

She is married to Dale C. Ruth and has two girls, Dale Lee Oke, age 47 and Kim Carol Hite, 37. She has three grandchildren: Bayo Oke, 9; Sam Oke, 8, and Maria Elizabeth Hite, 3.

MARY SAMUEL HARGIS RUTHERFORD, born in 1938 in Tell City, IN, began flying in 1983 to "slay a personal dragon," a fear of flying. Licensed in 1984, she joined the 99s and fear changed to passion.

She has held all chapter offices and is presently North Central Section vice-governor. In 1988 she received the Amelia Earhart Scholarship Award, is a commercial/instrument pilot and spends most of her time in a Cessna Skyhawk. Activities include flying air derbies, judging NIFA events, acting as Air Race Classic's ground crew.

She is an American Cancer Society board member, instituted Daffodil Days Airlift in Indiana, and in 1991, was awarded the Indiana Quality of Life Award.

Mary Sam is one of the founders of a community health center where she is a nurse practitioner and director of clinical services. She and husband, Charles, a CPA, have two sons, Christopher and Matthew and reside in Newburgh, IN.

HEATHER MARY RYAN, born Aug. 27, 1975, in Tarrytown, NY, has lived in Iran, Kuwait, Kentucky, Virginia and currently resides in Macungie, PA. Heather started her flying career at 14 at Solberg Airport in Solberg, NJ, in a Navy SNJ-5 Texan. She then progressed upward to a 450 hp Boeing Stearman before soloing on May 6, 1992 in a Cessna 152. Heather joined the Keystone Chapter of the 99s while still a student pilot. Upon completing her FAA private pilot checkride on Nov. 27, 1992 and taking on the position of treasurer of the Keystone Chapter of the 99s, Heather progressed rapidly forward and now has her instrument rating from Danville Regional Airport. Heather is currently a junior at Averett College in Danville, VA, where she is majoring in aviation/criminal justice and flying on a daily basis working on her commercial and certified flight instructor certificates in both single and multi-engine and instrument airplane. Heather has also passed her ground school and aircraft portion of the course to fly the Grumman (TBM) Avenger, one of the largest single-engine/single pilot WWII torpedo bombers still flying. Heather's ultimate goal is to become a special agent of the Federal Bureau of Investigation (FBI) in their aviation division and to fly aerobatics like her idol Patty Wagstaff.

She has been treasurer of the Keystone Chapter of the 99s for over two years, secretary of the Averett College flying team, member of the Aero Club of Pennsylvania.

LOUISE SACCHI, started flying in June 1939 at Nassau airport, Long Island, NY. Earned her private in September 1939, got flight instructor and ground instructor ratings in 1940, taught ground subjects at night to Grumman Aircraft Flying Club. Winter of 1941-42, persuaded Merchant Marine School, NYC, that women could take their navigation course.

In May 1942 went to Terrell, TX, to teach advanced navigation to RAF cadets. While there got a commercial on flight test waiver.

In 1944 returned to New York City, worked for then Socony-Vacuum Oil Co. on Navy fuel test program, spent weekends flight instructing at Sussex, NJ. In 1947 got SES rating in Pennsville, NJ. Sold an aircraft to a corporation with herself as the pilot in 1948, so she was first woman corporate pilot.

Sacchi did freelance international ferrying from 1963 until she started her own company Sacchi Air Ferry Enterprises (SAFE) in 1965. As owner/chief pilot of SAFE flew 333 of their 475 deliveries. She delivered 85 Bonanzas, Barons, Kingairs to Spanish government, was named honorary member Spanish air force and was first woman and first foreigner to be awarded Spain's Cruz del Merito Aeronautico Blanco in 1976.

In 1971 set SEL speed record between New York and London and entered London – Victoria Air Race with Bonanza 36. Was the only solo woman, came in second in SEL class and seventh of 53 contestants.

In 1978 eye problem forced retirement. Briefed neophyte ocean flyers, wrote *Ocean Flying – A Pilot's Guide,* McGraw-Hill publisher.

Honored by Beech Aircraft, Aero Club of Pennsylvania, Aero Club of New England, 732nd MASAFR, 99s, Silver Wings, for Distinguished Service to Aviation.

MARINA FRANCES SAETTONE, born Jan. 30, 1966, and raised in the Chicago area. Started flying in 1984 and graduated from Purdue University 1988. Became a member of AOPA in 1987, involved with the ROTC program at Purdue, and became a 99 in 1994. Moved from Chicago to San Francisco in 1994. Flown primarily Piper aircraft. Exposed to the passion to fly by father Jerry a veteran USAF pilot.

Future endeavors: helicopter, charter, aerobatics and racing. Growing up attending air shows and having a pilot at home encouraged her to go beyond commercial applications. The Reno Air Races of 1990 got her hooked on the challenge of personalizing her flying in a creative way. She has also taught with her CFI 1989-93 and currently has commercial, multi-engine, instrument and CFI.

KATHLEEN K. SAGE, Indiana Chapter, started flying lessons with her husband, Dr. Russell Sage, in 1960 at Shanks Airport. Daughter Jean preceded and inspired them. She had soloed while a student at Indiana University, 1959. She received her private license January 1962 in a Piper Tri-Pacer.

Joined the 99s in 1962. Served as vice-chairman 1964, 1965-66; membership chairman 1967-68; was awarded the Indianapolis Aero Club Dee Nicholas Trophy in 1969; and received commercial rating, 1969.

Flew as co-pilot in the 1966 Powder Puff Derby with Dorothy Smith as co-pilot. Has flown seven Indiana races as pilot. Placed fifth, 1963, Kokomo.

REIKO SAKURAI, born July 21, 1962, in Tokyo, Japan, started to fly gliders in 1981, then fixed wing and helicopter in California.

She joined the 99s in 1991. She has flown C-152, 172; BE-55, 20; R-22, AS350B, BELL206, and gliders. She was a corporate pilot (fixed wing and helicopter) from 1985 to 1994. She has logged 2,300 hours in glider, fixed wing and helicopters.

Especially interested in sports glider flight including challenging record; cross-country and glider aerobatics. Joined the world glider aerobatics championship as a member of Japan team in France, 1995.

BETTY LOU SALCEDO, born March 5, in Los Angeles, CA. Started flying January 1986 and joined Mount Diablo Chapter of 99s in 1990 (treasurer, 1992). Also, belongs to Alameda Contra Costa County Sheriff Posse Air Patrol (search and rescue); Aircraft Pilots of Bay Area (APBA) secretary three consecutive years; AOPA; Travis AFB Museum; Cessna Pilots Assoc.; California Pilots Assoc.; and Western Flyers Assoc.

She has flown Cessna 150s to 206s; Piper 140, 180, 200; and 10 hours helicopter simulator. Has the distinct pleasure of owning a Cessna 175 and loves flying non-pilots, along with family and fellow pilots.

She has four sons, one daughter and one grandson, Nathan Edward, 3 years old, who has a passion for airplanes and helicopters. Has been square dancing since 1978 and has held every office in the Northern CA Square Dancers Assoc. (started a blood bank for square dancers in 1980, still going strong), along with being active in the CA State Square Dance Council and United Square Dancers Assoc. (nationwide). She is pursuing her IFR rating and just completed FAA Phase V Wings Program. Every time she flies, she thanks God for making her "special" to allow her to enter His air space.

JACQUELINE M. SALISTEAN, born Sept. 8, 1934, in Villerupt, France, learned to fly in the early 50s in France. She was the youngest pilot of France in September 1952, but lost her title to Josette Aubertot shortly afterward. Went back to flying in the late 70s in the USA and received her private and SE land in 1983; got her instrument rating in 1986. She has flown the Piper Club, Stampe, Rockwell Commander on a regular basis and Cessna 152, also Dualson. She joined the 99s in 1986.

Salistean is the widow of Nicholas B. Salistean MD. She has three children: Daniel Albert, 40; Christian George, 38; and Annette Nicole, 33. She has two grandchildren, Daniel Allen, 11 and Jason Armando, 8.

CATHY SALVAIR, born Oct. 17, 1963, in Perth, Western Australia, took her first flight in a powered aircraft in 1979. She took commercial theory studies and gained her private license in 1981. She received her commercial rating in 1983, instructor rating in 1984 and began instructing at Jandakot. In 1987 she won the Lady Casey Scholarship and used it for command instrument rating at Bankstown, Sydney, Australia. Started full-time at Bankstown and studied senior commercial theory in 1988. She attended AWPA/NZAWA Rally, Christchurch, New Zealand and was captain of the winning AWPA team of Jean Batton Trophy. Competed in NZAWA competition and won Fitton Rose Bowl. Stopped full-time instructing. In 1993 Salvair started as first officer flying tours around Australia in a DH Heron. She was the winning pilot of the AWP Nav. Trail in Fremantle in 1994. She has been flight instructor/ferry pilot Cootamundra and is now chief flying instructor/chief pilot at Camden. Salvair has more than 4,000 hours and is holder of an airline transport license (first class), grade 1 instructor rating with IFR multi-approval with aerobatic, tailwheel, formation, and pressurization. Her ambition is be an old pilot.

HELEN M. SAMMON, a lifetime Ohio resident, learned to fly in 1948. She earned her license in 1950. In 1952 shortly after obtaining her private license, she joined the All-Ohio Chapter of the 99s Inc. where she was a very active member holding many offices including chairman from 1963-65.

In 1974 she became a charter member of the Lake Erie Chapter. She has attended chapter, section and many International conventions. Helen volunteered her services at many aviation functions including the For-

mula One Races (midget planes) held in Cleveland where she was a scorer and timer. Her timer expertise was also used at Toledo, OH in 1976; she was the chief timer for the terminus of the Air Race Classic in Cleveland.

In 1985-86 Helen received the Chapter Achievement Award, then in 1989 was awarded the Pilot of the Year. In 1986, 34 years later after attending her first meeting, Helen became a lifetime member.

In 1995 Lake Erie Chapter selected her as an honoree into the Forest of Friendship.

KATHY SAMUELSON, born Oct. 3, 1946 in Piqua, OH, joined the 99s in 1980 after receiving her private license and CFII ratings 1979-1984. She has flown the AA1B, AAS, AA5A, AA5B, C-150, C-152, C-172, C-172RG, C-177, C-182, Beech Sundowner, PA-28 140-180, PA Arrow, Commander 114B, Lance and Tomahawk. She was employed as a CFI Columbus pilot center from 1983-84 and is currently a CFII Columbus flight instructor since 1985.

She has two children, Michael, 25 and Amy, 22. Whether flying for fun on vacation or to see relatives or friends, flying has been great fun. Instructing accounts for most of her flying. Seeing students progress and then join the rank of other pilots has and will continue to be rewarding. She has 11,000 hours now and each hour has been fun.

DARLINE "DOTTIE" SANDERS, born April 23, 1918, in Buffalo, WY, joined the San Diego Chapter of 99s in September 1949 and served as secretary, vice-chairman and chairman of that chapter. In 1961 she formed and was the first chairman of the El Cajon Valley Chapter. She celebrated her 45th year as a life member of the 99s in 1994. In addition to chapter activities, she served one year each as governor of the Southwest Section and as International membership chairman, and attended many Southwest Section meetings and International Conventions until 1975. She has a private license with instrument rating.

She has flown Cessna 120, Patroller 140A, 182G and Piper Cherokee 140.

From 1950-1972, she flew the Powder Puff Derby 20 times, seven times as pilot, having placed in the top 10 four times and the rest as co-pilot, placing second in 1971 and first in 1972. When not racing she helped on the timing committee.

As co-pilot, Dottie was in the Angel Derby three times. Also as co-pilot, entered the Intercontinental Air Race from Mexico to the Bahamas, placing fourth, in the Palms to Pines Race, placing first and entered the Bahamas Treasure Hunt, and helped time contestants in the Angel Derby to Puerto Vallarta. She was a member of the Pacific Air Race board of directors, raced in it several times as pilot once as co-pilot and served as treasurer and chief timer from 1983 to the last one in 1990. She was chairman and chief timer of the stop at Loreto, Baja, CA, in the Baja-California Air Race that ended in La Paz, Baja, CA, in 1982.

In other aviation affiliations, she represented the El Cajon Valley Chapter on the aviation committee of the El Cajon Chamber of Commerce from 1964 until 1992; was a member of AOPA since 1956; Aviation Breakfast Club; was an original member of the San Diego Aerospace Museum and Silver Wings, pilots who soloed 25 or more years ago, in the San Diego Chapter. She also ferried airplanes from the Cessna and Piper factories to Gillespie Field. In 1978 she was awarded the Gillespie Field Achievement Award for service to that airport. Among other things, she helped dedicate the control tower at Gillespie Field in May 1962 by pulling the large ribbon and bow off of it with the tail of her airplane and was the first to be officially controlled by the tower for a takeoff and landing. She was inducted in the International Forest of Friendship in 1985.

Dottie is most thankful to the 99s for the many years of camaraderie with other women pilots.

DELIA V. SANDERS, Indiana Chapter, started flying in 1953, and received her private license July 1954 in a J-3 Cub N61964 and then joined Indiana Chapter of 99s. She served as secretary/treasurer 1955-56, and chapter chairman 1957-58, and 1958-59, when Indiana hosted the Fall Meeting of the North Central Section. She served on many committees and was always ready and willing to help.

She flew the first Indiana Air Race, coming in fifth and Powder Puff Derby in 1961 with Ethel Knuth.

Delia passed on to new horizons March 23, 1973.

SYLVIA SANDERSON, born March 24, 1945 in San Bernardino, CA, earned her airplane single-engine land, instrument rating. She has flown the Cessna 150, 152, 172, 172RG, 182 and various tailwheel aircraft. She joined the 99s in 1978.

While a student pilot, Sanderson flew from Santa Paula to Blakesburg, IA, with her friend in his Stinson 108-2 to attend the Antique Airplane Assoc. Annual Fly-In, where they camped out under the wing. In addition to the 99s, she is a member of AOPA, the Angeles Antiquers and the Vintage Aeroplane Assoc. As a 99, she has served as a member or chairman of various committees, as well as, a member of the chapter board of directors (treasurer one year, recording secretary two years).

SHERYL J. SANDHAGEN, born April 28, 1938 in Altadena, CA, earned her private pilot license and instrument rating. She has flown the Cherokee Archer II (owner). Sheryl learned to fly in 1974 at Brackett Airport in Southern California. She joined the 99s the same year and has participated in many activities since that time.

She and her husband, Robert, who is also a pilot, have flown on many cross-country trips, including to Alaska in their Cherokee Archer.

She is a stockbroker with a BA from the University of California, Santa Barbara. She has held many 99 board positions, and is also a member of Pomona Valley Pilots Assoc. and AOPA.

She has three sons, a stepson and daughter, two grandchildren, and is married to Robert Sandhagen, a college instructor.

PAULA BAZAR-SANDLING, a native of Southern California, learned to fly in 1980 and joined the 99s in 1981. Paula owns a Cessna 182. She has logged more than 1,500 hours in Cessna, 152, 172 and 182. She is instrument rated. Paula has participated in the Valley Air Derby, Pacific Air Race, Palms to Pines and Jim Long Air Rally.

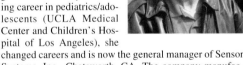

Paula earned a BS in nursing from Mount Saint Mary's College in Los Angeles. After a 20 year nursing career in pediatrics/adolescents (UCLA Medical Center and Children's Hospital of Los Angeles), she changed careers and is now the general manager of Sensor Systems, Inc., Chatsworth, CA. The company manufactures navigation, communications antennas for aircraft.

Paula is a member of the San Fernando Valley 99s and has been a past chapter chairman, treasurer and chaired numerous other committees. She served one term as a director on the 1992-94 Southwest Section Board and co-chaired the 1995 spring Southwest Section Convention. Other aviation affiliations: AOPA, Cessna Pilots Assoc., International Women's Air and Space Museum, National Counsel for Women in Aviation and Flying Samaritans.

She is married to Michael John Sandling, also a pilot. Michael is an orthopedic physician assistant. Their flying adventures have taken them all over the US and parts of Canada. They have also encouraged four other members of their family to take up flying and join the ranks of the aviation community.

GLORIA P. SANDS, born Aug. 22, 1939, in Seattle, WA, earned her private pilot license in 1991 at the age of 51. She has flown many years with her husband as a passenger in her husband's 1948 Luscombe as well as a Cessna 172 and a Beech Bonanza. She joined the 99s in 1991.

She learned to fly in a Tri-Pacer and she is fondly known as the *Pink Lady,* (Sands calls her Rosie). She has written a book of poems called, *The Pink Lady,* which is about learning to fly and some of the many exciting experiences all pilots share.

She is married to Paul F. Sands and has two children, Kimberly Bartlett, 35 and Yvonne Turney, 32. She has four grandchildren: Joshua Turney, 8; Stesha Turney, 6; Matthew Bartlett, 5; and Andrew Bartlett, 20 months.

BARBARA SANTAMARIA, born in 1951 in Toronto, joined the 99s in 1991 and obtained her private commercial Class IV instructor land and sea rating. She has flown the Cessna 150, 152 and 172; Supercub Piper Arrow; Archer; Warrior; and Dakota. She is a Class IV instructor for Island Air Flight School, Toronto.

She is married to Xavier and has three children: Alexander, 17; Stefan, 15 and André, 11. She hopes to own an airplane someday and has found a special challenge in flying that allows her to combine her professional life as a high school language teacher with her passion for aviation. She flies for Operation Skywatch, an aerial surveillance government project for the prosecution of environmental offenders.

GLORIA L. SANTUCCI, born July 10, 1921, was introduced to aviation in 1929 by meeting Amelia Earhart. Took first ride in 1934 in a Driggs Skylark bi-plane with a limited commercial rated pilot. Started flying lessons at Utica (NY) Municipal Airport in 1935. While in high school attended night classes in aircraft and engine mechanics from 1935 to 1938 at New York State Aviation School in Utica. Graduated from same school in 1939 as only woman in competitive pilot ground school in Civilian Pilot Training Program (CPT). Class took first CAA multiple choice written test. Received private pilot certificate under CPT at Utica Aviation school in 1940 and became the first woman in Utica area licensed to carry passengers. Immediately joined Utica Aviation staff as coordinator of CPT/WTS (War Training Service) programs in conjunction with Hamilton College, Clinton, NY.

From 1941-42 completed Syracuse University Army Air Corps Accelerated College Program. In 1943 was issued CAA ground instructor certificate with all ratings including radio navigation and Link trainer. In 1948 became the first person licensed by NY State Education Department to teach aviation subjects. Introduced concept to CAA/FAA of open-book pilot written testing. Worked continuously at Utica Aviation School for 14 years. From 1954-1967 managed Hylan Airport, Rochester, NY; also was ground instructor for Ray Hyland School of Aeronautics and Hylan Flying Service averaging 500 students per year. During same period supervised flight and ground pilot training for ROTC programs at University of Rochester, NY, and Hobart College, Geneva, NY. In 1967 returned to Utica to be a founder, corporate officer and director of Riverside School of Aeronautics, Utica/Riverside Airport, Marcy, NY, FAA Aviation Maintenance Technician School #1356. The aviation mechanic school placed graduates throughout the industry – nationally and worldwide – and the school had the distinction of being the first school of its kind to receive national accreditation by the Office of Education, Washington, DC. Served as FAA designated written test examiner from 1979 to June 1995. In 1991 awarded FAA Eastern Region plaque for working over 50 continuous years in aviation education. Resigned from mechanic school in 1989 to form her own company, Aviation Services. Serves as FAA accident prevention counselor and holds a lifetime membership in OX-5.

FRANCES ROHRER SARGENT, Florida Goldcoast Chapter, began flying in 1942. Member of first group of women chosen as Link instructors for Navy Flight Instructors School. Joined WASP in 1943 and assigned to first women's group in a tow-target squadron. With encouragement of her three children (now three grandchildren) completed bachelor's degree, Cum Laude, 1967 and MBA in 1968. Joined 99s in 1968 and has served as International and section NIFA committee chairman and held several chapter offices. ATP, flight instructor, ground instructor and commercial glider pilot. Recipient of the 99 Amelia Earhart Scholarship and twice recipient of Florida Air Pilot Annual Service Award. Appointed FAA accident prevention counselor.

Served as president of the University Aviation Assoc. and National Intercollegiate Flying Assoc. Board. Professor at Miami-Dade Community College Aviation Department for 25 years and awarded professor emeritus upon retirement.

FAA designated written test examiner for 17 years until replaced by computer testing. She is happy to be able to continue to teach flying as well as share the joy of flying with family and friends.

EVELYN LOIS SASSER, was born in Harriman, TN, in 1922. She now resides in Phoenix, AZ, with her husband of 50 years, Keith. She has a son Don, a daughter Nancy Eden and six grandchildren.

Evelyn received her private pilot license July 16, 1965, flying a Beechcraft Musketeer. She joined the 99s in October 1966. After getting her license, the rest of her flying was in her and Keith's Piper Twin Comanche. She learned to fly because of her husband's love of flying. Evelyn attended many of the Southwest Sectional meetings and most of the local fly-ins.

Even though she has not flown for many years, she continues to enjoy keeping up with the activities of the 99s.

DOROTHY CHAPMAN CARPENTER SAUNDERS, learned to fly an Aeronca and Great Lakes bi-plane at Ann Arbor while taking a Ph.D. in biology at the University of Michigan. Got her private pilot license #32234 in August 1935 and joined the 99s that year. Obtained Federation Aeronautique Internationale Certificate No. 8681. While professor of biology at Cedar Crest College, Allentown, PA, was governor of Middle Atlantic Section of 99s from 1939-41, and editor of *The 99s Newsletter*.

From 1941-44 lived in Lima, Peru where she flew the Stearman of the Aero Club del Peru, a Phillips, Luscombe and Rearwin Sportster. Was a research biologist during WWII for the office of Foreign Relations of the USDA in Peru, Ecuador, El Salvador, Cuba and Guatemala, and was later Fulbright Postdoctoral Research Scholar to Egypt for a year. Saunders is married to Dr. George Saunders, a retired research biologist.

EVIE SAUNDERS, born Oct. 21, 1921 in England had a lifelong ambition to learn to fly. Was introduced to airplanes at age 4. At 14 she went to Brooklands Flying School for lessons, but had no money. War came, joined WAAFs as administrator and PT instructor. Got married for a short while and bought a home. Saunders started tourist business still with flying in mind. At age 68 she finally had enough money. Received PPL(A) September 1990, IMC rating August 1991; night rating September 1991; and a multi-rating (B) June 1992. Experience on helicopters, float planes and Tiger Moth. In 1992 received the Jean Lennox Bird Trophy for noteworthy performance in aviation awarded by the British Women Pilots Assoc. Won timed circuit competition, bang on, in 1994. Spent many happy hours flying around Europe and Florida, Bahamas, etc. Trained on twin in Orlando. More projects in mind.

Saunders is delighted to be a member of the 99s, Inc. British Women Pilots Assoc., Popular Flying Assoc., Aircraft Owners and Pilots Assoc. and the Spitfire Society.

She has one daughter Pamela, (age 50) who married. She has three grandchildren: Lucy, age 14; John, age 16; and Karl age 28. She has two great-grandchildren, Samuel, age 3 and Stephanie, age 1.

JERELYN W. SAUR (JERE), born June 20, 1931 in Shreveport, LA, joined the Shreveport Chapter of 99s in 1962 and Shreveport 99, Helen Hewitt (ATP) was her instructor for all advanced ratings and licenses. She earned her private on Aug. 7, 1962; commercial Dec. 20, 1969; instrument June 23, 1971; basic ground instructor Aug. 21, 1974; CFI Nov. 12, 1973; and CFII Nov. 1, 1978. She has flown the Colt (PA-22-108), Cherokee (PA-28-140), Cessna 150, 172, 182, 140, 152 Tri-Pacer; Beech (T-34A), PA-28-161.

She has participated in Powder Puff Derbies from 1971-75. Was a member of CAP about 20 years and attained rank of captain. Instructed flying at Downtown Airport for several years, taught ground school for Louisiana Tech University Barksdale Air Force Base. Presently teaching music for Louisiana Tech University.

Saur married Henry C. Saur, CPA in 1957. Henry is also a CFII. She has a BM from Baylor University 1954; an MA in music from Louisiana Tech University in 1968; attended summer session at Eastman School of Music; and studied piano with Orazio Frugoni, 1955.

HILDA FOWLER SAVAGE, born Jan. 21, 1926 in St. Louis, MO, received flight training in Memphis, TN, and received an instrument rating with Mr. Jim Zuenichet at Memphis Aero. She has flown all Piper, all Cessna, Mooney Mark 21 and Bonanza. She is married to Doyle C. Savage and has three children: Linda S. Hammons, John Savage and Ralph Savage. She has four grandchildren: Lauren Pearcy-Melissa, 16; Meredith, 12; Elizabeth, 10; and Alexandra 7.

A wonderful and fulfilling educational experience. Her husband's commercial construction company kept him traveling around the world. Her children were 17, 15 and 13 and busy with their "things." She decided to take flying lessons. Three months later she had her Wings and life really began. She joined the 99s and met many flying ladies and their husbands. She became an active member of a most wonderful organization, the 99s. Her husband retired when he was 55 and having traveled so much wanted to stay home, so she sold her Bonanza and started a new phase in her life. May God bless and keep all pilots in the palm of "His" hands.

KAREN SCALLAN, has always been fascinated by airplanes and wanted to fly for many years but, for some reason or other, she didn't take the first step until December 1993. While living on Hilton Head Island, SC, she went up in a Cessna 172 to see if she would like to fly – she never looked back. From then on she was hooked. Scallan finally received her private pilot license on June 18, 1995 (Father's Day) and enjoyed taking up her first passenger that evening – the man she married on July 2 of that year. It has been so much fun flying the children to Cedar Point, taking her family up over Akron to watch fireworks from the sky and visiting friends in Ashtabula, OH, and back to Hilton Head, where she took her husband up for a "bird's eye" view of the island! She plans on pursuing her instrument rating and keep on going. Flying is something she plans on doing for the rest of her life! Her family is so supportive, and she is hoping her husband pursues his desire to learn to fly next year.

Scallan's husband, John, and their house full of children: Matt, Marcie and Rebecca, all enjoy flying. She has a son in the Navy who is anxious for his turn up in the sky with his "High Flying Mom." Blue skies to everyone!

MARION ROOKE SCANIO, a fifth generation south Texan, has had many memorable flying experiences. However, her least enjoyable were back to back flights across the Atlantic and Pacific Oceans in the early 70s. The anxiety of these trips occasioned her determination to overcome her flying phobia.

It was at this time that she met Pauline Glasson, who was her instructor, and the answer to her need. They became fast friends. Marion received her private license in 1973, the year she co-piloted for Pauline in the All-Women Transcontinental Air Race (Powder Puff Derby) from Carlsbad, CA, to Elmira-Corning, NY. Two years later, she co-piloted for Pauline in the International from Hamilton, Ontario to Titusville, FL.

Of the many wonderful flying memories, her two favorites are coming into Sedona just before sunset after a later afternoon shower, and flying over Niagara Falls on the approach to Elmira-Corning.

JETTA SCHANTZ, discovered aviation with her first flight in a hot air balloon in 1983. Her journey into ballooning started as volunteer ground crew and evolved into her completion of a commercial pilot license in 1991.

Born June 14, 1959, in Wichita Falls, TX, raised in Oklahoma, Jetta now lives in Jacksonville Beach, FL, with her husband Rob. They sell balloons, train students, operate corporate balloon programs worldwide and produce Balloon Tour America, a professional nationwide balloon racing circuit.

She first flew into the record books in 1993 setting nine national and world records for distance (292 miles) and again in 1994 setting nine national and world records for altitude (32,657 feet).

Jetta's altitude flight was recognized by the National Aeronautical Assoc. as one of The Ten Most Memorable Flights of 1994. For her altitude flight, she also was honored to receive the Montgolfiere Diplome, the highest International award in ballooning, presented by the Federation Aeronautique Internationale and the International Ballooning Commission.

DORIS DEGARMO-SCHELL, born April 10, 1941, in Los Angeles, CA, never got to fly until she was 35. Her three daughters saved up the money to pay for book and tuition for her to go to ground school for her 40th birthday. There she met the people who had the FBO Southern Oregon Aviation. She got a free first hour of flight training and was hooked. She got many of her student pilot hours by trading her services. She would work four hours to get one hour of flight time. She cleaned their offices, and homes, washed windows, shampooed carpeting, etc. She got her pilot certificate in 1984. When she married Leroy in 1988, they bought a little 1957 Tri-Champ. She joined the 99s last year, a new chapter, the Crater Lake Flyers. They are a wonderful group of women who all have fascinating flight stories.

She is married to Leroy Schell and has three children: Penny Hoskins, 35; Connie Ketarkus-Schade, 31; and Stephanie Leach, 25. She has two grandchildren, Rebecca, 12 and Hayden, 2.

LYNDA SCHELL, born in 1949 in Cody, WY, took her first flight lessons in Worland, WY, in 1980 along with her husband, Mike, however, stopped taking lessons shortly thereafter because of financial reasons. In 1991 she decided to resume flight training and received her license in September 1992 in Gabbs, NV. She flies a Cessna T-210 and a 172.

Schell's educational background includes a military tour of three years in the Air Force where received training in radiological technology. She pursued a 21 year career in X-ray and is a member of the American Registry of Radiological Technologists. She is also a member of the High Country Chapter of the 99s, Aircraft Owners and Pilot Assoc. and Cessna Pilot Assoc.

She is married to Mike Schell who is also a pilot (he

194

finished in 1980) and a successful business entrepreneur. She has a stepdaughter, Tannah, age 26 and a stepson, Tony, age 21.

DEEANN SCHIAPPACASSE, born Feb. 8, 1951, in Detroit, MI, joined the 99s in December 1988. She earned her private pilot certificate (ASEL) December 1988 and her instrument rating on October 1990. She has flown the Piper PA-28 140, 180, 183, 236; Piper Saratoga PA-32R-301.

She earned a BS in nursing in 1973 from Wayne State University; 23 graduate credits; past jobs: coronary care nurse, supervisor for public health and is now a radiology oncology nurse. She is married to Richard, MD (also a pilot) and has two children: Michael 13 and Angela 10.

An active member of Greater Detroit Area Chapter of 99s, including "pinch hitter" course annually, Air Bears programs, IFR-VFR Seminars, chapter office. Also belongs to AOPA, Michigan Aviation Education Foundation, flown IFR cross-country to Palms Springs and to Florida often, as well as the Midwest.

DAPHNE SCHIFF, is a professor of natural science at Glendon College, York University, teaching meteorology and aerodynamics. She has an airline transport pilot license with 4,000 hours and has worked as a commercial pilot. Her flying trips have taken her across North America and as far as Australia. Each year, she lands a light aircraft on the York campus as part of the Science of Flight course she has taught for the past 16 years. She was co-captain in recent Around the World Air Race 1994; co-captain in New York-Paris Race, 1985; co-captain Air Race Classic 1989-90; co-captain Great Southern Air Race 1992-93; and co-director, pilot for Air-O-Sols flying team to measure air pollution. When she is not flying government-sponsored pollution monitoring flights, she is busy producing award-winning documentary science films, which have included "The Northern Wetlands: Its Role in Global Warming." Daphne currently chairs the Amelia Earhart Scholarship Program for the 99s. Daphne is blessed with four grandchildren.

SUZANNE "SUSIE" M. SCHLERNITZAUER, born April 6, 1956, in Youngstown, OH, joined the 99s in 1992. She started flying in September 1991 and was licensed in June 1992. She has flown the Cessna 150, 152 and 172; 1946 Piper J-3 Cub (owns). She and her husband, Bob Schlernitzauer, were married in May 1993 and decided to delay their honeymoon until August to coincide with Sentimental Journey, the annual trek of all Piper Cubs back to Lock Haven, PA, where they were built. The members of her 99s chapter teased them before they left that "two pilots in a two place airplane for two weeks . . ." they were sure to be divorced before they made it home. Well, Bob and Susie did spend two weeks in their little J-3 Cub! From Homestead, FL to Lock Haven, PA, to Pittsburgh, PA, and back to Homestead. 48 flying hours, 31 fuel stops, seven sectionals and a wet compass . . .she learned the true meaning of "cross-country" and "pilotage" and in a few cases "dead reckoning" the hard way. Bob and Susie had a wonderful time and were married and still speaking when they made it home.

JESSIE SCHILLING, started flying in 1980 at Palo Alto, getting her license later that year with Mayetta Behringer at San Jose Airport. She is both a glider pilot and powered (ASEL) and has logged about 500 hours 350 of them in gliders. Her most memorable flight was in an open cockpit Schwitzer 126 glider in Maryland at sunset. Jessie joined the SCV 99s in 1980 and has served as corresponding secretary. She works in administration and development at Stanford University and has worked for the Soaring Society as a government liaison and safety and training officer. She belongs to the Soaring Society of America and AOPA. Jessie recently had a house built in Portola Valley, CA.

ELSIE MCBRIDE SCHMIDT, is a lifetime resident of Philadelphia, PA. While employed by the city of Philadelphia at PHL, she organized a flying club and was the first woman to solo and get a license. In 1959 she flew for radio station WIP and gave weekend traffic reports in a Comanche – possibly the only woman at the time to do this. She received her commercial and multi-engine ratings. Shortly after she joined the 99s in 1956, she was co-pilot in the International Air Race from Texas to Cuba – they won. Martin M. Decker, who became her employer, sponsored Elsie in all subsequent races.

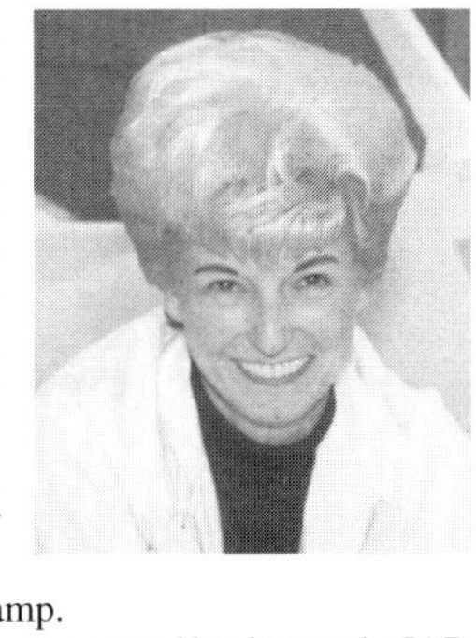

After joining the 99s, Elsie held chapter and section offices and progressed to and including the executive committee. In 1960 and 1962 she was terminus chairman of the TAR and flew for many Pennies-a-Pound Rides. In 1963 she arranged for her chapter of fly First Day Covers of the Amelia Earhart stamp.

The last race Mr. Decker sponsored her in was the IAR in 1968. Business started going downhill and he went bankrupt. Her flying days were over.

KIMBERLEY L. SCHMIDT, born Sept. 19, 1963, in Bellefonte, PA, began flight training with Meriden Aviation in Meriden, CT and Robinson Aviation in New Haven, CT. So far, the only rating she has acquired is her private license. She has flown the PA-28-161, PA-28-140 and C-152. Her goal this summer is to do some aerobatics training. She received a BS in economics from Allegheny College. She is currently employed at Sikorsky Aircraft of Stratford, CT, as a military pricing analyst, Army and Air Force Products. Her position requires her to price and negotiate contracts with the Army and Air Force. She is an active member of Junior League of Eastern Fairfield County and for the past three years has served as the Point of Contact for military aircraft for the Tweed-New Haven AirFest, a non-profit airshow that raises money for scholarships for inner-city high school students who want careers in aviation.

She has a special story about a very special friend of hers that was the one who finally convinced her that she could be a pilot. While working to organized the Tweed-New Haven AirFest, she was introduced to one of the test pilots at Sikorsky Aircraft. He arranged for her to fly as co-pilot on a test flight in an S-76. During the flight, he allowed her to take the controls' letting her take-off, hover and fly for nearly an hour. On the way back to the plant, he called "his plane" and pulled the helicopter into a 60 bank over a friend's house. One second she was looking out over the clouds, the next she was looking at the tree tops! It happened so incredibly fast and was the most exciting experience that she had ever had. After they landed, she knew she had to learn to fly!

NELDA JEAN SCHMIDT, born April 15, 1940, Paris, TN, joined the 99s in 1978 after receiving her ASEL and a ASES instrument rating. She has flown Cessnas, Cub, and Aeronca Chief. She is married to George J. Schmidt and has four children: Leren Wilson, 29; stepson, Dan, 35; Dennis Schmidt, 30; and Mary Beth Filer, 32. She has three grandchildren: Phillip Schmidt, 11; Megan Schmidt, 8; and Jimmy Filer, 7. She credits meeting her husband (a retired Air Force pilot) through aviation.

MARGARET GOODRICH SCHOCK, born Nov. 23, 1920, in Nogales, AZ; raised in Glendale, CA; attended Glendale Junior College and Immaculate Heart Women's College, Hollywood, CA. Married Raymond Schock in 1941. She has one daughter, Susan Schock Grinold and one granddaughter, Katie. Ray taught her to fly through her private and commercial ratings. He kept her current and on the mark until he was diagnosed with Parkinson's disease in 1973. He began flying in 1928. She joined the 99s in 1948, in San Antonio, TX. She learned to fly and earned her private SEL in San Antonio, TX, on Jan. 31, 1948 and commercial SEL July 25, 1964. She transferred to Phoenix, AZ, in 1950 and founded Tucson Chapter in 1951. She has held all chapter offices over the years.

The most heart-stopping moment she ever experienced was at the Powder Puff Derby stop in Tucson. In her rush to reach the time clock, one girl jumped off the leading edge of their Navion right through the slowly turning propeller. Soon after that, the ground timing was changed to a fly-by. Working on Pennies-a-Pound, many stops for the PPD and several SW Sectionals as well as flying their Tucson AW Tours and fly-ins, culminated in the PPD Commemorative Flight 1977.

She was co-owner/operator of a repair station and two GI flight schools, San Antonio and Sequin, TX, 1947-50. She continues a little bookkeeping and caring for the financial business of a couple of elderly clients and a non-profit organization.

ANNA SCHOLTEN, born Feb. 27, 1963, in Topeka, KS, learned to fly in 1988 and joined the Colorado Chapter of 99s. Received private pilot license, instrument rating, commercial pilot license, multi-engine rating, flight instructor certificate for single and multi-engine airplanes and airframe and powerplant mechanic license all by the end of 1992. In 1991 she was the recipient of the Amelia Earhart Memorial Scholarship which was used to obtain multi-engine instructor certificate.

Scholten currently works for Jepperen Sanderson as a NOTAMs analyst. Has been very active on the Colorado Chapter's Mile High Air Derby Air Race committee as an impound inspector and handicap chairperson, as well as actually participating as a racer when not on the committee. She has flown the Cessna 152, 172, 172RG, 182, 210 and 414; Beech Duchess; Stinson 108-2 Voyager; Stearman; Ercoupe; Bellanca Super Decathlon; Piper Cherokee; and Taylorcraft F-19.

ELIZABETH ANN SCHOPPAUL, was exposed to aviation at an early age. By the time she was 10, her father, Art, an airline captain, had his three children restoring a crop duster in their backyard which they later learned to fly. This early experience "bucking" rivets eventually led her to acquire both airframe and powerplant and flight engineer licenses.

Currently, she is a 727 check airman for Flight Engineers at American TransAir where she has worked since 1990. Of over 200 FEs and 800 flight officers at ATA, she is the only female FE and check airman. Previously, she worked at Rich International Airlines as a flight engineer on the DC-8. She initially entered aviation as a flight attendant at Arrow Air in order to see the world and work on her licenses. Four direct family members (her father, stepmother Willy, and both brothers) were also flight crew members at Arrow.

Her fellow mechanics (her brothers) also pursued careers in aviation. David is a pilot at America West, and Bob is a check airman for Flight Engineers at Carnival Airlines. She was raised and still lives in Miami where she recently married Ken Lindeman, a marine scientist (no, not a pilot). Besides helping him raise her wonderful stepsons, Bryan and Eric, she also assists her husband's corporation, Coastal Research & Education, Inc. She acquired her private pilot certificate in 1992 only to have Hurricane Andrew claim the Cessna 150 she and her father had recently restored. She plans to continue her flight training to further advance her aviation career.

SHARON ANN SCHORSCH, born April 24, 1942, received her private pilot in 1967, just a few weeks before their daughter, Ann Marie, was born. (with . . . instrument rating, ASEL and ASES) She learned to fly in a Cherokee 140 and now flies a Cessna 206 on amphibious floats. Joined the 99s at the Friendship Tent in Oshkosh. Served as Chicago Area Chapter chairman from 1988-90, safety seminar chair from 1991-93, and has participated in many other rewarding chapter activities. Especially enjoys working with NIFA. Served as chief judge for Region VIII in the fall of 1994.

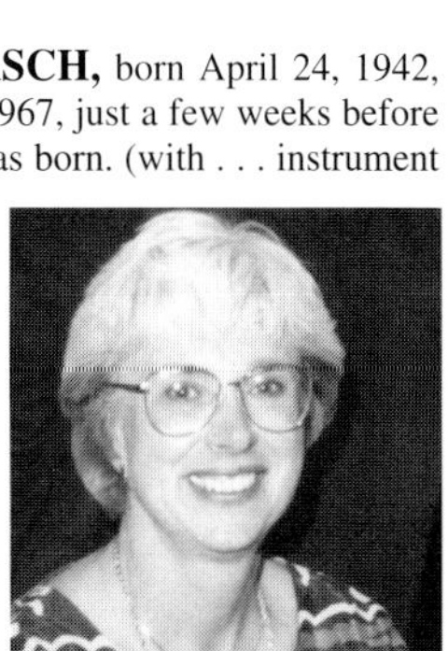

Schorsch is married to Frank and has three children: Cary, 28; Ann Marie, 27 and Jeanne, 25. Her educational background includes a BA in economics from Mundelein College of Loyola University, Chicago. She has taught elementary school in Chicago and currently works with her husband, Frank, in real estate management. When not flying

SHIRLEY SCHREIBER, born in Hinsdale, IL. When her husband proposed in 1987, she was teaching dance at the University of Massachusetts-Amherst. He promised she'd sit by the pool and eat bonbons in West Virginia. Shirley's pool is a T-hangar and her bonbon a Grumman AA5. She earned her private pilot license in it.

Schreiber's educational background includes: BS degree in dance Julliard School of Music; MA in dance management, Sangamon State University; and MA in counseling and guidance, West Virginia University. She is currently an adult therapist at a community mental health clinic and also teaches ballet.

She is a national certified counselor, a Ham (KB8SYC) and a certified first responder. She is a member of the 99s, Inc., Morgantown Pilot's Assoc., American Yankee Assoc., Aircraft Owners and Pilots Assoc., and aerospace education officer in the Morgantown Civil Air Patrol.

Her co-pilot/spouse, Richard, has a commercial pilot license and teaches at West Virginia University. She has one son, two step-children and one step-grandchild.

KAREN SCHREINER, is a lifelong Kansas City, MO, area resident. Karen started taking flying lessons at age 19 and received her private pilot license in July 1990. She then got her instrument rating in two months. During the next year, Karen built flight hours for her commercial rating by taking trips to several different locations. One of the most memorable was a spring break flight to the Bahamas in a Cherokee 6. In July 1991, she earned her commercial rating.

Karen is a 1991 graduate of Central Missouri State University in Warrensburg, MO, with a bachelor's degree in aviation. She has been a member of the Northeast Kansas Chapter 99s since 1992.

Karen is also an officer trainee in the Kansas Air National Guard 190th Air Refueling Group in Topeka, KS. She has been selected by the unit for Air Force pilot training. During the course of pilot training, she will fly the US Air Force trainers: T-3, T-37 and T-38 or T-1. After undergraduate pilot training, Karen will go to Castle AFB, CA to train in the KC-135. The KC-135 is the air refueling aircraft shown behind Karen in the picture.

The other picture shown is of Karen and Connie Johnson, a fellow 99 in the Northeast Kansas Chapter. This picture was taken of Connie and Karen on their 1994 flight to Oshkosh, WI, for the annual EAA Fly-In. They make the trip every year in Connie's Cessna 182G.

CATHERINE MILLER SCHREVE, a student pilot in 1986, Cathie met 99s in the Lake Charles area who mentored and encouraged her towards her private pilot license. Owning her own airplane, she decided to combine her dream of flying and working for herself and went on to obtain her CFI, commercial and multi-engine rating.

She briefly instructed in Dallas and upon returning to Lake Charles she operated her own flight school, Flightline. In 1991 she accepted a position as airport manager of Southland Field in Sulphur, LA. Managing fuel concessions and daily operations, she still found time to do flight instructing.

Seeing the need to nurture general aviation, she has worked to obtain a localizer, NDB approach, new rotating beacon, enhanced runway lighting system as well as an extended parallel taxiway. She also campaigned and procured funding for T-hangars. Best of all she recently married her student, Lee, and they fly a Cessna 210.

MARTHA SCHRINER, and her husband, Tom, have been flying their Cessna Turbo 210, N6048N, since September 1994. They previously own a Cessna 172, purchased when Martha got her license in 1992.

Martha joined the SCV Chapter that year and immediately was elected to the treasurer's office, and has served as chairman of the chapter's Flying Companion Seminar.

Martha and Tom run their own landscaping business and have worked with troubled youth for years, running a home and offering employment opportunities to disadvantaged teens.

She and Tom have made many long cross-country flights in their Cessnas, traveling as far south as Baja, Mexico and as far north as Fairbanks, AK. Thus far, they traveled over 11 states in the T-210. Martha has logged more than 600 hours and is working on her instrument rating.

JEAN FORSYTH SCHULTZ, began flying in 1967, following in the footsteps of her mother, Pamela Van Der Linden, of Coyote County Chapter, and two brothers. Holds private, commercial and instrument ratings. Charter member of the Santa Rosa 99s, organized in 1973.

Flew three PPDs as co-pilot for her mother (two races flown under former name, Jean Clyde), also three Palms to Pines.

Jean serves on the board of Canine Companions for Independence and the Sonoma County Community Foundation. Has produced two documentaries for Canine Companions, "What a Difference a Dog Makes," (1987) and "Heart of a Hero" (1994).

Jean was awarded an honorary Doctorate of Humane Letter from Sonoma State University in May 1994.

Jean has two children and five grandchildren. Married to Charles Schultz in 1973.

LINDA KAY SCHUMM, born Sept. 28, 1957 in Moline, IL, became interested in flying through skydiving. She felt the most dangerous part of the jump was the airplane ride, so she decided to learn to fly. In 1984 Linda earned her private pilot certificate.

Since that time, her focus has moved from skydiving to flying. She has earned her ATP certificate, flight instructor and multi-engine flight instructor ratings.

Linda was terrified during her first solo cross-country. She cried from the time she lost sight of the local airport until she returned there to land. She has come a long way since that trip. Her aviation interests are now in the areas of cross-country air racing and international travel.

Linda competes in air races with co-owner, Rosemary Emhoff, in their Cessna 177. They received a NAA Certificate of Recognition for placing third in 1992 Air Race Classic. They also hold a US National Record for Speed Over a Recognized course from Peoria, IL to Kitty Hawk, NC.

Linda has visited all 50 states and landed in most of them. Now her goal has expanded to landing in every continent. She has flown to Alaska, the Caribbean, South and Central America. She has even flown in New Zealand. Current plans include a trip south of the Equator and a trip to Japan via Alaska and Russia.

Linda hopes to move into an aviation career, full-time, as fixed base operator. Her goals are to expand awareness and interest in general aviation throughout the community – and the world!

AUDREY M. SCHUTTE, her first airplane ride was at the age of 6 and she vowed then that she'd learn to fly someday. She soloed in June 1956, in a Cessna 170, obtained her private certificate in December 1956 and her commercial certificate in October 1957. Her husband, Ray, was her reluctant but patient instructor, with their two kids in the back seat. She flew her first Powder Puff Derby (AWTAR) in 1958 in a Cessna 140 and joined 99s that same year. She flew seven more AWTARs as a contestant and the last one in 1976 as a member of the AWTAR Board. Audrey began instructing in 1961 and is commercially rated in aircraft single and multi-engine land, single engine sea, glider, rotorcraft and hot air balloon. She is proud to be Whirley Girl number 146.

Audrey won the Woman Pilot of the Year Award in the San Fernando Valley Chapter 99s three different years and was selected by the FAA Pacific Region as Flight Instructor of the Year in 1974 and 1975.

Audrey owned and operated Viking Aero Service, Inc. at Van Nuys Airport from 1969 to 1976, where she was chief pilot and director of operations as well as an FAA designated pilot examiner. She taught aeronautics at Glendale Community College. In 1979 she became an air safety investigator for the National Transportation Safety Board in Washington, DC, and Los Angeles, CA. In 1987 she transferred to the FAA at Van Nuys, CA, as an operations inspector. After retirement from the government in 1991, she moved to Redding, joined the Mount Shasta 99s, Redding Area Pilot's Assoc. and EAA and is currently director of operations at Hillside Aviation at Benton Field, and a member of the Redding Airports Commission.

MARILYNN L. SCHUYLER, wanted to a pilot ever since she was 10 years old. When she was 13, her father, Maj. Donald Lee Schuyler, died while piloting an RF-4 in England. The dream didn't fade; however, and Marilyn earned her private pilot license just after her 19th birthday in August 1984.

One of the most memorable excursions was flying to Catalina Island in 1985. While she lived in the San Francisco Bay area from 1985 through 1992, a favorite fly-in for lunch was Half Moon Bay. She also has fond memories of participating in an airmarking as a member of the Bay Cities Chapter.

Marilynn now resides in the Washington, DC, area. She works for the Department of Labor during the day and attends Georgetown University Law Center at night. She's even willing to provide legal service for the 99s after graduation in May 1996.

BONNIE NELSON SCHWARTZ, born May 12, 1942 in Madison, WI, is a Washington, DC based Broadway and television producer who began to fly at the age of 48. She has a private pilot license and instrument training rating. She has flown the Cessna 172 and the Bonanza BE36. Her constant commuting between Washington, DC and New York now supports a Bonanza BE36 instead of the air shuttle business.

Her recent productions include segments for the 1992 Barcelona Olympic and the Olympic Woman Project for the 1996 Atlanta Olympic Games. She founded Washington, DC's Helen Hayes Award and has taught for the Smithsonian Institution. Her love affair with flying began 15 years ago when she taught a class at the Smithsonian Air and Space Museum, called "To Fly."

She is married to Dr. David Abramson, fellow pilot, scuba diver and sailor. She has four children.

JEAN SCOGGIN, received a private pilot certificate in July 1971. A member of Nebraska 99s.

Her husband, Dean and son, Kenwood are both pilots. She enjoys hunting, fishing, sewing and watching their mules, besides teaching elementary grades in a rural school. She is a Nebraska 99s Amelia Earhart Scholarship chairman. They fly a Cessna 172.

CAROLE MARY SCOTT, has spent most of her life in Fairbanks, AK, living, working and flying. What a privilege to be in the land known as the Last Frontier with all its majesty.

She earned her private pilot license in 1965 and continued her education in aviation related fields. Most of her 2,000 hours were logged in a PA-18 and Aeronca Chief both on floats. She enjoyed flying many people to remote fishing and camping sites throughout the years.

Her ratings include that of an aircraft dispatcher,

commercial pilot airplane single and multi-engine land/ airplane single engine sea.

Her membership in the 99s, Midnight Sun Chapter as a charter member and being a Pylon Judge during the National Air Racing Championships in Reno, NV, during the past 19 years, brings many happy memories.

She currently works for a Canadian airline which still flies the legendary DC-3 and DC-4 on their main schedule. She is frequently reminded from other pilots in aviation that someone still flies "real airplanes."

PHYLLIS V. SCOTT, born Sept. 21, 1931, in Argonia, KS, has always had a love for aviation but did not have the opportunity to fly when she was younger. Took her training while children were in college and received her private in 1979. She has flown for enjoyment since and has had a wonderful time.

She joined the 99s in 1980 and has flown the Cessna 152 and 172. She is the widow of Albert Scott and has three children: Leonard Scott, 41; Gary Scott, 39; and Susan Scott Haggard, 37. She has six grandchildren.

SHEILA SCOTT, OBE, was a former actress who had trouble passing a driving test, took her first flying lesson in 1959 on a dare and flying became her life! She flew around the world solo three times, and held over 100 World Class Records. She was the first in the world to fly a light aircraft solo from Equator to Equator via a Pole, and first woman in the world to solo the Arctic Ocean and the True North Pole. During this flight in 1971 she and her aircraft were used for experiments in the biomedical, environmental and positioning fields via NASA's cooperation and the Satellite Nimbus. Her major awards include the Harmon Trophy, the Britannia Trophy and in 1968 she was decorated by H.M. Queen Elizabeth with the OBE. Sheila wrote several books and was a professional lecturer and broadcaster.

She was co-founder of the British Section of 99s and became the first governor. She is now deceased.

MARY SCRIBNER, born Jan. 26, 1928, in Bangor, ME. It might have started as far back as Aug. 12, 1934 when Amelia Earhart came to Bangor, ME, to give rides in a Boston-Maine Tri-Motored Airline or it could be the feeling of the closest place to God that started the first step to becoming a pilot and 99.

Started flying September 1968 and was a member of the Connecticut Chapter, New England Section. She has three sons: Charles, Joseph and Tim and three daughters: Maureen, Cynthia and Sally. She also has nine grandchildren: Christie,

Mike, Pat, Jim, Sara, Katy, Matthew, Will and Thomas. She is now a member of Florida Suncoast Chapter.

Scribner had a 112 Rockwell Commander in which she did traffic reporting for WPOP radio and also used to show clients property in Connecticut for Mary E. Scribner Agency (Real Estate). She enjoys Explorer Air-Scout Group, US Power Squadron, tennis, photography. Will graduate from University of South Florida with a mass communications degree in May 1996.

ERNA SCRIVEN, born in Clementsport Nova Scotia, Canada, on Dec. 18, 1933. In 1976 she enrolled in the private pilot training course at the Halifax Flying Club at age 43. Six months later, she received her private license. She continued on to a night endorsement and a commercial license and time on a twin-engine Apache. The commercial career was limited to several local charters. Erna joined the Montreal Chapter of the 99s in 1980. During the years with the Montreal Chapter, Erna organized the planting of the Forest of Friendship tree at Chebucto Road Park in Halifax on Sept. 18, 1983 and

then on July 25, 1987 planted a tree on the air strip in Harbour Grace, Newfoundland. Amelia Earhart stopped at Harbour Grace on her solo cross-Atlanta flight in May 1932.

Erna started the Atlantic Chapter in 1989 with nine members. The charter was granted on Sept. 9, 1989. The chapter became visible at Fly-Ins and Poker Runs. Erna, chapter chair, was always willing to speak to a group about flying and/or the 99s. She presented an award in aviation to the Science Fair recipient on behalf of the chapter and represented the Canadian 99s in the presentation of the 99s Award in Aviation. She served on the board of the Halifax Flying Club and published their newsletter. She also served on the board of the Atlantic Canada Aviation Museum. She served as APT chair for the East Canada Section.

Erna owns a 1967 Cherokee and flies an average of 50 to 60 hours a year. She is married to Jim Scriven, a professional engineer; and they have three children.

BERYL SCUDELLARI, Pickering, Ontario, Canada, learned to fly in 1968. She flew her Mooney for 26 years and has logged 1,150 hours. Beryl flew with her husband, Norm, for pleasure and travel enjoyment. Many of her trips were tied in with 99s conventions with added time for travel. The holiday trips took them from Arctic Canada to southern US as well as the east and west coasts.

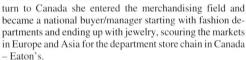

Beryl's educational background is a BA degree from Queen's University in Kingston Ontario. Following graduation she did volunteer work in Mexico. On her return to Canada she entered the merchandising field and became a national buyer/manager starting with fashion departments and ending up with jewelry, scouring the markets in Europe and Asia for the department store chain in Canada – Eaton's.

Beryl is a member of the First Canadian Chapter of 99s. She has served on the executive board of the chapter, participated in the early start up of the Operation Skywatch, as well as the annual Poker Runs. She is also a trustee on the board of the 99s Canadian Award in Aviation, and is a member on the Honor Roll of Canada's Aviation Hall of Fame.

She is past president of the Mooney Aircraft Club of Canada, a member of the Canadians Owners and Pilots Assoc. Beryl is also a fully licensed amateur radio operator and her call sign is VE3UHG.

LINDA SCULLY, a New Jersey resident, earned her private pilot certificate in 1981. Scully obtained her CFI and CFII in 1983 and began instructing part-time at Trenton-Robbinsville Airport while employed full-time as an accountant.

After returning to North Jersey and deciding to make a career change, Scully purchased a Cessna Aerobat Taildragger in December 1993 and incorporated Aero Safety Training, Ltd., specializing in spin training, unusual attitude recovery and tailwheel transition. She is based at Lincoln Park Airport.

In 1994 it was an exceptional year, launching Aero Safety Training, Ltd., participating in the Air Race Classic and Garden State 300, earning an IAC Basic Aerobatics Patch and serving as secretary of the North Jersey Chapter of the 99s. Scully also attended the Women in Aviation Conference and the 99 International Convention in Norfolk. She has logged more than 2,100 hours.

Scully is a member of the 99s, Inc., Aircraft Owners and Pilots Assoc., Experimental Aircraft Assoc., International Aerobatics Club, National Assoc. of Flight Instructors and Air Race Classic. Scully received her BS in business administration from the College of St. Elizabeth.

ERICA SCURR, a native of England, began flying gliders in 1965 and continued after immigrating to the United States in 1968. She participated in US Soaring Competitions in the 1970s and achieved the FAI Three-Diamond Soaring award in 1978. In 30 years of flying, Erica has logged more than 2,300 hours and now holds an ATP certificate airplane multi-engine land and a commercial pilot certificate airplane single-engine and glider. She is a CFI, airplane single and multi-engine, instrument and glider and also teaches weekend ground schools. Erica's chemistry degree led to a career in information science which has taken her around the world and provided opportunities to fly in the United Kingdom, France, Switzerland, Poland and Australia. A longtime All Ohio Chapter

member, Erica now enjoys touring North America, including Alaska and Canada, in her Piper Arrow and writing about aviation personalities and adventures.

SYLVIA SEARS, had an interest in flying since 13 years old. She started taking lessons in 1975 after finally being able to have time for herself. Her husband had been flying for sometime and had bought their Cessna 150 in 1975. Received a private pilot license in 1980. In 1993 she received an instrument rating. Sears has been a 99 member of All Ohio Chapter for two years. Of course this all came about because of her love of flying and her desire to learn more about it.

JEAN M. SEATON, born May 8, 1927, became involved in aviation in 1973. She received her private pilot certificate with commercial, multi-engine and instrument ratings.

Seaton joined the Reno Chapter of 99s in 1974 and was chairman from 1977-79.

She is married to Dunham, H. Seaton and has seven children.

SUE SECCIA, a New Jersey resident was born in 1963. Sue earned her private pilot license in September 1995. She began flying in 1993 after experiencing a paraplane ride during a vacation. After hearing the story, a pilot friend convinced her to explore flight training which she did promptly. She flies Cessna 152, 172 and a Grumman Traveler out of Essex County Airport in New Jersey. She has also flown in Cherokees, Pitts Specials, an Aerobat, Katana, Stinson and Luscombe.

Sue is an active member of the North Jersey Chapter of the 99s and has been since joining the chapter as a student pilot in 1993. "It has been tremendous to get to know these accomplished, talented women who are always willing to help and support you. It's an honor to join their ranks. I have many friends through the 99s and without their support, I don't know how I would have completed my primary training." Sue has participated in her chapter's Poker Run and Pennies-a-Pound fund-raisers, airmarking, section and International meetings. She flew in the Garden State 300 twice as a student pilot. She has also hosted a local cable TV show on aviation in New Jersey and is a member of AOPA. In 1996 she plans to start work on her instrument rating. A marketing communications professional for over 10 years, Sue currently works with Dun & Bradstreet Information Services.

CYD SELLERS, has been a member of the All-Ohio Chapter since July 1992, where she currently serves as secretary and newsletter editor. She started flying in March of the same year inspired by a friend who took her for a flight, and taught by a fellow 99. Flying out of Dayton General Airport for recreation and renting aircraft at business destinations, provides a lot of variety in this much loved sport. "The people I've met and the friendships that I've made are opportunities that I never would have had if I hadn't learned to fly."

GRETA LOREN SELLERS, born in Peoria, IL, now resides in Palm Beach Gardens, FL, with her husband, Jeff. She comes from a family of pilots and is proud to carry on the tradition having received her private license in 1989 and instrument rating in 1993.

Before her grandfather, "Charlie" Albert, passed away, she used to love to sit with him and listen to him tell her the story about one day in 1903 when his father came home and told his mother that "those two brothers from Ohio flew their airplane today." He had heard the news about Wilbur and Orville Wright's first flight come in over the Western Union telegraph wire to the telegraph station in Leonore, IL. Seventeen years later, at the age of 23, Greta's grandfather received his pilot license after training at Midway Airport.

Greta's parents, Don and Anita Albert, are also private pilots. As a youngster, Greta enjoyed flying with her mom and dad to breakfasts held at local airports in Illinois. She also used to coax her dad into "taking her up" after school on days when the weather looked good. The most memorable airplane ride Greta had was in an experimental airplane purchased by her dad and uncle that was a replica of a Woody Pusher. She liked the feel of the wind through the open cockpit. The most satisfying flight experience Greta remembers to date was when she and her mother flew in the Illi-Nines Air Derby. They placed somewhere in the middle but had a lot of fun planning their strategy and plotting the course. Greta is a member of Chandelle Flying Club at Palm Beach International Airport and plans to continue to have fun flying by venturing to various Florida airports on the weekends.

SHIRLEY MAE SENDRAK, was born in Detroit, MI, earned a master's degree, had four children, and taught in the South Redford School District. The family moved to Santa Barbara, CA, where she and her husband taught school. Shirley is a life member of the 99s and her new goal is to fly long enough to become a UFO – United Flying Octogenarian.

Shirley feels that she has lead a fantastic life and treasures her many adventures. Flying was always in her mind since she was 12. Her dad declared flying to be too dangerous for a girl (her mother didn't) and hence, gaining a pilot license was delayed. Shirley enjoyed life. She was named Miss AFL in 1941, Detroit, and there were parties, dances and boyfriends. She describes her boyfriends as clean-cut, handsome, gentlemen who drank cokes, danced beautifully and played baseball.

It wasn't until 1969 that Shirley discovered that husband, Ted, shared her interest in flight. As a youngster she had devoted many hours to building models of aircraft. Both Ted and Shirley became licensed pilots in 1969, purchased a Mooney and flew across the United States yearly, including flying the Oregon Trail. They attended most conventions of the 99s for over 20 years. Both are, also, members of the Silver Wings.

JEAN SERVAAS, born Jan. 11, 1919, began flight training in 1970 at Freeway Airport, and obtained a private license in 1972. She joined the 99s in 1972 and has been chairman twice. She is currently fly-in chairman. Her sister and husband are good friends of Chuck Yeager, and they had him come down and give their chapter and friends an interesting talk. He spent the night with them and went hiking with them. He brought photos to show Piet and her of hikes around Mount Whitney knowing that they had hiked up with their daughter, Tanya, who is also a pilot.

Aviation experiences include flights to the Bahamas, Mexico, Baja to tip (Cabo San Lucas), Canada and across the US to Niagara Falls, checking out air museums along the way.

MARION SERVOS, received her private pilot license in 1974. Husband, Gerald Servos, was a general aviation pilot. Becoming acquainted with the 99s Chicago Area Chapter was an incentive to learn to fly. Starting out in a Cessna 150 she became the proud owner of a Cessna 172, a Cherokee Warrior and now a Piper Cherokee 140.

The chapter has sponsored a safety seminar every year. Many of the Seminars were chaired by Marion. Also, wonderful memories include flying in the Illi-Nines Air Derbies. Came in first place once.

The year 1983 brought the untimely death of Gerald and the family business has been carried on by Marion. Instrumentation and Control Systems, Inc. designs and manufacturers industrial electronic controls.

Educational background includes a BA from Fontbonne College, St. Louis. She has a daughter, a son and four grandchildren. Being a mom, a grandma, a pilot, a 99, a president of a small company, has enriched her life and broadened horizons beyond imagination.

CONSTANCE M. SEWELL, was introduced to flying in 1980 with a few cross-country flights. After those flights, she realized that flying was something that she wanted to pursue and as a result, learning to fly became her goal. She started pilot training in 1982.

Her employment provided opportunities for affordable flying and she was able to earn her private pilot license in 1984. In 1988 she joined the Florida Suncoast 99s. She is the proud owner of a Cessna 152II, N25293. She has accumulated 422 hours, mostly in her Cessna 152, but has also had time in a Piper Arrow II, Piper Comanche PA-24-250, Mooney M20F and a Mooney M20K/Rocket Conversion.

She has been a member of the Suncoast Chapter since June 1988; the Suncoast Chapter 99s newsletter editor from June 1989 to June 1991; chapter treasurer, June 1991-June 1993; vice-chairman, June 1993-June 1995; and chairman from June 1995-June 1997. She has also been a timer for the 1990 and 1991 Florida 400 Air Rally (sponsored by the Suncoast Chapter); helped with the chapter Young Astronauts program; assisted the Tampa Air Traffic Control Center in their Operation Raincheck; worked at Sun 'N Fun/ EAA spring fly-in at Lakeland, FL, for 13 years.

She is a charter member of the Florida Aero Club/ Vandenburg Chapter since 1989 and has served as secretary/ treasurer for the chapter from July 1989 to January 1991. She also served as secretary for the state board of directors for the Florida Aero Club from January 1990 to December 1993.

ELIZABETH VIRGINIA SEWELL, best known in aviation as "Susie," a name acquired while she was learning to fly, was born in Tulsa, OK.

As a member of the 99s since 1946, Susie has served the organization from the chapter level to the top, including International treasurer, vice-president and president. She helped establish the first headquarters office for 99s in Oklahoma City in 1956. During her term as International president, 1972-74, Susie led the 99s to build their original International Headquarters building, contracting with the Oklahoma City Airport Trust for the facility on Will Rogers World Airport. Susie served as an elected trustee on the AE Memorial Scholarship board, coming full circle as a former scholarship winner. She was treasurer of AEMSF one of her two terms.

During WWII, Susie began flying lessons in Oklahoma City hoping she would qualify to join the WASPs. Before she reached the age and pilot qualifications required, the program was disbanded. She continued to fly, became employed in aviation, obtained her private pilot certificate in 1945. A commercial certificate and CFI rating followed with help from an Amelia Earhart Scholarship award received in 1947.

Later, attending the University of Oklahoma, Susie studied business law, business management and corporate finance at the senior level and completed insurance courses through OU and the Insurance Institute of America. She became secretary/treasurer of the FBO, Catlin Aviation Co., where she was employed and served as it operations manager from 1959 to 1972 on Will Rogers World Airport, Oklahoma City.

In 1972, Susie became president and general manager of Aviation Development Co., distributor for Piper Aircraft Corp., in the state of Oklahoma and western Arkansas, serving in this capacity until 1983. During this time, she established an aviation insurance agency to serve both aviation firms along with many private and corporate aircraft owners.

Appointed to the Women's Advisory Committee on Aviation in 1972, Susie served on this committee under various FAA administrators for several years, making numerous recommendations to the Federal Aviation Administration.

As chairman of the International Committee, Insurance – 99s events, Susie continues to work in the interest of promoting aviation safety, more flying and keeping in touch with members in chapters sponsoring and conducting 99 flying events.

BONNIE SEYMOUR, earned her private pilot license in July 1967 at San Jose Municipal Airport. She has flown the C-150, 152, 172, 182; PA-28-180, PA-28-235 (owned 55W); PA-28-236 (own 8298W). Seymour obtained an ASEL, commercial in 1974 at Van Nuys, CA, and joined the 99s in 1968 with Santa Clara Valley and the San Fernando Valley Chapter in 1971. She formed the Lake Tahoe Chapter in 1975. She has been a 27 year member of the 99s.

Attended most 99 International conventions and section meetings since 1969. She has participated in flying to Oshkosh, Watsonville, Baja Bush Pilots, Sacramento Valley and Santa Clara Valley Airmen fly-ins. Wrote monthly publication for *The Flying Country Club* of Reid Hillview Airport. Served two three-year terms as trustee of the Amelia Earhart Memorial Scholarship fund for International 99s. She is proud of the video produced about the AE Memorial Scholarship program. Wrote and published an accompanying handbook about the AEMSF. Air races: PPD (AWTAR), 1976 Sacramento to Wilmington, DE, several Palms to Pines races, Pacific Air Race and His and Hers Races.

Seymour is married to Milt, who is also a pilot. They have a daughter, Linda, a pilot, and late son-in-law Gregg Gojkovich, was a pilot. She is a retired speech and language specialist/short term special assignments. Community service as a member of the North Lake Tahoe Redevelopment Committee and the North Lake Tahoe Advisory Committee reporting to the Placer County board of supervisors. Formed homeowners association in Kings Beach. She is also a member of the American Speech, Language and Hearing Assoc. with national certification, state licensed in California and Oklahoma as a speech, language and hearing pathologist.

LINDA SEYMOUR, has flown since she was a teenager with her family all over the west, Baja and to the East Coast. She has many stories to pass on to her daughters about "Mom and Grandma's adventures in 55 Whiskey." And then, of course, there are the airline flight attendant tales. She earned her private pilot license in May 1976 at Palomar Airport and received her ASEL rating. She joined the 99s in July 1976. She has flown the C-152 and 172; PA-28-235 and PA-28-236.

Seymour is married to non-pilot Wayne Cowie and has two daughters, Laura and Sarah. They live in Oceanside, CA. Both parents are pilots. She is a flight attendant for Continental Airlines and a substitute teacher in elementary and secondary schools. Her previous occupations include: manager of Thrifty Rent-A-Car in Truckee and South Lake Tahoe, city manager for Hertz Rent-A-Car in San Diego and was research assistant at Scripps Institute for the development of artificial blood.

She flew co-pilot with her mother, Bonnie, on the last Powder Puff Derby in 1976. She had about 50 hours when they started the race. They also flew the Palms to Pines, Pacific Air Race and other races. She flew with her father in the His and Hers Race. She has attended section meetings and various aviation activities with her mother and family. She has helped a few times at the Truckee Airshow 99s hot-dog booth.

GWEN ELIZABETH SHAFER (FREEMAN), had her first ride in a small plane as a teenager, but didn't learn to fly until she was graduate student in California. There she enjoyed whale watching by air and flying to Lake Tahoe in the Sierras, gold-rush era towns, and wine country. She also tried gliding and skydiving. She has a private pilot license with a SEL rating.

After getting a Ph.D. in chemistry in 1985, Shafer moved to Baltimore, where she joined the 99s. She is

currently chairman of the Maryland Chapter of 99s. She worked at the National Cancer Institute in Bethesda, MD, for five years as a molecular biologist. Since 1994, she has worked for a small biotech company.

She was born in Boston, MA, and received her education at the Massachusetts Institute of Technology and the University of California, Berkeley. She enjoys the love and support of her husband, Rick, and son, Robert (born September 1993).

PAGE SHAMBURGER, died Dec. 7, 1991, after a valiant battle with lung cancer. Page started her 6,700 flying hours in the mid-40s while working as an apprentice mechanic to pay for her flying lessons. One of her first professional jobs was landing at every airport in the US during 1952-53 and reporting on it for *American Aviation Magazine*. She made it to nearly 4,000 airports.

Page wrote for *Cross Country News* and was the editor of *Southern Aviation Times* while also a radio script writer, announcer and TV writer. She contributed over 1,500 aviation articles to virtually all the aviation magazines, flying her Bonanza to where the stories were. Page authored seven aviation history books, including *Tracks Across the Sky, Command The Horizon,* and *Summon The Stars.*

She was the first woman to fly on an Air Force hurricane hunt in a Lockheed WC-130 into the eye of Hurricane Betsy. "It did right much damage on the first flight," Page said in her soft Carolina accent. "I counted 138 rivets popped out of the plane." Later, Page was the first woman to ride in an RF-4 Phantom Jet.

Page's awards include the 1971 Non-Fiction Aviation Book of the Year Award; the Lady Hay-Drummond-Hay/ Jessie R. Chamberlain Memorial Trophy; the Doris Mullen Scholarship for helicopter training (she was a Whirly-Girl #142); and membership on the FAA's Women's Advisory Committee.

She was a 43 year 99s member serving as Carolinas Chapter chairman, Southeast Section governor, the International board of directors and curator for the Resource Center. She served the Powder Puff Derby for eight years, was a National Intercollegiate Flying Assoc. Trustee and a member of the board of the University Aviation Assoc.

A member of the old Moore County family which founded Aberdeen, NC, Page participated in fox hunting throughout her life, riding to the hounds and showing thoroughbred dogs. She received the Madeline M. Coleman Perpetual Memorial Award, from the horse community just last year. She was deeply involved in the Moore County Hounds, the Horse Trials, the Hunter Trials and her great love, the Walthour-Moss Foundation of which she was a trustee and secretary. 99s may wish to honor her aviation life in 99s projects through headquarters or contribute in her name to the Walthour-Moss Foundation, PO Box 147 Southern Pines, NC 28388. Page had started a brand new business after "retiring" from her travel agency.

BEVERLEY SHARP, became fascinated with flying things (birds, kites, insects, planes, etc.) as a youngster. Early memories include Sunday afternoon family excursion to Smith-Reynolds Airport, Winston-Salem, NC. She'd climb halfway up the chain-link fence for a decent view of take-offs and landings. Dreams of herself flying were tempered by the notion that pilots were born in blue uniforms.

Fortunately, she met a "normal" person with a private license. "Hark," says she, "perhaps I can do this." Family circumstances and economics forced her to postpone the aspiration for 10 years; but, in late September 1981, she began flying lessons. Bev earned her license in May 1982 and joined the 99s (Washington, DC Chapter). She quickly earned her instrument rating. In 1992, after 10 years with a Skylane, she and her 49 1/2, Ed, acquired a Baron E55 and multi-engine certificates. Grown sons, Eddie and Sam and kitty, Squeek, round out the family.

Bev became membership chairman almost immediately and has been a working 99 since, holding chapter, section and International office. She serves as chapter membership chairman again, Mid-Atlantic Section treasurer (ending second term), and International director. She is a candidate for International treasurer. Bev edits and publishes *Briefing*, a forum for discussing member rights issues.

Favorite projects; Pennies-a-Pound (first flights are thrilling) and the Flunkbusters Program, which Bev has coordinated for six years. The chapter provides incentives for marginally performing high school students: tower tours (no unexcused absence), airplane rides (10 highest grades), lesson and log book (most improved).

Most memorable piloting experience – cross-country trip with Scott Crossfield in the right seat!

EVELYN SHARP, born in Melstone, MT, on Oct. 20, 1919. Evelyn began her flying when an itinerant flight instructor got behind on his board and room bill and offered to give 14-year-old Evelyn flight instruction in payment of his debt. Enough money was raised in those depression times to put a down payment on a brand new Taylor Cub. Evelyn repaid their kindnesses from profits earned while barnstorming county fairs, rodeos and special community celebrations in Nebraska, Iowa and Kansas. Between 1938 and 1940, her personal register recorded the names of more than 5,000 passengers she took up.

Evelyn earned her private license at 17, her commercial at 18 and was one of only 10 women flight instructors in the United States at the age of 20. Between 1940-42, she taught over 350 men and women to fly in the government's Civilian Pilot Training Program in South Dakota and California.

Evelyn was one of the original 23 WAFS (Women's Auxiliary Ferrying Squadron). She brought 2,968 hours of flying time, more than any other woman pilot, to the 2nd Ferrying Division at Wilmington, DE, on Oct. 20, 1942. She was transferred to the 6th Ferrying Group in Long Beach, CA, in February 1943. Over 150 deliveries in 18 different kinds of planes are recorded.

On April 3, 1944, Evelyn was ferrying a Lockheed P-38 out of New Cumberland Airport near Harrisburg, PA. As she rotated the powerful flight off the end of Runway 30, black smoke belched from the left engine. Not able to gain enough altitude to clear Beacon Hill, Evenly set the aluminum-skinned fighter down on a grassy knoll near a wooded ravine. At that moment, Nebraska's beloved and best known aviatrix lost her life.

In memory of Evelyn Sharp, the airport was dedicated as Sharp Field in September 1948. Today a war surplus P-38 propeller secured by her father, a Nebraska state historical marker designating the entrance, and her nearby grave site marked with a simple white, wooden cross commemorate her life.

ROSE SHARP, born Sept. 18, 1929, in British Columbia, grew up in a remote section of Canada. As a teen, she was instrumental in starting a flying Girl Scout Troop in high school and she soloed at 16. She received her flight training in Troutdale, OR, and Marin Co. She has flown the Cub, Citabria, Piper Arrow, Cessna 210 and anything "old."

Flying took a back seat to work and family and it was 20 years before her husband gave her flying lessons as a gift. Although not accumulating a lot of hours she flew two Powder Puff Derbies and several local races. Holding all the offices in Bay Cities Chapter she worked on conventions, section meetings and right seat seminars. Being a founding board member of the Western Aerospace Museum afforded the opportunity to learn about antique aircraft which became a favorite with Rose along with flying aerobatics.

Together with her late husband, Donald, she operated a book store in San Francisco. She has two stepdaughters and a son and a daughter, five grandchildren and six great-grandchildren.

ELEANOR SHARPE, was in the late 50s, age, not the year, when she learned to fly at Honolulu International Airport, Hawaii. She had always wanted to fly, but felt she couldn't (polio) until a friend at Civil Air Patrol (who became her instructor) found her an Ercoupe (no rudders or toe brakes). Learning to fly and flying in Hawaii was a special experience. There was always something new and beautiful to see. As soon as she received her license in 1975, she joined the Aloha Chapter of the 99s, actively participated and became chairman 1980-81. It was a special rewarding time, and her toy poodle, Iwa, flew with her to all the islands. In 1989 they moved to New Port Richey, FL, and she joined the Suncoast Chapter. Ercoupe N3665H joined them the next year and now they are exploring Florida.

ELIZABETH "LIZ" SHATTUCK, knew the day she soloed in 1963, her life would never be the same again. She has a commercial, ASMEL rating and has accumulated more than 2,000 hours.

She was Orange County Pilot of the Year in 1975; flew in 13 races placing first twice, second three times and fourth once; worked three Powder Puff Derbies starts in inspection; organized eight competitive flying events; participated in Mercy Flights; FAA accident prevention counselor for six years; served six years on City of Oceanside airport committee; Safety Education chairman for Orange County Chapter, writing a monthly safety column for chapter newsletter; second woman to qualify as a Baja Bush Pilots; 1994 Southwest Section "Hot Wings Honoree" and owned aviation equipment sales business for eight years, Skybuys.

Her special interest in mechanics has helped in the ownership and maintenance of a Mooney M20C for 25 years. Flights over the US, Bahamas and Mexico include many of the destinations that fill her log book, and here's a toast to another 30 years!

CHARLOTTE SHAWCROSS, never wanted to fly. Unlike her husband, she never dreamed of soaring with the eagles. It never even crossed her mind. Shawcross and her husband, Lee, are the same age. After an abortive attempt at flying lessons for him at age 40, the flying bug bit him again at age 50. They bought a Tomahawk and it suddenly occurred to her that if he ever got his ticket, he was going to expect her to go up in the plane with him. She thought to herself, "No way am I going to do that if I can't land the thing." So, she decided to take a few lessons.

She's sure you guessed the rest! The few lesson soon turned into an all-out effort (a bit of ego involved, she admits) and soon she, too, had her ticket. They sold the Tomahawk and bought a Warrior. They have taken several trips together and had many a $100 hamburger. Flying does not come naturally to Shawcross, but with the enthusiasm and encouragement of Lee, her instructors and a terrific bunch of 99s, she is still hanging in there with even a few hours of instrument in her log book. It's easier and more fun for both of them to have a licensed pilot sitting there and they have shared many an animated conversation about flying.

"We have yet to soar with the eagles – but we do call ourselves the Flying Turkey Airlines!"

TAMRA SHEFFMAN, learned how to fly when she turned 40. She has since pursued her new avocation with great vigor, and continued her growth and development by advancement with her commercial and her instrument rating.

She is a proud member of the 99s, Inc., AOPA and the Florida Aero Club.

Tamra is the 1995 president of the Miami Beach Assoc. of Realtors, and the owner of Royal Palm Realty. She has been involved in real estate ownership, management, operations and sales on Miami Beach since 1978.

Her professional actives include active participation in

the Miami Beach Assoc. of Realtors, several community committees, the Greater Miami Jewish Federation and Toastmasters. She is past president of both the South Pointe Assoc. and the Professional Women's Sales and Marketing Assoc.

Her personal interests and activities which she enjoys with Ron Mayer, her husband, include ballroom dancing, underwater photography, skiing and travel.

RUBY WINE SHELDON, Phoenix 99s, holds pilot ratings for CFII single and multi-engine airplane, seaplane and helicopter. Her career includes service as an instructor, chief pilot, charter pilot, corporate pilot and also as a pilot/remote sensing technician for the US Geological Survey. She retired from the USGS. She now owns and operates a business manufacturing products for agricultural use, and is still an active flight instructor. Many of her former students are career airline pilots and embrace her warmly when they happen to meet in far-flung places.

For the USGS, Ruby flew throughout the North American Continent, collecting water resources information by means of aerial photography, side-looking airborne radar and thermal imagery. She worked extensively in Arctic Alaska and was based on an ice island 400 miles north of the Alaska coast for four months providing helicopter support between ice floes for American and Canadian scientists. From the Arctic, she flew to the jungles of Panama for additional USGS work.

Also for the USGS, she piloted a Lockheed T-33 jet, a twin-turbine Grumman Mohawk OV1-B, a Douglas B-23, a DeHavilland Beaver (on floats or wheels), a Sikorsky H-19 helicopter and a Bell UH-1F (Huey) helicopter. The honor Ruby obviously treasurers most is that the Smithsonian hung her picture on their hallowed walls in honor of her being the "first person to win a helicopter instrument instructor rating in what was then called the 'free world.'" She got it in a Huey.

In addition to the 99s, Ruby is a member of the Whirly Girls and is on the board of the Air Race Classic. Her passion is air racing. She has flown three Powder Puff Derbies, 12 Air Race Classics and numerous small races, collecting 32 trophies. Her record includes two first places in the Air Race Classic, three firsts in the Mile-High Derby and one in the Kachina Air Rally.

TAUNI SHELDON, a Native Canadian (Inuit), born Feb. 7, 1970, in Thunder Bay, Ontario, has recently become the first-ever female Inuk to be employed as a commercial pilot. Upon completion of the multi/IFR ratings, Tauni traveled north to Iqaluit, Baffin Island to fly PA31 aircraft. She describes the overwhelming beauty of the Arctic from the air as she traveled north of the 75th parallel to spots such as Grise Fjord, Resolute and Cape Dorset.

After just a short stint on Baffin Island, Tauni left to resume training with Air Inuit on their Twin Otters, and is looking forward to a career flying with them to the remote regions of Northern Quebec.

Raised in Milton, Ontario in a flying family, Tauni has flown the C-150, 172; PA-16, PA-44 and PA-38, as well as commercially the PA-31 and the Twin Otter. Of her favorite, the PA-16, she says, "I learned the meaning of crosswind landings in this four-seater, tail-wheel while flogging in and out of grass strips."

Tauni is very proud of her membership with the 99s, and enjoys her time in Operation Skywatch flying C-172s with the MOE (Ministry of Environment) personnel involved in aerial surveillance of potential illegal landfill sites in Ontario.

Throughout her flight training days, she worked operations/passenger services at Toronto Pearson International Airport. Prior to that, her high school days saw her swimming to victory claiming the Ontario gold medal for butterfly. It seems that Tauni is happiest either in the air or in the water.

KAREN SHERMAN, born in Providence, RI in 1942 to George and Leonore Sherman. Karen learned to fly in 1968 as the mother of three children, Deborah, Paul and Michael.

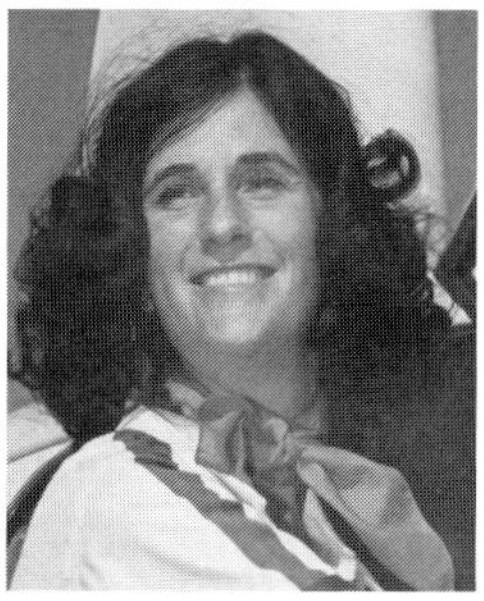

Logging more than 6,000 flight hours, commercial, single, multi-engine with instrument and rotorcraft ratings. In conjunction with her aircraft sales profession, flies all single/multi-engine aircraft, piston helicopters and turbine twins.

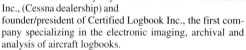

Member of 99s, Whirly Girls (#329), National Aircraft Appraiser Assoc., president of Southwest Skyways Inc., (Cessna dealership) and founder/president of Certified Logbook Inc., the first company specializing in the electronic imaging, archival and analysis of aircraft logbooks.

Grandchildren around the world include: Megan Nicole and Thomas George (Paul and Leslie's); Sophia Rose (Michael and Sarah's); Nimrod and Jonaeon (Deborah and David's).

Flown the Powder Puff Derby, Palms to Pines (with daughter Deborah as PIC took tenth place one year and placed third two years in a row with Jean Schiffmann) and the Pacific Air Race. Recipient of many aircraft sales awards from the Cessna factory. Flown aircraft in every corner of our country and recently completed an Atlantic crossing, Paris to Los Angeles flight in a Cessna Conquest 11. Has plans to be one of the pilots on an around-the-world flight in an Astra Jet in 1996.

VICKI LYNN SHERMAN, grew up and began flying in Minnesota/North Dakota and holds a bachelor of arts and science degree from North Dakota State University together with master's work in political science. She has been a resident of DeLand, FL, since 1974, is a registered real estate broker, with a Graduate Realtor Institute rating and a member of the Environmental Assessment Assoc. and is the owner of Rainbow Realty of DeLand, Inc. She holds ratings as ATP, MEL, SEL, CFI, CFII, glider instructor, BGI, AGI and IGI.

She is a member of the International Nominating Committee of the 99s and Southeast Section vice-governor and Liaison section chairman. She has served as chairman of the Wolf Aviation Fund, chairman and trustee of the Bonnie and Archie Gann Memorial Scholarship Fund, chairman of the Spaceport Chapter of the 99s, as well as, holding other chapter offices. She developed and continues to give the "Cockpit Cool" Seminar at numerous aviation safety meetings and the EAA Sun 'N Fun.

She as been involved with the FAA Accident Prevention Program since it was founded in 1991, and is an accident prevention counselor and serves on Speaker's Bureau and as liaison to the program for the 99s, Inc.

She is the deputy commander of the DeLand Composite Squadron of Civil Air Patrol, for 16 years, and serves as legal officer, check pilot, mission check pilot, flight safety officer and flight release officer. In 1989 she received the Charles E. "Chuck" Yeager Aerospace Education Achievement Award.

She has been president of the Florida Race Pilots Assoc., Inc. (sponsor of the Great Southern Air Race). For 15 years, she has been a member of the board of visitors of Embry-Riddle Aeronautical University. She is married to William E. Sherman who is also a pilot and a lawyer.

AKIKO SHIMOHIRA, born Aug. 11, 1968, in Japan, joined the 99s in 1994. She has been flying gliders for eight years and learned to fly power planes in 1991 in California. She has five Japanese National Soaring Records flown in Minden, NV, USA in 1994. She made out and return distance, free distance and distance around a triangular course. (These are multi-place, open class gliders DZ) and made an absolute altitude record, (single-place, feminine glider DIF class).

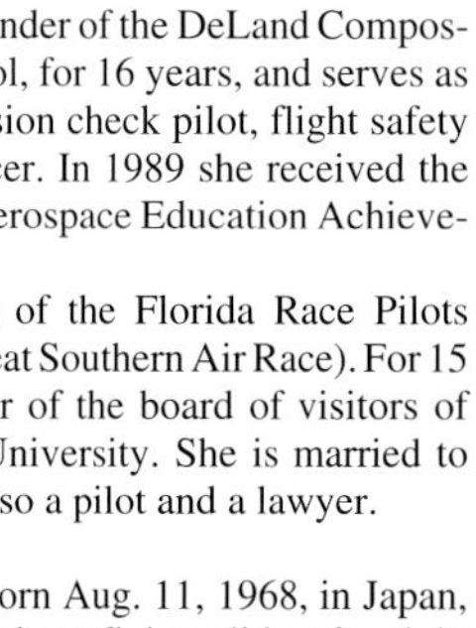

When she attempted another Japanese record the summer of 1994, she landed away from the airport and hitch-hiked because she had no choice. Fortunately, she was picked up by a kind couple who turned out to be pilots. The woman, Sandy Hart, from Fallon, NV, gave her some information about the 99s and suggested she join. She feels grateful to her as her savior and so glad she took her advice.

MILDRED "MILLIE" SHINN, 1940, CPTP allowed one female for every 40 men to qualify for 36 hours flight time. Millie Reid obtained her private license #54001 in a Rearview Sportster on Jan. 13, 1941 in Billings, MT, her commercial rating #3173 in 1946 in her Fairchild PT-19A.

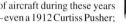

Millie flew with the Civil Air Patrol and State Aeronautics Search and Rescue from 1941-61 in Northwest as well as charter, ferry, private, cross-country. She soloed 25 to 30 different types of aircraft during these years – even a 1912 Curtiss Pusher; aerobatics was her dear "love."

She is a 1954 Montana 99 charter member, then 99s in Idaho and Washington State in Northwest Section. Held all chapter and section offices, including governor. Flew the Powder Puff Transcontinental Air Races 1961 and 1964. One of five NW pilots qualified for the WASP program 1942, then "frozen" to CAA office (now FAA).

Millie, now 76 years of age, continues flying with other intermountain pilots – keeping active old navigation skills as well as up to date and still loves every minute in the air.

DONNA LEE SHIRLEY, born July 27, 1941, in Oklahoma, was airplane struck since age 5, Donna built models and planned to be a pilot. Her father gave her flying lessons at 15. She soloed at 16 and was a flight instructor by the time she graduated from college. She was in the University of Oklahoma "Airknockers" with Gene Nora Stumbough (now Jessen) and earned a seaplane rating at Oklahoma State with Wally Funk (two of the Mercury 13 almost astronauts). She got a multi-engine rating one summer in the University of Illinois' ancient Bamboo Bomber and also soloed a sailplane at Oklahoma State. Donna was a 99 for a while in college and instructed a grand total of two students.

Graduating from Oklahoma University in 1963 and 64 with degrees in journalism and aerospace engineering, Donna worked at McDonnell Aircraft in St. Louis and then moved to Pasadena, CA, in 1966 to work on space projects at Cal Tech's Jet Propulsion Laboratory (JPL). She gave up flying after college because of money and the no-fun environment of big cities, and dropped out of 99s.

Since 1966 she has had a variety of jobs at JPL, the last 20 years in management. She currently leads the nation's Mars Exploration Program which is flying 10 robotic missions to Mars between 1995 and 2005. Presently recruited to speak at a 99 convention in Idaho, she renewed her membership, joining the Idaho Chapter as their most remote member.

JANNE SHIRRELL, Santa Rosa Chapter, native Californian, lived in the Yosemite area before moving to northern California. Worked in engineering at Standard Oil Co. and Douglas Aircraft. Put herself through college, graduating in 1975 with her bachelor's degree.

Began flying in 1971 after the death of her 14-year-old son, Christopher. She needed a diversion so she chose learning to fly. One day someone said, "Why not fly the Powder Puff Derby?" This would involve six months of preparation with charting, sewing, pre-flight and working eight hours a day. Flew San Carlos to Toms River, NJ, 1972, in a Comanche 240, placed fourteenth out of 245 planes that entered.

Has one son, Kevin, who developed the mobile kitchen for fast food chains, hotels and built 400 units for TCBY Yogurt who liked his company so well they bought it.

Two grandchildren, one step-grandchild. Married to Robert Shirrell, a health insurance actuary.

AVANIEL SHIVELY, learned to fly at San Jose Municipal from Dee Thurmond's Flight School in 1976. She has logged about 200 hours and is active with the Civil Air Patrol. She flew the Palms to Pines Race in 1977. Nell works for the post office now, after running her own bakery for many years and lives in Sedona, AZ.

ELDRIS B. SHOGREN, her 50th birthday present to herself was to learn to fly. She was licensed Sept. 8, 1973 and joined the Golden West Chapter of the 99s (San Carlos, CA) immediately thereafter. Has been active in chapter doings ever since. Attended the first ever Overseas International at Canberra, Australia in 1979, and hasn't missed many Internationals since then. Retired in 1985 after 37 years as a "school marm." Her proudest moment was in 1990 when her chapter sponsored her for induction into the Forest of Friendship in Atchison, KS. How elated she was to see her name in the walkway along with the flying greats of history. Today she teaches a two-week course called "Let's Learn About Flying" to elementary school children and conducts airport tours at San Carlos airport. Because of loss of hearing her flying days are numbered but she still is a 99, and in her heart she knows she can fly!

JOAN T. SHONNARD, born Feb. 5, 1927, first became involved in aviation in 1946 or 47 through some friends. She is a self-employed artist and the widow of Claiborne P. Shonnard. She has seven children (all adults). She has a BA in fine arts from the University of Louisville. She is past parish council secretary, Our Lady of Snows Church, president of auxiliary, Casa de Vida, board member of Reno Executive Club; projects and planning committee and past president of Auxiliary Washoe County Medical Society and past president Croesus Investment Club, etc.

VIRGINIA SHOWERS, born Dec. 6, 1925, in Leavenworth, KS, for most of her life she had a love affair with flying. Before the age of 5, she had a ride in a barnstormer's biplane, a night air tour of Washington, DC, and a dirigible flight. She was hooked! She earned her private license in 1958 and became a 99 in 1959. Two years as chapter chairman; two years as Southwest section treasurer; and three years USPFT treasurer. She has flown the C-150, 172, 182, 210; A-36 Bonanza, Piper Apache, Aztec, Warrior, PA-180, PA-260 Cessna Citation (two hours).

Worked in 1979 AWIAR start, Back to Basics Races, judged NIFA and USPFT Regional/National meets, Flying Companion Seminars, Mechanics Courses and Flight Instructors Clinics.

Flown extensively in US, Canada, Bahamas, Central and South America. Logged 2,000 hours with commercial, ASMEL, SES, BGI and instrument rating. Flown six AWTARs, nine ATWARs (second in 1976 and 1979) plus many others. Flown DFR, medical supplies to Mexican Clinic. In 1977 WPY (chapter) fourth in Section, Forest of Friendship Honoree, two Service Awards, retired from UCLA.

Two sons, David and Neal, and one grandson, Henry, 17 years.

MOHINI SHROFF, a 99 since 1965 courtesy of El Cajon Valley Chapter initially for five years. She became a charter member of India Section in 1976. Mohini took flying in 1959 as a government scholar and got her A License in 1961. She got her commercial from San Jose in 1973 with the help of Santa Clara Valley Chapter and late Mrs. Marion Barnick. She managed to attend two International Conventions, one in 1973 and the other in 1985 at Hawaii.

Mohini is the first lady to be on the Council of Aero Club of India and Aeronautical Society of India, Bombay.

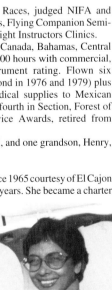

In 1973 she became the first lady Honorary Secretary of the Bombay Flying Club. She was the only lady participant flying her Piper Super Cub at India in Bangalore the same year. Mohini loves aviation and keeps motivating men and women equally to see them airborne. Mohini is the vice-governor of the 99s.

HELEN SHROPSHIRE, born May 7, 1909, in Nebraska, is a life member of the 99s since 1965, and has logged more than 2,000 hours. She has her commercial, multi/single engine land with instrument ratings. She has flown the Ercoupe, Tri-Pacer, Comanche 180, Comanche 260, Apache and Navajo.

She was co-owner/founder of Monarch Aviation, a fixed base on the Monterey Peninsula Airport. She helped to organize the Monterey Bay Chapter of 99s, served in all offices, including the first chairman. She also held offices in the Southwest Section.

She flew in air races, including the Powder Puff Derby and received a trophy for her work. She flew in treasure hunts in Mexico and the Bahamas. She received an award from the Bicentennial Commission of California for work as chairman of the Re-Enactment of the de Anza Expedition. She is on the board of directors of the Cannery Row Foundation and president and manager of the California Heritage Guides.

CONSTANCE HELEN SHUTT, born March 16, 1955, in Lexington, KY, learned to fly in Raleigh-Durham, NC, in the fall of 1978-spring of 1979. Trained in a Cessna 150 and 172. First solo on Nov. 14, 1978. Ferried new single-engine Cessnas from the factory in Wichita, KS, to North Carolina from 1979-1981.

Airman certificates include: commercial pilot, airplane single and multi-engine land, instrument airplane, ground instructor, advanced, control tower operator ratings.

She was employed as a line service technician from 1979-81; employed by FAA (air traffic control) 1981 to present; FAA accident prevention counselor (now aviation safety counselor) since 1991; and joined the 99s in 1994.

MINETTE (LEARNED) SICARD, born 1940, Yakima, WA, holds a BA music degree from UCLA, and was an established teacher and professional pianist throughout west Los Angeles. In 1966 she pursued a dual career in music and aviation, becoming ATP/CFI/AGI rated, airplane single/multi-engine land and sea, instrument and helicopter. Se was a flight/ground instructor and charter pilot for Santa Monica Flyers and Lease-A-Plane, Santa Monica and Monarch Aviation, Monterey, CA. She piloted *LA Times* helicopter photo mission, including the Sharon Tate murder scene. In 1968 she co-piloted a Cessna 337 from San Francisco to Honolulu, and also enjoyed air racing. She co-piloted the Powder Puff and Angel Derbies, and as pilot, placed fifth in the inaugural Palms to Pines Air Race. Minette holds a California teaching credential in aviation, and has taught both civilian and military classes. She is a Whirly-Girl #129, former member of Professional Helicopter Pilots Assoc. and in 1969 presented a paper "VTOL: Helicopters" to AIAA. In 1970 Minette was selected Outstanding Young Woman of California for her contributions to both music and aviation. She has 3,500 flight hours and her aircraft experience ranges from hot air balloons, blimp and gliders to B727s.

In 1973 Minette became the fourth woman FAA aviation safety inspector and first in the Pacific-Asia Region, serving at Guam IFO and Honolulu FSDO. In 1975 she became the first woman inspector and accident prevention specialist in FAA's Southwest Region, serving at Dallas FSDO. In 1976 she married FAA executive, Robert Sicard, and closed her aviation career.

AMBER SICHAU, born in Cleveland, OH, in 1967. Her first flight experience came while on spring break from Kent State University's School of Fashion Design. This inspiring hang gliding flight off the sand dunes of Kitty Hawk, NC, encouraged her to obtain her private pilot license in December 1993. She then joined Women with Wings Chapter of the 99s. She found a tremendous group of women from all walks of life and one main thing in common that bonds them together – flying

For Sichau, flying is a place where strength, incredible freedom, and inner peace can be found every time you leave the earth. Currently she has 121 hours in a Cessna 150, but she knows her flying adventures have just begun.

MARGARET E. SIEDSCHLAG, her flying experiences have always given her a feeling of accomplishment, as well as pleasure. After struggling to complete her private pilot license, she took a break from the training aspect of flying for about one year, then began training for her instrument ticket. After completing her instrument rating she moved into a complex higher performance airplane, the Mooney 201. She is presently working on her commercial ticket using the Mooney.

Her husband, Paul, is a captain with Southwest Airlines and retired Navy. When they met, she was working on the ground school training for her private. He turned out to be a real blessing! He told her that aviation was his life, which was fine with her. Now, aviation is "their" life! When Paul is working, she can fly the Mooney to meet him on his overnights. She enjoys flying with him as well, but it is to her advantage to fly cross-country alone as often as possible. This is what builds confidence and skill in a pilot. She cannot stress this enough to her sister flyers!

Siedschlag is co-founder of the Brazos River 99s, a member of the EAA, participant of the Falcon Flight Program, Communities in Schools, and overall, a supporter of women who have dreams of accomplishment.

BEATRICE SIEMON, took her first airplane ride in a Ford Tri-Motor at the 1939 World's Fair at a cost of $15. Thereafter, she went out to a nearby farmer's field to watch the barnstormers take up passengers and on Sundays, planes take-off and land at Rochester, (MN) Municipal Airport.

Later, after the arrival of three children, Bea and her husband Vern took lessons at Waukegan Airport, IL, dividing a one-hour lesson into one-half hour of instruction each. Then they bought a factory for their business adjacent to the airport, and in 1946 Bea received her private pilot license. One month later, she joined the 99s.

By 1948 Bea had earned her commercial and instrument and the couple bought two airplanes, a J-3 Cub and a Cessna 140, the latter equipped with a jump seat for their two older children. When the third child was old enough to accompany them, they purchased a Cessna 170 and then a 195.

A 99 for almost 50 years, Bea has served as chairman of airmarking and scholarship committees, newsletter reporter, and treasurer of the Chicago Area Chapter. Her vigorous flying resulted in numerous trophies, some from three Michigan Small Races, AWTARs of 1957, 1958, 1959 and the Commemorative Race of 1977. In 1991 she was inducted into the International Forest of Friendship, an honoree from her Chicago Chapter.

BARBARA LATTURE SIERCHIO, born March 8, 1932 in Abingdon, VA, attended Emory and Henry College, Emory, VA. She got her private pilot with instrument rating and soloed in 1979. She joined the 99s in 1980 and currently flies a BE-36 Bonanza.

She is a member of American Bonanza Society, AOPA, EAA, Florida Suncoast Chapter 99s and Women in Aviation, International. Judge for NIFA, USPFT and WPFC. Helps with FAA Safety meetings each month. Headed drive for donations for the 99s building at Sun 'N Fun, EAA Fly-In, Lakeland, FL, and participated in the annual convention. Participated in Air Bear in local schools and programs with Young Astronauts Chapters.

She is married to Gerald P. Sierchio, MD, and has three children: Michael, 37; Peter, 36; and Stephen, 34. She has one grandchild, Peter, 3.

HEATHER ANN SIFTON, First Canadian Chapter, learned to fly because of her husband's involvement with aviation and ownership of Toronto Buttonville Municipal Airport. She joined the 99s on attaining her pilot license in 1967 and went on to get a commercial rating with a multi-engine rating in 1971.

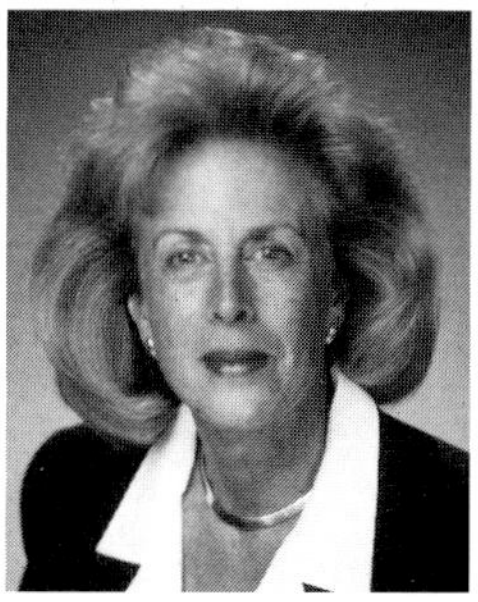

She flew a Piper Cherokee 180 to win the Governor General's Cup with Shirley Allen as co-pilot in 1971.

Heather was chairman of First Canadian Chapter in 1972 and 1973. A very active 99, she has held the offices of secretary and treasurer for the chapter, also vice-governor for East Canada Section in the 70s. She was involved with setting up the Aviation Safety Seminars and on the original committee for Operation Skywatch. She also chaired the FCC Poker Runs for three years.

When the 99s Canadian Award in Aviation was established in 1974, Heather became chairman of the board of trustees and has been in that office for the past 15 years.

Heather opened one of the first general aviation retail shops in Canada at Buttonville Airport – The Prop Shop was and is a must stop for pilots. She was also the owner of Aviation Unlimited, a Piper aircraft dealer and distributor for Eastern Canada.

Heather is presently president and CEO of Toronto Airways Limited at Toronto Buttonville Municipal airport. She is a board member of ATAC (Air Traffic Assoc. of Canada.)

BARBARA WEIBLER SILAGI, a teacher certified K-14 in special education and supervision when Sputnik appeared and with it many new training requirements for educators. Science credits beyond the graduate level without prerequisites were hard to come by, but she found an aerospace class that included going up in an aircraft. She had never flown before but set off bravely and fell in love with flying. Within the next 10 years, she amassed thousands of hours; more night than day, more instrument than visual, more multi-engine than single and also earned her ATR multi-engine land and sea. She holds commercial single-engine land and sea, multi-engine and instrument flight instructor, is a certified flight dispatcher who loves flying hot air balloons.

Finding flying more challenging than 30 years of teaching, she went to work for the airlines as a flight dispatcher. It wasn't long before she was flying C-47/DC-3Cs as the company's instructor-pilot and making countless flights, mostly cargo runs, into the snowy reaches of Canada and throughout the states. Deregulation brought much of the flying to a stop but she continued to work with experienced chief pilots writing aircraft/airline instruction manuals before reaching the FAA's mandatory age deadline. This mother of one and grandmother of four big boys has flown in Powder Puff Derbies and other races but hasn't retired. She manages a publishing business in Florida, lives beside a runway and maintains her membership in the Chicago Area Chapter, where she has many friends.

BARBARA ANN SILCOX, born Oct. 14, 1929, in Camden, NJ, had her first flight lesson in April 1973 and received her private certificate in October of that year. During WWII, she dreamed of learning to fly and she was sure she would have joined a service had she been a few years older. After she raised her family of five children and then divorced, she was fortunate to have a generous and enthusiastic pilot as her employer. He paid for her flight training and they have been together 25 years.

They have flown their B55 Baron extensively throughout the Caribbean and Central and South America. In 1977 they flew all over Alaska in their brand new plane, visiting Anchorage, Nome, Kodiak Island, the Diomedes, Barrow, Fairbanks, Skagway and many more.

In 1981 they flew down the west coast of South America to Punta Arenas, around Cape Horn and back north along the east coast to Rio, Brasilia, Belem and home. In many places, none had ever met a woman pilot! They made friends in many countries.

Between them, Tom Styer and Silcox have eight children: Ginny Silcox, 46; Dianne, 42; David, 40; Barbara, 38; Wendy, 34; James Stanley, 24; Jessie Myers, 21; and Justin Stanley, 18; 8 of her 16 grandchildren: Sarah, 15; Lianna, 13; Julia, 10; Grey; Kimberly Silcox, 6; Benjamin Silcox, 4; Lisa McPherson, 4; Kyra McPherson, 1; two great-grandchildren and another due momentarily. One daughter and one grandchild are pilots and five grandchildren are student pilots. She hopes flying opens their eyes to the world the way it has for her!

JEAN SHIELDS SILVERSTEIN, was born and still resides in Pittsburgh, PA. She began flying lessons in a Beechcraft Skipper at the Allegheny County Airport in Munhall, PA, at the same time her youngest son started flying early in 1984. She completed flight training in November of the same year flying a Cessna 152 and joined the 99s that same month. Jean taught the aviation ground school course at Community College for five years. She has held several offices in the Western Pennsylvania Chapter, as well as, editing the chapter newsletter. Wherever her travels take her, she rents a plane, takes pictures from the air and gets it in her logbook – Pearl Harbor; Kissimmee, FL; Milwaukee; Virginia; New Jersey, Scotland and more! She is married to Jerry and has three children: Amy, 32; Scott, 30; and Andrew, 28.

BARBARA J. SIMMONS, born in Tulsa, OK, on Nov. 18, 1923. She moved to Indianapolis at age 8. Barbara started flying lessons in July 1966 and bought a 1960 Cessna 175, as co-owner, one week before she received her pilot license (ASEL) in August 1967. After several years of flying the 175 she then purchased a 1964 Cessna 182 Skylane (again as co-owner) which she flew until she retired from flying. Barbara joined the 99s, Inc., in October 1968 and has held chapter offices as vice-chairman, secretary, treasurer, APT chairman, air activities and ways and means chairman. She was co-chairman of the 1978 FAIR (Fairladies Annual Indiana Race) as well as co-chairman of the North Central Section meeting held in Indianapolis in 1989. Barbara attended many section meetings as well as International meetings and enjoyed flying to many of them. She has been a member of AOPA, PIA and is a charter member of the Cessna Skylane Society. Even though she is no longer flying, she is still very active in her Indiana Chapter and the 99s, Inc.

GINGER JONES SIMMONS, Oregon Pines Chapter, Northwest Section, cannot decide if the emotional suction to the airport began when her dad drove her brothers and her to the airport to see the 4th of July fireworks when they were little, or if it was orientation ride taken in college to tag along with her cousin, Amy's latest wild hair idea. In either case, it planted the seed of flying in her heart and she found herself at a meeting of the local flying club in the spring of 1982.

She joined the 99s Chapter of Oregon Pines in 1984 and has experienced the greatest thrills and challenges of her life with these women. She had held several offices within the chapter, including four years as chapter chairman. She has enjoyed the opportunity to attend many Northwest Section conventions, met dozens of aviation enthusiasts and made wonderful friends throughout the United States. The 99s have provided the gold medal bonus to flying in her life.

The experience of a lifetime was to travel to Cape Canaveral with fellow 99, Lyn McGuire, to watch the morning launch of STS-54 with friend Susan Helms aboard. To see Orlando, a manatee, innumerable pelicans, too many alligators and beautiful beaches combined with Apollo rockets, moon buggy, space suit and shuttle displays, launch pads and a launch of Endeavor, is really an explosive feeling for a pilot.

Currently she holds a single-engine land private pilot certificate at 300 hours and enjoys the simple fun of general aviation with other pilots, or taking off with family to visit more family. Her daughter, Mary (3) makes an airplane by crossing two hairbrushes and daughter, Marina, (2), makes engine sounds whenever an airplane passes overhead. She looks forward to sharing the freedom, fun, challenge and reward of aviation to them.

LADELL CHARLENE SIMMONS, born Feb. 13, 1938, in Tacoma, WA, while in grade school, read all the books about Amelia Earhart, and her dream was to be a pilot. Her dream started coming true in 1985, when she started flight training at Rose Aviation, Inc., at Hawthorne Municipal Airport in Hawthorne, CA. She and her boyfriend purchased a 1947 Bonanza as he was a pilot also, while she was still working to get her ticket. She joined the Long Beach Chapter in 1986. She has flown the Cessna 150, 152, 172, 182RG; and Beechcraft Bonanza A-35.

In November 1987, she transferred her membership to the High Desert Chapter as she had moved to the Antelope Valley area. In November 1988 the Antelope Valley Chapter was formed of which she is a charter member. She has served as treasurer, vice-chairman and chairman and many committees. She flies just for pleasure and loves it so much and has met so many friends in the 99s. Never give up on your dreams.

Simmons' previous employment was with Boeing and Northrop Grumman Corp. She has one son, Duane Lynn Harpel, 37 an air traffic controller and pilot; and two grandchildren, Amanda Kay Harpel, 14 and Brandon Jay Harpel, 12.

SANDRA SALIBA SIMMONS, born July 14, 1941, in Prescott, AZ. Orphaned at 4 years old, Sandra was raised by maternal grandparents on a farm in Arkansas. She graduated from Alma, AR High School, attended one year at the University of Arkansas and graduated from Southern Methodist University in 1964, with a BA in art. Sandra learned to fly in 1966 at Waco, TX. She was the first female pilot to become type-rated in the CE-500 (Cessna Citation) receiving the rating in March 1973. She was also the first female pilot to receive the flight engineer turbojet certificate from Braniff Education Systems Incorporated in 1973. Along with other firsts, Sandra was the first female pilot hired by Braniff International Airways, March 19, 1974. After Braniff declared bankruptcy, Sandra was hired by Republic Airlines, flying Convair 580s, based in Minneapolis. She resigned six months later to return to the "New Braniff." Furloughed 10 months later she was hired by Alaska Airlines, April 8, 1985, and is currently based in Long Beach, CA.

Sandra resides in Crystal Beach, TX with her husband, Nellis Dye, a United Parcel Post Service captain. Aircraft flown are the C-150, 172, 177RG, 182, 310, 421; CE-500, Beech Musketeer, Beech Baron, Citabria, J3Cub, Moony 21, Grumman American Tiger, B-727, 737; DC-9, MD-82-83, Lake Buccaneer and CV-580, with type-ratings in the CE-500, B-727, B-737, and DC9. She has accumulated more than 17,000 flight hours. Sandra has been a 99 since 1971. She has participated in two Powder Puff Derbies (1972 and 1976) along with Dallas Doll Derbies (1971 and 1972) and the San Diego Bicentennial Race in 1969. Sandra is the proud mother of Andrea Simmons Swanson, 29 and Serena Simmons, 25.

SHARON SIMMONS, born April 19, 1933, joined the 99s in 1962. She has the SEL, SEA ratings and has 600 hours flying time. She has flown the Cessna 150, 172; Mooney, Piper Archer, Bonanza and Commander 112. She has been chapter chairman, section vice-chairman, section governor and AE Scholarship chairman.

Simmons is married to Bob Dorr and has four children: Deborah, Eugene, Scott and Tammie, all grown. She has six grandchildren: Sarah, Adam, Mark, David, Jonathan and Nicole.

MARGARET JUNE SIMPSON, born in Jersey City, NJ, began flying in 1952 at Walden-Newburgh Airport (now a golf course) in Aeroncas. She holds a commercial rating and has flown Cub, Cherokees, Cessnas and Grumman.

Other aviation experiences include hang gliding, ultralight, parachuting, seaplane and gliders. Part-time work at Aurora Aviation for a short time.

Simpson holds a master's degree in professional nursing. Through the years she has traveled to all continents, involved in health projects and field research on a number of subjects. She takes advantage of any flying opportunity when abroad.

As a member of the 99s, she has held offices of past president of International Flying Nurses Assoc.; a mission pilot in the Civil Air Patrol, rank of lieutenant colonel. Simpson has published articles in nursing and aviation journals and newspapers. Involved in aviation history in Orange County, NY, and in 1994 published book, *Orange County Airports, Past and Present*.

She is a widow with one daughter, Pamela L. Phelps, and three grandchildren: Megan, 19; David, 22; and Ryan, 20.

LINDA SINACORI, grew up and currently lives in Westchester County, NY. She is an associate director of marketing for a large investment firm in New York City. She started flying in 1986, earning her private certificate in 1988 and instrument rating in 1991. Her first passenger was her mother, who likes to recall the early days of aviation in Dutchess County, NY, and from whom Linda inherited her love of flying.

Linda met her husband, Stephen, also a pilot, when both were members of a local flying club. She has served as an officer of that club since 1992. She and her husband own a Cherokee 6 (whose N number on the open cargo door nicely displays a 99!). They fly for pleasure, especially enjoying long weekends all along the East coast and taking family and friends to the islands off of New England for day trips. They also take advantage of the six's hauling capabilities for charity flights through AIRLIfeLine.

Linda plans to continue earning ratings with the goal of instructing or operating a small Part 135 business.

BEVERLY J. SINCLAIR, born in Mineola, NY, Oct. 21, 1953. At 25, Beverly started flying for pleasure. Working as an interior designer, she was spending so much time at the airport, she went to work for the local FBO at Arapaho County Airport in Englewood, CO.

In 1979 Beverly got her private certificate, and joined the 99s in 1980. Beverly found a job selling aircraft parts, flying the company's C-182, in Colorado and Wyoming. When the economy declined, Beverly was laid off and returned to the interior design field to finance her commercial, instrument and flight instructor ratings.

While working as a first officer for a commuter airline, flying a Beechcraft 1900, arriving in Minneapolis after an extremely turbulent flight, she overheard one of the passengers tell his wife, "Honey, the flight was so bumpy, the flight attendant sat up front, next to the pilot the whole trip."

At 36, Beverly began to teach flying during the day and worked as a dispatcher for an air ambulance company at night. She is now a Beechcraft 1900 first officer for US Air Express, with over 3,800 hours, and hopes to work for a major airline. Beverly is married to Alan Stewart, an A&P technician for Learjet/Bombardier.

LORETTA SINCORA, "What took you so long?" Those were words of Loretta Sincora's jumpmaster in preparation for her first parachute jump. The jump, which took place in May 1989 when she was 61 years old, was the culmination of a desire she had held since she began flying in 1946. Loretta earned her private license in August 1947 while she was a member of Civil Air Patrol, 1945-1957. She joined the 99s in 1963 at the urging of her late husband, Russell. She works at many Chicago Area Chapter activities.

She also serves as a volunteer for the 99s and EAA at Oshkosh. This life member received the Chapter Service Award in 1993 and was inducted into the Forest of Friendship in 1995. Since retiring from General Electric in 1988, she has attended almost all chapter meetings, section meetings and all 99s International Conventions.

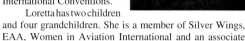

Loretta has two children and four grandchildren. She is a member of Silver Wings, EAA, Women in Aviation International and an associate member of the Chicago Flight Instructors Assoc. She received her first glider instruction in 1993 and her first helicopter instruction in 1994.

VIRGINIA BAKER SISSON, born in Boston, MA, in April 1957, and now resides in Houston, TX, where she teaches geology at Rice University. She has flying in her blood as grandfather, father and five uncles all flew for various branches of the US military. However, she began her flight training when her husband, Will Maze, enrolled her in ground school as a graduation present. After a slow start, she completed her private pilot certificate in 1986. Then, with the excuse of needing a plane for acquiring an instrument rating, she and husband purchased a N1941V, a Piper Archer. Now, she has accumulated more than 650 hours of flight time, most of them in a N1941V. She and her husband especially enjoy long cross-country trips up through Montana, North Dakota and Michigan. She uses her plane to fly students over their field research areas along the Texas coast to give them an aerial perspective of geologic processes.

RUTH LOWE SITLER, born on May 7, 1933, in Pittsburgh, PA, learned to fly 30 years ago at Kent State University Airport in Ohio and has 10,000 hours. A charter member of Lake Erie 99s since 1965, she participated in the Powder Puff Derby with Marsha Orton and with Ruby Mensching in the Angel Derby and Indiana Fair Ladies Race. She has flown Cessnas, Pipers, Seminoles, Senecas, Navajos, Aztecs and Apaches 411s.

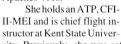

She holds an ATP, CFI-II-MEI and is chief flight instructor at Kent State University. Previously, she was safety program manager for the Cleveland FAA.

Ruth holds a BS in Ed studies, an MS in counseling, and is pursuing a Ph.D. in curriculum and instruction at Kent State. Her research involves gender differences in learning to fly

She has presented to Women in Aviation, Orlando, FL, the Minneapolis Magnet School Conference, Dowling College and Ohio State's Symposium on Aviation Psychology.

Ruth's family includes children: Bob, Susan and Kathleen and grandchildren: Annie, 18; Chaya, 16; John, 12; Luke, 8; and Daniel, 4.

EVELYN ADAMS SIZEMORE, was given a check for her first flying lesson as one of her 35th birthday presents from her husband, Bob, a USAF pilot. Bob flew combat in all the wars his country fought from 1944 through 1973. Bob and Evelyn had five children. Their first son, Robert Sizemore Jr., earned his private pilot license flying in Virginia in 1964 just a year after his mother was licensed while they lived in Tokyo, Japan. He went on to be rated commercial, instrument and instructor pilot.

Sport flying was Evelyn's love. Nothing was more thrilling to her than perfecting loops and rolls in the sky at the controls of a Beech T34 or an old J3 Cub. She joined the 99s in 1965 in Washington, DC. Following her husband's retirement from the USAF after 31 years active duty until his death in December 1993, Evelyn and Bob had the joy of adding hours to their logs flying together in everything from Aero Commanders to sailplanes.

With her active flying days brought to an end by physical disabilities, Evelyn now spends her days writing about her favorite subject, aviation, and the people who have brought it from cloth wings and single-engines to supersonic jets and spaceships.

SUZANNE LYNN SKEETERS, born in Santa Maria, CA, on Sept. 27, 1953, graduated from California Polytechnic State University in San Louis Obispo, CA. In March 1977 after seeing an ad for a $10 introductory flight at the airport, she signed up for it and it was "love at first flight." She received her private pilot license in January 1978, and her instrument, commercial, CFI and multi-engine rating in 1979. She joined the 99s in 1978 and has been an active member in various chapters since that time. She has worked as a flight instructor, charter pilot, air ambulance pilot, corporate pilot, commuter airline pilot and was hired by Northwest Airlines in 1984. She has flown as Boeing 727, DC-10 and Boeing 747 second officers and B-727 and B-757 and DC-10 first officers and will check out as captain in the near future. She has also served as an instructor for DC-10 second officers and B-757 captains and first officers. She is also a member of the International Society of Women and Airline Pilots. Suzanne has one son, Kevin Novotny, born Dec. 24, 1989.

DORA DAVIS SKINNER, born on July 25, 1884 in Niigata, Japan to a missionary family. A graduate of Oberlin College and the University of Pennsylvania Nursing School, she married Dr. Henry Harlow Skinner (1878-1953) and moved to Yakima, WA, in 1913. She served in WWI as a nurse and volunteered with the Visiting Nurse Service in Yakima.

Dora soloed on Nov. 5, 1930 and became the 385th woman pilot licensed in the United States. In the 1930s, *Dawn*, her Cardinal airplane with the Kinner engine, could be seen doing fly-overs along Yakima Boulevard during Armistice Day parades. Dora and her husband flew for many years and were said to be the oldest flying couple in the United States.

Dora was an active member of several aviation associations, including the 99s. She liked to recount the time she attended a 99s meeting in Portland, OR, at which Amelia Earhart was the guest speaker. When Earhart stated that her next speaking engagement was in Seattle, Dora told her she would be delighted to offer her a lift. Miss Earhart noted the inclement weather, declined and took the train.

When Dora's daughter, Ellouise (Beatty) learned to fly to 1944, they became one of the few mother-daughter pilot teams in the 99s. Then Ellouise's daughter Jenny Beatty carried forward the aviation torch by soloing on Dora's 97th birthday.

Dora Davis Skinner lived in Yakima, WA, until her death at the age of 98 on Aug. 11, 1982, survived by her two children, six grandchildren and two great-grandchildren.

NANCY SLIWA, born Oct. 5, 1955, in Monterey, CA, has bachelor's and master's degrees in computer science and has worked extensively in fields of artificial intelligence and robotics. From June 1981 through November 1994, she worked for NASA in Virginia, California and Florida, as a researcher and manager. She is currently continuing her career through private consulting in business process re-engineering, as well as dabbling in writing, doctoral education and other entrepreneurial activities.

Nancy learned to fly in 1990, and joined the Santa Clara Valley 99s soon thereafter. She transferred her membership to the Spaceport Chapter when her husband accepted the presidency of Embry-Riddle Aeronautical University in Daytona Beach, FL. She logs over 120 hours each year flying

for business and pleasure in her Mooney 201. She and her race partner, Marcie Smith from the SCV 99s, have done a number of races together, including the 1993 and 1995 Air Race Classics, the 1992 Great Southern Air Race and the Palms to Pines in 1990-93. In 1994, they also did a wonderful trip together across Canada, flying from Victoria, British Columbia to Halifax, Nova Scotia. Nancy is a regular at Southwest section meetings and at the annual International Conventions.

Nancy also belongs to the Aircraft Owners and Pilots Assoc., Mooney Aircraft Pilots Assoc. and Angel Flight. Her daughter, Tabitha, studies psychology at the University of California at Santa Cruz and is contemplating law school. In addition, Nancy and her husband, Steve, own two golden retrievers, Honey and Chandelle, who also love to fly.

SHANNON D'ANDREA SMALL, born in Irving, TX, on Dec. 9, 1968, started flying on Dec. 8, 1990 and soloed Feb. 21, 1991, while completing her bachelor's degree at Texas Christian University. She completed her private certificate in August 1991 and moved to Tulsa, OK, to complete her training.

She joined the 99s in 1991 while training in Tulsa, OK. She currently holds a commercial pilot certificate, single and multi-engine land, instrument airplane and certified flight instructor rating single and multi-engine land, instrument airplane. She completed her certified flight instructor training in June 1992. She earned her Gold Seal flight instructor certificate on Dec. 10, 1994. She currently works as a freelance flight instructor in the Tulsa area as well as work as a contract instructor for Tulsair Beechcraft.

Her goal in aviation currently is to fly Corporate Aircraft. She feels that the sky is the limit to future possibilities in the aviation/aerospace industry and she cannot say what the future may hold for her. She does know one absolute: She is unwilling to let her feet stay on the ground.

PATTY SMART, 46 flies a 1979 Cessna Skyhawk XP, out of the old Naval Air Station, now the Grosse Ile Municipal Airport. Her flying world started in 1974 after having moved from Yuma, AZ, where an observant instructor noticed her yen to try to do what her husband was accomplishing. Twenty years later, finds her with 915 hours (planning large party at 1,000 hours), an instrument ticket, experience in four aircraft (American Yankee, Cherokee 180, Mooney, Hawk XP) a new husband, Addison, who actually learned to fly and got an instrument rating due to the relationship (49 1/2), 3,000 elementary school children initiated into the world of flight, charter membership and past president of the Greater Detroit Area Chapter of the 99s, and current airmarking committee chairman. All the flying is pleasure flying and promotion of aerospace education. When she leaves public education she may teach ground school.

ALMA GALLAGHER SMITH, introduced many to the joy and benefits of aviation as a flight instructor and newspaper woman. She was born in Concord, NH, in 1918 and graduated from Saint Mary of the Woods College in Indiana in 1940.

Her interests in aviation was roused on an assignment for her father's newspaper. She became a flight instructor and taught Naval air students at Dartmouth College under the V-12 program during WWII.

She returned to the family newspaper instructing part-time, and traveled extensively with the Strategic Air Command as a newspaper woman. On one trip, she broke the sound barrier at the controls of a F101.

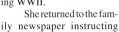

She was a charter member of the Northern New England Chapter of 99s, served as chapter chairman, North East Section governor, North East Air Race Board and Mahn Scholarship Committee. She held office in the Aviation Assoc. of New Hampshire and was a member of the Aviation/Space Writers Assoc. She passed away in 1991.

ANNETTE CATHARINE SMITH, born in Minneapolis, MN, and migrated to California in 1951. Bob, a retired systems analyst, and she have been married 41 years and have three children and four grandchildren. She has 18 years currently as an assistant administrator, with the state of California. Flying was always a deep desire of hers and even as a child, her mother couldn't stop her from climbing or jumping off high places. She was especially fond of the roof and determined to fly one way or another. Finally in 1982, her husband surprised her with a flight demonstration. Explaining to her that she would be doing the flying, butterflies quickly took flight in her stomach. Soon their one hour flight became one and a half hours because as her husband puts it, "The smile never left your face." Upon landing, they purchased a new Piper Warrior with lessons starting shortly thereafter. Her first solo cross-country found her enveloped in smoke for which she used the standard 180 to retreat and her second cross-country found her wrestling with the, then unfamiliar, auto pilot. Undaunted by these events, she received her license in April 1983, after a three month delay, due to rain, for her uneventful long cross-country. Priority one, after the license, was joining the 99s (Fullerton Chapter) the following month. She has acted as both vice-chairman and chairman since joining. Other memberships include AOPA, NCWA and past finance officer in CAP. To date, flying has been confined to pleasure trips.

BELINDA M. SMITH, born March 29, 1949, in Chicago, IL, joined the 99s in 1990. She has her CFI, commercial SE, commercial ME, instrument ratings.

She has flown the Beech Sport, Skipper, Sierra Duchess, Cessna 150, 152, 172; and Piper Arrow.

She won the Amelia Earhart Memorial Scholarship in 1994 and used it to obtain her CFI certificate. She is the owner of a 1975 B24R Beech Sierra.

BILLIE L. SMITH, Indiana Chapter, became a member of the Indiana Chapter of 99s in 1948. She is the mother of one and grandmother of four. She and her 49 1/2, "Smitty," own a Beech Musketeer N2316Q. Their flying has been for pleasure, which has taken them over the greater part of the USA.

Billie was chapter chairman in 1960, when the first Indiana Air Race was conceived. This annual race became known as FAIR. She was on the race board for many years and has served as its hospitality chairman for a number of years. She has served on many committees, attended many North Central Section meetings and International Conventions.

CAMELIA M. SMITH, born Sept. 8, 1936, in Fayetteville, AR, has been a lifetime resident of Northwest Arkansas, except for three years as a flight attendant for TWA, learned to fly in 1971. She has accumulated more than 4,000 hours flying time. With commercial and CFII ratings the next step was an air charter operation called Smith Air. Living on a beef and poultry farm plus an aviation business created a very busy lifestyle. She has flown many SEL, Cessna 310, 414, 421 and KingAir A90.

Becoming a 99 in 1975 has been one of the greatest benefits of being in aviation. The wonderful 99 network and the inspiration of other women pilots has been invaluable.

Other flying organization memberships and involvement's are AOPA, the Arkansas Pilots Assoc., Clipped Wings (TWA former flight attendants), an FAA accident prevention counselor and past chairman of the Arkansas Chapter of 99s. Air races and flying vacations have also brought many memorable adventures.

She is married to Bill Smith, a native of Hindsville, AR, who is also a pilot and farmer. They own a Cessna 150, 182 and T210. They live on their farm at Hindsville with their two children, Whitney Smith, 29 and J.B. Smith, 31, nearby.

DONNA JEAN SMITH, born June 27, 1945, in Norton, VA, learned to fly in 1975. Donna received her BA from the University of Kentucky and her MS from the University of Southern California. Her career led to work as a civil servant at Patuxent River Naval Air Station in Maryland where she worked on various test and evaluation projects for the Navy, Air Force and NASA.

As a mathematician, she worked on numerous projects such as the X-29, LRACCA, AH-1 helicopter night targeting system, P-3 Update IV, E-2C ARPS, VH-60 Presidential Helicopter and GPS. Her work concentrated on software engineering. She has a back seat ticket for Navy jets.

Donna is an instrument rated commercial pilot. Her love for aviation has led her to try everything from gliders, ultra-lights, balloons to Navy jets. She is an active member of the EAA. She has held numerous positions in the EAA, 99s, BPW, as well as professional organizations. She served on the St. Mary's County Airport Commission.

Donna now teaches mathematics at St. Mary's College of Maryland and Charles County Community College.

Donna is married to Dennis Smith. They own a restored 1947 Navion and are building a Lancair. They have three children, Belynda, Richard and Crystal and two grandchildren.

DOROTHY J. SMITH, born Nov. 12, 1925, in Decatur, IL, started pilot training in 1959 at Bob Shank Airport, Indianapolis, IN. Received her private license in 1960. Commercial in 1965, joined the 99s in 1961.

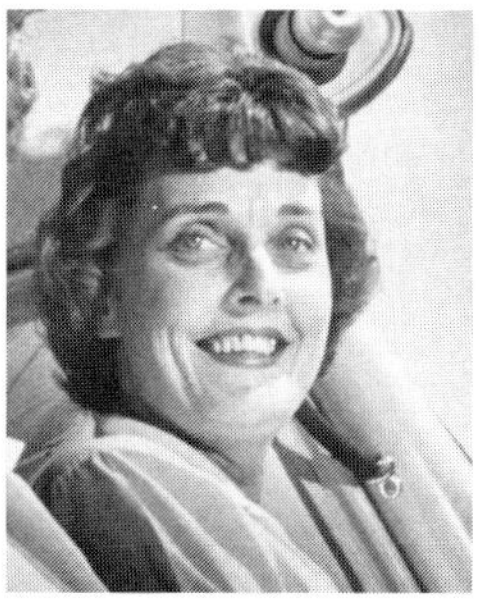

She has flown the Piper J-3, Aeronca Chief, Piper PA-20, PA-22, PA-28-180, PA-28-140, Cessna 172, 175, 180, 210. Has held all chapter offices. Has served as chairman of all chapter committees. Was section nominating chairman one year. Active in racing. Was chairman of the Fairladies Annual Indiana Race two consecutive years. Has served on the FAIR board eight years. Has flown the PPD, IAR, Michigan Small Race, Illi-Nines Air Race and the Fairladies Annual Indiana Race. Received the "Pilot of the Year" Award from the Indianapolis Aero Club in 1965 and 1992. North Central Section Governors Award in 1983. She has been married to William T. Smith for 49 years.

IDA VAN SMITH, was born to Mr. Theodore D. Larkin and Mrs. Martha Jane Keith Larkin on March 21, 1917 in Lumberton, NC. It was after college and beginning a career teaching math in a North Carolina high school, marriage and a move to New York that Ida began to think seriously again about flying. Then came her lovely family of four. After years had gone by, Ida resumed her teaching career in New York.

While sitting at her desk in New York University about to begin work on a doctorate degree, Ida left her desk and drove to LaGuardia Airport, where she had her first flight lesson. It did not take long for Ida to decide LaGuardia was just too big, so to North Carolina she went.

Ida felt she was fortunate indeed to have this excellent flight instructor, Ed McCorvey. She drove from Lumberton to the Fayetteville Airport, arriving every morning at 6:30 a.m. "rain or shine."

Suddenly there were groups of adults and children who stopped Ida as she approached her car to return home. They were excited and had many questions. This group as well as others in Lumberton were invited to meet with Ida in her parents' home. Basically that was the beginning of the Ida Van Smith Flight Clubs in 1967.

With her private pilot license in hand she returned to New York, continued her training toward her instrument rating and advanced ground instructor license at Farmingdale Airport, Long Island, NY.

In the meantime, the Ida Van Smith Flight Clubs were growing in numbers. Ida was asked to set up clubs not only in Jamaica, NY, but also in Manhattan and the Bronx, Dayton, Washington, DC, Long Island and other places as the years went by. The only International Club was in St. Lucia.

For several years Ida hosted "The Ida Van Smith Flights Clubs Cable TV Show" in New York City, assisted by pilot and ground instructor, George G. Daniel, the executive manager of the flight clubs.

In 1974 Ida Van taught a ground school aviation course for adults provided by York College of City University of New York. Of the 40 students, 39 passed the FAA written the first time taking it.

Ida Van is grateful to the staff and to the Long Island 99s, as well as 99s all over who "have always been there for us." Ida says thanks also to all the other organizations.

Ida Van's Long Island Chapter of 99s surprised her by placing her name in the International Forest of Friendship in 1984. "It was truly a wonderful experience!" said Ida. Ida has not missed sponsoring names for the Forest through the flight clubs continuously through 1994. In 1986 the Flight Clubs sponsored the name of the great Bessie Coleman – the world's first black woman pilot. Each year thereafter the Ida Van Smith Flight Clubs, Inc. has placed a name or names, including 1993 – Dr. Mae C. Jemison, the world's first black woman astronaut.

As the Ida Van Smith Flight Clubs, Inc. grew, they began to publish booklets entitled, *Historical Review*, along with the name and date, to help keep their history correct and up to date. There have been five such publications.

On Jan. 11, 1995, Ida Van Smith was one of eight 99s to be honored on the 20th Anniversary of International Women's Year in "A Salute to Women in Aviation and Aerospace," sponsored by the Washington, DC Chapter of the 99s, Inc. and the Zonta Club of Fairfax, VA.

Ida Van's picture and story lines appear in the Smithsonian's National Air and Space Museum, in the Pentagon and in the International Women's Air and Space Museum in Dayton. Ida Van's famous column, "Come Fly With Me" ran for many years in a Queens, NY, newspaper, *The New York Voice*. She also penned a column, "Flying Over Rockdale."

Ida Van appears in the 99s movie "Women in Aviation" in the FAA movie "The Ace Program" and others. In 1978 Ida Van held three workshops at the National Aerospace Congress held in Dallas. She was also a guest speaker on her IVS Flights Clubs in the General Assembly.

In 1990 the Borough of Queens dedicated May 2 as the Ida Van Smith Flight Club Day and has incorporated the Flight Clubs in the school program at Duke Ellington P.S. 150, Queens.

In 1986 Ida Van received the World Aero Space Education Organization Award voted to her by the International Women's Conference held in Nairobi and presented by NASA. In 1983 NASA flew Ida Van from Washington to Kennedy Space Center to witness the first black astronaut's launch into space. In 1983 Ida Van received the Tuskegee Airmen's National Community Service Award for outstanding aviation work and accomplishment with youngsters. Ida Van has received several awards from the FAA: 1988, 1986, 1982.

In 1984 she received the National Sojourner Truth Award (highest honor of the National Assoc. of Negro Business and Professional Women's Clubs, Inc.)

JANETTE SMITH, earned her private pilot license in Santa Maria in 1993. Invited by a friend to ride along on a sightseeing trip, she knew immediately that flying was something she would have to do. Janette has been a member of the Santa Maria Valley Chapter of 99s since earning her wings. Learning to fly in a Cessna 152, she immediately transitioned to a Mooney 201 which is her aircraft of choice.

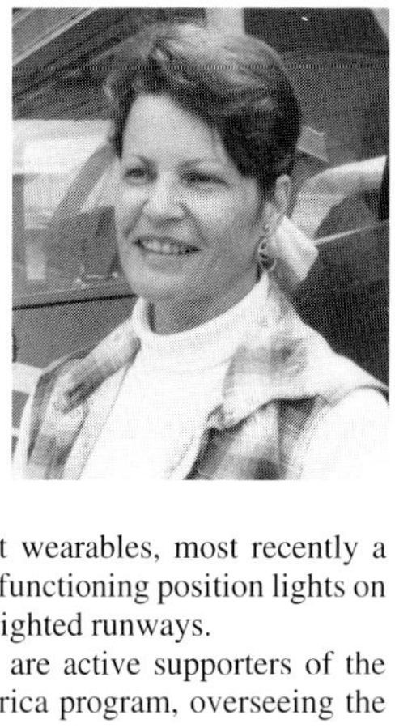

Janette is employed in the construction field in the Santa Maria area. Other interest include golf, bowling and handicapping horses. She also enjoys designing aircraft wearables, most recently a collection of sweatshirts with functioning position lights on planes which are flying over lighted runways.

She and husband, Don, are active supporters of the local Future Farmers of America program, overseeing the care of a variety of animals as students prepare the animals for local and statewide competitions and often accompanying them as chaperones to the events.

JOAN JUDEEN SMITH, born Aug. 4, 1930, in Duluth, MN, earned her pilot license in 1982 at the age of 52. She went on to get a float rating, commercial, instrument and multi-engine rating. She is a member of the Minnesota Chapter of the 99s.

She broke no speed or distance records, but counts her 12 years, so far, of flying as one of the most significant things in her life.

A flight from Superior, WI, to Santa Fe, NM, in her Super Cub, as well as several Canadian seaplane seminars are highlights. Float flying is her favorite. Other activities include talking to Girl Scouts, Red Cross blood flights and serving at 99 booths.

She has a married daughter, Darcy Smith 42; and a married son, Brad Smith 40, who is a bush pilot; and one grandchild, Emily Jo Hillmer 1 1/2 years old. Her husband, William Chas Smith, has been a pilot for 40 years.

Meeting 99s from Walla Walla, WA, to the Turks and Caicos remains the most delightful aspect of flying.

JOYCE SMITH, a native Texan, started flying in 1971 and holds an airline transport rating with the Beech 1900/300 type rating. She is a flight instructor (ASEL, AMEL, AI) and ground instructor (advanced, instrument). Smith has logged 5,000 hours of flying as a charter pilot, chief corporate pilot and chief flight instructor.

Smith's educational background includes a MS in vertebrate zoology from Texas Tech University and a teaching certificate issued by the state of Texas. Smith accumulated 15 years of high school teaching and a year of university instructing under a graduate teaching fellowship.

Smith worked for six months as an ATC data specialists, six months as FSS specialists (developmental) and seven and a half years as a FAA principal operations inspector. She has worked for the NTSB as an air safety investigator for three and a half years and has been an FAA designated examiner for three years.

LAURA SMITH, born July 28, 1967, in Long Beach, CA, completed private, instrument, commercial and CFI at Sanford Air, Sanford, ME, in 1992. She has logged 350 hours since then. Smith has flown the C-152, C-172 and Piper Warrior. She joined the 99s in 1992. Flying is the most rewarding activity she has undertaken, and therefore she has decided to pursue it as a career. She is currently chapter chairman of Katahdin Wings Chapter. Her most exciting flying adventure to date was a solo flight in a Warrior from Sanford, ME, to Omaha, NE, to visit her mom.

MAE ELAINE SMITH, born Dec. 30, 1935, in Camden, NJ, a life member of the 99s, joined the Long Island Chapter in 1977. She has held the offices of chairman, secretary, treasurer along with hospitality, aerospace education and many other chairmanships. She has as a delegate attended every annual convention since 1979.

As an active member of the Long Island Chapter, she has volunteered as a speaker for Air Bear, teachers, and flight companionship seminars, Painted Compass Roses, flew children for the Ida Van Smith Clubs and many first timers for their Pennies-a-Pound.

Currently she is also serving as secretary for New York/New Jersey section and has also been treasurer, director and nominating chairman.

She is a graduate of Rutgers University with a BA in biology and teaching. She has been an active church member and community volunteer for Scouts and Little League. Her husband, Robert Smith, got her into flying and they have two adult children, April Leslie, 35 and Robert C. III, 34.

Mrs. Smith owns an Archer II and has been a member of the AOPA since 1971. She has also flown the Champion Citabria, Cessna 150, 152, 172; Piper Cherokee 140, 180, 181; Cardinal, Comanche 180 and Musketeer.

MARILYN PATRICIA SMITH, born July 5, 1942, in Long Island, NY, is a recent "transplant" to Punta Gorda, FL, where she resides with her husband, Adrian, a 25,000 hour retired TWA captain, and their 1952 Cessna 195 – a classic with a 300 hp Jacobs radial engine – which they affectionately call *Red Rover*. Her husband claims that every great event in his life had some connection to flying; and so he felt it appropriate that their marriage in December 1991, take place in *Red Rover* – in the air – with each of them taking turns at the controls as the other recited their portion of the marital vows. Thanks to *Red Rover's* five-seat capacity, the back seat accommodated a minister who conducted the airborne ceremony, flanked by the best man on one side and the matron of honor on the other. Four other C-195s and a Stearman formed the "wedding party procession in the sky" flying in formation behind the nuptial aircraft. A keyed mike carried the entire ceremony on 122.9 to their aviator friends aloft and to those with hand-held receivers observing from the ground. While no marriage is truly made in heaven, this one came close to it!

Included among their 35 honored guests in attendance for the occasion were the FAA examiner who gave Marilyn her flight instructor and instrument flight instructor checkrides over 20 years ago, and Ellie (McCullough) Odorico, founder of the Long Island Chapter of the 99s and the "mentor" Marilyn adopted when first joining the 99s in 1969.

"Ellie was the one who taught me 'real world' flying – some pilots fly governed by the FARs; Ellie's flights are governed by EARs (Ellie's Aviating Regulations!)," Marilyn recalls, "Ellie was and still is one smart, no-nonsense, skillful pilot and a fiercely competitive air racer. I considered it an honor when Ellie began inviting her to be her co-pilot for several air races after we'd flown together in her Cherokee 140 (*Schatzy*) on several occasions years ago. Flying races with Ellie meant you passed all stages of Ellie's co-pilot test – in other words, you had a compatible personality, had the ability to be on the same brain waves and do what she needed before she asked, share her serious approach to racing as the means to winning, and didn't do anything that would cause her to view you as a liability rather than an asset when the race results were announced!" To this day, 25 years after first meeting through their 99s affiliation, Marilyn and Ellie share a mutual respect for one another's capabilities, achievements and long-standing friendship. As Marilyn puts it, "My biography would not be complete if it did not include Ellie's influence on the path I pursued in aviation."

For someone who got into aviation by accident (no pun intended) and never really had any prior desire to fly either as a passenger or pilot (a classic white-knuckler), Marilyn's love-of-air which began with the completion of her first flight lesson is one that has influenced many aspects of her life and provided additional career paths and interests to which she devotes times and effort.

After earning her private pilot certificate in 1968, she was "hooked" and went on to earn her commercial certificate and instrument rating for single and multi-engine airplane, her advanced and instrument ground instructor certificate and her Gold Seal flight and instrument flight instructor certificate. Upon joining 99s, she pursued active involvement in the organization, as well, holding a variety of committee chairs and offices on both chapter and section level, including chapter chair and election to the office of section governor. Additionally, she was the first chair of the International Safety Education Committee, has served as a facilitator for the council of governor's and board of director's and most recently, chaired the International membership committee. She has flown as pilot and co-pilot in a variety of regional 99s air races; and in 1973, she developed a syllabus

for and then conducted several seminars for people afraid to fly – SAFE (Seminar on Air travel for Everyone) – which was based on her own past fear-turned-love experience with flying. She was assisted in the conduct of these courses by members of her Long Island Chapter, as a chapter fund-raising project. A year later, at the 1974 International convention, Marilyn gave a slide presentation to the 99s attendees. Upon their adoption of it as an organization-wide fund-raiser and educational program, she donated the syllabus to the International organization for by and of the chapters. Known also in the FAA's Eastern Region as the creator of Courses Plotted for Safety (an FAA program which had record-breaking pilot participation), she earned the distinction of appearing on several television shows which highlighted her aviation involvement's – "To Tell The Truth" for founding the course for people afraid to fly; "American Adventure" series; "Weekend Pilot" as the FAA Eastern Region's selection to be filmed and interviewed while piloting her plane; and several local news and educational shows. In 1975 the Long Island Chapter presented her with the bronze Amelia Earhart Medal for her contributions to and efforts for the 99s and aviation.

Meanwhile, a career that started out in business management with a focus on architecture and construction and included ownership of construction and interior design companies, was cast aside in favor of an opportunity to be a part-owner and manager/chief pilot of a flight training and aircraft rental business. After four glorious years of flying and starvation, Marilyn learned that "the way to make a little money in aviation was to start with a lot!" And so she returned to business management and picked up some advanced academic training in the areas of human resources development, organizational behavior, corporate culture and the latest developments in management philosophies and processes. The application of this advanced knowledge during her 12 years with the leading cosmetic packaging company in the US, enabled her to become one of their top seven executive managers reporting directly to the president – the only woman at that point in the history of the corporation to be promoted into the top executive ranks. Still hungry for a challenge and no higher position to which she could climb, Marilyn eventually resigned her corporate executive post and started her own business as a management consultant and facilitator. However her marriage to husband, Adrian, and their relocation to Florida signaled a formal break for her from the business world and a return to increased flying opportunities and 99s involvement.

Marilyn describes herself as "A woman with a zest for life, a thirst for challenge, a love for flying, and at age 53 . . . who knows what I'll be when I grow up?!"

MARTHA SMITH, got her license with the Moffett Field Navy Flying Club in 1988, and joined the SCV 99s in 1990, where she is past chairman. Marcie works as an aerospace engineer on spacecraft operations for NASA-Ames on both the Pioneer and Galileo Probe Projects. Marcie has her commercial ASEL with her instrument rating, with more than 800 hours. Marcie has raced the Air Race Classic, the Great Southern and Palms to Pines with Nancy Sliwa. She won SCV WPOY in 1991. Marcie also enjoys volleyball and sewing and enjoys making airplane quilts to donate for auction every year. Marcie's husband, Chris McKay, is a space scientist at Ames. Marcie now flies out of West Valley Flying Club at Palo Alto.

MARY WARNER SMITH, First Canadian Chapter, decided to learn how to fly herself in 1970. At that time and for 30 years, she was a teacher and librarian at Bayview Junior High and Earl Haig Secondary School in North York, Toronto.

Mary joined the 99s in 1975 and held many offices in the 70s. She was chairman for the 25th Anniversary of the founding of East Canada Section. In 1977 Mary was co-chairman of FCC's Aviation Safety Seminars. She became chairman in 1978 through to 1980.

She organized the design and presentation of an illuminated scroll to Charles Kingsford Smith. It was flown with him to be given to the 99s in Australia in 1978, on the occasion of their hosting the International Convention. The route in a Cessna 340 was undertaken by Mr. Kingsford-Smith to commemorate the anniversary of his famous father's historic flight 50 years ago.

She was dedicated to teaching children and Charlie Brown flights were one of her special projects. Mary and her husband, Warner, flew many trips in their own plane across Canada and south of the border. Mary attended at least seven International Conventions and events in the US, including one memorable convention in Philadelphia that she flew herself.

In 1990, Mary retired from teaching. She has time now to visit her grandchildren and catch up with posting events and pictures in the chapter scrapbook. Still traveling extensively, Mary is presently the official photographer for First Canadian Chapter.

PEGGY DAWN SMITH, Maple Leaf Chapter, received her private pilot license in 1973 and joined the 99s the same year. Flying has always been a form of relaxation for her and she has really enjoyed exploring the countryside both here in Canada, the US and Australia. She recalls that when they were in Europe, they rented a car and one could usually find them at all the "little" airports and on the back roads everywhere. A big treat was renting a 172 in Australia and meeting some 99s "down under." She always takes the roster when on a cross-country and she has met so many interesting women by giving them a "local" call.

Smith acquired her first plane in 1976, an old 172, CF-SSH and it was known affectionately around London as *Sugar-Sugar-Honey*. That plane took her everywhere – good times and bad, met good friends, remember getting stuck in the middle of nowhere more than once, flying to convention with other 99s. It was a hoot! In 1986, upgraded to her current baby, C-GDZG – *Dizzy* for short. Its a Cessna Cardinal, C-177RG and that has proven to be another adventure, going places and doing things a little bit faster. Longer cross-countries, it has a six hour range which is a little bit longer than her range.

She has held several offices at the chapter level, including chapter chairman and went on to serve as governor of East Canada Section in 1982-84. Section committees included nominating chairman and aerospace education. She has been a trustee for the Canadian Award in Aviation since 1984. Currently, her time has been spent at chapter level participating in the Science Fairs and Poker runs. She is also a member of the Cardinal Club and COPA. In her spare time, she loves her job at Fanshawe College where she is involved with cooperative education in the business school.

WENDY ANNE HARE SMITH, born in western Queensland, the birthplace of Qantas. Her grandfather, Walter Cecil Miller, was an original shareholder. She learned to fly on the family sheep station, Coreena, an Aboriginal word meaning, "big waterhole," when her father built an airfield and it became an authorized training facility. Wendy won two flying scholarships to gain her commercial rating and an A Grade instructor rating. For five years she was constantly on the move in outback Australia instructing, before moving to Sydney to join Illawarra Flying School at Bankstown Airport.

In 1979 she began flying LearJets out of Sydney and Melbourne airports for Stillwell Aviation. Wendy the first Australian woman to hold a senior commercial pilot license.

Today, Wendy lives on the beautiful north shore of Oahu, HI, with her husband, Carlton Bruce Smith III, where she enjoys to the fullest her two other great loves, oil painting and golf.

YVONNE C. SMITH, born Oct. 25, 1943, became interested in aviation as a Unicom operator. Later became the first licensed female air traffic controller in the Bahamas, then earned her pilot license on Aug. 17, 1975. She joined the 99s in 1976. She has flown the Cherokee Warrior, Cessna 172 and glider. She is still actively flying today for her personal enjoyment. She and her husband own a C-172 and fly mostly from the Bahamas to the Florida mainland. Current member of the 99s and past governor of the Caribbean Section. Also a member of AOPA. She is married to Henderson M. Smith (pilot) and has three children: Eric V. Smith (pilot), Linda Forney and Barry Smith.

PATRICIA JO "PAT" SMITHSON, joined the 99s in December 1986 as a 66. She began flying in October 1986 after a friend, Shirley Gage, took her for a ride in her Luscombe; that fall she enrolled in the TMCC Ground School and she started flying lessons at General Aviation in Stead. She currently has a commercial, CFI and BGI ratings.

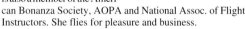

Smithson is a sales administrator assistant for Baker & Taylor Books. Smithson has two children, Chip and Raymond.

She is the SW Section chairman, 66 Program and has been chairman and vice-chairman and treasurer. She is also a member of the American Bonanza Society, AOPA and National Assoc. of Flight Instructors. She flies for pleasure and business.

BARBARA JEAN SCHOELEN SMOLA, born Aug. 11, 1941, in Kingfisher, OK, started flying lessons in the summer of 1993 at the age of 52! It proved to be quite a challenge. One of the proudest moments of her life was passing her checkride to get her private pilot license. It was a great day and one of the most fun things she has ever done! Her husband, Steve, is also a pilot and they really enjoy trips together in their Cessna Skylane.

Smola is a registered nurse, graduating from St. Anthony School of Nursing in Oklahoma City. She practiced nursing for several years, then retired to raise a large, active family. When the children were all in school, she opened the first of five different retail stores. She now owns and operates a wholesale supply business to retail stores throughout Oklahoma and Arkansas. She has been active in all phases of community and church volunteer work including two terms on her city council in Watonga.

Her husband, Steve, is president of Wheeler Brothers Grain Co. and they farm and ranch in this area. They have six children: Brenda, Marilyn, Shelly, Chris, Angela and Michael; one grandson, Daniel, 2; and numerous exchange students in their family.

KATHLEEN SNAPER, moved to Las Vegas, NV, in 1969 and flying was not a priority. Never having been too interested in aviation which was strange because her father was an aircraft mechanic. In fact, she was frightened of being in a "little airplane."

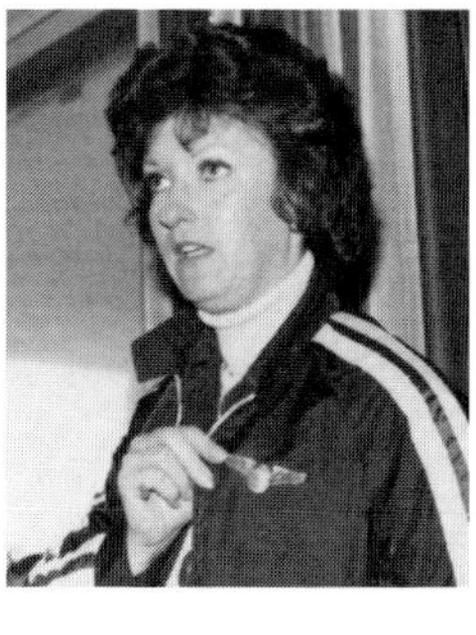

In April 1973 a friend mentioned he had started flying. She said that sounded like fun. For her wedding anniversary present, her husband said if she was interested, she should take a demo flight. She went, signed up for lessons, and four months later had a private certificate. That started a career she never anticipated.

This included instructor certificates (IFR, SEL and MEL), ATP, seaplane, more than 10,000 hours, first female designated flight examiner in Nevada, racing the final 1976 Powder Puff, many other races since 1979. Established two records: Flying Below Sea Level – distance and endurance; 14 years chief flight instructor and since 1974, a 99.

LINDA SNELL, member of the Florida Goldcoast Chapter for 16 1/2 years. Began flying in 1971, earning her private license. Her mother and her husband are also licensed pilots.

Recently retired after 21 years as an air traffic control specialist at the Miami Flight Service Station.

Presently serves as membership chairman with the Goldcoast Chapter and works as a volunteer at Fairchild Tropical Gardens.

EVELYN PEARCE SNOW, born Dec. 12, 1925 in Santa Rita, NM, soloed 1961 in Abilene, TX. Earned private license 1965; joined the 99s in Lincoln, NB. Degree in aviation from Louisiana Tech University; taught aviation at Louisiana Tech/Barksdale.

Received Amelia Earhart Scholarship in 1978 for multi-engine rating in BE-55. Flew three Powder Puff Derby Races, including 1976 race with daughter, Betty Heise of Abilene Chapter.

Snow married in 1944 to AAF Pilot Daniel W. Snow, has one daughter, Betty, and three sons: Daniel Jr., David and Steven; also five grandchildren and one great-grandchild.

ANASTASIA MARIE SNYDER, became a pilot in 1980 at the age of 19. She went on to earn an instrument rating in 1981, commercial in 1982, certified flight instructor in 1983; multi-engine in 1984, CFII in 1985 and multi-engine instructor in 1986. She joined the Bay Cities Chapter 99s in 1986.

Stacy attended College of Alameda where she became a certified airframe and powerplant mechanic. She then enrolled in San Jose State University's aeronautical operations program.

Stacy was an active member of the Alameda Aero Club from 1982-87. She taught dozens of student to fly. On March 31, 1987, Stacy and her student were killed in a midair collision, after taking off from Oakland Airport. Flying is what Stacy loved to do and always had time for. From competing in aerobatics to painting watercolors of aircraft, her interest in aviation was multifaceted and a lifetime commitment.

MARILYN SNYDER, born in Oakland, CA, bought a Cessna Skyhawk in 1990, and her instructor, David McConnell, taught her to fly it. She obtained private pilot license in June 1991 and joined the 99s the same year.

She now owns a Bonanza A36 which lives at Gnoss Field in Novato, CA. Snyder has also flown a Cessna 172. She belongs to the American Bonanza Society, 99s.

She is an active critical care nurse and also works as an administrative nurse and belongs to a number of nursing organizations. She has four children: Steve, Scott, Sandy and Spence.

CAROL SOKATCH, born Aug. 13, 1934, in Freeport, IL, was licensed on April 10, 1969, and her husband, Jack, was licensed on April 11, 1969. Their 200 hp Beechcraft Musketeer (fuel injected) is similar to the planes used in their original flight training. They have owned N469JS since 1982. Sokatch joined the 99s in 1982 and she has also flown the Cherokee 140, 160; and Cessna 172, 206.

Flying is basically for pleasure for both Sokatches. Carol currently is a volunteer with the National Cowboy Hall of Fame and Omniplex Sciences Museum, Oklahoma Pilots Assoc. and OUHSC Faculty Women's Club as well as 99 activities. She chaired the 1993 Okie Derby.

Carol has three children: David, born 1958; Barbara, born 1960; and Karen, born 1962. She has six grandchildren: John Sokatch, 9; Jill and Ellyn Sokatch, 6; Austin Buchan, 7; Colling Buchan, 4; and Dalton Buchan, 2.

DORRO CONVERSE SOKOL, began flight training in 1959 when they acquired a Comanche 250 shortly after the birth of their fourth child. Private SEL was issued in January 1960. This was upgraded to MEL when they traded for a Beechcraft Travelair in 1961. Instrument rating 1969, followed later with a commercial rating 1973. After moving to Oxbow Ranch, Prairie City, OR, and building an airstrip, they traded for an E50 model Twin Bonanza whose supercharged engines were better suited to their mountainous area. Later on in the mid-60s, they were fortunate to have Columbia Cascade 99 chapter (which she had joined) weekend Fly-In at the ranch. They replaced the E50 with an H50 Twin Bonanza in 1969 and it came with her to Sisters, OR (still ranching) in 1971.

Central Oregon 99s is her chapter now and she has assisted some with the Palms to Pines annual race. She participated in the 1966 Petticoat Derby put on by Oregon Chapter 99s and placed first, then flew the Northwest Section Derby 1967, ending in Montana where she was second. Her flying has covered the US northwest, southwest, south central to Florida and the Bahamas also Canada and Mexico. Over the years, she has logged in excess of 5,000 hours, has flown several models of Barons plus some time in a Citabria (on the ranch), Piper Tri-Pacer and Cessna 172 and 150.

Before leaving California, as a member of the Flying Farmers, she served as Flying Farmer Queen and was active in many of their flying activities.

NIKKI SOLPER, Phoenix 99s, joined the Phoenix 99s in 1993 after several years in the Monterey Bay Chapter. She had learned to fly in Watsonville, CA, in 1974. Nikki holds multi-engine ATP, FE turbojet, commercial SEL, advanced ground instructor and previously held SE and ME instructor ratings. She has worked as a flight instructor in California and Alaska, flown freight and charters out of Chicago, as first officer and captain with American Eagle at O'Hare and second officer on DC8s for UPS.

Nikki married husband, Mark, 1987, and they have a son Mark, born in 1988. Besides her heavy-duty 99 activities which have included aerospace education, air races, section meeting planning, oral history and Future Women Pilots, she volunteers a lot of time to her son's elementary school and considers herself a "full-time mom" for the present.

MARIE SPENCE, born in Kent County, Ontario, Canada, earned a private pilot license with training from Chatham, Ontario. She joined the First Canadian Chapter in 1966 and became a charter member of Maple Leaf Chapter when it formed in 1969. Spence and her husband, Bob, have flown all over Canada and the US in their Cessnas. Bob rebuilt a Fairy Swordfish, (Navy plane that torpedoed the Bismarck) the only one flying in North America. They now have about 60 hours flying time in it and will be at several airshows. Marie has a BA and master of education degrees.

She taught school for 23 years and is now an artist. She and Bob have two children, Peter, 37, has his commercial rating and Susanne, 36, has her private pilot license, and both can fly formation at airshows. She has two grandchildren, Kate Wilkins, 9 and Aaron Wilkins, 7.

PRISCILLA H. SPENCER, born Jan. 5, 1912, in Vermont, soloed Jan. 12, 1946; forced landing March 16, 1946; private license on Aug. 30, 1946. She joined the 99s in September 1947 and then joined the Civil Air Patrol in 1950 (flew as observer for downed aircraft). As a 99 member she flew in AWTAR – TAR #2 as co-pilot in July 1963. The pilot was Bertha Haycock (owner of Comanche 250) in Bakersfield, CA, to Atlantic City, NJ. She has flown the Piper J-3, J-5, Ercoupe, Stinson, Piper Comanche 250.

She has one child, Mrs. Coral W. Poole, 55. She has two grandchildren, Carter Poole, 34 and Jamie Poole, 25.

PAMELA GINDLESBERGER SPRANG, was looking for a new challenge in her life. She decided to take up flying since it was something she had thought about doing someday. Connie McConnell, from the All Ohio Chapter, was her first flight instructor. She did not receive her private pilot license until 1990 because a new job and moving got in the way. Pam has earned a rating each year since then. Her most recent rating being a flight instructor certificate in 1993. Other ratings

include instrument in 1991 and commercial in 1992. She is now taking a rest from earning ratings and enjoying traveling with her pilot husband, Phil, in their Cessna Skylane. Pam has belonged to the International 99s since February 1991, a charter member of the Women With Wings Chapter, and is also a member of AOPA and Skypark Plane Janes.

HELEN F. SQUIRES, born Oct. 3, 1952, in Australia, became involved in aviation in 1986. She had gone up with a friend who really wanted to learn but needed moral support. After a few flights, she was hooked and has flown ever since. She joined the Reno Chapter of 99s in 1986. Her current ratings are private pilot license with high performance, retract. and taildragger ratings. She flies a Citabria C-172 for pleasure.

Squires holds a (E)AS in computer science, BS majors in biology and secondary education; and minors in business; an MS educational leadership, administrative credential. She enjoys taildragger flying, hiking, fishing, camping, swimming, travel, black and white photography, travel to Australia, reading in all areas, research in alternative education and community projects. She is currently working on an educational specialist degree at the University of Nevada, Reno.

She is also member of the Kiwanis Club, Audubon Society, Sierra Club, AOPA and Stead Neighborhood Council (an advisory group to the Reno City Council), National Assoc. of Graduate and Professional Students and currently Human Diversity Coordination, board member; representative for the Educational Leadership College, University of Nevada, Reno Graduate Student Assoc.

JENNIFER L. ANDERSON-STACK, born June 27, 1956, in Menominee, MI, received solo training with the Civil Air Patrol in June 1973 in 172. Completed license in 1990, instrument rating in C-337 in May 1994 and commercial in April 1995. She has flown the C-152, 172, 177, 182, 337; BE24R and Cub J3. She is married to James G. Stack and has three children: Sandra Jean, 16; Wendy Lynn, 14; and Erik David, 12.

E. DIANE STAFFORD, born Jan. 25, 1939, in Louisville, KY, joined the 99s the year she received her private license, 1969. Her husband had learned to fly; had house built on backyard of airport (2,200 foot grass strip) known lovingly as Blue Lick International. She would fly anytime there was an empty seat. Decided she should know how to land an airplane. She found out that was the hardest part, you had to learn everything else to make a good landing and it sort of snowballed from there ... to getting her private license and instrument rating. She has flown the J3, Aeronca Champion, Rockwell Commander, DeHavilland Chipmunk, Cessna 150, 172, Great Lakes Bi-Plane, 2-22 Schweizer Sailplane and Piper Seminole Turbo. She is married to Ed and has three children: David, 37; Mark, 35; and Dale, 24.

VI BLOWERS-STAMM, soloed Oct. 20, 1967, and has logged more than 2,100 hours in various aircraft including Cessna 150, Citabria 150, Cessna 172 and currently owns a Cherokee 140. She has entered and flown in 41 99s sponsored races including the Angel Derby, Illi-Nines, Fair Ladies Indiana Race, Kentucky Air Derby, Michigan Small Race and the Buckeye Air Rally from 1970 to 1994. She has many racing trophies and won the Achievement Award (All Ohio) in 1977 and 1994. She recently participated in the relay flight carrying the ceremonial

shovel from Parkersburg, WV, to Dayton, OH, by flying the first leg.

Vi has been a 99 since 1968. Other organizations to which she belongs are: AOPA, EAA, Flying Angels, Inc., Ohio Cherokee Pilot's Assoc. and Silver Wings. She has taken 20 Young Eagles for their first flight. She is retired from Wright Patterson's AFB, passed her 75th birthday and flies every day, weather permitting.

CORALIE A. STAMP, on a beautiful October day in 1964 in the state of Ohio, Coralie began her flying career. In

July 1966, she joined the 99s. Over the years she has been busy with raising four sons, two of whom are in aviation (Randolph and James are helicopter mechanics and William and Norman are the only non-flying members of the Felger men). Coralie has been active with the CAP working with the cadets. She also lent her aviation knowledge to the Girl Scouts assisting them in acquiring aviation badges. Lake Erie Chapter recognized her accomplishments by naming her 1985 Pilot of the Year. Coralie and husband, Don, are restoring a 1944 SNJ and are members of EAA, attending Sun-N-Fun and Oshkosh each year. The couple own a Navajo, Cessna 182RG and a Sky Bolt.

JANICE STANFIELD, born Sept. 9, 1954, in Corpus Christi, TX, went out to local FBO in a small mountain town in Colorado and went on a friend's check-out ride. The thrill and challenge had her taking lessons a week later from a former Air Force Flying Tiger, Skip Stanfield. After three flight, she soloed and was hooked forever. She just kept going one step at a time, and as she looks back now, if she had known all she would have to do, she might have been discouraged. She earned her private license and instrument and commercial rating at Glenwood Springs, CO, in 1987-88 and her multi-engine and ATP in 1990 in Dallas, TX. She joined the 99s in 1992. She has flown the C-150/150, C-172/180, CP-210, C-206, C-207, C-402, C-414, DHC-6, BE-1900, SA-227, B-737, C-180, N25-3, and BN2A.

Stanfield always looked ahead and loved every minute of it. In 1988 she moved her family from Glenwood Springs, CO, down to Belize in Central America for her first flying job. She flew for Tropic Air and Wings of Hope from May 1988 to June 1990. Her children went to school on the island of Ambergus Caye and tolerated her quest for flight time. They all moved back to the states in 1990 and she went to work for Conquest Airlines in Austin, TX. Began there as an FO in 1900s and later they switched to SA227s which she upgraded in 1993. She has more than 5,000 hours and continues to help other females become interested and/or viable candidate for their first flying jobs. She has logged more than 200 hours in taildraggers.

Stanfield has three children: Qurisha Leigh, 19; Ben Robert, 17; and Destry Scott, 15. She is a member of the Fort Worth Branch.

PAT STAPLETON, a lifetime New Jersey resident earned her private pilot license in 1993.

She flies a Cessna 172 and has logged more than 200 hours to date.

Pat is an active participant in the North Jersey Chapter. She is currently working on her instrument rating.

Pat also likes to travel and has been to many domestic and foreign places.

BETTE PAT STARK, started flying at Bowling Green State University and Findlay, OH. After graduation and a private pilot license, she headed for the WASP at Sweetwater, TX. A B-25 bomber, which she flew, was named *Miss Pat*.

During the war she received her commercial and instructor ratings. She married Paul Stark and has three children: Roger, Sandra and Bette.

She was chairman of Lake Erie Chapter, Pilot of the Year and received the first Achievement Award for her extensive flying experience and contribution to avia-

tion. She was inducted into the Forest of Friendship at Atchison, KS.

She was the commander of Flotilla 716 of the US Coast Guard Auxiliary and founded the first air operations in 1979, the only air division in the auxiliaries seven state 9th Eastern District, that takes part in search and rescue over Lake Erie. She is a proud charter member of Women With Wings and is still flying.

ALICE-JEAN MAY STARR, born Norwood, NJ, in 1920. Joined the 99s in 1949. Served as WASP for 22 months during WWII. First lieutenant AFR. Learned to fly at Sky Harbor Seaplane Base, Carlstadt, NJ, as a member of Women Flyers of America, Inc. Has flown approximately 30 different types of planes; winding up WASP service ferrying pursuits. Prior to WASP, worked for Navy, Atlanta, GA, as Link trainer instructor. Ferried surplus planes after war. Worked for Navion sales agency, Teterboro Airport, NJ. Published WASP cartoon book, 1947.

Married Ted Starr; two children: Gerald W. and Julia M. Two grandchildren: Amelia, 5 1/2 and Susannah, 3. Won 11th Annual Reading Air Show Women's Race, and came in fourth in International Air Race, Montreal to West Palm Beach with Canadian co-pilot. Presently PR liaison for NY-NJ Section.

ANN RACHELLE (DOANE) STARRET, born in Brampton, Ontario, Canada, Nov. 9, 1950. Always wanted to fly but it wasn't practical. Did the practical . . . became a chartered accountant, a partner in a CA practice in Georgetown, Ontario, and had two children, a son, James and a daughter, Rachelle. She became quite involved with various community service organizations and the Institute of Chartered Accountants serving at various executive levels and on various committees.

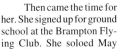

Then came the time for her. She signed up for ground school at the Brampton Flying Club. She soloed May 20, 1991, and became licensed Sept. 11, 1992. In September 1993 she and a partner purchased a Cessna 172.

As well as flying numerous hours in Ontario, she has flown from Ontario to Prince Edward Island, loves flying in British Columbia and northern Washington.

She joined the Civil Air Search and Rescue in St. Catharines, Ontario and COPA Flight 28 before becoming involved with the 99s First Canadian Chapter. She was the chapter treasurer for 1994-95, 1995-96 and coordinated the chapter's first Young Eagles Rally.

Motivated by the comradeship and enthusiasm of the fellow 99s, she is directing more of her energies to flying. She looks at this as a beginning and hopes to accomplish a lot more flying and plans to introduce many more people to the world of aviation.

WYVEMA STARTZ, a native Oklahoman, employed by Inovest, Inc., as an accountant. Married to Lawrence in 1954 and has two sons, David and Paul, and an 18 month-old grandson. Member of Oklahoma Chapter where she served as airmarking chairman and currently as treasurer.

She got her private license in 1989 just in time to participate in the Oklahoma Chapter's Okie Derby Race, where she placed fourth. She and her race partner have become known as the "Airheads" and are sought out every year to see what new and innovative costume they have come up with. One of the highlights to her flying career was renting a Mooney and flying to the Smokies. To date, she accumulated more than 540 hours, mostly in Mooneys, Arrows and C-172s.

JESSICA RENEE STEARNS, born Aug. 3, 1940, in Birmingham, AL, a resident of Princeton, NJ, and Bonita

Springs, FL, logged more than 14,000 hours as a USAF navigator and pilot, Continental Airlines flight engineer, pilot and now flies as a B-727 first officer. After receiving her private pilot certificate at age 17, she went on to complete a 20 year career in the USAF, retiring in 1980 as a major. She has flown most general aviation aircraft, instructing in many and now flies and tows gliders with the So. Jersey Soaring Society and C-172RGs to do aerial photography. She is also an active Daedalian Life member and has been a 99 member with the No. Jersey 99s since 1992 and enjoys the annual Poker Run. She is also an accomplished photographer and is president of the Princeton Photography Club. Jessica is a graduate of Golden Gate University, San Francisco, CA, with a BS and an MBA. She is presently certified as an ATP SMEL, commercial L-300, CV-240, 340, 440; CFII SMEL and private glider. Her true love is to be in the air. She has one daughter, Heather Anne Stearns, 24.

VIVAN STEERE, a California native, began her flying career as a grandmother in 1982. She flies her own Cherokee 140 and co-pilots with husband, Dick, in his Beech Debonair. Her educational background includes a BA from Stanford University where she met her future husband and then spent the next 32 1/2 years moving around the world as an Air Force wife. She has four children and five grandchildren.

In addition to flying, her interest in aviation resulted in being a guide at the Air Force Museum in Dayton, OH, as well as a Docent in training at the Smithsonian Institute's Air and Space Museum in Washington, DC. Upon completing the year long course at the Air and Space Museum, she was ready to conduct tours when her husband's orders moved the Steeres to Florida.

She is currently chairman of the Coyote Country Chapter of the 99s in Fallbrook, CA.

TERESA LEIGH STEIN, born in Salisbury, MD, and now resides on Long Island in New York. She has three children: Ethan, 5; Devin, 2; and Samantha, 1. While attending the University of Maryland, College Park, she learned to fly and attained her private pilot license in 1978. Shortly thereafter, she joined the 99s and has since been a member, first in the Potomac Chapter and currently in the Long Island Chapter.

Teresa's interest in aviation extends to her profession. She has been an air traffic controller at New York Center since 1981, and has flown several controllers in her Rockwell Commander to familiarize them with the world of the private instrument pilot. She appeared with her plane in the FAA film, "Nothing Left to Chance," which has been shown at the National Air and Space Museum in Washington, DC and at the EAA air show in Oshkosh, WI.

As her children grow older, Teresa hopes to spend more time flying with them and her husband David, as well as with her friends in the 99s.

JOAN STEINBERGER, born Nov. 25, 1929, joined the 99s in 1960 after starting flight training in 1953 on an Aeronca 7AC. She has flown most singles from 1946 on and has owned a PA-28 for 29 years. She has logged 3,800 hours and has flown in the Powder Puff Derby, Air Race Classic, Palms to Pines and Pacific Air Race. She helped talk down a fellow racer in the 1969 PPD. Presently working on becoming

an A & P mechanic. She helps with the local elementary school in the Future Flyers and is a member of AOPA, Silver Wings, EAA, etc. She is the mother of two children, Ron, 40 and Norma, 33. She has two grandchildren, Kenny, 7 and Ryan, 5.

LAUREEN MYRA STEINKE, was born in Vancouver, British Columbia and now resides in Calgary, Alberta. She took her first plane ride in a light float plane in 1961 but wasn't able to pursue her interest in flying until 1991. After completing her bachelor of commerce, Laureen started flight training in May and completed her private license on New Year's Eve, 1991.

Laureen has been actively involved with sport parachuting since 1986, when she did her first jump. She stopped skydiving in 1991 but maintains her rating as a judge. She has also been the Alberta director for the Canadian Sport Parachuting Assoc. since 1988 and has served as the association's president since 1992.

Laureen's love of aviation and involvement with skydiving came together with the purchase of a Cessna 182 in 1993. Laureen and a friend have formed their own company, Wind Dancer Air Services, and the plane is used to fly skydivers from April until October.

Laureen's plan is to complete her commercial rating and multi-engine and instrument ratings. She would also like to take aerobatics training. Laureen joined the 99s in 1995 but has been volunteering for the organization at a local level for the past three years.

SHARRON STEMLER, High Desert Chapter, fell in love with soaring in 1971, took lessons in Fremont, CA and traveled all the way to Yugoslavia for the World Competition in 1972. She flew as a guest pilot for the local government aviation official doing a test hop to test out the radio in the new Citabria that they had sold to the contest officials to be used as tow planes.

After living at airports and working on the 1975 Powder Puff Derby as publicity chairman, she counts her biggest claim to fame as being the wing runner for Bob Harris when he set the new high altitude record in a sailplane. This flight is in the *Guinness Book of Records* and serves as a permanent fond reminder of her days out on the runway, helping to launch sailplanes. She also enjoys the fellowship with her friends in the 99 Chapters wherever she happens to live and has done airmarking at Catalina, Byron CA, Apple Valley and Barstow Daggett airports.

ANNE STEPHENS, born June 30, 1944, in Seattle, WA, joined the 99s in 1989 after earning a private pilot license with instrument rating in 1984. She has flown the C-150, 152, 172, 182, 172RG and TB20. She is employed as assistant manager/administrator at International Flight Training Academy in Bakersfield, CA, since 1992. She is married to Keith Stephens and has two children, Christine Duckering, 27 and Scott Malone, 26. She has one grandchild, Casey Duckering, 6 months. She made the solo trip from Southern California to Seattle in 1989 – the most satisfying experience of her flying career.

BEVERLY SUE STEPHENS, born on a farm near Bowie, TX, Stephens watched airplanes fly high overhead but never thought she would be a "special" person who flew them. While living near Fort Worth, she and her husband, J.C., and two sons, Charles and Michael, often spent Sunday afternoons at Mangham Airport watching others fly. That eventually led to certificates for the entire family. Beverly received hers on her 37th birthday in 1975. She joined the Golden Triangle Chapter of 99s in 1976 and has been an enthusiastic, active member serving in many positions including four terms as chapter chairman. Flying has been 'for the fun of it," but it has opened new opportunities for friendship and understanding.

MYRNA M. STEPHENS, born Oct. 3, 1941, in Moose Lake, MN, began flight training 1972 while studying for her doctorate degree. She received her private pilot license in 1973. She has been a member of the Lake Michigan, Quad Cities and Iowa Chapters of 99s since 1973. She has flown the Cessna 172, 182; and Beech Bonanza.

Stephens, an audiologist, has a BS from the University of Minnesota and MS from the University of Wisconsin and a Ph.D. from Michigan State University. She is the owner and director of Audiology Consultants in Davenport, IA. Stephens is active in a number of professional organizations and is a past president of the Iowa Speech-Language-Hearing Assoc.

Stephens and her husband, Peter, have two children, Sarah, age 25 and John, age 22.

CONSTANCE STEVENS, joined the Lake Tahoe Chapter of 99s in August 1988. Stevens learned to fly with money earned from buying cars in need of repairs, paint, etc. then rehabbing them and selling them to pay for her flight lessons. She soloed in 1975 with eight hours but it took a few more cars and a few more years to complete her license! She received her private pilot certificate on June 17, 1978, has an ASEL rating and 250 hours flight time. She has flown the C-150, 152, 172, 182; Super Cub, C-195, B-35, PA18, Mooney, PA24, BE185 and MD-11, 767, S-80, DC-10, 727 simulators.

She is the founder and director of Wildlife Shelter, Inc. Her previous occupations include: flight attendant with a major carrier, director of cabin safety program for cabin crew of major airline from 1988-1990, EMT certificate, aircraft accident investigator's certificate from USC. She is a CPR instructor, SFA, cabin safety firefighting (instructor). Enjoys downhill and cross-country skiing, mountain biking, wildlife lectures and travel, domestic or foreign.

Stevens has taken part in the Truckee Airshow, International Society of Air Safety Investigators seminar coordinator and presenter, Flight Safety Foundation seminar, presenter and US delegate to China and the Soviet Union. She also belongs to the WSI, EAA local chapter member and CAP. She is married to Louis who is also a pilot.

ORA KING STEVENS, born Jan. 15, 1915, in E. Taunton, MA, flying education was all taken at King Aviation Service. Her father and mother owned and operated the Cow Pasture Airport. She worked in the office, got her commercial rating at 18, soloed at 16 and private pilot license at 17. She has flown the OX5 Travelaire, Hess, Argo, Cubs, Cessnas, Navion and Bonanza.

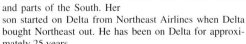

Her first airline ride was with Delta while in high school, Atlanta to Fort Worth, 1929, their first year. Now her rides are Boston to Miami and parts of the South. Her son started on Delta from Northeast Airlines when Delta bought Northeast out. He has been on Delta for approximately 25 years.

Her letter came from the 99s to join when she got her private. Whether it was from Amelia Earhart or not she doesn't remember as her mother who was the boss said, "No you can't join! Those women are all well-to-do and you can't afford it anyway. How do you think you are going to get to Long Island to the meetings once a month by yourself?" So that settled that, you better throw that letter away. When Jean Adams, the president of the Northeast Section and airport manager of Nantucket Airport asked her, it was O.K. for her to join and go to Boston for the meetings. She was the youngest member at 17 with her private license tucked in her billfold, she was off at night in the family Model A Ford by herself, her first trip to Boston. She got lost in Boston and forgot to look at what street she had parked the car on. Got a real bawling out when she got home at midnight. She needed more experience in the car, the only time she could use it was to go for the mail, never on a trip. She was made secretary of the Northeast Chapter in 1933.

She is the widow of Leroy M. Stevens, and has four children: Leroy, 50; Linda Michaud, 48; Don, 46; and Joanne, 40. She has four grandchildren: Scott Stevens, 25; Kelly Stevens, 23; Sherry Stevens, 22; and Kurt Stevens, 18.

THYRA STEVENSON, learned to fly at San Jose Airport in 1994 and already has her single and multi-engine instrument ratings with 600 hours logged. She owns a C-182, which she flies out of San Jose, and also occasionally flies a C-401. This summer she got her DC-3 type rating. Thyra joined the Santa Clara Valley 99s in 1995 and already serves as chapter vice-chairman. Thyra has her Ph.D. in Latin American literature and has spent time living in Spain, Mexico and Latin America. She now owns the Los Gatos Cyclery Shop. Thyra is a member of AOPA and volunteers with the San Jose FSDO. She is very interested in getting new people into flying and improving the relationship between GA pilots and the FAA. Thyra has a daughter, and her husband Walt Greenham is a design engineer at Amdahl. Her interests include skeet shooting and of course, flying.

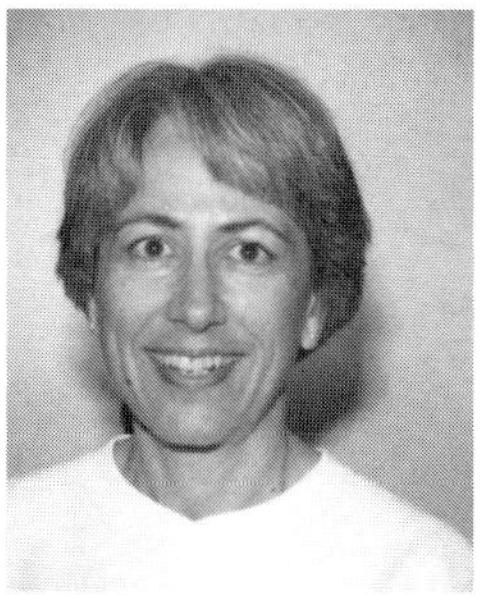

MARGARET C. STEWART, born May 27, 1923, in Oakboro, NC, received degrees as follows: BS Pfeiffer; MA from ASU; Ph.D. Georgia State University. She received flight training at the local CFI and is a member of the Carolinas Chapter. She has flown the Cessna 172 and flies mainly for the pleasure of it. She is married to Edward P. Stewart and has two children, Peggy Kluttz, 51 and Betsy Heath, 40. She has two grandchildren, Neil Kluttz, 24 and Jessica Heath, 15. She retired from NC public school system as a college administrator in 1983. Presently she is an entrepreneur in Oakboro.

STEPHANIE M. STEWART, Phoenix 99s, both got her private ASEL license and joined the 99s in Phoenix in 1993. She had grown up in Salt Lake City and moved away to start a career in law enforcement, serving four years as a city police officer in Wyoming. She is now a special agent with the US Department of Justice Drug Enforcement Administration.

Stephanie has started planning a career in aviation, is an active 99 Chapter member and enjoys running.

SAUNDRA KAY YOUNG STIENMIER, born April 27, 1938, in Abilene, KS, got her flight training at Ft. Bliss Flying Club, Texas. She has flown the C-152, 172, 182; Aztec, Baron, Warrior, Arrow, Cherokee, Lance, Bird Dog and T-41A, 41B, 41C.

She has participated on July 4, 1977 in the Commemorative Powder Puff Derby near Cross City, FL. Their T-41B was hit by lightening! What a 4th! Smoke filled the cockpit from a burned out 24 volt battery. They made an emergency landing on a field under several inches of rain with no electrical system/flaps, but completed the race. Also flew in two European Competitions (1978-79), first and second place and Grand Prix Air Race, Mile High Air Race plus local events.

She is married to Richard H. Stienmier and has four children: Richard Bruce, 33; Susan Irene, 27; Julia Teresa, 25; and Laura Saundra, 22. She has been in the USAF since 1977 and is the Aero Club manager.

MILDRED NATALIE STINAFF, born Sept. 13, 1911, soloed in 1929, the same year she graduated from Akron Ohio North High School. Milly held a limited commercial pilot license at the age of 17. It was also in 1929 that she became a charter member of the 99s.

In January 1930, Milly flew 42 (inside) loops and established herself as the Women's World Record holder. She worked at Akron Airport as a secretary and airport terminal hostess and in her spare time worked on regaining her "loop" title which had been bettered by 46.

On June 23, 1931, while flying her bi-plane Milly

started practicing loops and stalled. She recovered the aircraft, but it stalled again too close to the ground to recover. She was dead at 19.

Her grave marker is engraved with NAA Wings, pilot license number, the 99s emblem and across the bottom, "Jesus Savior Pilot Me." Mildred is an honorary member of the Women With Wings Chapter.

KATHARINE STINSON, has been a member of the 99s for the past 50 years. She knew at an early age she wanted to be an airplane pilot, and was inspired to enter the field of aeronautical engineering by Amelia Earhart.

In 1935 when Miss Earhart was flying around the country demonstrating a new auto-gyro, she stopped in Raleigh, NC, Katharine's home town, where she spent several days while her auto-gyro was being serviced.

Upon graduation from North Carolina State University, where she was the first woman graduate in engineering, Katharine pursued a career with the Civil Aeronautics Administration, now the Federal Aviation Administration, for 32 years. During her years with the agency she perfected a distribution system of air worthiness directives adopted by the aviation safety departments in all the developed nations.

Her honors include: Distinguished Women in Aerospace Industry Award, Women's Advisory Committee on Aviation (WACOA), Institute of Aeronautics and Astronautics Aerospace, Pioneer Award (first women to receive award), President Soroptimist International of the Americas, President Society of Women Engineers, and Who's Who of American Women.

BRENDA STOCKMAN, as a child in the midwest, her family owned a single-engine aircraft, in which she and her brother always rode in the back seat. Just another trip to grandma's house, except they all landed at Fowler Airport, walked across a field, ate dinner, walked back across the field and climbed into the airplane to go back home. She thought everyone traveled to relatives in this method of transportation.

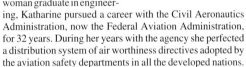

Her interest in airplanes happened when she was 21. While working at Bakers' Flying Service in Kansas City, MO, she was asked if she would like to ride along to ferry airplanes. She was a passenger when the pilot asked her to take the controls. Then he told her to turn to the right, she did . . . she was then hooked! She received her private pilot license in September 1973. She trained and took her flight test in a Grumman American Trainer.

Soon after receiving her license, Stockman flew a Cherokee 140, Citabria, Cessna 172 and Mooney. She was seven months pregnant and could not pull the yoke back far enough on the Mooney, so she started ground school for her instrument rating. Then the 18 year gap to raise three boys occurred.

Aviation is a constant learning experience, no matter how many hours are logged. The wrong attitude can kill.

Her children are grown now. She is an active member of 99s and Civil Air Patrol. Her aircraft insurance business keeps her busy, but she loves it! Stockman's experience includes being a flight secretary for an FBO with airframe and powerplant school, maintenance shop, flight school with VA training. She also managed a flight school with 11 aircraft and did charters.

Currently, Stockman dates a man who has logged more than 8,100 hours in various aircraft and helicopters, and he still takes the time to teach her.

SANDRA M. STOKES, remembers watching airplanes fly overhead while growing up in Indiana, wishing that she could be part of the sky. That dream came true in October 1984, when she earned her private pilot license. Hours of fun, challenge and continued learning lay ahead.

One of Stokes' first actions as a private pilot was to joined the 99s. She has been a member of the Lake Erie Chapter for most of the time since she joined in 1984. She also belongs to the Aircraft Owners and Pilots Assoc. and the Experimental Aviation Assoc.

Stokes is associate professor of reading and special education at the University of Wisconsin at Green Bay. She has a Ph.D. from Kent State University. She and a student are presently studying factors contributing to flight instructors' success in teaching students to fly. Daughter, Jenny and son, Andrew, complete the family.

CHRISTINE MARIA-LUCIA POSTER ST. ONGE, born Nov. 9, 1951 in Atlanta, GA. She began flying in June 1973, received private license on Dec. 7, 1972 and has logged more than 3,500 hours in 45 different aircraft, and has her commercial, instrument, single and multi-engine land; certified flight instructor-instrument single-engine and multi-engine; basic, advanced and instrument ground instructor certificates. Christine was also an FAA designated written test examiner 1991-95.

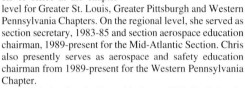

Christine joined the Greater St. Louis Chapter in November 1974, held all levels of office at the chapter level for Greater St. Louis, Greater Pittsburgh and Western Pennsylvania Chapters. On the regional level, she served as section secretary, 1983-85 and section aerospace education chairman, 1989-present for the Mid-Atlantic Section. Chris also presently serves as aerospace and safety education chairman from 1989-present for the Western Pennsylvania Chapter.

Christine flew for Trans Air Cargo, 1979-81, St. Louis, MO, in Beech 18s. Her other multi-time includes Beech King Air, Baron and Travelair, Cessna 310s, Piper Navajo, Chieftain and Turbo-Commander.

Christine is a graduate registered nurse from Maryville College, St. Louis, 1971 and worked all the ICU levels, specializing in cardiology. She now teaches CPR/first aid for Southwestern Pennsylvania Chapter of the American Red Cross. Chris also teaches private, commercial and instrument ground schools for CCAC-Center North, private and instrument ground schools at the Zelienople Airport, PA.

Christine is a member of AOPA since 1976; Staggerwing Museum Foundation and Staggerwing Club since 1974; Missouri Pilots Assoc.; Air Heritage; EAA; Condor Aero Club, served as board of trustee and senior pilot, since 1994; Civil Air Patrol, rank of major, serving as medical officer, Emergency Services Officer, Search and Rescue Mission Pilot, Special Ops and DEA Ops Pilot.

Christine is president and CEO of Wings Aviation, Inc. Bi-plane rides and flight instruction at Zelienople Airport, since 1992. She is also an FAA safety counselor, since 1985.

Christine is chairman and president of Wings Weekend Extravaganza – 76 pilots received their wings last year.

Christine's air races include the Allegheny Air Derby, which she won in 1989, and the Buckeye Air Rally, 1995 being her ninth consecutive year. She flew in her 1936 Beechcraft C17B Staggerwing, the first bi-plane entered in the rally and won Oldest Aircraft, Best 99 Team and Best Female Team.

Christine has flown more then 150 hours in her Beechcraft C17B Staggerwing. She has also accumulated 30 hours in a D17S Staggerwing. For the 1996 Bendix Commemorative Tour, Terry Von Thaden, Louise Thaden's granddaughter, will be her navigator. They will be the only women and 99s team. Her 1936 Beechcraft C17B Staggerwing is the same model aircraft that Louise Thaden and Blanche Noyes won in the 1936 Bendix Air Race.

Christine is married to Paul St. Onge, an electrical engineer for Westinghouse and has a son, Joseph, 13 and daughter, Laura, 5 and lives in Wexford, PA.

RUBY ST. ONGE, born Jan. 19, 1922, a Minnesota native, learned to fly following her husband's death in a float plane accident in 1961. She earned her private license in 1962. During this time she worked with Fred Van Dusen while he wrote and published, *The Instrument Rating.* She has flown the Cessna 150, 172, 182, 206; Mooney, Bonanza and Beach 18.

During the 60s, she spent summers at Great Bear Lake Lodge in Canada's Northwest Territory. She acquired her Canadian license and served as co-pilot for numerous fishing trips to the Arctic Ocean. She also served as executive secretary for the Condominium Owners Assoc. at Breezy Point Lodge in northern Minnesota. After six years she became the office manager at the New Arrowwood Lodge in Alexandra. In 1971 she also flew as co-pilot for Caroline Olson who flew her Mooney in the 25th Powder Puff Derby. Ruby received her commercial rating in 1964.

In 1965 Ruby won the Amelia Earhart Scholarship and pursued her instrument rating, receiving it at IFT in Minneapolis.

Ruby continued in property management until retirement in 1984. She then served 27 months in the Peace Corps as a tourism advisor for Costa Rica. After Peace Corps she lived in Oklahoma City and volunteered some time at the 99 headquarters during the transition from old to new building. She now resides in Des Moines, IA, near her daughter, Patricia, 48, and continues her world-wide travels.

SUE STORM, Phoenix 99s, got her private ASEL at Sky Harbor Airport, Phoenix in 1962 and joined the Phoenix Chapter 99s in 1964. She got her commercial in 1966. Sue was chapter chairman for two terms, and also held every other elective office. She especially enjoyed the Kachina Air Rally, airmarking, the Girl's Ranch Flight activities, airlifts and Fly Without Fear seminars. She recalls taking 70 relatives, many of whom had not flown before, up in one day in Pennsylvania.

Sue's chief occupation was as secretary/receptionist/ partner in her family business, but she had many interesting aviation experiences such as flying her family to Florida and Pennsylvania, and making a trip to Alaska. Joe and Sue Storm have been married 54 years, have four children and 10 grandchildren. They have visited all 50 states, and lived in all the Rocky Mountain states. When Joe worked for AT&T (1941-43) they moved 31 times in two years. Now retired, their hobbies are recycling and gardening.

MARY LOUISE STROH-STORY, born Aug. 23, 1929, in Chicago, earned her private license in February 1967, and she was the only person (other than her instructor) she knew who flew. It was very lonely. She heard about the 99s and someone at the airport put her in contact with Nita Fineman who sponsored her. Became a 99 in 1969, and what an experience and inspiration. Earned her instrument rating in 1973. Has always been active in her chapter, serving as chairman on most committees at one time or another. Has flown many proficiency races, with good results. She has flown the Cessna 150, 182; Piper 140, 180; and Beech V35A, trained in 150 and owned the others. Beyond the encouragement in flying, the 99s have been her caring buddies for the past 25 years. She is married to Bert and has four children: Bill Stroh, 45; Barb Dawson, 44; Linda Boydston, 40; and John Stroh, 37. She has four grandchildren: Amy Stroh, 17; Mat Dawson, 17; Paul Stroh, 14; and Mary Boydston, 4.

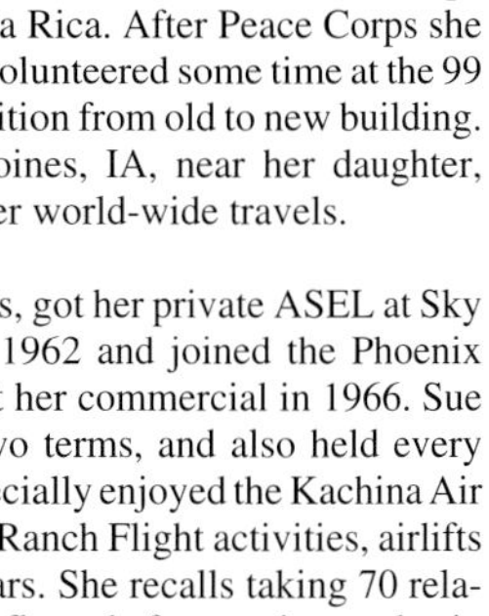

MARILYN J. STOWE, is a lifetime resident of Arkansas. She is married to Ken Stowe and they have two sons, Kevin, 14 and Adam, 10. Marilyn became interested in aviation in 1978 on her first date with Ken. Ken owned his own air charter service and flew her off to a nice dinner. They married in July 1979. Her family came first, so it was in 1985 when she began to sprout wings and learn to fly. She loves stalls, climbs, turns, takeoffs and descents. She was hooked and got her pilot license in September 1985.

Marilyn is active in her local chapter, Arkansas 99s. She has served as chapter chairman, membership chairman and is currently working as scrapbook chairman. She is also active in church and children's school PTA. She is treasurer of the local Chapter of Arkansas Aero Club and is on the board of directors of Arkansas Pilots Assoc. which is affili-

ated with the USPA. Marilyn is a member of Aircraft Owners and Pilots Assoc. She now owns and flies a Cessna Skyhawk 172.

Marilyn owns and operates a successful catering business in North Little Rock, AR.

GINGER STRANGE, born Oct. 28, 1925, in Wray, CO, received her license on Oct. 19, 1974. Lee Agnew was her instructor. She flew in nine Palms to Pines Air Races and is charter member of Mount Shasta Chapter of 99s. Flying has been one of the greatest experiences of her life. She is married to Chuck, and has three children. She has eight grandchildren and one great-grandchild. She has flown the C-150, 172 and 182.

WANDA WREN STRASSBURG, originally from Tulsa but met and married husband, Don, in the Ventura/Ojai area of California. Since 1955, San Louis Obispo has been home where they raised their two girls and her younger brother. All are now grown with their own families and are now expecting their first great-grandchild.

In 1961, Don, who was a pilot when they first met in school, renewed his interest in aviation and talked her into learning to fly to forestall her fears. They purchased their first plane, an Aeronca 7AC and within a few hours she was hooked BAD!

Couldn't afford a lot of flying time after they were both licensed and outgrew the Aeronca, so she figured the only way to satisfy her need was to get paid to do it. Began preparing for writtens – first, ground instructor certificates, then commercial, CFI, etc., paying for it by working at the local airport, once their girls were both in school all day. This led to her being a local FAA examiner designee and finally to the vice-presidency of a commuter airline. Four PPDs, a PAR and a Palms to Pines.

DOT STRATE, born Aug. 12, 1947 in Plentywood, MT, joined the 99s in 1989. She earned her single-engine land and IFR. She has flown the Cessna 150, 152, 172, 182, 340; Piper Warrior, Arrow, Saratoga, and Malibu. She has four children: Paul, 28; Dawn, 26; Dave, 25; and Carrie, 22 and three grandchildren: Chase, 3; Tamesha, 2; and Jake, 9 months.

She hasn't had too many harrowing flight experiences but she does fly all over the country to small scenic towns, sets up her art easel and oil paints. The combination of these two make for an exciting life. She flew the Classic Air Race in 1991 which was the experience of a life time. On one leg of the trip an airplane took off before her then she took off. She saw something burning in the trees below her and found out it was the aircraft in front of her. Luckily the girls were alive and rushed to hospital with burns. Another leg she encountered some fog, found the nearest airport which was a crop-duster field in Mountainview, MO. They landed and a big man with coveralls and a red beard came up to the plane and said, "Ya'll lost?" We laughed and said, "No, just had to land because of the weather." He said that we'd probably be stuck there for another week and went to town and brought us back some hamburgers. We still came in 32nd so, other girls had problems, too. All in all the most learning experience she had ever had.

CECELIA "CECI" M. STRATFORD, born March 2, 1943, in Los Angeles, CA, former nurse and Peace Corps volunteer (Guatemala 1966-1989), leaned to fly in 1975 at Whiteman Airport (near Los Angeles, CA). There she met her future husband, Malcolm, who wholeheartedly has supported her enthusiasm for flying. She joined the 99s in 1976 after earning her private in 1976, instrument rating in 1978, commercial in 1979, ASEL, CFI and seaplane in 1986.

Favorite flying activities: Flying Companion Seminars, Direct Relief Foundations flights, Future Women Pilots scholarship program, Pennies-a-Pound rides, restoring WWII aircraft, educating others about flying and its value and safety.

Memberships: San Fernando Valley 99s (chairman 1984-85), Confederate Air Force Southern California Wing, Toastmasters, Vintage Aeroplane Assoc. (president 1984), Angels Antiquers, AOPA, California Aviation Council, National Aeronautic Assoc., National Assoc. of Flight Instructors.

Awards: FAA Accident Prevention Counselor, Union Bank Volunteer of the Year (1986), 99s Amelia Earhart Memorial Scholarship (1985), Who's Who in California (1992), San Fernando Valley 99s Service Award (1982) and Woman Pilot of the Year (1983).

Aircraft owned: Piper Cherokee 140 (converted to a 180 hp with constant speed propeller). She has also flown the PA-28-140, 180; Cessna 150, 152, 172; Grumman Tiger, Cheetah and J3 Cub.

Longest flight: Los Angeles, CA, to Anchorage, AK. Many cross-country flights around the Pacific Northwest and the Southwest, including nine air races. She has flown 2,100 hours.

Education: BS in nursing, Mount St. Mary's College, Los Angeles; MS rehabilitation administration, University of Wisconsin, Madison. She is employed as risk analyst at Union Bank and a part-time flight instructor.

C.J. STRAWN, is an instrument rated private pilot and the newly elected president of the Palms Chapter of 99s. Her father was an Army Air Corps flight instructor in WWII and her memories include the Tri-city Airport, San Bernardino, CA, where he flew his BT-13 and knew Joe and Pinky Briar. She is currently chairman of the Palms Chapter.

C.J. completed the MFA at UC in theater design. Then spent 10 years designing Off Broadway theater, western and circus shows.

Currently she is a production designer for film and TV including notable films such as, "My Chauffeur", "The Hidden" and "Nightmare on Elm Street" films.

At the Santa Monica Airport, C.J. has a studio when she paints and exhibits her current passion, "European Coffee Houses," a collection of oil paintings. Her goals include designing period films and flying around the world.

SHARON STROBERG, born Feb. 20, 1920, a lifetime Kansas resident learned to fly in 1989. Sharon has been flying since "preconception!" Both her father and brother are pilots. Her family was active in the International Flying Farmers. In IFF, she won second place in an international Al Ward Essay Contest. She was also a Farmerette and Teen President for the state of Kansas. On an International level she became International Flying Farmerette in 1977 competing with over 40 girls for the title.

Stroberg's educational background includes a BS in dietetics and institutional management from Kansas State University. She is a registered dietitian; and is also a licensed dietitian for the state of Kansas.

Her community activities include giving a lot of time to her college sorority, Kappa Alpha Theta. The Hutchinson Alumnae Club won two national awards in 1990 while she was president of the group. The Thetas have donated over $18,000 to local charities (the money was raised with group's fund-raiser) in the past few years.

Stroberg Equipment Company manufactures livestock equipment. In the past 10 years, Sharon has spent a lot of her time at Agriculture Trade Shows promoting their portable panels, continuous/permanent fencing and numerous types of feeding equipment for a variety of livestock. Some of the sports she enjoys besides flying are racquetball, scuba diving and snow skiing.

JUNE STRUTHERS, First Canadian Chapter, the motivator for June to get her pilot license was her husband, Donald. When they attended the doctor for their medical examinations together, unfortunately Donald did not pass due to strict medical requirements. June, however, decided to go ahead anyway with her training. June and Donald both took the ground school course as an added measure of safety.

June received her private pilot license on her birthday, June 1, 1965. She joined the 99s the following year and was one of the original members when the First Canadian Section Chapter was reinstated. She obtained her night rating and also took 90 hours instrument training under the hood. Most of June's flying was cross-country in Ontario and Quebec. She has flown in many chapter Poker Runs and has enjoyed the challenge of many local air rallies. Since 1965, she has maintained membership with COPA (Canadian Owners and Pilots Assoc.), Ottawa, Ontario.

A graduate of the University of Western, London, Ontario, in nursing, June was a director of Public Health Nursing for the Waterloo County Health Unit, Waterloo, Ontario. Upon graduation as a registered nurse, she was a nurse with Victorian Order of Nurses, Hamilton, Ontario. For seven years June taught at the College of Education in Toronto and later high school with the Waterloo County Board of Education, Waterloo. During this time she did flight familiarization with the high school students at the Guelph Airport.

In 1987 June embarked on a new career, after obtaining a degree in theology, and joined the ministry. She is recently an ordained minister at the Britten Memorial Church of Canada in Toronto, Ontario with a congregation of 350 members.

DOROTHY C. STURMAN, a lifetime Ohio resident, fulfilled a long time dream of flying by getting her private pilot license in October 1978. In 1985, she purchased her own plane, a 1976 Cessna Cardinal 177B. Her longest flight was to Fort Lauderdale, FL, to visit her son. She has also enjoyed flights in a Christen Eagle, a hot air balloon and a Schweizer sailplane. She has a BA from Ohio University and was inducted into Phi Beta Kappa Honorary. Her varied work history includes periods as a manager for the Internal Revenue Service, as a real estate agent and as a high school teacher. She is married to John C. Sturman and they have four daughters and three sons as well as 11 grandchildren. She is a member of AOPA, EAA, CPA and the Cardinal Club.

VALERIE SUBERG, born Jan. 14, 1960, in San Antonio, TX, learned to fly at the age of 15 1/2 in Germany. On her 16th birthday, she soloed in Santa Rosa, CA. She got her private license at age 17, and joined the Santa Rose Chapter of the 99s. She graduated college from San Diego State University in 1982 with a BA in fashion merchandising. After college she decided to continue her education at Miramar College and went on to get her airframe and powerplant mechanics license.

While in A&P school, she continued her flying and acquired her instrument rating in 1984. In 1985 she embarked on a career in aviation maintenance. She worked in general aviation on all kinds of airplanes in Healdsburg, CA. In October 1985, she went to work for Flying Tiger Airline as the first female mechanic to work at their base in San Francisco.

She continued with her flying and in 1986, her commercial rating was added. In 1987 she got her multi-engine rating. And in 1987 she was awarded the AE Scholarship Award and used it to achieve her certified flight instructor rating. After achieving her CFI in 1989, she started working as a flight instructor at Sonoma County Airport for Redwood Aviation.

In 1989 Federal Express bought out Flying Tigers and

she continued to work for Federal Express. In 1990 she acquired her multi-engine instructor license. She continues to flight instruct. Suberg started a flying club with a Piper Arrow, and owns a partnership in a Beechcraft Baron. She is also a member of AOPA, Cherokee Pilots Assoc.

ELIZABETH MCGEORGE SULLIVAN, deceased 1993, Maryland Chapter 1965, a former WASP started flying in 1941, logged more than 1,000 hours, added commercial and SMEL ratings to her private pilot rating. She was checked out in Taylorcraft BC-11; Cessna 120, 140, 150, 172; UC-78; Cub Sport; J-3; PA-11; L4H; Aeronca 7AC; LC-3; Fairchild PT-19, 26; Vultee BT-15; N Am AT-6; B-25; Curtiss AT-9; Fleet PT-26; and Douglas C-47. She and her husband operated Doersom Airport near Gettysburg, PA, and an FAA approved ground school. She earned her BA from Wilson College in Chambersburg, PA, enabling her to be a self-employed IFG (insane fruit grower) at her apple orchard.

SANDRA HEWITT SULLIVAN, born July 27, 1946 in Jennings, LA, started flying at 11 months with her dad! Her dad was a pilot and also her three brothers. Flight training at 19 and license at 20 with her husband as instructor. She also received her commercial, instrument, ASEL, ASES and 2000II. She has been a courier and charter pilot in a C-185 and is now part-time jump plane pilot (C-182) in the Seattle, WA, area.

She joined the 99s in 1968 and has flown in two Powder Puff Derbies and local races. She has held all chapter offices and is active in Flying Companion Seminars and anything to promote women in aviation.

She is also a member of Christian and Missionary Alliance Church, a supporter of Mission Aviation Fellowship and high school girls soccer coach for three years, local club coach for 10 years. She and her husband, John Lian, a retired NWA pilot, both own small apple orchards. She has two children, Michael, born Aug. 5, 1969, a commercial pilot in Alaska; and Tanna, born June 16, 1972, soloed and a student pilot.

SUSAN SULLIVAN, is a licensed commercial pilot with instrument and multi-engine ratings. Susan was inspired to obtain her pilot license by her father, Noah W. Sullivan, who is also a commercial pilot, as well her five uncles who were the first Ace Air Force pilots to lead the first wing of Thunder Jets into the Korean Conflict. Her uncles were awarded many medals of honor as well as ticker-tape parades for their bravery and achievement.

She has been flying for over 20 years. Susan obtained her private pilot license and joined the 99s in 1986. She obtained her instrument, commercial and multi-engine ratings.

Susan now owns and operates a Piper Arrow based at Dutchess County Airport, Poughkeepsie, NY, where she also owns Direct Air, an aircraft charter, sight-seeing and aerial photography service. Susan is the chairman of Hudson Valley 99s, Inc. She is an aviation safety counselor with the Federal Aviation Administration, a member of the Lawyer Pilot Bar Assoc. and American Bar Assoc. Aviation Subcommittee. She is also active in scheduling aviation safety seminars with the Aircraft Owners and Pilots Assoc. as well as local FBOs. She currently attends New York Law School.

Susan's goal is to provide aviation education to the public and inspiration to aspiring pilots.

LESLIE SUMMERS, earned her private pilot license on May 14, 1977 at Livermore, CA. She also has glider, ASEL, (1985) and instrument (1992) ratings. She joined the Lake Tahoe Chapter of 99s in June 1994. She has flown the C-172. Her first opportunity to learn to fly was when she met her husband, who at the time was taking gliding lessons. One year later they got married and bought a glider, instead of buying a house like most people. After they got married, she finished her bachelor's degree, then decided she wanted to learn to fly power. Later they decided to buy a Cessna 172 to ease the commute from Livermore to Truckee every weekend. Summers then became chief pilot and Mark became "passenger in command." The fog on Monday mornings in Livermore prompted her get her instrument rating and she studied with Chuck Davidson at Truckee, and took her check-ride in Reno. She works at Lawrence Livermore Labs.

AVA C. SUMPTER, born May 12, 1969, in Los Angeles, CA, joined the 99s in 1993. She has a private pilot with instrument rating and has flown the C-152, 172, 172RG and Aeronca Champ. Her education consists of an AS in airframe and powerplant technology and is working on a BS in professional aeronautics from Embry-Riddle Aeronautical University.

She enjoys entering Poker Runs and going to airshows. She has a love for old WWII aircraft and aircraft prior to that time period. She is also is the Civil Air Patrol, Confederate Air Force, AirLifeLine of Texas and Aircraft Owners and Pilots Assoc. She is married to Bradley Sumpter. She has been in the US Army and Army Rescue since Nov. 1, 1987 and an A&P instructor.

MARILYN SUNNERGREN, received her private pilot license at Stockton, CA. She has ATP, CFI, CFIAI, CFIME, advanced and instrument ground instructor certificates and ASEL, AMEL and instrument ratings. She joined the Lake Tahoe Chapter in February 1991. She has more than 7,000 hours flight time. She has flown the C-182 (owned N6196B), A36TC and about everything that has one engine. She has been a farmer, corporate pilot, flight instructor, computer operations manager at Harrah's. She is a Realtor with Prudential California Realty. Married to Bill, a pilot, has a son, a helicopter pilot.

She has been involved Truckee Tahoe Plane Talkers, Tahoe Trekkers, 99s for four years and received Lake Tahoe 99s, "Woman Pilot of the Year" in 1992. Flew the Goodyear blimp *Columbia* and logged 1/2 hour, taught her son to fly. He soloed three aircraft on his 16th birthday and got his private pilot license on his 17th, rode rear seat in a P51 Ridge Runner owned by Dan Marten from Merced Gathering of Warbirds to Hollister for 40th birthday present. Non-flying activities include snow skiing and hiking.

GINI SUTHERLAND, waited until her children were in high school before learning to fly in 1972. In 1974 she joined the Michigan Chapter of the 99s and when the Greater Detroit Area Chapter was formed in 1975, she became a charter member of that chapter holding all of the offices becoming chairman in 1978.

North Central Section saw Gini progress through the various offices to governor from 1992-94. Aerospace Ed is a high priority for Gini, and she welcomes the chance to talk to children and have them tour her airport and examine her airplane.

Second to 99s, in Gini's heart, is her love of animals and hosts for section and International are sure to include information on the local zoo. She and her husband, Neal, fly their Cherokee N73GS from Oakland Pontiac Airport.

CAROLE ALICE SUTTON, began flying in 1968, received her private in 1972, her commercial in 1975 and was awarded an Amelia Earhart Scholarship in 1977 to be used for an instrument rating which she received in 1978.

She joined the Nebraska Chapter of 99s in 1973 and has served through the ranks of offices of the Nebraska Chapter of the South Central Section serving as governor 1994-96.

Her husband, Stuart and Sutton have two daughters, Sonja and Linda. Linda is a private pilot. Since 1962 they have owned and operated Chester Flying Service, Inc., an aerial application business. She was the first and only licensed female agricultural pilot in Nebraska receiving her license in 1977.

Sutton also serves as president of Belleville Flyers, Inc., a flying club, and as secretary of the board of trustees of the Amelia Earhart Birthplace Museum. She also enjoys crafts.

LINDA RANAE SUTTON, having come from a flying background she received her private in 1988 and joined the Nebraska Chapter of 99s that same year. When she was small, she would read the charts and navigate for her mother, Carol Alice Sutton, who is serving as governor of the South Central Section of 99s (1994-96).

Sutton's being a 99 has given her the ability to make new friends around the world. She has flown over London with Aileen Egan, British Governor of the 99s when in England on business. Also when she and her mother visited Australia they met many gracious 99s including Nancy Bird Walton, who autographed her book for them. Presently she is corresponding with Merce' Marti Inglada of Spain who was a 99 at large, and now with her help, has been adopted by the Nebraska Chapter of 99s.

GAIL A. SWAIN, started flying airplanes in 1980 at the age of 26. She is now a CFII and a seaplane instructor. She and her husband own and fly a Cessna T210. They frequently fly south to Baja, CA, and have flown all throughout the United States.

She is a member of the 99s, AOPA, Baja Bush Pilots and the Stick and Rudder Club.

She has raced in the Air Race Classic (1990) and in the Palms to Pines Air Race (1992-94) placing second in 1994.

She is proud of women's contributions to aviation and to be a member of the 99s.

KAREN CHRISTINE SWAIN, born July 24, 1951, in Richardson, TX, received her private license in 1989 and immediately joined the Dallas Chapter of 99s. She has held various offices and is the current chairperson (1995-96). Active at section level, Chris was elected to the office of director for the South Central Section (1995-97) and appointed chairperson of the membership committee (1994-96). Chris is currently working on her instrument rating and aspires to attain the commercial and seaplane ratings. Chris is devoted to antique and classic aircraft and proudly flies her 1948 Stinson 108-3 to attend as many aviation activities as possible. She has also flown the 68 Cessna, Skyhawk 172, Cessna 152, Stearman, Aeronca Champ, Navion, Piper Cub, Luscombe 8A, Stinson 108-3 and Piper 140.

Chris has been employed at Northern Telecom Inc. as a compensation analyst for the past 12 years. In her spare time, Chris enjoys scuba diving, any outdoor activity and spending quiet time reading a good book. Chris is a native Texan and is married to Garry Robert Swain who is also a pilot and a supervisor for the Texas Department of Criminal Justice. She has two daughters, Janna Michele, 21 and Shana Marie, 19.

DONNA FROST SWANK, born in Hollywood, CA, became a member of the 99s in Hawaii in 1988. Donna soloed when she was 16 years old at Renton Airport, Renton, WA, in a Cessna 150 aircraft. Her father, Don Frost, a WWII fighter pilot and Air Force flight instructor taught her to fly.

Donna served four years as a pilot in the US Army in Korea, flying CH-47 Chinook helicopters, assisting the Korean Army with their sling load flight operations. While on duty in Korea she married Will Swank, a US Army helicopter instructor.

Upon her discharge from the Army and her return to the US, she flew as a pilot for Aloha Airlines on inter-island flights from Honolulu.

Currently Donna is with the Federal Aviation Administration holding the position of principal operations inspector for Continental Micronesia and is a check pilot for Hawaiian Airlines in DC-9 aircraft.

Prior to working for the FAA Donna received a scholarship from the 99s which helped her earn her flight instructor certificate. She holds an airline transport pilot certificate with DC-9 rating.

EULA MAY HODGES "SKIPPY" SWEET, born Sept. 8, 1929, in Honolulu, HI, used her first paycheck for flying lessons a month later purchased an Aeronca Defender, only one of its kind in the islands; this plane was also shot at on Dec. 7, 1941 while on a training flight, was the only woman to enter the air races in Hawaii the summer of 1954, considered a "dark horse." She joined the 99s on April 11, 1989.

She received her flight training at Hawaii School of Aeronautics, Honolulu, HI, on Nov. 19, 1952. She has flown the Aeronca, Ercoupe, Cessna 172 and soloed five 1/2 hours at Bellows Air Field, Hawaii in 1953. She built and started CAVU Flying School in Kipapa, Wahiawa amidst canefields. She worked as a dispatcher at the Hawaii School of Aeronautics till CAVU (Ceiling and Visibility Unlimited) was started with two planes, a Defender and L-5.

She is married to Richard Bayliss Sweet and has three children: Richard Bayliss Sweet Jr., 40; Richie May Sweet, 39; and Robert Byron Sweet, 32. She has four grandchildren: Erika Shirley, 8; Shaun Robert, 7; Shannon Augusta, 6; and Joni Anne, 5.

AMELIA SYLVESTRI, learned to fly in 1960 at Half Moon Bay in a C-140. She and her husband, Frank, owned and operated West Coast Aviation there for many years, and still own a rebuilt Stearman, Super Cub and a C-185. Amy has logged more than 300 hours, and has many more right seat hours navigating with her husband on trips to Baja and British Columbia. She served as vice-chairman to the Santa Clara Valley 99s, which she joined in 1963. A mother, grandmother and photographer, she enjoys swimming, sewing and music.

JENNIFER ANN JEWETT SYME, born Dec. 15, 1964 in Painesville, OH. Love for flying started as a young pioneer. Flying was the best nap time known to her, buckled in the back seat with the soothing roar of the engine. Eventually, she worked her way up to the front and began taking lessons at the ripe old age of 8. She earned her private pilot license Aug. 11, 1986, 14 years later and a day before leaving for the International 99s Convention in Hawaii where she registered officially as a 99. Her love of flying has been nurtured by her family, her mother, Dodie, and father, Harlan, are pilots; brother, Bruce is a United Airlines pilot.

Met her husband, John, at Portage County Airport, married 1990. John and his family are pilots too.

She graduated in 1982 from Geneva High School; member National Honor Society, Thespians, varsity softball team; swam competitively for 10 years during elementary and high school; did competitive synchronized swimming. Bachelor's degree in nursing in 1986 at Kent State University; master's degree in critical care nursing 1990 from Case Western Reserve University. Taught critical care at Summa St. Thomas School of Nursing, Akron, OH. Member of Sigma Theta Tau, the International Honor Society of Nursing. Earned American Assoc. of Critical Care Nursing certification for CCRN; served as president-elect and president of the Akron-Canton Area Chapter of AACN. Currently a healthcare representative for Pfizer Pharmaceutical.

Captain in the Army Nurse Corps, United States Army Reserve and served on active duty during Operation Desert Storm. She earned the Army Commendation Medal, Army Achievement Medal with two Oak Leaf Clusters, Army Service Ribbon and the National Defense Service Medal and a lifetime member of the Reserve Office Assoc.

Member Medina Jaycees, AOPA, Lake Erie Chapter of 99s and served as treasurer.

She truly enjoys the freedom of flight and sharing those wonderful experiences with her family. She and her mother enjoy flying together and have flown into Oshkosh for the airshow. She and brother, Bruce, have flown to Great Exuma Island, Bahamas, in their mother's (Dodie) Cherokee 140 during Christmas break from college.

The first mother and daughter to be inducted at the same time into the International Forest of Friendship in 1994. Jennifer believes it is a great honor to be among these extraordinary contributors to the world of aviation, especially her mother.

BOZENA SYSKA, if you're looking for Bozena Syska at Brookhaven Airport, don't ask for her by name. Instead ask for the woman with the flying dog, anyone will point her out to you. For the past six years, Bozena's favorite co-pilot has been her dog, Czarna. In fact, the only time Bozena has had engine problems while flying her Warrior, which she has owned since getting her private in 1984, is when Czarna wasn't with her. Now she rarely flies without her dog.

Bozena learned to fly at Santa Monica Airport at age 29. But with the ink still fresh on her private license, her job relocated her to Long Island. Numerous requests to have her plane ferried to her new home, convinced her she must do it herself. With 117 hours under her belt, she soloed her craft to its new home. In the 11 years since that flight, she obtained her IFR rating and has made shorter cross-country flights up and down the east coast. As a 99, she is the one her chapter can rely on to participate in events where flying is needed. She an enthusiastic supporter of programs which give first flights to children and young adults.

Bozena's second love is writing, She's written and edited for local aviation newsletters, is currently the editor of the NY-NJ Section Newsletter and has published poems some of which feature flying and her favorite co-pilot. Bozena supports her aviation habit as a graphic designer at *Newsday*, a newspaper.

As for flying with a sheepdog, Bozena says, "Czarna's the perfect co-pilot. She's always available to fly when I am, she doesn't criticize my flying and she's a great bodyguard at unfamiliar airports."

MARY SHYNE TAIT, born Dec. 31, 1944, in Cohasset, MA, learned to fly in Fryeburg, ME, in 1992 after moving from Massachusetts in 1990. Prior to moving she worked in the supermarket industry for 14 years in New England developing and implementing efficient systems for purchasing and distribution of store supplies.

During the first month with her private pilot license, she flew into Logan Airport and continues to do so, not wanting any airspace to be a barrier. Other longer trips have included Maryland, Montreal and Nova Scotia. She has flown Cessna 150, Piper Warrior, Piper Archer II, Piper Arrow RG and Cessna 172 float plane.

Mary obtained her instrument rating in October 1994, commercial in July 1995, is presently working on her single engine sea rating and plans to continue on to CFII single-engine land and sea.

She is married to Lindsay Tait and has two children, Anna Eddy, 30 and Lisa Tait, 25. She has one grandchild, Brianna Eddy, 1.

MARY JANE SEERY TALBOT, learned to fly in Glendale, AZ, and earned her private pilot license on Aug. 1, 1993. "A little over 20 years ago, I was a flight attendant, and, although my job was incredibly boring, it piqued my interest in flying. Later as an undergraduate AFROTC cadet at Northern Arizona University, I investigated the possibility of competing for a slot in navigator training. (I already knew my age, 27, and glasses would preclude me from pilot training.) Unfortunately, I was too old for navigator training as well, but the interest never died."

Mary served in the Air Force as an enlisted person and as a Titan II missile maintenance officer. After earning a master's degree in higher education from the University of Arizona, she worked in planning and research for a number of years. Recently, Mary and her husband, Neil Talbott, and their daughter, Christina, moved to Colorado Springs. "I will continue to fly here just for the love of it, but some day, perhaps in three or four years, I will be able to fly medical supplies and personnel for a non-profit organization."

GUDRUN HENLE-TALIRZ, born July 16, 1967, in Innsbruck, Austria, got her first flying license in 1983 in Innsbruck, Austria. At that time she was 16 years old, and her first plane (an ancient K4 glider) was of more than double that age. Though she now likes soaring modern competition gliders like the DG-300 (especially in the Foehn storms in Innsbruck), her main interest is in aerobatics. For aerobatic training she went to Poland. (Gudrun's teacher there was Jerey Makula the gliding aerobatics world champion). Unable to afford a private single engine education in Austria, Gudrun came to Oklahoma and got her PPL there. Since then she has returned several times to the US and logged hours in C-152, 172, C-170A, Decathlon and Luscombe T8F.

Gudrun's profession is medical doctor at the university clinic of Innsbruck. In her flying club she works as a radio instructor and is responsible for repacking the parachutes. Her husband, Wolfgang Henle, is a physicist and earns his living as an airline pilot.

LURANA E. TALLY, born Oct. 21, 1918, in Mansfield, MA. She graduated as an RN from Massachusetts General Hospital in 1940. She had her first airplane ride in a Piper Cub at Wiggins Airfield which is now Logan Airport. Thirty years later, at the age of 52, she earned her private pilot license followed by an instrument rating in 1974. She has logged a total of 1,029 hours.

She has owned a Rally Minerva-220, flown a variety of fixed and retractable Piper products and the Grumman American Traveler.

She joined the 99s in 1970 and has been in the altitude chamber-Pease AFB (SAC) Portsmouth, NH; midair refueling-Pease AFB; flew two AWNEAR (All Women NE Air Race); ground crew for three AWNEAR races; Girl Scout aviation orientation seminars; three airmarkings; and attended chapter, section and International 99 meetings.

She is married to Sidney K. Tally, husband and first passenger and the best navigator around. She has three children: Joanne, daughter 52, almost soloed; Sidney Jr., son 50, commercial, instrument, multi-engine rating with 1,500 hours; and Dr. Taz, son 44, a private pilot. "My association with the 99s opened a new, exciting and rewarding life. I recommend it to women of all ages."

SHIRLEY TANNER, Orange County Chapter. Since her first flight in 1969, Shirley has earned her commercial, instrument and multi-engine ratings, and certified flight instructor certificate. Combining motivation, talent and the drive to excel, Shirley has filled her years of 99 membership working to advance the cause of women in aviation. She has served as chapter chairman and airmarking chairman, and won chapter "Pilot of the Year," but her first love is air racing. She has competed in many local air races and aerobatics events

in her Citabria, as well as, the IAR and AWTAR, usually bringing home a trophy; but best of all the AWTAR of 1976 with sister, Joan Paynter, when with a blown exhaust stack, still finished second.

ELAINE (HAWORTH) TANTON, originally from Saskatchewan, learned to fly during school holidays in

1974 at the Regina Flying Club. After graduating from Havergal College in Toronto in 1975, she attended the aviation technology program at Selkirk College in Castlegar, B.C.

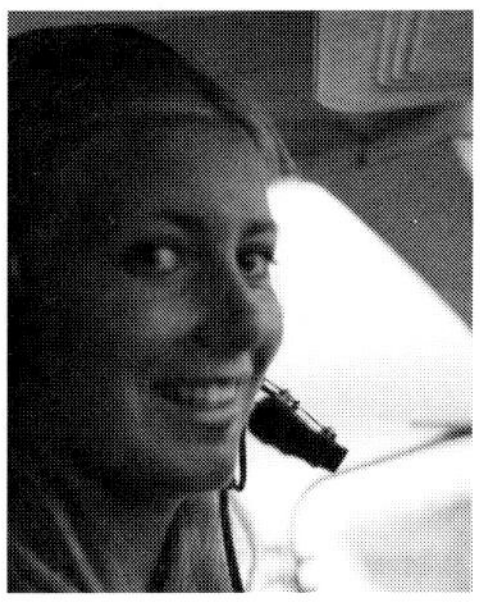

Between 1977 and 1986 she flew over 25 different types of single and light twin aircraft as a flight instructor, charter pilot and corporate pilot. This included three years flying on severe storm research and hail suppression programs in Canada and Greece. With more than 3,300 hours she currently holds an ATR license with a Class I multi-engine instrument rating and float endorsement.

An active member of the 99s in Calgary, Alberta since 1990, Elaine enjoys life by flying her PA 28-235 to a mountain hideaway on Kootenay Lake in British Columbia.

ALYCE STUART TAYLOR, born April 16, 1935, in Roanoke, VA, earned her commercial, ASMEL, instrument, CFI and CFII ratings. She has flown most single and light twin engine planes. She joined the 99s in 1969 and is a charter member of the Greater Pittsburgh Chapter. She has been designated pilot examiner from 1978-1992; participant in Angel Derby 1972; FAA safety counselor; various positions, chief instructor for Part 141 flight school; and chief pilot for Part 135 air taxi. She is married to Hameed Afzal who also has his ATP, FO, CFI and CFII ratings. She has one child, Ernest Adolph Conrads III and two grandchildren, Desmond Conrads, 5 and Julian Conrads, 1. She has been the owner of Alpha Tango Flying Services, Inc., since 1980.

ANNA TAYLOR, born on May 12, 1960, in Swindon, England. As the only daughter in an Air Force family, Anna followed the family tradition and joined the Air Force after high school. While in the Air Force, she met and married Kent Taylor. No longer on active duty, Anna is still an enthusiastic member of the Air Force Reserve.

Anna graduated from Oklahoma State University in 1989 with a degree in aviation management. During her time at OSU, she earned her private pilot license and multi-engine rating and joined the 99s. Most exciting hours have been in the right seat of a Beech 17 (owned by OSU) and a back seat ride in an A-7 attack jet with her former guard unit in Albuquerque.

Currently, Anna is treasurer of the South Central Section of the 99s and editor of the section newsletter, *The Approach.*

DONNA LOU TAYLOR, born Feb. 4, 1936, In Tillamook, OR, can only first remember that after working all summer (age 12 years) picking fruit and hops, she took a small portion of the money and went for a 15 minute plane ride in Newberg, OR. She then watched aircraft at various fields as she grew up – never knowing that she could be a pilot! She watched, fascinated at the Blimp base at Tillamook as blimps floated up and down the Oregon coast during WWI and the Coast Guard planes as they flew hourly, watching for Japanese subs. After marrying and having their four children, her husband felt that by flying he could go up and down the Sacramento Valley to meetings and be away from their business less. The minute she said she would not fly with him unless she knew what she was doing, he started her out with flying lessons. She got her PPASEL in 1974 and her instrument rating in 1994. She has flown the Cessna 150 and bought 1971 Piper Arrow. She had 20 minutes in an ultralight.

Then she found the 99s! Was a charter member of the Mt. Shasta Chapter since June 1972. Then discovered cross-country air racing and has participated in about 16 races. This has been great. Every time Taylor puts the plane, she thanks God and her husband that she can do this. She is not a great, well known pilot. She is not many things. But she has an ability to do something that is indescribable happiness! The icing on the cake is that her 99 friends understand this, and they feel the same. Other friends cannot begin to comprehend her inner joy at flying. She prays that this wonderful USA does not make it too difficult for the younger potential pilots to do the same.

Taylor is married to Howard L. "Bud" Taylor and has four children: Brent 32 years; Jana, 31 years; Dana and Dian, 29 years (twin girls). She has two grandchildren, Joshua, 4 years and Lauren, 2 years. Her daughter, Dana, is a pilot and relates to her special joy in flying and she to hers.

ELOISE ELIZABETH TAYLOR, Phoenix Chapter, as a cadet wife in 1943-44 watching husband Jim go from Stearmans to B-17s, little did she think that 34 years later on a dark, rainy and windy morning early in January he would pick up the phone in his office to hear, "Hi, honey, I'm a pilot." He couldn't believe she took the check-ride in those marginal conditions. But for most of the next 16 years it has been "severe clear" in the southwest. She has flown in Kachina Doll Air Rallies and other chapter fun and games. Jim and Eloise have been into many of the airports and dirt strips in Baja, CA in their Cessna 182, and mountain flying in the Rockies, the Sierra Nevadas; and the coast ranges have presented challenges and rewards. She has been chapter historian, delegate to the International Convention and coordinator of the Phoenix intermediate stopover for the Angel Derby. They recently celebrated their 53rd wedding anniversary with their two sons and daughter, their spouses and three grandchildren. Flying has enhanced her vocation as a Christian Science practitioner.

HELEN MARIE REEVES TAYLOR, Her darling husband, Gary, came home one day and asked if she would like to take flying lessons. One solitary year later they were the proud owners of a 1984 Piper Warrior II #4374N, and he with his very own private pilot license. Her first cross-country was a day of adventure to Bryce Canyon for a hike among the stunning hoo doos, return flight over the Salt Lake Valley with its glistening glow and an iridescent moon rising over the mountains, an incredible adventure in a Cessna 152 #5384P and a delightful day of friendship indeed!

Landing in Utah from the Washington, DC, area was another fun adventure. At the age of 28, she decided to finish her undergraduate BS. Choosing to major in logistics management she went to the library to find the available options. There was one, Weber State College, now University of Ogden, UT. Looked at a map to locate Utah, since easterners have the vague idea that it's out west somewhere. Loaded up her BMW 530 Taylor, with her "child from a previous marriage," a Pomeranian named Brandy, and arrived three days later. Little did she know the greatest adventure of her life was now in motion . . .

Taylor and her husband have five wonderful children: Howard, Wendy, Michael, John and James. They are the happiest married couple they know and have their own business selling copiers, faxes and computers. Gary is a brilliant marketeer whom she loves with all her heart.

What a great feeling it is to fly Gary down to Manti to fly the valley and surrounding mountains, for their mutual enjoyment. Both of them are very proud and pleased with the gifts of love and enjoyment of life that the Lord continues to bless them with, that they look to share with others. She knows that flying will test her and keep her in a growth mode for the rest of her life – what an exciting thought! What more could one ask from life?

SUZANNE JABLONSKI-TAYLOR, received her private pilot license in 1977, instrument and multi-engine rating in 1980, commercial single and multi-engine 1982, CFI 1983 and sea plane, 1989.

Taylor joined the San Fernando Valley Chapter 99s in 1977. Flew first air race in 1978, subsequently earning several first place and top ten trophies in Palms to Pines, Pacific Air Race, Valley Air Derby, Back to Basics, Air Race Classic and the first International California-Baja Air Race (a few of the many 99s sponsored events). Proud recipient of the 1981 Amelia Earhart Memorial Scholarship. Participated in more than 15 years of the Jim Hicklin Memorial Air Rallies.

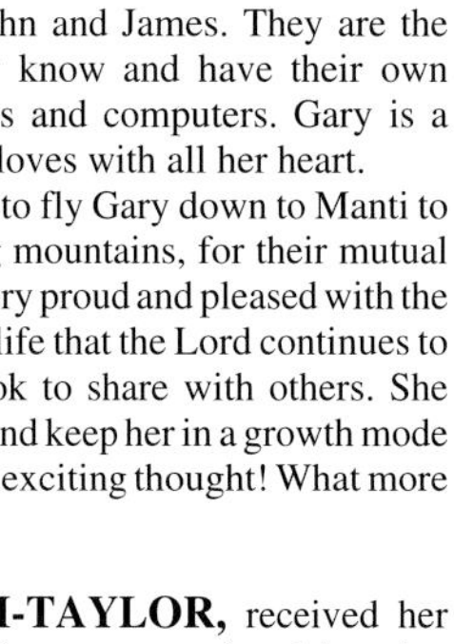

Aircraft owned: Piper Cherokee 140; Piper Dakota 236; with experience in Boeing Stearman, B25J Mitchell Bomber, T-6 Trainer, Piper Cub, P51 Mustang and Waco F5.

"Good friends and cherished memories are mine forever thanks to the 99s!"

JANE CHAMBERLAIN DUNBAR TEDESCHI, born Oct. 15, 19__, in Washington, DC., received flight training at Congressional Airport, Rockville MD, Stevens Airport, Frederick, MD and Avenger Field, Sweetwater, TX. She was a Woman Air Force Service Pilot during WWII. She has flown the Cub, Taylorcraft, Aeronca and Stinson as a civilian pilot; the PT-19, PT-17, BT-13 and AT-6 as a military pilot. She was employed during WWII in the engineering department, test pilot at Craig Army Airfield, Selma, AL.

Wearing an oversized, made-for-men winter WWII flight suit almost washed her out when the Army check pilot pulled the stick back for spin recovery. The stick caught under the folds of her heavy jacket and the heavy gloves were too clumsy to free it for quick recovery as required. Her instructor had faith and arranged another test checkride which she passed easily.

She is married to Romolo D. Tedeschi and has three sons, Craig, Alan and Jody. She has two grandchildren, Zachary, 7 and Matthew, 2.

NANCY TEEL, born June 21, 1932, in Alameda, CA. Became a professional accordionist at age 16. Musical career included USO Tours of Europe and Far East. Also had many jobs on West Coast, Alaska and Denver, CO by age 30.

Married Don on Aug. 5, 1953. Has two children, two stepsons, 10 grandchildren and three great-grandchildren.

Had first flying lesson December 1968 and earned private pilot license, April 1969. Joined Tulsa 99s in 1969. Founded Tri-State Chapter 99s (now defunct) May 1971. Chairman two years.

Co-produced 1971 Joplin, MO, air show featuring Blue Angels, Harold Crier etc.

By January 1976 had earned commercial, instrument, CFII, ground instructor and airline transport pilot. Flew charter from 1973 through 1981 for a Kansas corporation and logged 5,000 hours before retirement.

Member of AOPA, International Comanche Society, American Bonanza Society, EAA, Eastern Star and presently vice-chairman of Tucson 99s.

BEBE TEICHMAN, a native of Virginia, educated at Old Dominion University, Norfolk, VA, learned to fly in the sky over Virginia's countryside. She and her father learned to fly during the same period and soloed on the same day in 1985. She owns and operates a 1946 Aeronca Champ, which she received as a birthday present from her husband one year after she became a private pilot. She and her husband also own a Piper Twin Comanche. Building an Acro Sport II is a project that Bebe and her husband have been working on for six years and is now more than half-way completed.

She now lives in Florida with her husband, David; daughter, Amelia, age 7 and son, David II, age 4. She is a full-time mother. She checks on her children by flying the Aeronca over their schools almost every day. Being a part of a flying family she plans many trips in the twin Comanche. Recently the family has been to Yellowstone, MT, and on several trips to the Caribbean.

Before becoming a mother, Bebe was a professional sales person and has lived and worked in Washington, DC, and Philadelphia, PA.

She is a member of the 99s, EAA, AOPA and many other not so well known flying organizations.

DIA TERESE, Ah! To be able to fly! What dreams she had for as long as she can remember. And now, especially as she flies among the craggy peaks of the Sawtooth Mountains in Stanley, ID, she feels like she is still in her dream world. Even though she was born and raised, 1945-1960, in Los Angeles, CA, she has always been "a country girl" at heart. So at this particular juncture in her life – turning 50 – she feels extremely lucky to be living and flying in some of the most spectacular country in the world!

At age 43 in 1988, Terese, at long last, had the money and the time to get her pilot license. She knew she wanted a female pilot instructor and was thrilled to find Pam Penkoff in Sun Valley, ID, even though she was 75 miles from where she had chosen to live. Her passion to learn to fly fueled her to be able to drive over a mountain pass in the middle of winter, leaving her cozy, tiny cabin on the Salmon River in temperatures of 25 degrees below zero! And she hates to be cold, yet there she was, preflighting that Cessna 172! Pam's humor always prevailed – what a wonderful woman!

That spring of 1989, Terese fell in love with Robert Danner, cowboy and bush pilot, lifelong resident of the little frontier town of Stanley, ID (population of about 50 in the winter). Through his know-how and guidance she was able to purchase an old 1960 Cessna 182. He took over where Pam left off, working with her until she could take her check-ride in her own high performance plane. She highly recommends that learning to fly with your sweetheart as your instructor should be avoided if at all possible. The man deserves a medal (and so does she).

Her dear airplane 8690T has provided her with a way to make a living. Over the years, *Tango,* received a new engine (when pulling gliders finished off her old one) new Cleveland brakes, windshield, radios, transponder and everything else she needed (and continues to need) to be a little workhorse for Robert's charter service, Stanley Air Taxi. Doing scenic air tours, back-country flights, forest service and fish and game recon is all very interesting and satisfying for her. She's not yet pretty – a much needed paint job always takes a back seat to the more important items.

Terese finds herself in a love/hate relationship with airplanes and flying. Making money by the fruits of a passion can definitely eat away at the fun and glamour. Nevertheless, she feels very fortunate to have her airplane sitting out in her front yard, next to a dirt strip in the beautiful mountains of Idaho, and she is looking forward to the day she can say "I just play" in the skies.

SHIRLEY A. TEUTSCH, born Nov. 22, 1928, in Los Angeles, CA, joined the 99s in 1976. She has a private pilot license with a commercial and instrument rating. She has flown the C-150, AM5-A, PA-28-161-PA38112, B-F33T, M20J, B-7-36, PA28R-201T.

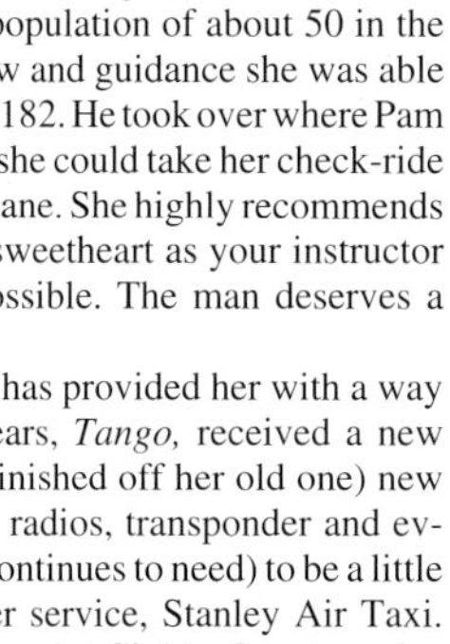

In the more than 2,200 flight hours, she has flown "North to Alaska" by way of the Alcan (twice). She flew to Albany, NY, via Chicago and Philadelphia for the 50th Anniversary of the 99s and to other 99 conventions. She has enjoyed many trips to Bermuda Dunes in Palm Desert. In 1993 she flew to Catalina Island which was challenging and fun. Most recently, her travels have taken her "South to Santa Ana" and the John Wayne Airport to visit family. Her flying has also been a wonderful mode of transportation to her many golf tournaments country-wide.

She is married to Thomas L. Teutsch, and has four children: Karen Henslee, 45; Marcia Dietz, 43; Theresa Jones, 41; and Thomas Teutsch, 28. She has four grandchildren: Jaime Henslee, 17; Angela Dietz, 17; David Dietz, 15; and Brittany Dietz, 8.

LOUISE MCPHETRIDGE THADEN, Carolinas Chapter, born in 1905 in Bentonville, AR, the airfield there is now named for her. She went from her 1927 solo in San Francisco to one of the outstanding pilots in American aviation history.

She was married to Herbert von Thaden, in 1928, and they have two children. She lives in High Point, NC, today and owns Thaden Engineering Co. She was a charter member of the 99s, secretary in 1929, vice-president in 1931-32 and co-chairman of the board of trustees of the Amelia Earhart Scholarship Program in 1962-65. Her professional experience and aviation history take two pages merely to list, and can be read in *Who's Who of American Women.* Highlights include:

1928-29: altitude record for women 20,200 feet — at 24 degrees F; solo endurance record, 22 hours, 3 minutes, 28 seconds; speed record, Los Angeles to Cleveland, Women's Air Derby; Transport License, fourth woman in the US to receive it (she also earned a commercial later). 1930-35: refueling endurance record, 196 hours, in a Curtiss Thrush; co-developer of the national airmarking program for the Dept. of Commerce; light plane speed record.

In 1936 she was awarded the Harmon Trophy as the world's outstanding flier. That year she also won the Bendix Trophy race with Blanche Noyes, the first woman to do so, in 14 hours, 55 minutes. And the same year she captured the east-west speed record in a Beech Staggerwing.

In 1938 she set the 100 km. speed record for women at 199 mph. That year she also published *High, Wide and Handsome* (new edition in 1973). She was active as a lieutenant colonel and command pilot in CAP, 1949-1970; on the advisory committee on women for the Defense Department, 1959-61; in the OX5 Hall of Fame; 1973 received the Silver Wings Aviation Achievement Award.

MARJORIE THAYER, born Wenatchee, WA, on July 20, 1942. Started flying in 1969. Moved to Arizona in 1972. Has logged approximately 3,000 hours. Holds a commercial pilot certificate with SEL, SES, MEL and instrument ratings. Joined the Phoenix 99s in 1973. Helped to charter the Arizona Sundance 99s in 1982. Has been involved in cross-country speed racing for the last 10 years. Has flown for Val Vista Management Corp. since 1981. She is chief pilot and also flies charter for SAS Executive Aviation, Ltd., Falcon Field, Mesa, AZ. Her 491/2 Ron and she have a combination of nine children and 15 grandchildren, none of them pilots. She is a member of the 99s, ARC and AOPA.

Thayer's hobbies are cross-country speed racing, deep sea fishing, traveling in their RV, their annual Santa Claus trip to Guaymas, Mexico and working at Phoenix 500.

PATRICIA THEBERGE, a Massachusetts resident has been flying since 1988 and currently flies a Cessna 172. Included in her flight training was a week long IFR training flight to the Bahamas.

In addition to being actively involved as an officer in the Eastern New England Chapter of the 99s, she is vice-chairman of her local airport commission, a member of the Aircraft Owners and Pilots Assoc. and Experimental Aircraft Assoc.

Pat is a graduate of the New England School of Art and holds a BS and a professional achievement certificate from Northeastern University. She is currently employed as an industrial engineer.

DOROTHY THEURER, joined the Santa Clara Valley Chapter in 1982, after coming to a Flying Companion Seminar and learning to fly. She served as membership chairman for many years and has been an active 99. Dottie was chairman of the Amelia Earhart Luncheon at the Hawaii International Convention in 1986.

BARBARA WALKER THISTED, born on June 25, 1927, in Los Angeles, CA, where she had her first airplane ride at age 10, in 1937. In high school, she completed a ground school course and subsequently obtained a pilot license, April 1947 and joined the 99s the following September. In May 1947 she married Dale Thisted, a former P-38 pilot and flight instructor. Their honeymoon was in their Vultee BT-13, the first of nine airplanes that they have owned. She has flown the Vultee BT-13A, Taylorcraft L2, Stearman A-75-220, Cessna 150, 172, 182, 210; Aeronca 7AC, 90; Sedan 145, L3, L16 and Tri Champ; Piper J3, PA-24-250; Beech 55 (Baron); Bonanza D-35, C85; Ercoupe, Stinson 165, OY2, Bellanca 90 and Mooney Mite 65.

Over the years, Barbara worked in experimental flight test, North American Aviation; with husband Dale, operated the Del Mar Airport, CA, 1949-51, and were partners in Jimsair Aviation Services, Lindbergh Field, San Diego (1968-78). Other activities included: organizer and first president of the Los Angeles Chapter of Women's International Aeronautical Assoc.; membership in National Aeronautical Assoc.; Washington Pilots Assoc.; Silver Wings and officer/pilot in Civil Air Patrol. Her 48 year membership in the 99s has resulted in many elected/appointed offices and innumerable activities, including working on several terminuses and flying in the Powder Puff Derby. The Thisteds have four children (all with pilot licenses): Ronald, 44; Blair, 42; Dana, 40; and Sarah Ellen, 35; and five grandchildren: Jennifer Thisted, 15; Walker Thisted, 11; Matt Grossman, 18; Aliza Grossman, 4; and O'Farrell (born September 1995).

SHIRLEY THOM, learned to fly in 1959 and has logged more than 1,600 hours. She loves air racing, especially with her daughter, Linda Johnston. The mother-daughter team has participated in the Palms to Pines, PAR and the Powder Puff Derby. She has flown the Piper Apache, Tomahawk 140, 180, 235; Beech F33A, Cessna 150, 172, 182RG, 182 and P210.

Shirley is very active in her chapter activities such as, FIRC, past chairman of the Jim Hicklin Memorial Air Race for men only, and other chapter projects.

Shirley sold aircraft for eight years, was partner in a ComputerLand and became a financial planner while in Dallas, TX. She is in the process of starting a video company with her daughter Linda.

She is married to Eugene Thom, also a pilot, and an electrical engineer. They have a daughter, Linda, and a son Jerry along with four grandchildren: Sandy Falley, 29; Kenneth, 10; James, 8; and Steven, 7; and two great-grandchildren.

LYNN THOMA, born in Connecticut in May 1965. She, at the time of this writing, has flown most of the well-known Cessna and Piper singles. In addition to these, she has flown a Decathlon and shamelessly lusts after all aerobatic airplanes, the more aerobatic the better. Lynn got bitten by the flying bug at age 9 at Waterford Airport where in addition to a model airplane show complete with a flying Snoopy doghouse, she and her family crammed into a 182 (seven total) for a 15 minute hop. She turned to flying models in her teens, as a less expensive means of enjoying aviation and was startled to find on the day of her first model flight, the same Snoop doghouse already there. A modeling friend also flew full scale and invited Lynn to share expenses. It was just a matter of time before the desire to hold the controls overwhelmed her and lessons began. She soloed at Chester Airport and got her remaining training at Groton Airport where she is now a flight instructor. One of the job "perks" is showing many people how fun flying is. Lynn is studying for instrument instructor and multi-engine ratings. She aspires to a successful airline carrier and hopes to have an aerobatic beasty of her own. Mrs. Thoma joined the 99s in 1994 and also belongs to AOPA and the National Assoc. of Flight Instructors. Prior to marrying in 1993, her sweetheart, George, a jet engine mechanic for Pratt and Whitney, Mrs. Thoma went by the name Michaud. She wants to ask, "Aerobatics, anyone?"

BETTY KIRK THOMAS, born Oct. 1, 1934, in Theo, AR. Flying was her husband's thing and a means for her to get from point A to point B in the shortest amount of time. Her husband, Gerald F. (Jerry) Thomas, and she were well into "middle age" when she realized that she had no idea how to land that Cherokee. After reading a couple of horror stories about women like herself crashing after spouse incapacitation, Thomas decided that she wanted to learn to land the plane and operate the radio. Her husband was delighted and encouraged this interest. She soon realized that if she could land the thing, she could fly it; and somewhere along the way

it became fun. She stuck with it, got her license, and joined the Golden Triangle Chapter of 99s as soon as she could.

Betty and Jerry have two children, Kirk F. Thomas, 40 and Jerry Kim Thomas, 38. They have three grandchildren: Morgan Yshin Thomas, 7; Darby Kim Thomas, 5; and Whitford Kirk Thomas, 3.

CAROLYN RUTH THOMAS, born May 10, 1946, in Ottawa, Ontario, Canada, learned to fly at the Ottawa Flying Club in 1974. She holds a pilot license with a commercial rating and has flown in many air rallies and the Powder Puff Derby in 1976 with Betty Jane Schermerhorn. She has flown the Grumman Cheetah and the Cessna 172.

Thomas' educational background consists of a BA in biology from Carleton University in Ottawa and a B.Ed. from the University of Toronto. She is a school teach in Ottawa.

She is married to Ron Neufeld, who is also a pilot, and they own a Grumman Cheetah. They fly all over Canada and the US and have flown from Ottawa to Seattle and Vancouver every summer for the past six years.

RUTH WOLFE THOMAS, born Oct. 28, 1918, in Jefferson County, TN, joined the 99s Aug. 9, 1940, as a charter member of Tennessee Chapter; married Ferris Thomas, flight instructor, who gave her flying lessons on their first date. First female hired as airway traffic controller in Detroit Airways Center during WWII; a public school teacher/librarian who combined her profession with aviation and taught Tennessee teachers in aerospace workshops from 1954 to 1965. She received NAA commendation in 1953 for aviation education activities in Knoxville school system. Attended week-long celebration of 50th Anniversary of Powered Flight in Washington, DC. Awarded Amelia Earhart Medal at National Aerospace Workshop, Boulder, CO in 1954. Flew in both AWTAR and IAR races; was instrumental in getting her hometown, Knoxville, TN, to terminate the 1954 AWTAR. She won the 1976 Tennessee Historical Air Photo Contest. Flew her Cessna 172 (N99RT) to International Conventions in Alaska and Vancouver and celebrated her 50th wedding anniversary plus 50th Anniversary of Charter of Tennessee 99s Chapter at Las Vegas Convention. In 1974 the Thomas' purchased a paved drag strip in Maryville, TN, renamed it Montvale Airpark (Atlanta Sectional Chart) and built there home there.

SANDRA LEA THOMAS, born Oct. 2, 1936, in Evart, MI, was a grandmother of three and over the age of 50 when she received her private pilot license: It is never too late to take flying lessons.

Her interest in flying came after her husband started at Detroit Metro Airport in 1961. Shortly after that she started at Bowman Field. She soloed at 12 hours on May 17, 1962 and had a total of 22 hours when she couldn't continue.

Back to flying lessons during the summer of 1988, soloed again at 15 hours on July 20, 1988. Received her license Nov. 26, 1989.

Has been a member of the 99s since receiving her license and has participated in several areas of her chapter activities including two years as chapter chairman. Has also been safety chairman for North Central Section.

Some community activities over the years have been as Girl Scout leader followed by Girl Scout Neighborhood chairman, president of the Jaycettes of Louisville, KY, member of CAP and volunteer driver for the American Cancer Society.

Other organizations she currently belongs to are the Aircraft Owners and Pilots Assoc., treasurer of the Michigan Aviation Assoc., and also treasurer of the Michigan Aviation Education Foundation.

Their recent trip to Florida via Cincinnati West Airport: Due to draining oil instead of adding oil and not wiping the stick clean, they had temperature, oil pressure drop, aircraft shaking and RPM problems upon their decent below cloud layer northwest of Cincinnati. But thanks to Cincinnati Approach and Cinn West Airport's IA mechanics, they learned a very valuable lesson. Always wipe the oil stick when the oil is clean even though you may have checked it three times. She also learned she did not panic.

She is married to Charles H. Thomas and has three children: Dawna Rennae, 38; John Charles, 36; and Kimberly Michelle, 29. She also has three grandchildren: Alisha Lynn, 19; Brent Aaron, 11; and Melissa Kay, 15.

KEETA J. THOMPSON, Nebraska Chapter, and all of this after her 46th birthday, January 1977, found her implementing a Christmas gift from her husband, Dick, the Cessna private pilot course. Her flight endeavors progressed as follows:

Private – May 1977, glider – August 1977, multi-engine March 1978. Currently she is devoting her free time, (she is employed full-time outside her home) toward finishing her commercial/instrument – then instructor. The continued support and encouragement of her husband and four sons enables her to reach for aviation's higher echelons. What an exciting, pleasurable and rewarding way to invest the rest of her life. Total hours are 241.

VIRGINIA THOMPSON, obtained her private pilot license in 1941; instrument rating, 1971.

She joined the 99s in 1954. Virginia was International historian for many years; past governor of the Middle-East Section (now Mid-Atlantic), past chapter chairman of the Washington, DC, Chapter and charter member and past chapter chairman of the Shenandoah Valley Chapter. She was the first historian of the Middle-East Section.

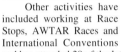

Other activities have included working at Race Stops, AWTAR Races and International Conventions and has attended 29 of the latter.

She has also served as secretary of the International Women's Air and Space Museum and is currently president of the New Market Garden Club and secretary of the New Market Family and Community Education Club and the Shenandoah Valley Chapter of 99s.

Her husband, Donald M. Thompson, is a retired aeronautical engineer who is currently an FAA flight examiner, a flight instructor and an A&P mechanic.

They have two sons and four grandchildren.

CHERYL LYNN THORNTON, born Feb. 25, 1971, in Pompton Plains, NJ, raised in Fairfax County, VA, and now works at Teterboro Airport in New Jersey as a flight instructor.

Cheryl started flight training in October 1989 while pursuing a BS in aeronautical science at Embry-Riddle Aeronautical University in Prescott, AZ. She graduated in December 1992, after obtaining her commercial certificate and instrument rating for both single and multi-engine land. She has a minor in aviation safety and was the first student member and student chapter president of the International Society of Air Safety Investigators.

After graduating, Cheryl returned to New Jersey and obtained her certified flight instructor and instrument instructor ratings through Air Fleet Training Systems, where she is currently employed. Having accumulated more than 1,600 flight hours, she is currently working on her ATP.

Cheryl finally decided to join the 99s in 1995 and has flown everything from Cessna 150s, Skyhawks, a Decathlon, a Grumman Tiger and a Bonanza to Seminoles, Duchesses, Barons, a King Air 90, a Merlin and a Falcon 50 right seat. She plans to keep flying as long as life permits and hopefully to pass on her love of flying to her future children.

GABRIELLE ANITA THORP (CRIPPES), began flying in 1977 at Falcon Field in Mesa, AZ. After earning her business degree from Arizona State University in 1978, she worked as a CFII/Multi I for four years. During the summers, she flew DC-4s and PV-2s involved with large scale spray programs throughout the US. In 1979 Gabi was honored with an Amelia Earhart Scholarship from the 99s. She used this award to earn her ATP license.

In 1982 Gabi was selected for a pilot position with the 161st Air Refueling Group/Arizona Air National Guard. She completed UPT Williams AFB, survival school (Fairchild AFB), and KC-135 training (Castle AFB) before becoming mission qualified in the KC-135E by the summer of 1984.

As a part-time pilot for the Arizona Air Guard, Gabi has had the wonderful opportunity to travel the world for the past 12 years. She upgraded to aircraft commander in 1990 and recently became an instructor pilot in the KC-135E. Gabi holds the rank of major.

Gabi married Joseph Crippes (a full-time Guard pilot) in 1987. There are the very proud parents to twin boys, Austin and Kyle, who were born in 1990. They reside in Phoenix, AZ. In addition to her flying duties at the Air Guard, Gabi manages Joe's Custom Home Construction business and is a licensed realtor. Gabi also recently finished her classroom requirements for her MBA and is currently completing her thesis at the University of Phoenix.

PATRICIA M. THRASHER, ENE Chapter, born March 3, 1928, in Needham, MA, this former Northeast Airline stewardess surprised the airline pilot one day in 1969 when he arrived at the small airport to see her taking-off alone in his small Cessna. It is not unusual that an ex-stewardess, who loves to fly, would earn her pilot license. It's just that the airline pilot, who had no inkling of her accomplishment, was her husband. She earned her commercial rating in their later acquired Bonanza, spent one summer giving scenic rides at the Chatham Airport on Cape Cod and went on as a sales representative for Piper Aircraft for four years there. Pat joined the 99s in 1973 and has served on many committees, including membership chairman, operations chairman of New England Air Race 1977, co-pilot in AWNEAR in 1975 and placing fourth and Powder Puff Derby in 1977. Offices held – treasurer, secretary, vice-chairman ENE Chapter; and section treasurer, secretary and vice-governor 1992-93. One very ironic incident worthy of mention is that having forever loving anything to do with flying, she was listed in her high school Who's Who as another Amelia Earhart. She is married to Thomas F. Thrasher and has three children.

NANCY HOPKINS TIER, born June 16, 1909, in Washington, DC, began her flight training at Hoover Field, Washington, DC, and Roosevelt Field, NY, from 1927 to 1929. She served 18 years in the military from 1942 to 1960. She has owned a Fleet, Kittyhawk, Aristocrat, OX Waco, Eaglerock, Taylorcraft, two Monocrafts, Cessna 140 and Cessna 170.

Flat spin (she recovered from); six forced landings; flew from her own airfield for 23 years; made all repairs; CAP requirements; taught navigation, aerodynamics, military training; Ford Reliability Tour; Woman's Dixie Derby; Connecticut Speed Champion; many local races; National and State Conventions; charter member of 99s; OX5, Silver Wings, UFO (United Flying Octogenarians); president IWASM (International Women's Air and Space Museum) Dayton, OH; New England Air Museum, Smithsonian membership. Flew to California in 1933 from Connecticut alone in an Aristocrat and was involved in Air and Space Research during the war. She has 66 years of flying and is in the National Hall of Fame, C.W. Post Certificate of Honor and in the Early Bird Assoc. "38."

She is married to Irving V. Tier (deceased) and has three children: Mary Anne Tier Oxford, 61; Benjamin I. Tier,

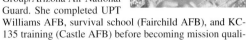

59; and David H. Tier, 45. She has six grandchildren: Peter B. Oxford, Lee Anne Oxford (Cunningham), Glenn Oxford, Julia S. Tier (Osborne), Stephen Tier, Christopher Tier and many great-grandchildren.

DOREEN TIGHE, a New Jersey resident, learned to fly in 1984. She flies Cherokee 140s and 180s out of Sussex Airport and has logged nearly 200 hours. Flying for recreation and pleasure as well as participating in local flying events keep her busy in her spare time. Doreen has placed in the Garden State 300 top 10 several times and serves on the Sussex Air Show planning committee.

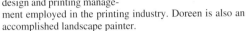

Doreen is a graduate of Georgia Southern University and the Fashion Institute of Technology with degrees in fashion illustration, graphic design and printing management employed in the printing industry. Doreen is also an accomplished landscape painter.

An active member of the North Jersey Chapter 99s, Doreen has served as secretary, vice-chair, director, advertising and promotions chairman on the Young Eagles committee. Additional memberships include AOPA, EAA Chapter 73.

ADELAIDE TINKER, born in Chicago, IL, her husband started the family interest in flying when he came home with his private license in 1949. She acquired hers in 1950 and a commercial in 1953. Son Robert soloed on his 16th birthday (1957) before he could legally drive a car. Flying vacations brought them over most of the US, into Canada and once to Alaska (1961) with all three taking turns as pilot, navigator and photographer. The only other relative who flies is brother, Phillip Camerano, helicopter test pilot.

Joining the 99s, Washington, DC, Chapter in 1951, activities included governor of the Middle East Section three times (1958-60, 1971-72 and 1976-77), PPD timer several times and presently incorporation resident for the International organization.

Lt. Col. Tinker has been a member of the CAP since 1951 where she has served on the Delaware Wing staff in various capacities and participated in search and rescue activities.

JEAN KAYE TINSLEY, born March 24, 1927, in San Francisco, CA, has 50 years of flying this year! It all began when her mother took her to watch Amelia fly out the Golden Gate. She started to fly in 1945 right after WWII when her dad had some left over gas coupons and took her to the valley to fly a T-Craft one of his patients had kept running. No surprise that it was love at first flight! Started flying balloons and helicopters in the 1960s and really fell in love with rotary wing flight. Constant speed prop gyroplane rating in 1976. Been in all World Helicopter Championships in which USA has participated: 1973 as pilot on first US Team, England; 1978 Russia; 1981 judge; Poland; 1986 senior judge England; 1989 chief judge France; 1992 head of US Delegation, England; 1994 judge, Moscow; founding member of Helicopter Club of America; member AOPA; AHS; NAA; FAI; HCA; California Pilots Assoc.; Silver Wings; Quad A; Twirly Birds; Guild of Air Pilots and Air Navigators of the City of London; Western Aerospace Museum Board; past International secretary; Whirly Girls; chairman Bay Cities 99s; president Helicopter Club of America; currently executive director of the Whirly Girls, International Organization of Licensed Women Helicopter Pilots; first female in world to fly the Tilt Rotor.

She has been busy working, raising six children: Margaret Tinsley Mulcahy, 46; William Barnett Tinsley, 42; Martha Tinsley Minot, 37; (deceased children, Christopher Tinsley at age 27; Sally Tinsley Sicard, at age 40 and Carolyn Tinsley, 28). She has eight grandchildren: Dawn Mulcahy Brown, 25; Victoria Mulcahy, 19; Lynecia Sicard, 19; Craig Tinsley, 18; Brian Tinsley, 16; Scott Tinsley, 15/ Katy Minot, 12 and Patrick Minot, 10. She couldn't have done all of her working and flying without the love and support of the family and her late husband, Dr. Clarence "Bud" Tinsley. Still working and flying. She is absolutely delighted that aviation really has opened up to females and is so very proud of their pilots and their accomplishments, in space and aviation.

LINDA MARIE "SUNNI" TODD, born March 30, 1948, in Los Angeles, CA, received her private pilot license on Oct. 15, 1990 at Truckee, CA. She has an ASEL rating with 130 hours. She has flown the C-172 which she owns. She joined the Lake Tahoe Chapter of 99s in February 1991. Todd has been involved in the Truckee Tahoe Airshow every year since 1984, Reno Air Races, Fallon Airshow, FAA Safety Seminars, airmarking, FAA Wings program. Elan Vital, co-founder of Church Davidson Memorial Scholarship (1993).

As a novice student pilot on her fourth solo flight, she went out to practice maneuvers northwest of TRK. It was a lovely morning, VFR conditions and light winds. Maneuver training went well and on returning to the airport she felt pleased and calm. The traffic was landing runway 28 left traffic as it had been when she left earlier. There seemed to be more action with the wind sock, but Unicom still recommended 28 and with several aircraft set up for 28, she didn't question it. The pattern, landing, set-up and rotation all felt fine, then as the main wheels came to touch, she got a stiff quartering tail wind from the left. The touch turned into a hard bounce, the wind pushed her Cessna to the right, nearly into the sagebrush. In a timeless instant her mind computed the danger and the inability to "save" the landing. She brought on full power, took off full flaps, kept the nose lowered and mushed the plane in slow flight at about six feet above the ground until it gained climb speed. As she climbed up her mind raced over past events. She realized she set up with full flaps expecting light winds. This time she monitored AWOS for accurate wind information, set up for 28, prepared for crosswinds, wing slightly lowered to the wind and in case of a last minute shift – no flaps. It was a good landing, a big relief and a lesson she will never forget.

She is married to Robert (a pilot) and has one son, Jesse. She has been a counselor, worked in communications and as a teacher. She us currently a Unicom at TRK and a financial planner.

UTE TOELKE, born Dec. 10, 1936, in Bremen, Germany. She learned to fly in 1973 at the airport Bremen "very, very" secret because her husband was against her wish to learn to fly. "A woman has only to be for the family and her husband," he always said. When she got her license in 1974, he was very proud and the best co-pilot she ever had and suddenly was not anxious any more. Ute participated in many air races including Deutschlandflug.

Ute has been a member of the 99s since 1978. All over the world she has met many flying-friends. Her most beloved friends are in India (Bombay). Since then, they meet each other every year. At the convention at Las Vegas in 1990 she was also heartily invited by many 99s pilots. The hospitality of her American friends was generous.

Her husband, Dieter, died unexpectedly in January 1990 during a holiday in Hawaii with friends. Now she flies only for fun and her two little white dogs are often her companions. They also like to fly. She has two children, Kai, 36 and Jska-Elena, 29. She has one grandchild, Jasna Toelke, 4 1/2.

KATHERINE PAT TOLSMA, born May 25, 1928, in Buffalo, NY, as a young girl was fascinated with flying. In seventh grade she wrote a composition on becoming a stewardess when she grew up — that was the only way she knew of that women could fly. She had her first flight as a sophomore in high school – a scenic tour from Buffalo Airport. WWII ended the year she graduated high school. She went to work, married and raised her boys. When they were 14 and 13, they joined the Civil Air Patrol. Then her husband joined and so did she. Through the aerospace education portion of the program Tolsma learned the ground school requirements. In the emergency services, she gained flying experience, becoming comfortable in a small aircraft. After her son, Paul received a flight scholarship through CAP, her husband decided to learn. He attained his certificate in 1977 and encouraged her to try, but she thought she was too old. Then in 1981 at a CAP Aerospace Education Congress in Las Vegas, Tolsma began talking to the 99s who had a booth there. When she came home the Western New York 99s had a pinch-hitter course where her husband and his partner kept their airplane. She took the course, and those lovely ladies also encouraged her and said they looked forward to her joining them. So she did. Got her certificate and joined the 99s. They are retired now but still active in CAP. It is very satisfying working with these young aviation interested cadets. The highlight of her 22 years in CAP came in 1985 when as commander of Southtowns Cadet Squadron they were not only squadron of the year for Niagara Frontier Group but also the National Squadron of Distinction. She is membership and aerospace chairman for the Western New York 99s. She is married to George J. and has two children, Paul Anthony, 37 and George Jr., 36.

NANCY THOMPSON TOON, born March 9, 1935, in Atlanta, GA. No one in Nancy's family had aviation leanings, except for her father. He would take his daughter to the observation deck of Atlanta's now Hartsfield Airport to watch the take-offs and landings of various aircraft.

After seeing the film, "The High and the Mighty," Nancy decided flying was what she wished to do. Thwarted by family and friends until 1980, when husband, Ralph, a non-pilot purchased an Aerostar for business, Nancy would ride shotgun with the company pilot. A few months later at age 45, Nancy enrolled in flight school. She continued to get advanced ratings, culminating in an ATP in 1988. She has flown Cherokees, Mooney, Cessna 172, 182, 340; Barons; Bonanza; Aerostar and Duchesses.

Nancy began teaching in 1986, and until recently would teach multi-engine to students in her Beech Duchess. This Duchess she flew in the 1985 New York to Paris Air Rally, having only 650 hours in her logbook. Since then racing an airplane has been her favorite sport!

Nancy, with racing partner, Susan Collen, a former student, has won an Air Race Classic twice, and has been in the top five several times. The Mile High Derby was won twice, as well as high place finishes in the Great Southern Air Race.

Nancy and Ralph have three grown daughters: Julie, 37; Libby, 35; and Jennifer, 34; none of whom have shown an interest in becoming pilots. She hopes one of the four grandchildren: Caroline, 6; Drew, 4; Katie, 3; and Brent, 2; will inherit the flying gene from their grandmother.

GLORIA TORNBOM, born Feb. 22, 1930, joined the San Fernando Valley Chapter in 1977, Intermountain 1979 and the Tucson in 1986. She has been chairman of the Tucson Chapter for two years, chairman; Tucson Treasure Hunt, five Cent-a-Pound rides and aerospace for two years. She learned to fly at Van Nuys, CA, and received her private pilot license in 1977. She got her instrument rating on July 12, 1987.

She is a registered nurse and was a flight nurse on call for Schaeffer Air Ambulance in Van Nuys, CA. She was flying her Ercoupe in a Civil Air Patrol practice search in Coeur d' Alene, ID, in the mountains east of Coeur d' Alene, when her engine quit. Her options were slim, when her passenger excitedly said there was a runway out in the distance. It was way too far away, but she headed for it. She kept going, never deviating from her best glide speed. She suddenly realized she was too high, so with no flaps, she had to do all sorts of maneuvers to slow down and lose altitude. She came fast and dove to the ground with a few bounces. They made it with a few bruises and a damaged plane. She found out a little later that she had a 20 to 40 knot tact wind. Without that she probably wouldn't be here.

She has a very supportive husband, three grown children, six grandchildren. She is a seamstress and a member of the CAP, Flying Samaritans and Wright Flight.

DEBRA TORRES, now residing in North Carolina, learned to fly in 1978. She owns and flies a Cessna 172. Torres had logged more than 300 hours and has 19 free-fall parachute jumps to her credit. Torres is currently a second lieutenant with the Civil Air Patrol and has participated in many CAP activities including Aviation Challenge at Space Camp in Huntsville, AL.

Torres' educational background includes a BS in business administration from Marywood College. She is currently enrolled at Embry-Riddle Aeronautical University in the master of aeronautical sciences/management program at Pope AFB, NC. Torres is a member of the 99s, Inc., and Aircraft Owners and Pilots Assoc.

She is married to Rolando Torres who is a student pilot and a command sergeant major in the US Army Special Forces. She has two daughters, Yolanda and Miriam.

FUMIKO TOYODA, born March 3, 1937 in Japan, received her flight training at Japan Flying Service K.K. She joined the 99s in 1985. She has flown the Piper Cherokee, Warrior, Archer and Navajo.

When she was a student pilot, she experienced continuous 26 times touch and go without any mistake, so that Japanese defense force took a film of her practice and showed at Japan Hall. She attended USA cross-country flights from Oct. 4, 1975; departed from Bello Beach, FL, via Panama, New Orleans, Austin, El Paso, Phoenix to Riverside, CA. Also she experienced International cross-country flight to China with two other pilots and an instructor.

She is married to Tsuneo and has three children: Mami, 37; Mari, 27; and Saeko, 33.

MARIAN TRACE, Bay Cities Chapter, learned to fly at Mills Field, where she was secretary to a flying school.

In 1935, she accepted a position with Condor Airways in Honduras.

AILEEN TROTTER, North Jersey Chapter, has resided mainly in New Jersey. She considers herself a late bloomer in the wonderful world of flying, starting in 1976 in a Cherokee 140, followed by a Mooney, then 12 years in a Beech Debonair, and now a Cessna 172 which is based at Lincoln Park (N07). She is especially thankful for the first nine years shared with her husband, Joseph. The North Jersey Chapter of the 99s became a part of her life in 1992 and she is active in all their activities including piloting for Pennies-a-Pound and the Young Eagles program and currently serves as chapter news reporter. She returned from her first attended International Convention, Norfolk 1994, sporting a 99s jacket, which several chapter members then purchased and proudly wear. Aileen is a member of AOPA and is active in volunteer work for the area homebound and handicapped.

EVELYN BOBBI TROUT, born Jan. 7, 1906, in Greenup, IL, and got her name when she had her hair bobbed. In her early years, she contracted Spanish influenza, needed air and was sent west to recuperate. At 16, Bobbi had her first airplane ride and five years later, she wrote a check for $250 to learn to fly. Her first lesson in the air was Jan. 1, 1928, at the Burdett Fuller Flying School in Los Angeles. She was the fifth woman in the USA to obtain her transport license # 2613.

In the fall of 1928, Bobbi was approached by R.O. Bone, owner of Golden Eagle Aircraft Manufacturing Co. He asked Bobbi to demonstrate his newly developed airplane, the 60 hp Golden Eagle.

On Jan. 2, 1929 Bobbi set her first record at LA Metropolitan. Others followed: Solo endurance record for women; was the first woman to fly all night long; a woman's altitude record.

In 1929 Bobbi was one of 20 women in the first Woman's Transcontinental Air Derby. Was at the end of the Derby that Bobbi, Amelia and a couple of other girls decided to form what was to become the 99s.

Bobbi and Elinor Smith established the first in-air refueling endurance record for women in 1929. Bobbi teamed up with actress Edna May Cooper and they set the second in-air refueling endurance record in 1931.

Bobbi then worked for the Cycloplane Co. and was active in the Women's Air Reserve. She later was a member of the Civil Air Patrol and during the first part of WWII, went on to start her first defense business, Aero Reclaiming Co., sorting rivets for the aircraft factories. Two years later she sold the business and co-founded De-Burring Service of Los Angeles, still in business today.

Bobbi keeps active with various aviation groups and was most recently honored May 30, 1995 at Burbank/Glendale Airport in California. In was 65 years earlier in 1930 that she won the women's pylon race flying a Kinner Fleet at opening day of the airport. You can read more about Bobbi and her early aviation friends of the 1920s and 1930s in her book, *Just Plane Crazy,* through her distributor Aviation Book Co.

Bobbi is a director of Aviation Archives, a California nonprofit corporation, to preserve aviation history. Aviation Archives is in the process of producing their first CD-ROM on "Women in Aviation" and invites fellow 99s and all women in aviation to contribute their stories for inclusion in a series of multimedia aviation productions. Lt. Col. Eileen Collins took Trout's international pilot license (endorsed by Orville Wright) into space as Eileen became the first woman to pilot the shuttle.

JANICE EMILY TERNES TROVER, became involved in aviation as a totally naïve person. She had no comparison to base her learning. Her instructor had previously taught her amateur radio classes. She became a licensed ham radio operator. Ground school amazed her and she struggled with it. Flight education followed after an introductory lesson to connect ground school theory with practice. Her husband became a pilot after her involvement. Never had they discussed or thought of doing this but now can't imagine having ever missed this interesting activity. One very calm patient person, Dan Whipple, a high school teacher in Minnesota and an adult evening education teacher, taught her all of these activities. She has been a member of the Minnesota Chapter of 99s since June 1992.

CAROLE TOSH TRUMP, born July 19, 1939, in Crumpler, WV, started her aviation training in October 1992. She received her private pilot license, ASEL in Williamsburg and Jamestown.

She joined the 99s in 1994. She has flown the Cessna 150, 152 and 172.

She has been an office manager for Maida Development Co. for 15 years.

She has one child, Kenneth L. Trump Jr., born Jan. 3, 1960.

MARILYN ELAINE TRUPIN, Champaign, IL, logged more than 3,000 hours, since 1971 and held SEL, MEL, IFR and CFI tickets. She flew her Seneca II in the Caribbean, Central America, Canada and Alaska. Currently she does private/IFR flying in her 1985 Turbo Saratoga SP and 1978 Warrior.

MBA, New York University in 1955. Her master's thesis, "Married Women in the Labor Force," predicted that 75 percent of married women would be working, eventually, including over 50 percent who had children. This was considered a "bizarre" theory in those days before anyone heard of the "women's movement."

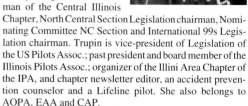

Positions in the 99s since joining in 1979, Chairman of the Central Illinois Chapter, North Central Section Legislation chairman, Nominating Committee NC Section and International 99s Legislation chairman. Trupin is vice-president of Legislation of the US Pilots Assoc.; past president and board member of the Illinois Pilots Assoc.; organizer of the Illini Area Chapter of the IPA, and chapter newsletter editor, an accident prevention counselor and a Lifeline pilot. She also belongs to AOPA, EAA and CAP.

She has two daughters who are physicians, a son who is a CPA and seven grandchildren.

SONJA TRUXEL, born Sept. 27, 1941, in St. Paul, MN, learned to fly in the Civil Air Patrol and obtained her license in 1991. She joined the 99s in 1994. She has flown the Aeronca Champ, Cessna 152, 172; Cherokee 140; Archer II; Dakota. She owns an Archer II with her husband. She has flown with CAP in search and rescue missions locating downed aircraft and flies a CAP C-172 plane for proficiency flights.

She is married to Jim Truxel who is also a pilot and a retired FAA air traffic supervisor. She has three daughters: Leandra Nelson, Lisa Harding and Lara Stang; a son, Bryan Johnson; stepson, Jim Truxel; and a stepdaughter, Kim Truxel. She has five grandchildren: Megan, 9; Emily 8; Becky, 6; Katie, 5; and Rachel, 5. She enjoys flying her children and granddaughters in the family Archer II.

She is also a member of the AOPA, Air Force Assoc. and is active in church. She enjoys snow skiing, boating, crocheting, sewing, silk flower arranging, bird watching and music.

CORALEE TUCKER, native of Los Angeles, has been flying since 1966, earning her ATP, CFI(A & II) ground instructor ratings. She has more than 5,000 hours, including 2,000 hours as flight instructor. She has flown several Powder Puff Derbies, the Powder Puff Commemorative, the Angel Derby, many Palms to Pines, Salinas His and Hers, Pacific Air Race, often placing in top 10 and winning the Palms to Pines.

She earned her BA at the University of Redlands; MA from California State University at Northridge, taught primary children for Los Angeles city schools for over 25 years; worked for over 17 years teaching ground school and flight training for a Part 141 flight school at Van Nuys.

She is married to Alfred Tucker, an engineer and pilot, who is also an avid backpacker (he has hiked the entire Pacific Crest Trail and Continental Divide). She served as flying transportation on portions of the Pacific Crest, dropping him and his backpacking dog off and picking them up again several weeks later. They have been married 45 years and have two children.

MARY ANN TURNEY, began flying in 1976, soloed that same year and in what was to become a foretelling of her strong desire to teach aviation, earned her ground instructor rating before her private in 1977. Within the next three years, at Republic Airport on Long Island, Mary Ann obtained her instrument, commercial and CFI. In 1979 she became the first female flight instructor at Republic, then continued on to earn her CFII and ATP. She has more than 3,000 hours.

As a 99, Mary Ann is a strong supporter of the 66 program. She is director of flight programs at Dowling College's School of Aviation and Transportation. She administers the Dowling Flight Training Program, teaches aviation courses and is the advisor for the College's Precision Flying Team.

She is a doctoral candidate at NOVA Southeastern University, holds a BS from Le Moyne College and a master's degree from Hofstra University. She has published research on gender-related issues and is currently working in this area with several other institutions.

She belongs to the University Aviation Assoc., AOPA, EAA, the American Assoc. of University Women and the Women in Transportation.

Before they were married, Mary Ann taught her husband, Robert Maxant, to fly. And when he purchased an airplane she taught him the finer points of maintaining it. Friendship blossomed into love and they became lifetime flying partners.

Mary Ann has owned three aircraft: C-150, C-172 and Piper Arrow. They have taken her to exciting places such as the Bahamas, Florida, Oshkosh and California.

"Flying," Mary Ann says, "has given me strength, courage, and above all great joy."

MARILYN JUNE TWITCHELL, her first flying lessons were interrupted by WWII. Her original logbook with five hours of flight time in a Taylorcraft, recorded in 1941, is a proud possession. Thirty years later, after serving in the WAC, attending undergraduate school and veterinary college on the GI Bill and fencing scholarships, and building a veterinary practice and hospital, this California native returned to the sky.

Marilyn holds a commercial certificate with an instrument rating. She is an enthusiastic air racer and co-owned a Cessna 182 with 99 Jean Pyatt. They flew N299PT in the Powder Puff Derby, Angel Derby, Palms to Pines, and other races. Marilyn is the 1977 All American Glamour Kitty Sissy Earhart's veterinarian. Sissy enjoyed air racing with Marilyn and Jean and is still a frequent flyer at 19.

Marilyn was born in Los Angeles, April 27, 1925. She graduated from Calumet, MI, high school and attended Adelphi College in New York before being accepted at the Veterinary College of Michigan State University in East Lansing, where she was the only woman in her class. She received her Doctor of Veterinary Medicine degree in 1952 and established a veterinary practice in Santa Monica, CA. In 1976-77, she was selected for a residency in veterinary ophthalmology at the University of Pennsylvania in Philadelphia and qualified as a Diplomate in the American College of Veterinary Ophthalmology. She practices veterinary ophthalmology at the Veterinary Centers of America hospital in West Los Angeles and at Cahuilla Veterinary Hospital in Palm Springs.

PATRICIA EMERY UDALL, born May 24, 1926, in Denver, CO. She received her private pilot license with commercial, instructor, CFI and CFII ratings. She joined the 99s in 1955. She has flown the Piper PA-18, PA-11, PA-22, PA-22-150, PA-22-160, Cherokee 180, Piper Arrow, Cessna 140, 172, 177RG, 182RG, 152; Citabria, Grumman AA5B, Piper ME20 and various ultralights.

She has been employed by Prescott Embry-Riddle as a flight instructor, in flight standards, responsible for safety standards, in the career center; AOPA ground school; technical research, Jeppesen Sanderson project editor. International experience: Royal Nepal Aviation Corp. ground school for cabin crews; ground school for pilots seeking US certification.

Udall has six children: Mark Emery Udall, 45; James Randolph Udall, 44; Judith King Udall, 42; Anne Jeannette Udall, 40; Bradley Hunt Udall, 38; and Katherine Lee Udall, 36. She has seven grandchildren: Loren Udall, 12; Catlin Udall, 9; Jed Udall, 7; Tess Udall, 5; Clay Harding, 5; Tor Udall, 5; and Luke Harding, 2.

ANNE UMPHREY, born Nov. 30, 1940, in Niagara Falls, NY, a newcomer to aviation, began flying at age 51. She has logged more than 300 hours, all in helicopters. She owns an R-22 which she flew from California to Boston, sightseeing along the way. She has also flown the Bell 206B3 (Jetranger).

She works as operations manager at the Boston Helicopter Co. in Bedford, MA. She is retired from the board of assessors in her hometown and has been active in other community service activities, including the local PBS auction. Hobbies include skiing, biking, knitting and other hand work. She is a member of 99s, Aircraft Owners and Pilots Assoc., Whirly Girls and the Aero Club of New England.

She has two children, Catherine, 23 and Will, 20.

PATTI LINDEMUTH UNCAPHER, born March 1, 1943, in Beechwoods, PA, started her aviation adventure with an introductory ride in a Piper 140 in 1975. She worked at the flight school as office manager and earned a private instrument rating. This was a rewarding and special time, giving her the opportunity to fly many types of aircraft with many instructors and learning from their different teaching techniques. She has flown the C-150, 152, 172, 182, Piper 140, 180; Arrow 200, Citabria, Piper Aztec and Apache.

She is treasurer of the Michigan Chapter of 99s, Inc., a member of the Experimental Aircraft Assoc. and a member of the Aircraft Owners and Pilots Assoc. She is a member of the PEO Sisterhood, an organization promoting education of women.

She now lives in Royal Oak, MI. Her husband, Chet Uncapher, is a pilot and financial consultant. They have two daughters, Andrea, 28 and Karen, 26, and a grandson, Maxfield Jacob, born Sept. 22, 1994. In 1989 they purchased their first airplane, a Piper Arrow 200, which they lovingly refer to as *Delta J*. Together, they renovated the Arrow and enjoy many happy flying hours visiting family and friends.

VIRGINIA LEE ELMGREN UNGER, born May 5, 1950 in Lake Forest, IL, graduated in 1973 from Southern Illinois University. While at SIU she earned private, commercial and flight instructor certificates and an instrument rating, all in airplanes. She was the first woman to receive a commercial or flight instructor certificate at SIU.

Lee and her husband, Tom, moved to Tucson, AZ, in 1973, the year Lee joined the Tucson 99s. Lee has enjoyed the 99s as chapter chairman, vice-chairman and in various other capacities. The Tucson Chapter awarded her with a flight scholarship in 1979. On the International level, Lee was a recipient of an Amelia Earhart Memorial Career Scholarship in 1980.

In Tucson, Lee has attained multi-engine, instrument instructor, multi-engine instructor, basic, advanced and instrument ground instructor certificates. Lee was a flight instructor at Roadrunner Aviation (1973-75), assistant chief flight instructor and airplane salesperson at Arizona Frontier Aviation (1975-82) and an instructor and salesperson at Hotton Aviation (1983-85). Following 1985, Lee has kept her certificates current while focusing on her husband and their children Alexander (1983) and Krista (1985).

Other aviation activities include membership in the Arizona Pilots Assoc. (APA) since 1980. During three years as an APA vice-president, Lee helped organize APA members in the Tucson area. Lee has also been an FAA aviation safety counselor since 1979.

EVELYN I. URBAN, born Oct. 20, 1933, in Marshland, OR, with the encouragement of her husband, learned to fly. Her husband, Tony, in his spare time, was buying and selling planes because he loved to learn about different planes. He had a TV repair business. He died in an auto accident. After that she bought and maintained rental properties and worked for the Census Bureau. She sold the properties and moved to Carlsbad, NM. She has flown the Musketeer, Cessna 150, Cherokee 260, Bonanza M and S Models.

She was married to Ralph "Tony" Urban and had four children: David and Margaret (whom they adopted); and Tony and Teresa.

JEAN E. URBAS, born in Chillicothe, OH, and currently resides in Fontana-on-Geneva Lake, WI, with her 49 1/2, retired TWA captain, Adolph M. Urbas. She soloed in her late 60s at the Belvidere, IL Airport, obtained a private license in 1994, and is presently studying for the instrument written test. Jean flies a Cessna 172, owned with her son, Reno Air MD-80 captain, Robert M. Urbas, who along with her husband, encouraged her to fly.

Her daughter, Susan K.B. Urbas, is an environmental lawyer in Chicago and was Jean's first passenger. Susan has taken flying lessons, instructed by Adolph and Robert Urbas.

Jean is a real estate broker in Wisconsin and an Illinois sales agent. She is a member of several Realtor professional organizations. Bridge, down-hill skiing, ballroom dancing and DAR and genealogy are other avid interests that she enjoys. Jean is a member of the Flying Farmers, Cessna Pilot's Assoc., and is on the membership committee for the Chicago Area Chapter of 99s. She has logged 315 hours of flight time. Jean is committed to sharing the joy of flying by helping women become aware of the career opportunities in, and the pleasures and challenges of flying.

NORMA WYNN VANDERGRIFF, born Sept. 17, 1928, in Piedmont, OK, learned to fly in 1967. She has a private pilot with commercial, instrument, advanced and instrument ground school instructor ratings. She spent most of her 2,000 hours in a C-210. She joined the 99s in 1968 and has flown the C-210, 182 and 172. Flew two Powder Puff derbies and local races. Had her own hangar and air strip in her backyard. Held all chapter offices. Flew PIC through Canada, Alaska, Mexico and Central America.

She has a BS in technical education/aviation and taught aviation for 10 years in two colleges. She was forced into early retirement for medical reasons. Norma now owns Wynn Flites. She is married to Charles and has three children: Gary Rybka, 43; Alex Wynn, 31; and Neal Wynn, 24. She has four grandchildren: Gray Rybka, 15; Alex Wynn, 5; Adriana Wynn, 2; and Sydney Wynn, 3.

PAM VAN DER LINDEN, born Oct. 18, 1912, in Southampton, England, died Oct. 6, 1990. She was educated at St. Margaret's Folkstone in England and the Sorbonne in Paris. She taught languages in Paris, Barcelona, Berlin and Mannheim from 1930-39. Married Archer Forsyth in 1933 and had three children: Ian, Allan and Jean.

Emigrated to the USA in 1939 and operated Berlitz Language School in San Francisco and ran Radio Good Neighbor Program for Sacramento Bee (Central and South America). Owned Forsyth Realty, Marin County from 1942-47. Moved to Fallbrook in 1947 and married Victor van der Linden. Bought Fallbrook Real Estate Co. in 1948.

Linden's flying career began when the board of realtors gave her a coupon for flying lessons as a thank-you for two years as board president. She was hooked! She had more than 4,000 hours; flew in nine Powder Puff derbies (10th place in one); five Pacific Air Races; two Baja Races; 10 Air Race Classics; circumnavigated Central and South America in a Bellanca Viking, 1978; chairman of Palomar Chapter of 99s; charter member of Coyote Chapter; Pilot of the Year of Southwest Section; and in 1986 inducted into International Forest of Friendship, Atchison, KS.

Pamela is featured in San Diego Aerospace Museum in PPD section. She was also an avid tennis player. Her ranking in 1989 in Southern California for the 75 years and over, was first in doubles, second in singles. She ranked sixth nationally in women's doubles, 1989.

EVELYN VIRGINIA "GINNY" ANTHONY VAN KESTEREN, Albany, GA, in 1943, married Henry Van Kesteren, an Army Air Corps pilot. Part of Ginny's marriage contract was that he teach her to fly. That agreement was fulfilled in 1946 at the close of WWII. She and her husband built an airport, Dallas Field, in Selma, AL, where she earned her private pilot license in an Aeronca Champion.

Through the years, Ginny earned her commercial, multi-engine, instrument rating, had three wonderful daughters, two grandchildren and then obtained a commercial helicopter pilot rating, becoming a Whirly Girl #339.

Always an enthusiastic pilot she flew the Powder Puff Derby in a tailwheel Bellanca in 1968; and, with her husband, the New York to Paris Air Race in a Golden Eagle; Beech Baron to Australia; an MU-2 to Uruguay, Argentina and Chili and a Bell JetRanger to Venezuela.

She presently flies her own A-36 Bonanza from their home in St. Petersburg, FL, to their island cottage on North Captiva island, FL. Having joined the 99s in the 1960s, Ginny is now a life-time member.

LIEVE VAN LINT, born and raised in Belgium, moved to Alaska in 1981. She always had a fascination for airplanes, but never thought she would ever become a pilot herself.

Became a private pilot in 1994. It was not so easy for her. Motto: "Never give up, just think about your dream and work harder."

At the moment she flies a Cessna 150. In May 1995 she started some aerobatics classes in a Pitts Special S2-B. She hopes to start her instrument rating. But, at the moment, she just enjoys flying Alaska, the beautiful.

CAROLYN J. VAN NEWKIRK, born Sept. 22, 1937, in Ephrata, PA, learned to fly in 1991 and holds a private license with commercial and instrument rating. She has logged more than 550 hours and has participated in the Air Race Classic.

Her educational background includes a doctorate in educational administration from the College of William and Mary in Virginia. She has served as a teacher, reading specialist, adjunct professor and administrator in schools and colleges on the East Coast. Van Newkirk is principal of the Lower School, Beth Tfiloh Community School in Baltimore, MD. She is a member of numerous educational associations.

She is active in York Twinning Assoc. which promotes people and cultural exchange with foreign countries. She and her husband have hosted many foreign visitors in their home.

Van Newkirk is a member of 99s, Inc., Aircraft Owners and Pilots Assoc. and Air Race Classic.

She is married to Dr. Jack C. Van Newkirk, a superintendent of schools. They have two children, Lauri, 34 and David, 29; and one grandchild, Jess Collins Altman.

JACQUELINE ANN VAN VALKENBURG, a private pilot from New Jersey who spends most of her time up in the air. For fun she flies her own classic taildragger; for a living she works as a corporate flight attendant for Jet Aviation in Teterboro, NJ.

Learning to fly was a dream since childhood when she would visit her grandparents in Lake Placid, NY. Her grandfather owned a gas station a half mile from the municipal airport, and she would often spend afternoons watching the planes fly overhead on their approaches and departures. In 1976 Jacqueline became a flight attendant for Overseas National Airways based at JFK. Two years later, she joined Continental Airlines, based at LAX, then Houston and finally Denver. After nine years with the airline, she moved back to the East Coast and joined the corporate aviation world.

While enjoying her job working in the back of airplanes, she never abandoned her desire to take the controls and finally earned her SEL rating in 1988 at Richmoor Aviation in Poughkeepsie, NY, in a Cessna 152. In 1989 she married Hank Van Valkenburg Jr., a corporate pilot. They built a house on Alexandria Field in Pittstown, NJ, and co-own a 1950 Piper Pacer. Hank, who also holds a CFII, helped her transition from tricycle gear to taildragger. Jacqueline joined the 99s in January 1995. She is actively pursuing her instrument and commercial ratings and has more than 600 hours of flight time.

SALLY VAN ZANDT, took ground school in the 1940s, but is was not until 1971 that she learned to fly when her husband bought her flying lessons for their 25th wedding anniversary. Her flying has chiefly been in Piper Cherokees. She has flown to the Bahamas twice in a Piper 180.

Active in the Nebraska Chapter, presently serving as chapter chairman, and South Central Section bylaws advisor, she has flown in local air races and Poker Runs and has been head timer at three ARC Must Stops.

She recently retired after 30 years as associate professor of human development and family at the University of Nebraska. Her husband is a retired air traffic controller. They have three children and four grandchildren.

Sally says that one of the best things about the 99s is meeting fantastic women at Sectional and International meetings, most memorable, Puerto Rico, Alaska, Hawaii and Tyler, TX.

LOIS R. (SWEENEY) VAN ZELF, born Oct. 8, 1931, in Santa Rosa, CA, joined the US Marine Corps in 1951 and the Mt. Shasta Chapter of 99s in 1977.

In 1977, she flew with Pam Vanderlinden through Central America, over the Panama Canal, stopping in the San Blas Islands and the Mayan ruins in Tikal, coming home determined to obtain a pilot license, which was earned Oct. 7. 1977. Her instrument rating was obtained Feb. 19, 1990 after much hard work and many hours of flying. In 1980, she flew with Pam Vanderlinden and Kay Brick around South America. Very exciting adventure with two wonderful ladies. She and Pam flew the Air Race Classic five years and attempted a sixth, 1985-87 (received eighth place) and 1988. They started the race in 1990 but had to quit due to engine trouble.

She and Suann Prigmore flew Suann's Grumman Tiger to Alaska during June 1994. They flew through Canada, Yukon Territory and Alcan Highway to Anchorage, AK. She flew the Palms to Pines Air Race with Suann, 1991, first place; 1992, third; 1993, seventh; 1994, sixth.

She belongs to the Mt. Shasta Chapter of the 99s, Shasta County Flying Posse, Redding Area Pilots Assoc. and AOPA. She participated in search and rescue training with the Shasta County Sheriff's Flying Posse, has flown Young Eagles with the EAA, participated in Partners in Aviation for the Mount Shasta 99s, along with other things they do such as painting runways, etc. She was chapter treasurer for four years (1988-91), chapter newsletter in 1992 to the current time.

GWEN VASENDEN, born Nov. 4, 1931, in Minnesota. Her husband soloed her in a J3 Cub on Aug. 6, 1954. She received her SEL private-instrument in March 1984. She has flown the Cub J3, Tri-Pacer C-135, Aeronca C-65, PA-32-300, Archer PA-28-181 and C-172.

The ARC Race from Theramo, CA to Elk City, OK with co-pilot, Beth Lucy, was definitely the highlight of her flight experiences. They were the best of the "First Time Racers." Flights to Alaska, Bahamas, Annapolis, Sun-N-Fun, Seattle, Jackson Hole – any cross-country flying for the first time is an adventure to her!

She is married to Arnold Vasenden, a pharmacist and aviator, and has five children: Kari; Arnold, Pilot Corps; Jon pilot for USAir; David, pilot and pharmacist; and Andrew, a student pilot. She has been self-employed with her husband for 40 years. They own a drugstore in Fertile, MN.

CLARICE J. VASOLD, born March 1, 1940, in Saginaw, MI, earned her private pilot license and joined the 99s in 1987. She has flown the Cessna 172 and 150. Most enjoyable flight experience was to Alaska. Her husband was her flight instructor which made for an interesting experience. She has flown in many Michigan Small Races. They enjoy flying to their daughter's home in Cincinnati.

She is married to Duane and has two children, Greg, 35 and Kelly, 33. She has five grandchildren: Bradley, 8; Jeffrey, 5; Jordan, 5; Courtney, 3; and Kristen, 2.

MARY MARGARET VERMEULEN, while working for RCA at BMEWS, Clear Air Force Station, Alaska, 39 engineer/tech., she attended FAA pilot written exam. She and 31 engineer/techs, passed the exam. She had not yet flown a plane nor had any instruction.

Her first cross-country was in a Colt from Haines via Skagway. Her last in that Colt was from Haines to Yakutat. Icebergs breaking up, severe up and down drafts and the churning Gulf waters astonished her. Upon landing, she realized the Alaska earthquake of March 1964, at 8.4 on the Richter scale, was the happening.

In 1964, she moved to Anchorage to take a position with Metcalf and Eddy Engineers. Worked with CAP as information officer and later became a 99.

She was elected chairman of Alaska State 99s in 1967. Acquired a 1957 Tri-Pacer, helped form Alaska Air Museum, elected chairman of the board Alaska Airmen Assoc., awarded Aviation Woman of the Year by the Anchorage, AK, Business and Professional Women's Clubs. She was re-elected chairman of Alaska 99s.

Enjoyed hunting and fishing jaunts with her Tri-Pacer. At -70 degrees Fahrenheit you keep your battery in the house and plug in electric warmers for the oil and engine, sometimes take-off and land on the highway and park the plane in the PDO auto parking lot.

Moved back to Anchorage to work with USDOL/RACAP as director of operations mainstream. She devised, developed, implemented more than 50 village needed projects. Also worked with the union and contractor personnel for her village employees to acquire professional training.

Moved to the lower 48 states, continued CAP and 99 activities. Elected Top of Texas 99s chairman, 1983-84. Attained rank of major, CAP, earned Proficiency Ribbons and Badges and Life Saving Award.

For eight years, she bred and raised registered Texas longhorn cattle; awarded 49 ribbons, two plaques, three grand champion trophies.

Elected chairman of Top of Texas 99s in 1991-92. Has been guest speaker concentrating on flying experiences, needs for women in the flying and related professions and CAP missions. Appointed aviation activities committee chairman 1993-94 by SCS Governor, Judy Rinehart.

She has been a realtor/broker for 20 years. She is married to Victor and has five grandchildren: David, 10; Sara, 6, Katie, 3; Christina, 6 months; and Allen, 5 months.

MARY B. VIAL, born in Three Rivers, MI, learned to fly at Phoenix in 1962, obtaining ASEL, commercial and instrument ratings. She joined the Phoenix 99s and held several chapter offices, including chairman. She was governor of the Southwest Section in 1972-73, and then went on to the International board where she served for several years.

She loved racing and flew in six Powder Puff derbies, numerous Kachina Air Rallies and about 1970, an International Race to Nicaragua. Whenever possible she took along some of her four children and eight grandchildren, but only interested one in becoming a pilot too. Now she lives in a Phoenix area retirement community, and says she thoroughly enjoys it.

BARBARA VICKERS, didn't start flying until her 40s, when a friend opened her horizons and told her she could. The friend, Ed Maxson, became her flight instructor right through an instrument rating and also introduced her to other

local women pilots and told her about the 99s. She formed the Sedona Red Rockettes 99s Chapter in 1988. She has been Southwest Section aerospace education chairman and delegate to the World Aerospace Education Congress in India.

Babycakes is the name of her 1979 Grumman Tiger used for business, pleasure and 99s projects.

Vickers earned a BS from the University of Minnesota and a MA in sculpture from George Washington University. She has been a teacher, writer, antiques dealer, sculptor and real estate broker, owning her own company, Buyer Brokers of Sedona. She is married to sculptor and former Spitfire pilot, Philip Vickers, who now lets her do the flying.

PATRICIA VIKER, her first flight was at 14 in the right seat of a float plane at Walker Lake, MN. Her husband, Dean Viker, became a private pilot in 1969; Pat became an enthusiastic cross-country co-pilot. They were frequently accompanied by their three children, all of whom are now pilots.

Pat earned her private pilot license in 1986, after the Vikers purchased their Cessna Skylane 182. She and Dean have logged many cross-country hours, sometimes joined by one of their nine grandchildren. In 1993 a memorable flight spanned the Panama peninsula. The Vikers' beautifully restored 1940 Porterfield, purchased in 1990, continues to take trophies at West Coast air shows.

Pat has recently discovered air racing, participating in the Palms to Pines in 1993 and 1994. She was Santa Maria Valley 99s treasurer (1988-90) and is currently vice-chairman. Pat is co-owner of Viker Tractor in Santa Maria.

ELIZABETH "BETTY" VINSON, an Alabama native, learned to fly in 1990. She flies Cessnas 150, 172 and 182. An instrument rated private pilot, she has logged 600 hours.

Her educational background includes a BS in nursing from the Medical College of Virginia, where she graduated in 1982. She and her husband, Laymon, have two sons, Darryn, 35 and Kevin, 33; two grandchildren, Lauren, 3 and Avery, 8 months.

Vinson has been a member of the 99s since 1991 and has served three terms as chairman of the Virginia Chapter. Her special interest is spreading her love of aviation to school children. A self-professed late bloomer, Betty tells children and adults alike her motto, "Never give up on your dreams."

Vinson is a member of Wing Nuts Flying Club, Aircraft Owners and Pilots Assoc. and International Flying Nurses Assoc.

DOLORES A. VITULLO, born on Jan. 6, 1942 in Long Island, NY, and raised a family there with little time for aviation. Her husband was a flight instructor and visited the fly-in community of Spruce Creek, and they moved there when retired. They brought an airplane (Tri-Pacer) with them that was half-restored on a glider trailer, with wings off, down I-95 with a tow. Many eyes watched in disbelief.

Since they wanted to fly immediately, they purchased a half share of a Cherokee 140. What a great time they had visiting friends and family in Florida that year. Her husband got ill and lost his medical which made her feel compelled to get her private certificate. She started lessons at 49 years old and got her certificate the day before her 50th birthday. They now look forward to Sun-N-Fun and their flight there to see all the aviation buffs and enjoy the day. Many thanks to a great instructor who made all this possible. He is a great friend of her husbands' and really never let her give up the quest to fly.

She is married to Edward and has four children and two stepchildren: Michael, Mary Jane, Billy, Kathy; Brooke Michelle and John Michael.

TIFFANY VLASEK, began flying at age 15 in 1982 at San Jose Airport in California. Since she got her license in 1984, she has gotten her CFI, CFII, MEI, ATP and FE and has logged about 2,000 hours. Tiffany attended Embry-Riddle in Daytona, joined the 99s there in 1987, where she served as treasurer and won a leadership award. When she returned to California, she joined the Santa Clara Valley Chapter and won their Professional Pilot of the Year award in 1993. She won an AE Scholarship which she used for her ATP, and won the Women's Airline Pilot's ISA +21 award for her flight engineer certificate with Northwest Airlines. Tiffany has flown traffic watch, skydivers, as a government contractor at Tindal AFB and has been a flight instructor at San Jose Airport. Just recently, she got a job flying for Air Vegas in Las Vegas, NV, where she was the right seat of Air Vegas' first all female crew. Her husband, Vance, is also a pilot; he flies a Lear 35 for Silicon Valley Express. Tiffany's interests include travel and music, and she is a member of AOPA.

ELEANOR MARY VOGT, R.Ph., Ph.D. learned to fly in a Piper Tomahawk over beautiful southeastern Wisconsin and earned her certificate on Jan. 4, 1984, on her mother's birthday. She progressed though Warriors and Archers and finally into a partnership in a 1968 Arrow, as well as, marriage to a fellow pilot.

Dr. Vogt is vice-president of public affairs at the National Pharmaceutical Council in Reston, VA (outside of Washington, DC). Currently she spends more time as a commercial business passenger than piloting herself. She misses the ease and freedom of flying the less crowded skies of the Midwest and particularly those low spectacular flights along the Lake Michigan shore where you could almost reach out and touch the Chicago skyline. Her current loves are hot air balloons and ultralights.

MARY VOIGTS, a flight attendant for US Airlines, started flight training in 1974 at Braniff International after she and her husband bought a Cherokee 140. They now own an Aero Commander 500 Twin. She obtained her private pilot license and multi-engine land rating. She joined the 99s in 1988, the Reno Chapter in October 1992 and has been a member of the Kittyhawk Chapter, North Carolina.

Voigts is married to Busch Voigts Jr. and has two children, Busch III and Christopher. She holds a BS from Oregon State University and enjoys snow skiing, swimming and horseback riding. She is also a member of the Junior League of Reno and Folded Wings of Reno.

NINA ANGELA VOLPE, born April 6, 1953, in Flushing, NY, when she was 16 years old, her dad took her to Pennies-Pound day at Warnington Airport. After her first flight she was enchanted with flying. She received flight training at Van Sant and Doylestown airports. She has flown the Stearman, Cub, C-172 and 182, Piper Archer and Warrior. She joined the 99s in 1981.

She is a clinical instructor and married to T. Bruce McKissock. She has three children: Casey, 13; Meredith, 10; and Garrett, 6.

JEWEL Y. VOM SAAL, the 99s lost a loyal and dedicated member, Julie vom Saal on June 17, 1990. Julie was born and educated in Oklahoma but learned to fly in 1935 with barnstormers in Texas. She married Dr. Frederick vom Saal in Enid, OK, in 1940 and after several years moved to the New York area, where she made her home until her death.

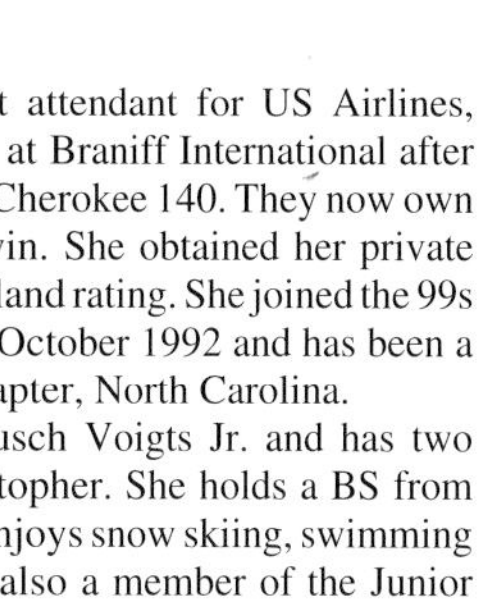

Julie had been a 99 for more than 30 years. She served in all chapter and section offices in the Greater New York Chapter and the New York/New Jersey Section. Her talent was recognized and she was elected to the 99s board of directors for two terms.

In 1969 she was co-chairman with Doris Renninger-Brell for the 99s 40th Anniversary Convention held at the famous Waldorf Astoria Hotel. It was a tremendous success, thanks to Julie's attention to the multitude of details.

Julie's many friends rallied around during her long illness and placed her name, along with that of her husband's, in the 1990 Forest of Friendship. Julie's last days were brightened with the knowledge that granddaughter, Jill, and her son's wife, Diane, would represent both her and Fred at the Induction Ceremony on June 16 in Atchison. That same day was also their 50th wedding anniversary.

Julie was always there when she was needed, a wonderful mother and wife, a tireless volunteer, a true friend to all who knew her. The world was a better place because Julie was in it, and we truly miss her.

MARY JO VOSS, born March 28, 1930, in Texarkana, AR, received flight training in Houston, TX, and Shreveport, LA. She trained in the Ercoupe, Luscombe and Cessna 170 and received her private license in 1967. She earned her instrument in 1976; commercial in 1976; flight instructor in 1981; ground instructor advanced in 1982; and dispatcher (DC9) in 1984.

She joined the 99s in 1971 and has been a timer for the Angel Derby. She was chairman for the Shreveport Chapter from 1974-76; flew co-pilot PPD Riverside, CA, to Boyne Falls, MI, 1975; participated in opening ceremonies of International Forest of Friendship, Atchison, KS, 1976; assisted in judging NIFA Contest at Monroe, LA, and designed, along with two other Shreveport 99s, the 70 foot compass rose now seen all over the US and served as assistant ground instructor for Southern Aviation, 1981; served as chief ground instructor Southern Aviation, 1982; served as administrative assistant to chief flight instructor, Royale Airlines, 1983; served as manager in training department, Royale Airlines, 1984-89; served as DC-9 dispatcher (part-time) Royale Airlines, 1988-89; FAA accident prevention counselor, 1983-present; and FAA designated written examiner assistant, 1983-present.

She was awarded the Amelia Earhart Scholarship, the Jimmie Kolp Award and FAA Safety Achievement Award. She has had numerous articles published in *The 99 News* on airmarking and items of interest to women pilots. She has been invited to speak at 99 Chapter Meetings, 99 Section Conventions, 99 International Conventions, Bicentennial Celebration at Downtown Airport, television interviews, Airport Authority, FAA Safety Seminars, airport meetings in various cities/states for Royale Airlines and church.

She is married to Benjamin Voss and has two children, Candice, 37 and Darryl, 41. She has seven grandchildren: Jo Marie, 21; Jason, 20; Micah, 19; Nikki, 13; Joey, 12; Julie, 9; and Matthew, 5. She has two great-grandchildren, Jordon, 2 and Kristina, 8 months.

ANDREA WAAS, Phoenix Chapter, born March 5, 1958, grew up around aviation, flying with her father in Missouri. Then in 1987, Andrea's father was killed in an aircraft accident. Shortly thereafter, she established the Willis A. Waas Memorial Scholarship at the University of Kansas (her father's alma mater). Each year, one student is chosen from a number of applicants and awarded the scholarship to go towards flight training.

After talking with others who have lost family members in aircraft accidents, Andrea established Wings of Light, Inc. to assist survivors, family members and those involved in the rescue and investigation of aircraft accidents. Financial support for Wings of Light, Inc. comes primarily from individual donations.

Andrea is a private pilot, working on additional ratings. She learned to fly at Gran-Aire, Timmerman Airport, Milwaukee, WI. She is a member of the 99s, Aircraft Owners and Pilots Assoc., Experimental Aircraft Assoc., National Aeronautic Assoc., Aircraft Rescue and Fire Fighting Working Group, American Society of Assoc. Executives and National Center for Nonprofit Boards.

Andrea has a BS in journalism from the University of Kansas and MBA from Cardinal Stritch College. She currently serves as president/CEO of Wings of Light, Inc.

LILLIAN I. WAGER, born Sept. 6, 1940, in Portland, OR, an instructor at Hillsboro, Ore City Airpark, Troutdale, Aurora. She has a private license with commercial, CFI and instrument ratings. She has flown the Cessna 152, 150, 172, 175, 182; Tri-Pacer, Colt, Aeronca, Cherokee 140, 180, 235; Apache, Goodyear Blimp Columbia.

Wager joined the 99s and has participated in many races including: three Petticoat derbies and Northwest

Women's Air Games. She is involved in the Civil Air Patrol and Red Cross Blood Flights. She belongs to CAP, Beaverton Chapter, Oregon Pilots Assoc., AOPA and Columbia Cascade 99s, Inc.

She is married to Edward J., also a private pilot. They have one child, Becky, 27 and two grandchildren, Tyler, 4 1/2 and Johnathan.

RUTH KITCHEL WAKEMAN, her first airplane ride was in an old OX bi-plane when a barnstormer came to her hometown of Coldwater, MI. A year later the same pilot returned, flying a six-place Stinson Detroiter. She decided that flying was not only fun, but remunerative as well. The affluence of the barnstormer was later explained when he was arrested for bootlegging. But the flying bug had bitten so in 1930, she enrolled in Curtiss-Wright School and eventually earned her transport license 18221. In March 1932, she joined the 99s and has been chairman of the Michigan Chapter, Bay Cities Chapter, Southwest Section Governor and editor of the newsletter.

DEBORAH BLACKWELL WALDROP, 37, is a native of Charlotte, NC. She has been an interpreter for the hearing-impaired in the Charlotte-Mecklenburg school system for the past 13 years. Debi took her first flying lesson in May 1991. She has more than 350 hours and is working towards obtaining an instrument rating. Debi is a mission pilot in the South Carolina Wing Civil Air Patrol. She was awarded the flight operations officer for the state through May 6, 1994 to May 6, 1995. She has combined her interest in flying and interpreting by joining the International Deaf Pilots Assoc. Debi lives up to her vanity tag, "AV84FUN." She enjoys jaunts to the beach and to fly-in restaurants. Her other hobbies include playing the piano and reading.

KYM ROSS WALDROP, born in Hattiesburg, MS, has had a life-long fascination with flying. Her first flight made an indelible impression: It was at age 4 in a DC-3. While growing up in Mississippi, she had occasions to fly in several types of private planes including: Cessna 150, 172; T-34 and Piper Arrow. However, it was only through the encouragement of an old and dear friend that she began flying lessons at the Norfolk Navy Flying Club. She received her private pilot license in 1974, thus fulfilling a long-standing dream.

She is a charter member of the Hampton Roads Chapter, and though inactive now, she took part in numerous chapter activities. During the years she lived in the Norfolk area, she logged more than 400 hours. Flying provided unforgettable experiences and opportunities and has forged many wonderful friendships.

Over the years she has traveled extensively and pursued many interests including cooking, gardening, music, needlework and writing. She has lived in Jacksonville, FL since 1979.

MAUDE MAXINE WALKER, life member, entered the Civilian Pilot Training Program in 1940 at Piedmont Aviation Flight School, Smith Reynolds Airport, Winston-Salem, NC. She soloed in the fall and earned her private pilot certificate in February 1941. During WWII she worked for Piedmont in fixed base operations, then when Piedmont started the airline division, she worked both. After obtaining a second class medical she obtained additional ratings such as commercial, instrument, flight instructor along with Link trainer and ground instructors. Max was fortunate to be working for a Piper and Beechcraft distributor. Not only did she have access to the newest and best but a variety of trade-ins, old and later models of all makes and sizes. There was also a good selection of company aircraft available for flight instruction, photography flights, air taxi, etc. After the airline start-up, there was endless opportunities to test hop and ferry DC-3s throughout the system.

In 1954 Max moved from home base to Richmond, VA. Five years later, in 1959, she joined the Federal Aviation Administration in the ATC control tower at Byrd Field, Richmond, VA. Flight instructing along with other types of flying was restricted but not completely prohibited due to the "conflict of interest" rule controllers had to abide by. She flew with tower personnel and their immediate families and area military flying clubs and owned a PA-12 with a fellow 99 for a number of years. After 55 years of flying she still loves it and never misses an opportunity.

RUTH WALTERS WALKER, born Feb. 25, 1921, in Denver, CO, was the only girl in a class of 10 chosen to take CPT training in 1940. Received her private pilot license in February 1941. Graduated from Purdue University and flew where Amelia Earhart was active. Her father used to drive her to an unplowed airstrip each morning for her lesson because she didn't have her driver's license yet.

She has been active in the 99s since 1941 and in the Utah Chapter for many years. Airmarking has been her specialty. Due to ill health she has not flown for many years.

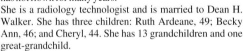

She is a radiology technologist and is married to Dean H. Walker. She has three children: Ruth Ardeane, 49; Becky Ann, 46; and Cheryl, 44. She has 13 grandchildren and one great-grandchild.

NINA M. WALLACE, a third generation San Diego native, learned to fly in Ramona, CA, in 1979. She flies a Cessna 210 and 182. Nina is a licensed airframe and powerplant mechanic and maintains her own Cessna 182.

As a member of the San Diego 99s, Nina wrote a monthly "Maintenance for Pilots" advisory column. Nina and her husband, David, also an A&P mechanic with inspection authorization, sponsored several clinics to teach 99 members how to perform preventative maintenance on their own airplanes.

Nina has taught aircraft systems, sheet metal techniques and composite inspection at San Diego Community College, Miramar Campus in the airframe and powerplant program.

Nina's other aviation memberships include Aircraft Owners and Pilots Assoc., Cessna Pilots Assoc., Professional Aviation Maintenance Assoc. and Experimental Aircraft Assoc.

Nina, David and their daughter, Whitney, live in Charlotte, NC, where Nina works as a maintenance controller and certified weather observer for Mountain Air Cargo, a cargo feeder operations for Federal Express.

BETTY M. WALSH, born Nov. 11, 1931, San Francisco, CA. Her fly! She never had a desire to fly. Her first flight was in a commercial airliner (age 23). Her husband always wanted to fly. When he decided to take lessons, she said she would too (the dutiful wife syndrome). His reaction was, "Who you?" Her reaction was, "Yes, Me!" They started lessons in their 40s. It was not easy for her, but thanks to the local 99s Chapter and all the encouragement from those gals the difficult times were made easier. She joined the 99s in 1973. She had more time to devote to lessons and was able to get her ticket first, even purchased their C-172 without any help. They both love flying and the many friends they have made through the 99s. She has also flown the C-150 and 152.

She has been married to Jim for 42 years and has three children: Steven, Sandra and Sheila (all adults). She has two grandchildren, Christopher, 18 and Michael, 5.

CLAIRE LEE WALTERS, born Feb. 20, 1924, in Santa Ana, CA, a twin of Betty Loufek. Started flying at age 17 in 1941. The war delayed flying until after graduation from Santa Ana Junior College, 1943. Private and commercial certificates, 1944. Joined the 99s in September 1944 and got instructor rating in 1946. She and Fran Bera won the Powder Puff Derby in 1951. Married Mike Walters, August 1951 and later divorced. Her two children are Michael and Susan. She has four grandchildren: Nicole, Michelle, Kathryn and Kent, and two great-grandchildren, Kayla and Mikala.

Owned and operated Claire Walters Flight Academy, Inc., 1960-1987 in Santa Monica, CA. Ferried Piper Aztec, Oakland to Australia 1966. FAA pilot examiner from 1967-1993. Instructed approximately 38,000 hours. She has flown many AWTAR, Air Race Classics, Pacific Air Race, Angel Derby, Shirts and Skirts, Back to Basics, Valley Air Derby. She is co-founder and chairman of the Palms to Pines Air Race, 1970 to present and has flown in all.

Helped charter many chapters. A charter member (1978), Palms Chapter, Southwest Section. A former governor of Southwest Section. She has her ASEL, AMEL, instrument and instructor ratings.

NANCY WALTERS, born Feb. 11, 1945, in Croswell, MI, took her first airplane ride in October 1978 and was hooked. She soloed on January 6 the following year and earned her wings six months later, along with her husband, Ron, the same day. They presently have a Beechcraft V-tail Bonanza and are building a Glasair Experimental Aircraft in their garage. Walters received her seaplane rating when they flew to Anchorage, AK. She has also flown the Cessna 150-152, 172, 182; Piper 140, 180; Beechcraft T-34, V-35; and Glasair Experimental.

She has participated in all the Michigan Small Races since she joined the 99s in 1980 and is past-president of the Michigan Chapter of 99s. She has organized many Michigan Air Tours, and is an active member of the Aircraft Owners and Pilots Assoc., Michigan Aviation Assoc., American Bonanza Society, Experimental Aircraft Assoc. Nancy is active with the Young Eagles Program and has taken numerous children for their first flight. She enjoys being active and also scuba dives. "And my sons told me I should take a sewing class, or something, when I told them I was going back to school."

She is married to Ron Walters and has three children: Maryan, 27; Kevin, 25; and Michael, 24. She has two grandchildren, Ashley, 4 and Kevin, 1 1/2. She is employed at Electronic Data Systems, Inc.

KATHY WALTON, born June 10, 1941, in Oakland, CA, grew up with her father, a private pilot, owning airplanes. She helped with the Powder Puff Derby in 1962 and met some of the 99s. It was then that she decided she had to learn to fly. She joined the Bay Cities Chapter of 99s a few days after earning her license in 1968. She has served on many 99s Chapter, Section and International committees and is a past governor of the Southwest Section. She has participated in many of the Palms to Pines Air Races and Air Race Classics. She has flown the Cessna 150, 172, 182 and Cherokee D.

Walton has spent her career in education as a teacher and administrator as well as owning and operating several businesses. Her fabric store supplied many pilots with "flying fabrics." She also found time to incorporate a city and serve on the Chamber of Commerce board for many years.

She is currently a member of the Santa Rosa Chapter. She is married to Richard Walton who is a former corporate pilot and air traffic controller.

MELINDA WALTON, born in 1963, in Maryland. She has been interested in flying since childhood, when her uncle took her on a tour of Delta's simulators. Watching that simulator flight made a lasting impression.

She learned to fly at 19 in Florida at Embry-Riddle Aeronautical University. While earning a BS in aeronautical science, she completed her private, instrument, commercial and CFI.

In 1987, Melinda began her career, flight instruction in Houston. Melinda enjoys instructing and did so for six years – producing some 40 solo and certified pilots. In that time she earned her CFII, multi-engine and MEI.

She joined the ERAU Chapter of 99s in 1984. Melinda has been secretary, newsletter editor, vice and chairperson for the Austin Chapter. In 1992 she won a 99 Amelia Earhart Scholarship for her ATP.

Melinda currently flies Metroliners for Berry Aviation, a Part 135 cargo company. Employed with Berry since 1993, she has flown to 28 states and Mexico. Melinda is presently based in Colorado Springs.

NANCY BIRD-WALTON, born Oct. 16, 1915, in Kew, New South Wales, took her first trial instruction flight at 15 and two years later started flying lessons with Australia's most famous airman, Sir Charles Kingsford Smith, obtaining her private license in 1933 and her commercial rating #474 in 1935.

There being no jobs in aviation, she planned barnstorming tours of the country in New South Wales landing in paddocks beside shows and race meetings hoping to obtain passengers. This led to the Far West Children's Health Scheme inviting her to fly their medical service in the "Outback," thus she became the first woman to be employed in commercial aviation in Australia.

Nancy joined the 99s in 1938. She visited the US in 1939 after spending 12 months studying aviation development in Europe. She was greatly impressed with the 99s. As a result and because of Australia's currency restriction, she founded the Australian Women Pilot's Assoc. in 1950, becoming their first president, after serving as Australian commandant of the Women's Air Training Corps during WWII.

Nancy flew in three Powder Puff derbies with American "Aces" Iris Critchell, Lauretta Foy and Betty Gillies, 1958, 1961 and 1977. After the war she engaged in charity work for the National Heart Campaign and helped put the first NSW Air Ambulance in the air by public subscription. She authored two books, *Born to Fly* and *My God! It's a Woman* in 1990.

Honors include the OBE (Officer of the British Empire) presented by King George VI in 1966; the Order of Australia (highest honor given to a woman in 1990); Honorary Master of Engineering, Sydney University; Honorary Doctorate of Science, Newcastle University.

She was married to John Charles Frederick Walton in 1939 (deceased 1991) and has one daughter and one son.

LOIS MERRITT WARD, born July 17, 1920, in Louisiana, received flight training in 1940s and 1950s. Became a commercial pilot in 1960s. Received a BA in aerospace education and taught ground school seven semesters at local college. Flew taildraggers Cessna 120 and 140. Her husband worked 23 years at local airport for Piper dealers and fixed-base operators. Flew their aircraft.

She worked for the US Dept. Interior's remote sensing unit and the Earth Resources Observation Satellite program enhancing satellite imagery with computer.

She participated in local proficiency rallies and flew in the last Powder Puff Race in 1977. Prior to joining the Phoenix Chapter in 1971, she belonged to the discontinued Northern Arizona Chapter. She is a life member in the 99s.

She has two children, six grandchildren and four great-grandchildren. Her husband is deceased.

She no longer flies but is active in helping form new 99 Chapters, encouraging young women to fly. She researches and records aviation history, people and events. She has also written a text book on ultralights.

MEREDITH O'KEENE WARD, born Jan. 26, 1910, in Tuscaloosa, AL. Charter member of Alabama Chapter of 99s and life member of the 99s. Student pilot certificate in February 1950; private pilot license #1199590 on Dec. 17, 1950 (40 years old).

Flew taildraggers mostly and owned a Cessna 120 most of the time, active until 1983, flying "for the fun of it."

Desire to fly directed her career from retail pharmacist to hospital pharmacist to obtain more flying time. Over 27 years was chief of pharmacy service at a VA Hospital. Active in Civil Air Patrol for years.

In 1950 was one of three delegates from Alabama to Pharmacopyeial Convention in Washington, DC; listed in *Library of Alabama Lives, 1961; The Dictionary of International Biography vol. 3; Who's Who of American Women 1968-69; Who's Who is the South and Southwest 10th Ed.*

Cracked up twice, unhurt. She flies a Cessna 120 #76826, built in 1948, still flies.

PAT WARD, an emergency unit RN, an instrument rated commercial pilot, owns a 1952 Cessna 195B. She is an FAA safety counselor, a member of NAA, BFA-Competition Division, Amelia Earhart Society of Researchers, a NIFA associate and a judge for USPFT and WPFT. She was competition events coordinator for the US National and the North American Hot Air Balloon Championships, 1989-91.

Pat first joined the 99s in 1963, chartered the South Louisiana Chapter in 1966, and has served as its chapter chairman for three terms over a period of 35 years. She was chairman of the Silver Anniversary Powder Puff Derby terminus in 1971, held in Baton Rouge, LA. She was instrumental in securing an exhibit and archives on Women in Louisiana Aviation at the state aviation museum, president and editor of *Louisiana Aviation Information Services* newsletter for airman and airport operators and co-founded the Aviation Assoc. of Louisiana. She is a 1992 honoree in the Forest of Friendship, a Jimmy Kolp Award honoree and a 1993 FAA Aviation Safety Award recipient.

Pat was elected secretary of the South Central Section in 1986, then vice-governor, and section governor 1990-92. She redesigned the format for the official publication of the South Central Section, the *SCS Approach*, to a 32 page magazine format and served as editor from 1990 through the fall 1993 issue.

Pat was one of the initiators of the International Council of Governors in 1990, served as coordinator for the Council of Governors through 1991, and correspondent to the governors in 1992. She served as International bylaws chairman 1992-94 and was elected to the International Board of Directors for the term 1994-96.

Pat is married to Roger Ward, a B-767 captain for Delta Airlines, and has two daughters who are also RNs and pilots. She has five grandchildren. Pat and Roger have recently built their retirement home at Aero Country Airport in north Texas. She is now a member of the Fort Worth Chapter.

JOYCE A. WARGER, born Feb. 5, 1939, in Michigan, began flight training at age 42 (her husband was 62). They took their training together, receiving their private certificates within six months of each other. Two years later, they again trained together to receive their instrument ratings. After moving to Florida, she commenced to get her commercial and multi-engine as well as go back to college, which resulted in graduation at age 52 with an associate's degree in flight technology and a bachelor's in aviation management. She has flown the Piper Warrior II, Arrow, Cherokee "6," Aztec and Simonel. She joined the 99s in 1983 and became Arkansas Chapter chairman in 1993.

They use their airplane to visit family scattered across the country. She is married to Arnold E. Warger and has three children: Deborah, 35; Kimberley, 34; and David, 32. She has six grandchildren: Annette, 14; Tyler, 3; Winston, 3; Taylor and Torrie, 1; and Wesley, 1 month.

LAURA WARMAN, born Oct. 9, 1963 in Trenton, NJ and reared in Virginia. She moved to Michigan in 1985 and joined the 99s in 1986. Her flight training includes Private, 1986; Instrument April 1987; Commercial November 1987; Instructor 1988; Instrument Instructor CFII 1990; and Multi-engine 1992. Warman has flown Cessna singles, Piper singles and Piper twins.

Warman is the recipient of two Amelia Earhart Scholarships, one in 1989 (CFII) and one in 1992 (multi-engine).

She remembers taking her first airplane ride in a 1941 Stearman Bi-plane at the Flying Circus Aerodrome, Bealeton, VA in August 1983. It was love at first flight. She was suppose to take her first lesson on her 21st birthday, but ended up riding along on a charter trip instead. They had microwave hot-dogs and cupcakes at 9000 feet over West Virginia – a birthday she will never forget.

Education: was graduated Summa Cum Laude in 1994 with as Associate Degree in Aviation Flight Technology.

Employment: Big Beaver Airport since May of 1988 after completing advanced flight training there in April. She also worked as office manager at another FBO while taking primary flight training.

She has flown nearly 5000 hours 90% of which is as a CFI. She joined the 99s three days after receiving her private pilots license. She was impressed by the encouragement and enthusiasm she received from a group of people she had never met before.

During here nine years in the group she has held many various offices within the chapter – from safety, aerospace ed. and flying activities committee chairman to secretary, vice chairman and most recently chairman for 1994-95.

ISABEL WARMOTH, born June 1, 1911, a retired clinical technologist, was first attracted to flying when her brother and sister-in-law became pilots. She first soloed on March 20, 1968, with a total of 270 hours flying time. Flying was something she did for herself after seeing her two older children through college and on their own. But having her younger child (who thinks the 99s should have a 49 1/2 badge for children of the 99s!) learn to fly at the same time was really spending quality time with him. He is now with the FAA in Puerto Rico (hopefully back in the US someday). He is AP, AI, ATP, instructor, multi-engine and helicopter rated.

She has a AB in biochemistry at the University of California, Berkeley 1932.

Most interesting trip: Flying commercial to Africa – but on one of these trips, she did fly a sight-seeing trip around Addis Ababa with daughter and son-in-law!

Most hair-raising: When a plane landing at Linds turned into them as they were leaving Stockton Airport. He dived as they climbed higher and they missed by about 30 feet! Not funny but exciting: as a student between the time she arrived and checked out the plane, the wind had risen considerably and when she took off – after – she knew she shouldn't have been allowed up in that wind. She changed her plan to fly 20 miles away and plotted her return at 3,000 feet and then she angled in downwind. With a 20 degree crosswind about 30 knots, she greased it down the middle of the runway. That was when she knew she could fly!

EMILY HANRAHAN WARNER, graduated from high school, Denver, CO, aspiring to become a flight attendant. After an airline trip in 1958 as a passenger, she became hooked on flying and began flying lessons the very next week.

Emily followed her career to become a flight instructor, air-taxi pilot and FAA pilot examiner in general aviation. Warner made aviation history when she was hired by Frontier Airlines as the first woman pilot in modern times for a US scheduled airline in January 1973. Emily flew as captain in the Boeing 737 and 727. She has logged more than 21,000 hours.

Emily joined the Federal Aviation Administration as an inspector in 1990.

Her honors include: the Amelia Earhart Award (1973), Colorado Aviation Hall of Fame (1983), Women's Aviation Hall of Fame (1992) and the International Forest of Friendship in 1993. Her first airline uniform is installed in the Smithsonian Flight Museum.

DOTTIE PRINCE WARREN, Waco Cen-Tex, private pilot license in 1967, Spaceland Airport, Houston; retired Texas teacher and administrator on all levels for 32 years. She received an M.Ed. from Baylor University, Waco, TX; BS from Sam Houston State, Huntsville, TX; involved as aerospace education consultant at the university teacher workshops since 1967. She is a colonel in Civil Air Patrol, USAF Auxiliary. Her various assignments: chairman National Aerospace Education Committee HQ CAP-USAF Maxwell AFB; vice-commander SWR; DC/S AE SWR; staff member of NCASE since 1978; director CAP's National Staff College CAP-USAF, Maxwell AFB, AL from 1987-88.

Recalling teaching-flying experiences: The most fun came as PIC giving orientation flights to "first time" flyers. Before approving aerospace education in a school district often both superintendent and high school principal needed their "first small aircraft" flight. Classes would be established after an air survey of surrounding areas, the administrator's ranches and much persuasion. Orientation flights created and maintained enthusiasm, and many students have become airline pilots or involved in today's aerospace world.

MARGARET THOMAS WARREN, born March 9, 1912, in Anson, TX, is a charter member of the 99s. She received her pilot license, age 17, at Fort Worth, TX, in 1929. She then joined the sales staff of Curtiss-Wright at Roosevelt Field, NY. At one time she was part of the aerobatics team of the Curtiss-Wright Exhibition Co., with Red Jackson and Freddy Lund.

She married Bayard Warren in 1936, has one son, Michael, and one daughter, Mary, and one granddaughter, Margaret, 8. She now lives in Ireland and London where she writes and paints. Her book, *Taking Off,* was published in 1993 in England, and is available in the US from Beverly Bookshop, Beverly Farms, MA 01915.

NANCY JANE WARREN, born Oct. 2, 1928, in Nashville, TN, has been fascinated with flying and the freedom it represents since early childhood. In 1983 at age 54, she signed up for private pilot ground school at her local airport. Her husband, a Corsair pilot in WWII, died the following year; but she returned to flight training and received her private pilot certificate in May 1985. In August 1985 she joined the 99s; and with the support and encouragement of her Indiana Chapter, she earned her instrument rating in 1988 and the multi-engine/multi-instrument rating and commercial rating in 1990. She flew air freight with a local DC-3 crew for over two years, plus joined them in Africa in 1989 to fly relief drops in the Sudan for UNICEF. She has logged almost 1,500 hours and loved every minute of it. She owns and regularly flies a 1968 Cessna Cardinal 177, her best friend. She has flown the Cessna 152, 172, 177; Citabria 7ECA, Beech Baron 55, DC-3 and Beech 18. She was married to Bruce Warren and has one child, Connie Mitchell, 43. She has one grandchild, Molly Mitchell.

EVIE L. WASHINGTON, born March 27, 1946, in Gordo, AL, is a FAA licensed ATP, a certified flight instructor for both multi-engine and single-engine airplanes, has earned six FAA Pilot Proficiency Wings Awards and is an FAA volunteer accident prevention counselor for #27 FSDO-Washington Dulles. Evie earned her pilot license while working a full-time job of handling family crises.

Evie L. Washington, lives in Washington, DC, and has a MA in supervision and management from Central Michigan University in Mount Pleasant, MI. Her undergraduate BS in psychology is from Howard University in Washington, DC.

Evie became interested in flying while watching military airplanes over her grandfather's farm in Alabama, but financial restraints kept her from taking lessons until April 1984. She wanted to join the Air Force, but none of the military services were accepting women into the flight program.

Evie has flown the Mooney, Apache, Navajo, Aztec, Cessna 310, Cessna 150, 152, 172; Warrior, Archer, Comanche, Bonanza A36, Cherokee 6, Lance, Cherokee 180 and the Travel Air.

She was the recipient of the Amelia Earhart Scholarship to help with her certified flight instructor license for single-engine airplanes and was one of 10 nominees submitted for Metro Three Airplane training.

Evie served as chapter chairman for the 99s, and has served in many elected offices at regional and national levels. She served many years as squadron commander for the Tuskegee Airmen Civil Air Patrol (CAP) Cadet Squadron, and is a CAP mission rated pilot and cadet orientation pilot. She is one of CAP's aerial surveillance pilots for Customs and DEA, a volunteer CFI for Virginia Aviation Safety Week activities, a member of the Capital Area Assoc. of Flight Instructors and volunteers annually to work the Aviation Career Day Program at Andrews AFB.

Evie served on the Board of Opportunity Skyway for minority and disadvantaged students, and has volunteered as the classroom and flight instructor for Aviation Career Education Program (ACE). She has received many honors, including a plaque for being a distinguished and outstanding "Woman in Aviation" from the National Aviation Club in October 1993.

VIVIAN BURGESS WATERS, born Jan. 26, 1934, in Oakdale, CA. At the age of 2, she came to St. Louis to live with friends of her dead mother and father. She became commercial hot air balloon rated. From 1980-1990 was the owner of Balloon Adventures, operating a balloon flight operation and balloon repair station at a small airport in St. Louis. During that time she logged more that 1,000 hours aloft flying in such races as the Great Forest Park Balloon Race, the VP Fair Races in St. Louis and Albuquerque Balloon Fiesta for over 10 years. She is the past chairman of the Greater St. Louis 99s. She has been married to Bernie Waters (also a commercial balloonist) for more than 40 years, and is a mother of five children and eight grandchildren.

AILEEN MARIE JOST WATKINS, since Aileen can remember, she wanted to fly. When traveling to the airport with her father, she knew that one day she would be a pilot. She grew up around airplanes and worked three jobs while going to college to save for flight lessons. At 19, Aileen was hired by Pratt & Whitney as a Senior Computer Graphic Illustrator, and by age 2 had saved enough to complete her Private Pilot Certificate in four months. In 1991, as a wedding gift from her husband, Bob, Aileen took an hour lesson in a Pitts S-2B, and with the help of Michele Thonney, she trained, competed and won/placed in several aerobatics competitions, while holding positions as Volunteer Coordinator and Judge.

She worked for a banner towing operation, freelanced aircraft renderings, and washed airplanes to finance her goal of becoming a professional pilot. In 1992, with the help of the Les and Martha Griner Scholarship, Aileen completed her Instrument rating, and shortly thereafter her Commercial certificate. In 1993 she was elected Vice-Chairman of the Florida Gulf Stream 99s and served until relocating with Bob to the St. Petersburg area. In 1994, Aileen was the proud recipient of the Amelia Earhart Memorial Scholarship for her CFI. She went on to receive her Multi-Engine and CFII, and was elected aerobatic/conventional gear instructor in the Tampa Bay area, is an International Sales Agent for Continental Airlines, and is attending Embry Riddle at MacDill AFB. She has logged PIC time in more than 35 different types of aircraft, and SIC in a DC3. She is married to Robert Watkins, as airport manager and CFI of ten years.

LUCILLE FRANCES (BRAUN) WATKINS, born in Hobart, OK, on July 12, 1922. After marrying John Watkins, widowed with three sons, she reared five more sons during which time she worked as a keyboard musician both as church organist and choir director as well as a night club entertainer. Having always wanted to fly, she began instruction after all the children were grown. Receiving her pilot license April 12, 1971, seemed the accomplishment of her life, so thrilling and intellectually rewarding! Actually, most of her activity was taking family, friends or any interested party for the first time on a pleasure ride in a Cessna 150 or 172.

Musical duties interfered with attendance of meetings after she joined the All-Ohio Chapter of 99s; since coming to Oklahoma, she has tried to be supportive of their activities while enjoying the company of such wonderful women in this organization!

LOIS WATSON, born March 30, 1929, in New York, started flying in 1953 in Miami at old Tamiami Airport. She has SEL, SES, MEL, glider, instrument, aerobatic instruction and helicopter instruction. She has flown the Piper J3 Cub, Aeronca Champ, Cherokee Arrow, Twin Comanche, Schwartzer 126 sailplane, Enstrom helicopter, Decathlon and Apache. She has flown into every state in the continental US, Canada, Alaska, Bahamas, Dominican Republic and eastern Caribbean. At Miami-Dade Community College in 1988 she established the Eig-Watson Scholarship endowment in memory of the late Saul Eig. This is awarded annually to outstanding students seeking a career in commercial aviation. To date, 67 scholarships have been awarded. She is married to J. Harvey Watson.

MARGARET WATT, began flying at Reid-Hillview Airport in 1977 and got her license later that year at San Jose Airport. She is an ATP rated CFII MEI with more than 8,000 hours. She owns a C-152 which she keeps and uses for instruction at Hollister, and she instructs at the Flying Country Club at Reid-Hillview. Margaret has been a corporate pilot for Hewlett Packard since 1989, where she currently flies an AstroJet. She has flown the Hayward-Las Vegas Race five times, placing third in 1994 and second in 1995. She is active with the Young Eagles Program and is a member of Flight Instructor's of America. She has been active in promoting women in aviation. Her husband, Eric, is a co-pilot with American Airlines; and they have two children. Margaret joined the SCV 99s in 1992. Her other interests include horses, golf, hiking and skiing.

NANCY WAYLETT, born Nov. 25, 1950 in Montana when cowpokes still wore six-shooters. Spent the balance of her childhood in the wild, wild Middle East. After living through two revolutions and a war, she developed a taste for adventure, but first had to endure four sleepy years in college. Entering real life, Nancy joined the Navy and did a tour as an Air Intelligence Officer, where flying was discovered.

After the Navy, she spent far too many years as a flight instructor as San Carlos Airport, CA, and Arapaho (Centennial) Airport, CO, before Lady Luck looked her way. A kind gentleman pilot realized that this women could fly and gave her a job as a Westwind co-pilot on Sept. 21, 1981. Mind you, it was temporary and part-time. Six months later she was full-time and in the left seat. Ye-Ha! Life was grand but the oil industry crash of the early 1980s precipitated a move to USAir on Sept. 21, 1983. Coincidentally, seven years later to the day, she flew her first trip as a Boeing 737-300 captain. She has flown almost everything with one engine, most light twins, Westwind I and II, Citation 500, BACI-III, Boeing 737-200, 300 and 400.

Nancy is currently raising three children: Chris Krahe, 14; Irina Berra, 7; and Andrew Berra, 5, with her husband, Bob Berra, while remaining active in the Potomac Chapter of the Mid-Atlantic Section having served as treasurer, chapter chairman and membership chairman.

CLEO O. WEBB, born March 10, 1932, in Polson, MT, is married to John Webb. She and John have two sons, Mark, age 35 and Vic, age 34. They have two granddaughters, Lindsey, 10 and Brandi, 7, and two grandsons, Jake, 8 and Bradley, 5. They have lived in Homer, AK, since 1979.

She had her first flight in a J-3 Cub with her friend's husband when she was only 19. Many years later those friends came to Alaska where she took them for a ride in their plane. She started flying lessons in 1984 and then received her license in June 1985. She received her training at the Alaska Flying Network in Homer, Kenai and Soldotna, AK. She became involved with the 99s in 1984 when the NW Section Meeting was held in Homer, AK, and International Convention was in Anchorage. What a great move that was for her, as she met women pilots from all over the world. She has flown the Cessna 150, 152, 172 and the Beech Duchess.

She has been chapter chairman, secretary and treasurer of the Cook Inlet Chapter and Achievement Awards chairman and Nominations Committee chairman of the Northwest Section.

PHYLLIS MARIE WEBB, born in Paw Paw, MI, always wanted to fly – since childhood. Finished graduate school, the University of Minnesota, before starting flying lessons. Activities include: those Indiana Dunes support, had 11 years Wing Program participation, FAA accident prevention program volunteer, EAA co-chairman women's activities, etc. She is IFR, commercial rated and a volunteer pilot, flew a patient to Mayo Clinic and has flown 371 University of Minnesota Blood Flights. She is co-owner of a C-150, lifetime member of 99s and a member of the Indiana Dunes Safety and Education. She has flown the C-150, 172 and Gulfstream Tiger.

DONNA TESSANDORI WEEKS, is from an old pioneer Italian family in the Southern San Joaquin Valley of California. A graduate of the University of California, she has been a junior high school teacher for over 20 years. She is active in several civic, philanthropic and educational organizations. She is an active pilot who averages about 170 hours a year. She has flown solo to Canada and Alaska. She is also planning a solo trip around the world. She is active in the International Women Pilots, the 99s and the Civil Air Patrol and teaches aviation-aerospace for the local squadron as well as various Air Force bases. She is a sought after presenter at numerous aviation and aerospace seminars and conferences. She has an active Aviation/Aerospace Club at Washington Junior High and also incorporates this area into her GATE classes. She speaks three languages fluently and knows eight more well enough to read, speak and understand.

She was selected: Outstanding Bilingual Teacher, 1995; Outstanding Teacher in 1994; Hot Wings Honoree for the Flames of Fame, 1994, a wall honoring members of the 99s; Outstanding Instructor of Cadets by the CAP in 1993; Outstanding Aviation/Aerospace Instructor for the Pacific Region in 1993; recipient of the Chuck Yeager Aerospace Education Achievement Award in 1993; Outstanding Social Studies Teacher in 1990; Outstanding Science Teacher in 1985; and Outstanding Teacher of the Year in 1976 and 1977.

She has held numerous offices for the 99s and other organizations and holds the rank of major in the CAP. She was placed in charge of the theme, Women in Aviation, for the 1990 Bakersfield, CA, Air Show. She developed aviation curriculum for the Kern County Airport that is used by all the schools in Kern County.

She has written two genealogical books; one on her mother's family line and a current book, which is being edited, on the Italians of Kern County. In her spare time, she quilts, golfs, paints and plays serious bridge.

SALLY WEICHERT, born Sept. 22, 1947, began her aviation career in 1986 through some friends. She earned her ATP, CFI, CFII, MEI, ground instructor (advanced and instrument) ratings. She joined the Reno Chapter of 99s in March 1990 and is currently vice-chairman of the air race committee. She has been chairman of the Reno Chapter.

She has an AA degree and has flown night cargo between Reno and Las Vegas. She teaches, flies charters and is currently a pilot for Reno Flying Service. She enjoys aerobatics, snow and water skiing, tennis, teaching scuba diving, gardening and music. She is also a member of the AOPA. Sally is married to Gary Weichert.

MONIQUE WEIL, born in 1928, in Paris, France, and learned to fly in 1966 after moving to California. Monique is an MSW and has worked 30 years as a clinical social worker, flying for fun and adding ratings: instrument, commercial, glider, multi-engine and ATP. She has been a self-employed flight instructor (instrument and glider) since 1984. She joined the 99s in 1989.

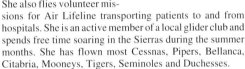

Weil has logged more than 5,500 hours, is an active search pilot and check pilot for the Civil Air Patrol and teaches High Altitude Search flying in the Sierras for CAP. She also flies volunteer missions for Air Lifeline transporting patients to and from hospitals. She is an active member of a local glider club and spends free time soaring in the Sierras during the summer months. She has flown most Cessnas, Pipers, Bellanca, Citabria, Mooneys, Tigers, Seminoles and Duchesses.

Monique Weil would like to encourage more women power pilots to learn to fly gliders for the pure joy of discovering soaring flights as well as for the safety benefits soaring skills can bring to the power pilot.

SUSAN WEINER, started flying in 1980 and joined the SCV 99s in 1981 after getting her license at San Jose Airport. She is a college instructor at West Valley College where she is currently chair of the chemistry department. Susan flies a Mooney, N3549N, out of San Jose and holds a private license with an instrument rating. She has logged more than 1,000 hours and her flights include trips to Canada, Texas and Oshkosh. She earned the Chapter Pilot of the Year Award in 1984, authored a textbook and has been a veteran speaker at the SCV Flying Companion Seminars.

RUTHIE WEISS, born in New York City and raised in Los Angeles. In 1985 she received her pilot certificate, joined the San Gabriel Valley Chapter of the 99s in 1988 and was named their 1990 Pilot of the Year. She is serving as vice-chairman this year and has twice won their annual Poker Run.

Ruthie graduated UCLA, 1977, with a BA in art history and is currently employed by the Bel-Air Assoc.

Through the years she has served as co-chairman UCLA Thieves Market, docent chairman UCLA's Wight Gallery, treasurer UCLA Art Counsel, membership chairman Planned Parenthood Guild LA, member LA Olympic Citizens Committee 1979-84. Currently, she is a sustaining member of the UCLA Art Council Board of Directors, member Planned Parenthood Guild LA, grant reviewer for the Los Angeles Educational Partnership, membership Los Angeles County Museum and AOPA.

Married to David Weiss, an intellectual property attorney who counsels the 99s. She has two sons, Jeffrey, a computer programmer, and Michael, a 1995 graduate of Santa Clara Law School.

SUE TAYLOR WEITZ, after raising three daughters, two married and one off to college, Sue decided to learn to fly. She earned her SEL license on Oct. 2, 1993. Soon thereafter, Sue began training for an instrument rating. After many hours of hard work, studying and training, Sue became instrument qualified on Aug. 24, 1994. She has been busy trying to stay current and helping with maintenance work around aircraft so that she can be more familiar with the aircraft she flies. Sue became a member of the Women with Wings Chapter in July 1994. She serves as newsletter editor, NIFA and aviation safety chairman. In September 1994, Sue flew her first air derby, the Michigan Small Race, with Debbie Downey.

CECILIA K.T. WELDON, has been actively flying since 1989. She sort of comes from a flying family, her father a career Army Airborne paratrooper. For Cecilia, the attraction of flying is a chance to participate in a dynamic and fulfilling pastime, one she shares with her fellow 99s, Palms Chapter.

Cecilia graduated from the University of Washington in Seattle with degrees in civil engineering and architecture. She is a professional civil engineer in California where she and husband, Jeff, a flight instructor, reside. The couple own a Cessna 182 for fun and travel and are also restoring a WWII Stearman biplane.

Besides the 99s, Cecilia is a member of the National Aeronautics Assoc., Aircraft Owners and Pilots Assoc., Cessna Owners Organization and the Stearman Restorers Assoc. Annually she participates in the 99s Palms to Pines Air Race, having won awards two out of the three years in which she has participated.

TINA M. WELLER, is a new member of the 99s, a new attorney and new member of the Lawyers-Pilots Bar Assoc. She belongs to a local missionary aviation group, Wings for Christ, and a local EAA Chapter. States she has lived in include: Ohio, California, Missouri, New Jersey, Massachusetts, Indiana, Tennessee, Illinois, Maryland, Virginia and Texas.

She started flying prior to law school at a small strip in Creve Coeur, MO, and received her private pilot license while taking lessons at Monarch Air Training at Addison Airport, northwest of Dallas. Her instructor, Jennifer Bankson, was the best instructor imaginable. Jennifer's last name may have changed, if she married recently. Patience and competence are a winning mixture. The 99s, she highly recommends her.

Currently, she is job hunting and not flying as much as she would like to; but that should change soon. She hopes to meet more 99s on the land, air and sea in the near future.

FAY GILLIS WELLS, earned her private license on Oct. 5, 1929, No. 9497. This outstanding lady became the first woman pilot member of the Caterpillar Club on Sept. 1, 1929. She is a charter member of the 99s and was the first American woman to pilot a Soviet civil aircraft and the first foreigner to own a glider in the Soviet Union (1932-34).

Fay Gillis Wells has been a free-lance correspondent, covering conflicts all over the globe; a buyer of strategic war materials in Portuguese West Africa; an interior designer; inventor; and White House correspondent. She was one of three women correspondents chosen by the White House to accompany President Nixon on his historic trip to the People's Republic of China in 1972 and to the Soviet Union in 1972.

She received the CBS Charlotte Friel Award for outstanding contributions to broadcasting in 1972, the Outstanding Woman of the year Award from OX-5 Aviation Pioneers in 1972, International Woman of the Year Award from the 99s in 1975 and many other awards and citations.

She has written a number of articles, reviews and a syndicated column. She is a member of Aviation/Space Writers Assoc., Aircraft Owners and Pilots Assoc., the 99s, OX-5 Aviation Pioneer's Overseas Press Club, Washington Press Club and numerous other professional organizations. She is listed in *Who's Who in American Women, Jane's Who's in Aviation and Aerospace, US Edition, Foremost Women in Communications, The World's Who's Who of Women,* and *Who's Who in the World,* 1989-90 and 1996.

A significant accomplishment has been the establishment of the International Forest of Friendship in 1976 in Atchison, KS (birthplace of Amelia Earhart). In 1991 a gazebo was built in this forest in honor of Ms. Wells.

JOYCE BRIZARD WELLS, of Larkspur, a third generation Californian, began flying in 1968 and has a SEL commercial/instrument rating with 3,000 hours. Learned to fly in Pipers and now flies a C-182 and C-T210. Aviation activities include air racing (Air Race Classic, Palms to Pines, Powder Puff Derby and Pacific Air Race), airmarking, Right Seat Seminar chairman and presenter, aviation speaker. A member of the Bay Cities Chapter of 99s since 1969, she held all offices as well as SW Section vice-president; International director, vice president and president (1994-96). She also serves on the National Aeronautic Assoc. Board of Directors and holds membership in the Aircraft Owners and Pilots Assoc., California Pilots Assoc., National Council for Women in Aviation/Aerospace, Air Race Classic, High Desert Museum, Sunriver Nature Center, Redwoods Presbyterian Church, Stanford Alumni Assoc., National Assoc. of School Nurses and Marin County Commission on the Status of Women.

Born in Eureka, CA, on Oct. 11, 1932, graduated from Porterville Union High School. A cum laude graduate of Stanford University with a BS in nursing, she has worked in surgery, public health and school nursing. Family includes: husband Harold Wells, a pilot and electrical engineer; daughters, Allison and Valerie; stepsons, Ross and Tim; and 10 grandchildren. Interests include: music, gardening, traveling, sports, and dancing. Community activities: Scout leader, first aid instructor, church choir and election board inspector.

PATRICIA WELLS, born Aug. 5, 1935, in Yakima, WA, had her first lesson on her 44th birthday and was so enthused that she went on to earn her instrument rating and commercial license.

She bought an Archer and flew 250 hours annually just playing. Her husband said that she should start a 135 Operation and have other people pay her to fly since she really didn't care where she went. At 49 she decided what she wanted to do when she grew up.

Working full-time when she founded Pegasus Air in 1984, she had to tell passengers that she must return by 2:30 and couldn't fly in the clouds. In 1985 they bought a Seneca so she could fly IFR, but she still had to be back early.

In 1989 she quit to fly full-time. Instruction, rental and seven more planes were added. Certificates/ratings include commercial instrument, SELS, MEL and 135 Charter. She still doesn't care where she goes. Every flight is fun and she still can't wait to get to the airport every morning.

She is married to David, and has two children, Susan, 38 and Donna, 34.

JANICE MARIE WELSH, born Jan. 22, 1955, in Brookville, PA, earned her private pilot license in November 1990 and joined the 99s before the ink had dried on the certificate. She got her instrument rating in May 1993 and IGI June 1993. She has flown the Cessna 172, Cherokee 180, Cessna 175, Piper Arrow.

She has a BS, M.Ed. in speech pathology. She is employed with a rehab company working in long term care. IFR flight to minimums. The fog so bad at LCL Airport couldn't see HIRL; ending up landing at control tower airport on east end as fog bank rolled in on west end. Had good mountain flying "experience" with husband and CFI back from Spokane, WA, to Indiana. Most of trip was VFR minimums, no snow capped mountain peaks or blue skies as anticipated.

She is married to Larry, who is also a pilot (learned at the same time), and has two children, Deana, 11 and Larry Jr., 8.

MARY WENHOLZ, learned to fly in 1945 in Miami, FL, with hopes of flying militarily but because at the time one must have "20/20 vision uncorrected," she could not qualify.

Mary arrived in Denver with an SES rating. Donna Meyers sponsored Mary into the 99s. She has held many chapter offices and served on several national committees. Mary has flown many races including the Powder Puff Derby, Palms to Pines, Pacific Air Races, Orange County and Kachina Doll, and placed in the top ten several times. After moving to the Los Angeles area Mary obtained her commercial and instrument ratings, joined the Long Beach Chapter and was treasurer of the Baja Air Race Board. She helped with several Fear of Flying seminars.

Community activities included Cub Scouts, church, Flying Samaritans and Meals on Wheels.

On Dec. 10, 1937, Mary married J.R. (Bob) Wenholz, a Continental Air Lines pilot, now retired after 30 years service. They have two sons and three grandsons.

CINDY WENK, 11-year Chicago Area Chapter member, flew with her dad growing up. In 1979 she wisely married into a flying family. Flying lessons in a Tomahawk started as a birthday gift, and she earned her license in 1984. One of three flying "Wenk Women," she enjoys flying Illi-Nines with mother-in-law, Gail. At an Iowa Derby, Cindy and equally mischievous sister-in-law, Wendy, convinced their trusting, non-pilot husbands that they had won first prize in the race! Busy jogging, the boys hadn't heard the race had been canceled due to marginal weather. Just as the proud husbands were purchasing champagne to celebrate, the only flyer who hadn't been "prepared" for the masquerade spilled the beans! Cindy works at home part-time for the family insurance business while raising two children, Sam, 8, and future 99 Lucy, 7, who love to critique mom's landings!

FELICIA M. WEST, born on Sept. 27, 1915, in Arkansas and moved to Florida in 1925. She graduated from Florida State College for Women in 1938. While teaching in St. Petersburg in 1941, she earned her private pilot license under the Civilian Pilot Training Program. For four years during the war she was an instructor for the US Navy in Atlanta, GA, teaching Link trainer operation, radio navigation and aircraft instruments to Navy enlisted personnel.

After the end of the war she went to Sao Paulo, Brazil, as an instructor in the Brazilian Army Aviation Technical School. On returning to the US she resumed her teaching in the Dade County Public schools and at Miami-Dade Community College. She retired in 1983 with 35 years experience and the rank of professor emeritus.

Before and after retiring she traveled extensively, including a five-month trip around the world and other special trips, some of which were with groups of 99s and WASPS. Some of the trips included: rafting down the Colorado River for five days, sailing in the Virgin Island, heli-hiking in the Canadian Rockies and travel in European countries, as well as in Australia, New Zealand, China, Japan, Africa and South America.

VERNA WEST, her first contact with aviation was to audit GI Flight Training books for the Veteran's Administration. Pat Gladney taught her to fly at Palo Alto after her husband, Harry, gave her lessons for Mother's Day. She joined the 99s the day she passed her checkride in 1965 (on her birthday). She flew Oceanic Conservation Patrols, Direct Relief Foundation, crewed for Hot Air Balloons, camped from airplanes in Alaska, Baja, most of Western USA. She worked at many races and flew the 1977 AWTAR. She was a member of the USPFT Council 1982-86 and traveled with the team to Norway in 1983. She has held many 99 offices including SW Section governor 1976-78. She was a ski racer, National Ski Patrolman and active in Scouts and PTA. A photographer, she has an extensive collection of 99s photos. She is currently section historian and resource center chairman at 99 Headquarters. Verna and her husband have three daughters and two grandchildren.

PHYLLIS M. WESTCOTT, born March 14, 1921, in Santa Ana, CA, earned her private pilot license – single-engine on July 3, 1963. She is a charter member of the Imperial So-Lo Chapter since April 1976. She has flown the Aeronca, PA18 and Navion. She was a co-pilot in 1965 Fallon Fun Race and got first place single-engine Navion.

She has three children: Kim, 35; Patricia, 34; and Scott, 26. She has five grandchildren: Robert, 9; Ryan, 8, Brian, 7; Lauren, 6; and Breann, 4.

VALERIE JALOVEC WESTEDT, born Aug. 28, 1948, in Muskegon, MI, is married to Craig Westedt and has a 23-year-old daughter, Andrea, who is currently a graduate student at Michigan State University. Val is employed as a math instructor and department chair for Muskegon public schools at their high school.

She started flying in October 1990; soloed on Jan. 7, 1991; received her private pilot license on May 4, 1991; and her instrument rating on Aug. 29, 1992. She also earned a ground school instructor certification on Nov. 1, 1994. Her current aircraft is a Piper Cherokee 140 and a Baby Great Lakes she is rebuilding. Her future plans are to learn aerobatics and earn a flight instructor license.

Val has been very active in aeronautical education. She has done several teacher workshops with the Lake Michigan 99s and the Michigan Bureau of Aeronautics. She received an Eisenhower grant to develop and implement an aviation program suitable for the middle and high school level. Materials from that program are available for other teacher's use. She is also teaching an aviation class at Muskegon High School. The class is designed for at-risk students to help motivate them to set personal goals and open up the many possibilities available in a variety of aviation related careers.

DOROTHY M. WESTLING, born May 27, 1920 in Aitkin County, MN, began flying intermittently in 1952, Cessna 170, at which time her husband, Donald, also began flying.

Her flying was diverted to flying floats. For some time, she trained in J3 Piper, Piper PA12, Lycoming 115. On Oct. 5, 1957, she received her private on floats, and some weeks later obtained her SEL. Soon after, she began flying Cessna 180 on 2600 Edo floats, retractable skis.

Around 1960, she joined the Minnesota Chapter of 99s,

spent two years as membership chairman and two years as Minnesota Chapter chairman. She is a life member of the 99s. She has two daughters, one son and seven grandchildren.

There is nothing as breathtaking as flying over Minnesota and its 10,000 lakes or into Canada, all the while viewing the panorama of water and forests.

BETTY M. WHARTON, born 1926, in Texas, a resident of San Diego, CA, since 1934, learned to fly at Gillespie Field, El Cajon, CA, in 1963. She flew all the single-engine Cessnas, and with her husband, owned several 210 models. After joining the 99s in 1964, she served as a member of the International Board of Directors for one term, was on the board of the Powder Puff Derby for nine years, held all offices in the San Diego Chapter and a couple of chairmanships on the SW Section board. Wharton was a contestant in several Powder Puff derbies, many Pacific Air Races and several Palms to Pines. (She won one of those.) She was part of the PAR Race committee for 25 races, always inspecting aircraft for safety and "stock" configuration.

Married to Claud for nearly 49 years, they flew all over the USA. She got up to Canada and down to Acapulco on air races. Those 27 years are among her fondest memories. Now she is active in the San Diego Air and Space Museum, working in the Women in Aviation exhibit room. She is also active in her church, and plays golf twice a week. They raised three sons: Don, 47; Larry, 45; and Clay, 42, and have one granddaughter, Jaime Wharton, 16.

DOROTHY WHEELER, lifetime 99, joined the then Seattle Northwest Chapter in 1942, a newly-arrived Boeing engineer's wife. A William Woods College graduate, she was among a select few women completing the federal Civilian Pilot Program, flying Pipers and Aeroncas.

Serving as Northwest Chapter governor, she was an early member of the newly-formed Civil Air Patrol, the young AOPA, and held a then required radio license. Dottie had earlier turned down a coveted airline stewardess offer to become a homemaker.

The mother of three sons, two who have flown privately, she boasts five grandchildren. Active in community projects and ever an ardent golfer, her club's woman champ, she has enjoyed competitive participation.

Dottie's father owned an Aeronca Chief; husband, Don flew, designed and built the Wheelair private airplane; and a brother is a retired Navy pilot and UAL captain.

MARY I. WHEELOCK, born Dec. 15, 1934, in Ada, OK, learned to fly at Shreveport Downtown Airport in 1971 while employed part-time for Shreveport Aviation. She credits her flight instructor, Helen Hewitt, and 99s Evelyn Snow and Jere Saur, for their encouragement and genuine friendship which enabled her to receive her private pilot license.

Mary joined the Shreveport 99s in 1971 and became a life member in 1981.

Mary served as squadron commander in the Civil Air Patrol in addition to being rated a search and rescue pilot. She is a designated FAA safety counselor.

Husband, T.W., has built a Steen Skybolt. He is a life member of the Experimental Aircraft Assoc. and holds an A&P mechanic's license. T.W. and their sons, Travis, Dean and Terry are strong supporters of Mary's 99 activities, especially airmarking and safety education.

Mary and T.W. are proud grandparents of grandson, Christopher and granddaughter, Michelle for whom flying accounts have been established.

EDDIE ELISABETH WHISTLE, born March 21, 1926, in Atlanta, GA, earned her SEL private pilot license and instrument rating. She has flown the Cessna 150, 172, 182, 210; Bellanca; Bonanza V35; Piper 235; and Comanche 180-250. She is married to L. Paul Whistle and has four children: Paulette, 47; Jill, 37; Lee, 36; and Wes, 34. She has eight grandchildren: Patrick, 24; Jennifer, 22; TJ, 18; Rechelle, 16; Ileah, 14; Josh, 14; Tabitha, 12; and Nichole, 10.

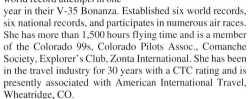

In 1980-81 she and her husband made two around the world record attempts in one year in their V-35 Bonanza. Established six world records, six national records, and participates in numerous air races. She has more than 1,500 hours flying time and is a member of the Colorado 99s, Colorado Pilots Assoc., Comanche Society, Explorer's Club, Zonta International. She has been in the travel industry for 30 years with a CTC rating and is presently associated with American International Travel, Wheatridge, CO.

EVA A. WHITE, born in Aurora, IL, in 1913. She attended Stephens College in Columbia, MO, the Chicago Academy of Fine Arts and North Central College. She taught art to handicapped children at Freeman School in Aurora and married Harold White of Naperville in 1937, becoming Eva Anderson White.

Eva earned her pilot license in 1954 at DuPage County Airport and became a member of the Chicago Area 99s in 1955. She served as chapter chairman for two years and twice won the chapter's top award for achievement in aviation education.

In 1958 and 1959 she was co-pilot with Beatrice Sieman in a Cessna 195 in the Powder Puff derbies; and in 1970 she was co-pilot with Geraldine Krause in the International Angel Derby race between Toronto, Canada and Nassau in the Bahamas.

Eva was a member of the Powder Puff Derby Assoc. board of trustees North Central College, and co-owner of *The Naperville Sun*. She is listed in *Who's Who in the Midwest*, in *Foremost Women in Communications*, and in *Who's Who of American Women*.

HARRIET URBAN WHITE, born in Buffalo, NY, on Aug. 19, 1921. Her first flight was in a Ford-Trimotor over Niagara Falls when she was 7. She soloed a J-3 Cub in Stonington, CT, in 1941; by spring 1942 she was instructing on a small field at Clarence, NY. Soon she flew for the company where she had taken her training. She reactivated the Western New York 99s and held workshops, participated in Civil Air Patrol.

She went into the WASP, in Sweetwater, TX. After graduation, Class 43-5, she was assigned to the USAF, 3rd Ferrying Group, ATC, Romulus, MI, mainly flying trainers, sometimes into Canada. When she was discharged she held a 0-3,000 hp, single and multi-engine rating, C-47, B-25.

She became a technical writer with North American Aviation in Grand Prairie, TX, working on pursuit and cargo handbooks for mechanics' use. She married her Air Force boyfriend, Chet White, and they moved to Albuquerque, NM, where she rejoined the CAP. She was a New Mexico state public information officer for a couple of years, while babies were little. Finally five children kept her on the ground, one girl and four boys: Landra, 49; Andy, 46; Carl, 45; Bill, 41; and Tomas, 39.

Harriet attended the University of New Mexico, getting degrees in Spanish, multicultural education and English as a second language, BA, MA, Ed. Sp., while running a boarding house to supplement funds. Scouting and PTA, etc., kept her in touch with the children's world. She taught various university level Native American programs in Panama, primarily in English and teacher education, for 25 years.

There are seven grandchildren: Jessica, 18; Jonathan, 17; Ruth Ashley, 12; Sandra, 12; Ross, 7; Christopher, 6; and Claire, 2; the eldest entered college in 1995. Harriet is an officer in the Albuquerque 99s, and Pan-American Round Table, while she learns to master a computer for writing articles.

LUETTA MARIE WHITE, born on a southern Iowa farm on March 19, 1931, (maiden name Stookesberry), remembers watching airplanes fly overhead and hearing stories of pilots such as Amelia Earhart and Charles Lindbergh when she was a young child.

Received ASEL certificate Oct. 21, 1967, and attended her first Iowa 99s meeting about that same time. Was an active pilot until the fall of 1978 when she sold her Cessna Skylane 3031F. She had not lost her love of flying, but she had lost her 49 1/2, James, in 1974. She was assisting their young son, Larry, in their 1,500 acre grain and livestock farming operation and also working nine to 12 hours a day at an off-farm job. She and Larry (also a pilot) had little time to use 3031F.

LuEtta remains a member of the 99s, IFF and AOPA, and loves to read the reports and activities. For the past several years she has been part of a timing team, working with the Air Race Classic.

There are many fond memories of flying activities she participated in with James and Larry, solo and with other flying friends. It's too difficult to select one special activity to write about – every flight and every flying activity was and is special to her.

MARY PARKER WHITE, born March 17, 1915, died Aug. 19, 1990, was born on property which later became the Oceana Naval Air Station in Virginia Beach, VA. She was associated with aviation as the spouse of an aviation medical examiner for most of her 75 years, but didn't learn to fly until age 50. Upon earning her private pilot certificate in a Bonanza, she joined the Virginia Chapter 99s in the mid-60s. In 1975 she became a charter member of the Hampton Roads Chapter. Mary was a mainstay of the chapter; she was quick to volunteer on projects,

but always avoided the limelight. Although personally unassuming, she delighted in other 99s achievements. In 1985 Hampton Roads Chapter honored Mary's outstanding service by awarding her a life membership. When Mary lost her battle with cancer, she left behind a daughter and son, plus many friends who sorely missed her quiet strength and caring nature.

VIVIAN "PENNY" WHITE, learned to fly a Cessna 172 in Manhattan, KS, at age 40, while teaching kindergarten and writing and composing songs and stories for a children's program called "Penny's Pardners" over KSAC.

Her first passenger was a 99, Helen Simmons, who upon landing on a ranch and being greeted by Marilyn Copeland, Gene Nora Jenson and Pat McEwen asked her to joined the Kansas Chapter. Meetings were held at the home of Olive Beech of Beechcraft as well as Bill Lear of Lear Jet.

Trips to follow were to Alaska, New Zealand, New Guinea and Australia. Remember, whenever you travel, a 99 is there to help – the camaraderie is most contagious!

Perhaps the highlights of White's career would be: Chartering Golden Triangle, South Central with Brenda Strickler in 1969; teaching first grade for 25 years with an abundant dose of aviation as well as basics; receiving "Outstanding Senior Citizen" of Arlington, TX, 1990; awarded the Rose of Honor by Sigma Alpha Iota (honorary music); climaxing with retirement luncheon on January 13, conducted by the regional commissioner, White being the age of 73 working for US Immigration.

Ladies, live each day to the fullest and travel when you can – don't wait for retirement. The 99s are always there for comfort and strength, for well in time of need have they taken time for a world of encouragement to husband, Donald, a double amputee, saying we care.

Remember, most of all, by being a good listener one can encourage young people, the future leaders.

CANDACE I. WHITFIELD, born Oct. 13, 1948, in Washington, DC. After much prodding and cajoling by her husband, for approximately seven years, she finally decided it "was my idea" and began her flight training in 1983 in Concord, CA. Five instructors later and many more than the minimum required hours, she passed her checkride in March 1985. She has often said to others that she felt obtaining her pilot license was a major accomplishment and one that has given her much confidence in other aspects of her life. In regard to the 99s, she joined the organization as a student pilot and has been an active member ever since; and she has held many positions at the chapter level. Today, you can find her husband, also a pilot, and her taking pleasure flights. They own a 1963 Cessna 150, which Gary fondly describes as, "it ain't much, but it's ours."

She has one child, Michelle, 22, and two stepsons, Bryan and Kevin. They have one grandchild, Derek, 3.

JACQUELINE (JACKIE) KAY WHITFORD, Phoenix 99s, who joined the 99s as soon as she qualified in 1991, has just gotten her instrument, and is now working on her commercial/multi-engine rating. She has been elected to her second term as Phoenix Chapter chairman. She got interested in aviation very early, flying out of Elgin, IL, with her father who flew for his business, from the time she was 2 until she was 8. She says she was about 7 before she realized that not all kids' fathers took them flying.

Jackie finally was able to realize her dream to be a pilot in 1990 when a friend who owned a Cessna 120 loaned her his plane and introduced her to the 99s. A co-worker gave her instruction, and it wasn't long until she had her private. She expects to take her commercial/multi in a Cessna 303, and then check out in several twins. She says that for someone who started flying for the fun to it, she has really taken it seriously and that she likes the feeling of independence and freedom that she experiences when flying.

Jackie's granddaughter and grandson love to fly with "Nana," and she spends some of her air time with a friend in his C-340. Before becoming chapter chairman, she served as airmarking chairman, NIFA chairman, co-chaired a Kachina Air Rally for the Phoenix Chapter of 99s and co-chaired PCIFA/NIFA for the Southwest Section. She became the first woman president for her employer's flying club, the Arizona Blue Sky Flyers.

WANDA WHITSITT, a resident of Illinois, learned to fly at the age of 48, in 1979. She continued her training by attaining her instrument rating, commercial license and ground school instructor rating. In 1980 she founded Life-line Pilots, a non-profit organization of pilots who fly people with special needs – primarily to medical centers for treatment and diagnosis. The organization grew to over 450 pilots from 17 Midwest states. She was a founding member of Air Care Alliance, a national federation of public flying groups. In 1986 she was inducted into the Illinois Aviation Hall of Fame. In 1992 she received a Certificate of Merit from the National Aeronautical Assoc. and was featured in a *Family Circle* magazine article entitled, "Women Who Make A Difference." She is married to Don Whitsitt, also a pilot, and has four children and nine grandchildren.

ANNE M. WIDGER, was born in Wiesbaden, German, where her father was stationed in the Air Force. She had always wanted to fly airplanes for as long as she remembered; her boyfriend told her to, "Go for it!" At the age of 27, on May 13, 1992, she started flight training. She soloed 10 hours later on June 2, 1992. She passed her private checkride July 22, 1992, with 49 hours. Her 84-year-old grandmother begged her to be her first passenger when she earned her private license, and she was. The next summer she passed her instrument checkride. The fall of 1994 she passed her multi-engine commercial checkride. She joined the 99s, December 1994. Most recently, July 1995, she passed her single-engine commercial checkride. Since 1992 she has logged more than 500 hours.

She flew across the US starting from Great Valley, NY, to Taos, NM, June 1994. The flight crossed the Sangria de Cristo Mountains of New Mexico with the highest elevation of Wheeler peak at 13,161 feet.

She graduated in 1987 from the University of New Hampshire with a BS in chemistry. Anne hopes to continue her flight training in the future and maybe add a seaplane rating or learn to land taildraggers. It is not often that people get their dreams fulfilled. Anne feels fortunate that she had the love and support of her boyfriend and family to encourage her to reach for her dreams.

LEW JANE RICE WIESE, born March 6, 1945, in York, NE, lived in Waco, NE, until marrying Verne Wiese. She learned to fly in 1978 in Hawaii. Once she overcame motion sickness and began the cross-country (in Hawaii, cross-water) phase, she became enthusiastic about flying. She continued on to get commercial and instrument ratings flying Cessna 150s, 172s and 177RGs. Lewie is part owner of a PA-28 Warrior and flies to 99s Section and International meetings as well as on vacations.

Lewie joined the Three Rivers Chapter in 1983. She has a BA in mathematics, BEE (electrical engineering) and a MS in management. She is a staff electrical engineer at Magnavox Electronic Systems Co. and a member of IEEE, NCMA, Greater Fort Wayne Aviation Museum, Assoc. of Old Crows and Magnavox Management Club of Indiana.

Lewie enjoys classical music and traveling. Verne is also an electrical engineer and a flight instructor.

LYNNETTE E. RENNEKE-WIEST, born Nov. 11, 1954, in Gaylord, MN, received her private pilot license in LeSueur; instrument/commercial at Mankato; flight instructor and MEL, Minneapolis; ATP at Albert Lea; and flight engineer with Sun Country Airlines, Minneapolis and Denver. She has flown the DC-10, second officer and just trained for first officer; ME turbo props; King Airs, Conquest HII, comprehensive list of single-engine and multi-engine aircraft.

She has been a pilot for Sun Country Airlines, Minneapolis, MN, since Nov. 1,1991; current flight instructor, SE, ME and instrument.

She is married to Donald Weist, a farmer and pilot and has two children, Trevor, 4 1/2, plans to be a warbird pilot and fire fighter. Trevor already flies left seat. Her second child was due on March 16, 1995, but at this writing the name unknown.

Enjoying experience as an airline pilot, though got into flying in 1977, just for fun. She was influenced and exposed to flying by then friend and husband since 1978, Don. Always gotten support for her flying from Don and her parents. Got into commercial flying in 1984. It was a major career change from being a land use planner. Instructed full-time at Mankato, MN, 1985 and 1986 then flew single pilot, Part 135 from 1987-1993. Flying single pilot in all weather from single engine cargo to multi-engine turbo props has given her lots of experience. She really enjoys flying their Archer with Don and Trevor, just for fun.

CHRISTINE E. WIETBROCK, born April 10, 1974, in Hammond, IN, began her flying career at the age of 16. She was influenced by her dad, who is a private pilot. She received her private pilot certificate in October 1991. The day before she started classes at Indiana State University she earned her instrument rating. She has flown the C-152, 172, 172RG; Piper Arrow; Warrior; Tomahawk and Seneca.

Since she has been in school, she has been able to obtain her commercial and multi-engine ratings, including her flight instructor (CFII and MEI) from American Trans Air Training Corp. She has also been fortunate for the opportunity to earn her seaplane rating. She graduated from college in December 1995 with a BS in aviation technology. She has been an active member and the vice-president of Alpha Eta Rho, aviation fraternity. She joined the Illiana Cardinals in January of 1992.

DELYLE RICE WIGGER, born Dec. 20, 1952, in Fairbanks, AK, is one of only a handful of women bush pilots with more than 5,000 hours of flying scheduled passenger service to the villages and towns of Alaska's vast interior. DeLyle was raised in a family rich in aviation background. Her father, Rober Rice, was a pioneering Alaska aviator who helped chart much of Alaska. She has an airline transport pilot and current certified flight instructor. She has flown the C-46, DC-3, C-150, PA-31, PA-32, BE-99, C-207, C-208 and Dornier 27. She is currently working for Larry's Flight Service, Inc.

A member of the 99s since 1983 and a member of the Aircraft Owners and Pilots Assoc., she is dedicated to flying safety and furthering education. DeLyle is presently attending the University of Alaska to further stretch her horizons. Among her other dedications are her husband, Walter Wigger, who is a longtime Alaska gold miner and her two sons, Kelly O'Neil, 22 and Jeremy O'Neil, 20.

One of DeLyle's favorite pastimes is her yearly visit to Mexico where she visits with warm Mexican families and brushes up on Spanish when she is not playing in the waves.

DeLyle enjoys the friendship of the 99s and welcomes anyone of the 99s who visits Fairbanks, AK, to give her a call. Maybe you'll go flying in her Dornier 27 or maybe you'll go gold mining.

EILEEN JUTSON WILD, born in 1944, in Brooklyn, NY, became interested in flying as a Civil Air Patrol cadet while in high school. However, went to college to become a nurse and kept working and going to school, earning a doctorate in health care administration in 1990 and works as a management consultant. She began flying again in 1987, when she met her future second husband, Fred. Fred was a private pilot, so he encouraged her to take lessons. After the first lesson, she was hooked! Eileen bought a Cessna 150 the next week. Shortly, she moved up to high performance (SEL) and began flying Fred's 1947, 98 percent original Beechcraft V-Tail Bonanza. Now, every weekend, weather permitting, they argue about who is flying which leg of their weekend excursions. Most of the time, Eileen will end up flying both legs. She hopes to start IFR training when she and Fred can agree on how to do the instrumentation on the Bonanza without disturbing its original antique panel. Meanwhile, they fly for pleasure, or as members of the US Coast Guard Auxiliary, doing safety patrol missions. Eileen and Fred live in Baldwin, NY, and fly out of Brookhaven Airport in Shirley.

She joined the 99s in 1994 and she has flown the Cessna 150 and 172; Beech 35 and Bonanza. She has two children from her first marriage, Thomas Alessandro, 22 and Eileen Alessandro, 20. She has one grandchild, Adrianna Alessandro, 3.

BARBARA ANN WILDER, born Nov. 22, 1936, in Kentucky, learned to fly and opened up a whole new world for herself. She not only joined the 99s but her state organization called the Grasshoppers. Through these organizations she was able to participate in numerous flying activities, along with the local Civil Air Patrol, serving as their medical officer.

She was in the USAFNC. Wilder received her flight training in Kissimmee, FL and got her license on March 16, 1971. She joined the 99s in 1970. She has flown the Tri Champ, Aeronca Champ on floats, DC-10, DC-3, Piper Cub, Cessna 140, Piper Cherokee 140 and 180; Lake Amphibian, J. Bonanza, Twin Comanche and 150.

She was married to J. Lloyd Wilder (deceased) and has

two children, Charles E. Lee, 34 and Tamelia D. Oliver, 32. She has one granddaughter, Jennifer Lee, 10.

CONSTANCE WILDS, while other 9 year olds were spending their Saturdays at the matinee, Connie was at the Goodyear Blimp Base on Watson Island in downtown Miami waiting for a vacant seat to view Miami Beach from the air. This was possible as her uncle was a pilot and dad was foreman of the ground crew. This began a long love affair with the wonders of flight that would lead her to attain her private certificate in 1958.

In 1960, a new challenge was brought, the All Women's International Air Race. This race was later dubbed the Angel Derby by Mexican controllers who were intrigued by all the angels communicating with them. With less than 100 hours to her credit, a 1960 Cessna 172 with fewer hours, the youngest contestant with the oldest contestant as her co-pilot, took off from Tamiami Airport to San Salvador, El Salvador. Although she captured the "Tail End Toni" Award, the experience gleaned from the race was invaluable.

She participated in several other races including Alabama's Petticoat Derby and Florida Suncoast's 500 Rally. She was involved in the organization of the Angel Derby – serving as president of Florida Women Pilots, sponsors of the race.

In 1989, Connie was elected vice-governor of Southeast Section of the 99s and became governor 1991 where she attended five board of directors meetings as a proud member of the Council of Governors. In 1993, at the annual convention in Portland, OR, Connie was elected to the International board of directors.

SHERYLE K. WILKERSON, born June 1922, joined the Reno Chapter of 99s in February 1993. She first became involved in aviation in 1976 when she joined the Air Guard. She started flying for personal reasons in 1986. She has a private pilot license.

She is married to Bradley N. Wilkerson and has been a photographer for 16 years and is currently the foreman for the PPIF imagery processing. She holds a BS in business and is seeking her masters in education. She enjoys photography, gardening, biking, flying and race walking.

She is also a member of the Silver State Striders, volunteer for the VA Hospital, volunteer for the READ Program.

BEVERLY J. WILKINSON, born Feb. 24, 1924, in Cadott, WI, got her flight training at Avenger Field, Sweetwater, TX. She flew general aviation planes from factories to El Monte Airport, CA, during summer vacations from the University of California, Berkeley. There were five WASPs in her group. She graduated as a WASP in October 1949, instructed and flew war surplus planes to Reno Sky Ranch. She has flown the PT-17, PT-19, BT-13, AT6, Otter, Piper Club, Aeronca, Luscombe and Waco. They drove to factories and had much more dangerous experiences being driven there than they ever did flying the planes back. She joined the 99s in 1991.

She is married to Jimmy E. Wilkinson, a retired lieutenant colonel in USAR (December) and has two children, Brad A. Wilkinson, 36 and Lesley Wilkinson Singletary, 32. She has three grandchildren: Joshua Wilkinson, 14; Marissa Wilkinson, 9; and Robert Singletary, 1.

BONNIE WILLIAMS, learned to fly in 1991 and lives in Virginia. She joined the 99s in 1994 and is presently a member of the Virginia Chapter. She is currently working on her instrument rating. She flies a Commander 114B. She and her husband, Dan, also a pilot own a Nanchang CJ6A. They have owned and operated a FAA certified repair facility for the last 15 years doing turbine engine repairs. Previously, she worked for Pan-American World Airways at JFK Airport for 13 years in operations and maintenance.

DIANA WILLIAMS, born May 18, 1947, learned to fly because she was a fearful flyer and got her ASEL. Although she has only 300 hours she has flown thousands of hours in their Aerostar with her husband. Business takes them all over the US, and flying has become a part of their life.

She has been an active member of the 99s working with flyers for 12 years, a program called Flight Without Fear.

Married to husband, Mark, for 25 years, they have four children: Diana, 34; Mark, 33; Jeff, 30; and Brian, 27. They own and operate Mark William Enterprises, Inc., a manufacturer of automotive racing components. A graduate of Oregon University with a BS in business, she has also had time in an Enstrom helicopter and has had some multi-engine training

ELOISE BALL WILLIAMS, born Jan. 5, 1925, in Eden, MD, at age 60, received her private pilot license. *The St. Petersburg Times* ran an article with pictures on her accomplishment. She was extremely proud! She received flight training at Sunshine Flying Club in St. Petersburg, FL, at SPG. She has flown the Cessna 152, PA-112 (Tomahawk) and Cessna 172.

Today, she and her husband, Tommie E. Williams, who is an instrument pilot, own and fly a Cessna 172M. They enjoy their flights.

JUDITH ANNE WILLIAMS, joined the SCV 99s in 1991 soon after learning to fly. She became active immediately by filling a sudden vacancy as vice-chairman and is now the chapter secretary; she has also served on the Marion Barnick Scholarship committee. Judy has logged about 200 hours and is working on an instrument rating. She flies at Trade Winds Aviation at San Jose; her favorite flying trips (after the SCV 99s monthly fly-ins) are long weekend trips to Baja with pilot husband, Dennis Stark. A native of Michigan, she has two daughters and works as a technical writer/engineer for a Silicon Valley Software company.

JUDY EMILY WILLIAMS, born May 25, 1943, founded the British Columbia 99s which had no official or practicing chapter and were the last province in Canada to become officially a chapter. She earned her wings at Pacific Flying Club, Vancouver, BC, (now relocated at Boundary Bay, Delta, BC in 1975). She has flown Cessna 150s, 152s, 172s; Skyhawks and Lances.

While practicing touch and go's on a snowy strip in the broad Squamish Valley, British Columbia, Judy had the privilege of literally flying with eagles! About 30 eagles were training their young to fish the river while soaring the thermals on this CAVU day in December 1978. A wise man once wrote that we remember moments, not days, this was her unforgettable spiritual "moment."

Since first thrilling to Peter Pan, spending countless hours with a mirror tucked under her chin to give her the illusion of walking upside down on the ceiling, and making her own wings out of plywood before leaping off her childhood clubhouse, Williams had hungered to fly.

Earning her single-engine land, day, wings in 1975, Williams' love of flying led her to found the British Columbia 99s in the fall of 1976 with only five members. Now, BC has three chapters and over 30 members! The 99s have run booths at the Abbotsford Airshow, conducted educational seminars, hosted a convention, established a Poker Run tradition, participated in numerous fly-ins and put the 99s name on an aviation emergency flight kit.

BC's coastal mountain ranges and miles of coastline have afforded Williams the kind of aviation challenges she loves; but more importantly, it has put her in the company of the fine men and women who share her deep and abiding love of flying. Her favorite flying, however, is when she can island jump in the San Juan Islands as she holds both Canadian and American pilot licenses.

MARGARET H. WILLIAMS, born July 18, 1926, at Ft. Lauderdale, FL, enlisted in the US Air Force in 1952. Graduated from Officer Candidate School at Lackland AFB in June 1953 as second lieutenant. Served with Air Training Command at Fort Sam Houston, TX; Dover AFB (MATS), Lackland AFB (ATC), Langley AFB (TAC). Retired in 1985 with the rank of lieutenant colonel.

She got her private license in 1956 and commercial rating in 1973. She has flown the Taylorcraft, Cessnas, Stinsons and many others, owned a Piper Tri-pacer (PA-22) and a Cessna 150. First flight lesson at Ft. Lauderdale, FL, in 1939 when there was only one paved runway on the field. Served with Civil Air Patrol during WWII in Savannah, GA. Holds rank of lieutenant colonel in CAP and has flown many missions over the years, with one "find" of a crashed aircraft. Wear the wings of a senior pilot, CAP.

She is a pilot with the Coast Guard Auxiliary and has flown many pollution and safety patrols and transported Coast Guard officers. Served as instructor for Coast Guard Auxiliary air observers and wrote curriculum for their ground school and flight instructions. A member of the Silver Wings (aviation pioneers). Holds rating of coxswain in Coast Guard Auxiliary and skipper Auxiliary Vessels on surface patrols. Also is a vessel examiner, instructor and lecturer for the Coast Guard Auxiliary. Holds a BS from Georgia State College for women and a master's from the College of William and Mary, Williamsburg, VA. Taught earth science and physical sciences in Norfolk, VA, city public schools. She is ham operator with an extra class license and the call of K14W.

She is single with two Doberman pinschers, Gus and Patsy, her "kids" and only family.

PEGGY WILLIAMS, began flying in Wisconsin in a 50 hp Cub, soloed Dec. 3, 1942, on "skis" (softest landing she ever made) and joined Civil Air Patrol where she met her husband, Roger, a pilot and chemical engineer. The war eventually grounded civilian flying . . . then came three children and when the last one left for college, she resumed flying, getting her license on May 24, 1976, and subsequently an instrument rating. She has accumulated more than 1,000 hours and has flown from California to Florida, Alaska to Baja.

Peggy has held several positions in her local 99s chapter; is a member of AOPA and Silver Wings Fraternity. Community services have included Boy Scouts, Girl Scouts, Campfire Girls, PTA, Meals on Wheels and hospital volunteer.

She loves racing and has flown in Palms to Pines, Pacific Air Race, Air Race Classic and chapter "mystery" races.

Peggy's children and four grandchildren all have been up in her favorite airplane, the C-182. She loves flying friends and golf clubs to golfing resorts. Her dream is to fly the Atlantic before packing up her wings.

KARIN WILLIAMSON, born April 14, 1940, in Frankfurt, Germany; moved to Canada in 1957 with her parents and brothers. She earned her private pilot license in 1979 after being introduced to flying by her husband.

They purchased their plane, CGKKW, a Cessna 172, in 1980 and have steadily upgraded it. The plane has been flown to conventions, section meetings and chapter fly-ins.

Karin has been a member of Maple Leaf Chapter 99s since 1980 and has held every position of the chapter executive once or twice and has chaired many chapter activities. She also flew for Operation Skywatch in the early 1980s.

She was a member of CASARA, Search and Rescue in London and acted as secretary treasurer for seven years.

She enjoys giving an annual introductory flight to Junior Science Fair winners. Karin is married to Ken and they have two sons, a daughter and two grandsons.

CLAIRE D. WILSON, born Feb. 17, 1938, in Westfield, NJ, moved to Massachusetts 21 years ago. Learned to fly in 1987 in a Cessna 172. Currently, owns and flies a Skylane 182 Cessna and has logged more than 995 hours. Joined the 99s in 1988 and is ENE Chapter treasurer this year and also aviation safety education chairman and insurance chairman, NE Section. Aviation orientated activities include operational officer of the Coast Guard Auxiliary Air Wing Division involving sunset and weekend patrols and SAR missions and also hold public affairs officer at division level.

Education background includes an associate in banking, insurance and real estate and paralegal certificate both from the Cape Cod Community College. Other affiliations include: AOPA, Aero Club of New England, Business and Professional Women's Club, National Assoc. of Insurance Women, American Society of Notaries (Notary Public). Occupation: broker and co-owner of Presidential Insurance Agency, East Dennis, MA.

She is married to George A. Wilson, who is also a pilot and the "other half-owner" of Presidential Insurance Agency. They have one daughter and three grandchildren.

JUDITH E. WILSON, born April 11, 1945, in Albany, CA, took up flying as a hobby in July 1992. She earned her private license in June 1993 in an Archer and her instrument rating in November 1994 in the family Bonanza F33A.

Judee has logged more than 400 hours and participated in Air Fairs, Poker Runs and Safety Seminars – currently at Phase II in the Wings Program. She is a member of the San Gabriel Valley 99s, Aircraft Owners and Pilots Assoc. and the American Bonanza Society.

She met her husband, Jack, a stockbroker, at California State University at Los Angeles. They have been married 30 years and have two daughters, Tamara and Janis. An executive secretary until their first daughter was born, she spent 19 years at home raising the girls and being active in the community. After taking computer classes and becoming proficient in WordPerfect, she now works as a temporary and teaches WordPerfect when she isn't flying.

KATHLEEN A. WILSON, born Dec. 6, 1949, in Pennsylvania, was always a little "flighty." As a small child, she would lie on her back watching the clouds go soaring past, wishing she could fly with the birds, pretending she was "Tinkerbell" of Peter Pan fame!

At age 33, within one week of her first small plane ride, Kathy started ground school at Shannon Memorial Airport, Downington, PA, taking flight lessons in a Cessna 150 from Bob Shannon Jr., obtaining her private pilot certificate one year later.

Rather than continue on into an instrument rating, she decided to "fly for fun," joining the 99s Eastern PA Chapter, AOPA and EAA.

Within two years, Kathy had fallen madly in love, purchasing a bright yellow and blue 1946 Ercoupe on the field, but after happily flying it for five months, discovered it was full of corrosion with a "bad" engine and not airworthy. Now called "parts" with a broken heart, she stopped flying. Four years later, another 1946 Ercoupe caught her eye . . . black, trimmed with red and gold and Kathy's love affair with the little two-place, Fred Weick designed, rudderless, two-control aircraft was now an addiction.

Flying with her classic "coupe" was the joy in Kathy's life, carrying her in 1992 to Sun-N-Fun, Lakeland, FL, and many Ercoupe owners club national conventions throughout the country, until finally in 1994, she became a five state regional director of the EOC.

Kathy's hopes for the future definitely include retaining her Coupe, and enjoying revisiting the many friends she has made across the USA, now solo.

RENEE M. WILSON, earned her private pilot SEL certificate while living in Fairbanks, AK, in 1990. While buying sectional charts one afternoon, the proprietor of the store asked if she might be interested in taking the air traffic controller exam that was to be held the following Saturday. Renee took the exam and a few months later was hired as a flight service station controller. She has worked at Fairbanks AFSS and Prescott AFSS and has been very active in aviation education for the past several years.

Renee is a member of the 99s, AOPA and the Professional Women Controllers Assoc. She has an associate's degree in air traffic control and is pursuing a BS in professional aeronautics.

TRACIE WILSON, was first introduced to flying when she had the opportunity to fly in a friend's Super Cub in Alaska. Upon her return from that trip in 1992, she began her flight training in Fredericksburg, TX, in a Cessna 172 and obtained her license in 1993. She is pictured here with instructor Bob Carter shortly after her "shirt tailing" ceremony.

She has been an active member in the Austin Chapter of 99s since 1993 and served as chapter news reporter/historian in 1994-95. Other aircraft flown include: Cessna 150, 152; Mooney, Luscombe, Scout and RV4.

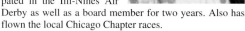

Tracie holds a BA of business administration and teacher certification in vocational education. She is an aspiring court reporter and is active in local, state and national court reporting associations. Her husband is Leonard Wilson, a public education administrator and fly-fishing instructor/guide, who also loves flying.

SUSAN WINDUS, learned to fly at San Jose Municipal Airport in 1972 with Janet Hitt. She now has both ASEL with instrument, and ASES, with more than 1,500 hours logged. She and her husband, Walt, own a Mooney and a C-172 on floats. They are very active promoting seaplane landing facilities and developed and built the Frazier Lake Airpark south of San Jose. Susan is by profession an interior designer, and she designed the buildings and facilities for the airport. She has given seminars on airport facility design at the AOPA conventions. Susan and her husband have been instrumental in coordinating the annual Clear Lake Seaplane Fly-Ins for the Sea Plane Assoc. She has been a member of the Santa Clara Valley 99s since 1974 where she has been active in airmarking and aerospace education.

SHIRLEY WINN, had her first two flying lessons at age 21, after finishing three years graduate work in psychology. The results were disastrous – she fainted both times. Blood pressure too low, they said. She raised it by getting married, working and having seven children. Finally 26 years later in 1971 started flying again and joined the Sacramento 99s. She has 4,000 hours, her commercial rating, seaplane, instrument, all ground instructor ratings.

She was Woman Pilot of the Year and governor of the Southwest Section, and elected chairman of the International nominating committee. She was member of the Yolo County Aviation Committee 17 years, the California Aviation Advisory Committee six years and the Yolo County Sheriff's Aerosquadron.

Shirley raced in over 30 air races, including the Powder Puff and International Angel derbies, Air Race Classic (board member six years), PAR and Palms to Pines. She is also a proud member of the Forest of Friendship.

KIMBERLEY DAWN WINSOR, born Jan. 29, 1975, at St. John's, Newfoundland, Canada, started flight instruction, December 1993. Soloed March 22, 1994, obtained private license Dec. 15, 1994. Joined the 99s in January 1995, attending first 99s convention in July 1995 in Halifax, Nova Scotia youngest member attending.

Presently enrolled in commercial pilot program while building time on family Cessna. Intends to land at Harbour Grace Airstrip Nfld. each May 20 to commemorate Amelia Earhart's departure for her solo transatlantic flight on May 20, 1932. Monument to Amelia placed and tree planted by 99s Eastern Chapter in 1986, located adjacent to airstrip. She has also flown the Piper Cherokee 140 and Cessna 172 and 150.

Kim's other interests include competitive equestrian show jumping since age 12. Kim and her horse, Popple, have won numerous awards in jumping, horsemanship and sportsmanship of the year awards. She is a therapeutic riding coach instructing physically disabled children in therapeutic riding programs. She is an avid sailor and downhill skier. Kim intends to become an airline pilot and establish her own therapeutic riding facility for disabled children.

Kim is a proud member of the 99s and wears her 99s pin with pride.

MARLENE BUERGI WINTERS, born March 26, 1936, in Rice Lake, WI, always wanted to fly, but did not have the opportunity until her children were in school. She received flight training at Palwaukee Airport, Wheeling, IL, for private and instrument rating. She has flown the Twin Comanche, C-172, Archer, Arrow, C-150, Lance, Tomahawk and C-152 Aerobat.

Learning to fly and joining the 99s has been the greatest experiences for her. As a 99 member she has participated in the Illi-Nines Air Derby as well as a board member for two years. Also has flown the local Chicago Chapter races.

She was the Forest of Friendship chairman for three years and on numerous committees for her chapter. She is currently the news reporter for the 99s news magazine.

Vacationing and seeing the country with a general aviation plane has been a great pastime.

She is a registered X-ray technician and is now a cardiac ultrasound diagnostic technician at a large suburban hospital. Her hobbies are oil painting, knitting, crocheting, reading and traveling. She has five children and seven grandchildren.

IRENE N. WIRTSCHAFTER, is a commercial instrument pilot, single-engine land and sea with more than 2,100 hours logged. She has participated in Powder Puff derbies, International Air Races, Section Air Races, as well as proficiency races. Irene is a retired Navy captain. Her other careers have included international banking specialist, Internal Revenue agent, insurance broker, real estate sales and appraisals and is currently a tax consultant. She has and is serving on many different boards, has attained national offices as well as much recognition. Current activities: senator, Florida Silver Haired Legislature; Cocoa Beach Code Enforcement Board; Florida Space Coast Philharmonic; Cap Canaveral Hospital Senior Advisory Board; Cocoa Beach Friends of the Library Board; International Forest of Friendship Board; Brevard County Cultural Grants Board; a trustee of the Assoc. of Naval Aviation, etc. She is a 99, a Grasshopper, Lady Bug, and a charter member of Women in Aviation, Int.; a Rotarian and a consultant for Junior Achievement.

NAOMI R. WITMER, born July 15, 1936, in Dexter, MO, received her private pilot license, single and twin-engine instrument ratings. Her husband, Loren A. Witmer, is a longtime commercial, instrument single and twin pilot. His love of flying sparked her interest, but he "created a monster." She loves to fly. She has flown the Cessna 152, Piper Warrior, Archer, Mooney 201, Bonanza A-36 and Twin Comanche.

In 1989 she was co-pilot with Margaret Ringenberg in the Air Race Classic taking second place, wonderful experience for a novice pilot. Loren and Naomi have made several trips to Fargo, ND, to visit relatives; Montana and Washington for vacation. Her son, John, and his 99 wife, Lois Collum, were their flying buddies until they bought their own Cessna 172. They still travel with them on the longer excursions. She retired from the Bank of America three years ago after 23 years, so she is free to do more flying. She is enjoying Comanche Fly-Ins and 99 trips.

She is married to Loren and has four children: Dennise Ann, 40; Sheryl (deceased 1994); John, 32; and Angela 30; three stepchildren: Jim, 47; Mike, 45; and Vickie, 43. She has six grandchildren: Jeff, 18; Joshua, 17; Leanne, 15; Aaron, 14; Crystal, 9 and Chase, 3. She has four step-grandchildren: Keith, 16; Lauren, 11; James 9; and Elizabeth, 7. These step-grandchildren are not considered as step. She is the only grandmother they have ever known.

BEVERLY LECHNER WITT, born on March 5, 1946, in San Luis Obispo, CA, currently residing in Gastonia, NC. Self-employed accountant. Proud owner of Cessna 172.

Witt started flying Aug. 17, 1993, earning her private pilot license Feb. 2, 1994. Joined the 99s, July 1994 in Oshkosh, WI. She is currently working on her instrument rating. Also she is taking aerobatic lessons in a Cessna A150 Aerobat. The lessons are for safety and a better understanding of unusual attitudes.

Never in her life had she ever done anything that has given her such rewards as learning to fly. Not only has she made new friends, but the joy of going flying with her husband is the best thing that has every happened to her. (He is a student pilot.)

She is married to Gary Gene Witt.

BETTY DEWITT WITTMER, a charter member and second chairman of Oregon Pines Chapter, represents the state's 99s on the mostly-male Oregon Aviation Alliance. She is an FAA written examiner, an AOPA member, a past president of the Salem Chapter of Oregon Pilots Assoc., past board member and secretary of NW Section and officer of many education and community organizations. She owns a Piper Dakota, which she and her late husband, Russ, have flown into a major number of the US states.

Wittmer earned a BA in English and speech at Iowa Wesleyan College and an MA in journalism at the University of Oregon. Her long career of high school teaching was interspersed with work as city editor of an Iowa daily paper and free-lance writing for *The Iowan* magazine. She now works as a volunteer with young adults at Salem Public Library.

The Wittmers, who celebrated their 50th wedding anniversary in 1993, moved to Salem in 1955. To their hobbies of music, gardening, house building and European travel, they added flying in 1976, bought an airplane at once and the sky has been home ever since.

GAYE WOHLIN, born in Holland, MI, and raised in the West Michigan area. She graduated from Central University of Iowa in 1964 and taught junior high and high school in Iowa, Michigan and Ohio. When she left the teaching field, she looked for another challenge in life and found it in flying. In 1989 she received her private rating and her instrument in 1993. She and her husband own an Archer II.

Gaye is a charter member of the Women With Wings Chapter of 99s, currently serving as treasurer of the chapter and student pilot chairman of the chapter and the North Central Section. She has organized the Flying Companion Seminars for the chapter and has conducted classes for grade school children.

Gaye is married to Don, also a pilot and owner of D/C Machinery Sales; they have a son, and a daughter and two granddaughters.

JEANE WOLCOTT, 1994-96 Women With Wings chairman, soloed March 1967 and received her ASEL, July 1967. She joined the 99s in December 1967. In 1975 she obtained commercial and instrument ratings. After 27 years in the Army she retired as a lieutenant colonel. Jean administered FAA written exams and flew pipeline. In 1976 she became a corporate pilot flying first for a trucking company and later, a law firm in an A-36.

From 1989-91, she resided in Texas and as a member of CAP, flew drug interdiction flights out over the Gulf. As a 99, she has served as news reporter, vice-chairman, chairman twice, Ways and Means and NIFA judge. Jeane has flown Powder Puff and Angel derbies and judged at countless race stops. She assisted in originating the Buckeye Air Rally and served on its board for over 10 years. Jeane has more than 6,300 hours, most in her Bonanza.

BETTY WOLFE, born May 11, 1927, Baltimore, MD, learned to fly at Dawn Aeronautics, New Castle County Airport, DE, receiving her private pilot license on Oct. 28, 1981. She flies a Piper Cherokee. Wolfe has logged about 465 hours out of New Castle County Airport, DE, and New Garden Airport in nearby Pennsylvania. She is a member of the Delaware Chapter of 99s, past secretary, vice-chairman and chairman. Wolfe's educational background includes a BS in elementary education from University of Delaware.

Some of her community activities include being a former Girl Scout leader, Pioneer Girls leader, unit director in Summer Camp at Sandy Cove in Maryland, Sunday School teacher, water safety instructor, and delivering Meals-on-Wheels to shut-ins.

Wolfe is married to Paul Wolfe who is a chemical engineer, a graduate of John Hopkins University in Baltimore, MD. She has a son, William, 47; two daughters, Winifred, 45 and Lois, 43; and four grandchildren: Cyrus, 24, Wendy, 22, Sandy, 18; and Edward Jr., 16.

KIMBERLY ANN WOLFE, born Nov. 7, 1963, in Columbus, OH. Her first experience with flying was in May 1984 when she joined the Air Force and headed to San Antonio, TX, for basic training. From there she went to the 376th Strategic Wing in Okinawa, Japan, where she was a personnel manager in support of the 909th Air Refueling Squadron. As part of tanker support, Wolfe frequently traveled to Korea, the Philippines and mainland Japan on a KC-135 (Boeing 707). While en route to such places, she witnessed air refuelings of various bomber, fighter, reconnaissance and attack planes.

Her real love for flying came when she was given the opportunity to ride back-seat in an F-15 over the islands of Japan. At the end of her tour in Japan, she was reassigned to Hawaii where she set a goal to get her private pilot license. Wolfe has since logged 160 hours and is pursuing an instrument rating and has earned a bachelor of general studies in aviation studies from the University of Nebraska at Omaha. While attending UNO, she was instrumental in starting a Nebraska Chapter of Alpha Eta Rho, an international aviation fraternity and designing an aviation program for local Omaha high schools.

She is married to Andrew Mikio Wolfe.

DORIS I. WOLFSTONE, born in Colorado, received her ASEL and joined the 99s in 1971-72. She has flown the Cessna 150, 152, Bonanza and Sierra. She has six children; two sons and four step-children. She has 14 grandchildren. Wolfstone loves to travel and has gone to Alaska, Mexico, Australia, New Zealand, Hawaii, Fiji, besides various parts of the US. She has been fortunate in being able to fly with other 99s to various events.

DOTTIE LYNN WOOD, born Oct. 29, 1940, in Houston, TX. Her aviation activities were born out of her honeymoon trip in 1974 with husband, Bob, to the EAA Fly-In at Oshkosh, WI. Encouraged to attend a "white knuckles" seminar in the "women's tent," she discovered a group of women excitedly revealing their experiences flying in without radios (experiencing their own white knuckles). This was the impetus to enroll in a flying school immediately.

She obtained her private license in 1975, 99s membership in 1978, instrument rating in 1984, completion of and flying their Thorp T-18 in 1995. She has also flown the Pipers, Grumman's, Mooneys, Citabria and Decathlon.

Her 99 activities include the Space City Chapter chairman, South Central Section bylaws advisor and currently vice-governor. She has served on the International bylaws, convention credentials and tellers committees.

She is married to Robert Leon (Bob) Wood and together they raised six children: Kris Wood, 26; Kelley Wood Corwin, 29; Kim Shaw, 37; Jody Trainer, 38; Danny Trainer, 39; and Stephen Trainer, 40.

JAN MARIE WOOD, born Nov. 6, 1921, in Los Angeles, CA, had 35 hours civilian training then a WASP in 1944. She had six months training in Sweetwater, TX, then was assigned to Stockton, CA. Her father loved to travel, rode the rails even and her mother went along. In 1946 she bought a Taylorcraft side-by-side wheel and in 1955 bought a 1953 Cessna 170B. She has flown Stearman BT-13 at WCLA and has had 20 hours in floatplanes two place, four place.

Education: 1943-1986, physical education teacher at Los Angeles school district in junior high and high school. She has had 20 years at LA Valley College. Since retirement she has traveled the world commercially and sailed the Caribbean, Sea of Cortez, Bahamas and Tahiti, Fiji Cruise, Panama, Costa Rico, East Coast inner-coastal waters.

From February 1956 to April 1957, flew plane around the world on sabbatical leave from teaching. She shipped the plane across Pacific on PAA Cargo plane. Reassembled and toured 10 months in Europe, the top of Norway to Mediterranean; from December 1956 to March 1957 East Middle East to Far East; she was jailed as a spy; had seven forced landings because of dirty fuel in Turkey; Singapore plane boxed in a crate and shipped Japanese freighter to Long Beach, CA; toured PAA to Philippines, Japan, Hawaii; cruised Hawaii to Long Beach. Reassembled plane.

She flew Powder Puff Derby; Alaska Point Barrow twice; Mexico to USA to Canada. She is a member of the Cessna 170 Club, AOPA, the 99s, Confederate Air Force, Float Plane Assoc. and Sea Pilots Assoc.

JOSEPHINE WALLINGFORD WOOD, a charter member and a native of Oklahoma, learned to fly in 1929 and received her private license in Santa Monica, CA, flying an OX5 Swallow. In 1931 she received her limited commercial rating.

JOANNA M. WOOD, born Jan. 17, 1940, in Pasadena, CA. Her interest in flying, along with her husband's began in 1979 when they purchased a 1964 Cessna 182, before they learned how to fly. The purpose was for her business (women's retail clothing stores) and plain pleasure. They still own the same plane, having upgraded equipment though the years. They have flown as far as Alaska, Canada, Mexico and Central America.

Probably the most memorable flight was in 1987, when they flew into Nicaragua during the tough political times. Recalling the difficulties of getting in and out of that country are hard to forget. They are both instrument rated pilots and will continue to fly their plane as much as possible as N2185R is truly a "magic carpet." She earned her private pilot license in 1980 and instrument rating in 1981.

She is married to Kenton Wood and has one daughter, Susannah M. Hough. She has one grandchild, Joseph Evan Hough, age 1.

MARGUERITE GAMBO WOOD, started professional flying in Honolulu, HI, in 1939 with her Gambo Flying Service. She was Hawaii's first woman commercial pilot, flight instructor and flight school operator.

Her activities halted abruptly Dec. 7, 1941. She was flying with a student when Japanese Zeros suddenly appeared. She saw smoke at Kaneohe NAS, became suspicious and returned to the Honolulu Airport just in time to take cover in a ditch while the field was strafed.

Soon after, Marguerite decided to join the Women Air Service Pilots, but was sidelined in California where flight instructors were needed. Later, she became chief flight instructor for Pierce Flying School in Arizona where she met her husband, Robert Wood.

After the war, they returned to Hawaii and Marguerite taught ground school at the University of Hawaii. Marguerite started her second flying school, The Hawaii School of Aeronautics.

In 1952, her Air Force husband was transferred to Germany. She sold her school and turned the hangars over to the Territory of Hawaii. When they returned to Honolulu, Marguerite was told she'd have to take over management of the hangars, so she established Hawaii's Country Club of the Air.

Marguerite died from cancer in 1984. Although her flying school no longer exists, her legacy lives on in the Marguerite Gambo Wood Foundation which provides annual scholarships for women pilots in Hawaii.

She was a charter member of Aloha Chapter which sponsored her plaque at the International Forest of Friendship in 1978.

INEZ WOODWARD WOODS, born Oct. 7, 1917, in St. Louis, MO, joined the 99s in 1943. She began flight training privately in 1940 and obtained her private license (74077-41) in 1941. She also had WASP training. She has flown single-engine from Cobs-Douglas A-24, multi-engines C-78, C-45, and C-60-C-47.

She ferried planes after the war for the Defense Plant Corp. and was a flight instructor for Executive Pilot Steel Co.

While with the Los Angeles school district, she had the opportunity to set up course (experimental) in aeronautical science in five of the 63 senior high schools. Wrote the course of study. Was asked to collaborate with four others to produce course of study for all California high schools. FAA liked it, reprinted it and made it available to high schools nationally through Supt. Of Documents, Government Printing Office, Publication No. GA-20-10 (printed twice).

She is married to Jack Woods, and has one child, Barbara Anne Woods, 39. She has one granddaughter, Amanda C. Woods, 5.

JESSIE MARTIN WOODS, said it was easier to wing walk in the air than on the ground! Jessie was a charter member of the Mount Tahoma Chapter, one of 10 members, whose charter was issued July 10, 1969, in Bellingham, WA.

She and her husband, Jimmy, entertained paid crowds with parachute jumping, wing walking and the rope ladder. They started in 1928 and continued for nine years. Jessie was truly a pioneer with many "firsts" to her credit.

The chapter bid her God speed February 1974 when she moved to Florida.

DIANE ELOISE WOODWARD, born Feb. 26, 1943, in St. Louis, MO, became interested in aviation because it was her husband's hobby. Her husband, John, a pilot himself, always loved airplanes and flying.

In 1985, they bought their airplane, a 1968 Cessna Cardinal, which is the plane Diane later learned to fly in. After purchasing the plane, Diane found she was terrified of flying. However, she would go along with her husband on all his flying excursions – white knuckling it all the way.

Finally she overcame this fear and began working towards her pilot license. In June 1993 she earned her license, and now she is so addicted to flying that she readily pushes her husband out of the left seat every time she gets a chance. She has also flown the Cessna 150. She joined the 99s in 1993.

She and John have two children, John Alden Woodward Jr., 31 and Clayton Michael Woodward, 29. They have four grandchildren: Heather Anne Woodward, 11; John Alden Woodward III, 8; Julia Shea Amos, 7; and Richard Skylar Woodward, 5.

CHARLENE WOOLSEY, born Jan. 24, 1942, in Sayre, OK, joined the 99s in 1980. She has her CFI, commercial, instrument ASEL/AMEL, advanced, instrument ground instructor certificate. She has flown the Piper 140, 150, 161, 180 R, 181, 31, 34-220T; Dakota; Cessna 172, 182, 195XP, 42; Lockheed C130E; and King Air B-90. She is a charter, photography and instructor pilot.

Service to the 99s includes: International Ways and Means chairman, judging USPFT national events, NIFA *SAFECONs*, section 99s news reporter and nominating committee and past chapter chairman of the Oklahoma Chapter. Received "Pilot of the Year" award by the National Race Pilots of America, Inc., in 1984, and the Oklahoma Chapter in 1985. Air Races flown include the Okie Derby, Kansas Sunflower Rally, Arkansas Aero Derby and Fairview Fly Lady Derby. Memberships in professional organizations are Oklahoma Pilots Assoc. and AOPA. She was a public school teacher and counselor for 25 years, holding a masters of education in counseling and an associate's in aviation education.

She is married to Corkey Woolsey and has three children: Susanne, 35; Linette, 32; and Davey, 27. She has nine grandchildren: Kyle, 15; Daniel, 14; Mandy, 9; Matt, 8; Landon, 5; Kaley, 5; Dillon, 3; Shawn, 3; and Bailey, 8 months.

ANITA CONLEY WOREL, born on June 17, 1922, in Vallejo, CA, and now resides in Fair Oaks, CA. She has two daughters, Dale Ball and Linda Kennard, six grandchildren: Lora Collins, Kristi Matheson, Steven Ball, Denise Barker, Kevin Kennard and Clint Ball; seven great-grandchildren: Daniel and Jeffrey Collins, Douglas and Jenica Barker, Whitney Matheson, Stephanie Ball and a Kennard-to-be.

In 1958 she went for a ride in a Swift. It was so much fun that she started taking lessons in a Cessna 120 in which she soloed. In 1959 she made her second solo in an Aeronca 7AC. In 1960 her third solo was in a Piper PA-12. She finally got her license in 1961 in the PA-12. She immediately bought an Aeronca 11AC. With 65 hp and no electrical system, it was pure fun. In 1964 she bought a Cessna 170. In 1966 she and her new husband bought a 35 Bonanza which she dearly loved. She has also flown the Culver Cadet Ercoupe, Travelaire, Cessna 120, 140, 172, 182; Beech, Musketeer, Piper-PA-16, Tri-Pacer, T-6, Navion, Fleet, Eagle and Bell Helicopter.

She joined the 99s in 1961 and held all chapter offices in the Redwood Empire Chapter. She is now a member of the Reno Area Chapter. She flew the 1962 and 1964 AWTARs with Myrtle Wright, the 1966 with June O'Donnell and the 1971 with Nina Rookaird. Placing 19th was the best they did, but they finished all of them.

She married Jack Worel at Fallon Airport right after the 1965 San Diego/Fallon Fun Race. She was late to her wedding as she had to throttle back due to low fuel and ran out of gas while taxiing, but she finished and wasn't last.

ELEANOR WORTZ, learned to fly in 1940 in the first Civil Pilot Training Program class at Catawba College, NC, and joined the 99s in 1949. She flew as a WASP in class 43-W-4, from February 1943 to December 1944. She has more than 1,000 hours of flying both single and multi-engine aircraft and holds an instrument rating. Eleanor taught aviation ground school in Brazil (she is proficient in Portuguese) and is a retired community college professor from Canada College. She has a master's degree in business education and enjoys reading, swimming, traveling and embroidery. She is currently recording her WASP experiences into a memoir.

MARIE WRAY, born on a farm near Mount Pleasant, IA, and now lives in Cedar Rapids, IA. She received her private pilot license, May 18, 1986, at the age of 69.

Her husband, Bob, flew for the Army Air Force during WWII in the Philippines. He was taken as a Japanese prisoner and after three and half years in the prisons camps, he retired as a major and became an executive pilot for Iowa Manufacturing Co. in Cedar Rapids.

Her son, David, retired as a colonel from the Air Force, and is now flying with United Airlines.

Her daughter, Janice, and husband, Perry, are both pilots also. They own and operate the Marion, IA, Airport, flight instructor, and fly charters. Perry is also an FAA examiner. It was her daughter who said, "Mom, why don't you learn to fly?" Since all her family were fliers, "Why not?"

Marie has been associated with flying since 1936 when her husband started courting her in an airplane. He really impressed her when he landed on the cow pasture on the farm. What a thrill!

However, learning to fly on her own, was the biggest thrill of all. Flying is a wonderful responsibility teacher. She says it taught her to be more alert, to think, to never assume anything; besides that, it was fun. Everyone should learn to fly.

NANCY VESTAL WRENN, native North Carolinian, graduated Duke University 1943, spent two years WWII in the Far East with American Red Cross, settled near Asheville, was co-owner of private summer camp and had a 34 year career in education.

Member of the Carolinas Chapter since starting to fly in 1966. Holds commercial, ASEL certificate. Served all levels chapter offices (some twice), several Southeast Section committee chairs, past section treasurer. Charter member Western North Carolina Pilots Assoc. Past president North Carolina Council for Social Studies. Served on governor's aerospace education advisory committee. Long time member Delta Kappa Gamma.

Active in St. James Episcopal Church, Altar Guild, Outreach Committee, historian/photographer. Spends retirement enjoying lifelong commitment to her horses and dogs. Flies occasionally.

When she was 10 years old in Southern Pines, relatives of George Palmer Putnam lived next door. George and his wife, Amelia Earhart, often spent time there. (Pity, Nancy was too young to fully appreciate it.)

ALICE M.A. WRIGHT, born May 10, 1938, in Hagerstown, MD, joined the 99s in 1981. She received her private pilot license with an instrument rating. She has 800 hours flight time. She has flown the Cessna 152 and 172; Beech Musketeer A23A; and Piper PA-28-140.

Wright has been a member of the 99s for 14 years and also a member of AOPA, Cherokee Pilots Assoc. and NJ Pilots Assoc.

She participated in Garden State 300 Proficiency Race for many years. Placed fifth, seventh, eighth, and tenth as co-pilot in various years.

Wright is married to Peter and has one child, David, 25. She has two grandchildren, Benjamin, 18 months and Margaree, 2 months.

ELAINE M. ATKINSON-WRIGHT, Alberta Chapter of 99s, received her private pilot license in the summer of 1968, the same year of Alberta 99s charter. She purchased a 1958 Aeronca Champion in 1968 and still owns and flies it.

Wright helped at the Powder Puff Derby held in Calgary; rode in the Calgary Stampede Parade as a 99; served on a number of executive positions at chapter and section level, 1970-74; Alberta 99s secretary/treasurer, vice-chairman and chairman and Western Canada's vice-governor. At the same time she served on the Alberta Flying Farmers executive, becoming Queen of the Alberta Chapter in March 1974. She served as International Duchess for the International Flying Farmers, chosen at New Orleans, August 1974. Wright later served as one of a team of IFF teen advisors. From 1982 she owned and operated a travel agency and a fashion business, but sold the travel business in 1989 and started college.

She has completed two diplomas the spring of 1995 in biological sciences technology, one an environmental diploma and the second a laboratory and medical research diploma. Wright presently will enter her second year of a bachelor of science degree with studies in molecular biology.

MARTHA E. WRIGHT, born June 27, 1921, in Childress, TX, soloed on April 30, 1946. She received her private pilot license on July 22, 1951. Wright has flown the Ryan, Vultee, Aeronca, Steerman, Cubs, Eaglet and Cessna.

While flying co-pilot for Edna Whyte, Wright left for Fort Myers, FL, in her Cessna 120 heading for Nassau, Bahamas, in the All Women's International Air Race. They won first place. It's always really fun to win. She still loves to fly, and now she has ultralights putting the original fun back into aviation.

She is married to Linley S. Wright and has four children: Danella, 48; Carol, 45; Toni, 43; and Patti, 35. She has 10 grandchildren; Lynnette, 28; Russell, 25; Jana, 26; Mirielle, 23; Dustin, 18; Joelle, 16; Lindsey, 12; Kara, 10; Danny, 8; and Bethany, 5.

NANCY LUCILE WRIGHT, learned to love flight at an early age. Her father was a pilot in WWII, and her brother was an Air Force officer for many years and now flies for American Airlines.

She learned to fly in the late 70s and early 80s. She was encouraged to learn by a very dear person, Jim Pappas. He was her right arm and her support through every certificate and rating she got. Without him her feet would still be on the ground with her heart in the air. Above all he is the reason she joined the 99s, because his aviation friends had told him about the organization. Wright didn't join women's club she told him he persevered and she joined him.

Wright learned to fly on a grass strip in Michigan where she earned her private license and her instrument rating. She earned her commercial in Florida. Her ground instructor certificates came next right through instrument ground instructor.

Life in the 99s is never dull as she found out in her early years with the Florida Suncoast Chapter. They let her go through most of the chairman duties along with all of the chapter offices. Wright's first love came when she was named aerospace chairman. She still holds this position and will keep it as long as the chapter wants her. Wright also held this position at the International level. She was instrumental in starting a space camp scholarship given by her chapter to a young lady from Florida, they now give two every year. Along with her service to the chapter, she entered the section level. Wright went though the offices and is presently serving as governor of the Southeast Section. As governor she attended many International BOD meetings and is now the chairman of the standing rules committee. Wright loves to attend section meetings and conventions and seldom misses any.

Another activity she loves to do is her time at Sun-N-Fun in Lakeland, FL. She spends at least a week there every year helping Barbara Sierchio man the building; recruiting new members and enjoying everyone who stops in to see them.

Currently Wright is on the board for the World Precision Flying Championship. They will work with teams from all over the world. It is very demanding and very satisfying work. She loves the 99s. She loves the friends she's made and the wonderful experiences she has enjoyed. Fly Safe 99s.

CHARI KAYE WROOLIE, born Aug. 31, 1969, in Fontana, CA, has been flying since she was 14 and has loved every minute of it. Wroolie has been a flight instructor for four years, three of which were full-time. Now, she just does it on the side. She has a CFI, multi-instrument rating. She joined the 99s in 1994. She has flown the Citabria, C-152, 172; Piper Cherokee, Arrow, Malibu, Grumman T-Cat, King Air 200 and Mooney.

Wroolie met her husband, Noel Eric Wroolie, at the airport. He worked down the hall as a Lear captain, and she worked at the flight school. He is still flying the Lear, but she took an office position working for one of her former students. How's that for ironic?

They bought a Citabria, so Wroolie figured she had better learn how to fly it. After much more instruction, she hopes to participate in airshows and aerobatic competitions.

Recently the Coachella Valley Chapter of 99s graciously gave her the position of membership chairman. It's been worth the hard work.

Her goal is to be as skilled as her aerobatics instructor, Denny Brown, and as humble as Jeana Yeager.

MARY MARGARET WUNDER, born Oct. 7, 1953, in Norristown, PA, while growing up, can remember many parties at the Pylon Club that her dad belonged to. Later in life she was dating a man who was taking flying lessons. Not wanting to be left on the ground, or up in the air alone – if something should happen to the man in the left seat – she started taking flying lesson. After being bit by the aviation bug, Wunder not only received her private pilot certificate, but her commercial certificate with multi-engine and instrument ratings. She has flown the PA-28-140, 180; C-150, 172; BE-95; and Aeronca Champ.

Wunder never dreamed of a job in aviation. Even after getting her pilot license, she never thought that it would be more than a hobby. But she became an air traffic controller in 1981 and has worked at the Wilkes-Barre, Allentown and Philadelphia Airports.

She is currently chapter chairman of the Eastern Pennsylvania Chapter of 99s. Wunder and her co-pilot, Barbara Strachan placed fifth in the Garden State 300 their second year of competition. She received the "Woman of Vision" award in 1993 from Montgomery County for her work in promoting women in aviation.

She is married to Arnold J. Wunder.

GEORGIA BEWLEY WURSTER, born March 4, 1935, first learned to fly while living in Uganda (1967-1972) and enjoyed flying to game parks in Uganda and Kenya. She and her ex-49 1/2 made dual flights to Ethiopia, Tanzania, Zanzibar and a loop from Entebbe to Cape Town, South Africa back through Mozambique landing in all the countries in between.

While living in Peru (1973-1981), she first flew on a converted East African license but decided she might as well go for her FAA ticket and actually passed her test when a certified FAA examiner passed through the country. Her most memorable flight was flying "in command" (She was usually co-pilot of their Cessna 337.) in their Cessna 182 from Lima to Santiago, Chile, Bariloche, Argentina and back. On the return leg, the winds were so strong in Antofagasta, Chile, that the plane wouldn't stick on the runway! After three hops, she finally slowed to a rolling stop. When she disembarked, the official wanted to talk to the "piloto." Wurster told him she was the pilot, but he wouldn't believe a woman could be flying her own plane. After that landing, he should have believed! There were numerous trips over the Nazca Lines, flying visiting friends over that curious wonder, and flights over other Indian ruins, such as Chan-Chan, near Trujillo. The ferry flight of the Cessna 182 from New Jersey to Lima was also very memorable and convinced her to get an instrument rating. (She flew with their co-owner, who was also SEL-no instrument, hugging the coastline in places about 150 feet due to cloud.) Wurster learned about high altitude landings in Arequipa, Peru and the strip near Huascaran (highest peak in the Peruvian Andes). Dual flights in their 337 were made annually to New Jersey and to Iquitos, Peru on the Amazon River, and Chile and Argentina, crisscrossing the backbone of the Andes.

She has flown very little since 1982, when her health and fortunes took a bad turn. She still keeps up 99 ties and hopes to feel well enough to climb back into the command seat.

RUTH A. WYATT, born Oct. 10, 1935, in Greeley, CO, became a licensed private pilot on March 11, 1989. Ruth is a recreational pilot and co-owner of a 1957 Cessna 172. By way of historical information, Ruth was the first student pilot to solo at the Huntsville-Madison County Regional Airport, Huntsville, AR, on June 6, 1988. Bill Smith, CFI, was her instructor.

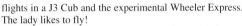

During the first nine months after receiving her license, Ruth logged more than 100 hours in the air. Types of aircraft she has flown include the Cessna 150, 172, 182 and 210. She has also had short flights in a J3 Cub and the experimental Wheeler Express. The lady likes to fly!

Twenty years of life as a military wife provided many wonderful experiences in four countries, but none can compare to the thrill of flying for Ruth. She became interested in piloting while serving as the navigator/observer for her husband, Thomas. Safety considerations caused Ruth to begin taking pinch-hitter training. The love of flying caused her to continue lessons culminating in her private pilot license.

Ruth's only regret concerning being a pilot is not having become a pilot at a much earlier time! She has four children and four grandchildren.

EILEEN HANSEN WYCKOFF, began flying lessons at Vandenberg AFB Aero Club in 1973, with veteran instructor, Jim Asel, out of self-defense. It was a life insurance policy for a family of four in case pilot-husband, Bob, became disabled while aloft. She has been a member of four 99 chapters and helped charter both her current Santa Maria Chapter and its predecessor, the Central California Coast Chapter in 1974. Her All Women Transcontinental Air Race Team (AWTAR 100) with Linda Schreck and Judy Sorce finished fifth in the 1977 Powder Puff Derby. Both kids grew up in their 1947 model-run Bonanza, *Chiquita* – one of Mr. Beech's original efforts – visiting grandparents in New Jersey and Florida from California annually.

Flying influenced their kids' lives also. Steve has checked out in *Chiquita* and is building an experimental GP-4. Lynne completed her master's in airport design and management, complementing her BA in architecture.

CARLA LEA YANCEY, born Sept. 15, 1964, in Winnemucca, NV, had her first lesson with the start of the new year in 1987, along with her husband. She received her private pilot certificate on July 22, 1987. Since then she has logged 300 more hours, most of which has been in a Stinson 108-3, a Chinese Nan Chang CJ-6 equipped with a nine-cylinder radial engine and pneumatic functions and currently a Beech P-35 Bonanza.

Yancey has commercial and instrument ratings and is a certified advanced ground instructor. In 1992 and 1993, she participated in the Palms to Pines Air Race along with Rhea Bastian in her Cherokee 140. She is an original charter member and past chairman of the Crater Lake Flyers Chapter.

She and her Ag pilot husband, Monty, have two children, Kurt and Alex Andra.

ROBERTA ANN YATES, born in Galveston, TX. She lived in Texas most of her life and moved to Baton Rouge, LA, in 1986. She began flying in January 1992, soloed July 11, 1992, and obtained her private pilot license in February 1993. Since that time she has been busy in the 99s, joining as soon as she took her checkride and attending the meetings as a 66 prior to that time. She flies "every time I get a chance," and you can usually find her at a local airport on the weekends.

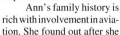

Ann's family history is rich with involvement in aviation. She found out after she began flying that both her mother and her father had a held a private pilot license! In 1994 she took her first ride in an aerobatic plane, and she was also the "poster person" for her 99 chapter ad to attract new members to the 99s. She became an FAA aviation safety counselor, receiving her appointment, June 1995. Ann is also currently serving as the chairman for the South Louisiana Chapter of 99s for the 1995-96 year while she actively works on her instrument rating. She is a registered nurse with a certification as a chemical dependency nurse specialist and works part-time in an emergency room for personal fulfillment and "extra flying funds."

Ann would like to complete her aviation career by obtaining her CFI. Her hope is to teach at the private pilot level and to be able to assist others in achieving the tremendous feeling of satisfaction that flying imparts.

MARION YEAGER, learned to fly at Reid-Hillview Airport in San Jose in a Citabria. She has logged 210 hours and flies at the Flying Country Club at Reid. She joined the Santa Clara Valley 99s in 1992 and is interested in aerospace education and has been active in fly-ins. Marion is learning to be a court reporter. She and her husband have two children and live in Saratoga, CA.

AUDREY YEANDLE, (nee Ashfield) was born in Brooklyn, NY, on May 3, 1916. As a girl, a ride in a biplane fired her interest. Lacking education, flying had to wait.

Time passed, marriage came, three sons before the opportunity to fly arrived in 1976 when a business venture made it practical.

When Audrey received her license at 61, her husband, Thomas, purchased a Piper Dakota which enabled her to supervise the landscaping of the project 100 miles to the northeast. Audrey joined the Alameda County 99s at the Hayward Airport. She enjoyed being active locally. The young women were so dedicated.

Audrey spent 900 hours flying around California in her Dakota. Also, accompanied her husband as co-pilot in his Cessna 414. They have flown all over the US, parts of Mexico, Canada and the Caribbean.

Difficulty with arthritis put an end to her personal flying at 78. She refused to use age as a reason not to learn and enjoy this wonderful activity.

DELORES PYNES YELLIN, born May 12, 1941, in Victor, MT, has an instrument rating, MEL and has logged more than 2,500 hours. She has checked out in over 30 aircraft (SEL and MEL) and currently owns a Debonair. She previously owned and operated an A/C maintenance shop at Van Nuys Airport. Yellin is now a bookkeeper/sales director at an Awards company in Los Angeles.

Active member of the San Fernando Valley 99s since 1970, Yellin won the San Fernando Valley Chapter "Woman Pilot of the Year." She has flown many rallies, treasure hunts, Poker Flights, air races (local and cross-country) including the Baja California Air Race, Angel Derby, Palms to Pines (won four times), Pacific Air Race (won two times).

She is married to Dale Yellin, an insurance executive, and has three children: Lori Brown, Danelle Bayne and Kurt Simington; and one stepson, Jeff Yellin. She has two grandchildren, Brian Bayne, 4 and Cody Bayne, 2. Her hobbies include flying to Baja, drinking Margaritas, water skiing, fishing and gardening.

LOIS K. YOUNG, born Jan. 2, 1928, in Franklin, VA, is a member of the 99s, Inc., Suffolk Christian Church, Order of the Eastern Star, Daughters of the American Revolution and Suffolk, Franklin, Grassy Creek Golf Assoc.

She and her husband, Harry, just celebrated their 50th wedding anniversary. They have two daughters, Carolyn Y. Cobb, 49 and Karen Y. Norako, 42; and five grandchildren: Trey Cobb, 17; Cristin Cobb, 14; Kathleen Norako, 12; Jennifer Norako, 10; and Vincent Norako, 8; who fly with her often. She had the opportunity to fly as co-pilot the world's oldest flying airplane, a 1928 Ford Tri-Motor, built the year she was born. While in Hawaii she was checked out and rented an airplane, which she flew, with hubby, around the island of Oahu.

Having always loved flying, the dream finally came true when she soloed at the age of 55.

LUCY B. YOUNG, born in Waterbury, CT, on Sept. 9, 1954, and attended Shepaug Valley High School in Washington, CT. She was selected for a Navy ROTC scholarship to Purdue University, graduating with a BS degree in 1976. Upon graduation, she was sent to Attack Squadron 42 (VA-42) at NAS Oceana, VA. She reported to Pensacola, FL, in October 1976 for flight training. Flying the T-28 Trojan and the T-44 King Air, ENS Young won her wings of gold in October 1977.

After qualifying in the TA-4J Skyhawk, she reported to Fleet Composite Squadron 1 (VC-1) NAS Barbers Point, HI, where she accumulated more than 1,000 hours. Lt. Young qualified as section leader, instructor pilot and air combat maneuvering pilot then went to Training Squadron 21 (VT-21), NAS Kingsville, TX. As a TA-4J flight instructor, she instructed student naval aviators and carrier qualified on the USS *Lexington* in May 1982.

Leaving active duty in July 1983, Lt. Young accepted a commission in the Naval Reserve and a position in Atlanta, GA, as the first female FAA test pilot. Lt. Young affiliated with VA-2267 in January 1984 then joined VR-58, NAS Jacksonville in December 1984. In November 1985, LCDR Young reported to VR-46, NAS Atlanta, GA.

In May 1986 Lucy began training as a Boeing 727 flight engineer with Piedmont Airlines in Winston-Salem, NC. In October 1987 she upgraded to first officer on the B-737. Promoted to commander in July 1991, CDR Young subsequently transferred to VTU-6767, NAS Atlanta in March 1992. CDR Young qualified as a transport aircraft commander with overwater/international qualification. She is a veteran of Operation Desert Shield and Operation Desert Storm.

Lucy is a member of the 99s, Tailhook Assoc. and Women Military Aviators. She is an air safety representative and accident investigator for the Air Line Pilots Assoc., and is facilitator for the USAir Crew Resource Management program. She holds the air transport pilot, flight engineer and certified flight instructor ratings and has 9,000 hours of pilot time in over 40 different aircraft. She is currently flying international flight as a B-767 first officer based in Philadelphia, PA.

RAMONA YOUNG, born Aug. 26, 1952, in Memphis, TN. She spent much of her life in Osyka and Pascagoula, MS. In 1977 she married aviator, Obie Young, who gave her a choice between a microwave oven or flight lessons. Naturally, she chose the flight lessons (she got the microwave later for Christmas). A few months later she earned her private pilot license and has flown many types of airplanes. The majority of her flight time has been in the Piper Warrior and the Cessna 172. She especially enjoys aerobatics and enjoys flying the Decathlon, Christian Eagle and Pitts Eagle. Ramona and Obie owned a Pitts Special for about five years, and Ramona emceed Obie as he flew airshows. For 14 years she assisted her husband in the direction of the Jackson County Airshow, Mississippi's Official Airshow. When her husband, Obie, made a career change to the FAA in 1990, the family moved to their present location, Orlando, FL.

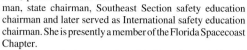

Ramona joined the 99s in 1981. In Mississippi, she was safety education chairman, public relations chairman, state chairman, Southeast Section safety education chairman and later served as International safety education chairman. She is presently a member of the Florida Spacecoast Chapter.

Ramona is also a member of the Experimental Aircraft Assoc., International Aero Club, Aircraft Owners and Pilots Assoc. and is a captain, pilot and former public affairs officer in the Civil Air Patrol (Auxiliary of the United States Air Force). She has received the Pilot Proficiency Wings Phase IV.

Her non-flying affiliations include American Business Women's Assoc., Assistant Scoutmaster with the Boy Scouts of America, Delta Gamma Alumni Assoc., University of Southern Mississippi Alumni Assoc., Epsilon Sigma Alpha International (women's service organization) and active member of St. Luke's United Methodist Church.

Young presently teaches sixth graders with specific learning disabilities where she incorporates flying activities into the curriculum.

Ramona's 15-year-old son, Obie II, has been flying since before he was born. He is very active in the Air Force ROTC and works with the Young Eagles Program.

LIBBY M. YUNGER, born Feb. 20, 1944, in Ohio has always been fascinated with airplanes. As a child she dreamed of flight, and as a young adult she was totally besotted by a college friend in his rented Cherokee. Alas, several "authorities" convinced her that a medical limitation would prevent her being licensed by the FAA. Grounded! Twenty-five years later a ride in a Stearman changed her life. "Of course you can fly," exclaimed pilot/owner Carl Cochran, "you'll just have to fly with the FAA for a medical waiver." Six years later her license reads private pilot SEL, SES, instrument-airplane. Bravo! Yunger's 1956 C-182, is her main mount; but once on a grass strip on a calm day, she managed to solo that Stearman. Pursue your dreams – it doesn't get any better than this!

TONI ZABALA, began flying at the Moffett Navy Flying Club in 1989 and got her license in 1990. She is about halfway through her instrument rating and has logged about 200 hours. She now flies at Reid-Hillview Airport with Wings over California. Toni joined the Santa Clara Valley 99s in 1995 and is interested in aerospace education. She works for Airborne Express at San Jose Airport and is a member of AOPA. Toni has been a member of the ROTC program at San Jose State and is considering a military career.

MARIE ANNE ZAREMBA, a week after she turned 18, she and her dad went skydiving. Since then, Zaremba joined Miami University's Parachute Dropout Club and became vice-president. During one of her free falls she got tangled up in her chute, and had to pull her reserve. Her dad was very upset and "grounded" her until she turned 21. It was then that Zaremba decided to learn how to fly. She always wanted to, but now she had an excuse. She signed up through Miami University and took lessons at the airport down the road. One day when her dad, who is also a pilot, picked her up from school, Zaremba saw a flier in the back of his Cherokee for Women With Wings. Her mom and he found it at Wadsworth Airport. After speaking with Gaye Wohlin, she joined the Women With Wings as a 66. On April 29, 1995, she passed her checkride and became a 99.

LOIS HOLLINGSWORTH ZILER, born Oct. 14, 1920, in Tulsa, OK, joined the Indiana Chapter of 99s in 1938 and in 1957 became a charter member of the El Paso Chapter. She was a WASP during WWII from January 1943 to Dec.

20, 1945. She received flight training at Tulsa, West Lafayette, IN; Hartford, CT; Military Houston in Sweetwater, TX. She has flown the Cub, Taylorcraft, etc., PT-19, BT-13, AT-6, AT-17, C-45, B-34, A-24, A-25 and Cessnas.

She was a flight instructor from 1944 to 1953. Husband Doyle was her student in 1945. They married in 1946, and he got his commercial rating in his Luscombe. She was a flight instructor also in 1976 till the present time. She and her husband, Doyle, have three children: Don, Jerry and Mary. They have five grandchildren: Doyle, Jerry, Danielle, Johnna and Ryan.

PATRICIA MICKEY ZIMMER, a native Californian began flying in 1978. She acquired her private license with instrument, commercial and CFI-A airplane ratings in short order. As an owner of a commercial interior design firm and general contractor in Orange County, CA, she had many opportunities to enjoy flying in conjunction with clients involved with development of the area.

When a client expressed an interest in helicopters, she researched the requirements for him. The addition of private, commercial and CFI rotorcraft soon followed as did the multi-engine rating and Citation-type rating.

Patricia (AKA "Nancy Narco") is married to Ed Zimmer, president and CEO of Narco Avionics where she is vice-president/director of communications/marketing. They met through mutual interest in aviation and helicopters (the helicopter was on its pad at a party at his house). They were married in their Lear Jet March 24, 1992, at 45,000 feet over Lovelock VOR. She has a Lear Jet type rating too!

MACSEEN ZIMMERMAN, born June 22, 1929, in Spalding, Greeley County, NB, began building balsa wood planes as a small child. Her first flight was in a Piper Cub for 15 minutes out of a cow pasture at a county fair at 14. After getting private license continued training to receive commercial and instrument rating. Flight training was received at Hood River, OR, from September 1965 to January 1966. She has flown the Ercoupe, Aeronca 7AC, C-150, 172, 180, 182, 206, 320, 310; and Piper Apache.

She joined the Civil Air Patrol in July 1966 and became a mission pilot for Search and Rescue and serves in that capacity today. Worked as co-pilot for a small charter company for a few years but never had the opportunity to fly professionally due to prior commitments and responsibilities. She just loves to fly, any time, any where.

Zimmerman is a charter member of the Columbia Gorge 99s.

She is married to James Zimmerman, and has one daughter, Judith Ann Pointer, 44, born Dec. 22, 1950. She has one granddaughter, Caroline Pointer, 10, born July 1, 1984.

CONNIE LYNN ZOOK, born Dec. 12, 1961, in Brattleboro, VT, lifetime resident of Chesapeake, VA. Received her first airplane ride the day she applied for a job at the Suffolk Airport, her 19th birthday. Shortly thereafter she began taking lessons receiving her license before her 21st birthday. Membership in the 99s and Aircraft Owners and Pilots Assoc. soon followed. As a 99 she has served in all four offices and chaired several committees in the Hampton Roads Chapter.

Other activities include her participation in the Capital Proficiency Race in 1984, taking second place and racing the Allegheny Air Derby in 1986, placing third.

Connie is part-owner of a Piper Cherokee 160. She serves as president and treasurer of the 08 Tango Owners Assoc. She now earns her flying money working for a periodontist. Other hobbies include: making crafts, snow skiing, clogging and country dancing.

Honored at the 1982 Forest of Friendship ceremonies as active pilots with over 50 years of flying time, respectively, are Alice Hammond, Betty Huyler Gillies, Connie Wolf, Edna Gardner Whyte, Evelyn Waldren, Melba Gorby Beard and Nancy Hopkins Tier.

NINETY-NINES ROSTER

Charter Members Top, L to R: Marjorie C. Stinson, Gladys O'Donnell, Bernice C. Blake Perry. **Middle, L to R:** Neva Paris, Evelyn Bobbi Trout, Achsa Peacock Donnels, Meta Rothholz. **Bottom, L to R:** Marion Clark Clendaniel, Josephine C. Wood Wallingford, *(oval)* Mary E. Von Mach, Mildred Stinaff.

Editor's Note: This is the most recent roster available from The Ninety-Nines International Headquarters at the time of this printing.

— A —

Aavang, Theresa K.
Abbate, Doris
Abbey, Barbara J.
Abbott, Faye V.
Abbott, Kathy Hamilton
Abbott, Mary Ann
Abel, Mary H.
Abernathy, Della Irene
Abitz, Brenda Louise
Aboytes, Charlene Willis
Abramson, Jeanne M.
Abresch, Nellie Joyce
Ackley, Deborah D.
Ackley, DeAnne Webb
Ackley, Dorothy Sue
Acton, Anneliese René
Adams, Bettye C.
Adams, Laurie Ann
Adams, Mary Sue
Adams, Meigs K.
Adams, Renee Annette
Adams, Rita V.
Adamson, Geraldine Y.
Adcock, Dorothy Mae
Adderson, Vanecia
Addison, Joan
Adeison, Irene
Adelman, Betty F.
Adelman, Gabrielle Ariadne
Ades, Bonita Joan
Adkins, Cheryl A.
Adler, Susan Straus
Adrien, Lolita Marie
Ady, Barbara C.
Affleck, Marilyn June
Agee-Housler, Jo Evelyn
Agiza, Charl
Aguilar, Katherine Yvonne
Ahrens, M. Lynn
Aivalotis, Barbara Ann
Akerlund, Frances J.
Akers, Jerre Lee
Akers, Paula Joyce
Alair, Betty Lou
Albert, Anita P.
Albiez, Michele C.
Albinger, Dorothy Ann
Albright, Dorothy F.
Albright, Esther Alicia
Albright, R. Maxcine
Albritton, Ellen P.
Albury, Katherine H.
Alderman, Marilyn M.
Aldrich, Nancy Welz
Alegre, Mary (Kitty) T.
Alexander, Elizabeth M.
Alexander, Genevieve Mahre
Alexander, Linda L.
Alexander, Mary Beth
Alford, Eleanor Carr
Alfson-Kerkof, Sheryl Ann
Allan, Jeanne
Allen, Angela Gail
Allen, Beverleymae
Allen, Jean Elenor
Allen, Judith E.
Allen, Linda Lou
Allen, Lynne Gay
Allen, Marie Barclay
Allen, Mary Clair
Allen, Shirley Kathleen
Allen, Verna B.
Alley, Donna J.
Alley, Kathryn Fuqua
Alley, Kay
Allinson, Gail Lynne
Allison, Betty Best
Allison, Brenda Carol
Allison, Josephine
Alofs, Josephine Enneking
Alper, Bernice "Chickie"
Alsbrook, Nancy A.
Alsworth, Cinimin Sea
Alter, Lavonna Scott
Alustiza, Manuela
Alvarado, Leslie Michele
Alvis, Ellie B.
Alwardt, Martha Alice
Alyson, Kathryn T.
Amaral, Julia R.
Ammons, Constance
Amundson, Janet B.
Amundson, Trudi
Anders, P. Diane
Andersen, Robin L.
Andersen-Gross, Stephanie Jean
Anderson, Aileen Roda
Anderson, Antoinette
Anderson, Carol K.
Anderson, Clare Louise
Anderson, Deborah Ann
Anderson, Dorothy J.
Anderson, Dorothy M.
Anderson, Eileen Burk
Anderson, Ellen E.
Anderson, Janet N.
Anderson, Judy Lynn
Anderson, Karen Marie
Anderson, Kate L.
Anderson, Leonora M.
Anderson, Lisa Lee Yadao
Anderson, Lori
Anderson, Marjorie R.
Anderson, Marjorie S.
Anderson, Mary Karen
Anderson, Marylin R.
Anderson, Meriem Roby
Anderson, Nancy Carol
Anderson, Ruth S.
Anderson, Vicky Joyce
Anderson-Stack, Jennifer L.
André, Babette Yvonne
Andreason, Jane Davis
Andress, Judy Anne
Andrew, Rose Sissons
Andrews, Carol Ann
Andrews, Marilyn Kay
Andrews, Marion
Andrews, Sandra
Andrews, Suzanne Lehman
Andrus, Laurie K.
Andtbacka, Helena Birgitta
Angelini, Claire Z.
Angell, Margaret A.
Angello, Debra Ann
Angeloro, Elise Ann
Angle, Gail E.
Anglin, Mary
Angott, Kelly Sue
Anthony, Leann D.
Apgar, Joy Pauline
Apperley, Lois
Aragon, Cecilia
Archibald, Jan
Archibald, Penny
Arendt, Karen Ann
Arentz, Zada F.
Ariza-Quinonero, Mar Marie
Armanino, June Alice
Armbrecht, Betty J.
Armington, Judie
Armstrong, Anne Constance
Armstrong, Lori Lee
Armstrong, Margaret Janet
Arnett, Dorothy K.
Arnold, Constance Taksel
Arnold, Dorothy L.
Arnold, Earline
Arnold, Ellen B.
Arnold, Vera F.
Arsenault, Camille Michelle
Arterberry, Martha Elizabeth
Ash, Ann C.
Ash, Nancy J.
Ashcraft, Linda Fitzpatrick
Ashe, Wilma E.
Asher-Young, Karen Sue
Ashley, Jeanette Louise
Askew, Susan E.
Atkeison, Agnes Ann
Atkinson, Patricia R.
Atkinson-Wright, Elaine M.
Atlan, Leslie
Attardo, Julie Lynne
Atteberry, Barbara B.
Attwood, Peggy A.
Auchterlonie, Lois Dobbin
Augustine, Angela Sue
Ault-Meyers, Elisabeth C.
 (Betty Jo)
Aurand, Cheryl E.
Ausman, Romaine J.
Austin, Alicia Irene
Austin, Diana
Austin, Louise
Austin, Ruth Campbell
Austin, Valorie M.
Averman, Lee
Avery, Dorothy Arline
Axell, Cynthia Ann
Axelrod, Beatrice
Axinn, Joan F.
Axton, Mildred Darline
Ayars, Albina R.
Ayers, Kathleen Marie
Ayers, Melody Diane
Azar, Suzanne S. (Suzie)

— B —

Babar, Laura Maureen
Babbitt, Carol Lynn
Bachman, Adele May
Bachman, Martha M.
Bachman, Ruth Ann
Backer, Carole A.
Backhaus, Christine Daniela
Bacon, Ann Liv
Baddour, Anne Bridge
Badger, M. Anne
Baehr, Reva M.
Baer, Deanna Dittnock
Baer, Dorothy H.
Baer, Phyllis M.
Bagg, Carolynn L.
Bahr, Sylvia A.
Bailey, Betty C.
Bailey, Dr. Karen Joan
Bailey, Eleanor Joan
Bailey, Ethel Marian
Bailey, Helen Jane
Bailey, Juanita Jo
Bailey, Kathryn Jo
Bailey, Susan P.
Bailey, Tamar Dawn
Baird, Audrey L.
Baker, Ann D.
Baker, Annette
Baker, Carolyn J.
Baker, Carolyn Riley
Baker, Cheryl Ann
Baker, Diana W.
Baker, Janet L.
Baker, Jill Anne
Baker, Joye Smith
Baker, Roberta F.
Baker, Tricia Ann
Bakti, Beverly Ann
Balazs, Marguerite
Bales, Katherine E.
Ball, Jeannie Lehir
Ball, Mary D.
Ball, Patricia Ann
Ball, Virginia B.
Ballard, Judith Chewning
Ballard, Karen Jean
Ballew, Susan L.
Balogh, Mary Lou
Baltzer, Marian
Bancroft, Barbara
Baney, Carol J.
Banks, Beverly R.
Banks, Nicole Darlene
Bann, Elizabeth A.
Bansemer, Denise L.
Barbas, Doris Oster
Barbee, Priscilla Lane
Barber, Susan Elizabeth
Barchie, Christine Ann
Barclay, Ruth
Barcus, Kate C.
Barden, Dorothy Marie
Bargerhuff, Courtney Deniese
Barker, Dorothy Louise
Barker, Judith H.

Barker, Mary Lynn
Barker, Nancy M.
Barklage, Linda Margaret
Barlia, Betty P.
Barnes, Bobbi
Barnes, Chris M.
Barnes, June Elizabeth
Barnes, Maureen Lund
Barnes, Suzanne Marie
Barnett, Beverly Ann
Barnett, Candace Denise
Barnett, Laura Ann
Barnett, Rebecca Louise
Barnick, Penny
Baroody, Leila J.
Barr, Agnes Ruth
Barr, Holly Ann
Barr, Judith Lynne
Barr, Katharine
Barr, Norma McElvain
Barras, Kim Cheryl
Barrera, Patricia Jean
Barrett, Barbara McConnell
Barrett, Charleen Sue
Barrett, Mildred
Barrett, Misti Lin
Barrett, Ruth Anne
Barris, Bernice M.
Barry, Michaela Marie
Bartels, Diane Ruth
Bartels, Laura Victoria
Barter, Sylvia B.
Bartholet, Betsy Jane
Bartik, Marjorie Ruth
Bartizal, Dolores
Bartlett, Lola M.
Bartling, Lois K.
Bartman, Kathleen Annette
Bartolet, Hazel E.
Barton, Irma Marie
Basamania, Cynthia Carol
Bascomb, Elsie Wahrer
Base, Elaine Beverly
Basham, Betty Sue
Basile-Bartelt, June
Baskin, Karen M.
Bass, Beverley Jean
Bassett, Kimberly Ann
Bassler, Letty Jane
Bastian, Rhea Reed
Batchelder, Jean Frances
Bate, Jennifer Patricia
Bates, Barbara L.
Bates, Irene S.
Bates, Mary Jo Ann
Batie, Jayne-Alice
Batson, Rhoda M.
Batt, Jacque
Batten, Marie Saunders
Batterby, Janet L.
Batto, Jeannie Lee
Batty, M. Ann
Batty, Mary Helen
Baty, (Peggy) Margaret J.
Batzel, Virginia Claire
Bauer, Catherine J.

Bauer, Kathryn Anne
Bauer, Sheila M.
Baugh, Anne T.
Baugh, Deborah F.
Baughman, Pauline M. "Polly"
Baumann, Sigrid M.
Bavis, Kathleen M.
Bawcom, Judith Marie
Baxter, Mary Pat
Bay, Constance J.
Bayers, Pauline L.
Bayes, Carol L.
Bayless, Lori Jeanene
Bazaire, Connie L.
Beail, Joyce D.
Beaird, Bettie Jean
Beasley, Angela Eller
Beattie, Janice E.
Beattie, Luz A.
Beatty, Jenny
Beauchamp, Patricia B.
Beaumont, Ann K.
Beauregard, J. Susan
Becherucci, Ruth J.
Beck-Pfeiffer, Sharon C.
Becker, Mary Jane "Terrie"
Becker, Penelope Ellis
Becker, Robin
Beckstrand, Lyle B.
Bedinger, Jeanne H.
Bedrossian, Adelle M.
Beeler, Judith C.
Beeler, Maycay
Beeman, Peggy
Beering, Donna Janine
Beers, Clydia L.
Beery, Merlene Ranee
Beeson, Glenda Sue
Beeuwkes, Nancy Jean
Begg, Susan Margaret
Behrend, May Pietz
Behringer, Mayetta
Belcher, Aleta R.
Belcher, Sherry Evans
Beliveau, Nadine Anne
Bell, Jeanette Marie
Bell, Jo A.
Bell, Nancy Carol
Bellafiore, Mary Ann
Bellino, Clarice Mae
Belliveau, Patricia Ann
Bencene, Gretchen B.
Benech, Sheila Esther
Benedict, Ruth Taksel
Benenati, Irene Joanne
Benham, Martha (Marty) A.
Beningfield, Theda M.
Benjamin, Judith A.
Benjamin, Sheila P.
Benker, Dorothy Swenson
Bennett, Caroline Leslie
Bennett, Elizabeth Cumming
Bennett, Jennifer Danette
Bennett, Susan Kirstein
Bennett, Vivian C.
Bennett-Williams, Shadlea

Benoit, Catherine Masterson
Benson, Karen Ann
Bentley, Christina Monthan
Benton-Walsh, Rita Kelly
Bera, Frances S.
Berdar, Jan Marie
Berentson, Judy Lyn
Berg, Kathleen Ann
Bergemann, Clarice I.
Berger, Susan Helene
Bergeron, France Jenny
Bergishagen, Jane Fisher
Bergman, Anna "Boo"
Bergman, Judith Sue
Berkley, Ester Gustafson
Berner, Esther
Berner, Mary G.
Bernet, Mary Jo
Berrong, Lynn Lambert
Berry, Betty Lou
Berry, Janice Eileen
Berry, Margaret A.
Berry, Phyllis A.
Bertagna, Hazel L.
Bertea, Hyla Holmes
Berthelet, Dorothy Arlene
Bertino, Amy Marie
Bertles, Joan B.
Bertrand, Jeanine M.
Bertucci, Mary Ann
Berwyn, Cyndhi K. Hughes
Beshears, Nancy Ellen
Bessler, Jennifer Ann
Betchce, Shirley Ann
Bethel, Norma A.
Betz, V. Lavelle
Bew, Deborah Ann
Bhajekar, Sunila Ashok
Biagetti, Marilyn Lois
Bianchi, Patricia Ann
Bickers, Margaret Aline
Biedron, Margaret Olive
Biele, Frances R.
Biggin, Sally W.
Biggs, Barbara A.
Bigham, Loretta J.
Bilbrey, Sherry Rennier
Billingham, Claire Elizabeth
Billington, Selena
Bills, Sandi
Biloff, Sharon R.
Bingley, Diane Norine
Binsfield, Adele
Birchmore, Doreen "Dee"
Bird, Dominique Deckers
Birdsong, Dorothy Lanelle
Birmingham, Margarete Cora
Biron, Helen B.
Bisanti, Bertha Margaret
Bishop, Cheryl A.
Bishop, Edna H.
Bishop, Elizabeth Pauline
Bishop, Janice Elaine
Bishop, Thea Nancy
Bissell, Marcie B.
Bissonnette, Alana Lynne

Bithel, Lillian
Bittermann, Sivika
Bixel, Beverly D.
Bjerre, Lise
Bjorklund, Maxine
Bjornson, Carolyn Helen
Bjornson, Rosella M.
Black, Barbara J.
Black, Joy Anna
Black, Linda Bess
Black, M. Anne
Black, Margaret M.
Black, Viki Kay
Blackburn, June F.
Blackman, Pamela S.
Blackwelder, Patricia P.
Blackwood, Joy Parker
Blair, Lara Gail
Blair, Marianne T.
Blais, Mary Lee
Blake, K. Jean
Blake, Stephanie Rene
Blakely, Mabel G.
Blakey, Deborah Renee
Blanc, Betty Gay
Blanchard, Sharon Kay
Bland, Jo Ann Hammond
Blank, Gloria Jean
Blanton, Phyllis Jean
Blasi, Patricia Barrett
Blazier, Ann
Blazyk, Nancy Suzanne
Blech, Lorrie
Bledsoe, Anna M.
Bledsoe, Lucile Mary
Blidberg, Madeline Ann
Blietz, Beverly Bolles
Blinco, Pegge Marguerite
Bliss, Betty J.
Blissert, Barbara F.
Bloch, Carol Alexandria
Block, Paula Kruse
Block, Rosemary A.
Blodgett, Linda S.
Blodinger, Sharon G.
Blondeau, Neita Loy
Bloomingdale, Gwen
Bluestein, Leah Anne
Bluhm, Thelma Nadine
Blum, Luellen
Blum, Priscilla H.
Board, Betty Blackley
Boatman, Julie Kristen
Bock, Linda K.
Bock, Madeleine Dupont
Bockhoff-Gattmann, Patricia Ann
Boehler, Lorraine J.
Bogden-Stubbs, Linda
Bohrer, Fran
Bolander, Dorothy Estelle
Bolen, Denise Ann
Bolinder, Trudi
Boling, Shirley Sweet
Bolkema, Judith Anne
Bollinger, Janice M.
Bollinger, Margaret (Peggy) Mary

Bolton, Joanne Gail
Bolton, Margaret Mason
Bolton, Marilyn Arlene
Bond, Miriam J.
Bones, Virginia Gail
Bonesteel, June
Bonnard, Debora Jean
Bonnard, Dolores Liegey
Bonomo, Rose
Bonzon, Rachel
Booker, Linda L.
Booker, Lisa Marie
Boom La, Chantal C.
Boot, Barbara
Booth, Lisa
Boothby, Sarah Lynn
Borbas, Linda Katalin
Borek, Margaret Mary
Borella, Ann Louise
Bormes, Barbara Nafis
Bornwasser, Louise Weisser
Borst, Evelyn Louise
Bosca, Caro Bayley
Bossio, Natalie
Bostic, Ruby L.
Bottum, Olivia Graye
Boucher, Roberta Ann
Boudreau, Rachelle
Boudreault-Carrera, Ginette
Bourdon, Diane Reina
Bouvier, Patricia Anne
Bovee-Chesnut, Mary
Bowden, Joy Ruth
Bowden, Sandra Hughes
Bowditch, Gladys Marie
Bowen, Sandra Waegner
Bower, Gray Gordon
Bowerman, Barbara
Bowers, Deanna M.
Bowersox, E. Lou
Bowles, Alice Marie
Bowser, Nicole Kimberly
Box, Ed.D., Ph.D., Gertrude Elaine
Boxmeyer, Carolyn Joyce
Boyd, Dr. Jacqueline B.
Boyd, Joan Marie
Boyd, Minnie M.
Boyd, Rosemary
Boyes, Madaline Marcia "Lindy"
Boyes, Wendy J.
Boylls, Virginia Wright
Boysen, Dorothy
Bozarth, Lois Marie
Brack, Annabelle SanSan
Brackett, Andrea Inman
Brackett, Jacqueline G.
Brackley, Carol Ann
Bradberry, Geraldine R.
Braddock, Joan Forshaug
Bradford, Maryann
Brady, Diane
Braese, Evelyn B.
Brafford, Marilyn Jean
Brainerd, Annemarie Leslie Sissi
Brake, Cynthia Celia
Braman, Helga Barbara

Branch, Doreen Gail
Brand, Lisa
Brand, Lori Ann
Brandell, Therese N.
Brandli, Madeleine D.
Brandon, Nancy Lafarge
Brandson, Elizabeth Bailey
Brantingham, Shirley A.
Brantley, Melba Lewis
Brantly, Nancy C.
Brasfield, Marion
Brassil, Maura A.
Brawley, Kristin Kendall
Brecher, Virginia Howard
Bredin, Lynn N.
Breed, Virginia
Breeden, Kristal Lynne
Breedon, Shelley Marie
Bregman, Harriet Anne
Breien, Constance J.
Breiner, Joyce Karen
Brell, Doris R.
Brennan, Carol L.
Brennan, Margaret B.
Brenneman, Holly Florence
Bresett, Elizabeth K.
Bresley, Heloise C.
Brewer, Barbara C.
Brewer, Sue Marguerite
Bridges, Angela Jane
Bridges, Mikelyn Nanette
Briggs, Doris J.
Briggs, Thelma Jean
Bright, Linda B.
Bright, Nell S.
Brin, Harriet R.
Brine, Marcia E.
Brinkley-Murphy, Michele
Brinkman, Annelie-Susanne
Brinnand, Sheila Patricia
Brito, Margarita C.
Britt, Virginia Janssen
Brittan, Carol Ann
Brittingham, Verda Eloise
Brockett, Joan L.
Brocksmith, Gisel L.
Brolan-Thomas, Margaret
Bromberek, Mary Denise
Bronson, Judy Ann
Bronson, Laurel Nanette
Brooks, Glee Ruth
Brooks, Laura Marie
Brooks, Margette E
Brooks, Margot Ann
Brooks, Theresa J.
Broomfield, Robin Elizabeth
Brossard, Betty Louise
Brotherton, Barbara S.
Brower, Elayne M.
Brown, Amy Lynne
Brown, Andrea Guerra
Brown, Betty J.
Brown, Carol Ann
Brown, Carol J.
Brown, Carolyn Olivia
Brown, Cheryl Marie

Brown, Claire Louise
Brown, Cynthia
Brown, Ena E.
Brown, F. Christine
Brown, Gail S.
Brown, Heather Ann
Brown, Helene G.
Brown, Janice Lee
Brown, Janis M.
Brown, Jewell Bailey
Brown, Krys M.
Brown, Mallory E.
Brown, Margaret L.
Brown, Marion S.
Brown, Marjory Fay
Brown, Marlene Sharon
Brown, Mary Louise
Brown, Nellie Irene
Brown, Sandra A.
Brown, Sandra Lea
Brown, Sarah Duke
Brown, Shirley Watts
Brown, Terry H.
Brown, Toni C.
Brown, Virginia Ruth
Brown-Bell, Elaine Lockhart
Brown-Dawson, Nora Lee
Browne, Kathleen Fagala
Browne, Lucy Williams
Browne, Sandra Raye
Brownlee, Gale
Broyles, Edna Bates
Brubaker, Susan Kay
Bruce, Paula Ruth
Bruce, Victoria Jean
Brugh, Jill Christine
Bruland, Leona Gail
Brumfield, Audre Hopson
Bruner, Cledith L.
Brunetto, Natalie
Brunton, Marilyn Kay
Brush, Barbara W.
Brusnahan, Kathleen
Brusseau, Barbara J.
Bryant, Margret Nelle
Bryant, Marjorie E.
Bryant, Mary Snell
Bubb, Joan M.
Buchanan, Gladys M.
Buchyns, Marcia J.
Buckley, C. Dawn
Buckley, Catherine Carlsen
Buckley, Ruth P.
Buckman, Kathryn L.
Buckwalter, Mary A.
Budell, Stacey Marie
Budhabhatti, Chanda Sawant
Buegeleisen, Sally P.
Buehl, Ed.D., Elizabeth Anne
Buehler, Barbara Ann
Buell, Linn M.
Buelna, Elizabeth K.
Bufka, Sandra K.
Buller, Marcia S.
Bullington, Marion Sandra
Bullock, Sandra L.

Bundgard, Olive Grainger
Bundy, Shirley Elaine
Bunn, Catherine Lynne
Bunnell, Eris Marie Bronander
Burch, Marilyn Taylor
Burch, Thelma Juanita
Burcham, Miriam Parsons
Burdick, Ginger Marie
Burford, Michelle W.
Burger, Renee Lin
Burgess, Barbara C.
Burgoyne, Bonnie E.
Burjak, Catherine Elizabeth
Burk, Ramona England
Burke, Mary Helen
Burkett, Candace
Burkett, Rose Marie
Burkholder, Heather Lynn
Burkig, Laura L.
Burklund, Jeanette Marie
Burnett, Mary Parham
Burns, Mary Lee
Burns, Robyn J.
Buroker, Gladys
Burrell, Patricia B.
Burriss, Madeline W.
Burrow, Catherine
Burrow, Shauna Mckamey
Burrows, Donna Maria
Burton, Diana Lynn
Burwell, Jacqueline D.
Burysz, Marilyn J.
Busby, Yvette Linda
Buschhorn, Sandra Everett
Bushko, Nancy Bailey
Buss, Rhonda K.
Bustos, Jean Roberta
Butcher, Beverly Jo Ann
Butler, Linda S.
Butler, Lyn
Butler, Peggy Carol
Buzzell, Elizabeth Anne
Byerly, Mary Louisa
Byers, Diane
Byers, Juliet Rosemarie
Byers-Jones, Charmian
Byous, Martha E.
Byrd, Iris Olive
Byrne, D. Lynn
Byrne, Maggie
Byrum, Beverly Ann
Bytwerk, Helen P. (Betty)

– C –

Cabassa, Nancy Jo
Cabot, Michelee Morgan
Cadmus, Judith Ann
Cadorette, Nancy C
Cagle, Myrtle K.
Cain, Linda R.
Cain, Patricia L. (Pat)
Caine, Jennifer Mary
Calatayud, Amparo Theresa
Caldwell, Janet L.
Caliri, Gloria J.
Callar, Donna Howe

Callaway, Margaret W.
Calvert, Colleen D.
Calwell, Bonnie June
Camden, Lillian Georgeana
Cameron, Carol Jean
Cameron, Elizabeth Starker
Campbell, Betty Jean
Campbell, Coleen Anne
Campbell, Dorothy B.
Campbell, Dorothy T.
Campbell, Elizabeth Ann
Campbell, Grace Norma
Campbell, Joan A.
Campbell, Joanne
Campbell, Joanne Angela
Campbell, Karin N.S.
Campbell, Margaret A.
Campbell, Marlene Ann
Campbell, Mary M.
Campbell, Patricia Ann
Campbell, Roberta S.
Campbell, Sue C.
Campbell, Susan Mary Irene
Campton, Debra Joan
Cangiano, Loreli B.
Cannell, Janice A.
Cannon, Janet Mansfield
Cansdale, Carol A.
Canulette, Linda Jane
Caolo, Erin M.
Capizzi, Jane G
Caples, Roxana S.
Capps, Christina Louise
Caputo, Teresa A.
Carastro, Marie F.
Carico, Altha Miller "Polly"
Carl, Ann B.
Carlin, Virginia T
Carlson, Diana L.
Carlton, Laverne Billingsley
Carmichael, Ava S.
Carmien, Amy M.
Carmine, Marguerite
Carnahan, Joyce Elaine
Carney, Melanie M.
Carney, Mildred Castle
Caron, Roxanne M.
Carpenter, Joann
Carpenter, Martha McKenzie
Carpenter, Reva Horne
Carpenter, Rikako Sugiura
Carpp, Carolyn Claire
Carr, Bonnie Lundquist
Carrano, Carmen Janel
Carroll, Aurelia Lou
Carroll, Gina Marie
Carroll, Joan Frances
Carroll, Karla
Carroll, Kathleen Brown
Carroll, Linda J.
Carroll, Mary Joanne
Carroll, Wanda Wilma
Carruthers, Sandra Kaye
Carson, Cindy
Carson, Linda Kelley
Carter, Anne Christine

Carter, Betty Kirbo
Carter, Beverley Ann
Carter, Brenda Joyce
Carter, Janet Gunn
Carter, Joyce Kim
Carter, Kimberly Lynn
Carter, Lucretia Mae
Carter, Marian Carol
Carter, Nancy Ann
Carter, Susan M.
Carts, Judith A.
Caruso, Helene P.
Carver, Julia
Case, Elizabeth Owen
Case, Kay Loree
Case, Lois H.
Case, Martha Anne
Case, Stephanie Dawn
Casey, Carol A.
Casillas, Cheryl Lorraine
Casper, Audrey Jane
Cassidy, Mildred L. "Mid"
Casteel, Angela Paprskar
Caston, Ashley Elizabeth
Castro, Carylon Marie
Caudle, Judith Ann
Causley, Susan Candler
Cavaliere, Lorraine Agnes
Cawley, Joy Ann
Centrella-Hooper, Michelle Jenae
Ceparano, Janet S.
Cerise, Barbara
Cessna, Karen Marie
Chabot-Fence, Dene
Chadwick, Patricia Louise
Chaffey, Kay Gott
Chagnot, Catherine Jane
Chalmers, Lois M.
Chambers, Amanda Faith
Chambers, M. Veronica
Chambers, Susan Coon
Chambers, Susan Hitt
Champlain, Eleanor Louise
Champlin, Marian D.
Chan, Patricia
Chandler, Ann L.
Chandler, Leanne Marie
Chandler-Law, Linda D.
Chapelson, Sherry L.
Chapman, Barbara Ann
Chappell, Frances B.
Chapple, Jody Hill
Chase, Dorothy Jean
Chase, Elaine F.
Chase, Roberta Sorensen
Chay, Andrea B.
Chelgren, Gwendolin Jo
Cheney, Donna Louise
Cherry, Charlotte F.
Chester, Patricia V.
Cheung, Katherine Sui Fun
Chichester, Catherine Joan
Childers, Linda
Childress, Mickey R.
Childs, Elsie F.
Choy, Nelwyn C.

Christensen, Beverly Jane
Christensen, Dorene M.
Christensen, Marie H.
Christian, Carol Ann
Christian, Elizabeth Dorrance
Christiansen, Marilyn Ann
Christie, L. Dawn
Christmas, Gayle Elaine
Christofferson, Mary Elizabeth
Christy, Martha M.
Church, Carol E.
Church, Elizabeth J.
Church, Nance A.
Church, Patricia B.
Churchill, Janet I.
Churchman, Priscilla Ann
Cichocki, Cheryl Lynn
Ciciora-Ewart, Karen Marie
Clapper, Connie E.
Clarey, Alison Ann
Clark, Elinor Ann
Clark, Elizabeth A.
Clark, Gail Hoce
Clark, Josephine M.
Clark, Julie E.
Clark, Katherine Hamby
Clark, Kathryn Irene
Clark, Leslee DeAnn
Clark, Maureen
Clark, Patience L.
Clark, Sandra Sprague
Clark, Valerie Ann
Clarke, C. Tessa
Clarke, Carolyn Elizabeth
Clarke, Peggy Ann
Clawson, Michelle Leigh
Claxton, Clara T.
Clay, Suzanne McRae
Clayberg, Linda Marcus R.
Claypoole, Susan Louise
Clayton, Ruth McKinney
Clayton-Jones, Mary Catherine
Cleary, Jill
Clelland, Julee Ann
Clemens, Frankie Bea
Clement, Kimberly R.
Clements, Winoma Alice
Cletsoway, Emily C.
Cleveland, Margarette W.
Cleveland, Marian L. Alexander
Cleveland, Nancy J.
Clever, Barbara Lucille
Clifford, Eileen Jean
Clifford, Royce Ellen
Clift, Chrystal Caren
Cline, Rita E.
Clinton, Margaret Louise
Clinton, Nancy Jean
Clothier, Cathy Colebrook
Cloud, Connie Lee
Cloud, Patricia P.
Coates, Annie Lena
Coates, Yvonne C.
Cobb, Jerrie
Cochrane, Barbara Jean
Cochrane, Shirley

Cockerill, Cora L.
Coco, Susan Deanne
Cody, Dianna Dickie
Coe, Bertha Hall
Coe, Diane Wilder
Coffeen, Virginia M.
Coffman, Jean T.
Coggins, Minnie Wade
Cohrs, Jennafer Louise
Coker, Sally Walker
Colbert, Virginia L.
Colburn, Carol
Cole, Diane Phyllis
Cole, Elizabeth Joanne
Cole, Kelly E. Margaret
Cole, Mary Ellen
Coleman, Barbara A.
Coleman, Carolyn J.
Coleman, Polly P.
Coleman, Stephanie A.
Coleman, Tweet Theresa
Coles, Dana Lynn
Collier, Jess Ann
Collier, Lorraine Albin
Collier, Patricia Ellen
Collinge, Julia Patricia
Collings, Cynthia Sue
Collings, Pamela Ann
Collins, Barbara Jean
Collins, Beverly Bell
Collins, Eileen Marie
Collins, Elizabeth Ann
Collins, Jean Elizabeth
Collins, Loretta Kay
Collins, Patricia
Collis, Mary Ross
Collum, Lois Diane
Colon, Linda Gail
Colony, Bonita G.
Colvin, Janice Knestout
Comat, Hella Marion
Combs, Georgia L.
Common, Elizabeth J
Compardo, Eloise E.
Compton, Patricia Anne
Condon, Christine Ann
Condon, Doreen Joyce
Conkle, Patricia G.
Conklin, Lynn Ann
Conley, Mary A.
Conlin, Anne S.
Conlin, Margaret M.
Connal, Jennifer Robin
Connell, Linda Janette
Connell, Lucille Hoffer
Connell, Mary J. "Joey"
Conner, Amy Correll
Conners, Gayle Denise
Connon, Kathleen A.
Connor, Catherine Erwin
Connor, Romy Merrit
Connors, Sheri Kaye
Conrad, Christine Kay
Conway, Patsy Ann
Cook, Audrey Melenyk
Cook, Constance

Cook, Ilse S.
Cook, Jeanne D.
Cook, Kelli Colleen
Cook, Loretta Ann
Cook, Lynelle E.
Cook, Margaret A.
Cook, Marian F.
Cook, Marlene E.
Cook, Priscilla O.
Cooke, Mary S.
Coon, Lorene Elsie
Coonce, Kimberly Joy
Cooney, Cheryl Jean
Cooper, Ann Lewis
Cooper, Carin L.
Cooper, Donna S.
Cooper, Joan Katherine
Cooper, Nadine Marcelle
Cooper, Sherry D.
Cope, Cheryl L.
Copeland, Constance Lee
Copeland, Marilyn F.
Copeland, Marsha E.
Copeland, Velma Lee Barnett
Copley, Gail
Corcoran, Suzanne Mary
Corless, Dorothy A.
Cornelius, Melinda G.
Cornell, Sally Ann
Cornell, Susan Denise
Cornette, Joy M.
Cornish, Christine Alyce
Cornwall, Penelope Jane
Correia, Kasey June
Corry, Hazel Marcella
Cosby, Margaret M.Z.
Cosner, Carol R.
Costa, Betty Lee
Cotham, Lisa Anne
Cotton, Adele
Cottrell, Nina Louise
Coughlin, Kim B.
Coulson, Kristin Ann
Councilman, Kathryn G.
Countiss, Carol A.
Courtnay, Debra Ann
Courtney, Cathy Ellen
Coussens, Ellen J.
Covington, Candace Jean
Cowan, Patti Blue
Cowden, Violet C.
Cowlard, Lisa Bernice
Cox, Barbara C.
Cox, Deborah Louise
Cox, Georgia Jean
Cox, Judy B.
Cox, Priscilla Rose
Cox, Victoria Hunt
Coy, Donna Claire
Coyle, Jean M.
Coyne, Lisa Michelle
Cozart, Anne
Cozzi, Diane Marie
Crader, Margaret Virginia
Craft, Joan
Cragin, Marilyn O.

Craig, Marilyn Kaye
Craig, Renee S.
Craighead, Joy
Craik, Evelyn
Crane, Mardo C.
Crane, Noreen Ann
Crane, Pamela H.
Crane, Sylvia Marlene
Crane-Bailey, Donna Gale
Cranford, Geneva Mae
Cranz, Helen
Crary, Phyllis Fleet
Crawford, B. Nicole
Crawford, Mary A.
Crawford, Sharon R.
Creamer, Christina
Creason, Mary Rawlinson
Creger, Charlyne
Crespi, Angelina B.
Crevani, Carole
Crist, Grace B.
Cristiani, Mary J.
Critchell, Iris C.
Critz, Sue Anne
Croak, Pamela Jacqueline
Crocker, Patricia Suzanne
Croff, Janice Lee
Croft, Kathy Anne
Crommelin, Jean Sadie
Cronan, Selma
Cronin-Gaffney, Mary Corky
Cronk, Lena Martha
Crooker, Barbara Louise
Crooks-Visgar, Dorothy Ann
Crosby, Janet Sue
Cross, Joye E.
Cross, Judith Hutter
Crotty, Sharon Ann
Crow, Peggy A.
Crowl, Marjorie
Crown, Lori D.
Croy, Barbara J.
Crum, Ellen Elizabeth
Crum, Ruth May
Cruse, Ann L.
Cruz, Stephanie Penkivich
Cuddie, Patricia Marie
Cue, Linda A.
Cull, Thelma J.
Cullere, Barbara Ann
Culp, Doris F.
Culp, Mara K.
Culver, Elsie E.
Cumming, Nancy Bottler
Cummings, Dodie
Cummings, Nancie J.
Cummock, Marguerite H.
Cunneen, Judith Ann
Cunningham, Barbara
Cunningham, Deborah Rose
Cunningham, Florence Jean
Cunningham, Tanya Marie
Curley, Juanita Dale
Currie, Jan Michele
Currin, Irene Mildred
Curry, Eleanor Neemann

Curtis, Diana Marie
Curtis, Ruth Ann
Curtis, Virginia L.
Cutrona, Alice A.
Cutter, Ginger
Cutting, Marian M.
Czech, Jeanine Marie

– D –
D"Alessandro, Sally Jo Ann
D"Alterio, Marie Murphy
D"Armand, Ann Savage
Dade, Diana
Daffin, Elizabeth Jane
Dahl, Lisa Buswell
Daisy, Jessica Stetson
Dale, Alison G.
Dalrymple, Tracy A.J.
Dalton, Fran W.
Dalton, Gloria B.
Daly, Anna Amelia
Dalzell, Cecilia Marie
Damer, Stacey K.
Damschroder, Lulu J.
Danford-Klein, Amy
Daniel, Violet
Daniels, Joanne S.
Daniels, Kari L.
Dant, Margaret Ellen
Dare, Shirley
Darling, Marilyn Stagner
Darling, Maureen R.
Darling, Valdene
Darlington, Diane Marie
Darnell, Lauren Mae
Darr, Barbara
Davenport, Tracy Dawn
Davidson, Cecile Elizabeth
Davidson, Kathryn Joan
Davidson, Margaret Sage
Davidson, Ursula Malluvius
Davis, Anne M.
Davis, Barbara June
Davis, Bonita J.
Davis, Donna M.
Davis, Dorothy F.
Davis, Dourelle Jay
Davis, Edna A.
Davis, Frances Perdue
Davis, Francesca S.
Davis, Glee Ann
Davis, Janet (Jan) L.
Davis, Janet M.
Davis, Joyce Luanne
Davis, Luana S.
Davis, Marjorie Faye
Davis, Miriam S.
Davis, Nancy
Davis, Sherry L.
Davis, Shirley Rogers
Davis, Susan Margaret
Davison, Helen Smith
Dawson, Emily Susan
Day, Errol Stephanie
Day, Jeanne
Day, Lenora Ellen

Day, Mary E.
De Canter, Mary Elizabeth
De Lisle, Jane Mccaffery
De Rosa, Addie
De Simone, Mary Louise
De Vries, Amy Janel
De-Blicquy, Lorna Vivian
De Bernardi, Fiorenza
de Castro, Pamela C.H.
deVeze, Nancy J.
Deaken, Donna Marie
Deal, Sarah Marie
Dealey, Elizabeth Young
Dearborn, Valerie Jean
Dearth, Julianne
Decker, Anette
Decker, Jane Frances
Dedera, Carla C.
Dederer, Susan I.
Deeds, Barbara Lucile
Deerman, Ruth
Deeter, Wendy Wenk
Degarmo-Schell, Doris T.
Degler, Anne S.
Deimler, Julia
Del Giorno, Velma B.
Delatorre, Konnie C.
Dell Olmo, Linda Susan
Dellacioppa, Caren Neila
Delles, Martha Puryear
Demcak, Deborah Dawn
Demetre, Georgia
Demko, Beverly Ann
Demmer, Maryanne
Demore, Louise H.
Dempsey, Valerie J.
Demuth, Joan S.
Denham, Melinda Murrel
Denler, Agnes J.
Dennis, Carole Jean
Denny, Edith L.
Denson, L. Dianne
Dent, Aggie Mary
Denton, Shirley Ray
Depra, Jane V.
Dermott, Jill Patricia
Derrick, Lisa Deniz
Deshmukh, Saudamini Madhav
Desio, Phyllis Anne
Desjardins-Smith, Joanne P.
Desper, Beatrice S.
Detmar-Pines, Gina Louise
Detombe, Suze
Dettmer, Barbara Jean
Devereux, Hilda Ethel
Dewey, Gwen E.
Dewitt, Joan Karsten
Dewulf, Suzanne Maloney
Dexter, R Marchine
Dey, Sharyn Elizabeth
DeBaun, Betty Eilene
DeCamp, Kristin J.
DeCourley, Claudette
DeFronzo, Helma
DeGraaff, Teresa M.
DeLay, Sharon Aileen

DeLong, Barbara Reibel
DeLuca, Janice
DeLuca, Rosanne
DeMars, Karen L.
DePue, Carole B.
DeVries, Julia Leanne
DeYoung, Esther May
Di Giovanni, Maris Lee
Di Misa, Christine Marie
Dick, Helen R.
Dick, Sheila Ann
Dickerhoof, Dorothy Kathleen
Dickerson, Nettie Durant
Dickerson, Ruby Smith
Dickeson, Tammy Diane
Dickey, Frances Caroline
Dickinson, Christine Margaret
Dickson, Marilyn I.
Dickson, Marilyn L.
Dierolf, Carolgene
Dieser, Anna Jo
Dieterich, Sharon Gay
Dietz, Heidi Martina
Dignum, E. Kim
Dilda, Mary Elizabeth
Dilley, Joan E.
Dillman, Lois
Dimentstein, Erella
Dinan, Elizabeth Crowley
Dingley, Lucille M.
Dirian, Kathryn Marie
Distaso, Pamela Anne
Distefano, Maria A.
Ditsch, Patricia Ruth
Dittman, Marion B.
Dittmer, Linda G.
Ditton, Delores E.
Diversey, Lori Louise
Dixon, Gwen C.
Dixon, Shirley Margaret
Djurklou, Alessandra Katarina
Dobbins, JoAnn H.
Dobbins, Vi
Dobrescu, Ruth S.
Dobrzeniecki, Michaline
Dodd, Debi
Dodson, Margaret Alice
Does, Joanne Maria
Dolan, Debra Tomlinson
Doll, Sally Maureen
Dombroskie, Ann Gibson
Don-Wheeler, Grace
Donaghy, Delinda
Donahue, Nelda McDowell
Donahue, Susan McDonnell
Donahue-Lynch, Margaret E.
Donn, Tara Anne
Donnelly, Mia Joanne
Donnels, Achsa B. Peacock
Donner, Teresa Marie
Donoghue, A. Maureen
Dooley, Florence C.
Doolittle, Barbara Ann
Dorsey, Katherine Gayle
Dostoler, Denise A.
Dougherty, Alberta Marie

Dougherty, Melody Ann
Douthitt, Iona Faith
Dover, Christie Celestia
Dowd, Gladys K.
Dowdy, Linda Margaret
Downey, Deborah Anne
Downing, Billie Marie
Downs, Daphene
Doyle, Margaret G.
Doyle, Mathea A.
Doyle, Victoria Suzann
Dragoo, Edna M.
Dragten, Marita Carol
Drake, Holly Anne
Drane, Jana Sharp
Draper, Linda Mae
Dratler, Cheryl Lynn
Drayster, Sheila Anne
Drebing, Mariana D.
Drescher, Margaret Judith
Drillette, Elysa Jean
Drilling, Betty Ann
Drizos, Margaret Irene
Drum, Allison Elizabeth
Drummond, Alma S.
Druskins, Linda Langrill
Dubbe, Connie Jean
Ducey, Mary Elizabeth
Dudek, Jeanette A.
Dudley, Jutta Siefert
Duerk, Valissa Lee
Duffey, Leslie Jo
Duffy, Bertie K.
Duffy, Marilyn Clare
Duffy, Sandra Lee
Dugan, Sonya Louise
Dugger, Carolyn D.
Dugger, June J.
Duguay, Toni D.
Duke, Jean Fraser
Duker, Jan D.
Dumont-Tribolet, Dana Louise
Dunbar, Martha C.
Duncan, Athalie
Duncan, Phyllis Anne
Dunfield, Frances J.
Dunlap, Jeannie Powell
Dunlap, Marion Elizabeth
Dunlop, Helen L.
Dunn, Betty Jean
Dunn, Lynne E.
Dunn, Myra A.
Dunn, Rosemary
Dunn, Susan S.
Dunn-Mangnall, Taya Ann
Dunnam, Mary Helen
Dunphy, Daniele Suzanne
Dunseth, Elizabeth Ann
Duperow, Winifred S.
Durand, Loubelle Marie
Durbin, Frances Margaret
Durden, Deborah Shadoan
Durham, Helen E.
Durkin, Carol Lesley
Durnal, Bonnie Ann
Dutcher, Billie F.

Dwight, Catherine Lynn
Dwinnell-Moore, Catherine Ann
Dwyer, Laura J.
Dykema, Dr. Deborah L.
Dykema, Muriel Muzzy
Dyvad, Deloris E.
Dziadulewicz, Melanie

– E –

Eargle, Dorothy Eleanor
Earley, Laura J.
Easom, Ruby Lee
Easterly, Ann Sandra
Eastman, Patricia Ann
Easton, Catherine
Eaton, Elizabeth Ashlee
Eaton, Lois Marie
Eaves, Rita C.
Eberhard, Christine Lucille
Ebert, Lynn Marie
Ebert, Mary Elizabeth
Ebert, Nancy Kay
Ebey, Ruth
Eby, Patricia Smith
Eby-Buck, Gloria Joyce
Echemann, Barbara B.
Eckrote, Debra Jean
Eddleman, Jocile Baggett
Edison, Betty Jean
Edmonson, Elinor Ruth
Edwards, Charlotte Jean
Edwards, June K. Reynolds
Edwards, Marion Lawry
Edwards, Terry Lynn
Edwards, Virginia Elaine
Egan, Aileen
Egan, Joanne Laura
Egan, Shari L.
Egbert, Elaine
Egge, Yvonne May
Eggert, Kimberly Louise
Eggleston, Judy L.
Egglestone, Denise
Ehle, Margaret MacDougall
Ehrich, Sharon Slechta
Ehrlander, Sue Ann
Eichenberger, Candace R.
Eide, Wynola Thornton
Eiff, Mary Ann
Eisemann, Mary Ellen
Ekin, Theresa Lee
Eklund, Jill B.
El Hajj, Nancy E.
El-Din, Donna Joy
El-Nadi, Lotfia Mahmoud
Elam, Lorraine
Eld, Margot J.
Eldredge, Barbara D.
Eldredge, DeAnna Grace
Eldridge, Linda Barker
Elistve, Mary Anne
Elkerton, Eveline C.
Elligott, Margaret Cecile
Ellingson, Jean T.
Elliott, Janet Lynn
Ellis, Mary E.

Ellis, Merle E.
Ellsworth, Amy Akemi
Elsbach, Ann Ensley
Emberg, Patricia Ann
Emerson, Lillian Ann
Emhoff, Rosemary Ann
Emmens, Carol A.
Emmons, Christine B.
Emry, Rose
Enders, Deirdre Jeannine
Engard, Irene P.
Engelmann, Marjorie L.
Engelmann, Ursula
England, Diane N.
Engle, Mary Edith
Engleman, Ann G.
English, Anne Louise
English, Kara
English, Mary Harden
Engstrom, Elizabeth Jean
Enniss, Marty L.
Eno, Minerva Ann
Enos, Lucille Camp
Erbach, Laura Lynn
Erbaugh, Lynne M.
Erickson, Alicia Ross
Erickson, Dorothy W.
Erickson, Elizabeth Murphy
Erickson, Lois Ann
Erickson, Patricia Ann
Eriksen, Anne C.
Erkes, Mary Grace
Ernst, Kim D.
Ernst, Patsy L.
Erotas, Lisa Denise
Erwin, Jane Mary
Esh, Lucinda Ann
Espino, S.J. (Sam)
Esposito, Anne Johnson
Essaye, Gloria Gustafson
Estep, Dorothy P.
Esterbrook, Jennie Jane
Estes, J. Alika
Estes, P.C.
Estrada, Darla E.
Eubanks, Lottie J.
Evans, Barbara J.
Evans, Blanche W.
Evans, Elaine Kay
Evans, June Marie
Evans, Linda Lee
Evans, Marijane
Everett, Nancy Louise
Everett, Shirley S.
Everling, Jeanette
Ewald, Nelda
Ewert, Margaret J. (Peggy)
Ewing, Barbara K.
Eychner, Joan Marie
Eytinge, Isabelle I.
Ezell, Nancy J.

– F –

Faber, Maria Johanna
Fabiszak, Delores E.
Fadner, Barbara D.

Fagan, Vivian H.
Fairbanks, Nancy Pat
Falco, Mary Louise
Falk, Donna L.
Falkenberg, Charlene H.
Falls, Gwenda Nell
Falstrom, Mary Treve
Falwell, Beth Parnell
Fanfera, Virginia G.
Farey, Gale Jeannette
Faries, Lindi Locke
Farmer, Constance Anne
Farnham, Evelyn V.
Farnsworth, Carol G.
Farnsworth, Karen Lynn
Farrell, Constance Gwynn
Farrell, Heidi Belle
Farrell, Joyce Marie
Farris, Anita
Farris, Laurie Mclean
Farris, Mary L.
Fasano-Ramos, Marie A.
Fast, Alison Faith
Faught, Paula Bea
Faulk, Barbara Lucille
Faulstick, Judith
Fautin, Daphne Gail
Faux, Betty J.
Fazzino, Kimberly Wright
Feader, Barbara Louellen
Feather, Bonny L.
Fechter, Charity
Fedor, Naomi N.
Fedorchak, Penny M.
Fee, Charlene
Feeney, Jennifer Kelly
Fehr, Lorraine R.
Feigenbaum, Lois
Feik, Mary S.
Fein, Edythe Gordon
Feldman, Arlene Butler
Feldstein, Lynda Sue
Felker, Kathleen Mary
Fenimore, Jeanne L.
Fenn, Lily Kar Bak
Fennimore, Charalene
Fenske, Donna Smith
Fenton, Karen W.
Fenwick, Beverly Ann Peters
Fera, Margaret Leanna
Ferch, Eileen Marilyn
Ferdinand, Rosemary
Ferency, Garrie Jo
Ferguson, Andrea Lyn
Ferguson, Barbara Ann
Ferguson, Claudia Kay
Ferguson, Helen Squires
Ferguson, Lori A.
Ferguson, Susan Teresa
Fernandes, Pansy Ivonne
Fernandez, Mary Elizabeth "Skipper"
Fernandez, Nohema
Ferrarese, Colleen Kay Bradley
Ferree, Mildred House
Ferrel, Lyla
Ferrell, Betty R.

Fetsch, Linda Lee
Fichter, Jill Elise
Fickett, Marion L.
Fielding, Nancy H.
Fields, Evelyn M.
Fields, Janet Booth
Fierro, Paola Isabella
Figley, Peggy Sue
Finch, Linda K.
Findlay, Linda Joyce
Fineman, Juanita Rae
Finley, Juanita M.
Finnegan, Iva L.
Fintak, Florence F.
Fioravanti, Barbara Kussmaul
Fiorelli, Tina Embich
Firminhac, Lajean Ardel Roy
Firth, Mary Louise
Fischbeck, Vicki Lynn
Fischer, Constance Mary
Fischer, Edith Copeland
Fischer, Ph.D., Margaret E.
Fish, Linda E.
Fish, Norma Gene
Fisher, Barbara
Fisher, Carolyn Wentworth
Fisher, Carolynn Mary
Fisher, Diane Tribble
Fisher, Elizabeth
Fisher, Nancy M.
Fishman, Pamela Sue
Fitch, Elaine E.
Fitzpatrick, Beryl Ethelene
FitzPatrick, Gene Teresa
Flahart, Mary
Flakker, Jean
Flanagan, Roberta Gail
Flanary, Virginia J.
Flashaar, Rilla Margaret
Flaspohler, Melissa (misti)
Fleck, Marcia Geralyn
Fleisher, Ruth Shafer
Fleites, Vicky
Fleming, Betty Jane
Fleming, Delryn R.
Fleming, Pamela Kay
Fleming, Sally Ann
Fletcher, Mary P.
Fletcher, Maybelle
Fletcher, Patricia L.
Flewellen, Irene W.
Flint, Diane Kaye
Flinterman-Scammell, Marielle
Floren, Natalie Louise
Flores, Patricia
Florey, Ruth Thomason
Florio, Claire
Flower, Jean Frances
Floyd, Anita Louise
Floyd, Catherine Merica
Floyd, Janie Davis
Flynn, Dorothy Llewellyn
Flynn, Kathryn Anne
Flynn, Margaret Anne
Fodge, Ellen S.
Foeh, Ruby Alice

Foellner, Lindalou W.
Fogle, Adele
Fogle, Beverly
Foley, Mary F.
Folk, Kathryn W.
Folk, Sharon Lynn
Folsom, Jane Taylor
Folsom, Karen A.
Fonseca, Laura Lynne
Fontenot, Rita Rae, MD
Foose, Betty S.
Foote, Vivian
Forbes, Lynne Donivan
Forbes, Patricia L.
Ford, Jiska Marlies
Fordham, Esther Brunen
Forman, Dorchen R.
Forster, Darlene
Forte, Laurie E.
Foster, Judith A.
Foster, Yvonne Karen Anne
Fouquet, Nancy D.
Fouts, Melissa S.
Fowler, Dorothy M.
Fowles, Trista J.
Fox, Kathleen Carol
Fox, Marie L.
Fox, Mary Ann
Foy, Carol M.
Foy, Judith L.
Frame, Mearl M.
France, Elisabeth Anne
France, Nancy D.
Franck, Darlene D.
Frank, Kathleen Marie
Frank, Sandra Kaye
Frank, Vicki Rose
Franke, April Lynn
Franklin, Alicia Jane
Franklin, Ellen A.
Franklin, Susan
Frantz, Etta Sue
Frantz, Ruth Ellen
Franz, Wendy
Fraser, Catherine Una
Fraser-Jones, Jeannette E.
Frasher, Cheryl Lynn
Frazee, Teresa Ann
Frazier, Jeannette
Frazier, Vicki S.
Frederick, Frances Maxine
Fredericks, Ruth May
Fredrick, Dianne Marie
Freeburg, B. Jean
Freed, Norma Jean
Freedman, Marilyn
Freeman, Arlee Ardeth
Freeman, Barbara Silva
Freeman, Gayle Gorman
Freeman, Katherine Elizabeth
Freeman, Linda R.
Freeman, Louciel B.
Freeman, Rachele Becker
Freese, Gustava (Gussie)
Freeze, Wendy Harshman
Freier, Norma J.

Frelin, A.J.
Fremont, Margrette H.
French, Donna Jean
French-Gatipon, Mary
Fricker, Constance Eddiline
Friede, Eleanor
Friedman, Blossom Ruth
Friedman, Holly M.
Friedman, Linda J.
Friedman, Patricia Claudette
Friedmann, Dorothy Louise
Friel, Kari L.
Frierson, Patricia B.
Fritsch, Christina A.
Frost, Jane K.
Frost, Shirley Mae
Frost-Swank, Donna Marie
Fry, Bonnie Michelle
Fry, Jolene
Fry, Nancy W.
Fry, Patricia S.
Frybarger, Gail Lenore
Fryer, Stephanie
Fuchs, Alice S.
Fudoli, Ruby S.
Fuentes, Ana
Fugiel, Jennifer Renee
Fugit, Pamela Hatley
Fujimura, Dorothy Etsu
Fulks, Donna L.
Fuller, Harriet Jean
Fuller, Kris R.
Fuller, Nell Wolfe
Fuller, Susan Olson
Fullerton, Carolyn Janet
Fullgraf, Rosalind R.
Fullington, Carlyn Dupin
Fulmer, Glenda West
Fulton, Laurie Paige
Fultz, Sara Spitler
Funk, Iona Eleanor
Funk, Wally
Funke, Barbara Joan
Futehally, Rabia
Futehally, Sumaira
Futterman, Norma Lamkin
Fydrych, Josie Joan

– G –
Gable, Jacklyn McKenna
Gacher, Phran Elizabeth
Gaddy, Janet Lynn
Gafford, April D.
Gagnon, Linda Kay
Gallagher, Denice Lynne
Gallant, Angela Victoria
Gallina, Pamela Lea
Gallo, Vanita Lou
Gamble, Carole A.
Gamble-Lerchner, Kathryn Mary
Gamertsfelder, Iona Inmon
Gammell, Jan Vawter
Gann, Doris Post (Dodie)
Gann, Wilma Minyard
Gant, Kelli
Gaona, Susan Aileen

Garber, Ethel Ritenour
Gardanier, Mary W.
Gardiner, Esther H.
Gardner, Mignon Roscher
Gardner, Sandra S.
Gardner, Willy H.
Gardy, Christine Stephanie
Garibay, Yvonne V.
Garin, Roberta V.
Garms, George Ann
Garner, Faye A.
Garner, Mary F. Sumrall
Garner, Patricia Newman
Garrett, Lisa Readhimer
Garrison, Ruth H.
Garrison, Terri Joanne
Garrymore, Anne Blake
Gartman, Robin J.
Gasker, Kathryn G.
Gaskill, Marla K.
Gassoway, Victoria Rask
Gatchel, Leta M.
Gates, Madi
Gatewood, Julie Lynn
Gauby, Karen Ann
Gauthier, Cindy Marie
Gawlik, Elizabeth Marie
Gay, Barbara A.
Gay, Eileen Ruth
Gay, Ruth Y.
Gay, Sarah Elizabeth
Gaydos, Jennifer Lynn
Gaylord, Nanette
Gaynor, Nancy A.
Geaney, Martha Mary
Geary, Gisele E.
Genaro, Marie Muccie
Gentry, Kristine Rene
Gentry, Marlaette (Molly) T.
George, Marilyn L.
George, Mary E.
Geraci, Jacqueline Ann
Gerfen, Sonja L.
Gerhardt, Margaret C.
Gerlach, Darla Jean
Geromi, Virginia E.
Gerren, Donna Sue
Gerritsen, Jeanne Louise
Gerritsen, Kristi Lea
Gerson, Gabrielle
Gesualdo, Clare J.
Gex, Maria R.
Giacomo, Louise Poirier
Giambattista, Susan I.
Gibbons, Nannette M.
Gibbs, Elizabeth L.
Gibford, Sabrina Marie
Gibson, Dana
Gibson, Diana Lynn
Gibson, Ethel Cook
Gibson, Julie Ann
Gibson, Louanne
Gibson, M. Rita
Gibson, Nancy Lynn
Gibson, Rhoberta H.
Gibson, Rose
Gibson, Rosemary Raynor
Giese, Deborah (Debbie) Jean
Giesen, Britta
Gietz, Marcia Kay
Giffin, Beverly Ann
Giggey, Erika Rae
Gignac, Suzanne
Gigray, Margaret E.
Giles, Jackie
Gilgulin, Ursula
Gilkison, Pauline
Gillcrist, Amie Jo
Gillespie, Chele Maree
Gillette, Kimberly Kraeszig
Gillies, Betty H.
Gillies, Glen Judith
Gilliland, Christina Leslie
Gilmore, Rae F.
Gilmore, Tammy Lynn
Gilroy, Carla Coggins
Gitelman, Marcia Kramer
Giustina, Verda Marie
Given, Jeanne
Gladen, Vera Virgene
Glasel, Judith M.
Glasser, Carma J.
Glassman, Mary Elyse
Glasson, Pauline
Glenn, Jamie L.
Glenn, Linda Marie
Gleszer, Ann Griffin
Glidewell, Tammy Sue
Glover, Cindy-Lee Jensen
Gluhushkin, Wendy Debra
Godfrey, Nancy Jane
Godwin, Linda M.
Goetsch, E. Lynn
Goforth, Janice Lorraine
Goins, Rebecca Stokes
Goldsberry, Laura Jane
Goldwater, Kathleen A.
Gomes, Martha Duponte
Gomez, Traude Elisabeth
Gonzales, Jaime Sharon
Goodall, Jean Ann
Gooding, Anne W.
Gooding, Karen Rose
Goodman, Betty Jeanne
Goodman, Peggy Myra
Goodman, Susan Annette
Goodrich, Lois Lauretta
Goodwin, Barbara Ann
Goodwin, Rhonda L.
Goppert, Martha Nye
Gordley, Sandra Lee
Gordon, Gwyn Voorhees
Gordon, Peggy L.
Gorham, Marie Elizabeth
Gorman, Marjorie N.
Gorra, Valerie Ann
Gorzell, Margie Comstock
Gosko-Sadowski, Laura Kay
Gosling, Carol Anne
Gosnell, Mariana E.
Gottdiener, Mari Lynn
Gottlieb, Kelly S.
Gougeon, Patricia A.
Goulding, Deidre Ann
Gouveia, Julie Marie
Govesky, Eleanor B.
Gowans, Mary Jane
Gowthrop, Janna Lee Imlay
Graddy, April Darice
Graham, Elizabeth Hegarty
Graham, Erica C.
Graham, Jennifer Jane
Graham, Lisa Diane
Graham, Patricia A.
Graham, Virginia (Ginger)
Grandy, Vicki Jo
Grange, Susan Elaine
Granger, Joan B.
Grant, Claire E. (Geni)
Grant, Cynthia S.
Grant, Fran
Grant, Kathy
Grant, Loretta Ida
Grant, Shirley
Grass, Susan Jane
Graul, Mary R.
Graves, Dale V.
Graves, Ginger Anne
Graves, Paula D.
Gray, Janice Marie
Gray, Jean Lisa
Gray, Kathleen M.
Gray, Linda M.
Gray, Marjorie M.
Gray, Michelle Stadler
Gray, Rita I.
Gray, Roberta Helen
Gray, Robin Michael Renee
Gray, Sue Bess
Greaves, Erika Margret
Greb, Greta O.
Green, Fara E.
Green, Janet
Green, Janet C.
Green, Lorraine
Green, Lynda Kay
Green, Ruth G.
Greene, Bambi Lee
Greene, Dianne
Greene, Janice C.
Greene, Leslie Hall
Greene, Lynda Sue
Greenfield, Karen Marie
Greenham, Marcia Ellen
Greenhill, Beverlee K
Greenwalt, Grace Astrid
Greenwell, Patricia W.
Greenwood, Krystina D.
Greer, Delia Frances
Greer, Mary Ann
Grein, Marie Elisabeth
Greiner, Robin E.
Greply, Elizabeth
Grey Eagle, Gwendoline
Grieco, Johannah
Griffin, Jean M.
Griffin, Lynda Lea
Griffing, Knansie Beth
Griffith, Carol Sue
Griffith, Irene
Griffith, Thon
Grimm, Wendy L.
Gronau-Fietz, Lois Ann
Gronewald, Mary Helen
Gross, Anja Rita
Gross, Gail
Grosser, Janet Mary
Grosskortenhaus, Birgit
Groth, Elizabeth Ruth
Grove, Doris F.
Groves, Kathy
Groves, Laura Margaret
Grubbs, Susan Dallas
Grunwell, Julie Marie
Grupenhagen, Esther
Guay, Helene Louise
Guernsey, Nancy Patricia "Red"
Guhsé, Ruth Nydine
Guiang, Stephanie
Guiberson, Marian Cook
Guilfoyle, Kimberly Lane
Guillet, Nancy
Guin, Sarah Cherry
Gulasy, Terri L.
Gullino, Candis Hall
Gumbert, Eva M. (Skip)
Gunderson, Cathy Delline
Gunderson, Lucinda Isabel
Gunther, Kathryn L.
Gurke, Sharon Mccue
Gustafson, Alice Jerome
Guthrie, Alyce Newberry
Guthrie, Deloris Jean
Guthrie, Sally Ann

– H –
Haag, Sandra Marlo
Haas, Del L.
Haase, Madeleine L.
Haber, R. Bonnie
Haberfelde, Beverly Jane
Hach-Darrow, Kathryn C.
Hackett, Monica F.
Hackler, Jeanette D.
Hadden, Isabel Maciver
Hadfield, Josephine B.
Hagan, Darlene Dolores
Hagan, Mary Lou R.
Hagans, Nancy Ferguson
Hahn, Eleanore M.
Hahn, Tacy Weeks
Haid, Beverly J.
Hailey, Lois Brooks
Hainline, Jean Sheffield
Hair, Christina A.
Haizlip, Mary
Hake, Dorothy Arline
Haldeman, Jane Gwynn
Hale, Isabelle G. McCrae
Haley, Carolyn Ann
Haley, Gayle
Haley, Melissa Courtright
Haley, Patricia Lynn
Hall, Auleen Katheryn

Hall, Barbara Lee
Hall, Carol Ann
Hall, Dolores Ellen
Hall, Elisha Spence
Hall, Emma Whittington
Hall, Harriett Monroe
Hall, Hopi L.
Hall, Julia "Judy" Corbett
Hall, Nancy M
Hall, Sylvia Sue
Hall-Canatsey, Shirley
Hallatt, Marion Estelle
Haller, Ann C.
Hallett, Brenda Diane
Halloran, Sherlyn Denise
Halpain, Sue
Ham, Gerry Shanelle
Hamann, Marie M.
Hamblin, Juanita Avita
Hamer, Linda Lee
Hamill, Doris
Hamilton, Carol Yates
Hamilton, Col. Kelly Sue
Hamilton, Helen Louise
Hamilton, Lori Ann
Hamilton, Nadine C. Rose
Hamilton, Onalee (Lee)
Hamilton, Sharon Kaye
Hamilton, Sharon Y.
Hamilton, Sylvia L.
Hamilton, Twila Marie
Hamlett, E. Ellen
Hamlin, Diane E.
Hammarback-Speer, Rene Wicks
Hammer, Betty Jo
Hammer, Jean Henry
Hammond, Tabitha Olivia
Hamner, Marvine P.
Hamzah-Braybrook, Noor Azizah
Hanan, Rose Rowland
Hancher, Jean Franklin
Hancock, Marilyn N.
Hancock, Mary Margaret F.
Hancock, Melanie Lee
Hancox, Eileen C.
Handrahan, Colleen Marie
Hane, Marcia L.
Hange, Patricia J
Hanley, Mary Anne
Hanlon, Evangelia
Hanna, Patricia Jane
Hansen, Coral Bloom
Hansen, Edna Lee
Hansen, Kathryn (Kathy) Lynn
Hansen, Mary L.
Hansen, Renee Jo
Hansen, Valli M.
Hansen, Wanda T.
Hansen (Dr.), JoAnn Brown
Hanshew, Donna Lynne
Hanson, Caryn J.
Hanson, Catherine A.
Hanson, Donna Lee
Hanson, June Glaser
Hanson, Marjorie Crawford
Hanson, Shirley Clark

Hapgood, Randy
Happy, Heather Amanda
Haraldson, Nancy L.
Haranka, Helen Marjorie
Harby, Ethelyne A.
Harclerode, Zoan Ruth
Hardcastle, Marjorie M.
Harder, Beverly J.
Hardey, Ray Prescott
Hardin, Sherry Brunson
Harding, Joyce Autry
Hardwick, Ellen
Hardy, Janice C.
Hare-Smith, Wendy Anne
Harker, Karen
Harman, Donna Lee
Harmer, Virginia Curtis
Harmon, Dalita
Harmon, Gloria Yost
Harmon, Kay Beverly
Harmon, Virginia M.
Harms, Kimberly
Harned, Anna Marie
Harper, Barbara Lee
Harper, Kimberly Anne
Harper, Rhonda Ann
Harper, Sue Mapp
Harper, Susan Kay
Harper, Sylvia Irene
Harrell, Bonnie Jean
Harrend, Barbara Ann
Harrington, Elizabeth
Harris, Carol Anne
Harris, Jettie June
Harris, Keitha A.
Harris, Leritha Marie
Harris, Mary Hart
Harris, Mary-Jane Frances
Harris, Sylvia Burleson
Harris, Terry Lallement
Harris-Cirillo, Jan Patricia
Harris-Para, Barbara A.
Harrison, Mary-Kathryn
Harrold, Rita Mary
Hart, Peggy Irene
Hart, Sandra Gail
Harte, Eileen Wadle
Hartley, Edweena D.
Hartley, Leora Chapin
Hartley, Marian Burke
Hartnett, Colleen I.
Hartnett, Maureen C.
Hartzler, Amanda Jean
Harvey, Debbie
Haschel, Wendy Catherine
Haselmann, Susan Keizer
Haskell, Charlene Dee
Haskell, May E.
Haskins, Mardell D.
Haslett-Mowrey, Laura E.
Hassell, Sally M. Flynn
Hastings, Trudy M.
Hasto, Kristine Ann
Hataway, Rosemary S.
Hatch, Joy Arden
Hatch, Karen Sue

Hatfield, Cecile
Hatfield, Patricia Marie
Hathaway, Diane A.
Hathcock, Carol Joyce
Hatton, Hope Renee
Haugarth, Nancy Kirby
Haught, Joan L.
Haupt-Spangler, Dorothy J.
Hausteen, Karen Grace
Havens, Barbara C.
Havens, Kathryn Anne
Havice, Lucy Thelma
Havice, Shirley Lucille
Hawbaker, Jody Lynn
Hawkins, Katherine
Hawkins, Sandra
Hawks, Ruth J.
Hawley, Lois Carol
Hayden, Helen D.
Hayden, Sara P.
Haydu, Bernice Falk
Hayes, Darlene R.
Hayes, Jane Anne
Hayes, Mary Evelyn
Hayes, Patricia Jean
Hayes, Tommy Jean
Haynes, Linda Campbell
Hays, Ruby C.
Hayward, Wyn P.
Hazlett, Margaret
Hazlett, Sandra Adams
Head, Billie Louise
Heale, Marian Dodge
Healey, Victoria M.
Heaps, Heather Frances
Hearn, Erika Lyn
Heathman, Lenora Ash
Heaverlo, Frances Marie
Hebden, Janice R.
Hecksel, Nancy E.
Hedges, Jessica Wimmers
Hedglon, Emily Leda
Heeding, Michelle Anne
Heesacker, Madelyn Mary Ann
Hefel, Lori Ann
Heffner, Jane Elizabeth
Hegedus, Alice I.
Heggland, Elin-Mari
Hegranes, Geils-Adoue Todd
Hegy, Roberta Ann
Heikkila, Dorothy A.
Heim, Patricia M.
Hein, Emmy J.
Heinemann, Rosalind Margaret
Heinonen, Laura Eveliina
Heinsohn, Mary A.
Heiser, Carol Jeanne
Heitsch, Betty Parsons
Heitzman, Mary M.
Helf, Lesley
Helgeson, Dorothy E.
Helleloid, Kristen Lou
Heller, Mary Ann A.
Hellmann, Amie Lynne
Hellsten, Gunilla Maria
Helly, Karen Diane

Helm, Alice M.
Helm, Deborah T.
Helm, Dorothy Davis
Helms, J. Lynne
Helms, Michelle Bigham
Helms, Susan Jane
Helquist, Joyce Esther
Helvey, Toni Lynn
Hembel, Caroline Etheredge
Hembel, Helen E.
Hempel, Becky
Hems, Gwen F.
Hems, Helen Mary
Henderson, Deborah Ruth
Hendricks, Barbara Jean
Hengesh, Charlene D.
Hengsteler, Pamela Chesley
Henig, Mary Elizabeth
Heninger, Vonne Anne
Henley, Shirley M.
Hennes, Carla Jean
Hennessey, Mary Ann
Henrotin, Maysie Morris
Henry, Alice K.
Henry, Betty J. (Janie)
Hensley, Tookie
Henze, Gayl I.
Hepner, Barbara W
Heraty, Melissa R.
Herman, Kathryn Frances
Herman, Wrenn Redford
Hermann, Mary B.
Heron, Sharon Muriel
Herr, Dianne Lodeen
Herr, Ellen Louise
Herring, Ellen
Herrington, Dorothy Means
Herrmann, Beth Ann
Herron, Lisa Ann
Herschelmann, Ph.D, Kathleen M.
Hershkowitz, Diane Ellen
Hertel, Jane V.
Herzog, Deborah Lee
Herzog, Dorinda Renee
Herzog, Lisa Diane
Hess, Virginia L.
Hessin, Anita B.
Heston, R. Elaine
Hetherington, Kathryn Jane
Hettenbach, Christine A.
Hettinger, Corinna
Hetw, Patricia Ann
Heuer, Nadine
Heuermann, Katie M.
Hewett, Nancy Ann
Hewette, Nena Jo
Hewgley, Linda Sue
Hewitt, Connie Ann
Hewitt, Helen Smith
Hewitt, Lois Jean
Hickey, Carol A.
Hicklin, Dolly
Hickman, Carole J.
Hickman, Sandra Annette
Hicks, Betty
Hicks, Roberta Ann

Hiebel, JoAnn H.
Hiern, Sara Smith
Hiestand, Sally Grace
Higgins, Clare Brooks
Higgins, Donna L.
Higgins, Leah A.
Higgins, Virginia Bond
Higgs, Mary Brock
Highleyman, Leslie Ann
Hightower, Ora Delle
Hijos, Rosie L.
Hilbert, Jane D.
Hilbrandt, Kathleen Ann
Hilburn, Helen M.
Hilchie, Joyce L.
Hill, Geraldine Masinter
Hill, Joan Marilyn
Hill, Karen A.
Hill, Kathryn D.
Hill, Lori Susan
Hill, Margaret Ann
Hill, Ruth Janette
Hill, Ruth V.
Hill, Sherrie L.
Hill, Susan Leigh
Hilliard, Patricia C.
Hillis, Jeanne B.
Hillmann, F. Sue
Hilsberg, Kerry Francis
Hilst, Katherine Louise
Hilton, Diana Rene
Hinchcliffe, Carolyn Anne
Hine, Peggy Jo
Hinman, Katherine Joan
Hinn, Betty Ardell
Hinneburg, Brig. Gen. (Ret.) Patricia Ann
Hinterberg, Kristine E.
Hinton, Jeanette Balderson
Hipfel, Ilse Erika
Hirahara, Maire Lani Kei
Hirsch, Katharine M. Stanley
Hirst, Jessica S.
Hirzel, Beverley Ann
Hisaw, Mary Jo (Jody)
Hissem, Jeanne K.
Hitchcock, Elaine Marie
Hitson, Amy Elizabeth
Hitt, Janet Russell
Hivick, Fonda Rose
Hixon, Denise Marie
Hoagland, Sheryl Ellen
Hobson, Barbara B.
Hockings, Dorothy Jane
Hodges, Bonnie Jean
Hodges, Joanne Platt
Hodson, Tonya Susann
Hoefer, Marie
Hoeffner, Marilyn
Hoelscher, Ute Maria
Hoelting, Kathryn Carpenter
Hoerle, Shirley Jean
Hof, Marion
Hofer, Susan L.
Hoffbeck, Kimberly Kathleen
Hoffman, Diana L.
Hofford, Dorothy Jean
Hofschneider, M. Susan
Hogan, Betsy L.
Hogan, Carol Knight
Hohn, Hazel M.
Hoiby-Griep, Kay L.
Hoit, Elise L.
Holbird, Helen
Holbrook, Allison B.
Holcombe, Katherine Elizabeth
Holden, Ruth Richter
Holden, Tammy Lou
Holder, Ashley Elizabeth
Holdsworth, Marjorie E.
Holforty, Wendy L.
Holifield, Sally Lynn
Holladay, Ann Sanders
Holland, Susan Townsend
Hollander, Lu
Holley, Robin Helene
Holliday, Patricia Ann
Holman, Ellen B.
Holman, Margaret Gale
Holmblad, Claire Marie
Holmes, Jane F.
Holmes, Joan Marie
Holmes, Sherry
Holmes, Virginia B.
Holt, Lilian Darling
Holtman, Elizabeth Dorothy
Holton, Helene M.
Holub, Shirley Ann
Homuth, Sharon Joan
Honacki, Sharrilyn Andrea
Honer, Anne S.
Honisett, Jillian
Honjo, Misaki
Hood, Carol Rebecca
Hoof, Eugenia L.
Hook, Mary M.
Hook, Paula G.
Hooker, Linda Mary
Hoopmann, Janet Elizabeth
Hoover, Amy Lynn
Hoover, Loraine Gay
Hoover, Mildred Ann
Hoover, Tracy Ann
Hope, Rosalie June
Hopfenmuller, Jill Ann
Hopkins, Evelyn Craig
Hoppe, Felicia R.
Hopson, Karen P.
Horn, Linda J.
Horner, Mary Royster
Hornsby, Dorothy
Horowitz, Susan Amy
Horsch, Pamela Kay
Horsey, Helen Louise
Horton, Carolyn
Horvath, Marilyn G.
Hosoya, Noriko
Hostler, Betty Hawkes
Houchin, Frances E.
Houck, Maurine Johanna
Houden, Patricia Ann
Hough, Mary
Houpt, Helen Eileen
House, Linda Katheryne
House, Robin
Householder, DVM, Laurie Robin
Houseknecht, Sethany Ann
Houston, Deborah Gaye
Houston, Lynn Elizabeth
Houston, Maura Mary
Houston, Winifred "Winnie"
Hovel, Mary Bernadette
Howar, Beth Gladys
Howard, Betty Jean
Howard, Phyllis Ann
Howard, Stacy Hamm
Howard-Phelan, Jean Ross
Howell, Jean M.
Howell, Teresa J. Caywood
Howerton, Constance Elizabeth
Howes, Jaye
Howren, Evelyn Greenblatt
Hrindak, Sue
Hsia, Lynne Kastel
Hubbard, Joan Johnston
Hubbell, Rosemary E.
Huber, Julia T.
Hubert, Ruth Clifford
Hucabee, John Dell
Huck, Betty
Hucke, Kara
Huddy, Charma L.
Hudgins, Jewel E. (Jaye)
Hudson, Alice A.
Hudson, Dorothy Lynn
Hudson, Joan Marie
Hudson, Justyna M.
Hudson, Marjorie K.
Hudson, Shelly J.
Huffman, Barbara Ann
Huffman, Beverly L.
Hughes, Dorothy "Dottie" M.
Hughes, Janice K.
Hughes, Kathryn G.
Hughes, Marcia B.
Hughes, Marjorie Ann
Hughes, Patricia M.
Hughes, Shirley G.
Huhndorff, Lisa Jane
Huie, Elizabeth A.
Hukill, Lorraine M. (Lorry)
Hulett, Kathleen A.
Hull, Mary Mason
Hull, Terri Lynn
Humphrey, Gloria Jean
Humphreys, Ann M.
Humphreys, Mary Jo Ann
Humphries, Judith Lynn
Hunsaker, Jana Jean
Hunt, Altha Ernesteen
Hunt, Cary
Hunt, Eve Yarbrough
Hunt, Jean S.
Hunt, Merilee Sue
Hunt, Rachel Snead
Hunter, Betty A.
Hunter, Crystal Starr
Hunter, Jayne
Hunter, Penelope A.
Huntsberger, Ruth Catharine
Hurd, Mary Jo
Huritz, Frances
Hurley, Marilouise
Hurwitz, Jolynn Edwards
Hussan, Cheryl Jean
Hustead, Dixie Lee
Huston, Barbara M.
Hutchins, Wendy
Hutchinson, Deanna Lynn
Hutchinson, Judi Daughtry
Hutchinson, Judith R.
Hutchinson, Louise Prugh
Hutson, Camilla D.
Hutson, Christina K.
Hyland, Margaret R. Watson
Hyles, Helen Marie
Hyson, Bonnie Leonard

– I –

Iglesias, Elizabeth Rose
Ilves, E. Joyce
Imperral, Joanne C.
Infusino, Joanie A.
Ingle, Lillian Jean
Ingold, Elaine Parker
Ingraham, Joeann J.
Ingram, Carolyn F.
Inman, Suzanne E.
Innes, M. Elizabeth "Betty"
Ioannou, Georgia I.
Iovine, Jane Ann
Iredell, Elizabeth A.
Ireland, Barbara A.
Ireland, Nancy Jane
Irvin, Kristine K.
Irvin, Paulette Diane
Irwin, Donna-Marie
Irwin, Elisabeth Barbara
Isbell, Eva M.
Iselin, Karen L.W.
Ishiyama, Yasuko
Israel, Anita Pearl
Israel, Karen Raye
Ito, Keiko
Ivany, Mary A.
Ivarsson, Eivor
Ives, Emily J.
Ivie, Neva Renae
Ivy, Beverly Reich

– J –

Jablonski, Mary Suzanne
Jackson, Anne Robbins
Jackson, Bette D.
Jackson, Brenda E.
Jackson, Donna J.
Jackson, Inez M.
Jackson, Kerry Marie
Jackson, Linda H.
Jackson, Melody Ann
Jacob, Marlene
Jacobs, Arlene M.
Jacobs, Barbara Cowan
Jacobs, Jeanie L.

Jacobs, Linda Ann
Jacobs, Ruth Estelle
Jacobson, Caroline A.
Jacobson, Doris
Jacquot, Ruth M.
Jaderborg, Jana Marie
Jaffe, Hedy
Jaffe, Kathleen Lynn
Jahn, Gretchen Lois
Jahner, Sharleen A.
James, Ellen Eisendrath
James, Kathleen A.
James, M. Joyce
James, Teresa D
Jameson, Cathleen Gail
Jamison, Mary Ann
Jamison, Myra S.
Jankord, Susan Elizabeth
Janov, Abbe J.
Janssen, Beth Ann
Janus, Priscilla Wallace
Jauch, Kayla L.
Jayne, Barbara Kibbee
Jayne, Marion P.
Jeffery, Doris Sumiko
Jeffery, Lynn Marie
Jeffrey, Nancy C.
Jeffries, Rhonda Detert
Jenison, Barbara W.
Jenkins, Bettina
Jenkins, Dorothy R.
Jenkins, Jeanette J.
Jenkins, Patricia E.
Jenkins, Ruth B.
Jennings, Barbara F.
Jennings, Denise Evelyn
Jennings, Sandra Elizabeth
Jensen, Jenette
Jensen, Karole K.
Jensen, Lydia L
Jensen, Marilyn Anne
Jensen, Nancy Kelley
Jensen-Coonrod, Rosemary
Jenson, Mary Goodrich
Jeschien, Alma Jean
Jessen, Gene Nora
Jessup, Helen E.
Jewett, Delores "Dodie" Grace
Jipsen, Shannon L.
Jogtich, Elizabeth Lundin
Johnson, Aleta M.
Johnson, Beverly Bond
Johnson, Beverly Boucher
Johnson, Bonnie Lorraine
Johnson, Brenda Marie
Johnson, Charlotte Beverly
Johnson, Charlotte Bloecher
Johnson, Cynthia Jean
Johnson, Elaine M.
Johnson, Elinor Reay
Johnson, Evelyn Bryan
Johnson, Harvella K.
Johnson, Jeannie E.
Johnson, Judith A.
Johnson, Judith E.
Johnson, Kay J.
Johnson, Kendra Ann
Johnson, Kendra Lea
Johnson, Lisa Nicole
Johnson, Lori K.
Johnson, Marcia Jeanne
Johnson, Marjorie
Johnson, Marjorie Ruth
Johnson, Nancy Ruth
Johnson, Nedra Dynelle
Johnson, Ruth "Roni" L.
Johnson, Shirley Curran
Johnson, Susan M.
Johnson, Una Rae
Johnson, Wendy Elizabeth
Johnston, Dianne
Johnston, Ella
Johnston, Linda G. Thom
Johnston, Loretta Adams
Johnston, Virginia B.
Johnstone, Karen Kay
Jones, Bertha Marie
Jones, Betty Era
Jones, Elly W.
Jones, Jennifer Elaine
Jones, Jo Carol
Jones, Joan Barbara
Jones, Jovita A.
Jones, Joyce Whitmore
Jones, Katherine S. Putnam
Jones, Kelly Lynn
Jones, Patricia Annette
Jones, Patricia L.
Jones, Ruth Craig
Jones, Ruth Owen
Jopson, Leanne T.
Jordan, Elizabeth Edna
Jordan, Jennifer Ann
Jordan, Jo Rita
Jordan, Margaret Jean
Jordan, Priscilla Bridget
Jorgensen, Lynn D.
Jorgensen, Trine
Jorgenson-Roach, Carol Anne
Joyner, Carol M.
Judd, Lorraine
Judges-Lemmin, Patricia Nora
Judkins, Elizabeth Marta
Juhasz, Margaret Mary
Juillerat, Mary Anne
Julian, Shauna D.
Jung, Cecilia
Junge, Karina E.
Junkins, Sonya Lin
Jurenka, Jerry Anne
Juricek, Susan Elaine
Justice, Nell
Jylanki, Leila Maria

– K –

Kaeder-Carpenter, Deborah Ann
Kaes, Mary Barbara
Kahak, Mary Olivia
Kahn, Karen M.
Kahonen, Jaana Kyllikki
Kaier, Lorna Marie
Kaiser, Ann Larue
Kaker, Virginia A.
Kalman, Caroline Rosemary
Kalthoff, Rosemary Helen
Kaminsky, Fran G.
Kamm, Patricia Lynn
Kamp, Marilyn
Kamps, Ronnie Diann
Kanao, Miyako
Kane, Sarah Ann
Kane, Theresa Rose
Kanzelmeyer, Evelyn V.
Karas, Deborah Chisholm
Kardatzke, Marcia L.
Kariolich, Cynthia Marie
Karp, Audrey J.
Kase, Gayle Louise
Kass, Gail Diane
Kastanas, Gaynelle R.
Katapodis, Mary Margaret
Katz, Maureen Block
Katz, Rochelle B. "Shelly"
Katzen, Debi
Kauffman, Erma M.
Kaufman, Adele
Kaufman, Ursula R.
Kawai, Reiko
Kazmark, Ruth Mycol
Kearns, Joann Mary
Keating, Ailine
Keating, Carla I. Irving
Keck, Janet A.
Keck, Susan Marie
Keefer, Patricia Jayne
Keefover, Kathleen Ann
Keffer, Hazel Snyder
Kehmeier, Louise Engblom
Keidel, Helen J.
Keinath, Carol M.
Keith, Ann H.
Keith, Janet Birt
Keith, Nancy
Keller, Linda Kay
Keller, Mary "Mitzi"
Kelley, Charlotte S.
Kellogg, Jane B.
Kelly, Anne N.
Kelly, Cynthia A.
Kelly, Diana S.
Kelly, Joey
Kelly, Leigh
Kelly, Marian Wallace
Kelly, Mary Gayle
Kelly, Mary Sutton
Kelly, Patricia A.
Kelman, Margaret Mary
Kelman, Naomi Gabrielle
Kelsey, Dorothy Ellen
Kemichick, Geraldine Joann
Kemp, Cheryl Ann
Kemp, Elicia M.
Kemp, Mitzi Lyn
Kempas, Mila Maarit
Kemper, Cynthia Gardner
Kempton, Doris W.
Kendrick, Mona
Kennard, Fredrica Lois
Kennard, Gaby P.J.
Kennedy, Catherine Bonnie
Kennedy, Doris J.
Kennedy, Evelyn Carole
Kennedy, Janice Duncan
Kennedy, Kristian R.
Kennedy, Madeline B.
Kennedy, Pam
Kennedy, Susan Jane
Kenner, Linda D.
Kenny, Lisa Irene
Kensett, Lenore B.
Kensey, Decki Joy
Kent, Annemarie Elisabeth
Kent, Jeanne
Kent, Shirley Y.
Kenyon, Emily Anne
Keon, Linda E.
Kerbaugh, Gale D.
Kern, Beverly J.
Kern, Hazel E. (Dwiggins)
Kerner, Carol M.
Kerr, Laura Taylor
Kerr, Susan K.
Kerr, Victoria Laureen
Kerscher, Helen Ann
Kerwin, Joan
Kestenbaum, Janice Kay
Kester, Elizabeth Louise
Kesterson, Roberta A.
Kesti, Susan Beth
Ketchum, Sharon Kay
Kettley, Dorothy B.
Keyes, Doreen Amy
Keys, Dell Avery
Kidd, Betty Jane
Kidder, Delores Christine
Kieffer, Ann Terese
Kiff, Linda Mary
Kilpatrick, Anne Burford
Kimball, Jane S.
Kime, Susan Esther
Kimura, Kazuyo K.
Kinard, Dawn Louise
Kincel, Katya Molochko
Kindberg, Opal Antonia
King, Janet Mary
King, Josette B.
King, Kirsten Maria
King, Leslie A.
King, Luanne Paul
King, Martha Ann
King, Mary Ann
King, Patricia L.
Kinnaman, Cheryl Anne
Kinnaw, Mary Margaret
Kinninger, Andrea Elizabeth
Kintop, Dale B.
Kinzy, Vicki S.
Kirchner, Heather Michelle
Kirhofer, Jeanne
Kirk, Faye L.
Kirkland, Jacqueline Cercek
Kirsch, Debi Dolores
Kirschke, Robin Jean
Kirschner, Deborah Ann

Kiser, Bobbi Kay
Kistner, Joan Carol
Kitson, Denise Valerie
Kivland, Jana B.
Kizziar, Veretta
Klabacha-Stevenson, Cheryl Ann
Klaus, Jacqueline Hunter
Kleihege, Nancy Buck
Klein, Carol A.
Klein, Maj. Paulette M.
Klein, Mary Lou
Klein, Sherralyn Jean
Kleist, Jo Marie
Klements, Mathilde M.
Klemm, Margaret Fae
Kleynhans, Joanna Elizabeth
Kliewer, Karol Sue
Kline, Carol Christine
Kline, Elinor J.
Kline, Emily I.
Kling, Julie Wells
Klingberg, Carlene
Klippert, Elizabeth Melissa
Kloos, Dianne Eleanor
Klopfer, Joy Lorraine
Kloth, Carolyn Marie
Klyn, Charlotte S.
Knaan, Alona
Knapp, Joan Elizabeth
Knapp, June D.
Knaute, Judith L.
Knepshield, Silver Grenoble
Knianicky, Christine Irene Obuch
Knickerbocker, Carol Beth
Knight, Linda C.
Knipmeyer, Laura Louise
Knolinski, Pamela J.
Knouff, Mary Jo
Knowles, Katrina Anne
Knox, Patsy Dawn
Knudson, Vicky Jean
Kochan, Janeen Adrion
Kocisko, Jo Nell
Kodis, Bonnie Hefte
Koehler, Alexis C.
Koehs, Kimberly Annette
Koen, Mildred Albert
Koen, Patrice C.
Koenig, Ann Embry
Koenig, Beth Ann
Koenig, Mary Lynn
Koepke, Yvonne P.
Koerner, Mary A.
Koerwitz, Janet Marie
Kofke, Julia Evans
Kogel, Ruth Pegeen
Kohen, Joanne S.
Kohl, Mary Coleen Finnegan
Kohler, Betty D.
Kohler, Katharine G.
Kokesh, Louise Rogene
Kokkola, K. Anneli
Koll, Karin
Konger, Julia A.
Konno, Naoko
Koonce, Janet Williams

Kopp, Claire B.
Korda, Lya
Koshan, Kathryn E.
Koslowsky, Pat M.
Kothawala, Ann Kramer
Kovacs, Angela Therese
Koval, Marjorie J.
Kovalchuk, Julianne Kathleen
Kovar, Catherine L.
Kowalewski, Deborah
Kowalski, Inez J.
Kozak, Anna Mary
Kozak, Elma
Kraemer, Norma Janelle
Kraeszig, Lana Edwards
Krafft, Helene Krumholz
Kragness, Gloria A.
Krajnik, Barbara Ellen
Kramer, Betty Lou
Kramer, Doris Jensen
Kramer, Elizabeth Anne
Kramer, Phyllis J.
Kras, Judith Ann (Judy)
Krass, Teresa Brayson
Kraus, Rita
Krause, Geraldine H.
Krauth, Esther F.
Krawchuk, Darlene Carol
Kreth, Evelyn Inman
Kreutzen, Sharon Louise
Krick, Alice C.
Kriss, Kathy Hinders
Kritz, Beverly J.
Krongold, Helene N.
Kropp, Evelyn
Krottinger, Starr S.
Krotzer, Jean
Krouse, Olive Leslie
Krueger, Amy Sue
Krueger, Patricia Ann
Krumwiede, Erin Francine
Kudiesy, Norma M.
Kudrna, Alice G.
Kuechenmeister, Janice Rose
Kuechle, Betty Reid
Kuehn, Geraldine M.
Kuhn, Maria Jean
Kuhn, Martha Ann
Kunica, Waltraud Maria
Kunichika, Heidi Norie
Kuortti, Marianna
Kuortti, Orvokki
Kupchuk, Dana A.
Kuprash, Aimee
Kurrasch, Madeline E.
Kurth-Weninger, Kristin Janet
Kurtz, Barbara Lee
Kuzenko, Edith Pearce
Kwarciany, Sheryl Ann
Kyle, Nancy R.

– L –

L'Herisson, Mary Sloan
L'Hoir, Michelle Ann
La Brie, Beverly A.
La Salle, Margaret Ann

La Vake, Margaret E.
Labadie, Jamie Marie
Labate, Catherine P.
Lacarrubba, Madeline Steiner
Lachance, Danie
Lack, Anne M.
Lacomette, Debra Ann
Lacrambe, Annie-Claire
Ladd, Barbara Jean
Lafia, Theresa Rose
Laflin, Danette Aimee
Laforge, Sally
Lafrinere, Cynthia Dianne
Lager, Kelli Lyn
Laine, Linda S.
Laing, Diann N.
Laird, Dorothy Pemberton
Laird, Sharon H.
Lake, Fern Lillian
Lake, Linda G.
Lamar, Jane Arnold
Lamb, Charli L.
Lambert, Georgia Butler
Lambrechtse, Sammy Lou
Lamermayer, Amy Elizabeth
Lamm, Edith Raymond
Lammers, Nancy Ross
Lamont, Sherrill Ott
Lancer, Dennice
Landfried, Janet Bal
Landis, Linda Sue
Landrum, Edna "Ozelle"
Lane, Alacia L.
Lane, Nancy E
Lane, Rosemary J
Lange, Marcy S.
Langner, V. Rudene
Lanke, DiAnn L.
Lankenau, Sandra C.
Lankford, Kathy
Lanning, Carol Ann
Lanning, Judy E.
Lansden, Evelyn E
Lanson, Susan
Lanzi, Patricia Anne
Lapis, Susan Fawley
Lapook, Gail S.
Lapsley, Norene M.
Lapsley, Saundra
Large, Nancy C.
Larmon, Virginia Sue
Larsen, Anne
Larsen, Diane S.
Larsen, Patricia Lynn
Larson, Betty J.
Larson, Deidre A.
Larson, Diane B.
Larson, Susan
Lashbrook, Gudrun
Lashchuk, Kathy M.
Lasher, Barbara J.
Laska, Pearl Bragg
Latshaw, Billie F.
Lauer, Denise A.
Lauer, Linda K.
Laughbaum, Nina Elaine

Laughlin, Vickie Kay
Laurila, Ritva Anneli
Lauro, Jan Elaine
Laux, Kathryn L.
Lavelle, Debbie
Lavin, Rosalind Sue
Law, Mary Jane
Lawrence, Charlene M.
Lawrence, Cody M.
Lawrence, Marguerite Marie
Lawrence, Rolinda
Laws, Amy Elizabeth
Lawson, Bette Irene
Lawson, Margaret S.
Lawson, Sharon Lee
Lawton, Peggy Sanders
Laxague, Marianne Jean
Laxson, Janice E.
Layman, Kathryn Stephanie
Layton, Linda Lee
Lazaro, Cassandra Colleen
Lazurenko, Lydia B.
LaFontaine, Suzanne
LaMoy, Mary M.
Leach, June Adell
Leach, Lucie B.
Leahy, Mary I.L.
Lear, Shanda
Leatherman, Linda Irene
Leatherwood, Shirley Jean
Lecklider, Nancy Carol
Leckrone, Mable M.
Ledbetter, Dorthy W
Ledbetter, Gwen Bjornson
Leder, Dr. Sandra J.
Leder, Maryrose Catherine
Lee, Barbara Jeanette
Lee, Carol Joan
Lee, Clararose
Lee, Dorothy Tutt
Lee, Elizabeth Lasater
Lee, Janette K.
Lee, Joyce
Lee, Kathey Ann
Lee, Marilyn Ann
Lee, Nelda Kaye
Lee, Patricia
Lee, Tracey Marie
Lee, Vicki Balfanz
Lee, Wayna Lea
Leferson, Susan R.
Lefgren, Helyn M.
Leftwich, Karen Holcombe
Leger-Miller, Lee
Leggett, Marjy
Legierski, Diane Marie
Lehman, Elizabeth Slade
Lehman, Helen B.
Lehman, Jan Wilson
Lehr, Shirley W.
Leiblie, Loy Anne
Leininger, Ann Marie
Leistikow, Frances Ferguson
Leiter, Barbara Ann
Leland, Rebecca G.K.
Lende, Andrea Susan

Lengyel, Mary Ann
Lenhard, MD, Parwin
Lenoch, Leslee Ann
Leon, Ines Perez Gavilan
Leon, Patricia
Leonard, Tracy Suzanne
Leone, Sharon Sheldon
Leoni, Marion Y.
Leota, Nancy Lee
Lepore, Ellen
Lepore, Marie C.
Lepore, Rose Marie
Leppiaho, Cathy Elizabeth
Leriche, Jeannine S.
Lesher, Cathy Ann
Leslie, Joy Kay
Lester, Judy A.
Letzring, Lois M.
Leudesdorf, Olga M.
Levandoski, Theresa L.
Leve, Fran Kantor
Leverentz, Patricia Lynn
Leverton, Irene H.
Levesque, Bettina Marie
Levesque, Patricia M.
Levick, Merlee Anna
Levinson-Adler, Rina
Levy, Helen Egan
Levy, Shana Gayle
Lewinski, Barbara Jean
Lewis, Anita Lorraine
Lewis, Barbara D.
Lewis, Bonnie Ann
Lewis, Carol E.
Lewis, Janet Marie
Lewis, Lauri Ellen
Lewis, Marva Jean
Lewis, Mary Rosso
Lewis, Nicole M.
Lewis, Shirley C.
Lewis, Sue Ann
Lewis-Minschwaner, Sharon Lee
Leyner, Carol J.
LeClaire, Briana Jessen
LeGrande, Pamela
LeMaitre, Susan Kay
Lian, Marcia Alice
Liberty, Janet Lee
Lichtiger, Barbara G.
Lichtle, Laura Lynn
Lienemann, Grace Helen
Liersch, Leah S.
Lilienthal, Eleanor Ann
Liljegren, Agnes B.
Limbach, Dorothy D.
Limmer, Rita G.
Lindauer, Jacqueline Jane
Lindell, Kristen Lee
Lindelof, Gretchen S.
Lindeman-Waingrow, Debora
Lindgren, Linda Lee
Lindley, Annabelle
Lindner, Kimberly I.
Lindsey, Dorothy Lafitte
Lindsey, Georgia Bea
Lindsey, Margaret Lee

Lindsey, Terri Ann
Lindstrom, Marikay
Lingo, Gail P.
Link, Jane L.
Lion, Nikki Sylvia
Lipman, Shelley L. Rosenbaum
Lippert, Laurel Hilde
Lisk, Jerrie Parker
Liss, Cheryl Virginia
Lissant, Bettye L.
Litsche, Mildred E.
Little, Diane
Littler, Marie Katherine
Littrell, Mary Ann
Litwin, Linda Joan
Livada, Irene Guertin
Liverman, June L.
Livingston, Mary M.
Livingston, Pamela Janelle
Livingston, Patricia R.
Llamido, Margaret Pirz
Llorens, Maribel
Lloyd, Grace M.
Lockhart, Maybelle Lillian
Locklear, Sara J
Lockness, Doris E.
Loebbaka, Bettie Anne
Loetscher, Ila Fox
Loewinger, JoAnna Lynn
Lofton, Laura Grossman
Logan, Deborah Joanne
Logan, Myrna Sue
Logan, Tricia Lee
Logue, Judith (Judy) Ann
Lohman, Lillie L.
Lollar, Dianne
Lombardo, Donna Lynn
London, Barbara Erickson
Long, Esther D.
Long, Katherine M.
Long, Maureen L.
Longenecker, Judy A.
Loob, Joan J.
Loomis, Mary P.
Loomis, Nancy Ellen
Lopresti, Arlene E.
Lor, Socheata Krystyne
Lor, Tania Danuta
Lore, Anna Marie
Loricchio, Susan
Louchheim, Valerie Pingree
Loufek, Betty McMillen
Lough, Dorathea Arline
Love, Ruth
Lowe, Gayle Conklin
Lowe, Mary A.
Lowe, Sylvia Jean
Lowenberg, Grace C.
Lowers, Patricia A.
Lowry, Esther P.
Loxley, Carolyn Lee
LoGiudice, Rosemary Jo Anne
Lu, Therese
Lubline, Lori O"Brien
Lucas, Christine Marie
Lucas, Jacqueline Ann

Lucas, Kelly Jean
Luce, Ann R.
Luchs, Linda Lee
Luck, Carolyn W.
Luck, Marceline Dorothy
Luckett, Charlotte Ann
Luckhart, Frances H.
Lucy, Beth L.
Ludington, Shirley Wood
Ludlow, Reba J.
Luedtke, Jacqueline Rae
Luehman, Constance
Luehring, Lois Weatherwax
Lueninghoener, Florence Marie
Lugo, Lissette
Luhta, Caroline N. "Connie"
Lukowitz, Ruth
Lum, Vada M.
Lummis, Ruth M.
Lund, Alice Lee
Lund, Jamie S.
Lund-Bell, Judith Lynne
Lundstrom, Evelyn C.
Lunnemann, Marion Wright
Lupina, Paula Williams
Lupton, Ann Orlitzki
Lusteg, Kathryn Ann
Lustig, Joelle
Luther, Karen Elaine
Luther, Rebecca Lynn
Lutley-Borland, Kathy
Lutte, Becky K.
Lyksett, Joyce R.
Lyn, Sally D. Patricia
Lynfoot, Ernestine Marie
Lynn, Johnnie M.
Lynum, D. Joan
Lyon, Melinda Mitchell
Lyon, Zula V.
Lyons, Dottie
Lyons, Evelyn B.
Lyons, Linda J.

– M –
Mabe, Janice L.
Maben, Georgia L.
Mac Leod, Martha Lozar
Macario, Katherine B.
Macario, R. Canivet
Macarthur, Irene Edith
Macdougall, Shirley
Mace, Joan Elizabeth
Machado, Anesia Pinheiro
Machinek, Dr. Angelika A.
Mack, Barbara Ann
Mack, Connie Lee
Mackoul, Anne Glynn
Macky, Carolyn June
MacDonald, June E.
MacDonald, Mary Margaret
MacIvor, Celeste B.
MacKinnon, Heather Lynn
MacLeod, Barbara
MacLure, Myra Ione
MacMillan, Mary Luella
Madgett, Ruth Ann

Madison, Carey Catherine
Madsen, Cynthia S.
Maestre, Ruth L.E.
Maffettone, Carol Ann
Magee, Mary E.
Magidson, Cynthia S.
Magon, Katherine M.
Magon, Patricia A.
Magouyrk, Nell Sellers
Maher, Sara R.
Mahonchak, Pamela
Mahoney, Laverne Hoyt
Mahoney, Minerva C.
Mahoney-Epstein, Beverly J.
Mahoney
Mairs, Pamela Harter
Maitland, Caro
Majneri, Muriel
Makarsky, Faye Susan
Malady, Susan Mary
Malan, Eileen C.
Malby, Vesta W.
Malcomson, Christine Sue
Malden, Joan Williams
Malek, Lucia Virginia
Malkmes, Joyce C.
Mallary, Pauline L.
Malm, Nancylee
Malmgren, Marie L.
Malone, Abby L.
Malone, Kathleen Ann
Maloney, Dorothy Jean
Maloney, Linda S.
Malpassi, Jill Ann
Manchip, Kristine Elizabeth
Mandel, Mary Catherine
Manheim, Leslie
Manley, Elizabeth A.
Manley, Gayle Margaret
Mann, Eva Fern
Manos, Patricia
Mansel, Kristen Michele
Mansfield, Karen L.
Mantello, Deborah Anne
Manuel, Margaret M.
Manwaring, Betty Lou
Mara, Robin Halley
Marble, Kay Lynn
Marcec, Mona R.
Marchbanks, Karen Rae
Marecek, Melanie
Marie, Dawn
Mariner, Rosemary Bryant
Markert, Lynn Adkins
Marland, Ph.D., Eileen T.
Marquis, Ruth Maureen
Marriott-Johnson, Marion B
Marsell, Wendy
Marsh, Jacqueline Hite
Marshall, Betty
Marshall, Laverna G.
Marshall, Linda H.
Marshall, Margaret Elizabeth
Marshall-Johnson, Tracy Lynn
Martell, Carolyn Glenda
Marti, Merce

Martin, Betty Jane
Martin, Betty Rockwood
Martin, Bonnie Ann
Martin, Dorothy A.
Martin, Eileen
Martin, Elaine J.
Martin, Gisele Elisabeth
Martin, Heidi Diane
Martin, Helen Tabor
Martin, Jane L.
Martin, Jane Marie
Martin, Jean Nicole
Martin, Lois Jean
Martin, Paula Rose
Martin, Regina Ann
Martin, Sabra Hassel
Martin, Stephanie Marie
Martin, Virginia Ann
Martirano, Diletta
Martlew, Glenda L.
Masching, Debra Sue
Mascorro, Marsha Purvis
Mason, Joan
Mason, Judy Ann
Massee, Janice Tuggle
Masson, Dr. Angela
Mastenbrook, Shirley C.
Masters, Nancy Robinson
Masterson, Angela Denise
Masura, Judith Elyse
Matarese, Elizabeth Ann
Matheis, C. Arline "Sue"
Matheny, Yvonne Ann
Mather, Cheryl Ann
Matheson, Gretchen C.
Matheson, Maxine Lynn
Mathews, M. Dolores
Mathews, Valeri Ann
Mathias, Linda B.
Mathieu, Ann Elizabeth
Mathys, Gertrud E.
Matlock, M. Sue
Matthews, Martha Ann
Matthis, Wilma "Diane"
Mattingly, Linda Marie
Mattison, Patricia Dorothy
Mattiza, Virginia Dare
Mattman, Joan Marie
Mattocks, Mary (Willy) Eugenie
Mattuch, Laura Lynne
Matut, Rachel
Mauldin, Linda T.F.
Mauldin, Sandra Ilene
Maule, Susan De Etta
Mauritson, Janet Elizabeth
Mauthe, Marci A.
Maxim, Constance K.
Maxim, Edythe Salo
Maxson, Mary Hass
Maxwell, Ione Hooker
Maxwell, Patricia G.
May, Dian Ward
May, Gloria Louise
May, Judith Ann
Maye, Letitia Teets
Mayer, Celeste Marie

Mayes, Nancy L.
Mayeur, Terri-Jo Annette
Mayfield, Barbara Jeanne
Mayle, Ruth O.
Mayr, Catherine E.
Maze, Donna Hale
Mazur, Jessica Natasha
McAdam, Bobbie Ann
McAllister, D. Gail
McArthur, Patience O"H.
McAteer, Carol Ann
McBurney, Mary Z.
McCall, Nancy Ann
McCallister, Terri Lynn
McCammond, Ila Leona
McCann, Jennifer Dawn
McCarrell, Norma "Jody"
McCartan, Beth A.
McCarthy, Andrea Deane
McCarthy, JoAnne H.
McCarthy, Virginia S.
McCaslin, Gladys I.
McCauley, Elizabeth D.
McChesney, Fiona Elizabeth
McChesney, Grace M.
McClaskey, Marlyn G.
McClellan, Alanna Marie
McClintock, Bonnie S.
McClister, Mary T.
McClung, Dena L.
McClure, Barbara Lee
McCollom, Anne Spalding
McCollom, Phyllis Maxine
McCollum, Erdine I.
McCombs, Joan Esther
McConnell, Constance Ann
McConnell, Georgiana T.
McConnell, Janet Ann
McConnell, Jean O.
McCormack, Diane Douglass
McCormack, Jan E.
McCormack, June
McCormick, Linda Elva
McCormick, Olive A.
McCoy, Mary Margaret
McCreery, Ardath B.
McCullough, Carol Charlene
McCune, Mary Lee
McCurry, Kathleen Gloria
McCurry, Nancy Minor
McCurry, Ruth Margaret (Maggie)
McCutcheon, Margo
McDaniel, Karen M.
McDaniel, Virginia Mae (Ginny)
McDermot, Jill E.
McDermott, Judith L.
McDermott, Yvonne K.
McDonald, Adele Marie
McDonald, Joyce Louise
McDonald, Kathy S.
McDonald-Waugh, Tonya Lee
McDuffee, Patricia E.
McEachern, Alice M.
McElhatton, Jeanne
McEniry, Paula P.
McFall, Shirley

McFarlin, Maj. Margaret R.
McGee, Helen McDonald
McGee, Renee Sharp
McGettigan, Sandra Lynn
McGinley-McCarty, Jill Anne
McGinnis, Nancy W.
McGlasson, Rosella J. (RJ)
McGowan, Suzanne P.
McGrady, Colleen Mary
McGraw, Elizabeth H.
McGuinness, Mary Catherine
McGuire, (Lyn) Linda L.
McGuire, Emma L.
McGuire, Grace
McHaffie, Natalie Eleanor
McHenry, Eva Laura
McIlveen, Evelyn L.
McInnes-Stine, Pamela Margaret
McIntire, Jane Lee
McIntosh, Barbara J.
McIntosh, Rosaly J.
McKay, Mary Catharine
McKeever, Helen
McKelvey, Jeanne Wolford
McKendry, Felicity Helen E.
McKenna, Barbara Louise
McKenna, Sandra Spradley
McKenzie, Jan
McKenzie, Sandra Sue
McKenzie, Susan D.
McKerracher, Lee
McKillip, Mary J.
McKillip, Pamelia Hopper
McKinley, Jan K.
McKissock, Nina Angela
McLaughlin, Evelyne
McLaughlin, Jean
McLaughlin, Renate
McLaughlin, C.S.J., Mary Loretta
McLeod, Sheila Kathleen
McMahan, Helen Frances
McMahon, Catherine M.
McMaster, Maureen Louise
McMillan, Jessica Antonik
McMillan, Marie E.
McMurtrey, Dori Ann
McNabb, Betty W.
McNamara, Kathryn M.
McNamee, Edith Geneva
McNeal, Heather Michelle
McNeil, Heather Anne
McNeil, Kathryn B.
McNeil, Mary
McNeil, Mary Jane
McNeil, Racquel Marilyn
McNiff, Marion Lillian
McNutt, Sue K.
McPherson, Kathrynne Alexine
McRae, Virginia M.
McReynolds, Elaine Harrison
McReynolds, Norma L.
McSheehy, Nancy Ellen
McSorley, Betty Violet
McTague, Linda K.
McWilliams, Janice Kathryn
McWilliams, Lynne F.

Meachem, Cilla Elizabeth
Mead, Barbara J.
Mead, Margaret A.
Mead, Mary Jo
Mead, Shelley Renee
Meadors, Sherry Ann
Meadows, Lynn Ann
Meeder, Dorothy F.
Meeks, Kimberly
Meengs, Ann Louise
Meese, Gabriele Linda
Megill, Shauna
Mehaffie, Charleen Ann
Meier, Joyce Rodgers
Meisenheimer, Alice J.
Mejia, Eva R.
Melby, Kathleen F.
Melchiorre, Robbin Dee
Meldrum, Deirdre Ruth
Melius, Helen K.
Mellinger, Kristin L.
Mellott, Aileen Saunders
Melroy, Pamela Ann
Mendenhall, Merry Kathryn
Mendonca, Eleonora C.
Menkveld, Anna Marie
Menkveld, Anna-Maria
Mennitto, Helen Mcchesney
Merchant, Linda Ann
Mercier, Dorothy L.
Mercker, Mary A.
Meredith, Barbara Laura
Mermelstein, Isabel Rosenberg
Merrell, Genie
Merritt, Carol B.
Merritt, Sylvia M.
Mersch, Ashlee Mariah
Mertz, Marilyn Ruth
Mertz, Susan Vankirk
Mertz, Teresa
Meschi, Janet Ruth
Messerrly, Julie Grace
Metayer, Estelle
Metzger, Carolyn "Lynn" D.
Meyer, Donna M.
Meyer, Doris L.
Meyer, Linda S.
Meyer, Mildred J.
Meyer, Patsy Lee
Meyer, Shari Sloan
Miceli, Heidi
Michaels, Barbara Kadish
Michaud, Karen Marie
Mickelson, Thelma Elizabeth
Miele, Janis M.
Miele, Ruth A.
Might, Carol Ann
Migis, Mary Jane
Mikesch, Gloria Jean
Miklozek, Beth Snyder
Milchanowski, Karen M.
Miles, Lisa A.
Milhausen, Mary Lou
Millan, Cinthia Nicole
Millar, Bernice G.
Millard, Joyce Janet

Miller, Bertha Louise
Miller, Betty J.
Miller, Bonnie Jean
Miller, Cheryl C.
Miller, Davette Dawn
Miller, Dixie Lee
Miller, Donna J.
Miller, Doris E.
Miller, Dorothy R.
Miller, Dr. Brenda Gale
Miller, Elizabeth G.
Miller, Grace Elizabeth
Miller, Heather Ann
Miller, Ilona M.
Miller, Juli
Miller, Lorelei Elizabeth
Miller, Margo Leslie
Miller, Marie T.
Miller, Marilyn Jean
Miller, Marilynn Lucille
Miller, Nancy A.
Miller, Nancy J.
Miller, Phyllis Anne
Miller, Phyllis J.
Miller, Ramona Lynn
Miller, Susan Riva
Miller, Tess S.
Miller, Vivian Grasby
Miller-Grubermann, Connie L.
Millmore, Kimberly Lynne
Mills, Dr. June
Milmine, Sylvia Fern
Milmont, Diane Leslie
Minichiello, Cheryl A.
Minner, Kathryn Joy
Minniear, Louise Kirby
Minnig, Ronni
Minor, Valerie Lynn
Minter, Doris Jane
Minton, Madge Rutherford
Mirabel, Susan
Misiowiec, Leda
Mitchell, Anne Elizabeth
Mitchell, Beverley Heineman
Mitchell, Judy K.
Mitchell, Marcia Elizabeth
Mitchell, Marge
Mitchell, Pamela
Mitchell, Patty L.
Mitchell, Susie
Mitrovich, Rona W.
Mixon, Anita
Mixon, Lisa Ann
Mlady, Patricia Ann
Mlnarik-Holt, Janice A.
Moberly, Mary Ann
Mock, Gail Lynne Steger
Modestino, Linda Marie
Moeller, Hildegard M.
Moffat, Ann
Mohandiss, Carol A.
Mohorovich, Janet L.
Mohr, Elizabeth Bright
Moller, Margaret Jane
Moloney, Mary Jo Gumbert
Monaco, Madeleine Joyce

Monahan, Linda J.
Monahan, Pamela A.
Monk, Ena Catherine
Monroe, Alice J.
Monroe, Cara L.
Monroe, Hazel Shirley
Montague, Jean B.
Monteith, Karen S.
Montgomery, Camille Penny
Montgomery, La Velle M.
Montoya, Nora Isela
Moody, Marilyn Jo
Moody, Patti W.
Moody, Patty Hartley
Moon, Sharon G.
Mooney, Roseann Theresa
Mooney, Tangerine W.
Moore, Aimee Lynn
Moore, Barbara Collester
Moore, Barbara J.
Moore, Donna Diane
Moore, Evelyn Janice
Moore, Evelyn R.
Moore, Greta L.
Moore, Jennifer Lee
Moore, Kathleen Mary
Moore, Kaye Combs
Moore, Martha Lucille
Moore, Mildred Jeannette
Moore, Nancy Cooper
Moore, Peggy Anne
Moore, Shirley L.
Moore, Verba A.
Moorman, Bridget Anne
Moote, Janette Lynnd
Moran, Deborah Marilyn
Moran, Myrna M.
Morchand-Holz, Denise
Morehead, Bonnie M.
Morfitt, Grace Louise
Morgan, Cathy A.
Morgan, Elizabeth J.
Morgan, Jennifer L.
Morgan, Karen Lee
Morgan, Lois Marilyn
Morgan, Velda Lucas
Morgenthal, Becky Holz
Morin, Colette Cecile
Morinaka, Reiko Sakurai
Morison, Janet Z.
Morris, Carol Craig
Morris, Carol Leslie June
Morris, Janice Garrity
Morris, Patricia Ann
Morrison, Beatrice "Betty"
Morrison, Gladys Mae
Morrison, Janet M.
Morrison, Linda Marie
Morrison, Pamela Jane
Morrison, Patricia A.
Morrison, Sheila Ann
Morrow, Elaine Ruth
Morse, Betty Jeanne
Morse, Teresa Elizabeth
Morshead, Catherine Anne
Mortensen, Jette

Mortensen, Linda A.
Morton, Marcia S.
Moseley, Betty H.
Mosher, Berneta Jean
Moshier, Grace Anne
Moskow, Kathleen May
Moss, Betty Sue
Moss, Cara Jean
Moss, Noriko Date
Motley, Elizabeth Marie
Motola, Maureen
Mouhot, Rebecca Ann
Moulder, Helen Marie
Moxley, Ngaire Lorraine
Moynihan, Penny E.
Muehlhausen, Barbara J.
Mueller, Mary Lou
Mullin, Ann
Mullis, Betty Lee
Mulvaney, Jana L.
Munck, Miriam Diane
Murakawa, Keiko
Muranko, Louise G
Murawski, Laura Alyce
Murayama, Mari Diane
Murdock, Christine
Murphey, Margaret D.
Murphy, Catherine Ann
Murphy, Florence Colanth
Murphy, Gertrude "Trudy" M.
Murphy, Katherine B.
Murphy, Linda
Murphy, Mary Margaret
Murphy, Micheal T.L.
Murray, Carol Anne
Murray, Jackie L.
Murray, Jean
Murray, Kimberly Anne
Murray, M. Susan
Murray, Patricia Jo
Murray-Demaree, Barbara Kay
Murren, Barbara Ann
Murry, Jean Emily
Murto, Ingrid Eileen
Music, Sydney Marie
Musser, Carolyn Ann
Musser, Patricia F.
Myers, Christina Holm
Myers, Deborah R.
Myers, Diane Rumble
Myers, Donna T.
Myers, Judith E.
Myers, Karen J.
Myers, Krystal Dawn
Myers, Margot Abbie
Myers, Mary Rose
Myers, Rena L.
Myhre, Deanna S.
Mynster, Velma Fay
Myshatyn, Maria

– N –

Naas, Bonnie S.
Nadig, Betty Lou
Nagy, Penelope (Penny) Jean
Nagy, Teresa Ann

Napoli, Diana J.W.
Narehood, Susan Stewart
Narezo, Sara E.
Nash, Andrea Camille
Nassimbene, Andrea H.
Nasypany-Downey, Mary Elizabeth
Natof, Margaret
Nave, Linda R.
Navia, Ana Maria
Nay, Nancy A.
Nayak, Geetantalee N.
Nead, Lisa Marie
Neal, Jennifer Rae
Neal, Mary Joanne (Jo)
Neale, Mary Lou Colbert
Nealey, Sue C.
Needham, Mary L.
Neel, Faye Wright
Neese, Laura Louise
Neil, Jean Morrell
Neil, Jeanne Marie
Nellans, Marcia A.
Nellis, Deborah Jo
Nelms, Renee Elizabeth
Nelson, Constance Lovatt
Nelson, Janice Rae
Nelson, Kirsten Anne
Nelson, Margaret M.
Nelson, Marlene Dee
Nelson, Shirley Ann
Nelson, Shirley Ann
Nelson-Boutet, Laureen
Nemhauser, Vivian G.
Nerroth, Barbara Sue
Netherton, Rebecca Tullis
Nettleblad, Jane H.
Neu, Donna E.
Neuman, Golda Maurine
Neumann, Linda H.
Neumann, Nancy Joan
Neumeier, Melanie Lorraine
Neville, Lois Elaine
Nevitt, Norma J.
Newbold, Jeannie A.
Newcomb, Anne Michael
Newcombe, D. Ann
Newell, Juanita
Newhouse, Kathleen M.
Newhouse, V. Lorraine
Newman, Alice Marie
Newman, Judith Marie
Newth, Aline Kay
Nicholas, Elizabeth P.
Nicholls, Judith Lee
Nichols, Ruth S.
Nichols, Wilma Joyce
Nicholson, Alberta H.
Nicholson, Carol M.
Nicholson, Constance Francis
Nicholson, Judith R. Panizian
Nicholson, Lauren Trent
Nicholson, Margaret
Nicholson, Margaret F.
Nickell, Bernita M.
Nickoles, Marion Corzine
Nicks, Betty L.

Nielander, Hope Hayden
Nielesky, Janice Gerber
Nielsen, Janice Ann
Niles, Angela Marie
Niles, Joan Bates
Niles, Sandra Marie
Nilson, Wendi M.
Niquette, Beverly J.
Nishihara, Junko
Nissen, Joanne
Nix, Elizabeth Ann
Nixon-Rios, Cynthia Elizabeth
Noble, Alice Faye
Noble, Merrilyn Martin
Nobles-Harris, Ellen Marie
Nobmann, Elizabeth D.
Noffke, Esther E.
Noland, Dodie Sue
Nolen, Janet Githens
Noll, Johanne
Noonan, Ellen Huffman
Noonan, Lynn P.
Noren, Patricia Ruth
Norkus, Dorothy A.
Norman, Arlene Janice
Norman, Debbie Travis
Norman, Martha Helene
Norman, May Marguerite
Norris, Kelli Kae
Norris, Nancy L.
Northam, Ellen R.
Northrop, Barbara Kay
Northrop, Karen Anne
Northrop, Kimberlei Anne
Northrop, Patricia Kay
Novaes, Nancy
Novosel, Lorna M.
Noyes, Hazel Jane
Nozoki, Yae
Nugent, Charli
Nunn, Elizabeth A.
Nunnery, Joyce Cutsinger
Nusbaum, Dolores Helen
Nutter, Zoe Dell
Nydegger, Margaret Josephine

– O –

O'Brien, Adrianne Lucyle
O'Brien, Anne T.
O'Brien, Blanche E.
O'Brien, Elizabeth A.
O'Brien, Mary
O'Brien, P.K. "Kathy"
O'Brien, Pamela Azar
O'Connell, Jane M.
O'Connell, Lee Eastin
O'Connell, Teresa Lynne
O'Connor, Connie Marie
O'Connor, Debra
O'Connor, Ruth Elkinton
O'Day, Valleta M.
O'Donnell, Lynn
O'Farrell, Kieran Katlin
O'Grady, Barbara Mary
O'Hara, Carol Anne
O'Hara, Ellen L.
O'Hara, Shannon Colleen
O'Hora-Webb, Denise B.
O'Kelley, Genie Rae
O'Laughlin, Nancy Carol
O'Neal, Katie Lee
O'Neil, Debra Divine
O'Riley, Amy Lee
O'Shaughnessy, Linda Marie
Oakes, Lianne Rene
Oakley, Teresa M.
Odell, Carol H.
Odenthal, Mary Ann
Odom, Mary Elizabeth
Odom, Nora Sue
Odorico, Eleanor Ellie
Oehler, Frances Jean
Oelschlager, Mary Lou
Ohl, Shirley T.
Ohlau, Margaret Ann
Ohlsson, Patricia Ann
Ohnemus, Jeanne K.
Oja, Hannah Elina
Ojala, Claire M.
Olberding, Debra L.
Oldaker, Joyce B.
Oldershaw, Maude H
Oldham, Candie
Oleson, Jennifer M.H.
Oliver, Mildred Diehl
Olivera, Olive May
Olivier, Cheryl Yvette
Olivolo, Elizabeth Jean
Olmsted, Marion Rose
Olovitch, Edith Adeline
Olsen, Barbara A.
Olson, Barbara Jane
Olson, Jennifer Eliene
Olson, Jimmie Clark
Olson, Patricia Mckennon
Olson, Ruth M.
Omstead, Barbara Marie-Piper
Orchard-Armitage, Nicola J.
Orcutt, Patricia M.
Orlando, Kathleen Yvonne
Orloff, Marilyn C.
Orloski, Elizabeth Jane
Orosz, Ginny
Orr, A. Lee
Orr, Janice Marie
Orr, Rita Ann
Orsini, Jean R.
Osbakken, Patricia Louise
Osborne, Allegra A.
Osborne, Susan H.
Osland, Linda R.
Osman, Catherine Howes
Osmon, Patricia A.
Oster, Diane M.
Oster, Eva Elisabeth
Oster, Julie Dawn
Oster, Maureen P.
Osterud, Joann
Oswald, Marie Elizabeth
Otiker Ph.D., S. Toni
Ottaway, Pauline Margaret
Owen, B.C. (Be Be)
Owen, Pamela Marie
Owens, Sally J.
Owsley, Leslie C.
Oxford, Janet Irene
Ozbun, Martha

– P –

Pace, Clara Pearce
Paddeck, Jeraldine L.
Paddock, Astarte
Paine, Katherine B.
Paine-DuPont, Melinda Anne
Paisley, Suzanne
Pal, Susan
Palazzolo, Valerie Geraldine
Palmer, Anita Joan
Palmer, Patricia Texter
Palmer, Rosemary Elizabeth
Palmer, Susan B.
Palmer, Zoann
Palombi, Joan M.
Paluzzi, Victoria Ann
Panczyszyn, Mary
Pandorf, Denice K.
Pangrazzi, Anna Marie
Pankalla, Bonnie C.
Pantages, Judy Lorraine
Paoli, Sylvia Lee
Papayans, Sheila R.
Papp, Lori Ann
Pappas, Georgia Elaine
Para, Carol
Parish, Suzanne D.
Park, Dolores D.
Park, Valerie Jean
Parke, Helen Marie
Parker, Adrienne
Parker, Alma B.
Parker, Heather Mary
Parker, Inga Corinne
Parker, Lynnell Page
Parker, Maxine M.
Parks, Eva M.
Parks, Mary Ellen
Parmenter, Sara Lester
Parmer, Carolyn Sue
Parr, Sharon E.
Parrish-Jones, Judith Totman
Parson, Jill Marie
Parsons, Lori Kae
Parsons, Mary E.
Paschket, Darlene Joy
Pasqualino, Carolyn Gaye
Pasten, Dr. Laura Jean
Pasternak, Bernice A. "Bunnie"
Patamapongs, Malawan
Pateman, Yvonne Celeste
Patierno, Marilyn
Patino, Toni N.
Patricia-Mars, Colleen Marie
Patrick, Carolyn D.
Patrick, Pamela S.
Pattavina, O. Frances
Patten, Marcia A.
Patterson, (Billie Jo) Pud
Patterson, Betty J.
Patterson, Camille Minor
Patterson, Carolyn Maria
Patterson, Kennie Ruth
Patterson, Marianne A.
Patterson, Marla
Patterson, Shirley A. Phillips
Pattison, Ella May
Patton, Elizabeth K.
Patton, Janet Lee
Paul, Janice W.
Paul, Sandy D.
Paul, Wendy Elizabeth
Paulet, Roberta Angela
Paulus, Barbara E.
Paver, Wendy S.
Pawlowski, Cynthia S.
Paynter, Joan T.
Payton, Sophia M.
Peake, Margaret Juliet
Pearce, Molly Martin
Pearcy, Donna Louise
Pearson, Jean H.
Pearson, Patricia Liebeler
Pearson, Penelope Kristina
Peck, Lois Ann
Peck, Sharon Lynn
Peckham, Andrea
Peckham, Judith E.
Pecora, Lisa Allyson
Pedersen, Ingrid Elisabeth
Peel, Deborah R.
Pence, Henrietta
Pendleton, Carole A.
Penney, Dorothy Montgomery
Penstone, Ann E.
Pentecost, June P.
Peper, Irene Iowa
Pepler, Anne M.
Peppler, Isabel L.
Perelman, Debra A.
Perez, Dr. Rosalind
Perez, Stephanie Lorraine
Perica, Analee Holden
Perkins, Valerie Jene
Perrigo, Susan F.
Perrin, Helen Margaret (Ryan)
Perrine, Lynn Marie
Perry, Evelyn Skeen
Perry, Helena Bruin
Perry, Jan
Perry, Jane Kathleen Hutton
Perry, Paulie M.
Pescatello, Ann Marie
Peters, Jeanne M.
Peters, Jo Ellen
Peters, Marilyn G.
Peters, Mary Kay
Petersen, Sandra Kae
Peterson, Diana Jane
Peterson, Dorothy Agnes
Peterson, F. Darlene
Peterson, Hazel I.
Peterson, Jan M.
Peterson, Laurel Rae
Peterson, Linda G.
Peterson, Penny W.

Peterson, Sandra Lynn
Petosky, Patricia Ann
Petroline, Nancy Joan
Pettigrew, Suzanne
Pevehouse, Jean O.
Pfeffer, Suzanne Ruth
Pfeifer, Maxine L.
Pfeiffer, Sharon Rose Groome
Pfender, Janet Louise
Pfister, Betty H.
Phelps, Carol Ann
Phelps, Janet Vee
Phillips, Ardith Derr
Phillips, Billie Kathryn
Phillips, Doris C.
Phillips, Elizabeth Tyler Davis
Phillips, Janet E
Phillips, Louise M.
Phillips, Martha W.
Phillips, Sylvette Elizabeth
Philpot, Susan
Philpott, Glenda
Phipps-Alden, Margo Lovendale
Pickering, Aileen L.
Pickering, Jean Allen
Pickett, Cindy Lynn
Pickle, Tiana Marie
Pierce, Virginia Marie
Pifer, Betty Ann
Piggott, Ann Tingley
Pilaar, Carolyn
Pilkinton, Amy L.
Pillows, Darlene
Pinaire, Marla Lynn
Pinciotti, Margaret L.
Pinkney, Mary Chernus
Pinneri, Danielle
Pinto, Donna M.
Piper, Marian Lucille
Piper, Sara Martin
Piperis, Janice (Jan)
Pippen, Elizabeth Jean
Piramoon, Mary Jo
Pirman, Linda Diane
Pistorius, Roberta Snyder
Pitt, Mary S. (Terry)
Pittman, Joyce S.
Pizzolato, Dora May
Plaia, Gayle Rene
Plante, Isabelle Virginia
Plantz, Patty Ann
Platt, Sharon S.
Plumley, Sally Ross
Plummer, Dorothy Jane
Plummer, Sally Lynne
Poarch, Jacquelyn Martha
Pobanz, Carolyn S.
Pochert, Patricia Best
Pocock, Janice B.
Podeweltz, Dana Annette
Podwys, Bertha U. Siegl
Poeling, Linda Lee
Poetzman, Kelly Ann
Pohl, Margie Lee
Pohl, Marylou A.
Poklar, Janeen Ann
Polen, Kateri Lee
Polen, Linda Louise
Poling, Patricia Ann
Polsky, Amy J.
Pomeroy, Christina R.
Pons, Angela Antoinette
Ponton, Charleen Sullivan
Pool,md, Marjorie K.
Poon, Tsui Ping
Poore, Marilou Soanes
Popa, Geri-Sue
Pope, Kitty Maynard
Popovich, Marina L.
Porter, Barbara Hamilton
Porter, Bonnie Andrew
Porter, Frances L.M.
Porter, Kathryn Lee
Porter, Louisa
Porterfield, Joyce Ann
Portnoy, Judith Ann
Ports, Dorothy L.
Portwood, Esther Mae
Posner, Patricia Ann
Post, Virginia A.
Postlethwaite, Jane Marie
Postma, Frances Huff
Potter, Ilovene N.
Potter, Jane Palmer
Potter, Kirstin Andrea
Potter, Linda Susan
Powell, Barbara Cohrssen
Powell, Brooks Mooney
Powell, Frances M.
Powers, Amy Elizabeth
Powers, Maureen Therese
Prakken, Betty L.
Prater, Jo-Ann
Pratt, Mary Lynne
Pratt, Sandra Lynn
Prentiss, Patricia Noyes
Presley, Marcia Lynn
Presley, Sandra D.
Presnell, Georgeanna
Preston, Theresa Marie
Prewitt, Katherine E.
Price, Beverly Franz
Price, Dallas P.
Price, Emmy Louise
Price, Jacqueline Lee
Price, Katherine A.
Price, Mary A. (Toni)
Price, Nora Jean
Priest, Lisa M.
Prigmore, Suann W.
Prochaska, Verna Mae
Proctor, Anne P.
Proctor, Joyce A.
Proctor, Virginia M.
Prokop, Adrienne Vollmer
Pronczuk, Danuta Veronica
Prophett, Marian E.
Provart, Joy Sybilla
Provencher, DeLaney
Province, Anita C.
Pruett, Rachel M.
Pryor, Joyce C.
Pucci, Susan J. Meeker
Pugh, Carolyn C.
Pugliese, M. Gabriela
Pulaski, Lori Jaye
Pulaski, Madine
Pulis, Dorothy
Pulley, Michele Lea
Pullins, Miriam H.
Pulver, Linda Rose
Purcell, Kim L.
Purcell, Mary C.
Purdy, Lulu Mae
Puri, Manisha Mohan
Pustmueller, Helen Maxson
Putman, Sandra C. Phillips
Pyatt, Mary Jean
Pyne, Mary Ethleen
Pynes-Yellin, Delores A.
Pyysalo, Satu Marika

– Q –
Quarles, Nona
Queen, Cynthia D.
Quinn, Mary F.

– R –
Rabung, Virginia
Racka, Ingrid Antanina
Rader, Karen Arnett
Rader, Melanie Mae
Radford, Dianna Kay
Radford-Price, Lindsey Helen Mora
Radzai, Karen Thelma
Radzin, Monica Therese
Raffaelli, Virginia Rae
Raftery, Deborah Jean
Ragaz, Margaret K.
Ragland, Carole C.
Ragsdale, Nancy
Rahn, Diane L.
Rahn, Jodi Elizabeth
Raiche, Voline L.
Rainwater, Virginia Alice
Rajnus, Sharon Lyn
Rakestraw, Christine Gail
Raleigh, Constance Mary
Raleigh, Maureen G.
Raley, Claire
Ralph, Sheila Rae
Ralston, Jane
Ramelli, Sigrid E.
Ramer, Helen D.
Ramirez, Nancy L.
Ramming, Deborah Anne
Ramos, MarieLouise A.
Ramsey, Dora Lee
Ramsey, Suzanne Pauline
Rancourt, Judie
Randall, Deborah Lynn
Randolph, Julia Ann
Rands, Amy M.
Rank, Patricia Wilson
Raphael, Beth Ann
Rapier, Bonnie Lee
Rappaport, Margaret M.
Ratley, Sarah Lee
Ratliff, Pamela S.
Rauth, Kathryn Anderson
Rawlings, Irene Marie
Rawlings, Mary
Rawson-Sweet, E. Gloria
Ray, Ava J.
Ray, Dorothy H.
Ray, Georgianne Gale
Ray, Hilda W.
Ray, Rose M.
Ray-Geier, Claire Louise
Rayburn, Carol S.
Raymond, Janice R.
Raymond, Lorraine R.
Raymond, Mary W.
Read, Andrea Beatrice
Read, Anita Ann Atwater
Read, Laura B.
Reading, Martha Ann
Reagan, Linda S.
Reagan, Sandra Linelle
Ream, Roberta L.
Reavis, Debbie Kay
Recken, Roberta Louise
Reckson, Paula Lynn
Rector, Diann
Redford, Charlotte Ann
Redlawsk, Judith A.
Reeb, Morgen Ann
Reece, Flora Belle
Reece, Lisa Rae
Reed, Dana Ann
Reed, Ginny Clarice
Reed, Karen Rae
Reed, Mary Lou
Reed, Patricia Ann
Reed, Tia Linn
Reed-Spera, Connie A.
Reeder, Deborah J.
Reel, Elizabeth Louise
Reep, Diane Michelle
Reep, Jennifer Belle
Reep, Kelli Dawn
Rees, Lorin Caren
Reeve, Faye Lynette
Reeves, Laurie S.
Reeves, Monica Manners
Reeves, Sandy A.
Regan, Dorothy Marie
Regan, Robyn E.
Regier, Elaine Roxanne
Rego, Jane Marie
Reichel, Diane Elizabeth
Reichenbach, Eleanore
Reid, Amelia Carman
Reid, Janet Robison
Reid, Joanne Wixon
Reid, Mary C.
Reid, Raquel Erin
Reider, Eileen M.
Reilich, Hialeah
Reinbold, Edna Marie
Reindl, M. "Joan"
Reinemer, Joy Lorraine
Reinhard, Antonia Wehmann
Reinhardt, Nancy Marie

Reinhart, Judith Anne
Reintjes, Penelope A.
Reis, Lee D.
Reisman, Brenda Gail
Reiss, Elizabeth Linn
Reiter, Betty A.
Reitmeyer, Dr. Janice Pachence
Reitz, Shirley
Rellihan, Paula Stringer
Remol, Carolyn Boatwright
Rempp-Tate, Cynthia
Renbeck, Andrea Lynn
Render, Shirley Linda
Renema, Arlene Diane
Renneisen, Carol L.
Renneke-Wiest, Lynnette Elizabeth
Renning, Rhonda Harriet
Rentfrow, Lorene P.
Repine, Phyliss Laverne
Retzer, Laurie Gail
Reukauf, Carol Ann
Revelle, Joyce S.
Reynolds, Constance Young
Reynolds, Jean Hart
Reynolds, Jean W.
Reynolds, Nellie Margaret
Rhea, Delores Maxine
Rhian, Dera Lorraine
Rhodes, Ginger
Riach, Dorothy Frances
Rials, Rhonda Jo
Ricci, Lola
Rice, Janice M.
Rice, Mary Jane
Richard, Christine Anslow
Richard, Lorraine
Richard-Jones, Carol A.
Richards, Faith B.
Richards, Judy Lee
Richards, Linda Stephens
Richardson, Betty
Richardson, Josephine
Richardson, Nancy Jean
Richied, Debra D.L.
Richland, Bonny L.
Richter, Darla
Richter, Doreen K.
Richter, Laura Jane
Rickabaugh, Elynore
Rickel, Joy Louise
Ricks, Laura Kay
Ricord, Deane F.
Riddle, Betty F. Martin
Riddle, Jean E.
Riddle, Margaret C.
Ridgeway, Sondra Joan
Riemers, Endora Aletha
Rieske, Betty Jane
Rifkin, Lois Ann
Rigden, Gail E.
Riggs, Barbara B.
Rihn-Harvey, Deborah Furstenberg
Riker, Claudette Marie
Riley, Heidi H.
Riley, Mary Virginia
Riley, Melissa Nell

Rinaldini, Julie S.
Rinck, Lois Lemay
Rinehart, Terry London
Ring, Marie-Christine
Ring, Mary Kathryn
Ringenberg, Margaret J.
Ringer, Lynn Diane
Rippelmeyer, Lynn Janet
Risberg, Irene O.
Ritchey, Doris Jean
Ritter, Diane Campbell
Ritter, Dorothea E.
Roach, Dorothy M.
Roach, Frederica S.
Roach, Jane Louise
Roam, Kay M.
Roark, Colleen Joye
Robbins, Joan D.
Robbins, Joy H.
Robbins, Rennie
Robens, Jane Florence
Roberts, Alice
Roberts, Anne W.
Roberts, Barbara
Roberts, Carol Ann
Roberts, Carol L.
Roberts, Diane C.
Roberts, Elizabeth N.
Roberts, Mona Elizabeth
Roberts, Patricia W.
Roberts, Sally J.
Roberts, Shirley Ann Brownfield
Roberts, Stephenie R.
Robertson, Alice Margaret
Robertson, Deanna Dale
Robertson, Doris
Robertson, Helen M.
Robertson, Julie
Robertson, Lorene E.
Robertson, Natha Jane
Robey, Senja Raymond
Robichaud, Dominique Marie
Robins, Elizabeth Rachel
Robinson, Beth Ann
Robinson, Brenda Lee
Robinson, Dorothy
Robinson, Dorothy Callahan
Robinson, Florence Abbie
Robinson, Suzanne Margaret
Robishaw, Lori L.
Robison, Glenys Anne
Roby, Bernice B.
Rochowiak, Pamela Joan
Rock, Lee C.
Rock, Molly Anne
Rockcastle, Ruth J.
Rockwell, Patricia
Rockwell, Patricia Thorpe
Rodd, June Evelyn
Roderick, Elaine K.
Rodewald, Ann E.
Rodewald, Rosemary L.
Rodgers, Nancy F.
Rodriguez, Valerie Rose
Rodriquez, Raquel Orietta
Roe, Holly Lee

Roe, Roberta G.
Roedel, Shelley Lee
Roehrig, Elaine C.
Roemer, Mary E.
Roethke, MD, Anne E.
Roettger, Dr. Belinda Faye
Rogers, Erna Higgins
Rogers, Fay N.
Rogers, Jacki-Lyn
Rogers, Margaret Nancy
Rogers, Martha S.
Rogers, N. Jean
Rogers, Nancy Jean
Rogers, Rita
Rogers, Sheryl
Rogus, Linda R.
Rohde, Barbara Jo
Rohde, Peggy Irene
Rohrberg, Eugenia Trapp
Rokos, Georgeanna
Rollen, Lucile Katherine
Rollo, Vera Foster
Romagno, Mary Lou
Romberg, Tanna May
Romero, Beverly Catherine
Romero, Lynn
Rooney, Dorothy M.
Roose, Linda Sue
Root, Lisa Ann
Rorick, Mary Jane
Rose, Diane F.
Rose, Jeanne C.
Rose, Myrt Anne
Rose, Shelley J.
Rosen, Nancy Elperin
Rosenau, Marlo J.
Rosenman, Bernice
Rosenthal-Metler, Karen
Rosevelt, Jane Margo
Ross, Dorothy A.
Ross, Linda Herron
Ross, Mary K. (Kitty)
Ross, Phyllis Marguerite
Ross, Polly
Ross, Ryan Paige
Ross, Victoria
Roth, Marilyn M.
Roth, Sylvia
Rotstein, Frances Jean
Rotzinger, Wadad "Poochie"
Round, Laurie-Lee
Rowe, Patricia L.
Rowland, Judy M.
Rowlett, Linda Irene
Rowley, Joyce Ann Marie
Roy, Florence E.
Roy, Jane E.
Royce, Marci
Royle, Madeleine Kelly
Rubin, Katherine Leah
Rubin, Melanie S.
Ruby, J. Eloise
Rucker, Cindy M.
Rucker, Joan Ellen
Ruckerbauer, Dr. Gerda M.
Rudischhauser, Barbara A.

Ruehle, Jane LeFever
Ruggles, Gloria Ruth
Ruhe, Gerda
Rulik, Betty H.
Ruller, Sandra L.
Rumbaugh, Paula K.
Runkle, Charla Ann
Ruokolainen, Outi Helena
Ruppert, Liliane
Rush, Catherine Maureen
Rusk, Benetta Lee
Russ, Joyce E.
Russell, Corneil Marcia
Russell, Jennifer Marie
Russell, Lois Roberta Langley
Russell, Martha F.
Russell, Maryellen
Rust, Martha Anne
Ruth, Anitra Doss
Ruth, Marion Babe
Rutherford, Mary H.
Ruzicka, Lyn Williams
Ryall, MD, Jo-Ellyn M.
Ryan, Barbara S.
Ryan, Dorothy Swanson
Rylee, Freda B.

– S –
Saba, Elizabeth Garrard
Sabels, Sonja Louise
Sabins, Lynda Joan
Sable, Karen Anne
Sacchi, Louise
Saettone, Marina F.
Sageser, Denise Jean
Sailer, Helen R.
Saito, Takako
Sakakihara, Lucille A.
Salcedo, Betty L.
Salerno, Alison Lea
Salisbury, Pat J.
Salkin, Janet Lynne
Salmans, Caroline Grubbs
Salsamendi, Kelly Kay
Salvair, Cathy M.
Salvo, Sandi
Sammis, Charlene
Sammon, Helen M.
Samuels, Mary Elizabeth
Samuelson, Kathleen Sue
Samuelson, Margaret Ann
Sanborn, Mary Lorraine
Sanchez-Eaton, Gail D.
Sandau, Sidney Wales
Sanders, Carolyn Coleman
Sanders, Christiana Lee
Sanders, Darline "Dottie"
Sanders, Diane Jane
Sanders, Kimberly Kay
Sandhagen, Sheryl Jean
Sandling, Paula Bazar
Sandman, Nurit
Sands, Gloria Pauline
Sands, Sherry Ann
Sani, Patricia Guichard
Sansone, Carolee A.

Santa Cruz, Susan M.
Santamaria, Barbara Gail
Santosuosso, Mary E.
Santucci, Gloria L.
Saperstein, Linda G.
Sargent, Frances Rohrer
Sarkisian, Jessica
Sarkison, Rhonda M.
Sartori, Grazia Serena
Sasser, Evelyn Lois
Satterfield, Jeanette Parker
Satz, Karen Christina
Sauder, Jacquelyn A. Luke
Sauerwein, C. Jane
Saunders, Evelyn Florence
Saur, Jerelyn Weyer
Sautter, Dawn Patterson
Sauvage, Janice J.
Savage, Hilda F.
Savage, Patricia
Savinsky, Tanya Lynn
Scanio, Marion Rooke
Scanlon, Carol Ann
Scavone, Jeanne Anne
Schaaff, Harrilyn A.
Schalk, Barbara A.
Schamber, Mary J.
Schantz, Jetta Denone
Scharr, Adela Riek
Scheer, Nancy Aedes
Scheinert, Paula J.
Scher, Helene Lenz
Schermeister, Janet Marilyn
Schermerhorn, Elizabeth J.
Schiappacasse, Deeann Lynn
Schick, Dorothy M.
Schiek, Jayne A.
Schiermbock, Jean M.
Schiff, Daphne
Schiff, Susan Eisner
Schiffmann, Jean M.
Schillen, Maria Cristina
Schlachter, Kathleen Ann
Schlafly, Laura L.
Schlieckau, Ellen Kay
Schlundt, Tannie W.
Schmauder, Sherie G.
Schmertzler, Patricia Anne
Schmid, Karen Gloria
Schmidt, Adeline E.
Schmidt, Elsie McBride
Schmidt, Frances Marie
Schmidt, Kimberley Lynne
Schmidt, Mary Orene
Schmidt, Nelda Jean
Schmidt, Renate
Schmidt, Valerie Lee
Schnaubelt, Loretta Cecelia
Schneeweis, Diane Carin
Schneider, Laura E.M.
Schnelker, Diane D.
Schock, Margaret Goodrich
Schoessler, Kathryn Laverna
Schofield, Cynthia Ann
Scholder, Beth Yvonne
Scholten, Anna Marie
Schoppaul, Elizabeth Anne
Schorer, Carolyn Jones
Schorsch, Ann Marie
Schorsch, Sharon Ann
Schottle, Ruth Neva
Schrader, Corinne Elizabeth
Schram, Velma Jane
Schramm-Ogne, Wanda Lee
Schrank, Vivienne D.
Schreiber, Shirley
Schreiner, Laura J.
Schreve, Catherine Miller
Schrick, Elizabeth Ann
Schriner, Martha Coffman
Schroder, Beverly A.
Schroeder, Joyce Ellen
Schroeder, Mary Kay
Schroeder, Patricia Cleary
Schu, Mary A.
Schubert, Nancy E.
Schuerman, Pamela Faye
Schuetze, Elizabeth Ann
Schuhmann, Josephine Claudia
Schultz, Lucinda Ann
Schultz, Paula Diane
Schulz, Jean Annette
Schulz, Jean Forsyth
Schulz, Sigrid
Schumacher, Juanita Jane
Schumm, Linda K.
Schurr, Connie Lee
Schutte, Audrey M.
Schuyler, Marilynn L.
Schwartz, Bonnie Lee
Schwartz, Inez E.
Schwartz, Merav
Schwarz, Pamela Bodie
Schweizer, Virginia M.
Scibetta, Jean E.
Sclair, Robyn
Scott, Anna Regnera
Scott, Beverly Rhonda
Scott, Carol Patricia
Scott, Carole Mary
Scott, Eleanore Hacking
Scott, Jacqueline Smith
Scott, Marykate J.
Scott, Michele Edwards
Scott, Phyllis V.
Scribner, Mary E.
Scriven, Erna Marie
Scroggs, Linda Peckham
Scudellari, H. Beryl
Scully, Linda Maria
Scurr, Erica Margaret
Sealy, Christina Marie
Seamans, Shirley Jean
Searle, Lynn M.
Sears, Susan C.
Sears, Sylvia Jean
Seaton, Jean M.
Seccia, Sue Ann
Seck, Etta "Ruth"
Seddon, Marilyn Carole
Seguine, Nancy Carolyn
Seidenberg, Faith April
Seisser, Cheryl Ann
Self, E. Anne Wallis
Sellers, Cyd A.
Sellers, Greta Loren
Sellers, Murray Mccowen
Sellinger, Laura
Selwitz, Barbara Jean
Semas, Betty
Sena, Lynnette Yuki
Sendrak, Shirley M.
Senft, Mary Ann
Senior, Anona V.
Senko, Katherine Ann
Serasio, Michaele Sharlene
Serfass, Janice Elizabeth
Servaas, Jean E.
Servos, Marion Louise
Seslar, Mary E.
Sesock-Miller, Donna Marie
Sestito, Barbara A.
Sewell, Elizabeth V. Susie
Sexton, Susan Joy
Seydoux (Nairobi), Fabienne
Seyferth, Joyce Margaret
Seymour, Bonnie Lee
Seymour, Linda Claire
Shadel, Ruth Arlene
Shafer, Gwen Elizabeth
Shaffer, Margaret
Shaffer, Marilyn Jean
Shakespeare, Jan M.
Shaneyfelt, Ann
Shannon, Elizabeth Ann
Shannon, Margaret B.
Shapiro, Evelyn R.
Shapiro, Sally Keck
Shapiro, Sara
Sharkey, Sabrina Therese
Sharp, Beverley D.
Sharp, Erika
Sharp, Jane Neal
Sharp, Rose Harrison
Sharpe, Eleanor Nettie
Shattuck, Elizabeth A.
Shaum, Tania J.
Shaw, Diane E.
Shawcross, Charlotte Yates
Shea, Mary A.
Shearer, Grace J.
Shechter, Smadar
Sheeks, Kathleen Sue
Sheets, Jeri
Sheets, Linda L.
Sheffman, Tamra H.
Shelby, Gloria D.
Sheldon, Nancy Ruth
Sheldon, Ruby M.
Shelrud, Constance Joan
Shelton, Frances Sears
Shenkman, Beverly Allyn
Shepherd, Carol A.
Shepherd, Lucille B.
Sherburne, Mariel B
Sherman, Karen Isabel
Sherman, Mildred (Deana) M.
Sherman, Tessa Louise
Sherman, Vicki Lynn
Sherman-Yonker, Clara J.
Shields, Anne M.
Shields, Sue Carol
Shigley, Carol Mann
Shilling, Martha Jo
Shimer, Julie Ann
Shimohira, Akiko
Shinn, Lyn C.
Shinn, Mildred Lorraine
Shirrell, Janne M.
Shively, Avaniel
Shofer, Cynthia Royce
Shogren, Eldris B.
Shonk, Sara E.
Shonk, Stephanie F.
Shonnard, Joan T .
Short, Pamela Joy
Shortreed, Mary Lou
Showers, Virginia
Shroff, Mohini Khubchand
Shropshire, Helen Mae
Shubel, Barbara A.
Shuhart, Nita L.
Shutt, Constance Helen
Sibenik, Lois H.
Sicard, Minette Marie
Sichau, Amber Rean
Sickler, Joan V.
Siderwicz, Margaret M.
Sieber, Delrose
Siedschlag, Margaret E.
Siegel, Jacqueline G.
Siemon, Beatrice C.
Sieracki, Rosemary S.
Sierchio, Barbara Latture
Siew, MD, Shirley
Sifton, Heather Ann
Sig-Hester, Hazel L.
Sikorsky-McCallum, Siegrid A.
Silagi, Barbara
Silberman, LeAnn S.A.
Silcox, Barbara Ann
Silver, Bonnie
Silverstein, Jean Shields
Simmons, Barbara J.
Simmons, Ginger Jones
Simmons, Jane J.
Simmons, LaDell Charlene
Simmons, Sharon K.
Simon, Marion I.
Simonson, Kimberly J.
Simpson, Betty D.
Simpson, Margaret June
Simpson, Susan Diane
Sims, Beverly Z.
Sims, Jane S.
Sims, Sharon L.
Simsarian, Arax
Sinacori, Linda
Sinclair, Beverly J.
Sinclair, Christine Aleece
Sincora, Loretta B.
Singer, Christine Anita
Singewald, Velzora
Singh, Meera

Siporin, Susan
Sisson, Sherry Elizabeth
Sisson, Virginia B.
Sitler, Ruth L.
Sizemore, Evelyn Adams
Skaggs, Barbara Jan
Skaggs, Sandra W.
Skalla, Susan L.
Skeeters, Suzanne Lynn
Skerman, Gabrielle
Skiber, Carol Ann
Skinner, Carol Joyce
Skinner, Mary J.
Skliar, Janis
Skolfield, Cynthia L.
Skomars, Michelle Ann
Skoos, Norma
Skywork, Angelee Alice
Slaton, Marie Rushing
Sleeper, Christine Fernald
Sleeper, Sara Fair
Slimmer, Fran
Sliwa, Nancy Ellen
Sloan, Alice Jean
Sloan, Susan Leigh
Slodowy, Rosemarie
Smagalski, Carolyn M.
Small, Shannon D'Andrea
Smallish, Judith Wilk
Smart, Patricia Gail
Smead, Pamela Kay
Smet, Laurie Ann
Smetana, Julie Ann
Smit, Marie-France Gisele
Smith, Angel Marie
Smith, Annette Catharine
Smith, Barbara "Bobby"
Smith, Becky K.
Smith, Belinda M.
Smith, Betty Jo
Smith, Camelia Morrow
Smith, Carey Ruth
Smith, Carlene Marie
Smith, Cheryl-Ann
Smith, Christy Anderson
Smith, Cynthia Aulbach
Smith, Deborah Ann
Smith, Donna Jean
Smith, Donna Jeanette
Smith, Doris Phillips
Smith, Dorothy Jean
Smith, Elizabeth W.
Smith, Eva Jane
Smith, Harriet Joy
Smith, Janette E.
Smith, Jill Dianne
Smith, Joan J.
Smith, Joyce Mize
Smith, Judith L.
Smith, Katherine M.
Smith, Katie Marie
Smith, Kim Renee
Smith, Kitty Sue
Smith, Laura
Smith, Lillie Irene
Smith, Louise McEwen

Smith, Madelyn T.
Smith, Mae Elaine
Smith, Margaret L.
Smith, Marianne M.
Smith, Marilyn Patricia
Smith, Marlene Cecile
Smith, Marlene Elizabeth
Smith, Martha Ann (Marcie)
Smith, Mary Elizabeth
Smith, Nanoya
Smith, Pamela Ann
Smith, Pamela S.
Smith, Patricia
Smith, Patsy Ruth
Smith, Peggy Dawn
Smith, Shari Radawn
Smith, Sheelah R.
Smith, Sheila J.
Smith, Tina Denise
Smith, Virginia M.
Smith, Virginia Mary
Smith, Yvonne C.
Smith-Capps, Susan K.
Smith-Lynch, Sandra Kay
Smith-Wright, Deborah Jan
Smither, Kathryn Mary
Smithson, Patricia Jo
Smolik, G. Virgene
Smotrich, Susan Lieberman
Smudin, Jane Farneth
Smythe, Sandra J.
Snaper, Kathleen M.
Snead, Nancy Ann
Snell, Linda Rice
Snider, Margaret M.
Snoj, Teresa Ines
Snook, Jayne M.
Snow, Evelyn L.
Snyder, Geraldine Brooks
Snyder, Marilyn A.
Sochacki, Jennifer Ann
Sokatch, Carol Mae
Sokol, Dorro Converse
Sollars, Linda Langenfeld
Solorio, Genevieve Ann
Soloway, Sheri
Solymosi, Tanya Lisa
Sommer, Theoclete B.
Sommerfeld, Joan Ellen
Sommers, Susan Aalbu
Song, Hyeran
Sonnenberg, Melanie Diane
Sorensen, Arthelle Schreiber
Sorensen, Jonna A.
Sorenson, Deborah Renee
Sorrell, Janet R.
Sothman, Maridee Jill
Soucy, Joanne Marie
Sousa, Doreen A.
Southworth, Heidi Lynn
Sowders, Sandra Gettelfinger
Spaniol, Pamela Cecilia
Sparagowski, Jane Lee
Sparks, Karen Michelle
Speaker, Margaret O.
Spector, Joan N.

Spees, Clarinda P.
Spellenberg, Marla Anna
Spells, Mary B.
Spence, E. Marie
Spencer, Anelladee
Spencer, Charlene J.
Spencer, Mary Katherine
Spencer, Priscilla H.
Spencer, Tamar Lish
Spicer, Billie Faye
Spier, Cheridah Frye
Spikes (Col. Ret.), Virginia M.S.
Spivey, Anna M.
Spoerry, Dr. Anne
Sposeto, Billie Ann
Sprague, Jacquie Bennett
Sprague, Jeannine Edith
Sprang, Pamela Gindlesberger
Sprauer, Constance D.
Spriggs, Thatch (Mary)
Springer, MD, Christiane
Sproul, Phyllis
Spry, Shirley Mae
Srour, M. Leila
St.onge, Christine Maria
St.onge, Ruby Jane
Stackhouse, Jeanne T.
Stacy, Nadine Fraser
Stafford, E. Diane
Stafford, Holli Anne
Stafford, Ruth Margaret
Stahlnecker, Naomi W.
Staiger, Maxine Schotte
Stalcup, Anna H.
Staley, Ellen Jean
Stalk, Joan Fleming
Stamford-Krause, Ph.D., Shari
Stamm, Violet Blowers
Stamp, Coralie Ann
Standifer, Harriet Ramsay
Stanfield, Janice S.
Stanford, Barbara A.
Stanger, Lynn Beth
Stanley, Mary D.
Stannah, Margaret E.
Stansbery, Eileen Katherine
Stapleton, Jean
Stapleton, Pat C.
Starck, Marnie L.
Starer, Kimberly Ann
Starer, Merle Ann
Stark, Bette P.
Starkey, Eleanor S.
Staroski, Roseanne J.
Starr, Alice-Jean M.
Starret, Anne Rachelle
Startz, Wyvema F.
Stasko, Nancy Grace
Staszak, Sandy
Statham, Sylvia R.
Staudt, Laura Callaway
Stauffer, Michele Sue
Steadman, Bernice T.
Stearns, Anne M.
Stearns, Jessica Renee
Stearns, Kathlyn

Stears, Maisie R.
Steel, Susanne Sandifer
Steele, Cathleen Moira
Steele, Helen H.
Steele, Lawanna Faye (Lonnie)
Steele, M. Leanne
Steely, Frances Yon (Dee)
Steenland, Pam
Steere, Vivian Johnson
Stehle-Deberry, Gisela (Gigi)
Steimle, Gail Theresa
Stein, Ellen M.
Stein, Teresa Leigh
Steinberger, Joan
Steinke, Laureen Myra
Stenger, Louanne Purvine
Stephens, Anne B.
Stephens, Beverly Sue
Stephens, Carol Lyn
Stephens, Marsha
Stephens, Myrna M.
Stephens, Sandra Lynn
Steuerwald, Mary Lee
Stevens, Betty McKay
Stevens, Constance I.
Stevens, Donna M.
Stevens, Ora K.
Stevenson, Bernadine E.
Stevenson, Claire Rhodes
Stevenson, Grace
Stevenson, Thyra Kay
Stewart, Carol C.
Stewart, Kristie L.
Stewart, Lou Ann
Stewart, Margaret Crowell
Stewart, Nancy
Stewart, Patricia Ellen
Stewart, Stephanie
Stienmier, Saundra Kay
Stierman, Paulette Marie
Stiles, Judith L.
Stilley, Sue York
Stilwell, Frances Beverly
Stinson, Katharine
Stites, Carol C.
Stivers, Margaret Garrison
Stock, Pamela Sue
Stocklin, Diane M.
Stockman, Brenda Jo
Stokes, Sandra Marie
Stokes-Roelofs, Clodagh Anna
Stoll, Doris H.
Stomberg, Cam
Stone, Ksena M.
Stoneking, Linda Kay
Storhok, Sandra Wilson
Storm, Susan L.
Story, Irma
Story, Mary Louise
Stotlar, Nada
Stott, Barbara Anne
Stouder, Judy G.
Stouffer, Patricia R.
Stovall, Linda Joy
Stovall, Marsha Lynn
Stover, Patricia Ann

Stowe, Marilyn J.
Strachan, Barbara A.
Strand, Deanna Rae
Strassburg, Wanda S.
Strate, Dot M.
Stratford, Cecelia M.
Stratford, Nancy J.
Stratford, Sonia A.
Strawn, C.J.
Stream, Judy A.
Streeter, Megann
Streeter-Kempinski, Jan
Strehle, Katherine Smith
Strickland, Corinne Sills
Strickland, Elaine Lila
Strickler, Brenda M.
Stroberg, Sharon K.
Strohfus, Elizabeth Bridget
Strom, Connie R.
Strother, Dora Jean D.
Strubeck, Frances F.
Struthers, June Pauline
Studen, Maureen
Stulik, Christine Marie
Stults, Ruth Celestine
Stump, Esther J.
Stumpf, Karen Linnea
Sturman, Dorothy Crafts
Stype, Joann J.
Subach, Marilyn Butler
Suberg, Valerie Gayle
Suchodolski, Jeanne Constance
Sudermann, Lynda Ellerbe
Sugden, Nadine R.
Suisman, Mary Jane
Sullivan, Gloria Steffen
Sullivan, Jan
Sullivan, Mary Lawrence
Sullivan, Sandra E.
Sullivan, Sharon Ann
Sullivan, Vicki C.
Sullivan-Bisceglia, Susan Marie
Sullivan-Walsh, Morgan T.
Sultan, Jane W.
Summers, Debora M.
Summers, Leslie J.
Sumpter, Ava Christine
Sunden, Annica Linnea
Sundmacher, Marjorie Anne
Sunnergren, Marilyn Harriet
Supplee, Anne Abbott
Suta, Nancy L.
Sutherland, Virginia
Sutliffe, Jane Elizabeth
Sutton, Beth Howell
Sutton, Carole A.
Sutton, Cheryl Dianne
Sutton, Linda Ranae
Svenson, Elizabeth
Swain, Chris
Swain, Gail A.
Swaine, Elizabeth
Swallow, Helen H.
Swann, Linda Kay
Swanner, Nancy Gail
Swanton, Karen Ruth

Swartz, Martha Maye
Sweeney, Barbara F.
Sweeney, Sandra Stoen
Sweet, Audrey M.S.
Sweet, Eula M. "Skippy"
Swengel, Julie Anne
Swink, Shirley Jeanne
Swoyer, Mary P.
Sylvestri, Amelia R.
Syme, Jennifer Jewett
Syran-Fox, Syd (Sidsel)
Syska, M. Bozena
Syverson, Margaret Marcella
Szatmarcy, Susan E.
Szydlowski, Charlene A.

– T –
Taber, Lovina
Taft, Lana J.
Tait, Mary S.
Takacs, Laura Katalin
Talbot, Sarah
Talbott, Mary Jane
Talley, Mary E.
Tally, Lurana E.
Tanner, Anne Marie
Tanner, Shirley Louise
Tanouye, Tracey Midori
Tanski, Tina Marie
Tanton, Elaine
Tarrio, Frances Baldwin
Tasker, Lynn Theresa
Tassa, Yvonne (Bonnie) B.
Tasseaux, Gloria
Tate, Phyllis M.
Taylor, Alyce Stuart
Taylor, Angie Dee
Taylor, Anita Hapka
Taylor, Anna Loraine
Taylor, Donna Lou
Taylor, Elizabeth Eyre
Taylor, Eloise Elizabeth
Taylor, Frances McLaurin
Taylor, Linda
Taylor, Linda Lou K.
Taylor, Lucille T.
Taylor, Martha C.
Taylor, Nancy Elizabeth
Taylor, Priscilla J.
Taylor, Roberta Ellen
Taylor, Sondra Lee
Taylor-Sanders, Susan
Teel, Nancy Jean
Teel, Ruth Woodruff
Teetor, Joyce
Teiber, Antoinette Marie
Teichman, BeBe H.
Teixeira, Sammy K.
Telfer, Maureen June
Temple, Dr. Alice D.
Temple, Jane Hallenbeck
Temple, Peggy Lynn
Templeton, Sandra Jean
Tenborg, Diane Lynn
Tennant, Nikki P.
Terese, Dia

Terrana, Judith Hahn
Terry, Beverly D.
Terry, Carolynn M.
Teufel, Katherine I.
Teutsch, Shirley A.
Thacker, Betty Jean
Thal-Slocum, Valerie Jean
Thayer, Marjorie Jean
Theberge, Patricia M.
Theile, Heidi Lorraine
Theilgaard, Dagmar
Theisen, Marie C.
Thern, Lucille E.
Theurer, Dorothy Priscilla
Theurer, Jane C.
Theurkauf, Susan Jane
Thibodeaux, Sophia Rose
Thies, Angelique Cheri
Thiessen, Gina Zinn
Thisted, Barbara W.
Thisted-O'Farrell, Sarah Ellen
Thom, Shirley R.
Thoma, Natalie Lynn
Thomas, Betty Kirk
Thomas, Carolyn
Thomas, Edith "Micki"
Thomas, Flora Lee
Thomas, Jacqueline Carey
Thomas, Juliet Mary
Thomas, Kristina (Tina) Jo
Thomas, Leota E.
Thomas, Linda Z.
Thomas, Margaret-Anne M
Thomas, Millie
Thomas, Pat M.
Thomas, Ruth W.
Thomas, Sandra Lea
Thomas, Susan Gene
Thompson, Darla Jill
Thompson, Deborah Pellett
Thompson, Dorothy Love
Thompson, Dorothy M.
Thompson, Edna Marie
Thompson, Evie B.
Thompson, Gail L.
Thompson, Jackie C.
Thompson, Linda R.
Thompson, Tracy Sue
Thompson, Virginia Louise
Thompson-Pudwill, Debra Ann
Thornberg, Dr. Hope Bouvette
Thorne-Davis, Nadene Leilani
Thornton, Cheryl Lynn
Thornton, Regina Ann
Thorp, Beryl M.
Thorp-Crippes, Gabrielle Anita
Thorpe, Brenda K.
Thrasher, Patricia M.
Thunich, Patricia A.
Thurmond, Dee M.
Thweatt, Suzanne Sherry
Tickner, Beverly Ann
Tidmore, Martha R.
Tidwell, Mary Alice
Tidwell, Suzanne
Tier, Nancy Hopkins

Tierney, Carol Jean
Tighe, Doreen Lynn
Tilden, Peggy Jane
Tilley, Heather Ann
Tilley, Susan
Tilvikas, Eleamor R.
Tinker, Adelaide
Tinsley, Jean J.K.
Titlow, Eleanor M.
Titterington-Machado, Diane Marie
Titus, Patricia Joan
Tobey, Carole Ann
Tobias, Elaine Susan
Todd, Eleanor Lucille
Todd, Judith Lynn
Todd, Linda
Toelke, Ute
Tokunaga, Ayako
Toland, Ada Rebecca
Tolsma, Katherine Patricia
Tomlinson, G. Lola
Tompos, Grace M.
Toney, Florence S.
Tonkinson, Eileen F.
Tonn, Jonalea Rose
Toolan, Deirdre Mary
Toon, Nancy Thompson
Topp, Janice Irene
Topper, Sylvia Jeanne
Torgerson-Watson, Delia-Marie
Tornbom, Gloria J.
Torres, Debra A.
Toskes, Jane M.
Totten, Julia Kay
Townsend, Patricia G.
Townsend, Serena Jane
Townsend, Virginia May
Toxen, Jean P.
Toyoda, Fumiko
Traulsen, Ilse Gamero
Travers, Margaret
Traynelis, Mary Sue
Treister, Ethel
Tremaine, Dorothee A.
Treml, Karen E.
Tretchick, Rachel Jane
Trimble, Evelyn
Trip, Christy Jo
Triplat, Bonnie Elizabeth
Tripp, Karen E.
Tripp, Lavonne M.
Trippe, Julia Elizabeth
Trippensee, Concha Elvira
Trischler, Juliann
Tropea, Daryl Anne
Trott, Dorothy E.
Trotter, M. Aileen
Trout, Bobbi E.
Trover, Janice Emily Ternes
Trudell, Nancy Anne
Trueman, Yvonne Margaret
Trump, Carole Tosh
Trunzo, Fran
Trupin, Marilyn Elaine
Truxel, Sonja Elaine
Trykall, Terry Lynn

Tschanz, Ute
Tschida, Suzanne M.
Tseu, Gale Louise
Tucker, Coralee
Tucker, Karen P.
Tucker, Pat Avon
Tucker, Shirley Kay
Tulane, Geraldine K.
Tuley, Babs
Tull, Etta Lorene
Tune, Ava Lee
Tunnell, C. Diane
Tunnell, Lorrie Kay
Turek, Alyce Ann
Turk, Bev I.
Turken, Dian R.
Turner, Carol Ann
Turner, Maxine
Turner, Michelle L.
Turney, Maryann
Turney, Patricia
Tuttle, Ethelyn G. "Tuckie"
Tuttle, Jacqueline S.
Twigg, Shirley Anita
Tyler, Kathryn J.

– U –
Udall, Patricia
Ulrey, Ann Pike
Ulrick, Lynda Marie
Umphress, Nancy J.
Umphrey, Anne R.
Uncapher, Patti M.
Unger, Lou Ann
Unger, Virginia Elmgren
Upfield, Romona Rolfson
Upton, Shirley A .
Urban, Ann Barille
Urbas, Jean E.
Ursin, Virginia Marie
Uusiaho, Marja Riitta
Uznys, Linda Joyce

– V –
Vaden, Henrietta Roberts
Vaessen, Janet Lyn
Vail, Janna Downs
Valdata, Patricia
Valdez, Pamela Patricia
Vallarino, Mary Emily Jackson
Valor, Joan Diona
Van Bloem, Patricia M.
Van Der Velden, Rose M.
Van Doren, Janet B.
Van Dusen, Hazel Lee
Van Duyne, Yuri K.
Van Dyke, Esther Patricia
Van Dyne, Jennifer D.
Van Etten, Lynn T.
Van Fleet, Susan Jo
Van Grunsven, Denise A.
Van Howd, Nancy Earlene
Van Lint, Godelieve Clothilde
Van Maare, Catrina W.
Van Newkirk, Carolyn Jane
Van Ovost, Jackie D.
Van Reeth, Joan D.
Van Valkenburg, Jacqueline A.
Van Velzer, Mary Louise
Van Zelf, Lois R.
Van-Kesteren, Ginny
Van-Zandt, Sally L.
Vance, Rosemary Jean
Vander Ley, Nichole Kegel
Vandergriff, Norma Wynn
Vanderpool, Celia C.
Vanecek, Barbara Joan
Vanscyoc, Mary E.
VanWormer, Jill A.
Vasenden, Gwen N.
Vasko, Donna M.
Vasold, Clarice Jean
Vaughan, Nancy Lee
Veen, Elisabeth T.
Vennell, Claudia Ruth
Verduzco, Lynda J.
Veris, Carol V.
Vermeulen, Mary Margaret
Vermoelen, Maria Joanna
Vernon, Marlyn F.
Viall, Patricia Lee
Vickers, Barbara G.
Vietz, Bridget L.
Vigeant-Langlois, Laurence N.
Vigneron, Joan E.
Vijayakumar, Vijayalakshmi
Viker, Patricia J.
Villaggio, Cheryl E.
Vinas, Cristina
Vincent, Denise M.
Vincent, Jamie K.
Vinson, Elizabeth "Betty"
Violet, Wilma Ruth
Vitullo, Dolores Ann
Vlasek, Tiffany Tokar
Vlieger, Faye A.
Vogel, Georgia I.
Vogel, Katherine C.
Vogt, Eleanor M.
Voigts, Mary L.
Volk, Kimberly Lynn
Volkerts, Patricia
Volkner, Gisela
Vollstad-McKean, Kristin
Von Gruben, Rebecca
Von Isser, Aldine
Von Thaden, Terry L.
Von Tobel, Beatrice S.
Von-Maltitz, Elizabeth Amalia
Voss, Carol A.
Voss, Mary Jo
Vredenburg, Judith M.

– W –
Waas, Andrea S.
Wacker, Mary Wills
Wackwitz, Linda Kay
Waddell, Anne
Waddell, Juanita R.
Wade, Dianne Lee
Wade, Joyce Ann
Wade, Norma Jean
Wadington, Barbara J.
Wadsworth, Katherine Elizabeth
Waehler, Christa Anna
Wageman, Becky Annette
Wagner, Eleanor Isabel
Wagner, Helen C.
Wagner, Rebecca Ann
Wagner, Ruth Keller
Wagner, Sheilagh Farmer
Wagner,MD, Marion Holmes
Wagstaff, Patty R.
Wahle, Marjorie Helen
Waide, Jacqueline Ann
Waite, Mary H.
Wakelin, Dee Bond
Walden, Debby T.
Waldrop, Dorothy G. (Dot)
Waldrop, Kym Ross
Walker, Helen
Walker, Janis P.
Walker, Leslie Ann
Walker, Maude M.
Walker, Ruth Walters
Walko, Ann-Marie
Wall, Hannelore W.
Wallace, Elizabeth Ann Utzig
Wallace, Jean Sharon
Wallace, Katherine Ellen
Wallace, Nina
Wallash, JoAnn Navik
Wallen, Vera S.
Waller-Binns, Carole B.
Wallingford, Josephine Wood
Wallis, Shirley Patricia
Wallis, Susan E.
Walsh, Ann Elizabeth
Walsh, Betty M.
Walsh, Mary Margaret
Walsh, Nancy Diane
Walsh, Sheila Louise
Walter, Jane Ann
Walters, Betty Frances
Walters, Claire Lee
Walters, Lynne J.
Walters, Mary Flowerree
Walters, Nancy
Walters-Phillips, Barbara Ellen
Walther, Ethelyn (Lyn)
Walton, Kathleen
Walton, Melinda A.
Walton, Nancy-Bird
Walton, Teresa A.
Waltz, Lucia Mariscal
Wanamaker, Beverly
Wappler, Catherine Garvey
Ward, Ann Cordes
Ward, Barbara Goode
Ward, Deborah M.
Ward, Diana Madama
Ward, Lois Merritt
Ward, Marilou
Ward, Meredith O"Keene
Ward, Pat M.
Ward, Rosalie Ann
Ward, Suzanne V.
Wardlow, Gail Gunter
Ware, Jennifer Kathleen
Warford, Ellen Elizabeth
Warford, Marilyn Junette
Warger, Joyce Annette
Warman, Laura Jean
Warmoth, Isabel Huntley
Warner, Emily Joyce
Warner, M. Margaret
Warner, Monica E.
Warner-Smith, Mary
Warren, Dorothy J.
Warren, Dottie Prince
Warren, Ellen Mary
Warren, Lynette Erb
Warren, M. Ellen
Warren, Margaret Thomas
Warren, Nancy Jane
Warren, Yvonne Morris
Wartman, Lois L.
Washburn, Patricia
Washington, Evie Lou
Waters, Denise A.
Waters, Mary Martha
Waters, Vivian Burgess
Watkins, Lucille Frances
Watkins, Margaret (Peggy)
Watkins, Rachel Alice
Watson, Elsie G.
Watson, Florene Miller
Watson, Lois
Watson, Margaret Harper
Watt, Margaret Bumb
Watts, Rebecca
Waugh, Cynthia Catherine
Waxmundsky, Carol Ann
Waylett, Nancy E.
Wayne, Janet W.
Weaver, Annette Lee
Weaver, Victoria L.
Webb, Cleo Olive
Webb, Patricia Thaden
Webb, Phyllis Marie
Webb, Shirley Ann
Webber, Lynn
Weber, Joyce A.
Weber, Katharine Kingman
Weber, Renee Megan
Webster, Audrey Ellen
Webster, Donna Marie
Weeks, Donna Tessandori
Weeks, Nancy
Weeks, Patricia M.
Wegner, Margaret Anne
Wehman, Frances Jetter
Wehr, Eileen
Weichert, Sara R.
Weightman, Joan S.
Weiman, Nancy Hunt
Weinbaum, Shirley Adell
Weiner, Susan Ambrus
Weiss, Margaret "Peg" R.
Weiss, Ruth S.
Weitz, Susan Taylor
Weitzel, Karen Ann
Welch, Diane Marie
Welch, Nikki C.

Weldon, Cecilia KT.
Weldon, Karen M.
Welisek, Terri Louise
Weller, Mary Doris
Wellington, Margaret T.
Wellman, Barbara Bigelow
Wellons, Chris
Wells, Dawn Louise
Wells, Fay Gillis
Wells, Joyce B.
Wells, Kay
Wells, Marjorie A.
Wells, Nancy
Wells, Nancy Elizabeth
Wells, Patricia Kay
Wells, Phyllis E.
Welpton, Mary Anne
Welsh, Barbara F.
Welsh, Debra Elaine
Welsh, Janice Marie
Welsh, Olivia Wilma
Wempe, Judy L.
Wenger, Niki Mason
Wenholz, Mary R.
Wenk, Cindy Craig
Wenk, Gail W.
Werlin, Claudia Jenny
Werly, Ann L.
Werner, Barbara
Werner, Karen B.
Werner, Penny H.
Wertheimer, Beverly H.
Wertz, Kathleen M.
Wertz-Thomas MD, Charity Marcy
Weser, Marcia Goren
West, Felicia M.
West, Frances
West, Jean S.
West, Mary Ellen
West, Verna Steele
Westaway, Nancy Louise
Westby, Dorothy Carlson
Westcott, Phyllis M.
Westedt, Valerie Jean
Westerheide, Judith Lee
Westerlin, Eleanor J.
Westermeyer, Rachel Mary
Westervelt, Susan Conley
Westling, Dorothy M.
Westmoreland, Mary Lou
Weston, Kim Renee
Wetherby, Mary Ann
Whale, Margo C.
Wharton, Betty M.
Whatley, Nancy Lee
Whatmough, Jocelyn E.
Wheatley, Cheryl Ruth
Wheeler, Carole Sue
Wheeler, Dorothy J.
Wheeler, Gary Anne
Wheeler, Linda Susan
Wheeler, Rebecca Warner
Wheelock, Mary Imogene
Wheless, Shelle Elizabeth
Whipple, Florence G.
Whirley, Loreen Carol

Whistle, Eddie Elizabeth
Whitaker, Gina Carolyn
White, Ann
White, Cheramy Lynn
White, Dorothy E.
White, Harriet Urban
White, Jacqueline Dianne
White, Karen Kay
White, Katherine R.
White, Louise G.
White, LuEtta Marie
White, N.C. Jeep
White, Nancy Colleen
White, Rozann Skozen
White, Shirley Maye
White, Susan Louise
White, Vivian Penny H.
Whiteman, Margaret A.
Whitfield, Candace (Candy) I.
Whitford, Jacqueline Kay
Whitington, Doris Jean
Whitman, Bonnie Chittenden
Whitman, Bonnie J.
Whitman, Denise J.
Whitsitt, Wanda I.
Whitson, Judy Maely
Whittaker, Phyllis Lynn
Whorton, Myrna Winrow L.
Wichmann, Monica M.
Wickman, Janet Ryan
Wicks, Louise C.
Widger, Anne M.
Wiemers, Karyn Danille
Wiener, Louisa M.
Wiersma, Martha Marie
Wiese, Lew Jane (Lewie)
Wietbrock, Christine Edna
Wigger, DeLyle Rice
Wikert, Alinda Hill
Wikle, Sandra J.
Wilbur, Anne B.
Wilcox, Joan Perry
Wild, Eileen
Wild, Janis Louise
Wild, Lindsey Anne
Wilder, Barbara Ann
Wilder, Barbara W.
Wilds, Constance
Wileman, Patricia A.
Wiley, Kimberly Kay
Wilford, Sonja
Wilke, Helen Louise
Wilkerson, Lee
Wilkerson, Sheryle K.
Wilkey, Bette A.
Wilkey, Terrie Hearns
Wilkins, Sharon Kay
Wilkinson, Berenice
Wilkinson, Beverly Jeanne
Wilkinson, Peggy Ann
Wilks, Heather Mary
Will, Carolyn Jane
Willcutt, Bettie
Wille, Rosanne Louise
Willerth, Jeanne Given
Williams, Ardyth Martin

Williams, Bonnie M.
Williams, Candace S.
Williams, Chiyomi Hupfield
Williams, Deborah West
Williams, Diana Carol
Williams, Dianne E.
Williams, Doris Jeanne
Williams, Dorothy Laird
Williams, Ellyn
Williams, Eloise Ball
Williams, Gay Zena
Williams, Jane C.
Williams, Judith Anne
Williams, Lindy Marie
Williams, Margaret H.
Williams, Margaret Kathleen
Williams, Patricia Landis
Williams, Peggy P.
Williams, Robyn Debra
Williams, Rosemary Arnold
Williams, Ruth A.
Williams, Scotty Harmon
Williams, Shirley Maxine
Williamson, Elaine Shirlann
Williamson, Karin
Willis, Judy M.
Willis, Patricia
Willmann, Janis Q.
Willmore, Betty Fay
Wills, Bonnie M.
Wills, Christine Margaret
Wills, Glorice Pool
Wilper, Barbara
Wilson, Alma Dorothy
Wilson, Betty P.
Wilson, Carolyn Munk
Wilson, Claire D.
Wilson, Cynthia J.
Wilson, Debra Sue
Wilson, Donna Ruth
Wilson, Elizabeth A.
Wilson, Evelyn F.
Wilson, J. Priscilla (Pat)
Wilson, Judith Elaine
Wilson, Marilee Dyer
Wilson, Marolyn Piersma
Wilson, Patti Kay
Wilson, Stacy Lea
Wilson, Susan Lynn
Wilson, Tracie L.
Wilson, Victoria Celeste
Wilt, Donna Forsyth
Wiltsie, Victoria Cromwell
Windus, Susan Paige
Winfrey, Onita L.
Wingett, Victoria Katherine
Winn, Shirley
Winnard, Martha M.
Winsor, Kimberley Dawn
Winter, Ann Catherine
Winter, Joan Lee
Winter, Kristin Amelia
Winters, Karen Lanford
Winters, Marjorie S.
Winters, Marlene Kay
Winthrop, Pauline Hunter

Winzer, Christine L.
Wirtschafter, Irene N.
Wiscaver, Mary Sue
Wister, Alberta Mary Ethel
Witbeck, Jerry Melton
Witmer, Naomi Ruth
Witt, Beverly
Wittman, Diana
Wittmer, Betty De Witt
Wohlgemuth, Mary Esther
Wohlin, Gaye
Wojcik, Catherine
Wolcott, Jeane M.
Wolf, Darlene Evonne
Wolf, Selina Mary
Wolfe, Kimberly Ann
Wolff, Barbara Lee
Wolff, June
Wolfstone, Doris Ione
Womack, H. Virginia
Wood, Ann
Wood, Donna Lee
Wood, Dottie Lynn
Wood, Dr. Patty
Wood, Jan Marie
Wood, Jane Parker
Wood, Joanna
Wood, Kathleen
Wood, Linda S.
Wood, Patricia Ann
Wood, Penelope Gayle
Wood, Shela Ann
Woodhouse, Mary Ellen
Woods, Inez Woodward
Woods, Jessie
Woods, Justine Fletcher
Woods, Marie L.
Woodson, Kathleen T.
Woodward, Beverly R.
Woodward, Diane Eloise
Woodward, Jane Weldon
Woodward, Linda Catherine
Woodward, Marlene J.
Woodward-Duncan, Christine
Woolever, Patricia
Wooley, Adrienne Barbara
Woolley, Christine Signy
Woolsey, Charlene
Worel, Anita Conley
Worland, Norma Lou
Wormsbacher, Joanne Alice
Wormsbacher, Nadine
Wortz, Eleanor Thompson
Wray, Helen Heath
Wray, Marie Lund
Wrenn, Mary Beth Hunt
Wrenn, Nancy Vestal
Wright, Aina G.
Wright, Alice M.A.
Wright, Barbara Brock
Wright, Edna H.
Wright, Edna Jeannette
Wright, Lauren
Wright, Nancy Lucile
Wright, Patricia Ann
Wright, Sheila King

Wright, Shirley Howes
Wright, Susan Rothrock
Wrigley, Susan C.
Wulczak, Marie Therese
Wunder, Mary Margaret
Wurster, Georgia B.
Wurts, Janet Lynn
Wyall, Marty
Wyandt, Esther A.
Wyatt, Alta Laura
Wyatt, Leanne Bernice
Wyatt, Ruth Ann
Wyckoff, Eileen H.
Wyers, Cheryle Ann
Wyland, Joan Eleanor
Wyse, Rosalie Bracht

– Y –
Yager, Ellen P.
Yager, Marion Elizabeth
Yamada, Motoyo
Yamagishi, Mariko

Yamamoto, Sue S.
Yancey, Carla Lea
Yarborough, Linda
Yarborough, Michelle Faith
Yarbrough, Deborah C.
Yarbrough, Michele Emma
Yarger, Peggy Lee
Yarworth, Sandy M.
Yates, Nina Burdg
Yeager, Darcell
Yeager, Susan Henley
Yeandle, Audrey A.
Years, Patricia Miller
Yeary, Elaine Carole
Yecny, Christine H.
Yendes, Patricia Cannon
Yenkelun, Bernice W.
Yerian, Judy B.
Yersavich, Amy Crum
Yetke, Kristin Nicole
Yi, Chong Hui
Yoder, Janet Sue

Yoder, Nancy Lee
Yodice, Kathleen Ann
York, Patricia
Yoshida, Maia
Young, A. Ruth
Young, Audrey "Willa"
Young, Barbara Kay
Young, Betty L.
Young, Florence English
Young, Lois Keeter
Young, Lucy Burwell
Young, Maureen Maggie
Young, Ramona O.
Young, Sharon K.
Youngblood, Angela
Yunger, Libby Marie

– Z –
Zabala, Antoinette Selena
Zagaria, Teresa L.
Zahner, Lynn E.
Zaitz, Donna M.

Zak, L. Edith
Zanetti, Sylvia Charlotte
Zapata-Harshbarger, Carolyn Ann
Zdunczyk, Mary Ann
Zehnacker, Gloria D.
Zeigler, Jennifer H.
Zeise, Joan Linda
Zeissner, Carol Moreno
Zellen, Carol Ann
Ziadie-Haddad, Maria Theresa
Ziegler, Angelika
Ziler, Lois Hollingsworth
Zilik, Elaine Pender
Zillig, Shirley M.
Zimlich, Margaret Smith
Zimmer, Patricia Mickey
Zimmerman, Kathleen
Zimmerman, Ruth E.
Zipfel, Ericka
Zoch, Syria Jane
Zook, Connie Lynn
Zopf, Peggy Jean

Oklahoma 99 Nema Masonhall, with husband Mace, was honored for her long-time support of The 99s at her 52nd consecutive section meeting in 1981 at Santa Fe, NM, when she received the Jimmy Kolp Award from the Wichita Falls Chapter. Before her death in 1993, she had attended 75 consecutive South Central Section meetings.

Moya Lear and Lois Erickson

Ruth Rueckert

Dell Hinn

THE NINETY-NINES PRAYER

Oh, Lord, it is good to be able to fly, to share with your winged creatures the joy of our feedom of flight.

We thank you for your wondrous natural laws that lift our wings skyward and we thank you for the courageous pioneers who first mastered those laws so that all of us could fly through your friendly skies.

But as we soar high above this beautifully jeweled island home we call earth and survey its glory as few others can, heighten, oh Lord, our appreciation for the beauty of your creation, and increase our dedication to care for it.

And we beseech your presence with each of us whenever we are airborne. Keep us steady when we find ourselves suddenly "on top" and our human balances begin to play tricks on us.

Keep us alert when our throttles are at cruise, but our minds become lulled into idle. And keep us safe as we gently return to the earth from which we came.

We believe you are with us, Lord, but help our disbelief. Stay close by us and show us the way for it is so good to be able to fly.

The Rev. Robert S. Layne
Aviation Leader Banquet
Aug. 14, 1971

1976 Sacramento Bicentennial line-up, 200 planes awaiting the signal to taxi.

INDEX

Editor's Note: This index does not include the names already listed in the biographies section since they appear in alphabetical order.

-A-

A-25 Helldiver 19
A-26 Bomber 26
Adam, Joan 36
Administrator's Championship Awards 35
Aerial Nurse Corps in Cleveland 16
Aero Club of India 40
Aero Club of Southern California 41
Aero Reclaiming Company 15
Aero-Valley Airport 16
Aeronautical Society of India 40
Air Bear 34, 37
Air Line Pilots Association 29
Air Marking Program 20
Air New Zealand 30
Air Race Classic 27, 30
Air Safety Investigators 29
Air Transport Auxiliary 16, 17
Air Transport Command 18
Airborne Express 41
Aircraft Owners and Pilots Association 23, 40
Alabama Aviation Hall of Fame 37
Alaska Airlines 35
Alaska Chapter 38
Alberta Chapter 41
Alexander, Mary 7, 12
Alexander, Mike 21
All Woman Air Show 3, 19, 22, 26
All Women Air Maneuvers 20
All Women International Air Race 22
All-Ohio Chapter 23
All-Woman Transcontinental Air Race 26
All-Woman Trophy Race 26
Allegheny County Airport 16
Allen, Belinda 21
Allison, Gail 21
Aloha Airlines 32
Ambrose, Evelyn 21
Amelia Earhart Award 29
Amelia Earhart Birthplace Museum 33, 36
Amelia Earhart Board of Trustees 20
Amelia Earhart Field 19
Amelia Earhart Medal 28
Amelia Earhart Memorial Race 26
Amelia Earhart Memorial Stadium 25
Amelia Earhart Scholarship 15, 20, 21, 25, 31, 34, 41
Amelia Earhart's Electra 24
America West 37, 40
American Airlines 29, 34
American Cancer Society 35
American Declaration of Independence 27
American Eagle 16
Andersen, Janet 21
Andersen, Robin 21
Anderson, Adeline Fisit 7
Anderson, Eileen 21
Anderson, Katherine 21
Andrews, Marion 26
Angel Derby 22, 31
Anglin, Mary Elaine 21
Annette Gipson All Women Air Races 16
Annual Achievement Award 30
Anson, Mark V 29
Arlene Davis Memorial Trophy 16, 23
Army Air Corps 15
Arnold, Gen. 19
Arnold, Pat 22
Arrow Sport 16
Arunachalam, V.S. 40
Asbury-Oliver, Suzanne 38
Aspen Chapter 39
AT-6 19, 35
AT-6 Race 20
Atchison County Commissioners 36
Atchison, Kansas 12
Atherton, Jean Kaye Tinsley 35
Australian Woman's Pilot Association 37
Aviation Country Clubs 14
Aviation Hall of Fame 31
Avid Flyer 36
Avro Avian 14
Award of Merit 25, 41
Axton, Mickey 39, 41
Azar, Suzie 35

-B-

B-17 19
B-25 40
B-29 Superfortress 19
B727 25
Backenstoe Leh, Dorothea 7, 42
Bacon, Mary Ellen 7
Bailey, Martha 21
Bailey, Susan 21
Baker, John 40, 41
Bakersfield Chapter 30
Bancroft, Barbara 7, 8
Banks, Marian 27
Barber, Susan 21
Barker, Linda 21
Barnes, Florence "Pancho" 11, 12, 15
Barnick, Marion 30
Barnstormers Trophy 30
Barnwell Peacock Donnels, Achsa 35
Barracuda torpedo bomber 18
Bartos, Janet 21
Basler, Lane Joan 21
Bass, Beverly 29
Baty, Dr. Peggy 35
Batz, Suzanne 21
Bauer, Katherine 21
Baumann, Joanruth 21
Bayley, Caro 19, 20
Beard, Hank 32
Beard, Melba Gorby 7, 10, 14, 15, 235
Beatty, Jenny 21
Beech 33A 30
Beech Aircraft 26, 28, 30
Beech, Olive Ann 26, 41
Beech Travel Air 13, 38
Beech, Walter 12
Beekman, Bitsy 26
Beeler, Rich Gritter & MayCay 37
Beery, Wallace 15
Beggs, Gene 30
Bell Helicopter Co. 19
Bellew, Gwen 38
Belt of Orion Award for Excellence 39, 41
Bendix Race 16, 19
Bendix Trophy 13
Benn, Velta 21
Bennett, Alverna Williams 31
Bergstrom, Cynthia 21
Berry, Phyllis 21
Bertram, Marion 19
Betsy Ross Corps 12
Bikle, Helen V. Cox Clohecy 7
Bill Barber Award 35
Bird Biplane 15
Bird Walton, Nancy 13, 40, 41
Bishop Wright Air Industry Award 29
Bixby, Dianna 26
Bizzoso, Pat 21
Bjornson, Rosella 28, 29, 32
Blair, Mary Frances 21
Blanchard, Wendy 21
Blankenship, Clyde D. 29
Blankenship, Denise 29
Bledsoe, Lucile 39
Blok, Manja 35, 40
Blood Flights 28
Blum, Patricia 32
Blumberg, Juanita 31
Boeing 707-720 29
Boeing 727 jet 29
Boeing 737 29, 32, 35
Boeing 747 34
Boeing Aircraft 40
Boeing's Flying Fortress 19
Bond, Langhorne 27
Bonesteel, June 34, 35
Borgerding, Carol 21
Boston Harbor 10
Bowditch, Gladys 21
Bower, Gray Gordon 21
Boyle, Jennifer 23
Braese, Evelyn 21
Brainerd, Annemaire 21
Brave Bessie 10
Breeden, Jacqueline 21
Breedon, Shelley 39, 41
Brenneman, Holly 21
Brewer, Frank G. Award 31
Brewer Trophy 40
Bribach, Dr. Eugene J. 36
Brick, Kay Menges 6, 19, 22, 24, 27
Bridewell, Ruth T. 7
Britannia Trophy 14
British Air Transport Auxiliary 18
Britton, Mabel 6
Broadhead, Jean 19
Broadwick, Tiny 10
Broken-back Cub 19
Brooklyn Bridge 16
Brown, Heather 21
Brown, Janice L. 30, 31
Brown, Margery 7, 11, 12
Brown, Vera 7
Brumlow, Nancy E. 21
Bucyrus Historical Society 16
Budhabhatti, Chanda 40
Buffington, Glenn 10, 36, 41
Bullington, Marion 21
Burchette, Rosalie 21
Burger, Chief Justice Warren 30
Burger, Stacy 21
Burke, Verna 19
Buroker, Gladys 17
Bush, Alan 30

-C-

C-130 31
C-141 Starlifter 32
Cagle, K. 33
Cagle, Myrtle 25
Caldwell, Duke 19
Caldwell, Myrtle R. 7
California National Guard 30
Camden, Conrad 27
Cameron, Laurel 21
Campbell, Coleen 21
Canada's Aviation Hall of Fame 39, 41
Canada's Webster Memorial Trophy Competition 39, 41
Canadian Armed Forces 31
Caniff, Milton 26, 27
Cannon, Barbara 38
Canyon, Steve 26
Carpenter, JoAnne 21
Carr, Bonnie 21
Carrigan, Joe 36
Carso, Margaret 22
Carter, Ann Shaw 25
Cary, Lynn Palmer 21
Case, Joyce 28, 30
Casey, Lady Marie 13
Caterpillar Club 12
Central Airlines 29
Cervero, Manuel 33
Cessna 13, 24, 28, 30
Cessna 170 16, 32
Cessna 172s 30
Cessna 177B 40
CH-47 Chinook 37
Challenger 33, 38, 40
Chamberlin, Candi 21
Champlin, Gail 27
Chappel, Harold 30
Cheetham, Lucile 21
Cheyenne Airport 40
Chicago Area Chapter 31
Christensen, Marie 6, 34, 36
Christy, Naomi 38
Church, Carol 21
Church, Pat 35
Citizens Advisory Committee on Aviation 24
Civil Air Patrol 12, 15, 28, 37
Civilian Pilot Training Program 18
Clark, Julie 35, 38
Clark, Patrice Francise 32
Clarke, Carolyn 21
Clendaniel, Marion Clark 7, 236
Cleveland Air Races 11
Cleveland Airport 16
Climbing Vines 11
Clinic, Lovelace 30
Clinton Aviation Co. 29
Clinton, Nancy 21
Clover Field, Santa Monica, CA 11
Cobb, Jerrie 24, 25, 29, 31, 33, 37
Cochran, Jacqueline 3, 6, 17, 18, 19, 26, 27
Coeur d'Alene Lake 17
Coleman, Bessie 39, 41
Collings, Pam 37
Collins, Eileen 13, 37, 38, 39, 40
Colorado Aviation Hall of Fame 39
Colt airplane 22
Commission Internationale d' Aviation Generale 34
Compton, Patricia 21
Conatser, Rosemary 25, 28
Consolidated Aircraft 16
Continental Airlines 29, 30
Cook, Audrey 21
Coonce, Kimberly 21
Cooper, Edna Mae 15
Copeland, Dr. John 32
Copeland, Marilyn 6, 30, 32, 36

-C- (cont.)

Corporate Angel Network 32
Corvair 580 29
Corvairs 28
Cotham, Lisa 21
Council of Governors 39
Courtney Turner Trust 36
Covington, Candace 21
Cox, Karen 31
Craik, Evelyn June 21
Crane, Mardo 26, 30
Crary, Nelson 8
Crary, Phyllis Fleet 7, 8, 16
Crawford, Marjorie 11
Cray, Dick 36
Cray, Evah 36
Cray Foundation of Atchison 36
Creason, Mary 21
Crittenden, Retha McCulloch 7, 8
Crockett, Anita B. 21
Cross, James 33
Crosson, Marvel 11
Crown Circle Award 32, 40
Cullen, Carolyn 19
Cummings, Wanda 27
Cummins, Dorothy 37
Cunha, Yvonne 29
Cunningham, Deborah 21
Currie, Jan 21
Curtiss Airport, Valley Stream, NY 11
Curtiss Challenger 11
Curtiss Field 16
Curtiss Flying Service 12
Curtiss, Glenn 10
Curtiss, Jenny 15
Curtiss Pusher 19
Curtiss School 12
Curtiss Thrush 12
Curtiss-Wright 11

-D-

Daffodil Day Flower Drop 35
Darcy, Susan 40
Darzniek, Morey 40
Davidson, Jean 7
Davidson, Ursula 21
Davis, Arlene 16, 23, 24, 26
Davis, Joan Faye Shankle 7
DC-7 28
DC-8 25, 29, 33, 41
DC-9 29, 38
DC-10 29, 30
de Bernardi, Fiorenza 25, 38
de Laroche, Raymonde 10
de St. Phalle, Brigitte 30
Deal, Sarah 39, 40
Deerman, Ruth 6
DeForest, Linda 21
DeGraaff, Teresa 21
deHaviland Twin Otter 29
deHavilland Beaver 32
Delta Air Lines 23
DeMore, Louise 21
Dennehy, Pat 32
DePue, Carol 21
Devereux, Hilda 36
DiAmico, Joan 21
Dickerhoof, Dorothy 21
Dickerson, Linda 34
Dietrich, Jan 21, 25, 28
Dietrich, Marion 33
Dinan, Elizabeth 21
Discovery 33, 40
Dobrescu, Charles 19
Dobrescu, Ruth 28
Dodge, Jane 7
Domas, Patricia 21
Donaldson, Jack 16
Donn, Tara Anne 21
Donnelly, Mia 21
Donnels, Achsa Peacock 7, 236
Donner, Theresa 21
Doolittle, Maj. James 17
Doyle, Margaret 21
Draper, Linda 21
Dubroff, Jessica 40
Dusenberry, Susan 35, 38
Dwelle, Diane 21

-E-

Earhart, Amelia 6, 7, 11, 12, 14, 15, 16, 20, 24, 25, 28, 29, 32, 34, 35, 36, 37, 39
Earhart, Amey Otis 3
East African Flying Doctors Service 29
Eastern Airlines 29
Edison, Mrs. Thomas A. 3
Edwards Air Force Base 15, 34
Egan, Eileen 38, 39, 40

Egan, Shari 21
Eggert, Kimberly 21
Eggleston, Denise 21
Ehrlander, Sue 39
El-Nadi, Lotfia 41
Elder, Ruth 7, 10, 11, 13, 15
Elder Statesman of Aviation Award 35, 39, 41
Ellington Air Force Base 31
Ellsworth, Amy A. 21
Embry-Riddle Aeronautical University 27
English Air Ministry 14
English, Anne 21
English Channel 10
Enola Gay 19
Ercoupe 19, 26
Erickson, Betty 39
Erickson, Lois 39, 260
Esselburne, Ann 21
Etheridge, Caroline 23
Evans, Broneta Davis 6, 32
Evans, Donna 21
Evans, Teresa 21
Ewanchew, Alexis 36
Experimental Aircraft Association 40

-F-

F-16 35, 39, 40
FA-18 40
FAA Certificate of Appreciation 35
FAA Certified Flight Instructor of the Year Award 35
FAA Gold Medal 24
FAI Rotorcraft Gold Metal Award 39
Fairbanks, Nancy 21
Fauth, Marjorie 20
Fear of Flying Clinic 35
Federal Aviation Administration 25
Federal Aviation Agency's Women's Advisory Committ 24
Federation Aeronautique Internationale 10, 12, 26, 34, 41
Feigenbaum, Lois 6, 23, 30, 34
Feik, Mary 39
Fellow Lady Astronaut Trainees 33
Fenno, Sarah S. 7
Ferguson, Janet 21
Fernandez, Nohema 21
Ferrell, Jean 30
Ferry Command 18
Figley, Peggy Sue 21
First All Women Air Show 19, 20
First Canadian Chapter 35, 39
First Women's Air Derby 35
Flaum, Donna 21
Fleet airplane 16
Fleet, Maj. R.H. 16
Flight Engineer Scholarship 21
Flight Instructor of the Year Award 34
Flight Standards Division 35
Florida All-Woman Air Show 26
Florida Chapter 26
Florida First Coast Chapter 39
Florida Ninety-Nines 19, 20
Florida Women Pilots Association 22
Flying Farmer 29
Flying Samaritans 28
Flying Seven 32
Flynn, 1st Lt Jeannie 39
Fokker F28 Twin Jet 29
Fokker Tri-motors 14
Foote, Jean LaRene 7
Ford Reliability Tour 16
Forest of Friendship 37, 38, 235
Formula 1 22
Fort, Cornelia 18
Fort Lauderdale Airport 16, 31
Foster, Karen 21
Fox, Christine 39
Foy, Lauretta 19
Frank Brewer Award 33
Fraser, Cathy 21
Fraser, Cora McDonald 21
Freedom Flight America 40
French Caudron C-3 13
Fricker, Connie 38
Friedman, Holly 21
Friedman, Linda 21
Friendship Through Flying 12
Frontier Airlines 33
Frye, Charlote 37
Funk, Mary Wallace 25
Funk, Wally 28, 33, 40

-G-

Gadflies 11
Gaffaney, Mary Tracy 20, 25
Gardella, Kathy 21

General Aviation Council 31
General Aviation Revitalization Act 39
Gentry, Viola 7, 11, 12, 16, 27
Georgia Hall of Fame 35, 37, 39
German Bucker Jungman 13
Gibson, Hoot 15
Giebe, Charlene 7
Giesen, Thelma Elliot 7, 9
Gilkison, Polly 31
Gladney, Pat 20, 21, 26, 27, 260
Glaserman, Marcy 21
Glasson, Pauline 26, 27, 41
Goetz, Barbara 21
Goldeman, Margaret 31
Golden Eagle Monoplane 15
Golden Pacific Airlines 28
Golden Triangle Chapter 29
Goodrich, Mary 12
Goodwin, Barbara 21
Goodwin, Dr. Linda 37
Goodwin, Kim 29
Gordley, Sandra 21
Gorelick, Sarah 25
Gossamer Condor 29
Grag, Loretta 34
Graham, Judy 21
Graham, Martha 21
Grandy, Vicki 21
Granger, Clema 15, 27
Grant, Fran 35
Graves, Dean 36
Gray, Marjorie M. 37
Greater New York Chapter 31, 32
Green, Anne Sawyer 32
Green, Janet 6, 28, 34, 36
Greenberg, Margery Louise 7
Greene, Irene J. Chassey 7, 9, 12
Greenhill, Bev 31
Greinke, Helen 19
Grider, Lesa K. 21
Grievance Committee 39
Griffith, Lori 34
Griffith, Thon 6, 36, 38, 41
Grimm, Wendy 21
Gross, Robert 16
Gullino, Candis I. Hall Hitzig 7
Gypsy Moth 11, 32

-H-

Haddaway, Abby Dill 16
Haigh, Henry 30
Haizlip, Mary 17
Halaby, N.E. 24
Haldeman, George 13
Haley, Patricia 21
Hallett, Carol 30
Halley's Comet 33
Halliburton, Ruth E. 7
Hamilton, Kelly Sue 21
Hammond, Alice 6, 20, 41, 235
Happy Bottom Riding Club 15
Happy Flyers 28
Harmon Trophy 17, 19, 24, 29
Harrell Marsalis, Frances E. 7, 12, 42
Harris, Georgina 21
Harris, Grace 38
Harris, Karen Walker 21
Hart, Jane 25, 33
Hartranft, J.B. "Doc" 23
Harwood, Alma 15
Haskell, Loretta 21
Haskins, Mardell 27
Hatfield, Jessica 21
Hawk HP 30
Hay, Lady Drummond 25
Hayward, L. Bernadette 21
Headquarters/Resource Center 5
Heart of Texas Chapter 32
Heath, Lady Mary 7, 14
Heilig, Donna 23
Helly, Karen Diane 21
Helse, Eugenia R. 6
Hembel, Caroline Etheridge 24
Hendricks Harwood, Alma 20
Hengsteler, Pamela 21
Henig, Mary 21
Hensley, Bessie 21
Henze, Gayl 39
Hester Stenzel, Dorthy 37
Hettenbach, Christine 21
Hetzel, Belle 6
Hickam Field 18
Highleyman, Leslie 21
Hinn, Dell 27, 34, 260
Hirsch, Mary 29
Hirth, Orene 21
Hixson, Jean 21, 25, 33
Hodges, Joanne 21
Hoelscher, Juanne L. 21

Hollander, Lu 2, 3, 6, 36
Hollowell, Linda 21
Homan, Alberta B. Worley 7
Homes, Gloria 21
Homing Pigeons 11
Honer, Anne 27
Honisett, Jill 38
Hooker, Linda 21
Hooper, Helen 26
Hoover, Amy 21
Hopkins Tier, Nancy 7, 15, 16, 38, 42, 235
Horstman, Susan 29
Houston Chapter 31
Houston, Lynn 39
Howard Hughes Memorial Award 41
Howard, Jean Ross 24, 25, 39
Howard, Marijane N. 21
Howell Warner, Emily 33
Howland Island 24, 25
Howren, Evelyn Greenblastt 39
Hoyt, Jean Davis 7, 11, 12
Hubert, Ruth 19
Huffman, Cynthia Jean 21
Hughes 300-C 33
Hurlburt, Marge 19
Hurrle Allison, Rhea 25
Hurst Jefford, Ruth 21
Husby, Peggy Bolton 21
Hussein, Saddam 35
Huyler Gillies, Betty 6, 7, 8, 12, 15, 18, 19, 26, 235

-I-

Iacocca, Lee 33
Ida Van Smith Flight Clubs 29, 30
Idaho Chapter 5
Illinois Department of Transportation Bureau of Aviation 31
Indian Airlines 35
Indiana Fairladies Air Races 22
International Aerobatic Club Championships 38
International Aerobatic Title for Women 20
International Biplane 15
International Board of Directors 27, 34, 36, 39
International Forest of Friendship 12, 25, 32, 34
International Harmon Trophy 30
International Headquarters 14, 20, 236
International Northwest Aviation Council 30
International Precision Championship 37
International Women's Air and Space Museum 22
Iris Harris, Melba 37
Izzo, Angela 21

-J-

Jablonski, Mary S. 21
James, Teresa 16
Jameson, Cathleen 21
Jayne, Marion 37
Jefford, Ruth 38
Jenick, Lorraine 28
Jenkins, Pat 33
Jensen, Mary H. Goodrich 7, 8
Jessen, Gene Nora 2, 3, 6, 25, 33
Jessie A. Chamberlin Memorial Trophy 25
Jetstar 28
Jetton, Pat 30
Jewett, Delores 21
Jewett, Dodie 39
Jimmy Kolp Award 260
Johnson, Elinor 30
Johnson, Evelyn Bryan 37, 38, 41
Johnson, Karen 21
Johnson, Katherine F. 7
Johnson Mollison, Amy 14
Johnson, Patti 30
Johnson, President Lyndon B. 24
Johnston, Thelma R. Burleigh 7
Jones, Claudia 29, 30
Jones, Connie Jo-Ellen 21
Jones, Hazel 6, 21, 23, 34, 41
Jones, Joan 21
Jones, Joyce 21
Jones, Lorraine 21
Jones, Rosemary S. 21
Joseph, Angela L. 7
Jubilee Race 26
Jurenka, Jerry Anne 34

-K-

Kachina Doll Air Race in Arizona 22

Kaeder, Deborah 21
Kahn, Karen 28, 29
Kansas Chapter 29, 39
Kansas Heritage Trust 36
Kansas Historical Society 35, 36, 39
Kaspar, Enid 21
Katharine B. Wright Memorial Award 35, 38, 39, 40
Kauffman Workman, Mildred E. 7
Kefauver, Sen. Estes 29
Keith-Miller, Jessie 10, 12, 13
Kelly, Mary 21
Kelly Royle, Madeleine B. 7, 9
Kelton, Helen 21
Kennard, Gaby 41
Kennedy Space Center 39
Kenny, Cecelia Roy 7
Kenyon, Cecil W. "Teddy" 7, 12, 15, 42
Kershner, William 41
Kim, Kyung O. 22, 24
King Air 30
Kingsley Field 39
Kirkpatrick, Daisy 6
Kirkpatrick, Estelle 21
Kitty Hawk Award 37
Kitty Hawk Chapter 37
Klingensmith, Florence E. 7, 43
Knapp, Brooke 32
Knight, Sherry 30
Knouff, Mary Jo Oliver 24, 28, 33
Kocelan, Janeen 41
Kochanek, Mary 21
Koepke, Lee 24
Koich, John 27
Koop, Claire B. 21
Korean War 24, 26
Kropp, Evelyn 21
Krupa, Marge 31
Kubeck, Candalyn 40
Kunz, Opal 3, 12, 16
Kuprash, Aimee 21

-L-

L-19 24
Lady Diana 30
Landefeld, Carol 21
Landry, Kaddy 19, 20
Lange, Eva Mae 7
Langher, Doris M. 29
Lanning, Judy 21
LaPook, Gail 21
Laurence P. Sharples Award 30
Law, Ruth 10
Lawerence, Laverne 37
Lay Ross, Eleanore B. 7, 43
Layne, Rev. Robert S. 261
Lear Jet Inc. 39
Lear, Moya 39, 260
Leebold, Nancy Ellis 13, 21, 29
Leh, Dorothea 12
Leiminger, Ann Marie 21
Leistikow, Frances Ferguson Leitch 7
Lempke Sovereign, Jeannette 19, 20
Lende, Andrea 21
Lennox, Peggy 19, 20
Leonard, Tracy 21
Lesser, Marjorie May 12
Lesser Van Antwerp, Marjorie May 7
Leverton, Irene 25
Lewis, Janet 21
Lindbergh, Ann Morrow 25
Lindbergh, Charles A. 25, 28
Lindbergh Memorial Fund Dinner 28
Linsley, Susan 21
Lippert, Laurel 21
Livingston, Nancy Miller 17, 18
Lockheed 10 24
Lockheed F-104G Starfighter 19
Lockheed Jetstar 19
Lockheed P-38 40
Lockness, Doris 37, 39, 41
Loening, Grover 23
Loening Trophy 23
Loetscher, Ila Fox 7, 8, 32
Loffredo, Geraldine Grey 7
Logan Kunz, Opal 7, 12
Logue, Judy 34
London, Barbara Erickson 19, 29, 38, 40
London Rinehart, Terry 29
Long Beach Hangar 40
Long Island Chapter 29
Loper, Rose 40
Loufek, Betty McMillen 27
Louth, Denise C. 21
Love, Nancy 16, 19, 20
Love, Nancy Harkness 15, 18
Lovelace Clinic 33

Lovelace, Ethel 7
Lucid, Shannon 40
Ludington, Shirley 39
Ludtke, Teresa 21
Luhta, Carolina 21
Lutz, Lola L. 7
Lynum, Joan 41
Lyons, France 10

-M-

MacCloskey, Helen 16, 17
MacDonald, Mildred H. Chase 7, 16
Machado, Anesia Pinheiro 13, 14
MacNei, Carol 21
Macy's Thanksgiving Day Parade 16
Mahan, Shirley 21
Mahonchak, Pam 36
Maloney, Linda 21
Manila Davis Talley Collection 14
Manser, Margaret Cooper 6, 7, 20
Mariner, Rosemary 41
Marsalis, Frances 29
Marshall, Linda 21, 36
Marshall, Sheri Coin 39
Martell, Carolyn 21
Martin, Glenn 10
Martin, Sasha Hall 7
Martin, Stephanie 21
Martlew, Glenda 21
Masonhall, Nema 260
Masson, Angela 29
Masson, Dr. Angela 34
Matthews, Helen Manning 7, 42
Mattingly, Linda 21
Maugham, Olivia "Keet" 7
Maule, Susan DeEtta 21
Maxwell, Jan 23, 34
Mayr, Cathie 39, 40
McBride, Helen 19, 20
McCann, Jennifer 21
McCarrell, Jody 23, 34
McCauliffe, Christa 40
McConnell, Constance 21
McConnell, Georgianna 21
McCormick, Jill 21
McCullough, Terry 32
McDonald, Adele 21
McDonald, CPT Kathy 39
McElhatton, Jeanne 31, 35
McEwen, Patricia Z. 6, 28
McGee, Helen 21
McGinnis, Nancy 21
McGonigal, Hallie 30
McMillan, Jessica 21
McMillian, Marie 35
McNabb, Betty 6, 28
McNamara, Ann 21
McNamee, Edith Geneva 36
Meacham Field, Fort Worth, TX 29, 34
Mead, Margaret 28
Medes, Bea 19, 26
Memphis Chapter 37
Mentzer, Ann Piggott 21
Mercury 13 33
Mercury Astronaut Program 24, 25
Merk, Ora 21
Metcalf, Lillian Porter 7
Meyers, Linda 21
Miami's All-American Air Maneuvers 20
Michigan Chapter 33
Michigan SMALL Race 22, 27
Mickelsen, Geraldine 6, 18
Mile High Derby 27
Miles Messenger 13
Military Pilots' Association 19
Miller, Betty 22, 24
Miller, Chuck 24
Miller, Donna Jeanne 21
Miller, Jessie Maude 7, 43
Miller, Marie 21
Miller, Michelle 21
Miller-Grubermann, Connie 31
Mills, Anges A. 7
Minnesota Chapter 28
Minnesota Lady Lifeguards 31
Mitchell, Virginia 21
Mix, Tom 15
Mock, Jerrie 22, 24
Mohler, Carolyn 36
Moisant, Mathilde 2, 4, 10
Moisant School 10
Monaco, Madeleine 21
Monahan, Dorothy Broadfield 21
Monteith, Karen 21
Monterey Chapter 30
Monterubio, Mary Murphy 21
Montrose Flood 12
Moore, Donna 21

Moore, Paula 30
Moorman, Dorothea 19
Morrissey, Muriel Earhart 30, 32, 41
Moser, Leah 31
Mount San Antonio College 23
Mullen, Doris 25
Murakami, Chiyoko 27
Murphy, Margaret Fzandee O'Mara 7
Museum of Flight 38
Myers, Donna T. 6

-N-

Nadi, Lotfia-El 38
Nancy Bird-Walton Trophy 37
NASA 25, 33, 37, 40
National Aeronautic Association 23, 26, 29, 35, 37, 39, 41
National Air and Space Museum 27
National Air Races 15
National Airlines 29
National Aviation Hall of Fame 33, 38
National Aviation Hall of Fame in Dayton, OH 32
National Collegiate Flying Club 23
National Congress on Aerospace Education 32
National Congress on Aviation and Space Education 40
National Intercollegiate Flying Assoc. 23, 24, 34
National Intercollegiate Flying Meet 16
National Park Service 29
National Transportation Safety Board 28, 29, 40
National Women's Air Derby 10, 11
Nealy, Sue 31
Nebraska Aviation Hall of Fame 38
Nelson, Sylvia A. 7, 12
Neuffer, Lt. Judith Ann 28
Nevado del Ruiz 33
New Castle Army Air Base 15
New England Air Race 22
New England Section's Woman of the Year 16
New Jersey Aviation Hall of Fame 37
New Orleans Air College 16
New Orleans Chapter 37
New Zealand Airwoman's Association 37
Nichols, Ruth 7, 11, 14, 20, 29
Nicholson, Mary Webb 7, 43
Niekamp, Dorothy 20, 21
Nielsen, Carol 21
Ninety-Nines Transcontinental Air Race 26
Nixon, Pat 26
Nobel Peace Prize 31
Noisy Birdwomen 11
Nolde, Fran 19, 20
North Central Section 31
North Georgia Chapter 39
Northrop P-61 40
Northrop T-38 jet 19
Noyes, Blanche 6, 7, 8, 11, 12, 13, 16, 17, 20, 27
Noyes, Dewey 16
Nozoki, Yae 27
Nutmeg Air Rally 27

-O-

Oases Aviation 29
Oberlander, Gertrude C. Ruland 7
O'Connor, Sandra Day 30
O'Donnell, Gladys 7, 11, 12, 15, 236
Oklahoma Aviation Hall of Fame 32, 37
Oklahoma City 5
Oldershaw, Maude H. 29
Oldfield, Barney 16
O'Mara, Margaret F. 12
Omlie, Phoebe 7, 11, 15, 20, 43
Omlie, Vernon 15
Opa Locka U.S. Naval Air Station 19
Operation Desert Storm 37
Operation Skywatch 35
Orr, A. Lee 21, 39
Osborne, Kathy 21
Osterud, Joann 37
Oswald, Marie 39
Oswald, Mary 41
OX Aviation Pioneers 37, 39, 41

-P-

P-38 19
P-47 19
P-51 40
Pabst, August 16
Page, Jane 19
Palisades Chapter 32

263

Palms Chapter 4
Palms to Pines Air Race 22, 27
Panther McClatchie Powerplant 15
Parachute Nurse 16
Parask, Pamela 21
Paris, Neva 7, 10, 11, 12, 236
Parker, Charlotte 21
Parker-Blackwood, Joy 39, 41
Patton, Janet 21
Paver, Wendy 21
Paxon, Peggy J. 7
Payne, Bill 24
Pearl Harbor 18
Pellegrino, Ann 24
Penfield, Phyllis M. Goddard 7
Pennington, Gladys 19
Perry, Bernice C. Blake 7, 236
Perry, June 21
Perry, Margaret 10, 11, 42
Persian Gulf 33, 35, 37
Peter O'Knight Airport 19
Peters, Jo Ellen 21
Peterson, Laurie 21
Pfile, Julie 30
Pfister, Betty 38, 39
Pfoutz, Louise 28
Phelps, Carol Ann 21
Phillips, Jane 21
Pickering, Jean 21
Pilaar, Carolyn 31, 37
Pilkinton, Amy L. 21
Piper Airplanes 28
Piper Comanche 260 14
Pirrung, Theresse 21
Pitts Special 20, 25
Place, Elizabeth F. 7
Planes of Fame Museum 40
Poberenzy, Audrey 39, 40
Poberenzy, Paul 39, 40
Pope, Yvonne 21, 29
Port Angeles 17
Post, Wiley 12
Postlethwaite, Janie 28
Potomac Chapter 40
Powder Puff Derby 15, 24, 26, 27, 28, 30
Powder Puff Derby Commemorative Album 26
President's Women's Advisory Committee on Aviation 29
Price, Katherine 21
Prichard, Gayle Conklin 21
Project Air Bear 31
Pronczuk, Danuta 21
Puckette, Margaret 21
Pulsar 37
Pyle, Ernie 16

-Q-
Queen Bessie 41
Quenzler, Bonnie 31
Questair Venture 37
Quimby, Harriet 10, 39

-R-
Radecki, Nicole 21
RAF Hazards 17
Raiche, Bessica 10
Raines, Hazel Jane 35
Ramachandran, Dee 21
Rasche, Thea 7, 10, 11, 13, 43
Ratley, Sarah 33, 39
Ray, Mathilda J. 7
Rayburn, Carol 21, 35
Reagan, Sandra 21
Red Cross 28, 32, 37
Reid, Joseph 40
Reindl, Mary 21
Reitsch, Hanna 13
Relief Wings 14
Render, Shirley 21
Renneisen, Carol 21
Renneke-West, Lynette 21
Renninger, Doris 22, 24
Reno Air Races 35
Reo, Rita 21
Research Scholar Grants 21
Resnick, Judith 40
Resource Center 18, 33
Reynolds, Jean 21
Rich, Melody 21

Richardson, Gini 26, 27
Richey, Helen 16, 17, 20, 29
Rickett, Sylvia 21
Ride, Sally K. 31, 32
Rihn, Debby 30
Roberts, Alice 6
Roberts, Pat 23, 34, 36, 41
Rockcastle, Ruth 31
Rockwell, Patricia 21
Roediger, Beverly 21
Rogato, Ada 13
Rogers, Betty 30
Rogers, Will 11, 12
Rolls Royce 13
Roosevelt Field, Long Island 16, 17, 19
Ross, Anitra Doss 21
Ross, Ann 20
Rossi, MAJ Marie 37
Rothholz, Meta 7, 12, 236
Rough, Helen 20
Royal Aero Club 14
Royal Netherlands Air Force 35
Rueckert, Ruth 10, 260
Russian Mir Space Station 40
Rutan, Dick 33
Rutherford, Mary 21
Rutledge, Karen 21
Ryan Aeronautical Co. 15
Ryan, Catherine 36
Ryan Fireball 15

-S-
S&H Green Stamps 22, 24
Sacchi Air Ferry Enterprises 24
Sacchi, Louise 24
Safe Air Taxi Inc. 31
Santa Monica Airport 27
Santa Rosa Chapter 30
Sargent, Frances 21
Saunders, Aileen 28
Saunders, Eve 38
Savage, Ann 27
Schantz, Jetta 39, 41
Scharr, Jack 30
Schermerhorn, Elizabeth 21
Schiffman, Jean 21
Schimmoler, Lauretta 16
Schirra, Wally 32
Schmitt, Julia 21
Scholten, Anna 21
Schorsch, Ann Marie 21
Schorsch, Sharon Ann 31
Schroeder, Gail 21
Schug, Lynn 21
Schulz, Charles 27
Schwartz, Merav 21
Scotia, Nova 21, 40
Scott, Anna 21
Scott, Blanche Stuart 10
Scott, Sheila 14, 40
Scully, Linda 21
Sea Turtle, Inc. 32
Second Annual All Women Air Show of the World 19
Seddon, Rhea 37
Sedivy, Evelyn 21
Senecal, Dick 36
Senior Executive Services (SES) 35
Sestito, Barbara 6, 34, 35
Sewell, Elizabeth 6, 21
Sewell, Susie 21
Shaffer, Margaret 21
Shamburger, Page 19
Shanahan, Catherine 21
Sheehy, Ethel 6, 26
Sheffer, Helen 21
Shields, Anne 19, 21
Shoemaker, Carolyn 39
Shoemaker-Levy 9 39
Shroff, Mohini 40
Shropshire, Helen 27
Siberia 12
Simons, Sandy 35
Skelton, Betty 19, 20
Skylarks 12
Sloan, Jerrie 25
Smith, Joan Merriam 22, 24
Smith, Laura 21
Smith, Louise 6, 24
Smith, Margaret Willis 7

Smithsonian Air and Space Museum 29, 31
Snaper, Kathleen 29
Snohomish Airport 17
Snow, Evelyn 21
Solar Challenger 31
Solar Penguin 30
Sopwith Camel 27
Southern Illinois University 23
Southwest Airlines 33, 35
Souza, Yvonne M. 21
Sovereign, Jeannete 6
Spanagle, Hazel Mark 7
Sperry Gyroscope Co 16
Spirit of Flight Award 32
Spoerry, Dr. Anne 29
St. Onge, Christine 21
St. Onge, Ruby 21
St. Paul's Cathedral 30
Staggerwing Beech 13
Standard Oil Co. 16
Stanford, Margaret 21
Stark, Anneliese Rene 21
Stationaire 6 32
Statue of Liberty 40
Steadman, B. 22, 25, 33
Steadman, Bernice T. 6
Stearns, Edith Foltz 7, 9, 10, 16, 29
Steel, Lawanna 21
Stevens, Donna 21
Stewart, Betty 30
Stewart, Ruth W. 7, 43
Stinaff, Mildred 7, 236
Stinson, Katherine 10
Stinson, Marjorie C. 7, 236
Stinson Monoplane 13
Stocker, Dorothy L. 7
Stone, Amalie Ward 21
Storer Broadcasting 12
Story, Irma "Babe" 26
Stott, Barbara 21
Stratford, Cecelia 21
Strawn, C.J. 4
Strike Eagle F-15E 39
Strother, Dora Dougherty 19
Studer, Clara Trenckmann 11, 25
Suberg, Valerie 21
Sullivan, Elinor Smith 15
Super Stallion CH53E 40
Susie Sewell-Research School 21
Sutton, Carole 21, 36
Swallow 17
Swartwood, Jean 21
Sweet, Virginia 21

-T-
T-34 38
T-38 31
Tanner, Lori 40
Taylor, Jim 36
Taylor, Roberta 21
Tennessee Bureau of Aeronautics 15
Tennessee Research Instructor School for Women 15
Tereshkova, Valenhtina 29
Texas Air National Guard 39
Thaden, Louise 7, 10, 11, 12, 13, 15, 16, 17, 20, 27, 35, 37, 38, 43
The 99 Memorial Project Fund 32
The American Girl 13
Thible, Madame 10
Thomas, Linda 21
Thomas, Patricia 21, 31
Thomas, Susan Gene 21
Thomas, Tina 21
Thomas, Virginia 10
Thorp, Gabrielle 21
Thro, Edwina McConnell 7, 8
Thurmond, Dee 26
Tibbets Jr., Lt. Col. Paul 19
Tiburzi, Bonnie 31
Tiger Moth 37
Tokar-Vlasek, Tiffany 21
Tompkins, Mimi 32, 41
Toney, Florence 21
Top Woman of the Year Award 37
Top Woman Pilot 23
Tormey, Patricia 21
Towers, Gordon 41
Transair Limited 29

Travel Air 11, 35, 38
Trenholm, Ardyth 21
Trout, Bobbi 7, 11, 15, 35, 38, 41, 236
Trubey, Verene 21
Truhill, Jerri 33
Trusler, Mary 21
Tuller, Dorothy 21
Turney, Patricia 21
Turtle Lady 32
Twin Beech model 18 31

-U-
U.S. Aerobatic Team 30, 33
U.S. Air Force 31, 33, 35, 39
U.S. Amy Air Corps 12
U.S. Coast Guard 25
U.S. Defense Department 38
U.S. Helicopter Team 25, 37
U.S. Marines 21
U.S. National Aerobatic Champion 37, 38
U.S. National Air and Space Museum 41
U.S. National Women's World Championship 25
U.S. Navy 27
U.S. Post Office 25, 39, 41
U.S. Precision Flight Team 31, 34, 37
U.S. Proficiency Flight Team 31
U.S. World Precision Flying Championship 34
Ultimate 10-300S 37
Unger, Virginia 21
United Airlines 21, 30, 32
United Flying Octogenarians 37
United Parcel Service 21, 29, 32
University of California 20
University of South Carolina 23, 24

-V-
Vail, Gail J. 21
ValuJet 40
van den Dool, Yvonne 21
Van Smith, Ida 29, 30, 39
Vance, Esther Combes 7
Vanderpool, Celia 21
Varig Brazilian Airlines 38
Vass, Gail 4
Voisin Biplane 10
Von Mach, Mary E. 7, 236
Voss, Mary Jo 21
Voyager 34
Voyager II 33
Vreeland, Melissa 21
Vultee BT13 28

-W-
Wagner, Judy 22, 31
Wagstaff, Patty 37, 38
Waldren, Evelyn 235
Walker, Vera Dawn 7, 10, 11, 15
Wallingford, Josephine C.
Walsh, Wilma 7, 12
Walters, Claire 4, 27
Walton, Melinda 21
Waltz, Jessica 21
Ward, Mary 21
Ward, Pat 23, 34, 39
Warman, Laura 21
Warner, Emily 25, 29, 30
Warren, Margaret Thomas 7
Washington 99 25
Washington D.C. Chapter 39
Washington, Evie 21
Watkins, Aileen 21
Watson Jr., Thomas 23
Way, Annette Gipson 16
Webb, E. Ruth 7, 12
Webster, Elizabeth 32
Wedell, James 17
Weeks, Betsy Kelly 7
Weeks, Kermit 30
Weigand, Don 30
Welch, Janice M. 21
Weldon, Jeffrey 27
Wells, Fay Gillis 7, 11, 12, 25, 38, 42
Wells, Joyce 6, 40
Wells, Linton 12

West, Caroline 26, 27
West, Carolyn 19
West, Verna 2, 3, 34
Western Airlines 29, 30
Westmoreland, Mary Lou 21
Weyl, John 2
Wheeler, Carole Sue 23, 34
Whirly Girls 24, 25, 29, 37, 39
White, Ann 21
White, Nora Alma 7
Whitman, Bonnie Chittenden 7, 9
Whitsitt, Wanda 11
Whyte, Edna Gardner 6, 16, 17, 22, 235
WIAC Aviation Pioneer Hall of Fame 38
Wichita Falls Chapter 260
Wicks, Rene 21
Wideroe, Turi 29
Wieben, Elizabeth 21
Wikander, Ruth 21
Wilcox, Blanche 16
Wilkerson, Angela 21
Wilkinsburg Airport 16
Will Rogers World Airport 34
Willhite, Nellie Zabel 7, 8
Williams, Ella D. 39
Williams, Gay Zena 21
Willis, Patty 11
Wingett, Victoria 33, 35
Wings for Direct Relief Foundation 34
Winkelmann, Laura 21
Winters, Karen 21
Winzer, Christine 21
Wladyke, Harriet 21
Woffteddies 18
Wolf, Connie 235
Women Fliers of America 12
Women's Achievement Award 23
Women's Advisory Committee on Aviation 22, 29
Women's Air Derby 11, 13, 16
Women's Air Reserve 12
Women's Airforce Service Pilots 18, 29, 40
Women's Auxiliary Ferry Service 15, 18, 38
Women's Flying Training Detachment 18
Women's International Association of Aeronautics 25
Women's Pylon Racing Association 22
Women's Transcontinental Air Race 19
Woodham, Dorothy 21
Workman, Shanon 39
World Aerobatic Championships 30
World Aerospace Education Organization 30, 38
World Aviation Education and Safety Congress 40
World Precision Flying Championship 33
World Speed Record for Women 17
World War II 18, 19, 23, 28, 29, 32, 40
Wright Brothers Memorial Trophy 29
Wright, Carol 21
Wright, Cynthia 21
Wright, Lucile M. 32
Wright, Wilbur 23
WWII Commemoration Committee 40
Wychelewski, Ann Marie 23
Wyoming Chapter 30

-X-
Xerox Corp. 28
XV-15 Tilt-Rotor 35

-Y-
Yarbrough, Michele 21
Yeager, Gen. Chuck 19
Yeager, Jeana 33, 41
Yeda Cox de Barros, Kalina 38
Yohoyama, Hideko 32
Young Astronaut Program 34
Young Eagles Program 39
Young, Janice Orr 21

-Z-
Zeidler, Terry 21
Zimbabwe Sun Air Rally 27
Zonta International 39

CHARTER MEMBERS Top, L to R: *Peggy J. Paxson, Nora Alma White, Ethel Lovelace, Helen V. Cox Clohecy Bickle, Myrtle R. Caldwell, unidentified.* **Middle, L to R:** *Betsy Kelly Weeks, Olive "Keet" Mathers Maugham, Wilma L. Walsh, Sasha Peggy Hall Martin, Angela L. Joseph.* **Bottom, L to R:** *Marjorie May Lesser Van Antwerp, Esther Combes Vance, Vera Dawn Walker, E. Ru Webb, Lady Mary Heath.*